The Transformation of the World

The Transformation of the World

A Global History of the Nineteenth Century

Jürgen Osterhammel

Translated by Patrick Camiller

PRINCETON UNIVERSITY PRESS

Princeton and Oxford

First published in Germany by C. H. Beck under the title *Die Verwandlung der Welt*
© Verlag C. H. Beck oHG, München 2009

English translation copyright © 2014 by Princeton University Press
Published by Princeton University Press, 41 William Street,
Princeton, New Jersey 08540
In the United Kingdom: Princeton University Press, 6 Oxford Street,
Woodstock, Oxfordshire OX20 1TW

press.princeton.edu

Jacket illustration: Harbor at Shanghai, China, 1875, © Getty Images. Cover design by Faceout
Studio, Charles Brock.

Paperback ISBN: 978-0-691-16980-4

Cloth ISBN: 978-0-691-14745-1

The Library of Congress has cataloged the cloth edition as follows

Osterhammel, Jürgen.
    [Verwandlung der Welt. English]
    The transformation of the world : a global history of the nineteenth century / Jürgen
Osterhammel.
        pages cm. — (America in the world)
    "First published in Germany by C.H. Beck under the title Die Verwandlung der Welt, Verlag
C.H. Beck oHG, Munchen 2009."
    Includes bibliographical references and indexes.
    ISBN 978-0-691-14745-1 (hardback : acid-free paper)   1. History, Modern—19th century.
I. Title.
    D358.O8813 2014
    909.81—dc23

                                        2013025754

British Library Cataloging-in-Publication Data is available

The translation of this work was funded by Geisteswissenschaften International - Translation
Funding for Humanities and Social Sciences from Germany, a joint initiative of the Fritz Thyssen
Foundation, the German Federal Foreign Office, the collecting society VG WORT and the
Börsenverein des Deutschen Buchhandels (German Publishers & Booksellers Association)

This book has been composed in Garamond Premier Pro

Printed on acid-free paper. ∞

Printed in the United States of America

10 9 8 7 6 5

For Sabine and Philipp Dabringhaus

CONTENTS

PART TWO: PANORAMAS

PREFACE

This book was first published as *Die Verwandlung der Welt. Eine Geschichte des 19. Jahrhunderts* by C. H. Beck publishers in Munich in January 2009. It rapidly went through five editions and two unnumbered special editions and is now being translated into Chinese, French, Polish, and Russian. For the American edition the manuscript was revised and brought up to date as far as that could be done without adding to the book's considerable length.

For a single author to tackle a one-volume global history of a "very long" nineteenth century borders on the foolhardy and may require if not an apology, then at least some explanation. Several of my previous books have been crisp and concise, and I fully appreciate the value of collaborative work, having the privilege of being, with Akira Iriye, one of the editors-in-chief of a multivolume "New History of the World" that is written by a distinguished group of scholars from several countries. Thus, *The Transformation of the World* should not be seen as a product of solipsism and conceit.

My own research experience has focused on two different fields: the final phase of British informal imperialism in China, and the role of Asia in the thinking of the European Enlightenment. I never wrote a source-based monograph on any aspect of the nineteenth century, but I have long been involved in teaching its history, and the present book draws on a lifetime of reading about the period. Two other ingredients went into the making of this book: One of them is a deep respect for historical sociology, especially the tradition going back to Max Weber, with whose works I was made familiar by two of my teachers: Wolfgang J. Mommsen and Wilhelm Hennis. Later, I had the chance to discuss issues of historical sociology with S. N. Eisenstadt on the occasion of his visits to the University of Konstanz, and today I enjoy the regular exchange of ideas with Wolfgang Knöbl at Göttingen, a sociologist with a deep understanding of how historians think. The second formative influence has been an interest in the history and theory of world history writing kindled by yet another of my teachers: Ernst Schulin at the University of Freiburg. A collection of my articles on historiographical topics was published in 2001. However, theorizing about world history can never be more than a preparation for historical analysis. In this sense, the present book is an attempt to put my own recipes into practice.

The book is an experiment in writing a rich and detailed but structured, nontrivial, and nonschematic account of a crucial period in the history of humanity. It was not commissioned by a publisher and has therefore been written oblivious to marketing constraints. Though easily accessible to students, it was never intended as a textbook. It does not disguise personal idiosyncrasies such as a special interest in animals, the opera, and the old-fashioned—though, as I hope to show, highly important—field of international relations. Uneven coverage would be an inexcusable sin in a textbook, whereas this work does not deny the fact that its author is more familiar with some parts of the world than with others. General and summary statements, frequent as they have to be in this particular kind of synthesis, derive from the logic of analysis and not from a pedagogical urge to simplify complex things for the benefit of the reader.

Work on the manuscript began in 2002 when I was a fellow of the Netherlands Institute of Advanced Study (NIAS) at Wassenaar, an excellent institution whose rector at the time, H. L. Wesseling, counts as one of the godfathers of the project. A first sketch of some of my emerging ideas was presented in November 2002 as the Sixteenth Annual Lecture at the German Historical Institute in Washington, DC (and later published in the institute's bulletin), given at the invitation of its then director, Christof Mauch. During the following years, regular teaching duties, relatively substantial as they are at German universities, slowed down work on the manuscript. Unsurprisingly, the publication of C. A. Bayly's magisterial *The Birth of the Modern World* early in 2004 caused me to reassess the project and threw its continuation into doubt. Ultimately, I wrote a review essay on Bayly and decided to carry on. There are already several world histories of the "age of extremes" (Eric Hobsbawm)—why not two of its predecessor, the nineteenth century? I was able to complete the manuscript when Heinrich Meier invited me to come to Munich for a year as a fellow of the Carl Friedrich von Siemens Foundation, whose far-sighted director he is.

The German edition owes its existence to the confidence and courage of the great publisher Wolfgang Beck and his editor-in-chief, Detlef Felken, both of whom learned about the unwieldy manuscript—any publisher's nightmare—at an advanced stage of writing. Contact with Princeton University Press had already been established on the occasion of a previous book with the help of Sven Beckert, and I am most grateful to him and Jeremi Suri for including *The Transformation of the World* in their prestigious series *America and the World*. At Princeton University Press, Brigitta van Rheinberg, Molan Goldstein, and Mark Bellis did everything in their power to turn the revised manuscript into an attractive volume. Patrick Camiller's translation was funded by the program Geisteswissenschaften International–Translation Funding for Humanities and Social Sciences from Germany.

As this book is based on secondary literature, my main debt is to the marvelous historians and social scientists in many countries who have, almost within one generation, hugely increased our knowledge, deepened our understanding, and thus radically transformed our view of the global nineteenth century. I only managed to sample a tiny fraction of their work, and in this I had to limit myself to the small number of languages that I am able to read. Among numerous reviews of the German edition, those by Steven Beller, Norbert Finzsch, Jonathan Sperber, Enzo Traverso, Peer Vries, and Tobias Werron were particularly useful in pointing out errors of fact and problems with the overall conception. Etienne François, Christian Jansen, and H. Glenn Penny provided critical comments that describe my methods and literary stratagems much better than I could have done it myself. Folker Reichert and Hans Schneider gave detailed advice on how to improve the accuracy of the book.

Not every suggestion could be heeded. A pervasive disregard of gender issues remains a serious drawback that will, hopefully, be remedied in a forthcoming attempt to expand chapter 15 of this book into a global social history of the period from the 1760s to the 1880s. A certain weakness of explanatory power may rest at the heart of the project, although in principle I disagree with a postmodernist aversion to causality. Readers who were—and are—vainly looking for insights into literature, music, the visual arts, and philosophical thinking may like to know that I am now doing some work on the social and cultural history of music. A more general response would be that world history should avoid the mirage of encyclopedic completeness and that the danger of superficiality never looms larger than when the historian is confronted with works of art and philosophy that require careful and elaborate interpretation.

At the University of Konstanz, the revision of the manuscript benefited enormously from the atmosphere of intellectual excitement created by the members of the Research Unit "Global Processes (18th to 20th Centuries)" that I was able to establish with generous funding from the Deutsche Forschungsgemeinschaft (German Research Foundation). I mention only Boris Barth, Franz L. Fillafer, Stefanie Gänger, Jan C. Jansen, and Martin Rempe. New work by these young scholars, by several PhD students, and also by myself, is emerging out of this stimulating context.

My family has been living with the book ever since our year at NIAS. It is a great joy to renew the original dedication to my son Philipp Dabringhaus and to add the dedicatee of a previous book, my wife Sabine Dabringhaus, an accomplished historian of China.

INTRODUCTION TO THE
FIRST GERMAN EDITION (2009)

All history inclines toward being world history. Sociological theories tell us that the world is the "environment of all environments," the ultimate possible context for what happens in history and the account we give of it. The tendency to transcend the local becomes stronger in the *longue durée* of historical development. A history of the Neolithic age does not report intensive contacts over long distances, but a history of the twentieth century confronts the basic fact of a densely knit web of global connections—a "human web," as John R. and William H. McNeill have called it, or better still, a multiplicity of such webs.[1]

For historians, the writing of world history has particular legitimacy when it can link up with human consciousness in the past. Even today, in the age of the Internet and boundless telecommunications, billions of people live in narrowly local conditions from which they can escape neither in reality nor in their imagination. Only privileged minorities think and act "globally." But contemporary historians on the lookout for early traces of "globalization" are not the first to have discovered transnational, transcontinental, or transcultural elements in the nineteenth century, often described as the century of nationalism and the nation-state. Many people living at the time already saw expanded horizons of thought and action as a distinguishing feature of their epoch, and dissatisfied members of the middle and lower strata of society in Europe and Asia turned their eyes and hopes toward distant lands. Many millions did not shrink from undertaking an actual journey into the unknown. Statesmen and military leaders learned to think in categories of "world politics." The British Empire became the first in history to span the entire globe, while other empires ambitiously measured themselves by its model. More than in the early modern period, trade and finance condensed into integrated and interconnected worldwide webs, so that by 1910, economic vibrations in Johannesburg, Buenos Aires, or Tokyo were immediately registered in Hamburg, London, or New York, and vice versa. Scholars collected information and objects from all over the world; they studied the languages, customs, and religions of the remotest peoples. Critics of the prevailing order—workers, women, peace activists, anti-racists, opponents of colonialism—began to organize internationally, often far beyond the confines of Europe. The nineteenth century reflected its own emergent globality.

As far as the nineteenth century is concerned, anything but a world-history approach is something of a makeshift solution. However, it is with the help of such makeshifts that history has developed into a science, gauged in terms of the methodological rationality of its procedures. This process of becoming a science, through the intensive and possibly exhaustive examination of sources, took place in the nineteenth century, so it is not surprising that the writing of world history receded into the background at that time. It appeared to be incompatible with the new professionalism that historians embraced. If this is beginning to change today, it certainly does not mean that all historians wish to, or should, take up the writing of world history.[2] Historical scholarship requires deep and careful study of clearly definable cases, the results of which form the material for broad syntheses that are indispensable for teaching and general orientation. The usual framework for such syntheses, at least in the modern age, is the history of one nation or nation-state, or perhaps of an individual continent such as Europe. World history remains a minority perspective, but no longer one that can be dismissed as esoteric or unserious. The fundamental questions are, of course, the same at every level of spatial scope or logical abstraction: "How does the historian, in interpreting a historical phenomenon, combine the individuality given by his sources with the general, abstract knowledge that makes it possible to interpret the individual in the first place? And how does the historian arrive at empirically secure statements about larger units and processes of history?"[3]

The professionalization of history, from which there is no going back, has entailed that history on a larger scale is now often left to the social sciences. Sociologists and political theorists who retain an interest in the depths of time and the vastness of space have assumed responsibility for engaging with major historical trends. Historians have an acquired predisposition to shy away from rash generalizations, monocausal explanations, and snappy all-embracing formulas. Under the influence of postmodern thinking, some consider it impossible and illegitimate to draw up "grand narratives" or interpretations of long-range processes. Nevertheless, the writing of world history involves an attempt to retrieve some interpretative competence and authority, visible in the public eye, from minutely detailed work in specialist fields. World history is one possible form of historiography—a register that should be tried out once in a while. The risk falls on the author's shoulders, not on that of the reading public, which is protected from spuriousness and charlatanry through the alertness of professional criticism. But the question remains of why it should be the work of a single hand. Why should we not be content with multivolume collective products from the "academic factory" (Ernst Troeltsch)? The answer is simple. Only a centralized organization of issues and viewpoints, of material and interpretations, can hope to meet the constructive requirements of the writing of world history.

To know all there is to know is not the key qualification of the world historian or global historian. No one has sufficient knowledge to verify the correctness of every detail, to do equal justice to every region of the world, or to draw fully adequate conclusions from the existing body of research in countless different areas. Two other qualities are the truly important ones: first, to have a feel for proportions, contradictions, and connections as well as a sense of what may be typical and representative; and second, to maintain a humble attitude of deference toward professional research. The historian who temporarily slips into the role of global historian—she or he must remain an expert in one or more special areas—cannot do other than "encapsulate" in a few sentences the arduous, time-consuming work of others. At the same time, the labors of global historians will be worthless if they do not try to keep abreast of the best research, which is not always necessarily the most recent. A world history that unwittingly and uncritically reproduces long-refuted legends with a pontifical sweep of the hand is nothing short of ridiculous. As a synthesis of syntheses, as "the story of everything,"[4] it would be crude and tiresome.

This book is the portrait of an epoch. Its modes of presentation may in principle be applied in the case of other historical periods. Without presuming to treat a century of world history in a complete and encyclopedic manner, it offers itself as an interpretative account rich in material. It shares this stance with Sir Christopher Bayly's *The Birth of the Modern World*, a work published in English in 2004 and in German two years later, which has rightly been praised as one of the few successful syntheses of world history in the *late* modern period.[5] The present volume is not an anti-Bayly but an alternative from a kindred spirit. Both books forgo a regional breakdown into nations, civilizations, or continents. Both regard colonialism and imperialism as a dimension so important that instead of dealing with it in a separate chapter, they keep it in view throughout. Both assume that there is no sharp distinction between what Bayly, in the subtitle of his book, calls "global connections" and "global comparisons";[6] these can and must be combined with each other, and not all comparisons need the protective backup of strict historical methodology. Controlled play with associations and analogies sometimes, though by no means always, yields more than comparisons overloaded with pedantry can do.

Our two books often place the emphasis differently: Professor Bayly's background is India, mine China, and this shows. Bayly is especially interested in nationalism, religion, and "bodily practices," which are the themes of superb sections of his work. In my book, migration, economics, the environment, international politics, and science are considered more broadly. I am perhaps a little more "Eurocentrically" inclined than Bayly: I see the nineteenth century even more sharply than he does as the "European century," and I also cannot conceal a fascination for the history of the United States, a topic I discovered in the course of writing. As regards our theoretical references, my closeness to historical sociology will become apparent.

But the two most important differences between Christopher Bayly and myself lie elsewhere. First, my book is even more open than Bayly's to the chronological margins of the period. It is not a compartmentalized history of a certain number of years sealed off from what went before and what came after. This is why there are no framing dates in the title, and why a special chapter is devoted to issues of periodization and temporal structure. The book anchors the nineteenth century variously "in history," allowing itself to look back far beyond 1800 or even 1780 as well as ahead to today's world. In this way, the significance of the nineteenth century is triangulated in longer periods of time. Sometimes the century is remote from us, sometimes it is very close; often as the prehistory of the present, but on occasion as deeply buried as Atlantis. The determination must be made on a case-by-case basis. The nineteenth century is viewed in terms not so much of sharply defined hiatuses as of an inner focal point, stretching roughly from the 1860s to the 1880s, when innovations with a worldwide impact came thick and fast, and many processes running independently of one another seemed to converge. The First World War does not therefore appear as a sudden, unexpected falling of the curtain, as it does in Bayly's historical staging.

Second, the narrative strategy I have chosen is different from Bayly's. There is a kind of historiography that might be described as time-convergent; and it has allowed some historians—operating with fine judgment, huge experience, and a lot of common sense—to present whole eras of world history in the main and secondary lines of their dynamic momentum. John M. Roberts's global history of the twentieth century, which he offers as an account of "what is general, what pulls the story together,"[7] is a perfect example of this. It is world history that seeks to identify what is important and characteristic in each age, shaping it into a continuous narrative without any preconceived schema or big guiding idea in the background. Eric Hobsbawm, with a pinch of Marxist rigor and therefore a compass that I cannot claim to possess, achieved something similar in his three-volume history of the nineteenth century, working his way back from each digression to the major trends of the age.[8] Bayly takes a different road, which may be described as space-divergent; it is a decentering approach, not so uninhibited in allowing the current of time to carry it forward. It does not make such nimble headway as a Roberts type, drifting along with the flows of history, but goes into the detail of simultaneity and cross section, searches for parallels and analogies, draws comparisons, and ferrets out hidden interdependences. This means that its chronology is deliberately left open and vague: it manages with few framing dates and keeps the narrative on course without too much explicit organization into subperiods. Whereas someone like Roberts—and in this sense he may represent the mainstream of older world-history writing—thinks within a dialectic of major and minor and constantly asks what of significance, whether good or bad, each period produced, Bayly concentrates on individual phenomena and examines them within a global perspective.

One case in point is nationalism. Again and again, we read that it was a European "invention" that the rest of the world took on in a cruder form and with many misunderstandings. Bayly takes a closer look at this "rest of the world" and arrives at the plausible idea of a polygenesis of forms of nationalist solidarity: that is, *before* nationalist doctrines were imported from Europe, "patriotic" identities had already taken shape in many parts of the world, which could then be reinterpreted in a nationalist sense in the late-nineteenth and twentieth centuries.[9] Bayly's historiography is primarily horizontal—or "lateral," as he aptly calls it[10]— and spatially determined, whereas that of John Roberts or Eric Hobsbawm is more "vertical" and temporal in its emphasis. All three authors would insist that they combine the horizontal and vertical dimensions, and that is certainly correct. But the relationship between the two approaches seems to display a kind of unavoidable fuzziness, rather like that which is found in the well-known tension between narrative and structural accounts: no attempt to marry the two achieves complete harmony.

The design of the present volume leans more in Bayly's direction, but it goes further than he does and may therefore be said to take a third road. I doubt that it is possible, with the historian's cognitive tools, to fix the dynamic of an epoch in a single schema. World system theory, historical materialism, or evolutionary sociology may contradict. But since it is the business of history to *describe* change before it ventures explanations, it soon runs up against remnants that stubbornly resist integration. Bayly is well aware of this, of course, yet he overcomes such scruples when he tries to define the distinguishing feature of an age. His main thesis is that between 1780 and 1914 the world became more uniform but also more internally differentiated;[11] the "birth of the modern world" was a slow process that only came to completion with the "great acceleration" after 1890, a process that one hopes Bayly will analyze more comprehensively in future work.[12] Since Bayly eschews any more-or-less clear dividing line between areas of historical reality, he cannot be really interested in the independent logic governing each of them. Only industrialization, state building, and religious revival feature in his account as discrete processes. A general "master narrative" for the world of the nineteenth century rises out of a cosmos of particular observations and interpretations, which are always stimulating and mostly convincing.

I experiment with a solution in which "grand narratives" are even more resolutely defended. Postmodern critiques have not rendered such overarching constructions obsolete but made us more conscious of the narrative strategies their authors deploy. To be sure, a grand narrative may establish itself at various levels: even a history of worldwide industrialization or urbanization in the nineteenth century would be passably "grand." This high level of generality, at which we are nevertheless still talking of *sub*systems of a scarcely discernible totality of communal life, gives the book its basic structure. It appears encyclopedic only at first sight but is actually made up of successive orbital paths. Fernand Braudel once described a similar procedure: "The historian first opens the door with which he is most familiar. But if he seeks to see as far as possible, he must necessarily find himself knocking at another door, and then another. Each time a new or slightly different landscape will be under examination. . . . But history gathers them all together; it is the sum total of all these neighbors, of these joint ownerships, of this endless interaction."[13] In each subarea, therefore, I look for the distinctive "dynamics" or "logics" and the relationship between general developments and regional variants. Each subarea has its own temporal structure: a particular beginning, a particular end, specific tempos, rhythms, and subperiods.

World history aims to surmount "Eurocentrism" and all other forms of naive cultural self-reference. It shuns the illusory neutrality of an omniscient narrator or a "global" observation point, and it plays consciously on the relativity of ways of seeing. This means that it must not be forgotten who is writing for whom. The fact that a European (German) author originally addressed European (German) readers cannot fail to have left its mark on the text, whatever the cosmopolitan intentions behind it; expectations, prior knowledge, and cultural assumptions are never location neutral. This relativity also leads to the conclusion that the centering of perception cannot be detached from core/periphery structures in historical reality. This has a methodological and an empirical side. Methodologically, a lack of adequate sources, and of historiography based on them, hampers many a well-intentioned effort to do historical justice to the voiceless, the marginal, and the victimized. Empirically, proportions between the various parts of the world shift with the long waves of historical development. Power, economic performance and cultural creativity are distributed differently from epoch to epoch. It would therefore be capricious to sketch a history of the nineteenth century, of all periods, that disregarded the centrality of Europe. No other century was even nearly as much Europe's century. It was an "age of overwhelming, and overwhelmingly European, initiatives," as the philosopher and sociologist Karl Acham aptly put it.[14] Never before had the western peninsula of Eurasia ruled and exploited larger areas of the globe. Never had changes originating in Europe achieved such impact on the rest of the world. And never had European culture been so eagerly soaked up by others, far beyond the sphere of colonial rule. The nineteenth century was a European one also in the sense that other continents took Europe as their yardstick. Europe's hold over them was threefold: it had power, which it often deployed with ruthlessness and violence; it had influence, which it knew how to spread through the countless channels of capitalist expansion; and it had the force of example, against which even many of its victims did not balk. This multiple superiority had not existed in the early modern phase of European expansion. Neither Portugal nor Spain nor the Netherlands nor England (before approximately 1760) had projected their power to the farthest corner of the earth and had such a powerful cultural impact on "the Others" as Britain and France did in the nineteenth century. The history of the nineteenth century was made in and by Europe, to an extent that cannot be said of either the eighteenth or twentieth century, not to speak of earlier periods. Never has

Europe released a comparable burst of innovativeness and initiative—or of conquering might and arrogance.

Nevertheless, "Why Europe?" is not the big question posed in this book, as it has been for so many authors, from the Enlightenment to Max Weber down to David S. Landes, Michael Mitterauer, and Kenneth Pomeranz. Two or three decades ago, a history of the modern world could still blithely proceed on the assumption of "Europe's special path." Today, historians are trying to break with European (or "Western") smugness and to remove the sting of "special path" notions by means of generalization and relativization. The nineteenth century deserves to be looked at again in the context of this debate, because a strong current among comparative historians now considers that socioeconomic differences between Europe and other parts of the world in the early modern period were less dramatic than previous generations used to think. The problem of the "great divergence" between rich and poor regions has thus been shifted forward to the nineteenth century.[15] Yet this is not the central issue of the book, and no novel interpretation will be added to the many that already try to account for Europe's ephemeral primacy. To approach the historical material through the lenses of exceptionalism would be to focus from the start more on what distinguishes Europe from other civilizations than on what civilizations and societies have in common with each other. There are dangers in both possible kinds of a priori assumption: namely, an a priori contrastive option that privileges difference in all possible ways but also, at the opposite extreme, an equally one-sided a priori ecumenism that rarely lowers its sight below the human condition in general. It makes more sense to find a way out from the well-worn "West against the rest" dichotomy and to measure again, on a case-by-case basis, the gap between "Europe" (whatever that may have been at the time) and other parts of the world. This can best be done in relation to particular areas of historical reality.

The book is divided into three parts. The three chapters of Part One ("Approaches") outline the presuppositions or general parameters for all that follows: self-reflection, time, and space. The equal treatment of time and space will counter the impression that the writing of world history is necessarily bound up with temporal dedifferentiation and a "spatial turn." The eight chapters of Part Two then unfurl a "panorama" of eight spheres of reality. The term "panorama" refers to the fact that although no pedantic claim is made to represent all parts of the world equally, an attempt is made to avoid major gaps in the field of vision. In the seven chapters of Part Three ("Themes"), this panoramic survey gives way to a more narrowly focused, essay-style discussion of discrete aspects, which deliberately refrains from trying to include everything and uses examples mainly to illustrate general arguments. If these themes had been developed in a "panoramic" scope, the requisite scale of the book would have made excessive demands on the reader's patience as much as on the author's stamina. Moving on from "panoramas" to "themes," the book shifts the weight from synthesis to analysis—two modes of investigation and presentation that do not stand in sharp opposition to each other. The chapters of the book are meant to hang together as a coherent whole, but they may also be read separately. Once readers have entered the book, they should not worry: they will easily find an emergency exit.

PART ONE

APPROACHES

CHAPTER I

Memory and Self-Observation

The Perpetuation of the Nineteenth Century

What does the nineteenth century mean today? How does it present itself to those who are not professionally involved with it as historians? Our approach to this age begins with the face it turns to posterity. This is not simply a question of *our* "image" of it, of how we would *like* to see it, of how we construct it. Such constructs are not entirely random, not unmediated products of contemporary preferences and interests. Today's perceptions of the nineteenth century are still strongly marked by its own self-perception. The reflexivity of the age, especially the new media world that it created, continues to shape how we see it.

It was only a short time ago that the nineteenth century, separated from the present by more than a full calendar saeculum, sank beneath the horizon of personal recollection. In June 2006 even Harriet, the giant tortoise that in 1835 may have made the acquaintance of the young Charles Darwin in the Galapagos Islands, finally departed this life in an Australian zoo.[1] No one remains to reminisce about the Chinese Boxer Rebellion of 1900–1901 or the solemn obsequies of Giuseppe Verdi and Queen Victoria, both of whom died in late January 1901. Neither the funeral procession for Japan's Meiji emperor in September 1912 nor the mood when the First World War broke out in August 1914 remains within the memory of anyone alive today. In 2009 the ultimate survivor of the *Titanic* shipwreck passed away; the last German veteran of the Great War died in May 2008.[2] Remembrance of the nineteenth century is no longer a matter of individual recall but rather of media information and book reading. The traces are to be found in academic and popular history, in the collections of historical museums, in novels and paintings, old photographs and musical sounds, cityscapes and landscapes. The nineteenth century is no longer actively remembered, only depicted. It has this in common with earlier ages. In the history of the representation of cultural life, however, it occupies a distinctive place that already sets it apart from the eighteenth century. Indeed, many of the forms and institutions of current cultural life are inventions of the nineteenth century: the museum, the national archive, the national library, statistical science, photography, the cinema, recorded sound. It was an era of organized memory, and also of increased self-observation.

The role of the nineteenth century in today's consciousness is by no means a matter of course, either for the aesthetic canon or for the formation of political traditions. China may serve as an example of this. The nineteenth century was disastrous for China politically and economically, and has remained so in the minds of most Chinese. They think back reluctantly to that painful age of weakness and humiliation, and official propagandistic history does nothing to raise their appreciation of it. At the same time, indictments of the West's "imperialism" have become more muted, since the newly rising nation does not recognize itself in that earlier role of victim. Culturally, too, the century counts for them as decadent and sterile: none of China's artworks or philosophical texts from that period can stand alongside the classical works of a more remote past. For today's Chinese, the nineteenth century is much more distant than the splendors of many a dynasty down to the great emperors of the eighteenth century, who are constantly evoked in popular histories and television serials.

The contrast between China and Japan could not be greater. In Japan the nineteenth century enjoys incomparably higher prestige. The Meiji Renewal (often known as the Meiji Restoration) that began in 1868 is conventionally seen there as the founding process not only of the Japanese nation-state but of a distinctive modernity. Its role in the consciousness of today's Japanese is comparable in many respects to that of the Revolution of 1789 for the French.[3] The aesthetic evaluation of the century is also different. Whereas in China a modern literature cannot be said to have begun before the 1920s, Japan's "1868 generation" was already producing modern works in the 1880s.

The historical memory of the nineteenth century casts a similar spell in the United States, where the Civil War of 1861–65 stands alongside the formation of the Union in the late 1700s as the constitutive event of the nation. The descendants of victorious white Northern settlers, defeated white Southerners, and newly emancipated slaves have each ascribed quite different meanings to the conflict and composed their own "useful past." But there is agreement that the Civil War represents a common "felt history," as the poet and literary critic Robert Penn Warren put it.[4] For a long time it operated as a collective trauma, which still has not been overcome everywhere in the South. As always with historical memory, we are dealing not simply with a quasi-natural formation of identity but also with an instrumentalization advantageous to identifiable interests. Southern propagandists, foregrounding "states' rights," made every effort to gloss over the fact that the war was centrally about slavery and emancipation, while the other side grouped around a mythologization of Abraham Lincoln, the president murdered in 1865. Not a single German, British, or French statesman—not even Bismarck, more respected than cherished, or the ever-controversial Napoleon I—enjoyed such veneration after his death.

In 1938 President Franklin D. Roosevelt could still publicly ask, "What would Lincoln do?"—the national hero as helper to posterity in its hour of need.[5]

1   Visibility and Audibility

The Nineteenth Century as Art Form: The Opera

A bygone age lives on in revivals, archives, and myth. Today the nineteenth century has vitality where its culture is staged and consumed. Its characteristic aesthetic form in Europe, the opera, is a good example of such revival. The European opera came into being around 1600 in Italy, only decades after the rise of the urban music theater in southern China, which marked the beginning of a development wholly independent of European influence that would reach its peak after 1790 in what we know today as Beijing opera (*jingxi*).[6] Despite the existence of a number of outstanding masterpieces, it was a long time before the cultural status of European opera became unassailable outside Italy. Only with the contributions of Christoph Willibald Gluck and Wolfgang Amadeus Mozart did it become the paramount genre in the theater. By the 1830s it was generally considered to be at the top of the artistic hierarchy.[7] This progression was paralleled in the Beijing opera, which at mid-century entered its period of artistic and organizational maturity. Since then European opera has triumphantly maintained its position, whereas its distant sister in Beijing, following radical breaks with tradition and the penetration of a Western-tinged media culture, has persisted only in folklore niches.

The opera houses that sprang up between Lisbon and Moscow in the nineteenth century are still in full swing, with a repertory that largely goes back to a "long" nineteenth century beginning with Mozart's masterpieces. Opera underwent globalization early on. In the mid-1800s it had a clear radial point: Paris. Around 1830, Parisian musical history was global musical history.[8] The Paris Opera was not only France's foremost stage. Paris paid composers the highest fees and outdid all rivals for the rank of music's leading magnet city.[9] Fame in Paris meant world fame; failure there—as happened in 1861 to Richard Wagner, already an established master, with his *Tannhäuser*—was a deeply wounding disgrace.

By the 1830s European operas were already being performed in the Ottoman Empire. In 1828 Giuseppe Donizetti, the brother of the celebrated composer Gaetano Donizetti, became the musical director of the sultan's court in Istanbul and built up a European-style orchestra there. In the independent empire of Brazil, especially after 1840 under Pedro II, opera became the official art form of the monarchy. Vincenzo Bellini's *Norma* was performed many times, and the major operas of Rossini and Verdi were staged. When Brazil became a republic, hugely wealthy rubber barons built a lavish opera house (inaugurated in 1896 and closed again after eleven years) in Manaus, then in the middle of the

Amazonian jungle, combining local precious woods with marble from Carrara in Tuscany, steel from Glasgow, and cast iron from Paris and featuring candlesticks from Murano.[10] Under colonial rule, too, opera spread far outside Europe, its sumptuous houses intended to display the superiority of French civilization. The theater inaugurated in 1911 in Hanoi, the capital of French Indochina, was especially massive, dwarfing many in the mother country with its 750 seats for the fewer than three thousand French in the city.[11] Like many others, it was modeled on the Opéra Garnier in Paris, whose 2,200 seats made it the world's largest theatrical space when it was completed in 1875.

Opera took root earlier in North America. The French Opera House that opened in New Orleans in 1859 was long considered one of the best in the world. In San Francisco, then a city of sixty thousand inhabitants, the passion for opera was so great that 217,000 tickets were sold in the year 1860. The Metropolitan Opera in New York, which opened its doors in 1883, went on to become one of the world's leading houses, a place at which American "high society" showed itself off in ways scarcely different from its counterparts in Europe. Architecturally and in terms of stage technology, the creators of the "Met" brought together elements from Covent Garden in London, La Scala in Milan, and of course the Opéra in Paris.[12] Almost its entire repertory came from the other side of the Atlantic. In the 1830s Chile was in the grip of Rossini fever.[13] In Japan, where the government had been encouraging the spread of European music since the 1870s, the first performance of European opera, a scene from Gounod's *Faust*, took place in 1894. Whereas in 1875, when an Italian prima donna had given an early guest performance in Tokyo, the event had been so poorly attended that the audience could hear the mice squeak, a steady interest in opera developed after the turn of the century and acquired a focus in 1911 with the opening of the first large Western-style theater.[14]

The figure of the itinerant stage celebrity active in various parts of the world also originated in the nineteenth century.[15] In 1850 Jenny Lind, the "Swedish nightingale," sang to an audience of seven thousand in New York, at the start of a tour comprising ninety-three performances. The soprano Helen Porter Mitchell, who called herself Nellie Melba after her native Melbourne, first appeared in Europe in 1887 and went on to become one of the first truly intercontinental divas, her voice reproduced after 1904 on gramophone discs; she was the icon of a new cultural self-confidence in her reputedly uncouth homeland. Nineteenth-century European opera was a global phenomenon, and so it has remained. The repertory of that age is still dominant today: Rossini, Bellini, Donizetti, Bizet, and above all Verdi, Wagner, and Puccini. But they are only some of the composers highly appreciated in the nineteenth century. Gaspare Spontini or Giacomo Meyerbeer, once celebrated masters, are rarely played nowadays, while others have entirely vanished into the archives. Who still knows any of the countless operas on Germanic or medieval subjects that saw the light of day alongside and after Wagner?

Similar points might be made about dramatic theater or another typical genre of the time—the novel—and a separation of the living from the dead in nineteenth-century high culture is possible for many countries. Nineteenth-century high culture is intensely present in the contemporary world, albeit with a strict selection that obeys the laws of taste and the culture industry.

Cityscapes[16]

The nineteenth century is visibly present to us in a quite different way in the cityscapes that often form the backdrop and arena of everyday life. London, Paris, Vienna, Budapest, and Munich are cities whose physiognomy is marked by nineteenth-century planners and architects, partly in neoclassical, neo-Romanesque, and neo-Gothic idioms that refer back to older models. From Washington, DC, to Calcutta, grand official buildings drew on and imitated this past, with the result that the architectural historicism of the nineteenth century offers us a global overview of European traditions. In many Asian metropolises, on the other hand, scarcely any old buildings have survived. In Tokyo, for instance, which for several centuries (at first under the name Edo) was the capital of Japan, earthquakes, fires, American bombs, and constant reconstruction have erased nearly all architectural traces older than a few decades and even cleared away many Meiji relics. The world's great cities range along a scale between the extremes of well-preserved urban ensembles (e.g., Vienna's Ringstrasse) and physical obliteration of the nineteenth century. The teeth of time gnaw selectively: the industrial architecture of the nineteenth century has worn away more quickly than many monuments from the Middle Ages. Scarcely anywhere is it still possible to gain a sensory impression of what the Industrial "Revolution" meant—of the sudden appearance of a huge factory in a narrow valley, or of tall smokestacks in a world where nothing had risen higher than the church tower.

2   Treasuries of Memory and Knowledge

Archives, libraries, museums, and other collections might be called *treasuries of memory*. Alongside the places of remembrance that crystallize the collective imagination of the past, these treasuries deserve our special attention. The boundaries between their various subcategories developed only gradually. Libraries were for a long time not clearly distinguishable from archives, especially if they held large numbers of manuscripts. In the eighteenth century, the term "museum" (especially in German) encompassed spaces for any kind of antiquarian study or exchange of ideas among private individuals, even journals whose declared aim was to present historical and aesthetic sources. The principle of universal accessibility to the public first appeared only in the nineteenth century. Treasuries of memory preserve the past as a virtual present. Yet the cultural past remains dead if it is nothing but treasured. Only in the act of appropriation, comprehension, and sometimes reenactment does it come alive.

Archives

The archive was more important in the nineteenth century than in any previous one. In Europe it was the time when the state everywhere took possession of memory. State archives were established as central depositories for the records governments left behind; and with them arose the professions and social types of the archivist and the public records historian. The latter could now have access to the collections of earlier princes and republics in Venice or Vienna or Simancas, Spain. In countries with constitutional rule, the government took over the archival task as one of the attributes of sovereignty. In September 1790 the French National Assembly renamed its still-modest collection the "Archives Nationales"; revolutionary confiscations, especially of church property, soon increased its holdings. Napoleon conducted his archival policy in grand style: he wanted the Archives Nationales to become a central depository—"la mémoire de l'Europe"—and had large quantities of documents brought to Paris from Italy and Germany. In 1838 Britain created the legal basis for the Public Record Office, and in 1883 the legendary archives of the Vatican were made accessible for the first time.

The "new history" that took shape from the 1820s in the work of Leopold Ranke and his disciples, first in Germany, then in many other countries, saw closeness to the text as its guiding principle. The past was to be reconstructed out of (mostly unpublished) written sources; history was to become more scientific, more verifiable, and more critical in its attitude to received myths. At the same time, historians made themselves rather more independent of the archival policy of governments that controlled access to the sources in which they were interested. The systematic organization of record keeping also helped to shape a new kind of scholar. Learning was uncoupled from the individual capacity to memorize facts and figures, the "polyhistor" became a mocked curiosity, and humanities scholars followed the natural scientists in seeing the investigation of causes as their main imperative.[17]

Archives were not, to be sure, a European invention, but nowhere else in the nineteenth century was there comparable interest in the preservation of documentary material. In China, the state had from early on monopolized the collection of handwritten material. There were few archives of nonstate institutions such as temples, monasteries, guilds, or clans. It was customary for a dynasty to destroy the records of its predecessor once the official history of it was complete. In 1921 the State Historical Museum in Beijing sold 60,000 kilograms of archive material to wastepaper dealers—only the intervention of the learned bibliophile Luo Zhenyu managed to save the collection, which is today kept at the Academia Sinica in Taiwan. Until the 1930s, official printed and handwritten material of the Qing Dynasty (1644–1911) was disposed of as trash. Despite a venerable tradition of historiography, there was still no archival awareness in nineteenth-century China. The documentation department of the Palace Museum, founded in 1925, was the first institution that brought the rule-driven conservation ethos of a modern archive to bear upon the relics of the imperial era.[18]

In the Ottoman Empire, where written records similarly contributed from an early date to the cohesion of a sprawling state, documents were produced and preserved on such a scale that research today can scarcely be imagined except as archive studies. Apart from the records of the sultan's court and the central government, tax registers and judicial proceedings (Kadi registers) are available from many parts of the empire.[19] Records, we conclude, were kept before the nineteenth century in Europe, the Ottoman Empire, and other parts of the world. Only during that period, however, did they begin to be systematically archived, safeguarded, and evaluated.

Libraries

Libraries, understood as managed collections of the printed cultural heritage, are also treasuries of memory. The seventeenth and eighteenth centuries made the first great advances in this respect in Europe. Between 1690 and 1716, Gottfried Wilhelm Leibniz, in his capacity as librarian, helped to place Duke August's magnificent collection in the small German town of Wolfenbüttel at the service of scholarship. Shortly afterward, the university in nearby Göttingen went a step further and for a time was reputed to possess the best-organized library in the world. The British Museum collection, initiated in 1753, was conceived from the outset as a national library; it incorporated the Royal Library in 1757 and was entitled to receive a copy of every book published in the United Kingdom. Antonio (later Sir Anthony) Panizzi, an Italian exile who joined the British Museum in 1831 and served as its chief librarian between 1856 and 1866, created the foundations of scientific librarianship: a systematic and comprehensive catalog, and a reading room organized to meet the needs of scholars, with a domed shape that made it the envy of the world.[20]

As the century wore on, national libraries in keeping with the British model were built on every continent. In the United States, Canada, and Australia, they arose out of parliamentary libraries.[21] Some were linked to academic institutions. They tended the nation's memory for the respectable public and all serious students, but they also collected knowledge in general. The most prestigious became known for their universal reach, as they gathered knowledge from all countries and all ages. Important prerequisites for this mission were a book trade with worldwide business links and the selling of private libraries on the antiques market. Newly founded Oriental departments collected books in rare languages, sometimes sending out special emissaries to acquire them. Libraries symbolized a country's pretension to equal or superior cultural status. In 1800 the young American republic staked its claim with the founding of the Library of Congress, and by the early 1930s its vast holdings, the largest anywhere in the world, completed the cultural emancipation of the New World from the Old. Countries that achieved unification late in the day had a harder time. The Prussian

State Library did not gain national status before 1919, and Italy has never established a single, all-embracing national library.

Municipal libraries served a public eager for education and were a mark of civic pride. But not until after mid-century did it become both legally possible and politically acceptable that taxpayers should foot the bill. Private sponsors were more important in the United States than elsewhere. The New York Public Library, built up after 1895 with funds from a private foundation, became the most famous of the numerous municipal libraries that set their sights high. The libraries of the West turned into temples of knowledge; Panizzi's British Museum, which housed the national library, made this architecturally palpable in its monumental neoclassical facade. In the 1890s the rebuilt Library of Congress took over this symbolic language and accentuated it by means of wall paintings, mosaics, and statues. The gigantic stores of knowledge were both national and cosmopolitan. Exiles conspired inside them: the Chinese revolutionary Sun Yat-sen, who in 1896–97 was in London forging plans to overthrow the Qing Dynasty, worked at the same British Museum where Karl Marx had developed a scientific basis for his struggle against the capitalist system.

The library is not a monopoly product of the West, as a look far back into history reveals. The first imperial library was founded in the palace of the Han emperor Wudi (r. 141–87 BC), and it was there that scholars developed a classification system that remained in use for a long time. Chinese libraries had a precarious existence, however: the imperial collections of books and manuscripts were destroyed and built up again at least twenty-four times between the third century BC and the nineteenth century. With the spread of xylography in the eleventh century, private academies (*shuyuan*), groups of scholars, and even individuals also developed large libraries. Details are known about more than five hundred collectors and their collections for the Qing period (1644–1911). The quantity of printed literature in private use was so great that the compilation of bibliographies became one of the scholar's principal tasks.[22] In China, then, libraries and catalogs were not a cultural import. What was of Western origin was the idea of a *public* library; the first opened in Changsha, the capital of the province of Hunan, in May 1905. The largest Chinese library today, the National Library of China, was founded in 1909, opened its doors to the public in 1912 and acquired national library status in 1928. The modern library in China was not the unbroken continuation of an indigenous tradition. The twin conception of the library as a public educational space and as an instrument of learning came from the West and took active root in early twentieth-century China, at a time when the country was facing difficult external conditions.

In traditional Japan the state acted much less often as a collector of documents. For a long time, Japanese holdings were mainly concerned with China. The nonpublic libraries built up from the eighteenth century by the shoguns of the Tokugawa family were mainly antiquarian and sinological, and they did not attempt to include the growing numbers of books produced in Japan. As in China,

Western book collectors appeared on the scene soon after the country opened up to the West (1853). The huge collections of Chinese and Japanese material in Europe and the United States were the result of a rising Western interest coinciding with a temporary neglect of indigenous cultural traditions in Asia and a lowering of book prices. After 1866 the journalist and educator Fukuzawa Yukichi, who had traveled to the West on a diplomatic assignment in 1862, familiarized the Japanese with the idea of a public library. But even amid the new enthusiasm for modernization, it took until the end of the century for the research library and publicly oriented publishing to become the accepted models.[23]

The Arab world was geographically closer to Europe but more distant in terms of the history of the book. China's long-standing use of xylographic text reproduction meant that the professions of calligrapher and copyist were less important there than in the Arab world, whose printing revolution did not take place until the early nineteenth century and which, until the early eighteenth century, had mainly relied on Christian Europe for the printing of books in Arabic and Turkish. Arab Christians and missionaries played a role alongside Muslims in the new industry. In the Ottoman Empire, there were private and semipublic libraries that also contained a number of European titles. But in the two centuries before the Turkish Republic switched to a Latin script, only twenty thousand books—many in very small editions—were published in the whole Ottoman and post-Ottoman territory. As a result of this small scale of publishing activity, public libraries developed there later and more slowly than in East Asia.[24]

Museums

The museum, too, owes its still-vital role to the nineteenth century. Despite many pedagogical innovations, there is a tendency for museums to keep returning to the dispositions and agendas of the nineteenth century. The whole range with which we are familiar today developed during that time: from art collections to ethnographic departments to science and technology museums. The prince's collection, which had already been accessible to his subjects at times, became the public museum in the age of revolution.

The *art museum* united a number of elements: the idea of the autonomy of art, first formulated by Johann Joachim Winckelmann; the "value" of the artwork over and above its material craft character; and the "ideal of an aesthetic community" in which artists participated, along with experts, knowledgeable laymen, and, in the best case, a princely sponsor (such as King Ludwig I of Bavaria).[25] The museum flourished as the public grew increasingly differentiated. Soon it was even being asked whether art should belong to the state or the prince—a sensitive issue in the nineteenth century, since the French Revolution had set a radical precedent by nationalizing private art treasures and making it possible for the Louvre to become Europe's first public museum.

Things looked different in the United States, where it was mainly the munificence of the rich and superrich, in what Mark Twain called the "gilded age," that impelled the construction of museums from the 1870s on. Many of the buildings received joint public-private funding, but the actual works of art were mostly purchased by private individuals. America had few older works on its soil, and so its collections took shape in close symbiosis with the developing art market on both sides of the Atlantic. It was the same market that fed the creation of *new* collections in Europe.

The monumental style of the museum buildings (Alte Pinakothek in Munich, Kunsthistorisches Museum in Vienna, Victoria and Albert in London) commanded ever greater attention in the cityscape. Since palaces were now rarely built in the cities, only opera houses, city halls, railway stations, and parliament buildings—for example, the neo-Gothic Houses of Parliament on the Thames (1836–52) or the parliament buildings in Budapest and Ottawa—could compete with the new museums. Nationalism, too, enlisted art for its cause. Many of the trophies that Napoleon had carried off to Paris were jubilantly repatriated after 1815—the Louvre lost roughly four-fifths of its holdings—and required prestigious places in their home countries for their display. Painters tackled historical subjects with a national resonance, and national galleries in many European countries are still adorned with huge canvases from the high point of this trend in the middle decades of the century.

The exterior and interior design of museums gave material form to an educational program that for the first time was in the hands of professionals—of art historians and learned curators. Connoisseurs had for centuries been devising such agendas for themselves and their circles in Europe, China, the Islamic world, and elsewhere: we need only think of Johann Wolfgang von Goethe and his private collections of art and natural objects. Now the rise of experts turned the museum into a place for guided walks through art history. Credibility, authority, and expertise facilitated the elevation of museums to unprecedented heights of prestige.[26] State museums for *contemporary* art, such as the Musée du Luxembourg in Paris, gave artists a further stimulus to earn public support and the fame that came with it. The museum did not only preserve and "museumize" objects, in the sense of separating art from life. It also presented something new.

Historical museums were based on a premise different from displaying collections of ancient relics. The first museum of this kind, the Musée des Monuments Français dating from 1791, grouped a chronological series of statues, tombs, and portraits of persons whom its founder, Alexandre Lenoir, considered to have been important in the life of the nation.[27] Beginning with the Napoleonic Wars, new collections with a historical focus were designated as national museums in many European countries, early on in Denmark, Sweden, and Hungary. In Norway and Finland, national collections predated independent statehood and contributed to the visibility of nationalist movements. In Britain, there was no national history museum; the British Museum was meant to encompass "civilization" in the widest possible sense. However, Parliament established the National Portrait Gallery in 1856, with the aim of strengthening national and imperial sentiment. When three imperial museums were simultaneously inaugurated in Japan in 1889, the problem resembled that of Hungary seventy years before: there were no ruler's collections, and objects had to be acquired from many different sources.[28] Painted scenes from the nation's heroic past could stand in for missing artifacts.

The historical museum proper rested upon a new understanding of "historical objects." It was not enough that they should be "old"; they had to have a significance that spontaneously communicated itself to the beholder, and they had to be both worthy and necessary objects to preserve. In Germany, where "fatherland heritage" associations were founded in numerous places after 1815, it took many years to advance toward a national museum. A decision to create one was finally made in 1852, and a Germanisches (not "Deutsches"!) National-museum subsequently came into being in Nuremberg, in a spirit of gushing patriotism and with a heavy emphasis on the Middle Ages.[29] No thought was ever given to a central museum in the capital, even after the founding of the German Reich in 1871.

In Asia and Africa, historical museums usually emerged only after a country won its political independence. Meanwhile, a large part of the indigenous art treasures, manuscripts, and archaeological remains often disappeared into the museums of the colonial metropolises.[30] In Egypt the outflow had already begun with the French invasion of 1798. Muhammad Ali, as the nominally Ottoman viceroy of Egypt from 1805 to 1848, did impose a ban on exports of antiquities in 1835, but he was himself extremely generous in giving them away. The Egyptian Museum in Cairo was essentially a private initiative on the part of the archaeologist Auguste Mariette, who had been appointed curator of antiquities in 1858. The Muslim potentates of the time were divided over the neo-pharaonic style that Mariette chose for the construction: the world of pagan mummies was alien to them, but they could see that the European enthusiasm for pre-Islamic antiquity was good for Egypt's reputation in the world.[31] For the museums in Istanbul (Constantinople),[32] it was important that in 1874 the Ottoman Empire established control over the division of finds from foreign-directed archaeological excavations. In China the huge decaying structures of the former imperial palace—the Forbidden City of hundreds of temples, halls, and pavilions—were designated a museum in 1925 and largely opened to the public. But only in 1958 did the state establish a national museum with a nationalist focus.

Ethnological museums were only intermittently associated with patriotic or nationalist strivings.[33] They first developed in the mid-nineteenth century, sometimes as the continuation of a princely cabinet of curiosities or a private scholar's collection. The Königliches Museum für Völkerkunde, founded in 1886 in Berlin, soon became known as the world's richest ethnological depository. German ethnological research was not a creation of colonialism but stemmed from an earlier, liberal-humanist tradition of cultural studies.[34] German travelers and ethnologists collected on every continent. The task of the museum was emphatically not to satisfy a crude appetite for the exotic and the sensational. The conversion of objects into scientific material was supposed to happen in the museum, which also served research purposes and helped to train new experts.[35] The ethnological museums displayed items that had come into the possession of Europeans by theft, or transactions akin to theft, and were not part of their national heritage.[36] The aim was to present the diversity of human life, but only in relation to "primitive peoples," as they were then known. Each museum was part of a newly developing world of collections and exhibitions. As in the case of art galleries, connoisseurs were soon able to survey items from all around the world. Museums competed with one another but were also elements in a global movement toward the representation of material culture. They had a subversive effect insofar as avant-garde artists were able to find inspiration in them. It was not necessary to travel to the South Seas, as Paul Gauguin did in 1891, to absorb the renewing energy of "the primitive."[37]

Not only objects but also human beings were dispatched to Europe and North America to demonstrate, for "scientific" as well as commercial purposes, the otherness and "savagery" of the non-Occidental. Toward the end of the nineteenth century, such human displays were an everyday entertainment in the metropolises of the West, and many smaller cities found space for mobile exhibitions. It was one peculiarity of this period of rapid cultural upheaval.[38] Such events had been very rare before 1850, and after the First World War they became subject to a slowly emerging humanitarian taboo. The commercial exhibition of nonwhites, and also of handicapped people, was everywhere outlawed in the twentieth century. Yet the *principle* of the ethnographic museum survived decolonization, its declared aim no longer being to spread knowledge of "primitive" lifestyles as objects but rather to preserve a common cultural heritage in a multiethnic world. The nineteenth-century type of museum was itself decolonized.

World Exhibitions

Another novelty of the nineteenth century was the world exhibition, the most salient combination of panoramic gaze with encyclopedic documentation.[39] It all started with the Great Exhibition of the Works of Industry of All Nations, in London's Hyde Park (1851), whose spectacular crystal palace, a glass-and-iron hall 600 meters long, has remained in the collective memory, although it burned to the ground in 1936 in its new location in the suburbs. The Great Exhibition was a creature of the railway age. Only the train made it possible to bring more than 100,000 exhibits and up to a million visitors from the provinces—a pointer to the "expo tourism" of the future. The rich symbolism of the event left a strong legacy: for some, it embodied the dawning age of world peace and social harmony; for others, Britain's economic and technological superiority; for others still, the triumph of imperial order over the chaos of barbarism. At the same time, the exhibition put forward an elaborate taxonomy of classes, divisions, and subdivisions that went far beyond the older classifications of natural history to unify nature, culture, and industry in one grand system. Ensconced in this was a dimension of temporal depth. For no opportunity was lost to demonstrate that humanity *as a whole* had not yet attained the same level of complete civilization.[40]

Numerous "world's fairs" and *expositions universelles* followed until 1914, each with its ideological agenda associated with a particular point in space and time: Paris (1855, 1867, 1878, 1889, 1900), Antwerp (1885, 1894), Barcelona (1888), Brussels (1897, 1910), Chicago (1893), Ghent (1913), London (1862, and the Colonial and Indian Exhibition of 1886), Liège (1905), Milan (1906), Melbourne (1880), Philadelphia (1876), Saint Louis (1904), Vienna (1873).

The two with the largest attendance were both in Paris: the Exposition Universelle of 1900 (fifty million visitors) and the Exposition Universelle of 1889, which left a landmark still visible today in the form of the Eiffel Tower. World exhibitions were events that conveyed a message; Philadelphia 1876, for instance, first alerted the world to the technological and industrial might of the United States. The aim was always to put the contemporary world on display: the most up-to-date achievements were the heart and soul of the exhibitions. This was not contradicted by the extensive spectacle of "alien" peoples and civilizations. These could be presented as exotica or as visible remnants of earlier stages of human development, at the same time providing evidence that the remotest areas and tribes in the world could be incorporated into the global knowledge-based order. The world's fairs symbolized more clearly than any other medium the universal pretensions of the Atlantic "West."

Encyclopedias

Great encyclopedias, as monumental shrines to what is known and worth knowing, are akin to archives, museums, and even world exhibitions; they are also memory hoards and cathedrals of knowledge. The *Encyclopaedia Britannica* (from 1771), the *Konversations-Lexikon* of Brockhaus in Germany (from 1796), and many similar publishing-house projects continued in new ways a rich tradition that had begun in the early modern period.[41] They grew over time, renewing themselves from edition to edition. Nationalists soon recognized the value of the encyclopedia as a harnessing of scientific energies, a cultural monument, and an international signal of self-confidence and cultural strength. With such reasons in mind, the historian and politician František Palacký publicly proposed the idea of a Czech encyclopedia; it came to fruition in a twenty-eight-volume work that appeared between 1888 and 1909, exceeded in size only by the *Encyclopaedia Britannica*.[42]

By the end of the century, all European countries plus the United States had at least one such multivolume encyclopedia claiming to be universal in scope—to gather the most up-to-date knowledge about all of the earth's regions, periods, and peoples. They were more than reference books or aids for middle-class people to hold their own in conversational and educational contexts. Their alphabetical listing dispensed with systematic coverage of a subject but allowed the material to be laid out in linear fashion. There must have been readers who spent years struggling to get through from A to Z. The most cohesive, and from today's vantage perhaps the most attractive, encyclopedic achievement of the nineteenth century was Pierre-Athanase Larousse's *Grand dictionnaire universel du XIXᵉ siècle*, which appeared in seventeen volumes between 1866 and 1876. Even though for years Larousse provided a small extra income to large sections of the indigent Parisian intelligentsia, he wrote many of the 24,146 pages in his own hand. He was a radical republican, a supporter of the Great Revolution, and an opponent of the Second Empire, but the authorities left him alone and no censor ventured to read the mammoth work. Larousse's aim was not to educate the bourgeoisie but to prepare "the people" for democracy; the volumes were printed on cheap paper and scantily illustrated to make them more affordable. No issue was too hot for him to handle.[43]

That encyclopedias could be perceived as subversive is apparent from the attempts of Sultan Abdülhamid II to keep them out of the Ottoman Empire. With a little skill, of course, it was possible to obtain one through the book trade, even in Turkey. Someone who managed this in the 1890s had previously translated 3,500 pages of crime novels—ironically for the pleasure of the sultan's court—in order to have the means to buy the seventeen-volume Larousse. Another enthusiast had a French encyclopedia smuggled bit by bit into the country in the regular letter mail.[44]

How does the other great encyclopedic tradition compare with these new European developments? Since the eleventh century at the latest, China had been putting together often quite extensive compilations of reprints and excerpts from older literature in every branch of knowledge; these encyclopedias (*leishu*) served not least to prepare candidates for the entrance examinations to qualify for the imperial civil service. Unlike in Europe, where a reference work organized alphabetically by keywords—the standard format after d'Alembert and Diderot's great collective *Encyclopédie* of 1751 to 1780—became the organon for public debate and a forum for scientific advancement, the Chinese encyclopedias served to codify a hallowed tradition of knowledge, adding no more than layers of supplementary notes. In the twentieth century, comprehensive Western-style works of reference began to be published in Chinese. The *leishu* genre disappeared.[45]

Only in the nineteenth century did European languages—which had often not been consciously appreciated until the Romantic period—acquire what had existed in China since the great dictionary commissioned by the Kangxi emperor around 1700: that is, a full inventory of all possibilities of written expression in a particular language. The brothers Jacob and Wilhelm Grimm, who embarked on such a project in 1838 with their *Deutsches Wörterbuch* (volume 1 appeared in 1854; the final volume in 1961), and James Murray, who did the same for English-speaking culture after taking over in 1879 as editor of the *Oxford English Dictionary*, were among the most admired cultural heroes of the age, and among those with the most lasting impact. Murray's network of readers and word collectors soon spanned the globe.[46]

How could these great stores of knowledge have such universal reach in what is often called the age of nationalism? The nineteenth century can be thought of today as global because that is how it thought of itself. The universality of libraries, exhibitions, and encyclopedias signaled a new phase in the development of the knowledge society in Europe. The most important theoretical currents of the time—positivism, historicism, evolutionism—shared a cumulative and critical conception of knowledge that went together with the idea of its public significance. Knowledge was supposed to be educative *and* useful. The new media made it possible to unite the traditional *and* the new. In no other civilization had the culture of scholarship developed in such a direction. In Japan and China among others, however, the educated elites were willing to play an active role in shaping the transfer of new European conceptions and the institutions associated with them. This transfer got under way in the last third of the nineteenth century, but in most places it became really noteworthy only after 1900. The nineteenth century was an age of well-nurtured memory. This is one of the reasons why it retains a strong presence in today's world. The collecting and exhibiting institutions that it created continue to prosper, without being tied to the goals set at the time when they were founded.

3   Observation, Description, Realism

Another obvious survival from the nineteenth century is the descriptions and analyses written by people living at the time. It is no privilege or peculiarity of the nineteenth century to have observed itself. Since Herodotus, Thucydides, and Aristotle, and since Confucius, Xunzi, and the old Indian state counselor Kautilya, thinkers in various civilizations have repeatedly attempted to understand their epoch in inner-worldly categories. The novelty in nineteenth-century Europe was that, over and above a normative political and social theory, branches of knowledge arose with the aim of *describing* the contemporary world and grasping the patterns and regularities beneath the surface of phenomena. Since Machiavelli, there had been no lack of attempts to investigate the true functioning of political and social life, and the best travel writers of the seventeenth century had already gained deep insights into non-European societies. In Europe itself, Montesquieu, Turgot, and the French physiocrats, as well as the eighteenth-century English, Scottish, and Italian economists and the German and Austrian cameralists and statisticians ("statistics" then included the compilation of nonnumerical facts), presented important accounts of real social conditions. They investigated state and society as they *were* (in their eyes), not as they thought they *ought to have been*.

"Factual investigation"—which Joseph A. Schumpeter contrasted to "theory" in his great history of economic thought—acquired new scope and significance in the nineteenth century,[47] when Europeans produced incomparably more self-observational and self-descriptive material than they had in previous centuries. New genres of social reportage and empirical inquiry came into being, as attention was directed at the living conditions of the lower classes. Both conservative and radical authors placed the bourgeoisie, from which they themselves often hailed, under a critical magnifying glass. For the most important analysts of political and social reality—one thinks of Thomas Robert Malthus, Georg Wilhelm Friedrich Hegel, Alexis de Tocqueville, John Stuart Mill, Karl Marx, Alfred Marshall, and the chief figures in the German "Historical School" of economics, including the early Max Weber—factual investigation was closely bound up with the theoretical quest for connections and correlations. The positivist bent typical of the philosophy of the period made a program out of just such a link.

Social Panorama and Social Reportage

A distinctive form in which precise observation found literary expression was the social panorama. On the eve of the French Revolution, Sébastien Mercier set the standard for this type of work with his *Tableau de Paris* (1782–88), a vast twelve-volume canvass of life in the metropolis. He does not philosophize about the city but, as he says, conducts *recherches* in and about it, looking behind the facades and self-conceptions. Mercier became "one of the greatest discoverers of a new field of attention."[48] Mercier's labor of differentiation brought the city to life as a gigantic social cosmos. Rétif de la Bretonne then took up Mercier's literary procedure in his *Nuits de Paris ou le spectateur nocturne* (1788), presenting the nocturnal counterworld of the capital in a narrative, fictional form.

In the following decades, social reportage shed many of its literary ambitions. Alexander von Humboldt's report on the slave island of Cuba, based on his trips there in 1800–1801 and 1804 and first published (in French) in 1831, was written in the detached tone of an academic researcher. He avoided any drama or sentimentality in his uncompromising critique of slavery, allowing the facts to speak all the more effectively for themselves.[49] In 1807 the medical doctor Francis Buchanan published an extremely detailed account of everyday life in the agrarian society of southern India, having been commissioned to do so by the East India Company, which ruled large parts of the Subcontinent at the time.[50] The first "modern" works of social reportage thus developed in the colonies, by combining the Enlightenment's sober "political report" (a genre with which Humboldt had been familiar as a student) with the ethnographer's gaze.

In 1845 the young manufacturer's son Friedrich Engels published his *The Condition of the Working-Class in England: From Personal Observation and Authentic Sources*, which, as he put it in the preface, described "proletarian conditions in their classical form."[51] For this he combined the features of a travel book about a distant land with those of the parliamentary "blue books," which are still today among the standard sources for nineteenth-century British social history. In particular, individual life stories add a graphic dimension to Engels's case for the prosecution. The writer and journalist Henry Mayhew followed this example for his four-volume encyclopedia of London life, based on twelve years of investigations and regular interviews, *London Labour and the London Poor* (1861–62). It "stood alone," the author proudly claimed, "as a photograph of life as actually spent by the lower classes of the Metropolis," a good part of it "from their own lips."[52] Frédéric Le Play, a mining engineer by training, began in the 1830s to study workers' living conditions in several European countries and vividly depicted a number of social groups ranging from Ural nomads to Sheffield cutlers to Austrian charcoal burners.[53] The wealthy Liverpool merchant and shipowner Charles Booth, driven by religious-philanthropic motives and a desire for political reforms, tried to achieve greater analytical clarity in his detailed descriptions of the London poor, which he published in 1889–91 after seventeen years of research. The third edition of his magnum opus, *Life and Labour of the People in London* (1902–3), stretched to seventeen volumes. Booth overwhelmed his readers with an abundance of precise data, abstaining from horror stories and sentimental effusion in his panorama of late-Victorian London. Unlike the impressionistic Mayhew, he employed statistical methods and a sophisticated model of social classes, distinguishing between types of poverty and coining the term "line of poverty" that is still current today. His work marked a step from social reportage toward empirical social survey.

Literary Realism

A close relative of reportage is the realist novel, one of the characteristic art forms of the nineteenth century. In its ambition to capture "real life," it does not simply reproduce it figuratively but probes for the social and psychological energies active within it.[54] Honoré de Balzac's *La Comédie humaine*, published between 1829 and 1854, undertook a sweeping dissection and diagnosis of French society at that time. Wolf Lepenies, in his great book on nineteenth-century sociology, saw "a little self-irony and a great deal of social awareness" in Balzac's description of himself as a "docteur ès sciences sociales"; and in the ninety-one novels and stories that make up the cycle, he found "a social system" and "an exact counterpart to that which Comte, the founder of the discipline, strove to achieve with his sociology."[55] Before there was a science of sociology (Comte coined the term in 1838) writers were the real specialists in the study of society, and later, too, they engaged in productive competition with sociologists. In the century from Jane Austen's *Sense and Sensibility* (1811) to Thomas Mann's *Buddenbrooks* (1901) and Maxim Gorki's *Mat'* (Mother, 1906–7), a chain of "social novels" tells us as much about moral standards, behavior, status distinctions, and material conditions as we know from the works of social scientists. James Fenimore Cooper and Henry James; Charles Dickens, George Eliot, and Anthony Trollope; Gustave Flaubert and Émile Zola; and Ivan Turgenev, Leo Tolstoy, and Theodor Fontane are among the most important witnesses to the history of nineteenth-century society, mores, and attitudes.

To what extent did the "realist" novel spread beyond its three main litera-tures—French, English, and Russian?[56] In some cultures it gained a foothold in the nineteenth century, in others only later or not at all. In the United States, after the end of the Civil War in 1865, it became the focus of opposition both to cultural conformism and to the destruction of social values by rampant in-dividualism. In Europe there are significant national literatures—the Italian or Hungarian, for example—in which social-realist narrative, as distinct from the historical or psychological novel, occupied a marginal position in the nineteenth century. On the other hand, lesser-known traditions contain novels in which the social problems of the time were given profound consideration. Directly influ-enced by Balzac, the Portuguese writer José Maria Eça de Queiros set out to offer a panorama of all layers of his society in the *Cenas de vida portuguesa* (Scenes from Portuguese life) cycle, but he completed only a little before his death, most notably a novel on salon life in Lisbon in the 1870s, *Os Maias* (The Maias, 1888). In Poland, Bolesław Prus's *Lalka* (The doll, 1887–89) drew an artistic portrait of social problems that was especially sharp on relations between the nobility and the bourgeoisie. Comparable for its place in Norwegian literature is Alexander Kielland's *Garmann og Worse* (1880)—a novel about a merchant family, laced with satirical touches, which influenced the young Thomas Mann when he was preparing to write *Buddenbrooks*. Alberto Blest Ganas's *Martín Rivas* (1862), the first Spanish American realist novel, followed the transformation of Chile from a patriarchal-agrarian order into a society shaped by capitalism. The novel *Max Havelaar*, a masterpiece in form and style, which Edouard Douwes Dekker pub-lished in 1860 under the pen name Multatuli, is considered the leading Dutch prose work of the nineteenth century. It is also of genuine importance for its un-flinching exposure of Dutch colonial policy in the East Indies, today's Indonesia. It had a great impact on the public and in Parliament, as a result of which some of the worst practices in the colony were discontinued.

In the dominions of the British Empire, a settler literature began to develop, but it was not until the twentieth century that the native population gained a hearing. The first description of South African conditions from within was Olive Schreiner's *Story of an African Farm* (1883). In Australia, nineteenth-century novels portrayed the lives of convicts: Marcus Clarke's *For the Term of His Natu-ral Life* (1870–72), based on actual events, is regarded as the classic work of social criticism in this field. Sara Jeanette Duncan took up the formation of a Canadian national consciousness in *The Imperialist* (1904).

Turning to China, we may say that the great Ming and early Qing tradition of the novel reached a climax in *Honglou meng* (Dream of the red chamber), a family saga that circulated only in manuscript during the lifetime of its au-thor, Cao Xueqin (1715–64). Since it first appeared in print in 1792, it has been one of China's most popular novels. The nineteenth century added little to it. The changes that came with the invasion by the West crystallized only later in novelistic forms. The great Chinese novel of the Taiping Revolution, or the one dealing with the Christian missionary challenge, was never written. The first one to face up to the new conditions was Han Bangqing's *Haishang hua liezhuan* (Exemplary biographies of flowers in Shanghai, 1894), set in the milieu of courtesans and their clients in the mixed Sino-Western society of Shanghai. Shortly after the turn of the century and the watershed of the Boxer Rebellion, novels began to appear that painted contemporary life in the darkest colors. The best-known of these, by Wu Woyao, the most productive novelist of the period, bears the eloquent title *Henhai* (Sea of woe, 1905).[57] On the whole, the Chinese novel of social criticism was not an import from the West but built on a prose tradition that had arisen independently of European influence in the sixteenth century. But it did not play a leading role among literary genres comparable to that of the realist novel in Europe until the thirties of the twentieth century.

The hierarchy of literary genres was different in Japan. Here, the prose novel reached an extraordinary perfection as far back as the eleventh century, in the works of court ladies, most notably, Murasaki Shikibu's *Genji monogatari* (Tale of Genji). During the Tokugawa period, however, lyrical verse and drama were more highly regarded. And with the opening to the West—especially after 1868, which is seen as the birth year of modern Japanese literature—national genres of narrative gave way to Western forms much more quickly than in China. The first modern Japanese novel, written in a colloquial style and thus also accessible to less-educated readers, was Futabatei Shimei's *Ukigomo* (Floating clouds, 1885–86). Despite, or because of, Japan's victory in its war of 1894–95 with China, the inner contradictions of modernization came increasingly to the fore. Many writers tackled socially critical themes but, for the most part, restricted themselves to the sphere of the family and private life. The panoramic vision of a Balzac, Zola, or Dickens was not in evidence among Japanese writers during the late Meiji period.[58]

Travel Writing

Alongside the realist novel, travel literature was an indispensable source of knowledge about the world for the nineteenth century, as it is today for historians of the period. Yet its importance was less than in the early modern age, when there had often been no other possibility of informing oneself about remote corners of the earth. In the nineteenth century, too, some travelogues achieved high status both in world literature and as factual sources. Outstanding examples are: Madame de Staël's hugely influential book on Germany (*De l'Allemagne*, 1810); Alexander von Humboldt's account of his travels in South America from 1799 to 1804; the journals of the expedition that President Jefferson commissioned Meriwether Lewis and William Clark to make across North America between May 1804 and September 1806; the report by the young French jurist Alexis de Tocqueville on his travels in the United States in 1831–32; Charles Darwin's book on his trip to the Galapagos Islands in 1831–36; Heinrich Barth's impressions from North and Central Africa during his period in British service from 1849 to 1855; Sir Richard Burton's narrative of his visit to Mecca and Medina in 1853; Franz

Junghuhn's encyclopedic account of the island of Java in the 1850s; the report by the Westphalian baron August von Haxthausen of a 10,000-kilometer trip he made through Russia on horseback, a book that, when published in 1847–52, opened the eyes of the country's urban intellectuals[59] for the first time to their peasant fellow-citizens; and Ferdinand Baron von Richthofen's five-volume work on China (1877–1912), based on his travels there in 1862–72, when few Europeans had yet seen the inland provinces.[60] What these texts have in common is the excitement of discovery, which would disappear in the next generation of travelers. All the authors (with the exception of the rather shady adventurer Burton) were united in their strong sense of duty to the cause of science. Not a few of their great journeys were youthful projects laying the basis for an academic or public career. More than ever before or since, in the century after Humboldt's emblematic trip to America, firsthand travels conferred an aura of scientific authority.

Unlike the early modern period, the nineteenth century witnessed a growing number of visitors to Europe from overseas who wrote back home about what they saw: Chinese emissaries, Japanese ministers, Indian and North African scholars, a king from what is today Botswana, even Oriental monarchs such as the sultan of the Ottoman Empire (Abdülaziz was the first Turkish head of state to visit Christian Europe, on the occasion of the Paris World Exhibition in 1867); Shah Nasir al-Din of Iran, who traveled three times to Europe (in 1872, 1878 and 1889) and kept a journal or had one kept; and the Siamese king Chulalongkorn, an unusually keen observer, who first visited Europe in 1897. Asian scholars such as Ram Mohan Roy from Bengal, who went to England in 1831 and died in Bristol in 1833, or the low-ranking official Li Gui, the first Chinese ever to make a trip around the world (in 1876–77), influenced how the West was perceived in their homeland.[61] A sizable literature of travel and observation also began to appear within East Asia itself. Fu Yunlong, who was sent by the Chinese government to Japan and America in 1887–89 and later headed a department at the war ministry, composed a country report on Japan in thirty volumes. Japanese reports from the East Asian mainland were no less thorough.[62]

The largest group of travelers to Europe were, of course, Americans: some, from both North and South America, were searching for the roots of their own culture; others, most prominently Mark Twain, went in the assurance of belonging to a younger and better world. In the second half of the nineteenth century, it was no longer necessary for Europeans to fabricate "foreign mirrors," in the manner of Montesquieu's *Lettres persanes* (1721), if they wanted to see themselves distorted beyond recognition or for the purpose of self-satire. The rest of the world began to articulate what it was absorbing from Europe. This was also true in the colonies—and earliest of all in British India, whose educated classes were the most influenced by Europe, and which had the most dynamic political and literary life.[63] In the nineteenth century, Asian reactions to Europe did not yet add up to a systematic "Occidentalism" that could be compared with Europe's budding "Orientalism." Only Japan had a basis for this in its "Dutch studies" (*rangaku*), which since the eighteenth century had involved observation of Dutch traders in Nagasaki and scrutiny of the literature they brought along with them.[64] When North American geographers began to concern themselves with Europe, they did so with the instruments of *European* science.

Measuring and Mapping

In the nineteenth century, research travelers, academic geographers, and other such writers still formed the largest group of European collectors of information about the wider world. Not surprisingly, their activity was ever more tightly linked to the imperial and colonial projects of the Great Powers.[65] One side of geography involved a global discourse that was increasingly imperial—although admittedly it could also be directed against European conquest, as in the writings of Carl Ritter and Alexander von Humboldt in the first half of the century. Its other side was a great success story of the eighteenth and nineteenth centuries, since the exact description of natural and social reality gave Europe one of its decisive advantages over other civilizations. However irrational or demented the ideas might have been that sometimes drove researchers in "the field," the sum of their activity brought a colossal gain in exact knowledge about the world.[66] Nowhere was this plainer than in cartography.[67] The measuring and mapping of vast areas of land and water was one of the great collective projects of modern science, closely bound up with European conquest of the oceans of the world. It began with the Spanish and Portuguese, continued after 1700 with the Dutch plan to map the whole earth, and profited later in the eighteenth century from the growing sophistication of measurement techniques and the global expansion of European sea travel. By the 1880s even "darkest Africa," south of the Sahara, could be represented in broad outline.

If the eighteenth century was a time of revolution in measurement and mapping techniques, the nineteenth was the age of their global application. As a result of these persistent efforts, it became possible to grasp the world in its entirety. The maps produced around the end of the century were scarcely surpassed until the advent of satellite cartography and computerized mapping. Non-Westerners were also involved in many European cartographic operations, as informants, helpers, advisers, and scientific partners. Most of them occupied a formally subordinate position, but without their local knowledge it would have been impossible to fill in all the gaps.

Outside the West, the Japanese were the first (and, for a long time, the only) nation to undertake measurement and mapping at European levels of precision. This was initially a private initiative, spurred by the alarming appearance of Russian ships off the coast in the 1790s. Only in the Meiji period after 1868 did cartography become a state-sponsored project on a grand scale.[68] Of all the non-European traditions, the Chinese might have seemed the likeliest to produce a "modern" geography. All district officials were required to give empirically detailed reports on the makeup of their area. In the same way that philologists developed a new precision in the verification of traditional texts, geographers fell in with the empirically oriented *kaozheng* scholarship that became dominant in the late seventeenth century.[69] However, nineteenth-century Chinese geography did not benefit from the large government commissions so characteristic of Europe;[70] it could not free itself from the narrowly practical goals of administration or from its subordination to the more prestigious discipline of historiography. Indeed, it forgot the innovations in measurement and mapping that had reached China with the Jesuits in the seventeenth century. More recently, from the 1920s on, Chinese geography was alive to older indigenous traditions, but at the same time it took in key elements from the scientific geography developed in the West. It was therefore from the beginning a hybrid discourse.[71]

Sociology

Geography was a globally sighted but locally rooted science. As economic geography it accompanied the industrialization process in America and North America; as colonial geography it consorted with the West's land-grabbing expansion. An even more important organ of self-observation was the newly emerging social sciences. Their theoretically grounded questioning took them beyond social reportage, but they never lost touch with the empirical description of reality—a reference that was already apparent in economics *before* Adam Smith's epoch-making work on the wealth of nations (1776). Tendencies toward abstract model building began to appear in 1817 with David Ricardo, but their influence became dominant only after 1870, as mathematical theories of subjective utility and market equilibrium developed more or less simultaneously in Austria, Switzerland, and Great Britain. At the same time, especially in Germany, *Nationalökonomie* continued to flourish as a largely descriptive study of economic patterns and changes past and present. This trend took organizational shape in 1872 with the foundation of the Verein für Socialpolitik (Social Policy Association); over the years it would make an enormous contribution to the knowledge of society.

Sociology, whose founding fathers were Auguste Comte in France and Herbert Spencer in Britain, thought of itself mainly as a theoretical discipline. In Germany, the bastion of historicism and source criticism, it had a less speculative and all-embracing cast than in France or Britain, with a particularly close relationship to history since the days of Lorenz von Stein, the author of a vast history of social and political movements in France (1842) and the first social scientist in the German-speaking world. Toward the end of the century, sociology everywhere, including in the United States, annexed the field of empirical social studies that had previously belonged to state-sponsored surveys and private reformers such as Charles Booth. In Britain the reform-oriented London School of Economics, founded in 1895, marked the breakthrough to a fusion of theory with factual research, even if "sociology" only acquired its separate professors in 1907, and the professionalization of the subject proceeded more slowly than on the Continent. In the United States, the creation of the first sociology department, at the University of Chicago in 1892, was a similar turning point.[72]

Only in the 1890s did academic sociology begin to contribute on a large scale to the empirical study of contemporary societies. Only then did the methodical self-observation of advanced societies enter a process of institutionalization that has continued to this day. Sociology spread rapidly, at least in East Asia, where influences converged from Europe and America. A chair in sociology was already created in 1893 at the Imperial University in Tokyo, just a few years after a Japanese equivalent had been found for the European term "society."[73] In China, sociology was at first taught by foreigners, who contemplated such topics as municipal guilds, relations within the ruling Manchu clan, and the structure of northern Chinese agrarian society. In 1915, when Émile Durkheim, Max Weber, and Georg Simmel were still flourishing, the first sociological account of Chinese society by Chinese authors appeared in print; and in the same year, the subject started to be taught by Chinese lecturers at a few universities. Chinese sociologists subsequently developed numerous analyses of contemporary society, with an increasingly Marxist orientation.[74]

Never before the nineteenth century had societies created such space for ongoing institutional self-observation. Many earlier civilizations might be said to have produced descriptions of their respective societies that were at the same time interpretations of them. Important insights into what would later be known as "sociological" contexts were already achieved in the eighteenth century—for example, the model of society as a process of circulation, developed by the French doctor François Quesnay, or the multifarious "science of man" in the Scottish, English, and French Enlightenment. Yet it was not until after 1830, in the context of accelerated social change in Europe, that a permanent social-scientific discourse developed among intellectuals and philanthropic reformers, and only in the closing years of the century that it took root in the universities. This was peculiar to Europe. The social sciences, however, soon proved to be a successful export. Political economy found much interest in Japan and India, and its pioneers—especially Adam Smith and John Stuart Mill—were among the European authors most widely translated in other parts of the world.[75] In its more radical variants, political economy could appear as a critique of colonialism: not just Indians opposed the forcible "drain of wealth" from the Subcontinent, as the civil servant and economic historian Romesh Chunder Dutt termed it, but European or Japanese analysts of imperialism were also drawing this conclusion around the turn of the century.

4 Numbers

Censuses

The nineteenth century was the founding age of *modern* statistics: no longer just the more or less haphazard compilation of data but their rigorously methodical collection and mathematical processing. The state increasingly took over these tasks, which were becoming so complex that only the state had the organizational capacity to handle them. In the second half of the nineteenth century, statistics became what it is today: the most important tool for the constant self-monitoring of society.

Its prototype was the census. Authorities began to count their subjects long ago. For military and fiscal reasons, the numbers of households, individuals, and livestock were recorded. Countries with a large area rarely accomplished this in full; the figures often have gaps, or have simply not survived the passage of time. Historical geographers, who rely on such sources, have a hard time of it, but they must decide in each case whether the data resulting from a census are usable. Europe or "the West" cannot simply claim to have been the first in this respect. The earliest data from China that are today considered usable come from the years 1368 to 1398, when the first emperor of the Ming dynasty ordered a census following the restoration of a central government.[76] In Japan, from 1671 all lords were obliged to compile annual population registers for their territory; the first countrywide census useful for demographic research dates from 1721, but the abundance of *local* data that have survived to this day tell us even more about premodern Japan.[77] The Ottoman authorities usually conducted a population survey of newly conquered territories: it was important to have an accurate picture, if only for fiscal and military reasons. Ethnicity was not recorded, but everyone did have to declare their religious affiliation, since non-Muslim inhabitants were subjected to a head tax until 1855. The first general census of the male population in the European and Anatolian provinces of the empire, which took place between 1828 and 1831, marked the beginning of the history of Ottoman-Turkish demography.[78] In the case of Egypt, then nominally a province of the Empire, the census of 1848 is considered reasonably reliable.

The pioneer in Europe was Sweden, where the first national census dates from 1755. In 1787 the great Enlightenment monarch Carlos III ordered one to be held in Spain, and its methods were so advanced that it has sometimes been described as Europe's first "modern" census.[79] Then, around the turn of the century, modernity came to population statistics in all the major countries of the continent.[80] This presupposed regularity, institutionalization, and verifiable procedures. Institutionally, four elements were involved: (1) a statistical office, usually under the interior ministry, which collected, evaluated, and published data; (2) a permanent statistical commission of senior civil servants, with the task of ensuring central coordination; (3) private associations of doctors, professors, engineers, and office-holders, operating as lobbies for improvements in statistics; and (4) municipal statistics offices (which became a normal feature only in the second half of the century).

These four components did not appear all at once: it took decades for them to be introduced throughout Europe. Britain (first census in 1801) and revolutionary-Napoleonic France got the ball rolling. In 1810, statistical offices were created simultaneously in Prussia and Austria. It was much harder to collect near-complete data in the multiethnic empires than in small countries such as Belgium or the Netherlands, whose statistical services were considered exemplary after 1830. By 1870 or thereabouts, modern statistical bureaus existed everywhere in Europe, and conferences of the International Statistical Congress (1853–78) set quality standards that no country could evade. In the United States, censuses of a reasonably modern character had been taking place since 1790. The sixth national count (1840), though full of gaps and other defects, was held up everywhere as one of the great achievements of the American nation.[81]

It was one of the most demanding tasks imaginable to produce population statistics for India. Unlike in China, Japan, or Burma, precolonial governments seem to have bothered little there about the number of their subjects, but the British soon turned to the work of empirical description. This meant first of all collecting information about the major cities: their living conditions, political significance, and number of inhabitants.[82] By 1820 the first, rather skimpy, gazetteer was available, containing neither an approximately correct figure for the total population nor an account of social structures in the Subcontinent. European measures of quantification could not be directly transferred. After all, what was meant in India by "family," "household," or "village"? At what age was the dividing line between "adult" and "child"? Was a "caste" always identical with a certain occupation? And, if not, how was caste membership to be understood? There were decades of experimentation, during which the population was counted in various provinces with different degrees of exactitude. Only after 1881 did regular and better-organized all-India censuses yield satisfactory results,[83] though at the price of more rigid categories; statistics did not simply reflect reality but imprinted its own rules on it. Thus, whereas a census in the British Isles never asked about religious affiliation, the colonial authorities treated it as central to the classification of society, thereby boosting the significance of "communities" that would later be so important in Indian politics. The demographers of British India and their ethnographic advisers were obsessed with the ranking of castes; race theories typical of the age also entered the picture, so that the 1901 census, reputed to be particularly scientific, rested on the assumption that India's social hierarchy reflected differing degrees of racial purity. An ambitious attempt at a fully integrated census of the entire British Empire was abandoned at the outbreak of the First World War.[84]

Modern censuses are not simply a matter of head counting. Scandinavia was the first country to include aspects that would eventually become a matter of course: births (divided into legitimate and illegitimate), age of marriage, and age of death. Whether such data were available depended on what the churches and secular authorities deemed worth registering. In the comparatively backward Catholic Philippines, for instance, patchy but revealing data may be found a long way back in parish registers. In general, the demographic data improve once marital status becomes a state-recognized civil matter. In a country such as China, where marriage remained a private affair, such information is lacking.

Statistics and National Politics

A census is public business, a matter for the authorities. As the state became
an organ through which society observed itself, the nineteenth century took
up and continued a number of older tendencies. In Central Europe it had been
the task of a special "science of public governance" (*Polizeywissenschaft*)—in
English-speaking countries, of "political arithmetic"—to gather data about the
present day. So what was new in the nineteenth century? Improved observation
techniques, institutions to preserve the results, a more objective approach. It was
the nineteenth century that first thought in terms of "populations." The "new"
mathematical statistics, which was fully developed by 1890, was an expression
of such thinking. As early as 1825, the Belgian astronomer and mathematician
Lambert-Adolphe Quetelet tried to identify averages and social regularities in
the numerical material, and to correlate various social facts with one another. He
was searching for a "social physics" beyond mere numbers and came up with the
statistical "average citizen" (*l'homme moyen*), one of the great mythical figures
of the modern age.[85] Quetelet was among the most influential thinkers of the
nineteenth century.

In the 1830s and 1840s, several European countries were gripped by a passion
for statistics. It made things visible that had previously been hidden or taken for
granted. The poor appeared as a social entity only when they were counted, and
the resulting emergence of "poverty" as an abstract concept helped to arouse a
moral commitment. Statistical societies and journals were founded, and government offices were called into being to gather, evaluate, and store social data. Politics rested more than ever before on exact information. In France, the systematic and regular collection of data was instituted at the prefecture level in 1801.
Seeking to make deep inroads into civil society, the Napoleonic state needed
as much accurate information as possible about it.[86] In Britain too, despite its
much less developed regional bureaucracy, the parliamentary government made
extensive use of empirical facts about all manner of things—from sanitation in
workers' districts to the medical condition of soldiers in the army.[87] The collection of these was entrusted to ad hoc royal commissions, whose conclusions were
publicly available both to the government of the day and to its critics. In *Hard
Times* (1854), Charles Dickens poked fun at the type of hard-boiled positivist
who collected such data, in the person of Thomas Grandgrind. However, such
positivism not only generated the knowledge base for control of society but also
provided grist for the analytic mill of an anti-positivist opponent of the system
such as Karl Marx.

In the United States too, statistics acquired a major place in public life, perhaps even more so than in Britain or France. Full-scale social integration was
conceivable only in a statistical perspective; only numbers could have brought
home the unparalleled, and otherwise elusive, dimensions of the United States.
For similar reasons, statistics played an important role in the unification of Italy,

both in the imagination directed to the nation's future and as special knowledge at the disposal of new elites. No sooner had political unity been achieved than statistical surveys spread like wildfire; even liberals were interested in recording the country's population and resources and in monitoring the performance of lower authorities from the vantage point of central government. Italy, in a sense, was a creature of statistics.[88]

The nineteenth century can be seen as the century of counting and measuring. The idea of an all-embracing taxonomy now grew into a belief that the power of number—of statistical processing or even "social mathematics," as the Marquis de Condorcet, a bright star of the late Enlightenment, put it—could open up truth itself to human reason. It was in the nineteenth century that societies measured themselves for the first time and archived the results.

There is much to suggest that they sometimes went too far. In some countries, more statistical knowledge was produced than could be scientifically and administratively handled. Statistics became what it still is today: a form of political rhetoric. The categories that statisticians had to develop were reified in the hands of government bureaucracies. Categories that statistics made technically necessary—classes, strata, castes, ethnic groups—acquired the power to mold reality for administrative departments and, indeed, in society's perception of itself. Statistics had two faces: a tool for sociological description and explanation, and a powerful mechanism for stereotyping and labeling people. In both respects, it became a central element of the social imaginary. Nowhere was the second face more apparent than in the colonial world. Where social relations were much more difficult to understand than in close and familiar surroundings, many European observers and administrators succumbed to the false allure of objectivity and exactitude—when they did not simply come to grief because of the practical obstacles involved in pinning down mobile populations.

5 News

The Press and Its Freedom

The nineteenth-century press ranged even wider than the realist novel, statistics, and empirical descriptions of society. Weekly or daily newspapers, as well as periodicals and magazines, opened communicative spaces of every conceivable dimension, from the local sheet to the London *Times*, which by the end of the century was bringing news from all around the world while delivering its papers to be read on every continent. The conditions for political communication changed as soon as the press took root. The demand for freedom of the press, hence for the opportunity to voice opinions without fear of punishment, became a transformative impulse in every country, creating for the first time something like a public space, where citizens exchanged ideas and asserted the right to be kept informed. The founding fathers of the United States thought that only well-informed members of the community would be capable of fulfilling their civic responsibility—a view whose optimistic assumptions about the rise of the popular press few would share.[89]

It is also possible to see the space opened up by the press in a different light, as a new level of society's reflection of itself. The distinctions between different types of printed media were fluid. In the early decades of the century, short "pamphlets" played an important role and evaded censorship more successfully than books or newspapers. The hazy boundaries were apparent in the fact that many novels, including most of those by Dickens, originally appeared in serial form in magazines.

The special characteristics of the newspaper were: (1) publication at regular intervals; (2) production by an editorial team; (3) division into separate departments and fields; (4) reporting that went outside the regional and social horizon of its readers; (5) a rise in topicality, which in Germany meant that the proportion of news less than a day old rose from 11 percent in 1856 to 95 percent in 1906;[90] (6) increasingly industrial production, based on the latest technology, which required considerable capital investment for a mass circulation press; and (7) a fluctuating market that depended on daily decisions by customers at the newsstand, except in the case of subscribers.

The newspaper established readers as politically mature subjects while at the same time mobilizing them for certain ends. The period from the middle of the nineteenth century until the end of the 1920s (when radio began to reach a wide public in Europe and America) was an age when the press had no rival in the world of media. Since the press was not as concentrated as it would soon become, it may be said of the United States, for example, that the number and variety of printed news sources was greater at the turn of the twentieth century than it has ever been before or since. By then, the press "tycoon" was a sui generis political force in countries such as the United States, Great Britain, and Australia.

The golden age of the press could begin only when there was freedom of the press. In countries like Germany, where censorship did not relax as production technology advanced, "family sheets" and illustrated magazines had an easier time of it politically than newspapers. The Karlsbad Decrees of 1819 established highly repressive press laws in the German Confederation, and although the censors often found it too difficult to apply them rigorously, they were a daily nuisance for publishers and journalists. But the Karlsbad system did not survive the revolution of 1848; pre-publication censorship was anyway no longer necessary, since the state apparatuses had other means to control the printed word. Police and the courts took over the job of the censor, whose passing began to be almost regretted. The first German state to introduce full freedom of the press was the Kingdom of Württemberg, in 1864, but it was another ten years before an imperial press law ended preventive censorship for good throughout the Reich. From then on, publications that offended the authorities had to fear harassment but no longer suppression. In his later battle against Catholics and especially

Social Democrats, however, Otto von Bismarck did not refrain from attacking press freedom.[91] Opposition journalists were never safe from prosecution, while behind the scenes the chancellor used sectors of the conservative press for his own ends. Only after 1890 did the bourgeois press—things were not so easy for the Socialists—enjoy the freedom that had long been taken for granted in the English-speaking world.[92]

The special place of countries marked by British culture is nowhere more evident than in respect to freedom of the press. John Milton's *Areopagitica*, which already called in 1644 for an end to the system of advance licensing of publications, would have a lasting influence. In the United States, the First Amendment (ratified in 1791) forbade Congress to pass laws restricting free speech or press freedom. Of course this remained open to interpretation, and from 1798 the question repeatedly arose as to when it was overridden by the common law on "seditious libel," notorious for the looseness with which it could be applied to protect "public figures."[93] But on the whole, nineteenth-century America was a country with a free press. The idea of the press as an institutional counterweight to the government, a "fourth estate," became ever more deeply rooted in its political culture. In Britain, after 1695, the state no longer had the legal right to act against critical publications, although a special stamp duty limited their distribution until its last vestiges were abolished in 1855.

A dynamic press followed quickly in Canada, Australia, and New Zealand. In Canada, a country with a population of 4.3 million, some thirty million newspapers were sent through the mail in 1880.[94] An English visitor to Melbourne in the late 1850s was amazed to see a newspaper in every doorway when she went for a morning stroll. With no real interference from the authorities, the press played an especially important role in the development of a democratic civil society in sparsely populated Australia. Newspapers were filled with news from the heart of the empire, but they also served to give voices from "down under" a presence in London. The press soon became a political force to be reckoned with in Australia.[95]

In each of these cases it is hard to say when press censorship was legally or constitutionally abolished, and harder still to determine when administrative obstruction in the form of sureties, police searches, confiscation, threats of prosecution, and so on actually fell below a minimum threshold and became merely sporadic. Punitive action *after* publication always disappeared later than preventive censorship. In countries like Spain, where the press stood on such shaky foundations that journalists could not operate without taking a second job provided by political sponsors, the most liberal press law was of little avail.[96] In continental Europe, Norway was the first country to have a free press (from 1814); Belgium and Switzerland joined it around 1830, and Sweden, Denmark, and the Netherlands by 1848.[97] It is true that in the Declaration of Human and Civil Rights (1789), the French revolutionaries proclaimed "the free communication of ideas and opinions" to be "one of the most precious of the rights of man" (article 11), but this meant little in practice. Napoleon III's Second Empire (1851–70) still went to great pains to control and depoliticize the press, although in the 1860s, as it moved toward a semi-parliamentary regime, it considerably loosened the reins.[98] After a period of state repression bordering on terror that followed the suppression of the Paris Commune (1871), the Third Republic finally drew a line in 1878 and made it possible to speak again of a functioning public sphere. In 1881 an exemplary press law ushered in a *belle époque*, in which the political press attained a quality and diversity never to be repeated after 1914, both flourishing economically and exercising great influence in the affairs of the republic.[99] Until the turn of 1881, a deeply divided France had witnessed a struggle over press freedom more severe than in any other country in Europe.

In the Habsburg Monarchy, a more liberal climate of opinion began to develop in the 1860s, but press confiscations were a regular recurrence down to the First World War. A further complication was the existence of a press in the many different languages of the empire. Being accused of high treason was always the risk one took for making statements that could be construed as separatist, the Czech press being especially exposed to this.[100] In the Tsarist Empire, more liberal legislation adopted in 1865 made it possible for a comparatively free press to develop, despite all the censorship and repression.[101] The measure of comparison is here the Russian situation before the reform, not the vitality and lack of restriction of the press in the United States, Britain, or Scandinavia at that time. But with this reform, Russia followed the Western European model of a transition from preventive censorship to legal and administrative control *after* publication. In the aftermath of the Revolution of 1905, the press was nominally as free as in Russia as in the West, but it remained subject to official harassment beyond what was normal in Germany or Austria. It was by no means the case that the whole of Europe was a haven of press freedom in an otherwise backward world.

Newspapers in Asia and Africa

The daily newspaper was a European-American invention that soon spread beyond the North Atlantic area. Where the colonial system offered the opportunities, indigenous educated classes soon took advantage of them to make their voices heard in both local languages and those of the colonial rulers. British India was again an especially clear case in point. Here the press developed in fairly close synchrony with Europe's, one difference being that the printing press appeared in India at the same time as the newspaper: a *double* communications revolution. The first English-language paper came out in 1780 in Calcutta; the first in an Indian language (Bengali) in 1818. The Gujarati-language *Bombay Samachar*, founded in 1822, is still published today (as *Mumbai Samachar*). Soon there appeared English-language papers produced by Indians. Lithographic technology, which soon spread to smaller cities, was common to all. Another reason why the new medium was taken up so quickly, eagerly, and successfully in India was that the country could build upon a rich culture of written reporting.[102] The years from 1835 to 1857 were a time of vibrant progress, in liberal conditions that people in the German Confederation could only dream of at that time. After the Great Rebellion of 1857–58, the colonial government reacted more heavy-handedly to Indian criticisms and tightened its control of the press, but this never escalated into a muzzling of public opinion. The viceroys valued the press both as a means of communicating with the population and as a source that relayed information and attitudes from Indian society. Together with the English legal tradition that generally tied the hands of the state, these pragmatic considerations explain nineteenth-century India's significance as a country with a highly developed press system. The same cannot be said of the colonies of other European powers. Although the Netherlands was at least as democratic a country as Britain, it was much more fearful of liberalizing press controls and public life in general in the East Indies than the British Raj was in India.[103]

Things were different again in China, whose old printing tradition led to the development of a nationally independent press. *Jingbao* (News from the capital, or the *Peking Gazette*, as it was known in the West) began to appear as early as 1730. In fact, there had been a precursor that for the past thousand years published palace reports, edicts, and petitions. This court gazette existed until the end of the empire in 1911, having adopted newspaper-like features in 1900 and naming itself *Guanbao* (News for officials). The modern newspaper was introduced by Protestant missionaries, who operated first from abroad (Malacca, Batavia/Jakarta) and, after China opened up in 1842, in Hong Kong, Canton (Guangzhou), and Shanghai, addressing their potential converts and protégés directly in Chinese. Their sheets, though very short on political news, brought not only Christian propaganda but also general cultural information about the West. In the treaty ports, subject to foreign law as they developed successively after the end of the Opium War in 1842—and especially in Shanghai, the largest of them by far—a foreign press soon took off and flourished. It reflected the views and interests of European and American merchants in the treaty ports but was generally very well informed about what was happening in China. A private Chinese press, outside the control of the Chinese authorities, developed after 1861, again in coastal cities such as Shanghai and Tianjin and in the British crown colony of Hong Kong.

A paper like *Shenbao* (Shanghai daily news), which appeared from 1872 until 1949 and was run until 1909 as a Sino-British joint venture, could compare favorably at the turn of the century with serious and highly regarded European papers such as the *Berliner Tageblatt* (also, as it happens, launched in 1872). Nevertheless, before the Revolution of 1911 its daily circulation was never higher than ten thousand. It attempted, with some degree of success, to provide accurate news reports along the lines of the London *Times*, and converted old Chinese forms of political discourse and criticism of rulers into the kind of front-page articles that reached their peak of importance in both China and Britain in the late nineteenth century. Its educated readership, which soon spread well beyond the treaty ports, saw the new approach to reporting by *Shenbao* not as an alien import but as a reshaping of older ways of treating the major issues of the day.[104] General trends in the press would make themselves felt in China too. Complaints about the "Americanization" of the press were to be heard there as in Europe after the First World War.

After China's defeat in the Sino-Japanese war of 1894–95, a climate of agitation gripped its intellectuals across the political spectrum as they analyzed the country's acute crisis and future prospects. In Japan, by contrast, the war led to the patriotic mobilization of the reading public, fueled by a press that justified overseas ambitions and had seen its sales permanently rise by at least a quarter in the wake of the conflict.[105] The *critical* Chinese press, much of it published abroad or in the treaty ports, sold fewer copies than the large dailies and used a demanding style that made it inaccessible to a mass readership. But it played an extremely important role in the politicization of new "middle strata"—reform-minded journalists actually spoke of a "middle level of society" (*zhongdeng shehui*)—in the cities of the interior too.[106] The Chinese press took on a new polemical tone. But the imperial government did not grant the room for maneuver that the semi-free press of colonial India was able to enjoy. Until 1911 both the Chinese and English-language papers could only prosper in the coastal enclaves, under the protection of foreign laws. Chinese and foreign journalists worked closely together there, sharing a common interest in the problems of reform in China.[107]

In the Ottoman Empire, too, the 1870s witnessed hesitant steps toward a private press independent of the state apparatus. The first semiofficial weekly (in Arabic) was founded in 1861 and kept going until 1883.[108] Censorship continued, of course, and was even placed on a legal footing in 1867. Under Sultan Abdülhamid II, controls on public opinion became more oppressive and the printed media had to be very cautious. There were no liberal enclaves such as Hong Kong or Shanghai in China. Opposition newspapers and periodicals were printed in Paris, London, or Geneva and smuggled into the country in private correspondence.[109]

An exception in this respect was Egypt, only nominally part of the Ottoman Empire, where Khedive Ismail (r. 1863–79) tried to cultivate good terms with the press and skillfully used it for his own purposes. Ismail understood that a docile journalism that relied on official handouts was worthless; what he wanted were papers that appeared independent and could be manipulated behind the scenes. Domestic as well as foreign journalists received lavish gifts, while the British and French news agencies were quietly subsidized.[110] However, such relative liberalism also encouraged genuinely private initiatives. The most important was the founding of *Al-Ahram* in 1876 by two brothers of Catholic Lebanese origin,

Salim and Bishara Taqla. A daily since 1881, *Al-Ahram* made reliable and up-to-date information available from all over the world, together with a measure of critical commentary; no one could fail to see that the Taqla brothers were in favor of greater liberalism and against foreign intervention. Between 1877 and 1882, thirty political papers were appearing in Cairo and Alexandria, with a total print run of 24,000 a day (in 1881).[111] Apart from articles of their own, they also contained translations of material from European papers such as the London *Times* or *Le Débat*. On the eve of the British occupation in 1882, the Egyptian press presented a varied landscape in both Arabic and European languages. This remained the case in the subsequent period of de facto British rule (1882–1922): the spread of printing technology, rising literacy, a more professional journalism, and the liberal attitude of the British authorities combined to make the country an island of freedom of opinion in the Middle East. During the last quarter of the nineteenth century, the potential readership for newspapers, though still tiny, grew in size and opened up a public sphere for political argument. Furthermore, oral forms of dissemination made it possible to satisfy a hunger for news that was increasing at a faster rate than literacy. In the Ottoman Empire, only the Young Turk Revolution of 1908, which put an end to the sultan's autocracy, unleashed the forces of a press system rooted in civil society.[112]

Birth of the Mass-Circulation Press

Most innovations in the press came from the United States, as did the major developments in print technology. The first rotary press was built in Philadelphia in 1846. Between 1886 and 1890 a German immigrant in Baltimore, the watchmaker Ottmar Mergenthaler, finally solved the problem of slow throughput by means of a keyboard-operated "hot metal" linotype machine that represented the most important advance since Gutenberg's movable-type printing press.[113] There were also organizational breakthroughs that began in the United States and then crossed the Atlantic. Unheard-of sales figures were achieved with the birth of the East Coast "penny press" in the 1830s—cheap newspapers for the masses, printed on poor-quality paper, with no stock market prices but teeming with crime reports and other sensational material. The same period also saw the growth of "investigative journalism," involving a house reporter who would probe suspicious deaths, immorality, and political scandals. For decades visitors from Europe would turn up their noses at these trends in the American press, until similar investigations became common in Britain and elsewhere.[114] This kind of press went together with the growth of democracy, a few decades ahead of Europe; the communications media took the workers seriously once the latter gained the right to vote. Such newspapers mirrored their age more than they analyzed it.

The penny press publishers introduced the mass-circulation press to the United States before any other country in the world. One of the convictions of the new age was that newspaper reading and a willingness to pay for news that used to be obtained for free by word of mouth were expressions of civic-mindedness.[115] Around the year 1860 the shrill *New York Herald* (founded in 1835), which was read by middle-class readers for its abundant news columns, had a daily circulation of 77,000, the highest in the world. Horace Greeley's *New York Tribune*, the first American paper that was able to combine seriousness with popularity—and that included Karl Marx in London among its special correspondents—was reaching a readership of 200,000 in 1860 with its weekly edition.[116] In the United States as elsewhere, all these developments required a railroad network to carry fresh editions overnight to distant corners. The first mass-circulation daily in France, Moïse Millaud's *Le Petit Journal*, appeared in 1863 at a quarter of the price of established newspapers.[117] In Britain, where highbrow papers dominated the scene much longer than in the United States, the turn came when Alfred Charles William Harmsworth launched the *Daily Mail* in 1896; he would later, as Lord Northcliffe, become the first of the legendary Fleet Street magnates. By 1900, when the South African War was fueling the need for information, the cheap new morning daily was selling an incredible total of 989,000 copies. Globally, only Joseph Pulitzer's *New York World* was riding higher, with a circulation of 1.5 million (in 1898).[118] The London *Times*, at the height of its prestige and political influence, had a readership of just 30,000—the establishment that it *wanted* to reach, and no more.[119]

Another statistic will serve to emphasize the upward trend. In 1870 a daily total of 2.6 million papers was being sold in the United States, but by 1900 this figure had risen to at least fifteen million.[120] A political "crusading press" came into being more or less simultaneously in the United States and Britain. Pulitzer, the Hungarian-born owner and chief editor of the *New York World*, built it up from the early 1880s into a financially successful paper specializing in investigative articles and social criticism. In Britain, W. T. Stead, the inventor of the interview, used a similar combination of information with political campaigning for his *Pall Mall Gazette*. But such papers did not merely react to events: they were soon trying to create them. They exerted public pressure on governments, forcing them to reverse old laws and pass new ones. This meant that, unlike in continental Europe, the press was not simply the mouthpiece of political parties and tendencies; its owners and chief editors were able to give free rein to their own convictions and obsessions. Paradoxically, the commercialization of news—which reached a new level with the growth of advertising and publicity—increased the independence of the newspaper founders. If they obtained half their funding from publicity, they had much greater leeway than if they were dependent on political patrons and parties.[121]

The quality press, as we know it today, and also, with fine gradations, the popular press, developed in the last quarter of the nineteenth century. The social type of the modern journalist came into being. Around 1900, countries with press freedom and a literate public spawned a large group of specialists in the collection and presentation of news. One of these countries was Japan, which already had an active record of publishing in premodern times. Closing the gap with developments in the West, it acquired a fully fledged press system in the course of the 1870s and 1880s, driven by new-style journalists and proprietors, employing cutting-edge technology, and adapting to the changes in society since the Meiji Restoration that began in 1868: rising literacy within a state education system, a nationwide mail network, and structural transformation of the public sphere thanks to a parliamentary system and the formation of political parties. The first major newspapers were not, as in China, founded by foreigners. Japan took in cultural elements from the West and gave them its own distinctive imprint. Characteristically, journalism remained close to the leading institutions of higher education; it was only a short step from ranks of the top universities to the chief editorial positions. A long-simmering rivalry between Tokyo and Osaka added a lively note of tension in an otherwise centralized and rather uniform country.[122]

Global Communications

One of the features of the nineteenth-century press was the global character of its leading organizations. The major newspapers felt they had a responsibility to print news from all over the world—indeed, only if they were capable of providing international coverage could they "hit the big time." The foreign correspondent was a new breed. At first he was scarcely distinguishable from the war reporter; the first man who rushed between locations to write about uprisings, sieges, and battles for the reader back home was William Howard Russell of the London *Times*. He relayed his impressions from India, South Africa, and Egypt; from the Crimean War, the American Civil War, and the Franco-Prussian War of 1871. Russell, who was no militarist and no friend of imperialist adventures, carried the genre of war journalism to literary heights that had rarely been seen before.[123] The kind of reporter he invented was here to stay, and the *Times* made a special effort to cultivate it.

When Russell began his career he still had to send his reports to London by mail, but the cabling of the world by telegraph changed the conditions for long-distance reporting within the space of twenty-five years. The electrical overland telegraph came into use in 1844. The first durable underwater cable was laid across the English Channel in 1851, and a permanent transatlantic link was established in 1866.[124] By 1862 the worldwide terrestrial telegraph network was 150,000 miles in length; by 1876 India and all the settler colonies of the British Empire were linked to the home country and one another; and by 1885 Europe could be reached from nearly all large cities by underwater cable. The telegraph network was much too cumbersome, overloaded, and expensive to be described as a "Victorian Internet"—it absorbed 15 percent of the *Times*' expenditure for 1898—but the basic model was there for a historically unprecedented world wide web.[125] It was much more centralized than today's Internet. The telegraph lines themselves, as well as the financial threads of a global cable business that served the needs of commerce more than those of the press, converged in London.

The new technology laid the basis for the news agencies. Julius Reuter from Kassel, Germany, opened his office in London in 1851—the same year that transmission time across the Channel was shortened to a couple of hours. Two other Jewish entrepreneurs had already founded news agencies or "telegraph offices": Charles Havas in Paris and Bernhard Wolff in Berlin. The Associated Press came into being in the United States in 1848. The agencies supplied reports to newspapers but also to governments and private individuals—including Queen Victoria from 1865. Reuter was so successful that he, the nobody from Germany, was introduced to the Queen of England in 1860. The Crimean War (1853–56) was the last major international event that was *not* mainly reported on by cable. By 1861, Reuter's firm—still the only news agency with a global reach—had built a network of correspondents that covered the whole of Europe as well as India, China, Australia, New Zealand, and South Africa. Where the telegraph did not reach, it used the express mail service provided by steamships. Reuter's war correspondents covered the American Civil War (1861–65) from start to finish for readers in Europe. Increasingly, the agencies also reported on developments in science, the arts, and sports. As Julius Reuter built up his news empire, his agency became an "institution of the British Empire."[126]

The agencies contributed to the globalized production and dissemination of news, passing it along without additional comment in a powerful expression of the ideology of "objectivity." On the other hand, their standardized reports promoted a uniform kind of journalism, now that all print media were more or less in the same boat. Only a few major papers, headed by the London *Times*, sustained their own networks of foreign correspondents and kept their dependence on the agencies to a minimum. For the *Times* it was a matter of principle to have its own coverage at least of British imperial interests.[127]

Not until four centuries after Gutenberg did the printed news media enter the daily lives of more than a tiny educated stratum of society. The basic structures of the press, as we know it today, were created in the second half of the nineteenth century. The press used advanced technologies. It obeyed market laws and operated within a certain legal-political framework. Freedom of the press was a basic demand of liberals all around the world. The distinction between West and East here is, as so often, of little relevance. In many colonies of the British Empire the press was freer than in parts of central and eastern Europe. The new breed of journalist also embodied an important facet of the "intellectual." Journalists exerted political influence as far away as India and China; they gave the public a face. The best of them contributed to the transition from the classical written language of the elite to more flexible idioms that broader, often newly literate, sections of the public found more accessible. Alongside "realist" art, statistics, and descriptive social science, the press was a further means of social self-observation in a world in which media-supported communication was dramatically extending its reach. It still had a monopoly by virtue of the technology it used. The achievements of the young Italian engineer Guglielmo Marconi, who built on the discoveries of his Serb-American colleague Nikola Tesla to transmit wireless messages across the Channel (in 1899) and then the Atlantic (in 1901), had not yet made the radio a *mass* medium. It would become that only after the First World War—and the boost given to the new technology by its naval and military applications.[128]

6 Photography

The Birth of Authenticity

Finally, the nineteenth century discovered how to use optical and chemical processes to record phenomena from the external world.[129] A watershed in the century is the moment when the first recognizably genuine pictorial documents were produced. No one knows what Ludwig van Beethoven (1770–1827) really looked like, but we do know how Frédéric Chopin (1810–49) appeared. Only paintings of Franz Schubert exist, but Gioachino Rossini, five years his elder, lived long enough to be photographed in the studio of the great portraitist, Félix Nadar. A few other heroes from the age of Romanticism and Idealism lived to see the age of photography, which dawned in 1838–39 with the invention of the daguerreotype, followed by the opening of the first studios two years later. There are photographic images of Friedrich Wilhelm Joseph Schelling and Alexander von Humboldt as old men, but not of Hegel, Goethe, or Wilhelm von Humboldt (Alexander's brother), who all died before the advent of the new epoch. When King Friedrich Wilhelm IV asked Hermann Biow, the first German photographer, to come from Hamburg to Berlin in 1847 to make portraits of the royal family with the new technology, the famous Humboldt—who had recognized the revolutionary significance of Daguerre's invention a few months after it was made public—also sat for his picture to be taken.[130]

Once photographs became reproducible, in the early 1850s, personal "prominence" acquired a new meaning. Portraits of rulers and political leaders—Lincoln, Bismarck, Emperor Wilhelm I—found their way into countless living rooms. But so long as they did not appear in print on a large scale—which was economically infeasible until the early eighties—their individual features were known only to a limited number of people. When Ulysses S. Grant, the Civil War hero and highest-ranking general of the Union army, arrived at the railroad station in New York, reporters were unable to pick him out in the crowd.[131]

Biow also made a large number of daguerreotypes of Hamburg's Alster district after the great fire of May 1842 had left it in ruins—one of the first photographic records of a disaster.[132] After the Crimean conflict, all wars involving Europeans or North Americans were captured on photographic material. There are no photographs and virtually no graphic representations of the great

Taiping Rebellion in China (1850–64), whereas the American Civil War (1861–65) is abundantly preserved in pictorial memory for later generations. A single photographer, Matthew B. Brady, took more than seven thousand chemically prepared glass-plate images at and between the battlefields.[133] Although in other respects, painting and photography often peacefully competed with each other, the vivid photographic reproduction of battlefields and living or dead soldiers spelled the end of the heroic war canvas. The cheap, easily transportable, hand-operated Kodak roll-film camera, which was invented in 1888, opened new possibilities for visual documentation. Few photographs of the Great Indian Famine of 1876–78 reached the international public, but when the catastrophe was repeated two decades later, every traveler or missionary was a witness potentially capable of documenting it.[134] In its early days, photography was little appreciated as an artistic achievement on the photographer's part;[135] its fascination was that it offered an objectivity and lifelikeness never seen before. Especially important was its use in the natural sciences—first in astronomy, then soon afterward in medicine (X-ray photography opened up a previously invisible realm).[136] From the sixties on, pictures from the world of work were increasingly common. Not long before, travel photography and related applications in geography and ethnography had greatly increased in importance.

The Closing of Distance

Photographic expeditions to archaeological sites (primarily Egypt) and the habitats of exotic peoples became more numerous.[137] In Britain, whose overseas possessions were far larger than those of other Western countries, the public only now realized who and what had been gathered beneath the imperial roof. In comparison with the illustrated travel books that for centuries had provided the only visual impressions, photography brought a great increase in knowledge and atmospheric detail about distant lands. As far as India was concerned, nothing before had come close to the eighteen-volume gem *The Peoples of India* (1868–75), which made 460 new photographs available.[138] Yet for many years the camera remained a tool in the hands of Europeans and Americans alone, who quickly discovered its usefulness in imperial war.[139] Subversive gazes directed back at the metropolis would only come later. But many photographers trained in remote places, enabling them later to focus more clearly on things closer to home. John Thomson, the author of the four-volume *Illustrations of China and Its People* (1873), came back and pointed his camera next at the poor of London—those whom Henry Mayhew had described in journalistic prose a few years earlier.

The camera had less of an exotic effect than the pen or paintbrush. As early as 1842, Joseph-Philibert Girault de Prangey took some wonderful daguerreotypes of both medieval European and Islamic architecture and established a strong aesthetic affinity between them.[140] The place of "the alien" in the European imagination of the second half of the nineteenth century is almost inconceivable without photographic representation. The idea of a "photo-graphic museum of the human races" was obsessively pursued, with highly diverse results. On the one hand, images of poverty—for example, Chinese opium dens or the devastated sites of the Indian revolt of 1857–58—served finally to strip the "fairy-tale" Orient of its enchantment. On the other hand, the alien became more palpable than in conventional depictions of the noble or not-so-noble savage, and colonialists could more easily illustrate scenes of their rule for the benefit of the public in the home country.

Photography was adopted more readily in parts of Latin America—Peru, for example—than in Europe and North America. The 1840s were a boom age for Peru, and the new medium fit perfectly into the boisterous atmosphere.[141] The first non-Western country in which photography gained a foothold was the Ottoman Empire. Studios began to spring up in its large cities in the 1850s, only a little later than in western and central Europe. At first they were run by Europeans and members of non-Muslim minorities for customers who were also mostly European. In the last two decades before 1900, however, the family portrait and the workplace picture became a basic part of the culture of the Muslim upper and middle classes. The state soon saw that photography could be advantageous, especially for military purposes. The autocratic Sultan Abdül-hamid II used it to check up on officials in the provinces—for example, on whether building projects were going ahead as planned—and to project an image of his country in Europe. He is said to have presented one of his daughters with snapshots of suitable marriage candidates.[142]

By the end of the century, photography had become part of ordinary life in many societies. All its branches that are familiar to us today have their roots in the nineteenth century, including advertising, propaganda, and picture postcards. Photography was a widely practiced trade; even small towns had their own studios and laboratories. The Kodak camera of 1888, which required no training or technical knowledge, democratized the medium and lowered its artistic pretensions. Easier and cheaper devices, as well as the invention of roll film, made the production of images for private use technically accessible to the lay public. Scarcely any middle-class home was without professionally made pictures of special occasions on display, or without an album containing photos taken by family members.

Of the observation systems that the nineteenth century perfected or devised, photography was the one that brought the greatest advance in objectification. This remains true even if we bear in mind the malleability and "subjectivity," and hence the artistic adaptability, of the medium. Of course large numbers of photos were "set up," and many show the inhibitions and prejudices of the age; photographic images have proved to be rewarding objects of such deconstruction.[143] Nevertheless, the technology afforded a novel kind of visual access to the world, created new concepts of truth and authenticity, and placed tools of image creation in the hands of those without artistic talent or training.

Moving Images

Film was born in the year 1895.[144] On March 22 in Paris, the "cinématographe"
of the manufacturer's sons Louis and Auguste Lumière and their engineer Jules
Carpentier showed moving images for the first time. The Lumières could immediately offer everything for sale: camera, projection apparatus, and film. Unlike
in photography, the new technology was ready for industrial production from
the outset; public performances for an entrance fee were already taking place
in December. The Lumière family also trained a squad of operators in the new
machines and sent them out into the wider world. By 1896–97, Lumière films
were being shown all over Europe, from Madrid to Kazan and from Belgrade to
Uppsala, as well as in a number of cities on the American East Coast. The coronation of Tsar Nicholas II on May 26, 1896, was an especially popular subject.
The triumphal progress of the cinema continued: Lumière operators appeared
in 1896 in Istanbul, Damascus, Jerusalem, Cairo, Bombay, Mexico City, Rio de
Janeiro, Buenos Aires, and Sydney; and by 1899 it was possible to see motion pictures in Shanghai, Beijing, Tokyo, and Yokohama.[145] Nearly everywhere, motion
pictures also started to be made at the same time.

From 1896, filmed documentaries were being made of royal appearances, military maneuvers, and everyday life on every continent. Spanish bullfights, Ni-
agara Falls, Japanese dancers, and all manner of street scenes were among the
earliest themes. Film started out as a medium of reportage, swept up by globalization. The first known demonstrator of the new French technology in a Shanghai teahouse was James Ricalton from Maplewood, New Jersey, who offered
his Chinese audience moving pictures of the Russian tsar's visit to Paris and an
Egyptian belly dancer's performance at the world's fair in Chicago.[146] A great
success in many countries was the film that Auguste Lumière shot of workers at
his own factory.[147] The new medium soon revealed its dual nature as staged production and as documentary. Since cameramen were not present at the main international event of summer 1900—the Boxer Rebellion in northern China—a
number of horrific scenes were reenacted in English meadows and French parks
and presented as authentic testimony; a special favorite was the reconstruction
of an attack by Chinese rebels on a Christian mission post. More traditional documentary footage came out of liberated/defeated Beijing only in 1901.[148] But it is
difficult to tell here what is genuine and what is illusory. Georges Méliès, who is
considered the inventor of the art film, shot his *Coronation of King Edward VII*
(1902) in a studio, after a careful study of the previous year's event and with the
help of a British master of ceremonies. His film about the Dreyfus affair (1899)
had been a kind of animated conversion of photographic material from newspapers and magazines.[149]

Media studies have recently tended to emphasize perspectival and subjective factors, casting doubt on claims to truth or objectivity. In light of present-
day experiences of the technical and material malleability of the media, such

mistrust does indeed seem justified. The arts have long distanced themselves from a model of "realism," and even documentary currents in literature and cinema, which first arose in the nineteenth century and have never disappeared, have lost much of their original naïveté. It is therefore no longer easy to grasp the emotive connotations of objectivism or the high regard for "positive" knowledge that were characteristic of the nineteenth century. Its quest for reality, not without roots in the early-modern empiricism of Francis Bacon, makes the world seem alien to us—although there was no lack of voices at the time, from the Romantics to Friedrich Nietzsche, who warned against the illusions of positivism and realism.

On the other hand, the nineteenth century belongs to the prehistory of the present day. It gave rise to institutions and cognitive forms of social self-observation that did not change fundamentally down to the spread of television in affluent societies, and even to the late-twentieth-century digital revolution. *Mass* communications media stretching far beyond small elite circles, public investment in the preservation of knowledge and objects of general interest, the monitoring of social processes through statistics and social research, the technical reproduction of texts and artifacts by means of fast printing presses, photography, and phonographic recording—the latter having been technically feasible from about 1888 and used already a year later to document Bismarck's voice[150]— all this was still a long way off and unimaginable in 1800 but was treated as a matter of course by 1910.

The nineteenth century developed an ambivalent relationship to the past that is not alien to us even today. Optimistic openness to the future, awareness of innovation, and faith in technological and moral progress had seldom been so great, and the old had rarely appeared so obsolete, yet the century was also the zenith of a historicism that was not only imitative and reconstructive but also conservationist. The age of museums and archives, of archaeology, and of textual criticism built gateways to the distant past through its work of collection, preservation, and classification—gateways that we still make use of today. Written knowledge about the earlier history of humanity piled up in the leap from 1800 to 1900 as it had never done in any previous century.

Strictly speaking, this only applies to the West. It was in Europe, and in its fast-growing offshoot across the Atlantic, that technological and cultural innovations started their journey around the globe, supported in some cases (the telegraph) by imperial power and imperial capital, and in other cases (the press, the opera, other Western-style musical entertainment) by complex, non-imperial processes involving both the export of taste and indigenous adaptations. No one forced the Egyptians to found newspapers, or the Japanese to listen to Gounod and Verdi. There was cultural mobility from east to west, as we can tell from the riveting effect that Japanese or African art had in Europe.[151]

But the new thinking, technologies, institutions, and "dispositives" that were supposed to achieve universality over time, and that by 1930 at the latest appeared as hallmarks of global "modernity," all came into being in the West in the nineteenth century and began their various global careers from there. In the main, the contents of memory and observation were and remained locally and "culturally" specific. But the frames and forms of their media everywhere came under Western influence, albeit in widely varying degrees and with distinctive mixes of adaptation and resistance to a partly feared, partly welcomed Europeanization.

CHAPTER II

Time

When Was the Nineteenth Century?

1 Chronology and the Coherence of the Age

Calendar Centuries

When was the nineteenth century? We speak of a century as if it were a self-explanatory term, implying that everyone connects it with a precise, perhaps the same, meaning. What is it if not the time span that is contained between the years 1801 and 1900, for example? Yet that time span does not correspond to a tangible experience: the senses do not perceive when a new century begins, as they do the daily cycle or the seasons of the year. The century is a creature of the calendar, a calculated quantity, which was introduced for the first time in the 1500s. For historians it is, as John M. Roberts put it, "only a convenience."[1] The less they believe in the "objective" coherence of an age, and the more they see dividing lines between epochs as pure convention, the fewer objections there can be to a simple chronology that operates with chunks of a hundred years. In the case of the nineteenth century, however, the lusterless boundary dates underscore the formal character of this procedure: neither the beginning year nor the end year of the calendar century coincided with a major turning point. Years with two or three zeros are often not the watershed that remains fixed in the memory of a nation. It is not 2000 but 2001 that is engraved in the mind.

All this can be an advantage for the writer of history. A tight border means that there is less of a distraction from the picture itself, and the whole problem of periodization can be solved in one decisionist swoop. Blind justice marks out a spatially and culturally neutral frame of reference, capable of encompassing all kinds of change around the world, which frees the historian from difficult debates about the major landmarks. Only this kind of photographic "frame" takes in various histories without treating one as a yardstick for the others. Books have been written about what took place in a certain year—1688 or 1800, for example—in the world's diverse theaters,[2] producing a panoramic effect whose *formal* simultaneity brings out the *substantive* nonsimultaneity of many phenomena.

Synchrony spread over a whole century can have the same result. But, of course, change becomes visible in the span of a hundred years. Snapshots at the beginning and the end of a calendar century reveal processes at different stages of maturity in different parts of the world. Other temporalities emerge alongside the familiar narrative of Western progress.

Nevertheless, such formalism does not satisfy so easily: content-blind periodization achieves its clarity of focus only at the price of contributing little to historical knowledge. That is why historians shy away from it. Some regard periodization as "the core of the form that historiography gives to the past" and therefore as a central problem for historical theory.[3] Those who would not go so far readily join in discussions about "long" and "short" centuries. Many historians are partial to the idea of a long nineteenth century, stretching from the French Revolution in 1789 to the outbreak of the First World War in 1914. Others prefer to operate with a short century—one, for example, that embraces the period in international politics from the European new order of 1814–15 (the Congress of Vienna) to America's entrance into the global arena in the Spanish-American War of 1898. The choice of a content-based temporal framework always involves a particular interpretive emphasis. The length and shape of a century is therefore by no means a pedantic question. Since every historian *must* answer it willy-nilly, he or she might as well do so explicitly right at the start. So, how should the nineteenth century be situated within the temporal continuum? The question is all the more pressing if it cannot be assumed that Europe's political events, economic cycles, and intellectual trends are the only ones that structure the continuum.

A century is a slice of time. It is given meaning only by posterity. Memory structures time, arranging it deep down into echelons, sometimes bringing it close to the present, stretching, shrinking, or occasionally dissolving it. Religious immediacy often leaps across time: the founder, the prophet, or the martyr may be fully present here and now. Nineteenth-century historicism locked them up in the past. A linear chronology is an abstraction, which seldom corresponds to how time is perceived. In many non-Western civilizations, the problem of the precise dating of past events first presented itself only when a time continuum made up of years following one after the other gained general recognition. Linearity arranges historical knowledge into a "before" and an "after," making a narrative possible by the standards of historicism.

Issues of dating were everywhere central for "modern" history and archaeology. In Japan, an extra-European pioneer in this respect too, it was only after the turn of the twentieth century that a satisfactory national chronology was developed for remote periods in the past;[4] whereas in China, whose rich historiographical tradition went as far back as Europe's, the necessary work of source criticism began in the 1920s, and it took decades before a reasonably dependable chronology of ancient times was established.[5] In many other countries, especially in Africa and the South Pacific, archaeological finds confirmed a wide range of human activity but did not enable precise dating even for the modern age. In the case of Hawaii, scholars posit a "proto-historical" period that lasted until 1795, the date of the first written records.[6]

In this book I have opted for the following solution. My nineteenth century is not conceived as a temporal continuum stretching from point A to point B. The histories that interest me do not involve a linear, "and then came such and such" narrative spread over a hundred or more years; rather, they consist of transitions and transformations. Each of these has a distinctive temporal structure and dynamic, distinctive turning points and spatial locations—what might be called regional times. One important aim of this book is to disclose these time structures. It will therefore contain many dates and repeatedly call attention to finer points of chronology. The individual transformations begin and end at particular moments, with continuities in both directions on the arrow of time. On the one hand, they continue developments from the past—let us say, from the "early modern age." Even the great revolutions cannot be understood without the premises that led to them. On the other hand, the nineteenth century is the prehistory of the present day; characteristic transformations that began then rarely came to a complete stop in 1900 or 1914. I shall therefore, with a deliberate lack of discipline, repeatedly look far ahead into the twentieth century or even to the present day. What I wish to conjure up and comment on is not a sealed-off, self-sufficient history *of* the nineteenth century but the insertion of an age within longer timelines: the nineteenth century *in* history.

What does this mean for the temporal framework of the account? If continuities are emphasized more than sharp breaks between epochs, it will not be possible to base definitions on precise years. Instead, I shall move nimbly between two modes of macro-periodization. Sometimes I shall refer to the bare segment of time, approximately from 1801 to 1900, without specifying content: that is, the *calendar century*. Elsewhere I shall have a *long* nineteenth century in mind, one beginning perhaps in the 1770s, that emerges only through contextual analysis. If I were to select a single "world-historical" event as emblematic of the period, it would be the revolution that led to the founding of the United States of America. At the other end, it would be convenient, dramatically effective, and conventionally acceptable to close the long nineteenth century with the sudden fall of the curtain in August 1914. This makes sense for certain transformations—in the world economy, for example—but not for others. The First World War was itself a time of colossal transition and greatly extended chains of effects. It began as a military confrontation in the space between northeastern France and the Baltic, but soon spread to West and East Africa and subsequently turned into a *world* war.[7] Conditions within almost all the countries involved changed dramatically only in 1916–17. Nineteen nineteen became the year of political restructuring in Europe, the Middle East, and Africa, and of revolutionary or anticolonial upheavals from Ireland to Egypt and India to China and Korea. Disappointment that the peace did not live up to its promise was widely shared around the world.[8]

Or, to put it more pointedly: only when the war was over did humankind realize that it was no longer living in the nineteenth century. In many respects, then, the long century that began in the 1770s should be thought of as having ended in the 1920s, with the transition to a world in which new technologies and ideologies established a deep gulf between the postwar present and the pre-1914 past.

Constructing Epochs

One of several ways of shaping historical time is to condense it into epochs. To the modern European mind, at least, the past appears as a succession of blocks of time. But the terms used to describe epochs are seldom crystallizations of raw memory; they are the result of historical reflection and construction. Not infrequently it is a major historical work that first calls an epoch into being: whether it be "Hellenism" (Droysen), the "Renaissance" (Michelet, Burckhardt), the "late Middle Ages" (Huizinga) or "late antiquity" (Peter Brown). In many cases, academic neologisms have scarcely trickled through to a wider public: "early modern age" is a good example. This was first proposed as the name for an epoch in the early 1950s. The term soon won recognition in the historians' lexicon, being seen almost as the fourth epoch of world history on a par with the previous three—and thus fulfilling the apocalyptic fourfold vision of world empires in the Old Testament.[9] Confusion reigns when it comes to "modernity," a concept applied indiscriminately and with a host of arguments to every century in Europe since the sixteenth, and even to "medieval" China in the eleventh: social history has employed it for the period since the 1830s; cultural-aesthetic theory limits it to one not earlier than Baudelaire, Debussy, and Cézanne.[10] The ubiquitous talk of modernity, postmodernity, and "multiple modernities," nearly always without even an approximate chronological definition, naturally indicates that the sense of epochs has been steadily weakening. It may be that "early modern age" is the last construction of its kind that commands general acceptance within university faculties.[11]

Whatever its precise dates, the nineteenth century appears to almost all historians as a freestanding epoch that resists naming. Whereas for earlier times, several centuries are readily grouped together into an epoch (as many as ten in the "Middle Ages," or three in the "early modern age"), the nineteenth century remains alone. No one has ever seriously proposed using the obvious term, "late modern age." German historians are not even sure whether the nineteenth century should be classified under "modern" (*neuere*) or "recent" (*neueste*) history: the former would define it as the culmination of developments that began before 1800; the latter as the prehistory of an age that began with the First World War.[12] Eric Hobsbawm, the author of one of the best general histories of Europe since the French Revolution, does not give the nineteenth century (which for him is "long") a single overarching name but divides it into three: the Age of Revolution (1789–1848), the Age of Capital (1848–75) and the Age of Empire (1875–1914).[13] Nor has the history of ideas yet managed to come up with a single appellation, along the lines of "the Age of Enlightenment" that is sometimes used for the eighteenth century. So, we are left with a nameless and fragmented century, a long transition between two ages that seem easier to identify. Perhaps a quandary.

2 Calendar and Periodization

In large parts of the world, people did not notice in 1800 or 1801 that a "new century" had begun. Official France did not want to know, because it dated the years from the beginning of the Republic (1792 = year I), and in 1793 it had arbitrarily introduced a new organization of the year that was observed with diminishing enthusiasm until the restoration of the Gregorian calendar in 1806. A new counting of the months meant that on January 1, 1801, the French people found themselves on the eleventh day of the fourth month (*Nivose* = "snow month") of year IX. Muslims, for their part, woke up on an ordinary day in the middle of the eighth month of the year 1215, in a calendar that went back to the flight (*hijra*) of the Prophet Muhammad to Medina on July 16, 622; the new century, the thirteenth, had already begun in 1786. In Siam and other Buddhist countries, people were living in the 2343rd year of the Buddhist era, which was the year 5561 in the Jewish calendar. In China, January 1, 1801, was the day of the second of the Ten Heavenly Stems and the eighth of the Twelve Earthly Branches, in the fifth year of the rule of Emperor Jiaqing; and other calendars were also in use within the vast Chinese empire, Muslims, Tibetans, and the Yi and Dai minorities each having one of their own. In China the turn of 1801 did not mark an epochal change; the only event that counted had taken place on February 9, 1796, when the glorious Emperor Qianlong, after sixty years on the throne, had handed over to his son Yongyan, who as ruler had taken the name Jiaqing. In Vietnam, earlier than in other Asian countries, the unification of the country in 1802 brought a switch to the Western calendar for certain official purposes, although people continued to use the calendar of the Chinese Ming dynasty (which had fallen in 1644).[14] These and other possible examples add up to a colorful picture of calendar pluralism. Their message is clear: the magic of the turn of the century was limited to the areas where Christianity had spread. The West was to be found wherever people noted the passing of the old century and the coming of the new. "Our" nineteenth century began only in the West.

Pope Gregory's Calendar and the Alternatives

Anyone who finds this surprising should consider that even in Europe a uniform calendar was achieved only slowly and in stages. It took all of 170 years for England, and with it the whole British empire, to adopt the Gregorian calendar that had been introduced in 1582–84 in the Catholic countries of Europe, soon afterward in Spain's overseas territories, in 1600 in Scotland, and in 1752 in Great Britain.[15] In Romania it became official only in 1917, in Russia in 1918, and in Turkey in 1927. The Gregorian calendar—not a radical innovation but a technical refinement of Julius Caesar's calendar—was one of modern Europe's most successful cultural exports. Initiated by a Counter-Reformation pope, Gregory XIII (r. 1575–85), it reached the farthest corners of the planet along the routes of Britain's Protestant world empire. Outside the colonies, it was imported voluntarily rather than being foisted upon "other" civilizations through the dictates of cultural imperialism. Where it remained controversial, it was often for scientific or pragmatic reasons. Auguste Comte, the positivist philosopher, made a great effort in 1849 to secure the adoption of his alternative calendar, which divided the year into thirteen months, each of 28 days, resulting in a total of 364 days plus a kind of bonus day outside the system. In this proposal, the conventional names of the months would have been replaced with dedications to the benefactors of mankind: Moses, Archimedes, Charlemagne, Dante, Shakespeare, and so on.[16] In terms of calendar technology, it was not devoid of refinement. Different variants would later often be suggested.

The Russian Orthodox Church still uses the unreformed Julian calendar from 46 BC, which Julius Caesar, in his capacity as pontifex maximus, created against a rich backdrop of thinking about time among Greek and Egyptian astronomers—an instrument tried and tested over the centuries, but one that had eventually accumulated a few extra days. The situation in the Ottoman Empire (and later Turkey) was especially complicated. Although the Prophet Muhammad had made the moon the measure of time and declared that only the lunar calendar should be considered valid, relics of the Julian solar calendar remained from the Byzantine period. The Ottoman state accepted that this was more practical for its purposes and geared its financial year to the four seasons. This was important in order to establish the point in time when the harvest would be taxed. There was no direct correspondence between the solar and the lunar calendar; overlaps, desynchronization, and time differences were inevitable. In many Muslim countries, the rural population continued for a long time to observe the lunar calendar, while the cities used the international (Gregorian) calendar.[17] Chinese all over the world, even the pioneers of globalization, continue to celebrate the New Year in accordance with the lunisolar calendar. And lastly, apart from "traditional" and "modern" calendars, there were and are specially created festive calendars that mark national holidays, commemorations of national heroes, and so on, or in some cases an entire separate system for the arrangement of time. The Bahai religion, for instance, has a calendar made up of nineteen months with nineteen days in each, and calculates the years from the divine inspiration received by its founder in 1844.[18]

Nor is historical time everywhere reckoned in terms of "Anno Domini" (or today's "Common Era"). Our linear dating system, capable of situating any point in time from a year 1 (*annus domini*), was originally conceived in the sixth century, further elaborated by the Jesuit Dionysus Petavius (Denis Pétau) in 1627, and propagated soon afterward by the great Descartes.[19] It spread worldwide in the nineteenth century but has never completely supplanted the alternatives. In Taiwan, one of the most modern societies in the world, the dating system begins with the revolutionary year of 1912, when the Chinese Republic (Minguo)— which the present regime claims to embody—put an end to the imperial era; the year 2000 was therefore "Minguo 89" on the island. Just as the counting of years in Imperial China began anew with each change on the throne until the new Republican government adopted the Western calendar in 1912, so too in Japan each new ruler ushered in a new sequence (1873 was thus "Meiji 6"). But in 1869—in a fine example of invented tradition—a succession of emperors uninterrupted by dynastic change was enshrined in a parallel dating system, so that 1873 became year 2533 since the mythical first emperor, Jimmu, ascended to the throne. This was thought to link Japan to the Western linear conception of time.[20] Despite the cautiously worded objections of many historians, this archaic reference to 660 BC as the fictitious year 1 of imperial rule remained a founding myth of Japanese nationalism after 1945 and received unmistakable endorsement in 1989 with the coronation of Emperor Akihito.[21]

The new chronology served the manifest political purpose of entrenching the emperor at the center of the mental world of the new nation-state. It is true that in 1873—nearly half a century before Russia—Japan introduced the Gregorian calendar and the previously unknown seven-day week. On day 9 of month 11 in the lunar calendar, an imperial edict decreed that day 3 of month 12 would be redefined in accordance with the solar calendar as January 1, 1873. Yet for all the modernization rhetoric, which attacked the lunar calendar as a sign of superstition and backwardness, the abrupt reform of 1873 had the main function at the time of preserving the state treasury from bankruptcy. For an intercalary month had been due under the old system, and the extra month's pay that this would have meant for all officials could not have been supported in the alarming budgetary situation of the time. So, the New Year was suddenly brought forward by twenty-nine days, allowing desperate housewives no time to prepare for it with the traditional house-cleaning. Japan's alignment with the most influential global calendar meant that court astronomers would no longer be necessary for working out the correct date.[22]

Epochal Chronologies

The relativity of chronology is even clearer if we consider the various appellations given to historical epochs. Nothing like the triad of antiquity, Middle Ages, and modern times—which Europe had gradually adopted since the 1680s—came into use in any other civilization that could look back at a continuous and comparably documented past. There were periods of renewal and rebirth, but before contacts with Europe it rarely occurred to anyone that they were living in an age superior to the past. Only the Meiji system change, promoted by Ōkubo Toshimichi and other energetic young nobles, brought that future-oriented rhetoric of new beginnings that is an essential component of any "modern" consciousness.[23] However, it immediately became trapped in traditionalism, as Japanese officialdom reemphasized the sacredness of imperial rule (even though there was no past model for Meiji political practice) and strove to invent an indigenous "middle ages" that would link the country to Europe's prestigious history.[24] The idea of a "medieval period" played a certain role in traditional Muslim historiography, but not in its Chinese equivalent. Nor did the importing of Western conceptions change this in any way: the term "medieval" is used neither in the People's Republic nor in Taiwan to refer to China's own history. Not only traditionalist historians preferred dynastic periodization; those working today in the People's Republic follow the same principle, as does the *Cambridge History of China*, the flagship of Western China studies published in a series of volumes since 1978.

The first variations from this rule emerged in relation to the nineteenth century. Marxist orthodoxy dates the beginning of China's "modern history" (*jindai shi*) to the Anglo-Chinese treaty of Nanjing in 1842, and of its "recent history" (*xiandai shi*) to the anti-imperialist protest movement of 1919. Thus, a nineteenth century with a content-based definition starts only in the 1840s. In today's China studies, however, both in the United States and increasingly in China itself, historians have begun to use the term "late imperial age," referring not simply to the final decades of the empire (usually known as the "late Qing period") but to the period between the mid-sixteenth century and the end of the long nineteenth century; some even go as far back as the eleventh century, which in China was an age of political consolidation, social regeneration, and cultural blossoming. This "late imperial China" stretches until the end of the monarchy in 1911. It has a formal affinity with the European concept of an "early modern age" or, in updated variants, with the idea of an "Old Europe" that began in the Middle Ages. But it does not share the emphasis on the period around 1800 as the end of a historical formation. The general view now is that, despite a number of innovative elements, the calendar nineteenth century represented a decadent final phase of an incomparably stable ancien régime. But China represents only one possible variant in the quest for a substantive definition of the nineteenth century.

3   Breaks and Transitions

National and Global Turning Points

Unless one accepts the mystical notion that a single zeitgeist expresses all aspects of life in an epoch, historical periodization must face the problem of the "temporal diversity of cultural domains."[25] In most cases, a break in political history does not also mark a turning point in economic history; stylistic periods in art history do not generally begin or end at points when new developments are thought to be emerging in social history. Whereas social history is often free of periodization debates because it tacitly takes over the usual division into political epochs, other writers warn against placing too much value on the history of events. Ernst Troeltsch, himself an important German theologian and intellectual historian of the early twentieth century, was unable to make much out of it. From a discussion of the non-event-based models of historical epochs to be found in Hegel, Comte, or Marx or in Kurt Breysig, Werner Sombart, or Max Weber, he drew the conclusion that "a truly objective periodization" was possible "only on social, economic, political, and legal foundations," only on the prior basis of "the great basic forces."[26] Troeltsch did not believe, however, that such basic forces enabled a sequencing of history that was unambiguous and clear-cut.

Troeltsch was concerned with the history of Europe as a whole, not of individual nations. Whereas a particular national history still finds safety in a general consensus about its key dates, it becomes all the more difficult for Europe as a whole to agree on the epochal shifts of common importance. The political trajectory of Britain, for instance, where not even the Revolutions of 1848 played a major role, has been such that popular historians are not alone in using the term "Victorian" for the period between 1837 and 1901—a term derived from the reign of a constitutional monarch. England experienced its profound political break considerably earlier, in the two revolutions of the seventeenth century, and the shock waves of the French Revolution after 1789 were by no means as powerful there as they were on the Continent. British history books today tend to situate the decisive turning point not in 1789 but in 1783—the year when the North American colonies were finally lost—and therefore attach much less significance to 1800 than one commonly finds in France, Germany, or Poland. In Britain the passage from the eighteenth to the nineteenth century is less dramatic, sidelining the rupture of the Napoleonic Wars that were fought on the other side of the Channel.

If the temporal shape to be given to European history is far from uncontroversial, how much more difficult is it to agree on a periodization for the world![27] Political dates scarcely help. Before the twentieth century, not a single year can be regarded as epoch-making for the whole of humanity. The French Revolution may be seen in retrospect as fraught with significance for world history, but the deposal and execution of the ruler of a medium-sized European country did not have the shattering impact of a world-historical event. In East Asia, the Pacific, and southern Africa it went largely unnoticed. The French philosopher and cultural historian Louis Bourdeau remarked in 1888 that the French Revolution did not exist in the minds of 400 million Chinese—hence his doubts about its true significance.[28] It was not the revolutionary program and its application within the borders of France but its expansion abroad by military means that had radiating consequences. The revolution had no impact in India or the Americas—with the exception of the French colonies in the Caribbean—until war broke out between Napoleon and the British. Even the First World War initially left large parts of the globe untouched; only its end in 1918 triggered a worldwide crisis, including a deadly influenza pandemic.[29] The Wall Street crash of 1929 was the first *economic* event of truly global weight; producers and consumers on every continent were reeling from its consequences within a few months. Not long afterward, the Second World War began in stages: in July 1937 for China and Japan; in September 1939 for Europe west of Russia; and in 1941 for the rest of the world, with the German invasion of the Soviet Union and the Japanese attack on the United States. Latin America and sub-Saharan Africa, however, were less affected by it than they had been by the First World War. So we may say that before 1945 not a single date in world *political* history had an immediate or near-immediate impact on the whole of humanity. Only in the second postwar period did a shared "event history" begin for the entire world.

Let us now consider what historians (and, following them, probably the public at large) in individual nation-states have seen as the key moments of *domestic* politics in a "long" nineteenth century. The years around the turn of the century had an epochal impact wherever Napoleon's armies toppled or irrevocably weakened the ancien régime. This was the case in the mosaic of Germany's western statelets, in Spain and Portugal, in colonial Saint-Domingue (soon to become Haiti), and in Egypt—but not in the Tsarist Empire, for example. There were also indirect effects. Had the Spanish monarchy not collapsed in 1808, the revolutions that won independence for Hispanic America would have begun *later*, not in 1810. The French occupation of the Ottoman province of Egypt in 1798—a short-lived affair that ended within three years—delivered a shock to the ruling elite in Istanbul that triggered a many-sided modernization drive. In a longer time perspective, however, the defeat of 1878 in the war with Russia represented a more serious blow to the sultan, since it led to the loss of some of the richest areas of the empire: the Balkan peninsula was 76 percent Ottoman in 1876, but only 37 percent in 1879. *That* was the great political turning point for the Ottomans, the key moment in their slide into decline. The deposal of the autocratic sultan by the "Young Turk" officers in 1908 was the almost inevitable revolutionary sequel. Lastly, indirect effects of the Napoleonic Wars were also felt in areas where Britain intervened militarily. The Cape of Good Hope and Ceylon (Sri Lanka) were severed from the Dutch state, then part of the Napoleonic empire, and subsequently remained under British rule. In Indonesia, a brief British occupation (1811–16) led to deep changes in the restored system of Dutch rule. In India the British made a bid for supremacy in 1798, under the most successful of the colonial conquistadors, the Marquess of Wellesley, and by 1818 at the latest it was securely in their hands.

In other countries, the main political breaks—more important than those around the year 1800—came well into the new century. Many states came into being only in the calendar nineteenth century: the Republic of Haiti in 1804, the republics of Hispanic America between 1810 and 1826, the kingdoms of Belgium and Greece in 1830 and 1832 respectively, the Kingdom of Italy in 1861, the German Reich in 1871, and the Principality of Bulgaria in 1878. Today's New Zealand began its existence as a state with the Treaty of Waitangi, which representatives of the British Crown concluded with Maori chieftains in 1840. Canada and Australia were converted by acts of federation (in 1867 and 1901, respectively) from groups of adjacent colonies into national states. Norway severed its union with Sweden only in 1905. In all these cases, the foundational date divides the nineteenth century into a time *before* and a time *after* the achievement of unity and independence. The structuring power of these slices through time is greater than that of the approximate calendar periods that we happen to call "centuries."

There is no lack of further examples. Britain's internal politics was unsettled but not thrown off course by the age of revolution, and the country entered the nineteenth century with a highly oligarchic political order. The Reform Act of 1832 eventually expanded the number of active citizens entitled to vote and brought to an end the peculiar British form of an ancien régime. The year 1832 thus marks the most extensive change within post-1688 British constitutional history, perhaps even greater at a symbolic level than in reality. Hungary, which remained off the campaign routes of the Napoleonic armies, underwent its first major political crisis in 1848–49—but then it was more intense than anywhere else in Europe. For China, the Taiping Rebellion of 1850–64 represented an epochal challenge of revolutionary dimensions, the first internal crisis on such a scale for more than two hundred years. Political system changes in the world became more frequent in the 1860s. The two most important—each revolutionary in its essence—were the collapse of the Southern Confederacy and the restoration of national unity at the end of the American Civil War (1865), and the fall of the shogunate and the beginning of Japan's intensive state-building effort in 1868 (the Meiji Renewal). In both cases, system crisis and a reform drive swept away structures of rule and political practices that had survived from the eighteenth century: the feudal federalism of the Tokugawa dynasty (in power since 1603) and the slave system of the Southern states of North America. In both Japan and the United States, the transition from one political world to another took place *in midcentury*.

"Early Modern Age"—Worldwide?

The *political* beginning of the nineteenth century can therefore scarcely be identified chronologically. To equate it with the French Revolution would be to think too narrowly, with France, Germany, or Saint-Domingue in mind. Ancien régimes tumbled down all through the nineteenth century. In a large and important country such as Japan, the modern age began politically as late as 1868. What should we make, then, of periodizations that have Troeltsch's social and cultural "basic forces" as their criterion? This question takes us back to the category of the "early modern age." The more convincingly we manage to define the early modern age as a rounded epoch, the more solid is the foundation on which the nineteenth century can be inaugurated. Here the signals are contradictory, however. On the one hand, a combination of specialist research, intellectual originality and academic politics has led to a situation in which many historians simply take the existence of an early modern age for granted and adjust their own thinking to a framework that extends from 1500 to 1800. The result is what inevitably occurs when routine use gives period schemas the appearance of a life of their own: transitional phenomena drop out of sight. It may therefore be not unwise to place major events—"1789," "1871," or "1914"—at the middle rather than the edge of their period, so that they are seen from a temporal periphery both before *and* after.[30]

On the other hand, it seems more and more compelling that both outer dates of the customarily defined early modern age should be left more open, if only for the sake of the continuities with previous and subsequent periods.[31] The only break that long went undisputed, at least for European history, is the one of 1500—although many historians insert it into a transitional period from roughly 1450 to 1520. It is obvious that a number of far-reaching innovative processes occurred together at this time: (late) Renaissance, Reformation, beginnings of early capitalism, emergence of the early modern state, discovering of maritime routes to America and tropical Asia; even, going back to the 1450s, the invention of book printing with movable type. Numerous authors of world histories have taken 1500 as the key orientation date.[32] But even the momentousness of 1500 is now in dispute: an alternative approach speaks of a very long and gradual passage from the medieval to the modern world, so that the boundary between the Middle Ages and the early modern age falls away. The German historian Heinz Schilling has emphasized the *slow* emergence of early modernity in Europe and has downplayed 1500 in comparison with the turning points around 1250 and 1750. He attributes the vision of a sudden dawn of the modern age to the nineteenth-century cults of Columbus and Luther.[33] Earlier, in an account of Europe's institutional structures between 1000 and 1800, Dietrich Gerhard held back from the categories "Middle Ages" and "modern age" and employed the term "Old Europe" for the entire period.[34] Analogies with the concept of "late imperial China" are easy to detect.

Paradoxically, historians of non-European civilizations have recently taken up and experimented with the classical Eurocentric designation "early modern age." Few of them actually intended to force alien concepts onto the history of Asia, Africa, and the Americas; most were looking for ways to incorporate these parts of the world into a general history of modernization and to translate the experiences of each into a language intelligible to a European readership. The historian who departed most from prevailing dogma was Fernand Braudel, who in his history of capitalism and material life from the fifteenth to the eighteenth century actually treated the *whole* world as if this were a matter of course.[35] Braudel was careful not to be drawn into a debate about the periodization of world history. What interested him were not so much the great transformations in technology, trade or worldviews as the functioning of societies and intersocietal networks within a given time frame.

Braudel's panoramic vision has found surprisingly few imitators. Recent discussions on the applicability of the term "early modern age" have tended to focus on particular regions. In the cases of Russia, China, Japan, the Ottoman Empire, India, Iran, Southeast Asia, and, of course, colonial South and North America, historians have looked for similarities and dissimilarities with contemporaneous West European forms of political and social organization. There is certainly much scope for comparing England and Japan, and there are striking parallels between the processes that Braudel described for the Mediterranean in the age of Philip II and those that Anthony Reid analyzed for the similarly multicultural world of Southeast Asia during the same period: growth of trade, deployment of new military technologies, centralization of the state, and widespread religious unrest (though introduced to Southeast Asia from outside, by Christianity and Islam).[36]

Insofar as the discussion is also about chronology, some agreement has been reached that the period from 1450 to 1600 was one of especially big changes in large areas of Eurasia and the Americas. There is much to be said for an approximately simultaneous transition to an early modern age in many different parts of the world. With the exception of Mexico, Peru, and certain Caribbean islands, incipient European expansion was not yet a major determining factor. Only in a "long" eighteenth century, whose beginning may be dated to the 1680s, did European influence become plainly visible *worldwide* and not simply in the Atlantic area. Then even China, still closed off and resistant to any attempt at colonization, was drawn into global economic flows of silk, tea, and silver.[37]

Up to now there have been no comparable reflections on the *end* of a possibly worldwide early modern age. For some regions the evidence seems clear-cut: in Hispanic America, the national independence gained by some regions by the late 1820s marked the end of the early modern era. Bonaparte's invasion of Egypt in 1798 not only toppled the Mameluk regime dating from the Middle Ages but shook the political system and culture of the suzerain Ottoman power; the French body blow became the trigger for the early reforms under Sultan Mahmud II (r. 1808–39). It has therefore been suggested that we should speak of an Ottoman "long nineteenth century" (1798–1922) or a "reform century from 1808 to 1908."[38] Things stand quite differently in Japan, which experienced much social turbulence between 1600 and 1850 but no sweeping changes comparable to those that followed the opening of the country in the mid-nineteenth century. If the term "early modern Japan" has any meaning, it must stretch well into the 1850s.[39]

The beginning of European colonization, at very different points in time, represented an epochal break in nearly all parts of Asia and everywhere in Africa, although it is not always easy to establish when the European presence really became tangible; overall, certainly not before 1890. Since the British conquest of India unfolded in stages between 1757 and 1848, while the French took from 1858 to 1895 to establish control of Indochina, a political-military periodization would have little relevance. In the case of Africa, leading specialists extend the Middle Ages as far as the period around 1800 and avoid using the term "early modern age" to characterize the first three quarters of the nineteenth century.[40] The decades until the European invasion remain without a name.

4 The Age of Revolution, Victorianism, Fin de Siècle

It is thus even more difficult in a global perspective than a European one alone to date the beginning of the nineteenth century in terms of content rather than formal calendar. There is much to be said for conceding an epochal character to what the great German historian and theorist of history Reinhart Koselleck once termed the *Sattelzeit* ("saddle period"), a time of transition to modernity from roughly 1750 to roughly 1850 (sometimes 1770–1830) when, in Koselleck's words, "our past becomes our present."[41] That period of dissolution and renewal may be variously seen as involving a forward extension of the eighteenth century or a backward extension of the nineteenth. It led into a middle period that, at least for Europe, articulated in a condensed way the cultural phenomena that are considered with hindsight as most characteristic of the nineteenth century. Then in the 1880s and 1890s such a jolt passed through the world that it is appropriate to describe those decades as the beginning of a further subperiod. We might call it the fin de siècle, as it was known at the time: not *a* termination of any given century but *the* fin de siècle.[42] Its end has traditionally been identified with the outbreak of the First World War, but, as we argued earlier in this chapter, 1918–19 seems a more appropriate date, since the war itself realized certain potentials of the prewar period. There is also much to be said for the even longer "turn of the century" suggested by a group of German historians, whose richly illustrated work has focused on the years from 1880 to 1930.[43] In many ways 1930 makes sense as a terminal date for such a protracted turn of the century. Particularly strong support for such a periodization comes from economic history.[44] One might also take it as far as 1945 and characterize the whole period from the 1880s to the end of the Second World War as "the age of empires and imperialism," since at root both world wars were clashes of empires.[45]

At the risk of an inadmissible Anglocentrism, the decorative word "Victorianism" might be considered for the nameless years between Koselleck's *Sattelzeit* and the fin de siècle: that is, for the "real" nineteenth century. It would relieve one of the embarrassment of having to choose from a variety of narrower, content-based terms: "the age of the first capitalist globalization," "the golden age of capital," or perhaps "the age of nationalism and reform." Why Victorianism?[46] The name reflects the remarkable economic and military—and to some extent, also cultural—supremacy that Britain exercised in the world during those decades (not before or after). It is also a relatively well established category, which in most uses does not coincide precisely with Queen Victoria's years on the throne. G. M. Young, in his famous portrait *Victorian England* (1936), referred only to the years from 1832 until the point when "the dark shadow of the eighties" descended.[47] Many others have followed his lead and treated the years from the mid-1880s until the First World War as a sui generis period—a transmogrification of "High Victorianism."[48]

A Global *Sattelzeit*?

Which factors are the most important for giving coherence to a global *Sattelzeit*? What follows from Rudolf Vierhaus's suggestion that the eighteenth century should be freed from its narrow association with the "classical" early modern age and opened up as "the threshold to the modern world."[49] Which aspects of that period of world history permit us to consider the roughly six decades around 1800 as an epoch in its own right?[50]

First, as C. A. Bayly in particular has shown, the global relationship of forces changed dramatically during this period. The sixteenth and seventeenth centuries were an age in which the most successful large organizations of European origin (Spain's colonial empire and the intercontinental trading networks of the Dutch and English chartered companies) were unable to gain clear superiority over China and the "gunpowder empires" of the Islamic world (Ottoman Empire, Mogul India, and the Iranian empire of the Safavid shahs). Only the advent of the fiscal-military state in England and elsewhere, organized for conquest on the basis of rational resource use, gave Europe a significantly greater punch in the world. This conqueror state appeared in various guises in Britain, in the Russia of Catherine II and her two successors, and in revolutionary-Napoleonic France. All three empires expanded with such force and on such a scale that the period between 1760–70 and 1830 may be described as a "first age of global imperialism."[51] The Seven Years' War (1756–63), fought in both hemispheres, had already been a war for hegemony between England and France, in which North American tribes and Indian princes had played a significant role on either side.[52] The great conflict of empires between 1793 and 1815 ranged even farther beyond Europe. Fought on four continents, it was a genuine world war that had a direct impact as far away as Southeast Asia, and in 1793 even affected China, when Lord Macartney traveled to Beijing to put out the first diplomatic feelers to the imperial court.

After 1780 two new factors joined the "mix" of the Seven Years' War: on the one hand, the struggle for independence on the part of settlers in British North America and (later) Spanish South and Central America, as well as of black slaves in Haiti; on the other hand, a weakening of the Asiatic empires, partly for reasons specific to each one, which for the first time caused them to fall behind Europe in military capability and in the game of power politics. The interplay of these forces changed the political geography of the world. Spain, Portugal, and France disappeared from the American landmass. The expansion of the Asiatic empires finally ground to a halt. Britain built a position of supremacy in India as a springboard for further assaults, established itself securely in Australia, and covered the globe with a network of naval bases.

Whereas earlier historians spoke of an "Atlantic revolution" all the way from Geneva to Lima, thereby correcting a fixation on the European twins (political revolution in France and industrial revolution in England),[53] we can go a step further and grasp the European "Age of Revolution" as only part of a *general* crisis and shifting of power that also made itself felt in the American settler colonies and the Islamic world from the Balkans to India.[54] The general crisis of the decades around 1800 was at the same time a crisis of the Bourbon Monarchy; of British, Spanish, and French colonial rule in the New World; and of such once-mighty Asiatic powers as the Ottoman and Chinese empires, the Crimean Tatar Federation, and the Mogul empire's successor states in the South Asian subcontinent. The French invasion of Algiers in 1830, when that "pirates' nest" was still *de jure* a part of the Ottoman Empire, and the defeat of China in the Opium War of 1839–42—the Qing Dynasty's first military setback in two hundred years—dramatically illuminated the new relations that had taken shape during the *Sattelzeit*.

Second, the provisional political emancipation of settler societies in the Western hemisphere around 1830 (with the major exception of Canada, which remained in the British Empire), together with the colonization of Australia around the same time, led to a general strengthening of the "white" position in the world.[55] While the American republics remained tied to Europe economically and culturally and assumed functional roles within the world economic system, they acted more aggressively than in colonial times toward the hunting and pastoral societies in their midst. In the United States, this reached a point in the 1820s when "native Americans" were no longer treated as negotiating partners but regarded as objects of military and administrative compulsion.[56] Australia, New Zealand, and Russia too, and in some respects South Africa, fit into this picture of repressive, land-grabbing colonization.[57]

Third, one of the major novelties of the *Sattelzeit* was the emergence of inclusive forms of social solidarity and a new ideal of civil equality. This "nationalism" stabilized the collective identity and demarcated it from that of neighboring countries and distant "barbarians." In its early period, until around 1830, this nationalist spirit was especially successful where it could serve as an integrative ideology of an existing territorial state and where it coincided with a missionary sense of cultural superiority. This was the case in France, Britain, and—at the latest by the time of the victorious war against Mexico (1846–49)—in the United States. Everywhere else in the world, nationalism was initially—things would change later—a reactive force: first in the German and Spanish resistance to Napoleon and the Spanish-American liberation movements; then, after 1830, in other continents too.

Fourth. It was only in the United States that the ideal of civil equality translated into broad popular involvement in political decision making—albeit with the exclusion of women, Indians, and black slaves—and a system of checks on the country's rulers. The presidency of Thomas Jefferson (1801–1809) had given a particular impetus in this direction. When President Andrew Jackson took office in 1829, the United States found its way to the form of anti-oligarchic democracy that would be the distinctive feature of its civilization. Elsewhere democratic modernity was in a sorry state before 1830. To be sure, the French Revolution was not as innocuous, conservative, or downright irrelevant as a "revisionist" historiography fixated on continuity claims it to have been. But neither did it lead to Europe-wide democratization, let alone world revolution. Napoleon, its executor, ruled at least as despotically as Louis XV, and the restored Bourbon Monarchy (1815–30) was a caricature of bygone times. Until 1832, aristocratic magnates ruled Britain unchallenged. Absolutist reaction held sway in large parts of southern and central Europe and Russia. Not until 1830 did a constitutionalist trend gradually begin to take shape, although even that halted at the "colored" colonies of the European powers. Politically, the *Sattelzeit* did not witness the breakthrough of democracy in either Europe or Asia; rather, it was the last fling of aristocratic rule and autocracy.[58] The political nineteenth century began after the *Sattelzeit* was over.

Fifth. Periodization is more difficult in social history than in political history. The transition from a society dominated by estates to a class society is clearly discernible in countries such as France, the Netherlands, Prussia, and, a few decades later, Japan. But it is not easy to find estates in eighteenth-century Britain, and they existed only at a rudimentary level in the United States and the British dominions, and a fortiori in India, Africa, and China. The model "from estates to classes" therefore lacks universal validity. For several countries, or even continents, the end of the Atlantic slave trade and the emancipation of the slaves in the British Empire in 1834 were of at least similar importance. Over the next five decades, slavery slowly disappeared from Western civilization and the overseas regions under its control. A different way of putting this would be to say that this relic of extreme coercion from early modern times went virtually unchallenged until at least the 1830s.

In terms of social history, a distinctive feature of the *Sattelzeit* was the growing contestation and subversion of traditional hierarchies. It remains to be proved whether the years around 1800 were also a period of agrarian change and rural unrest outside western and central Europe; there is a lot of evidence that they were.[59] Notwithstanding the revolutions in France and Haiti, this was a period when social traditionalism was shaken but not yet overthrown. With a handful of exceptions the "rise of the bourgeoisie," and more generally the emergence of new social forces, would be a feature only of the subsequent period. Fully fledged "bourgeois societies" remained in a minority throughout the nineteenth century. A growing tendency toward class formation was a direct consequence or accompaniment of the gradual spread of industrial capitalism around the world, which did not begin until 1830 and reached the most advanced country in Asia—Japan—only after 1870.

Sixth. Economic historians must address the question of when the dynamic of England's "industrial revolution" spilled over into one of general growth beyond British borders. Angus Maddison, a leading statistician of world history, gives a forthright answer: he sees the 1820s as the decade when worldwide stagnation gave way to more dynamic and "intensive" (in the economic sense) development.[60] The little reliable evidence we have about income trends supports the thesis that even in England, early industrialization led to a noteworthy economic upturn only *after* 1820. So, the years between 1770 and 1820 do indeed count as a period of transition from the slow income growth of the first half of the eighteenth century to the faster rates of the 1820s and beyond.[61] Almost nowhere other than in northwestern Europe did the industrial mode of production take root before 1830. Historians of technology and the environment point to a similar break when they suggest that the "fossil fuel age" began around 1820; it was then that the use of stored fossil energy (coal) in place of wood, peat, and human or animal muscle power became a visible option in production processes throughout the economy.[62] Coal gets steam engines moving, and steam engines drive spindles and pumps on ships and railroads. The fossil fuel age that dawned in the first third of the nineteenth century not only made possible the production of goods on an unprecedented scale but also greatly boosted the formation of networks, speed, national integration, and imperial control. Until the 1820s, however, an ancient regime still prevailed in the energy sector.

Seventh. The smallest degree of worldwide synchronization was to be found in the realm of culture. Contacts and exchange between civilizations, though not negligible, were not yet sufficiently strong to impart a general rhythm to the development of "global culture." As regards the exchange of experience among articulate minorities—which underlies Koselleck's concept of a *Sattelzeit*—we know little from non-Western settings for the period around 1800. So far it has been difficult to demonstrate such phenomena as a greater awareness of time in worldviews and cultural semantics, or a general experience of the speeding up of human existence, except in relation to Europe and its settler offshoots. The evidence for this starts to come in thick and fast only in the second half of the nineteenth century. Similarly, the discovery of previously hidden depths and causalities—which Michel Foucault highlighted in the natural sciences, linguistics, and economic theory around 1800—was probably peculiar to Europe.[63] In any event, 1830 marks one of the clearest watersheds in the entire history of European philosophy and arts: the end of the heyday of philosophical idealism (Hegel succumbed in 1831 to the global spread of cholera) and strict utilitarianism (Bentham died in 1832) as well as of the "Age of Goethe" in the arts; the weakening of Romantic currents in German, English, and French literature; the end of the classical style in music (when Beethoven and Schubert fell silent in 1827–28) and the shaping of the "Romantic generation" (Schumann, Chopin, Berlioz, Liszt);[64] and the transition to realism and historicism in West European painting.

All in all, there are good reasons to consider the "true" or "Victorian" nineteenth century as a shortened trunk: that is—as it has been said about German history—"a relatively brief, dynamic, period of transition between the 1830s and the 1890s."[65]

The 1880s Threshold

The 1880s were a time of especially radical change, a hinge period linking Victorianism and the fin de siècle. Of course, in terms of political and military history, the turn of the century also brought profound upheavals for many parts of the world. It may not have marked a striking break in most European national histories, but the final years before 1900 were certainly momentous for China: its unexpected defeat at the hands of Japan in 1895 resulted in a massive loss of sovereignty, and rivalry among the Great Powers had flung the country's doors wide open and triggered an unprecedented crisis that culminated in the Boxer Rebellion of 1900. In Spain, military failure in its war against the United States caused similar reactions in 1898, and today it is still regarded as a low point in the country's history. In both cases the victorious power—Japan, the United States—felt its path of imperial expansion to have been vindicated. The whole of Africa had been in turmoil ever since Britain's occupation of Egypt in 1882. Its conquest of Sudan in 1898 and the South African War of 1899–1902 basically concluded the "division of Africa" and were followed by a less stormy, less traumatic period of systematic exploitation. In the early years of the new century a wave of revolutions swept across the world: Russia in 1905, Iran in 1905–6, the Ottoman Empire in 1908, Portugal in 1910, Mexico in 1910 (the bloodiest of all, which lasted until 1920), and China in 1911. By the eve of the assassination of Archduke Franz Ferdinand in Sarajevo, all these upheavals had given a new impetus to political democratization; the world war would add little new of substance to it. When monarchies started to collapse east of the Rhine River toward the end of the First World War, they had already disappeared or lost much of their power in parts of the world that Europe considered "backward."

These processes added up to a cluster of crises in the age that we have called the fin de siècle. The transition to this age in the course of the 1880s may be characterized by a set of further traits.

First. As in the 1820s, a new threshold was crossed in the history of the environment. Around 1890, minerals (coal and petroleum) moved ahead of biomass in estimates of global energy use—even if most of the world's population still did not directly consume such fuels. The fossil fuel age began after 1820 only in the sense that these became the cutting edge in energy production. Around 1890, however, this tendency gained the upper hand quantitatively on a world scale.[66]

Second. Global industrialization entered a new phase. Japan and Russia experienced what economic historians used to call a "takeoff," that is, a transition to self-sustaining growth. Things were not yet so advanced in India or in South Africa (where large gold deposits were discovered in 1886), but a core of industrial and mining capitalism began to take shape in both countries, for the first time outside the West and Japan.[67] At the same time, the organization of the economy changed in the early industrializing countries of Europe and North America, as a "second industrial revolution" took them beyond steam-engine technology. One can dispute which were the most important inventions, and therefore those most fraught with consequences, but any list would have to include the incandescent lamp (1876), the Maxim gun (1884), the automobile (1885–86), cinematography (1895), wireless transmission (1895), and radiological diagnosis (1895). The most significant for economic history was the technological-industrial application of discoveries in the fields of electricity (dynamo, electric motor, power-plant technology) and chemistry, in both of which the 1880s were the decisive years. The serial production of electric motors alone revolutionized whole branches of industry and commerce that had been little served by the steam engine.[68] Science and industry drew closer together; the age of large-scale industrial research was beginning. This was associated in the United States and several European countries with a transition to large capital concentrations ("monopoly capitalism," critical contemporaries called it) and the spread of limited liability companies that placed managerial employees alongside family entrepreneurs ("corporate capitalism"). New bureaucracies appeared in the private sector, and ever more finely graded hierarchies developed within the growing ranks of the salaried classes.[69]

Third. This reorganization within advanced capitalism produced worldwide effects as large European and American companies increasingly opened up overseas markets. The age of multinational corporations was nigh. Steamship ocean travel and the telegraphic cabling of all continents greatly increased the density of world economic links. European global banks, joined around the turn of the century by US institutions, began to export capital on a massive scale across the Atlantic, as well as from western Europe to eastern Europe, to colonies such as South Africa or India, and to nominally independent countries like China and the Ottoman Empire.[70] Also in the 1880s, the flow of European immigrants to the United States suddenly shot up,[71] and new intercontinental systems of contract labor were developed to transfer Asian manpower to North and South America. The fin de siècle would be the most intense period of migration in world history. All in all, the 1880s brought a surge in globalization that for the first time linked all continents into economic and communications networks.[72] The great expansion of international trade lasted until 1914—or for some regions (e.g., Latin America) until 1930.

Fourth. After the British occupation of Egypt in 1882, a new climate of intense imperialist expansion became perceptible. While the instruments of financial control were perfected and the collaboration between European governments and private capital became ever closer, claims to occupy and, as far as possible, to rule overseas territories came increasingly to the fore. This was the quintessence of the "new imperialism" or "high imperialism." Indirect influence and access to bases and coastal enclaves were no longer enough. Africa became divided on paper and then soon on the ground, and Southeast Asia, with the sole exception of Siam (Thailand), was also incorporated into the European colonial empires.

Fifth. After a time of persistent unrest, new political orders were consolidated in a number of large countries around the world. The process differed in both its character and its causes: the provisional conclusion of nation building (Germany, Japan), a retreat from earlier reforms (the United States after the end of Reconstruction in 1877; the return to strict autocracy in Russia under Tsar Alexander III in 1881 and in the Ottoman Empire under Sultan Abdülhamid II in 1881); a transition to regimes geared to top-down reform (Mexico under Porfirio Díaz, Siam under King Chulalongkorn, China under the Tongzhi Restoration, Egypt under the proconsul Lord Cromer); or a refounding of parliamentary democracy (France in 1880 after the internal pacification of the Third Republic, Britain after the electoral reform of 1884). The results, however, were astonishingly similar: until the new outbreak of revolutionary unrest in 1905, the systems of rule around the world were more stable than they had been in the preceding decades. It is possible to view this negatively, as a hardening of state apparatuses, but also positively, as a revival of the state's capacity for action and a safeguarding of internal peace. It was this period, too, that witnessed the first attempts at state provision of essential services, over and above mere crisis management. The roots were being laid for the welfare state in Germany and Britain, and even in the United States, where the long-term humanitarian consequences of the Civil War had to be grappled with.

Sixth. The standing of the 1880s as a decade of cultural renewal in Europe is probably undisputed. The transition to "classical modernism" was not an all-European but a Western European, or indeed French, phenomenon. It began in painting with the late work of Vincent van Gogh and Paul Cézanne, in literature with the poetry of Stéphane Mallarmé, and in music, a little later with Claude Debussy's *Prélude à l'après-midi d'un faune*.[73] In philosophy, German authors such as Friedrich Nietzsche (especially his major works of the 1880s) and Gottlob Frege (his *Begriffsschrift*, first published in 1879, is the foundation of modern mathematical logic) offered new approaches that were as varied in content as they were influential in their impact. In economic theory, the Austrian Carl Menger (1871), the Englishman William Stanley Jevons (1871), and above all the Swiss Léon Walras (1874) had a worldwide impact in the 1880s that laid the foundations for twentieth-century thinking. Outside the West there seem to have been no artistic or philosophical innovations of comparable radicalism and impact. But meanwhile the press grew in weight—palpably in Europe, North America, Australia, Japan, China, India, Egypt, and elsewhere—while tending to disseminate the latest cultural trends around the world.

Seventh. What was most striking in the non-Occidental world around 1880 was a new critical self-assertiveness, which may be regarded as an early form of anticolonialism or a renewed attempt to draw on indigenous resources in the encounter with the West. Breaking from the sometimes uncritical fascination with which local elites greeted European expansion in the Victorian age, this reflective attitude differed from spontaneous xenophobic resistance, but it would be too simple to describe it as "nationalist" at that time. It expressed itself most clearly in India—where the Indian National Congress (founded in 1885), though remaining loyal to the Raj, campaigned in support of a series of grievances, reminiscent in many respects of the Italian Risorgimento—and in Vietnam, where 1885 is still commemorated as the year that saw the birth of a coherent national resistance to the French.[74] In the Muslim world, individual scholars and activists—for example, Sayyid Jamal al-Din ("al-Afghani")—advocated an up-to-date Islam as the basis for self-assertion vis-à-vis Europe.[75] And in China a young literatus by the name of Kang Youwei formulated in 1888 a kind of reformed Confucianism, thoroughly cosmopolitan and not at all defensive toward the West, that was meant to revitalize the Chinese Empire. Ten years later it would acquire political significance within the ambitious, though ultimately futile, imperial initiative called the Hundred Days of Reform.[76]

Such anticolonial stirrings occurred simultaneously with new forms and levels of protest that emerged among the laboring classes and women in many parts of the world. Obsessions with authority faded, new objectives were set for the protest movements, and more efficient organizational forms were devised. This was equally true of the great waves of strikes of the 1880s and 1890s in the United States and of the contemporaneous movement for freedom and political rights in Japan.[77] The forms of agrarian protest also began to change. In many peasant societies—the entire Middle East, for example—this period witnessed a shift from the premodern militancy of spontaneous uprisings (*jacqueries*) to peasant leagues or organized rent strikes that mounted a strong defense of economic interests.

Subtle Processes

Nevertheless, one must not be too naive or one-sided in looking for watersheds or historic shifts. World history is even less amenable to precisely defined time frames than the history of a nation or continent. An ability to recognize epochal changes comes not from deep insight into a essentialized "meaning" of the age but from study of a number of superimposed time grids. Epochal thresholds are condensations of such fine dividing lines, or, to use another image, derive from a coincidence of clusters of intensified change. At least as interesting as the crude division into epochs are those subtler periodizations that have to be developed anew for each spatial entity, each human society, and each sphere of existence from climate history to the history of art. All these structures help to provide bearings for the layperson's sense of history as well as analytic instruments for the historian.

In his theory of temporalities, Fernand Braudel shows that overlapping histories develop at quite different tempos—from the hourly precision of *l'histoire événementielle* in a battle or a coup d'état to the slow, glacier-like changes of climate or agrarian history.[78] Whether a process is faster or slower is a question of judgment: the answer depends on the purpose behind the observer's argument. Historical sociology and conceptually kindred ways of writing history often proceed very freely with time. In a typical example of this habit, the sociologist Jack Goldstone writes that "within a very short time," between 1750 and 1850, most countries of Western Europe arrived at economic modernity.[79] However, such off-the-cuff statements should not lead analysts of world history to dismiss as pedantic the meticulous chronology of years and months. They need to keep their temporal parameters flexible and, above all, to account for the different speeds and directions of change.

Historical processes do not only unfold within different time frames—short, medium, and long term. They also vary according to whether they are continuous or discontinuous, additive or cumulative, reversible or irreversible, decelerating or accelerating. There are repetitive processes,[80] and there are unique processes with a transformative character. One interesting class of the latter are those that unfold causatively between different fields that are usually kept apart. Here historians, for example, refer to environmental effects on social structures or to effects of mentalities on economic behavior.[81] If processes unfold in parallel, they often relate to one another in a nonsimultaneous manner; they are classified and evaluated differently within the same natural chronology by the measure of non-chronological phase models.[82] When compared with the challenges of describing such finer temporal structures, the division of history into "centuries" is no more than a necessary evil.

5   Clocks and Acceleration

Cyclical and Linear History

The temporal structures that historians enlist as aids are never created entirely out of the perception of time that historical subjects can be shown to have had. If that were the case, there would be not a binding chronology but a chaos of different cultures of time, each one self-sufficient in relation to the others. Only when astronomical-mathematical reconstruction plus the linear succession of narratives provide the twin bases for a secure chronology can the perception of time contribute to internal differentiation within history and histories. Temporal regularity is necessary to experience acceleration.

World history often involves unusually long chains of consecutive effects. Industrialization, for example, can be dated to a period of several decades in each individual European country, but as a global process it still has not come to an end. Despite many national peculiarities, the impetus of England's "Industrial Revolution" is still detectable in a number of Asian countries; China today displays some of the side effects of Europe's early industrialization, such as ecological depredation and untrammeled exploitation of human labor.

The idea of historical movement as "round-shaped" rather than linear-progressive should by no means be written off as the expression of a premodern vision. Nor is it analytically worthless. Economic historians work with models of production and trade cycles of varying length, the discovery of which was an important theoretical event in the nineteenth century.[83] And "long waves" of imperial control and hegemonic supremacy have proved an illuminating idea in studies of the global distribution of power.[84] The West has known both linear and cyclical historical movement, but since the eighteenth century it has adopted progress, no matter howoften blocked or even reversed, as its guiding temporal template.[85] Other civilizations only later took this over from Europe. Some—like the Islamic world—stuck to their own ideas of linearity: history not as constant development but as an interrupted succession of moments.[86] It should at least be considered whether the modern science of history can accept such conceptions as appropriate for the reconstruction of historical reality.

Let us take the example of Michael Aung-Thwin, an American expert in Burmese history who postulates a spiral shape for the social history of Southeast Asia until the second third of the nineteenth century. What led him to this hypothesis—for him it is no more than that—was the conflict between the historian's assumption of evolution, progress, and cause-effect relations and the anthropologist's reliance on structure, analogy, homology, and reciprocity. Historians are liable to conclude prematurely that the changes they observe in a particular period of time are *permanent* transformations. Aung-Thwin's account, by contrast, sees the history of Southeast Asia in terms of "oscillation" between an "agrarian-demographic" cycle (in countries focused on their internal economy) and a "commercial" cycle in coastal cities and political entities. Burmese society, for instance, after many changes in the middle of the eighteenth century, returned to a situation very similar to that which had obtained in the glorious Pagan dynasty of the thirteenth century. This was possible because of the strength of Burmese institutions.[87] British colonization, which subjugated Burma in stages between 1824 and 1886, undermined this strength, but it was only the coming of revolution and national independence in 1948 that invalidated the old model of historical movement for good.

We do not need to form an opinion about how much of this stands up as a general interpretation of Burmese and Southeast Asian history. Another example would have served the same illustrative purpose. What is at stake is a general argument: from around the 1760s onward, European philosophers agreed on the idea that Asia was "stagnant" or "stationary" in comparison with the dynamic societies of Western Europe.[88] Hegel elaborated this view at considerable length and with a great deal of sophistication in his lectures on the philosophy of history delivered in Berlin in the 1820s. Not long afterward a cruder version gained currency, and European authors routinely spoke of "peoples without history," among which some of them included not only "savages" without a written language or a state but even the Asiatic high cultures and the Slavs. This refusal to accept that different cultures can participate simultaneously in a common space-time has been rightly criticized as a crude instance of "binary simplification,"[89] which sees in Asia's past only the eternal return of the same or merely superficial dynastic-military complications. But the other side of this view is no less problematic: to bathe the whole of history—if only "modern" history—in the uniform glow of European concepts of progress. The sociological modernization theory of the 1960s fell into this trap with its vision of history as a competitive race, with the efficient North Atlantic out ahead and other regions as stragglers or late developers. Keeping open at least the possibility of nonlinear historical movement frees us from the false alternative of binary simplification or Eurocentric homogenization.

Reforming Time

We get closer to a history of nineteenth-century mentalities if we consider which experiences of time may have been characteristic of the age. This is a case of cultural construction and is one of the favorite criteria used by anthropologists and cultural theorists to distinguish civilizations from one another.[90] Indeed, there is scarcely a more demanding or productive starting point for a comparative approach to cultures.[91] Conceptions of time vary greatly both on the level of philosophical or religious discourse and in everyday behavior. Can anything sufficiently general be said about images and experiences of time in the nineteenth century?

No previous age had developed such uniformity in its measurement of time. At the beginning of the century there were myriad times and temporal cultures specific to particular locations or milieux. By its end the order of world time had settled over this reduced, but not entirely vanished, multiplicity. Around 1800 no country in the world had a synchronized time signal beyond the limits of a particular city; every place, or at least every region, adjusted its clocks by its estimation of the solar noon. By 1890 the measurement of time had been coordinated within national frontiers, and not only in the advanced industrial countries. This would not have been possible without technological innovations. The standardization of clock time was a challenge that occupied many engineers and technicians—even the young Albert Einstein. Only the invention and introduction of telegraphic electrical impulses made a solution practicable.[92]

In 1884 an international conference met in Washington, with delegates from twenty-five countries, and approved a single "world time" (the one we still use today), dividing the globe into twenty-four time zones each of 15 degrees of longitude. The driving force behind this historic agreement was a private individual, Sandford Fleming, a railway engineer who emigrated from Scotland to Canada, and who may safely be described as one of the most successful "globalizers" of the nineteenth century.[93] Advocates of time reform had been proposing similar plans since the beginning of the century, but governments had shown little interest until the 1880s. The logic of train timetables had cried out for coordination, but the actual work of reform had dragged on and on. As late as 1874, railroad time in Germany was calculated on the basis of local times in big cities, each of which had to be precisely measured and officially monitored.[94] Passengers had to calculate for themselves the hour at which they would reach their destination. In 1870 the United States had more than four hundred railroad companies and seventy-five different "railroad times"; each passenger had to report to the counter in accordance with the time in use for his or her journey. A first step toward standardization was the electrical synchronization of clocks for the reckoning of time within a single railroad company.[95] But where was the measure to be taken from? Since the eighteenth century, sailors had largely agreed on a standard time that took the longitude of the Royal Observatory at Greenwich as the zero meridian, and since 1855 some 98 percent of all public clocks in the United Kingdom had used Greenwich Mean Time (GMT), even though this became compulsory only 1880.[96] In 1868 New Zealand became the first country in the world to make GMT official. In the United States, where the coordination problems were of a different magnitude, a GMT-based national standard time was introduced in 1883 with four geographical time zones. This was the idea that caught on at an international level the following year, with adjustments in the many cases where the national territory was spread out over a wide area.[97]

Standardization occurred at two levels: within and between nations. Not infrequently the international coordination came first. In the German Reich, which was small enough to dispense with separate eastern and western time zones, an official standard time came into effect only in 1893, after the aged Field Marshal Helmuth von Moltke, the nation's foremost military authority, had made a moving plea in the Reichstag five weeks before his death. France adopted GMT as late as 1911. What were the reasons for its revealing hesitation?[98]

It is a remarkable paradox that the major moves toward international standardization—the same is true of weights and measures, postal and telegraphic communications, railroad gauges, etc.—went hand in hand with the strengthening of nationalism and nation-states. For this reason Sandford Fleming's plans met with fierce resistance in France. When the Washington conference of 1884 was considering the proposal to accept Britain's imperial observatory on the Thames as the zero meridian, Paris much preferred to see its own "older" meridian observatory play that role (there were also any number of other suggestions, from Jerusalem to Tahiti). However, not only had the Greenwich meridian long been in use in ocean navigation; the American railroads had already set their clocks to GMT, no doubt in acknowledgment of British hegemony that was freely given, not imposed. The French objections therefore had no practical chance of acceptance.

In 1884, relations between France and Britain were not particularly bad, but each of them had staked a claim to represent the peak of Western civilization. It was therefore no trivial matter whether Britain or France was the reference country dominating global standard time. France even offered a deal: it would accept that the zero meridian should run through a district of London if the British agreed to adopt the metric system of weights and measures. As everyone knows, that did not happen; an attempt to decimalize time, back in year II of the Revolution, had likewise been an utter failure.[99] Of course, no one could force the French to join an international time system. In the mid-1880s every city in France still had its local time adjusted to the height of the sun; the country's railroads ran to Paris time, which was 9 minutes and 20 seconds ahead of GMT. In 1891 a defiant law made this Paris time the *heure légale* throughout the country. In 1911 France finally adopted the universal time standard, essentially dispelling the anarchy in European time. The French example shows that national uniformity did not necessarily precede international standardization, and that global regulations did not automatically cancel national specificities. Tendencies to the nationalization of time were also present during the period of its universalization. But at least in this case the tendency to standardization was victorious in the end.

Chronometrization

All this took place in societies that were already wedded to precise timekeeping. The ubiquity of clocks and the obedience of their owners and users to the dictates of mechanical time struck many Asian or African visitors to countries like Britain and the United States as notable. A standardized time was possible only in societies that had agreed to measure time and grown used to doing so—that is, in clock societies. It is hard to say when not only academics, priests, and princes but whole societies became subject to chronometrization. Probably the threshold was reached only with the industrial mass production of cheap timepieces for the private living room, bedside table, and waistcoat pocket in the second half of the nineteenth century. This "democratization of the pocket watch," as David Landes described it, put punctuality within the reach of all. Annual world output of pocket watches climbed from 350,000–400,000 units at the end of the eighteenth century to more than 2.5 million in 1875, at which point the manufacturing of *cheap* timepieces had been extant for only a few years.[100] The main producer countries were then Switzerland, France, Britain, and the United States. It is not known how many watches found their way into non-Western pockets. In any event, like the commanding heights of world time, the devices for its measurement were mainly in the hands of white males; the world divided into the watch owners and the watchless. Missionaries and colonial rulers made new time resources available, but in doing so established their monopoly control of time. Lewis Mumford's observation that the clock, not the steam engine, was the most important mechanism of the industrial age is applicable at least for the non-Western world.[101] The clock was incomparably more widespread than the steam engine. It ordered and disciplined societies in a way in which production technology alone could not have done. There were clocks in parts of the world where people had never seen a coal-fired machine or locomotive. Yet the problem of making the prestigious device meaningful to them remained an ongoing challenge.

The watch became an emblem of Western civilization. In Japan, for want of pockets, it was initially worn around the neck or the waist. The Meiji Emperor awarded pocket watches—made in the United States at first—to the best students of the year.[102] By 1880, along with the top hat, laced corset, and false teeth, the watch was considered in Latin America to be a status symbol of the Western-oriented upper classes. In the Ottoman Empire, nothing more clearly exhibited the resolve of the state and social elites to introduce Western-style modernization than the public clock towers that Sultan Abdülhamid II ordered to be built in large cities in the last quarter of the nineteenth century.[103] The British did much the same in their world empire—for example, on the occasion of Queen Victoria's Diamond Jubilee in 1897. Such towers, a secular and culturally neutral offshoot of the clock-bearing church tower, made time publicly visible and in most cases audible. China, for its part, remained largely content with drum towers and their purely acoustic time signal until well into the twentieth century.

The spread of mechanical chronometry contributed to the quantification and continuation of labor processes. In the preindustrial world, E. P. Thompson argued in a famous essay, labor followed an irregular and uneven course. In the nineteenth century, however, as the division of labor intensified and production was organized within ever larger and more capital-intensive firms, entrepreneurs and market forces enforced a stricter time regime and a longer workday. Workers who moved from agriculture or handicrafts into the early factories found themselves subject to a strange new concept of abstract time represented by clocks, bells, and penalties.[104] This sounds a plausible account, all the more attractive in that it places English factory workers in a situation of social discipline and cultural alienation similar to that of workers in later-industrializing countries or subjugated colonies. Thompson's thesis, with its critique of modernity, thus appears to be universalizable. The clock everywhere became a weapon of modernization. Yet this seems to have happened later than Thompson claimed. For, even in Britain, clocks that told the precise time in accordance with standard norms came into widespread daily use only toward the end of the nineteenth century.[105]

It is a good idea to keep the quantitative and qualitative sides of this argument separate. Karl Marx already believed that the workday had been appreciably lengthened, and many other contemporary witnesses confirm that the beginning of industrial factory production was often, or almost always, associated with an increase in the number of hours worked by individuals; workdays as long as sixteen hours appear to have been normal in the early period of cotton-spinning machines. It is true that the full picture is difficult to uncover, even with the precise and detailed techniques and quantitative procedures available to the historical sciences, but meticulous studies have established a clear rise in the length of the workday, at least for England's early industrialization up to 1830.[106] This upward trend, over a period of roughly eight decades, was accompanied with increased ownership of clocks and watches, which made factory workers more aware of the quantitative demands being made upon them.[107] The struggle for a shorter working week presupposed that workers had an idea of their actual performance. With watch in hand, they could check the extent of the capitalist's impositions.

Qualitatively, therefore, it is questionable whether the clock was really nothing but an instrument of compulsion in the service of the factory owner. And if technological developments are not to be seen as an independent variable, we must ask whether the invention of the mechanical timepiece created a need for precise measurement in the first place, or whether the need had already been present and kindled a demand for the technical means to satisfy it.[108] Wherever precise timekeeping was introduced, it was an instrument of mechanization, and even of the more intensive form that involved the strict metronomization of production and numerous other processes in everyday life. This was emblematic of a time regime more uniform than that experienced in a close-to-nature peasant lifestyle.[109] In the nineteenth century, peasants and nomads were confronted on all sides with this regulation of time that radiated out from the cities.

Those who have learned from experience that the same strict standards of punctuality still do not apply everywhere in the world, will not underestimate the capacity of human beings to resist time and to live simultaneously in more than one temporal order: that is, to cope with discontinuous mundane experiences of time as well as with the abstract time of the clock and calendar.[110] Anthropologists have found many instances of societies without astronomy or clocks that are able to distinguish between "points in time" and ongoing processes and to coordinate their activities precisely in time.[111] E. P. Thompson's appealing hypothesis that the perception of time was a battlefield in the cultural conflicts of early industrial England seems to be of only limited applicability to other regions and other epochs. Its validity has been openly contested in the case of Japan. Japanese peasants of the late Togukawa period (up to 1867), who competed with one another in small economic units overwhelmingly geared to intensive agriculture and craft production for the market, did not by any means live in idyllic harmony with the rhythms of nature but related to time as a precious resource to be used in accordance with a well-thought-out plan. A bad economy of time would spell ruin for the family. When industrialization began around 1880, laborers were already occupied continuously in a flow of work through all seasons of the year. The new discipline of the factory—which in Japan was actually quite lax for a long time—did not feel too oppressive. Unlike their working-class comrades in Europe or the United States, Japanese laborers complained little about the intensity of exploitation and did not make a shorter workday one of their central demands. More important to them was the moral issue that management should recognize them as partners within the enterprise hierarchy.[112]

Things were different on the cotton plantations of the American South before the Civil War, where overseers had long imposed an intense rhythm on "gangs" of slaves and backed it up with extreme violence. Slave owners soon got their hands on the newfangled mechanical timepieces, which they made available as part of the arsenal of labor discipline. Unlike factory workers—whether in England, Japan, or the slave-free Northern states of the United States—slaves were in no position to argue with their bosses over working hours. Here the clock was much more plainly a one-sided instrument of compulsion, although in the end it changed the life of the slave owner too; master and slave shared the new world of pitilessly ticking hands. The clock also served another, quite different purpose, insofar as the plantation oligarchy tried to use it to link up with cultural practices in the more developed North. As in countless other situations around the world, a privately owned timepiece became one of the most potent symbols of modernity.[113]

On closer examination, it is necessary to make a number of further distinctions: between village and city time, men's and women's time, old people's and young people's time, military and civilian time, musicians' and master builders' time. Between the objective time of the chronometer and subjectively experienced time stands the *social time* of "typical" life cycles in the family and work. This in turn exhibits various mixes of cultural norms, economic tasks, and emotional needs. One question especially worthy of consideration is whether and under which circumstances social time was also experienced collectively, for example, as the cycle of a generation.

Acceleration

Was acceleration the characteristic experience that exceptionally large numbers of people shared as they moved into the nineteenth century?[114] In the wake of the steam engine and its mechanical combination with wheels and ship's propellers, the nineteenth century became the age of the speed revolution. Although the dramatic increases in speed made possible by air travel and greatly improved road transport would come only in the next century, the railroad and the telegraph marked a decisive break with all previous history. They were faster than the fastest horse and carriage or the fastest dispatch rider. The conveyance of people, goods, and news was released from the shackles of the bio-motor system. This development had no causes other than technological ones. However different the cultural reactions and modes of employment, the effects of rail travel were in principle the same all over the world.[115] The experience of *physical* acceleration was a direct consequence of new technological opportunities.

The fact that the railroad had been invented in Europe was less significant than that it spread across whole continents. The railroad was culturally neutral in its *potential* uses. But the same was not true of its *actual* use; there were many different ways to deploy it. It has even been claimed that the Russian public showed little enthusiasm for the fast speed of rail travel (which was anyway more measured than in the West), because of a cultural preference for slowness that faded only when the observation of other countries showed how backward Russia was becoming.[116] Trains were not only faster but also more comfortable than older forms of land transport. In 1847, en route from Tauroggen (today's Tauragé in Latvia) to Saint Petersburg, the French composer Hector Berlioz spent four days and four nights in an ice-cold sledge, which he describes as a "hermetically sealed metal box," enduring "torments I had never suspected in my most lurid dreams."[117] On the other hand, there was the new calamity of the train crash: in England, where Charles Dickens barely survived one in 1865 on a journey from the south coast to London; in Russia, where Tsar Alexander III suffered the same experience in 1888; as well as in India and Canada. By 1910 at the latest, mechanical acceleration and denaturation of the experience of time was in principle, though not necessarily in fact, a reality for most of the world's population.[118]

This can be stated less confidently about the new temporal categories used in interpreting the world, which Koselleck has analyzed in relation to the *Sattelzeit* around 1800 in Western Europe. The accelerated experience of history was only loosely connected with the greater physical speed of travel and communication. Nor did it attain the same universality. We have already seen how small was the radius of the *direct* influence of the French Revolution. But it also raises the question whether the philosophical-historical model that Koselleck detects in the epochal changes in Europe around 1800—that is, the forcible "breaking open" of a time continuum through revolutionary action in the present—can be found anywhere else in the world.[119] Was there anything comparable in those parts of the world that were not shaken by 1789—and if so, when? Did they doze on in the slumber of premodernity? Or were their "breaking open" experiences different? England, which had beheaded a king way back in 1649, was agitated but not convulsed by the events in Paris. By 1789 the United States had already codified its revolution into a written constitution and was directing it into safe institutional channels.

Where else in the nineteenth century do we find the perception that something totally new has irrupted into familiar life cycles and conventional expectations of the future? Millenarian movements and apocalyptic preachers lived on this effect. They did exist in various regions, from China via North America (among Native Americans as well as whites such as the Mormons) to Africa. As many testimonies show, African Americans experienced the end of slavery as the sudden dawning of a new age, even if the actual "death of slavery" was often arduous, protracted, and disappointing.[120] From the French Revolution to the Chinese Taiping movement of the 1850s, the vision of the new was often bound up with a resolve to link it with a reorganization of time. A calendar that breaks with tradition is itself part of what a revolution is about. However, it should by no means be seen as always involving messianic spiritualization or resistance to the logocentrism of a previously hegemonic culture.

More characteristic for the age since the late eighteenth century is an urge to rationalize the recording of time, to make it more in keeping with the modern world. This was the case in France in 1792, in Japan after the Meiji Renewal of

1868, or in Russia in February 1918 when the Bolshevik regime moved without delay to introduce the Gregorian calendar. The same impetus is evident in the counterstate that the Chinese Taiping rebels sought to construct, whose calendar had eschatological as well as thoroughly practical references. The "new heaven and new earth," we read in the Taiping documents, shall overcome the false teachings and superstitions of the past and enable the peasantry to distribute their labor time in a rational manner.[121] Time was supposed to be simple, transparent, and devoid of magic.

CHAPTER III

Space

Where Was the Nineteenth Century?

1 Space and Time

The relationship between time and space is a major theme in philosophy. Historians can be more modest in dealing with it. A point made by Reinhart Koselleck may be enough for them: "Any historical space constitutes itself by virtue of the time by which it can be traversed, the time that makes it politically or economically controllable. Temporal and spatial questions are always intertwined with each other, even if the metaphorical power of all images of time initially stems from experiences of space."[1] The geographer David Harvey, approaching the issue from a different angle, speaks of "time-space compression."[2] The separation of the two is therefore in a sense artificial. Despite the multiple intertwining, three important differences between space and time should not be overlooked in a historical perspective.

First, space is more directly perceptible to the senses than time. It can be experienced by each of them. In the form of "nature," it is the material foundation of humanity's struggle to gain a livelihood: earth, water, air, plants, and animals. Time limits human life by exerting wear and tear on the organism; space may confront it in particular situations as hostile, overpowering, and deadly. Human communities are therefore arranged within very clearly defined spaces, experienced as natural environments, but not within *specific* times. Time is a cultural construct, beyond the astronomical day-night cycle, the climatic yearly cycle, and the regularities of the ocean tides. Space, however, is first of all a prerequisite of human existence, which is interpreted culturally only at a later point in time.

Second, outside mathematics—the domain of rare specialists—space can scarcely be thought about at all *in abstracto*. It lacks the schematic regularity of chronologically structured and numbered time. Is there pure space, or only relational space that depends on the forms of life that exist within it? Is space a theme for historians at all until human beings start trying to shape it, to invest it with myths, to assign it a value? Can space be anything other than a set of places?

Third, time may be arbitrarily defined in terms of astronomical regularities, but it cannot be materially changed in a way that has an impact on later generations. Labor takes material shape in terrestrial space. Space is more malleable than time: it is the result of its own "production" (Henri Lefebvre). It is also easier to overcome, to subjugate, to destroy: through conquest or material exhaustion, but also through pulverization into myriad allotments. Space is the prerequisite of the formation of states. States draw resources from space. To be sure, space varies in importance from one epoch to another. As "territory," it becomes an intrinsically political value only in modern Europe.

Where was the nineteenth century located? An epoch is defined essentially by time, but its spatial configurations may also be described. The most important model for such configurations is the core-periphery relationship. Cores are places within a larger context where people and power, creativity, and symbolic capital are concentrated together. Cores radiate out and draw in. Peripheries are the weaker poles in asymmetrical relations with cores; they are receivers rather than transmitters of impulses. On the other hand, new things keep appearing on peripheries. Great empires have been formed from the periphery, religions have been founded there, and major histories written. In favorable circumstances, such dynamic peripheries may even become cores. The weight is constantly shifting between core and periphery. Often several cores will either cooperate or compete with one another. The map of the world looks different according to the place you take as your systematic observation point. Political geography does not coincide with economic geography, and the global distribution of cultural cores is different from that of concentrations of military power.

2   Metageography: Naming Spaces

The nineteenth century was transitional in a dual sense for the development of geographical knowledge.[3] First, it was the era when European geography came to dominate the pictures that other civilizations had of the world. By 1900 it had fully taken shape as an independent discipline, with its own research methods, taxonomy, and terminology; its own career paths, academic establishments, textbooks, and specialist journals. Professional geographers thought of themselves partly as natural scientists closely linked to exact disciplines, such as geology, geophysics, or hydrology, and partly as human scientists akin to anthropologists, but in any event no longer as helpmates to the historian. With every manual, schoolbook or map, especially if it had the official stamp of approval, they exercised the "power to name."[4] They became sought-after advisers to governments that were looking to establish new colonies or to "valorize" (that is, exploit) existing ones more scientifically. This model, which first arose in Germany and France, soon found adherents and imitators in other European countries and overseas, its popularity boosted by geographical societies in which amateur enthusiasts rubbed shoulders with representatives of interest groups. Wherever geography established itself as an academic subject, it did so along these European lines; it made little difference whether the importing country was still independent or had been colonized by Europe. By 1920 or thereabouts geography had become a uniform worldwide discourse, even if scientific hybrids developed in countries such as China that had a geo-scientific tradition of their own.[5] The nineteenth century was the age when the outstanding contributions of individual geographers were welded into an academic discipline, an institutionally safeguarded collective undertaking.

The Last Age of European Discoveries

But while the nineteenth century was the *first* phase in the conversion of geography into a science, it was also the *last* age of discoveries. There were still heroic travelers who ventured into regions where no European had previously set foot, still blank spaces on the map to be filled in, and still journeys that could prove highly dangerous for those who embarked on them. In 1847 Sir John Franklin vanished on an expedition to find the Northwest Passage, together with some of the Royal Navy's ablest officers and a set of the best instruments of the time. Only in 1857–59 did a search party discover skeletons and other remains of the Franklin mission, which had set off from England with a crew of 133.[6] The last age of discoveries began in 1768 with James Cook's first circumnavigation, which took the captain and his scientific companions to Tahiti, New Zealand, and Australia. The Franklin debacle cast a cloud over the period when the Royal Navy was the most active force in global exploration.[7] Unveiling exploration of entirely unknown parts of the planet came to an end with the Norwegian Roald Amundsen's dash for the South Pole in December 1911. Afterward, heroic feats were still possible in high mountain ranges, deserts, and deep seas, but there was little more to be discovered.

In the course of the century, travelers journeyed to various parts of the world for the first time and wrote up their accounts. The main regions of discovery were

- sub-Saharan Africa beyond long-familiar coastal strips (visited by the South African doctor Andrew Smith, the British-commissioned German geographer Heinrich Barth, and the Scottish missionary David Livingstone);
- the whole west of the North American continent (to which Thomas Jefferson sent the famous expedition of 1804-6 under Meriwether Lewis and William Clark, but which was first explored by Sir Alexander Mackenzie in 1793 and only fully mapped later in the nineteenth century);
- the interior of Australia (where the Prussian explorer Ludwig Leichhardt vanished without trace in 1848, and which long remained completely unmapped); and
- large parts of Central Asia (about which Chinese geographers had been better informed than Europeans since the eighteenth century, and which after 1860 became a growing field for Russian, British, French, and, in the next century, German travelers and researchers).

Otherwise, people in Europe had had a reasonably good knowledge of world geography since early modern times. This was true not only of Mexico (an old core area of Spanish expansion) and India (about which much had been known even before the colonial period), but also of countries that had never been colonized by Europeans, such as Siam, Iran, or Turkish Asia Minor. Large areas of Asia were so familiar that Carl Ritter, who ranks along with Alexander von Humboldt as the founder of scientific geography, began in 1817 to publish a vast work that eventually ran to twenty-one volumes and 17,000 pages (*Die Erdkunde im Verhältniß zur Natur und zur Geschichte des Menschen*): a summa of several centuries of European reports about the continent. But many of the sources of information were out of date, and Ritter, who was by no means gullible, had great difficulty in extracting the serviceable material. Thus in 1830, European knowledge of China's inland provinces still relied on reports by Jesuits from the seventeenth or eighteenth century; and as for Japan, still tightly sealed from foreigners, not much advance had been made on the classical report of a trip made in the 1690s by the Westphalian doctor Engelbert Kaempfer.[8] In all these cases a fresh pair of eyes were necessary. New expeditions were therefore undertaken, many of them inspired by science managers such as Ritter and Humboldt, Joseph Banks or John Barrow (strategically placed at the British Admiralty), and later with the growing support of organizations such as the African Association or the Royal Geographical Society (founded in 1830).[9] Alexander von Humboldt himself set the standard with his trip to America from June 1799 to August 1804; over the next quarter of a century, he evaluated the results in a series of works centered on his travel report—a key document of the nineteenth century.[10] By 1900, geographical accounts existed for most regions of the world, recognized as standard works representing the best research available at the time.

The geographical exploration of *Europe* paralleled these overseas enterprises; it did not necessarily precede them. In September 1799, a few months after Alexander von Humboldt boarded a ship for Havana, his elder brother Wilhelm set off for Spain. He was breaking new ground there almost as much as Alexander did in the New World. Seen from Berlin or Paris, Spain's Basque provinces were no less exotic than its American empire, and the same was true of other peripheral areas of Europe.[11] Throughout the nineteenth century there continued to be individual travelers driven by a lust for adventure and scientific curiosity. The category also includes a number of women, such as the English globetrotter Isabella Bird, who, though no scientific researcher, keenly observed foreign mores and customs.[12] Two other important figures representative of the age were the imperial pioneer, whose purpose was to "occupy" territories on behalf of his government, and the colonial geographer, who kept a look out for precious minerals, possible farmland and transport links.

The vision of geographers varies in range. Travelers and land surveyors see their immediate surroundings; only in the scholar's study does the larger picture emerge from the mass of descriptions and measurements. Like Carl Ritter, the path-breaking eighteenth-century French cartographers of Asia had never set foot on the continent whose form they drew in such accurate detail. The nineteenth century naturally based itself on the ball shape of the earth, for which recent circumnavigations had provided further clear evidence. But it should not be forgotten that before aerial photography, the ball shape could be seen only from the ground—the perspective of traveling or seafaring contemporaries. The bird's-eye view, not to speak of the view of the globe from the cosmos, was the stuff of fiction, for which the ball offered a mere approximation. In the case of a geological oddity such as the Grand Canyon, the techniques of conventional landscape drawing, which could easily cope with Alpine valleys, were stymied; there was no angle from which the drama of the precipitous gorge could be depicted in a naturalist manner. The graphic artist who accompanied the first scientific expedition to the Colorado River in 1857–58 tackled this limitation by means of an *imaginary* aerial view from a point a mile above the earth.[13]

Names of Continents

Geographers and cartographers have always been the ones who give names to places and localities.[14] However a name came about, it entered the public domain as soon as it appeared on a globe or a well-made map with scientific or political authority. If it was just a question of a single topographical feature—a mountain, river, or town—local names had a chance of being adopted by Europeans. Under the surveyors of British India, it was the rule in the nineteenth century that a place requiring an official name should, after consultation with knowledgeable local people, retain the one in customary use. A famous exception was "Peak XV" in the Himalayas, which in 1856 was named after the retired surveyor-general of India, George Everest—overriding his modest objection that Indians found it difficult to pronounce.[15] In other parts of the world, the names of European monarchs, statesmen, and discoverers were liberally sprinkled around: Lake Victoria, Albertville, Melbourne, Wellington, Rhodesia, Brazzaville, the Bismarck Archipelago, and the Caprivi Strip (in today's Namibia) are just a few examples from a long list.

Even more arbitrary and ideological than these local instances, however, was the choosing of names for large areas. Some have spoken of "metageography" to refer to this spatial schematization of the world, which everyone carries around in their head, usually without being aware of what is involved.[16] Metageographical categories are among the large variety of "mental maps" those that divide the globe into continents and other "world regions." In the nineteenth century the main geographical categories were still in flux, and one needs to beware of anachronism when employing names from a later time. Even the term "Latin America" is less straightforward than it sounds. To the present day, there is still disagreement as to whether the "West Indies" or Caribbean (where English or French, or Creole, is spoken) should be included in it. Alexander von Humboldt, and those who followed in his tracks, did not know the term "Latin America";

his America was the "midnight" or tropical regions of the Spanish empire in the New World, which evidently included Cuba. Simón Bolívar's generation spoke of "southern America." The name "Latin America" was coined in 1861, amid the "pan-Latinism" of the French Saint-Simonians, and soon afterward taken up by politicians. At the time, Napoleon III was seeking to build a French empire in the region—an ambition that came to a shabby end in 1867 with the expulsion of French troops from Mexico and the execution of Maximilian, the French-backed Habsburg emperor of the country. The strategic attraction of the "Latin" tag was that it promised to construct "natural" bonds between the Romance-speaking peoples of France and the Americas.[17]

"Latin America," though, is a comparatively old regional concept. Many other "world regions" are much younger. "Southeast Asia," for instance, came into being in Japan during the First World War, and its wider adoption was due to the fact that in 1943, in the middle of the Pacific War, it became politically necessary to define the position of Lord Mountbatten, at the head of a "South East Asia Command" distinct from the American-dominated military theater.[18] Until then the West had lacked a generic name for this topographically and culturally most heterogeneous region. When Europeans had not referred indiscriminately to the "East Indies," their terminology at a level higher than individual kingdoms and colonial domains had distinguished between a mainland "Further India" (today's Burma, Thailand, Vietnam, Cambodia, and Laos) and the "Malay Archipelago." Until a few decades ago, "Southeast Asians" felt little or no common identity, and the first history of the region as a whole appeared no earlier than 1955.[19]

The picture was similar farther to the north. Early modern maps featured a seldom clearly defined area in the middle of the Asian land mass: "Tartary." This corresponded vaguely to the terms "Inner Asia" and "Central Asia," which even today have not achieved conceptual stability. Russian authors employ them only for the mainly Muslim-populated areas of the former Russian Turkestan, whereas some other usages include Mongolia, Tibet, and the present-day Mongol regions of the People's Republic of China ("Inner Mongolia.") Tibet is often excluded—in which case it does not belong anywhere, since it is also not part of "South Asia." Southern Siberia and Manchuria—which in the eighteenth century were still mostly placed in "Tartary"—have disappeared from any concept of Central Asia. The boundaries between Central Asia and "East Asia" and the "Middle East" have long been controversial, and some authors have proposed a neologism such as "Central Eurasia."[20] An alternative approach to a region of such uncertain shape is a functionalist one that sees it as a pulsating network of exchange, expanding and contracting over the centuries. "Central Asia" is then coterminous with the scope of trade and conquest conducted by peoples of the steppe.[21]

While "Tartary" and "Central Asia" conjured up the mysterious wonderland, scarcely accessible to the ordinary traveler, which Halford Mackinder portrayed in his oft-quoted lecture of 1904 "The Geographical Pivot of History,"[22] the territorial referents of the "Orient" were even less developed.[23] It was essentially a culturally defined term for the lands—including the Ottoman Balkans—inhabited by Arab, Turkish, and Iranian Muslims that European commentators had covered over the centuries with various layers of meaning. It was never clear whether more distant Muslim regions such as the Mogul empire, Malaya, or Java were part of it; and in the nineteenth century, "Orientals" was often used for Indians and Chinese. Nevertheless, this was the only collective term available to Western observers at the time. The expression "Near East" (German *Vorderer Orient*, Russian *Blizkii Vostok*, French *Proche-Orient*) entered diplomatic usage toward the end of the century, when it designated the Ottoman Empire and areas of North Africa (such as Egypt and Algeria) that had once been, but were effectively no longer, part of it. "Fertile Crescent," coined in 1916, was a favorite among archaeologists and had a ring of pre-Islamic antiquity. "Middle East," on the other hand, though of older origin, was popularized around the turn of the twentieth century by the British journalist Valenine Chirol and the American naval officer and military theorist Alfred Thayer Mahan, had no cultural and few historical connotations; it referred to a zone of perceived political instability caused by the weakness of the erstwhile hegemon, the Ottoman Empire. In spatial terns, it designated the zone north of the Persian Gulf, then seen as a major theater in the conflict between Britain and the Tsarist Empire, but some geopolitical commentators included Asia Minor, Afghanistan, or even Nepal and Tibet (which others allocated to "Central Asia"). From a British point of view, the main focus was the strategically vulnerable borderlands of India. Geographical terms that specialists and laypeople alike now use as a matter of course, and which have in many cases been taken over by indigenous elites with varying degrees of enthusiasm, often rooted in geopolitical perceptions during the age of high imperialism.

"Far East"/"East Asia"

The metamorphoses of European spatial semantics are best illustrated by the region known as East Asia. The term is more common in geography and sociological "area studies" than among philologists, since there is no obvious linguistic case for bracketing China, Japan, and Korea together; the three languages are constructed quite differently. Sinology, Japanology, and Korean studies are still separate disciplines, often jealously protective of their independence. But since their origins in the nineteenth century, they themselves have had no difficulty in employing the common term East (or Eastern) Asia. In fact, as a vague, mainly topographical, designation, it first appeared in English, French, and German in the late eighteenth century. But it only became universally accepted in the 1930s, after the rise of the United States as a Pacific power made it seem absurd to continue using the Eurocentric "Far East"; logically, only the term "Russian Far East" (meaning Siberia) would have been plausible. Since then, more outside the region than in the countries directly concerned, attempts have been made to agree on a term such as "Sinosphere" or "Sino-Japanese cultural realm": a historical construct that mainly draws on the virtue of a shared "Confucian" link, though a conceptualization in terms of interaction offers an alternative vision.[24]

As for "Far East," which is still in occasional use, it stems like "Near East" and "Middle East" from the vocabulary of imperialism—and thus from the meta-geographical division of the world according to geopolitical-strategic criteria, which was so popular with geographers and politicians in the fin de siècle. Many statesmen—for example, the viceroy of India and later foreign secretary Lord Curzon—fancied themselves at the time as amateur geographers and indulged in speculations about the rise and fall of various world regions. When the term "Far East" was coined toward the end of the nineteenth century, it had a dual meaning. On the one hand, it took clichés about the Muslim "Orient" and extended them farther east, so that China, Japan, and Korea now appeared as those parts of a generalized "Orient" where the "yellow races" lived. But on the other hand, far more importantly, it operated as a geopolitical-strategic concept. This could appear only after the traditional Sinocentric world order had disappeared. Under European eyes, then, the "Far East" was a subsystem of world politics in which European influence was significant but did not, as in India or Africa, have the solid underpinning of colonial rule. The cultural specificities of the countries in question played only a secondary role; the main point of the concept was to define operational areas for the Great Powers. The geostrategic center of gravity of this Far East lay in the Yellow Sea and increasingly in Manchuria—regions that were more and more considered to be "pivots" (in Mackinder's sense) of Great Power rivalry. The crucial issue was the future of China as an imperial state. Unlike the analogous "Eastern Question" (which referred to the fate of another multinational entity, the Ottoman Empire) however, the "question of the Far East" also concerned the rise of a second, independent military power in the region: Japan.

Although, in terms of power politics, Japan was a major player alongside Britain and Russia in the Far Eastern arena, its relationship to other parts of the "East Asian" region was ambivalent. Korea, historically China's main tributary state, had had few, mostly unhappy, experiences with Japan, but in the Meiji period it came to be seen as a potential addition to the Japanese sphere of influence, and when a favorable opportunity arose in 1910 it was formally annexed. Since the last third of the nineteenth century, and especially since 1890, Japan had been mentally distancing itself from the Asian mainland. As Fukuzawa Yukichi put it in 1885, in his essay "Farewell to Asia" (*Datsu-a*), Japan was geographically but no longer culturally part of Asia, oriented politically and materially to the successful "West" and ever more inclined to belittle its former Chinese model.[25] In contrast to this, however, there was a tendency toward the end of the century to proclaim Japan as the head of a "pan-Asiatic" resistance to the might of the West. This basic contradiction was also present in Japanese attitudes to "East Asia" (*Tōa*): a wish to be a peaceful part of it, but also an urge to dominate and "civilize" the other countries.[26]

Metageographic Alternatives

In the age of Ritter and Humboldt, geographers worked with finer regional grids than in the period *after* the consolidation of a map taking in all "the regions of the world." By the first decade of the nineteenth century, they had left behind the schematism of "compendium geography" and "statistics" prevalent in eighteenth-century Germany in particular and were looking for new spatial entities to take as the basis for study. Carl Ritter was decisive in this regard. Rejecting the fixation on political states, he challenged the existing taxonomies and condemned the unmethodical collection of data in the old manuals.[27] His new, physical classification of the earth's surface featured "countries" and "landscapes" instead of static kingdoms. But this did not prevent him from investigating the material lives and actions of human societies, understood as the theaters of history. In his view, the task of geography was to follow the development of nations—hence of the "individualities" that were important to him—in connection with the "nature of the land." On the other hand, he avoided reducing the life of societies and the "movement of history" to natural constants such as climate. He was no geodeterminist. Ritter saw nature as the "school of the human race," the source of collective identities and particular social types;[28] there were correspondences rather than causal relations between nature and history. He took the descriptive vocabulary that geography had developed in the seventeenth and eighteenth centuries[29] and complemented it with "dynamic" metaphors of growth and activity. Starting from his concept of an integrated "regional geography," Ritter attempted to relate natural features such as mountain ranges or "water systems" to the theaters of history. Again and again he wrestled with the problem of "classifying the parts of the world"[30] with a seriousness that placed him above earlier and many later geographers. In this way he arrived, for example, at a concept of "Upper Asia" (*Hoch-Asien*) that was not shallowly geopolitical but included the specificities of the natural relief as well as the lifestyles of its inhabitants.[31] Instead of lumping everything into "the Orient" or "the Near and Middle East," he differentiated among West Asia (including the Iranian world), Arabia, and the "escarpment land" of the Euphrates and Tigris systems.

Ritter's original names did not become established. But his terminological ingenuity continued in the work of two important geographers of the last third of the century, who, though otherwise having little in common, each resisted the metageographical tendency of the age toward oversimplification. Both the French anarchist freethinker Elisée Reclus, who worked in exile in Switzerland and later Belgium, and the politically conservative but methodically path-breaking Leipzig geographer and ethnographer Friedrich Ratzel strove incessantly to use language in new ways to describe the world. Ratzel's *Anthropogeographie* (1882–91) and *Politische Geographie* (1897) spurned the newly fashionable megacategories in favor of a sophisticated study of landscape types and spatial "locations" in relation to political formations—for example, in the discussion of islands.[32] Reclus, in his last (partly posthumous) work, looked at the world situation shortly after the turn of the century through geographer's eyes, experimenting with an unusual macro-classification that used neither conventional divisions of the world nor geopolitical neologisms. Comparable only to Ritter in his knowledge of the geographical and political literature, he steered clear of a closed concept of Europe and divided the continent into three transgressive zones, each politically and economically related in distinctive ways to the extra-European world: (1) the Latin and German nations, including the whole Mediterranean basin as well as the Ottoman Empire, which in his view was "completely dependent on capitalists";[33] (2) land-based Eurasia from Poland to the Yellow Sea; and (3) maritime Britain and its associates and dependencies (*cortège*) including the whole empire headed by India.[34] Finally, the two Americas and the Pacific Basin (except for the British possessions) constituted an entity that was newly taking shape. Reclus was a relational thinker, not someone who thought in essentialized regional categories. For that reason his work—more than that of Ratzel, who was rather inclined to schematic theorizing—may be seen today as a geographical summa of the nineteenth century, even if it is not representative of nineteenth-century academic geography.

Ratzel and, a fortiori, Reclus were far removed from the theories of "cultural arenas" (*Kulturkreise*) fashionable around the turn of the century in Germany and Austria. Reclus's left-wing political temperament made him especially averse to any geopolitical definition of regional zones. The *Kulturkreis* theories used the steady flow of ethnographic material to construct a series of extensive cultural arenas or civilizations, not merely as methodological aids but as entities to which they ascribed an objective existence. "Cultural arena" thus became the key postliberal concept, supplanting the "individual" in the idealist geography and history of Carl Ritter's generation.[35] Those ideas, later to resurface in the work of Samuel P. Huntington, were a typical fin de siècle phenomenon, expressing an oversimplistic view of the world such as was also to be found in the terminology used by followers of geopolitics.

3 Mental Maps: The Relativity of Spatial Perspective

In order to reconstruct nineteenth-century conceptions of space, we need to keep questioning things that we today consider self-evident. The category of "the West" or "the Western world," for instance—the "community of values" influenced by Christianity that has been counterposed first to the Muslim "Orient," then to Soviet-style atheistic communism, and now again to "Islam"—does not appear as a dominant figure of thought before the 1890s.[36] The opposition between Orient and Occident, the lands of the rising sun and the setting sun, goes back to ancient cosmology and the Greek-Persian wars. But "the Western world" first arose out of the idea of an overarching *Atlantic* model of civilization. To speak of the West presupposes that Europeans and North Americans rank equally in global culture and politics. Such symmetry was not assured in European eyes until the turn of the twentieth century. The coupling of "Judeo-Christian civilization," now a widely used synonym for "the West," is an even more recent development, which had little public resonance before the 1950s.[37]

From the beginning, the idea of "the West" was even less bound to a particular territory than that of "the East." Should it extend to the neo-European settler colonies of the British Empire: Canada, Australia, and New Zealand? How could it fail to include Latin America, especially those countries with a high percentage of people of European origin? Shouldn't we follow the Italian historian Marcello Carmagnani in speaking of "the other West"?[38] In the long nineteenth century it was much more common to speak of "the civilized world" than of "the West"; it was a highly flexible, almost placeless designation. Its persuasiveness depended on whether those who described themselves as "civilized" could explain to others that that was what they really were. Conversely, after the middle of the century, elites all around the world made great efforts to satisfy the demands of civilized Europe. In Japan it even became the goal of national policy to be accepted as a civilized country. Westernization therefore meant not only to adopt certain elements of European and North American culture but, in the most ambitious cases, to gain recognition as an integral part of the "civilized world." This was not something that could be given tangible form or represented spatially on world maps. The "civilized world" and its approximate synonym, "the West," were not so much spatial categories as benchmarks within an international hierarchy.[39]

Europe

Even the category "Europe" was less clear at the edges than one nowadays likes to suppose. Elisée Reclus never tired of reminding his readers of that. To be sure, Europe was seen as being *in some way* a single historical entity and (internally differentiated) living space. A general "European consciousness," over and above the religious self-definition of Christendom, emerged here and there among the elites in the course of the Enlightenment, and for Europe as a whole by the Napoleonic period at the latest.[40] In the *first* half of the nineteenth century, however, a number of contradictory Europes appeared on the drawing board, each one linked to a particular vision of space:[41]

- the Europe of Napoleonic imperialism, conceived and organized around a core area from Tours to Munich, and Amsterdam to Milan, everywhere else being an "intermediate zone" or part of the outer ring of the empire[42]
- the Europa Christiana of postrevolutionary Romanticism, including, as a special variant of limited practical relevance; Tsar Alexander I's Holy Alliance of 1815, in which Orthodoxy, Roman Catholicism, and Protestantism came together in the high-flying rhetoric of a religious renewal under Slav leadership

- the power system of the Congress of Vienna, designed to create stable, peace-preserving balances without an all-embracing ideology referring to common European norms and values[43]
- the Europe of western European liberals (with the historian and statesman François Guizot as its most influential proponent), which, in contrast to the Holy Alliance, sharply differentiated western and eastern Europe and regarded western European solidarity, and especially the Franco-British axis, as more important than any Eurasian commonality
- the Europe of the democrats, who discovered the people as the subject of history (with great literary effect in Jules Michelet's *Le peuple* [1846] and his *Histoire de la Révolution française* [1847–53]) and who emphasized the national idea and a federation of European nations and liked to hark back to the Greek ideal of freedom[44]
- the revolutionary counter-Europe of Marx and Engels's *Manifest der Kommunistischen Partei* (1848), evoking an international workers' solidarity that was at first also European in its core

The British had their own conceptions of Europe. A minority of the political elite—such men as Richard Cobden, the indefatigable advocate of free trade, or John Stuart Mill, the liberal philosopher and economist—were internationalist and in some cases outspokenly Francophile; while a majority did not think of the British Isles as part of the European continent, rejected it as a model, and favored remaining outside a continental balance of power. When racial doctrines began to proliferate in Europe in the 1880s, a British equivalent glorified the global dominance and civilizing diffusion of "the Anglo-Saxon race," by no means limiting itself to the European continent.[45]

Whoever believed in the 1870s that Europe was no more than a geographical concept reflected a general feeling of disgruntlement in an age when older revolutionary, liberal, and even conservative solidarities had vanished and Europeans had again been waging war with one another. He or she was not only making a political diagnosis but also expressing a particular understanding of space: a kind of great-power Darwinism. The Great Powers were locked in rivalry with one another and looked down on smaller European states as potential troublemakers. Countries such as Spain, Belgium, or Sweden were of little concern to, and were not taken very seriously by, Britons, French, or Germans. Ireland, Norway, Poland, or the Czech lands did not even exist as independent states. The idea of a European pluralistic order consisting of states of every shape and size, such as lay at the basis of Enlightenment peace projects or the plan for European unification since the 1950s, would have been unthinkable in the late nineteenth century.

Furthermore, in the so-called age of nation-states, the largest and most important players were actually empires. This set up "Eurofugal" tendencies, and not only in Britain's external relations and its associated view of space. France, for example, had closer links to the Algerian coast than to Spain and perceived the

Mediterranean as a less forbidding barrier than the Pyrenees. Spain and Portugal clung to the remnants of their overseas empires, and throughout the century the Netherlands retained in what is now Indonesia a colony that was in many respects the most important European possession after British India. People at the time always saw the Europe of nation-states within a wider imperial framework.

In contemporary perceptions, Europe lacked not only internal homogeneity but even clear external boundaries. The eastern frontier at the Urals was (and remains) an arbitrary, academic construct with little political or cultural significance.[46] In the nineteenth century, it lay hidden in the middle of the Tsarist Empire. This influenced discussion of whether Russia was part of Europe or not—still a question of great moment, including for Western Europe's understanding of itself. Russia's official ideologies sought to minimize the opposition between Europe and Asia. How Russia saw "Asia" was always partly a result of its position vis-à-vis Western Europe. A neo-Petrine push westward during the Napoleonic wars was followed, under Tsar Nicholas I, by a mental withdrawal into the ancestral Slavic lands after 1825. From the time of Peter the Great until the Congress of Vienna, Western Europe had thought of the Tsarist Empire as increasingly "civilized." But after the suppression in 1825 of the moderately constitutionalist Decembrist movement, followed five years later by the defeat of the Polish November Uprising and the beginning of the "Great Emigration" of persecuted popular heroes, Russia became the great bogey of West European liberalism.[47] The despotic rule of Nicholas I was a setback from which Russia's reputation in the West took a long time to recover, if it ever did. Public opinion there tended to see it as a sui generis civilization on the margins of Europe, and many Russians internalized this view.

The Crimean War, which it lost, and resistance to its great-power pretensions at the Congress of Berlin in 1878, drove the Tsarist Empire to look farther eastward. Siberia acquired a new luster in official propaganda and the national imagination, and a major scientific effort was made to "appropriate" it. Great tasks seemed to lie ahead for this redeployment of national forces. The conviction that Russia was expanding into Asia as a representative of *Western* civilization[48]—an idea that had originated in the first half of the century—was now turned in an anti-Western direction by currents inside the country. Theorists of Pan-Slavism or Eurasianism sought to create a new national or imperial identity and to convert Russia's geographical position as a bridge between Europe and Asia into a spiritual advantage.[49] The Pan-Slavists, unlike the milder, Romantically introverted Slavophiles of the previous generation, did not shrink from a more aggressive foreign policy and the associated risks of tension with Western European powers. That was one tendency. But the 1860s, after the Crimean War, also witnessed the strengthening of the "Westernizers," who made some gains in their efforts to make Russia a "normal" and, by the standards of the day, successful European country. Reforms introduced by Alexander II seemed to restore this link with "the civilized world."[50] But the ambiguity between the "search for Europe" and the "flight from Europe" was never dissolved.

"La Turquie en Europe"

While the endless expansion across Siberia meant that Christian Europe saw its northeastern flank as open, both mentally and in reality, an old antagonism governed its attitudes to the southeast. Even after the much-discussed and often-dramatized "decline" of the Ottoman Empire could no longer be ignored in world politics[51]—that is, at the latest after its defeat by the Tsarist Empire in 1774 (Treaty of Küçük Kaynarca)—the Habsburgs thought it necessary to maintain a buffer zone (the so-called Military Frontier) between themselves and their southern neighbor. This area of military settlement, which stretched all the way from the Adriatic coast to Transylvania and survived in some degree until 1881, changed its purpose over time from a defending against Turkish armies to incorporating territories and population groups gradually wrested from the Ottomans. On the eve of its final dissolution this special zone was still an autonomous military state, with an area of 35,000 square kilometers (larger than Belgium and Luxemburg combined).[52] In the nineteenth century the Habsburg Monarchy no longer had any expansionist goals or extra-European ambitions, but it did remain a kind of "frontline state" against the Ottoman Empire. On the other hand, Vienna was very cautious throughout the century in the support it gave to anti-Turkish national movements, since these might easily acquire a pro-Russian and anti-Austrian coloration. In 1815, Ottoman rule still extended as far as Moldavia, and Belgrade, Bucharest, and Sofia were all in Ottoman territory. The war with Russia in 1877–78 lost it roughly one-half of its Balkan possessions, but until the second Balkan War of 1913, "La Turquie en Europe" was still a significant force and appeared under that name on most maps of the time.[53] For centuries the European Great Powers had had diplomatic relations with the Sublime Porte and entered into various treaties with it; in 1856 they formally admitted it into the Concert of Europe, which, though no longer effective in preserving peace, involved a circle of participants comparable to today's "G8" roundtables.[54]

Although history books influenced by Orientalist clichés and "cultural arena" theories long regarded the Ottoman Empire as an alien presence in nineteenth-century Europe, many people living at the time saw things differently.[55] Even someone who, following old Turkophobic traditions and the wave of aggressive philhellenism in the 1820s, condemned Ottoman rule in Europe as illegitimate could not avoid recognizing its de facto sovereignty over a large, if shrinking, area of the Balkans. So long as nation-states had not taken shape in the region, there was no nomenclature with which to visualize the political geography of southeastern Europe. In 1830 "Romania" and "Bulgaria" were ideas that stirred only a handful of activists and intellectuals. The British public discovered the South Slavs for the first time only when a travel report was published in 1867.[56] Scarcely anyone in the North had heard of "Albania" or "Macedonia." And even Greece, which by the grace of the Great Powers had been founded in 1832 as a kingdom of destitute peasants covering only a half of its present territory, played little or no role in the geographic imaginary of "civilized" Europe; it soon fell into oblivion after the great pro-Greece agitation of the 1820s has died down.

All descriptive spatial categories need to be situated historically. The insights of social geography seem to confirm the historian's belief that it would be wrong to treat areas or regions as so many givens. A historical (or "deconstructive") approach must pay close attention to academic studies and school textbooks, to journalistic coverage of world politics, to maps with a contemporary or historical reference, and to the compilation of maps in the atlases of the day. For maps are particularly effective bearers of geographical terminology and instruments of spatial awareness. The most diverse aims could lie behind the nineteenth-century need for precise cartography: not only the familiar ones of transport, warfare, or colonial control but also the urge to make one's nation visible. By now this close link between national awareness and cartographic representation has been extensively studied and documented.[57] Even more than compact nation-states, empires strung out across the world required their possessions to be made visually present. Indeed, there is much to be said for the view that only the publication of world maps, with their famous imperial red from the 1830s on, generated a sense of empire in the British public.

Chinese Spatial Horizons

Mental maps are part of everyone's basic cognitive equipment. The spatial images that individuals and groups have of the world stand in a complicated two-way relationship with each other.[58] Spatial perceptions should not be interpreted only as static world pictures and fixed codes; it is simplistic to speak of *the* Chinese or *the* Islamic vision of space. Images of space are always open to the new; they have to assimilate things that were literally unheard of. The historian Daniel K. Richter once tried to imagine how the original inhabitants of North America came to know of the arrival of Europeans on the East Coast: first a series of dramatic (perhaps contradictory) rumors would have spread with great effect; then strange objects might have begun to appear in villages by various convoluted routes; and finally, at a later stage, the Indians would have came face to face with white men.[59] In this way a completely new native American cosmology was built up over time. Many peoples around the world have had similar experiences.

None of the non-European world pictures could compete with European cosmology in the nineteenth century. Nowhere else did an alternative metageography arise that systematically divided continents and major regions from one another. Three central features of the modern European discourse of geography were: (1) the natural (not cultural or political) equivalence of different spaces; (2) the foundation in precise surveying and measurement; and (3) the reference to large inclusive entities up to the level of the world or, to put it the other way around, the general hypothesis of the earth as a global structure. A fourth characteristic was the autonomy of geographical discourse and its institutional crystallization in a separate branch of science. Premodern maps, for example, are often illustrations of *other* narratives: a religious history of human salvation, a series of travels, a military campaign, and so on. Modern geographical discourse, by contrast, is self-sufficient in both its text and its imagery.

A considerable amount is known about China, which may therefore serve as an example. Official scholars of the Qing period, who acted as administrators and bearers of culture, placed great value on news-gathering from the four corners of the empire. They used cartographic methods to perfect the internal ordering of space. They showed great interest in the boundaries between various provinces and districts, as the territorial organization of government, justice, and military affairs made geographical knowledge indispensable as a means of central control.[60] Surveying and mapmaking were geared to the same foreign-policy objectives that European monarchs were pursuing in the eighteenth century: to stake territorial claims in relation to neighboring states, especially the Tsarist Empire. However, the fully developed Qing Dynasty had no interest in the spatial form of the world beyond its own borderlands. Before the end of the Opium War in 1842, China sent no official travelers to distant countries, did not encourage private journeys, and made less and less use of Jesuits present at the imperial court as a source of information about Europe. The earliest firsthand accounts from overseas came in only after China opened up to the world. In 1847 the young Lin Qian set off from Xiamen (Amoy) to New York as an interpreter on a trade mission, and a year and a half later he returned and wrote a little book of "travel sketches of the Far West" (*Xihai jiyoucao*). So, it was not Europe but America that provided the first impressions of "the West" (as it was already known in China too). As far as we know, the book is the first account published by a Chinese about a Western country: rather skimpy in comparison with the voluminous works by European authors, but surprisingly open-minded about America's material culture and technology, which Lin Qian thought it would be good to transfer to China.[61] Though not cast in the form of a country report, it remains close to reality and by no means speaks of things foreign with dismissive incomprehension. But Lin Qian was a nobody in the Confucian system of scholar-statesmen; his text was not representative of how Chinese saw the world at that time, nor did it reach enough readers to have any real impact.

Much more influential was *Haiguo tuzhi* (Illustrated treatise on the overseas kingdoms), which the scholar and official Wei Yuan published in 1844. The versatile author developed an interest in foreign countries only as a result of the recent defeat in the Opium War. But although he collected much information about Europe and America, his chief focus was on China's long-neglected relations with maritime Southeast Asia; his conservative policy goal was to create (or re-create) supremacy over the hierarchically structured tribute system in the South China Sea, as a means of defending China against the European colonial powers.[62] Wei did not found a scientific tradition of world geography, nor did his successor in the study of foreign countries, the official Xu Jiyu, whose *Yinghuan zhilüe* (Short report on the maritime districts [1848]) was the first comprehensive account of the world political situation from the viewpoint of Confucian realism. Xu knew no foreign languages and had to rely for his source material on the little that had already been translated into Chinese.[63] It was only after 1866 that Xu's book won recognition and a wider readership in officialdom. By then China had had to endure a second war with Britain (this time joined by France) and had made knowledge of the West an urgent priority. In the nineteenth century, the Chinese did not explore global spaces intellectually but only tried to find their bearings in them when this became unavoidable in the mid-1890s.[64]

Japan began earlier to focus on events in the outside world and on their spatial aspects. In the mid-seventeenth century, when Japan sealed itself off from Europeans, the Tokugawa shogunate constructed a kind of foreign secret service to gather intelligence about events in mainland Asia, especially the dramatic conquest of China by the Manchurian Qing Dynasty between the 1640s and 1680s.[65] There were fears that the "barbarian" Manchus would stage a repetition of the attempted thirteenth-century Mongol invasion of Japan. The eighteenth century saw the development of "Holland studies" (*rangaku*), when a small number of European employees of the Dutch East India Company were allowed to reside in the country under strict conditions and close supervision. In the port city of Nagasaki, where they were allocated a special trading post, a whole hierarchy of translators busied itself with the evaluation of literature in Dutch (and later in English and Russian) for the use of politicians and scholars. Consequently in 1800 the Japanese were much better informed than the Chinese about the West and its colonial activities in Asia.

The real "discovery" of the West, however, had to wait until the opening up of the country in the 1850s, when Western geography began to receive widespread attention and methodical attempts were made to collect information and impressions from abroad. In 1871 forty-nine Japanese dignitaries and senior officials, comprising half of the ruling oligarchy, set off on a journey of discovery to the United States and Europe that was planned to last one and a half years. Some things were already known from books, and from nearly two centuries of diplomatic contacts. But much else surprised those who took part in this "Iwakura mission" (named after its leader): not only the strange lifestyle habits of foreigners but also Japan's backwardness in many fields, the differences between Europe and America, the decline of the level of civilization within Europe as one moved farther east from Paris and London, and above all the fact that Europe's spectacular successes had been achieved only within the past few decades.[66]

Two concurrent and in many ways related processes unfolded in the second half of the nineteenth century. *First*, European professional and amateur geographers pursued their program of discovery more systematically than ever before, increasingly competing with one another along national lines. Blank patches on the world map were gradually filled in, and travelers and geographers produced a growing body of knowledge that was of direct use to colonial and imperial rulers. At the same time, *local* cartography became more sophisticated. After all, the first map of Paris that accurately reflected the lay of its buildings dated only from the beginning of the 1780s, not as a service to tourists but as a tool for the resolution of property issues.[67] The result was a new standard in the objective, nonperspectival, geodetically precise depiction of the world—a scientific representation of the earth's surface, not a mental image of it tied to a particular place. The completion of this endeavor before the First World War contributed to the worldwide prestige of Euro-American geoscience. Military leaders were grateful for this material, and the better quality of maps served the Japanese well in their wars against China (1894–95) and Russia (1904–05).

Second, this greater objectivity went hand in hand with a general rearrangement of subjective spatial images. Horizons widened, centers lost their centrality. Many observers suddenly realized that they were no longer in the middle of a world of their own but on the periphery of newly developing larger contexts, such as the international system of states or trade and finance networks. New centers and reference points made their appearance. For example, after 1868, Japan changed its orientation away from nearby China toward the faraway, but militarily and economically closer, "West"—until it rediscovered mainland Asia thirty years later as a space for its own imperial expansion. Societies whose eyes had been turned inland realized that they faced new and unprecedented threats from overseas, but also that new opportunities seemed to be opening up from the same direction. New prospects beckoned to established imperial centers: the Ottoman leadership, for one, as it was gradually being pushed out of the Balkans, began to discover the potential value of Arabia.

4 Spaces of Interaction: Land and Sea

Historical geography works with various concepts of "space" that can also be used for questions relating to world history. Five concepts are especially important; they lead to distinctly different types of narrative.[68]

(a) *Space as a distribution of places—histories of localization.* How are phenomena from different times distributed in space, and is it possible to detect any regularities in the study of their distribution? Such questions suggest themselves in the history of population settlement, for example, including the spatial form of urbanization in the nineteenth century. They also arise in agrarian history with regard to the distribution of land use and enterprise types, or in the history of industrialization in areas close to abundant natural resources.[69] This approach is helpful not least because it can address the spread of institutions, technologies, and practices beyond national boundaries—for example, the printing press, the steam engine, or the agricultural cooperative. It also includes spatial analysis of epidemics or the use of particular languages. All this can be graphically presented on maps in cross sections over time.

(b) *Space as environment—histories of* Natura naturans *and* Natura naturata. How do human communities interact with their natural environment? Whereas the spaces of localized histories are rather empty and formal areas on which relations, proportions, and classifications are projected, those of environmental history may be understood as action spaces. The life of society rests upon natural premises: climate, soil quality, access to water and natural resources. Distance from the sea is also an important variable. The fact that Britain and Japan are both archipelagoes, for instance, cannot be totally disregarded.[70] As far as *world* history is concerned, Felipe Fernández-Armesto has suggested a sweeping environmental approach: he looks for correspondences between environmental conditions and forms of civilization, developing a typology of natural forms as they put their imprint on the evolution of societies: desert, uncultivable grassland, alluvial soil, temperate woodland, tropical lowland, highland, mountain, coastland, and so on[71] The early nineteenth century was the last period when such habitats had an inescapable impact on social life in many parts of the world. In the industrial age, which for most of the world began only after mid-century, intervention in nature was greater than ever before. Industrialization signified a huge increase in the capacity of societies to reshape nature; major technological changes to environmental space as a result of transport, mining, or land reclamation became a hallmark of the times. They were machine-driven operations. Later, the twentieth century became the age of chemistry (use of artificial fertilizer to raise agricultural output, exploitation of oil and rubber, development of synthetic materials).

(c) *Space as landscape—histories of the experience of nature.*[72] The concept of landscape opens up the question of cultural specificity. Societies—or rather, parts of societies—differ according to whether they are conscious of the landscape and, if so, to what degree. Paul Cézanne once remarked that the peasants of Provence had never "seen" the Montagne Saint-Victoire—the mountain near Aix that he painted numerous times.[73] What this implies, more generally, is that agrarian societies labored "naively" in and with natural environments, but did not gaze in admiration at landscapes. Of course, a word of warning about unhistorical, "culturalist" ascriptions is in order here. The Chinese, for example, had no "typical" attitude to the environment: *everything*, from ruthless exploitation and destruction to careful resource husbandry and delicate landscape poetry and painting, could and did appear at various times and in various social constellations.[74] From a transnational point of view, the most interesting processes are transfers—for example, the reception of the Asian garden aesthetic in Europe or the export of certain ideal landscapes by European settlers.[75] The reading of landscapes also has a history, as does the judgment of what constitutes a threat to, or destruction of, nature.

(d) *Space as region—histories of localized identities.* In any space, a central question concerns the factors that underlie its unity and make it possible to speak of an integrated context. In the optic of global history, regions are spaces of interaction constituted by dense networks of transport and migration, trade and communications. But they may also be understood as subnational units, since actual historical interactions, even over large distances, take place most often between territories that are smaller in size than nation-states. Networks are formed between regions. One region dispatches migrants, while another receives them; one region produces raw materials while another, on a remote continent, consumes or processes them. The economic center of the British Empire was not "Great Britain" but London and southern England.[76] Even comparisons are often meaningful or permissible only between regions. Thus the results are different if we compare the whole of Britain with the whole of China or only central and southern England with the regions around Shanghai and Nanjing (which have been economic powerhouses for centuries[77]). Of course, it is not always easy to establish what constitutes a region. Galicia, for example, in east-central Europe, was generally recognized in the nineteenth century as a small distinct region with a multiplicity of sharply divided nations, languages, and religions—one defined more by contrasts than by unity, whose main function was that of a bridge.[78] There are many similar cases of an in-between zone characterized by a high degree of ambiguity and instability.

(e) *Space as arena of contact—histories of interaction.* Spaces of interaction are spheres in which more than one civilization is in ongoing contact with another, and in which, despite manifold tensions and incompatibilities, new hybrid formations repeatedly come about. Since, in the age before air travel, ships were especially important in ensuring multicultural diversity and interaction, the oceans have been among the favorite spaces of global historians.[79] But their main focus of attention has been the early modern period; many interactive contexts are waiting to be explored for the nineteenth century.

The Mediterranean and the Indian Ocean

Ever since Fernand Braudel published his classic work in 1949 (a thoroughly revised edition came out in 1966), the Mediterranean and the "Mediterranean world" have been the prototype of a space of maritime interaction.[80] Despite the successive rise and fall of Roman, Arab, Christian-Italian, and Ottoman dominance, the Mediterranean area was characterized over the centuries by "dense fragmentation complemented by a striving toward control of communications."[81] In the nineteenth century we see contradictory developments. On the one hand, the North established an unparalleled maritime and colonial presence in the form of the riparian French state (with interests in North Africa), the Russian Black Sea fleet (rebuilt after the Crimean War), and above all the external power of Great Britain, which occupied the key strategic points from Gibraltar through Malta and Egypt to Cyprus; meanwhile the once respectable Ottoman navy disappeared as a force, as did the Algerian pirates. On the other hand, the *entire* Mediterranean region, including the Balkans and the French, British, and Italian colonies to the south, fell ever further behind economically as industry progressed north of the Alps. While Black Sea links forged by medieval Genoa were strengthened, Odessa developed into a major port, and the Suez Canal, opened in 1869, transformed the Mediterranean into one of the main transit routes in the world.[82] Historically minded anthropologists have long debated whether, over and above the geographical distances and the opposition between Islam and Latin or Greek Orthodox Christendom, it is possible to speak of a cultural unity at a more fundamental level, expressed for example in the traditional value of "honor."[83] The fact that the question can be posed with even a minimum of justification testifies to the *relatively* high degree of integration of the Mediterranean region.

Concentration on the oceans has long distracted attention from all the Mediterranean-type areas of water that were easier than the high seas for a sailing ship to navigate, and whose clear layout facilitated a high frequency of contacts. The Baltic and the North Sea are such "medi-terranean" seas or secondary arms of the oceans; so too are the Gulf of Guinea, the Persian Gulf, the Bay of Bengal, the South China Sea, and even the North American Great Lakes, around which several Indian civilizations grew up.

A Braudelian approach—which also involves inserting coastal hinterlands and port cities into the picture—was first transferred to the Indian Ocean. The most imaginative author to try this out was K. N. Chaudhuri, who moved from a fairly conventional history of interaction centered on long-distance trade to a grand canvas of four civilizations that developed on the ocean shores.[84] Unlike in Braudel's Mediterranean, where sixteenth-century Christians and Muslims had at least the inkling of a common destiny, historical subjects in the arc stretching from East Africa to Java—and in Chaudhuri's later vision, even to China—lacked any sense that they belonged together.[85] The early strong positing of culturally "alien" agents in trade was a peculiarity of this interaction space. The old notion that Europe's East India trading companies dominated trade in the Indian Ocean before the nineteenth century may have become untenable, but rigorous quantitative research has also corrected the opposite view that early modern European trade in Asia dealt only in unimportant luxury goods.[86]

In the nineteenth century, British rule was the cardinal political fact in South Asia. India was the center of an extensive political-military and economic field of force. It served as the military base for control of the entire Orient; Indian troops (*sepoys*) were deployed in Egypt as early as 1801. The Government of India had a say in everything to do with the security of sea routes and also felt responsible for the British presence east of Calcutta. Trade and migration, each supported by the introduction of steamships and the opening of the Suez Canal, became the most important forces in integration. One peculiarity of the Indian Ocean, in comparison with other oceans, was the absence of neo-European settler colonies—if we leave aside South Africa, which, though a staging post on the route to and from Europe, did not have a strong maritime orientation in its economic structure. Thus, despite the unbroken European presence and control on the

coasts and major islands, the Indian Ocean remained Afro-Asian demographically. It was also constantly crossed by travelers, pilgrims, and migrant workers,
who, in the decades around 1900, formed a transnational arena with a character
in many ways as distinctive as that of the Atlantic.[87]

The Pacific and the Atlantic

Things were different in the Pacific, the largest ocean and the one with the
most islands. Here the nineteenth century brought substantially greater changes
than in the Indian Ocean. The Pacific had from early times been the habitat
of genuine maritime civilizations that had mastered the skills of sea travel—a
kind of classical Aegean on a gigantic scale. The half-millennium before 1650
must have been a long period of island-hopping migration, in which extensive
communication networks were constructed.[88] In 1571 the founding of Manila—
which, with a population of 50,000, would be as large as Vienna by the mid-
seventeenth century—had boosted the role of the Pacific in world trade, one of
the main driving forces being China's demand for silver from the mines of the
Andes and Japan. For a time in the eighteenth century, the European imagination was attracted by no distant place more than by Tahiti and similar tropical
"island paradises."[89] In contrast, Japan, today such a key country on the "Pacific
Rim," was completely uninterested in the ocean, neither sending travelers across
it nor making active use of its commercial potential—its educated classes sensitive only to their own coastal areas. The nineteenth century then brought revolutionary changes that left none of the Pacific countries untouched: the migration from Europe to Australia and New Zealand; the settlement of California
and eventually the whole West Coast of the United States; the opening up of
China and Japan to overseas goods and ideas and their involvement in migration
flows; and not least the attachment of formerly isolated islands to international
networks, with often fatal consequences for populations that lacked the biological and cultural capacity for resistance.[90]

In the case of the Pacific, historians have until now asked fewer questions about
interaction than about mirrored economic development in the coastal regions on
both sides of the ocean. One reason for this is the absence, with the exception of
Chinese workers heading for America, of intensive migratory movements across
the Pacific. Even private journeys by Europeans were unusual. The emphasis on
economic development also reflects the experience of the second half of the twentieth century, when California, Australia, and Japan together, though not primarily as a result of a Pacific division of labor, became growth engines of the world
economy.[91] The Pacific moved up into the "first world," while the Indian Ocean,
once the trading sea of spices, tea, and silk, fell into third-world status. Way back
in 1890 the Japanese economist Inagaki Manjirō predicted the coming of a "Pacific
Age."[92] No such glorious future was foreseen for the Indian Ocean.

The countries bordering on the Pacific were culturally even less cohesive than
those on the Indian Ocean, where Islam was a powerful cement everywhere

(even as far as southern Chinese coastal enclaves), though not in southern India, Ceylon, or the Buddhist lands of Southeast Asia. China and the American West faced each other as cultural extremes: the oldest and the youngest of the major civilizations; two powers with a claim to primacy in their part of the world, which China never gave up even in the decades of its greatest weakness. Politically, the Pacific was never as clearly dominated by a single great power as was the Indian Ocean, which for a time was virtually a British lake. Australia soon became a self-confident part of the British Empire, not at all a flunkey of London. No foreign power could wrest the kind of supremacy that the United States would achieve in the region after the Pacific War of 1941 to 1945.[93]

Apart from the Mediterranean, no space of maritime interaction has been as extensively studied as the Atlantic. Large volumes have been written about its history *before* Columbus, whole libraries about the period since then. A new epoch began in 1492, and no one has doubted the intensity of the two-way traffic that developed between the Old and the New World. However, the forces driving this interaction and the effects resulting from it, as well as the respective shares of action and reaction, have long been the subject of debate. The European use of the word "discovery" has itself been sharply controversial in the case of the Americas; Creole "patriots" were already polemicizing in the eighteenth century against Eurocentric constructions of history.[94] Since Frederick Jackson Turner in 1893 interpreted North America's distinctive polity and society as a gradual advance of the *frontier* of settlement and "civilization," the prehistory and history of the United States have no longer been described only from the viewpoint of the Atlantic coast. Yet another perspective appeared when the Trinidad-born historian and cricket expert C.L.R. James published *The Black Jacobins* in 1938—a book that made the Haitian revolution of 1791–1804 known to a wide public. Since then, histories of the slave trade and Atlantic slavery have moved away from a pure discourse of victimhood. A lively, pulsating "Black Atlantic" has come to light.[95]

As a space of interaction the Atlantic, too, has been more intensively studied and more vividly portrayed for the early modern period than for the nineteenth and twentieth centuries.[96] The historical trade in human beings and commodities has become visible in the square formed by the two Americas, Europe, and Africa, and so too has the context of coercive relations and ideas of liberty, revolutions, and new colonial identities. Whole national histories have been interpreted anew in an Atlantic and imperial framework; the Irish, for example, a self-sufficient island nation, provided the (often reluctant) pioneers of globalization.[97] It remains a major challenge for historians to integrate the British, Iberian, and African Atlantic: what is distinctive about each of these partial systems? How can they be linked up and understood in a higher unity?[98] What would such a unity be, given the fact that the Atlantic—like the other oceans, but unlike the ecologically quite uniform rim of the little Mediterranean—does not form a *natural* arena of history, a "theater" in Carl Ritter's sense of the term? This raises a stream of other questions. How far does the "Atlantic space" stretch into the continental hinterlands? Does it reach as far as the Mississippi, where the Pacific region seamlessly begins? (In the case of the Seven Years' War, which in America, and in a British imperial perspective, was called "the French and Indian War," it has been shown how closely events in the heart of Europe and events deep inside America were bound up with each other.) Or should we stick to the idea of broad coastal strips and draw a clear distinction between "maritime" and "continental," so that there is an outward-looking and an inward-looking France (Nantes vs. Lyons) or Spain (Cadiz or Barcelona vs. Madrid), or a cosmopolitan New England and an introverted Midwest? Is not Sicily closer to North America than to Africa in terms of migration history? Should not Italy be seen as part of an Atlantic space of migration and socialization, at least for the period between 1876 and 1914 when fourteen million Italians left for North America, Argentina, and Brazil?[99]

In the nineteenth century, the Atlantic and the Pacific were subject to different tendencies. The "peaceful" ocean experienced a phase of integration in every domain; the two sides of the Atlantic drifted apart in reality and in people's minds. The slave trade, which involved the most important transactions across the early modern Atlantic, reached a peak in the 1780s and then began to decline, gradually at first, more abruptly in the 1840s. After approximately 1810 the flow of slaves headed mostly toward Brazil and Cuba; the United States and the British Caribbean withdrew from the trade.[100] Ira Berlin has shown that, by the mid-eighteenth century, the growth of plantations had narrowed the lifeworld of North American slaves and increasingly disconnected them from a wider Atlantic world, which he calls "cosmopolitan."[101] A second watershed was the independence of Hispanic America from Spain by 1826, and of Brazil from Portugal in 1823 (under the rule of a son of the Portuguese king), which severed a host of old imperial ties. In December 1823, US President James Monroe declared the eponymous doctrine that, though born out of specific problems in foreign policy, signaled a turning away from the Atlantic and a reorientation westward toward the interior of the continent. Subsequent trends down to the 1890s give the impression that, after a falling out that climaxed in the 1860s with the US Civil War and the French intervention in Mexico, Europeans and Americans drew closer again but with much hesitation. Only the mass emigration from the 1870s on, together with the innovations in transportation technology, makes it necessary to qualify the view that the Atlantic in the nineteenth century was by no means narrower than it had been in the densely entangled Age of Revolutions.

Continental Spaces

Continental land masses lend themselves less readily than bodies of water to fast and intensive contacts. Under preindustrial conditions it was quicker and more comfortable, though not necessarily safer, to travel long distances on a ship than on the back of a horse or camel, in a coach or a sledge, on one's own two feet or those of sedan-chair carriers. Europe was an exception in this respect. Thanks to its structured coasts, abundant harbors, and navigable rivers, travel by ship here played a much greater role than in other parts of the world. But it was possible to combine the technical advantages of land and water transport—in a way that happened elsewhere only in Japan, with its 28,000 kilometers of coastline.[102] The inexhaustible, and easily ideologized, question as to what Europe does and does not have in common with other (supposedly quite different) civilizations should be of less interest to historians than the division of the continent into regions whose boundaries rarely coincide with those of political entities. Another commonplace image that Europe has of itself is that more than any other part of the world, it combines unity with diversity. But how is this unity organized, and how should its elements be called? From Johann Gottfried Herder and his followers in the early nineteenth century comes the Romantic triad that is applied to the history of nations: "Latin—Germanic—Slav." It still echoed strongly in the propaganda of the First World War, and the Nazis later revived it in an extreme form.

Regional groupings of nation-states seem unproblematic in comparison. But even for the innocuous-sounding "Scandinavia," which Pliny the Elder already mentioned in his *Historia naturalis*, it is doubtful whether its regional coherence can be taken for granted for the nineteenth century. The conceptual division between northern and eastern Europe did not exist before the nineteenth century, when Russia was shifted from "the North" to a "semi-Asiatic" East. The prerequisite for a Scandinavian identity was the final collapse of Swedish great-power ambitions with the disappearance of the Polish-Lithuanian dual state in 1795 and the loss of the Grand Duchy of Finland to the Tsarist Empire in 1809. The "Scandinavianism" that appeared around 1848 in small political and intellectual circles was incapable of overarching the nascent nationalisms of the Swedes, Danes, and Norwegians. In 1864 Sweden did not practice Scandinavian solidarity in relation to the German-Danish war. And Norway, which the Swedes had taken from the Danes in 1814, strove for statehood that it finally achieved in 1905. Finland—which, though linguistically separate from the other three countries, has Swedish as a second language—has existed as an independent state only since 1917. A Scandinavian self-image became widespread in the region only after the Second World War. Today the four countries refer to themselves collectively as "Nordic," whereas observers from outside the region usually include Finland in "Scandinavia."[103]

If the right word for a rather clearly demarcated region like Scandinavia causes such difficulties, what is to be said about the conceptual precision and stability of other everyday names? "Western Europe," with the inclusion of (West) Germany, owes its designation to the post-1945 Cold War. As a term for Europe *west* of Germany, it was meaningless *before* the unification of the Reich in 1871 and the sharp clash between German and French nationalism. It presupposes an Anglo-French solidarity that did not exist before the First World War. In foreign policy France and Britain began to move closer to each other only in 1904, but it would be wrong to say that in the long run they shared the same constitutional-democratic values. The British political class still viewed the "despotism" of Napoleon III with grave mistrust. "Western Europe" is therefore a problematic entity as far as the nineteenth century is concerned. "Central Europe," at first a politically innocuous term that geographers dreamed up for a federated economic area rather than a Germanic imperial space, was later usurped by German hegemonism and wheeled into service during the First World War for the pursuit of maximum aims.[104] Only after the end of the Cold War did it again enter discourse as a term encompassing the Czech Republic, Hungary, Poland, and Slovakia. And today further versions, untouched by the lure of a Greater Germany, propose that it should also include Germany and Austria.[105] What has gained most ground, however, is "East-Central Europe"—with a strong anti-Russian note.

For the nineteenth century, which was characterized by the outwardly radiating model of "the West," the Hungarian-American economic historian Iván T. Berend has suggested applying the term "Central and Eastern Europe" to the entire region stretching from the Baltic to the northern frontier of the Ottoman Empire, including the whole of the Habsburg Monarchy and European Russia. He bases his history of this region between 1789 and 1914 on its possession of a distinctive identity and a number of characteristic features that set it apart from Western Europe and other parts of the world.[106] In this imagined cartography, the German Empire belongs in *Western Europe*.

Berend's dichotomy cuts across an older tendency to shun a binary opposition between East and West and to include eastern Europe within an all-European outline of history. The Polish historian Oskar Halecki, for instance, began in the 1920s to consider organizing the whole of Europe geographically and culturally along an East-West axis.[107] The Hungarian medievalist Jenö Szücs gave a major impetus to the "Central Europe" discussion of the 1980s by distinguishing three "historic regions" of Europe.[108] New conceptions of "historical regions" have also followed the model of "East-Central Europe." A stringent historical geography of nineteenth-century Europe based on nonnational regional categories is in the making.

Eurasia

Finally, there are spatial names that are pure constructs: "Eurasia," for example. "Asia" is itself a European invention, and the same is doubly true of the continental amalgam. The usage of "Eurasia" in Russia has been strongly ideological ever since the 1920s (there were precursors in the nineteenth century too), partly in the hope that Russia might play an "Asian card" to trump the West, but partly from fear of the disadvantages of being caught between Western Europe and China.[109] The term can be useful, however, for two reasons.

First, there are human groups who have intensely experienced the connection between the continents, and who may therefore be said to have Eurasian biographies. Among these are the "mixed ancestry" groups in Asia (known in India as the "Eurasian community"), most notably those of Portuguese-Asian and later British-Asian descent. In the early nineteenth century, many Indian Eurasian children of British soldiers had poor chances in the country's European marriage market on account of their low pay and social esteem. But in the early modern period and up to the 1830s, the ability of Eurasians to move and communicate between the two cultures made them essential to the functioning of the colonial system, accepted by Asians and Europeans alike. Predominately Christians, their status was comparable to that of Armenians or Jews. In the second third of the century, however, such European identities became more precarious. No one had had such a fast-track career as Lieutenant-Colonel James Skinner (1778–1841), a highly regarded cavalry commander and Knight Commander of the Bath. But now the "hybridity" of such men and their intermediate social existence was looked upon with disdain. Their upward mobility in the colonial civil service became more restricted than that of Indians and decreased still further as the century wore on. Their poverty, itself a result of limited opportunities, excluded them from the ruling stratum and placed them even below the "poor whites"; European racial theories considered them to be of lesser value. On the other side, they found themselves deprecated by the nationalism emerging in the various Asian countries.[110] Also categorized as biographically Euro-Asian were the colonial families who were linked to Asia for generations as settlers or officials, especially in the Dutch East Indies and British India.[111]

While that was a social and ethnic concept of "Eurasia," the term has been revived for a space of interaction, though mainly in the early modern period.[112] Europeans then felt more closely tied to Asia than they did in the nineteenth century. An Occident-Orient dichotomy with hierarchical intent came into being only in the 1830s.[113] The temporary unification of the Eurasian world from China to Hungary in the Mongol empire and its successors has meanwhile become a standard theme in the writing of world history. In the centuries after those Asian "middle ages," however, a world of plural states existed in continental Asia. An especially important factor here was the persistent integrative power of Islam, borne primarily by Turkic peoples.[114] "Inner Asia," the old heartland of initiatives in world history, was gradually colonized by the three advancing imperial powers: Tsarist Russia, the Sino-Manchurian empire of the Qing Dynasty, and the British hegemon in India. The military power of the Mongols, which outlasted the collapse of their empire in the mid-fourteenth century, was broken once and for all in the 1750s by the Qing armies. By 1860 the Muslim khanates had been incorporated into either the Chinese or the Russian empire. As a result of imperial conquests and interventions, as well as of incipient nationalisms and of the dynamism of Western Europe and Japan, Eurasia became more and more centrifugal and heterogeneous. By the end of the nineteenth century, it was scarcely possible to speak of it any longer as a space of interaction between the empires. Such episodes as Japan's mainland conquests between 1931 and 1945, which barely affected central Asia outside Inner Mongolia, or the construction of a Communist bloc from the Elbe to the Yellow Sea were able to change little in the overall picture. The Eurasian age—if one does not shy at such a pompous formula—began with Genghis Khan and ended sometime before 1800. For the nineteenth century, "Eurasia" is not a spatial category of prime importance.[115]

5 Ordering and Governing Space

The ordering of space is an old responsibility of the state. But not all states order space. Feudal and patrimonial systems, in which local power and customs protect landowners against regulation from above, are unable to achieve this. Only despotic and constitutional states can impose top-down planning targets. The ordering of space requires a central drive for rationalization and the instruments to carry it through. These conditions are found first and foremost in the modern world, but not only there. Three examples should serve to illustrate the range of variation in the nineteenth century: China, the United States, and Russia.

In the case of China, one is struck by a stability of spatial schemas that has no match elsewhere in the world. The division of the empire into provinces goes back to the thirteenth century. The basic template of spatial organization created at that time is still visible today.[116] Since China is equal to Western Europe in size, it would be as if the territorial structure of Europe had not changed appreciably since the Middle Ages. China's provinces are not organically evolved "landscapes" in the sense of European constitutional history; they are administrative constructs. Over many generations, the extraordinary normative strength of this territorial ordering has left its mark on human lifestyles. Even today, strong provincial identities shape the self-image of Chinese people and perceptions of people from other parts of the country, in much the same way that national stereotypes operate within Europe. Sometimes, though not always, the provinces are analytically meaningful units of economic and social geography. But in historical and geographical research, they are now usually combined into eight or nine (mainly physical) "macroregions" such as the Northwest, Lower Yangtze, or Upper Yangtze, each the size of a large European nation-state.[117] In any event, the classical regional names already covered supraprovincial areas, which in the Qing period were often assigned to a governor-general responsible for two or three provinces.

China's stable imperial ordering of space is an exception rather than the historical norm. The only comparable case is the United States of America, whose interstate boundaries have also changed less than those of many European or Latin American nation-states. But whereas the Chinese ordering of space remained the same in the nineteenth century—the empire reordered its peripheries, but the provincial boundaries were unchanged—the United States continually expanded. When it was founded, it was already one of the largest political entities in the world. By 1850 it had tripled in size, and there was still no end in sight.[118] New territories were incorporated in various ways: through straightforward purchase (Louisiana from France, New Mexico and southern Arizona from Mexico, Alaska from Russia), through a treaty with indigenous tribes, and through occupation by settlers or cession following a successful war (Texas). In each case, entry into the Union involved political difficulties of one kind or another. The question as to whether slavery should be permitted in a new territory was extraordinarily explosive, and of course it was the constitutional issue that eventually led to the US Civil War.

The westward movement of white settlers may seem at first sight to have been unplanned and spontaneous. But the United States was the first country in the world—even before the thorough reorganization of space and cadastral registration in Napoleonic France—to apply one simple ordering principle to the whole of its national territory. The American landscape is still today marked by a square planar grid to which state boundary lines as well as the layout of townships and private landholdings often conform. Complaints are often heard that frontiers in Africa were artificially drawn by the colonial powers, but it should be considered that the political geography of the United States was formed with equally deliberate artificiality. This grid, which covers roughly two-thirds of the country, goes back to the land ordinances that US congressional committees worked on and approved in 1784, 1787, and 1796. Its inspiration was the geometrical linear projection of navigational cartography associated with the sixteenth-century cosmographer Gerhard Mercator. A set pattern that could have only a fictitious astronomical character on the high seas was literally engraved on the "ocean-wide," untouched wilderness of North America. In sharp contrast to the confusion that reigned in England, the grid served the purpose of administrative rationalization and legal uniformity. To prevent anarchic appropriation, Thomas Jefferson and other architects of the system aimed to ensure that land was measured first before being sold to private individuals.

In the wake of westward expansion across the American landmass, the grid functioned as "a machine that translated sovereignty claims into property issues, territorial interests into economic interests, and in so doing bound together public and private interests in the acquisition of land." It meant that both the grand politics of nation building and the life choices of individual settlers became capable of planning.[119] It also brought in revenue to the state that allocated land to individuals. In the same way, the imperial government of China began in 1902 to sell off state lands to settlers in Manchuria in order to plug holes in the budget.[120] The policy aims in the United States went beyond mere mapping. Official surveys in the nineteenth century always involved conceiving of large tracts of land as uniform geometric surfaces to be recorded and registered *once and for all*; such was the case in India after 1814, when definitive surveys at every level were supposed to end the cartographic disorder and to bring geographical knowledge to completion. A scarcely older, partly contemporaneous model in Europe was the British-sponsored land survey in Ireland, which went far beyond the one conducted in England itself.[121] In the United States, on the contrary, the point was not (only) to describe the existing lie of the land as accurately as possible. The "grid system" was the outline of a plan for the future.

A third type of central ordering of space occurred in Russia: namely, the top-down founding of cities, which was very rare in modern China or the early United States. For that there needed to be a single will capable of imposing a decision, such as was lacking in American democracy (the founding of Washington as the new capital was an exception), and a capacity to see it through, such as the autocratic Chinese state could no longer summon up after 1800. A Tsarist administrative reform between 1775 and 1785, under Catherine II, divided the empire into forty-four governorships (the later *guberniyas*) and further subdivided these into 481 *uyezds*. Administrative entities with a population of 300,000 to 400,000 now sprang up in place of the historical provinces and oblasts, and since there were not enough cities to go around, a number of village settlements were converted into new ones by fiat. Special care was taken to found cities in the eastern and southeastern borderlands. However, by no means could all the new entities live up to the status of a city, and the promotion process was discontinued in the nineteenth century.[122] But although it did not come to fruition—unlike the American grid system—the Russian reform of territorial administration left permanent traces in the historical geography of the Tsarist space.

The Chinese ordering of space in the medieval period, as well as the later Russian and American equivalents, gave their names to the spaces of the nineteenth century. In other parts of the world the situation is more complicated. The norm is a mixture of indigenous names for regions and designations introduced from outside, the two origins standing in a highly varied relationship with each other. If today's atlas of nation-states is not to be uncritically projected into the past, historians must make an effort to ascertain the geographical nomenclature in the period they are studying. This applies especially to India, Africa, and western Asia. Not infrequently the present names for countries are at odds with nineteenth-century usage. By "West Sudan," for instance, an almost vanished term, people understood the whole gigantic stretch of savannah immediately south of the Sahara, from the Atlantic to Darfur in the country now known as Sudan. Before 1920, "Syria" denoted a geographical region roughly encompassing the territories of today's Syria, Lebanon, Israel, and Jordan. As to India, there used to be four by no means coextensive nomenclatures: (a) the pre-British political geography that survived in the princely states; (b) the British presidencies (Calcutta, Bombay, Madras) and provinces of the colonial period; (c) the federal states of the present-day Republic of India; and (d) the natural divisions of the area used by geographers.

The term "Islamic world" raises special problems of its own, since, as a reference to religious affiliation, it can never be given a precise territorial definition. As far as the modern age is concerned, it should include parts of South Asia, Afghanistan, and numerous islands of the Malay Archipelago. But this evidently does not accord with the conventions. Cultural geographers have proposed various subdivisions of a narrowly defined "Islamic world": for example, a "Turkic-Iranian world" spanning linguistic boundaries, alongside an "Arab world" further divided into "Middle East," North Africa, and the Sahara.[123] Unlike in East Asia and eastern Europe/North Asia, there was no all-encompassing political framework in the nineteenth-century Near and Middle East, even if the power of the Ottoman Empire to shape the administration of the region should not be underestimated.

The ordering of space operates at various levels—from the political restructuring of large regions (as at the Paris Peace Conference in 1919) to the regional planning of railroads down to the microorganization of agrarian property relations. The dissolution and privatization of common lands sometimes took place chaotically, without government regulation, while in other cases it was subject to planning and strict official instructions. Wherever the state levied taxes on land, it became essential to know *who* owed *what* to the revenue, whether from individual owners or occupiers (and no longer from village communities). All around the world, this was the strongest motive for the spread of government activity at the local level. Later came a further drive to disentangle jumbled land-ownership and to consolidate existing plots in a rational manner. Scarcely any of the land reforms of the nineteenth or twentieth century failed to make provisions in this regard.[124] The organization of landholdings is a basic operation of the modern age. It was plainly visible in the huge collectivizations of the twentieth century, in the Soviet Union, East Germany, or China, but otherwise it has mostly remained hidden to historians. There is a rule, however: no state is "modern" without a land registry and the legal right to dispose freely of real estate.

6 Territoriality, Diaspora, Borders

Territoriality

Until now all the considerations in this chapter have presupposed a seamless two-dimensionality. Spaces in the nineteenth century were indeed highly uniform and continuous; they became so as a result of government intervention. Whether in the US land ordinances, in the systematic mapping and recording of landownership from the Netherlands to India, or in the colonial administration of hitherto weakly governed regions, the activity of the state had a thoroughly homogenizing effect. It was a tendency of the age, especially after 1860, to conceive of governance not merely as control of strategic centers but as ongoing activity on the part of regional authorities. This may be described as a progressive "territorialization" or "production of territoriality"—a process that had deep roots in the early modern period, not only in Europe.[125] This territorialization was bound up with the projection of imagined shapes of the nation onto mappable space, with the formation of nation-states, and also with the reform of empires and the consolidation of colonial rule, which was understood for the first time as control over countries rather than simply over trading bases. In line with this revaluation of viable territories, there was a dramatic reduction in the world total of independent political entities—in Europe from five hundred in 1500 to twenty-five in 1900.[126] The Reichsdeputationshauptschluss of 1803 (a law of the Imperial Diet that secularized a large number of clerical territories and licensed medium-sized states in Germany to swallow up their smaller neighbors), the founding of the German Reich in 1871, the abolition of the traditional system of princely domains in Japan in 1871, and the colonial conquest of India and Africa involved the elimination of hundreds of semi-autonomous rulerships. Outside Europe this was not only a consequence of European expansion. In mainland Southeast Asia, for example, the precolonial eighteenth century had already witnessed a fall in the number of independent entities from twenty-two to three: Burma, Thailand, and Vietnam.[127] Diversified dynastic holdings were rounded off. Large states came into being—huge entities such as the United States, Canada (federated in 1867), and the Tsarist Empire, which only now really took possession of Siberia and expanded into southern Central Asia. The sober Friedrich Ratzel was not merely engaging in social Darwinist reverie when he elaborated a "law of the spatial growth of states."[128]

Territoriality was not only an attribute of the modern state but also a kind of monarchical politics. In nineteenth-century Iran, for instance, a country still hardly touched by Western influence, it was an important criterion of the ruler's success that he gained additional land or at least successfully defended the existing borders. Had he proved incapable of this, it would have been a signal for other princes to rise up in arms and seek to overthrow him. Control of the country was the basis of the kingdom (*mulk*), as it was later of the nation (*millat*).[129] In view of Iran's weakness vis-à-vis its imperial neighbors, this was not an enviable situation for a shah.

Discontinuous Social Spaces

One should not think of all spaces as continuous. In the nineteenth century too, the life of a society did not always unfold on a joined-together territory. The most important type of discontinuous social space is a diaspora: that is, a community that lives outside its real or imagined land of origin yet still feels loyalty and emotional attachment to it. It has its roots in forced dispersion from such a "homeland" or migration away from it in search of work, in business activity, or in colonial ambitions. An idealized myth of this (purported) homeland is cultivated down the generations, sometimes including plans to revive or rebuild it. Individual decisions to return there meet with collective approval. The relationship to the destination society is never completely untroubled; it always involves a sense of being tolerated as a minority, and may sometimes evoke fears that a new misfortune will befall the community. Also characteristic are empathy and solidarity with members of one's ethnic group who live in other (third) countries.[130]

Each diaspora differs from others in its origin and historical experience. The following categories may be identified: a victim diaspora (Africans in the Americas, Armenians, Jews), a labor diaspora (Indians, Chinese), a trade diaspora (Chinese, Lebanese, Parsi), an imperial diaspora (Europeans in settler colonies), and a cultural diaspora.[131] Those whose origins went far back *still* existed in the nineteenth century; others came into being in that period, for example, the Armenian diaspora after the beginning of anti-Armenian violence in 1895. Diaspora situations also vary according to the understanding of core and periphery: there may be no spatial core, as with the Jews before the *aliya* (the emigration from Europe to Palestine); a dominant core country that behaves protectively toward the diaspora (China); a colonized core (Ireland); or a foreign-ruled core that gives the diaspora the character of political exiles (nineteenth-century Poland, present-day Tibet). Diaspora groups vary according to their degree of acculturation in the host society. Limited adaptation, often a source of trouble, may sometimes be advantageous. The segregated Chinatowns that sprang up in the United States and elsewhere in the nineteenth century provided a measure of mental and physical comfort and protection for those living in them.

Diaspora formation as a result of mass migration was ubiquitous in the nineteenth century. Only the French stayed at home. China, the epitome of a rounded civilization that no one might be expected to leave, became the source of overseas communities. After a first wave of emigration in the Ming period, the foundations were now laid for a "Greater China." Even the travel-shy Japanese, who had never before left their islands, now asked their government for permission to start a new life in North America. Between 1885 and 1924, a total of 200,000 headed for Hawaii and 180,000 to the North American mainland.[132] The number of Japanese in the United States only became noticeable when they started to be interned after the attack on Pearl Harbor in December 1941. Nations were formed in order to unify those who felt they belonged together ethnically and culturally. Paradoxically, however, the readiness to recognize far-flung diasporas as part of the nation increased at the same time—even if no claims to foreign territory could be derived from the existence of such communities.

Diasporas led to the formation of discontinuous social spaces. For some this was a transitional stage on the road to integration into the society that received them. In many large American cities—New York, for example—Germans formed a compact community but in the long run were not resistant to New World assimilation.[133] In other cases the diaspora existence took forms that went far beyond nostalgia and folklore. "Lateral" networks between the destination society and the society of origin became indispensable sources of support for the overseas "homeland": parts of southern China, India, Sicily, Ireland, and (in the early twentieth century) Greece became downright dependent on financial transfers from compatriots living abroad. In the nineteenth century, the discontinuous social space of the diaspora acquired proportions never seen before—which puts into perspective the thesis that territorialization was generally on the rise. The formation of nation-states in Europe made the lot of minorities more difficult, so that they were more willing to emigrate at moments when overseas labor markets were thrown open. At the same time, improved communications systems made it easier for emigrants to remain in contact with their homeland. The rounding off of national spaces, where government control and emotional attachment centered on a single unambiguously defined territory, went hand in hand with the development of transnational spaces whose territorial moorings were weaker but by no means nonexistent.[134]

Borders[135]

Spaces end at borders. There are many different kinds of borders: those of soldiers, economists, lawyers, or geographers.[136] They seldom overlap. More concepts of borders appeared in the nineteenth century and found ardent champions. Linguistic borders, for example, were not much considered in the early modern period, but postrevolutionary France compiled statistics about languages and was soon entering them on maps; similar maps objects began to appear in Germany in the 1840s.[137] Still, the old military meaning of "borders" remained relevant throughout the nineteenth century: conquered lands were demarcated, borders would again and again become a casus belli. The history of relations with a neighboring country takes material shape in borders. The limits of sovereignty are nearly always expressed in symbols: frontier posts, watchtowers, border architecture. Political boundaries are therefore concrete: physical reifications of the state, symbolic and material condensations of political rule (since the state is constantly tangible there on a day-to-day basis).[138] On the other hand, there are also almost invisible symbolic borders that are sometimes much more stable, and much more difficult to move, than national boundaries.

The idea of political borders presupposes an "egocentric conception of the state" in which might is right.[139] Agreed borders come later—the more peaceful conception of legal theorists. In the nineteenth century there were both: imposed and negotiated borders. For the creation of the state of Belgium in 1830, the Great Powers reactivated the provincial boundaries of 1790.[140] The new German-French border of 1871 was dictated to the side that had lost the war. The political map of the Balkans was redrawn in 1878 at the Congress of Berlin, without any say from representatives of the Balkan countries. In Africa, borders were set by various protocols and conventions among the colonial powers; European commissioners had a good look at the place in question and put up signposts in the landscape. When the high-ranking Conference on West Africa met in Berlin in 1884, chaired by Bismarck, the territorial markers had already been laid down "on the spot" by the governments active in the region (Britain, France, Germany, Portugal, and Liberia). At first it was only a question of customs boundaries, but in the 1890s these hardened into international borders between the respective colonies (plus the independent state of Liberia). The Conference of 1884 also approved borders for territories in which no European had ever set foot, most notably in the Congo Free State belonging to King Leopold II of the Belgians.[141] On the other hand, the borders between the republics of Latin America were largely drawn without any outside intervention.[142]

The traditional view is that in modern times and especially the nineteenth century, borders became more entrenched and border*lands* were reduced to boundary *lines*. But this does not bear the weight of the evidence, given that sovereign territories with borders already existed at a time when personal jurisdiction was the norm. Besides, "linear" frontiers between countries were by no means a European invention carried by imperialism into the non-European world. Two treaties of 1689 and 1727, negotiated when there was an approximate balance of power in the region, bound the Qing empire and the Tsarist empire to a precise demarcation of their sovereignties in northern Central Asia. That such frontiers followed a geometrical line was by no means the rule. It was true of Africa, where roughly three-quarters of the total length of borders (including those through the Sahara) ran in a straight line, but was far less applicable to Asia.[143] There Europeans sometimes followed their ideology of "natural" borders, a dogma from the age of the French Revolution, and tried to establish "meaningful" frontiers.[144]

Efforts to grasp the actual power relations in a region were by no means ruled out. Between 1843 and 1847, a commission made up of Iranian, Ottoman, Russian, and British representatives struggled to come up with a border between Iranian and Ottoman jurisdictions that would be acceptable to all sides. The basis of the negotiations was that only states, not nomadic tribes, would be recognized as having sovereignty over a land, and both parties submitted reams of historical documentation in support of their claims. In practice, of course, the Iranian state could not force all the tribes in its borderlands to submit to its authority.[145] New measuring instruments and geodetic procedures made it possible to fix the borders with unprecedented precision. The border commissions—a second one followed in the 1850s—were unable to solve the problem entirely, but they made both parties more attentive than before to the value of their lands, thereby speeding up the process of territorialization independently of any "nationalism." It became quite common to call in mediators, often representing the British hegemonic power, as in the border demarcation dispute between Iran and Afghanistan.

In Asia and Africa, when the colonial powers introduced their fixed linear boundaries (which they automatically took to be the mark of superior civilization), the prevailing conception was still one of porous and malleable intermediate zones that not only defined spheres of sovereignty but also separated linguistic groups and ethnic communities from one another. These different ideas clashed on the ground more often than around the negotiating table. Usually it was the locally stronger side that prevailed. In 1862, when the Russian-Chinese frontier was redrawn, the Russians imposed a topographical solution even though it often separated tribes belonging to a single ethnic group, such as the Kirghiz. Russian experts arrogantly dismissed Chinese arguments on the grounds that they could not take seriously the representatives of a nation that had not yet mastered the rudiments of cartography.[146]

When a European conception of borders conflicted with another approach, the European one would prevail, and not only because of the power asymmetry. The Siamese state, with which the British more than once negotiated to fix the border with colonial Burma, was a respectable partner that could not be simply duped. But since the Siamese thought of a border as an area within effective reach of a guarded watchtower, they failed to understand for a long time why the British insisted on the drawing of a boundary line. So Siam lost more territory than was necessary.[147] On the other hand, in Siam as in many other places, repeated efforts had to be made to find criteria for the definition of borders. The imperial powers rarely appeared with intricate maps in the areas that required demarcation; "border making" was often an improvised and pragmatic activity, albeit one with consequences that were hard to reverse.

In extreme cases, the razor-sharp borders that the nineteenth century inaugurated had entirely destructive effects. This was especially true in areas with a nomadic population, such as the Sahara, where such a frontier might suddenly block access to pastureland, watering holes, or sacred places. Most often, however—there are good examples from sub-Saharan Africa and Southeast Asia—distinctive societies grew up on both sides of the border membrane, in which the location was used in productive ways appropriate to people's life circumstances. This might mean using the border as a defense against persecution: for example, Tunisian tribes sought refuge with the French-Algerian colonial army; people from Dahomey ran away to neighboring British Nigeria to escape from French tax collectors; and persecuted Sioux followed their chief Sitting Bull into Canada. The actual border dynamic, in which local traders, smugglers, and migrant workers also played an important role, often bore only a loose relationship with what the maps showed. New opportunities for making money offered themselves in local border traffic.[148] Borders had yet another meaning in high-level imperial strategies: a frontier "violation" again and again served as a welcome pretext for military intervention.

The nineteenth century saw the birth and spread of the clearly marked territorial limit as a "peripheral organ" (Friedrich Ratzel) of the sovereign state, equipped with symbols of majesty and guarded by policemen, soldiers, and customs officials. It was at once a by-product and marker of the territorialization of power as control over land became more important than control over people. Sovereign authority was no longer invested in a personal ruler but in "the state." Its territories had to be contiguous and rounded off: scattered holdings, enclaves, city-states (Geneva became a canton of Switzerland in 1813), or political "patchwork quilts" were now seen as anachronisms. In 1780 no one thought it strange that Neuchâtel in Switzerland should be subject to the king of Prussia, but by the eve of its accession to the Swiss Confederation in 1857 this had become a historical curiosity. Europe and the Americas were the first continents where the territorial principle and the state border gained general acceptance. Things were less clear within both the old and the new empires, where borders were partly administrative divisions without deeper territorial roots, and partly (especially under conditions of "indirect rule") reaffirmations of precolonial domains. Borders between empires were seldom marked with an unbroken line in the terrain, and it was scarcely possible to guard them as closely as a European national border. Every empire had its open flanks: France in the Algerian Sahara, Britain on the North-West Frontier of India, the Tsarist Empire in the Caucasus. The historic moment for the state frontier therefore came only in the post-1945 age of decolonization, with the formation of a plethora of new sovereign states. The same era saw the division of Europe and Korea by an "iron curtain," a frontier militarized as never before, whose integrity was guaranteed by nuclear missiles as well as barbed wire. It was thus in the 1960s that the obsession of the nineteenth century with borders came to full fruition.

PART TWO

PANORAMAS

CHAPTER IV

Mobilities

1 Magnitudes and Tendencies

Between 1890 and 1920, a third of the farming population emigrated from Lebanon, mostly to the United States and Egypt. The reasons for this had to do with an internal situation bordering on civil war, the discrepancy between a stagnant economy and high levels of education, the restrictions on freedom of opinion under Sultan Abdülhamid II, and the attractiveness of the destination countries.[1] Even in these extreme circumstances, however, two-thirds stayed at home. The older style of national history had little feel for cross-border mobility; global historians sometimes see *only* mobility, networking, and cosmopolitanism. Yet both groups should be of interest to us: the migrant minorities and the settled majorities visible in all nineteenth-century societies.

This cannot be discussed without numbers. In the nineteenth century, however, population statistics were often highly unreliable. Late eighteenth-century travelers to Tahiti, an earthly paradise that then aroused special "philosophical" interest, varied in their estimates between 15,000 and 240,000; a recalculation on the basis of available clues yields a figure slightly above 70,000.[2] When a national movement arose in Korea in the 1890s, its early activists were outraged that no one had ever taken the trouble to count the number of subjects in the kingdom. Estimates ranged widely between 5 million and 20 million. Only the Japanese colonial authorities established a figure: 15 million in 1913.[3] Meanwhile, in China the quality of statistics deteriorated as the central state grew weaker. The figures most often used today for 1750 and 1850—215 million and 320 million respectively—are cited with greater conviction than the usual later total of 437–450 million for 1900.[4]

The Weight of the Continents

Asia has always been the most populous region of the world, although the size of its lead has varied considerably. Around 1800, some 66 percent of mankind lived in Asia. During the seventeenth and eighteenth centuries, the relative

Table 1: World Population by Continent (in percentages)

	Asia	Europe	Russia	Africa	America	Oceania	World
1800	66.2	15.1	5.0	11.0	2.5	0.2	100
1900	55.3	18.0	7.8	8.4	10.1	0.4	100

Source: Calculated from Livi-Bacci, *World Population*, p. 31 (Tab. 1.3).

demographic weight of Asia had been increasing. This is reflected in the surprise that European travelers expressed about the "teeming human mass" in countries such as China and India. At that time a high population was considered a sign of prosperity; the Asiatic monarchs, we repeatedly hear, could count themselves fortunate that they had so many subjects. Then, in the nineteenth century, Asia's share of the world population fell dramatically, down to 55 percent around 1900.[5] Did Europeans, often unaware of such estimates, suspect this when they had the impression of Asia's "stagnation"? In any event, there was a lack of demographic dynamism. Even today Asia has not regained the share it had in 1800. Who was chipping away at Asia's leading position (see table 1)?

The estimates show that Asia's loss of quantitative share is correlated with the rise of Europe and, more generally, the Western hemisphere.[6] Africa, which probably had a larger population than Europe between 600 and 1700, was afterward rapidly overtaken as Europe's demographic growth accelerated. The population of Europe (not including Russia) soared between 1700 and 1900 from 95 million to 295 million, while that of Africa crept up from 107 million to 138 million.[7] At least demographically, the "rise of the West"—which should include the European immigration to Argentina, Uruguay, and Brazil—was an incontrovertible fact in the nineteenth century. Population growth rates differed within a world total that was increasing more slowly than it has since the latter part of the twentieth century. Between 1800 and 1850, the number of people living on planet Earth rose by a yearly average of 0.43 percent. In the second half of the century, the rate of increase accelerated to 0.51 percent—which is still little in comparison with the 1.94 percent growth rate reached in the 1970s.[8]

Major Countries

In the nineteenth century there were still many countries with a very small population. Greece, at the time of its founding in 1832, had fewer than 800,000 inhabitants, half as many as London. In 1900, Switzerland, with its 3.3 million citizens, equaled present-day Berlin. At the beginning of the nineteenth century the Canadian giant had 332,000 inhabitants of European origin; that number passed one million by 1830. Australia had its first big expansion with the midcentury gold fever, reaching the one-million mark in 1858.[9] Which were the populous countries at the other end of the spectrum? The best data we have are for 1913. For a world ruled by empires it is somewhat anachronistic to take today's

Table 2: The World's Most Populous Political Units in 1913
(in millions of inhabitants)

British Empire	441 (of which UK: 10.4%)
Chinese Empire	437–450 (of which Han Chinese: 95%)
Russian Empire	163 (of which ethnic Russians: 67%)[a]
United States Empire	108 (of which the 50 states: 91%)
French Empire	89 (of which France: 46%)
German Reich (with colonies)	79 (of which Germany: 84%)
Japanese Empire	61 (of which Japanese archipelago: 85%)
Netherlands Empire	56 (of which the Netherlands: 11%)
Habsburg Monarchy	52[b]
Italy	39 (of which "the Boot": 95%)
Ottoman Empire	21[c]
Mexico	15

[a] Census of 1897, including 44% Great Russians, 18% Little Russians, 5% White Russians.
[b] 1910.
[c] Without Egypt, prior to Balkan wars of 1912–13.
Sources: Maddison, Contours, p. 376 (Tab. A.1); Etemad, Possessing the World, p. 167 (Tab. 10.1), 171 (Tab. 10.2), 174 (Tab. 10.3), pp. 223–26 (App. D); Bardet and Dupâquier, Histoire des populations de l'Europe, p. 493; Bérenger, Habsburg Empire, p. 234; Karpat, Ottoman Population, p. 169 (Tab. I.16.B); Meyers Großes Konversations-Lexikon, vol. 17, 6th ed., Leipzig 1907, p. 295.

kind of nation-state as the reference. It is therefore best to be more flexible and to inquire about the major composite polities of the day (see table 2).

What stands out in these statistics? *All* major states were constituted as "empires." Most called themselves such. The only one that did not refer to itself officially in this way—the United States—should nevertheless be counted among the empires; the Philippines, over which the United States took sovereign control in 1898, was one of the most populous colonies anywhere in the world. Although it could not compete with the two giant possessions of British India and the Dutch East Indies (today's Indonesia), its population of 8.5 million was only slightly smaller than that of Egypt and larger than those of Australia, Algeria, or German East Africa. The most populous sovereign country that neither possessed overseas colonies nor constituted a spatially contiguous multinational empire was Mexico; its fifteen million inhabitants put it on a par with a sizable colony such as Nigeria or Vietnam. But Mexico too, torn by revolution and civil war in 1913, was no model of a compact and stable nation-state. In Europe, Sweden with six million inhabitants was the most populous nonimperial country.

Demographic size did not translate directly into world power status. In the age of industrial rearmament, absolute population figures were for the first time in history no guarantee of political weight. China, the strongest military power in Asia around 1750, was by 1913 scarcely capable of foreign policy activity and militarily inferior to the much smaller Japan (with 12 percent of China's population). The British Empire too, which India boosted to number one in the world in terms of population size, was not in reality the all-dominating superpower of the fin de siècle. But it did hold immense human and economic resources, and the First World War would show that it knew how to mobilize them in case of necessity. Table 2 reflects the overall relationship of forces on the international stage, though not exactly in their ranking order. Britain, Russia, the United States, France, Germany, Japan, and to some extent also the Habsburg Monarchy were the only Great Powers in 1913—that is, the only countries with the capacity and the will to intervene beyond their own immediate region.

Certain cases are particularly striking. The Netherlands was a very small European country with a very large colony. Indonesia had a population of fifty million, considerably larger than that of the British Isles and only slightly below that of the entire Habsburg Empire. It was demographically eight times the size of the mother country. The Ottoman Empire's humble ranking in the table may seem surprising, but it is the result of continual territorial shrinkage and a low natural rate of demographic reproduction; the loss of the Balkans should not be given undue importance in view of its sparse population. So, if we leave aside Egypt—which nominally belonged to the Ottoman Empire throughout the nineteenth century (until the British declared a protectorate in 1914) but was never actually ruled from Istanbul—then the population total even before its great loss of territory at the Congress of Berlin in 1878 was no more than twenty-nine million.[10] For demographic reasons alone, the early modern Mediterranean and West Asian superpower could barely stay in the race in the age of imperialism.

Paths of Growth

Asia's high absolute population concealed, as we have seen, a *relative* demographic weakness. Nowhere in the nineteenth century did it attain the extraordinarily high growth-rates that have molded our image of the twentieth-century "third world."

The most astonishing figure in table 3 has to be China's negative population growth in the "Victorian" age, coming as this did after an early modern period when its rate of increase had been higher than the average in Europe or other parts of Asia. The explanation lies not in anomalous Chinese reproductive behavior but in violence on a huge scale. Between 1850 and 1873, unrest of a destructiveness not seen elsewhere in the nineteenth century raged over large parts of the country: the revolution of the Taiping movement, the guerrilla war of the Nian rebels against the Qing government, and the Muslim revolts in the Northwest and in the southwestern province of Yunnan. In the five eastern and central provinces most affected by the Taiping revolution (Anhui, Zhejiang, Hubei, Jiangxi, Jiangsu), the population declined from 154 million to 102 million between

Table 3: Population Growth Rate in the Major World Regions
(annual average percentages in period)

	1500–1820	1820–1870	1870–1913
Western Europe	0.26	0.69	0.77
Russian Empire[a]	0.37	0.97	1.33
United States	0.50	2.83	2.08
Latin America	0.07	1.26	1.63
India	0.20	0.38	0.43
Japan	0.22	0.21	0.95
China	0.41	−0.12	0.47

[a] Within frontiers of USSR (without Poland, etc.).
Source: Simplified from Maddison, Contours, p. 377 (Tab. A.2).

1819 and 1893; a figure of 145 million was not reached again until the census of 1953. In the three northwestern provinces where the Muslim unrest was concentrated (Gansu, Shanxi, Shaanxi), the population fell from 41 million in 1819 to 27 million in 1893.[11] Grand totals of the numbers killed in the Taiping Revolution and its bloody suppression should be treated with great caution—partly because it is hard to distinguish between direct victims of violence and those who died as a result of the mass starvation bound up with revolution and civil war. However, figures as high as 30 million have received the endorsement of leading experts in the field.[12] The most recent estimate, based on research by Chinese historians, even arrives at a total death toll of 66 million.[13] The difference is not really relevant; what counts is the unrivaled scope of this man-made disaster.

The comparatively low Asian growth rates are surprising not only from the vantage point of the early twenty-first century but also against the background of deeply rooted European stereotypes of Asia. The great theorist of population Thomas Robert Malthus, whose analysis of trends in Western Europe and especially England before the nineteenth century has essentially stood the test of time, claimed that Asian peoples, and particularly the Chinese, differed from Europeans in being incapable of "preventive checks" on their fertility that would spare them the extreme poverty resulting from food shortages. At regular intervals, unrestricted population growth had run ahead of a constant level of agricultural production, until "positive checks" in the form of deadly famines had restored equilibrium. The Chinese had not managed to escape this vicious circle by planning their reproductive behavior (e.g., by marrying later). Malthus's account, however, was based on the anthropological premise that "Asiatic man," being less rational and closer to nature than Europeans, had been unable to achieve the leap in civilization from the realm of necessity to the realm of freedom. For two hundred years after its publication in 1798, his thesis was repeatedly left unexamined.

Even Chinese scholars perpetuated the image of China as a country in the grip of mechanisms of poverty and hunger.[14]

Things look different today. The fact that China had unusually low demographic growth in the nineteenth century is not disputed, but the reasons for it are. It is not at all the case that the Chinese reproduced in a blind instinctual manner and were then regularly decimated by ruthless natural forces. New research has shown that China's population was perfectly capable of making reproductive decisions; the chief method was the killing of new-born babies and neglect at later stages of infancy. Evidently Chinese farmers did not regard such practices as "murder"; they assumed that human life began around the sixth month after birth.[15] Infanticide, a low rate of male marriage, low fertility in marriage, and the popularity of adoption added up to a characteristic demographic pattern in the nineteenth century, which was the Chinese response to their straitened circumstances. The low "normal" rate of population growth, which various calamities turned into negative growth in the third quarter of the century, involved conscious adjustment to a falloff in resources. The contrast between a rational, provident Europe and an irrational, instinctual China gone to ruin does not stand up to scrutiny.

Similar points have been made about Japan. A century and a half of population growth under conditions of internal peace came to an end in the first half of the eighteenth century. This slowing was not due mainly to food shortage or natural disasters but rather to a widespread desire on the part of individual families to maintain or improve their living standards—and thus to preserve their status within the village.[16] As in China, infanticide was a common means of population control, but here it served more optimistic goals than a mere adaptation to scarcity. Shortly before the onset of industrialization in the 1870s, Japan left the demographic plateau of its "long" early modern period and entered a period of constant growth that (with the exception of the years from 1943 to 1945) lasted until the 1990s. In its early stages this was driven by higher birthrates, lower infant mortality, and increased life expectancy. The background factors were an increase in domestic rice production and grain imports, together with advances in hygiene and medical care. The demographic stability of Japan in the late Tokugawa era had not been an expression of Malthusian hardship, but had resulted from the achievement of a frugal yet, in global terms, respectable degree of prosperity. The new upward trend after 1870 was a concomitant of modernization.[17]

The most remarkable development in Europe was the biological spurt in British society. In 1750, England (without Scotland!) was demographically the weakest of the leading nations of Europe, with a total population of 5.9 million. The France of Louis XV was more than four times larger (25 million), and even Spain was considerably more populous (8.4 million). Over the next hundred years England rapidly caught up and overtook Spain, and narrowed the gap with France to less than 1:2 (20.8 million for England, Wales, and Scotland in 1850, against 35.8 million for France). By 1900 Britain (37 million) was all but level with France (39 million).[18]

Throughout the nineteenth century it had averaged by far the highest rate of population growth (1.23 percent per annum) of all major European countries. Even the lead over the second-placed Netherlands (0.84 percent) was immense.[19]

The population of the United States grew constantly, in the most exciting demographic story of the nineteenth century. Whereas the German growth-rate in 1870 was still a touch ahead, the United States by 1890 had left *all* European countries (except Russia) far behind. Between 1861 and 1914, the population of Russia more than doubled, keeping pace with England during the same period. The same trend was apparent in the Tsarist Empire as a whole; colonial expansion into Inner and East Asia did not play a major role in this, since the newly acquired territories were sparsely populated. So, at almost the same time as Japan, Russia entered a phase of rapid population growth, especially in the countryside. The Russian peasantry in the last half-century of the ancien régime was among the fastest-growing social groups in the world. Russia offers a rare example for the period of a country whose rural population grew faster than its city-dwellers.[20]

If an attempt is made to organize the quantitative country statistics into a qualitative picture for the century between circa 1820 and 1913, then three categories appear across the continents:[21]

(1) Temperate zones where frontiers could to a large extent be opened up underwent *explosive population growth*, even given the fact that the low starting point makes it appear particularly striking in the statistics. The population of the United States increased tenfold, and similarly extreme trends were apparent in the "neo-Europes" (the "Western offshoots" previously often known as "white settler colonies") of Australia, Canada, and Argentina.

(2) The other extreme of *slow growth bordering on stagnation* was present not only in northern and central India and China (and Japan before roughly 1870) but also in the middle of Europe. Nowhere was this as marked as in France, which in 1750 had the largest number of inhabitants in Europe yet by 1900 had been almost overtaken even by Italy. This slowdown was not due only to dramatic external influences. At the time of the Franco-Prussian War in 1870–71, France experienced an acute demographic crisis graver than any other European country had to face in the nineteenth century. War, civil war, and epidemics meant that there were half a million more deaths than live births—a deficit scarcely exceeded in the years from 1939 to 1945.[22] However, rather than the expression of a permanent crisis tendency, this was an atypical interlude mainly brought on by an earlier decline in fertility that is difficult to explain. Such a decline, nearly always accompanied with higher living standards, appeared in France before 1800 and in Britain and Germany only after 1870. "Depopulation" became an increasingly important issue in public debate in France, especially after the military defeat of 1871.[23] In Spain, Portugal, and Italy too, the pace of population growth was unusually slow, but unlike France, those three countries were not in the vanguard of social modernization. Demographic inertia is therefore not a particularly good indicator of modernity.

(3) There was *very high growth* in Europe (Britain and European Russia after 1860), as well as in parts of Africa (especially Algeria after 1870) and Asia (Java, Philippines, post-1870 Japan), and *fairly high growth*, though never at English levels, in Germany and the Netherlands. The demographic vicissitudes of humanity—this must be our chief conclusion—did not correspond to a simple East-West opposition, still less to the macrogeography of the continents. Dynamic Europe versus the stagnating rest? At least in terms of population history, things are not quite so simple.

2 Population Disasters and the Demographic Transition

The population disasters of the nineteenth century were not confined to one area of the world, but they did spare Europe more than other continents. Ireland was *the* disaster of the century in Europe, the only instance of negative growth. The Great Famine of 1846–52, following a period of rapid population increase that had begun in 1780, canceled the old demographic pattern. Triggered by a fungus that wiped out the potato crop, the famine caused the death of at least a million people—an eighth of the population of Ireland.[24] Emigration, already under way, turned into a flood. Between 1847 and 1854, 200,000 people a year left the island; the total population plummeted from 8.2 million in 1841 to 4.5 million in 1901, with the raising of the age of marriage, promoted by the clergy and landowners, as another important factor. The Irish economy recovered in the second half of the century, thanks in no small part to emigration. While real wages rose for agricultural laborers, Ireland—like Italy and southern China—benefited from overseas remittances.[25] In many respects, therefore, the consequences of the tragedy were overcome within a few decades.

In Europe, after the end of the Napoleonic era, wars and civil wars were a less important source of population loss than they had been in the eighteenth century or would be again in the twentieth. The major excesses of collective violence took place in other parts of the world:

- revolutionary civil wars, as in China between 1850 and 1876 or Mexico between 1910 and 1920;
- wars of secession, as in the United States, where the Civil War of 1861–65 alone cost the lives of 620,000 soldiers, or in South Africa at the turn of the century;[26]
- colonial wars of conquest, as in 1825 and 1830 in Java (probably more than 200,000 killed),[27] after 1830 in Algeria and later in many other parts of Africa, and all century long in the wars of repression and extermination that white settlers and their government bodies waged against the indigenous peoples of the Americas; and finally,
- the only Great Power conflict that took place outside Europe—the momentous Russo-Japanese war of 1904–5.

Meanwhile peace reigned in Europe. No war was fought there between 1815 and the beginning of the Crimean War in 1853, and the latter, like the Wars of German Unification, trailed in violence behind many conflicts outside Europe, not to speak of the great wars of the early modern period or those that lay ahead in the twentieth century. Of the ten deadliest wars between great powers since 1500, not a single one occurred between 1815 and 1914. There was no parallel to the War of the Spanish Succession (1710–14), which is thought to have left 1.2 million dead on numerous fields of battle, nor a fortiori to the wars between 1792 and 1815, which probably led to 2.5 million deaths among the armies alone.[28] All told, in proportion to the total population of Europe, there were seven times more war-related deaths in the eighteenth century than in the nineteenth.[29]

Microbe Shocks and Violent Excesses

Outside Europe it was still possible in the nineteenth century for "microbe shocks" to take whole populations to the brink of extinction. In 1881, after a series of diseases were introduced into Tahiti, the population fell to a low point of 6,000, less than a tenth of the total at the time of Bougainville's and Cook's famous visits to the island in the 1760s. For similar reasons, the number of Kanaks in French New Caledonia fell by 70 percent in the second half of the nineteenth century. In Fiji, in the year 1875 alone, more than a quarter of the population of 200,000–250,000 died as a result of a flu epidemic.[30] Several indigenous peoples in North America were wiped out by smallpox, cholera, or tuberculosis; most of the global pandemics of the nineteenth century also reached the native peoples in the New World. After the beginning of the Gold Rush, it was not so much disease as a frontal assault on their entire way of life that reduced the population of indigenous Californian peoples from about 100,000–250,000 in 1848 to 25,000–35,000 in 1860. Behind these figures lurk terror and mass murder up to the point of genocide.[31] Between 1803 and 1876 the native population of Tasmania fell from approximately 2,000 to zero. Before 1850, when the lawless years gradually came to an end, hunts for Aborigines were a regular occurrence in Australia; the killing of them went unpunished, and since resistance was not uncommon some whites also died in skirmishes and ambushes. Probably every tenth "unnatural" death among the Aborigines resulted directly from an act of violence. Outbreaks of smallpox (one was recorded as early as 1789, a few months after the arrival of the first Europeans), together with cultural stress and a general worsening of the material conditions of life, were responsible for a dramatic decline in the indigenous population.[32] It is likely that just before 1788 some 1.1 million Aborigines were living in all parts of Australia; by 1860 there were no more than 340,000.[33]

It is hard to put a serious figure on the total loss of life claimed by European imperial expansion.[34] Nevertheless, an attempt must be made to give some estimate of these human costs of colonization, which also include losses on the Western side, mainly among the military proletariat sent to fight in the tropics. The Genevan historian Bouda Etemad concludes that between 1750 and 1913 as many as 280,000 to 300,000 European and (in the Philippines) North American soldiers died in overseas colonial wars, either in battle or as a result of disease; India and Algeria were the two deadliest theaters for European troops.[35] Indigenous troops in the service of the colonial powers suffered a further 120,000 casualties, while Etemad calculates that the number of Asian and African warriors who died resisting the whites was between 800,000 and one million. All other losses among non-Europeans are difficult to quantify. Etemad includes the atypically high mortality in India between 1860 and 1921 among the consequences of the *choc colonial,* and he follows estimates of 28 million for the total casualties of famine and a new "ecology of disease" due to external factors. The high mortality in India is not primarily explained by colonial bloodletting and other misdeeds on the part of the British. The unusually severe famines of the 1860s to the 1890s, according to Etemad, accounted for only 5 percent of additional deaths during the period. More important were the concomitants of modernization (railroad construction, creation of large irrigation systems, increased mobility, urbanization under poor hygienic conditions), which afforded new opportunities for the spread of malaria and other indigenous, nonimported diseases. Only the focus on India and on a wide range of *indirect* effects justifies Etemad's high figure of 50–60 million for non-European deaths as a result of colonial conquest.[36]

In contrast to the post-1492 Americas, early modern Ceylon (Sri Lanka), or the aforementioned cases in Oceania and Australia, a "microbe shock" resulting from imported diseases did not play a large role in the nineteenth-century European conquests in Africa and Asia. In fact the shock operated there in reverse, since Europeans had no immunity from many endemic diseases. But colonization did lead everywhere to political, social, and biological destabilization. The often bloody wars of conquest and the ensuing "pacification" campaigns against resistance movements went hand in hand with disturbances to local production, drove large numbers of people from their ancestral homes, and opened new doors to diseases endemic in the area. European invasions therefore almost inevitably resulted in population loss, especially in sub-Saharan Africa, where they were concentrated during the period from 1882 to 1896. In a second phase, beginning in Africa after the turn of the century, the end of major fighting and the first results of a colonial health-care policy meant that conditions were generally favorable to population growth.

The scale of the invasion crisis varied greatly. The worst conditions prevailed in the Congo Free State, which was assigned to King Leopold II of Belgium as a kind of private colony at the Berlin Conference of 1884–85. Here an extremely brutal colonial regime, showing no concern for the natives and treating them as mere objects of exploitation, may have halved the total population between 1876 and 1920—although there is no reliable basis for the figure of ten million murdered Congolese that is today bandied around in the media.[37] In Algeria, brutally "pacified" over three decades, the indigenous population is thought to have declined by 0.8 percent a year between 1830 and 1856, and those who remained were threatened with drought, disease, and locusts in the especially harsh years from 1866 to 1870. A demographic recovery began after 1870 and continued without interruption.[38] Other especially grim and bloody theaters of war were Sudan, Ivory Coast, and East Africa. Where local resistance held its ground, the fighting could drag on for years. Thus, as many as 20,000 British soldiers waged a brutal war in Uganda between 1893 and 1899, and despite their possession of machine guns it was no easy triumph. Scorched-earth tactics were used to deprive civilians of their livelihood, especially the all-important livestock.[39] In South West Africa (today's Namibia), the local German "defense force" and a special marine corps sent out from Germany crushed the resistance of the Herero and Nama peoples between 1904 and 1907, using methods of extreme cruelty. The war of extermination continued against noncombatants and prisoners-of-war after the Africans had laid down their arms, either by driving them into the desert or by forcing them to work under conditions that led to an early death. Although reliable figures are lacking, the numbers killed must have been in the tens of thousands. "Genocide" is the appropriate term for what happened. However, the war of extermination in South West Africa was not one of many such episodes; the unbridled nature of the German actions and the scale of their impact make it an extreme case. It was not in "the logic of colonialism" to murder the colonial subjects. They could be and were used for labor.[40]

Demographic Transition

Did population trends follow a single pattern that eventually asserted itself everywhere in the world? The science of demography offers the theoretical model of a "demographic transition"[41]—that is, a transformation process leading from a "premodern" to a "modern" system of reproductive behavior. The starting point is a situation of high and closely matched rates of birth and death: many people are born, and most of them die early. In the "post-transformation" equilibrium situation, birth rates and death rates are also close to each other but are lower than before; life expectancy is high. The model postulates a multiphase transition between these opening and concluding equilibria. Birth rates and death rates move in opposite directions. At first mortality declines, without an immediate corresponding shift in fertility; more people are born, but they also live longer. There is a rapid increase in population. This model is not plucked out of thin air: it comes from observation of England, Australia, and the Scandinavian countries and has been tried out on other cases. Historically it means that a series of national societies realized at various points in time that families were growing larger, fewer children were dying, and existential horizons were lengthening as life expectancy increased. These experiences must have been similar *in principle*, but the causes would have been bundled together differently in each individual case. Fertility and mortality do not fit together mechanically; the factors determining them are in some degree independent of each other.

In particular, the transformation process that began with the decline in death-rates lasted for different lengths of time: 200 years in England (1740–1940), 160 years in Denmark (1780-1940), 90 years in the Netherlands (1850–1940), 70 years in Germany (1870–1940), and 40 years in Japan (1920–60).[42] So, only in a few European countries and overseas neo-Europes did the process begin before 1900. In the United States it got going in 1790 and lasted until the end of a "long" demographic nineteenth century. But it was a peculiarity of the United States that fertility continually declined during this period, even before mortality fell. The US pattern is therefore similar to that of the European special case, France.[43] Globally, the "Victorian" nineteenth century either still exhibited a premodern demographic structure or was caught up in the process of demographic transition. If we look for the turning point when fertility adjusted to declining mortality, we find surprising confirmation of an epochal shift in the fin de siècle. With the exception of France, this turning point shows up in the statistics only in or after the 1870s.[44] By the eve of the First World War, most European societies had adapted to the idea of individual family planning. The reasons for this are complex and controversial. Suffice it to say that the process was a fundamental one in the history of human experience: a "passage from disorder to order and from waste to economy."[45]

3   The Legacy of Early Modern Migrations: Creoles and Slaves

We like to think of a population, even a society, as something rooted to the soil, something stationary, clearly demarcated, capable of being shown on a map. At first sight this seems to apply particularly well to the nineteenth century, in which governance became territorialized and people rooted themselves in the soil by means of technological infrastructure. They lay railroad tracks and drove mine shafts to unheard-of depths. At the same time, however, it was an age of increased mobility. One characteristic form of this was *long-distance migration*: a definitive or long-term shift in the location of one's existence across great distances to a different social environment. It should be distinguished from frontier migration, in which pioneers were the spearhead of a march into wild, uncharted territory.[46] In the nineteenth century, long-distance migration gripped most parts of Europe and a number of countries in Asia; it was everywhere a factor marking the life of society. The engine driving it was the labor requirements of an expanding capitalist world economy. Migration affected many professions, many social layers, both men and women. It combined material and nonmaterial motives. No country of embarkation and no destination country remained unchanged.

In the nineteenth century, historians, especially in Europe, became fascinated by the role of migration in the origin of nations. A frequent inspiration was the story of Aeneas, the Trojan hero who finally settled down after a long odyssey in Italy. Germanic tribes in the era of the great migrations, Dorians in ancient

Greece, Normans in England after 1066: all found a place of honor in newly written national histories. Asian peoples, too, developed ideas about their past and imagined the arrival of their forebears, mostly from the North. The settled societies of the nineteenth century assured themselves of their mobile origins, and new societies, such as Australia, arose out of mobility then and there. The "immigration society" so often talked about today was in fact one of the great innovations of the nineteenth century, with mobility as its cornerstone. Migration has three closely related aspects: exodus and creation of the new community (the Mayflower motif), survival by means of further intakes of immigrants, and expansive occupation of new spaces. The nineteenth-century migrations represented three different time layers. They might be the sequel to *completed* processes of the early modern period; or they might rest upon movements stretching back into a previous period, such as the forced transfer of slaves; or they might involve a flow of forces that had newly appeared in the nineteenth century itself with the transportation revolution and the capitalist creation of employment opportunities. These flows do not always follow the political chronology: 1914 was for many of them a key turning point, but even more decisive was the Great Depression that began in 1929.

Early Modern Roots of European Emigration

Overseas emigration was already a distinctive feature of early modern Europe. At a time when the rulers of China and Japan made it virtually impossible for their subjects to leave the country, Europeans were spreading themselves around the world. England and the Netherlands were the two European countries that sent the largest proportion of their population overseas—the former overwhelmingly to the New World, the latter to Asia. Spain lagged behind in third place, while emigration from France, the most populous country west of the Tsarist Empire, scarcely featured at all. Many emigrants later returned, and their experiences enriched social and cultural life in the mother country. Of the 973,000 people (half of them German or Scandinavian) who went to Asia between 1602 and 1795 in the service of the Dutch East India Company, more than a third were repatriated to Europe.[47] Not everyone who stayed away lived to start a family.

Actually there were no self-reproducing European core settlements in the tropics. The 750,000 Spanish who remained in the New World mostly settled in nontropical highland regions, where they were not exposed to major health threats. They formed a Spanish society that successfully established itself through natural growth, achieved by *métissage* with indigenous women plus a certain influx from the home country that increased over time. The Portuguese experience was quite different. Portugal was a much smaller country, with a population that never rose above three million before 1800. Yet its emigration between 1500 and 1760 has been estimated at a maximum of 1.5 million—twice as large as the Spanish. In its golden sixteenth century, Portugal had numerous bases in Asia, Africa, and coastal Brazil, but all of them offered a worse environment than that of the Mexican or Peruvian highlands. Portugal—and in this it resembled the Netherlands—was much more likely than Spain to export unskilled labor; it was not a basis on which Creole societies could be constituted. The Netherlands also pursued a strategy of sending foreigners into the unhealthiest parts of the tropics. Generally in colonial history we often find "third" population groups in addition to the colonized and the members of the colonizing nation. At the end of the nineteenth century, for example, more Spanish than French lived in certain *départements* of Algeria.[48]

English emigration in the eighteenth century was equally selective. The unwholesome tropical islands attracted only a small number of plantation managers. The work was done there by African slaves, as it was in the southern colonies of North America, and it was mainly Scots and Irish who opened up the American frontier lands. The typical English settlers in America between 1660 and 1800 were quite highly skilled and gravitated toward the core settlements and cities. In India before 1800 the British need for personnel was much lower than that of the Dutch in Indonesia. Whereas the Dutch recruited their colonial soldiery in northern Germany and Saxony, the British soon began to enlist Indian troops (*sepoys*) on the spot. All in all, only the Spanish emigration was a great success from the outset—and was seen as such throughout Europe. For the other migration-inclined Western Europeans—English, Irish, Scottish, German—North America became an attractive destination only around the middle of the eighteenth century.[49] The prerequisite was finding ways to pass on the most unpleasant work to non-Europeans. But there were some special cases that deviated from the pattern of *ongoing* migration from Europe: the Boers of South Africa, for example, after their initial emigration from the Netherlands in the mid-seventeenth century, were replenished only by local propagation. The French Canadians, too, numbering 1.36 million in 1881, received few new intakes and were mainly descended from the immigrants who arrived toward the end of French rule in 1763.

The social history of the nineteenth century must therefore centrally address the consequences of migration *immediately* prior to it. It was not in the ancient times of the "great migrations" but in the seventeenth and eighteenth centuries that new foundations were laid for numerous societies. In a nineteenth-century perspective these were *young* societies, the virtual opposite of historically rooted social landscapes such as those of the Mediterranean or China. No other large region in the world witnessed the frequency of migration-driven ethnogenesis that was characteristic of Latin America and the Caribbean.[50] The societies of Latin America developed out of three elements: indigenous habitants who survived the conquest and the ensuing microbe shock, European immigrants, and enslaved newcomers from Africa. This mix, varying in its proportions, explains why the early-modern Atlantic slave trade helped to mold the four different types of society that had emerged in the Western hemisphere by the early nineteenth century.

The Slave Trade and the Formation of New World Societies

The *first* type of society developed in Brazil. Here a Luso-Brazilian society developed out of the descendants of Portuguese conquerors or immigrants and a half-African, half-native slave population. Between these two groups there were a number of intermediate layers. A wide spectrum of skin coloration, with various shades of mestizo and mulatto, corresponded to a relatively loose division among the legally free social classes. Although the Indians of the interior were enslaved throughout the eighteenth century by brutal bands living outside the law (the *bandeirantes*), the country's plantation and mining economy remained geared to slave labor imported from Africa. The gender imbalance among the slaves, most of whom came from present-day Angola and the Zaire River basin, as well as a high mortality rate due to harsh working conditions, meant that the African slave population in Brazil was unable to reproduce itself. Between the beginning of the trade around 1600 and the closing of the Atlantic slave importation to Brazil in the mid-nineteenth century nearly 4.8 million Africans were transported to Brazil. The peak was reached only in the four decades after 1810 when roughly 37,400 were arriving each year.[51] The trade continued until 1851, well after it had ceased in other parts of the Latin America. In Brazil it was easier than in other New World slave societies to buy one's way to freedom or to be granted personal emancipation.[52] Free blacks and mulattoes displayed the strongest population growth among all the groups in Brazilian society. Brazil remained marked by slavery until its abolition in 1888—a consequence of early-modern forced migration.

Slavery persisted everywhere for a while after the ending of the slave trade. In the United States it was declared illegal only in 1865, but the importation of slaves had ceased in 1808, having reached a record of 156,000 new arrivals over the preceding seven years.[53] The United States was exceptional in having high rates of slave self-reproduction even *before* the end of the international trade. Thus, after 1808 it had a self-perpetuating slave population in which those born in Africa soon constituted a minority.[54] Imports were no longer necessary to satisfy the demand for nonfree labor. All the more did the slave trade develop inside the United States, enabling special firms of "speculators" or "soul drivers" to make a fortune. Free blacks were captured and sold; slave families were brutally torn apart. Plantation owners from the Deep South, the realm of cotton, made trips to Virginia or Maryland to replenish their supply; it is likely that as many as a million blacks crossed interstate frontiers under compulsion between 1790 and 1860.[55] This internal commerce became the most visible and scandalous side of slavery, and the one most open to attack. At almost the same time, the end of the transatlantic trade boosted the circulation of slaves within the African continent.

A third pattern of correlation between migration and society building was found in Mexico. New Spain (Mexico), the administrative center of the Spanish empire, naturally shared the experience of slavery with the rest of the New

World, but unlike in Brazil or the Southern United States slavery never became an all-pervasive institution that marked every sphere of life. This was not because of any special aversion of the Spanish to human enslavement: Spanish Cuba remained a fully fledged slave colony right into the 1870s. But for mainly ecological reasons a large-scale plantation economy could not gain a firm footing in Mexico. In 1800, in contrast to Brazil or the United States, it was not a country of immigration. It is probable that from the beginning of the eighteenth century until the prohibition of the slave trade to Mexico in 1817, no more than 20,000 Africans were exported there.[56] The indigenous population slowly recovered after 1750 from various demographic setbacks. According to the 1793 census, blacks constituted at most 0.2 percent of the total population. The second smallest group, at 1.5 percent, were the 70,000 European-born Spanish (*peninsulares*). The majority of the Mexican population was made up of autochthonous *indios* (52 percent), followed by *criollos* (that is, people of Spanish origin born in Mexico).[57] In 1800 Mexico was a society cut off from intercontinental migration flows, whose population renewal was based on its own biological resources.

A fourth pattern developed in the British and French Caribbean. On most of the West Indian islands, the indigenous population had been killed during the first wave of European invasions. In the seventeenth century, on this tabula rasa, the dynamic of early capitalist production for the world market then created new kinds of society consisting wholly of nonindigenous outsiders. These out-and-out immigrant societies, totally lacking in local traditions, could fulfill their mission of producing plantation sugar only with an uninterrupted supply of slaves from Africa; the plantation system consumed human beings at a staggering rate. Those societies never progressed to self-reproduction of the black population, which in the Southern States of the United States had overcome the need for a constant intake of new slaves from overseas. The European share of the population stagnated after a wave of English, French, and Dutch settlement in the early seventeenth century. Although it was not upper-class planters but specialist workers and plantation overseers who later moved out from Europe, whites remained a small minority throughout the eighteenth century; black slaves accounted for 70 to 90 percent of the population on sugar islands such as Saint-Domingue or the British possessions of Jamaica and Barbados.[58]

It was much harder for a slave to buy freedom or to win emancipation in the Caribbean than in Brazil, and so the intermediate class of "free persons of color" remained comparatively thin until the ending of slavery. In Brazil roughly two-thirds of the population was legally free in 1800, while in the United States free men and women *always* constituted a majority. This differentiated both countries from the Caribbean sugar islands (although, of course, most free people were black or "mixed" in Brazil but white in the United States).

A further characteristic makes the special path of the Caribbean even clearer. The slave system was destroyed earlier in the Caribbean than in Brazil or the United States: partly as a result of a slave revolution (Saint-Domingue/Haiti,

1791–1804), partly as an effect of legislation in the metropolitan countries (Britain, 1833; France, 1848; Netherlands, 1863). These societies entered their own post-emancipation "nineteenth century" only with the abolition of slavery. Free immigration played only a very small role after the end of the slave trade, and numerous whites fled the region during the period of revolution and emancipation. Only Cuba continued to attract those who wanted a stake in the sugar boom: it drew 300,000 new settlers, overwhelmingly from Spain, in the years between 1830 and 1880. Elsewhere whites were unwelcome (Haiti) or saw few prospects in the stagnant island economies. In general, population growth in the Caribbean did not vary much between 1770 and 1870, while the demographic composition of the population underwent radical change. At the end of the eighteenth century, first-generation immigrants set the tone in Caribbean societies, whereas by 1870, native-born populations predominated.[59]

The transatlantic slave trade bridged the early modern period and the nineteenth century. It reached a peak during the decades around 1800, ensuring that the institution of slavery would survive the abolition of the trade by several decades. The formation of immigrant societies in the Western hemisphere entered a new phase in the second half of the nineteenth century, when forced migration across the Atlantic played a much lesser role than before. However, a visitor to the West Indies, Brazil, or the United States did not take long to realize that the nineteenth-century Americas were also a piece of Africa.

4 Penal Colony and Exile

Siberia—Australia—New Caledonia

What new elements in migration history are observable in the nineteenth century? Let us leave aside for the moment the opening of new frontiers, which will be discussed in chapter 7, as well as migration within individual countries, about which it is hard to say anything general. A newly popular institution was the penal colony, where malefactors and political opponents were exposed to isolation, privation, and the rigors of a harsh climate. Siberia had been used as a penal colony since 1648, and under Peter the Great, also as a location for prisoners-of-war. A growing number of offenses came to be punished with banishment. Rebellious serfs (until 1857), prostitutes, troublesome outsiders who were a burden to villagers, vagrants (sometimes the majority of deportees in the nineteenth century), and after 1800, Jews who had not paid their taxes three years in a row all found themselves being shipped off to Siberia. In the eighteenth century, compulsory hard labor (*katorga*) on state building sites became widespread. Only after the abortive Decembrist rising of 1825 did northern Asia start to be used on a large scale as a place of *political* exile. One wave of anti-Tsarist radicals followed another into the wastelands of Siberia. In 1880 there were still many who had been banished there since the Polish uprising of 1863;

soon they were joined by the first Marxists and anarchists. Few found conditions there as pleasant as the famous anarchist Mikhail Bakunin did, a relative of the governor who was to some extent allowed to share the social life of the local upper class. Many others had to perform hard labor in the coal or gold mines. Usually exiles were not kept behind bars and took some part in the life of society; some even had a family with them.

In the last three decades of the nineteenth century, Russian courts sentenced an average of 3,300 to 3,500 persons a year to deportation. In January 1898, official statistics revealed the presence of 298,600 deportees in Siberia, and if family members are included the total must have been around 400,000, or nearly 7 percent of the total population of Siberia. Shortly before 1900 the number of banishments to Siberia began gradually to fall off, but then it rose again after the 1905 Revolution.[60] Banishment to Siberia was repeatedly denounced in Western Europe as a sign of the "barbarous" nature of the Tsarist Empire. On the other hand, a statistical comparison shows that at the end of the nineteenth century—to take a generally applicable indicator– the death penalty was carried out more rarely, in proportion to total population, in the Russian empire than in the United States (where it was ten times more frequent), Prussia, England, or France.[61] Even mortality among prisoners was below the level in the tropical penal colonies of the French Republic. In the nineteenth century, the thinking behind the Siberian system was that it would provide a "prison without a roof" for political opponents and marginal social groups, while at the same time providing a labor pool for the giant state projects of colonizing and "civilizing" the region. It was a colonial development program that had much greater affinities with the colonial corvée system than with the pioneering advance into the American West driven mainly by market forces and voluntary decision.

At the time of the Russian Revolution of 1905, Western public opinion had long regarded deportation and forced labor as anachronistic and extremely hard to justify. In China, too, it had lost its usefulness to the state, having reached its peak in the eighteenth century. In 1759 the Qianlong Emperor completed the conquest of large tracts of land in Central Asia and immediately began to explore the possibility of using inhospitable borderlands as places of banishment. In the following decades tens of thousands of people, among them adherents of "evil" creeds that the state disapproved of, were exiled to what is now the province of Xinjiang, where they were subject to a regime that may be described as an exile *system* akin to the one that developed in Russia. Here, too, the goal of punishment was combined with the colonization of border areas. The Qing state continued the experiment until approximately 1820, but although it lingered on until the fall of the dynasty in 1911, the authorities lost interest as the problems multiplied and the conditions for new settlements became increasingly difficult. In China, government officials and army officers made up a high proportion of those punished with internal exile; the system generally allowed families to accompany the deportee, and a high value was placed on the aspect of moral reeducation. It was not unusual for an official to resume his career in the emperor's service after he had spent three to ten years in exile. Imperial China was more restrained than many parts of ancien régime Europe in its use of capital punishment; banishment was a common way of commuting death sentences. The transportation of prisoners and deportees to Xinjiang was painstakingly organized and was one of the great logistical achievements of the Qing state. Figures are not available.[62]

The French state deported political troublemakers after the unrest of 1848 and 1851. Following the defeat of the Paris Commune in 1871, more than 3,800 insurgents were sent in nineteen convoys of ships to the Pacific archipelago of New Caledonia, a colony under French rule since 1853; the deportation was conceived as a means of "civilizing" both the indigenous kanaks and the Communard revolutionaries, and that was the spirit in which it was carried out.[63] Previous attempts to settle ordinary French people there had fallen afoul of the climate. Until 1898 a yearly average of 300 to 400 convicts were sent to New Caledonia.[64] The other French place of banishment was the climatically even harsher colony of Guyana, in the northeast of South America—one of the most inhospitable lands in the world, which came to world public attention at the latest in 1895, when Captain Alfred Dreyfus (later found to be the victim of a conspiracy) was sent in an iron cage to the offshore Devil's Island. At the beginning of the twentieth century, French Guyana had a system of prisons and forced labor that encompassed roughly a fifth of its total population (not including indigenous tribes and gold prospectors). Banishment to the "pepper islands" was abolished only in 1936.[65]

Australia served on a grand scale as a penal colony: in fact, it owed its existence as a colony of any kind to the sending of the "first fleet," whose eleven ships and 759 convicts sailed into Botany Bay (close to today's Sydney) on January 18, 1788. The loss of the North American colonies had put the British state in the position of having to find somewhere else to send convicts. After a number of extreme alternatives—such as an island in the Gambia River in West Africa— were rejected on humanitarian grounds, someone remembered Captain Cook's discovery of Botany Bay in 1770. Although other motives, such as the maritime rivalry with France, should not be excluded, this spectacular solution would probably not have been adopted had it not been for the acute convict crisis of the mid-1780s. In any event, Australia was little more than a huge penal colony during the early decades of its colonial history. The first settlers were forced immigrants, dispatched by an English judge to faraway Oceania.

By the time of the last convict ship in 1868, 162,000 people had been transported as prisoners to Australia. Most of them were products of the growing criminal subculture in Britain's early industrial cities: burglars, pickpockets, swindlers, and so on, along with a small number who had been sentenced on political grounds. The government began to encourage free emigration to the Antipodes in the late 1820s, but that did not bring any slackening of the transports. On the contrary: 88 percent of the convicts left England for Australia after 1815. The peak was reached in the 1830s, when 133 ships with an average of

209 convicts arrived between 1831 and 1835 alone, after a sea voyage lasting four months or more.[66] Most of them still enjoyed at least the basic rights of a British citizen. From the beginning, the convicts were able to represent their interests in a court of law and were not totally at anyone's mercy in their choice of work. This was an important reason why Australia gradually developed a civil society without experiencing dramatic revolts.

The penal colony, indelibly imprinted on the mind through Franz Kafka's eponymous short story (written in 1914, first published in 1919), was a worldwide institution characteristic of the imperial nineteenth century, though even today it has not completely disappeared. In the flow of emigration from Europe, deportation remained an important element. Spain shipped delinquents off to Cuba or North Africa; Portugal, to Brazil, Goa, and above all Angola. British citizens might find themselves headed for Bermuda or Gibraltar. Convicted colonial subjects, too, were deported in convict ships: Indians, for example, to Burma, Aden, Mauritius, Bencoolen, the Andaman Isles, and the Malay Straits Settlements. The deportations did not always achieve their intended purpose; the deterrent effect was as questionable as the "civilizing" of the prisoners. Their forced labor did generally contribute to economic development in the region to which they were sent, but the colonial administrations in Burma or Mauritius, for instance, were interested only in strong and youthful work crews, not in the average Indian convict population.[67] Convict labor was rational only so long as no other labor pool was available.

Exile

Political exile, as a fate for individuals or small groups, was nothing new in the nineteenth century. There had always been refugees from war, epidemics, and famine, and in the modern age, especially in Europe, these had been joined by religious refugees (Muslims and Jews from Spain, Protestant Huguenots from France, Nonconformists from orthodox Protestant England). Figures are here very difficult to find. What is clear is that in comparison with the scale of the problem during and after the First World War, collective displacement was not a major form of migration in the nineteenth century. Nonetheless, the phenomenon did become more significant. There were several reasons for this: (1) more intense persecution of political opponents in the ideological atmosphere of a nonreligious civil war that first became manifest during the French Revolution and its repercussions in the whole of Europe; (2) a liberalism gap between states, which meant that some of them aspired to become bulwarks of liberty and were prepared within limits to give sanctuary to freedom fighters from other countries, thus contributing to the emergence of a transnational civil society;[68] (3) the greater scope for wealthier societies to offer foreigners at least a temporary living.

The refugees who differentiated the nineteenth century from others—anyway until the 1860s—were not so much the ones who came anonymously in large numbers as the individually conspicuous ones, often from a prosperous and well-educated background. The waves of revolution brought forth such exiles: the 60,000 empire loyalists who in 1776 fled the North American colonies to Canada and the Caribbean; the émigrés of 1789 and the subsequent years who remained loyal to the Bourbons; the victims of the repression of 1848–49 following the failed uprisings in many parts of Europe. Switzerland, for example, took in 15,000 exiles after 1848, most of them Germans and Italians, while 4,000 Germans ended up in the United States.[69] The Karlsbad Decrees of 1819 and the German Anti-Socialist Law of 1878 unleashed smaller waves. The most important legal watershed was the July Revolution of 1830, as a result of which the right to political asylum—and hence protection from politically motivated extradition—was firmly rooted in the legal systems of Western Europe, especially France, Belgium, and Switzerland. In the European revolutions of 1848–49, this principle found practical application. It was associated with public welfare support for political refugees, and also with the possibility of indirectly influencing their conduct.[70]

The links between exile and revolution are complicated. In 1830 the revolution in France awakened hopes for freedom in other nations and encouraged them to rise up in revolt—and at the same time it created political conditions that made France itself become a coveted place of refuge. In 1831, following the collapse of the November Revolution of 1830 in the Russian-ruled Kingdom of Poland, a large part of the Polish political elite—some nine thousand, more than two-thirds of them from the (very extensive) Polish nobility—marched in triumph through Germany to France. This Great Emigration (*Wielka emigracja*), most of whose participants settled in Paris, took cultural creativity and political initiative abroad with it. It came to be seen as a "metaphysical mission," whose sacrificial bearers represented all of Europe's oppressed.[71] In order to occupy the more unruly elements among the revolutionary refugees, the French government founded the Foreign Legion in 1831.

Never before the nineteenth century had so much politics been conducted from exile. Prince Adam Czartoryski in Paris, the "uncrowned king of Poland" whom people also called the "one-man Great Power," organized Europe-wide agitation against Tsar Nicholas I and tried to swear his divided compatriots to a common strategy and objectives.[72] Alexander Herzen, Giuseppe Mazzini, and the oft-exiled firebrand Giuseppe Garibaldi also operated from abroad. The Greek revolt against Ottoman rule was planned by exiles. At the same time, the Ottoman Empire was not simply a bastion of despotism but could itself become a place of refuge for defeated freedom fighters. In 1849, after a Tsarist interventionist force helped to crush the Hungarian independence movement, Lajos Kossuth and thousands of his supporters found sanctuary in the sultan's realms. British and French diplomats strengthened the resolve of the Sublime Porte to reject Russian extradition requests by referring to customary practice in the "civilized world" (in which, exceptionally, they were prepared to include Istanbul).[73]

Later in the century, exile activity also undermined the Asiatic empires—
something that had rarely happened before. In the case of China, the remaining
Ming loyalists in the seventeenth century had not known how to create an operational base outside the country, nor did any remnants of the Taiping Revolution of 1850–64 linger on abroad. In the nineteenth century, the Ottoman
Empire was heavily criticized by Turkish exiles, but only by individual dissidents
at first. Even before Sultan Abdülhamid II turned to autocratic rule in 1878, critical intellectuals such as the poet and journalist Namık Kemal had been sent into
exile, either internal (e.g., to Cyprus) or external. In the early 1890s an opposition
movement bearing the name *Jeunes Turcs* was formed in Paris against Abdülhamid. Its work with groups of conspirators inside the military eventually paved
the way for the Young Turk Revolution of 1908.[74] The Armenian revolutionary-
nationalist organization worked out of Geneva and Tiflis from the 1880s on.[75]
The Western-oriented opponents of the Qing Dynasty in China had the advantage of being able to prepare their revolutionary operations directly on the
doorstep of the empire. The revolutionary leader Sun Yat-sen and his followers
based themselves in the British crown colony of Hong Kong in 1895 and later
lived in overseas Chinese communities in the United States and Japan.[76] In the
1890s Tokyo became for a few decades the hub of various, and interconnected,
networks of exiled political activists from several Asian countries.[77]

The International Settlement in Shanghai, which was under international
(read: Western) control, served as another base for plans and operations against
the regime. When the young and politically weak Guangxu Emperor ventured
in 1898 to support an attempted constitutional reform (the "Hundred Days'
Reform"), only to suffer defeat at the hands of his aunt, the conservative Empress
Dowager Cixi, the leaders of the movement found safety abroad under British
protection. The most important of them, Kang Youwei, wrote in Darjeeling his
Datongshu (Book of the great unity), one of the major texts of utopian world literature.[78] The Americas, too, offer instances of an exile movement that managed
to push out a stable regime. The fall of the aged dictator Porfirio Díaz, who ruled
Mexico from 1876, was organized from San Antonio in Texas, where his main
adversary, Francisco Madero, rallied his supporters in 1910.[79] All these persons
and movements profited from the liberalism gap without directly becoming instruments of great power intervention.

Exile provided a degree of security (but not complete protection) from
henchmen of the regime under attack, making it possible to form articulate
circles of intellectuals who understood the uses of modern media, and it also
opened doors to private sympathizers and financial backers. In all these respects
exile politics was "modern"; it was premised on the emergence of advanced
communication techniques and a global public sphere. Opportunities for active exiles discontented with a life on the margins were concentrated in a small
number of places. Whereas émigrés after the French Revolution had first gathered in Koblenz, it was London, Paris, Zurich, Geneva, and Brussels that later

became the main bases for exile politics in the nineteenth century. Looking back today, one is amazed at the freedom that many exiled politicians enjoyed despite growing surveillance by the authorities (e.g., in France). In Britain not a single political refugee from the Continent was prevented from entering the country, or subsequently deported, throughout the nineteenth century.[80] No one thought that Karl Marx in London or Heinrich Heine in Paris should be subject to a gag order. No extradition treaties existed with other countries. Requests for legal action to be taken against regime opponents living in London were invariably rejected and sometimes not even answered. Nor was criticism of British imperialism legally barred in any way. Politically active exiles generally were regarded neither as saboteurs of British foreign policy nor as a danger to internal security.

Exile brought together not only revolutionaries and anticolonial resistance leaders (men like Abd al-Qadir from Algeria or Shamil from the Caucasus) but also rulers who had been toppled from power. A non-place like the island of Saint Helena entered history only because Napoleon was forced into exile there. In 1833, three years after the July Revolution, Chateaubriand came across the former Bourbon king Charles X wandering like a ghost through the empty Hradschin castle in Prague. Charles's successor on the throne, Louis-Philippe, ended his days in 1850 on a country estate in Surrey, and the Argentine dictator Juan Manuel Rosas breathed his last in Southampton in 1877. But the most curious spectacle of monarchical emigration occurred in 1807, when the Portuguese prince regent Dom João, hard pressed by Napoleon's invading army, assembled his whole court and much of the state bureaucracy (a total of 15,000 persons) and betook himself with a fleet of thirty-six ships to the colony of Brazil. Over the next thirteen years the viceregal capital, Rio de Janeiro, became the center of the Lusitanian world. It was a dual premiere: it was not only the first exodus overseas by a whole system of rule but the first time in the history of European maritime expansion that a ruling monarch had paid a visit to one of his colonies. In an age of revolution, a late-absolutist court took the risk of transplanting itself to a completely different political context, in a curious blend of evident self-interest and serious patriotism. Such an exile, bedecked with tragedy and legitimacy, fueled visions of renewal and rejuvenation, and of a prosperous empire with Brazil at its center. In 1815 an attempt was indeed made to form a tightly integrated Portuguese-Brazilian empire, but it came to nothing.[81]

5 Ethnic Cleansing

The Caucasus, the Balkans, and Other Arenas of Expulsion

Whereas political emigration and a heroic exile were a characteristic feature of the nineteenth century, at first in Europe but later also elsewhere, the image of impoverished refugees eking out a bare existence abroad is more associated with the age of "total war" and homogenizing, racially charged ultranationalism. Yet cross-border refugee flows triggered by government actions were not unknown in the nineteenth century. The Greek independence struggle, for example, was less a heroic enterprise—as it anticipated international solidarity with 1930s Spain—involving high-minded northern philhellenes à la Lord Byron and brave descendants of the freedom-loving ancient Greeks than a harbinger of later ethnic cleansing in the region. The population of Greece fell from 939,000 in 1821 to 753,000 in 1828, overwhelmingly because of the flight and expulsion of ethnic Turks.[82] In 1822 the Turks themselves had gone on the rampage on the Aegean island of Chios, massacring part of the Christian population, selling another part into slavery, and driving thousands more into exile. Delacroix immortalized the horror as early as two years after the event. New Chiotic communities began to appear in London, Trieste, and Marseille.

The Tatars who in the eighteenth century left their homes in the Crimean peninsula to settle inside the Ottoman Empire did so because of Russian contempt for their way of life, loss of land to Russian settlers, and the growth of Russian anti-Islamism. The emigration began during the Russian-Ottoman war of 1768–74 and intensified after the annexation of the Crimean Khanate in 1783. At least 100,000 Crimean Tatars, including nearly the entire upper stratum (the notables), moved to Anatolia over the following decade and became the core of what Tatars themselves call the "first exile" (sürgün). The Crimean War (1853–56) then sealed the fate of those still in the peninsula, whom the Russians regarded as a fifth column of the hated Ottomans. By the end of the conflict 20,000 Crimean Tatars had been given asylum and evacuated on board Allied ships, while the same number again had fled by other routes. In the early 1860s another 200,000 Tatars are said to have left the Crimea under wretched conditions.[83] It is true that in the *late* nineteenth century the Tsarist government tried to keep the Tatars and other Muslims in the country; it cannot be charged with a policy of systematic expulsions.[84]

The exodus of Muslim peoples from the Caucasus was much greater after the Russian army in 1859 crushed the resistance of armed highlanders under their leader, Shamil. In the conquest and "pacification" of the High Caucasus, the Russians resorted to all the methods of ethnic cleansing. At least 450,000, perhaps as many as one million, Muslims were driven from their mountain homelands between 1859 and 1864; tens of thousands died from starvation, disease, or accidents en route to the realms of the sultan. In 1860, 40,000 Chechens fled the region, and only a small minority of Muslims decided to brave it out in Georgia.[85] In the midst of disaster, the Tatars had the good fortune to be welcomed by a neighboring country, which they increasingly saw as a religious homeland. The impact of the expulsions on them was compounded by the attractiveness of the hallowed "Land of the Caliph." Messianic currents in the diaspora glorified the flight as a return home.

Such a refuge was unavailable to other persecuted ethnic groups. In early May 1877, after years of rearguard struggle and a victory the year before over the US

Army at Little Big Horn, the surviving Lakota Sioux under Chief Sitting Bull crossed into the land of the Great White Mother (Queen Victoria), who seemed a kinder ruler than the Great Father in Washington, DC, and in whose realm there were laws that applied to all. For the first time in his life, the chief met whites in Canada who treated him with respect; he thought he could trust them. But diplomacy soon thwarted his hopes. The United States, which considered itself at war with the now weak and impoverished Lakota, called for the Canadian authorities to intern the Indians. Hunger and relentless American pressure eventually forced the small Lakota community, a mere shadow of the once great Sioux nation, to make its way back to the United States, where its members were held as prisoners of the state.[86]

In an increasingly nationalistic Europe, cross-border refugee flows were the result of frontier changes imposed by force of arms or by political agreement. France expelled 80,000 ethnic Germans after the outbreak of war with Germany in 1870, and when the Reich annexed Alsace-Lorraine in 1871, under the terms of the Frankfurt peace treaty, 130,000 refugees who had no wish to live under German rule packed their bags and left.[87] On Germany's eastern borders, Bismarck's *Kulturkampf* against Catholicism spread to the already delicate sphere of German-Polish relations, and once the conflict died down the chauvinist character of the "struggle over language and soil" became plain to see. Pursuing a "Germanization" policy, itself supposedly a defense against "Polonization" of the eastern territories of the Reich (or swamping with Poles, as it was called), the German authorities did not shrink from using the instrument of expulsion. In 1885–86 a total of 22,000 Poles and 10,000 Jews with Russian or Austrian citizenship were driven from the eastern provinces of the Reich, many of them into the Russian-controlled "Kingdom of Poland," where they had no chance of making a living.[88] In the opposite direction, Germans were leaving a Tsarist Empire that defined itself more and more strongly in Russian national terms. Between 1900 and 1914, 50,000 Volga Germans abandoned their homes. Wherever new nation-states appeared in the decades before the First World War, and wherever a "nationalities policy" was pursued within multinational empires, the danger arose of an "unmixing of peoples."

Throughout the nineteenth century, the Balkans were one of the regions of the world with the most troubled ethnic politics. During the Russian-Ottoman war, Russian troops came within fifteen kilometers of Istanbul. The Tsarist government had begun the war in April 1877, utilizing ever stronger anti-Turkish sentiments after Ottoman troops had savagely crushed rebellions in Herzegovina, Bosnia, and Bulgaria: the "Bulgarian horrors," which stirred British opposition leader William E. Gladstone to glittering heights of moral rhetoric.[89] During their advance, Russian troops and Bulgarian mobs killed 200,000 to 300,000 Muslims and rendered an even greater number homeless;[90] when the war was over, roughly half a million Muslim refugees settled in Ottoman territory.[91] In 1878 the Congress of Berlin tried to put some order into the political map of Southeast Europe, but that very order would have grave consequences for religious and ethnic minorities. Refugees took to the road to escape the revenge of conquerors from a different religion or nationality, or to avoid being ruled by infidels. Christians sought refuge in the newly autonomous states or—borders being loose here—in areas under Russian or Austrian protection, while Muslims reached safety behind the shrinking frontiers of the Ottoman Empire. It is hard to see what difference there was between straightforward expulsion and unavoidable flight. By the mid-1890s, some 100,000 Bulgarian speakers had left Ottoman Macedonia for Bulgaria. Conversely, Muslim settlers and Turkish officials, but also Orthodox peasants, withdrew from a Bosnia that the Congress of Berlin had placed under Habsburg (therefore Catholic) occupation.[92] The total number of people uprooted by the Russian-Turkish war of 1877–78 may have been in the region of 800,000.

The refugee flows in Southeast Europe reached a peak during the Balkan Wars of 1912–13. The massacres and ethnic cleansing of those years already presaged what lay ahead in the wars of the Yugoslav succession in the 1990s. Population movements on such a scale had not been seen for centuries within such a small area of Europe. Muslims of every description (Turks and other Turkic peoples, Albanians, Islamized Bulgarians, etc.) fled from all the former Ottoman territories now occupied by Balkan states. Greeks abandoned the newly enlarged Serbia, the expanded Bulgaria, Thrace, and also Asia Minor (where many ethnic Greeks spoke only Turkish). Salonica—Ottoman since the fifteenth century, with a long history as a peaceful ethnic mosaic—turned into a Greek city in which Turks, Jews, and Bulgarians had to recognize the primacy of the Greek conquerors; by 1925 the Muslim population had abandoned the native city of Kemal Atatürk.[93] According to estimates made at the time by the British authorities, approximately 740,000 civilians were uprooted between 1912 and the outbreak of the First World War just in the rectangular area formed by Macedonia, western Thrace, eastern Thrace, and Turkey.[94] After the First World War and the Greek-Turkish war of 1919–1922, ethnic "unmixing" continued in the eastern Mediterranean and again led to the problem associated with all expulsions: the need to integrate people arriving in a new society. After 1919 the attempts of the League of Nations, particularly its Refugee Settlement Commission, to establish a modicum of order amid the chaos represented a small step forward.

The actual or threatened violence that lay behind these population movements did not stem simply from a religious clash between Christians and Muslims. The frontlines were more complex, and the Second Balkan War saw Christian states fight one another. Muslims too knew how to differentiate. Until relations between Greeks and Turks took a further turn for the worse, they could expect slightly less appalling treatment from Greeks than from the Slav peasantry who filled the Bulgarian and Serb armies. New, often hastily improvised visions of nation-states established the criterion for inclusion and exclusion. The authorities generally tolerated, sometimes even promoted, the refugee flows; the emigration was matched by immigration of the new citizens they wanted. To be sure, most governments refrained from encouraging too large an influx—after all, irredentist minorities in other countries might one day buttress annexation claims and perform useful services for a nationalist foreign policy.

Jewish Flight and Emigration

A new and especially important source of politically inspired cross-border emigration was the new anti-Semitism in the Russian Empire and elsewhere in eastern Europe.[95] Between the early 1880s and 1914, some 2.5 million Jews left eastern Europe and headed west. Care must be taken not to treat this exodus— probably the largest population movement in postbiblical Jewish history—as just another politically driven flow of refugees. The Jews in question formed part of a broader movement of people who wanted to improve their lives by em- igrating to the economically more advanced West, but they also had to deal with rising official hostility in the countries of their birth. In the 1870s approximately 5.6 million Jews were living east of the German Reich: four million under the tsar in a special "Pale of Settlement," 750,000 in the Habsburg lands of Galicia and Bukovina, almost 700,000 in Hungary, and 200,000 in Romania. In the Russian Empire, after Alexander II's accession to the throne in 1855, the hope had arisen that the authorities would encourage the integration of Jews into society. But the reverse occurred after the suppression of the Polish uprising of 1863; only a few discriminatory laws were repealed. The final years of Alexander's rule—he was assassinated in March 1881—were marked by a further autocratic clamp- down and a growing accommodation with conservative Russian nationalists, who saw their main adversary in the Jews. Nevertheless, although in the 1870s large sections of once-liberal public opinion also shunned the cause of Jewish emancipation, emigration did not for the moment reach dramatic proportions.

The picture changed with the first series of pogroms in that same year of 1881.[96] The involvement of a terrorist of Jewish origin in the tsar's assassination became the pretext for large-scale anti-Jewish violence, first in Ukraine, then also in Warsaw. To what extent the authorities unleashed the pogroms and to what extent they were "spontaneous" outbreaks among the mainly urban lower classes is still being debated. In any event, in addition to general poverty, a high number of children per family, a lack of job prospects, and a growing vulnera- bility to street violence, the Jewish population now had to face an official policy that denied it a place in the country's national life. In the 1890s nearly all Jewish craftsmen and merchants were brutally driven westward from Moscow into the Pale of Settlement, while the state placed great obstacles in the way of Jews (and others) wishing to emigrate. For many it thus became an illegal adventure to flee the empire, often by bribing corrupt officials, border guards, and policemen. The figures for Jewish emigration can be reconstructed with some degree of accuracy only from statistics in the destination countries. Whereas in the 1880s an average of 20,000 Jews a year left the Tsarist Empire for the United States (by far the first choice), the corresponding figure for the years between 1906 and 1910 was 82,000. The increase was due partly to the palpable attractiveness of the new life overseas and partly to competition among the shipping companies that had considerably reduced the cost of a transatlantic passage around the turn of the century. The fact that persecution was not the only factor fueling Jewish emigration is borne out by the not-insignificant numbers who *re*migrated to eastern Europe—perhaps as many as 15 to 20 percent in the 1880s and 1890s.[97]

Jewish emigration from Habsburg Galicia around this time was driven mainly by extreme poverty. After the legal emancipation in 1867, Galician Jews enjoyed full civil liberties and made certain advances in social integration, although the lack of social-economic opportunities meant that these did not lead to much. In Galicia, too, there were anti-Jewish stirrings in the 1890s, but the Habsburg government never officially engaged in action against the Jews. In Romania, which the Congress of Berlin recognized as an independent state in 1878, widespread poverty combined with an early and intense anti-Semitism. The state defined the Jewish minority as antinational, made its economic life as difficult as possible, and did not protect it from "spontaneous" violence. The Western Great Powers tried but failed to make the authorities in Bucharest comply with the clauses in the Treaty of Berlin that had provided for Jewish civil rights. It is therefore unsurprising that no other region of eastern Europe saw such a high proportion emigrate. Between 1871 and 1914 Romania lost a third of its Jewish population.[98]

Eastern European Jews were the first new-style refugees that people in Western Europe could identify as such. Most of them spoke Yiddish, wore traditional Jewish dress, and cut a wretched figure at ports and railroad stations and in city centers. Jews already living in the West viewed them with mixed feelings, as both "brothers" and "strangers" who, though deserving support, threatened the success of their own precarious integration. Most of the new arrivals saw Western Europe only as a stopover on the road to the New World. Craftsmen were more likely to stay on, but it was not made easy for them. In Germany, government policy created obstacles (though not so many as to spoil good business for the shipping companies that brought them there), and the public mood was unfavorable to their presence. Nevertheless, by 1910 a good tenth of German Jews were of eastern European origin.[99]

6 Internal Migration and the Changing Slave Trade

Although the nineteenth century was not yet the "century of the refugee," it was an age of labor migration across continents on a greater scale than anything seen before in history. This was not always entirely voluntary—though quite apart from the slave trade while that still existed—but on the whole it did involve a life choice that individuals made voluntarily. Its prerequisites were population growth, improved transportation, new job opportunities resulting from industrialization and the opening of frontier lands for agriculture, and postmer-cantilist government policies in both source and destination countries.

Transnational Migration in Europe and East Asia

A "new topography of cross-border migration" thus emerged on all conti-nents.[100] Historical research has afforded a fairly precise picture of this in the case of Europe but less so for other parts of the world. In central Europe the "Dutch" or "North Sea system," the only one of the early modern transnational migration systems still functioning in 1800, had given way by midcentury to the partly overlapping "Ruhr system."[101] Instead of Dutch trading and colonial activ-ity, the industrial development of mining regions now became the chief magnet for prospective migrants. The high spatial mobility of the early modern period increased still further and began to fall back again only in the course of the twen-tieth century. But it is clear that no other European country reached British or German levels of industrially driven mobility, and that in some it played scarcely any role. Areas in southern, southeastern, and eastern Europe (Italy, Russian-ruled central Poland, Habsburg Galicia) and to a lesser extent Belgium, the Netherlands, and Sweden were especially important sources in the new cross-border topography of migration, while the most attractive destination countries were Germany, France, Denmark, and Switzerland. In this complex pattern, the movement of Poles to the Ruhr and of Italians to France were of special impor-tance, occurring on a large scale from the early 1870s on. Those between two central host areas may be termed "secondary flows": for example, the migration to Paris of economically active Germans, ranging from subproletarian to petit bourgeois. In 1850 approximately 100,000 Germans were living in the French capital, some of them under wretched conditions. This "colony," as the French mistrustfully referred to it, began to disperse after the Franco-German war of 1870–71 and vanished entirely amid the economic crisis of the eighties.[102]

In Asia and Africa, the new migrations of the nineteenth century differed both from the chaotic mobility of crisis periods and from older patterns of sea-sonal labor movement. Europeans long cultivated the myth of Asia's sedentary small-plot farmers and overlooked the mobility that could be triggered by wars and natural calamities. In Java during the war of 1825–1830[103] and in many Chi-nese provinces during the turmoil of the Taiping Revolution, a quarter of the population found itself uprooted and homeless. Farmers everywhere remain "rooted" only so long as the fruits of their labor, or what they manage to retain of them, provide a livelihood—otherwise they look for different ways to make a living. Growing peasant communities also send young people who are unneeded in the fields off to distant parts. In the nineteenth century this gave rise to clear patterns whenever labor-intensive sectors such as mining or new agrarian devel-opment created a steadily increasing need demand for manpower.

China saw the continuation of a tendency in farming that had started to de-velop in the eighteenth century—that is, the move up from the lowlands into

hill and mountain country. The Qing state encouraged this with direct initiatives, tax relief, and military support for new settlers against hostile tribal populations. It did not bring traditional rice and wheat crops up from the plains but
introduced plants that had first been imported from the Americas in the Ming
period: above all, corn and potatoes. These were less demanding, allowed for
slash-and-burn clearing, and required less attention to soil management, fertil-
izers, and irrigation.[104] The nineteenth century also opened new avenues for migration, as the Qing government permitted Han Chinese to conduct trade and
own land in Mongolia. In 1858 it even became possible for both seasonal laborers
and permanent emigrants to cross the frontier into far eastern Russia. By the end
of the century some 200,000 Chinese had availed themselves of these opportunities. When Russian settlers after 1860 increasingly pushed north of the Amur
River, they often found Chinese farmers already there. Over the following years
the Chinese took to planting rye, wheat, and poppy, while traders used the free-
trade zones on either side of the border and carried on all manner of business in
the cities. From 1886 onward the Russian authorities took their own fear of the
"Yellow Peril" more seriously and repeatedly took action against the Chinese in
eastern Siberia, as well as the Koreans who, though somewhat less numerous,
were for that very reason more inclined to assimilation. The significance of the
Asian "diaspora" did not diminish as a result, however, and by the time the First
World War broke out in 1914, Chinese workers were indispensable in far eastern
Russia.[105] Today's economic dominance by Chinese in Russian territories north
of the Amur has a long prehistory.

By far the largest mainland migration of Han Chinese was not formally
"transnational," nor did it involve typically internal migration. The destination
was Manchuria, the ancestral homeland of the Qing dynasty, which had for a
long time been barred to Han settlers. It was partly opened up in 1878, but only
the combination of persistent or worsening poverty in northern China with
new opportunities in the huge expanses north of the Great Wall—soybean cultivation for export, railroad construction, mining, and logging—brought about
a real *wave* of migration. Cheap rail and steamer transportation created the
logistical foundations. Between 1891 and 1895 barely 40,000 northern Chinese
crossed the border per year. But at its peak in the late 1920s, the annual figure
was close to one million. Between 1890 and 1937, roughly twenty-five million
Chinese set out for the Northeast; two-thirds returned, but eight million settled
there for good. It was one of the largest population movements in modern history, exceeded only by the great transatlantic migration from Europe.[106]

Significant flows of peasant migration also occurred in mainland Southeast
Asia. Despite the tropical climate, the geographical pattern here was the reverse
of the Chinese: not from lowlands to uplands, but from the healthier highlands of ancient habitation down into the river deltas. Some of this migration
completed a tendency that had already been under way for some time. After the
British annexed Lower Burma in 1852, for example, the opening of the Burma

delta "frontier" for rice growing attracted hundreds of thousands of peasants from Upper Burma, and later from India too. In 1901 a tenth of Lower Burma's four million inhabitants originated in the first generation from Upper Burma, and another 7 percent in India.[107] Similarly, large numbers of peasants from the Northeast took part in the settling of Siam's central plain. In Vietnam the vast Mekong delta was opened up for the first time only in the French colonial period after 1866, when settlers moved down from the North. Major investment in canal construction subsequently transformed Cochin China into one the largest rice-exporting regions in the world, where immigrants performed most of the labor in latifundia under Vietnamese, French, or Chinese ownership.[108] During the same period, tens of thousands of Vietnamese peasants moved to Laos and Cambodia.

Internal migrants in South Asia were a small share of the total population, as compared with Europe. The state also intervened to restrict mobility. Much as attempts were made to control vagabonds and traveling people in ancien régime Europe, measures were taken in India against the nonsedentary population. The British colonial authorities sang the praises of the sedentary, taxpaying peasantry and persecuted mobile sections of the population, viewing them as bandit-like disturbers of peace and order, or sometimes even as anti-British guerrillas. In 1826, scarcely a decade after the end of the war against the Marathas and in a situation of major unrest in India, the British launched a campaign (within the law, admittedly) to wipe out the wandering cult of Thugs, who were feared and demonized as ritual murderers. In the 1870s herdsmen in northern India came under suspicion of constituting "criminal tribes" and were vigorously prosecuted.[109] The emergence of a new demand for labor set up powerful migratory patterns, which the state had to tolerate. Apart from migration to the urban magnets of Bombay, Calcutta, Delhi, and Madras, which had already grown considerably in the eighteenth century, the main flows were toward the newly established plantations, especially the tea-growing ones of Assam. Between 1860 and 1890, Chinese tea, once the dominant product, was driven out of the world market by tea from Assam and Ceylon. Local farmers, to whom the new-style plantations were alien, refused to work for a wage in Assam or Darjeeling, and there was no landless proletariat in the villages. Therefore workers were brought in from outside to work at cheap rates on long-term contracts—often whole families, who were expected to return to their home village for at least two months during the quiet season.[110]

For Russia and the whole of northern Asia, which came under *effective* Russian control until the 1890s, Dirk Hoerder speaks of a "Russian-Siberian" migration system. Unlike the two other extensive ones—the Atlantic system and the Asian contract-labor system[111]—this was not maritime but inland-continental. Free peasants, runaway serfs, landowners, criminals, and even, between 1762 and the 1830s, people deliberately recruited from Germany were the pioneers in this large-scale process of farm settlement. From 1801 to 1850, a yearly average of no more than 7,500 (including exiles and prisoners) moved to Siberia, but then the yearly figure rose to between 19,000 and 42,000 in the period from 1851 to 1890. The total number of immigrants to Siberia for the years from 1851 to 1914 is estimated at six million. In addition four million settlers moved to Kazakhstan and the regions beyond the Caspian and the Aral Sea. By 1911 the share of indigenous inhabitants in the population of Siberia, themselves split into numerous ethnic groups, had fallen to a tenth of its previous level. In the east they were caught between the hammer of Russian colonization and the anvil of Chinese.[112]

Nationalism and Migrant Labor

It is crucially important to distinguish between the migration of workers or agricultural settlers and the mobility of herdsmen. Pastoral existence is a special case of nomadism, of a collective nonsedentary way of life.[113] It has been of widely varying importance in different parts of the world. Nor was it absent in Europe: after all, in eighteenth-century France those who for one reason or another led a "nomadic" existence still accounted for 5 percent of the population. Yet pastoral peoples do not generally feature in written history. The urban civilizations in which historians reside have always seen them as "barbarian" others. This might be associated with either negative or positive values: the patriarchs of the Old Testament enjoyed high cultural esteem in the Jewish and Christian world, and here and there in the nineteenth century a kind of Bedouin romanticism saw the "sons of the desert" or the native populace of the American West as the rough but kind-hearted embodiment of an otherwise lost proximity to nature. They were "noble savages," at times more highly regarded in the West than in the city-dominated Islamic civilization. Realistic insights into their lives were extremely rare, however. Until the 1770s there were no European accounts of the internal "functioning" of nomadic societies. Only modern ethnology has systematically investigated the inner logic of nomadic lifestyles.

There were mobile livestock breeders in every continent. Europe's specificity was that livestock breeding was a branch of the division of labor within society, and with the exception of the Sinti and Roma, no ethnic groups were entirely nomadic. Europe had no pastoral *peoples*, although it did have small communities of shepherds and herdsmen who (sometimes accompanied by their families) moved from place to place with their animals. Today transhumance—the grazing of livestock at mountain pastures in summer, with winters spent in the lowlands—is an increasingly rare and marginal phenomenon in the Alps, the Pyrenees, the Carpathians, and Wallachia. Long treks with cattle, as a kind of living meat transport, used to take place in the American West in the nineteenth century, but no longer in Europe. The huge ox herds that once traveled across Hungary and as far as central Germany and Alsace became unnecessary as animal farming improved, slaughtering became industrialized, the railroad network expanded, and freezing technology developed in the 1880s. Nowhere else in the world was there anything like the herds of 150,000 to 400,000, occasionally as high as 600,000, that a couple of thousand cowboys used to drive north from Texas for three months at a time between the 1860s and 1880s.[114]

In no other part of the world did pastoral nomadism remain such an important way of life as in West Asia (the region between Afghanistan and the Mediterranean), Mongolia, and Africa. It is impossible to give a full survey of this here. An arc of pastoralism took in areas from the Hindu Kush through the Anatolian highlands to Sinai and Yemen. In Iran, the share of nomads in the total population fell from a third to a quarter in the second half of the nineteenth century.[115] All through the century, however, livestock breeding remained one of the most important sectors of the economy. The fact that a large part of the population lived a mobile existence created problems that had not been seen for a long time in Europe: three-way conflict and cooperation in the triangle of city-dwellers, sedentary tillers of the soil, and pastoralists; disputes over pasturage and transit rights; ecological destruction; and intertribal conflicts. Nomads also remained a power factor with which every ruler had to contend. Finally the dictator and later shah Reza Khan (r. 1925–41 as Reza Shah) brutally subjugated the nomadic tribes, which he regarded as unruly savages unworthy of a modern nation-state.[116]

In the Ottoman Empire, the sultanic center always had to negotiate with powerful tribes, and migrant labor played a significant role in numerous sectors of the economy. After the onset of reform in the late 1830s a newly assertive state broke the power of the tribes or drove them into marginal regions of the empire, thereby enhancing its internal security and improving mobility for nonnomads. This increased the area of land under cultivation and encouraged the formation of large, commercially run estates, but did so at the expense of nomads.[117] Yet, as Reşat Kasaba has pointed out, this strategy of intensified sedentarization did not meet all its aims and failed to comply with the Ottoman self-image of modernizing state-building. "[T]he Ottoman officials had to cooperate with some tribal chiefs in order to subdue others."[118] The Ottoman state always had to reckon with the tribal factor in ever-changing constellations of power.

In Africa pastoralism was widespread almost everywhere outside the tropics and the immediate coastal areas: from the Atlas Mountains to the highlands of South Africa. It existed in Sudan (which then encompassed the whole African savannah region south of the Sahara), in the Ethiopian highlands, in East Africa, and in Namibia.[119] As always in nomadism, the mobility radius varied greatly from group to group: it could involve the surroundings of a village or, in the *grand nomadisme* of North Africa, cover vast areas of desert.[120] Beginning in the last quarter of the eighteenth century, at the Cape of Good Hope, further along the coast, and later in the interior too, there developed a society of white nomadic pastoralists, the Trekboers, whose conflicts with their indigenous Xhosa neighbors centered mainly on pastureland.[121] Nineteenth-century Africa was a continent in constant nomadic movement.

Nomadism is not the same as migration, which implies that individuals, not whole societies or "peoples," are on the road either voluntarily or because they have been forced into it. Migrants leave behind a home society. Sometimes they will return to it, whether in a seasonal cycle that offers them employment elsewhere for part of the year or after a long stay in a distant land that may have disappointed their hopes of it. In Africa this kind of migration had two different origins. On the one hand, farmers and rural laborers moved of their own free will to new "cash crop" centers, such as the groundnut and cocoa areas in Senegambia and the Gold Coast (Ghana). The production of these goods was in the hands of Africans; foreigners provided only the link to world markets.[122] On the other hand, a directly colonial economy, in which foreigners also controlled the means of production, brought new opportunities for wage labor in mining and labor-intensive settler farms (which could often compete against African agriculture only with support from the colonial authorities). The change happened in such a short space of time that the term "mineral revolution" has been used for southern and central Africa between 1865 and 1900, especially for the years after 1880.[123] In the systems of diamond, gold, copper, and coal mining that developed from southern Congo (Katanga) to the Witwatersrand, entrepreneurs initially brought in trained Europeans to work alongside untrained African migrants. There inevitably came a point, rarely before the 1920s, when cost arguments spoke in favor of using *African* skilled workers. But until then, new seasonal patterns of unskilled employment were the norm. A topography of migration structured by new capitalist growth centers established itself on top of the traditional mobility of pastoralist societies.

Slave Exports from Africa

The Atlantic slave trade involved many areas of the African west coast in one of the principal migration systems, whose indirect consequences reached far into the interior. Sudan was furthermore the catchment area for the trans-Saharan and "Oriental" slave trade. As the African slave trade slowly contracted in the course of the nineteenth century, the continent became a less substantial part of intercontinental migration flows. In 1900 Africa was quantitatively less important in global migration networks than it had been a hundred years earlier: a case of *de*globalization. What was the scale of the nineteenth-century slave trade from an African perspective? This question, with its high moral and political charge, is all the more controversial because of the lack of hard data. Serious estimates of the total volume of the slave trade from Africa to America after 1500 vary by a wide margin. An especially thorough examination of the evidence has arrived at a figure of 12.5 million for the slaves who embarked from Africa; the horrors of the transatlantic "middle passage" meant that the number of arrivals was 10 to 20 percent lower (compared with a maximum loss of approximately 5 percent on ships carrying European emigrants).[124] The best estimate of the slaves who arrived between 1501 and 1867 at the major ports in the Atlantic world puts the number at 10,705,805.[125]

In the destination countries of the "Oriental" trade, slaves were put to work on plantations or in the households and harems of the well-to-do. Muhammad Ali and the rulers who succeeded him in Egypt needed to keep replenishing the great slave army (an old Islamic tradition) that they built up from the 1820s onward, at a rate that peaked around 1838 at 10,000 to 12,000 a year. At that time the initiative for the capture and recruitment of military slaves passed to private traders—the privatization of a growth sector in Sudan.[126] As to Ethiopia, the Arab North preferred child-slaves, especially girls, taking 6,000 to 7,000 a year in the second quarter of the century.[127] Europeans dealers did not participate in the Oriental slave trade, but its consequences for the affected regions of Africa were no less grave than those of the Atlantic trade. It is much harder to quantify, but we can be sure that—contrary to what is sometimes claimed—it was *not* significantly larger than the slave trade conducted by Europeans. If we accept an estimate of 11.5 million for the total number of African slaves who crossed the Sahara, the Red Sea, and the Indian Ocean, then it would be on the same scale as the transatlantic trade throughout its history—not including the slaves who ended up in Egypt.[128] Whereas the ceiling of the "Oriental" trade remained fairly stable in the eighteenth century, at around 15,000 slaves a year, it climbed to more than 40,000 a year by 1830.[129] This was the great age of the Arab slave hunts in eastern Sudan, the Horn, and East Africa. Brutal Muslim troops would make sorties from Khartoum or Darfur into "infidel" areas that were powerless to defend themselves. Deadly caravans of captives sometimes marched thousands of kilometers until they reached the Red Sea.

It depends on one's perspective whether one sees a slight decline in the slave trade between the eighteenth century and the period after 1800, or whether one emphasizes the continuity in an age when, at least in Europe, the forcible trade in human beings was beginning to be taken less for granted. In any event, post-1800 slave exports were down 1.6 million on totals for the eighteenth century, yet it has been cautiously estimated that 5.6 million people were still affected by it in the nineteenth century.[130]

East Africa was the only region on the continent that serviced both American and Afro-Asian markets. In the late eighteenth century, European slave traders scoured the French Indian Ocean islands, Mauritius (known as Île de France until 1810, then a British possession), and Réunion (known as Île Bourbon until 1793). Then came Brazilian traders who could not gain a foothold in Angola, followed by Spanish and North Americans seeking supplies to send to Cuba. The buyers included the Merina kingdom in Madagascar, which curiously also lost inhabitants as slaves. Portugal, the colonial power in both Angola and Mozambique, passed a decree in 1836 under British pressure that "completely abolished" the slave trade. But in reality nothing was abolished. In 1842, flexing its muscles more, Britain forced a treaty on Portugal that declared the slave trade to be piracy and gave the Royal Navy the right to conduct searches; British warships then began to patrol off the East African coast. But market forces, spurred on by Britain's global policy of free trade, proved to be stronger. Rising sugar and coffee prices in the 1840s increased the demand for African labor, and traders found ways to meet it. Human trafficking behinds the backs of British naval officers and missionaries was a mere trifle for experienced trading networks. In the 1860s the "illegal" trade from Mozambique was no less brisk than the "legal" trade had been before 1842, which had run to a yearly average of more than 10,000 slaves.[131]

Only after 1860 is it possible to speak of a real end to slave exports across the Atlantic, or anyway to trade movements that were recorded in some way and can be assessed by historians.[132] The slave traffic ended at different moments in different areas (again the actual circumstances need to be investigated). It first disappeared from the coasts of West Africa, where it had begun at an early date and notched up the highest turnover, and where it was all but over by the end of the 1840s.[133] West Africa—the arc stretching from Sierra Leone to the Bight of Biafra—was the first part of Africa to recover from the population drain, before it was caught up in the 1880s maelstrom of colonial conquest. Western central Africa—the Congo and Angola—enjoyed at best a brief respite of one generation, while throughout the east of the continent, from Somaliland to Mozambique, the European colonial conquerors of the 1880s arrived at a time when the slave trade was still in full swing.

South Africa—this should not be forgotten—was like the rest of the continent in experiencing the institution of slavery (on the eve of its prohibition throughout the British Empire, slaves made up a quarter of the population of the Cape Colony),[134] but it was never significantly involved in the slave trade. Only rarely did the new "legal" trade in agrarian exports and the old slave trade change places with each other as neatly they did in certain regions of West Africa (where palm oil products, for example, came to the fore). And if one looks more closely at particular localities, the juxtaposition of different systems becomes apparent. There might have been a local slave economy with its own institutionalized interests, but alongside it free African traders flowed into the cities and pressed on the markets.[135] The old slave-hunting routes were by no means everywhere a thing of the past when the European colonial presence created its new topography of migration.

The most important legacy of the slave trade was slavery itself. It had existed before the appearance of European slave traders in the sixteenth century, but the trade then generalized the institution and gave rise to societies based entirely on enslavement in military campaigns. Between 1750 and 1850, as much as a tenth of the population of Africa may have had the status of slaves—whatever that actually meant in practice.[136] And it was an upward trend. New internal slave markets came into being. The city of Banamba in today's Mali, soon after its foundation in the 1840s, was functioning as the center for a far-reaching slave trade network; it was surrounded by a band of slave plantations fifty kilometers wide.[137] Early colonial censuses often recorded a high percentage of the population with slave status, and the colonial authorities partly justified their rule by the claim to be "civilizing" the region in which they had intervened.

There is much to be said for the view that, far from being an archaic remnant of the premodern age, a slave mode of production fit well with the new possibilities that opened up in the nineteenth century. Whereas the colonial authorities, especially at first, used African labor in a corvée system, many African regimes continued to deploy slaves in production as the foundation of their economy. These might be prisoners-of-war, purchased slaves, tribute objects, debtors, victims of kidnapping, human beings captured specially for oracles, and so on. In West Africa, states such as the Sokoto Caliphate, Asante, and Dahomey often imported slaves from far away to work on plantations or in handicrafts. It is said that in the 1850s, just before it became a British protectorate (1861), nine-tenths of the population of Lagos consisted of slaves.[138]

In some parts of Africa slavery gained a fresh vitality in the nineteenth century, fueled by new economic opportunities and by Muslim revivalist movements whose state-building jihads depopulated whole areas as they swept through the sub-Saharan savannah belt from today's Mali to Lake Chad.[139] So, in addition to what was left of the maritime slave trade, impulses developed in the African interior toward that high mobility that is always related to slavery. It had inevitably to cover wide areas, because societies are disinclined to enslave their own lower classes en masse. The "weapons revolution" that began in the 1850s—that is, the availability of discarded rifles from European arsenals and their appropriation by of Africans—reinforced this process by making it possible to construct new kinds of armed forces.

Although Africa, after the dismantling of the slave trade, no longer served as the basis for a transcontinental system of migration—that is, unlike fin-de-siècle Europe, South Asia, and China, it no longer provided a long-term regular flow of labor in distinct geographical patterns—colonial *immigration* to the continent should not be overlooked. On the eve of the First World War, it was not Asia, with its ancient and populous colonies, but Africa that hosted the largest number of overseas Europeans in the Old World.[140] Algeria's 760,000 Europeans (two-thirds of them French) made it the largest settler colony outside the British Empire, well ahead of India's maximum of 175,000 (all categories included). At the same time South Africa had approximately 1.3 million white inhabitants, a large influx having begun after the mining revolution of the 1880s. More than 140,000 Europeans lived in British-ruled Egypt, almost exclusively in the cities, with Greeks as the largest single group. Most of the 150,000 Europeans in the French protectorate of Tunisia were Italians. As to the colonies south of the Sahara, there were a total of approximately 120,000 long-term-resident Europeans in 1913. All in all, roughly 2.4 million "whites" or people of European extraction then lived in Africa, most of whom had arrived there after 1880. The equivalent figure in Asia was no higher than 379,000—plus 11,000 Americans in the Philippines.

A European-organized labor migration from Africa to Asia did not exist in the nineteenth century. When the Dutch took slaves from the Cape to Batavia two centuries earlier, it was not the start of an ongoing large-scale export trade, any more than was the movement of slaves from India or Indonesia to the Cape Colony. The reason for the transfer was that the Dutch East India Company prohibited the enslavement of local subjects in its possessions. In the nineteenth century, after a long break, Asians began to migrate again to Africa in considerably larger numbers. Between 1860 and 1911, a total of 153,000 contract workers were shipped from India to work on the sugar plantations of Natal; some shopkeepers went of their own free will. In Kenya, 20,000 Indians were employed for the construction of the Kenya-Uganda railroad, and many of them stayed on after the end of their contract.[141] In Mauritius, too, there were many Indians. In precolonial times, a small community of Indian tradesmen already lived in the territory of what is now Tanzania. By 1912 German East Africa contained 8,700 Indians—indispensable middlemen, who kept the colonial economy going, but who were suspected by the authorities because the great majority of them remained British subjects.[142] All in all, perhaps 200,000 Asians arrived in Africa between 1800 and 1900.[143] At the intersection of several major systems, Africa in the nineteenth century was the continent with the greatest variety of migration.

7 Migration and Capitalism

No other epoch in history was an age of long-distance migration on such a massive scale. Between 1815 and 1914 at least 82 million people moved *voluntarily* from one country to another, at a yearly rate of 660 migrants per million of the world population. The comparable rate between 1945 and 1980, for example, was only 215 per million.[144] The migration of tens of millions of Europeans to America, an especially striking instance fraught with consequences, has been considered in many different ways:

- as emigration that partly developed out of migration within Europe
- as immigration that was part of the centuries-long settlement of America
- as a hostile invasion of land belonging to Native Americans
- in the perspective of social history, as the creation of new, and expansion of existing, immigrant (diaspora) societies
- sociologically, as a collection of phenomena of acculturation
- economically, as the opening up of new resources and the raising of the possible global level of productivity
- politically, as flight from a repressive Old World—from a monarchical old order to the realm of egalitarian liberty
- culturally, as a stage in the long-term Westernization of the world

Here it is enough to sketch the overall demographic picture.

Destination America

A sharp break, not uncommon in the history of population, here permits a fairly exact periodization of the nineteenth century. The break occurred around the year 1820, with the rapid and almost total disappearance of the "redemption system" under which new male and female immigrants had undertaken to pay back the cost of their passage soon after their arrival in America.[145] The system was a legal and humanitarian improvement on the old form of indentured service, customary first in the Caribbean and later in North America, which had always involved a period of bonded labor in a private relationship. Under the redemption system it was possible to cover the debt in other ways—if someone could be found to stand surety, for example—but even then the last resort was for the immigrant, or sometimes his children, to discharge it through their labor. The core of the redemption system therefore still meant voluntarily entering into a relationship of bondage.[146] It remained legal until the early twentieth century but soon dwindled in significance after 1820. Immigrants—Germans earlier than Irish, for example—were less and less prepared to accept such forms of service, and the American public, itself often not long in the country, increasingly viewed this "white slavery" as degrading. In 1821 the Indiana Supreme Court passed a landmark judgment against the debt bondage of white immigrants. Conditions in Europe would still force millions to move across the Atlantic, but in the eyes of the law that emigration was now free.

Meanwhile, processes were under way on both sides of the Atlantic that would lead to an "integrated hemispheric system,"[147] combining the various older patterns of migration within Europe and overseas. This subsystem of an emerging international labor market filled a vast space from the Jewish Pale of Settlement in western Russia to Chicago, New Orleans, and Buenos Aires, making contact at its margins with the Siberian and the Asian migration system. Mobility within the system was generated by imbalances—between poor and rich regions, between low-wage and high-wage economies, between agrarian societies and early industrial centers, between societies with steep hierarchies and few opportunities for upward mobility and societies of which the opposite was true, and between repressive and liberal political orders. All these dimensions shaped the changing rhythms of movement within the system, so that different parts of Europe channeled their surplus population into it at different times. With few exceptions, the migratory flows were mainly proletarian in character. Ordinary people looking for a better life were more characteristic than adventure seekers of genteel birth.

Net immigration into the United States for the whole period from its founding until 1821 has been estimated at 366,000,[148] more than half of whom (54 percent) came from Ireland, and just under a quarter from England, Scotland, and Wales. In 1820 the annual volume of slave exports to Brazil was still more than twice as large as that of free emigrants to the United States! Before 1820 the migration to the United States was a trickle. After 1820 it became a steady flow. The 1840s, 1850s, 1880s, and 1900s were the decades when it turned into a flood.[149]

The number of new immigrants in the United States grew from 14,000 a year in the 1820s to 260,000 in the 1850s and a peak of approximately a million in 1911. The main driving force all through the century was the robust growth of the American economy, whose curve roughly paralleled that of immigration, and the continual decline in transportation costs. After 1870 the proportion of immigrants from northern and western Europe receded, while that from east-central, eastern, and southeastern Europe increased. It was a dramatic trend. In 1861–70 east-central and eastern Europeans accounted for only 0.5 percent of immigrants, and southern Europeans for only 0.9 percent; in the 1901–10 decade, the shares were 44.5 percent and 26.3 percent respectively.[150] This had colossal effects on the cultural and especially the religious composition of US society.

The national shares of European migration across the Atlantic are a revealing indicator. Of the western and southern European countries, Ireland ranked first during the last three decades of the nineteenth century, followed by Britain and Norway in joint second place; third was a group made up of Italy, Portugal, Spain, and Sweden; while Germany followed some distance behind. Conversely, how important was transatlantic emigration for the individual European countries? In the decade after 1870, 661 per 100,000 of the local population emigrated from Ireland, against 504 from Britain, 473 from Norway, 289 from Portugal, and 147 from Germany.[151] In absolute figures, British, Italians, Germans, and Habsburg subjects were the most numerous among those who crossed the Atlantic, although until about 1880, Italians migrated more within Europe than overseas. Only one large European country did not take part: France. Of course, national averages give only a bird's-eye view; emigration was actually concentrated in certain regions within countries—for example, Calabria, western England, western and southern Ireland, eastern Sweden, or Pomerania.

Unforced migration across the Atlantic can also be estimated only approximately. Informed conjectures hover around 55 million for the whole period between 1820 and 1920,[152] 60 percent of it (33 million) to the United States. The second most important destination was Argentina, to which roughly 5.5 million (10 percent) emigrated between 1857 and 1924, followed by Canada and Brazil.[153] These figures do not take returnees into account. Although European emigration, unlike Indian or Chinese labor emigration, was normally considered definitive, there were always a certain percentage who went back or who moved on to another country. Canada had vast unpopulated territories but did not fulfill the expectation that part of the huge migratory flow to the United States would turn northward. Indeed, around the end of the century Canada was sending *more* people to the United States than it was receiving. Canada was a classic way station, a demographic sieve.[154]

Argentina is an extreme case in the history of emigration. Nowhere else in the world, including the United States, did immigrants constitute such a high proportion of the population by the end of the nineteenth century. In 1914, of the eight million people living in a country five times larger than France, approximately 58 percent had either been born abroad or were children of first-generation immigrants.[155] For decades, one-half of the inhabitants of the capital, Buenos Aires, had not been born in Argentina. Immigration from Spain to the River Plate region, other than of officials or military men, began only in the middle of the century; it had little to do with the fact that the country had once been ruled by Spain, and should therefore not be regarded as a postcolonial phenomenon.[156] By 1914 Buenos Aires was the third-largest Spanish city in the world, after Madrid and Barcelona, but it was Italians who made up the principal group of immigrants by numbers. Many of them moved there only temporarily, the journey from Italy being so easy to arrange that it was possible even on a seasonal basis. Italian musicians from rustic choirs to celebrated primadonnas sojourned there during the slack season of the peninsular opera year, turning Buenos Aires into a major center of Italian opera.[157] Since it had scarcely any continuity with the colonial past, the immigration to Argentina was not marked, as in North America, by the old practice of indenture, and unlike in Brazil African slavery was almost insignificant. It was therefore "modern," in the sense of not being burdened with unfree labor relationships. For lack of an adequate internal market, the economy was from the beginning geared to production for international demand: at first sheep farming (cattle played little role before 1900), and then an agricultural revolution that, within a few years after 1875, turned the former grain importer into one of the world's largest exporters of wheat. Immigrants were used here as farm laborers and sharecroppers; only few managed to acquire land on a significant scale. Argentina became much less of a melting pot than the United States. The Spanish-Creole upper classes did little for the integration of newcomers, while more than 90 percent of these spurned Argentine citizenship in order to escape military service.[158] Italians in Buenos Aires were renowned for their patriotic passions. Mazzini and Garibaldi found much support there, and conflicts between secular forces and church loyalists were pursued with passion.[159]

Contract Labor

In the nineteenth century there were also new migrations that did not originate in Europe. They were driven by the "pull factor" of labor shortage and occurred very largely, though not exclusively, within the British Empire and other areas under British control. The economic engine was not so much the processing industries as three other kinds of capitalist novelty: plantations, mechanized mining, and the railroads. Quantitatively the most important was the plantation, which combined agrarian and industrial revolutions by applying industrial mechanization and work organization to the production and processing of agricultural raw materials. Those who moved were, without exception, colored. And the extent of this migration was still greater than that of the transcontinental movement of Europeans. Indians arrived in East and South Africa, the east coast of South America, the Caribbean islands, and the Pacific island of Fiji; Chinese relocated to Southeast Asia, South Africa, the United States, and western South America. The geographic spread is apparent from table 4, whose figures are actually on the low side, since unrecorded migration and human trafficking must also have been considerable.

Table 4: Main Destinations for Contract Labor, 1831–1920

British Caribbean (Trinidad, Guyana)	529,000
Mauritius	453,000
Africa (mostly South Africa)	255,000
Cuba	122,000
Peru	118,000
Hawaii	115,000
Réunion	111,000
French Caribbean (Guadeloupe, Martinique)	101,000
Fiji	82,000
Queensland (Australia)	68,000

Source: Simplified from David Northrup, *Indentured Labour*, pp. 159–60 (Tab. A.2).

Before 1860 this migration served to fill gaps left by the ending of slavery in the sugar plantations of the British Caribbean and the island of Mauritius (which in mid-century was the largest supplier of sugar to the United Kingdom). Ex-slaves invariably turned their back on the plantations and tried to scratch out a living on land of their own, which could not often have been much more lavish than a slave's existence. The reduction in the supply of local labor accompanied an expansion of the global demand for sugar and a long-term fall in sugar prices, which was due in part to the fact that the production of beet sugar was growing faster than the production of cane sugar.[160] A labor supply was required, and it needed to be as cost effective as possible. New centers such as Trinidad, Peru, and Fiji entered the market. It was in this competitive atmosphere that the demand grew for cheap and acquiescent labor.[161]

Later, Asian migration flowed toward new plantations in colonies that had never known slavery, and into mining and railroad construction. The foundations of this "Asian contract labor system" were laid in the 1840s. It rested upon a generalizable and easy-to-manage form of cheap employment: indenture. The contractual compulsion to work thus reemerged in Asia soon after it had disappeared from new immigration to the United States. The distinctiveness of this new Asian system should not be underestimated, however. Although contract laborers were often kidnapped and cheated in the manner of slaves, and although they were often subjected to harsh discipline in the manner of early European factory workers, they were free persons in the eyes of the law, with no social stigma and no "lord" who systematically interfered in their private lives. They were employed for a specific period, and their children, unlike those of slaves, were legally unaffected by the relationship of dependence. On the other hand, they were often exposed to a racism in the new country that white indentured servants did not have to endure.

The sea journey itself was frequently a horrific experience; Joseph Conrad, in his novel *Typhoon* (1902), described this in the case of some Chinese "coolies" returning to their mother country. Conditions were especially bad on ships bound for Latin America and the Caribbean, which even after the introduction of steamers from southern China took 170 days to reach Cuba or 120 to reach Peru. Sailing ships remained in use here longer than in any other crossing. Laborers were crammed together on plank beds below decks, sometimes in chains, while troublemakers were confined in cages and pillories on deck. Nevertheless, the conditions were not comparable to the horror of the slave ships, whose human cargo had often been as much as six times larger for the same space.[162]

In many respects contract laborers were better off than European indentured servants of the early modern period, since they had not only room and board but a regular wage, usually free accommodation, and a modicum of health care.[163] Contract labor was not the continuation of slavery by other means, and therefore not an archaic practice, but rather an old system of (in principle) free labor migration adapted to imperial requirements in a capitalist age. It should not be seen as an exotic "tropical" form but is to be bracketed together with transatlantic migration. Wherever the wages on offer overseas were so low that they could attract only the poorest of the poor, the cost of the sea crossing had to be met either by someone else or through an advance on future pay.

In reality, colored migrants did not differ so sharply from white settlers: insofar as they were not repatriated for political reasons (as the Chinese were from Transvaal), Asians remained in their destination country as much as Europeans did—a virtual totality in the case of South Asians in the Caribbean. By 1900 Indians had overtaken Africans as the largest group in Mauritius, making up 70 percent of the island's population; they were more numerous than Europeans in Natal and formed a third of the population in Trinidad and British Guyana. Forty percent of the inhabitants of Hawaii were of Japanese origin, and another 17 percent were Chinese.[164] In the most disparate parts of the world, Asian minorities became a stable element in the local society, often constituting a kind of middle class. Asian contract labor essentially consisted of Indians and Chinese. Of the two million non-European contract laborers who figure in the statistics between 1831 and 1920, 66 percent originated in India and 20 percent in China.[165]

The Indian migration was the only one that continued on a significant scale throughout the period.[166] It began in the 1820s, climbed rapidly to a peak in the 1850s, and remained at an average of 150,000 to 160,000 a decade until 1910. This export of labor was a spillover from accelerated flows within India and a side effect of the extensive migration to Burma and other parts of Southeast Asia. The figures show a link not only with demand inside the British Empire but also with the chronology of famines in various regions of India. The chaos and repression following the defeat of the 1857/58 Great Rebellion were reflected in a sharp increase, but long-range factors also played a role. The unusually large number of weavers is attributable to the destruction of India's rural textile industry. The emigration was by no means restricted to the poor: members of higher castes also packed up and left. The especially detailed statistics that we have for Calcutta show that the movement involved a cross-section of the rural population of northern India.

Once the relevant statutory provisions were created in 1844, the Indian emigration became less arbitrary and prone to abuse than the export of labor from China. Deception and abduction were rarer occurrences, and it was to a larger degree voluntary.[167] At first it encountered much resistance from abolitionists, humanitarians, and colonial administrators,[168] but later the interests of imperial planters prevailed. The principles of liberal political economy decreed that no one should be hindered from freely searching for work. Governments within the empire also concluded agreements with one another to overcome labor bottlenecks. In Natal it was difficult to recruit the local population for work on the new sugar plantations, and so in the 1860s the government of the Colony of Natal (which Britain had annexed in 1845) negotiated a supply of contract laborers from India. The idea was that they would return to India at the end of the contract period, but most of them stayed on and helped to build the local Asian community.[169]

Criticisms of the system never died away in either Britain or India; they were part of a discussion on the acceptable restrictions of freedom that lasted throughout the century. The fate of Indians overseas was a permanent issue for early nationalist writers on the Subcontinent, and the lawyer Mohandas K. Gandhi waged a powerful campaign against limitations on the rights of Indians in Natal. In 1915 indenture even became the central issue of Indian politics, leading to its effective abolition the following year.[170]

In one respect the contract labor system differed fundamentally from the transatlantic migration: it was much more subject to political steering and could therefore be brought to an end by an executive decision. This happened not only to contain the rising tide of public criticism but also to protect "white" labor from colored competition. The end of the contract system was therefore a victory for humanitarianism and peripheral nationalism, but it was also the logical consequence of an increasingly racist response to the "brown" or "yellow" threat. No one asked the migrants what they thought. Whereas slaves fully supported the abolition of slavery, the picture was not quite so clear for Indian indentured labor. In any event there were few protests, and the migration from India continued after the system came to an end. The decisive factor on the Indian side was the wounded national pride of the Indian middle class, who were appalled that the Canadian and Australian dominions should close their borders on racial grounds to Indian labor in order to sustain the high wages of white workers.

The Chinese "Coolie" Trade

When the contract system was abolished in India, the Chinese "coolie" trade was already more or less over. It had started timidly after the end of the Opium War in 1842, flourished between 1850 and 1880, and then underwent rapid decline. A last flickering occurred with the export of 62,000 laborers from northern China to the gold mines of the Transvaal, where they undercut the wages of local Africans. Their role and their treatment became major issues in 1906 in parliamentary and electoral politics in both Britain and South Africa. In London the new Liberal government came out against the practice, while in South Africa the mining industry opted for a return to the use of native labor.[171] The heyday of Chinese labor emigration was literally a "golden age." It began with the California Gold Rush of 1848–49, continued with the flow of Chinese laborers to the goldfields of Australia between 1854 and 1877, and ended with the repatriation of the last Chinese from South Africa in 1910.[172]

The departure points for the Chinese labor migration were the southern coastal provinces of Guangdong and Fujian, where the introduction of the sweet potato and groundnut in the early seventeenth century had triggered a sharp rise in population. The accessibility of Southeast Asia by junk made it the natural focus for coastal inhabitants who wanted to establish contacts abroad. But the Chinese had traditionally been no keener than the Indians to migrate, the state acting as a major obstacle to overseas travel in the same way that religion had done in India. The Emperor's subjects were generally restricted in their freedom of movement; the state repeatedly took the liberty of ordering resettlement in frontier regions but was distrustful of spontaneous mobility. Imperial governments again and again prohibited people from leaving the country, or from returning once they had left. The social system, with its Confucian values, also made movement difficult, since the duty of piety toward parents and forefathers could not be discharged elsewhere than in the ancestral town or village. Between the fifteenth and eighteenth centuries, sporadic merchant emigration to various parts of Southeast Asia (the Philippines, Java, the Malay Peninsula) gradually crystallized into local Chinese communities with their own traditions and a distinctive cultural mix.[173]

The "opening" of China happened in 1842, at a time when new migratory structures were already emerging. There was a complementarity between southern China, marked by insecurity, overpopulation, and impoverishment, and the land of Siam, peaceful and prosperous but with a low population density. As rice production for export began to integrate the two countries into larger markets (somewhat earlier than in nearby Burma), an ethnic division of labor sprang up between Siamese agriculturalists and Chinese involved in milling, transport, and trade. By the middle of the century, Siam had the largest foreign Chinese community anywhere in the world.[174] Whereas emigrants to Siam—the main destination just before the beginning of the coolie trade—mostly had their passage paid by kith and kin already living there, the coolie emigration to Malaya, Indonesia, Australia, and the Caribbean was mainly organized on an indenture basis that marked a radical break from former practices. A second novelty was at the level of transportation technology. The traditional junk trade was partly supplanted and partly complemented by European steamships, whose spread in the second half of the century increased the flow of migrants to Southeast Asia and the Americas. Family ties and travel to join relatives constituted important contexts for the movement of Chinese labor across the seas. Operating from Southeast Asia, Chinese traders organized the transfers and sent agents along the coast of southeastern China in search of workers. The coolie trade was a multinational business. It was increasingly conducted by British, American, French, Spanish, German, Dutch, Portuguese, and Peruvian intermediaries, who often received a price per head from dubious Chinese partners. Naive farmers' children would be lured away from their parents with various tricks and tales of fabulous riches. Abduction was the simplest way for the recruiting agents to obtain exportable manpower.[175] It was not an abhorrent practice characteristic of Asian "barbarians." Until 1814, the Royal Navy had frequently used impressment to crew its warships.

The export of labor, like the contemporaneous opium trade, was against Chinese law and caused an uproar in the Chinese public. In 1852, revolts against abduction broke out in the city of Amoy (Xiamen); in 1855 there were protests throughout southern China; and in 1859 the spread of the practice led to panic and attacks on foreigners in the Shanghai region. In 1859, following the British occupation of Canton in the so-called Second Opium War, the Chinese authorities were forced into "cooperative" tolerance of the coolie trade, although it proved impossible to prevent them from sentencing kidnappers to death.[176] From the start it was a serious law-and-order problem—and it remained so until it came to an end. In 1866, after an incident in which no less than the son of a regional provincial governor was kidnapped, the Chinese government forced through an international regulation that made abduction illegal under Western law too. But Portuguese Macau continued to function as a loophole, where Chinese were promised work in California and shipped off to the harsher conditions of Peru's guano islands or Cuba's sugarcane fields. When Spain and Peru attempted to reach trade agreements with China in the 1870s, commissions of inquiry were sent out and the protection of coolies was made a prerequisite. After 1874, however, the Qing government took a tougher line, imposing a general ban on the coolie trade and dispatching consular officials to look after the welfare of emigrants.[177]

The struggle against Indian and Chinese indenture differed from earlier campaigns against the transatlantic slave trade in that the exporting countries also brought political pressure to bear. The colonial government of India never unanimously or wholeheartedly approved of contract migration; it was eventually prepared to remove this issue as a rallying point for early Indian nationalists. The Chinese government represented an independent country that was in a weak position vis-à-vis the imperial powers. The stubborn patriotic efforts of its diplomats on behalf of Chinese coolies were not without effect. But although it helped to put an end to the system, it did not play the decisive role. More important was the fact that Chinese labor ceased to be necessary to the economies of the host countries.

Classical indenture supplied plantations, where most Indians worked overseas. A large percentage of Chinese migrants headed elsewhere: they were not condemned to compulsory labor, even if they often had to raise a loan to cover the costs of their passage. In other words, most Chinese who left their homeland were not "coolies." The Chinese emigration to the United States that began with the California Gold Rush of 1848–49 was "free" and therefore more akin to the migration from Europe. The same is largely true of the exodus to Southeast Asia and Australia. Between 1854 and 1880, at the height of emigration from China, more than half a million nonindentured Chinese left from the port of Hong Kong alone.[178]

In no other mass migration of the nineteenth century was the quota of returnees so high. The ties of Chinese migrants to their place of origin were so strong that their residence abroad was sometimes regarded as temporary even after several generations. Europeans were much more inclined to think of migration as a permanent break with their past, much more prepared for assimilation and new life plans. In fact, Chinese emigration may be best understood as an overseas extension of the economy of southern China. Perhaps as many as 80 percent of the Chinese who left their country by sea in the nineteenth century returned at some point in their lives, whereas the corresponding figure for Europeans was probably around one-quarter.[179] The high circularity and fluctuation of migration also means that the absolute figures at the dates when statistics were compiled are surprisingly low. The US census of 1870 recorded only 63,000 Chinese, and the census of 1880 (when Chinese immigration was already beginning to slacken) no more than 105,465.[180]

The only part of the world where Chinese emigrants settled in large numbers was Southeast Asia—the oldest destination by ship (the main mode of travel to Siam, Vietnam, and Burma). Here the European colonial authorities generally encouraged the immigration of Chinese, who took on the roles of merchants, entrepreneurs, or miners that neither local people nor Europeans could fill adequately. Above all, they were good taxpayers. Their industriousness and business acumen enabled them to organize themselves under the leadership of local notables and secret societies, in well-functioning communities that caused little trouble. Despite their ties to the mother country, the Chinese minorities in Southeast Asia were loyal subjects of the European colonial powers. In the long run it mattered little how they had come there. In China, as in India, certain coastal areas were completely geared to emigration and relied upon it economically. Whole families, villages, and regions developed a transnational character; many people felt closer ties with relatives or former neighbors living in Idaho (whose population at one point was 30 percent Chinese) or Peru than with their fellow countrymen in the next village.[181]

Migration via contract labor was more controlled legally and politically, and better recorded statistically, than unregulated forms of emigration. If the latter is also taken into account, the figures for East and Southeast Asian long-distance migrants become considerably higher. Nor should the number of those who left as traders rather than workers be underestimated. According to some calculations now widely accepted, more than 29 million Indians and 19 million Chinese emigrated between 1846 and 1940 to countries around the Indian Ocean and the South Pacific—a population movement as large as that from Europe to the Americas. However, of the 29 million Indians, no more than 6 to 7 million had permanently settled abroad; with the Chinese, as we saw before, the proportion of returnees was also much higher than with European emigrants to the New World.[182] These migrations were quintessentially circular. Only a tenth of the Asian migrants were subject to indenture mechanisms, although private or public loans played a role in many other cases.[183] The First World War marked less of a break for this migration in the global "South" than for the human flow across the Atlantic. Only the Great Depression and the Pacific War seriously affected emigration by Indians and Chinese.

8 Global Motives

In the nineteenth century, more people than ever before were traveling across large distances and for long periods of time. In 1882 the Buddhist master Xuyun set off on a journey to Wutaishan, a holy mountain in the Chinese province of Shandong. Since he prostrated himself fully after every three paces, he needed two years to cover the roughly fifteen hundred kilometers.[184] Xuyun was a pilgrim, and any discussion of movement on a large scale should not neglect to consider pilgrimages. In Europe, Asia, and Africa religious centers continued to attract hundreds of thousands of travelers. The largest single movement was the hajj to Mecca, which was usually an expensive collective undertaking by ship and/or caravan, later made easier by the Suez Canal and the Hejaz Railway, though never safe from robbers, tricksters, and rapacious pilgrim guides, and always a major health hazard, especially since the first cholera epidemic in Mecca in 1865. The number of pilgrims (which nowadays may exceed two million) fluctuated sharply from year to year but tripled in the course of the nineteenth century to reach a total of 300,000. By far the most typical pilgrim from distant lands—Malaya, for example—was an aging member of the local elite, affluent enough to pay for the trip out of his own resources.[185] Further routes opened from the Balkans and Central Asia, and after the turn of the century there was a "push east" to the holy sites from the new Islamic realms of West Africa, which partly accounts for the intra-African migration by groups of followers or even whole ethnic groups. Millenarian expectations focused on a messiah or *mahdi*, to whom people wanted to be close, especially in the times of stress following colonial invasions.[186] By the nineteenth century, worldwide networks had come into being: Chinese Muslims traveled to Mecca and Cairo, while tombs of Sufi saints in the Chinese Empire became important destinations in their own right. Religiously motivated pilgrimage may shade imperceptibly into proto-tourism.

In eighteenth-century Japan visits to faraway temples and shrines came to be or-ganized in a way that has prompted historians to speak of a "tourist industry."[187]

A novelty of the nineteenth century was large-scale migration beyond and after the slave trade. It gradually developed after 1820 and dramatically increased after the middle of the century, at a rate significantly faster than that of the world population. Migration studies has long ceased to see it as an undifferentiated flow of "masses." The picture is now more of a mosaic, made up of local situa-tions in which the village community and its partial transplantation are often the focus of a microstudy. Up to a point the components of such migration his-tories were the same across different cultures: there were pioneers, organizers, and group solidarity. The decision to migrate was more often collectively made by the family than by isolated individuals. The transportation revolution im-proved the logistical possibilities, while the more extensive organization and faster speeds of capitalism required a more mobile labor force. Most emigrants, whether from Europe, India, or China, came from the lower classes; they aspired to join the middle classes in the host country, and they achieved this more often than former slaves or their descendants.[188] The link between internal and external ("transnational") mobility was variable. It would be superficial to claim that life in the modern world is always and everywhere faster or more mobile than in the past. Indeed, studies on Sweden and Germany long ago showed that horizontal mobility within societies, and not only the intensity of emigration, tended to diminish in the twentieth century, at least in times of peace.[189] In the case of Europe, the high level of mobility in the late nineteenth century was exceptional.

Up to the 1880s, governments usually put little in the way of cross-border migration, although in individual cases migrants might be subject to supervi-sion and harassment. This administrative restraint was an important prerequisite for the emergence of the vast systems of migration. The age of state-sponsored emigration began only after the turn of the century—not least with an eye to the beneficial effects of remittances sent by the emigrants to their kinsfolk back home.[190] The Japanese government embarked on active encouragement (and even funding) of emigration to Latin America. The Australasian colonies also pur-sued an active immigration policy, which in their case was especially necessary on account of the high costs of "unassisted passages." Australia needed people badly and had to compete with North America for them. Mass emigration there was possible only because the government began to offer financial inducements after 1831. Nearly half of the 1.5 million Britons who moved to Australia in the nineteenth century received official payments to cover their costs—not loans but grants, which came overwhelmingly out of the public purse. For decades the Colonial Land and Emigration Commission in London was one of the most im-portant and successful branches of the Australian state. This also facilitated the tasks of control and selection. In the conflict between the British government, which wanted to offload its "plebs," and colonies interested in a "higher class" of settler, the receiving countries prevailed in the end. The Australian case confirms the economic rule that governments of democratic states prefer an immigration policy that can be expected to maintain or increase the income of their electorate. The next question is when immigrants are to be offered naturalization on equal terms with the rest of the population.

The motives of individual migrants were, of course, culturally shaped. People from very hot regions do not like to work in very cold countries, and vice versa, and the tendency is to go where others from one's own homeland or members of one's own social group already have a reasonably contented life and can provide crucial advice and information. At the extreme—for example, among the Irish after the Great Famine—a magnet effect was set up that made emigration seem the only sensible thing to do. When people who left Scotland to supply an astonishing share of the workforce needed for conquering and running the British Empire responded to meager chances at home, they did this at various levels. Some were destitute peasants or younger sons of noble families, others alumni of Scotland's excellent universities, which produced more lawyers or medical doctors than the domestic labor market could absorb.[191] In other respects, however, where the scope for decision was fairly wide, attitudes might follow a culturally neutral rationality. One central element was the differential in real wages between the Old and the New World. The gap narrowed over time as a result of emigration, and this was a major reason why emigration itself slowly diminished.[192] But the wage motive was present everywhere in the world. In the last third of the nineteenth century, Indian workers migrated more to Burma than to the Straits Settlement, since wage levels were significantly higher there until the beginning of the Malayan rubber boom. Many other considerations pointed ahead to the future. Small agricultural producers frequently accepted temporary proletarianization in order to be spared permanent misery. Prognoses as to the outcome of emigration were always uncertain. The ill-informed or credulous allowed themselves to be lured into risky adventures by tales of fabulous wealth or by false promises of marriage. The topic becomes exciting for historians when it is a question of explaining micro-differences, for example, why one region produced more emigrants than others. Large interlinked systems must be granted a certain life of their own, although they form, persist, and change only through the interaction of countless personal decisions in particular life situations—in short, through human practice.

CHAPTER V

Living Standards

Risk and Security in Material Life

---

1 Standard of Living and Quality of Life

Quality and Standard of Material Life

A history of the nineteenth century cannot omit the material level of human existence, and we shall bring together the little that research can tell us about this at a *general* level. First, a distinction needs to be drawn between "standard of living" and "quality of life": the former is a category from social history, the latter from historical anthropology.[1] Quality of life includes the subjective impression of well-being—indeed, of happiness. Happiness is bound up with individuals or small groups; its quality cannot be measured and is difficult to compare. Even today it is nearly impossible to decide whether people in society A are more content with their lives than people in society B. As for the past, it is scarcely ever possible to reconstruct such appreciations. Furthermore, we need to differentiate between poverty and misery. Many societies in the past were poor in market goods yet enabled people to live a happy life; they based themselves not only on the market but also on community economics and the economics of nature. Personal or collective unhappiness affected not so much those without property as those who lacked *access*—to a community, to reliable protection, to land or forest.

"Standard of living" is a touch more palpable than "quality of life." But it involves a tension between the "hard" economic magnitude of income and the "soft" criterion of the utility that an individual or group derives from its income.[2] Recently it has been suggested that "standard of living" should be defined in terms of the capacity to master short sharp crises, such as a sudden drop in income due to unemployment, higher prices, or the death of a family breadwinner. Those who manage to pull through such crises and to plan their lives long into the future may be said to have a high standard of living. More specifically, under premodern conditions this was mainly a question of the strategies that individuals and groups applied to avoid an early death, and of their degree of success in doing so.[3]

Economists are rather more robust than social historians in their approach to the history of living standards. They attempt to measure the income of distinct economies (which in the late modern age are mostly national economies) and divide them according to their population level. In this way we obtain the famous per capita GDP (gross domestic product). A second question that economic historians like to ask concerns the ability of economies to save, hence to preserve values for the future, and perhaps also to invest part of what is saved so that it creates values in turn. However, there is no univocal positive correlation between statistical economic growth and the actually experienced standard of living. Growth of any degree, even high, does not necessarily translate into a better life. For a number of European countries, it has been shown that real wages moved downward in the early modern age, yet the material wealth of their societies increased overall; a massive long-term polarization must have occurred, whereby the rich grew richer and the poor poorer.[4] So, there is by no means a direct correlation between income and other aspects of an improved quality of life. When Japanese incomes gradually rose in the nineteenth century, growing numbers of consumers could afford the more expensive (and prestigious) polished white rice. But this created a problem since the vitamins present in rice husks were now lacking. Even members of the emperor's family died of beriberi, the vitamin B1 deficiency disease that is a risk associated with prosperity. The same link is observable between sugar consumption and poor dental health. History does not provide enough evidence that economic prosperity automatically translates into a higher biological quality of life.

The Geography of Income

However uncertain the income levels in the age before global economic statistics, the most plausible quantifications must serve as the basis for discussion (see table 5, which draws on Angus Maddison's work).

For want of statistical data, Maddison's estimates can be used only with considerable qualification. In particular it has been objected that they set Asia's economic performance too low. They are inherently "impossible," even if Maddison has attempted, by also making broad use of qualitative sources, to create an approximate impression that roughly reflects the true proportions. Nevertheless, if we take his figures as an at least plausible account of the relations of magnitude and accept that the GDP estimates have some degree of validity, then the following points stand out:

- Between 1820 and 1913, the richest and poorest regions in the world moved wide apart in their material living standards. The difference was 3:1 or perhaps 4:1 in 1820 but had climbed to at least 8:1 by 1913.[5] Even if such figures are not trusted, it is indisputable that the prosperity and income gap in the world increased considerably during this period, probably more than in any other epoch, though in the context of an overall rise in global

Table 5: Estimated per Capita Gross Domestic Product in Selected Countries, 1820 to 1913 (in 1990 $)

	1820	1870	1913	Factor 1870–1913
Europe				
Great Britain	1,700	3,200	4,900	1.5
Netherlands	1,800	2,700	4,000	1.5
France	1,200	1,900	3,600	1.9
Germany	1,000	1,800	3,600	2.0
Spain	1,000	1,400	2,300	1.6
Americas, Australasia				
Australia	—	3,600	5,700	1.6
USA	1,200	2,400	5,300	2.2
Argentina	—	1,300	3,800	2.9
Mexico	760	670	1,700	2.5
Asia				
Japan	670	740	1,400	1.9
Thailand (Siam)	—	700	830	1.2
Vietnam	540	520	750	1.4
India	530	530	670	1.3
China	600	530	552	1.04
Africa				
South Africa	—	1,600	—	—
Egypt	—	700	—	—
Gold Coast (Ghana)	—	—	700	—

Source: Maddison: *World Economy*, pp. 185, 195, 215, 224 (rounded up or down; factor calculated).

wealth. Only after 1950 did this trend subside, and even then there was a stable group of "ultra-poor" countries that benefited from neither industrialization nor the export of raw materials.[6]

· Alongside the industrial heartlands of northern and western Europe, the countries that Maddison calls "Western offshoots" (the neo-European settler societies of North America, Australasia, and the River Plate) achieved the highest income growth.

· The United States and Australia overtook the European frontrunners *before* the First World War, but the differences within the group of "developed" countries were much smaller than those that divided them from the rest of the world.[7]

- The formation of a statistical "third world," consisting of countries that made little progress from their low starting point, was already a feature of the nineteenth century, especially its last few decades.

- There was one exception in Asia and one in Africa: Japan began to industrialize in the 1880s, and around the same time South Africa discovered the largest gold deposits in the world.

- In many countries it is possible to identify an approximate turning point, when average prosperity, and therefore consumption potential, began to increase markedly. This point came in the second quarter of the nineteenth century for Britain and France, around midcentury for Germany and Sweden, in the 1880s for Japan, after 1900 for Brazil, and sometime after 1950 for India, China, and (South) Korea.[8]

2   Life Expectancy and *"Homo hygienicus"*

The limited value of Maddison's income estimates for the question of living standards becomes apparent when we look through his chapter on life expectancy. Here the "poverty" of Asia in comparison with Europe is not clearly reflected in the average length of human life, which is in turn a fairly reliable indicator of health. The lives of Japanese, the healthiest people in Asia, were scarcely shorter than those of Western Europeans, despite their lower per capita income. In fact, most people had the same life span everywhere in the early modern age. Before 1800 only small elites such as the English nobility or the Genevan bourgeoisie attained a male life expectancy above forty years. In Asia the figure was somewhat lower, but not dramatically so. In the case of the Manchurian Qing nobility, life expectancy hovered around thirty-seven years for those born in 1800 or thereabouts and thirty-two years for the generation born around 1830—a deterioration that mirrored the general trend of Chinese society.[9] As to Western Europe, life expectancy at birth averaged thirty-six years in 1820, with a peak in Sweden and a trough in Spain, while the corresponding figure for Japan was thirty-four years. By 1900 it had risen to forty-six to forty-eight years in Western Europe and the United States; Japan was almost level at forty-four years, with the rest of Asia behind it.[10] Considering that Japan's economy was then at least a generation behind those of the United States and the advanced European countries, we can see that, under conditions of *early* industrialization, it managed to achieve health standards that were elsewhere characteristic of *high* industrialization. However much weight one attaches to income estimates, the fact is that the notional average Japanese in 1800 led a more frugal existence than a "typical" Western European, without having a significantly shorter life expectancy. A hundred years later, after societies in both parts of the world had multiplied their wealth, the differential had not noticeably shrunk. Probably, though, national wealth was more evenly distributed in Japan, and the Japanese—who *today* have the highest life expectancy in the world—were unusually healthy. In the seventeenth and eighteenth centuries, they had diets, house-building techniques, dress habits, and public and private hygienic customs that reduced their susceptibility to disease, and they were exceptionally resource effective.[11] The Japanese were "poorer" than Western Europeans, but it cannot be said that their lives were therefore "worse."

Gaining Lifetime

In 1800 the average life expectancy at birth for the world population was at most thirty years; only exceptionally did it rise to thirty-five or a little higher. More than a half of all people died before reaching adulthood. Few enjoyed a life after work: either at the end of the day or in retirement following years of occupational activity. Death typically came as a result of infections: it came more swiftly than it does today, when protracted degenerative disease is the main cause of death in the rich countries.[12] By the year 2000, amid fast-increasing world population totals, the average life expectancy had risen to sixty-seven years, with a much greater leveling both within and between societies than in the case of incomes. In other words, people's ages increased faster than their material riches.

This "democratization" of a long life is one of the most important experiences of modern history. But there are exceptions to the rule. In the poorest countries of sub-Saharan Africa, many of which have also been hit hardest by AIDS, the average life expectancy for young adults aged twenty (*not* for the newly born) is today lower than it was in preindustrial England, China, and Japan or than it was in the Stone Age.[13] Why the human life span "exploded" in the nineteenth century is a controversial question: the decisive factor is variously considered to have been advances in medicine and sanitation, better nourishment, or new public health measures. Some experts adopt multicausal models in which all these elements play a role.

A reasonably precise dating of the processes that led to this life expectancy revolution is of great interest for any characterization of the nineteenth century. Robert W. Fogel has concluded from what is known to us today that the decisive leap occurred in "the West" (by which he means Western Europe, North America, and Japan) in the first half of the calendrical *twentieth* century, beginning with the period from 1890 to 1920.[14] There was by no means a constantly rising trend throughout the nineteenth century. During the early industrial age in Britain (c. 1780–1850), life expectancy initially went into decline and deviated from the high levels that England had first reached in the age of Shakespeare;[15] only after 1850 did wages catch up and overtake prices, and average life spans gradually began to increase.[16] In Germany, where industrialization began only around 1820, discussions were taking place a few years later about what would soon become known as "pauperism"—a new and disastrous mass impoverishment, affecting town and country alike.[17] This process, similar to that which England had undergone previously, may be attributed to two causes. *First*, the quantity and above all the quality of food did not keep pace with the physical demands of early-industrial factory labor, so that, according to Robert Fogel, the growth of real incomes registered in the statistics must be reduced by as much as 40 percent before it can be converted into physical well-being.[18] In the early nineteenth century, the United States was alone among "Western" societies in guaranteeing its citizens more than the minimum degree of nourishment. *Second*, the fast-growing cities, which brought in people from far and wide, were a breeding ground for health risks. Closely packed housing, without the necessary hygienic provisions, allowed deadly pathogens to spread, the most deaths resulting not from concentrated epidemics but from "normal" diseases present in everyday surroundings. This was essentially true of all European societies that entered the phase of industrialization. And it was true only of the cities. Life in the country was healthy in comparison—a differential that closed in northwest Europe only around the turn of the twentieth century.[19]

The worldwide trend for the increase in longevity, which began in Europe, North America, and Japan around 1890, manifested itself elsewhere at different times.

- Latin America's great advance came between 1930 and 1960.
- The Soviet Union caught up between 1945 and 1965 (but its successor states fell back dramatically in the 1990s).
- China pursued a successful health policy under the Communist regime, and its life expectancy soared from less than thirty years before 1949 to nearly seventy in 1980.[20]
- A number of African countries made advances in the two decades following independence, from approximately 1960 to 1980.
- Japan experienced a new surge between 1947 and 1980.[21]

Clean Water

Many of the foundations for the gains of the twentieth century were laid in the nineteenth. But it took time for them to spread more widely. Two especially important impetuses were new knowledge about disease prevention and the development of public health care. With regard to the latter, governments began to realize the need for a systematic policy sometime after 1850. In Western Europe, their range of measures to control and separate the sick and potential disease-carriers (e.g., the kind of port quarantines long practiced in the Mediterranean and the Black Sea[22]) were now expanded through infrastructural investment to remove the breeding grounds of disease. For the first time, mass health care was not entrusted to private philanthropists and religious institutions alone but was declared to be a task of the state. The "environmentalist" theories of the age showed that a start should be made with the clearance of urban garbage and wastewater and the provision of clean drinking water. England, the world leader in this "sanitary movement," had already begun in the 1830s to develop the basic principles and to take various pioneering initiatives. Thus, the collateral damage of the Industrial Revolution did not go unnoticed. Other countries followed suit—most comprehensively the United States, but soon also in continental Europe.[23]

The first step was civic and governmental initiatives to improve the water supply. The emergence of anything like a water policy presupposed recognition of water as a public good; water rights had to be defined, and public and private claims separated from each other. It was a long and complicated process to work out all the legal provisions for the ownership and use of water, including its industrial use. Even in centralized France this was not completed until 1964, and in many parts of the world it is still going on. For the creation of a modern water supply, not only political will and legal requirements but also an appropriate technology were necessary. In 1842, in one of the city's grandest festivals, New York celebrated the inauguration of a system of aqueducts, pipes, and reservoirs that supplied public wells, private households, and the fire brigade.[24] The value of clean water became especially apparent after an English doctor, John Snow, established in 1849 that cholera was not transmitted in the air or by bodily contact but was a water-borne disease. It took more than fifty years, however, for his findings to become generally accepted. The fact that London's water supply was in the hands of several private corporations stood in the way of change. In 1866 cholera entered the city once again along the pipes of one of these firms, claiming more than 4,000 lives in the East End alone. Water quality improved after that, however, and private wells gradually disappeared from the scene. Cholera and typhus epidemics were no longer seen in London after 1866.[25]

The importance of local scientific opinion is demonstrated by the example of Munich, where the doctor and pharmacist Max von Pettenkofer was the great authority in matters of hygiene. Like John Snow, he reacted to the threat from cholera, a second epidemic of which struck the city in 1854. But in his view, to prevent the spread of the disease, the main task was to ensure that the subsoil was kept pure and that the disposal of organic refuse was improved. Since he had ruled out poor drinking water as the cause, improvement of the water supply was pursued much less energetically than in London. Only in 1874 did Munich begin to draw up plans for its modernization, but there was still opposition to the contaminated-water theory even after the outbreak of a third cholera epidemic. In 1881, the city finally pressed ahead with the construction of new water installations.[26] Pettenkofer's error must have been costly to the capital of the Kingdom of Bavaria.

Munich, despite Pettenkofer's advice, also delayed the upgrading of its wastewater disposal until the 1880s. London had earlier been successful in developing a sewage system—a second prerequisite for the elimination of water-borne diseases such as typhoid, dysentery, and cholera from the British metropolis. It was known there that a clean water supply and a proper drainage system were twin sanitary requirements. This was not self-evident, though, and Napoleon had treated Parisians to public wells and aqueducts without concerning himself with other improvements. In London a Metropolitan Board of Works was founded in 1855—the first authority with powers covering the whole city. At first, its work was impeded by confusion over precise areas of responsibility and by resistance from supporters of a radical free-market liberalism. Then came the "Great Stink." Back in 1800 it had still been possible to fish for salmon in the Thames near London, and a few years later Lord Byron had enjoyed swimming in it. But in June 1858 such a stench rose from the river that the House of Commons, having tried coating protective curtains with chloride of lime, eventually had to suspend its sessions. The honorable members of Parliament were in a panic, realizing as they did that the exhalations of Old Father Thames were not only unpleasant but dangerous to the health. The chief engineer of the Metropolitan Board of Works, Sir Joseph Bazalgette, one of the pioneering modernizers of Europe's largest city, was commissioned to build a mostly underground system of sewers. Rumors that typhoid fever had caused the death of Queen Victoria's beloved forty-two-year-old consort Prince Albert, in December 1861, underlined the urgency of remedial action.[27]

By 1868 a total of 1,300 miles of sewers had been laid, of which eighty-two miles consisted of huge tunnels containing a total of 318 million bricks: one of the largest and most expensive public investments of the nineteenth century. Also part of this were the installations along the embankment, which included an underground railroad as well as all the pipes and cables of a modern capital city. The building work beneath London aroused great public enthusiasm.[28] The technology used for this monument of modernity was curiously preindustrial, if one leaves aside the magnificent Florentine or Moorish pumping stations equipped with steam engines. Brick-lined sewers and glazed ceramic pipes were nothing new; the movement of the water was simply down to their angle of incline. Technically speaking, the Victorian drainage system could have been built at any time in the previous hundred years. It was all a question of perception, political will, and a new attitude to dirt.[29] Whether the much-praised new installations really met all the requirements is another matter. When a pleasure steamer collided with a barge in September 1878 close to the effluent from the London sewers, there was a flurry of official speculation as to how many of the numerous casualties drowned in the Thames and how many were poisoned by its water.[30]

No comprehensive studies yet exist about urban hygiene on other continents. For the time being, we have to make do with a few impressions. Muslim West Asia was repeatedly praised by travelers for the high quality of its urban water supply; no report from Isfahan before its sacking by Afghans in 1722 failed to mention this point. Indeed, it was frequently remarked that nothing comparable was to be found in Europe. Western eyewitnesses condemned the barbarism of the Russians' destruction of Tatar water pipes, after their annexation of the

Crimea in the early 1780s. And in 1872 a German traveler to Syria, otherwise little impressed by the Levant, was still amazed that in Damascus, a city with 150,000 inhabitants, "every street, every mosque, every public and private house, and every garden" were provided "to overflowing" with channels and "fountains."[31] The origins of water modernization in Bombay lay not so much in public health considerations as in the inadequacy of supply for a fast-growing large city. After vigorous resistance from Indian notables, who not incorrectly feared higher taxes, a municipal water supply came on stream here in 1859, earlier than in many European cities. It also provided water for the booming cotton industry in the West Indian metropolis, and reduced the danger that owners of private cisterns would exploit periods of drought for their own profit.[32] In Calcutta a sewage system was opened in 1865 and water-filtering installations in 1869.[33] The first Chinese to encounter tap water were imperial emissaries on ocean steamers of the 1860s. Shanghai, where the quality of water had previously been better than in many large European cities of the time, acquired a modern waterworks and piping system in 1883; it was financed by private investors and initially served only prosperous Europeans and a few wealthy Chinese in the International Settlement, a colonial-style enclave governed by foreigners. The owners of the water plant tried to increase its operational radius and by no means wished to deprive the Chinese of clean water out of "colonial" motives. But the Chinese population remained skeptical: they had survived for generations, more or less, on water from the Huangpu River. Also the guilds representing more than three thousand water carriers protested against the new competition.[34]

Decline and Revival of Public Health

At first the age of modernity was an unhealthy one. In the first five or six decades of the nineteenth century, industrialization meant poverty, hardship, cultural decline, and reduced physical well-being for the working population of English cities. The country paid a price for having begun to industrialize before modern sanitary principles were understood and solutions attempted. Many people nevertheless weighed the risks of city living and accepted them of their own free will. The big cities and the new factory towns were unhealthier than the countryside—and they remained so throughout the century,[35] but the wages that could be earned in them were higher. The work discipline in factories was stricter, yet many preferred to escape the tight control of country squires and clergymen, and to have the freedom to found independent clubs and church communities.[36] The level of health declined in the United States too—historians like to use body size as the indicator—during the early phase of industrialization (c. 1820–50) that followed unusually favorable conditions at the beginning of the century. In Germany there were sharp oscillations in the standard of living, but with a long-term upward trend. A similar tendency was apparent in the Netherlands and Sweden, two countries that did not industrialize for a long time but experienced similar economic development centered on trade, finance, and modern agriculture.[37] In

France, the onset of industrialization in the 1820s was generally associated with clear and constant improvements in every area. This was an exceptional case, in which a second-generation industrializer (unlike the United States in the same period) did not have to contend with major losses in physical well-being. Two complementary reasons have been suggested for this: first, that France urbanized much more slowly than England, thereby avoiding the health risks of overcrowded slums; and second, that the urban population ate more meat in France than in England (the opposite had still been true in the eighteenth century) and therefore developed a higher resistance to disease. Furthermore, the French Revolution had helped to foster a slightly greater equality of income distribution. That, too, seems always to be a factor promoting good health.[38]

In general, late developers had to bear lower biological costs. As soon as new knowledge about epidemics and ways of combating them became available, big cities shed their "excess mortality" and became healthier places to live in than the countryside. It has been possible to demonstrate this for Germany as well as for colonies such as India, where Calcutta, Bombay, and Madras, for example, acquired at least some of the sanitary improvements of British cities. In both cases the new trend began in the 1870s.[39] The spread of medical and hygienic knowledge and of sewer and water supply technology was, at least in Europe, a "transnational" process; innovations took only a few years to pass across frontiers. For example, a modern water supply was being constructed by British firms in Berlin from 1853 and in Warsaw from 1880. Britain pioneered legislation on public health but took quite a long time to implement it. Germany, on the other hand, the industrial latecomer, swiftly adopted new sanitary measures, even before adequate legal provision had been made for them. Here the authorities applied their traditional right to intervene. The high administrative competence of Prussian governments proved to be an advantage, whereas in England powerful middle-class ratepayers were reluctant to take on extra costs, and weak municipal authorities were for a long time unable to stand up to them.[40]

The introduction of health systems had a profound impact all around the world. The new turn was palpable even in countries where indigenous arts of healing were well tried and recognized and enjoyed the confidence of the majority of the population. Traditional medicine—in Africa or Latin America, for instance—was strongly individualist, in the sense that it was bound up with the virtues and capacities of particular charismatic healers. There were three prerequisites for the introduction of public health systems: (1) a new definition of the tasks of the state and the will to commit resources to them; (2) the presence of biomedical knowledge, including its practical implications; and (3) an expectation on the part of citizens that the state should concern itself with health matters.

Intellectually the microbe theory developed by Louis Pasteur, which gained acceptance throughout Europe from the 1880s on, gave a scientific foundation to the observations of practical men such as John Snow, raising policies to promote public hygiene above the party-political fray. The earliest initiatives, though "well meant," rested on shaky premises and did not lead to generalizable conclusions. Only the theory of microbes established cleanliness as the highest priority, making *Homo hygienicus* the creation of bacteriology. Scientists such as Pasteur and Robert Koch became cultural heroes of the age. Disease was detached from its familiar ecological, social, political, and religious contexts, and health was proclaimed to be a supreme value. The middle classes, and more and more people from other strata of society, internalized this attitude.[41] Improved sanitation probably played a greater role in reducing mortality in Europe and North America than elsewhere in the world, where attempts are still being made to achieve comparable results with simpler and cheaper technology. The universality of ends was not matched by a universalization of means. The influence of the West, then, was differentiated.

Major public investment in hospital coverage became worldwide only in the twentieth century. The Allgemeine Krankenhaus in Vienna, founded in 1784 on the orders of Emperor Joseph II, was the first great modern hospital. In Britain the eighteenth century was the breakthrough age: hospitals were to be found by 1800 in all the large cities of England and Scotland, with a whole series of specialist centers already operating in London. Britain was the world pioneer; things took considerably longer to develop in the United States. All these early hospitals were private foundations—unlike in continental Europe.[42] In the German Reich, a growing number of hospitals were built after 1870, with the result that there was a surplus of beds on the eve of the First World War. The hospitals of the late nineteenth century were rather different from the care institutions of the early modern age. Geared to the new knowledge of hygiene, they mainly served the purposes of short-term medical treatment, the training of doctors, and the development of the art and science of medicine. The importance of these tasks increased with the advance of specialization (in Germany from the 1880s onward).[43] So long as there was a fear of epidemic outbreaks, it was a major task of hospitals to care for patients with acute illnesses—but for a long time no one could be sure that they increased rather than lowered the chances of survival.[44] The universalization of the Western-style clinic is a phenomenon of more recent times, closely bound up with new types of health funding.

The (Relatively) Healthy Slaves of Jamaica

The average state of health of a social group depends on numerous factors: adaptation to the local climate, quantity and quality of food, physical and mental stresses of work, risk-lowering behavior (such as personal hygiene), access to medical care, and so on. The information available for the nineteenth century allows a reasonably complete health profile to be drawn up for only a few groups, most of them in Europe. We still know little, for example, about the situation in the most populous country in the world, China. But there are exceptions to this rule. One is the slave population of the British Caribbean, between the end of the African trade in 1808 and the abolition of slavery in the British Empire in 1833. During that period it would have been foolish for even an unscrupulous and sadistic plantation owner to work his slaves to death; black laborers had become a commodity that was no longer so easy to replace. Most planters employed European doctors, or Creoles who had studied medicine in England or Scotland. Medical stations were not a rare sight on the large plantations. Of course it was in the logic of the exploitative system to care quite well for young and strong slaves, while neglecting older ones or even driving them from the plantation. All in all, however, medical facilities for slaves were not much worse than for English industrial workers at that time. The main limits to health care—in Europe as in the Caribbean—lay in the defective state of knowledge, which in the early nineteenth century still had not identified the causes of many diseases, especially those prevalent in the tropics. Many slaves wisely refrained from placing their trust in European medicine, often preferring to consult black healers who practiced a folk medicine unavailable to the European industrial proletariat.[45]

3   Medical Fears and Prevention

Major Trends

A second factor that helped to lower mortality wherever theory found practical application was the new knowledge of disease prevention. Like the "demographic transition," an epidemiological transition made itself felt at different times in various parts of the world. Generally speaking, the chances of succumbing to a mass outbreak of disease—what demographers call a mortality crisis—decreased over the course of the nineteenth century. For northwestern Europe the following sequence has been described: In a first phase that began in 1600 and reached its peak between 1670 and 1750, diseases such as bubonic plague and typhus lost their importance. In a second phase deadly infectious diseases such as scarlet fever, diphtheria, and whooping cough receded. In a third phase that began around 1850, respiratory diseases apart from tuberculosis gradually declined in significance. Finally, the twentieth century saw the gradual emergence of the mortality profile that is familiar today in all European societies: heart and circulation disorders and cancer as the main causes of death.[46] For each region of the world, a particular balance sheet of old and new diseases might be drawn up.

Tuberculosis was among the afflictions of the epoch that was thought of as new. Since it was recognized as a uniform disease pattern only in the early nineteenth century, little that is precise can be said about its appearance in earlier times. It was undoubtedly more common than the historical documents suggest. We can be sure that it was endemic in various parts of Eurasia and North Africa, and probably also in the "pre-Columbian" Americas. But its spectacular spread in the nineteenth century made it a token of the age, not only in the new working-class suburbs but also in the drawing-rooms of high society. The courtesan Marie Duplessis, immortalized as the "Dame aux Camélias" in Alexandre Dumas's eponymous novel (1848) and as Violetta in Giuseppe Verdi's opera *La Traviata* (1853), was one of its most famous victims. In the first half of the century, it doubled in frequency as a cause of death in France. It was still one of the great social calamities after the First World War, against which health policies fought with disappointing results. There were no drugs to treat it until 1944, and the truly effective ones became available only in 1966. Since tuberculosis was thought to be hereditary, it was often covered up in the families of the bourgeoisie. But silence was not possible in the case of prominent figures who succumbed to it—from John Keats (1821) to Frédéric Chopin (1849), from Robert Louis Stevenson (1894) to Anton Chekhov (1904) and Franz Kafka (1924).[47]

The cures that the rich began to seek in the 1880s, in a new archipelago of mountain sanatoriums, resulted in a special kind of international semipublic sphere. Here they were by themselves, but not alone, as they rested, ate healthily, shed the stresses of the big city, and willingly subjected themselves to the tyranny of the staff.[48] Thomas Mann's novel *The Magic Mountain* (1924), set in an Alpine sanatorium in the years before the First World War, depicts one of these characteristic institutions that sprang up even as far as Korea, where a fifth of the population was infected.[49] In Japan too, the number of tubercular patients rose dramatically after the turn of the century, to fall again only after 1919. Japanese scientists thoroughly studied new Western discoveries about the disease, but for that reason it sometimes took them a long time to act on them. Not until several decades after Robert Koch's simple and empirical identification of the tuberculosis bacillus (1882)—an effective vaccine followed in the 1890s—was the Japanese medical profession prepared to accept a clinical picture of it as a single infectious disease. But that was not the end of the story, since, as in Europe, there continued to be a divergence between popular and scientific perceptions. The majority of the Japanese population held on to the belief that "TB" was a hereditary disease that should be concealed as much as possible, whereas medical officials wanted to record as many cases as they could. Factory owners were also fond of the inheritance theory, since it relieved them of the need to improve conditions at the workplace. For the largest group of carriers in Japan were female workers in the silk and cotton industry, who subsequently spread the disease to their native villages.[50]

Some completely new diseases also appeared in the nineteenth century. One of these, first recorded among young people in Geneva in 1805, was meningitis, which in one out of two cases led to death within a few days. Soldiers on the move from one garrison to another seem to have been the most frequent carriers in France. Eventually the whole of France and Algeria were affected. At the peak of its virulence, between 1837 and 1857, the disease claimed several tens of thousands of lives, almost exclusively of people under the age of thirty. Poliomyelitis was another scourge of the nineteenth century. For a long time medical knowledge of it had been extremely vague, but in the last quarter of the century new conditions in France and other European countries caused it to assume epidemic proportions. A vaccine did not become available until 1953. Polio has never been a disease of poverty attributable to unhygienic surroundings: indeed, it first appeared in countries such as Sweden that had the most developed hygiene in the world. Other illnesses were rife among clearly defined risk groups: for example, the dreadful and incurable distemper, in principle an equine disease, spread to consumers of infected horsemeat and to coachmen or soldiers who had to deal with horses professionally.

In terms of global history, the nineteenth century saw a tension develop between easier transmission of diseases and more successful campaigns against them. On the one hand, migration and modern means of transportation proved effective conduits for the global spread of infections. The Black Death of the fourteenth century had already gripped most of the known world, by no means only Europe, and killed a third of the population of Egypt.[51] Now epidemics spread much more quickly across regions. The worst by far was the global influenza pandemic of 1918, which struck even remote islands in the South Seas, and is estimated to have killed between 50 and 100 million people—more than the total number of deaths in the recently ended First World War. Especially hard hit were Italy, which lost 1 percent of its population, and Mexico, where the figure reached 4 percent.[52] On the other hand, advances in medicine and disease control made it possible to combat some of the greatest epidemics that history had yet seen, not eliminating them altogether but breaking their power. The chronologies and spatial patterns of this counteroffensive provide information about global processes. The nineteenth century was the first epoch in which worldwide campaigns were systematically waged against medical scourges. In order to be successful they had to combine adequate biomedical knowledge with the idea of a public health policy. Here are a few examples.

The Preventive War against Smallpox

The primal story, later repeated elsewhere in modified forms, was the war against smallpox. It began, at least in Europe, with the English country-doctor Edward Jenner's successful vaccination trials in 1796, but there had been a prehistory to the campaign outside Europe. China had been practicing inoculation or "variolation" since the late seventeenth century, and the practice was common in India and the Ottoman Empire too. In this method, pathogens from a smallpox patient were directly applied to the skin of a healthy person to trigger an immunizing reaction. At the beginning of the eighteenth century, Lady Mary Wortley Montagu, a diplomat's wife and well-known travel writer, observed this immunizing effect among both peasant women and the wealthy upper classes of Turkey, and she reported it to her learned friends in London. In fact, inoculation had many advocates in England, Germany, and France in the last third of the eighteenth century, but failure to isolate the subjects properly at the stage when they were highly infectious often resulted in an epidemic outbreak. Before

Edward Jenner, who discovered the protective effect for humans of the much weaker cowpox pathogen, no one had found a risk-free way of guarding whole populations against smallpox. In 1798, after two years of experiments, Jenner presented his pathbreaking results to the public. A safe and inexpensive alternative to inoculation had been found in the shape of vaccination.

It soon became clear that vaccination would wipe out the disease only if the entire population was compelled to undergo it. Countries with centralist traditions or modernizing authoritarian systems of rule were particularly quick to act. In 1800 Napoleon gave the go-ahead for the first vaccinations, and between 1808 and 1811 nearly 1.7 million people in France were immunized.[53] Egypt under Muhammad Ali made vaccination compulsory, at least on paper, as early as 1818; the pasha sent teams of French doctors into the villages to vaccinate children and to instruct barbers in the necessary techniques. But the most important breakthrough came with the creation of a permanent health service in 1842, covering both the capital and the provinces.[54] Things moved faster in Egypt than in Britain, where immunization became obligatory only in 1853 (more effectively in 1867)—until libertarian MPs opposed to any state compulsion managed to prevail on the issue in 1909, at a time when public debate was still raging in the United States about its advantages and disadvantages.[55]

Jenner's discovery soon traveled around the world, and Jenner himself received news about this from remote corners of the globe, including letters of gratitude from Thomas Jefferson and from the chief of the Five Nations in Upper Canada.[56] European ships, previously notorious as vehicles of disease, carried cowpox lymph to many overseas countries, in an early example of the global diffusion of knowledge and problem-solving strategies. How was the vaccine transported? The best method was via infected human agents, and for this it was necessary to have a group of nonimmune individuals (often taken from an orphanage). A member of the group was infected, then the lymph pus was passed on to the next member, and so on; this ensured that there would be at least one virulent case on board when the ship reached its destination.

In 1803 the Spanish king Charles IV, an admirer of Jenner's, sent out an expedition with vaccine material to all the Crown's colonies. On its way from Buenos Aires, Chile, and the Philippines it put into southern China, where vaccine had arrived almost simultaneously from Bombay. In 1805 doctors at the East India Company settlement in Canton began to work with the vaccine, and in the same year literature on the subject was translated into Chinese. In Japan news of Jenner's discovery arrived in 1803; more was learned in 1812 from a Russian medical treatise that a Japanese prisoner-of-war had brought home with him. But vaccine was still lacking. The first batch reached Japan from Dutch Batavia only in 1849—an astonishingly late date in comparison with other countries.[57]

One should be wary, however, of a linear success story. For a long time the need to keep immune protection up to date was not understood. Unsuitable human carriers passed on other pathogens together with the vaccine; and many governments failed to recognize the importance of *mass* vaccination. All this gave rise to major unevenness. German soldiers who marched off to fight in France in 1870 had almost complete protection from a dual vaccine, whereas a large part of the French army had none. Around the same time smallpox was flaring up again in various parts of the country. The Franco-Prussian War thus took place in the midst of an epidemic crisis, and the asymmetry of protection contributed to the eventual French defeat. The French army lost eight times more soldiers than the German to smallpox, and as many as 200,000 civilians may have died of it in France between 1869 and 1871. Moreover, French prisoners-of-war carried the disease with them to Germany, where the general population was much less protected from it than soldiers. A severe epidemic in the years from 1871 to 1874 cost more than 180,000 people their lives.[58]

The degree of smallpox protection did not at all reflect the level of economic development. Impoverished Jamaica, for example, was free of smallpox decades before wealthy France; inoculation had been practiced there since the 1770s, and Jenner-style vaccination since the turn of the century, making the largest and earliest of the British "sugar islands" a model in this respect. The colonial authorities created a special Vaccine Establishment, and by the mid-1820s smallpox had disappeared from Jamaica, to be followed a few years later by most of the other British Caribbean islands ahead of most other parts of the world.[59] Ceylon, also an island under British control, would be smallpox-free by 1821 after a mass vaccination campaign. This was by no means the rule in Asia. In the giant subcontinent, outbreaks of smallpox occurred somewhere or other in every year of the century, the most dramatic being in 1883–84. In Kashmir vaccination only began in 1894. In Indochina, where the French colonial rulers showed less concern than the British did in India, smallpox proved especially stubborn.[60] In Taiwan, which the Japanese annexed as a colony in 1895, the authorities carried out an effective mass vaccination campaign, and by the end of the century the island was more or less clear of smallpox.[61] In Korea, the first Europeans who arrived in the formerly closed country in the 1880s found few people untouched by the disease; it had not been introduced to the peninsula from outside, and it was eventually eliminated under Japanese colonial rule in the second and third decades of the twentieth century.[62]

Although it was only in 1980 that the World Health Organization declared the world free of smallpox (the last natural case had occurred in Somalia in 1977), the breakthrough had been achieved in the nineteenth century. Where the disease lingered until the Second World War—and very rarely afterward—it was the result of government neglect, corrupt health administrations, or special epidemiological situations. The last epidemic in the West was recorded in 1901–3 in the United States. Sweden was the first country in the world to free itself even of endemic smallpox, in 1895. The disease was still deeply implanted in Africa and the Middle East on the eve of the First World War; only a small minority of those populations enjoyed vaccine protection.[63] The great advances in immunization occurred there in the twentieth century.

The problems that had to be solved before whole populations enjoyed immunity were in principle the same throughout the world: it was necessary to overcome opposition, in Britain as in Africa (where people distrusted the colonial authorities); governments had to make vaccination compulsory and to carry out checks; and high-grade vaccine had to be available in sufficient quantity. These were tasks that required complex organization, and they were not always fulfilled better in Europe than in Asia. Disciplined societies were the most successful, but even among them there were differences. Hesse and Bavaria were the first German states to introduce smallpox vaccination, under Napoleonic influence in 1807, but Prussia—which protected its army so well—otherwise put its trust in the commitment of local doctors.[64]

Western and Indigenous Medicine

Colonial regions seemed to have at least a theoretical advantage insofar as new vaccination techniques were made directly available to them. In Africa, Ethiopia—the only noncolonized country apart from Liberia on the eve of the First World War—was the last to introduce Jenner's methods. Elsewhere vaccine arrived early on, but for a long time it was restricted to the ruling circles. In Madagascar, for example, where smallpox victims had traditionally been buried alive, the king had the royal family vaccinated as early as 1818, but he could not effectively protect the whole island, a nodal point of the slave trade.[65] The procurement of vaccine from abroad was also a weak point in the otherwise successful reform policies of the kings of Siam. Only at the end of the century, later than in modest European colonies in Asia or the Caribbean, did government vaccination programs begin to get a grip in this independent country.[66] Colonies—at least those considered important—therefore had relatively good chances. The authorities understood that they could kill several birds with one stone: strengthening the labor capacity of the colonial population while also gaining a reputation as colonial benefactors and helping to protect the mother country from infection.[67]

What role did scientific knowledge play in this? Here, too, we need to pay attention to chronology. The important breakthroughs happened only after the middle of the century. From the late 1850s onward, Louis Pasteur and Robert Koch discovered that certain diseases were caused by microbes, and in a number of cases they developed medical therapies. The first post-Jenner vaccine, however, appeared only in 1881, when Pasteur isolated the anthrax bacillus; then Koch found an antitoxin against diphtheria in 1890.[68] Around 1900 medical science had only a few reliable drugs at its disposal—among them quinine, digitalis, and opium. Aspirin appeared on the market in July 1899. The twentieth century would be the great age of mass immunization against infectious diseases and of successes against bacterial illnesses with the help of sulfanomides and antibiotics. But one of the major achievements of the nineteenth century was a new insight into the underlying causes of inflammatory processes. From about 1880, the general use of antisepsis and disinfection reduced the incidence of mortality in childbirth, but only in Western countries.[69] The main contribution to the overall quality of life was in disease prevention rather than treatment—a trend reversal that set in with the new century. The generation that grew up in the West after the Second World War was the first in history not to live beneath the Damoclean sword of infection. In the United States, for example, the risk of dying from an infectious disease was twenty times lower in 1980 than in 1900.

Even for Europe one should not overestimate the speed at which the new advances in medical practice took hold. On other continents, the spread of Western medicine came up against systems of indigenous knowledge and practice; where these did not exist in written form, as in Africa, they commanded little respect from either native or European representatives of modern medicine and were relegated to a trivial everyday level.[70] Things were different, however, where "great traditions" met up. In Japan, where European medicine had been known even in premodern times, it began to be practiced after the middle of the century. In the Meiji period it officially replaced the Chinese medicine that had previously been dominant. In March 1868, in one of its first decrees, the new Meiji government—which contained an unusually large number of politicians with a medical background—proclaimed that Western medicine should be the only compulsory element in the training of doctors in Japan. After 1870, with the help of numerous German doctors, medical education was completely reshaped in accordance with the German model. The "old" (that is, Chinese) medicine was supposed to wither away gradually. Anyone who wished to become a licensed doctor had to pass an examination in Western medicine, but traditional doctors put up resistance. In the treatment of the commonly seen beriberi, indigenous medicine proved itself superior, partly because the disease was not a major health risk in Europe. In practice the two systems continued to coexist in a complementary relationship. Around the turn of the century, two thirds of statistically recorded doctors in Japan belonged to the traditional Chinese school.[71]

A knowledge transfer in the opposite direction, from Asia to Europe, had already occurred in the early modern period. Jesuit missionaries collected Chinese medical texts and herbals. Publicly disseminated reports by individual Jesuits, and especially the account published in 1727 of the Westphalian doctor Engelbert Kaempfer's trip to Japan in 1692–94, meant that Asian practices such as acupuncture or moxibustion were made known in the West. A number of Western textbooks tried to make sense of Chinese healing theories. Yet East Asian medicine did not find large-scale application in the West until the second half of the twentieth century. Unorthodox medical knowledge scarcely gains acceptance by itself. It requires a measure of intellectual receptiveness, a body of healers able to apply the new methods, patients ready to accept them, and sometimes an institutional underpinning in something like a "health system." Even failing such tough requirements, East Asian techniques of healing never ceased to fascinate Western medical experts. The ups and downs of that fascination plot a curve of Western openness toward alternative traditions of knowledge.[72]

4 Mobile Perils, Old and New

The End of the Plague in the Mediterranean

Any epidemic disease poses specific challenges to a society. Each develops at its own speed and has its own victim profile and pattern of spatial distribution. Each also has its own "image," a special significance that people attach to it. And each has its own mode of transmission, a distinctive moment of infection. Bubonic plague, a disease carried by rat fleas that was more deeply engraved than any other in the European imagination, was an Asian phenomenon in the nineteenth century. It receded from western Europe after the great surge of 1663–79, which gripped England, northern France, the Low Countries, the Rhine Valley, and Austria. The penultimate outbreak was unleashed in 1720 by a French ship returning from plague-stricken Syria; more than 100,000 people died of the disease in Provence over the next two years.[73] The last major epidemic in Europe outside the Ottoman-ruled Balkans overwhelmed Hungary, Croatia, and Transylvania in 1738–42. Improved checks at major ports, as well as the Austrian military *cordon sanitaire* in the Balkans completed in the 1770s, shielded Europe from further plague imports from Asia.[74] France and the Habsburg Monarchy were Europe's frontline states and therefore had the most experience; the continent owes them a major debt of gratitude for keeping it free of plague in the late modern period. An additional factor was the transition everywhere in eighteenth-century European cities from wooden and half-timbered construction to stone architecture, which meant that rats, the main carrier of plague, lost some of their habitat.[75]

A new plague cycle began in Central Asia in the middle of the eighteenth century—the third, after those of the sixth through eighth and fourteen through seventeenth centuries. In the Ottoman Empire this new wave joined up with stable plague centers in Kurdistan and Mesopotamia. Istanbul was considered the kingdom of rats and a dangerous focus of infection, while Ottoman troops ensured that the disease was transmitted all over the empire. The plague traveled by ship from ports such as Istanbul, Smyrna, Salonica, and Acre, as well as by land along the great highways.[76] Bonaparte's troops became infected in 1799 during their advance from Egypt to Syria; their commander tried to raise morale with a staged visit to the plague house of Jaffa. Half of his army died of plague, dysentery, or malaria in the siege of Acre.[77] Subsequent outbreaks were reported from Istanbul (in 1812, with 150,000 deaths), Syria (in 1812), Belgrade (in 1814), and Sarajevo (on several occasions). Helmuth von Moltke, then a young Prussian military adviser to the sultan, witnessed an epidemic in Istanbul in 1836 in which 80,000 people lost their lives, and on his return journey he had to endure the usual ten-day "detention" at the Austrian *cordon*-frontier.[78] Moltke had observed the last fling of the plague. Within the space of twenty years—between 1824 and 1845—it rapidly disappeared from the Ottoman Empire, with the exception of endemic areas in Kurdistan and Iraq. Tighter quarantines and new official health authorities played a key role in this, but the end of the plague in the Ottoman Empire, a turning point in the history of the disease, has not yet been fully explained. There remains an element of mystery.[79] Despite Europe's successful protective measures, it continued to live in the shadow of the plague until 1845, when the last outbreak was recorded in the eastern Mediterranean. It could not drop its guard any earlier.[80]

The New Plague from China

The last great wave of plague spread from southwestern China in 1892. It reached the southern metropolis of Canton in 1893 and the nearby British colony of Hong Kong in 1894, unleashing a panic reaction in the international public. Ships carried the pathogen to India in 1896, to Vietnam in 1898, and to the Philippines in 1899. By 1900, ports as far away as San Francisco and Glasgow were affected. In Cape Town one-half of those infected died in 1901: a total of 371 fatalities.[81] The most surprising exception was Australia, where the plague struck ports a number of times but never grew into an epidemic, because the authorities instinctively targeted rats with the utmost energy.[82] The pandemic continued to rage in the first decade of the new century—indeed, some medical historians argued that it burned itself out only around 1950. A later surge came in 1910, when a passenger ship carried the plague from Burma to Java, where it had never taken hold before; more than 215,000 Javanese died of it between 1911 and 1939. The long-term result was a major improvement in living conditions and health care in the colony.[83]

As in other epidemics of the age, experts set to work immediately on the spot. At first they were puzzled, because no one had been expecting the plague to reappear in Asia either. Japan had never been in contact with it. In India it was so little known that there had never even been a plague god (as there had in China). Soon British Hong Kong became the main focus of internationally competitive research: the worried government in Tokyo promptly sent the celebrated bacteriologist Kitasato Shibasaburō, who had been Robert Koch's assistant. Pasteur's Swiss disciple Alexandre Yersin hurried over from the Saigon branch of the Pasteur Institute. It was Yersin who in 1894 discovered both the plague pathogen and the essential role played by rats; soon afterward the flea was identified as the carrier.[84] Rats were now in for a hard time. The Hanoi city authorities paid 0.20 piasters for each one caught during the epidemic of 1903—a successful measure that also served as an incentive to private rat catchers.[85] In Japan isolated cases appeared in 1899, but they did not lead to an epidemic. The novelty of the disease there is shown by the lack of a term for it other than the phonetic loan word *pesuto*.[86]

Contrary to what was thought at the time, the turn-of-the-century pandemic did not appear out of the blue, nor did it burst out of the still mysterious "central Asia." The plague was already described in 1772 in Yunnan, a home of the yellow-breasted rat (*Rattus flavipectus*). It must have been present there for a long time, but only the economic development of the region created the conditions for it to spread. The promotion of copper mining by the Qing Dynasty made the province a magnet for workers within a radius of several hundred kilometers. Between 1750 and 1800 a quarter of a million migrants turned the remote wilderness into a region of work camps and growing urban settlements. With mining came trade and transportation, and the demand for food stimulated rice production in neighboring Burma.[87] The plague could spread only as a result of this greatly increased movement, which at first was entirely confined to China—or, more precisely, southwestern China, since there was little integration of the province into a countrywide market. For a time the problem therefore remained within China—out of sight for Westerners. An economic depression in the first half of the century had a dampening effect, but then the Muslim revolts that shook southwestern China between 1856 and 1873 rekindled the disease. Rebel forces and their Qing adversaries were the main carriers. At the same time, the opium trade from the coastal ports bound the province more than ever before to extensive international networks. Detailed reports in local Chinese chronicles allow us to follow the course of the plague from district to district.

Chinese medicine was not unprepared. One school of thought emphasized the importance of personal hygiene, while another focused on environmental factors, both the natural and social, in ways strongly reminiscent of the "miasma" theories that were common in Europe until mid-century. Neither school, however, considered that the disease was transmitted by infection. Collective efforts to combat it concentrated on ritual exorcism, public displays of atonement, and other symbolic acts. As in early modern Europe and the Muslim world, the plague was seen as a divine visitation or punishment, and here, too, people swept the streets, cleaned wells, and burned the possessions of plague victims. The big difference with premodern Europe was that neither leading doctors nor state officials believed in infection as the cause, and therefore in isolation of those suffering from or exposed to the disease. The West had been the first to demonstrate the effectiveness of such methods in the quarantining of affected ports. In 1894 the colonial authorities in Hong Kong applied another strategy. On the assumption that the plague bred amid the squalor of poverty, they intervened forcefully to keep Chinese and Europeans apart and to raze a number of districts inhabited by the poor. This provoked vigorous, sometimes violent, protests from the Chinese—not only among "the poor" but also among philanthropically inclined dignitaries.

What this resistance expressed was not premodern "Asiatic" superstition but a rational view that ruthless methods were of little avail. Western medicine was equally unable to offer a cure for the disease, and despite Yersin's discovery the word had not yet got about that rats and fleas should be the target of attack. In 1910–11 the plague reappeared in Manchuria with greater virulence, transmitted from Mongolia rather than southwestern China, in the last major outbreak to be seen in East Asia. Chinese authorities and doctors managed to bring it under control without foreign help, using Western-style quarantines and health checks. In 1894 the Cantonese authorities had done little to face up to the problem, but now perceptions had changed and the imperial government recognized the fight against the plague to be an important task. The late Qing state advertised its successes in public health as a patriotic achievement, which among other things forestalled any new intervention by foreigners against the country's "backwardness." China had dramatically narrowed its gap with Europe in the domain of plague control.

Nowhere was the plague more devastating than in India,[88] where it appeared with epidemic force in 1896, first of all in Bombay. Of the 13.2 million deaths from the disease recorded worldwide between 1894 and 1934, 12.5 million were in India. Hunger and plague were mutually reinforcing. The British authorities acted at least as harshly as they had in 1894 in Hong Kong, and more so than in previous epidemics of smallpox and cholera. Victims were locked up in camps or forced into special hospices, where the mortality rate was as high as 90 percent. Houses were searched for the dead and infected, travelers were subjected to physical examination, roofs and walls were removed to let in air and light, and huge quantities of disinfectant were sprayed around.[89] This heavy-handed approach was a result of international pressure to halt the spread of the disease and of a determination to prevent the complete breakdown of life in the big cities, but it also reflected the scientific self-confidence and image building of the medical profession. In any event, it proved as ineffectual in India as in Hong Kong. People ran away to escape the draconian measures and took the pathogen with them. The colonial authorities were flexible enough to correct their course in the end: whereas their main concern at first had been to protect the health of foreigners, they now—like the late Qing bureaucracy—took responsibility for the creation of a public health system.

The great fin-de-siècle Asian epidemic triggered a debate about how best to protect Europe. Earlier international health conferences that had been held since 1851 had been mainly concerned with cholera.[90] The one that gathered in Venice in 1897, with the participation of Chinese and Japanese experts, looked at measures to ward off the plague. Several European countries also sent health officials to study the situation in Bombay, and the health organization of the League of Nations—the precursor of today's World Health Organization—had its ultimate origins in these efforts at plague control.

The international outbreak of the plague that first became evident in the early 1890s was scarcely more "global" than other epidemics of the nineteenth century and less so than the Black Death of the fourteenth century (which was most probably a different disease). Most of the victims were recorded in India, China, and Indonesia (Dutch East Indies), with 7,000 deaths in Europe, 500 in the United States, and approximately 30,000 in Central and South America. The fact that it more or less spared the West was not due only to better medical provision in the "developed countries"; the contrast between "first" and "third" worlds, core and periphery, does not exhaust the subject. The new epidemic would not have been possible without the development of extensive international networks, without the linkup of southwestern China with overseas markets. When the rate of spread accelerated, "modern" cities such as Hong Kong and Bombay, accessible by either ship or rail, became for a time the most dangerous places on earth. Low standards of hygiene plus more tightly meshed networking created the basic conditions of which the plague could take advantage.

The official reactions did not vary along an east-west axis; the microbiological revolution and laboratory-based medical science were still so new and unfamiliar in their applications to health policy that Western authorities were no cleverer than their Asian counterparts. In a city like San Francisco people shut their eyes to the peril, while in Honolulu, newly annexed by the United States, districts inhabited by Chinese and Japanese were burned to the ground in a scapegoating reflex.[91] In a number of countries, foreign minorities, often with skin of a different color, were treated as carriers of infectious diseases and subjected to more intense health checks. One of the most rational approaches was that of the moribund imperial state in China, which avoided the pointless excesses of the British in India.

The Blue Death from Asia

At the end of the nineteenth century, Europe was by no means an island secure from epidemic disease. Just when the plague was spreading like wildfire in Hong Kong, the German port of Hamburg was hit hard by an outbreak of cholera. No other disease threw Europe into such fear and panic in the nineteenth century: it was not a passing shock, here today and gone tomorrow, but a constant threat to the quality of life in large parts of the world. Although Robert Koch discovered the bacillus responsible for it on a trip to Calcutta funded by the German government in 1884, thereby dispelling old speculative theories about its cause, another twenty years would pass before it was understood that replacement of the water and salt lost by the patient constituted a simple, cheap, and effective treatment. Until then people suffering from cholera, in Europe and elsewhere, had to endure often quite pointless and brutal medical procedures. Those who escaped the attention of doctors tried to make do with household items such as camphor, garlic, vinegar fumes, or burning pitch.[92] In terms of medical knowledge, Europe before Koch had no decisive lead over China. The Shanghai doctor Wang Shixiong, in his "treatise on cholera" (*Huoluan lun* [1838; 2nd ed. 1862]), stressed the importance of clean drinking water quite independently of John Snow and other European or Anglo-Indian luminaries.[93] People in Europe were as helpless as elsewhere in the face of cholera; no "all clear" signal could be sounded at any time in the nineteenth century. Any disease has a distinctive chronology that differs according to location. This shows itself in the polarity of India and Europe. Over the centuries Europe had grown used to the *plague*, never ceasing to fear it yet gradually learning how to keep it in check. In India it was something new in 1892; the only ones there who took countermeasures were Europeans. On the other hand, *cholera* came as an unpleasant surprise to both India and Europe in the nineteenth century. For decades European medicine was not much wiser than Indian when it came to explaining the disease and developing strategies to combat it.

Unlike dysentery, typhoid, or malaria, cholera is an itinerant disease; it travels from one continent to another and through village after village, it is borne on ships and in caravans. Like the plague, it came from Asia and was often described by people at the time as "Asiatic cholera." It therefore conjured up old fears of an invasion from the East, an Oriental menace. Its symptomatology underlined its horrifying nature: it appeared suddenly and could theoretically strike anyone, leading with plague-like probability (more than 50 percent of cases) to death in a time that might be as short as a few hours. Unlike smallpox, which causes a high fever, cholera is always described as a "cold" illness; unlike tuberculosis or "consumption," it is ill suited to any romanticism. Patients neither become delirious nor slip into a coma; they remain fully aware of what is happening to them. Diarrhea, vomiting, a bluing of the face and limbs: the symptoms resemble those of acute arsenic poisoning. Cholera, says the medical historian Christopher Hamlin, "was not a disease that a person lived with."[94]

The distribution of cholera can be clearly plotted.[95] European visitors to India drew a picture of the disease as long ago as the early sixteenth century. In 1814 it became more common in several parts of the country, and from 1817 there was a spectacular rise in the number of reported deaths in Bengal. With a speed unparalleled in people's experience, it then left the geographical confines of South Asia to become a global phenomenon. Medical historians identify a number of pandemics: six between 1817 and 1923, and a seventh after 1961. Their abrupt end is striking in each case. Cholera vanished as suddenly as it had appeared, and it might be another half-generation before it became visible again. In 1819 it arrived in Ceylon, and from there much-traveled shipping routes carried it west to Mauritius and East Africa and east to Southeast Asia and China. In 1820 it struck Siam and Batavia, and shortly afterward, moving simultaneously by sea via the Philippines and by land via Burma, it reached mainland China; by the following year it had moved two thousand kilometers north to Beijing. In 1821 it marched to Baghdad with an Iranian army and had already reached Zanzibar off the East African coast. In 1823, cases were reported in Syria, Egypt, and the shores of the Caspian Sea. Siberia was infected from China. It reached Orenburg in 1829, Kharkiv (Ukraine) and Moscow in September 1830, Warsaw and Riga in spring 1831.[96] Summer 1831 saw it reach Istanbul, Vienna, and Berlin; and in October it appeared in Hamburg, from which it spread to England and four months later to Edinburgh. In June 1832 it leaped across the Atlantic, probably in an immigrant ship from Ireland to Quebec, and by the twenty-third of the month it was in New York. In spring 1833 Havana lost 12 percent of its population. In Mexico City 15,000 people died in the space of a few weeks.

Later waves gave fresh vigor to local epidemics and added new localities to the list. Aggressive though this first wave certainly was, its devastating impact was later exceeded on several occasions. The third cholera pandemic (1841–62) raged during the Opium War in China, where British troops carried it from Bengal. In Paris, where the first attack occurred in 1832, as many as 19,000 people lost their lives in 1849. At the same time (1848–49) a million died of the disease in the Tsarist Empire.[97] Further outbreaks, each one weaker than the last, followed in Paris in 1854, 1865–66, 1873, 1884, and 1892. After 1910 France was free of cholera.[98] London had no more instances after 1866—doubtless because of the exemplary measures taken to improve sanitation. New York too escaped the epidemic of 1866 thanks to sensible preventive action, while other parts of the United States were severely affected. The last time that cholera invaded the country was in 1876.[99]

In the Crimean War (especially during the winter of 1854–55), the ravages of cholera among unprotected troops living in catastrophic hygienic conditions were the main impetus that led reformers such as Florence Nightingale—not only a ministering nurse but one of the great political and administrative talents of her age[100]—to call for radical changes in army health policy. Of the 155,000 British, French, Sardinian, and Ottoman soldiers who perished in the war, more than 95,000 succumbed to cholera and other diseases. In 1850 Mexico again suffered terribly, as did East Africa from 1865 to 1871; there were particularly severe outbreaks in Japan in 1861 and in China in 1862.[101] In Munich, an ill-famed hotbed of disease, the epidemic of 1854–55 was worse than that of 1836–37, and another major visitation would follow in 1873–74.[102] In Vienna cholera claimed nearly 3,000 lives during the world's fair of summer 1873. Hamburg was to some extent spared by the early pandemics, but in the 1892–93 outbreak (which was more severe than anywhere else in western Europe) more of its citizens died than in all previous ones combined. Since this happened at a time when statistical techniques had already made great advances, the records make it possible to analyze its social impact in greater detail than in the case of any other late-nineteenth-century public health crisis.[103] The Philippines suffered epidemics in 1882 and 1888; in 1902–4 (when vegetables from Hong Kong and Canton probably imported the bacillus) it saw as many as 200,000 deaths from cholera in a population weakened by the American war of conquest.[104] In Naples, three decades after the outbreak of 1884, cholera arrived again in 1910 from Russia (where it had claimed 101,000 lives), and US officials kept a close eye on the large numbers of Italian emigrants who were arriving at the time. Uniquely in the European history of the disease, the Italian authorities (under pressure from Neapolitan shipping interests) made a major effort to cover it up.[105]

The total number of people who died from cholera cannot be even approximately calculated. In India, probably the most seriously affected region, a figure of 15 million has been suggested for the period from 1817 to 1865 (when reasonably useful statistics began), with a further 23 million for 1865 through 1947.[106]

The suddenness of a cholera outbreak, which in one day can infect thousands of people in a large city by means of contaminated water, added to the drama. In 1831–32, and again in 1872–73, Hungary was hit harder than almost any other European country; its mortality rate in the 1870s was 4 percent higher than in the decades before and after. More generally, deaths from the disease varied from an upper limit of 6.6 per thousand in London to more than 40 per thousand in Stockholm or Saint Petersburg and 74 per thousand in Montreal (in 1832).[107]

The great pandemic of 1830–32, in which Georg Wilhelm Friedrich Hegel lost his life, made a particularly deep impression on people's minds in Europe. The speed with which it spread from Asia, suggestive of a Mongol-style microbial invasion, and the helplessness of its victims led to a veritable demonization of the "new plague." Among the rich it fueled fears of the lower classes as carriers of death, while among the poor it aroused fears that the authorities were poisoning them to solve the problem of unemployment. The "primitive Orient," to which the "civilized world" had felt so superior for decades, seemed to be providing proof of its continuing subversive power.[108] In Britain, France, and Germany, medical people tried to prepare for the future after the first disturbing reports came in from Russia, at a time when nothing was known about the likely extent or conduits of the disease or the efficacy of any countermeasures. The most precise descriptions of cholera came from British doctors in India, but these had received little or no attention in continental Europe.

Many sources tell of the first appearance of cholera in France and its social impact on the capital. The first cases, on 14 March 1832, afflicted doctors who had recently returned from Poland; cholera, unlike the plague, did not enter via Mediterranean ports but through the Rhineland or across the Channel. There were ninety deaths in March, but already 12,733 in April. Public places emptied, as anyone able to flee the city lost no time in doing so—a perennial type of response (the viceroy of Egypt in 1848 fled as far as Istanbul).[109] The problem of corpse disposal was almost insoluble. Rumors, reminiscent of a previous age, spread about the causes of the epidemic.[110] Revolts broke out, claiming at least 140 lives. On October 1 it was established that the outbreak had come to an end. As in all epidemics, the lower classes were hit disproportionately hard. The first waves of cholera rolled over societies that, in some cases, were passing through a stormy period of their political history. France had just experienced the Revolution of 1830 and had not yet adjusted to the new routines of the July Monarchy; the newly "emancipated" bourgeoisie was seeking fresh tasks for the state apparatus it had taken under its control. Cholera thus became a test for new forms of state regulation of civil life.[111]

Cholera appeared in India in 1817, at a point when the British had militarily defeated their strongest rival in the region, the Maratha Federation, and were moving to consolidate their own rule; the recent troop movements connected with this contributed to the spread of the bacillus. Moreover, India had just been opened up for the first time to Protestant missionaries. A link between conquest and epidemic therefore suggested itself to ordinary Indians: there was a widespread view that the British, in violating Hindu taboos, had called down the wrath of the gods. So, in their different ways, both British officials and Indian peasants saw cholera as more than a health crisis but as a danger to "order" in general.[112] All through the century, the British authorities adopted a laissez-faire attitude to the disease. The kind of massive health measures taken in the 1890s to combat the plague never applied to cholera; there was scarcely any quarantine, isolation, or even a slight tightening of controls on Hindu pilgrim flows. The events of 1865 in Mecca, when pilgrims from Java introduced cholera and triggered a global domino effect that began in Egyptian ports, had confirmed that pilgrimages could be a factor in the spread of the disease.[113] So long as the nature of cholera was unexplained, doing nothing could seem as good as any other response. A doctrinaire liberalism and the penchant of the colonial state for cheap solutions thus bolstered the dominant medical opinion in both British India and London: that expensive health measures were not warranted, because there was no proof that cholera was infectious.

In continental Europe the main reflexes were those associated with earlier battles against the plague, so that sealing off affected areas seemed to be the most promising course of action. Russia, Austria, and Prussia established *cordons sanitaires* around themselves: the Tsarist Empire in Kazan against Asia, Prussia on the Polish frontier against everywhere to the east of it. Prussia alone deployed some 60,000 soldiers along a line of 200 kilometers, subjecting travelers to a rigorous quarantine and new cleansing measures, and even washing banknotes or fumigating letters they had on their person.[114] Here, too, there were medical authorities and lobbies that represented various theories concerning the transmission of cholera—by air, water, or direct contact. States such as Pettenkofer's Bavaria that did not share such views did not impose *cordons* or quarantines either. The effectiveness of such measures was, of course, called strongly into question by the almost unstoppable dynamic of the various outbreaks. Indeed, one wonders whether the ritual incantations to ward off evil spirits, which the king of Siam ordered to be chanted, were essentially less appropriate. Yet the whole of Europe, pulled this way and that by the competing theories, again gave itself over to a quarantine approach in the 1890s.[115] Quarantines remained a feature of international travel during the great age of the steamship: ports reassured passengers and merchants when they built functioning, but not too irksome, quarantine facilities. The rise of Beirut as "gateway to the Levant," for example, began in the 1830s with the opening of a modern sick bay and quarantine station.[116] Countries unable or unwilling to halt the flow of immigrants faced special problems, but they had to adopt protective measures even if a strict quarantine had proved early on to be of little use.[117]

Smallpox, plague, cholera, and yellow fever are mobile diseases suited to globalization, enemies of human beings with truly military properties: they attack, conquer, then withdraw. Sometimes physical defenses such as quarantines and barriers remain the last hope. The growth of world trade and shipping in the nineteenth century increased the speed of transmission; humans and animals, but also goods, could become infected and disseminate deadly pathogens.[118] It should be added, however, that other, more localized epidemics also brought suffering and death.

In the nineteenth century the main one was typhoid or enteric fever, a good indicator of special historical problems. The classic description of this disease, which strikes an undernourished population living in conditions of "appalling misery," has come down to us from Rudolf Virchow, who in February and March 1848 was sent by the Prussian Ministry of Religious, Educational and Medical Affairs to Upper Silesia and sketched a powerful social panorama of one of the poorest regions in central Europe.[119] Industrialization and urbanization turned many large European cities into breeding grounds for typhoid. But it was also a soldier's disease, pointing to a failure to reform conditions in the army. It accompanied the Napoleonic armies, after they were infected by the waters of the Nile in 1798. It was especially grave during the Peninsular War in 1808, and even worse during the Russian campaign. In 1870–71 it was endemic in the Metz region during the Franco-Prussian War, and some of its worst ravages occurred in the Russian-Turkish war of 1877–78. At the turn of the century, a typhoid crisis could still bring the army medical service of any state to the brink of collapse.[120]

Finally, there was epidemic typhus, sometimes known as jail fever, quite devoid of glamour, or even of the frisson caused by the "democratic" horseman of the apocalypse who levels the highest and the lowest in society. It was a disease of poverty in a cold climate, the complete opposite of a tropical disease. Carried by lice, it tended to appear where poor sanitary conditions and fuel poverty meant that people living closely together did not change and wash their clothes often enough. Typhus, together with typhoid fever and dysentery, is a classic disease of war. Until the First World War it accompanied every modern conflict in Europe. The decimation of Napoleon's Grande Armée resulted more from dysentery and typhus than from the operations of all its other adversaries.

The Beginning of the End of the Medical Ancien Régime

In many respects, the medical history of the nineteenth century belongs to the ancien régime. There were still distinctive risk groups, the chief one being soldiers of every nation. The wars to conquer New Zealand were possibly the only ones in the century in which more European soldiers died in battle or from accidents than as a result of disease. The opposite extreme was the campaign in Madagascar in 1895, when some 6,000 French soldiers died of malaria and only 20 in military action.[121] A new era dawned outside Europe with the Russo-Japanese war of 1904–5, when the Japanese, thanks to meticulous vaccination and medical facilities, managed to keep their losses through disease to a quarter of the numbers killed in battle.[122] From a position of weakness, the emergent military state could hope for victory only if it carefully husbanded and deployed its scarce resources in personnel and material. But the nineteenth century also witnessed the beginning of the end for the medical ancien régime—something that, despite all the jolts and discontinuities, should not be denied the name progress. This transition had, roughly speaking, three aspects, which may be arranged in sequence.

The first aspect covers the global retreat of smallpox in the face of Jenner-inspired vaccination and the prevention and treatment of malaria with alkaloids obtained and developed from cinchona bark. After 1840 or thereabouts, and especially after 1854, deaths from malaria began to decline at least among Europeans in the tropics—an essential for military conquests in southern latitudes.[123] These were the only two effective medical breakthroughs until the emergence of microbiology.

The second aspect was the rise of laboratory medicine, associated with the names of Louis Pasteur and Robert Koch, which was one of the great innovations of the age. After its first major successes in the 1870s, it established itself in the following decade as an independent field of science, although it took a while before preventive strategies or even mass treatments could be deployed against the various diseases whose causes were now identified. Moreover, the idea that medical research had to take place in the laboratory remained controversial for a long time for the Western public. Such doubts were often expressed in the form of opposition to experiments with animals ("vivisection").[124]

Between these two breakthroughs (the Jenner and Pasteur moments in medical history, as it were), an intermediary aspect or third phase involved a triumph for practice rather than for theory. It is associated more with the names of social reformers and medical-sanitary practitioners than with researchers bent over a microscope.[125] The movement for improved sanitation that began in mid-century in Western Europe and North America soon had at least a sporadic impact in many other parts of the world. Long before causalities had been scientifically established, experience showed that it was healthier to live in cities with clean water, proper sewers, and organized garbage disposal and street cleaning (which, unlike today, was mainly a question of removing organic matter such as ash and horse dung). Medical people knew this even before they were in a position to classify clean water bacteriologically.

This third aspect concerns a change in attitudes, which in principle was possible on various cultural foundations and did not depend on a correct understanding of the latest scientific theories from Europe. Societies that could find the will and resources to make their cities healthier and to care better for their soldiers gained a mortality dividend, enhanced their military capability, and raised their general energy level. Experiences in handling epidemic disease could translate into a changed international weight for the countries concerned. The global "hygiene revolution" was one of the great breakthroughs of the nineteenth century. It began after 1850 in western and northern Europe and has continued down to the present day. It was soon taken up in parts of India, later in east-central

Europe and Russia, and from the 1930s in countries such as Brazil, Iran, and Egypt.[126] It would be too simple to interpret this global process as a straightforward result of the Industrial Revolution, or even of the new scientific discoveries of the age. National income growth and new expertise did not *directly* translate into gains right across society in health, life expectancy, and the quality of life. There also had to be a certain normative change, so that epidemics were no longer seen as divine retribution or a consequence of evil individual or collective behavior; morality had to be taken out of the medical understanding of the world. As it became clear that epidemics responded to social intervention, support grew for state-run programs to construct public health systems. The decisive innovation, in which cities such as London and New York took the lead, was probably the creation of local health authorities under central control but with the leeway to respond to conditions in their area. People now expected clean tap water and regular collection of the garbage they had recently learned to fear and loathe. And consumers were ready to pay for facilities that were beneficial to their health.

In the nineteenth century, tropical diseases endemic in latitudes close to the equator were less successfully combated than some of the great scourges that affected Europe.[127] Nonurban environments were often more difficult and more costly than cities to keep clean, especially in tropical climes. The disparity was due to a number of factors: to the fairly limited reach of colonial medicine, which, despite many successes (e.g., in the fight against sleeping sickness), did not have the means to root out endemic diseases at the source; to the fact that neither the regions concerned nor the colonial tax system could meet the exceptionally high cost of removing contributory causes such as swamps (insect bites were definitely established as a conduit of infection only in 1879); and to a vicious circle of malnutrition and defective resistance to disease, which Europe and North America mostly escaped. There is much evidence that in the worldwide retreat of fatal diseases, the biological and economic pressures declined faster in the temperate zones of the earth than in the tropics. Climate does not explain economic performance directly or override social and political factors, but it should not be overlooked that the health burdens in tropical zones were and are greater than those in temperate latitudes. This has contributed to an environmental fatalism in hot countries that acts as a dampener on hopes of development.[128] Whether tropical medicine was a tool of medical imperialism is a question that does not admit of a single straightforward answer. In some respects (e.g., malaria) it gave Europeans and North Americans the medical assurance with which to conduct further conquests, but it did not do this in other respects (e.g., yellow fever). On the one hand, important medical discoveries were made in the colonies; on the other hand, experiments were conducted with new treatments and drugs that could not be tried out on Europeans. The main goal of colonial medicine and sanitary services was to improve living conditions for the colonizers. But in many colonies efforts were also made to raise the working capacity of the colonized and to strengthen the legitimacy of colonial rule by means of reforms. Confronting potentially global scourges such as the plague in their non-European places of origin was a new approach that complemented the older strategies of protective shielding. The fight against disease was recognized in the nineteenth century as an international task. In the twentieth century it became one of the main areas of coordinated crisis control and prevention.

5 Natural Disasters

Apart from epidemics, there was no lack of other apocalyptic horsemen in the nineteenth century. Natural disasters seem to break into history from the outside; they are antihistorical free agents and independent variables. The most disturbing are those for which people are unprepared and against which human action is ineffectual. These include earthquakes. There is a history of earthquakes—as there is of spring floods or volcanic eruptions—but it can never be a history of progress. Only in the second half of the twentieth century did geology and meteorology, together with new measurement techniques, create some scope for disaster prophylaxis. Warnings are possible, and there is also a minimum, but nothing more than that, of preparation for the worst. Natural disasters are no peculiarity of the nineteenth century, but a portrait of the age would be incomplete without this ever-present menace to the routines of ordinary life. At times, certain spots of the earth were afflicted by a whole array of calamities. "In the first decade of the nineteenth century," reports a historian of Oceania, "Fiji experienced a total eclipse of the sun in 1803, the passage of a comet across the heavens in either 1805 or 1807, an epidemic of dysentery, a hurricane, and the inundation of many coastal areas as a result of either a tsunami or cyclonic storm waves."[129]

Earthquakes and Volcanoes

No event in nineteenth-century Europe had an impact on people's minds comparable to that of the Lisbon earthquake in 1755, whose horror still resounded thirty years later in the *terremoto* at the end of Joseph Haydn's *Seven Last Words of Christ on the Cross*. Heinrich von Kleist used a real case from 1647 as the basis for his novella *The Earthquake in Chile* (1807). But if any earthquake comes close to the one in Lisbon, it is the great tremor that shook San Francisco on 18 April 1906 at five o'clock in the morning. Many of the Victorian houses in the city collapsed, no thought having been given in their construction to the possibility that the earth would one day move. The social order itself was stretched to the limits as looters roamed the streets and the mayor called in the army to help. Fires blazed for several days and destroyed a large part of the city. Tens of thousands were rescued from the sea at the height of the crisis, in what was probably the largest maritime evacuation before Dunkirk in 1940. The most pessimistic estimates put the total loss of life at 3,000 and the number rendered homeless at 225,000;[130] early concrete structures, which were more resilient than masonry, kept those figures from being even higher. The quake of 1906 was exceptional not because of the scale of the losses (far below the 100,000 deaths or more in Japan following the Kanto earthquake of 1923) but for a different reason: like the earthquake in 1891 on the main Japanese island of Honshu, which had left 7,300 dead, destroyed buildings with a mainly European design, and fueled criticism of exaggerated Westernization—it seemed to embody a new type of "national" disaster, in which nature attacked the nation at its weak point but at the same time gave it an opportunity to display solidarity and ingenuity in the work of relief and reconstruction. This was a general trend in response to natural disasters. In the 1870s, when huge swarms of Rocky Mountain locusts devastated large areas in the American Midwest, the creatures were declared a national enemy and the army was mobilized, under the leadership of an old Civil War general and Indian campaigner, to get aid through to small farmers. In the winter of 1874–75 two million food rations were distributed in the states of Colorado, Dakota, Iowa, Kansas, Minnesota, and Nebraska. It was one of the logistically most elaborate operations conducted by the government since the end of the Civil War in 1865.[131]

Volcanic events too are sudden and localized, but their effects may stretch over a wide geographical area. The eruption of Krakatau on August 27, 1883, in the Sunda Strait in what is now Indonesia, threw up an ash cloud that spread all around the world. A tsunami triggered by the eruption claimed approximately 36,000 lives along the coasts of Southeast Asia, and the already quite advanced instruments of the time measured seismic waves on every continent. A local natural disaster thus became a global scientific event.[132]

Back in April 1815 the eruption of Tambora on the small Indonesian island of Sumbawa, more powerful and more devastating in its consequences (117,000 killed in the area), had not yet caught the attention of the international public. A large part of the Indonesian archipelago was covered in darkness for three whole days; people heard the volcanic explosions at a distance of several hundred kilometers, often mistaking them for cannon fire, and troops were put on a war footing in Makassar and Jogjakarta. A thick deposit of ash and rock settled over the export-oriented island, which lost most of its forest and saw its rice fields along the coast flooded with seawater. The eruption reduced the height of Mount Tambora from 4,200 to 2,800 meters. Sumbawa became virtually uninhabitable. There was no medical care for the often seriously injured survivors; food supplies were destroyed and drinking water contaminated; the island became completely dependent on imports. This situation lasted for several months until the colonial authorities and the outside world realized the full extent of what had happened. There could be no talk of speedy emergency relief. The neighboring islands of Bali and Lombok were covered with twenty to thirty centimeters of ash, and there too, the destruction of the standing rice crop led to outbreaks of famine. Agriculture in Bali—which suffered 25,000 deaths—was still seriously affected in 1821, but in the late 1820s the island began to reap the benefit of the fertile volcanic deposits. This was one of the reasons for the modern rise in its farm output.

The eruption of Tambora had global consequences. In many parts of Europe and North America, 1815 was the coldest and wettest year since records began, and 1816 went down in the annals as the "year without a summer." The impact was most severe in New England and western Canada. But Germany, France, the Netherlands, Britain, and Ireland also recorded abnormal weather conditions and poor harvests. For several more years particles in the stratosphere blocked the sun's rays, causing average temperatures to fall by three to four degrees Celsius. Nowhere did the crisis bite harder than in the southern Rhineland and Switzerland in the winter of 1816–17. Even the basic supply of imported grain broke down, since early frosts and harsh weather delayed shipments from Baltic ports. All of the old syndrome of food shortages, rising prices, and depressed demand for nonagrarian products established its hold. People flocked from crisis areas toward Russia and the Habsburg Empire, or via Dutch ports to the New World. Captains refused to accept penniless refugees, and many who were turned away had to make their way back home as beggars. The acute central European agrarian crisis of 1815–17 has often been seen as one of the last of "the old type," and quite a few historians have even thought that it destabilized European governments. Historians and climate researchers finally came to recognize in the twentieth century that it had been triggered by events in faraway Indonesia.[133]

Hydraulics

Water disasters lie at one extreme on the scale of events in which human activity is a contributory fact. They depend on the amount of periodic rainfall and snowmelt and are therefore difficult to predict even today, yet many societies learned early on to regulate the flow of water. Although few Asiatic societies can be said to have had a fully "hydraulic" character, it remains true that in many parts of the world, agriculture and other types of cultivation are possible only on the basis of irrigation and flood-defense technologies that go back a long way in time. The nineteenth century gave a new impetus to hydraulic engineering: it permitted major projects such as those regulating the upper and lower Rhine, or the great canals in North America and central Europe, and later in Egypt and Central America. In some cases, technological breakthroughs allowed new irrigation systems to be created out of ancient installations: for example, the massive projects initiated in the 1860s in the Bombay hinterland.[134] From 1885 on, in another project that took years to complete, the government of British India modernized and expanded a system of hydraulic installations in the Punjab (in today's Pakistan) going back to the time of the Mogul rulers. In this way, even the high plains of northwestern India were turned into wheat fields. Laborers were recruited from far and wide, and shepherds were replaced with taxpaying farmers reliable in their political loyalty to the colonial power.[135]

Sensitive irrigation systems—which require constant attention to work at their peak of efficiency—can be slowly degraded if private interests get out of hand and prevail over regulation in the common good.[136] War can destroy them in next to no time, as it did in Mesopotamia in the thirteenth century. The worst disasters occur where dams or dikes collapse—a constant danger not only in protected coastal areas but also on a number of great rivers. Such incidents were likeliest in China, the classical country of premodern water taming. Researchers have used the ample documentation on tax exemptions for flood victims to estimate the scale of the damage along the Yellow River (Huanghe), China's most difficult. For centuries a system of ever higher dikes guided the Yellow River through the provinces of Henan and Shandong, but the dangers of collapse also grew over time. In 1855 the northern dam in Henan gave way. The backwaters of gigantic floods could be seen three hundred kilometers away. And although the authorities deployed more than 100,000 men at the point of fracture, they were unable to hold the river again. After 361 years China's second-largest river altered its course for the sixth time in recorded history, now flowing northeast instead of southeast, so that its new mouth lay three hundred kilometers from the previous one.

In comparison with the catastrophe of 1938, when the Chinese high command blew up the Yellow River dikes in the face of advancing Japanese troops, the floods of the nineteenth century claimed surprisingly few lives. This was because the Qing state was then still capable of operating a kind of early warning system and, at many places, of maintaining protective dikes below the level of the main dams. Nevertheless, it was not unusual for many people to drown or lose their home in the escaping waters of the Yellow River, and floods often brought famine and disease in their wake. In some cases as many as 2.7 million people—7 percent of the population in the province of Shandong—received official disaster aid after dike breaches of the 1880s and 1890s. Social tensions, looting, and unrest were frequent consequences. In one region notorious for its banditry, in which the Taiping and Nian rebels had been active and sections of the population had been formed into armed militias, it did not take long for law and order to break down. Natural disasters alone seldom trigger social protest directly, but they were invariably a contributory factor in drought-prone northern China.[137] Floods there were not "manmade disasters" in any platitudinous sense of the term. The engineering challenges were enormous by any conceivable measure, as were those relating to work organization and project funding. The dike bureaucracy, the largest branch of the Qing state in the nineteenth century, concentrated many skills and discharged many tasks competently, but it was hobbled by its growing corruption, fiscal weakness, lack of planning, a tendency to act reactively rather than preventively, and resistance to new technologies.[138]

All in all, the old basic patterns changed little in the nineteenth century. In principle they still apply today. Owing to the bounty of nature, everyday life held fewer dangers for Europeans than for people in many parts of Asia. Although the capacity for government regulation was not noticeably different (no state in the world had as much experience as China in dealing with natural disasters), and although a massive impulse was required even in the West to galvanize the state (as the example of the American locusts showed), things were easier for Europeans when push came to shove: more resources could be concentrated on a small number of less serious cases. Nevertheless, the victims of a disaster generally had to fend for themselves or to rely on help from the narrow circle of people around them. Neither medical/humanitarian assistance nor international support entered the picture in the nineteenth century. Both have developed in the period since 1950. They presuppose the deployment of airlifts and a conception of international aid as an ethical principle within a nascent global society—one of the greatest advances of civilization in the contemporary world.

6 Famine

The extent to which famines are "man-made" is not something that can be determined in general. Nor is it easy to say what the "starvation" associated with a famine actually is. The difficulty is twofold: on the one hand, starvation is "culturally constructed," so that the word does not mean the same at every time and place; on the other hand, the question arises as to what must be taken into account, apart from human physiology and culturally specific "semantics," in order to reach a reasonably complete understanding of the existential state of "starvation." One big question therefore turns into a number of subquestions concerning: (1) the *quantity* of food—that is, the minimum of calories—necessary for people differentiated by age and gender; (2) the *quality* of nourishment required to ward off dangerous deficiencies; (3) the regularity and *dependability* of food grown at home or supplied through public distribution or the market; (4) the actual form and level of *distribution* according to social stratum; (5) the claims and *entitlements* to food associated with various positions in society; and (6) the *famine relief* institutions, whether governmental or private-philanthropic, that can be mobilized in an emergency.

The Last Famines (for the Time Being) in Europe

One simple distinction is the one between chronic starvation (long-term shortage of food) and acute famine with a high level of mortality.[139] Famine crises were more characteristic of the twentieth century than the nineteenth. The century of great medical advances and the doubling of life expectancy was also the one of the greatest famines known to man: in the Soviet Union in 1921–22 and 1932–34, Bengal in 1943, the Warsaw Ghetto in 1941–42, Leningrad during the siege by German troops in 1941–44, the Netherlands in the winter of 1944–45, China in 1959–61, and Sudan in 1984–85. The effects of starvation are the same across cultures: people of all age groups—but first the very young and very old—eat ever-smaller quantities of less and less nourishing food: grass, tree bark, unclean animals. They become "all skin and bones." Secondary effects such as scurvy are almost inevitable, especially where people (as in Ireland) are used to a vitamin-rich diet. The struggle for survival destroys social or even family ties, pitting neighbor against neighbor. Men and women commit suicide, children are sold, defenseless people are attacked by animals; cannibalism itself—however unreliable the reports always are—lies in a straight line from despair. Survivors are traumatized, children suffer lasting physical damage, and governments, bearing the original guilt of having failed to provide relief, are often discredited for decades. Memories stick in the collective mind.

Were there such famines in the nineteenth century, and if so, where? The question is rarely mentioned in the history textbooks. The German texts recall the terrible times of the Thirty Years' War, especially 1637 and 1638, as well as the great famine of 1771–72. Hunger again stalked the country in 1816 and 1817. After the subsistence crisis of 1846–47, the classic famine—brought on by harvest failure, grain profiteering, and inadequate government action—disappeared from the history of central Europe and Italy (where things were especially grim in 1846–47).[140] Of course, this needs to be seen in a broader framework: famine had marked many parts of Europe in the age of the Napoleonic wars; and hunger riots had broken out in England during the 1790s, even though it was then the richest country in Europe and had the best system of poor relief (the Poor Law) supported by religious and philanthropic private initiative. Few actually starved to death in England, but many of the things to which people were accustomed became prohibitively expensive. Those who could no longer afford wheat turned to barley, while those who found even that too expensive had to make do with potatoes and turnips. Women and children went short more than others, in order to maintain the laboring power of the head of the family. Household goods were pawned, and the number of thefts shot up. Such was the face of hunger in a country that after 1800, thanks to its wealth and its global connections, would be able to ensure its food supply from overseas.[141]

On the Continent, the specter of subsistence crises retreated after 1816–17. In some parts of Europe where famine had been a regular occurrence, it became much more of an exception—in the Balkans after the 1780s, for example. Spain remained vulnerable and in 1856–57 experienced another major crisis. And Finland lost 100,000 of its 1.6 million inhabitants after the harvest failure of 1867—the last true subsistence crisis in Europe west of Russia.[142] At the same time, and in similar weather conditions, Sweden's northernmost province, Norbotten, suffered a serious food bottleneck, although its much better organized disaster relief meant that the loss of life was much smaller than in Finland.[143] Scotland—unlike France, for instance—came through the eighteenth century rather well. But between 1846 and 1855 it endured hardship unparalleled since 1690, with year after year of poor potato crops in the western highlands and islands. The loss of life was not especially large, but it fueled massive emigration and was therefore of great demographic significance. It was the last great subsistence crisis in the British Isles.[144]

Europe's Exceptions: Ireland and the Tsarist Empire

In Ireland, the poorest part of the United Kingdom, the Great Famine of 1845–49 was caused by several years of potato crop failure resulting from the mysterious fungus *Phytophthora infectans*.[145] The potato blight hit a society in which the poor lacked not so much food as adequate clothing, housing, and education. English visitors described in dark hues the impoverishment of the island before the famine; they could hardly have failed to do so, given that they came as aristocrats and bourgeois from a country where living standards were twice as high. But, to keep a sense of perspective, we should bear in mind that Ireland's real per capita income in 1840 was equivalent to that of Finland in the same year, Greece in 1870, Russia in 1890—or Zaire in 1970.

The size of the potato harvest in 1845 was one-third smaller than normal, and in 1846 three-quarters smaller. The situation was a little better in 1847, but in 1848 it was scarcely possible to speak of a crop at all. The Irish famine, more than many others, was unleashed by the direct physical failure of the food supply. High prices and speculation, the usual triggers of early modern hunger revolts, played no significant role. The scale of the disaster becomes clearer by the criterion of land acreage of potatoes: two million acres before the famine, a mere quarter of a million in 1847. The death toll peaked in 1847–48, when dysentery and typhus ravaged an already weakened population and tens of thousands were dying in poorhouses, while at the same time the birthrate plummeted. Not only the poor were affected, since no one was safe from infectious diseases. As so often was the case in nineteenth-century epidemics, doctors succumbed too, in droves. Present-day research confirms the old figure of one million excess deaths in a total population of 8.5 million before the onset of the crisis. Perhaps a further 100,000 died of the consequences of starvation, either during or immediately after emigration.

It is still not altogether clear how the destructive fungus reached Ireland; one plausible theory is that it came in shiploads of guano fertilizer from South America. Relief measures, at first involving private initiatives, began shortly after the first crop failure became apparent, as reports aroused sympathy and support in many countries. The Catholic Church and the Quakers were especially active in the work of organization; even the Chocktaw nation sent donations from Oklahoma. As a reasonably good experience in 1822 had already shown, massive government aid at the beginning of the crisis might have been successful in controlling it; wheat could have been imported from the United States, for example, which unlike Europe had had a record harvest in 1846. But several factors determined the actual response of the British government. The ruling ideology of laissez-faire excluded any interference in the "free play" of market forces, because that would have been damaging to the landowning and commercial interests. Also influential was the view that the collapse of the potato economy would create opportunities for the modernization and reorganization of agriculture and allow it to achieve a "natural equilibrium." Some Protestants even believed that the crisis was a gift from the Almighty, making it possible to root out the evils in Ireland's Catholic society. Another element was British hostility to Irish landowners (whose greed and neglect of agricultural improvements were held responsible for the problems in the country), so that it saw little reason even to repair the damage.

In 1845–46, the first year of the famine, the Tory government of Sir Robert Peel bought emergency supplies of Indian meal (a cheap, coarsely ground corn-meal) from the United States and had it distributed at various official sales points; at the same time it inaugurated a program of public works. The Whig government of Lord John Russell that came to power in June 1846 continued with this approach but refrained from any involvement in the trade. Soup kitch-ens were set up in 1847 but soon were discontinued. It has often been asked how three million people could have been so dependent on the potato. The answer is probably that it had proved its worth for decades and that people did not think it left them open to excessive or incalculable risks. One theory is that the disas-ter of 1845–49 brought the long decline of the Irish economy to a head, while another school of historians sees the fungus invasion as an exogenous blow to a process of slow economic modernization. But a purely naturalistic explanation will not do. The Irish famine does not invalidate the general insight that from the beginning of the seventeenth century, European agriculture was productive enough to satisfy the basic needs of the population and that "famines were man-made rather than natural disasters."[146]

The famine of 1891–92 in the Tsarist Empire, which claimed approximately 800,000 lives, mostly in the Volga region, had quite different causes. It was not due to an absolute shortage of food: the harvest of 1891 was very small but no more so than those of 1880 or 1885, when Russia had pulled through without any major relief effort. A number of other factors came into play at the beginning of the 1890s, however. In the preceding years, farmers in the black soil region in particular had tried to raise output by redoubling their labor and putting a relentless strain on the earth. Then bad weather came on top of the exhaustion of people, animals, and soil; soon all reserves kept for a rainy day were used up. The famine of 1891–92 was a turning point in the history of Russia. It brought to an end the "reactionary" period following the assassination of Tsar Alexander II and introduced a phase of social unrest that issued in the Revolution of 1905. In general the Tsarist government did not perform badly in disaster relief, but this counted for little in the realm of symbolic politics. It seemed to the public of the time that famines happened only in "uncivilized" colonial or semicolonial coun-tries such as Ireland, India, and China. The anachronistic famine of 1890–92 ap-peared to demonstrate once again the growing gap between the Tsarist Empire and the progressive, prosperous countries of the West.[147]

The New World was also one of these "civilized" areas of the globe. North America was free of famine in the nineteenth century: only small communities of Indians may have been temporarily reduced to extreme subsistence levels. The fact that people in the Western hemisphere were not undernourished made a favorable impression on many poverty-stricken Europeans during the great crisis years of 1816–17 and 1846–47. An immigrant from northern Italy, where the rural population suffered from the vitamin deficiency disease pellagra and had meat on the table only on the main feast days, found a surplus of meat in Argentina. Even in Mexico, which was not a classic country of immigration, the age of famines lay in the past; the last one had occurred in 1786. The food situation improved markedly during the first half of the nineteenth century, as grain production increased twice as fast as the population. The new republic also took better precautionary measures than the Spanish colonial state had done, and on several occasions after 1845 it bought cereals from the United States in time of need.[148] In Australia and New Zealand, too, there was no longer any reason to fear an outbreak of famine.

Africa and Asia

Things looked different in the Middle East and Africa. In Iran, a great famine between 1869 and 1872 claimed approximately 1.5 million lives.[149] In sub-Saharan Africa, the 1830s, 1860s, and 1880s were marked by especially severe drought, and after 1880 the colonial wars of conquest everywhere exacerbated the food supply problem. In perhaps the worst known famine before the First World War, 25 to 30 percent of the population perished in 1913–14 in the Sahel region, not long after another famine in 1900–1903.[150] Drought does not automatically result in famine. African societies had a lot of experience in averting food shortages and starvation and in cushioning their impact. The mechanisms of crisis prevention and management included a change in production methods, the mobilization of social networks, and the use of ecological reserves. Supply maintenance techniques were highly developed. But it is true that in persistent drought, often followed by scarcely less dangerous periods of monsoon-like rainfall that brought diseases such as malaria in their wake, social orders might fall apart. People then dispersed into the bush to increase their chances of survival. Violence was more widely practiced by warrior groups in such situations. In southern West Africa (Angola), for instance, there was also a long-standing connection with the slave trade: drought victims would flock toward populated centers and become subjugated as "slaves"—a pattern still apparent in the generation affected by the extended drought of 1810–30.[151]

Even before the colonial invasions of the 1880s, however, two new developments made it more difficult to apply such tried-and-tested strategies. First, the spread of the caravan trade and the "Oriental" slave trade in the savannah belt south of the Sahara led to a new kind of commercialization from the 1830s on; long-distance trade started to bring in food supplies through regional distribution networks. Second, a new factor both in the Mediterranean North and in South Africa was the vigorous competition for land between African societies and European settlers. An additional complication was that colonial ideas about natural conservation often corresponded more to European fancies of a "savage" Africa than to the survival needs of the indigenous population.[152]

In Asia, which in the second half of the twentieth century left starvation behind faster than Africa did, the nineteenth century witnessed the most devastating famines. They seem to have been particularly deadly where, in conditions of low agricultural productivity and meager surpluses, societies found themselves temporarily trapped between growing marketization of the food supply and an underdeveloped structure of disaster relief. Despite its relatively productive agriculture and exceptionally good health conditions, Tokugawa Japan was not spared the visitation of famine. Like Europe, it had repeatedly witnessed hunger crises in the early modern period—for example, in 1732–33 and again in the 1780s, when the eruption of the Asama volcano in August 1783 added to the ecological and economic difficulties facing the country. The Tempō famine, the last great tragedy of its kind to strike Japan, broke out in 1833 as a result of crop failures and aggravated by infectious diseases; the next two harvests were not much better, and the one of 1836 was a disaster.

There are indications that between 1834 and 1840 Japan suffered a drop in population of about 4 percent.[153] A sharp rise in social protest was directly linked to the food crisis, but, as in large parts of Europe around the same time, it signaled the end of the recurrent threat of famine. The size of this threat should not be exaggerated. It had always been lower than in many parts of mainland Asia: Japan was not susceptible to climate-induced harvest failure (except in the far North), nor did its agriculture perform badly. The Tokugawa economy kept the growing cities fed, and the average food situation in the eighteenth century was probably not essentially different from that which prevailed in Europe. The second quarter of the nineteenth century followed a period of relative prosperity that had begun around 1790. The Tempō famine, comparable in scale to the European crisis of 1846–47, was felt as a great shock and a symptom of a broader social crisis precisely because it was *uncharacteristic*. Though the Japanese were by no means generally protected from hunger, they were no longer accustomed to the kind recurrent food shortage that haunted other societies in Asia.[154]

The Asian famines of the nineteenth century that caused the most deaths and attracted the greatest attention in the rest of the world were those in India and China. These countries experienced unusually severe weather conditions at almost the same time, from 1876 to 1879 and from 1896 to 1900–1902. Also Brazil, Java, the Philippines, and northern and southern Africa suffered poor harvests that have since been blamed on the meteorological phenomenon known as El Niño (although this is still disputed). For India and China together, the excess mortality during these years has been estimated at a total of 31 to 59 million.[155] In both cases, unlike Russia in the 1890s or Japan in the 1830s, it is questionable whether the famines triggered major historical changes. In China the famine of the seventies, which was considerably graver than the one at the end of the century, led to no really significant increase in political or social protest. The Qing Dynasty, which shortly before had withstood the far greater challenge of the Taiping Revolution, was not seriously destabilized and eventually collapsed in 1911 for quite different reasons. British rule similarly held firm in India—as it had in Ireland after the Great Famine. But the famous naturalist Alfred Russel Wallace, in the assessment of the Victorian age that he wrote in 1898, included both these famines among the "most terrible and most disastrous failures of the nineteenth century."[156]

But although famines are not always turning points in history, they invariably tell us something about the society in which they occur. In neither India nor China was the whole country affected. In India, where monsoon failure was the trigger, the worst famine of the nineteenth century was concentrated in the south, mainly in the provinces of Madras, Mysore, and Hyderabad, with a second center in the north-central region south of Delhi.[157] In China, only the northern parts of the country between Shanghai and Beijing were affected, especially the provinces of Shanxi, Henan, and Jiangsu. Undoubtedly the actions of the colonial government made the situation worse in India; contemporary critics already blamed the severity of the famine on doctrinaire adherence to free-market principles. It took some time before the administration was willing to acknowledge the scale of the disaster and to suspend the collection of taxes.[158] In northern India, where the harvest failure had been relatively minor, high prices in the British market sucked away so much grain that not enough was left to cover the subsistence minimum of the peasantry. Despite many initiatives by lower-level authorities to relieve the disaster, the policy of the Raj was to place nothing in the way of the private grain trade and to avoid as far as possible any additional public expenditure. The results were the same in 1896–98: grain could be bought at high prices even in areas where the harvest had suffered the worst damage.[159]

Commissions put in place by the government in London were among the critics of the British authorities, but they found no fault with the principle of "colonialism on the cheap." The great famines of the last quarter of the century were less an expression of primitive Indian resistance to progress than, on the contrary, the symptom of an early crisis of modernization. Railroads and canals, which made it easier to transport aid to crisis-hit areas, were at the same time the logistical basis for engaging in speculation with the harvest yield; they facilitated both an inflow and an outflow of grain. Poor harvests were inescapably reflected in high prices.[160] Hoarding and speculation had always been a possibility in premodern conditions. What was new was that traditional village reserves of food were also caught up in the flow of all-Indian and international trade, so that even small changes in harvest yield led to exceptional price increases. The severity of the impact on the rural population—the cities remained fairly well supplied—was ultimately due to the fact that incipient modernization made certain social groups more vulnerable, especially small leaseholders, landless laborers, and home weavers. The decline of home weaving in the countryside and of many social institutions that had formerly offered some protection against disasters (castes, the family, village communities) was an intensifying factor.

In many parts of India, farmers drove agriculture to the limits of the possible, mostly by using poorer soils that required a greater input of labor and reliable irrigation. Often these conditions were not present. The race to produce for export markets resulted in large-scale privatization of common land; shepherds were driven into mountainous country with their animals; trees and bushes were cleared away. Ecological stress on soil reserves was therefore part of the fateful modernization crisis. The growing economic vulnerability of families and individuals led to an upward spiral of debt, and urban moneylenders and their village agents, alongside grain speculators, were a great threat to the existence of the peasantry. The lack of adequate communal or government-controlled credit for small landowners fueled the debt spiral, which the colonial regime shrugged off as a consequence of the free play of market forces. The landless seem to have been the hardest-hit by famine, neither having their own means of production nor being able to assert ancient rights, however rudimentary, to the moral economy of mutual aid. The evolution from harvest problems to a full-blown famine did not depend only on the "free play" of market forces and self-interested policies on the part of the colonial rulers. Peasant producers were mostly cut off from the market and exposed to the machinations of landowners, merchants, and moneylenders, many of whom tried to profit from the crisis. The distribution of power in rural societies was one of the causes of starvation.[161]

In northern China, nightmare scenarios similar to those in India played themselves out between 1876 and 1879.[162] The Great North China Famine, which claimed 9 to 13 million lives (most of them from typhoid), was the most serious and geographically most widespread human disaster in any time of peace during the Qing era; the region had seen nothing like it since 1786. The only Westerners who observed it were not colonial officials but individual missionaries and consuls. It is therefore little documented in Western sources, whereas Chinese sources are filled with detail. The sense of horror that the Indian famine aroused abroad was due not least to the spectacular photographs of its victims, the first of their kind to be published anywhere in the world. Very few similar pictures exist from northern China; the famine there was in media terms the last of the "old type." Nearly a year passed before foreigners in Shanghai or Hong Kong became aware of its scale in a remote province such as Shaanxi. But then a private China Famine Relief Fund was soon set up in Britain, which transferred funds to China in an early charitable application of telegraph technology.[163]

Unlike India, northern China had not yet been opened up by the railroad and was all but untouched by capitalism. The province of Shanxi, for example, was linked to the coast only by narrow, frequently impassable roads that wound across high mountains. Aid from other regions of the country was more difficult to organize than in India, especially as the Great Canal, which for centuries had supplied the capital Beijing with rice from the lower Yangtze region, had silted up and fallen into disrepair. The famine-stricken regions had long been among the most precarious economically and the least productive in their agriculture. China's real granaries—the lower Yangtze and the southern coastal strip—were not hit by the natural disaster that lay at the origin of the famine. In the end the Chinese state did undertake considerable relief efforts, but the results were paltry in comparison with the size of the challenge, or indeed with some of the great relief campaigns of the eighteenth century. But this discrepancy had less to do with a doctrine of cheap government plus free markets than with the fact that the Qing Dynasty had been financially drained by the suppression of the Taiping and Muslim rebels. In contrast to the Indian famine, the one in North China was more a crisis of production than a crisis of distribution. It broke out in an ecologically precarious niche, where for centuries state intervention had been able to ward off the worst consequences of disastrous weather conditions. The limits to such intervention were now greater than in the past.

A "Land Stalked by Hunger"?

The famine of 1876–79 leads us on to the general standard of living in nineteenth-century China. Had it really become a "land stalked by hunger"? The question is so interesting partly because recent research, in both China and the West, has painted an extremely rosy picture of the eighteenth-century Chinese economy, confirming the favorable reports of missionaries at the time. The variants of agriculture in the Qing Empire ranged from pasture farming in the grasslands of Mongolia to the highly productive mix of rice terraces and fish ponds in the south to the export of products such as tea and sugar. But however hard it may be to make a generally applicable statement in this regard, there is now agreement among experts that until the last quarter of the eighteenth century, Chinese agriculture kept a fast-growing population adequately fed. The claim that eighteenth-century Chinese peasants lived at least as well as and probably better than their counterparts in the France of Louis XV—a claim that people in the West long found beyond belief—has something to be said for it in the state of our present knowledge.

The comparison with eastern Europe is certainly favorable. Almost constantly one district official or another would report food supply problems from a part of the vast empire and ask the imperial court for help. The Chinese state responded to such appeals on a scale that had no parallel in Europe at that time; the care and maintenance of its famed system of public grain reserves, which reached its peak of efficiency under the Qianlong Emperor (r. 1737–96), was one of the principal duties of local officials, and the relief it gave in an emergency was several times greater than the tax yield in a normal year. The emperor and provincial governors personally concerned themselves with the functioning of this system. The dynasty of the Qing conquerors from Manchuria derived some of their legitimacy as rulers of China from their success in ensuring internal peace and public welfare. When urban leaders other than government officials began in the 1790s to take on philanthropic commitments, the first goal they set themselves was to build up private grain reserves.[164] The state granaries also had ongoing responsibilities. Especially in Beijing and its surroundings, they took delivery of taxes and tribute in grain and even sold it in normal times at below the market price, keeping a close eye on private traders to prevent hoarding. The mixed state-private grain market that developed in this way had to be repeatedly kept on a middle course, and in general this was successfully achieved. In the last two decades of the eighteenth century, 5 percent of China's total grain harvest was being stored in public granaries. The system proved its worth under the rule of the Qianlong Emperor. Despite numerous droughts and floods, no famine remotely comparable to that of the 1870s is known to have occurred in the eighteenth century.[165]

It is not yet fully understood how Chinese agriculture fared in the nineteenth century. The climate seems to have worsened after the turn of the century, and there was a rise in the number of natural disasters. At the same time, the capacity of the state to intervene proactively in society gradually declined. Little use was made of the usual method of tax deferrals or exemptions, while fewer and fewer disaster areas received old-style direct support from the government. The general plight of the Qing Dynasty was palpable in the lower ethical standards of public officials and the spread of corruption, which must have negatively affected the complicated system of grain storage. Grain rotted away in poorly maintained storehouses, and there was a failure to keep the reserves regularly replenished. When the Opium War then opened a long series of conflicts with the Great Powers, and the Taiping Revolution shortly afterward started a chain of internal revolts, the Qing state began to set new priorities for its dwindling resources. The supply of food to the army would now take precedence over civilian disaster relief. This reorientation contributed to the virtual disappearance of the granary system in the 1860s, a hundred years after the height of its functioning.[166] Yet famine on the scale of the 1870s was a unique event. It may well be that until the 1920s Chinese agriculture was still capable of providing a reasonably tolerable average supply of food to the population.

Eurasia as a whole differed from North America in that its western and far eastern (Japanese) extremities left the constant threat of famine behind only in the second half of the nineteenth century, and the rest of the continent followed much later. This did not mean that all sections of society in Japan and Western Europe were now free of undernourishment or malnutrition, or that individuals were protected from extreme poverty, but it did mean that the specter of inescapable collective famine and widespread deaths from starvation was a thing of the past. Another ancient phenomenon also became unusual in nineteenth-century Europe: the starving of cities into submission through siege warfare. One notable exception was the siege of Paris in 1870–71, when the German blockade of food and fuel was partly responsible for a higher-than-usual number of civilian deaths, especially among the very young and the very old.[167] Ten years earlier, in the winter of 1861–62, there had been a similar episode in China, when imperial troops had besieged the city of Hangzhou in the hands of Taiping rebels, and two months of economic blockade had produced 30,000 to 40,000 deaths from starvation among the civilian population.[168] During the First World War, one of the few examples of such a blockade was that which Ottoman troops mounted in 1915–16 against the British garrison at Kut on the Tigris, although there were more soldiers than civilians inside the fortress. During the Second World War, this form of warfare was practiced against the city of Leningrad, with the new impetus of an ideological war of annihilation. Of a different order was the blockading of whole countries and regions—a strategy twice implemented on a large scale, each time with grave consequences for the civilian population. In 1806 Napoleon imposed the so-called Continental System against Britain, which retaliated by taking up the idea in an escalating spiral. And between August 1914 (a fortiori 1916) and April 1919, Britain maintained a blockade against Germany.

7 Agricultural Revolutions

The nineteenth-century changes in the geography of shortage and surplus must be seen against the wider background of a global development of agriculture.[169] The importance of agriculture everywhere at that time cannot be overestimated: most countries were still agrarian on the eve of the First World War; the world was still a world of tillers of the soil. This did not mean that the societies in question were mired in that general stagnation that city dwellers liked to ascribe to the alien world of the peasantry. After the middle of the nineteenth century, world agriculture experienced an extraordinary boom, most evident in the land area under cultivation. In the rice economies of East and Southeast Asia, there was literally no space for such expansion. But in Europe, Russia, and the neo-European overseas societies, total arable land rose by a factor of 1.7—from 255 million hectares in 1860 to 439 million hectares in 1910—which was a rate of growth without precedent in history over a period of five decades. Western Europe had only a minor share in this expansion, and the settlement and agricultural utilization of the vast Canadian prairie began only after 1900. The decisive advances were in the United States and Russia.[170] Only in a few countries for which estimates are possible—above all, Britain and France—did the total area of land given over to field and bush crops decline between 1800 and 1910. But there is not a direct correlation between industrial growth and a decline in agricultural acreage, since in the United States, Germany, Russia, and Japan (which all had industrial structures at the latest by 1880) the extensive development of agriculture continued.[171]

In the years from 1870 to 1913, world agricultural output grew by an estimated annual average of 1.06 percent—a rate far higher than any achieved between the two world wars. The per capita increase was smaller, of course. But annual growth of 0.26 percent meant that, by the eve of the First World War, more food and agrarian raw materials were available per capita of the world's population than there had been in the middle of the previous century. This outcome was made up of very different trends in individual countries. But the advances were by no means concentrated only in the North Atlantic space: output growth was higher in Russia than in the United States, and countries as different in their agrarian structure as Argentina and Indonesia occupied positions at the top of the league table.[172] The expansion of production concealed huge differences in productivity, and hence in the ratio of resource inputs to outputs. The yield per hectare in American wheat production and in Indian rice-growing, for example, was roughly comparable at the end of the nineteenth century, but productivity in the United States was fifty times higher than in India.[173]

The international trade in agricultural goods rose even more sharply than production, although at a somewhat slower pace than world trade as a whole. New export regions emerged for wheat, rice, and cotton, and challenged the position of traditional producers. Agrarian frontiers opened up in the American Midwest and in Kazakhstan, but also in West Africa, Burma, and Vietnam. In Cochin China—that is, the Mekong delta and its hinterland, which were scarcely populated before the arrival of the French—a dynamic rice-exporting sector geared itself mainly to southern China, while Burmese rice was sold chiefly to India. Between 1880 and 1900 the area used for rice nearly doubled and the volume of exports tripled.[174] New tropical products such as coffee, cocoa, and palm oil won overseas markets for themselves. "Developed" and "backward" countries alike offered agrarian products for sale on the world market; Britain obtained its wheat from the United States and Russia as well as from India.[175]

What did this mean for social history, in Europe, for example? Although the relative proportions of the three sectors—(a) agriculture and fisheries, (b) industry and mechanized mining, and (c) services—gradually changed in Europe, employment in the primary sector remained for a long time the highest in absolute terms. In 1910 the numbers working in agriculture were below the level of 1870 only in Britain, Belgium, Denmark, and Switzerland (plus Ireland, for altogether special reasons). At some point the share of those employed in agriculture fell below 50 percent in Europe: it happened before 1750 in England, between 1850 and 1880 almost everywhere in western and northern Europe, and only after 1900 in Italy, Portugal, and Spain.[176] The typical cause was more the emigration of rural laborers to urban industrial centers than a reduction in the number of family farms. All societies of Europe with the exception of England (much less Wales and Scotland) retained a strongly agrarian character throughout the nineteenth century. And even in England, with its towering (and tiny) landowning aristocracy, cultural ideals of a preindustrial country life continued to be dominant. The great contraction of agriculture, together with the social and cultural marginalization of the world of the peasantry, began in continental Europe after 1945 and is only today reaching a climax in countries such as China.

Statistically, then, the global food situation improved spectacularly between 1800 or 1850 and 1913. Engel's Law (so named after the Prussian statistician Ernst Engel), which is one of the few empirically rock-solid laws in social science, states that since the share spent on food decreases as total income increases, the rich are not the only ones able to profit from a growth in per capita production. Attempts have been made to demonstrate this by reference to an "agricultural revolution."[177] It is a concept about which there has long been intense debate, especially in relation to the economic history of England—that is, to the prehistory of the Industrial Revolution. The classic question is whether an "agricultural revolution" really did precede the Industrial Revolution and was perhaps even its necessary prerequisite. Suffice to recall a simple rule: "For industrialization to occur, it had to be possible to produce more food with fewer people."[178] There is no need to pronounce on the matter here. The relative proportions are the main interest for global history, and about those it is possible to say the following.

First, historians of England or Europe define agricultural revolution in general as the beginning of a long and steady increase in agricultural efficiency, measured both by rising yields per hectare (resulting in Europe mainly from new systems of crop rotation and preindustrial technological innovations)[179] and by a growth of labor productivity caused by mechanization and so-called economies of scale. Similar phenomena have already been recorded in the fourteenth-century Netherlands. The true agricultural revolution, however, took place in England in the late eighteenth century and continued in the first half of the nineteenth.[180] By 1800 an English rural laborer was producing twice as much as a Russian, and wheat output per hectare in England and the Netherlands was more than twice as high as almost anywhere else in the world. England was able to become a leading grain exporter to the Continent over the course of the eighteenth century, before its fast-growing population turned it into an even larger net importer and, beginning with the first Corn Law of 1815, made grain tariffs a central bone of contention in British politics.[181]

Second, England's special developmental path does not allow the conclusion that European or "Western" agriculture was unambiguously leading the world at the end of the eighteenth century. In large parts of Europe, agriculture was no more able to sustain the local population than in Indian, Chinese, Japanese, or Javanese regions of intensive farming. It was a long time before even the more dynamic European regions clearly benefited from advances in mechanization. The age-old sickle still cut 90 percent of the wheat harvest in southern England in 1790, being only slowly replaced by the scythe; and around 1900, when the sheaf-binding harvester was the leading technology in England, the scythe was still reaping most of the cereals on the Continent.[182] Steam-powered machines threshed most of the harvest from the 1880s in England, but only much later elsewhere. In 1892 the first tractor went into batch production in the United States, though no more than one thousand were in use in 1914 (one million by 1930). In 1950 horses still accounted for 85 percent of traction in European agriculture.[183] Artificial fertilizer, first used on a large scale in Germany and the Netherlands, came to be taken for granted all over Europe only in the 1930s, a full century after Justus von Liebig's trailblazing discoveries. Full mechanization and rationalization of agriculture was a *twentieth*-century development also in Europe and the United States, and relics of premodern ways of using the soil still lingered there. From Scandinavia to southern Italy, many farmers practiced technologically simple forms of subsistence agriculture, sometimes even slash-and-burn techniques as in Africa. Wherever, as around the Mediterranean, the use of horses came up against a lack of pasture and winter fodder, the energetic input into agricultural production faced narrow constraints. And also in Europe there were cases of agricultural "decay." Spanish agriculture had never recovered from the fact that the agrarian expertise of the Jews and Muslims (the last of whom were expelled in 1609) had been treated with contempt and their irrigation systems allowed to fall into disrepair.[184]

Third, labor-intensive rice cultivation on irrigated fields in tropical and subtropical latitudes had for millennia been among the most productive forms of agriculture. It too acquired its finished shape in a long process, which came to an end only in twelfth-century southern China: in the words of Fernand Braudel, "the most important event in the history of mankind in the Far East."[185] Since a transcendence of the given limits of agriculture is possible only at the highest "traditional" level, some regions of Asia were candidates for such a leap forward. Agricultural revolution presupposes a high population density, a functioning market system, reasonably free labor, and a high level and wide dissemination of know-how. These conditions were present also in parts of southern and central China in the mid-eighteenth century. Other factors, however, were working against an independent Chinese or Asiatic agricultural revolution: paddy cultivation was able to absorb ever increasing labor inputs on a given area; there were scarcely any reserves of land that could be opened up, given proper incentives; the ecological costs of intensive agriculture were more plainly visible in China (or Japan and India) than in Europe; alternative job opportunities were lacking outside the villages; absentee landowners living in the cities had few motives to improve production on their leased-out lands; and in the nineteenth and early twentieth centuries there was only limited access to industrially produced fertilizer. In northern China, where the ecological conditions were less favorable than in the south, and where producers tended to prefer millet and wheat over rice, the extreme parcelization of landownership and minimal "economies of scale" presented major additional difficulties.[186] Large estates or farms serve no purpose in the cultivation of rice: centralized management brings little benefit; and power-driven machinery—apart from small diesel and electric pumps, which first began to raise productivity in Japan in the 1910s—has very limited application in rice fields or tea gardens.[187] There is also little scope for soil-preserving crop rotation, since terraced pools can scarcely be used for anything other than rice and carp.

All this means that it would be unrealistic and inappropriate to use the Dutch-English "agricultural revolution" as a yardstick for an ecologically and socially quite different form of agriculture. In various parts of Asia at various times, that model reached a point at which it became difficult to feed an often-growing population. When that critical point was reached, however, also depended on *external* circumstances. Paddy cultivation in southern China, for instance, was part of a wider production complex, which included fish farming, tea growing, and silkworm breeding. From the early eighteenth century on, tea and silk were highly dependent upon the export trade, and the collapse of China's foreign markets (first for tea, then for silk) when Indian and Japanese competition emerged in the late nineteenth century was a decisive factor in the acute crisis of Chinese agriculture that many Western observers described in the 1920s and 1930s.

Fourth, the model of the English agricultural revolution did not spread through the West in the same way as the industrial mode of production, which could find a niche in the most diverse contexts. Agriculture is more bound up than industry with particular ecological conditions and much more tied to traditional social structures that are not easy to overcome. Certainly agricultural performance displayed wide variations. Only a few countries in continental Europe made spectacular gains in crop yield and productivity—by Germany first of all (where grain yields per hectare rose by 27 percent in the first half of the nineteenth century[188]), then by Denmark, the Netherlands, and Austria-Hungary, but scarcely at all by France, the largest agrarian economy of Western Europe. The absolute figures for output show a similar pattern: the grain harvest grew between 1845 and 1914 by a factor of 3.7 in Germany but only 1.2 in France.[189] One peculiarity of Europe and North America in comparison with Asia and large parts of Africa is the mixed economy of agriculture and livestock farming. In nineteenth-century Asia the distance between agriculture and (often nomadic) livestock breeding was still greater than in Europe—an important point, given that Europe's better integration of the two helped it to attain especially high productivity increases.[190] A country like Denmark managed to achieve its own quite distinctive agricultural revolution by specializing in animal farming. Butter, cheese, and bacon can also be a road to riches.

Fifth, the "pure" model of agricultural revolution, with the intensification of production at its core, assumes that improved performance is due primarily to increases in labor productivity and only secondarily to the expansion of arable land. In England and Wales, the acreage of farmland and pasture increased by nearly 50 percent between 1700 and 1800, but only by an insignificant amount in the following hundred years.[191] The great gains of the nineteenth century came rather from the *extensive* growth of production in frontier areas of the Tsarist Empire, the United States, Argentina, and Canada, as well as in India.[192] This expanding production of food staples had consequences that intruded far into political history. Two are particularly worth mentioning here. On the one hand, the adversaries of the Central Powers gained a decisive advantage in the First

World War from their ability to mobilize the far superior agricultural potential of North America and Australasia.[193] A lack of global political judgment led German leaders of the time to overlook this key factor.

On the other hand, agriculture had already become a central field of political conflict in a number of countries. This had only partly to do—as it has long been claimed for Germany—with the preponderance of authoritarian aristocratic elites, since the problem was similar in countries like the United States or the Netherlands where such elites were not a significant feature. By the turn of the century at the latest (gradually after the Emancipation of 1861 in the case of Russia), the intensive *and* extensive advances of agriculture had led on both sides of the Atlantic to the rise of an agrarian capitalism that employed wage labor and was highly export oriented. The crisis of world agriculture that began in 1873 and lasted for two decades was exemplified by the falling prices for agricultural goods and the less sharply falling, or even slightly rising, wages for farm laborers in line with pay increases in the cities. In this situation, large estates were often less capable of survival than smaller production units consisting essentially of members of one family. As the landowners' income fell, they asserted their interests ever more vociferously within the political system by calling, above all, for protective tariffs on agricultural imports—a campaign that was especially successful in Germany but less so in Britain or the United States. The prominence of agrarian issues in public debate, and of agrarian-romantic themes in cultural life, concealed the slow decline in the weight of the rural sector in several growing national economies of the West.[194] Elsewhere, in countries that had not seen the development of agrarian capitalism and where rural interests were represented in the political system by urban rentiers remote from village life, agrarian issues remained more or less out of sight. This was the case in the Ottoman Empire and Japan. But the most surprising silence was in the world's largest peasant country: China. It is a striking fact that in the whole discussion on reform, which began after the end of the Taiping Revolution in 1864 and grew more intense after the Sino-Japanese war of 1894–95, there was almost never any talk of the peasantry. China's public discussion was blind to one of its most pressing problems.

8 Poverty and Wealth

Poverty and Modernity

With the exception of the kind of utopian visions that exist in many civilizations, people before the nineteenth century never doubted that poverty was part of the natural, divinely ordained scheme of things. Classical political economy from Thomas Robert Malthus to John Stuart Mill, pessimistic in its basic mood, was not confident that modern capitalism was bringing a qualitative rise in productivity, or that the "uplifting" of the poor was possible except as the result of individual effort. But there was also a more optimistic school of thought that did not take poverty for granted and insisted that it could be overcome. The pioneers were two late Enlightenment thinkers: Tom Paine and the Marquis de Condorcet. Writing independently of each other in the 1790s, both formulated the idea that poverty was unacceptable in the modern world; that it should not be alleviated with alms but conquered through redistribution and development of the productive forces; and that society should help those who were unable to help themselves. Ever since Paine and Condorcet, two revolutionaries eventually killed or forgotten by their revolutions, the Western world has in principle regarded poverty as a scandal.[195]

Poverty and starvation are closely related to each other, without quite being mutually reinforcing. Although the poor may lack everything else, the last thing left to them is enough food to keep body and soul together. Not all poor people go hungry, and not all starving people are poor. Poverty as a concept embraces more. Societies have their own definitions of "the poor"; people who are not poor engage in discourse about those who are and make them recipients of their charity. In comparison with developed industrial societies, *all* premodern societies, whatever their cultural characteristics, were poor. But modern economies have not ended poverty—which is one reason why the achievements of "modernity" should not be celebrated too smugly. In the early twenty-first century there are still famines and hunger revolts in Africa and Asia; every sixth person on the earth is persistently undernourished. The increase in the productive forces of society in the nineteenth century—mainly the increases in agricultural productivity and the opening up of cheaper sources of fossil energy—did not as a rule go hand in hand with more equal opportunities in life.

Poverty and affluence are relative terms, both within one single society and between different societies. For example, an individual country—say, mid-nineteenth-century England—may become richer as a whole, but at the same time, if differences in income, consumption, and educational opportunity between the top and bottom layers of society become greater rather than smaller, relative poverty will become more evident than before. Long-term tendencies of income distribution are hard to identify in Western Europe (in spite of particularly good data), and even harder for the rest of the world. "Optimists" and "pessimists" have long stood in irreconcilable opposition to each other. There is much to suggest that, at least in England and France, the income and assets gap opened wider around the 1740s and gradually began to close again only a century later. In particular, the gulf between the big bourgeoisie and the class of manual workers broadened during that period. In many countries, the last third of the nineteenth century was a new era of narrowing differences. This is also consistent with the simple theoretical observation that the growth processes of "high industrialization" were driven not by working-class "underconsumption" but only by an expansion of mass demand.[196] This did not mean, of course, that the rich grew poorer.

The Rich and the Superrich

The richest were not immune from illness or misfortune. They ate better, enjoyed better clothing and housing, freed themselves of physical work, were able to travel more easily, and had unobstructed access to high culture. They lived in a world of luxury and, through their public conduct, set the norms of consumption to which others aspired. Whether in Europe, North America, or South Africa, the capitalist process created a degree of private wealth that in earlier ages had been attainable only by political and military rulers and by very small numbers of patrician merchants. In the year 1900 the rich were nowhere as rich, in absolute or relative terms, as they were under the conditions of capitalism. In some European countries, landowning aristocracies preserved their wealth from former times. At the end of the nineteenth century, the highest echelons of the English and Russian nobility (including many ennobled merchants) were still among the wealthiest people in the world. The Austrian, Hungarian, and Prussian (mainly Upper Silesian) nobility followed some distance behind, whereas the French had never really recovered from the Revolution of 1789–94.[197] Such wealth could be best preserved if it was invested not only in well-run landed estates but also in more modern sectors such as banking, mining, and urban real estate. At the same time, huge new fortunes had been amassed in finance and industry, and especially in Britain these nouveaux riches aped the lifestyles and symbolic displays of an aristocracy that did not constitute a closed caste but was separated from lower strata by fine shades of status. Old and new money, lords, knights (who had to be addressed as "Sir"), and untitled millionaires shared a world of sumptuous townhouses and country estates, a world inhabited by no more than four thousand people.[198]

In the pioneer societies of the New World and Australasia, nearly all the great fortunes were of capitalist origin. There were no feudal roots, even though many landowners in British North America could effortlessly mimic the grand lifestyle of wealthy English gentry. A distinction must be drawn, however, among the various "new Europes." In the antipodes few became spectacularly and lastingly rich as a result of either the gold rush or sheep farming. Although in 1913 Australia had a per capita income appreciably higher than Britain's and was even slightly ahead of the United States, it had few huge fortunes, and even the largest among them were considerably more modest than in Britain or the United States. There were more superrich people in Canada, but the real historical exception was the United States. When Alexis de Tocqueville traveled there in 1831–32 and felt himself to be fundamentally in a society of equals, he underestimated not only the ongoing formation of *very* large fortunes but also the widening of income differentials within American society—a process, it is true, that only later historical research uncovered. The growth and concentration of wealth was giving rise to prosperous oligarchies, both in the Northern states and among the planters of the South. The "self-made men" of the middle of the century, who used to be readily thought of as anti-oligarchic levelers, inserted themselves into this elite cosmos.

After the end of the Civil War in 1865, the elite divisions between North and South disappeared over time, while the transition began to a mature, high-growth industrial economy able to benefit from nationwide economies of scale and unparalleled corporate opportunities for capital accumulation. The richest tenth of the population owned one-half of the national wealth in 1860, but two-thirds by 1900; the top 1 percent of families held 40 percent.[199] Income inequality reached a peak between the turn of the century and 1914. The conviction of the founding fathers, especially Thomas Jefferson, that republican virtue required limits to material inequality continued to have some resonance into the 1880s, but then a new free-market ideology bestowed on boundless capital accumulation a legitimacy that would be occasionally questioned but never radically combated in the politics of the United States.[200] Extreme wealth even became one of the symbols of America's nascent world supremacy. The Astors, Vanderbilts, Dukes, and Rockefellers put Europeans in the shadow with their fabulous riches, staging a degree of luxury consumption that became known throughout the world.

Re-creations of English country estates, French châteaux, or Italian palazzi, filled with priceless works of art from the Old World, were the most visible display of the new superwealth; nor was there any problem endowing universities, thereby helping them achieve a position among the most highly regarded in the world. The first rank, and even the second rank, of American property owners were able to marry effortlessly into the European upper nobility: Consuelo Vanderbilt, for instance, having a share in an inheritance worth $14 billion, married the financially tarnished Ninth Duke of Marlborough and became the mistress of Blenheim Palace, one of the largest palaces in Europe. Around the turn of the century, a generation that had inherited wealth from their industrialist parents also appeared on the scene: those champions of luxury consumption who were the subject of the sociologist Thorstein Veblen's *The Theory of the Leisure Class* (1899). Nevertheless, ancestry was not altogether unimportant in the United States. The cream of the cream—assuming they had managed to preserve and multiply their riches—came from old families that went back to colonial times in cities such as Charleston, Philadelphia, Boston, and New York. People referred to them as "aristocratic," without implying that they occupied a fixed position at the top of a hierarchy. The US Constitution of 1787 made no provision for noble titles to be conferred on American citizens, and public officeholders, at least, did not accept foreign titles. "Aristocracy" was a metaphor for high prestige maintained across generations, and for a lifestyle expressing unshakable confidence in good taste that need fear no comparison with the summits of European *noblesse*. The American aristocracy, perhaps four hundred strong in late-nineteenth-century New York, could exude an exclusiveness and self-confidence that made even wealthier tycoons and captains of industry mildly aware of their parvenu status. Old and new money sometimes competed for political power in a city, but in the game of distinction even a weakened patriciate was usually able to keep its nose in front.[201]

The size of the top American fortunes was without precedent in the history of the world. Never before had private individuals accumulated such wealth. The money that could be made from oil, railroads, and steel in the late-nineteenth-century United States was several times greater than that which even the most successful European cotton industrialist could achieve; in fact, very few pioneers of the English Industrial Revolution had become truly rich.[202] The megarich looked down on those who were merely superrich. Thus, when the banker John Pierpont Morgan left a fortune of $68 million in 1914, the steel magnate Andrew Carnegie is supposed to have remarked pityingly that he had by no means been "a rich man."[203] Carnegie's own fortune and those of industrialists like John D. Rockefeller, Henry Ford, and Andrew W. Mellon were over half a billion dollars. The rapidity of the concentration of wealth may be gauged from the fact that the largest American private fortunes grew from about $25 million in 1860 to $100 million twenty years later and $1 billion two decades after that. By 1900 the richest man in the United States had assets worth twelve times more than those of the richest European (who was a member of the English aristocracy); not even the Rothschilds (finance), the Krupps (steel, machinery, weapons), or the Beits (British/South African gold and diamond capital) were in the same league.

The unique megafortunes in the United States are explained partly by factors such as the size of the internal market, the relatively high starting point for the economy, the wealth of natural resources, and the absence of political or legal obstacles to capitalist development. In addition there were synergistic effects within the industrial system. Rockefeller became a very rich man only after the emergence of the US car industry had presented his oil company with golden opportunities. Agricultural property did not stand behind any of the top American plutocrats, and in Britain too, by the 1880s, it was no longer land but finance, press ownership, or the gold and diamond trade that accounted for the largest fortunes. On the other hand, urban real estate was much in demand as a capital investment.[204]

Throughout the "West" (with the possible exception of Russia), the 1870s witnessed the birth of a "new," hierarchically differentiated, wealth. Beneath the megafortunes lay a stratum consisting of mere millionaires or half-millionaires. This elite had a different cultural style: Old Money began to complain about the New Rich, who flaunted their wealth in a vulgar or unthinking fashion and hollowed out aristocratic manners by dint of imitation. And something else was new. In the 1830s or 1840s, rich people with democratic or even radical political views had existed in the United States under President Jackson, in France during the July Monarchy, in England after the Reform Bill of 1832, and in pre-1848 Germany. But now, by the 1880s at the latest, the classic plutocracy of the fin de siècle had come into being. Political liberalism was largely divided within itself, as wealth virtually implied the representation of moneyed interests by conservative and right-leaning liberal parties. By no means were all the rich and superrich, in either Europe or the United States, vociferous propagandists for conservative values. But "radical plutocrat" had become a contradiction in terms.

Wealth in Asia

As in the United States, scarcely any large fortunes in Asia went back further than two centuries at the outmost. The conditions for the formation of private wealth were different from those in Europe and the neo-Europes. In China there had been no hereditary landowning aristocracy before the Manchu conquest of 1644, and large estates had generally been atypical; education more than property had been the qualification for elite membership. One could become prosperous in the state service, but not spectacularly wealthy, and few managed to keep their wealth in the family for many generations. The richest people in Qing China in the eighteenth and early nineteenth century were either members of the high Manchu nobility (e.g., princes living in Beijing in palaces arranged around a progression of inner courtyards[205]), merchants holding a monopoly from the state (salt, the Canton trade), or bankers from Shanxi province. They were joined in the nineteenth century by middlemen trading within the treaty-port system, the so-called compradors. The social prestige of merchants was far below that of scholar-officials, but they could indulge in luxury consumption that the latter deprecated as parvenu behavior, while also taking care to use their money to acquire landed property, purchase titles, and educate their sons. The dynastic accumulation of great fortunes was uncommon; it was more likely to occur among Chinese merchants, tax farmers, and mine owners in colonial Southeast Asia—in Batavia, for example, where ethnic Chinese had been active in the economy since the early seventeenth century. In 1880 the Khouw family, whose forebears had migrated there from China in the eighteenth century, were one of the largest landowners in and around Batavia and lived grandly at one of the best addresses in the city.[206] In China itself, wealth tended to be kept secret so that the envy of the authorities would not be aroused; manorial architecture—the most conspicuous expenditure of the European aristocracy and their American imitators—played scarcely any role. In late imperial China, "rich people" were not a model for the rest of society. Moreover, although the Manchurian imperial family occupied the largest palace complex in the world, the riches "belonged" to an imperial clan of several thousands rather than to a royal family of ten to twenty people.

In Japan, with its very different social structure, the outcome was similar. The aristocratic samurai, though sharply differentiated from "commoners," were seldom rich in a European sense: most lived on hereditary stipends awarded by their feudal prince (*daimyō*), who alone was entitled to raise taxes in his domain, and on low salaries for administrative duties. The objective impoverishment of many samurai, and even more their subjective experience of it, fostered discontent with the Edo ancien régime that found political expression in the 1860s in the Meiji Restoration.[207] Yet, in a way unknown in the rather austere Chinese empire, the Edo period was, until the end, one of conspicuous consumption. In a Japanese variant of the "royal mechanism," which Norbert Elias analyzed for the court of the Sun King, the real rulers of early modern Japan, the House of Tokugawa, tamed the territorial princes by compelling them to spend regular periods at the shogun's court in Edo (Tokyo). Edo was a great stage on which the princes and their entourages competed with one another to display the most glamorous buildings, festivals, gifts, and concubines. Many a thrifty prince, though aware of the impact on his finances back home, was driven to the brink of ruin by this contest of competitive splendor. Most of their treasuries had little left in them once the samurai stipends and the costs of running a court had been paid out.[208] Few large aristocratic fortunes therefore survived into the Meiji period. The feudal princes disempowered after 1868 lost their lands in return for a degree of compensation, while samurai status was abolished within just a few years. After 1870 Japan was a much more "bourgeois" country than Prussia, England, or Russia. Fortunes acquired through industrialization (some on the basis of merchant wealth from the Tokugawa period) did not constitute an upper class of "the rich," and private ostentation was also discreetly limited. It was considered improper to show off one's wealth in the shape of ostentatious private buildings, for example.

In South and Southeast Asia, wealth was traditionally in the hands of princes. The European colonial invasion narrowed the scope for enrichment, both in their case and in the case of court aristocracies. At the same time, it opened up new opportunities in commerce. Some Bengali merchant families, for example, amassed large fortunes after 1815, as did a number of cotton manufacturers in western India after 1870. In many places in Asia and North Africa, corporate assets had an importance similar to that of church property in Europe before the Reformation and the French Revolution. Clans and lineages, temples of various kinds, Buddhist monasteries, Muslim holy shrines, and pious foundations (*waqf*) owned and leased out land that was safe from state exactions, or controlled and multiplied large sums of money.[209] In the eighteenth and nineteenth centuries, private accumulation often occurred in the hands of religious or ethnic minorities that possessed extensive business networks: Jews, Parsis, Armenians, Greeks in the Ottoman Empire, and Chinese in Southeast Asia.

We know as yet too little about the financial circumstances of such merchant dynasties—or of Indian maharajahs, Malayan sultans, Philippine landowners, or Tibetan monasteries—to draw a substantive comparison with Europe or the United States. One thing is clear: these elites lived a life that was between comfortable and luxurious. But nowhere in Asia was Western-style aristocratic or upper-bourgeois wealth taken as a model, and apart from Indian courts and Japanese princely homes in Edo before the mid-nineteenth century, displays of luxury consumption were of less significance. This was not simply because Asian societies were poorer; material success in general had less of a function in guiding their cultures.

Types of Poverty

At the bottom end of the social ladder, the differences among the poor appear at first sight not to have been very great. On closer examination, however, all possible distinctions open up. In 1900, the pioneer social researcher Charles Booth identified five categories in London alone among the less "well-to-do." The decisive qualification for prosperity was the regular employment of one or more domestic servants, even in rented accommodation. From there it was a long way, through gradations of "shabby gentility," to outright poverty. If the rise of rich and superrich capitalists gave the nineteenth century a special place in the history of wealth, how does it appear in the history of poverty?

Poverty and wealth are relative, culturally specific categories. In sub-Saharan Africa, for example, the ownership of land was a far less important criterion than control over dependent persons. Many rulers in precolonial Africa had scarcely more storable wealth than their subjects. They stood out by the number of their wives, slaves, and animals, and by the size of their granaries. Wealth meant access to manpower that allowed the leap into conspicuous consumption and lavish hospitality. In Africa the poor were people whose situation in life made them especially vulnerable, and who had little or no access to other people's labor. The poorest of all were the unmarried and childless, especially if some physical disability made them unable to work, and doubtless also slaves (even if they were often well fed). Some African societies had institutions that provided a poverty net, but others (Christian Ethiopia among them) lacked anything that could be described as such. A precolonial "caring Africa," with a comprehensive community life, is a romantic myth.[210] The higher value given to control over people rather than ownership of land was not a peculiarity of Africa, since wealth is generally seen in terms of access to scarce resources. Thus, the status of Russian magnates before the emancipation of the peasantry in 1861 was measured more by their serfs or "souls" than by the size of their estate, and around the same time in Brazil the importance of a landowner depended on the number of his slaves. In early nineteenth-century Batavia, no European who wanted to count as somebody could afford to arouse the suspicion that he was skimping on the number of his black slaves.[211]

In societies of herdsmen—not only in Africa but also in West Asia, from Anatolia to Afghanistan or Mongolia—wealth was measured by herd size. The mobile way of life excluded the amassing of treasure as well as investment in buildings made to last. European conceptions of poverty and wealth apply to no one less than they do to nomads. This continually gave rise to the cliché that they were especially deprived, as many travelers reported from trips they made among African herdsmen, Mongols, or Bedouins. What is true is that a nomadic existence was (and is) especially prone to risk. It came increasingly into conflict with the interests of farmers and was exposed to the hazards of drought and food shortage. Herdsmen were the first to suffer in lean times: those who lost their herd no longer had any means of subsistence and were unable to pick up again after the end of a drought.[212]

In southern Africa, already before the First World War, poverty began to take on a form familiar from the densely populated societies of Europe and Asia: landlessness more than physical disability became the main cause of material deprivation, typically resulting from the state-supported takeover of land by settlers. Cities played a rather different role here, though. Whereas in Europe, at least during the first half of the nineteenth century, poverty was more visible and perhaps also greater in the town than in the country, African poverty was (and still is today) "made" above all in rural areas. It is likely that slumdwellers in Johannesburg felt better off in comparison with their relatives in the country. Extremes of structural poverty were found less among physically capable male migrant workers in the cities than among family members who remained behind in areas that, until the 1920s, were often still difficult to reach with famine relief. Nevertheless, there was an advantage in maintaining links with relatives in the country: the poorest sections of the population in Africa's growing cities were those for whom it was no longer an option to return to their village in times of crisis. There is little evidence, in large parts of the world such as Africa and China, that the lives of "the poor" improved to any noticeable degree in the course of the nineteenth century.

Poverty became most firmly entrenched in cities that displayed the full spectrum of income groups—from beggars to ultrarich manufacturers, bankers, or landowners. In any case, social research was still in its early days, and profiles of income and living standards were developed only for urban areas. In the English cities a turning point was reached around 1860 when the diet of the lower classes gradually improved and the proportion of people in the worst housing situation (statistically, more than two adults per bedroom) began to fall, partly as a result of the development of new working-class suburbs. But even in one of the richest countries in the world, destitution among the urban lower classes by no means disappeared. The number of males fit for work living in British workhouses is a good indicator of the scale of *extreme* urban poverty—and between 1860 and the First World War there was no significant drop in this total. The same is true of the figure for those classified as "vagrants."[213]

It is impossible to quantify global poverty for the nineteenth century. We rarely have any insight into proportions between Europe and other civilizations. Measuring income is scarcely ever possible in the case of the very poor, even in the cities. A minimum of data exists only where wages were paid, and actually recorded, at the bottom of the income ladder. Then we learn, for example, that between 1500 and 1850 the *real* wages of unskilled construction workers in Istanbul, the Muslim metropolis on European soil, followed the general trend in big cities to the north of the Mediterranean. They fell behind it only after 1850.

According to another estimate, shortly before 1800 the real wages (measured in wheat equivalent per day) of workers in Istanbul and Cairo exceeded those paid in Leipzig or Vienna and were significantly higher than in southern India or the Yangtze delta.[214] It is important not to assume a *general* superiority of "Europe" over "Asia." One has to differentiate according to region, type of work, social position, and gender. Toward the end of the eighteenth century, living standards of unskilled male workers in London or Amsterdam were already significantly higher than in the big Chinese cities, and that gap widened enormously during the nineteenth century. The contrast is less stark when we compare the more developed parts of China to those regions of southern and eastern Europe that remained untouched by industrialization.[215]

Begging and Charity

The gradual emergence of a welfare state in Germany and certain other European countries toward the end of the nineteenth century should not obscure the fact that in many parts of the world this was also an age of continuing, and freshly motivated, philanthropic efforts on behalf of the poor. There are many cases in Europe where poor relief funded by local authorities went hand in hand with private charity; the mix of the two varied, as did the motives behind them. In the Tsarist Empire, for instance, there was nothing that might be described as a public *system* of poor relief (such as existed in England under the Poor Laws, until their abolition in 1834); the altruism of large landowners and state officials, hardly on a large scale, stemmed partly from a wish to emulate Western European models of social commitment.[216] Contrasting examples outside Europe come mainly from philanthropic orientations in the Muslim world. In Egypt an ancient tradition of munificence persisted, not in ostentatious displays (which Islam prohibited) but out of the public view. It was a moral obligation that was often taken over by charitable institutions. This distinctively Muslim practice caused many European observers to tell stories about rich beggars. But in Egypt too, the nineteenth century saw the state increasingly assume the task of helping the poor.

One should not exaggerate the differences between Western Europe and North America, on the one hand, and the Muslim world on the other. In neither was there a linear development of a welfare state; family or community forms of aid coexisted alongside new state institutions. The greater failure of the Egyptian state, compared to "the West," to stem begging in the cities had to do with the public tolerance shown toward beggars (as in Tsarist Russia). Of course, Egypt differed in many respects from northern Europe: (1) its lower level of economic development meant that fewer resources were available to the state for poor relief; (2) its poorhouses were used as temporary accommodations, never as English-style workhouses; (3) poor relief acquired a colonial dimension when missionaries appeared on the scene, and when the British, after the occupation of 1882, started up some rather meager initiatives; and (4) the poor never disappeared from the public arena but vigorously asserted their claims—unlike the urban lower classes in England, for example, which from the 1860s on regarded poor relief and especially begging as shameful and demeaning.[217]

An absence of begging is very rare in history, and it was probably never attained before the twentieth century. We should bear in mind that in the nineteenth century begging was still seen as a normal part of social existence. It has always been a fairly precise indicator of poverty or even destitution, but also something else: a special kind of parasitic economy, often with a complex (in China even guild-like) organization and usually tolerated within limits by the authorities. The Victorian term "underworld" is here seldom apposite. In nineteenth-century Europe too, the social type of the penniless outcast, halfway between Franz Schubert's "hurdy-gurdy man" from his *Winterreise* (1828) and Charlie Chaplin's déclassé tramp (created in 1914), had not yet been rationalized away or pinned down in the categories of public welfare services. The struggle for existence at the lower depths was still visible.

9 Globalized Consumption

In both town and country, extreme poverty may be defined as a state of constant undernourishment. Beyond the threshold of a hunger that does not kill but does not abate, the range of variation is not as great as in other areas of consumption. The rich man whose monthly income is a hundred times greater than the poor man's is not a hundred times better nourished. As Fernand Braudel has shown, the differences between the culinary systems of various civilizations have greater importance than the vertical ones running within their respective societies.[218] The tables of the well-off were more diversified, fresher and more nutritious, and usually supplied by professional cooks but as a rule existed within one and the same culinary system. From the point of view of global history, therefore, only a few generalizations can be made.

The greatest interaction between the eating habits of continents occurred as long ago as the sixteenth century, when a "Columbian exchange" introduced European crops and animals into the New World and American crops into Asia and Europe.[219] Nor did this early modern transfer concern only rare luxuries: it changed the agricultural and garden economy, with huge effects on productivity and consumption habits in many parts of the world. The potato, which arrived in Europe shortly before 1600, took roughly two hundred years to become the main food staple in countries such as Germany, the Netherlands, or Britain. Much earlier still, the appearance of rice strains with a higher yield had considerably increased production in Southeast Asia and China. At the same time that the potato crossed the Atlantic, the sweet potato traveled from Manila to China and immediately became a tool of famine relief, while corn, tobacco, and groundnuts were introduced into the Middle Kingdom, and the chili pepper, today central to the cuisine of Sichuan and Hunan, was brought from the New

World. Once all these novelties had been absorbed in the space of a few decades, China's culinary system underwent no further major changes.[220]

The American manioc root became native in areas of Africa under Portuguese influence, and in the last third of the nineteenth century both indigenous and colonial initiatives helped to spread it to many other parts of the continent. Today it is by far the most widespread edible plant in the tropical countries of Africa. Centuries after plants of American origin first crossed the oceans, they were driven by new needs and applications to enter common use in the Old World. One example of this is the groundnut, probably first domesticated in Brazil and widely used in Inca Peru. It was introduced into China and soon became the main source of frying oil there. Then, in the nineteenth century, it was grown in the United States as animal fodder, before people realized that it could take the place of cotton in plantations devastated by pests. Nowadays the groundnut is firmly integrated into a number of Asian and West African culinary traditions, and over time groundnut oil has come to be appreciated in Europe too for its ability to withstand high temperatures. All in all, the use of tropical oils was one of the most important acquisitions of the nineteenth century, not only in cooking but also for soap and cosmetics.[221] The huge expansion of the international agrarian trade made tropical produce available even where it could not be acclimatized to local conditions.

Culinary Mobility

Culinary systems differ in respect to the innovations they acquire. The situation was clearest in countries such as the United States, where virtually all eating habits had to be imported. New tastes arrived on its shores with the great migrations of the nineteenth century: Italians were present in California from the midcentury gold rush on and were soon migrating from Italy to other parts of the United States. They brought with them durum wheat, the basis for pasta dishes. The international spread of Italian cuisine thus began long before the worldwide triumph of the pizza.[222]

The geography of dietary influences does not coincide with the distribution of political and economic power. The Chinese, for example, who in the sixteenth century had already demonstrated their willingness to learn from others, and who were politically much weakened by the forcible opening of their country in the Opium War, did not lose confidence in their own culture. At first they saw no reason to adopt Western influences in their cuisine. This changed slightly after 1900, when three "white" products from the West (produced in China, and often by Chinese companies) gained considerable popularity in the cities: white flour, white rice, and white sugar. A few European restaurants opened in the 1860s in the big cities, and beginning in the 1880s, a visit to one of them in Shanghai—complete with white tablecloths, silver cutlery, and "Western-style Chinese cooking"—became a demonstrative statement on the part of wealthy Chinese families. In general, however, affluent Chinese continued to show unusually scant interest in Western food and Western consumer goods in general.[223] Japan, which in many other respects proved extremely receptive to the West, adopted few culinary loans in the nineteenth century; one major exception was the increased consumption of meat.

On the other hand, since the time of Marco Polo numerous European travelers, missionaries, and Canton-based merchants had become familiar with Chinese cuisine and written reports about it. After the opening of China in the 1840s hundreds of foreigners made its acquaintance in the restaurants of the treaty ports and in the offerings of their private cooks. Those unable or unwilling to eat it regularly spared no cost or effort to keep themselves supplied with European foods and delicacies. Outside China there was for a long time no opportunity to taste Chinese dishes. Scarcely any Europeans or Americans ever ventured into their local Chinatown to try out the fast-food booths or diners used by émigré workers. Mark Twain, in his time as a journalist, was one of the first Westerners to describe the experience of eating with chopsticks outside Asia. The first Chinese restaurant to appear in Europe, in 1884, could be visited as part of a health fair in South Kensington; Sir Robert Hart, the powerful Irish inspector-general of the Chinese Imperial Maritime Customs, was responsible for the attraction. But China's gastronomic success with Western consumers still lay in the future. It began gradually in 1920s California and did not become a global phenomenon until after 1945.[224] As to Western food, it was only in the last third of the twentieth century that it began to have a marked influence on eating habits outside the luxury hotels and Western enclaves of East Asia, and then in the form of mass-produced industrial items.

At the end of the nineteenth century, "colonial goods" were appearing more often in European food stores. In London and the large provincial cities of England it had been possible throughout the eighteenth century to buy cane sugar, tea, and other exotic produce in a number of specialist locations;[225] nowhere else in Europe did food and delicacies from overseas play such an important role. The East India Company had made the British a nation of tea drinkers, especially after the duty on tea was sharply reduced in 1784. By 1820 they were consuming thirty million pounds of tea per annum.[226] The only other exotic import that changed habits outside the narrow circle of luxury food and drink was sugar. Already in the seventeenth and eighteenth centuries, the demand for cane sugar had set in motion the dynamic of the Caribbean and Brazilian plantation economy and the transatlantic slave trade. But only in the late eighteenth century, not least as a sweetener in tea, did it reach the level of mass consumption. The real expansion, however, took place in the nineteenth century: world sugar production doubled between 1880 and 1900, and doubled again between 1900 and 1914.[227] The share of sugar in the average caloric intake of Britons is thought to have increased from 2 percent to 14 percent in the course of the century. As the anthropologist Sydney W. Mintz has argued in an influential book, sugar actually became a food for the poor, a quick energy boost for the flagging labor force of industrial Britain.[228] This popularity of sugar was possible only because its real price was continually falling in retail outlets.[229]

Sugar can be produced only as cane in the tropics and only as beet at temperate latitudes. Salt, by contrast, can be extracted by various methods and is therefore more closely associated with particular localities. The same applies to livestock breeding, which like slaughtering was a local trade; the limited durability of meat in its fresh state was enough to ensure this. One of the major food trends of the nineteenth century was the industrialization of meat production, soon turned into a transcontinental business. Average meat consumption had slowly declined in early-modern Western Europe, and this trend persisted here and there into the nineteenth century, sometimes disguised as falling standards and expectations: in really hard times the poor of Paris ate cats.[230] By midcentury at the latest, however, meat consumption was rising among the lower classes of Europe: English working-class families doubled their intake between the 1860s and 1890s to more than one pound per person per week.[231] The Japanese, who otherwise stuck to the Tokugawa cuisine, were converted in the Meiji period to the eating of meat. Although certain groups such as samurai and sumo wrestlers had indulged in it before 1866, it was only in the final third of the century that people more generally became convinced that the imposing strength of the West was due in part to meat consumption and that a vegetarian diet was unworthy of a "civilized" nation.[232]

Expanding demand caused cattle stocks in Europe to grow faster than the human population between 1865 and 1892, while at the same time cattle breeding developed in the western United States, Canada, Argentina, Paraguay, Uruguay, Australia, and New Zealand. In 1876 beef was sent by refrigerated ship to Europe for the first time, and in the 1880s the new technology made it possible for Argentina and Australasia to export meat in large quantities.[233] After 1900, as more and more of the colossal US output was absorbed by its internal market, Argentina became the world's largest meat exporter.[234] The immediate reason, however, was the wish of the British government to supply canned and frozen meat to its troops fighting in the South African War. The real and lasting boom in Argentine exports to Europe began only in 1907, when American meatpacking companies with better deep-freeze technology took over the trade. It was the first important investment linkup between the United States and Argentina, which until then had belonged more to the British sphere of economic influence. On the other hand, access to the US market continued to be denied to Argentine producers.[235] Romantic social types such as the American cowboy or the Argentine gaucho were the mobile proletariat of a global meat industry.

The ranchers of the "Wild West" increasingly became suppliers of the giant Chicago slaughterhouses. The south of the city saw the rise of something that came to be one of its tourist sights: an industrial hell on earth for cattle and swine that flourished once the railroad was in full operation. Only the slaughterhouse districts of Buenos Aires, with their vast heaps of skulls and bones, were a shade more dramatic as animal necropolises. The industrialization of food production began during the American Civil War, when demand soared for the new powdered milk and canned meat. Chicago filled the gap for the Northern states, a second "porkopolis" next to Cincinatti. Its slaughterhouse complex could process 21,000 cattle and 75,000 pigs at the same time, so that by 1905 it had dispatched a total of 17 million animals.[236] It is no accident that one of the sharpest literary attacks on American capitalism, Upton Sinclair's *The Jungle* (1906), is situated in the Chicago slaughterhouses, which the author, using Zola's naturalist techniques, depicted as a Dantesque inferno. Quickly becoming a bestseller, the novel caused many readers to lose their appetite for meat, and demand took a temporary dip. It is possible that the average American in the Midwest was consuming 4,000 calories a day around the turn of the century, at a time when the intake per head in English working-class families was around 2,400 calories.[237] That age of meat surpluses, a new departure in the second half of the nineteenth century, gave rise to the American glorification of the steak, which had no parallel in any food culture other than that of Argentina.

Department Store and Restaurant

The industrialization of food production in the Western world—its beginnings can be dated to the 1870s in the case of Germany[238]—was correlated with other changes in society. The growing employment outside the home of working-class and lower middle-class women reduced the time available for household labor and increased the need for ready-made food. Such products could reach the final consumer only via translocal distribution systems. This presupposed—in addition to farm sales, periodic markets, and local butchers and bakers—the existence of grocery stores that, in turn, required wholesale dealers to keep them supplied with produce. But this new trend spread through Europe only right at the end of the century, with many gaps and much unevenness. In many rural areas, the supply of nonlocal produce remained throughout the period in the hands of peddlers and traveling dealers. In this respect the distribution mechanisms were not essentially different from those in China at that time, where periodic district markets operated alongside elaborate chains of middlemen. The passage from market to store (or sometimes consumer cooperative) was a necessary concomitant of the industrialization and internationalization of food production.[239]

The most spectacular innovation of nineteenth-century commerce was the department store. More than any other form of retailing it relied on standardized mass production of many of the goods on offer. Department stores opened up a novel commercial and social space, providing a stage for the world of commodities and enchanting the public with a kind of world's fair in miniature. The first such stores appeared in Paris in the 1850s. The philosopher and cultural historian Walter Benjamin made them a central theme (along with the famous arcades) in his analysis of the culture of French capitalism.[240] Paris was not a port city or international transshipment center like London or Hamburg or a center of industry like New York or Berlin. In France industrial mass production had not yet supplanted artisanal production to the extent it had in the United States; industry and crafts met up in the Parisian culture of consumption.[241] The great era of the Parisian arcades was the 1830s and 1840s, whereas the golden age of the department store lay in the Belle Époque between 1880 and 1914. The London department stores, which sprang up a few years after those in Paris, were even more uncompromising in their program of gathering all the necessities of life under one roof: not even a funeral department was lacking. Charles Digby Harrod built his store in the 1880s as a cross between a business and a club.[242] In New York the first department stores opened earlier than those in London, so early that a Parisian influence can be discounted. It was in 1851 that Alexander T. Stewart built a five-story Renaissance-style marble palace on Broadway and started an architectural rivalry in which newly founded cities such as Chicago soon took part.[243] The universal store did not, however, catch on at once everywhere in the developed world. The years between 1875 and 1885 were the launch period in Germany, when the Wertheim, Tietz, Karstadt, and Althoff families entered the fray, and architectural masterpieces such as the art deco Kaufhaus in the Saxonian town of Görlitz could hold up to comparison with stores in the largest cities. In Vienna, another great European center of consumption, it was only around the turn of the century that the department store overshadowed large stores with a more specialized range of goods.[244]

In Tokyo, department stores appeared toward the end of the Meiji period. A start was made in 1886, when for the first time one of the old silk stores also began to sell Western clothing. Subsequently, large stores witnessed many innovations: the city's first telephones were installed in them, and female assistants made their debut (traditionally only men stood at market stalls or behind counters). The first great Western-style shopping palace opened its doors in 1908. But there was also another novelty: the multistore covered market known as *kankōba* ("place for the encouragement of industry"), which fused the principle of the Oriental bazaar with that of the Parisian arcades, pointing ahead to the global "mall" of the present day. In the second decade of the twentieth century, however, fully fledged department stores supplanted these covered markets bazaars in Tokyo.[245]

Another innovation that one associates with the nineteenth century, the restaurant, is not actually a European invention. Rather, the evidence points to polygenesis of this type of commercial catering. Two attributes distinguish the restaurant from the manifold inns, taverns, and guesthouses that have existed since early times in numerous countries. On the one hand, it produced high-quality cuisine—previously a feature only of courts and elite private residences—and made it available to anyone who could afford to pay; it democratized fine dining. On the other hand, the restaurateur was an independent businessman, who offered a product and a service without the ties of a guild or corporation. A world in which food was not a biological need but an artistic passion came into being where it still has its center today: in Paris. But behind complicated issues of cultural history, there lies a relatively mundane process. The French Revolution destroyed the royal court with all its culinary splendor and threw out of work the private cooks of dispossessed aristocrats who had fled the country. A new supply thus became available to cater to a new market, as an urban bourgeoisie with sufficient purchasing power discovered the culinary arts. In the course of the nineteenth century, this public became increasingly international: one of the great attractions of the French metropolis for the new luxury tourism was its unrivaled gastronomy.[246] The rise of outdoor eating did not stop with expensive high-class restaurants but stretched all the way down to diners in working-class districts. Peculiarities of national culture also played a role. Across the Channel there were 26,000 fish-and-chip shops, which used a thousand tons of frying oil a week. The occasion when cod was first combined with potato strips in this way is lost in the mists of time, but it must have been at some point in the 1860s. The meal then developed into the favorite of the British working class, helping to shape its identity and symbolizing the national virtues on a plate.[247]

Good-quality commercial eateries certainly existed at an earlier date in China, and so the French claim to have "invented" the restaurant rests on shaky foundations. Private gastronomy blossomed in the late Ming period, essentially in the sixteenth century, when new mercantile wealth, together with a boom in foreign trade, led to a kind of embourgeoisement of large parts of urban culture. The burgeoning food culture managed to survive the upheavals of the seventeenth century, and reports and literary sources from the subsequent period testify to a varied culinary landscape that included public restaurants at any level of quality and price, from simple street grills to teahouses and specialized guesthouses to large banqueting halls. In the early modern period, China was much less segregated by estate or hierarchy than European or Japanese society; the boundaries between popular and elite culture were more permeable. Moreover, the urban residences of the rich, with their pavilions and inner courtyards, were more modest than the *hôtels* and mansions of the nobility in Paris or London. Top chefs could therefore enter the public domain earlier than in the West. What happened in France after the Revolution was by then a matter of course in China.

And Japan? There the beginnings of the restaurant date back to the eighteenth century. Until the nineteenth century Japanese society and culture remained strongly marked by status distinctions. The various kinds of restaurant therefore served more blatantly than in Europe, and a fortiori China, as social markers and upholders of distinctness. The first Chinese restaurant, an exotic creature altogether, opened in Japan in 1883, and Western ones were very few and far between. "Fine distinctions" were thus plainly visible in the gastronomic world.[248] In sum: the restaurant was a parallel invention in East Asia and Europe; the former was clearly in the lead, but there is no evidence that Europe took to the restaurant from China in the same way that it was inspired by Chinese horticulture in the eighteenth century.

Changed eating and consumption habits went together with new forms of marketing—a field in which the United States was the world leader, closely followed by Germany. The 1880s saw the birth and marketing of the branded product, with strategies planned like military operations. Singer's sewing machine and Underberg's herb liqueur in its characteristic bottle were present at the dawn of brand-centered marketing. It could develop because the serial production of articles of mass consumption was now a technical possibility. Whereas most consumers had previously been in the dark about where a product came from (unless they bought it directly from the producer), they were now surrounded by the names and logos of cigarette firms, soap producers, or canned soup manufacturers. Branding and the patent law were part of the new era of organized mass consumption.[249] No commodity embodied this watershed in cultural and economic history more strikingly than the sticky brown liquid that the chemist John Styth Pemberton launched in Atlanta on May 8, 1886 as a cure for hangovers and headaches: Coca-Cola. Sales rocketed from 1,500 gallons in 1887 to 6,750,000 gallons in 1913.[250]

Coca-Cola belonged to the first generation of industrial food and drink, which emerged in the 1880s in the United States and soon led to the founding of corporations in Europe too. The key products, from Heinz Ketchup to Kellogg's Corn Flakes to Lever's margarine, were all created in the laboratory. Branded goods rapidly spread around the world, so that by the early years of the new century the petroleum lamp burning oil from Rockefeller's Standard Oil Company, along with Western artificial fertilizer and cigarettes, could be found in remote Chinese villages. A further element in the new marketing complex, which was decisive in extending its reach, was the mail-order business. This, too, was an American invention: it seemed an obvious idea, given the size of the country and the isolation of many of its farms. Also essential was the expansion of the railroad, while the delivery of heavy packages by the United States Postal Service after 1913 made things simpler still.[251]

Does all this add up to a new "consumer society"? In the early 1980s historians rediscovered the consumer and thereby corrected, or fleshed out, a view of history that had focused too narrowly on the *productive* achievements of industrialization. The flywheels of human action, they showed, were oiled by needs and competition, hedonism and fashion. This is not only interesting for cultural history but also important for the explanation of economic progress. For only a sufficient level of demand could (and can) translate impulses to rationalize production into macroeconomic processes of growth. When did the consumer society begin? If by that we do not mean the same as the affluent society (in which nearly *everyone* pursues consumption as an end in itself), if we have in mind only the existence of consumption-oriented social strata beyond a tiny traditional elite, then eighteenth-century England undoubtedly qualifies as a consumer society.[252] Again, to be sure, we might ask whether China in the period from roughly 1550 to 1640 might not already be described as such a consumer society, and if the term might not be appropriate also for early modern Istanbul.[253] Clearly there was purchasing power among broad sections of the population outside the imperial court and officialdom. And contrary to the cliché that fashion was an eighteenth-century European fancy unknown in Asia at that time, it might be pointed out that the frequency of conservative-traditionalist complaints about the breakdown of morals is evidence of the extent to which official sartorial codes were eroded time and again.[254]

Hannes Siegrist has defined the ideal type of "consumer society" as follows: "Relatively high prosperity is not concentrated in a small elite. There is a minimum degree of civil equality and political rights, a broad middle class, social mobility and competition. A certain pluralism of values, diligence, a work ethic, and a striving for goods out of worldly but also partly religious motives are generally customary and understood to be legitimate. There is a division of labor and a degree of rationalization in agriculture, industry, and trade. The family is outwardly oriented in relation to work, professional life, and profit making; there is a well-differentiated institutional and legal system, rational knowledge that permits and fosters calculable and calculating behavior, a cultural apparatus that fosters understanding among the producers, procurers, and consumers of goods and that guides how buying and consuming are interpreted. Money functions as the general means of exchange."[255]

Most of the elements in this definition probably apply to late Ming China, although the country did not develop further in that direction and, like so many others, was overtaken by Europe in the nineteenth century. In Europe and North America, on the other hand, a long-term dynamic toward Siegrist's ideal type took shape. The extent to which this accentuated or flattened national cultural differences is an issue that was widely discussed in the twentieth century in relation to so-called Americanization. The most interesting aspect for global history is how far the rest of the world had already adopted Euro-American consumption goals and models in the nineteenth century. The answer to this question cannot be general but must proceed by way of examples.

The Creole elites of the new Latin American republics developed perhaps the strongest consumer orientation to Europe. British textiles flooded the region immediately after independence, and long before the arrival of the railroad, mule trains were carrying British cotton goods from port cities into the tablelands and high valleys of Mexico and Peru. Twenty or thirty years were enough to saturate the Latin American markets with British goods. Few imports passed through the cities to the haciendas and mines of the interior. The affluent elites, however, grew more and more accustomed to a European lifestyle. In the absence of local production, the prestigious symbols of Western progress had to be imported from England and Germany, Italy and France, and increasingly from the United States. The assortment ranged from machinery to French wine and English beer to coaches, spectacles, bicycles, and marble for the magnificent buildings of the rich. Gilberto Freyre considers that in the early nineteenth century, the rich in

Brazil tried to emulate the formerly despised Protestant heretics of Britain by wearing artificial dentures.[256] A small minority of Latin American consumers cultivated an ostentatiously European lifestyle, in which Spanish models usually played little or no role. From mid-century on, it became noticeable also in the appearance of a city such as Buenos Aires, with its shopping boulevards, grand hotels, *salons de thé*, and patisseries. The reorientation to European models went hand in hand with a new kind of racism: one switched custom from a baker of African origin to a genuine French pâtissier, and one's piano teacher, hitherto often black, was now brought out from Europe.[257] Meanwhile, social modernization passed by the majority of the population. Demand was funded increasingly out of the proceeds of Latin American exports to Europe (coffee, copper, guano, and so on).

Dress is always a good indicator of consumption preferences. In Latin America, especially in countries with a large indigenous population, society split into the peasantry who dressed as in colonial times and city dwellers for whom it was important to demarcate themselves from "uncivilized" fellow citizens. Mestizos, too, placed stress on sartorial markers, such as the polished leather shoe. Also in other spheres, the material cultures of town and country rapidly drifted apart. The identification of the Latin American upper classes with the civilization and commodities of England or France reached its peak in the Belle Époque, around the turn of the century. Equating progress with Europe, they were unreservedly prepared to interpret foreign goods as symbols of modernity. Their export *economies* were at the same time import *societies*, in either way occupying a peripheral position in the international order. Since the increasing prosperity did not rest upon domestic industrial production, the whole urban life of Latin America acquired a European stamp: not only clothing and furniture had to be imported but also the emblematic cultural institutions of contemporary Europe: the restaurant, the theater, the opera, the ball. Top chefs were enticed away from France, and in 1910, not a single indigenous dish was served at the official celebrations in Mexico to mark the anniversary of independence. In Lima golf and horseracing became an obsession. Railroad stations were built as exact copies of models in Paris or London.

The epitome of imitation was the wearing of heavy English men's clothing in tropical and subtropical zones. The British had already concluded that it was necessary in India. Around 1790 the governor-general Lord Cornwallis permitted himself to dine in his shirt sleeves, but two decades later it went without saying that members of the colonial elite should dress correctly for dinner when natives were present, even in intense heat, and in 1830 officials of the East India Company were forbidden to wear Indian clothes in public.[258] Such customs soon spread to Latin America. Whatever the temperature or degree of humidity, gentlemen in Rio de Janeiro and many other cities had to appear in penguin costume: black cutaway, starched white shirt and white waistcoat, tie, white gloves, and top hat; the disappearance of color and ornament from the fashion of the male European upper class between circa 1780 and 1820 had earlier led to a new vestural style of generalized functionality where clothes were no longer permitted to express social rank and personal identity.[259] Ladies forced themselves into corsets and wrapped themselves in layer upon layer of heavy material. Until the end of the 1860s crinoline was de rigueur in good Brazilian society. Such martyrdom was the price of being civilized.

Tropical cultures in which not even the upper classes had been accustomed to wearing covering clothes of a European or Middle Eastern description had a long road to travel before they reached what was considered as "civilization." Invariably, Christian missionaries insisted on a proper covering of the body and instilled in their charges Victorian notions of shame. In vast parts of the planet, such as the Pacific islands, this resulted in "a fairly total reclothing of the region."[260] King Chulalongkorn, the reformer of Siam, made every effort to get his subjects to wear buttoned-up garments, and by the beginning of the twentieth century the urban population was fully dressed.[261] In Lagos, in the 1870s and 1880s, a small group of Western-oriented Africans in frock coats and lavish women's costumes created a social life centered on churchgoing, balls, concerts, and cricket.[262] Gandhi, the great virtuoso of symbolic politics and friend of frugality, later reversed the process: the late-Victorian dandy we see in his early photos turned into the charismatic "naked fakir," as Churchill reviled him.[263] Nowhere else outside Europe, however, were the trappings of its civilization so faithfully and uncritically adopted as in Latin America; nowhere else, except perhaps in the Egypt of Khedive Ismail (r. 1863–79), was the imitative fetishism of consumption so great.[264]

Cultural resistance was stronger in West and East Asia. Sultan Mahmud II prescribed Western clothing for the senior Ottoman bureaucracy, and the military likewise switched to Western uniforms. This did not at all involve internalizing a European attitude to fashion but rather an outward change in public dress that scarcely reached beyond the court and the top administration. On the streets of Istanbul, men continued for a long time to wear traditional costumes, and no women were photographed before the 1870s in European dress; foreign influence showed itself, as it had for centuries, only in the use of new materials such as French or Chinese silk. European clothes became popular and culturally acceptable as late as in the last quarter of the nineteenth century.[265] Foreign fabrics should not be thought of as a conscious loan from another culture. Where European imports had largely destroyed indigenous textile production, there was often no other option. In the 1880s it was reported from Morocco—not yet a colony—that nearly everyone was wearing cotton goods from abroad.[266]

Japan, unlike Latin America, did not share a colonial past with Europe. Before 1853 there were few contacts with foreigners, and they did not radiate out to Japanese society as a whole. Later—especially after the Meiji Renovation of 1868 brought systemic change to the polity—the country opened up to the West and launched a modernization drive that took directly from Europe, and secondarily from the United States, new organizational forms for the state, the justice system, and the economy. But this far-reaching structural Europeanization was not matched by a de-Japanization of private life; people did not give up their traditional clothing, for example. It is true that following a decree of the State Council in 1872, top figures in the Meiji state, including the emperor himself, dressed in frock coat, top hat, or uniform, and that from the 1880s on lower officials fell in with the change. But traditional clothing kept its place in the home, as an early and expensive flurry of sartorial Westernization gave way to a moderate "improvement" of the *kimono*. Attachment to the familiar was even more self-assertive in other spheres of material culture. On the other hand, a fondness for leather shoes seems to have developed quite early, especially if they squeaked and "sang" as one walked. Those who wished to marry tradition with progress wore traditional dress plus leather shoes—a combination still popular today with Buddhist monks in various parts of Asia.[267] The hat became a universal symbol of bourgeois manners, civil servants wearing it for show in much the same way as a lawyer in Africa or India or a well-off worker on Sundays in the Polish industrial city of Lodz.[268] In the 1920s Kemal Atatürk ruthlessly forced hats onto the heads of Turks, banning the fez that had been introduced in 1836, in an earlier age of attempted modernization, as a symbol of the state's eagerness for reform. Before the hat became compulsory—having been prohibited to non-Muslim minorities in the Ottoman Empire—the Young Turk revolutionaries opted for the decidedly anti-Ottoman "Caucasian" cap.[269]

In China the resistance to foreign consumption models was even greater than in Japan, and Western clothing gained acceptance for the first time only through the military reforms of the Qing dynasty in the early 1900s. Photographs and moving pictures from the time of the nationalist protests in 1919, known as the "May Fourth Movement," show professors and students in Beijing, who were politically radical and often familiar with European culture, marching in the floor-length costumes of traditional scholars. Trousers and jackets, which finally won over these same circles in the 1920s, had traditionally been worn by peasants and ordinary soldiers only.[270] Groups of Chinese merchants who since the mid-nineteenth century had had close ties with Western business partners in Hong Kong, Shanghai, or other ports remained largely faithful to older models in their private life and were poor customers for European luxury items. Only in the 1920s did the appeal of these items increase in the cities, though even then with a bad conscience that regarded the display of "imperialist" appurtenances as national betrayal. The great opening of urban consumers to European and North American patterns of taste, fashion, and behavior occurred in mainland China only in the mid-1980s, a whole century after Latin America's, but now fueled by domestic industrialization and extensive brand piracy.

There are also examples of a reverse effect: of European acculturation to Asian customs. In China and especially in India, this was condemned with increasing severity as "going native"—as crossing a racial status barrier. Adaptation in the opposite direction was also frowned upon. Much as the "trousered Negro" was later an object of ridicule in Africa, many British in the nineteenth century refused to accept Indians in shoes and suits, seeing such sartorial behavior as an insolent aping of Europeans. The Indian middle classes were expected to dress in Indian style, and the symbol designers of British India concocted especially "exotic" costumes for the princes they liked to regard as feudal museum pieces. It caused a huge scandal when one maharajah, the reform-minded Sayaji Rao Gaekwad III of Baroda, arrived to greet the King-Emperor George V at the Imperial Durbar in Delhi—a sumptuous assemblage of Indian dignitaries—in December 1911, wearing a plain white European suit instead of the Oriental costumes and jewelry sported by the other princes, and with a walking stick instead of the prescribed sword.[271]

Acculturation in reverse had been on the agenda in eighteenth-century India, when the adoption of an Indian lifestyle had been a frequent and acceptable occurrence.[272] In the nineteenth century such things were still possible in the Dutch East Indies. Whites there had become so orientalized in the previous century that the British—who occupied Java during the Napoleonic wars and held it until 1816—sought to stem their fall from civilization, requiring the men to give up brazen cohabitation with female natives, and the women to forgo idleness, Oriental dress, and the chewing of betel nuts. It cannot be said that they were very successful. If anything, the lifestyle of both Europeans and Chinese in Batavia became even more Asiatic or perhaps hybrid: they ate *rijstafel*, wore sarongs (at least at home), and indulged in endless midday breaks.[273]

It cannot be stressed enough that adaptation to European culture was very often a voluntary process; colonial authorities and missionaries occasionally helped things along, but that was by no means the rule. A whole series of cases shows that European architecture was embraced in Asia and Africa even in contexts where there was no colonial or quasi-colonial dependence. In the eighteenth century, the Qing emperor had Jesuit architects build him a rococo-style summer palace on the outskirts of Beijing. The Vietnamese ruler Nguyen Anh (after 1806 Emperor Gia Long), who reunified Vietnam following many years of turmoil, built citadels inspired by the famous military engineer and architect Vauban—not only in his new capital, Hanoi, but in all large provincial cities. The building plans stemmed from French officers who, without an official contract from Paris, worked for the emperor in return for a salary. Gia Long preferred European architecture to Vietnam's traditional Chinese styles because he recognized its superiority for his purposes. French influence, or even a reflection of French prestige, played no role in the decision. Gia Long was not an imitator of the West but an early "free shopper" of what was on offer abroad. Good relations with Catholic missionaries did not prevent him from swearing his mandarins and officers first and foremost to the cult of Confucius.[274]

One final example: On Madagascar, which became a (French) colony only in 1896, amateur European master builders had been developing an imaginative architecture since the 1820s. A start was made with some modest buildings to house missionaries, but Jean Laborde, an adventurer shipwrecked on the island in 1831, had greater ambitions. In 1839 he built a new palace for the queen, skillfully combining local stylistic elements with neo-Gothic ones and stabilizing everything with European construction techniques. On other public buildings he put up Hindu quotations that he had learned in India. Later architects introduced granite facades, balconies, and Romanesque round arches. The resulting official style lent an unmistakable aspect to the capital, Antananarivo, where court ladies wore the latest fashions from Paris and London. In spite of all that, the Merina Monarchy did not belong among the zealous self-Westernizers of the age; the country was closed after being opened to the outside world several times, and deep suspicions remained about European intentions.[275]

Living standards, understood as a set of material circumstances or a measure of physical well-being, may be in part essentially the same for large differentiated societies but may also vary to a huge degree socially and regionally, and according to gender and skin color, within such societies. The epidemiological situation, for instance, may be very similar for all members of a society even if there are large income differences among them; the rich were no safer than the poor in the face of smallpox and cholera. On the one hand, then, the living standards of countries may be roughly quantified and ranked in a league table: "life" today is undoubtedly better in Switzerland than in Haiti. On the other hand, different societies and *types* of society operate by different yardsticks: wealth among rice farmers is not the same as wealth among Bedouins or among storekeepers. Societies, as well as social groups within them, differ in their perceptions of "illness" and in the language they use to speak about it. Some diseases are characteristic of particular epochs. Around the end of the nineteenth century, people in Central Europe complained of "neurasthenia"—a condition and a term that has all but disappeared in present-day medicine.[276] Yet the nineteenth century did not yet know the term "stress," which was borrowed in the 1930s from the realm of physics, from material science. This does not mean, of course, that people in the nineteenth century had "stress-free lives" by today's standards. But, whether it is a question of poverty and wealth, sickness and health, or hunger and adequate nourishment, the categories that describe such conditions are relative or—to use a trendy expression—"culturally constructed." They do, however, refer to tangible realities of bodily and material existence.

The nineteenth century, seen globally and in its full time span, was undoubtedly an age in which the material circumstances of life improved for a large part of the world's population. Today it seems perfectly natural to us to be skeptical about progress—the underlying ideology of the Atlantic West since the Enlightenment—but this should not be taken so far that it erases the idea altogether. Such a general statement suffers from a degree of triviality, however. A more interesting observation is that by no means do all tendencies lead in the same direction, that as a matter of fact they often contradict one another. There are numerous examples of this. In the early nineteenth century, many people in the big cities had a higher income than they would have had in the country, even though they often lived in worse environmental conditions. In one and the same society, living standards did not differ only on the scale from less to more; they often reflected different economic logics. Many working-class households lived only just above the survival threshold and could therefore not escape from a narrow time horizon; the property-owning and educated middle classes were able to make long-term plans, basing them on various sources of income.[277] Or, with regard to nutrition: Europe's "long" eighteenth century, which in terms of welfare sometimes lasted into the 1840s, was a lean century, but from the 1850s on, there was a visible "relocation" of hunger, as the capability of transporting food over longer distances was combined with improvements in preservation and storage and the beginnings of a processing industry.[278] The example of the Indian famines demonstrates, however, that this expanded circulation could have a deadly impact on economically weak food-producing regions. The victims of progress are therefore not to be found only among those who are "left behind" or untouched by innovation. The unfettered and uninterrupted invasion of "modernity" could also have baneful consequences.

Many aspects of the standard of living have not been broached in this chapter. For example, few things reveal the character of a society better than the way in which it treats its weaker members: children, old people, the disabled, and the chronically sick.[279] Histories of childhood and old age therefore need to be narrated. The best of them would show whether, in and since the nineteenth century, not only various curves of economic growth but also the survival chances of infants and the physically and mentally handicapped have gone up—whether, that is, the world has become more humane.

CHAPTER VI

Cities

European Models and Worldwide Creativity

1 The City as Norm and Exception

A "city" is a way of socially organizing space. It is hard to distinguish it clearly from other ways. The city always stands in a tension with something else, with non-city. This may take a number of forms: "the country" with its villages of settled farmers, the deserts and steppes of nomads, the world of large estates and plantations where the landowners' power is concentrated, or another city in the same region, with which there may be peaceful rivalry or sometimes—as in the case of Athens and Sparta, or Rome and Carthage—irreconcilable hostility.[1] A city is easy to recognize when it is taken in its specific polarity with non-city. But it is difficult to say which conditions a settlement must fulfill in order to be recognized as a city. Wall plus market plus city charter: nothing as clear-cut as in premodern Western Europe existed either in other civilizations or in the nineteenth century. Number of inhabitants is not a reliable guide: two, five, ten thousand—what is the starting figure? Not even national statistical bureaus have yet agreed on international criteria for what constitutes a city; statistical comparisons are therefore often a rather tall order. There are also other problems of definition. Many historians go so far as to question whether "urban history" can be distinguished at all from other fields of research: after all, is not almost every facet of history somehow reflected in cities? Nor is there agreement as to whether cities should be seen as social fields with a distinctive individual profile and a characteristic "spirit" or rather as interchangeable articulations of an overarching process of urbanization.[2] Urban history and the history of urbanization stand alongside each other as two different optics. The one focuses on the physiognomy of particular cities, the other on a major tendency of the modern age or even of human settlement in general.[3]

Models

Each civilization that has formed cities has its own idea of a model and its own terminology for different kinds of city. A Chinese *dushi* is not the same as a Greek *polis* or an English *township*, and over a long period of time—that which

saw Byzantium turn into Constantinople, then into Istanbul, for example—
quite different visions might follow on top of each other. Particular cultures have
developed their own understanding of "city" and "urban life." A city is thus a
concentrated expression of a particular civilization—a place where the creativity of a society is expressed most clearly. No one in the eighteenth or even the
nineteenth century could have confused Beijing with Agra, Edo (Tokyo) with
Lisbon, or Isfahan with Timbuktu. It is easier to know where you are in a city
than in a village. Urban architecture reveals more clearly than almost anything
else what is distinctive about a civilization; cultural traits become stone. Only
the growth of "megacities"—one of the most important trends in social history
in the second half of the twentieth century—has stripped away this personality
specific to a civilization.

On the other hand, even for earlier times, we should be wary of taking at
face value the city models that geographers and sociologists have construed. To
speak of *the* Chinese, Indian, or Latin American city, as if certain basic features
recurred in every instance, has a point only if one understands that this radically
abstracts from many particular cases. Such types are major simplifications; they
can only very incompletely embrace change over time—nineteenth-century urbanization, for example—and therefore give us an excessively static and "essentialized" picture. They also leave out the fact that, whichever the civilization,
cities with the same function (e.g., ports or capitals) have common features that
are often greater than their differences. It is particularly questionable to consider
civilizations as uniform spheres of social order that are sharply separated from
one another. It is by no means the case that "the Indian city" could be found in
every corner of South Asia, or that wherever the Chinese went they founded
the same kind of settlement. City forms are not latent "cultural codes" that automatically find expression amid changing circumstances. No doubt there are
preferences for certain kinds of urban life: Europeans seek out the center of a
city, North Americans are less likely to be drawn to it. But it is more interesting
to ask how the aims of a city were defined and achieved under a certain set of
circumstances than simply to take its distinctive morphology for granted. In "the
Chinese city," for instance, we will be on the lookout for what is *not* Chinese.

Cities are nodal points of relations and networks. They organize the area surrounding them. Either the market, an overarching state apparatus, or the initiative of local authorities creates trade networks, administrative hierarchies, and
federative associations between cities. No city is an island; influences from "the
outside world" penetrate through the city gates. In a strong tradition of Western,
as well as Middle Eastern/Islamic, thought, cities are the *fons et origo* of all civilization. The premodern traveler headed straight for them; they were his salvation
from the perils of the wild. As a stranger or outsider, he was in less danger there
than in a village. Knowledge, wealth, and power were concentrated in cities. They
offered opportunities in life for the ambitious, the curious, and the desperate. In
contrast to rural communities, cities were always "melting pots." Empires were

ruled, or global systems steered, from them: international finance from London, the Catholic Church from Rome, the fashion world from Milan or Paris. After the fall of a civilization, its cities are often what remains in the memory of a mythopoetic posterity: Babylon, Athens, the Jerusalem of Herod's Temple, the Baghdad of the caliphs, the Venice of the doges. The city is of premodern origin, but it is also the birthplace of modernity. Cities stand out from their surroundings by taking the lead, by wielding power, by being *relatively* progressive. This has been true at all times. What was new in the nineteenth century?

A World of Stone

The range of forms of urban life should not be underestimated. It stretches from the first skyscraper cities—Chicago nosed ahead in 1885, with its unprecedented seventeen-story building[4]—to the most evanescent settlements. In the nineteenth century there were still mobile cities that resembled the itinerant seats of power in medieval Europe. Only the founding of Addis Ababa in 1886 by Emperor Menelik II ended a centuries-long period of mobile capitals in Ethiopia, during which a huge herd of livestock and up to 6,000 slaves followed the ruler and aristocracy around, carrying their household goods and cult objects. The construction of the new capital was supposed to symbolize as clearly as possible the country's entry into modernity. After Menelik unexpectedly wiped out an Italian expeditionary force at Adwa in 1896, the Great Powers ratified the status of Addis Ababa and set about building embassies there in the European style.[5] Until the end of the nineteenth century, the kings of Morocco, too, spent more time in the saddle than in any of their urban residences; Sultan Mulay Hassan, actually an industrious master builder, is said to have journeyed in 1893 with a retinue and staff numbering 40,000.[6] Should such practices be regarded as inherently archaic? In any event, in China as much as in the Tsarist Empire or Great Britain, monarchs still had summer and winter palaces. From 1860 on, one of the largest countries in the world was governed for several months of the year from a health resort: Simla (now Shimla) in the Himalayan foothills. The whole apparatus of the British viceroy would travel there each summer by caravan and set up shop in a dramatic landscape—although it is true that from 1888, the representative of the Queen-Empress Victoria had his own permanent Viceregal Lodge, a castle built in the English late-Renaissance style.[7]

Nevertheless, the nineteenth century was in general an age when rulers settled down and cities turned into stone. Even in Europe stone constructions were by no means universal in 1800; a peripheral country like Iceland switched to them only after 1915.[8] The transition was especially apparent in the colonies, where the authorities sought to literally solidify the fluidity of local politics in the interests of a more manageable order. At the same time, this underlined their claim to have established themselves overseas for all eternity; they achieved their civilizing mission through the triumph of stone over clay and wood. But there was an ironic result. A lightweight house can easily disappear, be consumed by fire, or simply be replaced if political and economic conditions change. Stone buildings remain standing, so that today they are the most conspicuous testimony to the death of colonialism: despised ruins, villas converted into slum dwellings, power centers of postcolonial politics, or relics spruced up for tourists in parts of the world where they may be the oldest surviving monuments.

Sometimes forest depletion made the change to stone especially advisable. Wooden buildings were increasingly thought of as primitive and old-fashioned or as too reminiscent of pre-bourgeois grandeur; the timber facing of Victorian mock-Tudor houses was only an ornamental addition to the neoclassical solidity of their stone construction. Wooden or clay cities held out where ecology or economics excluded other options: then they could be a rational adaptation to circumstances. As in the West, the American zoologist and art collector Edward S. Morse noted in 1885, very few people in Japan could afford fire-resistant houses; it made sense to build only simple, collapsible buildings with ordinary inflammable material, in order to limit the expense and to make it easy for planks and floorboards to be quickly rescued in the event of a fire.[9] The fatalism of this way of thinking disappeared when houses in Japanese cities began to be built in stone and cement. The beauty of aging wood and tightly thatched roofs was sacrificed to the fireproof banality of concrete.[10]

The city is a nearly universal phenomenon. It has been said that the state was a European invention, but that is not true of the city. Urban cultures arose independently on all continents, with the exception of North America and Australasia. Usually with close links to agriculture, they developed in the Middle East, on the banks of the Nile, in the eastern Mediterranean, in China and India, and considerably later in Japan, Central America, and sub-Saharan Africa. The city as a physical form and a mode of social life is not a transplant from Europe. Although the "modern" city of European origin spread around the world, it encountered indigenous urban cultures that usually did not give way before it. Tenochtitlán was destroyed in the 1520s, so that colonial Mexico City could be built in its place. But old Beijing, with its gigantic walls (in three concentric rectangles) and sixteen city gates, survived European and Japanese invaders, until city planners and Mao Zedong's Red Guards tore down the "relics of feudalism" in the 1950s and 1960s. These were the two extremes—disappearance and persistence—in the face of the aggressive forces of the West. Everything else lay in between. Elements of architecture and urban organization were combined, overlaid, mingled, and juxtaposed in narrow spaces, often in sharp contradiction with one another. The general tendency to urban modernity broke through everywhere at different points in time, but seldom entirely on Western terms.

Tendencies in the Nineteenth Century

What happened with the city in the nineteenth century? The second half, in particular, was a period of intensive urbanization.[11] No other age had experienced such a spatial densification of social existence. The growth of the urban population accelerated in comparison with earlier centuries. For the first time, the city-dweller's way of life became economically and culturally dominant in a number of large countries. This had previously happened, if at all, only in core areas of the ancient Mediterranean, in East China during the Song period (960– 1279), and in early modern northern Italy. None of the established urban systems, whether in Europe, China, or India, was prepared for the huge influx into the cities. The adjustment, especially in the early stages, therefore led to crises. Part of the growth was channeled off into new cities outside the existing systems. Socially, if not always aesthetically, the most successful instances were in regions where no cities had existed before, especially in the American Midwest and Pacific West and in Australia. There urbanization started from scratch in the 1820s, although sometimes this meant taking over well-selected sites from their native inhabitants. The question of continuity and discontinuity was irrelevant.

In other parts of the world, the development was rarely continuous. Many people living at the time in Europe had the impression that the modern metropolis, as it existed from midcentury onward in nearly every country of the continent, represented a fundamental break with the past. Late-eighteenth-century French economists, evidently with Paris in mind, had been the first to observe that the big city was where "society" came together, and where the prevailing social norms took shape. The big city acted as the powerhouse of economic circulation and the multiplier of social mobility. Value increased not only through production—that happened in the country too—but also through the sheer force of human interaction. Rapid turnover created wealth.[12] Circulation was regarded as the essence of the modern big city: that is, the ever-faster movement of people, animals, vehicles, and goods within the city, as well as its speedier exchanges with surrounding areas both near and distant. Critics complained of the pace of life in the metropolis, while urban reformers wanted to adapt its physical aspect to its modern essence and to unblock its vital flow (to improve transportation by building railroad tracks, wider streets, and boulevards; to manage the water and wastewater with a systems of drains and underground sewers; and to purify the air by clearing slums and developing more evenly spaced housing). This was the basic impulse behind a large number of municipal programs, from English promoters of public health to Baron Haussmann, the creator of postmedieval Paris.[13]

The European metropolis of the late nineteenth century was socially more differentiated than the early modern city. Its oligarchies were less homogeneous. The simple threefold division into a patrician elite that made the political decisions, an intermediate stratum of artisans and tradespeople, and a mass of urban poor had become obsolete. Even the elite consensus on taste had lost much of its strength. City complexes were only rarely designed as a single whole—which would earlier have been the case not only for princely residences but also for many nonaristocratic towns. Aesthetically as well as socially and politically, the Victorian city was "a battlefield."[14] But it was more robustly built: less stucco, more solid brickwork, more iron. A city for eternity. And it was larger in volume. The average city hall and railroad station was of a size that only cathedrals or Versailles-style palaces would have reached in the past. Paradoxically, grand civic architecture made people smaller than princely ostentation had ever done.

Apart from its sheer growth of spatial size, number of inhabitants, and share in the total national population, the nineteenth-century big city underwent several other major transformations:

(1) Urbanization and the growth of cities took place at varying speeds in different parts of the world. In few other respects are regional discrepancies in the pace of social development—a fundamental characteristic of modernity—so clearly visible.

(2) Cities became increasingly varied around the world. Few old types of city disappeared, but many new types came to join them. This diversification stemmed from the appearance of further special functions: the railroad created the junction city, while greater leisure time and a middle-class need for relaxation led to development of the coastal resort.

(3) Since the days of Babylon and ancient Rome, there had been metropolises that stretched out and dominated large areas. The nineteenth century brought networking on a scale that permanently linked the world's largest cities with one another. This global city system is still with us today, even more interconnected and with a different weight distribution of its component parts.

(4) City infrastructure was built in ways that had no historical precedent. For millennia the "built environment" had consisted essentially of buildings. Now streets were paved, harbors lined with brick, railroad and street-car tracks laid, street lighting installed, and clinker-clad underground tunnels dug for sewage and subway trains. New structures went downward as well as upward. By the end of the century cities were cleaner and brighter. At the same time, the great metropolises added a mysterious underworld, which gave birth to a all kinds of fears and escape fantasies.[15] The new infrastructure absorbed huge private and public investments— along with industrial plants, the greatest employment of capital during industrialization.[16]

(5) Closely bound up with this new material solidity were the commercialization and steadily increasing value of urban real estate and the growing importance of the rental market. Only now did *urban* land become an investment and an object of speculation, valued not for its agricultural uses but simply because of its location. The "skyscraper" was emblematic of this trend.[17] Land values could soar at a speed unimaginable in productive sectors of the economy. A plot of land that changed hands in 1832 for $100 in the newly founded city of Chicago was sold in 1834 for $3,000 and was valued twelve months later at $15,000.[18] In an old

city such as Paris, real estate speculation began in earnest in the 1820s.[19] The same market mechanisms were at work in the boom years of Asian cities such as Tokyo and Shanghai. Under these conditions, land registers attained a new precision and economic significance; chapters on landownership, construction, and landlord-tenant relations were added to the law books; it was no longer possible to imagine the financial sector without mortgages. New social types appeared on the scene, such as the estate agent, the property speculator, the contractor or "developer" (who built standardized accommodation units for the middle and lower classes), and the tenant.[20]

(6) Cities have always been planned. They projected cosmic geometries onto the earth below. Princes laid out ideal cities; it was one of their favorite occupations during the European Baroque era. Only in the nineteenth century, however, did city planning come to be understood as an ongoing task of central or local government. Continually struggling against unbridled expansion, and often losing the struggle, city councils nevertheless continued to plan—and this impetus became an essential part of municipal politics and administration. If a city wished to be "modern," it outlined visions of its future complete with technical know-how.

(7) New conceptions of an urban public and community politics took shape and became more widespread. An oligarchy and an undifferentiated, unpredictable "people" were no longer perceived as the only actors in public space. A slackening of absolutist regimentation, together with wider electoral representation, new mass media, and the organization of interest groups and political parties in the municipal arena, changed the character of local politics. At least in constitutional states, the capital city was also the seat of a parliament where national politics was conducted: the electorate followed events there with an unprecedented level of involvement. A rich and lively world of clubs, associations, church communities, and religious sects, as it has been described with particular thoroughness for early modern England and Germany, also emerged in embryo under very different political conditions, for example, in the provincial cities of late imperial China.[21]

(8) New "urbanist" discourses and new critiques of urban life placed the city at the center of struggles over the interpretation of the world. Cities had always been something special, and those who lived in them—or at least in ones around the Mediterranean—had always tended to look on *rustici* with deprecation. But only the dynamic historical thinking of the nineteenth century elevated the big city into the pioneer of progress and real locus of cultural and political creativity. Jules Michelet even constructed a myth of Paris as the universal city of Planet Earth—a trope that later took root in the vision of the French metropolis as "the capital of the nineteenth century."[22] From now on, anyone who praised

rural life courted the suspicion that he was a simpleton or a reactionary; anyone who defended it no longer did so to strike a judicious balance between "court" and "country" but to sustain a robust critique
of civilization, in the spirit of either agrarian romanticism or militant
Junkertum. By the end of the century, even old pastoral ideals had been
redefined in the urban context of the "garden city." The new sociology,
from Henri de Saint-Simon to Georg Simmel, was fundamentally a science of the life of city dwellers, more of *Gesellschaft* than *Gemeinschaft*,
more of speed and edginess than of village placidity. Political economy
no longer saw the land as the source of social wealth, as the eighteenth-
century physiocrats still had. As one "production factor" among others, land was now viewed with skepticism as a stagnant obstacle to economic development. Value creation, for the generation of Karl Marx
and John Stuart Mill, took place in an urban-industrial space. This new
cultural preponderance of city over country mirrored the declining political importance of the peasantry. Between the Pugachev Rebellion in
southeastern Russia (1773–75) and the wave of protests at the turn of
the twentieth century (the Chinese Boxer Rebellion in 1900, the Romanian peasant revolt in 1907, the beginning of the Zapatista movement
in Mexico in 1910), there were few great peasant upheavals anywhere in
the world that mounted a challenge to the existing order. Many great
rebellions that might come to mind—especially the Indian Mutiny in
1857/58 and the almost contemporaneous Taiping uprising in China—
had a social base that extended beyond the peasantry. They were more
than spontaneous outbreaks of peasant rage.

In the nineteenth century, it is often said, the city became "modern," and "modernity" came into being in the city. If we wish to define urban modernity, and
perhaps even to gain some chronological bearings about modernity in general,
then we must take into account all of the above processes. The usual concepts of
urban modernity as it emerged in the second half of the nineteenth century tend
to be one-sided. Attempts have been made to define it as a combination of rational planning and cultural pluralism (David Ward and Olivier Zunz), as order
in compression (David Harvey), or as a space of experimentation and "fractured
subjectivity" (Marshall Berman).[23] Early Victorian London; Second Empire
Paris; post-1890 New York, Saint Petersburg, or Vienna; 1920s Berlin; and 1930s
Shanghai have been described as loci of such modernity. This has nothing to do
with sheer magnitude. No one has ever thought of describing Lagos or Mexico
City, two present-day metropolises, as epitomes of modernity. The heroic modernity of cities is a fleeting moment that sometimes lasts just a few decades:
an equipoise of order and chaos, a conjunction of immigration and functioning
technical structures, an opening of unstructured public spaces, a flow of energy
in experimental niches. The moment of modernity presupposes a certain form

of the city, which was still discernible at the end of its classical era, and an opposition to what is non-city. Inner and outer boundaries are lacking in the present-day megalopolis of endless, diffuse, polycentric "conurbations" with middling degrees of compression. There is not even a "countryside" that can be exploited or consumed as a local recreation area. The urban nineteenth century ends with the big cities' loss of shape.

2   Urbanization and Urban Systems

Urbanization used to be understood in a narrow sense as the rapid growth of cities in conjunction with the spread of mechanized factory production; urbanization and industrialization appeared as two sides of the same coin. This view can no longer be upheld. The definition that is common today takes urbanization to be a process of social acceleration, compression, and reorganization, which may occur under a range of very different circumstances.[24] The most important outcome of this process was the formation of spaces of increased human interaction in which information was swiftly exchanged and optimally employed, and new knowledge could be created under favorable institutional conditions. Cities—especially large cities—were concentrations of knowledge; sometimes that is why people headed to them.[25] Some historians distinguish between the growth of cities, seen as a *quantitative* process of spatial compression triggered by the concentration of new job opportunities, and urbanization proper, seen as the *qualitative* emergence of new spaces of action and experience, or in other words, the development of specific urban lifestyles.[26] This distinction calls attention to the wealth of aspects involved in the phenomenon, but it is a little schematic and difficult to sustain in practice.

City and Industry

Since the nineteenth century witnessed urban development almost everywhere in the world, urbanization was a much more widespread process than industrialization: cities grew and became more dense even where industry was not the driving force. Urbanization follows a logic of its own. It is not a by-product of other processes, such as industrialization, demographic growth, and nation-state building; its relationship with them is variable.[27] A higher level of urbanization at the end of the premodern era was not at all a basis for the success of industrialization. If it had been, northern Italy would have been among the trailblazers of the Industrial Revolution.[28]

Industrialization imparted a new quality to the concentration of people in urban settlements. As E. A. Wrigley has shown in a classic essay on London, we should assume that there was a two-way relationship with urbanization. On the eve of the Industrial Revolution, London had grown into a metropolis in which more than a tenth of the population of England lived (in 1750). Its commercial wealth, its purchasing power (especially for food, which in turn stimulated a rationalization of agriculture), and its concentration of labor and skills ("human capital") offered the best chances of a multiplier effect for the new production technologies.[29] Complemented and counterbalanced by an urban renaissance in English and Scottish provincial cities, the development of London was part of a more general increase in social efficiency and capacity. Powerhouses of the Industrial Revolution such as Manchester, Birmingham, and Liverpool became huge cities, but in the second half of the nineteenth century the fastest developers were ones with a large service sector and an exceptional ability to process information in relationships of direct contact.[30] In continental Europe and other parts of the world, and also in Britain itself, rapid urbanization occurred where local industry *could not have been* the ultimate cause.

Many examples show that nineteenth-century cities grew in the absence of a noteworthy industrial base. Brighton, on the south coast of England, was one of the fastest-growing cities in the country, but it had no industry. The dynamic of Budapest was due less to industry than to the interplay of agrarian modernization and key functions in trade and finance.[31] Also cities in the Tsarist Empire such as Saint Petersburg and Riga owed their constant population growth to commercial expansion, which was bound up with an extensive and productive crafts sector; industry played a subordinate role there.[32] In a particularly dynamic field of economic development, cities might let opportunities slip: Saint Louis, until the mid-1840s, grew with breathtaking speed into the leading city of the Mississippi Valley and the center of the whole American West. But it passed up the chance to acquire an industrial base. It soon faced economic collapse and had to surrender its lead to Chicago, having allowed the window of opportunity to close.[33] A trip around the downtowns of London, Paris, or Vienna reveals that they were never industrial; indeed, their cityscapes testify to a struggle in the past to prevent industry from destroying their distinctive culture. The emblematic metropolises of the nineteenth century created their enduring appearance more by fending off industrialization than by surrendering to its consequences.[34] And the twentieth-century growth of giant cities without an industrial base (Lagos, Bangkok, Mexico City, etc.) should make us further aware that there is only a loose association between urbanization and industrialization. Urbanization is a truly global process, industrialization a sporadic and uneven formation of growth centers.

Top Cities

Only if urbanization is considered outside the time frame of the nineteenth century and a narrow association with "modernity" can the place of that century in the *longue durée* of urban development be properly determined.[35] This also calls into question Europe's claim to a monopoly on urbanization. The big city was no more invented in Europe than the city in general was. During the longest part of recorded history, the world's most populous cities were in Asia and North Africa. Babylon had probably passed the 300,000 population mark by 1700 BC.

Rome under the emperors had a larger population than even the leading Chinese cities at that time, but it was a unique case. It embodied itself, not "Europe." In the second century AD more than a million people lived in Rome—a figure not reached by Beijing until the late eighteenth century or by London until shortly after 1800.[36] Imperial Rome was a one-off in the history of human settlement. It did not stand at the top of a finely tapering pyramid of cities; it hovered, as it were, above a world of scattered settlements. Only Byzantium at its height (before the catastrophe of the First Crusade, in 1204), which also did not rest upon a graded hierarchy of cities, came close to the dimensions of such a world city.

In general, population figures for non-Western cities—even more than for European or American ones—rely on often insecure premises until well into the nineteenth century. In 1899 Adna Ferrin Weber, the father of comparative urban statistics, laconically observed that the Ottoman Empire was full of cities but that only the larger ones were "known to the statisticians, and these imperfectly."[37] The following data should therefore not be regarded as anything more than informed guesses. The larger the city, the more likely it was to have made travelers and commentators feel that they had to judge its exact size. At least this allows us to gauge the order of magnitude of the world's largest cities at selected points in time.

In 1300, Paris was the only European city among the top ten in the world, taking sixth place after Hangzhou, Beijing, Cairo, Canton (Guangzhou), and Nanjing, but ahead of Fez, Kamakura (in Japan), Suzhou, and Xi'an.[38] Six of these cities were in China, Marco Polo's reports about which were beginning to reach Europe. By 1700 the picture had changed. As a result of development in the early modern Muslim empires, Istanbul was then number one, Isfahan three, Delhi seven, and Ahmadabad, also in the Indian Mogul Empire, number eight. Paris (5) had slipped a little behind London (4) and would never catch up to it again; they were the only two European cities on the list. Beijing remained the second-largest city in the world. The others in the top ten—Edo, Osaka, and Kyoto—were in Japan, which under the Pax Tokugawa had just bid farewell to a century of stormy urban development.[39]

By 1800 the picture had changed again, but only slightly:[40]

1. Beijing   1,100,000
2. London   950,000
3. Canton   800,000
4. Istanbul   570,000
5. Paris   550,000
6. Hangzhou   500,000
7. Edo (Tokyo)   492,000
8. Naples   430,000
9. Suzhou   392,000
10. Osaka   380,000

Six of these cities were in Asia—or seven, if we include Istanbul. After London, Paris, and Naples, the next European cities to figure are Moscow (fifteenth, with 238,000 inhabitants), Lisbon (sixteenth, with approximately the same number) and Vienna (seventeenth, 231,000). Of the twenty-five largest cities in the world in 1800—if we follow the estimates of Chandler and Fox, which are supported by other sources, though inevitably not based on a uniform concept of the city—only six were in Western Christendom; Berlin had a population of 172,000, which made it roughly the same size as Bombay (Mumbai) or Benares (Varanasi). The most populous city in the Americas was Mexico City (128,000), followed by Rio de Janeiro (100,000), the key center of Portuguese America. Even in 1800 North America was lagging in this respect: its largest city was still Philadelphia (69,000), the first capital of the United States. But New York was preparing to take the lead. Thanks to an extraordinary surge in immigration and an economic boom, it was already the main Atlantic port and in the new century would also become the largest city in the United States.[41] Australia, soon to join North America as an area of explosive demographic growth, had scarcely any urban history in 1800. Its entire population of European origin would easily have fit into a small German princely residence.[42]

These numerical impressions would suggest that in 1800, China, India, and Japan were still the dominant urban cultures in the world. It is true that what was meant by "city" varied enormously. European visitors found the walled cities to which they were accustomed most often in China, but even there not all throughout the country; and travel reports repeatedly speak of Asia's shapeless, "nonurban" cities. Sometimes the distinction between city and country seemed to lose all of its sharpness. The island of Java, for example, very densely populated in the nineteenth century, was not centralized in a few large cities, nor did it have the isolated, largely autarkic villages that people liked to imagine in Asia: it was one large intermediate area of settlement between city and country, essentially neither the one nor the other.[43] Nevertheless, *every* city was a sphere of dense communication and a consumer of surpluses produced in the country; it was in some way a nodal point of trade or migration. Each had to contend with supply and public order problems that were different from those in "the countryside." Asia's big cities somehow managed to solve these problems: otherwise they would not have existed. Even the most blinkered traveler could tell when he was in a city; the grammar of urban life was comprehensible across cultural frontiers.

Urban Populations: East Asia and Europe

Urbanization, understood as a measurable state of society, is a relative and obviously artificial indicator concocted by nineteenth-century statisticians. It implies that the growth of particular cities is related to their surroundings, the key yardstick being their share of the total population of a country. This share is not necessarily highest in regions with the largest cities. It is therefore

Table 6: Percentage of Population in Centers with More than 10,000
Inhabitants: 1820–1900

	China	Japan	W. Europe
1820–25	11.7	12.3	—
1840s	3.7	—	—
1875	—	10.4	—
1890s	4.4	—	31.0

Sources: China and Japan: Gilbert Rozman, "East Asian Urbanization in the Nineteenth Century: Comparisons with Europe," in: Woude et al., *Urbanization in History*, p. 65 (Tab. 4.2a, 4.2b); Western Europe: Maddison, *World Economy*, p. 40 (Tab. 1-8c).

illuminating to compare Europe with the countries of East Asia where the largest urban concentrations were found in the early modern period. In 1600, Europe had already reached a slightly higher level of urbanization than China, where the urban share of the proportion had remained roughly the same for a thousand years. But *on average* Chinese cities were larger than European ones. Two regions—one on the Lower Yangtze (Shanghai, Nanjing, Hangzhou, Suzhou, etc.), the other in the Southeast, around and inland from the port city of Canton—continually amazed early modern European travelers with the density of their population and the size of their cities. In 1820 there were 310 cities in China with more than 10,000 inhabitants; in Europe, outside Russia, there had been 364 at the beginning of the century. The average was 48,000 inhabitants in China, and 34,000 in Europe.[44]

Table 6 gives comparative population percentages for selected times in the nineteenth century. It indicates both Japan's constant middle position between China and Western Europe and the extraordinary speed of city formation in Europe after the first quarter of the century. Shortly before the West opened up the two largest East Asian countries, the share of cities in Japan's total population was more than three times greater than in China's. Is this, however, a methodologically legitimate comparison? Was Japan already—by the criterion of urbanization—more "modern" than China? The gap narrows if we break down the averages: that is, if we compare Japan not with the immense territory of China as a whole but with its economically most developed macroregion, the Lower Yangtze. In that case the demographic figures are roughly similar. In the Lower Yangtze in the 1840s, city dwellers made up 5.8 percent of the total population. By 1890 this rose to 8.3 percent, not far short of Japan's 10.1 percent in its early industrialization period.[45] This means that in these two densely populated regions, the absolute numbers were as follows: 3.7 million Japanese lived in cities with a population of 10,000 or more in 1825, but only 3.3 million did so in 1875; while the figures in China were 15.1 million at the time of the

Opium War and 16.9 million in the 1890s.[46] For Europe we have the estimates of Paul Bairoch and his collaborators, who define any settlement with more than 5,000 inhabitants as a city; this gives 24.4 million city dwellers in continental Europe in 1830, and 76.1 million in 1890.[47] The orders of magnitude can here only be approximate, but in 1830 it was not the case that an urban Europe faced a rural East Asia, whereas by 1890 the the gap between them had widened dramatically.

The nineteenth-century urbanization experiences of China and Japan were, to be sure, so different from each other that it would be misleading to speak of a common East Asian pattern. In Japan, paradoxically, state-initiated modernization actually led to a temporary *de*concentration: the abolition of the feudal principalities (daimyates, or *han*), the downgrading of castle towns as administrative centers, and the ending of the samurai's obligation to reside in them or at the shogun's court in Edo (Tokyo) increased horizontal mobility in the country, mainly to the advantage of medium-sized cities. In the transition from the Tokugawa era to the Meiji era, the population of Tokyo fell from more than a million to 860,000 in 1875—although this contained the seeds of future expansion, since much *daimyō* land in the area around Tokyo fell into the hands of the new government and would be used for urban development. In China, the negligible rise in the rate of urbanization may also be attributed to a modernization effect: that is, to the insertion of coastal regions into the world economy, and the rapid growth of a number of port cities, especially Shanghai. Urban population increases were visible almost exclusively in the Lower Yangtze and around Canton and Hong Kong. All in all, though, China remained as it had been at the beginning of the century: a markedly less urbanized country than Japan.

In the long term, the comparison with Europe is illuminating. Early modern Europe never reached the absolute urban population levels that China and Japan together displayed; East Asia also had many more *very large* cities. Europe experienced a first urbanization surge after 1550, and a second after 1750;[48] the share of its cities in the total population doubled between 1500 and 1800. But between 1650 and 1750 the degree of urbanization in Europe was lower than in Japan, the same as in the Lower Yangtze region, and above that of China as a whole. Europe's leap ahead in the nineteenth century was not entirely due to industrialization and the accompanying emergence of factory cities. It also had somewhat earlier roots in what Jan de Vries calls the "new urbanization" after 1750, which began in England and spread after the turn of the century to southern Europe, especially in small and medium-sized cities. The growth of *very large* cities was less spectacular, corresponding more or less to rises in the overall population; only the railroad increased their weight disproportionately, without ever resulting in "top-heavy urbanization" and generating the kind of megacities that would become common outside Europe in the twentieth century.

Hierarchies

In eighteenth-century Europe (apart from Russia and Spain), then, a finely graded hierarchy of cities gradually took shape, in which each size category was well represented. Jan de Vries, the cautious empiricist who generally much prefers to speak of "microregions" than of whole countries or Europe as a whole, thinks that the evenness of the rank/size distribution justifies the idea of a character-istic European pattern of urbanization.[49] The cities of Europe (west of Russia) formed a geographically linked, vertically differentiated system with a high de-gree of interaction, to which urban centers in the colonies also belonged in ways that are still imperfectly understood. De Vries points out that some countries in late nineteenth-century Europe—perhaps for the first time in history—crossed the threshold beyond which the main source of urbanization was not migration from the country or abroad but the process of natural reproduction within their own boundaries. By contrast, although the large North American centers of im-migration were at a level of economic development comparable to that of north-western Europe, they did not become self-reproductive until the First World War.[50] Considerable skepticism is doubtless warranted regarding the (often ideologically motivated) talk of Europe's distinctive historical trajectory. But there appears to be solid empirical evidence for the distinctiveness of its path of urbanization.

Scholars who study urbanization tend to make comparative assessments of city structures; they check whether the relationship among large, medium, and small cities "looks right." From this point of view, not only Britain, France, the Netherlands, and Germany but also the United States had a "mature" urban hi-erarchy in the nineteenth century. This was not true of Denmark or Sweden, given the dominance of Copenhagen and Stockholm, nor was it true of Russia, where there were no really large cities apart from Saint Petersburg and Moscow: the third in size, Saratov, had barely one-tenth of the population of Saint Peters-burg. The typical governor's capital, never amassing a population higher than 50,000, never outgrowing the administrative and military functions that it de-rived from the central state, was only marginally affected by the dynamic forces active in late Tsarist Russia.[51] The lack of a finely graded hierarchy of cities was a major obstacle in the modernization of Russia.

Japan, on the other hand, came close to the ideal of a city system with a wide unbroken spectrum of population size. So too had China in former times, although in the nineteenth century it lacked small cities in the 10,000–20,000 range, and rapid growth of large cities was limited to a few metropolises, nearly all of them on or near the coast. The suspicion that such gaps and disproportions on the size axis point to weak trade links among cities is nevertheless contradicted by the findings of Chinese historians, who have been able to demonstrate the increasing integration of a "national" market. In other words, it is problematic to start from the "aesthetic," Western-inspired norm of a uniform hierarchy of cities without clarifying precisely how different structures operate economically.

In China, apart from the few coastal metropolises, the cities that grew in population and average size were mostly those that were not administrative centers and could engage in commerce with little state regulation (experts in the field speak of "nonadministrative market centers"). A "nonideal" hierarchy might therefore perfectly well have had a certain functional point.

3   Between Deurbanization and Hypergrowth

Contractions

We have to be careful with the evaluations we make. Fast quantitative growth of cities is not per se a sign of impetuous modernization, and deurbanization is not always, though it is often, the expression of crisis and stagnation. In Japan as in Europe, the so-called proto-industrialization of the eighteenth century went hand in hand with emigration from the large cities. In fact, deurbanization was a feature of various parts of Europe before 1800—for example, Portugal, Spain, Italy, and the Netherlands.[52] However, the impoverished life of cities in southern Europe expressed a general tendency for the focus of European urban culture to shift toward the North and the Atlantic. Only around 1840 did the decline of older cities in the South come to a halt.

One exception, though, was the Balkans. It was highly urbanized in comparison with other regions at a similar level of economic development. This was not the result of a specifically nineteenth-century dynamic but the legacy of earlier developments: above all, the high value that the Ottomans attached to urban culture, and the importance of fortified garrison cities. After the end of Ottoman rule, a number of Balkan countries went through a phase of deurbanization. A particularly dramatic instance was Serbia, during the turmoil that lasted from 1789 to 1815. In Belgrade, which had approximately 6,000 houses in 1777, only 769 were recorded in 1834.[53] The Serbian revolution destroyed the institutions of the Ottomans with such thoroughness that even their urban structure was dispensable. A similar process occurred after 1878 in Montenegro, and in Bulgaria there was at least a long urban recession.

Deurbanization had other causes in Southeast Asia, where a trade boom after 1750 had led to strong city growth. By the early nineteenth century Bangkok, for example, held one-tenth of the Siamese population,[54] and the picture was similar in the multiple states of Malaya. As rice growing spread in the 1850s, however, a new "peasantization" began to appear, and with it an increase in the relative size of the rural population. Between 1815 and 1890 the share of the Javanese population living in cities with more than two thousand inhabitants fell from 7 percent to 3 percent, as a direct result of the growing export orientation of the economy. By 1930 Southeast Asia was one of the least urbanized regions in the world: only noncolonized Siam retained the traditional dominance that its capital had had since 1767; all the colonial capitals were functionally less important than the metropolises of the dynastic past had been.[55] Only in the Philippines, with its highly decentralized political conditions, had no cities functioned before the colonial period as sites of compressed power; this was why the Spanish founding of Manila in 1565 concentrated administrative, military, ecclesiastical, and economic functions to an extent never previously seen. The Philippines was an early and lasting example of the top-heavy structure later characteristic also of such diverse countries as Siam and Hungary.[56] The Dutch presence on Java began only a few decades later than the Spanish presence on Luzon, the main island of the Philippines, yet Batavia—whose economy, like Manila's, relied on an active Chinese population—never achieved complete hegemony over the princely seats of the native rulers. In the Philippines secondary centers, all relatively weak, took shape only toward the end of the nineteenth century.[57]

It depended on circumstances, then, whether colonial rule promoted, obstructed, or reversed urbanization. In India the urban population probably did not grow between 1800 and 1872. Nearly all the large cities from the pre-British period lost inhabitants: Agra, Delhi, Varanasi, Patna, and many more. In conquering the Subcontinent between 1765 and 1818, the British had taken over highly developed urban systems, but the fighting had destroyed a great deal of urban and interurban infrastructure, often including famed long-distance roads. The British introduced new taxes and monopolies, many of which made indigenous trade more difficult, so that merchants often abandoned the cities and retreated to the countryside. The disarming of indigenous troops, the decline of urban industries such as weapons production, and the dismantling of princely administrations contributed to the process of deurbanization. The tendency turned round in the early 1870s, but only slowly. In 1900 the degree of urbanization in India was not significantly higher than it had been a hundred years earlier.[58]

Deurbanization of a society entails shrinkage of individual cities.[59] As we have seen, Tokyo temporarily experienced this, while other Asian cities did not recover from earlier destruction by the end of the century. Isfahan, the glittering capital of the Safavid shahs, which had a population of 600,000 in 1700, remained a shadow of its former self (with only 50,000 inhabitants in 1800) after Afghan invaders laid it waste in 1722. Agra, the capital of the Mogul emperors, declined after the fall of the empire and regained only around 1950 the population of half a million that it had had in 1600. Its central political role was irretrievably lost. Many cities in Asia or Africa collapsed when the states with which they had grown were destroyed by colonialism, or when new trade routes passed them by. In early modern Europe, the decline of a city had been nothing unusual. Fast-growing cities such as London, Paris, or Naples coexisted with stagnating or shrinking ones. Many medium-sized German cities expanded little between the Reformation and the middle of the nineteenth century: Nuremberg, Regensburg, Mainz, and Lübeck, to name but a few. Venice, Antwerp, Seville, Leiden, or Tours had fewer inhabitants in 1850 than in 1600. Rome in 1913, with a total of

600,000, had climbed above one-half of the dimensions it had in antiquity. If we assume that roughly 150,000 people lived in Periclean Athens, the Greek capital did not regain its ancient size before 1900. For almost the whole of Europe an upward trend had set in by the 1850s; urbanization spread to every country on the continent—even to bottom-ranking Portugal. Not a single one of Europe's major cities would subsequently lose population. The phenomenon of urban decline was for the time being a thing of the past.

Supergrowth

If we take individual cities, growth was especially spectacular where the statistics started from zero. It is hardly surprising to find that cities grew nowhere as fast as in Australia or the United States. In 1841 Melbourne, the capital of the colony of Victoria (and of its successor, the present-day federal state), was a large village with a population of 3,500. Then came the gold rush and the rapid growth of Victoria's economy in general, so that by 1901 the city had passed the 500,000 mark.[60] At the turn of the century Australia had a top-heavy hierarchy of cities, with a number of large cities that functioned at once as state capitals, international ports, and economic centers, and a series of little-developed medium-sized cities. It was a "third world pattern," but in this case it did not prevent considerable dynamism. Statistically, Australia was one of the most highly urbanized regions in the world.[61]

Colonial North America was a rural world, in which the towns were so small that there could be no question of urban anonymity. Only a few—Boston, Philadelphia, New York, Newport, and Charleston—reached the size of an English provincial city. The big urbanization push in the United States came after 1830 and lasted a hundred years; the share of the population living in cities with more than 100,000 inhabitants has not appreciably increased since 1930.[62] Even more than in Europe, urbanization in the United States was geared to the new forms of transport: canal traffic and the railroad. A city like Denver now became possible without any link to a waterway—a pure creation of the railroad, which alone joined up individual cities into a system.[63] Even in the Atlantic Northeast, with its old cities from the colonial period, the railroad brought new centers into being and generally achieved their horizontal and vertical compression into an urban system.

In the West such a system suddenly appeared after the middle of the century. Its largest city, Chicago, exploded from a population of 30,000 in 1850 to 1.1 million forty years later.[64] Chicago and other Midwestern cities, like the cities of Australia around the same time, literally arose out of nothing. As the frontier moved westward, cities did not spring up from the land in keeping with the European model but anticipated an expansion of trade before their surrounding country had been opened up for farming.[65] On the Pacific coast, the loose network of Spanish mission stations had already laid the geographical foundation for the development of cities. California was never a stomping ground of cowboys and Indians: in the absence of village structures, its cities already accounted for 50 percent of the population in 1885, at a time when the average for the United States was around 32 percent.[66] But the real population boom, in absolute figures, began afterward. In the 1870s Los Angeles still had some features of a Mexican pueblo; only later did it become "Anglo" (or anyway mainly English-speaking), Protestant, and white.[67]

Alongside the growth of cities in the industrializing English Midlands, the urbanization of the American Midwest and of the Australian southeastern coast were the most spectacular cases of the sudden formation of an urban archipelago. In special circumstances, a city could blossom forth even in relative isolation. Driven by the rapid development of an export economy on an agricultural frontier, a city like Buenos Aires—which in Spanish colonial times had been of no great importance—could rocket from a population of 64,000 in 1836 to one of 1,576,000 in 1914.[68]

Such rapid multiplication was rare in Europe. Throughout the period from 1800 to 1890, Berlin, Leipzig, Glasgow, Budapest, and Munich were among the fastest-growing large cities, averaging 8 to 11 percent per annum. The others, including London, Paris, and Moscow, expanded more slowly. But no city in Europe matched the growth rates in the New World, even of older cities with a colonial past: New York (47 percent), Philadelphia and Boston (19 percent).[69] The picture changes somewhat if the two halves of the century are taken separately. In the second half, the growth pattern in the largest East Coast cities was not fundamentally different from what was happening in Europe. Indeed, it was then that the influx into big cities was at its height worldwide, and new land and groups of people were being incorporated into them from surrounding areas. Only in England and Scotland, and to a lesser extent in Belgium, Saxony, and France, was the change smaller than in the first half of the century. The period between 1850 and 1910 witnessed the highest annual rate of growth of the urban population in the whole history of Europe.[70] In 1850 there were two cities in Europe with a population over one million—London and Paris—and then a large gap before a group with 300,000 to 500,000. By 1913 the distribution was more even, with thirteen cities above the one-million mark: London, Paris, Berlin, Saint Petersburg, Vienna, Moscow, Manchester, Birmingham, Glasgow, Istanbul, Hamburg, Budapest, and Liverpool.[71]

Which forces were driving the growth of cities? Unlike in earlier history, political will was not the primary factor. There was little in the nineteenth century to compare with such titanic founding acts as the establishment of Edo (Tokyo) in 1590 by the warlord Tokugawa Ieyasu, the elevation of Madrid to capital status in 1561 (though it was built up into a metropolis only after 1850), the resolution of Tsar Peter I in 1703 to build the fortress of Saint Petersburg on an island in the Neva, or the decision of the young United States in 1790 to create a brand new capital on the Potomac. At best, the transfer of the viceregal government of British India from Calcutta to Delhi in 1911 and the construction of a high-prestige capital in the new location belong in the same category of events. Cities did not grow *because* they were seats of government or official residences. Only a few colonial capitals in Africa (Lagos, which in 1900 was only a one-third of the size of the old Nigerian metropolis Ibadan, or Lourenço Marques, the capital of Portuguese East Africa) and some advance posts of Russia's eastward expansion such as Blagoveshchensk (1858), Vladivostok (1860), and Khabarovsk (1880), constituted—demographically not very successful—exceptions. On the other hand, after the formation of nation-states in Germany, Italy, and Japan, many small residential seats lost their importance and often many of their inhabitants.

In the nineteenth century, the growth of cities was driven more than ever before by market forces and private initiative. The rise of some of the largest and most dynamic cities in the world was the result of private "civic" action; they were not so much seats of power and prestigious high culture as business centers that competed keenly with places of higher political status. Chicago, Moscow, and Osaka are good illustrations of this.[72] The advantages that really counted now were better organization of the social division of labor, better availability of sophisticated services (especially in the finance sector), more complex market mechanisms, and faster communications. Thanks to new technologies (steamship travel, canal construction, railroads, telegraphs, and so on), big cities could continually increase their radius of operations. The opportunities for rapid growth were especially great in cities such as Buenos Aires, Shanghai, Chicago, Sydney, and Melbourne, which developed the resources of a vast hinterland for the world market without initially being industrial centers in their own right. In both colonial and noncolonial contexts, it was generally port cities that were able to record the highest growth. In Japan, for example, not industrialization but the opening up of foreign trade has been seen as the main source of the growth of big cities.[73]

City Systems

Although viable cities were only seldom founded by administrative fiat, central state coordination usually had a favorable effect on the formation and construction of city systems, by creating a large degree of uniformity in legal and monetary matters, imposing exchange and communication standards, and planning and funding the city infrastructure for the common good. The last of these points was especially important. Even before the age of the railroad, the construction of river-canal systems in England and the United States made a major contribution to interurban communication. At the beginning of the nineteenth century, goods could be sent by waterway to London from all parts of Britain, and people in the United States had good reason to celebrate with pride the opening of the Erie Canal in 1825.[74] Elsewhere in the world, such things were a geographical possibility only in the Ganges Valley, the hinterlands of Canton, or the "macroregion" of Jiangnan. In China, however, the various city systems were never (not even in China proper) integrated into a single national system, and new technological possibilities were scarcely exploited. The horizontal integration and vertical differentiation of city systems was therefore bound up not only with fundamental social-economic processes such as industrialization but also with the building of nation-states. Economic success in the nineteenth century went to countries with an internally integrated and differentiated city system that was also open to the outside. Whereas nation-states required city systems, cities were themselves often not dependent on a functioning nation-state framework. The absence, or relative weakness, of integration into a national territory did not hinder the development of a large colonial port like Hong Kong or that of a noncolonial maritime city such as Beirut on the periphery of the Ottoman Empire.[75]

Most of the national city systems were permeable to the outside. Whereas the nation-state (if one existed) increasingly became the framework for the organization of economies in which urban industrialization played a growing role, *very large* cities were directly linked into international networks of trade, migration, and communications. In other words, even in the "age of the nation-state," individual countries were not necessarily "stronger" than big cities, which served to gather and distribute capital (not only national in origin) and provided a base for "transnational" connections. The development of cities is no more a direct consequence of state formation than it is an epiphenomenon of industrialization.[76]

For the early modern period, networking across great distances is already a dimension without which the history of cities cannot be written. There were regular trade links within Europe (e.g., in the shape of fairs), as well as maritime activity, first in the Mediterranean, then on a much larger scale between Atlantic ports such as Lisbon, Seville, Amsterdam, London, Nantes, and Bristol and their counterparts across the seas. These might be either colonial ports (Cape Town, Bombay, Macau, Batavia, Rio de Janeiro, Havana) or ports under indigenous control (Istanbul, Zanzibar, Surat, Canton, Nagasaki). Colonial cities such as Batavia or the ports of Spanish America were often slightly modified copies of European town patterns. At least one colonial city, however, saw itself not as a satellite or bridgehead of Europe but as a center with political and cultural functions in its own right: this was Philadelphia, which in 1760—barely eighty years old and, with 20,000 inhabitants, a little larger than New York—was one of the most dynamic cities in the English-speaking world. Its core, where trade, politics, and culture were concentrated, was the nexus between the colony of Pennsylvania and the rest of the Atlantic space.[77]

In the nineteenth century—and this was new—the geography of colonial urban growth, when seen on an international level, came to obey laws of the market more than it did political guidelines. Until the 1840s the Navigation Laws governed many overseas commercial relationships in the British Empire, stipulating, inter alia, that an export producer such as Jamaica, in return for a monopoly over its products in the British market, had to obtain its imports from the United Kingdom. These laws were a sharp weapon in the rivalry among

European trading empires and were one of the reasons why the influence of Amsterdam in the world economy fell behind that of London in the eighteenth century. At exactly the same time that the Navigation Laws ceased to operate, British pressure in China ended the monopoly privileges of Canton as a transshipment port for the whole of the overseas trade with Europe. Such mercantilist regulations, quite independently of one another in Asia and Europe and its colonies, had led directly or indirectly to a preference for certain cities and the impeding of the rise of others. Path dependencies created in the early modern period had continued to function in later centuries as established structural facts. But from the 1840s onward, the global imposition of free trade and the free movement of persons strengthened the market (nonstate) features of the changing city systems.

Networks and Hubs

City systems may be conceived in two different ways, vertically and horizontally. On the *vertical* plane, a size-graded hierarchy of settlements stretches up pyramid-like, in shorter or longer steps, from a multiplicity of villages at the bottom to a central location at the top. On the *horizontal* plane, it is a question of relations among cities, and thus of the networks in which they are inserted and to whose development and functioning they contribute. If the first model may be visualized as a structure of subordination and superordination, the second may be seen in terms of interaction between an urban center and its periphery, or with another urban center of a similar kind. The farther one moves up the hierarchy, the more easily the two models may be linked to each other, for many cities, especially large ones, have intensive vertical and horizontal affiliations. The horizontal networking model is more productive for a global historical approach: it lays greater emphasis on cities as hubs than on their dominant position within a regionally defined hierarchy, directing our attention to the fact that control over an immediate hinterland may be much less important for a city than the control it exercises over distant markets or sources of supply. Thus, for example, the textile cities of Lancashire had at least as close relations with the Russian Black Sea ports that supplied them with grain, or with the cotton latifundia of Egypt, as they did with the hinterland of the county of Suffolk. This kind of economic topography, which is invisible on conventional maps, also had political implications. For cities such as Manchester or Bradford, the consequences of the American Civil War were far more direct than those of the revolutions of 1848/49 in nearby continental Europe. But cities were also inserted into wider contexts within the same country. The boom cities of the Industrial Revolution may have been able to organize on their own the tasks of production, raw materials procurement, and marketing, but they were still dependent on political and financial decisions made in London.

The networking approach has the further advantage that it can elucidate city formation in the periphery. Many new cities of the nineteenth century did not so much grow out of their rural surroundings as they expanded because of their attractions to *external* interested parties.[78] This was true of numerous cities in the colonies and the American West, but also of Dar es Salaam—which in the late 1860s, before its colonial period, was created ex nihilo by Sultan Seyyid Majid of Zanzibar as the terminus of the caravan trade[79]—and of fast-emerging metropolises such as Beirut. At the beginning of the century Beirut had a population of just 6,000; by the end it was over 100,000. Its rise would have been impossible without the old urban tradition of Syria, but the real driving force was the general revival of Mediterranean trade originating in Europe.[80]

The openness of city systems to the outside world is a direct result of constant circulation. Networks are the product of human action; they have no "objective" existence. Historians, too, must try to see them within the perspective of their creators and users. Networks are also reshaped internally: the relations among their various hubs, the cities, are constantly shifting. If a particular city stagnates or "goes into decline," this also must be evaluated in the context of the city system of which it is part. City systems often display much persistence in change: thus, no completely new city has broken through to real preeminence anywhere in Europe during the last century and a half. It may also happen that the overall level of urbanization remains the same even though the system undergoes tectonic shifts internally; shrinkage and loss of function on the part of one city may be offset by growth elsewhere. In India many have nostalgically mourned the decline of old seats of residence, failing to see that the economic, and to some extent cultural, dynamic often switched to smaller market towns at a lower level in the hierarchy of functions and prestige. New patterns may, as it were, emerge in the shadows behind "official" urban geographies.[81]

Cities that are dominant in both models—that function, in other words, both as important hubs in horizontal networks and as the summits of vertical hierarchies—may be called "metropolises." Furthermore, a metropolis is a large city that (1) gives widely recognized expression to a certain culture, (2) controls an extensive hinterland, and (3) attracts large numbers of people from other areas to come and live in it. If, in addition, a metropolis forms part of a global network, it deserves the title of a "world city." Were there "world cities" in the early modern period and in the nineteenth century? It is difficult to give an answer, because the term has several meanings in current usage. It would be tautological and much too simple to define them as cities "of actual or potential global importance"; Uruk in ancient Sumeria would have counted as the first "world city" by that criterion.[82] Fernand Braudel defines a world city, rather more precisely, as one that dominates its own circumscribed "world economy," as Venice or Amsterdam did for a time.[83] Only in the nineteenth century, he argues, did a *globally hegemonic* city emerge in a single specimen: London. After 1920 its place was taken by New York. Yet this, too, is a highly simplified view of things: if there was anything like a "cultural capital of the world" in the nineteenth century, it was Paris rather than London, with strong competition around 1800 and around 1900 from Vienna (which carried scarcely any weight in world trade or finance). Nor was the "changeover" from London to New York so neat that it can be dated to a precise year; London remained the heart of a world empire, and it kept its central financial position even after its relative importance in trade and industry receded.

Nowadays it is more usual to speak of "world cities" or "global cities" in the plural, meaning that a global city is one among several nationally rooted "global players," rather than just one highly influential metropolis or heart of a great empire.[84] They are part of a global system, in which the links *among* world cities in different countries are stronger than their integration with a national or imperial hinterland. Such detachment from a territorial base is possible only as a result of today's information and communications technologies.[85] Many of the parameters that permit statements to be made about a city's ranking in the global hierarchy first took shape toward the end of the nineteenth century: for example, the presence of transnational corporations, with their own internal hierarchy of headquarters and branches, or of international organizations, or insertion into global media networks.

Empirical studies of the manner and frequency of contact among the largest world cities have not yet been undertaken for the nineteenth century. If they were, they would probably lead to the conclusion that only late-twentieth-century technologies brought about a special system embracing the metropolises, a true system of world cities. Before intercontinental telephony, radio communication, and airline links became normal and regular parts of life, it cannot be said that the largest and most important cities in various continents formed a permanent fabric of interaction and communication. Later, of course, satellite technology and the Internet brought a further quantum leap in networking. In this respect the nineteenth century—when crossing the Atlantic was still an expensive adventure, not an affordable routine—appears as the dull prehistory of the present day. This even includes the age of the zeppelin and the heyday of the fast, comfortable ocean liner able to complete the transatlantic journey in four to five days, which began in 1897 with the introduction of the first superliner, Norddeutscher Lloyd's 14,000-ton *Kaiser Wilhelm der Große*. The continuous linking of London, Zurich, New York, Tokyo, Sydney, and a few other top metropolises is an innovation dating from around 1960, which became possible only with swift and frequent airline travel.

4 Specialized Cities, Universal Cities

Pilgrimage Sites, Spas, Mining Towns

From a certain size up, it is not easy to classify cities in terms of a single function; they play several roles at once. Cities are mostly pluralist. In every age, however, this does not apply to ones that concentrate labor of a highly specialized kind. In the mid-seventeenth century Potosí, situated 4,000 meters above sea level in an extremely inhospitable part of what is now Bolivia, had a population of around 200,000; this made it the largest city in the Americas—a position due entirely to the fact that the most extensive silver deposits in the New World were to be found there. Significantly larger still, in the early eighteenth century, was Jingdezhen in the central Chinese province of Jiangxi, which produced pottery for the domestic and international market and, until the advent of the machine age, was probably the largest manufacturing center anywhere in the world. In the nineteenth century there were also single-function cities of an older kind: the religious pilgrimage sites, which, though with a highly mobile and fluctuating population, are themselves often stable over a long period of time. In addition to ancient cities such as Mecca and Benares, many new sites sprang up in Hindu and Buddhist, Muslim and Christian countries, such as Lourdes on the northern edge of the Pyrenees, which shot to fame in the early 1860s. Pilgrimages to such places were big business, never more so than in the late nineteenth century. The Dutch Orientalist Christiaan Snouck Hurgronje, who spent a year in 1884–85 in Arabia studying Muslim scholarship, noted that rampant commercialism was changing the character of the population of Mecca and had caused much disappointment among pious pilgrims;[86] things must have been similar in Lourdes. Charismatic movements can concentrate large numbers of people in a brief space of time. Not long after Omdurman was founded in 1883, the capital of the Mahdi movement in Sudan constantly had up to 150,000 people within its boundaries: religious devotees and soldiers; it was hard to tell the two apart.[87] Open at its back to the desert, where the Mahdi recruited most of his followers, the city was fortified on the Nile side—the opposite of the situation at Khartoum. It was at once a religious center and a military camp. Nothing remained of it after British troops crushed the movement in 1898.

Other kinds of single-function localities first emerged in the nineteenth century. The railroad created the junction city, where different lines crossed: good examples are Clapham Junction in South London, Kansas City, Roanoke in Virginia, and Changchun in Manchuria (a Chinese backwater astride the eastern railroad that the Russians built to China in 1898). Similarly, Nairobi grew out of a settlement that the British had built to serve as the logistical center for the construction of the Uganda railroad.[88] Railroad workshops, too, were usually located in such places. If these cities also provided the main connection between river and railroad, their opportunities for growth were especially favorable.

A further nineteenth-century novelty was the leisure and bathing resort. This must be distinguished from the spa town of the eighteenth century, where members of the upper classes traveled to fortify themselves by "taking the waters," and to mix in high society: Karlsbad in Bohemia, Spa in Belgium, Vichy in France, Yalta in Crimea, Wiesbaden and Baden-Baden in Germany were celebrated examples. They were also Western outposts of eastern European aristocracies and increasingly—in various degrees of exclusiveness and expense—magnets for the middle-class families of bankers and senior officials, considered slightly disreputable because of the gambling associated with them. Bad Ems, where Wilhelm I of Prussia took his cures, was the scene of the diplomatic imbroglio in 1870 that made it easier for Bismarck (kept up to date by telegraph) to "provoke a war in defense of the German nation."[89] Emperor Franz Joseph of Austria opted many times for Bad Ischl—when he did not head straight for Nizza where, in 1895, he shared the same grand hotel with the former British prime minister William Ewart Gladstone. The two aging gentlemen did not, however, exchange a word with each other.[90]

Democratization of the seaside vacation originated in England and Wales, and it was there, too, that a "holiday industry" first began to develop as an increasingly important factor in the economy. In 1881 there were 106 recognized coastal resorts in England and Wales; in 1911 there were already 145, with 1.6 million people living in them (roughly 4.5 percent of the total population). Demand grew in the sector, trickling down from the upper classes to other parts of an increasingly prosperous society, and the supply adjusted more and more smoothly to the needs of the different strata. In the same way that the older-style spas specialized in the treatment of certain disorders, the various coastal resorts were each geared to a clientele with a particular social profile. There had already been such a hierarchy in eighteenth-century England, with the aristocratic and upper-middle-class towns of Bath and Tunbridge Wells at the top. By midcentury, to the north in Lancashire, some members of the "lower classes" had discovered for themselves the joys of sea bathing.

The bathing resort was a special kind of urban environment, not centered on parks, cure facilities, and thermal baths but altogether geared to the beaches along the open shore. The social climate here was less formal than in the inland spas; life was more relaxed, status distinctions had to be displayed less often, and children found the latitude they were otherwise denied. The average sojourn was far shorter than in the spa resorts: one stayed for a week or two, not several months. By 1840 the bathing resort had taken shape in England and Wales, with most of the characteristic features that we still see today. The prototype was Blackpool on the West Coast, whose 47,000 permanent residents catered (in 1900) for more than 100,000 vacationers. On offer were the early achievements of a special "fun architecture," originally developed for various world exhibitions, and here presented—together with a circus, opera, and ballroom—in the imposing form of an imitation Eiffel Tower and a walk-in old English village.[91] Subsequently the seaside resort owed its growth to increased leisure time, greater affordability, and good railway and highway connections. By the turn of the century there were coastal resorts of more or less the same kind all around the central Atlantic and the Mediterranean, on the shorelines and islands of the Pacific, on the Baltic Sea, in the Crimea, and in South Africa. In China, people had traditionally gone to the mountains for relaxation; hot springs, not the sea, were the places for bathing. The opening of the Beidaihe resort on the Gulf of Zhili was mainly for the sake of the Europeans who, by the end of the nineteenth century, were living in large numbers in the nearby cities of Beijing and Tianjin. Today its hundreds of hotels attract droves of tourists, the best beaches being reserved, of course, for members of the party and state leadership. The seaside town is unambiguously an early nineteenth-century Western invention, whose origins went back to pre-industrial times, and which has continued to spread around the globe until the present age of the postindustrial service society.[92]

Another new form to be found on every continent was the mining town, already exemplified by Potosí in the early modern period. In the nineteenth century, societies dug deeper underground than ever before. Coal mining provided the energy source for industrialization and was, in turn, made more effective by a number of technical improvements. The specialized mining town became emblematic of the epoch. There were instances in Silesia and the Ruhr, in Lorraine, in the English Midlands, in the Ukrainian Donbass, and in the Appalachians. Soon after 1900, coalfields also began to be opened up in northern China and Manchuria, where it was partly British and partly Japanese businesses that introduced the latest technology. Industrialization also generated demand for other mining products, while the science of geology and advances in mine construction and extraction made it possible to work new deposits. Not only the technically and financially straightforward panning of gold in California and Australia, but also the opening of new mines that required considerable investment, led to outbreaks of gold fever and ultrarapid concentrations of laborers. In Chile, copper was already mined in colonial times alongside gold and silver, and in the 1840s there was a sharp rise in output and exports of the metal. For several more decades, however, copper mining remained in most cases a small-scale craft operation; steam engines were rarely deployed. Even after modern technology became the norm around the turn of the century, no real mining towns sprang up in Chile. Miners' camps tended to be isolated enclaves on the margins of the local economy.[93]

An example of a real mining city was Aspen, Colorado, where silver deposits were discovered in 1879 and urban developers followed hard on the heels of the first "prospectors." By 1893 two isolated log cabins had grown into the third-largest city in Colorado, with paved streets, gas lighting, two kilometers of streetcar tracks, a municipal water supply, three banks, a post office, a city hall, a prison, a hotel, three newspapers, and an opera house. But, also in 1893, what was called "the finest mining city in the world" lost the economic basis for its existence when the price of silver fell through the floor.[94]

Capitals

The opposite of such specialized cities were the metropolises, which, in addition to many particular tasks, carried out the *central* functions of the city: (a) civil and religious and administration; (b) overseas trade; (c) industrial production; and (d) services.[95] While a large number of cities have services constantly on offer, the importance of the other three functions may be said to define three different kinds of city: the capital, the industrial city, and the port. Of course, it is possible for the same city to be all three at once, but there are surprisingly few examples of this. New York, Amsterdam, and Zurich are not national capitals; Paris, Vienna, and Berlin are not ports; Beijing, a long way from the sea, had scarcely any industry until a few decades ago. At most London and Tokyo are seats of government *and* seaports *and* industrial centers. Nevertheless, the functional emphases diverge so much that it is not entirely arbitrary to isolate the three distinct types.

A capital, however large or small in terms of population size, stands out from other cities in being the center of political and military power. Other distinctive features follow from this. A capital is also a residential location—the seat of a court and of a central bureaucracy. The labor market of a capital is more geared to services here than in other cities—services that range from supplying members of the ruling apparatus to an especially active, artistically demanding, construction industry. Rulers must attend especially well to the population of the capital, since even in the most repressive political systems it is the stage of mass politics. In premodern societies the grain supply to the capital was a hugely important political issue—in imperial or papal Rome no less than in Beijing, which obtained most of its food by canal from central China. The Ottoman sultan was directly responsible for the population of Istanbul, and he was expected not only to ensure its basic food supply but also to protect it from usury and other abuses; this had not changed by the early nineteenth century.[96] The urban "mob," especially feared in London but also active elsewhere, contained revolutionary dynamite; it could be manipulated or repressed but not always reliably kept under control. The nineteenth-century capital was a place where sovereigns were crowned, and often buried, with pomp and circumstance. It was also a symbolic terrain, on which conceptions of political order were converted into geometry and stone. No other cities are as charged as capitals with layers of historical meaning; their prestige architecture expresses in visual terms the sovereign will of their past rulers.

With the notable exception of Rome, capitals have rarely been religious centers of the first order. Places such as Mecca, Geneva, and Canterbury never functioned as capitals within the framework of a nation-state. Yet, by virtue of the sacralized monarchy, the capital was automatically an arena of religious ritual. The Chinese emperors of the Qing Dynasty performed the prescribed rites in the course of the year; and the Ottoman sultan, in his capacity as caliph, was the supreme head of Sunni believers. In Catholic Vienna the alliance of throne and altar consolidated itself after 1848. Emperor Franz Joseph never missed an opportunity to take part in the magnificent Corpus Christi processions or to perform the Maundy Thursday ritual foot washing on twelve carefully chosen residents of municipal retirement homes.[97]

Finally, capital cities always strove to be independent in their exercise of key cultural functions. But true cultural capitals are not selected by governments or commissions; their decisive magnet effect can arise only through communicative compression and the development of culture markets, neither of which is really susceptible to planning. The outcome is not always successful, however. In the eighteenth century Philadelphia was for a time what it wished to be: the "Athens of the New World." But its successor as capital of the United States, Washington, DC, was never able to establish such a degree of cultural hegemony vis-à-vis other American cities. Nor did Berlin, at least before 1918, acquire the cultural weight of a dominant national metropolis, in the manner of London, Vienna, or Paris.

Few new capitals appeared on the scene in the nineteenth century, apart from those of the Spanish American republics, which had already been the main administrative centers in colonial times. Exceptions were Addis Ababa, Freetown in Liberia (a "real" European-style capital[98]), and Rio de Janeiro, which, as the seat of the Portuguese monarchy after 1808 and then the capital of the independent Empire of Brazil after 1822, was built up into a "tropical Versailles."[99] In Europe the most important new national capitals were Berlin, Rome (which followed Turin and Florence for the honor in 1871), Bern (since 1848 the "federal city" of the Swiss confederation), and Brussels (which could look back to a past as capital but only in 1830 concentrated all the central functions of the Kingdom of Belgium). Another interesting case is Budapest, which became the second capital of the Danubian monarchy after the "Compromise" of 1867. In the competition with Prague, it was a factor of utmost importance that the Czech metropolis never obtained the status of capital within the Habsburg Monarchy. Budapest, formed as such through the fusion of Buda and Pest in 1872, became one of the great showcases of urban modernization in Europe, and by the end of the century its gradual Magyarization had also given it a markedly national character in cultural as well as ethnic terms. The tension between Vienna and Budapest nevertheless continued within the new imperial context.

The Austro-Hungarian Dual Monarchy was an expression of a broader nineteenth-century trend toward twin metropolises, which often involved a deliberate separation of political and industrial functions. Not only Washington, DC, but also Canberra and Ottawa were tranquil provincial centers in comparison with the towering commercial, industrial, and service-providing cities of New York, Melbourne, Sydney, Montreal, and Toronto. Many other regimes encouraged such forms of competition. The pasha of Egypt, Muhammad Ali, stuck with Cairo as his capital, but he did more to raise Alexandria out of the decay into which it had fallen.[100] Elsewhere, "second cities" came forward with the strength of bourgeois assertiveness. Moscow got over its loss of capital status in 1712 and became the main center of early industrialization in Russia. Osaka, which received little support from the central government after the Meiji turnaround in 1868, strengthened its position as a port and industrial city; a modern rivalry between Osaka as business center and Tokyo as seat of government replaced the old antagonism between the shogun in Edo and the emperor in Kyoto. In China, the rise of Shanghai from the 1850s onward was a serious challenge to Beijing as the seat of government such as the centralized system of rule had not experienced since the fifteenth century. The tension between bureaucratic-conservative Beijing and commercial-liberal Shanghai persists to this day. A similar dualism, not at all politically planned, took shape in colonial urban geography, especially in the older colonies. Economic centers such as Johannesburg, Rabat, and Surabaya gained ground at the expense of capital cities such as Cape Town (replaced in 1910 by Pretoria), Fez, or Batavia/Jakarta. In Vietnam the roles were similarly distributed between the political capital in the North, Hanoi (which had been the ruler's residence before 1806 and again became the seat of government in 1889 under the French) and the economic center in the South, Saigon. In the new Italian nation-state, an opposition developed between Rome and Milan. In India the conflicts sharpened in 1911, when the government apparatus was transferred from the economic center, Calcutta, to the recently built capital of New Delhi. It is a striking fact that few nineteenth-century cities in the world followed the model of London or Paris to become metropolises with all-embracing functions. Even in dynamic counterexamples such as Tokyo and Vienna, which rested on foundations going back hundreds of years, the challenge of a "second city" was not far away. In Rome itself a dualism persisted between the secular regime and the Vatican.

Princely and Republican Residences

None of the top five European metropolises (and population centers) in 1900—London, Paris, Berlin, Saint Petersburg, and Vienna—was a creation of industry like Manchester, which since 1800 had risen from twenty-fourth to seventh place among the cities of Europe and pushed up close to the front-runners. But these were also too big to be purely political capitals, or to allow themselves to be dominated by a royal court. In France, Napoleon and Joséphine had created a new-style court of parvenus and winners from the revolution, but since the emperor was often away, no physical center of rule established itself in Paris before his final demise in 1815. Subsequently, the restored Bourbons and even more the "bourgeois monarch," Louis Philippe, cultivated a rather modest style of self-presentation, which Ahmed Bey of Tunis liked so much that he faithfully copied it to mark his distance from the Ottoman rulers in Istanbul.[101] In London the monarchy projected itself even more soberly: Prince Pückler-Muskau, the penetrating observer of things British, wrote in 1826 that it was only thanks to John Nash and his lavish work on Regent Street that the English capital had retained the aspect of a seat of government.[102] But the conversion of the ruined Buckingham House into Buckingham Palace between 1825 and 1850 was no architectural masterstroke, and Queen Victoria preferred her other palaces at Windsor, Balmoral, and the Isle of Wight. In Vienna the imperial Hofburg residence looked positively unassuming beside the pomp of the Ringstrasse.

Nowhere did court overshadow city to such a degree as in the late-absolutist imperial centers of Istanbul and Beijing, where whole districts were reserved for the use of the rulers and their household. Over the course of the century, however, many imperial properties in Istanbul—often gardens or sites of wooden palaces—were converted for public use as arsenals or port or railroad installations.[103] Beijing, a much older city in appearance, remained untouched by the railroad until 1897 and by modern industry even longer. At the turn of the century, the court and central government offices were still grouped together behind the walls of the Forbidden City, but they had already lost much of their power to representatives of the Great Powers in the diplomatic quarter, to governors in provincial capitals, and to the capitalists of Shanghai. Beijing was an architectural shell, a densely populated symbolic landscape with little political substance. When it was invaded in 1900 by peasant bands from the countryside and by troops of the Great Powers, an era came to an end. Army boots marched through the halls of the Forbidden City, horses were stabled in its temples, and officials had burned the state papers and fled. Beijing remained the capital of China until 1927 and became capital once again after the Second World War. The Christian churches sacked during the Boxer Rebellion were rebuilt, but scarcely any of the damaged temples were. Imperial Beijing never recovered from the shock of 1900, its dignity and ritual aura dispelled forever.[104] A few years later Beijing, now equipped with modern hotels, beckoned alongside Rome, the Giza pyramids, and the Taj Mahal as one of the great attractions of the dawning age of international tourism.[105]

The core of the American republic in Washington, DC, also had a war behind it. In August 1814 the British set fire to the Capitol and the White House. The city on the Potomac was the prototype of a planned capital. The first design was approved by Congress as early as 1790, and in 1800 it became the seat of the presidency. The approximate location was a compromise between the Northern and Southern states, while the exact site was personally chosen by George Washington, who had engaged the architect Major Pierre Charles L'Enfant. The very first plan, like so much else, stemmed from Thomas Jefferson, who opted for a chessboard schema. L'Enfant then worked this out on a grand scale, with wide boulevards, "magnificent distances," and splendid open spaces. The master builder, who had grown up as a boy at Le Nôtre's Versailles (where his father had served as a court painter), had learned there to think in terms of axial planes. His design for the American capital was therefore ultimately inspired by a late Baroque vision. It is striking that around the same time (between 1800 and 1840), but without any demonstrable connection, the Russian capital Saint Petersburg was redesigned as a neoclassical ideal city in a similar *esprit mégalomane*—at much greater expense and with more clearly defined results.[106] In this case, the initiative came from a man who was the polar opposite of the republican George Washington: Tsar Paul I, one of the worst despots of the age. The Kazan Cathedral in Saint Petersburg was intended as a Russian match for the dome of Saint Peter's in

Rome, and Saint Isaac's Cathedral as a synthesis of the whole European cathedral tradition. In Washington, DC, sacred architecture played no role at all.

For a long time, the original plans were inconsistently applied in both spirit and detail. As early as 1792, following a violent argument with the good-natured President Washington, L'Enfant was dismissed and took many of his plans away with him.[107] The urban area of "Washington" then became a field of experimentation, albeit on a smaller scale than L'Enfant had imagined; his official residence for the president would have been six times larger than today's White House. What first went up showed little sign of L'Enfant's sense of grandeur. Charles Dickens, who passed through in the spring of 1842, was distinctly unimpressed: it was a city not of magnificent distances but of "magnificent intentions": "spacious avenues, that begin in nothing, and lead nowhere; streets, mile-long, that only want houses, roads and inhabitants; public buildings that need but a public to be complete."[108] Capitol Hill, the site of both Houses of Congress, acquired its domes and side wings only at the end of the 1860s. The final design of the Mall followed only in the 1920s. The Lincoln Memorial was finally inaugurated in 1922, the funds for the Jefferson Memorial approved only in 1934. The eclectically conceived classical complex, adorned with the late nineteenth-century neo-Romanesque Smithsonian Castle, is essentially a creation of the architect John Russell Pope from the period between the two world wars. Washington belies its own youthfulness.

Manchester, a "Shock City"

Washington occupied a marginal place in the city system and was a long way from the industrial powerhouses of the nineteenth-century economy. Of the capital cities, it was Berlin—no match for the history of London, Paris, and Vienna or for their central location in the city system—that most closely corresponded to an industrial city. No other place during the industrialization of Germany concentrated so much cutting-edge technology, especially in the electrical industry. Berlin was not a center of the first phase of industrialization based on steam power; it found its character with the systematic application of science to industrial production. Corporate research and development, closely linked to state-organized science and large customers, had never before been so important for economic innovation. The Berlin of the Kaiserreich became the first "technopolis" or, as Peter Hall put it, "the first Silicon Valley."[109] Paris, in particular, was by comparison a city of services and small businesses; the two were, from an economic point of view, a metropolis of the past and a metropolis of the future.

Not only Paris but some of the fastest-growing and economically most modern cities in the world never became really prominent centers of industry. The dominant pattern in London, unlike late nineteenth-century Berlin or Moscow, was a combination of small and medium-sized industry with a large service sector, including the international financial services of the city. New York, around 1890, was still essentially a mercantile city and a port.[110] Both New York and London drew much of their dynamism from their own internal needs; both had a construction industry that was an important engine of growth. London did not have large iron and steel plants on the scale of Krupps in Essen (where one entrepreneurial family dominated an entire city), and its textile sector (like those in Paris and Berlin) consisted more of tailors and ready-to-wear manufacturers than of mechanized cotton spinners and weavers. London and the Lower Thames led the field in shipbuilding at the beginning of the century, but by its end Glasgow and Liverpool had moved into first and second place.[111] London's locational advantages pointed not toward large corporate specialization but to a wide variety of branches of production; they ensured that it did not look like a typical center of the textile, steel, or chemicals industry. The impression that large firms are generally more modern than small businesses is misleading. As the nineteenth century progressed, the economic modernity of a big city lay increasingly in its capacity for innovation—which is possible in many different forms of enterprise.[112]

Where does one find "typical" industrial cities—a nineteenth-century type doubtlessly without historical precedent?[113] At first they were to be seen only in England. People from France or Germany who visited the English Midlands before 1850 were used to the old-style towns of the early modern age and did not understand industry-driven urbanization. They might have experienced the damp basement dwellings of Manchester as an intensification of the urban poverty familiar to them back home, but they were unprepared for the smokestack landscapes and giant factories. Manchester, in particular, because of its new physical dimensions, seemed in the 1830s and 1840s like the "shock city" of the age.[114] Here were seven-story factory buildings that had been built without a thought for aesthetics or how they fit into their urban surroundings; this was clearest not so much in the inner city as in small localities where industry had shaken everything up in the briefest space of time. In the first generation of industrialization—roughly from 1760 to 1790 in England—the new factories already towered over most of the settlements in which they were built. Two or three might turn a village into a small town, and later a single company created many an industrial center. Chimneys became hallmarks of a new kind of economy, defining a cityscape even when they were disguised as Italian campaniles.[115] Other cities were completely refounded as localities that for a long time would have industry as their basic reason for existence: from Sheffield to Oberhausen, from Katowice to Pittsburgh. Others might have had a really significant preindustrial past, but were transformed by industry for the first time into large cities.

The much-maligned Manchester, which observers such as Charles Dickens, Friedrich Engels, and Alexis de Tocqueville saw as an apparatus to transform civilization into barbarism, was the best-known example of such a single-function metropolis.[116] Here new industrial concentrations and an influx of labor ran ahead of any possible development of infrastructure. The population of Birmingham more than tripled between 1800 and 1850, from 71,000 to 230,000, while

Manchester grew in the same period from 81,000 to 400,000, and the port city of Liverpool from 76,000 to 422,000.[117] Manchester and cities like it shocked people at the time by their dirt, noise, and smells, but also because they seemed to lack a clear urban form; they grew very quickly, without the institutions and distinguishing features regarded as essential to a city. Economic functionality created spaces and social environments for itself, whereas earlier it would not have occurred to anyone to see the economy as the ultimate basis for city life.[118] This was reflected in the architecture, since the factory could not be inserted smoothly into the general design. In such cities, the focus of urban planning changed from comprehensive design to local problem-solving. Factories whose location had been chosen purely to maximize profits inevitably had a centrifugal effect, whereas the European city had traditionally always tended to build up the downtown area.[119] Perhaps the reason why new city halls, from Manchester and Leeds to Hamburg and Vienna, were so much larger than their early modern predecessors was that those responsible for them wanted to balance the symbolism of capital (and in Vienna, the court) with a symbol of the public spirit.

To be sure, the Manchester model was not the only possible way of linking industry and city. Birmingham, for example, as Tocqueville recognized after a visit to both cities, used a different formula that corresponded to its more diversified economic structure, and Manchester was not as typical as the young Friedrich Engels would claim not long afterward.[120] The Ruhr region, too, arose out of a pure combination of economic factors, yet it came up with quite different solutions. Its recipe for success was found at the moment when four elements came together: coal extraction, coke technology, the railroad, and the influx of labor from farther east. At first, however, there were no urban structures in the Ruhr Valley, only sprawling workers' settlements with up to 100,000 inhabitants, which initially had the legal status of villages. The Ruhr did not develop a single urban core throughout the nineteenth century. It was an early example of a "conurbation," a multipolar urban space, as radically new in its way as the concentrated industrial city of the Manchester type.[121]

Today some historians doubt whether even Manchester corresponded to the stereotype of a pure industrial city grinding down its human population. They stress that its early economy was much more varied than an exclusive focus on its cotton industry would suggest. Manchester, too, was part of a city system and a division of labor that eventually took in the whole of central England. Large industrial cities could continue developing only if they played their special role within such systems and if they managed to organize their insertion into a number of environments, from the immediate vicinity to the world market. Industrialists—the pioneering generation as well as those who came later—were more than slave-driving factory masters; they had to form "networks," to keep in mind both advances in technology and the general economic and political situation, and to concern themselves with the collective representation of their interests.[122]

The industrial city should therefore not be viewed only from the point of view of the factory. At least in the larger cities, which were not dominated by a handful of enterprises, a cultural climate took shape in which innovation was possible. Cities such as Manchester, Birmingham, and Leeds were able to rise above the chaos of their industrial takeoff and to draw largely on their own resources of civic involvement. They improved the community infrastructure, founded museums and municipal universities (as opposed to the medieval institutions in Oxford and Cambridge), and adorned their centers with prestigious buildings—above all, a theater and a magnificent city hall, in which a giant organ had pride of place in the main assembly room.[123] The spectrum of human settlements shaped by industry was very broad. It encompassed primitive barracks (as in Russia and Japan), where conditions were at least as bad as in the slums of large industrial cities, but also model instances of entrepreneurial patriarchy, where the factory owner lived beside his factory and ensured that the workplace and his workers' housing conditions were tolerable.[124]

5 The Golden Age of Port Cities

London was, apart from everything else, a port city. Indeed, its whole history from at least the seventeenth century, when overseas trade with the West and East Indies began in earnest, might be described from an ocean vantage. If we were to distinguish between a maritime-mercantile and a continental-political model of a capital city, no place represented them both as perfectly as London did.[125] At first sight, port cities appear to be archaic; industrial cities, modern. But this is deceptive. Not only did some large cities—Antwerp is a good example—convert economy from a preindustrial production economy to an international port/service economy;[126] the nineteenth century also witnessed a transportation revolution that radically altered the nature of port cities. In some parts of the world, urbanization actually began in ports and is still largely confined to them; in the Caribbean, *all* cities that remain important today were founded in the seventeenth century as export-oriented ports. A world of small colonial ports thus came into existence, with Kingston and Havana as the most important; it was tightly woven together by trade and (before 1730 or thereabouts) by piracy.[127]

Rise of the Port City

The nineteenth century was the golden age of ports and port cities—or more precisely, of *large* ports, since only a few could handle the huge quantities involved in the expansion of world trade. In Britain, exports in 1914 were concentrated in twelve port cities, whereas at the beginning of the nineteenth century a large number of cities had been involved in shipping and overseas trade. On the East Coast of the United States, New York constantly strengthened its leading position. After 1820 it became the main port for America's most important export good: cotton. At first cotton ships sailed from Charleston or New Orleans to Liverpool or Le Havre, then stopped in New York, loaded with immigrants and European exports, on the return trip. Increasingly, however, cotton was shipped directly north from the Southern plantations to New York. Until the Civil War, New York middlemen, shipowners, insurers, and bankers dominated the international trade of the Southern states.[128] In China a series of treaty ports opened for overseas commerce between 1842 and 1861, to be joined later by many more. Toward the end of the century, only Shanghai and the British crown colony of Hong Kong had kept pace with the demands of ocean transport, and to a lesser extent Tianjin, the main port in northern China; and Dalian at the southern tip of Manchuria grew fully into the role.

Seaports were what airports became in the second half of the twentieth century: the key transaction points between countries and continents. The first things that arriving travelers saw from the sea were the quays and buildings of a harbor front; the first local people they encountered were pilots, longshoremen, and customs officials. As steamships, freight loads, and crowds of intercontinental migrants multiplied in size and number, sea travel acquired a significance it had never previously had. Of course, not every nation in history with an opening to the sea has shown a liking for salt water; many island dwellers forgot the nautical techniques that had brought their ancestors there in the first place. Tasmanians even lost the habit of eating fish.[129] As Alain Corbin has shown, continental Europeans—or at least the French, who are his main focus of interest—developed an open-minded attitude to the sea only around the middle of the eighteenth century. Amsterdam, which in 1607 was brilliantly conceived as a cityscape between land and water, was an early exception.[130] Outside the Netherlands, coasts and harbors did not become popular themes in painting until the eighteenth century—the period when ports also came to be thought of as worthy of architectural expenditure and top-level feats of engineering. Promenades were built for the first time on the shores of many coastal cities; even in Britain it was only after the 1820s that such an addition was considered de rigueur.[131] On the other hand, the Ottoman upper classes, leaving continental Asia behind, discovered way back in the fourteenth century the delights of a life by the sea. Istanbul, which they conquered in 1453, offered ideal conditions for the construction of palaces, pavilions, and villas with a view over the Bosphorus and the Golden Horn.[132] The idea of declaring a stretch of bare sand as a beach on which to enjoy the pleasures of the sea occurred to people in Europe only in the late nineteenth century.

That turn to the sea was by no means everywhere a "natural" tendency is also shown by the fact that farsighted early modern governments (as in France under Louis XIV or in Russia under Peter the Great) had to make special efforts to construct trading stations and naval bases. It is probably the case that in every historical era before the nineteenth century, most of the largest cities and main centers of power or cultural splendor were *not* situated on the coast: Kaifeng, Nanjing, and Beijing; Ayudhya and Kyoto; Baghdad, Agra, Isfahan, and Cairo;

Rome, Paris, Madrid, Vienna, and Moscow; and not least, Mexico City. North America was the only conspicuous exception to the rule: *all* major cities in the early United States were ports or had easy access to the sea. The great Japanese historian Amino Yoshihiko, who took a close interest in people living on the coast, came to the conclusion that even insular Japan, with a total coastline approaching 28,000 kilometers, always defined itself as an agrarian society and never made sea travel, fishing, and maritime trade central elements of its collective identity.[133] Here a clear distinction should be drawn, however, between fishing villages and port cities. In all civilizations fisherfolk live in small, often isolated, communities that preserve their special way of life for an unusually long time. Port cities, on the other hand, are plugged into wider and more up-to-date social trends, in which world market fluctuations determine economic conditions. A port city has a denser web of relations with its counterparts across the water than with fishing villages in its vicinity.

Most written history has treated port cities rather shabbily.[134] They are by definition on the periphery, far from inland centers, their populations turbulent and uncontrollable, cosmopolitan and therefore suspect for upholders of cultural, religious, and national orthodoxies. Even the Hansa remained on the fringes of the newly emerging German national context. Hamburg became part of the German customs area only in 1883, having previously been treated for such purposes as a foreign territory cut off from its natural hinterland. Scarcely ever have port cities housed important sanctuaries or places of top scholarship. Major temples, churches, and shrines, as well as leading universities and academies, have usually been located inland. All this applies as much to Europe as it does to North Africa and the whole of Asia.

A Special World

In the nineteenth century, two general trends enhanced the role of port cities and changed their character: the growing differentiation of maritime activity, and the replacement of wooden ships with metal ones.

With the growth of overseas trade and naval power, shipping involved an ever more intricate mosaic of activities. Functions that had been united in large overseas companies (e.g., the East India Company) now became separate from one another, most particularly the civilian and military sides of sea travel. In the eighteenth century, naval warfare required special facilities under the exclusive control of state apparatuses. Ports such as those at Plymouth, Portsmouth, and Chatham in England, or Brest and Toulon in France, or Kronstadt in Russia now acquired great significance as bases and shipyards for huge war fleets. German examples followed later: Wilhelmshaven was founded in 1856 as the military port of Prussia. In the nineteenth century, such naval bases spread around the world. The British Empire maintained large military shipyards in Malta (which became even more important after the opening of the Suez Canal in 1869), Bermuda, and Singapore.[135] The rise of the steamship at first necessitated more frequent shore stops and therefore gave birth to a new kind of port: the coaling station. Many of the seemingly absurd imperial disputes of the nineteenth century—in the Pacific, for example—become understandable once one realizes that the main issue was the supply of coal to warships.[136]

Similar to the military-civilian cleavage was the one that opened between freight and passenger transport, as one can see from the growing complexity of harbor layouts. The processing of passengers took place as close as possible to the city center, whereas the railroad made it possible to load and unload freight in more out-of-the-way areas. Marseille offers a good example of this bifurcation of port space. Around the middle of the nineteenth century, its Vieux Port—which had scarcely changed since Roman times—was superseded by a Port Moderne not far away. The old ports had been closely integrated into the life of the city, and mighty ships had similarly dominated the interior of cities such as Boston and Liverpool. The new-style ports became self-enclosed organisms with their own administration, conceived as a technical whole and both spatially and mentally remote from the city.[137] The first separate "docklands" emerged in London, Hull, and Liverpool. The model for the modernization of Marseille was the West India Docks in London, whose construction had begun in 1799, but new installations were added to the English port throughout the nineteenth century to handle the increase in tonnage. The total volume of shipping entering London from abroad rose almost thirteenfold between 1820 and 1901, from 778,000 to 10 million tons, while the largest ships were ten times larger than before.[138] Unlike the open quays on the Thames that they replaced, the West India Docks were a closely guarded space enclosed by a wall eight meters high; they were like deep artificial lakes, surrounded by towers and fortifications reminiscent of the Middle Ages. At the very moment when external city walls were coming down all over Europe, the new port enclosures were reaching skyward. The activities inside them were based on an increasing division of labor. The London docks were regarded as a miracle of engineering, and Karl Baedeker's famous tourist guide spoke of them as a sight not to be missed.[139]

The modern port in Marseille, the second-largest city of nineteenth-century France, was meant to surpass even its London model. Very large ships were able to enter the docks, whose construction had been considerably simplified by the use of concrete, and it was possible to anchor quite close to the warehouses. Iron and steel technology produced ever stronger steam-driven and hydraulic cranes. The pressure to modernize compelled ports all over Europe to follow the lead of Marseille and London, and the middle of the century marked the greatest turning point in port history since the Middle Ages. In Hamburg the old natural harbor gave way in 1866 to newly built installations,[140] and here, as in many other cities, there was resettlement on a large scale.

The innovations spread to Asia. After much wrangling over finance, Bombay—which had profited from the opening of the Suez Canal in 1869—obtained an up-to-date port in 1875. In Japan the city council of Osaka, with no help from the government in Tokyo, put up the money for a very expensive harbor, the largest urban construction project in the country in the late nineteenth century. In Batavia it was only in 1886 that it became possible for ships to load and unload directly at the quayside—too late for the old colonial capital to keep its lead over the rising port of Surabaya. The year 1888, when the first modern quay facility opened in Hong Kong, may be thought of as the beginning of port modernization in China.[141] But the process crept only slowly along the Chinese coast, since the surplus of ultracheap labor was a barrier to mechanization. What need was there for cranes if porters could be hired for next to nothing?

The new ports formed a special world of mass freight, hard manual labor, and a little mechanization, increasingly separate from the areas where upper-class passengers and herds of migrants boarded ship. The ocean liner finally disappeared in the 1950s, while at the same time container ports and petroleum depots were located far out on river estuaries. The "modern" ports of the nineteenth century gradually fell silent—demolished, filled in, and used for skyscraper development.

In the nineteenth century, inadequate port facilities sometimes proved a serious hindrance to the blossoming of trade. In Buenos Aires, which lacked a serviceable natural harbor, ocean steamers continued through the 1880s to anchor out at sea, loading and unloading by means of barges. Then the wing of the Argentine oligarchy that was prepared to shoulder the cost of a modern port won the day. In 1898 the project on the River Plate reached completion, providing the city with nine kilometers of cement quays, deep sea basins, and modern loading equipment.[142] In Cape Town only the South African War triggered such modernization; it was financially and technically the most formidable task the municipality had ever taken on.[143]

The Iron Ship and the Iron Horse

The switch from timber to metal hulls, and the related, though slightly later, transition from sailing ships to fuel-powered vessels, was the second major new trend. It became generally visible around 1870 and reached a conclusion around 1890. The consequences were higher transport capacities, lower freight and passenger charges, greater speeds, less dependence on the weather, and the possibility of keeping to regular schedules. Speed was not just a question of journey time. Steamers did not have to spend as long as sailing ships waiting at a port. The pace of life and work in the docks therefore accelerated dramatically.

A further result of the advent of steamships was that the barriers between sea and river transport were partly overcome. It is difficult to travel upriver on a sailing ship, but gunboats or small trading vessels have no trouble moving on engine power into previously inaccessible areas. China was "opened up" twice: once on paper, by the so-called unequal treaties that began in 1842, and once by the arrival of steamships, after 1860. Decades before railroads penetrated the Chinese interior, Western and Chinese steamships were already getting the process under way. Between 1863 and 1901 it became possible for ocean steamers of any size to reach Hankou (today's Wuhan), the great city right in the middle of China, when the waters of the Yangtze were high. Only after the turn of the century did major port improvements make Shanghai a final destination that could beat any competition for the ocean giants. From then on, it would be the transshipment point for goods to and from Hankou.[144]

The building of railways also had a great influence on the functioning of port cities, as we may see again from the example of East and Southeast Asia. It is true that some optimally located ports—above all, Hong Kong and Singapore—could function for a long time without an effective rail link to the interior. But in this they were exceptions. The general rule, valid on all continents, was that port cities without an adequate rail link had no future. The great port cities of the modern age are points at which land and water transport meet up and interact with each other.

Many, though not all, of these great ports were centers of shipbuilding. Often—in Barcelona and Bergen, for instance—this was the first industry they developed: one of the hardest and technologically most demanding branches of the engineering industry, especially at a time when ships' hulls were riveted and could not yet be welded. In China industrialization began—before any cotton factories—with the big shipyards in Shanghai, Hong Kong, and Fuzhou, which at first were all under state control. But the Chinese government was not alone in grasping the importance of shipbuilding for national development, both economic and military. Some port cities, such as Glasgow and Kiel, had greater importance in shipbuilding than in overseas trade. After 1850 Glasgow successfully switched from cotton spinning—which was then in decline—to the building of ships and machines; at their height, in the 1880s and 1890s, its yards were the most productive in the world.[145]

Port Societies

For social historians the most important aspect of port cities, especially those undergoing industrialization, is the diversity and flexibility of their labor markets. There was a need for sailors and transport hands, for skilled shipyard workers and unskilled labor in local light industry, and for captains, officers, pilots, and port engineers. Services of all kinds were in demand and in supply—from trade finance to red-light districts. One might go so far as to define a port city not by its geographical location but by the peculiarities of its job structure.[146] What crucially distinguished a port city from an inland city was the importance of short-term employment in its economy; laborers were hired from one day to the next, and there were a large number of people looking for work. The labor force in port cities was almost entirely male, whereas in light industry during the early period of industrialization the female share might be as high as three-quarters. Dockworkers in Europe were far down in the jobs hierarchy, whereas in early twentieth-century China they were leading figures in anti-imperialist strikes or boycotts and usually belonged to the political vanguard. In Europe they were paid badly and treated harshly and were rarely more than casual laborers; even when day wages receded in other areas of employment, they remained the dominant form in the docks. Besides, the mechanization of transport labor reduced demand for the mass of workers. Sharp seasonal fluctuations in employment often meant that women and children had to pitch in to augment family income. Children did not work in the docks, but dock labor indirectly brought child labor in its train.[147]

The unstable, fluctuating character of the population of port cities was not a nineteenth-century novelty; earlier, too, they had been magnets for commercial diasporas Nor should they be seen simply as conglomerations of foreigners. People who migrated there from the hinterland were strangers only to a lesser degree. In Chinese port cities, for example, they often lived alongside others in the same line of business, constituting distinctive social milieux, guild organizations, and recruitment networks. Shanghai, in particular, was a patchwork quilt of such communities based on a solidarity of origin. Attempts in the early twentieth century to organize a harbor proletariat into unions and political parties had to contend with such particularism.[148]

Groups based on place of origin were not peculiar to Asian cities. The transcontinental networking of port cities invariably tended to produce a differentiated ethnic structure. In Trieste, for instance, Armenians, Greeks, Jews, and Serbs lived alongside one another; Odessa grew after 1805 through the targeted recruitment of Jews, Swiss, Germans, Greeks, and others.[149] Following the Great Famine, Irish workers emigrated to British port cities such as Liverpool, Glasgow, and Cardiff, living there in tightly knit and fairly closed communities. In 1851, Irish people constituted more than one-fifth of the population of Liverpool, but the relative lack of segregation seen in Hamburg was not reproduced there. Immigrants often had the worst housing, and their children had the lowest chances of climbing the social ladder.

Security forces of every kind view port cities as breeding grounds for crime and civil commotion—a reputation that was borne out in the twentieth century even more than in the nineteenth. In Germany the revolution of 1918 started with a naval mutiny; in Russia, sailors rose up in 1921 against a revolution that had betrayed its principles. Dockworkers stood in the forefront of the struggle against colonialism and foreign interests, whether in China (Hong Kong and Canton), India (Madras), Vietnam (Haiphong), or Kenya (Mombasa). Port cities were and are more open than inland cities, not only to people from abroad but also to foreign ideas. In Germany, Prussian authoritarianism was counterbalanced by the bourgeois liberalism of Hanseatic ports such as Hamburg and Bremen. Similar oppositions may be found elsewhere in the world; port cities have tended to be places of deviance and innovation. The state was represented by people it seldom needed in other parts of the country: customs officials. Piracy and naval warfare were sources of vulnerability, and special courts administered a special law of the sea. Ever since the era of Elizabethan corsairs, the British Empire had been aware of how much "naval pressure" could be exercised through the blockading and bombardment of port cities. One celebrated episode was the Royal Navy's destruction of the old town of Copenhagen in 1807, an unprovoked attack on a neutral country that severely damaged Britain's reputation in continental Europe. In 1815 the United States declared war on the "pirate nest" of Algiers, for which no one in Europe had any sympathy, and fought successful naval actions against Algerian frigates.[150] In 1863, out of revenge for the murder of a merchant, British warships destroyed large parts of the Japanese fortified city of Kagoshima.[151]

Overseas trade was an important engine of urbanization, not only in the colonies but also in Europe. In 1850, 40 percent of cities with a population above 100,000 were ports; it was not until the mid-twentieth century that they lost their first place to industrial centers.[152] In some European countries, urbanization was essentially a coastal phenomenon: in Spain all the large cities (Barcelona, Cadiz, Malaga, Seville, Valencia—though not Madrid) lay on or near the sea, and the same was true of the Netherlands and Norway. Even in France, some of the great provincial centers (Bordeaux, Marseille, Nantes, Rouen) were on or near the coast. The industrial structure of port cities, except the very largest, was different from that of inland population centers. Typical port industries were cereal or cooking-oil processing, sugar refining, fish packing, coffee roasting, and (later) petroleum refining; heavy industry, or heavier branches of light industry, rarely entered the picture. In cities such as New York and Hamburg, it was not the industrial districts but the ports that were the main zones of innovation.[153] Only in fairly rare cases did a port city later branch out into industry on a large scale: Genoa, whose development at the end of the century owed more to industry than to foreign trade, was one example, as were Barcelona and (after the First World War) Shanghai.

Port cities were often governed by small oligarchies of merchants, bankers, and shipowners, a *grande bourgeoisie* that created many a chamber of commerce to represent its interests and to ensure social exclusiveness. This was no different in Rotterdam or Bremen than in Shanghai or Izmir. Landowners had less political influence than in large cities inland. However, the oligarchies were not always united among themselves: tensions could arise between commercial and industrial interests, or between supporters and opponents of free trade. In general, the dominant ideology of the commercial capitalist oligarchies involved a preference for the night-watchman state, one that intervened little and was satisfied with low taxes; the highest priority was to ensure the smooth flow of trade. Merchants tended to regard city planning with skepticism and to recoil from investment in infrastructure other than port facilities. Such cities seldom came up with administrative innovations, nor were they often in the vanguard of measures to improve public hygiene. Longer than elsewhere, they relied more on paternalist benevolence and ad hoc philanthropy than on the regular provision of social support. Sharp class conflicts were therefore a characteristic result of the polarized structure of port cities such as Liverpool and Genoa; the middle classes were less important than in strongly industrial inland cities such as Birmingham, Berlin, and Turin.

6 Colonial Cities, Treaty Ports, Imperial Metropolises

Is it meaningful to refer to port cities or administrative centers in the colonies as "colonial cities"?[154] At the end of the nineteenth century, such a large part of the earth was under colonial rule that it seems reasonable to assume that the "colonial city" was a typical form of the age. Right from the start the Spanish had exported Iberian city forms to the New World, though not always the same standard pattern. Then at the end of the sixteenth century, the Spanish-American colonial city was transferred to the Philippines: Manila differed in no way—except for the presence of Chinese—from a city in Mexico. Unique among the bridgeheads of early European expansion in Asia, it was not simply a trading port but also a center of secular and religious control.[155] On a more modest scale, the Dutch—also from a highly urbanized background—followed the example of the Spanish in Asia, or at least in the city of Batavia, which they founded in 1619 with visible success.

Calcutta and Hanoi

Once the British were firmly in the saddle in India, they made of their main base, Calcutta, a city of palaces. From 1798 on, after more than four decades in which the East India Company had exercised supreme political authority, the Bengali capital mutated into a splendid neoclassical ensemble almost unrivaled anywhere in the world. The function of the city did not change fundamentally. What it had lacked, despite brisk construction activity since the 1760s, had been a fitting architectural garb. At the core of the new design was the gigantic New Government House, which, unlike its modest predecessors, no longer drew sneers from critical Indians or envious Frenchmen. The new governor's residence, inaugurated in 1803, outshone every dwelling place or official seat available to English monarchs. They also erected a whole new series of public buildings (city hall, law courts, customs office, etc.), churches, and private villas belonging to East India Company officials or merchants—and high above them all towered the watchful Fort William.[156]

The Calcutta of porticoes and Doric colonnades did not simply transplant an English city to India. It was the stone utopia of a new Imperial Rome, conceived less as a functioning city than as a power landscape in which Indians, too, were meant to find their place. Architecturally, it is not difficult to follow the colonial traces of Europe around the globe, but they do not often appear as compactly and forcefully as they do in Calcutta. Few other colonies were loaded with such symbolic weight. Few were so rich and so easily exploitable that colonial splendor could be funded locally (colonies, after all, were not supposed to become loss-making businesses, unless this was unavoidable for reasons of international prestige). A set of European-style buildings, then, did not add up to a fully self-contained colonial cityscape. The minimum that even the poorest colonial capital required was a governor's palace, an army barracks, and a church; a hospital and a couple of villas for European officials and merchants completed the core. Whether whole districts sprang up in the European style depended on the size of the foreign presence in the city.

The will to plan and fund a whole new model city was quite exceptional. Dakar, founded in 1857 and later rising to be the capital for the whole of French West Africa, is a particularly impressive example.[157] Dublin may be thought of as a special case: not a planned colonial cityscape, but an opulent symbolic field with an imperial character. The capital of Ireland was lavishly provided with statues of English kings and queens, who expressed London's will to rule the country and served as the departure point for Protestant ceremonial occasions. But since the British never had the municipal government of Dublin fully under their control, national memorial sites were gradually established as symbols of resistance in opposition to the imperial monuments.[158]

Early-twentieth-century Hanoi was a fully developed colonial metropolis, at once center of the protectorate of Tonkin and, since 1902, capital of the Indochinese Union (comprising the three French *pays* of Vietnam plus Cambodia and Laos). Vietnam had posed thorny problems of imperial control right from the beginning, and the France of the Third Republic felt a special need to impress the natives and to convince the world of its colonizing abilities. Hanoi, the main city in Tonkin and since 1802 no longer the seat of the Vietnamese emperor, came under de facto French control in 1889 and immediately began to turn into a French city. The city walls and, also the Vauban-style citadel dating from early in the century were pulled down; new streets and boulevards were laid in a grid pattern and provided with a paved surface. Government buildings and an ugly cathedral rose up alongside a railroad station, an opera house (a smaller version of the Garnier Opera in Paris), a *lycée*, a prison, a technically remarkable bridge over the Red River, monasteries and convents, numerous official buildings, glass-domed department stores in the Parisian style, villas for top bureaucrats and merchants (two hundred individually designed luxury houses by the end of the colonial period), and standardized suburban dwellings for lower-ranking French personnel. The crassest monuments of colonialism, the governor's palace and the cathedral, were erected with brutal symbolism on the sites left vacant by demolished pagodas and Confucian examination halls. Whereas the British in Calcutta built their colonial city *beside* the indigenous old town, the French colonial authorities put theirs in its place. Streets and squares were named after "heroes" of the French conquest or great historical or contemporary Frenchmen.

The architectural style of this early colonial period made no concessions to Asian forms; indeed, settlers in Saigon set themselves quite consciously against such references. The glitter of France was meant to radiate its civilizing effect in all its original brightness. Corinthian, neo-Gothic, early Baroque: everything was jumbled together. British India did not shrink from historical allusions either, but at least—as in Victoria Station in Bombay—its designers dared to combine English, French, or Venetian Gothic with elements of what counted as the "Indo-Saracenic" style.[159]

Only after the turn of the century was there growing discontent in Vietnam and Paris with the bombast of the nineties. Scholars discovered a politically less explosive "old" Indochina behind the Sino-Vietnamese traditions, and after the First World War some art deco designs were also introduced in Hanoi.[160] Politically too, Europeanization was carried as far as it was possible to go. Hanoi with its up to four thousand (1908) French residents was endowed, like a good old French provincial capital, with a mayor, a city council, a budget, and heated factional struggles.[161] The main difference with Tours or Lyons was that although locally born people, as well as non-European immigrants from China or India, enjoyed some legal protection and a degree of informal participation (rich Chinese merchants even belonged to the chamber of commerce), they had no say when it came to politics.

The Ideal Type of a "Colonial City"

Hanoi looked more amazingly European than anywhere else in the colonies, and so it might be taken as the basis for the ideal type of a "modern" (as opposed to early modern) colonial city. Like the global city of the late twentieth century, the colonial city has as its most general characteristic a primary orientation to the outside world. Its other features are[162]

- a monopoly of political, military, and police control in the hands of rulers from abroad whose legitimacy derives solely from conquest;
- exclusion of the indigenous population, even its elite, from decisions about how the local authorities should regulate the life of the city;
- the introduction from Europe of secular or religious architecture, usually in the latest or next to latest style, or in one reflecting the supposedly "national" style of the colony in question;
- spatial dualism and horizontal segregation between a district for foreigners, lavishly and healthily designed in accordance with European principles, and a halfheartedly modernized (at best) "native city" that was regarded as backward;
- a fragmented urban society, with rigid compartmentalization along racial lines, and the relegation of locally born people to badly paid and dependent service jobs; and
- an orientation to the opening up, reshaping, and exploitation of the hinterland, in accordance with foreign interests and the requirements of international markets.

Such a list of distinct features has the advantage of avoiding hasty labels: the colonial city cannot be defined *only* in terms of its architecture or its economic function. On the other hand, the list mixes together form and function, for example; and the sum of the characteristics yields such a narrow definition that few cases are likely to correspond to it in the real world. Hanoi, for instance, was not unimportant economically, but it was neither a port city nor a typical colonial "vacuum pump" for the extraction of resources. Its functions can be adequately described only in the context of a *city system*, which in the case of Indochina would also have to include the port city of Haiphong and the southern metropolis of Saigon, as well as Hong Kong, Batavia, and ultimately Marseille or Nantes.

As an ideal type, the "colonial city" may help to bring the observed reality into sharper focus and to draw out its distinctive characteristics; it therefore also rules out a number of things in advance. If a colonial city is understood as a place of ongoing contact between different cultures,[163] then all large multicultural ports had a colonial element, whether or not they were in the colonies, London, New Orleans, Istanbul, or Shanghai among them. All had plural social structures. That characteristic alone would therefore not be sufficiently specific. If, on the other hand, "colonial city" is understood entirely politically, so that its decisive criterion is incapacitation of the local elite by an autocratic ruling apparatus implanted from outside, then Warsaw (as part of the Tsarist Empire) would fulfill that condition. At the end of the nineteenth century, a city that was not allowed to become the capital of a Polish national state had a permanent garrison of 40,000 Russian soldiers. An intimidating citadel towered over the populace, cossacks patrolled the streets, and ultimate authority lay with a Russian police chief answerable directly to Moscow. By comparison, a "normal" European metropolis such as Vienna had a regular garrison of 15,000 troops, mostly of local origin.[164]

Many characteristics of a colonial city need to be defined dynamically, not in a "binary" grid of presence and absence. Some historians are especially inclined to detect strict segregation or "urban apartheid," while others have a sharper eye for "hybridity" and admire the "cosmopolitanism" of many large colonial cities. But in between there are many different gradations. The social composition of colonial cities was marked by shades, transitions, and overlapping, against the background of a dichotomy between colonizer and colonized that operated *in principle* but did not take effect in each and every sphere of life. Social and ethnic hierarchies were superimposed on one another in complex ways. Even at the high-water mark of racist thinking, the solidarity of skin color and nationality by no means universally cancelled the solidarity of class. Wealthy Indian merchants or Malayan aristocrats were as a rule barred from British clubs in the large colonial cities, but so were "poor whites." In case of doubt, the social distance between a British official in the Indian Civil Service and the white inmate of a workhouse in India was greater than the ethnic distance between the same official and a prosperous, well-educated Indian lawyer—unless the relationship was clouded by politics (as it came to be after the First World War). The typical "colonial city" society was not organized simply in accordance with a two-class or two-race stratification.

Segregation

It is easy to identify the spatial dualism between a privileged foreigners' district, well protected and often climatically more agreeable, and areas of the city inhabited by locally born people. But this binary opposition is also a model construct. Power relations and social stratification were not consistently reflected in a rigid division of the city layout. And even when they were, the dependence of European colonials on teams of local domestics stood in the way of a sharp separation between the areas in which people lived. The colonizers were rarely alone among themselves. They acted in everyday life on a semiofficial stage, before an indigenous public that had its eyes trained on them. The segregation of housing did not always entail a univocal relationship of subordination and superordination. Kazan on the Volga, for example, had a Russian district and a Tatar district, even though it was impossible without qualification to describe conditions there as colonial.[165] In very large cities, at least in Asia, special minority communities had been tolerated since early modern times; often these were located in the same part of the city, as in Istanbul, where at least 130,000 non-Muslims were permanently resident in 1886.[166] Also many South and Southeast Asian cities offered to Europeans a picture of coexistence within integrated communities that was mostly, though not always, peaceful. They were—like cities in the Ottoman Empire—*villes plurielles*, where religion and language were the most important sorting criteria.[167] European colonialism overlaid such mosaic structures without actually erasing them.

Segregation did not at all derive from some "essence" of the colonial city; it had a history of its own. In Delhi, conquered by the British in 1803, there was no special British district until the Indian Mutiny of 1857/58. Lord Palmerston and many others then called for the city to be razed to the ground in punishment, but despite major destruction (e.g., of the Red Fort of the Mogul emperors) things were never taken that far.[168] After the horrors of 1857, many Britons wanted to live away from the "native city." Yet Indian landownership continued to be permitted in the new foreigners' district, and the police were never able to guarantee the complete security of the colonial "masters" from Indian robbers. Many English people rented accommodations from Indians in their "own" district, while continuing to work (and to enjoy themselves) in the old city. After the outbreak of plague in 1903, the advantages of suburban housing construction were proved, and more and more Indian landowners moved into the "civil lines" (as the British district was called).[169] In Bombay, however, the fortresslike "factory" of the East India Company formed the nucleus of urban development, to which a "native town" was attached only in the early nineteenth century. Later still, garden suburbs were created as a third element for well-to-do Europeans.[170]

What actually distinguishes *colonial* segregation within a city from other kinds of spatial separation? In European cities there were (and are) micropatterns of segregation, sometimes from one street to the next or vertically within the same apartment block (with the bourgeois on the *bel étage* and the impoverished writer in the garret).[171] Segregation, loosely defined, is a widespread phenomenon, an elementary form of social differentiation that manifests itself in many different ways. "Colonial" may here mean no more than urban apartheid along ethnic lines, enforced by the ruling apparatus of a regime consisting of minority foreigners. However, there are few examples of this. Some of the toughest segregation practices known in modern history were completely without ethnic overtones: for example, the separation of warriors from commoners in Edo during the Tokugawa period. Conversely, it is hard to decide whether the Irish in early Victorian industrial and port cities, and a little later in North America, remained in lower positions of the social hierarchy for social or for "ethnic" (which here also means religious) reasons.[172] The Irish were "white," but there were many shades of whiteness.[173]

On closer inspection, then, the ideal type of the colonial city loses its sharp contours. Not every city in the territory of a colony becomes a typical colonial city, and the distinction between a colonial city and a noncolonial one with similar functions should not be overstated. The fact that both Madras and Marseille are port cities probably means that what they have in common more than outweighs the colonial/noncolonial distinction. On the other hand, there was something like a colonial transition period in the global development of cities, stretching from the middle of the nineteenth to the middle of the twentieth century. Whereas the "frontier city" as typified by Boston, New York, Rio de Janeiro, and Cape Town had been an early modern innovation in previously nonurban settings, the "modern" colonial city of European origin imprinted itself on the old urban cultures of North Africa and Asia and sometimes provoked their resistance. Never before in history had European urban patterns had such an impact in the rest of the world. The colonial city, in the strict sense of the term, disappeared along with the colonial empires. Today it appears as a stopover point on the way to the postcolonial megacity of the present, whose evolution has departed from earlier European models and is fueled by partly local, partly global sources—a dynamic that is not specifically European or Western.

Colonial Westernization

Colonial past and later evolution into a megacity are so variably related to each other that general statements are hard to justify. Of the ten largest cities in the world in the year 2000, only one was a former imperial metropolis, Tokyo; or two if New York is considered the center of American world hegemony.[174] The most important imperial metropolises of the period between 1850 and 1960—London and Paris—have long ceased to figure among the top cities in population size, but they have ensured themselves the status of "global cities," that is, as nodal points at the highest level of the global city system and multiple concentrations of worldwide steering capacity. With the exception of London, today's global cities (the frontrunners are Tokyo, New York, London, and Paris) do not have this status *because* they used to be colonial metropolises. Apart from Tokyo, all ten *leading* cities (even New York!) were once "colonial cities," albeit in different ways and at different points in time. When Seoul fell under Japanese colonial rule in 1905, Mexico City already had nearly a century of postcolonial history behind it. Cairo was formally colonized for only thirty-six years (1882–1918); Batavia/Jakarta for 330 years (1619–1949). Other formerly spectacular colonial cities have not gone the way of megapolization; Cape Town, Hanoi, and Dakar, to name but a few, now lead a relatively modest existence. Centers of once great colonial empires, such as Madrid or Amsterdam, have become middle-ranking tourist destinations. Cities that, by anything other than a purely statistical yardstick, might also count as megalopolises—Bangkok and Moscow, for example—were never colonized; and Shanghai was, but only in a quite special, limited degree.

The era of colonial cities was a *nonspecific* preparation for the age of globally networked megacities, and it is all too easy to ask whether a colonial past has proved to be an advantage or a disadvantage for the present day. A negative formulation would be safest: a past as a colonized city has been neither a necessary condition nor a main cause of the urban explosion since the middle of the twentieth century, and previous "possession" of a colonial empire has been no guarantee of a leading place as a city in the postcolonial world.

Neo-European frontier cities in the British settler colonies (dominions), most strikingly in Australia but also in Canada (especially the west) and New Zealand, constitute a type of their own. They are a direct product of European colonization and have little in them that is "hybrid." Since they were not inserted into a preexisting cityscape but took shape under frontier conditions, they do not correspond to the ideal type of the colonial city, as defined above. Nor were Australian cities mere copies of British ones in the way that Spanish settlements in the Americas, for all their local differences, essentially reproduced a Spanish model. What they resemble most closely are the cities of the American Midwest, whose key advance took place around the same period. Unlike the colonial cities of Asia or North Africa, Australian cities have experienced continuous development. There was no sudden decolonization but rather a slow, constant, and peaceful process of political emancipation within a British constitutional framework. Economically, the Australian cities remained "colonial" so long as they were dependent on the London financial market (which gradually changed after 1860),[175] so long as they represented markets in the British Empire for which there were no alternatives, and so long as their own external trade was handled largely by agencies of British firms.[176]

One striking novelty of nineteenth-century colonialism was the treaty port.[177] In Asia and Africa, rulers had normally restricted trading activity by foreigners to special zones and tried to control them as tightly as possible. The traders were granted certain residence rights.[178] After 1840, when China, Japan, and Korea opened up one after the other to international trade, it was clear even to the most fanatical free trader that these economic spaces could not be "penetrated" through the unfettered operation of market forces alone. Special institutional forms were required, with the threat of military force ultimately behind them. A series of "unequal treaties" gave Westerners unilateral privileges, especially immunity from legal action in Asian courts, and trade regimes were set up that denied local governments control over customs policy. In some of the cities opened to foreigners under the treaty provisions (not all of which were treaty *ports*), small downtown areas were even withdrawn from the jurisdiction of the local state and placed under either foreign consuls (the concessions) or self-governing foreign trading oligarchies (the settlements).

The general significance of these extraterritorial enclaves or port colonies, as they were appositely known, should not be exaggerated.[179] In Japan they were for some years the main gateway for Western influences, but after 1868 they soon lost importance as the modernization policies of the Meiji state turned to unreserved cooperation with the West. They did not play a large role in the urbanization of Japan. Yokohama was the only one of the country's major cities to be founded as a treaty port. The first foreigners settled there in 1859, and thirty years later the port city had a population of 120,000 (mostly Japanese, of course)—a growth as swift as that of Vladivostok, founded nine years after Yokohama.[180]

In China the treaty ports were much more significant. Nevertheless, of the ninety-two ports that had at some point acquired this status by 1915, only seven ever had European minicolonies within their boundaries. And of those seven, only two were deeply marked by their foreign enclaves: Shanghai with its International Settlement and French Concession, and the northern city of Tianjin, where nine concessions, much smaller than those in Shanghai, came into being. The rapid growth of the two cities after 1860 was due mainly to an increasing orientation of the Chinese economy to the world market, which was in turn encouraged by the presence of foreigners in the protected treaty ports.

Some of the smaller concessions (e.g., in Canton and Amoy) were akin to insular "ghettoes," but such a word is not appropriate to the special areas in Shanghai and Tianjin. As late as the 1920s, the International Settlement in Shanghai was governed by representatives of the large Western corporations in China, with no formal Chinese involvement. But 99 percent of its population was Chinese, who were allowed to own real estate and could engage in many kinds of economic activity. The scope for radical politics was also greater there than in the part of the city under Chinese jurisdiction, its theoretically law-based polity enabling the formation of a critical Chinese public.[181]

Beyond its many other locational advantages, Shanghai grew up around its colonial core. The concessions and settlements in the treaty ports became entry points for the transfer of Western models of the city. Instead of a pompous palace architecture, the buildings were designed mainly to express an openness to the world market, although it was only in the 1930s that large corporate headquarters gave the Bund its well-known skyline. From time to time a Disneyland fantasy would appear: for example, Gordon Hall, the administrative center of the British Concession in Tianjin, whose towers and battlements made it look for all the world like a medieval fortress; or the reproduction of a German small town, complete with half-timbered buildings and bull's-eye windows, in Qingdao, the main city of the German "leased area" (i.e., colony) in the northeastern province of Shandong. More important, a new image of the city emerged in places where the settlements were able to expand: wide streets, somewhat less dense housing, stone and mortar materials even in Chinese-style houses, and above all a greater openness to the street (in contrast to the windowless walls that had traditionally sealed off the houses, so that only storefronts looked out at passersby).[182]

Urban Self-Westernization

"Colonial cities" did not exist only in colonies. Some of the most striking "colonial cities" originated not in an initiative by a colonial authority but in acts of preventive self-Westernization. In the twentieth century such things were no longer surprising. By the 1920s at the latest, everyone was agreed on what should be part of a "modern civilized city": paved streets, potable water on tap, drains and sewers, garbage removal, public toilets, fire-resistant buildings, lighting in the main streets and squares, some elements of a public transportation system, extensive rail links, public schools for some if not all, a health service with a hospital, a mayor, a police force, and a reasonably professional municipal administration. Even when external conditions were unfavorable—for example, in China of the 1920s and 1930s, torn apart by civil war—local elites and potentates tried at least to approximate to these goals.[183] It troubled no one that the model was of Western origin. But local circumstances imposed the most varied adaptations and omissions.

Before the First World War, when Europe was at the height of its prestige, urban self-Westernization was not only a practical demand but also a political signal. Cairo offers a good illustration of this, even before the colonial period that began in 1882 with the British occupation. Within the space of a few years, between 1865 and 1869, an urban dualism arose in the pure form that one finds elsewhere at most in some French colonial capitals. After the French under Bonaparte had caused severe damage to the city in 1798 and 1800, Egypt's first modernizer, Pasha Muhammad Ali (r. 1805–48), did surprisingly little for his demographically stagnant capital. The preferred architectural style began to change slowly: glass windows came into use, the space inside houses was redivided, house numbers were introduced, and the pasha commissioned a French architect to build a "neo-Mamluk" monumental mosque, which he declared to be in the Egyptian national style. But otherwise the aspect of Old Cairo remained mostly unchanged under Muhammad Ali and his two successors.[184] A major break in the history of the city came only with the reign of Pasha Ismail (r. 1863–79, after 1867 with the viceregal title of khedive), who dreamed "the dream of Westernization."[185] Between the river and the labyrinthine old city, in whose narrow lanes there had been no room even for Muhammad Ali's coach, Ismail had a new city built in accordance with a geometric plan, brightly lit boulevards instead of dark streets accessible only on foot, green parks instead of swirling dust, fresh air instead of lingering odors, a drainage system instead of waste tanks and open sewers, the railroad instead of long-distance caravans. In Cairo, as in Istanbul around the same time, the introduction of dead-straight avenues with long lines of sight amounted to an aesthetic revolution.

In 1867 the world's fair in Paris convinced the khedive of the advantages of European-style city planning, and he let himself be guided by the master of the re-design of Paris, Baron Haussmann. Upon his return he sent his minister for public works, the capable and energetic Ali Pasha Mubarak, on a study tour to the French capital. The opening of the Suez Canal, planned for 1869, became the focus of hectic building activity in Cairo, which was expected to gleam forth as the modern pearl of the East. The khedive spared no expense for the construction of a theater, an opera house, city parks, a new palace for himself, and the first two bridges over the Nile.[186] That all this helped to bankrupt the Egyptian state was another story. In part Ismail had a tactical goal: to demonstrate that Egypt was determined to modernize and to gain entry to the magical circle of Europe. In part he was deeply convinced of the superiority of the modern world that he saw taking shape north of the Mediterranean. City planning seemed to him the ideal instrument to achieve modernity and to make it perceptible at a symbolic level. Ismail did not spare the Old City and—in a decision that would have been unthinkable under Muhammad Ali—had some straight roads driven through it. He understood that it was essential to improve sanitary conditions in all parts of the city, but the advances made in this respect—installations to provide a supply of drinking water, and a conduit system—were by their nature scarcely visible in the cityscape.[187] The stark contrast between the old and the new city was scarcely softened as a result.[188] The British colonial period after 1882 took over the basic structures of Ismail's and Ali Pasha Mubarak's Cairo and added only a few things that were new, above all a concern for the preservation and even fanciful conjuring up of "medieval" or "Mamluk" elements of the city. "Colonial" Cairo was the creation of an Egyptian ruler who believed in progress and who attempted (in the long run unsuccessfully) not to become a politically dependent client of the Great Powers.

Similar stories of self-Westernization might be told about other cities in Asia and North Africa, such as[189]

- Beirut—unlike Cairo, a new city—which became a showcase of Ottoman modernity unencumbered by tradition, a bourgeois mirror to the admired Marseille across the sea;

- Istanbul, less violent but also more thorough than Cairo in adapting European city forms, in a way that avoided crass dualism and took infrastructural improvements more seriously;
- Tokyo, where, by 1880, centrifugal forces had made parts of old Edo look like suburbs of Chicago or Melbourne and generally created an architecturally ugly appearance, while at the same time helping a self-confident neo-traditionalism to assert itself in the conduct of everyday life; or
- Seoul, opened up very late (1876) and formally colonized only in 1910, which in the intervening period remodeled itself as a capital city in the international-Western architectural language of the time.

The story of Hankou sounds different again.

Noncolonial Dynamism: Hankou

The opening of more and more links to global trade networks gave a major advantage to coastal areas. Almost the whole of Australia's urban development was ocean oriented, and in noncolonial countries with an old city system (e.g., China or Morocco) the demographic, economic, and political center of gravity shifted from the interior to the coast.[190] Shanghai and Hong Kong, Casablanca and Rabat profited from this spatial shift, but inland cities also successfully linked into a dynamic in which world market forces combined with domestic trade flows. Had such economic centers been situated in the colonies, they would have been described without hesitation as "typically colonial"—which they were only in the sense that they counteracted the long-term trend toward a structural disparity between more dynamic ("developed," "Western") and more static ("backward," "Oriental") economic environments.

Cities of this type could be of different sizes and exist at various levels of city systems. One example was Kano in the Sahel region, the metropolis of the North Nigerian Sokoto caliphate. Hugh Clapperton visited the area in 1824–26, and the German traveler Heinrich Barth (on a British assignment) in 1851 and 1854. Both men saw an imposing walled city, which at the height of the spring caravan season in the Sahara had 60,000 to 80,000 people living in it at any one time. On the eve of the British intervention of 1894 it held approximately 100,000 inhabitants, one-half of them slaves. Kano was a dynamic economic center, with an efficient craft sector and a large catchment area for trading operations. Leather goods were exported to North Africa, textile material and tailored cloth to western Sudan. Cotton, tobacco, and indigo thrived in the surrounding area, much of it also being sent abroad. The slave trade remained important; slaves served as soldiers or worked in production. As a base for jihads, Kano had control of its own slave supply. It was a city that had grasped its economic opportunities, the most important of a number of commercial hubs in the Sahel.[191]

The location of Hankou, today part of the triple city of Wuhan, is an economic geographer's dream: at the center of a densely populated and fertile countryside, with a system of waterways leading via the Yangtze in all directions, including to Shanghai and overseas.[192] Unlike Hong Kong on the southern coast, which the British slowly developed into a major port from 1842 onward, Hankou in the late nineteenth century was an inland center with overseas connections, not an entrepôt and organizational center with relatively weak land links.[193] Jesuits already described Hankou in the eighteenth century as one of the liveliest cities of the empire, and a Chinese merchants' handbook called it the most important transshipment center in the country.[194] Hankou was a huge city, its population at least one million shortly after 1850—before the devastation of the Taiping Revolution. Indeed, it was one of the largest cities in the world, in the same league as London, of which it reminded many travelers on account of its high-density housing. Its growth had not resulted from any foreign presence or any links with the world market. In 1861 it was declared a treaty port. The British and French at once established small concessions, in which Chinese were accepted only as domestics, and starting in 1895 the Germans, Russians, and Japanese followed suit. Immediately after its "opening," Hankou attracted foreign consuls, merchants, and missionaries. This sudden appearance of Europeans, the palatial mansions they built for themselves, and the political demands they made with the backing of gunboats on the Yangtze, marked a dramatic change in the history of the city.

But none of this made Hankou "colonial" in character. The concessions did not dominate the inner city in the way that they did in Shanghai already at that time; the largest of them, the British Concession, had just 110 resident foreigners in 1870. A truly "European city" did not rise up as in Shanghai, Hong Kong, Cairo, or Hanoi. Above all, Hankou's extensive trade did not fall under the hegemony of foreign interests; its economic rationale was not transformed, in accordance with imperialist trends, from that of a "national" trading city into a vacuum pump for European and North American capitalism. As William T. Rowe has shown in a masterful analysis, Hankou before 1861 had been anything but the "Oriental city" familiar from Western sociology: static, geometrically designed, subject to an overbearing municipal authority. Nor, after 1861, was it a typical "colonial city." Rowe avoids choosing a label. His account describes a quite "bourgeois" urban world, in which a highly differentiated and specialized merchant class developed existing trade networks and branched out into new lines of business. Guilds—which, in the light of Rowe's urban history, should no longer be tagged "premodern"—adapted flexibly to changing circumstances, making tried-and-tested credit institutions more effective instead of discarding them in favor of Western-style banks. Hankou society accepted newcomers, becoming more pluralist and, under the leadership of local notables, developing into a community in which the lower orders, by no means always deferential, found a place for themselves. After the end of the Taiping terror, which came to the city from outside, massive reconstruction work became necessary and was carried out. The people of Hankou did not allow themselves to be passively colonized. Only the onset of industrialization in the 1890s changed the social climate and structures of the city. Some of the early factories were founded by foreigners, but the larger ones—such as the vast, technologically modern iron and steelworks at Hanyang—had their origin in Chinese initiative. Early industrial Hankou did not become a colonial city either. The great urban center of the Middle Yangtze Valley is a particularly good illustration of the fact that not every contact with the world market leads a weak economy into colonial dependence.

The first *post*colonial cities also emerged in the nineteenth century: cities that, turning away more sharply than those of the young United States from their colonial past, attempted in some sense to "reinvent" themselves. Mexico City was in such a situation. Here the first step in decolonization went back to 1810 when, even before the gaining of formal independence, the so-called Indian republic—above all, the fiscal exactions of the "Indian Tribute"—was abolished. But the belt of corn-growing Indian villages around the city remained part of the landscape for a number of decades, after which Indian land fell increasingly into the hands of private speculators. In November 1812, under the terms of the Constitution of Cadiz, Mexicans were called upon to vote for the first time in local elections, and from April 1813 on—still eight years before independence— Mexico City was governed by an elected council that consisted only of *Americanos* and included a number of Indian notables. This was a veritable anticolonial revolution; people wanted to wipe out the last traces of the ancien régime. The actual changes, however, were less dramatic than those envisaged in the program. Mexico City did not play a major role in the liberation movement, and in the new republican state it lost the aura and power it had enjoyed in colonial times. The cityscape remained essentially unaltered until the middle of the century. In particular, since Catholicism retained its position as the state religion, the city continued to resemble a huge cloister; it counted seven monasteries and twenty-one nunneries in 1850. Mexico City remained "Baroque," with state and church firmly yoked together. Only in the second half of the century did it undergo major changes.[195]

Imperial Cities

Ultimately, the colonial city was the counterpart to the imperial city, to the ruling metropolis that was the source of the colonizer's power. The imperial city is easily defined: it is a political command center, a collection point for information, an economically parasitic beneficiary of asymmetrical relations with its various peripheries, and a showplace for emblems of the dominant ideology. The Rome of Augustus and of the two centuries that followed was such an imperial city in its pure form, as were Lisbon and Istanbul in the sixteenth century and Vienna in the nineteenth. In modern times, the various criteria are otherwise not easy to match up. The cityscape of Berlin bears many traces of its past as a colonial metropolis (between 1884 and 1914), but economically Berlin never had an appreciable dependence on Germany's comparatively meager colonial empire in Africa, China, and the South Pacific. Conversely, the prosperity of the

Netherlands in the nineteenth century would have been unimaginable without the exploitation of Indonesia. This rather major dependence was hardly discernible to the casual visitor; only weak attempts were made to fit Amsterdam to the phenotype of an imperial city. The Royal Museum of the Tropics is today the most visible reminder of erstwhile colonial luxury. Rome, by contrast, whose colonial empire was rather insignificant, adorned itself after 1870 with imperial monuments—no problem, in view of the stage scenery inherited from the Caesars.[196] In Paris too, the conditions were favorable. The overseas colonialism of the Second Empire and the Third Republic could insert itself into the imperial cityscape shaped by Napoleon I. Marseille played the role of a second imperial city, rather as Seville had done in relation to Madrid. Glasgow, in many respects the center of a distinctively Scottish empire, convinced itself that it was "the second city of the Empire," even if that was not immediately obvious to visitors.

Even London, the center of the only world empire of the age, did not display its imperial side too obtrusively. In 1870 Calcutta looked more "imperial" than the great metropolis itself. For a long time London refrained from imperial monumentality, and in the architectural contest with Paris it often ended up the loser. John Nash's Regent Street was a feeble answer to the Arc de Triomphe (on which work lasted from 1806 to 1836), and the reshaping of the French capital under Napoleon III elicited nothing comparable on the other side of the Channel. Over time the French also learned better how to glamorize their world's fairs and colonial exhibitions. London remained the ugly duckling of Europe's metropolises, always looking poorer than it really was, although throughout the nineteenth century it had better drains and street lighting than its immodest sister capital.

When reasons were sought for London's imperial reticence, many pointed to traditions of private and public parsimony or the antipathy toward absolutist pomp in a constitutional monarchy. Besides, there was not a unified city administration with sufficient planning powers. Complaints grew louder that the great capital city had to hide its face in shame when confronted with the glamour of Vienna or Munich, and that tourism was suffering from the lack of sights and well-run hotels—yet still nothing was done. Only Queen Victoria's Golden Jubilee in 1887, followed by her Diamond Jubilee ten years later, finally roused the nation from its imperial slumber; she was after all empress of India as well as the reigning monarch. Admiralty Arch was erected on the southwest corner of Trafalgar Square, but little else happened architecturally.[197] Apart from a couple of statues of conquering heroes and the plethora of imperial imagery on the Albert Memorial, London before 1914 did not look very imperial—considerably less so than Chengde (Jehol) in Inner Mongolia, for example, the summer residence of the Chinese emperor, where claims to power over Central Asia were subtly represented in the architecture. Buildings like Australia House or India House appeared only after the First World War, functionally defined as high commissions (or de facto embassies). Nevertheless, in matters other than city planning or architecture, London was truly an imperial metropolis: in its docks and constant inflow of people from Asia and Africa, in its dark-skinned visitors from overseas, in the ornamentation and lifestyle displayed by colonial officials on their return home, or in the exotic subjects of music-hall merriment. The imperial nexus had its maximum impact away from the limelight.[198] London did not need to lay the symbols on thick.

7   Internal Spaces and Undergrounds

Walls

The premodern city was a walled space protected by defensive installations. Even when walls no longer fulfilled a military purpose, they continued to operate as customs boundaries. When they lost that function too, they served as symbolic markers of space. Whole empires expressed their superiority over the "barbarians" around them by the sheer force of their technological, organizational, and financial capacity to build walls. Barbarians might destroy walls—they could not put them up. Walls and gates separate city from country, compression from dispersion. The "typical" city in Europe, Asia, and Africa was walled, but not every single one was. Damascus and Aleppo had walls; Cairo, though crisscrossed by inner walls between districts, was never protected by a closed outer ring of fortifications. For military reasons the French removed many of those inner walls when they briefly occupied the city in 1798–1801. They were promptly replaced when the French left, but after the 1820s were guarded by policemen instead of the older private militia.[199] In the New World city walls were a rare sight—visible in Quebec or Montreal, for example. Australian and US cities never had them. On the other hand, since the 1980s Americans have enjoyed putting up new walls: the "gating" of prosperous apartment complexes and city districts, combined with protective walls, tall fences, and watchtowers, is still a growing trend. This colonial practice spreads whenever income differences and socially segregated housing reach a certain threshold. It has become common even in the big cities of (still officially socialist) China.

In 1800 the average European city still took it for granted that it should have outer walls. People did not always live in the area enclosed by them: many Russian cities were wide and sprawling. Sometimes suburbs would spill out and overrun the masonry, but the actual structures remained intact. Their eventual disappearance was not a linear process and should not be taken as a measure of modernity. In a place like Hamburg, certainly modern in many respects, the city gates were closed at night until the end of the 1860s, and in Rabat in 1912, not long before it became the capital of French Morocco, every sundown witnessed the locking of the gates and the handing over of the keys to the governor.[200]

The "defortification" of cities was not just a question of removing walls, filling in ditches, and developing bare slopes. Such changes always had a colossal impact on the real estate market. Different interests often stood sharply opposed to one another. The municipal authorities had to weigh not only the costs and benefits of demolition work but also of the development of newly released land. Often its incorporation became inevitable as the physical city limits fell away, and that, too, was associated with numerous conflicts.[201] Defortification usually began in the big cities and later spread to small and medium-sized ones. In Bordeaux the city walls succumbed as early as the mid-eighteenth century to an extensive modernization program that replaced them with squares and avenues. Nîmes likewise converted its walls into promenades.[202] But not all French cities followed so quickly; Grenoble left its walls intact until 1832, and even then they were initially not demolished but widened.[203] In Germany a number of large cities had removed their walls by 1800: Berlin, Hanover, Munich, Mannheim, Düsseldorf. During the Napoleonic Wars, many cities were compelled to pull down their fortifications: for example, Ulm, Frankfurt am Main, and Breslau (today Wrocław). If the land was converted into green spaces or promenades, the former perimeter of the walls remained recognizable in the cityscape. In the decades following the Congress of Vienna, the general stagnation of society in Germany was reflected in a slower pace of defortification compared to other parts of Europe. The last city walls vanished in the second half of the century, by 1881 in Cologne and 1895 in Danzig (today Gdańsk). None of Europe's major cities clung to its city walls more firmly than Prague; it had reinvented itself only in the 1830s as a Romantic-medieval-magical city, in opposition to the resolute modernism of Budapest.[204] In Britain, by midcentury there were no more city walls to cater to the aesthetic nostalgia of others; in the Netherlands they were all gradually removed between 1795 and 1840.[205] It took rather longer where conservative patricians ran things and indulged in a dream of the enclosed city—until 1859 in the case of Basel, whereas in Zurich and Bern the rural population and urban radicals had together ensured the walls' disappearance in the 1830s.[206]

In Spain the the dynamic city of Barcelona had been hemmed in until 1860, when its walls were demolished. In Italy only the port cities of Genoa and Naples gave up their walls early; most Italian cities remained until century's end "in the wall garb with which the late Middle Ages or the early modern period had fitted them."[207] When the demolition occurred in the age of intensive road building, planners recommended using the freed strips of land for prestigious rings that would unclog the city center, which is what happened in Vienna, Milan, and Florence. In 1857 Emperor Franz Joseph ordered the removal of old fortifications left in Vienna since the time of the Turkish wars, with the express aim of creating a new stage for the imperial court to display its splendor.[208]

Especially in smaller cities, entrance gates were sometimes left standing for decorative reasons, and now and again city walls were even rebuilt in the nineteenth century. In Paris, which still vividly recalled the Russian and German occupation of 1814/15, it was decided in 1840—when the threat of war loomed again—to construct a new defensive perimeter. Between 1841 and 1845, a city wall thirty-six kilometers long was put in place, with ninety-four bastions and a fifteen-meter moat that even encompassed areas not yet administratively part of the capital. The remains of this long-obsolete wall were finally removed in 1920 and partly replaced with parks and sports grounds.[209] In India the British scored an architectural "own goal": an earthquake in 1720 had severely damaged the Delhi city walls, and so between 1804 and 1811 the British rebuilt them so thoroughly that it would take four months and a great deal of effort to capture the city in the Rebellion of 1857/58. In response to those events, they then tore down fortifications wherever they still existed. In Delhi, where it would have been too expensive to blow up seven kilometers of thick masonry, the "walled city" remained in place with its perforated bastions, but the gates were no longer shut.[210]

After the land and sea walls gradually came down in Istanbul, the walls of Beijing endured into the new century as the last monuments of bygone days, a kind of urban mirror image to the Great Wall a few dozen kilometers north of the capital. During the Boxer Rebellion, photographs made the huge city walls from the Qing period familiar all over the world. They seemed to symbolize the medieval character of the Chinese empire, especially as they followed classical Chinese models and, unlike the bastions of Istanbul, showed no influence of European fortress architecture. The assault on Beijing by the armed forces of eight powers, which began in early August 1900, resembled the medieval storming of a fortified city, the main initial targets being the gates in the east. After breaching the levees, the attackers placed ladders against the walls and engaged in hand-to-hand fighting on the top. For the last time the Chinese double perimeter served its purpose, as the space between the outer and inner walls became a death trap especially for the Russian soldiers. The Chinese capital was the largest walled area in the world, holding within it the imperial palace, the so-called Forbidden City, itself surrounded by the walls of the Imperial City, which also contained lakes, parks, and official and business institutions. The two even more extensive outer walls around the northern city (dubbed "Tatar City" by nineteenth-century Europeans) and the southern city (or "Chinese City"), with their thirteen well-guarded monumental gates, mostly dated from the fifteenth and sixteenth centuries and had been enlarged in the mid-eighteenth century. Following their victory against the boxers, the eight powers spared the empire further humiliation and expense by not insisting that the walls be pulled down. Only in 1915 was a short stretch removed near one of the city gates to ease the flow of traffic.[211]

All Chinese cities were girded by walls—the *cheng* character may be translated as either "wall" or "city"—and followed approximately, though not schematically, the same cosmologically derived pattern. Local considerations also played a part in planning decisions. The walls of Shanghai were built in the 1550s, when pirate attacks all along the coast made life insecure for the population of the city, but a few decades later the danger receded and the defensive perimeter no longer served a purpose. By the mid-nineteenth century the fortifications, mostly built of clay and unfired brick, were in an advanced state of dilapidation, with the ditches and watercourses clogged up inside the city. In the late 1850s China's aspiring southern metropolis began to develop a new image for itself; the crumbling walls survived in unsalutary neglect.

In the early twentieth century, the fate of these walls became the object of a heated dispute between modernizing "demolition men" and their traditionalist opponents. Bustling suburbs with narrow winding streets had sprung up outside the walls, and alongside this "southern city," as foreigners called it, a "northern city" had come into being within the space of a few years. After the end of the Opium War, Shanghai had been opened to foreigners by treaty, and in subsequent years the British and French had brought large areas of the city under their control. A European-style city had then taken shape, with a grid system of streets and squares, a park, a racecourse, and a riverside boulevard where major European corporations gradually opened their Chinese headquarters.[212] As foreigners in Shanghai built themselves a kind of counter-city (they would later do the same in Tianjin and Saigon), the Ming-era walls reversed their function, serving not to repel attackers but to shut out an old "walled city" that symbolized in foreign eyes the filth and decay of native China. In the crown colony of Hong Kong too, the little "walled city" remained an enclave where British police and officials did not care to meddle—almost up to the end of the colonial period. British maps of Shanghai in the late-nineteenth century often left blank the area inside the city walls. Foreigners did not surround themselves in Shanghai with physical walls of their own making, but elsewhere they did retreat behind protective installations. The diplomatic quarter in Beijing had a wall around it, and this was further strengthened after the Boxer Rebellion. In Canton, in the far South, foreigners resided on an artificial island in the Pearl River.

The Railroad Invasion

If anything made city walls obsolete it was the railroad (they can coexist more easily with the automobile).[213] No other infrastructural innovation has ever cut so deeply into the social organism of the city; it brought about "the first great laceration of the traditional urban fabric."[214] One thinks primarily of new links: the first intercity line in Britain opened in 1838 between London and Birmingham; the first in India in 1853 between Bombay and the small town of Thana. Proximity to a river or the sea was no longer decisive for the development of a locality. Cities became enmeshed in national and later in cross-border networks. This happened in Europe and the East Coast of North America within the space of two or three decades, mostly in the 1850s and 1860s. More interesting than the chronology of individual lines, however, is the threshold beyond which it becomes possible to speak of a railroad *system*. This is not only a question of the number and distribution of lines in a network; there must also be a certain mastery of equipment and organization, a basic level of safety, regularity, profitability, and passenger comfort. France and the non-Habsburg German lands achieved such a degree of systematic cohesion in the 1850s; the New England states had done so some years earlier. By 1880 Europe all the way to the Urals, excluding only the Balkans and northern Scandinavia, was covered with a railroad network that met the requirements of a system.[215] By 1910 the same was true of India, Japan, North China, and Argentina.

What did the arrival of the railroad mean for a city? Everywhere, the early "railroad manias" not only involved money and technology but also affected the future shape of cities. Heated debates broke out over the relationship between private interests and public utility, and over the location and design of stations. The great pioneering age of railroad construction, which brought with it a novel technology and aesthetic, was the 1840s in the case of Britain and central Europe. The last of the great Paris stations, the Gare de Lyon, was in full operation by 1857. One reason for this speed of construction was that railroads and stations devoured huge quantities of urban land as they carved their way into the inner city, sending property prices through the roof. By the end of the urban railroad transformation, railroad companies possessed between 5 percent (London) and 9 percent (Liverpool) of the land in British cities and indirectly influenced the use of another 10 percent.[216] The argument that they were clearing slums in the process rarely caught on, since little concern was shown about the rehousing of displaced families. The problem was literally shunted aside. Hundreds of thousands of people in Britain lost their home as a result of railroad construction. It might take just a couple of weeks to rip out the heart of a district and to form a new neighborhood on either side of the tracks. Viaducts, much loved at first, did not solve the problem. Railroad lines and stations were loud and dirty. The expectation that they would breathe life into surrounding areas was sometimes fulfilled, but not often. In high-immigration cities such as Moscow, there was also a danger that new slums would spring up around the stations.[217] Passenger travel was the first to develop on the long-distance routes in Britain, followed in a second phase by freight transport, which often required the construction of an additional land-hungry station. Only in a third phase, after 1880, did local commuter services begin to appear—a lower priority for railroad companies, and one that sometimes relied on government subsidies.[218] In the countries that had pioneered the process of construction, the physical marking of inner cities by railroad stations was generally complete by the beginning of the 1870s.

Railroad stations altered cityscapes: they could sometimes revolutionize the whole character of a city. The main station in Amsterdam, which opened in 1889, was built on three artificial islands and a total of 8,687 piles, driving a huge wedge between the inner city and the harbor front. The much-admired contrast between the cramped city and the open maritime vista disappeared, and Amsterdam changed in its perceptions and lifestyle from a city by the sea to an inland city. At the same time, one canal after another—sixteen in all—were filled in. The aim of the planners was to "modernize" Amsterdam, to make it conform to the model of other metropolises. Only protests by local conservationists ensured that the canal destruction came to an end in 1901, so that Amsterdam was able to keep at least its basic early modern design.[219]

Railroad stations posed some of the greatest architectural problems of the age—or anyway they did once the rail companies or relevant public authorities were prepared to spend the necessary money, since the earliest stations (such as Euston in London) were built with thrift in mind. Never before had roofed spaces been designed for circulation on such a scale. The station had to organize movement, to direct machines and people, and to satisfy the requirements of the timetable. New iron and glass materials, tried out shortly before in the Parisian arcades, created a potential for easy construction that was expertly utilized in stations such as Newcastle (1847–50). The facades, on the other hand, had to be weighty and to accentuate strong visual features; many of them were at the end of a street, capable of being seen from a long way around. Stations were often admired as the ultimate artworks, which combined the latest technology with comfort and a pleasing external appearance, Jakob Ignaz Hittorff's Gare du Nord in Paris (completed in 1846) being a shining example.[220] Their architects were influential people, with a broad range of skills, who had to make any number of decisions on technical and stylistic matters.[221] Nothing went untried: Renaissance (Amsterdam, 1881–91), Romanesque combined with Gothic (Madras, 1868), wild European eclecticism plus Indian handicrafts (Bombay, 1888), the station as fortress (Lahore, 1861), glazed neo-Gothic extravagance with masterly wrought-iron details (Saint Pancras in London, 1864–73), a huge round arch facade (Gare du Nord, 1861–66; Frankfurt am Main, 1883–88), a mishmash of everything (Antwerp, 1895–99), "Moorish" fantasy (Kuala Lumpur, 1894–97), beaux-arts style (Gare d'Orsay in Paris, 1898–1900), allusions to ancient Rome (Pennsylvania Station in New York, 1910), and Nordic neo-Romanticism (Helsinki, 1910–14).[222] As these examples show, India also was a stomping ground for early station architects. Istanbul, with two stations built by German engineers (1887 and 1909), allowed itself the nice touch of greeting travelers from Europe with architecture of Islamic inspiration, and visitors from Asia Minor with a classical Greek exterior.

Horses and Pedestrians

People arriving by train in a European city around 1870 used a technology that is essentially still in use today—and a few moments later found themselves in an archaic world of horse transport. In 1800 all cities in the world were still filled with pedestrians and were therefore, in this respect, at the same evolutionary stage.[223] Their main outward difference was in the degree of their use of horses, which was not possible everywhere or available to everyone without restriction. In Chinese cities those who did not go about on foot had porters carry them in sedan chairs; horses were uncommon. In Istanbul non-Muslims were forbidden to ride a horse within the city limits, and until the nineteenth century even donkey or mule carts were less used than human traction to carry goods.[224] In Japan, until the end of the Tokugawa period, only samurai nobles were allowed to travel by horse; all others dragged themselves, often barefoot, through streets that were either muddy or dusty. After the opening of the country in midcentury, people were forbidden to walk in bare feet, on the grounds that foreigners would think this shameful.[225]

In the pedestrian city, the way from home to work could not take too long. This was a major reason why slums tended to concentrate in inner-city areas and why they were cleared so slowly. First there had to be mass transportation affordable even to the low-paid. Preindustrial technologies survived long after the beginning of the "industrial age." Horse-drawn omnibuses, the first significant innovation in the inner cities, rested upon the same operational basis as private coach travel and involved little technological advance. Later the horse-drawn bus developed as a form of public transport, operating a regular timetable on fixed routes and for a set price. This was an American invention, first introduced in 1832 in New York. It took twenty-four years for it to appear in Paris.[226] Such vehicles were inevitably expensive due to their high running costs. A large number of horses had to be kept in reserve; each horse usually worked for only five to six years; fodder and maintenance did not come cheap. Besides, a horse-drawn bus could at best travel only twice as fast as an average pedestrian; it was not a solution for the journey from home to work. Horses also created a lot of muck. Around 1900 the Chicago garbage disposal service was collecting from the streets an incredible 600,000 tons of horse dung a year.[227] Smelly manure heaps continued to be a feature of urban landscapes even in ambitiously modernizing countries, and stables were a ubiquitous element of the built environment up to the end of the century.[228] The clatter of horses' hooves on pavement and the cracking of whips made a noise about which the philosopher Arthur Schopenhauer in Frankfurt was not the only one to complain.[229] Congestion and accidents were part of everyday life. The disposal of dead horses was itself a major sanitary problem.

The horse-drawn tram or streetcar, which made its debut in 1859 in Liverpool and spread in the 1870s to continental Europe, did not solve these problems, but it did mark a certain advance since the use of rails doubled the weight a horse was able to draw. Costs and fares went down, though not dramatically. Nowhere was the "horsecar" more popular than in the United States. By 1860 New York had 142 miles of rails, and 100,000 people a day used the service. In the 1880s there were 415 "street railway companies" in the United States, which carried 188 million passengers annually.[230] In Istanbul, where water transport has remained important up to the present day, tramlines were laid on the existing broad, Western-style streets. The trams made the Turkish metropolis look like a great European city, even though stick-wielding men walked in front of the horses to shoo the infamous Istanbul dogs away from the rails.[231]

In Britain (though not in the United States) tram companies were forbidden by law to speculate in land, and so they had little incentive to open new routes in

the suburbs. But horse-drawn buses and trams did contribute to the differentiation of social space. They enabled the middle classes—who could afford the fare
as well as the rising price of real estate along the tram routes—to live far from
their place of work, thereby triggering the disintegration of what sociologists call
"workplace communities."[232] To come into their own, horse-drawn buses and
trams needed the railroad, since their main strength was as a feeder for intercity
and suburban trains. In turn, the train made the wider use of horses indispensable, since it increased the total circulation of people within the city. It is a curious paradox that literally until the end of the century, there was no improvement
in inner-city transit that came close to matching the most advanced transportation of the age. In 1890 people still moved with 1820 technology through the
streets of Europe and America.

In 1890 a total of some 280,000 horses were deployed on buses and trams in
Great Britain.[233] For no other city do we know as much about the use of horses
as we do for Paris. It is estimated that in 1862 there were 2.9 million horses in the
whole of France (a large part of them in agriculture and the army); Paris had at
least 78,000 in 1878, and approximately 56,000 in 1912.[234] Hackney carriages had
been in use since the seventeenth century, and in 1828 they were introduced for
the first time in a kind of regular service. Horse-drawn buses spread only after
the founding of the Compagnie Générale des Omnibus in 1855, while at the same
time new types of demand for transportation began to appear. The new Bon
Marché department store, for example, had perfectly run underground stables
with more than 150 horses and a large fleet of vehicles that could take customers
home. The post office, the fire brigade, and the police also required horses. Well-
off individuals kept saddle and carriage horses until well into the automobile
age; there were more than 23,000 private carriages in London alone in 1891.[235] In
the French Second Empire, under English influence, promenades on horseback
became more popular than ever before. Riding lessons, racecourses, and the hiring of horses were a feature of middle-class leisure. The key social distinction ran
between those who could and those could not afford to have a carriage with a
private driver. The lower classes profited from this last golden age of the horse by
having access to cheap horsemeat.[236]

In the long run, of course, overland coach travel was no match for the railroad. But it did not disappear overnight. Indeed, in the early nineteenth century, mail coaches reached their height of efficiency and elegance in Europe, in
accordance with the policy—first developed in France—of carrying passengers
as fast as letters. In England cross-country coaches had never been used as much
as they had in the transition to the railway age. At the beginning of the 1830s,
the London-based Chaplin & Company maintained a fleet of sixty-four passenger carriages and 1,500 horses. In 1835, every day saw fifty coaches leave the
capital for Brighton, twenty-two for Birmingham, sixteen for Portsmouth, and
fifteen for the ferry port at Dover. Altogether, the long-distance coach business
in London had a capacity of 58,000 seats for passengers. Like the sailing ship, it

reached its peak of technical perfection right at the end of its heyday. Improved vehicles and road services (tarring), deriving from both private economic initiative and political decisions, meant that, under favorable weather conditions, the 530-kilometer journey from London to Edinburgh could be completed in two days, as opposed to the ten it had required around 1750. Whereas a traveler needed four to five days to go from Moscow to Saint Petersburg, the Frankfurt to Stuttgart run, after the "express coach" was introduced in 1822, took twenty-five instead of forty hours. The ideals of smoothness and punctuality had never seemed so close.[237] On level roads, the best coaches could reach speeds of twenty kilometers per hour or even slightly more. At the other extreme were the heavy coaches used by American settlers, whose teams of four or six horses struck west across the continent at no more than three or four kilometers an hour. The railroad would render them obsolete by the 1880s.[238] Elsewhere in the world, suitably modernized, the horse kept a place in long-distance travel until the end of the century and beyond. In 1863 a good road opened between Beirut and Damascus, and an express coach could complete the trip in twelve to fifteen hours; as many as a thousand horses were kept available for it. It is true that a rail line opened in 1895 and cut the time to nine hours, but only in the 1920s did the train finally knock the horse out of the race.[239]

Streetcar, Subway, Automobile

Many problems of city transportation were eventually solved with the introduction of the electric streetcar: 1888 in the United States, 1891 in Leeds and Prague, 1896 as a Tsarist prestige project in Nizhni Novgorod, 1901 in London, 1903 in the small German town Freiburg im Breisgau (where I live). Technically, it involved the conversion of energy from an electrical drive into rolling movement. The streetcar brought a real revolution to the modern city: it was twice as fast and only half as expensive as a horse-drawn tram and finally made it possible for ordinary workers to commute to work. Tumbling fares had the same consequences as in the case of transatlantic steamships decades earlier. In Britain the number of trips per capita via public transportation soared from a mere eight in 1870 to 130 in 1906. On the eve of the First World War, nearly all large European cities had a streetcar network; the end of horse traction there was truly imminent. New York withdrew all its horse-drawn buses in 1897, and by 1913 there were none left in Paris either.[240] For the very poor, however, the streetcar was still prohibitively expensive. It was a boon mainly for workers with regular employment.

In Asia it was not horses but men who supplied the energy to carry people to the railroad. The Japanese rickshaw (also called *kuruma*), a kind of sedan chair on two wheels, was invented in 1870 and soon entered mass production; by the 1880s it was being exported to China, Korea, and Southeast Asia.[241] Large firms moved quickly to organize the rickshaw trade in Japan's big cities, engaging in sharp price wars with one another. In 1898 more than 500 rickshaws waited for customers outside the Osaka train station. In 1900 Tokyo had a force of 50,000 pullers. To ride in a human-powered single-axis vehicle was at first a luxury; it later became a necessity for many, before the spread of streetcars toward the end of the Meiji made it once more a service with a high-class profile.[242] In Japan too, horse traction was displaced by the electric streetcar soon after the turn of the century.

At the end of our era, the age of the automobile had not yet begun. This technological innovation enabled the real explosion of cities, first in the United States, then after the Second World War in Europe too. In 1914 there were 2.5 million personal motor vehicles in the world. By 1930 there would be 35 million. Around the turn of the century in continental Europe, it was still a sensational experience for many to come face to face with one. Those who did not own such a rare and costly machine might perhaps have the chance to ride in a motor cab: the number of horse-drawn hackney carriages in Berlin fell dramatically after 1907, and the fleet of motorized "taxis" (including some with an electric motor) had drawn almost level by 1914. In 1913 there was one passenger car per 1,567 inhabitants in Germany, one per 437 in France, and already one per 81 in the United States; in southern and eastern Europe there were scarcely any at all in private ownership. Outside the large cities, the automobile was not part of everyday life before the First World War. The United States, where the technically best were produced, was the only country on earth to which that statement did not apply. From the point of view of transportation technology, the twentieth century began in the United States. It was only there that the car was, by 1920, more than a curiosity but the technical basis for new kinds of mass transportation system.[243]

The largest pioneering enterprise in urban public transport was the London Underground, the first in the world, which combined a railroad system with tunneling techniques tested in the construction of sewers. It was a private initiative, originating not in farsighted urban planning but in the vision of one man, Charles Pearson. Throughout the nineteenth century it remained a profit-oriented project in the spirit of capitalist entrepreneurship. Work on the Underground began in 1860. Three years later, the first six-kilometer stretch of the Metropolitan Line ("Metro" would become a standard name all over the world) went into operation. The train lines were fifteen to thirty-five meters below ground, but it was possible to speak of a real "Tube" only when new techniques in the 1890s allowed the tunneling to go deeper still. Electrification work began straightaway. Until then the (originally windowless) carriages had been drawn by steam trains—which presented special problems in a closed tunnel—and dimly lit by oil or gas lamps. Locomotives had difficulty climbing underground slopes; they would often grind to a halt and roll backward.

Many property owners did not allow construction work to take place on or under their land—which explains the frequent bends and the generally awkward layout of the line. But since the resistance to overground rail lines was even greater, the Underground in effect owes its existence to being considered the lesser of two evils. At first, like the railroad, it had to convince numerous skeptics. In 1863 Lord Palmerston, the seventy-nine-year-old prime minister, refused to take part in the opening celebrations: at his age one should be happy to remain above ground as long as possible. The public, however, had no such reservations. On its very first day of business, January 10, 1863, the new line carried 30,000 passengers. Uncomfortable and dirty though it was—a retired colonial official from the Sudan later compared the noise to the breathing of a crocodile—it proved itself to be a relatively fast and accident-free means of transport. The gradual enlargement of the network crucially assisted the integration of the metropolis with the newly developing suburbs. It was both affordable for a wide circle of users and profitable for the business that ran it. The other underground systems that followed the London model were: Budapest (1896), Glasgow (1896), Boston (1897), Paris (1900), New York (1904), and Buenos Aires (1913). In Asia the first metro launched in 1927, in Tokyo.[244] The old brainchild of British engineers is today a reality all over the world. Never have so many subways been built as in the period since 1970.

Slums and Suburbs

In the pedestrian city, the best private addresses were also the most central. From premodern Paris to Edo, localities outside the city walls were regarded as distinctly inferior. Mexico City is a good illustration of such a concentric order: the Spanish occupied the center with their offices, churches, monasteries, colleges, and business premises, including many dark-skinned servants in their midst. The next circle consisted of new immigrants, ranked according to their place of origin. Finally, the outside circle was made up of Indian villages.[245] In 1900 Moscow looked much the same: the best places were in the center, and conditions grew worse the farther out one went. Outer areas were wild and rough—poorly lit streets, wooden huts still without kerosene lamps, a lot of overgrown land, barefooted people—the end of civilization for bourgeois and aristocratic Muscovites.[246] In many of today's megacities, the shantytowns of the jobless poor lie similarly on the outer periphery, cut off from the center.

It was therefore not a matter of course that the values of core and periphery should be reversed. Where this happened, where it became desirable to live far from downtown, it became the third major revolution in the urban history of the nineteenth century, after the railroad invasion and the general cleanup. Suburbanization, understood as a process whereby outlying areas grew faster than the inner core and commuting became a normal part of life, began in Britain and the United States around 1815. It would eventually be taken to extremes in the United States and Australia, whereas Europeans would never develop such a fondness for living outside the city center.[247] Even before private automobiles became widespread in the 1920s, the ideal of the spatially isolated household took solid root in the United States. Few things are as characteristic of the US model of civilization as the preference for home ownership and for detached housing in low-density areas far from the workplace. This trend reached its apogee in the post-1945 "metropolitan sprawl," nowhere more so than in Los Angeles, which has been described as "the rejection of the metropolis in favor of its suburbs."[248]

National styles of suburbanization differ from one another in many respects: a French *banlieue* is not the same as the German or Scandinavian type of garden-plot settlement (*Schrebergarten*) that first became popular in the 1880s. Nevertheless, there are basic mechanisms of European suburbanization, which are well exemplified by trends in England. In London, the birthplace of suburbia, and elsewhere in southeastern England, it had long been an upper-class custom to retreat to the country, to enjoy a well-cushioned retirement on a landed estate or in a villa (modestly referred to as a "cottage"). Suburbanization was something new and different. People who still had a regular job in the city center gave up their residence there and commuted on a daily basis. As early as the 1820s the upper middle classes, who could afford to commute by carriage, began to move into mansions and semidetached houses in gentrified areas in the vicinity of central London. John Nash's Regent's Park created an attractive combination of city and country, which would be the model for parklike abodes all over England. When the Parisian townsman Hippolyte Taine visited such districts in Manchester and Liverpool in the 1860s, he was startled by the calm that prevailed there.[249] The central areas of Manchester were abandoned by "swells" even earlier than those of London; one would have lunch at one's club, and in the evening be driven home in a carriage. The pattern was similar in any, the second major country of suburban villas and "fine residential areas" set apart from the city center. But was the "villa," with its ancient Roman connotations, really a European specialty? When the Moroccan sultanate liberalized in the last third of the nineteenth century, so that well-to-do people no longer had to make themselves small in the eyes of the ruler, the heights above Fez, an old Islamic city with a medieval feel, were soon built over with magnificent houses.[250]

The suburban living of ever larger sections of the middle classes presupposed higher incomes, more convenient transportation links, more time available for traveling, and a greater supply of commercially built housing. Early suburbanization should not, of course, be considered in isolation; it was closely bound up with another process in the Victorian city—the rise and fall of slums.[251] Industrialization driven by the middle class led to more densely packed low-grade housing in the inner cities, as a result of which the middle class fled the insalubrious poverty and moved into the suburbs. But it still continued to draw an income from the slums, either in the form of rent or from the proceeds of selling the land on which they were built. "Slumification" and suburbanization thus appear as two aspects of the same capitalist process, which for a long time acted itself out under the conditions of a politically unregulated market. Not before 1880 did the view spread in Europe that a free housing market might not provide minimum standards for all, and only after the First World War did an effective public housing policy begin to operate in a number of countries, such as Great Britain.[252]

This gradual politicization of the housing question presupposed that it was actually defined as a problem. So long as policymakers regarded extreme poverty and slum conditions as "normal," or even, in a moral twist, as brought on through the fault of those who suffered them, there seemed to be little need for action—as if it was evident that slums were only a first step toward the integration of immigrants into urban society.[253] In the United States, rather exceptionally, slums arose in high-density city centers (that is, in multistory tenements, most commonly in New York and Cincinnati), where the composition of their population was more ethnically mixed, and they were seen more and more as an abyss of misery and deviant behavior. Under these conditions, the limits to the assimilation of lower-class immigrants were widely discussed around the turn of the century. The continuing slumification in certain European cities—for example, Glasgow, Liverpool, Dublin, Lisbon, or the twelfth and thirteenth arrondissements of Paris—was present as a constant warning.[254] In Britain, slums were feared and abhorred more as a breeding ground for physical or moral disease, as a nagging reminder of the limits of modernity, and not least as an unproductive misuse of valuable land.[255]

The middle classes seldom returned to cleared slum areas, which usually became commercial districts. But the flight to spacious and healthy green belts and villa districts was not always the rule in Europe, or even in the United States. The bourgeoisie of Paris, Budapest, or Vienna hung on in its large urban dwellings with sumptuous reception rooms and fairly modest private quarters; in 1890 the centers of those cities were accordingly twice as densely populated as inner London.[256] They too saw a trend away from center to suburbs, but it never reached the scale or speed of the process in London. In this respect New York was an exception to the characteristic American pattern. Between the 1860s and the 1880s, the city dwellings of the upper middle classes—hitherto mostly narrow tenements raised above street level—became more and more voluminous; this happened, at the expense of private gardens, in a context of rising land prices. In the end, apart from the superrich, *all* New Yorkers had hardly any private land at their disposal. One alternative to a move away from the center was the fashion for "French flats" that developed in the 1880s. With the proliferation of hydraulic or electric elevators, it became possible to market high-rise luxury apartments in the inner city. Someone who did not have their heart set on an expensive villa was able, from the turn of the century on, to move into a mid-Manhattan unit complete with telephone, pneumatic mail, and hot and cold running water in a building with a swimming pool and a basement laundry[257]

Why did most cities in nineteenth-century North America and Australia manage to avoid the emergence of slums? Why did the detached single-family house become the core of suburbanization, affordable for a large section of the population, including skilled workers? Why did conditions not develop as in

Paris under Louis Philippe (the "bourgeois monarch") where one-quarter to one-fifth of workers lived in rundown "bed and breakfast" hotels: five-story houses with damp low-ceilinged rooms, often lacking a fireplace or wallpaper and provided with only the bare essentials in the way of furniture?[258] In other words, why did housing patterns in the neo-Europes generally develop quite differently from those in the Old World?

This is perhaps the most interesting question of nineteenth-century urban history. An explanation in terms of the greater supply of land, though convincing at first sight, is inadequate; the answer appears to lie elsewhere. Space-taking urbanization is considerably more expensive than more-compact forms, because it requires greater investment in infrastructure: longer suburban train lines with more stops, more extensive sewers, and so on. Three factors must come together if dispersed housing is to be possible in practice: (1) new and cheaper construction techniques (prefabricated structures), (2) mechanized transportation in the shape of electric streetcars and steam-driven subway and suburban trains, and (3) a high average level and fairly even distribution of income. This combination of elements, perfectly achieved in a city such as Melbourne, was lacking in European countries at the time when each began its intensive urbanization.[259] An "Anglo-Saxon" or "American" cultural preference for single-family houses cannot therefore be treated as an independent variable; it also had to be a feasible proposition.

The new cities in Australia and the American Midwest may have looked dull or even ugly by the aesthetic standards of European city planning, but they made the petit bourgeois dream of a protected private domain, with a wholesome family life in a house of their own, accessible to a large section of the working population. From the early nineteenth century on, capitalist serial production was able to turn out standardized housing of the most diverse kinds, from the small brick terraces, often standing back to back, that characterized English cities to the apartment blocks of Glasgow, Paris, or Berlin. Nine-tenths of the housing in Victorian London was not built out of need but consisted— just like today—of "speculative" projects in anticipation of future demand. But the mode of production did not yet guarantee production quality. Only the democratized suburbs of the New World solved the problem of overcrowded residential districts. But in so doing it bequeathed to the twentieth century the problem of deserted inner cities.

The technically advanced suburb of 1910 still feels close to us today: we describe it without hesitation as "modern." In comparison, the pedestrian city of the early nineteenth century was positively medieval: a place where executions were still one of the favorite popular amusements (140 people were publicly hanged in London between 1816 and 1820).[260] It was also a dark place.[261] House lights were put out early, and one could walk in the streets only with the help of torches and lanterns. Gas lighting was first installed in cotton factories, to lengthen the working day. In 1807 the first gaslights came on in London streets, and by 1860 some 250 German cities had followed suit.[262] In Japan, kerosene and gas lighting were introduced simultaneously in the mid-1870s. If kerosene kept its place, this was because it required scarcely any fixed installations: the railroad distributed it to large and small consumers all over the country. In 1912, the year of the Meiji Emperor's death, Japan was a land of kerosene lamps.[263] Interior spaces were fitted with modern alternatives at a later date than central squares and streets.

From the 1880s the average British working-class home had access to gas for lighting, cooking, and heating. The relevant technology had close links to industry; gas stoves, in particular, which came into use around this time in Western Europe, consumed large quantities of iron. In 1875 electricity was made publicly available in Paris, and there was constant electric street lighting by 1879 in Cleveland, 1882 in Nuremberg (the first German city), 1884 in Berlin, and 1897 in Mexico City (where the whole system had to be imported). At first it was difficult to break into the gas market. Gas lighting served its purpose, and it took a while for the advantages of electricity to become apparent. Perhaps the most spectacular was stage lighting in theaters. The dimming of gas lamps in the auditorium had already proved effective in Paris in the late 1830s, but only full illumination of the scenic space laid the basis for the modern dramatic arts, with their sharp focus on the body.[264] As soon as the new technology was operating on a mass scale, it led to a veritable light mania. European cities competed with one another for the title "City of Light."[265] The consequences were enormous in the inner cities: the evening was democratized, since it was not only people with coachmen or torchbearers who ventured onto the streets. At the same time, the state could keep a closer check on the nighttime pursuits of its subjects and citizens. Nothing created such a disparity between city and country as the transformation of light from a glow emitted by candles and lamps into a glare produced by technical systems.

8   Symbolism, Aesthetics, Planning

Punishment and Exoticism

In a certain sense, the specificity of urban spaces defies formal analysis; literary description can do greater justice to local color, to the genius loci.[266] There is no need to assume that the "spirit" of a society is expressed in built-up cityscapes. It is simpler to investigate what contemporaries thought about the essence of the city. In nineteenth-century Europe we often hear them say that the city is a natural organism—an idea that is at the origins of sociology. To take "modernity" as an external standard for a city is problematic. Historians too easily share either the enthusiasm of new "city people" or the aversion of old elites—the landed nobility or mandarin class—for the ascent of mercantile and industrial urbanites. A discourse of backwardness is hard to disentangle. What does it mean to say of a city that it is "a big village?" Western Europeans in Moscow or Beijing used to sneer at the appearance of simple countryfolk in the cityscape, wondering at the social mix and implying that these urban societies might be of a different kind from their own.

Images and evaluations of a city can change abruptly. Lucknow, the nawab of Awadh's capital with a population of 400,000, was the glittering residential seat of one of the richest princes of India, probably the most prosperous inland city of the Subcontinent in the mid-nineteenth century and the cultural center of a sophisticated Persianized elite. Yet in 1857, in the eyes of the British, it changed overnight into a hotbed of rebellion and wickedness. Admiration for the old Muslim India disappeared from one day to the next. The British garrison in Lucknow was besieged for 140 days during the Great Rebellion—and the cramped and crooked layout of the old city was given as the explanation of how this was possible. After 1857 the British therefore rebuilt it to make it more secure and also to improve public hygiene (disease had claimed more European lives than the actual fighting). The reshaping continued for two decades, until 1877. Other major precolonial cities that had been battlegrounds in the rebellion—Agra, Meerut, Jhansi—received similar treatment. Most of their older districts were demolished, and their symbolism systematically degraded. One of the main Islamic holy sites in Lucknow, the mausoleum of a nawab venerated by the people, was turned into a barracks where British soldiers marched around in hobnailed boots, drank alcohol, and ate pork. The great Friday Mosque, until then the religious heart of the city, was closed down and left to decay—a violent intrusion into the social space of the city, after which all that remained was small local mosques. Wide avenues suitable for military use were driven through the city, destroying streets and alleys in their path. The British, again for military reasons, defined traditional monuments as cases for slum clearance. Lucknow was radically de-exoticized.[267]

Model cities and architectural styles interacted with each other in different ways. The latter could be more easily copied than the former, but the cultural "spirit" of a city almost not at all. In the nineteenth century, there was a tendency to oscillate between eclecticism and a quest for cultural authenticity. Nor was this true only in Europe: the architects Nahouchi Magoichi and Hidala Yitaka introduced art nouveau and the latest designs to Osaka; the newly built city of Yokohama became a hodgepodge of the most diverse influences, with domes and colonnades, Gothic spires and Moorish round arches.[268] Another uncolonized country, Siam, made a conscious effort in the early twentieth century to develop a "national" Thai style, which, as the architectural expression of an emerging nation, had first to be created out of previously existing elements.[269]

On the other hand, following the previous experience in the mid-eighteenth century, the period from roughly 1805 on witnessed in Europe a second (and in America, a first) wave of architectural exoticism. The Royal Pavilion in Brighton sported "Indian" domes and minarets, while the nearby stables of the Prince of Wales's thoroughbreds (today used as a concert hall) were given a pseudo-Oriental splendor.[270] The American showman and entrepreneur Phineas Taylor Barnum tried to outdo the Brighton Pavilion with his three-story "Iranistan" fantasy in the "Mogul style." The fragile structure, completed in 1848, succumbed to a fire nine years later. Other "Oriental villas" across the Atlantic had a longer life and were aesthetically more pleasing.[271] In fact, one finds Oriental interiors more often than entire buildings: tiles, open woodwork and metalwork, rugs, and tapestry. Technically avant-garde places such as railroad stations and pumping stations were decorated with "Moorish" touches, and cemeteries embellished with exotica. Chinese pagodas and Japanese wooden door arches were even featured in city parks.[272] (Conversely, the typical European equestrian statue never had any resonance among Asians.) World exhibitions became displays of architecture from all around the world, or of what was thought to count as such.[273] Two "Oriental" elements—the bazaar and the obelisk—went beyond the effect of individual buildings. From the first shopping arcade to be called a bazaar in the West (in 1816) to the shopping malls of the present day, the Oriental form of the covered market has enjoyed persistent popularity. There was no haggling in European "bazaars," however. On the contrary, they were pioneers of the fixed and marked price.[274]

Obelisks have a special history. In Renaissance Europe they were the aesthetic symbol for the profound wisdom that was supposed to have been attained in ancient Egypt; they stood less for the contemporary Orient than for the early perfection of civilization in the depths of time. What was new was the idea of adorning optically central locations in European metropolises with such culturally remote objects. Later, in 1885, the Americans would take the simpler approach of building their own fifty-meter high obelisk and erecting it in their capital city as the Washington Memorial, but the imperial powers of the nineteenth century became fixed on the idea of shipping home lapidarian monuments. A "Cleopatra's Needle" was installed in 1880 on the Thames embankment, and another one the following year in New York's Central Park. But the ultimate example was the unveiling of a giant obelisk in the middle of the Place de la Concorde, on October 25, 1836—a gift to the French king from Muhammad Ali. In fact, the pasha of Egypt was personally indifferent to the art treasures of pre-Islamic antiquity. The aim of his gift-dispensing diplomacy was to delight the French public, who since Bonaparte's Egyptian campaign of 1798 had shown much enthusiasm for ancient times on the Nile. He would eventually need French support in his efforts to shake off his overlord, the sultan in Istanbul.

The only problem for the French was that they had to ship the 220-ton colossus themselves. No less a person than Jean-François Champollion, the decipherer of hieroglyphs honored in France as much as in Egypt, traveled out in 1828 to confirm the offer and recommended trying to obtain the obelisks at Luxor. In 1831 the new government of the July Monarchy sent a special ship and a team of engineers to Upper Egypt, but it took more than five years for the obelisk to be taken down, loaded, shipped north on the Nile, Mediterranean, and Seine, and finally erected in a spectacular public ceremony. As a result of this extremely costly venture, the "capital of the nineteenth century" furnished one of its most animated public spaces, the domain of the guillotine, with a Near Eastern monument dating back thirty-three centuries.[275] The tranquil solemnity of the huge stone stood in stark contrast to the bloody spectacles that had been enacted on the square during the Revolution. The obelisk, covered with markings obscure to the layman and thus politically neutral, had the great advantage that no one was likely to take umbrage at it. An integrative symbol, not a divisive one, it was quite unlike that other postrevolutionary monument in France: the penitential Sacré-Cœur on the hill of Montmartre (built between 1875 and 1914), which would be erected to affirm the triumph of law and church after the suppression of the insurrectionary Commune—a provocation to many.

North American and Australian cities adapted various European models to their own environments and social needs. The suburb—an English invention, as we have seen—became thoroughly naturalized in the United States and Australia. Some details even migrated in the opposite direction. The Panopticon—that model prison with cells radiating like spokes from a central observation hall, which the English philosopher, political theorist, and social reformer Jeremy Bentham conceived in 1791—actually was first built in the United States and later exported back to the old country. The Americans also pioneered the construction of giant hotels, which in the 1820s initially struck Europeans as graceless imitations of army barracks. Structures of comparable size and splendor reached Europe only in 1855, when the Grand Hôtel du Louvre opened with an unprecedented seven hundred rooms on offer. So successful was the new model that by 1870, Edmond de Goncourt was already bewailing the "Americanization" of Paris.[276] Every city in the world that claimed to be modern and sophisticated—and that, more prosaically, had to accommodate travelers—was now in need of hotels. The three decades before the First World War witnessed the birth of legendary luxury accommodations in Europe, Japan, and parts of the colonial world: the Mena House outside Cairo (1886), the Raffles in Singapore (1887), the Savoy in London (1889), the Imperial Hotel in Tokyo (1890), the Ritz in Paris 1898, the Taj Mahal Palace in Mumbai/Bombay (1903), the Esplanade in Berlin (1908), and so on. By 1849 Beirut already had a luxury hotel that Europeans could identify as such. North Africa and West Asia sometimes drew on the formula of the caravansery, which easily lent itself to modernization: often an enclosed courtyard where traders spent the night with their animals and conducted business.[277]

Regulatory Planning and Development Planning

Were nineteenth-century cities planned?[278] There probably has seldom been so much *and* so little planning as in that age. In the emblematic fast-growing "shock cities," from Manchester to Chicago and Osaka, any will to plan gave way before spontaneous forces of social change. There could not be planning unless political bodies made it their task. London, for instance, was more or less without a government; its first central body, the Metropolitan Board of Works, was not provided with adequate funding until 1869. Only in 1885 was the metropolis represented in Parliament in accordance with its position in the country, so that it could have a due influence on national policy; and only four years later was a directly elected council set up, the London County Council. Visitors to Manchester such as Alexis de Tocqueville and Charles Dickens were appalled at how little what was new in the city resulted from comprehensive planning. But critics easily overlooked the fact that precisely in Manchester an administration sensitive to social issues had begun to take shape just a few years after the opinion-changing reports of the 1830s and 1840s.[279] A further necessary clarification—following suggestions by Josef Konvitz—concerns the two distinct kinds of urban planning: *development planning*, which constructs the outline and the general aesthetic image of the city; and *regulatory planning*, which conceives of the city as a space requiring permanent technical and social management. Common to both was the rise of professional city planners, who in some cases might exercise major influence.

Regulatory city planning arose in Europe and North America in the 1880s. Urban elites saw then that it was necessary to move beyond ad hoc palliatives, such as most of the measures involved in the early cleanups, and to take charge of the whole urban environment on an ongoing basis. Infrastructures were now understood as regulatory systems. A systems viewpoint in technical matters and social policy gained the upper hand over uncoordinated private economic motives (of which the anarchic construction of London's railway stations was a striking example). This implied not least that landownership interests would command less respect. The rise of regulatory city planning can be clearly seen in the lack of concern about compulsory land purchase in the public interest.[280]

Development planning was an ancient practice, not a recent European invention. At least in China and India, the geometry of rule and the geometry of religion had older and stronger roots than in medieval and early modern western Europe, where often little more was required than the correct orientation of church axes. A uniform spatial alignment was one simple and effective form of planning; it may be found in the rectangular layout of ancient Chinese cities as well as in European geometric patterns (e.g., Mannheim, Glasgow, Valetta, Bari) and the grid pattern that was imprinted on both the land and the cities of the United States. With few exceptions (e.g., Boston and Lower Manhattan), these followed a logic of rectangular cell proliferation. Boston in the early nineteenth century constantly reminded travelers of an early medieval European city, but Philadelphia already faced them with an urban Enlightenment rationalism geared to the future; first the land was divided up in a grid and assigned to owners, then the grid was filled in.[281] Again and again, however, land speculation meant that attempts at orderly urban development spun out of control.[282]

Nineteenth-century urban planning attracts so much attention because it was not the norm. Many cities on every continent expanded without restraint: planning in Osaka, for instance, began only in 1899.[283] Whether anything was planned depended on special circumstances. A large fire might provide a stimulus—or it might not. After the Great Fire of 1812, Moscow was rebuilt in accordance with a plan of 1770; the reality looked less orderly. Another conflagration, in 1790, robbed Madrid of part of the Rococo charm that redevelopment had given it back in the age of Charles III; its golden times were never to return.[284] Hamburg, on the other hand, obtained and used an opportunity for planning after the fire of 1842. In Chicago the whole business district (but not the factory area) went up in smoke in 1871, after which the city rose again as the world's first skyscraper metropolis.[285]

The sheer speed of expansion of the most dynamic metropolises condemned Baroque-minded authorities to failure, while making it all the more necessary to establish order amid the rampant growth. Moscow's accretion of houses, gardens, and streets, for example, created a picture in which foreign visitors could see nothing but a confused jumble. The reality of urbanization here clashed with all visions of city planning, whether traditional or modern, west European or Russian.[286] It was similar in many other cities around the world. The contradictions could be especially blatant where a late absolutist regime with ambitions to shape the whole cityscape was replaced by one that gave free rein to private interests. A dramatic case in point was Mexico City. Under the liberal government of Benito Juárez, a brief transitional period in mid-century was succeeded by ruthless destruction of the Baroque cityscape—a process which, after the removal of ecclesiastical privileges, could roll on without meeting any resistance. The year 1861 saw the great demolition, when dozens of religious buildings were cleared within the space of a few months. Soldiers would burst into churches and rip images from altars with horses. Some were saved by allocation to other purposes, the National Library itself finding accommodation in a former church. Large-scale iconoclasm corresponded to a political program: the liberal intellectuals of an independent nation were rejecting its colonial past and an art they considered to be a cheap imitation of European models. Like a half century earlier in France, public space underwent violent secularization.[287]

Haussmann's Paris and Luytens's New Delhi

Development planning sought to make a new start, and it did this in three different ways. The first was surgical interventions in city centers that sacrificed them to a broad aesthetic vision: the Haussmann model. At first it was a Parisian specialty, stemming from the resolve of the president and later emperor, Louis Napoleon, to modernize France so thoroughly that it would regain the hegemonic position in Europe that it had occupied under the first Napoleon. In 1853 Baron Georges Haussmann, the prefect of the Seine *département*, was appointed director of public works and provided with sweeping powers and lavish funding.

For a long time his goals and methods were the subject of intense controversy in France, but the results eventually proved him right, and his ideas on city planning set the tone for the rest of Europe.

Few other cities were capable of planning on such a scale; first among them perhaps Barcelona.[288] Often a city would take over individual elements, as Nottingham did early on with the Haussmann boulevard. The adoption of this by Buenos Aires in the 1880s heralded its general switch from English to French cultural models, which were now perceived as more comprehensive in their modernizing ambitions; the *salons de thé* built around this time would survive until the McDonald's invasion of the 1980s.[289] As soon as the Parisian model was there for every visitor to behold, others could do with it what they wished. In Budapest, they decided to build the finest opera house in the world and cast their eyes around selectively: at Paris, but also at Gottfried Semper's splendid opera house in Dresden and the Burgtheater in Vienna. In one respect, the result in the Hungarian capital surpassed all others when it opened in 1881: the Budapest opera house had all the latest equipment and was considered one of the most fireproof in the world.[290] As a late developer, which had made the transition from timber to stone only in the final years of the eighteenth century, Budapest generally showed a sure hand in choosing its models, especially at the height of the construction and development boom between 1872 and 1886. From London it took the organization of projects by a central committee, the building of embankment roads, and the design of its parliament; from Vienna, much of the Ringstrasse conception; from Paris, the boulevard. By the turn of the century, Budapest had become a pearl studied with interest by German and American architects.[291]

The immediate impulse for the redevelopment of French cities was the need to create space for new railway stations and their access roads. Other factors were the removal of slums from city centers and a nostalgia for the grand planning of the empire. Not least, a construction boom promised to have spin-offs for the whole of the economy, providing a stimulus both locally and nationally. Politically initiated, though increasingly driven by private investment, the dynamic was of greatest profit to Paris, where many attempts at redevelopment had already been undertaken in the 1840s but fallen afoul of the lack of legal provision for massive state intervention. Now a government decree created the necessary framework, making it much easier for the municipality to buy up land in the inner city. Haussmann took advantage of a period in which the courts, infected by the construction boom, were prepared to interpret the new legal tools to the authorities' advantage. But he was by no means omnipotent, and many of his plans for street widening were thwarted by real estate interests. The fact that most of his visions became reality was due both to political will and to the calculations of many small investors that they would gain from rising land prices. Haussmann, as Peter Hall put it, "was gambling on the future."[292]

The prefect was driven by three passions: a love of geometry; a wish to create spaces that were both useful and pleasant, such as the boulevards on which

traffic could flow and walkers stroll for relaxation; and an ambition to place
Paris at the pinnacle of metropolises. The city was to be a wonder of the world,
and after 1870 that was indeed how it was perceived. Huge as the technical
effort was for this redevelopment of a whole city core, Haussmann and his colleagues also played close attention to aesthetic detail in their successful adaptation of seventeenth- and eighteenth-century Parisian classicism to the dimensions of a mass city. Stylistic unity held the project together; local variations
and the high quality of architectural execution prevented monotony. The basic
element was the five-story apartment house, whose facades formed integrated
horizontal lines along the new boulevards, their ubiquitous limestone brought
en masse to Paris by the new railroads. Squares and monuments lent a characteristic structure to the cityscape.[293]

The second form of urban planning bears a German signature. In Germany
a certain tradition of planning came together with one of strong local authorities. The later onset of industrialization, in comparison with Britain and some
other parts of Western Europe, made it possible to become familiar with the
problems of fast-growing modern big cities and to look for solutions in time.
The German model of urban planning focused less on the grand reshaping of
city centers than on growth in the periphery; it was essentially a question of expansion. This began in the mid-1870s and developed into comprehensive urban
planning in the early 1890s.[294] Around the turn of the century Germany was
widely regarded as a model of orderly urban expansion and holistic planning
of the city as social space, traffic system, aesthetic ensemble, and collection of
privately owned real estate.[295] In other words, development planning was coordinated at an early date and in an exemplary manner, with an awareness of the
need for regulatory planning.

In comparison with France and Germany, Britain had no really distinctive
model—unless its early and strong public concern with urban hygiene is regarded as such. London had been rather conservatively rebuilt in the wake of the
Great Fire of 1666, and after the work on Regent Street in the 1820s, linking the
palace of the prince regent (Carlton House) with the new Regent's Park to the
north, no further intervention took such a radical character. Regent Street was
the first new main street, after centuries of uncompleted projects, to be driven
through a densely populated European city core.

There was much building and transformation in London, but nothing comparable to Haussmann's great achievement. To find another example of such energy, we must look to the empire and the building of a new capital for India.
Work on it began shortly before the First World War and was not completed
until the 1930s: for this reason, and also because of its basic modernizing impulse
despite many Orientalist touches, it goes beyond the limits (however defined)
of the nineteenth century. Yet the imperial political will to launch and fund
the project (or, more precisely, to get taxpayers to fund it) bears the hallmarks
of the prewar period, when the British liked to think that colonial rule would

last forever, or almost. In New Delhi the architects Edwin Luytens and Herbert Baker, assisted by a large planning department and an Indian workforce of up to 30,000, could implement grand visions for which the conditions were present neither in the mother country nor anywhere else in the empire. The outcome was not so much a smoothly functioning, "livable" city as a prestigious urban complex, but—unlike 1880s Hanoi or Albert Speer's remorselessly vulgar plan for the capital of the "Greater Germanic world empire"—it was not one in which an imperial aesthetic brutally proclaimed its superiority. The Viceroy's House, government offices, and missions of larger princedoms were intended to form a harmonious ensemble together with the public archives, gardens, fountains, and avenues.

The New Delhi of Luytens and Baker was to be a stylistic synthesis, in which long-imported architectural idioms fused with Indian elements of Muslim or Hindu origin. Luytens had closely studied the work of early city planners, especially Haussmann's Paris and L'Enfant's Washington, DC. Being as familiar with the sketches of garden cities (an old Islamic idea recently revived in Europe) as with the latest currents of architectural modernism, he nurtured a deep aversion against the kind of Victorian bombast that he had seen at the railroad station in Bombay. It was not in Europe, or even in Washington or Canberra (Australia's new capital since 1911), but in India, on the soil of an ancient architectural tradition, that the greatest extravaganza of urban planning was launched at the end of the age that is the object of our study.[296] In the surfaces and straight lines designed by Luytens and Baker, we find a "de-kitschified" Orient combined with a modernist distaste for ornamentation, epitomized by Luytens's exact contemporary, the Austrian architect Adolf Loos. This gave their post-Victorian architecture a degree of timelessness, bringing it remarkably close to a cultural synthesis in stone.

The New Delhi project was unique, and so it would remain. The modernism that became the universal language of architecture in the twentieth century had originated on the other side of the world, in the 1880s, when the first skyscrapers expressing the new style in their external appearance grew upward in Chicago. The Monadnock Building Complex (1889–93) is perhaps the first building that an observer spontaneously ascribes to a new era in architecture.[297] Until the 1910s it was not technically possible to build skyscrapers with more than fifty stories. For a long time, this modernism remained American: the fact that planners and architects increasingly formed a kind of international—studying one another's work, making trips, and exchanging experiences—or the fact that stylistic borrowing and technology transfers became perfectly normal did not at all imply a global homogenization of tastes. The most spectacular new buildings in nineteenth-century Madrid were huge bullfighting arenas—not necessarily an export hit.[298] Europeans did not eagerly adopt the skyscraper, any more than they did the largely American vision of suburbia. City planners in the Old World fought against the disproportionate height and the obstruction of the view of churches and public buildings.[299]

The nineteenth century was one of the most important in the multimillennial history of the city as a material structure and a way of life. From the vantage point of 1900, even more so from that of the 1920s, it appeared as the founding age of urban modernism. The backward continuities with the early modern period are weaker than the forward ones with the twentieth century. Until the growth of "megacities" and the annihilation of distance by telecommunications and information technology, all the features of contemporary urbanism originated in the nineteenth century. Even the automobile age was looming on the horizon, if not yet the tyranny of the car over *all* cities in the world.

What remains of the neat cultural types that the old urban sociology, but also today's urban geography, have been so fond of demarcating? Even for the premodern period, the differences among "European," "Chinese," and "Islamic" cities have become less sharp and vivid for the contemporary student of global urban history; functional similarities appear at least as clearly as cultural specificities. But it would be superficial to go to the other extreme of seeing only crossovers and hybridity.[300] Many tendencies spread worldwide, supported by Europe's demographic, military, and economic expansion, without being byproducts of imperialism and colonialism. Examination of cities in noncolonized countries outside Europe (Argentina, Mexico, Japan, Ottoman Empire) has repeatedly demonstrated this. Designs for the city of the future were increasingly elaborated in a broad Atlantic, Mediterranean, Pacific, or Eurasian spatial context. The "colonial city" immediately evaporates as a sharply defined type; the bald dichotomy of "Western" and "Eastern" is unsustainable.

Thus, within the European and neo-European West, completely new cityscapes appeared in North America and Australia, not at all mere reproductions of Old World models. There was no direct European inspiration for the Chicago or Los Angeles of 1900. Types such as the "American" or "Australian" city are also difficult to construct, since cross connections leap to the eye in a global historical perspective. Melbourne was sparsely built over a wide area, like the cities of the American West Coast, whereas Sydney was dense and compact, like New York, Philadelphia, and the big cities of Europe.[301]

The modernization of urban infrastructure was a worldwide process, which required political will and a high degree of administrative capability, money, and technologies, and also the agency of philanthropic institutions as well as profit-oriented private interests. There were temporal disparities, but the process was generally complete by the 1930s in the major metropolises. In China, for example, then a very poor country with a weak state, the sanitization and physical development of cities was not confined to the cosmopolitan showcase of Shanghai. After 1900 urban modernization was also found deep in the interior, far from any strong foreign influence, where upper classes at the provincial or municipal level often encouraged and accomplished projects out of nationalist motives.[302]

New construction materials, techniques, and organization did not, however, automatically engender a corresponding change in urban society. A city is both a distinctive social cosmos and a mirror of the wider society around it. In various settings, therefore, specific mechanisms and institutions of social integration were at work. Thus, Western models of social stratification fail to uncover the logic of cities in the Islamic Middle East if we do not also recognize the tremendously important functions of religious foundations (*waqf*) as centers of political authority, religious and secular scholarship, exchange, and spirituality. They had a stabilizing effect, protecting property and defining its significance in space; they offered mechanisms for mediation between individual or private-corporate interests and the general requirements of urban society.[303] Similar examples may be found all over the world. Special societal institutions, often going back many centuries, resisted external adaptive pressure and remained woven into the social fabric of a fast-changing city.

CHAPTER VII

Frontiers

Subjugation of Space and Challenges to Nomadic Life

1   Invasions and Frontier Processes

In the nineteenth century, the opposite extreme of "city" is no longer "country," the realm of farming, but rather "frontier": the moving boundary of resource development. It advances into spaces that are rarely as empty as the agents of expansion talk themselves and others into believing. For those who see the frontier approaching them, it is the spearhead of an invasion; it will leave little as it was before. People flow into the city and to the frontier; these are the two great magnets for nineteenth-century migration. As spaces of boundless possibility, they attract migrants like nothing else in the age. The city and the frontier share a permeability and malleability of social conditions. Those who *have* nothing but are *capable* of something can achieve it here. The opportunities are greater, but so are the risks. At the frontier, the cards are reshuffled to produce winners and losers.

In relation to the city, the frontier is "periphery." It is in the city that frontier rule is ultimately organized, there that the weapons and instruments for its subjugation are literally forged. If cities are founded at the frontier, then the pre-frontier area moves farther out; newly established trading posts become the bases for further expansion. But the frontier is not a passive periphery. It brings forth special interests, identities, ideals, values, and character types interact with the core. The city can see its counterpart in the periphery. To a patrician from Boston, backwoodsmen in log cabins were scarcely less savage and exotic than Indian tribal warriors. The societies that take shape in frontier areas live within wider and widening contexts. Sometimes they break free; sometimes they succumb to the pressure of the city or to the consequences of their own exhaustion.

Land Acquisition and Resource Use

Archaeological and historical records are filled with processes of colonial land acquisition, in which communities open up new areas as a source of livelihood. The nineteenth century brought such tendencies to a climax but, in a sense, also to an end. In no previous age had so much land been used for agriculture. This expansion was a result of demographic growth in many parts of the world. It is true that the total population would increase even faster in the twentieth century, but the *extensive* use of resources would not grow at the same speed; the twentieth century as a whole is characterized by more *intensive* exploitation of existing potential (which by definition consumes less additional space). Destruction of tropical rainforest and overfishing of the oceans do, however, perpetuate the earlier pattern of extensive exploitation, in an age that in other respects has reached new heights of intensive development as a result of nanotechnology or real-time communications.

In nineteenth-century Europe, especially outside Russia, colonial landgrabs on a large scale became a rarity; it mainly took the form of settlement elsewhere in the world. Here all the dramas of European history seemed to repeat themselves while, at the same time, comparable processes were unleashed by Chinese and by peoples in tropical Africa. Migratory movements to the Burmese "rice frontier," or to the "plantation frontier" in other parts of Southeast Asia, were the result of new export opportunities in international markets. Land-grabbing settlement was associated with highly diverse experiences, which find their reflection in historical writing. On the one hand, active settlers drove into the "wilderness" on their heroic wagon treks, claiming "ownerless" land for themselves and their livestock and introducing the appurtenances of "civilization." The older historiography tended to glorify these pioneering deeds, depicting them as contributions both to modern nationhood and to the progress of humanity as a whole. Few authors put themselves in the place of the peoples who had lived for centuries or even millennia in the supposed "wilderness." James Fenimore Cooper, a patrician's son whose family occupied frontier land in New York State, already evoked the tragedy of the Indians in his *Leatherstocking Tales*, a series of novels published between 1824 and 1841 and soon widely read in Europe. But it was only in the early twentieth century that this bleak vision gained occasional entry into the work of American historians.[1]

After the Second World War and especially with the onset of decolonization, when doubts emerged about the white man's role in spreading good in the world, historians began to take an interest in ethnology and concerned themselves with the fate of the victims of colonial expansion; both academia and the wider public became aware of the injustices done to indigenous peoples in the Americas or Australasia, and the heroic pioneers of old became brutal and cynical imperialists.[2] Then, in a third stage, the one we are still in today, this black-and-white picture was refined into various shades of grey. Historians discovered what the American historian Richard White has famously called "the middle ground," that is, spaces of long-term contact in which the roles of perpetrator and victim were not always clear-cut, and in which negotiated compromises, temporary equilibria, and intertwined economic interests—sometimes also cultural or biological "hybridity"—developed between "natives" and "newcomers."[3] Regional variations have also come in for closer scrutiny; the view on frontiers has become pluralized and polycentric; the role of "third parties" in frontier expansion—that of the Chinese, for example, in the American Northwest—has been given as much attention as the fact that many (though not all) of these processes were driven by families and not by vigorous males on their own. There were cowgirls alongside the cowboys.[4] An especially rich literature now exists on the mythology of colonization and its representation in the media, from early illustrated travel reports to Hollywood Westerns.

For all the nuances, it remains of fundamental importance that the winners and losers of colonial landgrabs can be easily distinguished from each other. Although some non-European peoples, such as the Maoris in New Zealand, put up more successful resistance than others, the global offensive against tribal ways of life led almost everywhere to the defeat of the indigenous population. Whole societies lost their traditional sources of livelihood without being offered a place in the new order of their homeland. Those who escaped merciless persecution were subjected to "civilizing" procedures that involved complete devaluation of the traditional native culture. In this sense, the nineteenth century already witnessed the *tristes tropiques* about which Claude Lévi-Strauss wrote so poignantly in 1955. The massive assaults on those whom Europeans and North Americans regarded as "primitive peoples" left even deeper traces than the subjugation—at first sight more dramatic—of those non-Europeans who might at least become economically useful in systems of colonial exploitation. Sir Christopher Bayly has identified this as one of the key processes of nineteenth-century world history and has justly discussed it in close conjunction with ecological depredation.[5]

Colonial rule formally came to an end in the third quarter of the twentieth century. But almost nowhere was there a change in the subordinate position of "ethnic minorities" who had once been masters in their own land. The process of their subordination was quite swift. In the eighteenth century there were still semi-stable areas of "middle ground" in many parts of the world. But such zones of precarious coexistence were unable to survive in the second half of the nineteenth century. Only with the general delegitimation of colonial rule and racism after 1945 was fresh thought given to original injustice, "aboriginal rights," and the question of reparations, including compensation for slavery and the slave trade. The beginnings of recognition by the outside world also created a new scope for the affected minorities to build an identity. The fundamental marginalization of their way of life, however, is tragically irreversible and irreparable.

Frederick Jackson Turner and the Consequences

Land-grabbing colonization is one of the ways in which empires come into being. The legionnaire does not always have to go first; often the great invasion begins with the merchant, settler, or missionary. In many cases, however, it is a nation-state itself that "fills" a predefined territory. There is something that resembles inner frontiers and internal colonization. The most striking and

generally successful example of pioneer development was the European settlement of North America from the Atlantic coast westward, which the older
American historical tradition celebrated as "the winning of the West" (Theodore Roosevelt). The name for this gigantic process is itself of American origin.
The young Frederick Jackson Turner coined it in 1893 in a lecture that is still
probably the most influential text to have been written by an American historian.[6] Turner spoke of a "frontier" that had been driven ever farther from east
to west until it reached a state of "closure." Here civilization and barbarism met
each other in an asymmetrical distribution of power and historical right; the
efforts of the pioneers had formed a special national character; the peculiar egalitarianism of American democracy had its roots in the common experience of
life in the forests and prairies of the West. "Frontier" was thus the keyword that
made possible a new grand narrative of US national history, and that would later
be generalized into a category applicable to other settings.[7]

In hundreds of books and essays, Frederick Jackson Turner's original concept of the frontier has been situated within the history of ideas, refined by
neo-Turnerians, condemned by critics, and rather pragmatically adapted to the
standpoint of various historians. The associated vision of the national past has
deeply marked America's understanding of itself even in places where Turner's
name is unknown. The myth of the frontier has a history of its own.[8] Turner's
originality lay in his elaboration of a concept that was at once clearly defined as a
scientific category and intended as a master vision for understanding the special
historical destiny of the United States. For Turner, the settlers' opening up of the
thinly populated regions of the West was the key to nineteenth-century American history; the ever-shifting frontier had carried "civilization" into a realm of
untouched nature. Insofar as the "wilderness" was inhabited by natives, it was
a place where humans at different "stages" of social evolution encountered one
another. Not only was the frontier geographically mobile; it also opened up a
space of social mobility. The "transfrontiersmen" and their families were able
to achieve material success through hard work and a constant struggle against
nature and "native peoples." They forged their own happiness and in the process
created a new type of society. This new society involved an unusual degree of
sameness and coherence in its underlying assumptions and attitudes—in comparison not only with Europe but also with the less volatile and more hierarchical society of the American East Coast.

At once visionary and painstaking researcher, Turner identified several different kinds of frontier. As always happens with model building, though, his followers took the labor of classification to excessive lengths in trying to fit the rather
general basic concept to an endless variety of historical phenomena. Ray Allen
Billington, for example, the most influential of the neo-Turnerians, differentiated six successive "zones" and "thrusts" in the push to the west: first came the fur
traders, then the cattle drivers, the miners, the "pioneer farmers," the "equipped
farmers," and finally the "urban pioneers" (who closed the frontier and built

stable urban societies).[9] Critics objected to this overly abstract sequence, point-ing out that it neglected the political-military dimension and wondering what was meant by terms such as the "opening" or "closure" of a particular frontier. Turner himself had explicitly refrained from offering a precise definition. But the new frontier studies that took over his impetus replaced a sharp dividing line between "civilization" and "wilderness" with the concept of "zones of encoun-ter," without ever providing a widely accepted definition of what this meant.

One tradition diverging from Turner—Walter Prescott Webb is here the key author—looked back at world history and emphasized the active and "organic" side of the frontier, its ability to change those things with which it comes into contact.[10] This idea later inspired Immanuel Wallerstein's concept of the incorpo-ration of peripheral regions into a dynamic world system. Alistair Hennessy, in a remarkable comparative essay, presented the sum of all frontier processes as quite simply "the history of the expansion of European capitalism into non-European areas," as the irreversible spread of the commodities and money economy and European conceptions of property into overseas expanses of endless grassland: the prairies of Canada and the Great Plains of the United States, the Argen-tine pampas, the South African veldt, the Russian/Central Asian steppes, and the Australian outback.[11] And William H. McNeill, the great master of world-historical analysis, applied Turner's concept to Eurasia, with a stress on the theme of freedom that had already been supremely important to Turner himself. McNeill sees the frontier as ambivalent: a clear political and cultural dividing line, but also an opening up of free and empowering spaces no longer to be found in the more highly structured core zones of stable settlement. For example, the position of the Jews was markedly better in frontier areas, where they often set-tled, than under less fluid conditions.[12]

Should the frontier be regarded as a space that can be demarcated on a map? There is much to be said for the alternative view of it as a special social constella-tion. This would give us the following definition, sufficiently broad but not too woolly:[13] a frontier is an extensive (not simply local) situation or process where, in a given territory, at least two collectives of different ethnic origin and cultural orientation, usually under the threat or use of force, maintain contacts with each other that are not regulated by a single overarching political and legal order. One of these collectives plays the role of the invader, whose primary interest is in ap-propriating and exploiting land and/or other natural resources.

A specific frontier is the product of a push from outside, which mainly orig-inates in private initiative and only secondarily enjoys state or imperial support or rests upon conscious instrumentalization by a particular government. The settler is neither a soldier nor an official. The frontier is a sometimes persistent but theoretically fluid state of affairs marked by high social volatility. In the beginning, at least two "frontier societies" stand opposed to each other, each inserted into externally driven processes of change. In a minority of cases (the "inclusive frontier") they merge together into one (always ethnically stratified) hybrid society, whose *métissage* existed, above all in North America, as an "underground" beneath the respectable society of white Protestant heads of families.[14] As a rule, unstable equilibria break down in a way that disadvantages one of the sides, which is then excluded, separated, or even physically expelled from the ever more solid ("modernizing") social context of the stronger collective. An intermediate stage on the way to this is the situation in which the weaker side becomes dependent on the stronger. While the frontier opens up space for communication—in new pidgin languages, for example—and for the enhancement of special types of cultural self-understanding, the most important lines of conflict are in noncultural spheres: on the one hand, the struggle over land and the elaboration of ownership concepts, and on the other hand, various forms of work organization and labor-market structures.

The invaders marshal three self-justificatory patterns, separately or together as need dictates:

1. the right of the conqueror, which may simply declare existing occupation rights to be null and void
2. the seventeenth-century Puritan doctrine of *terra nullius*, which regards land populated by hunter-gatherers or herdsmen as "ownerless," freely acquirable, and in need of cultivation
3. the missionary duty to civilize "savages," often added afterward as a secondary ideology or *post festum* legitimation of coercive dispossession

Although the frontier concept is today used in everyday speech for any conceivable case in which profit can be made in a spirit of entrepreneurship and innovation, historical frontiers have the aura of transitions from premodern conditions. Once a region in question has linked up with the main technological macrosystems of the modern world, it soon loses its frontier character. The taming of nature then also passes quickly into corporate exploitation of resources. Thus, the coming of the railroad—not only in the American West—destroyed the precarious balances already in existence. The frontier is a social constellation that essentially belongs to an intermediate period, on the eve of the steam engine and the machine gun.

Frontier and Empire

How are frontier and empire related to each other?[15] Here the argument has to be mainly spatial. Nation-states never have frontier spaces on their borders. Frontiers, in the sense in which we are using the word here, can persist after the initial invasion only when no clear territorial borders are defined and when the process of state organization is still patchy or rudimentary. In the frontier perspective, "the state" is relatively far away. The borders of empires are typically, but not always, frontiers. As soon as empires stop expanding, frontiers stop being zones for potential incorporation and are transformed into exposed flanks in the struggle against external threats. They become uncontrollable spaces beyond what is perceived as the empire's defensive perimeter—menacing voids beyond the last watchtower, from which guerrillas or mounted warriors might suddenly appear. In the nineteenth-century British Empire, the North-West Frontier of India was one such neuralgic zone, which required special techniques of mountain warfare (traveling light in unfamiliar terrain); the Russians in the Caucasus and the French in Algeria fought similar border wars.[16] In contrast, the Northern Frontier of British India toward Tibet evinced no vulnerabilities of that kind; it was not really a "frontier" but an international *border*, defined through complicated negotiations between states.[17] This was also true of the borders that European colonial powers agreed to among themselves in Africa or Southeast Asia, although locally they were often of such little practical effect that the paper geography of political sovereignty was overshadowed by a more real geography of "lived" frontiers, not infrequently in the interaction between plainsmen and highlanders.

Where two or more colonial powers disputed a region with modern concepts of territorial statehood, we should speak not of frontiers but of *borderlands*, which, according to a student of Turner, Herbert Eugene Bolton, are "contested boundaries between colonial domains."[18] Here the possibilities of action are different from those in a frontier zone: indigenous people may to some extent play rival invaders off each other, continually crossing the various border lines. But once an intercolonial agreement is reached, it always operates to the disadvantage of local people. In an extreme case, whole peoples may be deported across borders, or transfers may be negotiated, as they were as long ago as the eighteenth century between the Tsarist and the Qing empires.

The attitude of imperial powers to frontiers is structurally ambivalent. Frontiers are constantly turbulent and therefore threaten what any empire must regard as the highest good following a period of conquest: namely, peace and order. Armed and unruly pioneers vitiate the monopoly of force that the modern state, including the colonial state, seeks to wield. A frontier on the edge of a colony can therefore seldom be more than a temporary state of affairs, a region that is "not yet" or "soon no longer" imperial. Nation-states are less able than empires to tolerate special "frontier societies," except where the natural environment compels them to do so. Frontier areas, then, do not realize the idea of empire in a pure form; they are at best countenanced as an anomaly. Or more generally, settler colonialism and empire are two quite different things. Unless settlers are actually dispatched as "armed farmers" to an insecure border zone, the imperial center regards them as inherently contradictory: "ideal collaborators" (Ronald Robinson[19]) but also a source of endless political trouble (from intractable Spanish *conquistadores* to the white elite of Southern Rhodesia, who declared independence unilaterally in 1965).

Recently the most interesting new meaning given to the frontier has been ecological. Turner already mentioned the "mining frontier" as perhaps second in importance to the settler frontier: it usually brought about more complex societies than a purely agrarian constellation and was capable of being completely independent. More generally, one might speak of resource-extraction frontiers—an economic but at the same time also an ecological concept. In fact, "ecology" also played a major role in the classic frontier, where settlers had to attune their farming methods to new environmental conditions. They lived with wild animals, bred livestock, and—to simplify greatly—drove their cattle-based civilization into regions where Native American civilizations had depended on the bison. One cannot speak of frontiers and be silent about the environment.

Another approach, independent of Turner, leads in a similar direction. In 1940 the American traveler, journalist, and Central Asia expert Owen Lattimore published his pathbreaking work, *Inner Asian Frontiers of China*. This interpreted the history of China in terms of a permanent conflict (symbolized by the Great Wall) between farming and pastoral cultures, two ways of life explicable mainly by their different natural foundations.[20] Lattimore's method was not at all geodeterminist, however, since in his view the basic contradiction between farmland and steppe was to be understood politically. Wide scope for manipulative action developed both in China and in the steppe empires that repeatedly emerged on its peripheries, the ultimate antagonism being the clash between nomadic herdsmen and settled farmers.

In light of the new interest in ecological aspects, Turner's insights may be usefully combined with the much broader viewpoint developed by Lattimore. Historians who range the whole of man's extensive intervention in nature under the "frontier" category directly link up with the idea of resource-extraction frontiers. In John F. Richards's environmental history of the early modern world, for example, the "frontier of settlement" appears again and again as the guiding thread. The process that reached a climax and a conclusion in the nineteenth century can be traced back to the early modern period, when technically better-equipped settlers occupied land that had previously been used (but not "deeply tilled" in an agricultural sense) by pastoralists and hunter-gatherers. The pioneers everywhere invoked their more productive land use to justify displacement of existing types of farming and hunting. They cleared forest, reclaimed marshes, irrigated dry land, and massacred the part of the fauna that they regarded as useless. At the same time, they had to adapt their methods to new environmental conditions.[21] In the argument of Richards's great work, the social, political, economic, and ecological aspects of frontiers cannot be separated from one another; he himself investigates frontier constellations around the world and is therefore able to present the phenomenon from each of these angles. Since this chapter cannot aspire to such regional completeness, the resource-development frontier will be treated only briefly in a later section below.

Transgression and Statization

One of the merits of the ecological approach is that it sharpens one's feel for frontier *processes*. It is hardly possible to describe frontiers statically. They are spaces where effects occur that it would be an understatement to describe as "social change." These processes are varied in kind. Two are especially widespread:

-   The "transfrontier process," that is, the movement of groups across ecological boundaries. A good example of this is the Boer treks in South Africa as they began in the last third of the eighteenth century. When fertile, easily irrigable land became scarce in the Cape Colony, many Afrikaans-speaking whites abandoned European-style intensive agriculture and took up a seminomadic way of life. Some of these—estimated at a tenth of the total—attached themselves to African communities. At the beginning of the nineteenth century, people of mixed origin (*griquas*) formed social organizations, townships, and even parastate structures (East and West Griqualand). Such "transfrontiersmen" appeared in South America too, though not in the midst of scarcity but in conditions where an abundance of wild animals made it possible to hunt for livestock and horses. Still, there were great similarities with Africa: in particular, transfrontier communities in the interior were virtually ungovernable from outside. Ethnic-biological mixing typically occurred, and it was only in the nineteenth century that racial doctrines produced attempts to draw clear lines of separation. Further examples were the Caribbean "buccaneers" and the Australian "bushrangers": quasi-military bands, consisting mostly of former convicts, who were suppressed by government action after 1820.[22]

-   The seizing of frontiers by the state. Even if colonization and frontier violence could initially proceed without constant military support, and even if the justice system by no means unambiguously differentiated between criminal and law-abiding behavior, the state was always on hand where it had to guarantee landownership. Already in the early modern period, the most general contribution that governments made to frontier settlement was the sweeping legalization of land occupations and the flat rejection of indigenous peoples' property rights. Frontier regimes differ in the thoroughness with which the state assumes the tasks of measuring, allocating, and registering land. And it was precisely in the "Wild West," so anarchic in the popular imagination, that land-ownership was tightly regulated from very early on. Governments rarely went so far as to influence the concentration of ownership, however. American frontiers, with their seemingly limitless supply of land, meant that the utopia of relatively equal distribution and general prosperity was theoretically achievable. This had been Thomas Jefferson's grand vision: a society without an underclass, where the bonds of scarcity were shattered. It is revealing here to compare the United States with Canada and Argentina, where frontier land was initially treated as a public good. In Canada it was mainly small farmers, highly mobile and venturesome, who took up the state's offer of land—and speculation began to appear at an early date. In Argentina the land fell into the hands of big land-owners; they often leased it to tenants on favorable terms, but in the long run everyone who had believed in the egalitarian frontier spirit fell prey to disappointment. If despite similar environmental conditions and similar links to the world market, the two countries developed opposite structures of landownership, this had to do with the fact that in Argentina government policy was aimed toward export-led growth, whereas in Canada it attached greater importance to a balanced social order. The ruling oligarchy was itself interested in landownership in Argentina but not in Canada.[23]

2 The North American West

An Exceptional Case

The frontier in the United States, especially between 1840 and 1890, stands out from all others for a number of reasons.

First, no other settler movement in the nineteenth century involved such large numbers; it filled a continent with people, to a far greater extent than in Australia. This was true of the overall long-term process but also of particular episodes of dramatic acceleration. The Californian gold rush, for instance, was the largest continuous migration in the history of the United States: 80,000 people flooded into the state in 1849 alone, and by 1854 some 300,000 whites were living there. The "gold rush" to Colorado in 1858 was of comparable dimensions.[24] Structurally similar experiences occurred in the Witwatersrand (South Africa), New South Wales (Australia), and Alaska, but they were more localized and less associated with a settlement push across a vast land mass.

Second, no other frontier had such an impact on society beyond its immediate areas. Nowhere else were structures of frontier society so successfully integrated into a national context. The American West did not develop into a backward and marginal "internal colony," partly because of a geographical peculiarity of the United States. After the midcentury gold rush, the region of extraordinary economic dynamism that took shape along the Pacific coast was not mainly the result of mechanisms of extensive land acquisition. The true frontier thus lay *between* the long-dynamic East Coast and an economically up-and-coming region on the other side of the continent; it was topographically a genuine "middle ground." From the point of view of social history, the basic distinction that should be drawn is between two kinds of frontier society: (1) a West of farms and small towns, inhabited by middle-class people and marked by families, religion, and closely knit communities; and (2) a much more turbulent pioneer West, defined by livestock herds, gold prospecting, and army outposts, where the characteristic social type was the young single male, often employed seasonally, highly mobile, and exposed to dangerous working conditions. In addition, as a special regional form, there was (3) the society that sprang up in the wake of the gold rush in California. It clashed so sharply with many features of the conventional West that there has long been controversy as to whether, or in which sense, Pacific California should be included at all in "the West."

Third, the nineteenth-century American frontier operated without exception as a mechanism to exclude the local native population. The picture was much the same in South America, whereas in Asia and Africa greater scope remained here and there for indigenous people. Earlier, on North American soil, there had definitely been instances of assimilation between "Indians" and Europeans; the French, much more than the English or Scottish, had reached a kind of modus vivendi with the Indians in the eighteenth century. In relations between the Spanish and native peoples in today's New Mexico, a stable "frontier of inclusion" had developed in conditions of approximate equilibrium.[25] This was not repeated in the US sphere of control, where the reservation gradually developed as the characteristic way of treating native people. The more the center of the landmass filled up with settlers, the less it was possible to drive the Indians into an open "wilderness." After the Civil War, and a fortiori after the end of the Indian wars in the 1880s, the system of scattered special areas became the norm. In no other frontier—although there are similarities with the homelands in twentieth-century South Africa—did this encircling isolation of the indigenous population occur on such a scale.[26]

Fourth, as a scholarly concept and a popular myth (little affected by academic "de-heroization"), the frontier has been the great integrative theme of national history since long before Turner gave it a name. Around 1800 Jefferson was in no doubt that the future of the United States would lie in the western continent, and in the 1840s the ideological motive of the Manifest Destiny was repeatedly used to justify an aggressive foreign policy. In this sense, some historians have interpreted the US maritime expansion in the Pacific—spearheaded by whaling—as the carrying of the frontier beyond the country's land borders.[27] The opening up of the West was and is seen as the distinctive North American form of nation building. The integrative force of the theme is due also to the fact that at some point in its history, nearly every region in North America has been a "West."

The vast research on the question cannot be summarized here.[28] At one extreme, the history of the West has been completely divorced from the frontier concept: this was to some extent inevitable when the focus shifted almost entirely to particular regions and localities, since that meant giving up Turner's basic idea that the various geographical and sectoral frontiers were ultimately interrelated parts of a single process. Another direction in American studies, to which our present treatment is closer, rejects the tendency to reify the West, seeing it not as a region describable in terms of objective geographical features, but as the outcome of relations of dependency. In this optic, "West" denotes a special kind of force field rather than a place that can be marked on a map. Other perspectival changes refer to the multiplicity of social actors—which cannot be reduced to a simple opposition of ranchers and Indians—and to the increasingly urban character of the West in the twentieth century. Cities never figure in the classic Western movies of the 1930s and 1940s, although at the time when they were made parts of the West were already among the most urbanized areas in the United States. Revised historical interpretations rarely feed only on advances in empirical knowledge. Consequently, the debate between neo-Turnerians and their opponents cannot be decided only by reference to advances in research. Each revisionism has a political backdrop, and attempts to dismantle the Turner orthodoxy may also, for example, involve a critique of American "exceptional-ism." If the frontier evaporates, then at least that claim to a special American way goes by the board.

As far as nineteenth-century world history is concerned, however, one cannot fail to be struck by the distinctiveness of the United States. We have already seen that its patterns of urbanization did not simply reproduce those of the Old World, and that its suburban sprawl defined a neo-European path that drew it closer to Australia typologically. If Europeans had not regarded the conquest and settlement of the West as such a unique phenomenon, they would not have described and commented on it with such fascination, or taken it as the starting point for fantasies and fictions of their own. America's striving to acquire a "normal" national history finds itself confronted with Europe's astonishment at the special development of the American frontier. Europeans will therefore not criticize American "exceptionalism" as vigorously as some American historians have done. In South Asian or East Asian eyes, America's peculiarities are even more apparent: in crowded spots of the planet they inspire ceaseless wonder at the abundance of fertile land. In many parts of Asia, nearly all highly productive areas were settled and cultivated by 1800; virtually all reserves of land were in use. America could not but appear as a land of plenty and waste.

Indians

A consideration of what is distinctive about the North American frontier must look first of all at the relationship between Euro-Americans and American Indians, taking account of the fact that any generalization about these extremely heterogenous groups is reckless to the extreme. As happened earlier in the Caribbean and Central and South America, the size of the indigenous population here fell sharply in the wake of the European invasion. A *general* accusation of genocide on the part of whites is exaggerated. But some American ethnic groups were certainly wiped out, and there were some dramatic regional irruptions. In California, where approximately 300,000 Indians had lived at the start of the Spanish settlement in 1769, only 200,000 remained by the end of the Spanish period in 1821. After the gold rush, a mere 30,000 survived until 1860. Disease, starvation, and sometimes murder—one leading historian has spoken of "a program of systematic slaughter"[29]—were the causes of this decline. It was a catastrophe for the survivors too, since the white society of California made no offers to integrate them.[30]

American Indians exhibited great diversity, with neither a uniform way of life nor a common language, and so it was difficult for them to coordinate armed resistance against the whites. The spectrum went all the way from bison hunters on the western plains to settled Pueblo farming communities to the sheep-breeding and jewelry-producing Navajo and the very loosely organized fisherfolk of the Northwest. Often they had little or no contact with one another; there was no unified Indian consciousness and solidarity, no united front against the invasion; and often cruel warfare arose even among related or neighboring tribes. So long as Indians were in demand as allies of the whites, they were sometimes able to play off British, French, Spanish, and rebellious settlers against one another. But this was no longer an option after the British-American war of 1812; the opportunity had passed for a pan-Indian resistance, organized from the North in a spirit of militarized religious fervor.[31] In all future Indian wars, renegade Indians would fight on the Euro-American side and provide logistic support.

One thing common to most of the Indians of the Great Plains was the impact of a technological revolution. There is no other way to describe the use of horses for riding and carrying, which had first been introduced in the early seventeenth century in the Spanish-held south of North America.[32] With the horse came firearms, which the French deployed to strengthen their Indian allies against the Spanish. Horses and muskets radically changed the lives of tens of thousands who had not set eyes on a white man before. As early as the 1740s there were reports of horse herds, horse trading, horse theft, and horseback combat, and by 1800 virtually all Indians west of the Mississippi had adapted their lifestyle to some degree to the animal. Whole peoples reinvented themselves as centaurs. This was not true only in ancestral lands on the edge of the plains. Sometimes following chosen migration routes, sometimes pushed westward by the Euro-Americans, Indian peoples from the Northwest such as the Lakota Sioux settled on the Great Plains and came into conflict there with farmers or rival horse-riding nomads. Whereas a relatively stable peace was negotiated in 1840 among the mounted hunters and warriors (Sioux, Comanche, and Apache, for instance), fighting continued between nomadic and settled Indian peoples: the bloodiest source of conflict in North America during the four decades before the Civil War.[33] On the other hand, the horsemen relied on farmers and vegetable growers to keep them supplied with carbohydrates and to exchange objects from the East for their hunting produce (mainly dried meat and skins).[34] This was perfectly possible, because low-tech Indian agriculture (no ploughs, no fertilizer) had achieved a high productivity from which Euro-Americans also initially profited. In 1830 the Great Plains were more densely populated than ever before. It has been estimated that 60,000 Indians then shared the vast habitat with up to 900,000 domesticated horses, 2 million wild horses, 1.5 million wolves, and up to 30 million bison.[35]

Only the horse permitted the complete opening up of the plains between the Mississippi and the Rocky Mountains—800 kilometers from east to west, and more than 3,000 kilometers from north to south. It functioned as an energy transformer, converting the energy stored in grassland into muscle power obedient to human command.[36] Now humans could keep up with the bison. It was no longer necessary for the whole population to take part in driving them to the edge of an abyss; mobile groups of young men could take potshots at them from the back of a horse. At the same time a new exchange economy evolved around the horse, and some tribes, foremost among them the Comanche, acquired "prodigious animal wealth" and became suppliers of horses for all sorts of customers far and near.[37]

The new hunting techniques revolutionized Indian communities. Women's labor was devalued, since their main activity was no longer to produce food on their own but to process the animals killed in the hunt. On the other hand, an increasing demand for bison skins meant that more women were needed to prepare them, so that one man could do with several wives. Women were bought with horses, and the resulting pressure to accumulate was a factor encouraging horse theft.[38] The allocation of men to hunting groups led to social fragmentation and an erosion of hierarchy, but it also created new demands for cooperation and coordination. At the same time, the Indian communities and tribes became more mobile than ever, as they had to follow in the tracks of the vast herds of bison.[39]

It was this horse-and-bison culture that turned the Indians of the Great Plains into genuine nomads. Packhorses made it possible to transport heavy loads such as tents. Anyone who had personal property needed horses, and horses in turn were regarded as prestige objects. They also created advantages in time of war. Here too, creativity and adaptability were required of the Indians. For there was no indigenous tradition of mounted warfare, and the Spanish heavy cavalry, which became known in the South in the seventeenth century, was no model. Since the horse had to serve for hunting *and* battle, it was essential that the techniques for both should be as close as possible to each other. The Indians therefore developed light cavalry tactics, in some cases reaching unsurpassed heights of mastery. The stereotype of the expert rider applies only to the last period in which the Indians led a free existence; it took them three or four generations to perfect their skills. Best of all were the Comanche, who, having expelled previously settled groups, controlled the area east of the southern Rockies and south of the Arkansas River and even constructed a formidable system of dependencies that has been described as a "Comanche Empire" and a powerful player in the imperial game on the North American continent.[40]

The new horse-and-bison culture of the eighteenth century may be seen as a superb adaptation to a dry climate that was unsuitable for agriculture. The image of ecofriendly Indians, however, living in caring harmony with nature is a sentimental idealization remote from reality;[41] the new integration into wider commercial circuits set up many pressures of its own. The first regular contacts between Indians and whites came about through the fur trade—which for two centuries had linked hunters and trappers in the North American interior as well as Siberia to the world market—and were stabilized thanks to the great adaptability of Euro-American "backwoodsmen" and to marriages across ethnic boundaries. Through the fur trade Indians developed a familiarity with alcohol—a drug which, like opium a few decades later in China, would greatly weaken the cohesion and power of resistance of their communities. The horse-and-bison culture strengthened the ties with external markets. In one direction, the Indians had to cover a growing part of their needs through the buying and selling of goods. Even the most implacable opponent of the whites did not refuse the knives and cooking pots, rugs and materials that could be bought via agents from the factories and workshops of the east. In addition, many Indians acquired firearms that they did not know how to produce or repair themselves. This pushed them further into the web of trade, as did the growing dependence of their bison specialism on uncontrollable market factors. After 1830, for example, bison hides became more important than meat produce in cross-frontier trade, and it was around then that the problem of excessive herd depletion set in. An annual "yield" of six to seven animals per person was manageable (as we know today), but anything above that meant a dangerous level of overexploitation.[42]

The livelihood of the Plains Indians, whose reaction to the demand stimulus was rational economically but not ecologically, faded before their eyes. As Pekka Hämäläinen has demonstrated for the Comanche, the very success of the Southern Plains horse economy was its undoing: an overabundance of horses and overgrazing "proved too heavy for the grassland ecology, triggering a steep decline in bison numbers."[43] White hunters also muscled in and organized the slaughter of bison on a scale unknown to the Indians, averaging as many as twenty-five *daily* per hunter. Between the end of the Civil War and the late 1870s, the number of bison on the Great Plains fell from 15 million to just a few hundred.[44] Profit seeking was cynically dressed up as a wish to remove the "savage" bison herds in favor of a "civilized" economy centered on well-behaving cattle, at the same time forcing the Indians to give up their "barbarian" way of life. By 1880 the horse-and-bison culture of the Great Plains had been wiped out: the Indians no longer had subsistence resources under their control. Only the reservation remained for the erstwhile masters of the prairie.

Was it good or bad for the Indians that the settlers did not systematically need their labor? Perhaps, at the cost of social marginalization, it spared them the fate of forced labor or enslavement. Here and there we come across Indian cowboys, but not an Indian proletariat. As early as the seventeenth century, there were unsuccessful attempts to incorporate Indians into colonial society as a toiling underclass. The Indians of California became the most integrated into the market economy, although this did not open up a stable perspective for them. Adaptation was seldom an effective resistance strategy, and the advance of the increasingly dominant whites everywhere limited the Indians' room for maneuver.

From the beginning there were two different reactions. Sometimes close neighbors might be miles apart in terms of their behavior: the Illinois Indians preferred a strategy of assimilation and near-total abandonment of their own culture; the nearby Kickapoo put up some of the fiercest resistance to intruders of any kind, whether Europeans or other Indian tribes, earning the reputation as the bitterest foes of the whites. Broken militarily by 1812 and eventually driven from their homelands, they nevertheless managed more than most to preserve their culture.[45]

Settlers

There were two sides to the American frontier: suppression of the Indians and official or private occupation of land that increased the national territory. Each side had its particular demography. The evolution of the Indian population can be calculated only approximately. There are very different estimates of its size on the eve of the first contacts with Europeans, but a figure of 1.15 million would appear to be well founded; the total of their descendants in 1900 was around 300,000.[46] On the other hand, official statistics exist for the inhabitants of what they refer to as the "West" of the United States—that is, the entire national territory apart from New England and the Atlantic states down as far as Florida (also excluding Alaska and Hawaii). Since the 1860s, more than half the US population lived in the West thus defined.[47] The settlement of the West did not proceed only in Turner's sense of the inexorable filling of empty spaces. There were also sudden leaps: when the Oregon Trail opened up the Pacific coast and, a few years later, the gold frontier appeared in California. The Oregon Trail cut through land where there had previously been no roads, from the Missouri River to the mouth of the Columbia River in Oregon (declared the thirty-third state of the Union only in 1859). It was along the 3,200-kilometer trail that the first settler wagons and cattle herds reached the Far West in 1842, and within a few years the old route of trappers and traders had become a busy transcontinental link. It remained in use until the railroad made it redundant in the 1890s.[48]

While the reality of the westward movement was shaped by millions of individual decisions, they were all part of a sweeping political vision. For the founding generation, whose spokesman in this respect was Thomas Jefferson, the country's turn to the West created the possibility of achieving a grand spatial utopia; the United States had the chance to avoid the alleged decline of the exhausted and corrupted societies of Europe by developing mainly in space rather than time. This was associated with the further idea that the space could and should be used, indeed exploited, for the general good as well as for personal enrichment.[49] Jefferson's ideal for both the eastern and western United States was the farmer as small businessman, who lived with his family in a self-sufficient community and participated in the democratic governance of its affairs.

This was also the model for the settlement of the West in the nineteenth century; the government repeatedly supported it with measures such as Abraham

Lincoln's Homestead Act of 1862, which was intended as a social and political alternative to the slave system of the Southern states. This law gave every adult head of family the right to own 160 acres of public land in the West, at almost no cost, after five years of continuous work on it. The reality not infrequently looked different, as numerous families from the urban East that took up the offer eventually sold their homestead to investors with ready cash. The realtor and the speculator were as characteristic of the frontier as was the rugged and frugal pioneer.

The settler's mobility, so often celebrated in the mythology of the frontier, was in many cases a bitter necessity. People had to seek out land where it was available and affordable, moving on to keep out of trouble and repeatedly abandoning unsustainable positions. Alongside the many success stories are lesser-known experiences of failure. Settlers from the eastern cities were not prepared for a hard life in a world almost without infrastructure, where the state could often provide no effective protection. Many feared they would slide into savagery and revert to a low stage of culture long since left behind.[50] The developing myth of the frontier could not entirely dispel such anxieties: the contempt of city people for nomads was transferred to the mobile pioneers, and commentaries of the time underlined the affinity with mass migration in other parts of the world.

Until small-town communities stabilized, male pioneers had to find brides from the "civilized" hinterland—which involved a constant to-ing and fro-ing. It was not like the days of the fur traders: marriage across ethnic boundaries was decidedly frowned upon. At least in theory, the frontier had to remain white and to reproduce the Christian family with its clear division of roles. The husband would conquer the world outside, while the wife ensured civility within the home. Almost nowhere else in the world was the ideal of the nuclear family, independent but woven into a web of neighborly relations, as resolutely upheld as it was in the North American West.[51] But individualist gold diggers and panners were not the only deviants from the norm of the autonomous pioneer household-cum-business. In California, where the land fell into the hands of large owners, agriculture was soon being conducted along aggressively capitalist lines, and the great majority of immigrants had a future only as landless wage laborers.[52] Those who joined the system as farmhands or tenants rarely worked their way up. Second-generation immigrants, too, were in a relatively unfavorable situation; Irish or continental Europeans, for example, who did not manage to acquire land of their own ended in dependent positions.

In the Southwest, the underclass of rural laborers and miners was recruited mainly from Mexicans, who were often discriminated against and overexploited. This was chiefly a result of the offensive war against Mexico, which overnight turned 100,000 Mexicans into inhabitants of the United States. Racist mentalities also played a role.[53] Alongside classical "Turnerian" settlers who had headed west as patriotic Americans, the frontier held all manner of other ethnic groups: immigrants from communities in Europe (e.g., Scandinavia) who had come without first acclimatizing themselves in East Coast cities; blacks both free and servile (some even as slave labor for Indian tribes); and considerable numbers of Chinese, in the wake of the gold rush and especially the beginning of railroad construction. In the second half of the nineteenth century, the frontier was even more ethnically mixed than the urban societies of the East, and just as little an all-devouring "melting pot."[54] It therefore cannot be reduced to a "binary" opposition between "whiteskins" and "redskins." The settlers had color hierarchies as perceptible as those in the cities.

In comparison with many parts of Europe, it was relatively easy on the North American frontier to obtain land cheaply—in most cases by buying it from the government or at an auction. The minimum price per unit of area, as well as a minimum business size, was usually fixed by law. Since land was not always (as under the Homestead Act) given away free, and since there were few legal obstacles to speculative abuses, funding proved to be a problem for many settlers.[55] The pioneer in his log cabin is by no means the whole of the picture. The degree of insertion into market relations at particular times and places has long been an issue of debate. No doubt there was a general trend toward commercialization. By midcentury the dominant social type on the agricultural frontier was no longer the countryman living off the land but the entrepreneurial farmer. Land was by no means as freely available as the official ideology claimed. There was always competition for *good* land, and the costs of acquiring and developing it had to make economic sense. After the Great Plains were "cleansed" of bison and Indians, the Big Business of the cattle barons spread out from Texas, largely funded by city sources or by British capital; it was a "big man economy," as in frontier lands of other continents.[56]

The variety of frontier experiences was also reflected in the problems that came to the fore; there were asynchronies such as those already theorized in Turner's conception of social evolution through stages. Whereas, after the end of the Indian threat, the Great Plains farmers from Texas to North Dakota had to solve typical nineteenth-century problems—mortgages, railroad charges, cash flow—people in California were already debating issues that would be characteristic of the twentieth century, such as water supplies, fruit growing, transpacific trade, or urban real estate markets. Water was not by chance a keyword: none of the West's other ecological problems was more threatening. The myth of the frontier waxed lyrical about its "limitless" natural resources, but we need to remind ourselves that one resource was scarce from the very beginning: water.

Indian Wars and Pistol Terror

A frontier nearly always has violence as part of it, but the North American West is the paradigm. From the First Anglo-Powhatan War of 1609–14 in Virginia to the end of the last Apache war in 1886 in the Southwest, the relationship between whites and Indians was marked by one conflict after another.[57] All in all, the Eastern Indian peoples—often joined together in a brittle confederation—held out longer and were comparatively stronger opponents. The last of them were eliminated militarily only when the remaining Seminole warriors were deported from the Florida swamps in 1842. The battles in the East had lasted for roughly 240 years. West of the Mississippi, on the other hand, they were packed into just forty.

The invasion of the Great Plains by Euro-American settlers began in the 1840s. The first deadly Indian attacks on overland wagons were recorded in 1845, but the tribes were often content to extract a toll and to exchange provisions on terms that they considered fair; some of the most brutal raids on wagon trains were staged by white bandits in Indian dress.[58] In the 1850s the number of incidents increased, and the 1860s witnessed the outbreak of the classic Indian wars so deeply rooted in the national memory and immortalized by Hollywood. In 1862, when Sioux warriors killed several hundred white settlers in the greatest massacre since the founding of the United States, fears were even rife of a major uprising in the rear of the Civil War armies.[59] No more than a minority of tribes, however, were involved in the Indian wars. Only the Apache, Sioux, Comanche, Cheyenne, and Kiowa offered lasting resistance. Other tribes (Pawnee, Osage, Crow, Hopi, etc.) fought on the side of the federal troops.[60] A military frontier against hostile Indian tribes came into being after 1850, when New Mexico was added to the Union as war booty and the Southwest was strewn with army camps to keep the "savages" under control.[61] Although they found it hard at first to beat off Apache and Comanche attacks, the forts later became bases for an effective "pacification" of the region. Troops that had fought on the Union side in the Civil War were sent to the South to break the independence of the Indians.

Modern European thinking was by no means inapplicable to many of the Indian wars. Excellent strategists appeared on the Indian side and, given the approximate material balance, were able to inflict many a defeat on the whites. The Indians of the Great Plains were probably the best light cavalry in the world, extremely effective against an enemy that was inadequately trained and equipped. Their often undermotivated adversaries suffered from the harsh conditions in the forts and on the battlefield. Apart from young elite cavalrymen, the motley crew included often overage Irish veterans of the British army, Hungarian hussars, and in the early years even some survivors of the Napoleonic wars. The weaknesses of the Indians were, of course, their inferior weaponry (they were ultimately powerless against the dreaded mountain howitzer), but also their inadequate discipline, lack of a proper command structure, and poor protection of camps and villages. The asymmetries that favored Europeans militarily in many Asian and African theaters were repeated here too.[62]

The transition from war to massacres and attacks on defenseless settlements was fluid enough. Both sides were armed, and lawless violence was part of everyday life in large parts of the frontier; it was a legacy that all had inherited from the colonial wars of the late eighteenth century.[63] The use of force between civilizations was interconnected with the general violence of civilian life on the Euro-American side of the frontier. The pioneers of the "Wild West," who settled their everyday disputes with a handgun or rifle, were among the most heavily armed populations in the world. The readiness to "shoot it out" marked social life in peacetime, in a way that is usually characteristic only of civil war situations. Extreme standards of male honor, unknown in the cities of the East, meant that it was more normal to sharpen a conflict than to soften it ("No Duty to Retreat"). People took their own initiative in defending their interests, sometimes with a suicidal cult of "valor." Typical was the vigilante band, operating in situations where the law held no sway, as a kind of revolutionary force, as it were, taking the place of the locally absent state. Behind this stood the idea of a right to self-defense and a highly muscular interpretation of popular sovereignty. Richard Maxwell Brown surmises that despite the high human costs, this practice preserved order more cheaply than a regular judicial system.

The reign of terror exercised by pistol-packing heroes reached its maximum intensity and compass in the four decades or so after the end of the Civil War. Brown actually describes it as a kind of mini civil war in its own right: most of the two hundred or three hundred most famous or infamous killers (plus a large number who are less well known) were acting on the orders of big landowners and enforcing their interests against those of small ranchers and homesteaders. They were not social bandits with a sense of justice and a sympathy for ordinary people but rather agents in a class war directed from above. In contrast, the great massacres of Indians—such as the Sand Creek Massacre of 1864 in eastern Colorado, where some two hundred Cheyenne men, women, and children were slaughtered—tended to be organized by regular troops rather than by militias or vigilantes. The fact that in many other cases the army protected Indians against private white violence makes the complexity of the situation apparent.[64]

Deportations

Indian policy was overwhelmingly made in Washington but put into practice at the frontier. At the time of the founding of the United States, most Indian communities already had considerable experience of external challenges. They had undergone medical, ecological, and military shocks and repeatedly found themselves in the situation of having to react and reinvent themselves. Around 1800, it was by no means the case that cunning "civilized people" stood face to face with dimwitted "savages."[65] At times the Indians had been fairly treated, especially by Quakers in Pennsylvania, but much more often vile behavior toward them had clashed profoundly with their sense of justice. The attitude of the US government was contradictory. On the one hand, it recognized their de facto nationhood, by entering into treaties that were by no means always a one-sided diktat. On the other hand, the old Puritan belief in the superiority of Christians over pagans passed into the Enlightenment idea of a civilizing mission: the "Great Father" in Washington would watch strictly and benevolently over his Indian "children";[66] the civilizing influence would initially come from outside. Until midcentury there was no legal provision for intervention in the *internal* affairs of the tribes, but they were subjected to a special kind of indirect rule. Only after 1870 did it become accepted that the Indians, too, should obey the general laws of the land.[67]

In 1831 the aged Chief Justice John Marshall, for thirty-five years one of the most influential figures in the country, declared that the Cherokee nation was "a distinct political society, separated from others, [and] capable of managing its own affairs and governing itself." The "tribes" were therefore not sovereign states on American soil but, as Marshall put it, "domestic dependent nations."[68] On paper this influential formulation seemed to give the Indians protection. But the executive had long since taken a different course, ignoring the judgment of the constitutional court. General Andrew Jackson, who took office in 1829 as the seventh president of the United States, had already shown himself to be an energetic fighter against the British, the Spanish, and the Indians. He thought nothing of breaking treaties with the Indians, and he did not share Marshall's view that any expropriation of Indian land should at least have a solid legal basis. Jackson's popular and effective policy of deportation ("Indian removal") has occasionally been explained in individual psychological terms, as if the president's unhappy childhood had made him envious of the Indians as "eternal children" and, at the same time, aroused in him a desire to exercise overpowering paternal authority over them.[69] That may well be. What matters more are the results of his policy.

In Jackson's eyes the civilizing mission of the Jefferson generation was a failure. He took his cue instead from the mentality of the so-called Paxton Boys, who in the 1760s had perpetrated horrific massacres of Indians in Pennsylvania.[70] He thought there was no point in tolerating Indian enclaves. His aim—with methods that one would today describe as "ethnic cleansing"—was to drive the Indians beyond the Mississippi. During the 1830s, a cataclysmic decade second only to the 1870s, some 70,000 Indians were deported, mainly from the Southeast. The expulsion drive stretched right up to the Great Lakes; only the Iroquois in New York State put up successful resistance. Concentration camps were built, and whole Indian communities were force-marched with few personal belongings (and in sometimes extreme weather conditions) to the so-called Indian Territory. The great efforts that some tribes had made to "civilize themselves" gave them no protection. On the endless long marches, thousands of Indians died of disease, malnutrition, and hypothermia. But the horror of it all should not make us forget that Jackson's "Indian removal" only intensified an older process. As early as 1814 people had been induced of their own free will to leave the Creek homelands for the West. For many enterprising Indians, the "open" West held the same kind of attraction that it had for white settlers.[71]

The worst episode was the deportation of the Seminole people from Florida, in which Jackson's campaign was intertwined with the issue of slavery. The whites of Florida were less interested in the swampland abode of the Seminole than in the Afro-Americans, some of them runaway slaves, who lived there either in separate villages or as part of the Indian community. But the Seminole fought back, and in several years of war many white soldiers also lost their lives.[72] Some of the deported tribes kept up their (disregarded) adaptation to Euro-American ways in the new areas to which they were sent. The "Five Civilized Tribes"—Cherokee, Creek, Choctaw, Chickasaw, and Seminole—fared relatively well between 1850 and the beginning of the Civil War. They got over the effects of the removal, found their way to a new unity, adopted constitutions of their own, and built political institutions combining the old Indian democracy with the institutional forms of US democracy. Many ran family farms, others worked black slaves on plantations. They developed a bond with their new lands, in the same way that white farmers did with theirs. In the 1850s they created a school system that could have been the envy of whites in the nearby states of Missouri and Arkansas. Missionaries were warmly welcomed and accepted into the community. In all these ways the five tribes followed the prescribed path to civilization and moved ever farther apart from their Indian neighbors.[73]

If the Indian peoples had received firm guarantees that they could remain in the new areas allocated to them, Andrew Jackson's brutal policy might have foretokened the final phase in the development of the Indian frontier. But no such security was forthcoming.[74] The land hunger of the settlers and railroad companies, together with encroachments by undisciplined miners, prevented the consolidation of viable communities. The general brutalization of American society during the Civil War carried over into new assaults on the Indians and a discourse of extermination like that heard a century before. The notorious saying that "the only good Indian is a dead Indian" first appeared in 1860, and it represented the spirit of the age.[75] It proved fatal for the Five Civilized Tribes in their so-called Indian Territory (in present-day Oklahoma) that they sided with the Southern states, since federal government policy after the end of the Civil War punished them for disloyalty and treated them as vanquished Confederate troops. The Indian peoples lost large tracts of their land and had to let the railroad companies in. Within twenty years they became minorities in the very territory that they had been forced to exchange for their homelands under President Jackson.[76]

The major Indian wars of the 1860s and 1870s should be seen in this light. Following the war in the East, a new influx of settlers, and a series of local provocations, Indian resistance became more intense throughout the Great Plains. Previously the US Army had cultivated a neutral relationship with the Indian tribes and repeatedly defended them against acts of violence, but now it became a tool of the government's policy of finally resolving the "Indian question." The resistance eventually collapsed in the early 1880s, as the famous Lakota chief Sitting Bull capitulated in 1881, and the Apache wars in the Southwest came to an end.[77]

A rough pattern can be discerned behind the Indian wars. Long before whites and Indians became locked together in military hostilities, most of the contacts between them had been marked by growing distrust on both sides. The federal government played a major role in this, since it had responsibility for Indian affairs, and its civilian or military representatives often claimed to stand above local parties (therefore to some extent above Euro-Americans) and to bring the wisdom of statecraft to bear on problem solving. The result was often confusion all around—a situation that could easily give rise to military conflict. Early hostilities were rarely due to calculated aggression; it was more typical for spontaneous clashes to escalate into something more serious. The Euro-American side did not generally see itself as the agency of a great historical trend to expansion, and local developments often sufficed for it to consider itself in the right. Whereas whites seldom differentiated between Indian fighters and civilians, they invariably cited Indian attacks on settlers as proof of their own moral and legal superiority. Any atrocities were used to underline the justice of their position.

Until the final phase of the wars, the Indians pulled off surprising tactical victories even against the Federal Army. The white side tended to overestimate its own strength and to underestimate the enemy's prowess in battle, considering them to be primitive and inflexible. It is indeed amazing how such arrogance prevented the learning of lessons. Yet despite their tactical successes, the Indians had no way of avoiding defeat in the end. Hostilities rarely ended in the conventions of the time for "civilized" warfare. Once the Indians' resistance was broken, they appeared not as an enemy army to be honored in defeat, but as a mass of impoverished, half-starved, and half-frozen people struggling to survive in makeshift accommodations or on the road of flight. Mighty warriors could instill fear; defeated Indians were a pathetic sight to behold. At the end of the wars, so much bitterness remained among victors and vanquished that no one ever imagined the transfiguration they would later undergo in literary and cinematic romanticization. The brutality on both sides often left behind such traumas that anything like reconciliation or even peaceful coexistence seemed scarcely possible.[78]

If the legendary West of the cowboys-and-Indians movie ever existed, then it was limited temporally to the period from 1840 to 1870 and spatially to the Great Plains at the foot of the Rocky Mountains. What had "closed" by 1890, when Frederick Jackson Turner formulated his theory, was not the settlement frontier—many of today's historians think that that remained open until the 1920s—but the military and economic-ecological dimensions of Indian resistance. At the same time, the commercial carve-up of the great expanses of the Midwest had made great advances. After barbed wire was patented in 1874 and produced in massive quantities, the consolidation of private ownership drew a line under the "open West."[79] The "wilderness" was divided up and colonized, until no space was left over for "wandering savages" (to use the language of the time). A single measurement grid was now applied in practice to the whole territory of the United States, making cross-boundary ways of life impossible.[80] The age of the reservation was dawning. Even the last Indians became "captive peoples under relentless pressures to make themselves into something that seemed to contradict all they had ever been."[81]

In the 1880s, the last combatant peoples had been disarmed and turned into dependent charges of the state. The Indian "nations" were no longer regarded even nominally as negotiating partners, as the decision of 1871 to sign no further treaties with them had clearly shown. The old ceremonies, usually prepared by advance by both sides, had reached a climax at the Treaty Council of September 1851, which Thomas Fitzpatrick had staged at Fort Laramie as the Indian agent of the federal government. Some 10,000 Indians from various peoples and 270 white envoys and soldiers had come together to negotiate and to exchange gifts.[82] Although the event went off peacefully, it had been made clear to the government negotiators that very few of the Indians were willing to be cooped up in reservations. By the 1880s the repetition of such a scene would have been unimaginable. Indians in California and the coastal Northwest had long ago been driven into reservations, and the same had happened in Texas, New Mexico, and the Great Plains after the Civil War. From the Indian point of view, it made a difference whether a reservation was in an area they considered ancestral land, or whether it counted as a permanent exile. It was mainly for this reason that in March 1850 some 350 Cheyenne, under their chiefs Dull Knife and Little Wolf, embarked on an adventurous journey of more than 2,000 kilometers—a kind of parallel to the Long March of the Torghut Mongols in 1770–71 from the Volga back to their homeland.[83] The impetus was not only sentimental, since the authorities had not been providing them with sufficient food. Suffering unprovoked attacks from the army, few of them would reach their destination. In any event, a commission of inquiry came to the conclusion that it made no sense to "civilize" Indians if they interpreted their situation as captivity.[84]

Property

Agrarian land use was not everywhere the kernel of the frontier constellation. In Canada, where there was no counterpart to the fertile Mississippi Plain and even the prairies were inhospitable, the assault on the wilderness and its inhabitants did not mainly involve agricultural colonization by settler families. The old Canadian frontier was a "middle ground" of hunters, trappers, and fur traders. The nineteenth century preserved its commercial character but gave it a new capitalist form. The fur trade, logging, and livestock farming were organized by large corporations on an industrial, capital-intensive basis; not independent pioneers but wage laborers bore the physical burden of the exploitation of nature.[85] The US frontier, however, involved a permanent conflict over agricultural land. It was this, rather than racism or a belief in Christian superiority, that gave such a sharp edge to the clashes between indigenous people and newcomers. Trade contacts are "intercultural," whereas control over land is an either-or question. European concepts of property armed the settlers ideologically and left little room for compromise.

The formula that European property concepts are individualist and exchange related whereas Indian ones are collectivist and use related is not entirely inappropriate, although it greatly simplifies complicated matters. The American Indians, like many other hunter-gatherers and farmers around the world, were perfectly familiar with private property, but for them it referred not to the land itself but to things *on* the land. In principle, those who produced the crops also had them at their disposal.[86] The idea of dividing up the land into fixed plots was as alien to the Indians as the idea that individual persons, households, or clans might take permanent possession of more land than they were able to cultivate. Claims to control land had to be justified over and over again by actual labor. Those who made due use of their land were allowed to continue doing so without hindrance. Communal control or "ownership" of the land, which nineteenth-century Europeans all over the world regarded as archaic, was paradoxically strengthened in response to the white invasion.[87] Thus, for example, when the Cherokee realized in the late eighteenth century that they were being continually cheated in land deals, they forbade individuals to sell land to whites and made communal rights over the land stricter still.[88] The exercise of such rights was a complicated business, especially in the British Empire with its sophisticated legal tradition.

The French never recognized Indian land rights in North America and appealed to rights stemming from conquest and effective occupation, as did the British in Australia. The English colonial authorities in America, however, claimed all land for the "sovereignty" of the Crown, while accepting the existence of "private" Indian land rights. Only this made it possible for Indian land to be directly assigned and sold. The US courts followed this practice. With the Northwest Ordinance of 1787, one of the founding documents of the new republic (adopted even before the Constitution and known mainly for the limits it set on the spread of slavery), the United States committed itself to the principle of contractual land disposal—not a good solution for the Indians, but not the worst possible either.[89] In practice, however, the state did little to protect the Indians from the aggressiveness of frontiersmen. In this light, President Andrew Jackson's policy of deportation may indeed be seen as an adjustment to the reality on the ground. Around 1830 the position of the East Coast Indians was already unsustainable.[90]

The history of the North American frontier may therefore be written as one of continual and irreversible loss of land by the Indians.[91] Even impressive innovations such as the horse-and-bison culture of the eighteenth century offered no alternative in the long run. The native inhabitants of North America were separated from their natural means of production, in a classic example of what Karl Marx called the "primitive accumulation of capital." Since Indians were neither tolerated as owners of land nor indispensable as a source of labor, and since their role as suppliers of pelts and leather was over within a few decades, they were left with no dignified way of fitting into the social order created by European immigrants. The wilderness turned into a series of national parks, empty of residents or garnished with folkloristic trappings.[92]

3 South America and South Africa

Argentina

Did South America, with its even older European colonies, also have a frontier?[93] Two countries in particular might be thought to have had a pioneer West: Argentina and Brazil. A third case is Chile where the military *pacificación de la Araucanía* was conducted in close coordination with the Argentine subjugation of the desert and its peoples (*conquista del desierto*).[94] The earliest frontiers in South America appeared with the mining of gold and silver; agricultural ones came later. The greatest similarity with the United States was in Argentina, where the pampas stretched from the Gran Chaco region in the North to the Rio Colorado in the South, as well as a thousand kilometers westward from the Atlantic. There was a lack of rivers corresponding to the Mississippi, however, to carry immigrants into the heart of the continent. Until about 1860, unlike in the North American West, no changes were observable to the natural environment of wild vegetation, with a theoretically fertile soil. In the 1820s the pampas began to be "opened up" as land was acquired on a large scale.[95] In contrast to the United States, the land in Argentina was not divided into small units; governments sold it off wholesale or donated it in the form of political gifts. Large cattle ranches therefore came into being, and sometimes their land was leased out to smaller ranchers. Only hides were produced at first; grains played no role and actually had to be brought in from outside.[96] It was decidedly a "big man's frontier." Legal regulations favoring small autonomous settlers could never be pushed through, and property rights in general took shape only slowly and patchily.[97] The Italians who flooded into the country in the late nineteenth century were incorporated into the system as tenant farmers rather than as owners of land of their own. Few even became Argentine citizens. They therefore carried little political clout against the big *latifundistas*. There was no basis for the formation of a stable agrarian middle stratum, such as that which gave social coherence to the whole of the American Midwest. The small rural town with service functions and a gradually developing infrastructure, so typical in the United States, was absent from the scene.

Thus, in Argentine conceptions of the *frontera*, the opposition between civilized city and barbarian country, was not very sharply drawn. The lack of a credit system for small farmers and the failure to compile a land register made it even more difficult to gain a foothold. Strictly speaking, Argentina had no settlement frontier and no real frontier society that carried weight politically or could form the stuff of legend. The periphery never became—like the cities on the Mississippi or Missouri—a core area in its own right. When the railroad arrived, it facilitated the influx into coastal cities rather than settlement of the interior. In Buenos Aires, people feared that ill-bred migrants from the pampas would bring their uncouth ways into the city. The railroad led at least as much to a contraction of the frontier as to an expansion.[98]

A characteristic social type in Argentina was the gaucho: migrant worker, ranch hand, and horseman of the pampas.[99] (The cowboy was essentially a *Latin* American invention, spreading from the huge ranches of northern Mexico to Texas and from there to the rest of the Wild West. The cowboy's first and last appearance on the *political* stage also took place outside the United States, in the shape of Pancho Villa's armed campaign after 1910 in the Mexican Revolution.[100]) As a conspicuous social group, the gauchos were squeezed out in the last third of the nineteenth century by an alliance between the powerful landed elite and the state bureaucracy; this was a central process in the nineteenth-century history of Argentina.

The gauchos—a term apparently first coined in 1774—emerged in the eighteenth century out of big-game hunters, who were usually of mixed Spanish-Indian extraction and therefore subject to the racism virulent in both colonial and postcolonial Argentina. They earned a reputation as fighters in the War of Independence (1810–16), but they were unable to preserve the esteem that came with it. By 1820 the age of hunting game and wild horses was over, as was the uncontrolled slaughter of cattle for their hides and tallow. Ordinary firms now took up the processing of salted and dried meat, selling a large part of it to slave plantations in Brazil and Cuba.

Sheep became another factor in the Argentine economy over the next two or three decades; it was a robust and undemanding animal that did not have to be killed to yield a profit. Stockades and special livestock farms transformed the economy from pastoral to mixed. By 1870 in the province of Buenos Aires, the most populous in Argentina, perhaps as much as one- quarter of the rural population could be described as gauchos. Thereafter the proportion rapidly declined, as fences reduced the need for horseback riders. In 1900 the *frigorífico*, with modern techniques of packing and freezing, acquired great significance, and the industrialization of meat production led to the rapid shedding of labor. The gaucho, less valued than ever, was robbed of the last vestiges of his independence.

When General Julio A. Roca in 1879 attacked and largely annihilated the Mapuche (the "Araucanos" of the Spanish sources), the largest Indian people in the country, the more turbulent elements among the gauchos were also reined in. The social elite saw them as (potential) criminals and forced many of them into leaseholds, dependent labor, or military service; drastic new laws curbed their mobility. As it so often happens, urban intellectuals began to romanticize the figure of the gaucho at the very moment when he was disappearing as a social type. In his demise, the gaucho would become the embodiment of the Argentine nation.[101]

In Argentina, unlike in Brazil, the Indians did not give way for a long time; they were still carrying out repeated raids in the 1830s in the province of Buenos Aires. Hundreds of Euro-Argentine women and children were abducted. The advance inland required both a more extensive definition of the state territory and a discourse that devalued the indigenous peoples and excluded them from the national community. The struggle no longer concerned particular natives but raised the whole question of civilization versus barbarism. The "Desert War" against the Indians, which dragged on from 1879 until 1885, was eventually settled only when the government introduced the breech-loading rifle on a wide scale. In almost the same year as the last of the great Indian wars in the United States, the vast expanse of the Argentine interior was opened up for agricultural development. The Indians were not allowed even the wretched future of life on a reservation.

Brazil

In Brazil, whose land reserves were at least as large as those of the United States, the development of the frontier was quite different—and also varied from the pattern seen in Argentina.[102] It is the only country in the world where some of the post-1492 frontier processes of exploitation and settlement persisted right through the twentieth century. In addition to the mining frontier, there was a kind of slave-operated sugar-plantation frontier, similar to Alabama's or Mississippi's before the American Civil War, while a patchy farming frontier developed late. Even today the social life of Brazil is concentrated in a narrow coastal strip. The interior (sertão), originally the whole country beyond the reach of Portuguese cannons, was (and to some extent still is) a symbolically inferior place that attracted few explorers. The Amazon jungle—until the assaults on the rainforest in the final decades of the twentieth century—was something like a "frontier beyond the frontier."[103] In Brazilian literature, the frontier is theorized in explicitly spatial terms, hardly at all as a process. So the spatial category sertão is the closest equivalent to Turner's concept, while fronteira denotes the state boundary line.

In Brazil many objective prerequisites were missing for the opening up of the interior. In particular, there was no serviceable network of waterways remotely comparable to the Ohio-Missouri-Mississippi system in the United States; nor were minerals that might have been useful for industrialization (like the coal and iron in the North American West) present in the ground. Only when Brazil blazed a trail for itself in the world coffee market did something like an agrarian development frontier see the light of day. In the mid-1830s coffee overtook sugar for the first time in the country's exports, and Brazil became the world's leading producer.[104] But with the technology of the time it took only a generation for soil exhaustion to set in, compelling planters to move farther west. After slavery was abolished in 1888, a demand for plantation labor drew many Italians to Brazil, but the conditions there were as unfavorable as in Argentina—so appalling, in fact, that in 1902 the Italian government banned publicity for further emigration. The power structures were similar in Argentina and Brazil, latifundistas in the former corresponding to big coffee planters in the latter. In neither was there a policy of land allocation or redistribution to small farmers.[105] The Brazilian fronteira was essentially a land of coffee monoculture, run by large businesses with or without slaves; it was not a place where independent pioneers (in

Frederick Jackson Turner's sense) and a home-centered middle stratum could take shape or an open-air school of democracy establish itself. As John Hemming has depicted in a moving trilogy, the Brazilian Indians trapped before 1910 in the Amazonian rubber business (not, it is true, in the coffee and sugar economy) were not even afforded the protection of reservations.[106] The rainforests, unlike the enclosed Great Plains settled by Euro-Americans in the United States, were an open frontier. The *indios* therefore retreated farther and farther into the remotest regions, their resistance to the colonists finally exhausted by the turn of the century.

South Africa

The lack of any real interaction between the frontier processes in South America and South Africa makes their exact coincidence in time all the more striking. The last Indian wars in North and South America took place in the 1870s and 1880s, just as the white (British) conquest of the South African interior was being completed. For South Africa the year 1879 saw the closing act, when the Zulus, the most important African counterpower, suffered military defeat. It was the last in a series of wars between the colonial power and African armies. The Zulu king Cetchwayo, provoked by British demands that were impossible to meet, was able to mobilize more than 20,000 men (a figure quite out of reach for North American Indians), but in the end he, too, had to bow to the superior might of the British.[107] Both Sioux and Zulus were significant regional powers, having reduced their indigenous neighbors to subjection and dependence, but they knew all about the whites' military strength from decades of contact with them. Both had taken only small steps to assimilate with the invaders and to adopt their way of life. Both had complex political structures and belief systems, which remained alien to Europeans and Euro-Americans and provided material for their propaganda concerning the savage's imperviousness to reason and civilization. By 1880, in the United States as in colonial South Africa, the supremacy of the whites had become unshakable.[108]

These common features contrast with differences in the fate of the Sioux and Zulus. The two peoples did not have the same capacity to resist economic pressures: the Sioux were nomadic bison-hunters, organized in bands, lacking a pronounced political or military hierarchy, and completely devoid of an economic role in the expanding internal market of the United States; the Zulus were sedentary and had a much stronger mixed economy based on livestock breeding and agriculture, with a centralized monarchy and a socially integrated system of well-defined age groups. Despite military defeat and occupation, it was therefore not as simple to break up and demoralize the Zulus as it was the Sioux. Moreover, Zululand was not marginalized within the wider South African economy but transformed into a reservoir of cheap labor. The gradual proletarianization of the Zulus thus played an important role in the division of labor inside the country.

The earlier frontier chronologies of South Africa and North America also displayed striking parallels. The first contacts between European immigrants and the indigenous population occurred in the seventeenth century, and in both countries the 1830s proved to be an important watershed: with Andrew Jackson's policy of deporting southern Indians in the United States and the beginning of the Boers' Great Trek in South Africa. One peculiarity of South Africa was the division within the white population after the British takeover of the Cape of Good Hope in 1806. From that point on, the Boer population dating back to the seventeenth-century Dutch immigration existed alongside a smaller British community, which was linked to a wealthy and powerful imperial metropolis and took over the key decisions in Cape Province.

At first the Trekboers, living entirely from farming, were forced into mobility by the shortage of land. Then in the 1830s the wind of slave emancipation blew down from London, finding application in the cape and becoming a central structural element for Boer society. The Boers found the British policy of legal equality for the races unacceptable. But the presence of well-armed African forces, especially the Xhosa to the east, meant that the pioneers' ox wagons had only one direction open to them: into the more lightly defended High Veld to the north. The Boers profited from the disintegration of many African communities, itself the result of a period of military conflict among African peoples that became known as the Mfecane. Between 1816 and 1828, the lightning advances of the Zulu state under its war leader Shaka had depopulated large areas of the grasslands, while at the same time handing allies to the white settlers from the anti-Zulu camp.[109]

The Great Trek was a militarily and logistically successful maneuver on the part of one of the ethnic groups competing for land in South Africa. It became a campaign of conquest, initially "private" in character. A process of state formation followed only later, as a kind of "by-product" (Jörg Fisch) of private land appropriation, when the Boers created two republics of their own: the Transvaal Republic in 1852, and the Orange Free State in 1854. These two entities were breakaways from the Cape Colony, but the British officially recognized them and exercised a certain influence over their economic life. So, nineteenth-century South Africa did not have a unified state that could have mapped out a general "black policy" analogous to the federal "Indian policy" in the United States.[110] Militarily, the Boers had no central army to give them support. As armed settlers, they had to fend for themselves and to prove their capacity to form a state of their own. They were reasonably successful in the Orange Free State, but much less so in the Transvaal (which the British temporarily annexed in 1877). In both cases, the state apparatus was rudimentary and the financial situation precarious, and there was a lack of "civil society" integration outside the church.[111] Since the South African frontier of the 1880s was "closed," in the sense that there was no more "free land" to distribute, the Boer republics were not essentially states on a settlement frontier.

Any frontier exhibits special demographic features, and in this respect South Africa differed importantly from North America. Before the 1880s there was no mass immigration into South Africa, and even subsequently the influx into its gold and diamond fields cannot be compared to the gigantic flows across the North Atlantic. By midcentury the Indians constituted a tiny share of the US population, whereas Africans made up the vast majority of the total in southern Africa. Black Africans were much less devastated than North American Indians by diseases brought in from Europe; nor was their cultural trauma so deep as to produce a steep demographic decline. In South Africa, then, the precolonial inhabitants did not become a minority in their own country.[112]

In South Africa as in North America, the armed pioneer providing for himself and his family was at first the principal frontier type. In America, however, the frontier was penetrated early on by large firms producing for export markets. In the eighteenth century, tobacco and cotton plantations—many of them situated at the frontier—formed part of extensive commercial networks, while in the nineteenth century the frontier increasingly became the site of capitalist development processes. In South Africa, after their partial exodus to the interior, the Boers were initially more remote than before from world markets. Only the discovery of diamonds in the 1860s, and of gold deposits two decades later, established in the Boer republics a mining frontier (largely aimed toward the world market) alongside subsistence farming.[113]

At the end of the nineteenth century, the Bantu-speaking population of South Africa managed to occupy a place in the social order relatively more favorable than that of the Indians of North America. Whereas the Khoisan peoples in the Southern Cape lost nearly all access to farmland early in the colonial period, the Bantu speakers in the interior, despite the advance of the settlement frontier, were able to make effective use of considerable land resources. In large areas of Lesotho (Basutoland) and Swaziland, and in eastern parts of today's Republic of South Africa, African small farmers worked their own land. This was partly the outcome of their resistance, and partly thanks to ad hoc decisions by various governments against the complete expropriation of the Africans. In North America no such concessions were forthcoming; the nomadism of the bison-hunters came into direct conflict with the expansion of farmland and the exploitation of the prairies for capitalist stock breeding. Neither of these economic forms had any need for Indian wage labor. In South Africa, farms and mines did require native wage labor, and so black Africans were not shunted off into subsistence niches but often integrated, at the lowest level of a racially defined hierarchy, into dynamic sectors of the economy. The rulers of South Africa tried to prevent the spread of a black proletariat throughout the country, creating instead a series of ghetto-like separate territories in some ways reminiscent of the North American Indian reservations. But the South African reservations, which came into their own only much later (after 1951) under the appellation "homelands," were not so much an open-air prison to isolate an economically functionless population as an attempt to control the black labor force politically and to channel it economically. They rested on the principle that families should feed themselves in the homelands through subsistence agriculture, while the male workers—whose reproduction costs were thus kept to a minimum—found employment in the dynamic sectors.

The attitude of whites to the black majority was largely marked by brutality and cynicism. One side effect of this was that, with the exception of some missionaries, no one took the trouble to "civilize" the Africans and hence to undermine their cultural autonomy. In contrast, that is precisely what happened in America, in the last third of the nineteenth century, through the well-meaning attentions of "friends of the Indians." All told, Bantu-speaking Africans in South Africa did not suffer a total defeat. They remained demographically the majority, were allowed a minimum of cultural autonomy, and played an indispensable role in the economy. When the United States in the 1930s switched to a kind of humane Indian policy, it was too late for a genuine "Indian revival." In South Africa at that time, full-scale repression of the black majority population still lay in the future. Only the overthrow of the coercive state apparatus at the end of the twentieth century would create the conditions for popular self-determination. The frontier had deeply marked South African statehood, but after a long delay it finally issued into a "normal" nation-state development. In the United States there are still reservations. In South Africa the homelands have disappeared on paper, but their imprint remains in the distribution of landownership.

Turner in South Africa

Apart from the United States, the frontier thesis is applied to no other country more often but also more controversially than it is to South Africa. All those who sustain one of its many variations are essentially agreed that social tensions and racist attitudes grew sharper as the distance increased from a colonial-cosmopolitan atmosphere. The Trekboers in the interior are generally seen as the epitome of uncouth pioneers, but whereas for some this means freedom-loving outdoorsmen, for others it means feral racists. What such interpretations have in common is an emphasis on the isolation of frontiersmen from "Western civilization," or anyway from the urban Europe that had its African outposts in Cape Colony. A rigidly Calvinist sense of mission is part of this image. Critical histories argue that South Africa's racist system, which came to a climax after 1947–48, first began to take shape precisely on that frontier, and that nineteenth-century experiences therefore marked the whole social order in the second half of the twentieth century. This idea of a long-term continuity in racist attitudes, from the 1830s until the heyday of apartheid, forms the kernel of the frontier interpretation of South African history.

In 1991 a book much read in South Africa repeated the claim that the Enlightenment and liberalism completely passed the Boers by, that these were "the simplest and most backward fragment of Western civilization in modern times."[114] Critics of this thesis were unwilling to go quite that far and found elements of racist thinking in the late-eighteenth-century Cape; while others came close to the "middle ground" interpretation of American history by pointing to numerous instances of contact or cooperation between whites and Africans. The historian Leonard Guelke, in particular, has sought a way out of such sharp counterpositions, identifying both an orthodox "frontier of exclusion" and a liberal "frontier of inclusion." Another proposal is to distinguish between a phase when the frontier was still open and another when it was closed, and to show that the situation hardened in the extreme only during the latter period. Nowadays a strict continuity thesis has few supporters among South African historians: neither the nineteenth-century frontier nor slavery in Cape Colony (before its abolition in the British Empire in 1833–34) is seen as the direct source of apartheid; rather, both the one and the other contributed to the fact that a (partly religious) sense of white cultural superiority, together with practices of sharp segregation, was already developing in the late nineteenth century. The frontier thesis does not provide a key to South African history, but it does emphasize the importance of geography and environmental factors for the crystallization of social attitudes.[115]

Turner's theme of the emergence of freedom at the frontier is only intermittently applicable to South Africa. The Boer exodus to the interior was, among other things, a response to the social revolution brought about by the liberation of the Cape slaves in 1834, and by the Governor's decree of 1828 that any person not having the status of a slave was equal in the eyes of the law and enjoyed its full protection.[116] In their own republics, founded at a time when such a polity was rare even in Europe, the Boers created a quasi-Hellenic form of democratic self-government involving all male citizens but excluding a section of the population regarded as immature (although slavery itself was not permitted there).

This frontier democracy calls to mind not so much a modern constitutional state as the egalitarianism of frontiersmen all over the world. In Argentina Juan Manuel de Rosas, a prototypical caudillo, first created a power base for himself by fighting the Indians at the frontier, then won support as a strongman from the Buenos Aires oligarchy, and in a startling volte-face, turned against his erstwhile gaucho followers. In South Africa, British colonial rule in the cape was too firmly entrenched to be threatened by a Boer liberation movement, while the Boers themselves cared only to be left in peace in their isolated republics. But the gold rush that began in 1886 in the Witwatersrand disturbed this self-sufficiency. Eager to profit to the full from the new riches, the Boers gave British capitalists a free hand but ensured that they kept political control, asserting their frontier democracy against not only the black underclass but also white newcomers (*uitlanders*). The South African or Boer War of 1899–1902 developed out of this tangled situation. It ended with the victory of an imperial power which, having had had to make extraordinary efforts to overcome a seemingly insignificant enemy, began to doubt whether colonial domination—especially over other whites—was worth imposing at such a high price.

The war deeply wounded Boer society on the High Veld; one-tenth of the population lost their lives. But Afrikaners still formed the great majority of the white population of South Africa, and they remained in control of agriculture. There were no other allies to whom the British could turn. Since a regime based on permanent occupation was not an option, some arrangement had to be made with the subordinate Boers. A younger and relatively more liberal Afrikaner leadership saw things in much the same way, and this provided the basis for a compromise. The founding of the Union of South Africa in 1910, as a self-governing dominion within the Empire, represented a triumph for the Afrikaners, a defeat for black Africans, and a safeguarding of basic economic and strategic interests for the British—at least until the Statute of Westminster was passed in 1931.[117]

Subsequently, older elements of racial discrimination came together in a fully fledged system. The political and cultural values of the Boer frontier took hold of the entire state, first gradually and then more dramatically with the 1948 electoral victory of the National Party. Unlike in Argentina, where the power of the gaucho frontier soon waned, the frontier periphery here conquered the political core and stamped it for almost the entire twentieth century. Nothing like it had been seen before, even in the United States. In 1829, with Jackson's presidency, a representative of the frontier had for the first time dislodged the urban East Coast oligarchy from the highest office of state. From then right down to the Texan oil dynasty of George Bush, father and son, "Western" attitudes repeatedly marked American politics. But in the nineteenth century a greater challenge came from the slave-owning South. The Civil War was for the United States what the Boer War was for South Africa, albeit in a more compressed time frame. The secession of the Southern states of the United States in 1860–61 was an equivalent of the Great Trek, and their planter democracy before the secession displayed great similarities with the master-race republicanism of the Boer pioneers (which justified itself, however, less in terms of an elaborate racial ideology than through a muffled, barely articulated, sense of superiority).

The defeat of the South in 1865 prevented the ideology and practice of white supremacy from engulfing the American state as a whole. Nevertheless, from the late 1870s on, blacks were again deprived of many of the rights that had been granted, or at least promised, to them during and after the Civil War; the ending of slavery by no means made them into citizens with equal rights. In the great compromises that followed the Civil War in 1865 and the Boer War in 1902, the vanquished whites were to a great extent able to maintain their own interests and values—in each case at the expense of the blacks. Evidently, however, the frontier did not triumph in the United States as it did in South Africa: the values and symbols of the true "Wild West" made themselves visible not at the level of the political order but as components of America's collective consciousness and "national character." In the United States, the North-South opposition complicated the political geography. It became the equivalent of a rebellious frontier in other parts of the world.[118]

4 Eurasia

At the beginning of this chapter, a frontier was defined as a special kind of contact situation, where two collectives of different origin and cultural orientation encountered each other in exchange processes combining conflict and cooperation in varying proportions. Turner's old premise that these collectives represent societies at "different stages of development" has turned out to be not generally sustainable. At the time of the Great Trek—to take just one example—the pastoralist Boers were by no means at a different stage of social evolution from their Bantu neighbors. Nor was it at all evident—to take another of Turner's themes—who were the "barbarians" and who the "civilized." In North America it was only fairly late, with the advent of Indian bison hunting, that a sharp opposition developed between different economic forms: on the one side, sedentary pioneers supplementing agriculture with fenced-in livestock breeding; on the other side, pastoralist nomads with the additional mobility of mounted hunters. Such clear-cut contrasts were rarely found in Africa, with its numerous gradations of nomadism. But as Owen Lattimore pointed out long ago, they were characteristic of the whole of northern Asia.

At the beginning of the nineteenth century, mobile lifestyles based on the breeding and exploitation of animal herds stretched all the way from the southern boundary of the Scandinavian-Siberian-Manchurian forest belt to the Himalayas, the highlands of Iran and Anatolia, and the Arabian Peninsula, and eastward from the Volga almost to the gates of Beijing: an area far larger than the "Central Asia" to be found on today's maps. Sedentary agriculture was concentrated on the margins of the Eurasian landmass, from northern China to the Punjab, and in Europe west of the Volga, which rounded off the world of grasslands and steppe.[119] Such an ideal-typical opposition between static and mobile should not, of course, make us forget that in the nineteenth century wandering population groups also existed in Europe and South Asia.[120]

Nomadism on the Steppe Frontier

Within these huge spheres of mobile lifestyles, ethnologists identify the following variants: (1) the camel nomads of the desert, also found throughout North Africa; (2) the tenders of sheep and goats in Afghanistan, Iran, and Anatolia; (3) the horseback nomads of the Eurasian steppe, the best known being the Mongols and Kazakhs; and (4) the yak herders of the Tibetan plateau.[121] These all have certain features in common: a detachment from, and often violent rejection of, urban existence; a social organization in lineage groups with elected chiefs; and a great stress on proximity to animals in the formation of cultural identity. Traversed by countless ecological boundaries, nomadic Asia was divided into numerous linguistic groups and at least three major religious orientations (Islam, Buddhism, and shamanism, each with a range of subvariants). On the frontier of this world, which in area constituted the largest part of Eurasia, conditions were relatively straightforward. Where nomadism did not—as in Arabia and the Persian Gulf—stretch all the way to the sea, it always encountered sedentary farmers in its path. This was true for millennia in both Europe and East Asia: both had a steppe frontier.

History has seldom been written from the viewpoint of nomads. European, Chinese, and Iranian historians saw, and still see, them as the Other—an aggressive threat from outside, against which any means (usually forward defense) was justified. Although Edward Gibbon, the greatest of Enlightenment historians, already asked what made the mounted warriors of early Islam or the Mongols of Genghis Khan into such an elemental force, sedentary societies found nomads almost beyond comprehension. Conversely, nomads often felt at a loss when confronted with representatives of nonmobile urban cultures. This did not, however, prevent both sides from developing a wide range of strategies in their dealings with each other. Methods of handling barbarian peoples from central Asia were always one of the most well-developed fields of Chinese statecraft. And in the late fourteenth century, Ibn Khaldun made the opposition between city dwellers and Bedouins the cornerstone of his theory of (Islamic) civilization.

The life of nomads is riskier than that of farmers, and this leaves its mark on their view of the world. Herds can multiply exponentially and lead to sudden wealth, but they are biologically more vulnerable than cultivated plants. Mobile ways of life constantly require decisions about how to manage herds and how to behave with neighbors or strangers; they therefore involve a quite distinctive kind of rationality. As the Russian anthropologist Anatoly M. Khazanov has emphasized, nomadic societies—unlike subsistence farmers—are never autarkic; they cannot function in isolation. The more socially differentiated a nomadic society is, the more actively it seeks contact and interaction with the outside world. Khazanov mentions four broad strategies available to nomads:[122]

1. a voluntary shift to a sedentary way of life
2. exchange with complementary societies or trading by means of well-developed forms of transport (such as the camel)
3. voluntary or unresisting subordination to sedentary societies, in a relation of growing dependence
4. domination of sedentary societies and development of long-term asymmetrical relations with them

The fourth of these strategies reached its peak of success in the Middle Ages, when peasant societies from Spain to China fell under the control of nomadic horsemen. Similarly, the great dynasties that ruled Asia in the early modern period had a nonfarming, though not necessarily nomadic, origin in Central Asia; the Manchurian Qing rulers of China (1644–1911) were the most notable, but also the last, example of this type of empire building, which in their case took more than a century to complete.[123] In various parts of Eurasia, however, nomadic societies remained strong enough to plunder their sedentary neighbors and to reduce them to tributary dependence; even Russia continued until quite late in the seventeenth century to pay astronomic sums to the Crimean Tatars. Thus, for very long stretches of time, the most diverse frontier processes were part of the historical reality of Eurasia, and the need to ward off threats from nomads was a significant factor in the formation of centralized states along either Russian or Sino-Manchurian lines.

Frontiers of this kind run like a thread through particular histories of power and exchange relations. Since farmers and nomads each had access to resources that the other needed, cooperation was much more characteristic than outright confrontation. Even if a middle ground of cultural hybridity, crossovers, and multiple loyalties failed to materialize in a lot of cases, the frontier nevertheless joined people as often as it divided them. This remained so until the eighteenth century.

It has long been a commonplace of world-historical interpretation that the Mongol conquests of the early thirteenth century opened up an unparalleled space of interaction and communication; some go so far as to speak of a "medieval world system." Subsequently, it is usually argued, the states and civilizations of Asia withdrew again into themselves—Ming China (1368–1644), sheltering behind its Great Wall, is given as an example—and put an end to Eurasia's medieval "ecumenism." The latest research suggests, however, that open channels and a multiplicity of cross-frontier relations persisted until the threshold of the nineteenth century, and that for this period too, it makes sense to speak of Eurasia as a continuous entity. A crude dichotomy of Europe versus Asia is only an ideological construct of the early nineteenth century.

Imperial Peripheries

One peculiarity of frontiers in Eurasia is that they were molded by empires. Unlike in America and sub-Saharan Africa, the centralized and hierarchically structured empire was here the dominant polity. Roughly speaking, it took one of two forms: either a steppe empire supported by nomadic horsemen and parasitic on a sedentary world of farmers; or an empire whose principal resources came directly from taxation of its own peasantry.[124] Transitional forms were also a possibility. The Ottoman Empire, for example, came into being as a loose military entity, structurally similar to the Mongol Empire, but over time mutated into an empire of the second type. With the general consolidation of this type—also (less happily) known as the "gunpowder empire"—the empires of Eurasia drew closer to one another, until in many places they had contiguous borders. In particular, the growth of the Qing Empire, unstoppable until the 1760s, and then the beginning of the real expansion of the Tsarist Empire meant that inter-imperial "borderlands" (in Herbert Bolton's sense of the term) often took shape out of open frontiers. In the early modern period, therefore, the nomads of Central Asia were already encircled by empires. They themselves (especially Mongols, Kazakhs, and Afghans) were sometimes capable of great military efforts, but they never built a new empire in the manner of a Genghis Khan or a Timur.

An event of world-historical importance was the final extension of the Chinese Empire toward Central Asia. It was, of all things, the non-Chinese conquering dynasty of the Manchurian Qing that succeeded between 1680 and 1760 in partly subjugating (Inner Mongolia) and partly reducing to dependence (Outer Mongolia) the Mongol tribes, and in integrating the Islamic oasis societies of East Turkestan (today's Xinjiang) into the Chinese imperial system. Thus, at the end of the eighteenth century, the heartlands of the old dynamic of redoubtable herdsmen were divided among the empires. This would remain the situation until the end of the Soviet Union in 1991 and the founding of the Central Asian states.

The imperial molding of the frontiers means that the frontier theme blends into the related one of empire building. What interests us for the moment is the fate of the nomads in the nineteenth century, and the great importance of the empires allows the question of the frontier to be posed in this framework.

In its expansion after 1680, the Qing Empire encountered a number of peoples—in southern China, on the newly conquered island of Taiwan, and in Mongolia—who, not being ethnically Chinese ("non-Han"), were classified as needing to be ruled over and civilized.[125] Once conquered, these peoples were subject to a finely graded system of imperial rule or control. They did not form semiautonomous tributary states, like Korea or Siam, but lived as colonized populations within the empire. From the mid-eighteenth century on, this was also true for the Tibetans, governable only indirectly from the distance of Beijing—geographically almost on a different planet. In the Chinese case, the primacy of politics was maintained; movements of settlers outside state control occurred only toward the inhospitable mountainous heartlands of the interior. In the non-Han periphery, which was regarded primarily as a buffer against the Tsarist Empire in the north, the Ottoman Empire in the west, and the emergent British Empire in India, the central authorities had no interest in a dangerous destabilization of the existing social order. The ideal solution was therefore a kind of indirect rule, although the Sino-Manchurian military still had to be present in sufficient number to ensure loyalty to the empire. Until late in the nineteenth century, the Qing state did all in its power to hinder the flow of Han Chinese settlers to Xinjiang, Mongolia, and especially the well-protected dynastic homeland of Manchuria that might one day serve as a place of refuge for an unlucky imperial house.

Nevertheless, Chinese traders could not be prevented from spreading to all these regions and often reducing the commercially inexperienced Mongols to ruinous levels of debt dependence. The settler movement became demographically significant in the early twentieth century, mostly concentrated in nearby Manchuria. But in the 1930s there were loud complaints about the neglect of the inner periphery, especially the Mongolian provinces, as a source of power for the nation; millions of Han Chinese eventually expanded into the periphery after 1949, under Communist rule. Thus, it was only in the twentieth century that a Chinese development frontier opened up and led to the predictable loss of land by the original population—without the emergence, however, of American-style reservations. The Muslim inhabitants of Xinjiang were able to preserve an especially high degree of cultural and political autonomy, enjoying the advantages more than the disadvantages of an inter-imperial borderland until the consolidation of Communist power after the middle of the century.[126]

Despite its growing relative weakness, the Qing Empire maintained its land borders surprisingly well until 1911 (with the exception of southern Manchuria).[127] It did not lose nearly as many regions as the Ottoman Empire, nor were the ones it did surrender nearly as important economically or demographically. The gradual retreat of Ottoman power from the Balkans repeatedly made existing borders and frontier guards obsolete, allowing new Balkan states to replace them largely under the direction and with the guarantees of the European Great Powers. Internal resettlement, such as that which took place in the Qing Empire and on a larger scale in the Tsarist Empire, did not happen in areas under Ottoman domination. No traditional models were applicable there, since in the early modern period Ottoman armies had pushed into regions with a stable peasantry, such as the Balkans and Egypt, where no virgin land was available for Turkish settlers to open up. Besides, Anatolian peasants were much less familiar than their Chinese or Russian counterparts with the techniques needed for agricultural development. The ecology set limits as well, since the Ottoman Empire contained scarcely any large areas that could be brought back into cultivation through a fresh input of labor. Yet there were some forms of frontier expansion. As the Ottoman state came under pressure from Southeast European national movements and Tsarist armies, and as it lost control of North Africa between Egypt and Algeria, its attention was directed to the remaining tribal regions in Eastern Anatolia.

In the early nineteenth century, the population there consisted mainly of Kurds and their khans. Even at the height of its power, the Ottoman state had feared the Kurds and contented itself with a loose form of sovereignty over them, and so the shift in policy after 1831 was the result of a new self-image that the Ottoman elite was beginning to develop. Seeing itself as the modern reform-minded administration of an empire that ought to have more and more elements of a nation-state, the government in Istanbul thought it necessary to eliminate semiautonomous domains and to absorb marginal areas such as Kurdistan, situated on the border with Iran, into an increasingly homogeneous polity. In order to achieve this, the government of the early Tanzimat period resorted to military force. A series of campaigns in the 1830s broke up the principal Kurdish khanates, and by 1845 Kurdistan was treated for the first time as a region under direct rule. However, no constructive policy followed the military success. Kurdistan became an occupation zone: devastated and partly depopulated, its remaining inhabitants bitterly anti-Turkish. This put great strain on the central budget, without stimulating economic growth in a way that increased tax revenue. Members of Kurdish tribes could not be turned into loyal Ottoman citizens by force. While the Balkan boundary contracted ever farther south, the eastern periphery of the empire was becoming more and more expensive to secure militarily; nor could there be any talk of linking Kurdistan to wider markets.[128] The expansion did not go hand in hand with settler colonization, except at best through the resettlement of Muslim refugees from the Balkans and the Caucasus, thousands of whom were directed toward Syria and Transjordan.

If there was a fully articulated frontier in nineteenth-century Eurasia, the places to look for it are the South and East of the Tsarist Empire.[129] The Russian state came into being as a frontline state, a concentration of forces against the Golden Horde of the Mongols. Scarcely was the "Mongol yoke" shaken off when the economic and cultural superiority of Western Europe made itself painfully felt. Peter the Great finally set out to raise the country from its second-rank position, but it was only under Catherine the Great that it became an imperial power of the first order, capable of routing the mighty khanate of the Crimean Tatars and gaining access to the Black Sea. Russia established military superiority over the Ottoman Empire and would never again lose it, although the Turks fought back on a number of occasions. After 1780 began the conquest of the Caucasus; it proved a long haul and was completed only in 1865, but the climax came in the 1830s in the drive to crush the newly unified Chechens.[130] By the end of Catherine's reign, representatives of the Russian state had established relations with a wide range of peoples and states in eastern Eurasia—from Siberian ethnic groups (who had previously had contact only with trappers or explorers) to various Tatar groups and the Kazakh hordes to the emperor of Georgia.[131] As for other empires, it had links not only with the Ottomans but also with China (a long-standing border treaty had already been signed in 1689 in Nerchinsk), Iran (which until the war with Russia of 1826–28 had been addicted to expansionism, laying waste to large parts of Georgia in 1795 and carrying off tens of thousands of its inhabitants), and of course with Great Britain (with which it entered a coalition in 1798 against revolutionary France).

Despite these foundations, the real building of the multinational Tsarist Empire and its expansion to the other end of the Asian continent took place in the nineteenth century. The exact time frame for this drive, without parallel in Eurasia, may be said to stretch from the (partly only nominal) incorporation of Georgia in 1801 to the Russian defeat in the war with Japan in 1905.

Although Frederick Jackson Turner himself warned in his later writings against the oversimplistic idea of a single, uninterrupted pioneer front moving ever westward across North America, conditions in the New World were incomparably easier to grasp than those on the multiform frontiers of Russian-influenced Eurasia. The great variety there was a result of the geography and ecology, the social and political forms of many different ethnic groups, the character of Tsarist policy, and the local decisions of Russian commanders. It is possible to speak of a frontier *policy* at least from the moment in 1655 when the tsar concluded a border treaty with the Kalmyks—not one of subjugation but more or less an agreement between equals.[132] So, early on, the Russian state adopted a policy instrument that the United States would from the beginning use in its relations with Indian peoples. Contractual agreements, even when unequal, presuppose that both sides are able to exercise a minimum of freedom of movement and negotiating skill. They are therefore typical not of fully fledged colonialism but at best of its preliminary stages. Treaties such as the one of 1655 originally served to pacify militarily stronger neighbors across the frontier, but subsequent Tsarist policy developed a wide range of options, from appeasement to genocide.[133] Behind them there was never a uniformly conceived policy of imperial expansion and internal colonial rule. Each of the frontiers should therefore be treated separately, as is usually the case in historical research today.[134]

The Tsarist Empire and North America Compared

The Eurasian-frontier problematic cannot be reduced here to the putting together of the multiethnic Tsarist Empire as seen through Russian eyes. Rather, we shall consider the specificities of Eurasian frontiers in comparison with North America.

First, until the founding of the United States—indeed, until the British-American War of 1812—the most powerful Indian nations remained to some extent foreign-policy partners of the white settlers, in a relationship roughly similar to that which existed between the Tsarist state and the Tatars, Kyrgyz, and Kazakhs. In both parts of the world, the great shift in the balance of power occurred only in the years around 1800. In North America, the Indians were never integrated into society on the settler side of the frontier, but this very exclusiveness of the frontier made possible the formation of a "middle ground," a mixed or transitional zone of contact. In contrast, as Andreas Kappeler put it in his standard account, the Tsarist Empire had "ancient traditions of multiethnic symbiosis which went back to the Middle Ages."[135] The non-Russian peoples included within it were not completely unarmed, and their elites were to some extent recognized by the Russians as aristocracies in their own right. Moreover, ill-defined zones on the margins of the empire (in what is now Ukraine and elsewhere) had been home since the late fifteenth century to semiautonomous, militarized Cossack societies—a form without an equivalent in North America, though similar in many respects to the *bandeirantes* in Brazil. The Cossacks were typical people of the frontier, who scarcely differed in lifestyle or military tactics from nearby steppe nomads such as the Nogai Tatars or the Kalmyks. Long feared by the tsar, they were not at all willing tools of central government in the early modern period.

Such special societies were transitory by nature, because at some point they became an obstacle to the development of solid imperial or national structures.

Much as the British state, around 1720, took energetic action against Caribbean pirates who had previously served it in wars against the French and Spanish, the Cossacks' position grew ever weaker as they lost their usefulness as a buffer against steppe nomads and as the Tsarist state took direct charge of its security requirements. It would be wrong to imagine the Cossacks as "European" fighters against rampaging Asiatic hordes. In many ways their social organization and cultural models made them closer to their nomadic neighbors than to the core Russian population. This was especially so in the Caucasus, where Terek Cossacks and Caucasian mountain peoples lived in close contact, each mirroring the other's martial culture. For the Cossacks, Russian merchants and caravans were easier prey than their armed neighbors. In the first half of the nineteenth century, when the Tsarist state put pressure on the Terek Cossacks to fight against the Caucasian peoples, many were divided in their loyalties or even defected to the other side and converted to Islam. Only in 1824 were they officially accepted into the realm of the Russian state, thereby obligating them to perform services and to pay taxes.[136]

Second, the US Army's role in the hostilities with the Indians should not be underestimated. Apart from the interlude of the Civil War, the frontier saw the largest deployment of troops in the period between the war with Mexico (1846–48) and the Spanish-American War (1898). The high point of army activity in the American West happened to coincide with the Tsarist offensives in the Caucasus and against the emirates of Central Asia (above all Khiva and Bukhara). The most notable difference was that the US army gave flanking protection to private settlers, engaging in what were ultimately major police operations rather than campaigns of conquest, whereas the Tsarist army became an instrument of conquest neither preceded nor followed by agricultural settlements. Continuing an earlier pattern, the Russian state exhibited a greater talent for military action than for the systematic organization of new settlement. Economic motives were not altogether absent from this army-led expansion of the empire: the conquest of Central Asia entered its decisive phase in 1864, when the American Civil War was interfering with cotton supplies to the Russian textile industry and Moscow was looking for alternative sources.[137] Moreover, strategic objectives in the confrontations with the Ottoman Empire, Iran, and the British Empire were at least as important as aggressive decisions on the part of army commanders on the spot. Such military imperialism did not lead to the development of a frontier. It was a state matter, which invariably rocked the foundations of the non-Russian societies under attack, without resulting in the construction of new kinds of society.

Third, unlike the Indians of North and South America, the embattled Central Asian peoples had the opportunity (often only minimal) to enlist the support of external allies, or at least to be welcomed as exiles in a third country. At best the North American Indians could escape to Canada; few were offered a safe haven there. The peoples of the Caucasus, inserted as they were into a web of Islamic solidarity, could at least count on acceptance in the Ottoman Empire. Caught in a pincer between Russian tsardom and Sino-Manchurian imperialism, the peoples of Muslim Central Asia had little room for maneuver toward the end of the eighteenth century—although some were able to jockey for a while longer between the two great empires. Several paid tribute until 1864 both to Russia and to China. From 1820, when China's hold on Xinjiang began to slacken, uprisings of the Muslim population broke out there and just across the border in Kokand. Repeated attempts were made until 1878 to create independent Muslim states in the space between the empires.[138] With the exception of a number of Siberian peoples, the victims of Tsarist expansion were able to retain a leeway that was denied to the Indians of North America.

Fourth, it is possible to speak of a frontier-like invasion of settlers in two large regions: Western Siberia and the Kazakh steppes. Since its beginning in the seventeenth century, the Russian conquest of the vast Siberian expanse east of the Ural Mountains was driven by the demand for animal skins and furs, so that the region joined the broader nexus linking the forests of the northern hemisphere with European and Chinese markets.[139] But the resources tapped by hunters and trappers were too widespread for a real "fur frontier" to come into being. Much as in North America, the indigenous peoples were at first able to take great advantage of the new market opportunities. But their situation worsened as the agricultural colonization of western Siberia, first made possible after 1763 by the construction of a road from the Urals to Irkutsk on Lake Baikal, gathered speed in the late eighteenth century; it was then not much farther to the Chinese border. A ribbon thousands of kilometers long had to be cut through the forest, and a surface laid that was capable of bearing wagons and sledges. This was a major technical accomplishment, carried out several decades *before* the work on the Oregon Trail in the United States and more than a century before the construction of the Trans-Siberian railroad. The so-called *trakt'* went so far south that the need for dangerous river crossings could be kept to a minimum. It stimulated the growth of existing localities en route, especially Omsk, which in 1824 became the seat of the governor of Siberia. But it also facilitated the exploitation of nature and profoundly affected the living conditions of indigenous Siberian peoples.

A second watershed was the emancipation of the serfs in 1861. It is true that this still did not give them full mobility—the legal restriction tying them to their village community was not lifted until 1906—but hundreds of thousands managed to get around this. In the 1880s an average of 35,000 people a year were emigrating to Siberia from European Russia; in the 1890s the figure was close to 96,000, and the flood after 1906 peaked two years later with an annual total of 759,000.[140] Multiple tensions appeared between the newcomers and earlier migrants to Siberia, the *starozhily*, who had largely adapted to the subsistence way of life of the Siberian peoples and sometimes even unlearned their Russian.[141]

The consequences of the new colonization were disastrous for the indigenous peoples. Their social capacity for resistance was as low as that of the

North American Indians vis-à-vis the Euro-Americans, or that of the Mongols in relation to the Han Chinese. The growing difficulty of hunting and fishing, debt burdens, and alcohol undermined traditional ways of life and cultural orientations. As far as the Sea of Okhotsk—and in the East there was additional pressure from Chinese settlers—native Siberians either tried unsuccessfully to adapt to the new conditions or retreated still deeper into the forests. Like the *indios* of South America, they were not even afforded the protective shelter of reservations.[142]

The most important region for agricultural settlement was the Kazakh steppe: that is, the area between the Lower Volga and the foot of the Altai Mountains near Semipalatinsk (today's Semey).[143] In order to defend itself from nomadic Kazakh horsemen (organized in great "hordes") and from steppe peoples such as the Bashkirs, the Russian state began in the 1730s to build a chain of forts, of which the chief at first was Orenburg. From these, representatives of the tsar conducted a policy mixing negotiation with division and intimidation, but despite their many successes this steppe frontier was not brought under control until the nineteenth century. As late as 1829, when Alexander von Humboldt visited the region at the Tsar's invitation, he was given a large Cossack escort for the route between Orenburg and Orsk, which was considered especially dangerous. Nomadic horsemen often raided Russian territory and carried off humans and livestock; some people were sold as slaves to Khiva, where they were apparently much prized for irrigation works. Russian soldiers watched events in the steppe from their wooden towers. The Kazakh absorption into the Russian Empire happened not through rapid conquest but through a slow process involving both ad hoc military expeditions and a gradual replacement of feudal allegiances with imperial subjection. The aim was not only to secure the region but also to convert the nomadic horsemen into farmers and to "civilize" them within overarching imperial structures.[144]

An even deeper impact was made by the settlement of Russian and Ukrainian peasants, who set about cultivating marginal areas of steppe more energetically than the Cossacks had done with their seminomadic mixed economy. As earlier in Siberia, the emancipation of the peasantry created the initial impetus, but once again the state lent a powerful helping hand. The Steppe Statute of 1891 drastically curtailed the ownership of land by Kazakhs. The nomadic herdsmen, few of whom could be induced to settle down, were driven farther south and cut off from the wetter pastureland essential to grazing cycles. The comparative chronology is striking here. Not until the 1890s, when there was no more "ownerless" land left in the Midwest or High Veld frontier areas, was the South Russian steppe frontier being *opened*. Here too, this happened at the expense of the indigenous peoples, although they did not disappear into enclosed enclaves but continued their nomadic existence on marginal land. The Kazakh settlement frontier was the most striking instance of its kind anywhere in the Tsarist Empire; a nomadic lifestyle was displaced by the farmer's plough. The conflict was less between populations at different "stages of development" than between different types of society or ethnic groups. The region where the frontier process unfolded was transformed "from a frontier zone of nomads and Cossacks to an imperial realm of farmers and bureaucrats," and from a Turkic-Mongol world into a multiethnic sphere under Slav domination.[145] It matters little whether one calls the outcome an "internal colony" or a "borderland." However, since it did not come under a special administration but was incorporated into the Russian state, there is much to be said against the term "colony."

A similar sequence may be found in other frontier regions of the Tsarist Empire: first came the Cossacks, then garrison towns and frontier fortresses, and finally settlements of farmers. The state tried to steer this process, and indeed every aspect of the opening of frontiers, much more forcefully than in the United States or South Africa. The main contribution of the American state was to make cheap land available to settlers in an orderly manner. The pioneers were completely free individuals: no one could send them anywhere. In Tsarist Russia, by contrast, until the liberalization of agrarian policy under Prime Minister Stolypin, the state intervened to guide the process of settlement. This posed no problem in the case of "state peasants," but even with other categories, whether dependent or "freed," the state presumed to act in a guardian-like capacity. Although many settlers eventually shaped their own lives, the settlement frontier was not, as in the United States, theoretically formed by their free decisions.[146] A further difference with the United States was the small weight of urban settlements. The North American frontier was everywhere associated with the formation of small towns, some of which profited from a favorable transportation location to develop rapidly into major cities. At the western end of the continent, the frontier ended in a densely settled urban zone that did not actually owe its formation to the frontier. No Russian California would ever emerge; Vladivostok did not blossom as a second Los Angeles. But neither did frontier urbanization in the strict sense become a large-scale phenomenon.

Fifth, all forms of the eighteenth- and nineteenth-century expansion of Russia were highly ideologized. Public rhetoric in the United States toward the Indians also went through phases in which the task of "civilizing" them was seen either as futile or as important for mankind. But the fantasies in the eastern Tsarist Empire were far more extravagant; nowhere in the whole history of European expansion was a "civilizing mission" taken so seriously.[147] Since many Russians at the time believed that civilization should mainly follow colonization, interpretations of history appeared—for example, in the influential Moscow historian Sergei M. Solovyev—which in many ways anticipated the frontier thesis of Frederick Jackson Turner. In the early nineteenth century, the view began to spread that Russia should act in Asia on behalf of progressive Europe. The space between the Arctic and the Caucasus seemed to be one where the enlightened Russian upper stratum could prove itself as a promoter of European civilization; conquest and colonization proceeded, as it were, with a look over the shoulder toward Western Europe. At the same time, it was intended to distance Russia from all the ill-famed aspects of colonialism and imperialism; indeed, Russian and Soviet historians have always shied away from admitting the imperial character of Russian policies. This shamefaced urge to camouflage reality, similar in a way to the American aversion to admitting the colonial side of US expansion, echoes in the much-loved talk of "assimilation" (*osvoenie*) of non-Russian regions and their inhabitants. But—and this is another important difference— whereas Turner's frontier involved a turning away from Europe and the birth of the distinctively American pioneer, Solovyev and his followers continued to regard Western Europe as the measure of all things. The Europeanization of Russia was supposed to advance farther, in the form of a Russification of other nationalities within the Empire.[148]

An American "wilderness" concept does not seem to have played a major role in Russia. On the other hand, a particularly high degree of ideologization was reached when expansion was dressed up as a struggle against Islam. Propagandists attuned to a philosophy of history argued that the "historical decline" of Christendom in relation to Islam could and should be reversed. Archaeologists went in search of "pure" (that is, pre-Islamic) cultural forms in the conquered periphery. Islam was discursively defined as a foreign import, and Christian outposts such as Georgia were incorporated into God's plan for salvation;[149] the purifying effect of the frontier experience would stand Russians in good stead. Similarly after 1830, partly to protect its heartlands from heretical contamination, the Russian state preferred to populate the periphery with religious dissenters—Old Believers, for example, whose persuasions had distanced them from the Orthodox Church since the mid-seventeenth century. By the 1890s heterodox Christians made up the overwhelming majority of ethnic Russians in Transcaucasia.[150] As usual, however, the imperial discourse was shot through with contradictions. The same Islamic fighters who in Dagestan were demonized as enemies of Christian civilization might appear in a different context as mountain warriors or "noble savages." Such romantic-Orientalist themes linked Russian thinking about "the alien" with the ideologies of other empires, such as the glorification of the Berber in French North Africa or the British admiration for martial races in India and East Africa.[151]

Sixth, unlike with the North American Indians, a few success stories can be reported in the case of the Tsarist Empire. Under pressure from the forces of expansion, many peoples displayed a high degree of cultural resistance as well as adaptability. One of these, the Siberian Bukharans, stood out among the inhabitants of eighteenth-century Central Asia by virtue of their urbanity, their relative loyalty to the Russian government, and their widespread literacy in Arabic and Persian; they formed the core of a merchant stratum and maintained intra-Islamic contacts between Bukhara and the Tsarist Empire. Other examples are the Yakuts and the Buryats. As one of only two Mongol peoples in the Russian Empire (the other was the Kalmyks), the Buryats were seen by Russians as representing a higher stage of development than that of the "primitive" shamanistic peoples of Siberia, especially since they had a differentiated social structure with a clearly recognizable aristocracy inclined to act as colonial "collaborators." Despite all manner of importunities from state officials and missionaries, the Buryats were able to command respect and to maintain a freedom of action that no Indian people in the Americas enjoyed. In particular, they set out to develop a modern, educated middle-class elite alongside the traditional political and ecclesiastical hierarchies—one that would articulate their interests both publicly and within the bureaucracy.[152] All over the world, the worst-placed ethnicities and societies were those unable to fulfill at least one of three long-term criteria: to be feared militarily, to be useful economically, and to gain representation in the forums of modern politics.

5 Settler Colonialism

State Settlement Projects in the Twentieth Century

Frontiers can be places of annihilation and places of regeneration. Destruction and construction are often dialectically intertwined; Joseph Alois Schumpeter called this, in a different context, "creative destruction." In the nineteenth century, whole peoples in frontier regions were decimated or reduced to poverty while constitutional democracies were taking shape there for the first time. Frontiers may thus be sites of archaic violence as well as birthplaces of political and social modernity.

Let us first cast a glance beyond World War I. There were still frontiers in the twentieth century, some of which continued processes from the previous century. But it would appear that they lost their ambiguity. Constructive developments were few and far between, as frontiers turned into peripheral zones of tightly controlled empires far removed from the internal pluralism of the British Empire.

The period after 1918 brought an intensification of ideology and state intervention in the opening up of new farming settlements. In general, the settlers in question were not enterprising private individuals, such as those who emigrated around the same time to Canada or Kenya, but people from the lower depths of poverty, sent out in the wake of conquering armies to secure "boundary markers" under harsh conditions. The idea that strong nations needed living space to escape the danger of resource shortage that came with overpopulation, and that they had a right and duty to take inadequately "cultivated" land from less efficient or even racially inferior peoples, can be found among numerous far-right movements and opinion makers in the early twentieth century. It became official policy in the new empires that appeared in the 1930s: fascist Italy in the case of Libya (and to a lesser extent Ethiopia), post-1931 Japan in Manchuria, and Nazi Germany in its short-lived *Drang nach Osten*. All three combined visions of a nation tested in frontier warfare with a special emphasis on the soil. Hitler, an admirer of the exotic adventure novels of Karl May, drew direct parallels between the Wild West of May's resourceful hero Old Shatterhand and the Wild East that he began to create in the early 1940s.[153] Frontiers were stylized as experimental spaces where new men and new types of society could develop without hindrance from tradition: a utopian military order in Manchuria, an Aryan racial tyranny in conquered Eastern Europe. Germany's "blood and soil" ideology, in which ethnic cleansing and mass killing were preprogrammed, represented the extreme form of such thinking. The settlers were not meant to carry out these murderous objectives themselves, but in each case they served as instruments of policy. It was the state that recruited and dispatched them, providing marginal land in foreign colonies and convincing them of their sacred duty to endure the inevitable rigors for "the good of the nation." The settlers of fascist imperial dreams—whether in Africa, Manchuria, or on the Volga—were guinea pigs for a state-directed *Volkstumspolitik*. They lacked the essential features of Turner's pioneers: freedom and self-reliance.

A further dimension that appeared in the twentieth century, and not only in fascist or (in Japan) ultranationalist systems, was what the sociologist James C. Scott termed the "social engineering" of rural settlement and production. Nature, it was widely believed, could be rationally exploited to the maximum through planned labor inputs and uniform conditions of agrarian production.[154] One side effect of this was always greater state control over the rural population. The collectivizations in the Soviet Union and the People's Republic of China, each associated with programs to bring "new land under the plough," had this momentum, as did many projects (fundamentally less illiberal in design) of the Tennessee Valley Authority under Franklin D. Roosevelt's New Deal. In the Communist version, the element of settler freedom totally disappeared, and the actual clearing of land was often undertaken by soldiers or state farms. But the idea that the molding of space could be taken beyond ecological or "civilizational" limits was common to all twentieth-century variants of state-initiated land clearance and to older forms of settler colonialism.

The key term "settler colonialism" is usually found in the context of empires and imperialism. There it is mostly treated—at least for the nineteenth and twentieth centuries—as a special case, since before 1930 there were not very many colonies in which European settlers made up a considerable part of the total population and where political processes occupied a dominant role. The only instances—apart from the British dominions, which had long resembled nation-states in their forms of government—were Algeria, Kenya, Southern Rhodesia, Angola, and Mozambique. There were no European settler colonies anywhere in Asia, and Northern Ireland was a particular exception in Europe. Histories of colonialism therefore focused little on settler colonies; only Algeria, the most important component of France's overseas empire, attracted greater attention. To discuss settler colonialism under the theme of the frontier involves a reference shift, so that it appears not as a special type of colonial rule but as an outcome and expression of special forms of expansion.

Settler Colonialism: The Congealed Frontier[155]

Not all frontier expansion by nonstate players leads to a permanent and recognizable line of divide between types of economy and society. The early Canadian frontier was an undemarcated zone of contact between Indians and white fur hunters and traders, all highly mobile people poles apart from settlers, and the Amazonian frontier was never anything more than a space of plunder and overexploitation. Frontier *colonization* is therefore a subcategory of frontier *expansion*,[156] a phenomenon known in most civilizations, which denotes a push into the "wilderness" beyond the existing cultivation boundary to develop land for agriculture or mineral extraction. Such colonization is by its very nature coupled with settlement; the economic objective is to bring the mobile production factors—labor and capital—closer to location-dependent natural resources.[157] It does not necessarily have to involve a new political entity, since the colony is often founded at the edge of an existing area of settlement: for instance, the gradual extension of the Han Chinese agricultural zone at the expense of the pastoralist economy of Central Asia, which reached a peak in the nineteenth and early twentieth century. But such colonization may take place in a secondary relationship to core overseas areas of new settlement; the best-known example of this is the opening up of the North American continent from its east coast outward. Industrial technology enormously increased the extent—and the environmentally destructive impact—of colonization. The railroad, in particular, strengthened the role of the state in a process that was in most cases historically organized by nonstate communities. The most extensive state-driven railroad colonization was the opening up of Asiatic Russia from the late nineteenth century on.[158]

Settler colonies are a special form of frontier colonization that first appeared in Europe in Greek antiquity (and before that in Phoenicia): a city would plant offshoots across the sea, in regions where only a relatively small commitment of military power was possible and necessary. In both ancient and modern times, this involved a decisive logistical difference with other kinds of frontier colonization. The sea, but also forbidding distances on terra firma (Gulja in Xinjiang, in preindustrial times, took longer to reach from Beijing than Philadelphia did from London), stood in the way of the regular links that alone permit social continuity.

Under such conditions it was possible that colonization would give rise to genuine colonies, in the sense not only of frontier settlements but of communities with their own distinct political structures. The classic example is the early English settlement of North America. The groups that founded settler colonies sought to create bridgeheads with a large degree of economic independence, reliant for supplies neither on the mother country nor on their local surroundings. Unlike the Romans in Egypt, the English in India, or the Spanish in Central and South America, European settlers in North America, Argentina, and Australia did not find efficient systems of agriculture capable of generating a surplus to sustain a militarily protected apparatus of colonial rule. It was therefore not possible to divert a structurally existing tribute from the coffers of the old rulers into those of the new; nor were Indian peoples or Australian aborigines suitable for forced labor in European-style agriculture. These were the circumstances that gave rise to the *first*, "New England" type of settler colonization: that is, to the growth of an agrarian population that filled its labor requirement out of its own families and indentured servants while ruthlessly driving off the land a small indigenous population of no use to it economically. Around 1750, regions had arisen in North America—and *only* there in the world outside Europe—that had a high degree of social and ethnic homogeneity and the potential to become the core of a neo-European national state. The British followed the same model of colonization in Australia, in the special conditions resulting from forced migration of convicts, as well as later in New Zealand (despite especially strong resistance from the native Maori).

A *second* type of settler colonialism emerged where a politically dominant minority, with the help of the colonial state, was able to drive the majority population off the best land, yet remained dependent on its labor and constantly competed with it for resources. Unlike in the New England model, settlers in this second type—which we may call "African" because of its main modern locations (Algeria, Rhodesia, Kenya, South Africa)—were economically dependent on the indigenous population.[159] This also explains the instability of the second type. Only the European colonization of North America, Australia, and New Zealand became irreversible, whereas powerful decolonization struggles eventually developed in the African settler colonies.

A *third* type of settler colonialism solved the labor supply problem due to expulsion or elimination of the indigenous population by importing slaves and putting them to work on medium-sized to large plantations. This may be called the "Caribbean type," after the region where it was most in evidence, but it was also found as a less dominant form in British North America. Demographic proportions were an important variable. In the British Caribbean in 1770, blacks made up approximately 90 percent of the total population, whereas in the northern colonies of the future United States they accounted for only 22 percent, and in the future Southern states for no more than 40 percent.[160] Type 3 is a limiting case, however. With the exception of the American South in the half-century before the Civil War, no coherent planter oligarchies with an independent political vision and capacity for action developed anywhere on the basis of modern slavery. This was anyway virtually impossible in places such as Jamaica or Saint-Domingue, where many large plantation owners resided in Europe. Hence plantation owners may be only loosely described as settlers at all.

What then, in the long history of settler colonialism, was specific about the nineteenth century, when type 1 became so pronounced as a culmination of older trends and a model for the future? The answer falls into five parts.

First, as Adam Smith already saw in 1776, this type of colonialism corresponded to the principle of voluntary settlement and therefore to an individualist market logic. Settlers, as small entrepreneurs, flowed to places where they saw opportunities for optimum use of their own resources (labor power and sometimes capital) in conjunction with extremely cheap land. They were not officially sponsored colonists or imperial agents. Their form of economy rested on family businesses, but did not, after the early pioneering days, aim at total self-sufficiency. Settler agriculture, based on a division of labor, produced staples for internal and domestic markets and obtained its own provisions through trade.[161] It employed wage laborers and abstained from forms of extra-economic compulsion. In many cases in the nineteenth century—from Argentine cereals to Australian wool—it achieved above-average productivity and was cost effective and internationally competitive. In short, frontiers in the nineteenth century, or at least those run on capitalist lines, became global granaries. This process of putting grassland under plough and incorporating it into the capitalist world economy reached a climax around the turn of the century. In 1870, Canada and Argentina were still relatively poor countries with little attraction for immigrants. But between 1890 and 1914 they made immense strides, achieving prosperity not through industrialization but as leading suppliers of wheat. In the period from 1909 to 1914, Argentina produced 12.6 percent and Canada 14.2 percent of the world's wheat exports.[162] What made this possible was the development of an open frontier—a process concluded by the outbreak of the First World War.

Second, classic settler colonialism rested on a surplus of cheap land, which settlers made their exclusive possession by a variety of means, ranging from purchase to deception to violent expulsion.[163] It would not be quite correct to say that it was always "stolen" from its previous owners, since in many cases mixed use and unclear property relations prevailed *before* the settler invasion. The decisive point is that the previous users—very often mobile tribal societies—were denied further access to the land. The producers were separated from their means of production or driven into marginal areas; nomads lost their best grazing land to agriculture or the settlers' fenced-in pens, and so on.

Settler colonialism everywhere ushered in a modern European conception of "property," in which the individual owner had exclusive disposal over precisely measured and delimited pieces of land. Clashes between different ideas of ownership were an ubiquitous accompaniment of *European* frontier expansion.[164] The dispossession of indigenous overseas communities followed from processes in Europe, earlier or contemporary, especially those involving the privatization of common land. On the European side too, however, a distinction must be drawn between various legal concepts in play. Of cardinal importance was the freedom to buy and sell land. In the British Empire and its successor states (e.g., the United States), land became a freely tradable or pledgeable commodity, whereas in the Spanish legal tradition, family links played a much greater role and, even after the end of the colonial period, latifundia could not be simply divided up and sold. This was a crucial element in the stabilization of rural oligarchies in Spanish America and may have been an obstacle to economic development there.

Third, classic settler colonialism, unlike the twentieth-century fascist variant, stood in an ambiguous relationship with the colonial state. The Spanish monarchy of the early modern period already made it difficult for land to be permanently accumulated in private hands, thereby preventing the early conquistadores from crystallizing into a landowning class resistant to state control. In the nineteenth century, the British Crown by no means always acted as a cat's paw for settler interests. In New Zealand, for example, in the early decades after colonization began in 1840, the authorities went to great lengths to protect Maoris from land-grabbers, prohibiting direct transfers of ownership to private individuals of British origin. Like the North American Indians, the Maoris did not have a conception of land as independent of tribal communities and the authority of their chiefs; use rights could be surrendered or even sold, but not the earth itself. The European legal framework was therefore at first completely incomprehensible to them. The colonial state stuck to royal prerogatives over the disposal of all land, exercizing a kind of right of first refusal and using grants of Crown land as a way to stem the anarchy of private interests. Such grants were, of course, a stepping stone to permanent land transfers, and in principle the courts gave "security of tenure" precedence over what were regarded as fictitious "aboriginal rights." However, Crown grants could be withdrawn if the land was not "improved" through use. In all British colonies (and many others) the authorities took action at some point to protect indigenous people from settlers' exactions, though naturally within the framework of a general affinity between state and settlers. One major common interest was the curbing of mobile population groups. But the motives were often different: settlers viewed "wandering tribes" as competitors for land, while the state saw them as a threat to order and an untapped source of tax revenue.[165]

Fourth, classic settler colonialism had an inherent tendency toward *semi*-autonomous state building. Settlers want to govern themselves and strive for a democratic, or at least oligarchic, political system. The abrupt secession on which the majority of North America's British settlers embarked in 1776–83, and the declarations of independence by the South African Boer republics in 1852–54, remained exceptions. Not until 1965, in Southern Rhodesia (later Zimbabwe), was there another settler revolt concerning the political form of the state. Most settlers needed the protective umbrella of an empire: the mother country was supposed to let them get on with things but to make its instruments of power available to them in an emergency. For this reason the position of settlers—especially in African-type colonies with native majorities—could only be one of semiautonomy. Under no circumstances were they mere tools of the metropolis; indeed, they often sought to gain influence over the political process there. The Algerian *colons* were especially good at this: their representation in Parliament in Paris was a source of strength, but their dependence on the colonial military served as a constant reminder that their position might one day be threatened. The British dominions chose another path. In Canada, Australia, New Zealand, and, in its special way, South Africa, settlers took over the colonial state and its main instruments of coercion in the course of the nineteenth century without tying their hands through formal incorporation into the British political system. No British colony ever sent MPs to Westminster, and time and again the dominions opposed plans for greater integration within the empire. Long before national liberation movements appeared on the scene, settlers were the main source of unrest in the European overseas empires. From the point of view of the colonial state, these "ideal collaborators" were also a headstrong and unmanageable clientele. "Settler democracy" was an objective that invariably pointed beyond empire.

Fifth, classic settler colonialism was a historical force with huge transformative energy. The natural realm experienced this more than any other. Seldom in history have relatively small groups of people made such radical changes to the environment in such a short space of time as the settlers did in neo-European regions of world. This took place before the great technological revolutions that came with the tractor, artificial fertilizer, and the motor-driven chainsaw. For a long time European and Euro-American settlers knew very little about nature in the regions where they sought to make a new life for themselves, and so their first reflex was to carve out familiar kinds of agrarian landscape.[166] Their main initial successes were in areas where the natural conditions resembled those of Europe. But over time they came to recognize the potential of uninhabited spaces, as well as the natural limits of all options open to colonizers. The Rocky Mountains, the Australian outback, the Canadian Far North, the swamplands of western Siberia, the Saharan South of Algeria: these all presented challenges on a scale beyond what Europeans had been anticipating. Settlers destroyed ecosystems and created new ones in their place. They wiped out animal species and introduced new ones—sometimes intentionally, sometimes as unknowing bearers of an "ecological imperialism" that spread life forms, from the microbe up, across the planet. New Zealand, a territory so distant that Europeans did not travel there expecting to return soon or at all, had its biological setup revolutionized. Captain Cook's ships had seemed like Noah's Ark when they put ashore there in 1769, in a country lacking mammals apart from dogs, bats, and a small species of rat. Decades before the first settlers made their appearance, tiny pathogens and splendid hogs arrived with Cook and his men—and stayed behind. Then the settlers brought horses, cattle, sheep, rabbits, sparrows, trout, and frogs, as well as game that English gentlemen, even in the colonies, depended on for their favorite sport. The Maoris, seeing this invasion not only as a threat but also as an opportunity, took up pig farming with considerable success. Wool became the colony's main export item: the two islands had 1.5 million sheep by 1858, and 13 million twenty years later.[167] New Zealand was only a particularly dramatic example of environmental changes that settler colonialism triggered everywhere. In the nineteenth century, the "Columbian Exchange" of plants and animals developed from a transatlantic into a global phenomenon, and the encroachment of settler agriculture went wider and deeper than ever before.

6 The Conquest of Nature: Invasions of the Biosphere

Frontiers interact with one another. Certain kinds of experience that occur in one can be subsequently transferred to similar general frameworks elsewhere. The frontier war of the medieval Spanish nobility against the Islamic kingdoms and the later assaults on the indigenous population of the Canary Isles formed a character type well equipped for the conquest of America. And men who in the seventeenth century had served the English Crown in Ireland could later be made good use of overseas. Linked up by international trade, frontiers became subject to adaptive pressures from the world market; those producing the same export goods—such as wheat, rice, or wool—were locked in sharp competition. Often they adopted similar strategies to secure their interests. For example, in the late nineteenth century, both California and Australia saw horticulture and fruit farming as a protection against world cereal price fluctuations.[168] Frontiers also stood in ecological relationships with one another. Exchanges among them were increasingly planned, so that Californians, for example, imported Australian eucalyptus as the key to afforestation of arid landscapes, while Australia adopted the Monterey pine from California as its favorite plantation tree.[169] Political visions lay behind the apparent innocence of botanical experiments: many in Australia dreamed that the Fifth Continent might become a second America.

At least since Owen Lattimore's work we have known that the frontier has ecological as well as demographic, ethnic, economic, and political dimensions. Large parts of environmental history could even be written as a history of frontier expansion. This is true especially of the nineteenth century, the most important but also the last phase of *extensive* development, before the last remaining frontiers (save the Arctic and the rainforest) closed in the first third of the twentieth century. This book does not have a special chapter on environmental history. There is one chief reason for this: "environment" and "nature" are virtually ubiquitous factors making themselves felt in many of the various fields covered by our survey: migration, cities, industrialization, and so on.[170] This section will look at some ecological frontier processes, each involving a dramatic expansion of human control over natural resources, and each continuing trends from earlier periods.

Of course, industrialization created unprecedented levels of pollution, generated totally new demand for farm produce, and developed technologies that made human intervention in nature incomparably more effective than before. But very often it was only modifying processes of more ancient origin. Frontiers to nature also emerged where no extension of arable land was involved. Mountains, for example, came into human purview to an extent unknown in the past;

demographically driven migration to ever higher valleys and slopes, together with new associated forms of land cultivation, was observable in many places on the planet, from the Alps to the Himalayas and the mountains of southwestern China (where a special kind of anarchic frontier, more or less outside state control, arose in the eighteenth century in sharp contrast to the agrarian-bureaucratic order on the plains).[171] Only in Europe, however, did the eighteenth-century aesthetic admiration of the high mountains—at first a special interest of intellectual circles in Geneva and Zurich—mutate into a sport of mountaineering that united foreign gentlemen-climbers with rustic local guides.[172] Alpinism began around 1800 on Mont Blanc and the Grossglockner, at exactly the same time as Alexander von Humboldt's extraordinary feats in the Andes, which took the German naturalist to heights at which no European had ever before stayed for long. In the nineteenth century, mountains on every continent were climbed, surveyed, and named. This, too, involved the opening and closing of a frontier, which symbolically concluded with Edmund Hillary and Tenzing Norgay's ascent of Mount Everest in 1953. The sporting challenge would intensify with the selection of new and more difficult peaks or the refusal of oxygen equipment; but the extensive conquest of the high mountains closed in much the same way as that of Antarctica had done in 1911.

Deforestation

In the long history of planned deforestation and of protests against it (which in Europe and China began between the 1850s and 1900), a precise place cannot be easily assigned to the nineteenth century: certainly the most destructive age yet seen for the earth's primeval forests, but still innocuous in comparison with what was to come. It has been estimated that of the major clearances since the dawn of agriculture, roughly one-half occurred in the twentieth century.[173] The pace of deforestation had accelerated over the previous century. Between 1850 and 1920 probably as much primeval forest was lost worldwide as in the period double that length from 1700 to 1850. By far the most affected region was North America (36 percent), followed by the Russian Empire (20 percent) and South Asia (11 percent).[174] The great forest clearance in the earth's *temperate* zone then came a halt almost everywhere around 1920, marking an important break in the history of the environment. (In some countries this turn in favor of the forest had actually begun much earlier—as far back as the early nineteenth century in France and Germany—and even in the United States individual activists had initiated a gradual rethink in the last third of the century.) Thereafter, many of the forest stocks in temperate zones stabilized or regenerated.[175] The two main reasons for this were the end of extensive land clearance at the expense of the forest; and the raising of tropical production to cover the requirement for wood in the North.

Even today it is difficult to cut through a host of opinions on desertification and wood shortage to get to verifiable facts. Moreover, assuming that such facts are available for certain times and places, there is the additional problem of gauging the short-term and long-term consequences of forest loss. Shrinkage may continue for a long time in a particular region before its deleterious effects enter the picture. And when does a crisis become "general"; when does it acquire supraregional significance? A number of histories might be narrated to show that different development paths exist within a general worldwide trend of destruction and unsustainable forest use.[176]

In China, forest destruction has been taking place for close to twenty-five hundred years, but one should not speak of a general wood crisis before the eighteenth century. Since then wood has been in short supply as a fuel and a construction material, not only in densely populated provinces with intensive forms of agriculture but in most of the core areas of the country. Non-Han communities in remote peripheries organized themselves for the first time in the eighteenth century to defend their remaining forest from Han Chinese, who often appeared as large-scale commercial raiding parties. Wood theft became a widespread crime in the Chinese heartlands. If new trees were planted for commercial purposes, they were of fast-growing varieties—and even those were not given enough time to grow.[177] A general deforestation crisis ensued in the nineteenth century, but neither the state nor private individuals did anything to combat it; little has changed today in that respect. There was no tradition of official forest conservation, such as began to develop in Europe in the sixteenth century. Today's environmental crisis in China has its roots in the nineteenth century. This cannot be fully explained by the weakness of the nineteenth-century Chinese state and its relative lack of concern for the common good, or by the fact that control over the forest (as in the Mediterranean, but not as in India, where it was in various ways a point of departure for state-building[178]) never served as a power base, or by cultural indifference to the myth and beauty of the forest. At least one economic factor also needs to be acknowledged: namely, a kind of path dependence of natural calamities. The crisis reached a point where the costs of overcoming its causes would have been greater than society could bear.[179]

External factors played no role in this development. China was not traditionally a wood exporter, nor did foreign businessmen show an interest in its forests in the post-1840 era of Western aggression. In any event, China found itself heading for a *homemade* forest crisis without the means to correct it. No supposed inadequacies of "Asiatic" societies in general can account for this. Japan—which had undergone a great deforestation crisis since the late sixteenth century, mainly as a result of fortress and ship construction during the period of unification around 1600—halted the tree loss in the late eighteenth century and launched new planting initiatives. This happened under the political ancien régime of the Tokugawa period, without any help from European forestry. The industrialization of Japan, beginning in the 1880s, then had a major adverse effect on forest resources, and the state did not see it as a priority to protect them. With scarcely any fossil fuels of its own, the country derived a large part of its industrial energy from charcoal (plus water power). It was only after 1950 that trends began to favor the forest again.[180] Japan, like China, never became a significant exporter of wood. Siam/Thailand, on the other hand, the only country in Southeast Asia to remain politically independent, granted concessions to European firms interested in its teak. There, forest conservation was not on the agenda.

Another history may be told about the Indonesian island of Java, one of the oldest and most deeply penetrated colonies in the world. In Southeast Asia large-scale deforestation got under way long before the age of voracious plantation forestry dawned in the nineteenth century; many areas had already laid out pepper gardens by 1400, before any contact with colonialism. European consumers were reached via the Mediterranean and later by way of the Portuguese monopoly trade. This replacement of primeval forest with monoculture spread more widely over the next few centuries, especially on Sumatra.[181] The 330-year Dutch presence on Java passed through a number of phases.[182] In the 1670s the Dutch East India Company took control of the areas of Javanese teak forest regarded as most valuable and, unlike dense jungle, easy to exploit for export. Over time the destructive impact of its logging methods became apparent. In 1797 the "sustained yield" principle generalized and made permanent an originally temporary ban on felling that had been introduced in 1722 for certain areas of forest. The basic idea of conservation, now seen explicitly as an alternative policy, was first applied against harmful indigenous methods, especially the burning of teak forest (completely prohibited in 1857). In 1808 a forestry department was created, all private use of the forest was forbidden, and the rationale of conservation was spelled out in greater detail. This was also the period when a science of forest maintenance emerged in Germany; it was not long before it came to notice in other European countries, the British Empire, and North America.

In 1830 the introduction of the so-called Culture System, a system of colonial exploitation based on compulsion, soon swept away the whole previous tradition of Dutch operations on Java, as wood and land requirements—for agriculture, especially new coffee plantations, as well as for roads and (after 1860) the railroad—suddenly shot up. This phase of unregulated, predatory cultivation by mainly private interests lasted until 1870. Between 1840 and 1870 Java lost roughly a third of its teak forest, with no thought given to reforestation. Then began another phase of conservationist reforms, involving the re-creation of a forestry department, a ban on private exploitation, and the regeneration of stands by means of tree nurseries. By 1897 the teak economy was definitively under state control; forward planning now ensured that requirements for wood were covered without inflicting the damage of the earlier period.

The example of Java shows that colonialism—in many respects, a "watershed in environmental history"[183]—could have various effects on the forest at the resource-development frontier, ranging from extreme overexploitation for short-term profit to rational planning in the interests of long-term conservation. It would be too sweeping a judgment to blame all the destruction of Indonesian

forest reserves on the colonial state. As in India or the Caribbean, it also introduced new ways of seeing and new methods of conservation.[184]

The effects of colonial rule in India were similarly ambiguous. The British
extracted wood (primarily the costliest kinds) on a large scale from the Himalayan forests; their most pressing needs were associated with shipbuilding, once
the East India Company and the Royal Navy began to farm out large orders to
Indian yards during the Napoleonic Wars. After the age of the sailing ship, deforestation in India received a fresh impetus in the 1850s from the conjunction of
railroad construction (which here as everywhere drove wide corridors through
the country), population growth, and the progressive commercialization of agriculture.[185] But while the colonial authorities countenanced and pursued "modernization," they also promoted reforestation and absorbed traditions of conservation from Indian rulers (more than from local farmers). Where colonial
representatives showed some respect for the demands of local people, as the British sometimes did in India, they had to grapple with a host of old rights of use to
the forest and to engage in protracted negotiations in search of a compromise.[186]
Protective measures were easiest to enforce where officials acting in a competitive bureaucratic framework knew how to make them seem fiscally useful in the
long term. A possible downside of conservationism, however—not only under
colonial conditions—was that communities traditionally living in and off the
forest might become objects of state intervention: "quiescent serfs of the Forest
department."[187] Analogous to forest ordinances and game laws in early modern
Europe, the conservationist measures of an environmentally aware government
created new boundaries between legality and illegality.[188] Again and again they
provoked resistance on the part of peasant communities.[189]

India illustrates a paradox of the colonial state with unusual clarity. During
the second quarter of the nineteenth century, the British built there, with the
help of expert German advisers, a forestry department and a body of legal regulations that would have no match elsewhere for decades to come. The Forest
Department designed and operated a rational system of maintenance that finally
brought the chaotic destruction of Indian forests under control. It was a model
copied all over the world, not least in England and Scotland, partly because it
proved both efficient and profitable in business terms. At the same time, however, it appeared to many Indians as an especially ugly face of the colonial state, a
ruthless alien intrusion into the lives of millions who, whether to preserve or to
clear it, had to have dealings of one kind or another with the forest.[190]

In the nineteenth century, India and Indonesia participated in a worldwide
tendency to clear forest land for monoculture plantations (tea, coffee, cotton,
rubber, bananas, etc.). The disposal of wood was of secondary concern; the main
aim was the ancient one, now fanned by capitalist forces, to extend the area of
land under cultivation. It was this powerful motive that impelled the destruction
of forest in the coastal regions of Brazil. Coffee-growing began to spread as early
as 1770, and by the 1830s coffee had replaced sugarcane as the main commercial

crop—a position it maintained until the beginning of the 1960s. It was mostly hill country that made way for the coffee shrub, but without its former protection the land underwent rapid erosion and soon had to be abandoned. This mobile economy was predicated on the belief that coffee shrubs needed the "virgin" soil of freshly cleared forest. Thus, well into the second half of the nineteenth century, coffee growing developed as a peculiar mix of modern and archaic forms of agrarian plunder: a plainly visible frontier pushing irresistibly into the interior. From the 1860s on, railroad construction made it possible to exploit highlands at some distance from the coast, while at the same time immigrants began to pour in from southern Europe and to take the place of black slaves in production. By 1900 the country had 6,000 kilometers of railroad, and the laying of track had everywhere led to major deforestation along with the advance of coffee. Cultivation methods did not change: fires continued to play a major role, often spreading out of control, and the freedom of livestock to graze on unfenced land blocked any natural regeneration of the forest. Land use in Brazil thus disregarded both the future of the forest as a resource and the long-term sustainability of farming. Often what was left behind was no more than steppe or inferior scrubland. No one had an interest in high-quality forest. It was simpler and cheaper to import ship timber from the United States or railroad sleepers from Australia.

Brazil represents an extreme example of wasteful forest use unchecked by official supervision. Unlike the colonial state, which in the best of cases aimed at long-term resource maintenance, the independent Brazilian state allowed free rein to private interests. The destruction of the Atlantic rainforest, which began in the Portuguese colonial period but really took off only under the postcolonial empire (1822–89) and the subsequent republic, was among the most savage and thorough processes of its kind anywhere in the modern world, all the worse because it was of no benefit to the economy as a whole and met with no political or scholarly opposition that might have at least slowed the work of devastation.[191]

There is not just *one* history of the European forest in the nineteenth century, if only because the whole peninsular and insular part of the continent (Iberia and Italy, Denmark and the British Isles) had little or virtually no remaining forest by the turn of the century (nor did the Netherlands as well). The other extreme was Scandinavia, especially Sweden and Finland. Here a cultural closeness to the forest, its sheer immensity in comparison with the size of the population, the ongoing incorporation of forest into the farming economy, and clearly defined government policies added up to a set of motives that have kept the Scandinavian forests in existence up to the present day. The picture was very different in England, where the Royal Navy's insatiable needs led first to extensive tree felling and then to inevitable laments about the strategic dangers of dependence on foreign sources of wood. After all, at least 2,000 fully grown oaks of the best quality were required for the construction of a single large ship of the line. Wood shortages forced the Admiralty early on (under pressure from the House of Commons) to employ iron technology; it became noticeable everywhere after 1870 that this made large ships lighter than comparable ones built with timber, and the effect was further strengthened by the replacement of iron with steel. In France too, the navy made an almost complete switch from wood to iron between 1855 and 1870. This reduced the dual pressure, from ships and railroads, to which the European forests had been subject. And at the very same moment, around 1870, the chronic crisis of British agriculture was creating new scope for land to be used for forest. Fast-growing lumber was planted again, and for the first time woodland areas supplied the recreational needs of the city population. The little that remained of the English forest now received conservationist attentions.[192]

Sometimes these particular histories were closely interconnected. Commercially, for example, Napoleon's Continental Blockade of 1807 had the effect of diverting British timber interests from the Baltic and Russia toward Canada, so that by the 1840s the province of New Brunswick alone was exporting 200,000 tons a year to Europe.[193] The late nineteenth century saw the emergence of a veritable global wood market, boosted by mass-circulation dailies hungry for newsprint. Meanwhile, the transfer and "acclimatization" of tree species continued on a larger scale than in the eighteenth century. Whereas 110 tree species were introduced into Britain before 1800, the figure for the next hundred years was over 200. Nevertheless, although local histories can and must be linked together, it cannot easily be argued that they add up to an overarching history of unrelieved environmental degradation. Deforestation did not always continue until the last tree had been felled. In many countries, it was up against the logic of energy use and a rudimentary conservationism whose motives might vary from Romantic nature worship to a sober appreciation of the effects of unchecked exploitation. It would be mistaken to imagine that industrialization continually displaced the wood economy as part of an archaic "primary sector." But of course it originally raised wood consumption in the form of charcoal for early steam engines and ironworks, both in economies such as Japan's that were starved of fuel resources and in areas such as Pennsylvania and Ohio where cheap wood was plentiful and where charcoal long remained an energy input for heavy industry.

A further major source of demand was private heating. A warm home soon came to be taken for granted as an accompaniment of material progress. In 1860 wood was still the most important fuel in the United States (80 percent), to be overtaken by coal only in the 1880s.[194] Even where industrialization had little impact on the economy in general, the transport industry devoured huge quantities of wood in the form of railroad sleepers—in India, for example, where the material first had to be procured from distant places. Early locomotives were driven by firewood, to the tune of 80 percent in the 1860s in India; a changeover to coal became evident only around the turn of the century.[195] In Canada's "modern" economy, and even in the United States, the timber business (which included large sawmills) continued to be one of the sectors with the highest creation of value added. Some of the world's largest fortunes were made out of wood.

Let us now finally consider another kind of ecological frontier, one that resulted not so much from man-made destruction as from gradual climate change. For this we must turn to the Sahel zone, a desert frontier roughly three hundred kilometers wide, which stretches along the southern edge of the Sahara. Life there was affected by increasingly arid conditions from the beginning of the seventeenth century on. Livestock breeding was pushed ever farther south, while the camel, capable of surviving for eight to ten days without water or grass and of moving firmly across sand, assumed greater importance. By the mid-nineteenth century a Great Camel Zone had come into being, extending from the Maghreb to the Adrar Plateau in present-day Mauritania. Growing aridity also imposed new patterns of transhumance within the southward-moving livestock zone, where a mixed economy of cattle, goats, and camels prevailed. These conditions gave rise to a desert frontier, in which Arabs, Berbers, and also black Africans coexisted with one another and took on a kind of "white" identity distinct from that of blacks farther to the south. The lifestyles of nomadic pastoralism and settled agriculture became ever more sharply defined. They also expressed themselves in differential mobility: camel and horse riders could easily make raids against which black communities or villages had little or no defense. Complex tributary relationships stretched across borders in both directions: the dependence of southern farmers being all the greater, the less the "whites" had to do with agricultural production in their own sphere. In the end, however, many common features at the level of social hierarchy—above all, a clear division between warriors and priests or into caste groups—bound the frontier zone together. Islam spread throughout the Sahel zone, by means both martial and peaceful, creating exceptionally deep roots for the slavery it brought from the North. The remnants of slavery in Mauritania in the second half of the twentieth century are clear evidence of this.[196]

Big-Game Hunting

Another ecological variant is the game frontier. In the nineteenth century the world was still full of human communities who lived from hunting, not only in the American Midwest but also in the Arctic, in Siberia, and in the rainforests of Amazonia and Central Africa.[197] At the same time, Europeans and Euro-Americans discovered new dimensions of the old pursuit. What had once been an aristocratic privilege and a training of belligerent masculinity became *embourgeoisé* in the thoroughly middle-class societies of the New World, as well as in parts of Europe where the bourgeoisie sought and found a way of linking up with the lifestyle of the nobility. The hunt served as a symbolic setting for status convergence. A nobleman hunted, although not everyone who took it up as a hobby thereby became a nobleman; it was a favorite subject for satirists.

A new aspect was the assault on exotic big game, the largest and most organized since the bloodbaths in the arenas of the Roman Empire, which for an unconventional commentator such as Lewis Mumford made Roman civilization especially repugnant.[198] In Africa, Southeast Asia, and Siberia, early travel reports expressed wonder at the paradisiacal abundance of the large fauna, but all that changed as soon as a struggle for "civilization" got under way against the beasts of the wild. In the name of upholding the colonial order, for which a figure such as the tiger could only be a rebel both real and symbolic, wild animals were killed and abducted on a grand scale to satisfy the curiosity of visitors to menageries and circuses in the capitals of the North and to provide spectacles for the greater prestige of their rulers. The technical prerequisite for this was the dissemination of the rifle, which made it possible for Asians and Africans to imitate the exterminatory practices of Europeans. The profession of big-game hunter appeared only after the repeater rifle became widely available, since this reduced the likelihood of having to face a bellicose tiger or elephant with one's last bullet spent.

In many Asian societies big game hunting had been a royal prerogative, but now, in keeping with the European model, lower ranks of the aristocracy began to join in. In India, the tiger hunt served to cement the British alliance with native princes that was essential to the stability of the Raj. A maharajah and a high official of the colonial government might have little to say to each other, but they could always find common ground in the hunter's lifestyle. European penchants often had a trickle-down effect. In the early twentieth century the sultan of Johor—a prince in the hinterland of Singapore dependent on the British—was considered a great tiger hunter: thirty-five stuffed trophies were on display in his palace. But he was not following in the footsteps of any ancestors; there was no such tradition. The sultan, for reasons of prestige, simply copied the behavior of Indian maharajahs, who in turn imitated the British rulers.

Villagers, too, had no tradition of ferocity toward wild animals. Of course, an ingenuous harmony had never prevailed between the two. Tigers were capable of terrorizing whole districts; and villages would be abandoned if the livestock (their most valuable possession) could no longer be protected, if the gathering of fruit and firewood (a task for young girls and old women) became impossible, or if an excessive number of children fell into the clutches of wild beasts. There are harrowing stories about such things, but the water buffalo who defends a child from the tiger is also a popular literary theme. Some regions could be crossed only at great peril. People undertaking such a journey often positioned an old horse at the rear of their column, as a sacrifice to a stalking predator. In West Sumatra, as late as 1911, a tiger attacked a mail coach and dragged its driver into the jungle.[199]

The tiger hunt was not only a luxury but often an actual necessity, having existed since before the arrival of European colonizers. In many cases it mobilized whole villages, under the leadership of an elder or a low-ranking colonial official, for a full-scale punitive expedition. Especially on Java the tiger was straightforwardly defined as a military enemy, liable for revenge and annihilation; Muslim Javanese knew no bounds in this, since their monotheistic religion excluded any superstitious notion that a spirit (good or evil) dwelled within the tiger.

Nevertheless, the idea that tigers should be wiped out seems to have remained fairly uncommon. There was a tendency to leave "innocent" ones in peace, and in general the non-Muslim population of Asia—as well as Muslims marked by popular culture—had an awkward feeling when they went after a tiger. Often they asked pardon of the slain beast, even blaming themselves for its (practically necessary) killing as if it were a case of regicide, or else it would be honored like a fallen war chief on the village square, with dancing and weapon play.[200] The European custom of the hunting gallery, arranged according to a hierarchy of the animal kingdom, or the use of distinctive horn signals for different species, shows a certain mental affinity with such practices.

The dead tiger was scarcely ever sold on the market until the early twentieth century, and, although it is reported that tiger meat was a delicacy among the Javanese aristocracy, ordinary people never ate it. At least in Southeast Asia there is almost no evidence that the animal was killed for its skin, which had no particular value attached to it. To decorate houses with tiger skins was unusual even among the nobility. The hunting trophy seems to have been invented in Europe, where it sometimes degenerated into a bedspread. In the early twentieth century there was a substantial tourist demand in Indian port cities for animal skins or even stuffed bodies. Traders and taxidermists often ordered a supply from native hunters. Tiger remains were especially sought after in the United States.[201]

Some hunters specialized in acquiring big cats for European or North American zoos and circuses. The first modern zoo opened in London in 1828; Berlin followed in 1844 (with the addition of a large predator house in 1865), and there were zoos in the United States after 1890. They were supplied by a small number of internationally linked dealers. Johann Hagenbeck, the half-brother of the Hamburg dealer and circus pioneer Carl Hagenbeck who eventually opened his own zoo in 1907, set himself up in 1885 as an animal procurer in Ceylon, buying specimens from local people and undertaking expeditions of his own to India, the Malay Peninsula, and Indonesia. Such people did, of course, employ methods that were somewhat more merciful than those of other hunters, but the effect was the same: a decline in the animal population. The business itself was risky; many animals did not survive the trip. But huge markups amply compensated for this. In the 1870s a rhinoceros purchased in East Africa for 160 to 400 German marks could be sold in Europe for 6,000 to 12,000 marks. By 1887 the Hagenbeck company had traded more than 1,000 lions and 300 to 400 tigers.[202]

The tiger was the most spectacular victim of deforestation and the hunting passion imported from Europe. Specialists in India, Siberia, or Sumatra might shoot 200 or more in the course of a career; the king of Nepal and his hunting guests totted up a combined score of 433 between 1933 and 1940.[203] After timid beginnings in the colonial period, the effective protection of tigers began only after 1947 in the Republic of India. Elephants gained legal protection earlier, in 1873 in Ceylon, and the times when a single hunter could claim to have killed 1,300 specimens did not last. The deployment of elephants as working animals does not seem to have promoted the biological stability of the species in Asia. On the other hand, the colonial authorities ended their use in warfare—traditionally a cause of major losses.

In the nineteenth century, some hunting was big business in the world economy. This was not entirely new. The fur trade, by no means a totally "premodern" sector, had been spanning continents since the seventeenth century, and in 1808 Johann Jacob Astor founded his American Fur Company, soon to become the largest of any kind in the United States. The advance of a commercial hunting frontier had an especially deleterious effect on the African elephant. In the Boer Republic of Transvaal, until the gold and diamond boom, ivory was by far the most important export item. Elephants were slaughtered en masse to keep Europe supplied with knife handles, billiard balls, and piano keys. In the 1860s alone, Britain imported 550 tons of ivory a year from all parts of (not yet colonized) Africa and India; exports from Africa peaked between 1870 and 1890 year, at the height of the rivalry among colonial powers to grab territorial possessions. In those years, 60,000 to 70,000 elephants were killed per annum. In 1900 Europe still imported 380 tons of ivory, representing the "yield" from approximately 40,000 elephants otherwise of no commercial value.[204] After the elephant population slumped in a number of colonies, resulting in the first timid measures (in the British Empire) to protect it, the Belgian-ruled Congo Free State remained the last source of tusks—not only a place of extreme human exploitation but also a gigantic cemetery for elephants. Between the beginning of the nineteenth century and the middle of the twentieth, the regal animal disappeared from large parts of Africa, from the northern savannah belt as well as Ethiopia and the entire South. Until after the First World War, more elephants were killed in Africa than were born. Only in the period between the wars did something like an effective species protection strategy get off the ground.

Similar stories could be told about many other animals. The nineteenth century was for all of them—as it was for the North American bison—an age of defenselessness and mass slaughter. The rhinoceros was seen as a special challenge by European big-game hunters. But until very recently it was demand in Asia rather than Europe that proved its undoing, since both the Muslim East and the Far East valued the substance of its horn and were prepared to pay astronomical prices to obtain it. The popularity of ostrich-feather hats meant that this African wild bird began to be raised on farms; this at least saved it from extinction. The pattern was the same all over the world: ruthless violence against wild animals in the nineteenth century, then a gradual change of mind among early ecologists, followed by British colonial bureaucrats. In the perspective of human history, the twentieth century is rightly considered the century of violence. From the point of view of tigers and leopards, elephants and eagles, it looks rather more favorable—as the age when humans tried to reach a modus vivendi with creatures that for millennia in the past, before the invention of firearms, they had faced in a relationship of approximately equal chances.

Naturally there were other reasons for hunting, apart from the pursuit of profit. Big-game hunters became cultural heroes. The ability to tackle a grizzly bear in the wild seemed to concentrate the highest qualities of the North American character. Around the turn of the century, President Theodore Roosevelt went to great trouble to present himself as its embodiment; big-game hunts for the benefit of the media took him all the way to Kilimanjaro. Gentlemen hunted, but settlers could profit from their natural surroundings and were nearly always farmers and hunters rolled into one. At least in the early nineteenth century, large predators were still so common in all of the world's settler zones that pioneers were well advised to protect their property.[205]

Moby-Dick

The fishing of cod or herring was more like sea harvesting than a clever stalking operation, but whaling was one maritime labor in the nineteenth century that did not lack the character of the hunt. One of the epic feats of the age, it was also a kind of industry. The Basques had hunted whales as far back as the Middle Ages, honing special techniques that the Dutch and English adopted in the seventeenth century. By the beginning of the nineteenth century the seas near Spitzbergen were so empty of aquatic fauna that whaling had become unprofitable there, and so the attention shifted to Greenland.[206] As for the North Americans, they entered the fray in 1715 from the port of Nantucket in Massachusetts, concentrating at first on the great sperm whale in the Atlantic. In 1798 American whalers appeared for the first time in the Pacific, and over the next three decades they pushed into nearly every important whaling ground in the world.[207] Whaling reached its peak internationally between 1820 and 1860, with the United States as the leading nation after the War of 1812. By 1846 the US whaling fleet, mostly based in New England ports that vigorously competed with one another for precedence, consisted of no fewer than 722 vessels. Half of these hunted the great sperm whale, whose blubber (*spermaceti*) inside its giant head was needed to produce oil for the world's best and most expensive candles.

Whaling was a global business, with a complex geography and chronology determined by, among other things, the large number of whale species. In the South Seas, sperm whaling grounds were found off the coast of Chile, where the great white whale Mocha Dick (the inspiration for Herman Melville's literary monster) sowed terror in the years around 1810.[208] At that time, international whaling centered on a stretch of ocean between Chile and New Zealand and in the seas near Hawaii. The discovery of new grounds triggered "oil wars," reminiscent of the Californian or Australian gold rush, between individual ships or whole fleets. Australia was especially successful for a time in 1830.[209] In the western Arctic (Alaska, Bering Straits, etc.) in 1848, the location of the now almost vanished Greenland whale was one of the most important finds of the century, since no other species produces such high-quality whalebone. It led to the commercial entry of the United States into the maritime North, mainly from New

Bedford, Massachusetts (Nantucket's rival), and the backup port of San Francisco; the American territorial interest in Alaska would scarcely have developed without this background. A turning point came in 1871, when the greater part of the US Arctic whaling fleet was lost in pack ice.[210] At the same time, the main grounds were already approaching exhaustion, and the 1870s were generally a decade of crisis for American whalers. Temporary relief—though not for the whales—was provided on the demand side by the new wasp's tail ideal of feminine beauty, which popularized elastic corsets stiffened with whalebone stays. This made it worthwhile to sail even farther out to sea.[211]

Whaling was not an Anglo-American specialism. New Englanders did certainly hunt in the South Pacific, to keep Parisian ladies supplied with candles and girdles. But until the late 1860s Frenchmen pitched in too, operating mainly from the port of Le Havre. Their hunting grounds stretched as far as Australia, Tasmania, and New Zealand—regions where as late as the 1840s whalers lying at anchor were sometimes set upon and killed by the locals. Nor was that the only danger. Between 1817 and 1868, French whaling expeditions ended nearly 6 percent of cases with the loss of a ship, mostly during a storm, yet caught no more than 12,000 to 13,000 whales (a fairly modest total if one considers that, before the Second World War, 50,000 whales a year were being slaughtered).[212]

The Moby-Dick age of the duel between man and whale, in which the animal opponent still stood at least a minimal chance, ended with the introduction of harpoon guns and rockets. By the 1880s open-boat lancing was a thing of the past. Kept up by only a few romantics, it had become especially difficult because the clever sperm whale avoided coming too close to the boat. The Norwegian Svend Foyn ushered in this post-Ahab whaling era in 1860, when he invented the onboard harpoon gun capable of firing 104-millimeter shots that exploded in the body of the whale—more an artillery weapon than a hunting device.[213] The steamships deployed after 1880, though initially doubling construction costs, added a further element to the unequal contest. But even from the whaler's point of view, the new killing techniques were a dubious advance, since numerous grounds were totally depleted by 1900.[214] Many whale species were close to extinction, while others had withdrawn to more remote parts of the oceans. In any case, new plant and fossil oils had come into use, making obsolete the demand for many whale products. (As early as 1858, farsighted people in the New Bedford whaling business had founded a factory for the distillation of petroleum.[215]) How whaling soon recovered from this trough is another story.

The only non-Western nation that pursued whaling independently of Western influences was Japan. Activity began there at more or less the same time as in the Atlantic, and toward the end of the sixteenth century many coastal villages were literally feeding themselves off it. From the late seventeenth century, there was a switch from harpooning to the method of catching whales (mostly of the smaller and faster species) in large nets off the side of boats. The processing of the whale, none of which went to waste, took place ashore rather than onboard (as it did in the United States). After American and British whalers in 1820 discovered rich hunting grounds between Hawaii and Japan, hundreds of Japanese whalers were soon putting out to sea, and in 1823 it was reported that Japanese officials had boarded a foreign whaling ship. In 1841, in a case that became known all over the country, a shipwrecked fisherman's son, Nakahama Manjirō, was rescued by an American whaling ship; the captain took the boy home with him and looked after his education. This first Japanese student in the United States excelled at college, specializing in navigation and eventually (in 1848) becoming an officer on a whaling ship. After various adventures, homesickness led him back in 1851 to Japan, where the authorities, eagerly taking the rare opportunity to learn more about the outside world, questioned him for months on end. Nakahama became a teacher at the clan school in Tosa, and some of his students would later become leaders of the Meiji Renewal. In 1854 the shogun used him as a translator in the negotiations with Commodore Perry, the commander of the American flotilla that "opened up" Japan. Nakahama also translated a number of foreign books on navigation, astronomy, and shipbuilding and acted as a government adviser for the construction of a modern Japanese navy.[216]

The rapid expansion of whaling was a key element in the opening of Japan in 1853–54, after centuries of self-imposed isolation. The US government was eager to protect American whalers stranded there from official sanctions, as well as to provide for the bunkering of its ships in Japanese waters.[217] The Japanese were among the first to adopt Svend Foyn's unsporting methods of shooting whales, but it was the Russians, not the Americans or Norwegians, who brought them to the attention of people in Japan. This, too, would eventually have foreign policy implications, since it was only Japan's victory in the 1905 war with Russia that drove this major rival out of its territorial waters and handed Japanese whalers a monopoly in the grounds between Taiwan in the South and Sakhalin in the North.[218]

Herman Melville's *Moby-Dick* (1851), one of the greatest novels of the nineteenth century, rooted the world of whaling deep in the minds of Western readers at the time, and even deeper in those of posterity. The work contains long, exhaustively detailed passages about whales; Melville knew them inside out. Having spent four years on whalers as a young man, he had firsthand experience of their social world, and there were real-life models for the White Whale, Captain Ahab, and various whaling tragedies. The most famous case, which Melville had closely studied, was that of the Nantucket-based *Essex* that was rammed and sunk by a raging sperm whale on November 20, 1820, thousands of miles from home in the South Pacific. Twenty crew members managed to escape in three small boats, and eight of them survived for ninety long days by eating the flesh of seven comrades. In 1980 a newly found report by one of the men in question confirmed and supplemented the eye-witness account of Owen Chase that Melville had used in *Moby Dick*.[219] The drama occurred four years after a similar incident of cannibalism, when only fifteen men had survived out of the 149 shipwrecked off West Africa from the French frigate *Méduse*, immortalized by the painter Theodore Géricault who finished his famous *The Raft of the Medusa* only three years after the event.

Faust: Land Reclamation

If whaling and deep-sea fishing involve an aggressive relationship to the ocean and its animal inhabitants, as well as representing a maritime way of life centered on fish and whales, the opposite extreme of a defensive attitude to the sea may be found in land reclamation projects. The taming of great rivers, such as the Upper Rhine beginning in 1818[220] or the Mississippi a century later, was spectacular enough. A source of even greater fascination was the "Faustian" project of wresting land from the sea for permanent settlement. It caught the attention of one of the world's greatest poets. Goethe, who had already made a study of hydraulic engineering in Venice in 1786, kept himself *au fait* with the Bremen port works in 1826–29 and turned the aged Faust into a land reclamation entrepreneur on a grand scale:

Kluger Herren kühne Knechte
Gruben Gräben, dämmten ein,
Schmälerten des Meeres Rechte
Herrn an seiner Statt zu sein.

Clever Lords set their bold servants
Digging ditches, building dikes,
To gain the mastery of ocean,
Diminishing its natural rights.[221]

The poet also saw that such projects required the sacrifice of workers' lives ("Human blood was forced to flow, / At night rose the sound of pain"). Dike construction, the draining of swamps, and the digging of canals were among the harshest exertions of the early modern period, usually organized by government departments and often performed by armies of convicts or prisoners of war (Turks, for example, in some of the German lands). The twentieth century had a special passion for dam construction and drained as much as a sixth of the wetlands on the earth's surface.[222] It also witnessed a continuation of major coastal projects, such as the land reclamation in Tokyo Bay (begun in 1870) and the mouth of the Yangtze, as well as the enclosing of the Zuiderzee, planned in 1890 but tackled only after 1920, which would eventually expand the territory of the Netherlands by more than tenth.

In the nineteenth century too, people were active in many parts of the world on this ecological frontier. In France, for instance, all of the major fens had been drained and converted to pasture by 1860—a prerequisite for the rise in meat consumption as its society grew more prosperous. Flood defenses and land reclamation remained an existential necessity, especially in the case of the Netherlands, where drainage had been organized since the Middle Ages and a protection system had been in place since the early sixteenth century. Here farmers were required to pay taxes, not to perform labor services. This promoted the commercialization of agriculture, while also helping to form a mobile proletariat of dike laborers. The decisive technological advances dated back to the sixteenth, not the nineteenth, century; the high point of drainage activity between 1610 and 1640 was rarely surpassed. Between 1500 and 1815 a total of 250,000 hectares were obtained in the Netherlands—one-third of the land under cultivation.[223] Windmill improvements raised pumping efficiency. Whereas efforts in the eighteenth century focused on controlling the flow of the Rhine and the Waal, the nineteenth century saw a new burst of land reclamation. All in all, 350,000 hectares were brought under cultivation between 1833 and 1911, of which 100,000 hectares were gained through dike construction and drainage.[224] In 1825, as a result of devastating floods, coastal defenses and the upkeep of dikes gained priority for the first time over land reclamation.[225] Another novelty was that—as in China for the past two thousand years—hydraulic engineering became a central government responsibility, instead of being left to provincial authorities and private individuals.

The main project in the nineteenth century was the draining of the 18,000-hectare Haarlemmermeer between 1836 and 1852. This low-lying lake, in the middle of Holland province, had taken shape during the storm floods of autumn 1836 and wreaked havoc with the road system, in particular the technically advanced interurban *straatwegen* (made of brick and natural stone) of which the Dutch were especially proud. There were also fears that the ever-expanding Haarlemmermeer would endanger Amsterdam and Leiden, while a new concern with economic policy focused on the employment effects. The drainage was organized along modern lines that are still customary in infrastructural projects. Precise scientific calculations preceded and accompanied the work; legal experts were hired to reconcile the numerous interests of people living on the shores of the lake. The project was put out to tender and entrusted to private firms. The laborers, known as *polderjongens*, worked in teams of eight to twelve under an overseer. Most of them were single, but some brought along their family and lived with them in reed and straw crofts near the construction site. In summer, at the height of activity, several thousand workers would be employed at the same time. Like other projects on this scale, it did not fail to pose health risks as well as problems to do with crime and the supply of drinking water. From 1848 on, the project was able to employ British steam pumps and three large pumping stations—another example of the diverse application of steam engines outside industrial production.[226] By 1852 the Haarlemmermeer was dried out and could be slowly converted into farmland. Today's Schiphol Airport lies on part of this reclaimed land.[227]

All frontiers have an ecological dimension. They are both social and natural spaces. This does not mean that social relations should be naturalized frontier-style: the expulsion of hunting peoples is something different from the evacuation of seawater; nomads and steppe are not indistinguishable elements of one and the same "wilderness."[228] However, the rollback of steppe, desert, or rainforest always entails that habitat is destroyed and that the people living there lose their livelihood. The nineteenth century was the period in world history when resource development reached its maximum extent, and when frontiers acquired a social and even political significance they have never had before or since. In today's zones of rainforest destruction, or in outer space, no new societies are being formed as they were in the nineteenth-century United States, Argentina, Australia, or Kazakhstan. Many frontiers—not only the one in the United States—were "closed" around 1930. Often they had originated in the early modern period, but it was the nineteenth century that founded a new era of mass migration, settler economies, capitalism, and colonial warfare. Many frontiers had a "posthistory" in the twentieth century, as we can see in the state-organized colonial subjugation of "living space" between 1930 and 1945, or in the giant social and environmental projects conducted under the banner of socialism, or in the politically driven expansion of the Han Chinese, who in the late twentieth century turned Tibetans into a minority in their own land.

Frontiers were many things in the nineteenth century: spaces of cultivation and increased production, magnets for migration, disputed zones where empires came into contact with one another, focal points in the formation of classes, spheres of ethnic conflict and violence, birthplaces of settler democracy and racial domination, breeding grounds of phantasms and ideologies. For a time frontiers became major foci of historical dynamics. Only a conception of the epoch narrowly centered on industrialization will limit this dynamic to the factories and furnaces of Manchester, Essen, or Pittsburgh. As far as its consequences are concerned, we should not overlook an important distinction. Industrial workers in Europe, the United States, and Japan became increasingly integrated into society, creating organizations to represent their interests and improving their material lot from generation to generation. But the victims of frontier expansion were excluded, dispossessed, and disenfranchised. Only in recent years have courts in the United States, Australia, New Zealand, Canada and a few other countries begun to recognize many of their legal claims, while governments have accepted moral responsibility and apologized for past misdeeds.

CHAPTER VIII

Imperial Systems and Nation-States

The Persistence of Empires

1 Great-Power Politics and Imperial Expansion

All the chapters in this book have something to say on empire and colonialism. That aspect of the nineteenth century is omnipresent, as it has to be in any attempt to employ a world-historical perspective. Thus, there is no need to provide a comprehensive overview of the various empires and to cover the standard topics of imperial history.[1] Nor is it necessary to join the debate about the peculiar position of the nineteenth century in the long sweep of global power politics and economic dynamism, a debate that leads invariably to a probing of the roots and causes of the "great divergence" that made Europe and the United States—usually bracketed as "the West"—for some time the masters of the world. How this "virtuous circle of incessant growth" (John Darwin)[2] of wealth and power came about and how it is connected to empire has intrigued the greatest minds for the better part of two centuries. Recent attempts to solve this mystery of mysteries, formerly labeled "the rise of the West," have been made (among others) by Daron Acemoglu, Robert C. Allen, John Darwin, Jared Diamond, Niall Ferguson. Jack A. Goldstone, David S. Landes, Ian Morris, Prasannan Parthasarathi, Kenneth Pomeranz, and Jeffrey G. Williamson; the debate has been monitored by supreme critical spirits such as Patrick K. O'Brien or Peer Vries. In spite of all these efforts and a long tradition of reflection on the "European miracle" from Adam Smith via Karl Marx and Max Weber to Immanuel Wallerstein, E. L. Jones, and Douglass C. North, agreement is nowhere on the horizon, and even basic methodological issues—do all those great historians and social scientists address the same questions and do they agree on a strategy and logic of explanation?—remain to be solved. In this bewildering situation, the present essay sets itself a decidedly more modest task: it sees empire as a special type of *polity*[3] and as a framework for social life and individual experience, and it simply argues that the nineteenth century was much more an age of empire than, as many European historians continue to believe and to teach, an age of nations and nation-states.

In the nineteenth century, empires and nation-states were the largest political units in which human beings led a common existence. By 1900 they were also the only ones with real weight in the world: nearly everybody lived under the rule of one or the other. There was no sign yet of world government or of supranational regulatory institutions. Only deep in rainforests, steppes, or polar regions did small ethnic groups live without paying tribute to a higher authority. Autonomous city-states no longer played any role: Venice, for centuries the epitome of a civic community well capable of defending itself, had lost its independence in 1797; the Republic of Geneva, after an interlude under French rule (1798–1813), had joined the Swiss Confederation in 1815 as a yet another canton.[4] Empires and nation-states provided the framework for the life of society. Only the communities of a few "world" religions—the Societas Christiana or the Muslim *umma*—had an even wider scope, but no political entity of similar extent corresponded to them. Empires and nation-states also had a second side to them. They were players on the special stage of "international relations."

Driving Forces of International Politics

International politics is essentially about questions of war and peace. Until the state-organized mass murders of the twentieth century, war was the worst of man-made evils; its avoidance was therefore especially valued. Although the fame of conquerors might be more dazzling for a time, all civilizations have—at least in retrospect—thought more highly of rulers who created and preserved peace. Those who both won an empire and subsequently brought peace to it have enjoyed the highest esteem of all: Augustus or the Kangxi Emperor, for example. Like the apocalyptic horsemen that bring pestilence and famine, war attacks a society as a whole. Peace—the inconspicuous absence of war—is the basic prerequisite for civil life and material existence. Hence international politics is never an isolated sphere: it has a close interrelationship with all other aspects of reality. War is never without implications for economics, culture, or the environment, and other dramatic moments in history are usually associated with it. Revolutions often arise out of war (as in seventeenth-century England, the Paris Commune of 1871, or the Russian revolutions of 1905 and 1917) or flow into it (like the French Revolution of 1789). Only a few revolutions, such as those of 1989–91 in the Soviet sphere of hegemony, remained free of military consequences,[5] although the events of 1989–91 had indirect military causes too (the arms race of the "Cold War," about which no one could ever be sure that it would not escalate into a hot confrontation).

This multiple interweaving with the life of society should not make us forget, however, that in modern Europe international politics has partly followed a logic of its own. There have been specialists in interstate relations ever since the emergence of (European) diplomacy in Renaissance Italy, and their thinking and values—for example, concepts of reasons of state, dynastic or national interests, or the prestige and honor of a ruler or state—have often been alien to the ordinary subject or citizen. They constitute distinctive "codes," rhetorics, and sets of rules. And it is precisely this ambiguity of autonomy *plus* social insertion which makes international politics such an intellectually appealing field for historians.

The nineteenth century saw the birth of international relations as we know it today. This has become especially apparent in recent years, because the end of the "bipolar" nuclear stand-off between the United States and the Soviet Union brought to the fore many patterns of warfare and international behavior that remind one of the period before the Cold War or even the two world wars. But there is a major difference. Since 1945 it has no longer been self-evident that states wage war in order to impose their political objectives. By international agreement, offensive war has lost its legitimacy as a means of politics. The capacity to engage in it is no longer considered—as it still was in the nineteenth century—to be a proof of modernity, if we leave aside the symbolic importance of nuclear weapons for certain countries in Asia today. Five major nineteenth-century trends are identifiable.

First. The American War of Independence (1775–81) represented a transitional form between the old duel led by officer castes and the role of patriotic militias. But it was the wars accompanying the French Revolution that established the principle of arming the people. The starting point was the decree of the National Convention on the *levée en masse* (August 23, 1793), which, coming after a four-year preparatory period, made all Frenchmen subject to permanent conscription.[6] The nineteenth century would be the first age in which mass armies were conceivable, and constant improvements soon appeared in their organization. Compulsory military service was introduced at various times in Europe (in Britain only in 1916), and there were wide variations in its practical effect and public acceptance. If, after the fall of the Napoleonic Empire in 1815, such armies were seldom deployed over the next hundred years in international wars, the reasons were not only countervailing forces such as deterrence, the balance of power, and rational circumspection, but also the rulers' fear of the uncontrollable tiger of an armed people. Nevertheless, the instrument of the conscript army now existed. Especially where the armed forces were seen as an embodiment of the national will, not merely as a tool of the government, a new kind of war became a latent factor that could always be deployed.

Second. In the nineteenth century it is possible to speak for the first time of an *international* politics that sets aside dynastic considerations and obeys an abstract concept of raison d'état. It presupposes that the normal unit of political and military action is not a princely ruler's arbitrary patrimonium but a state that defines and defends its own borders, with an institutional existence not dependent on any particular leadership personnel. This is, again in theory, a nation-state. But it is a special kind of state organization, which first emerged in the nineteenth century and began to spread hesitantly and unevenly around the world. International politics in the nineteenth century was acted out between "powers" organized partly as nation-states, partly as empires. Practice conformed most closely to this model after other players had quit the stage: pirates and partisans, semiprivate military operators and warlords, transnational churches, multinational corporations, cross-border lobbies, and all other forces on a medium level of activity, forces that can be understood by the term *communauté intermédiaire*.[7] Parliaments and democratic public opinion muddied the waters in new and unpredictable ways, and "foreign policy experts" went to great pains to restrict their influence. In this sense, the period from 1815 to the 1880s was the classical age of craftsmanship in interstate affairs, shielded to a higher degree than before or after from other intervening factors, and largely in the professional (though not necessarily capable) hands of diplomats and military men.[8] This by no means ruled out populist actions for public effect; we find them even in a traditionalist-authoritarian system like the Tsarist Empire.[9] The discovery that public opinion was not merely a pliant sounding board for official foreign policy but one of its driving and elementary forces pointed beyond the nineteenth-century understanding of politics. An early and dramatic example was the Spanish-American War of 1898, in which a jingoist mass-circulation press egged on the initially reluctant President William McKinley to confront the forces of (a by no means innocent) Spain.[10]

Third. The development of technology gave the new-style nation-state a destructive capacity previously unknown in history. Crucial innovations were the advanced bolt-action rifle, the machine gun, more powerful artillery and chemical explosives, the iron-hulled warship, new forms of engine-propelled locomotion (the submarine became technically feasible shortly before the First World War), troop trains, and signal systems that replaced dispatch riders, semaphores, and light telegraphy with electrical telegraphy, telephony, and eventually the radio.[11] Technology as such does not breed violence, but the effects of violence do increase as a result of it. Until the second half of the twentieth century, when ABC (atomic, biological, and chemical) weapons raised the threshold of horror, every military invention was applauded by the apostles of progress and actually employed in war.

Fourth. At the latest by the closing third of the nineteenth century, these new instruments of power were directly related to industrial capacity. The widening economic disparity between countries went hand in hand with the gap in military technology. A country like the Netherlands, for example, lacking an industrial base of its own, could no longer claim the international supremacy it had once enjoyed as a maritime power. A new kind of great power came into existence, defined not so much by population size, maritime presence, or potential revenue as by its industrial production and its capacity to organize and finance an arms drive. In 1890, before it began to strike out overseas, the United States had a troop strength of no more than 39,000, yet its position as the leading industrial power assured it of as much international respect as Russia enjoyed with an army seventeen times larger.[12] Size still mattered—more than in the post-1945 "nuclear age"—but it was no longer the key criterion for success. Outside Europe, the Japanese elite soon appreciated this once it set its sights after 1868 on making Japan both "rich and strong"; it was to be an industrial country with a military capability, which in the 1930s would develop into an industrialized military state. Over little more than a hundred years—from the 1870s until the arms race of the 1980s that crippled the USSR—industrial might was the factor of decisive significance for world politics. Since then, terrorism and guerrilla warfare (the old weapon of the weak) have again reduced its importance; nuclear weapons are now in the hands of industrial midgets such as Pakistan or Israel but not of substantial industrial nations such as Japan, Germany, or Canada.

Fifth. The European system of states, created essentially in the seventeenth century, expanded in the nineteenth into a global system. This happened both through the rise of the United States and Japan as great powers and through the forcible incorporation of large parts of the world into the European empires. The two processes were closely bound up with each other. The colonial empires were a transitional form on the way to a mature international community of states. It can be argued whether they speeded up the transition or slowed it down, but in any event the global plurality of the international system was still in a kind of imperial latency before the First World War. Only later in the twentieth century did the present-day system take shape in two distinct stages: the creation of the League of Nations immediately after the First World War, which made it possible for countries such as China, South Africa, Iran, Siam/Thailand, and the Latin American republics to establish permanent, institutionalized contact with the Great Powers; and the decolonization that occurred during the two decades following the Second World War. Imperialism, it is now recognized, became the opposite of what its protagonists had sought to bring about—that is, the great realigner of political relations in the world, and hence the midwife of a postimperial international order, albeit one still burdened in many ways with an imperial legacy.

Narrative I: Rise and Fall of the European System of States

In history textbooks dealing with the nineteenth century, one finds two master narratives that are nearly always kept separate from each other: a history of great-power diplomacy in Europe and a history of imperial expansion. Generations of historians have worked on each. An initial, highly simplified overview might summarize them as follows.

The first story tells of the rise and fall of the European system of states.[13] It could open with the Peace of Westphalia in 1648, or with the Treaty of Utrecht in 1713, but it is sufficient to begin in 1760. The dispute at the time concerned which countries were and which were not the European "Great Powers." Older hegemons such as Spain and the Netherlands, large but weakly organized territories such as Poland-Lithuania, and temporarily hyperactive but middle-ranking military powers such as Sweden were unable to maintain their position. The rise of Russia and Prussia sealed the formation of a "pentarchy" of five Great Powers: France, Britain, Austria, Russia, and Prussia.[14] After the Treaty of Karlowitz (1699) it was not necessary to reckon with external pressure from the Ottoman Empire, an aggressive and once even superior adversary. Special mechanisms of unstable equilibrium now took shape within the five power constellations, based on the principle of the egoism of individual states. There were no overarching visions of peace, and in case of doubt a smaller country could be sacrificed (as Poland was more than once to its larger neighbors). The attempt of postrevolutionary France, under Napoleon, to change this balance of power into a continental empire exercising hegemony over its neighbors collapsed in October 1813 on the battlefields near Leipzig. Until 1939 no country would risk another such grab for supremacy (if we leave aside certain German extremists in the First World War). The pentarchy was restored at the Congress of Vienna in 1814–15, with respect for France despite its two defeats (one in 1814, one in 1815 after Napoleon's return from Elba), but now the political elites shared a common will to secure peace and to avoid revolution. The system was stabilized and reinforced by a set of explicit rules, basic consultative mechanisms, and a conscious, socially conservative aversion from the new techniques of military mass mobilization. In a considerable advance over the eighteenth century, this new order preserved the European peace for several decades. It was shaken, though not entirely annulled, by the revolutions of 1848–49. But the Vienna system did not guarantee the "perpetual peace" for which many longed, and which Immanuel Kant, for one, had considered possible in 1795. In the second half of the nineteenth century, it was dismantled piece by piece.

The Congress System, whose true architect and deftest operator was the Austrian statesman Prince Metternich, involved a kind of freezing of the situation as it existed in 1815 (or more precisely in 1818, when France was again received into the circle of the Great Powers). Thus, insofar as the respective governments opposed liberalism, constitutionalism, and any form of social change centered on citizenship, the system stood as a bulwark against newly developing historical trends and, above all, against nationalist programs and political movements. In the multiethnic Romanov and Habsburg empires (and in the Ottoman Empire, which after 1850 also belonged pro forma to the "Concert of Europe"), smaller national groups began to stir against their perceived repression and to strive for either autonomy or full political independence. At the same time, a nationalism originating mainly in bourgeois middle strata called for the creation of larger economic spaces and rationalization of the state apparatus. This tendency was especially strong in Italy and in northern and central Germany, but the various regime changes in France were also largely motivated by the quest for a more *effective* national politics.

Another new factor was the major regional differentiation associated with industrialization. But the potential that this created for power politics, in the period roughly up to 1860, should not be overestimated. The old idea that the

Congress System was undermined by the independent variables and irresistible forces of nationalism and industrialization falls rather wide of the mark. The Crimean War, which from 1853 to 1856 pitted Russia against France, Britain, and eventually Piedmont-Sardinia (the core state of the later Kingdom of Italy), is good evidence that this is so, since it was the first military conflict for nearly forty years among the European Great Powers, fought out in a region on the periphery of Western Europe's mental maps. It showed that it was a disadvantage of the Congress System not to have settled the position of the Ottoman Empire in relation to Christian Europe. The Crimean War did not solve the "Eastern Question"—the future of the multinational Ottoman Empire—or any other problem of European politics.[15] Most crucially, however, it was neither a clash between industrialized war machines nor an ideologically heated contest between rival nationalisms. It was therefore by no means the expression of "modern" trends of the age.

At the end of the Crimean War, an opportunity was lost for a timely renewal of the Congress System. It was no longer possible to speak of a "concert of powers," and into the normative vacuum stepped Machiavellian realists (the term *Realpolitik* was coined in 1853) who risked international tensions or even war to impose their plans for new and larger nation-states. The big names here are Camillo Benso di Cavour in Italy and Otto von Bismarck in Germany.[16] They achieved their objectives amid the ruins of the Vienna peace. After Prussian-led Germany had prevailed against the Habsburg Monarchy and the Second Empire of Napoleon III (a disturber of the peace in his own way), in 1866 and 1871 respectively, it became a great power that carried much heavier weight internationally than Prussia had done. As German chancellor between 1871 and 1890, Bismarck dominated politics in continental Europe with a system of finely graduated treaties and alliances, whose chief aim was to secure the Reich, newly created in 1871, and to shield it from French revanchist ambitions. But the Bismarckian order, which passed through a number of phases, did not involve a pan-European peace settlement in succession to that of Congress of Vienna.[17] Although its core was meant to be defensive and served in the short term to preserve a given equilibrium, it produced no impulses toward a constructive European policy. By the end of Bismarck's time in office, the overly complex "balancing act" between various antagonisms was already scarcely functional.[18]

As for Bismarck's successors, they abandoned the relative restraint shown by the founder of the Reich. In the name of a new *Weltpolitik*, partly based on Germany's economic strength, partly driven by ideological hypernationalism, and partly responding to similar ambitions of other powers, Germany gave up any claim to be building peace for Europe. Moreover, its foreign policy induced the other Great Powers to bury their mutual antagonisms (which Bismarck had resourcefully fomented) and to regroup in a way that excluded Germany. By 1891, just a year after his dismissal by Wilhelm II, one of Bismarck's worst nightmares—a rapprochement between France and Russia—was beginning to come true.[19] At the same time, almost unnoticed by European politicians, a transatlantic rapprochement was taking place between Britain and the United States. By 1907 at the latest, a new power configuration was visible in international politics, though not yet at the level of alliances. France had found a way out from the isolation in which Bismarck had constantly sought to surround it, drawing closer first to Russia, then in 1904 (laying aside contentious issues in the colonies) to Britain. In 1907, London and Saint Petersburg defused their decades-long conflict in many parts of Asia.[20] A split also opened up between London and Berlin, exacerbated by a provocative German naval program. Germany—which, for all its economic strength, scarcely concealed its lack of means for a true *Weltpolitik*—eventually fell back on its only ally, Austria-Hungary, whose Balkan policies zigzagged ever more irresponsibly between aggressiveness and hysteria. The outbreak of the First World War in August 1914 was by no means foreordained. But all sides would have had to deploy exceptional statecraft, military restraint, and curbs on nationalist sentiment in order to ward off open conflict among at least some of the European Great Powers.[21] The First World War completely destroyed the European international system of the previous century and a half. In 1919 it could no longer be rebuilt as it had been in 1814–15.

The new Great Powers, the United States and Japan, played only minor roles in this scenario. But Russia's surprising defeat in 1905 at the hands of Japan, in a war fought mainly on Chinese territory, triggered a Russian policy crisis that was not without implications for Europe and the "Eastern Question." America's part in brokering a peace between the belligerents—the not always irenic President Theodore Roosevelt even won the Nobel Peace Prize for it—staked its claim to a great-power role for the third time in less than a decade, after the Spanish-American War of 1898 (in which the United States had been unbridled in its aggression) and Washington's involvement in the Eight Power expeditionary force against the Yihetuan ("Boxer") Uprising in China in 1900. Such a role was recognized for Japan as early as 1902, when the leading world power, Great Britain, concluded a treaty alliance with the archipelagic empire.[22] In 1905 the step from a European system of states to a global one became irrevocable. However, neither the United States nor Japan was directly involved in the outbreak of the First World War; it was a European conflict in its genesis. The European interstate system was destroyed from within.

Narrative II: Metamorphosis of Empires

Alongside this grand narrative of renewal, erosion, and catastrophe of the European interstate system, there is a second story of overseas expansion and imperialism. Although earlier versions of this history have been more strongly challenged in recent years than the standard narrative of the European interstate system, it is possible to reconstruct a sequential pattern more or less as follows. The end of the early modern period of European expansion and colonialism began in the early 1780s, with the British defeat in the American War of Independence and the formation of a new United States of America.[23] France, having lost its North American possessions in 1763, suffered a further sharp setback in 1804, when its economically most important colony, the sugar-producing Saint-Domingue portion of the Caribbean island of Hispaniola, renamed itself Haiti and declared independence. The revolution and the Napoleonic Empire, which led to supremacy in Europe, were paradoxically associated with France's withdrawal from overseas positions, since Napoleon conquered no new colonies. Egypt, invaded by Bonaparte in 1798, had to be given up three years later, and nothing came of projects to challenge England in Asia. With their successful campaigns in India between 1799 and 1818, the British were able to offset their defeat in America more easily than the French could recover from their colonial debacle. It is true that the British had been present in the Subcontinent as traders since the seventeenth century, and as territorial rulers of the province of Bengal since the 1760s, but it was in their global contest with France (which sought allies among the Indian princes) that they first managed to vanquish, or at least neutralize, the remaining indigenous military forces. As for the Spanish, their rule in mainland South and Central America was at an end by the mid-1820s. All that remained of the Spanish world empire were the Philippines, Cuba, and Puerto Rico.

European interest in colonies was not very great during the middle decades of the nineteenth century, although individual politicians (Napoleon III in France or Benjamin Disraeli in Britain) tried to stoke it up for domestic political reasons. Where political control already existed over a colony (India, Dutch East Indies, Philippines, Cuba), the aim was to make better use of them economically. There were several new additions: Algeria, first invaded by France in 1830 but not really conquered until the end of the 1850s; Sind (1843) and Punjab (1845–49) in an expanding British India; New Zealand, where the Maoris kept on fighting until 1872; inland extensions of colonies at the Cape of Good Hope and in Senegal; the Caucasus and the Khanate of Inner Asia. Britain and France, alone at midcentury in continuing aggressive expansion, established bases in Asia and Africa (e.g., Lagos and Saigon) that later served as springboards for territorial conquest, at the same time forcing Asian governments to grant concessions to European traders. The typical imperialist instrument was then not so much the expeditionary force as the cheap but effective gunboat, able to appear suddenly in a port and issue threats. But the two military conflicts with China (the First Opium War of 1839–42 and the Second Opium War, or "Arrow War," of 1856–60) also involved operations on land and were by no means walkovers. Some imperial enterprises ended in failure: for example, the first British intervention in Afghanistan (1839–42) and Napoleon III's expedition to Mexico after it became unable to repay its external debt. This bizarre episode, costly in both French and Mexican lives (approximately 50,000!), saw the Habsburg Archduke Maximilian crowned "Emperor of Mexico," only to be court-martialed and executed by firing squad in 1867. That France initially had British and Spanish support for its adventure has often been overlooked.[24]

In the 1870s, a change in the procedures and aggressiveness of European Great Powers was already looming. The Ottoman Empire and Egypt, deeply in debt to Western creditors, came under financial pressure that the Great Powers were able to exploit to their advantage. At the same time, a number of spectacular and widely publicized research expeditions made Africa once again an object of public attention in Europe. In 1881 the bey of Tunis had to accept a French "resident-general" as the power behind the throne; it was the beginning of the colonial "division of Africa." The race began in earnest the following year, when Britain occupied Egypt in response to the rise of a nationalist movement in a country that the opening of the Suez Canal in 1869 had made hugely important for the empire. Within a few years claims were staked throughout the continent and soon enforced by military conquest. Between 1881 and 1898 (the year of British victory over the Mahdi movement in Sudan), nearly the whole of Africa was partitioned among the various colonial powers: France, Britain, Belgium (with King Leopold II rather than the Belgian state as "owner" of a colony), Germany, and Portugal (a few old settlements on the coasts of Angola and Mozambique). In a final phase Morocco became a French possession (1912), and the Libyan desert, scarcely governable but viewed with new interest in Istanbul, came under Italian control (1911–12).[25] Only Ethiopia and Liberia (founded by former American slaves) remained independent. This "scramble for Africa," as it was known, though often chaotic, opportunistic, and unplanned in its finer details, should be seen as a single process. Such an occupation of a vast continent within just a few years was without parallel in world history.[26]

Between 1895 and 1905, a similar scramble developed in China, although not all the imperial powers had their eyes on territorial acquisition. Some—especially Britain, France, and Belgium—were more interested in railroad or mining concessions and in staking out informal spheres of commercial influence. The United States proclaimed an "open door" principle for all countries in the Chinese market. At that time only Japan, Russia, and Germany appropriated quasi-colonial territories of any significance on the periphery of China: Taiwan (Formosa), southern Manchuria, and Qingdao with its hinterland on the Shandong peninsula. But the Chinese state remained in place, and the great majority of Chinese never became colonial subjects. The consequences of the "mini scramble" in China were thus much less grave than those of the "maxi scramble" in Africa.

In Southeast Asia, however, the British established themselves in Burma and Malaya, while the French took control in Indochina (Vietnam, Laos, and Cambodia). Between 1898 and 1902 the United States conquered the Philippines, first from Spain, then from the Filipino independence movement. In 1900, Siam was the only nominally independent (if weak and therefore cautious) country in this politically and culturally diverse part of the world. The same justifications were given everywhere for European (or American) conquests in Asia and Africa between 1881 and 1912: a "might is right" ideology, mostly suffused with racism; the supposed incapacity of native peoples to govern themselves in an orderly manner; and an (often preventive) protection of national interests in the contest with rival European powers.

This second grand narrative does not flow as directly as the first into the 1914–18 war. The colonial world had been stabilized for some years before the outbreak of the war, and to some extent tensions among the colonial powers were even regulated by treaty. Occasionally non-European locations provided the setting for power games directed at the European public: for example, in the Morocco crisis of 1905–6 and 1911, the German Reich staged military exercises in North Africa as a bluff, while the press demonstrated its fateful power to stoke up conflicts. But genuine colonial rivalries were rarely at issue. Since the First World War was not unleashed primarily by the clash of imperialisms in Asia and Africa, History No. 2 is often understood as a branch line of the History No. 1, which in turn points straight toward the summer of 1914. Quite a few general accounts of nineteenth-century Europe mention colonialism and imperialism only in the briefest possible way, creating the impression that Europe's expansion in the world was not an essential part of its history but only a by-product of events in its various countries.[27]

Consequently, diplomatic history and colonial history have seldom really converged. A global historical approach cannot be content with this but must find a bridge between Eurocentric and Asia- or Africa-centric perspectives. It thus faces two demanding tasks: to relate the history of the European interstate system (which toward the end of the nineteenth century became a global system) to the history of colonial and imperial expansion; and to resist allowing the international history of the nineteenth century to run teleologically toward the outbreak of war in 1914. We know the war began on August 4, 1914, but just a few years earlier not many had suspected that things would go so far so soon. A genuine world war was virtually unthinkable for policymakers and the general public at the time, and it would unduly restrict our understanding of the nineteenth century if we were to view it simply as a long prehistory of the great conflagration.

A third challenge we face is to take into account the diversity of imperial phenomena. It would, of course, be superficial to lump together everything that describes itself as an "empire." The imperial vocabulary had quite different shades of meaning in different countries and languages. On the other hand, the consideration of frontiers in various contexts (chapter 7, above) has already brought to light great similarities between cases that are usually seen as unrelated. The same is true of empires. We must therefore seek to question the common distinction between the maritime empires of Western European powers and the land empires ruled from Vienna, Saint Petersburg, Istanbul, and Beijing. First of all, however, a glance is necessary at the nation-state.

2 Paths to the Nation-State

The Semantics of Empire

German and French historians in particular consider the
as the age of nationalism and nation-states.[28] The Franco-I
conflict between one of the oldest European nation-states an   ..ck-
ing to measure up to and outmaneuver the land of revolutions. ⊥ neirs were "entangled histories," if any in Europe can be described as such—not between fundamentally unequal partners but within a constellation that would in the very
long run lead to the post-1945 equilibrium. But can the Franco-German perspective sustain an interpretation of Europe or even the world in the nineteenth
century? British historiography, without the resonance that *Reichsgründung*
had for German historians, seldom placed so much emphasis on the process of
nation-state formation, seeing the founding of the Reich as a German affair with
implications for the rest of Europe. The British Empire, by contrast, did not owe
its existence to any "founding" event, except perhaps for those who wished to
glorify a couple of buccaneers from the Elizabethan age. It had not come about
in a Big Bang but had developed through a complex and lengthy process in many
world theaters, with no overall direction from the center. Britain, unlike Germany, did not need to found an empire in the nineteenth century, because for
a long time it had already possessed one whose origins could not be precisely
established. In fact, before the middle of the nineteenth century, it occurred to
scarcely anyone that the scattered possessions of the Crown plus various settlements and colonies added up to something as definite as an empire. Until the
1870s the settler colonies, whose "mother country" Britain claimed to be, were
seen as different in kind from the other colonies, where it was not maternal relations but a strict paternalism that prevailed.[29] Later too, there would be much
heated debate about the nature of the empire.

Also, in other cases the semantics of empire is multilayered, even contradictory. In 1900 the German word *Reich* had at least three different referents:
(1) a young nation-state in the middle of Europe, which had endowed itself with
a parvenu emperor (reminiscent of Peter the Great's self-elevation in 1721) and
called itself the German Empire (*Deutsches Reich*); (2) a small overseas trading
and colonial empire, to which the *Deutsches Reich* under Bismarck had gradually
added a few colonial acquisitions in Africa after 1884; and (3) a Romantic fantasy
of a sprawling land empire (for which Bismarck's petty German arrangement
came as a huge letdown), a revived Holy Roman Empire, a gathering-in of all
Germans or "Germanic peoples," a German *Lebensraum*, or even a German-
dominated *Mitteleuropa*—an empire, then, which would beckon in early 1918
with the diktat imposing the Treaty of Brest-Litovsk on Russia, and after 1939
would briefly become a reality under the Nazis.[30] It could further be shown that
concepts of empire have existed at all times and in many cultures, and that major

antic differences existed in late modern Europe, even within individual countries. An empire cannot therefore be adequately grasped by how it defines itself, and it is not a convincing solution to regard *everything* as an empire that calls itself by that name. It must be possible to describe an empire structurally, in terms of certain observable features.

Nation-State and Nationalism

Empires are a pan-Eurasian phenomenon of ancient pedigree, going back to the third millennium BC, and are therefore charged with meanings from many different cultural contexts. Nation-states, on the other hand, are a relatively recent Western European invention, whose emergence can be studied under laboratory conditions, as it were, in the history of the nineteenth century. It has proved difficult, however, to give a definition of the nation-state. "The modern nation-state," we read, "is a state in which the nation qua totality of citizens is sovereign, both determining and supervising the exercise of political rule. The equal right of all citizens to participate in the institutions, services, and projects of the state is its guiding principle."[31] This definition, plausible as it is at first sight, contains such high participatory requirements that it excludes too much. Poland under communist rule, Spain under Franco, South Africa until the end of apartheid: there would have been no nation-state in any of these cases. And if the word "citizens" is taken as gender-neutral, how should one classify Great Britain, which adopted universal female suffrage only in 1928, or the France of the Third Republic, which followed suit only in 1944? In the nineteenth century there was scarcely any country in the world that would have qualified as a nation-state by such criteria: at most Australia (but only after 1906) and New Zealand, where the right of all women to vote was recognized in 1893, though to run for office only in 1919, and where the indigenous Maoris also had the franchise.[32]

An alternative way of approaching the nation-state would be via *nationalism*.[33] This may be understood as a sense of belonging to a large collective that conceives of itself as a political actor with a common language and destiny. This attitude became operative in Europe from the 1790s on, resting on a number of simple general ideas: nations are the world's natural units, in comparison with which empires are artificial constructs; the nation—not the region or a supranational religious community—is the primary loyalty for individuals and the main framework for ties of solidarity; a nation must therefore formulate clear membership criteria and categorize minorities as such, with discrimination as a possible but not inevitable outcome; a nation strives for political autonomy within a certain territory and requires a state of its own to guarantee this.

The link between nation and state is not easy to grasp. Hagen Schulze has outlined how in a second phase "state-nations" and then "people-nations" took shape or even defined themselves as such, and how in the period after the French Revolution a nationalism with a broad social base—he calls it "mass nationalism"—took on the form of the state. Schulze avoids explicitly defining the nation-state, but he explains what he has in mind by offering neat successive periodizations of the "revolutionary" (1815–71), "imperial" (1871–1914), and "total" (1914–45) nation-state.[34] In any event, the nation-state appears here as a composite or synthesis transcending both state and nation: a *mobilized*, not a virtual, nation.

Pointing the historical discussion in a different direction, Wolfgang Reinhard claims, in agreement with such theorists as John Breuilly or Eric Hobsbawm: "The nation was the dependent variable of historical development, but state power was its independent variable."[35] In this view, the nation-state—which Reinhard, too, first locates in the nineteenth century[36]—is not the almost inevitable result of the formation of mass consciousness and identity "from below" but rather the outcome of a will to concentrate political power "from above."[37] A nation-state is thus not the state casing of a given nation; it is a project of state apparatuses and power elites, as well as of revolutionary or anticolonial *counterelites*. The nation-state usually attaches itself to an existing sense of nationhood and instrumentalizes it for a *policy* of nation building, whose aims are to create a viable economic space, an effective player in international politics, and sometimes also a homogeneous culture with its own symbols and values.[38] So, there are not only nations looking for a nation-state of their own, but also nation-states looking for the perfect nation with which to align themselves. As Reinhard convincingly observes, most states that are today designated as nation-states are in reality multinational states, with sizable minorities organized at least at the prepolitical level of social space.[39] These minorities differ from one another mainly according to whether their political leaders mount a separatist challenge to the wider state (until very recently Basques or Tamils, for instance), or whether they are content with partial autonomy (Scots, Catalans, or French Canadians). The "national groups" or (in a premodern sense of the word) "nationalities" of the great empires were such minorities. Some of the multiethnicity of *all* empires was preserved in the young nation-states of the nineteenth century, even if they constantly tried to conceal this behind discourses of homogeneity.

Where then are the nation-states that are the supposed hallmark of the nineteenth century? A glance at maps of the world shows empires, rather,[40] and in 1900 no one predicted the coming end of the imperial age. After the First World War, which irrevocably destroyed three empires (Ottoman, Hohenzollern, and Habsburg), the imperial era lingered on. The Western European colonial empires, as well as the US colony in the Philippines, reached the zenith of their significance for the metropolitan economies and mentalities only in 1920s and 1930s. The new Soviet regime managed within a few years to reconstitute the Caucasian and Central Asian *cordon* of the late Tsarist Empire. Japan, Italy, and—very briefly—Nazi Germany built new empires that imitated and caricatured the old. The imperial age came to an end only with the great wave of decolonization between the Suez crisis of 1956 and the end of the Algerian war in 1962.

Although the nineteenth century was not an "age of nation-states," two things are nevertheless true of it. *First*, it was the era in which *nationalism* emerged as a way of thinking and a political mythology, finding expression in doctrines and programs, and mobilizing sentiments with a capacity to arouse the masses. From the outset nationalism had had a strongly anti-imperial component. It was the experience of French "foreign rule" under Napoleon that first radicalized nationalism in Germany, and everywhere else—in the Tsarist Empire, the Habsburg Monarchy, the Ottoman Empire, and Ireland—resistance stirred in the name of new national conceptions. By no means was it always associated with the goal of an independent state, however. Often the initial aim was only to protect the nation from physical attack or discrimination, to achieve stronger representation of national interests within the imperial polity, or to widen the scope for the national language and other forms of cultural expression. The early, "primary resistance" to colonial conquest in Asia and Africa also seldom set its sights on an independent national state. "Secondary resistance" followed only in the twentieth century, when new elites familiar with the West warmed to the nation-state model and recognized the mobilizing power of a rhetoric of national emancipation.

Nevertheless, however hazy it remained in the nineteenth century, the idea of the nation-state as a framework for political leaderships to form and develop became ever more attractive in Poland, Hungary, Serbia, and other parts of Europe, as well as in a handful of extra-European contexts, such as the Egyptian Urabi movement of 1881–82 (so called after its main leader, it opposed an extremely pro-Western government with the slogan "Egypt for the Egyptians!") and the early stirrings of Vietnamese anticolonialism from 1907 on.[41]

Second, the nineteenth century was an age of nation-state *formation*. Despite many a spectacular founding act, this was invariably a lengthy process—and it is not always easy to indicate when national statehood was actually accomplished, when the "external" and "internal" building of the nation-state was sufficiently matured. The internal aspect is the more difficult to judge. One must decide when a certain territorial polity, usually undergoing evolutionary change, attained a degree of structural integration and homogeneous thinking that made it qualitatively different from the princedom, empire, old-style city republic, or colony that had preceded it. Even for the French nation-state, the usual model in this respect, it is no simple matter to say when such a point was reached. Already with the Revolution of 1789 and its national rhetoric and legislature? With Napoleon's centralizing reforms? Or with the transformation of "peasants into Frenchmen"—a decades-long process that its foremost historian sees getting under way as late as the 1870s?[42] If it is so difficult to give an answer for France, what can be said of more complex cases?

Less problematic is the question of when a polity became capable of international action and acquired the *external* form of a nation-state. Under the systems and conventions of the nineteenth and twentieth centuries, a country *counted* as a nation-state only if the great majority of the international community recognized it as an independent player. This Western concept of sovereignty is not a sufficient criterion—otherwise the external point of view would be absolute, and an entity such as Bavaria would have been a nation-state in 1850. But external recognition is a necessary condition: there is no nation-state that does not have its own army and diplomatic corps and that is not accepted as a signatory to international agreements. In the nineteenth century, the number of international players was smaller than the number of polities with some certifiable success in social and cultural nation building. Although around the year 1900 Russian-controlled Poland, Habsburg Hungary, and Ireland within the United Kingdom exhibited many features of nation building, it cannot be said that they were nation-states. They attained that status only after the end of the First World War—in a flurry of national emancipation outpacing all that the "century of the nation-state" had offered. The second half of the twentieth century witnessed the reverse: many states outwardly recognized as independent remained unstable quasi-states without institutional or cultural coherence.

In the nineteenth century nation-states came into being in one of three ways: (1) through the revolutionary breakaway of a colony; (2) through hegemonic unification; or (3) through evolution toward autonomy.[43] To these corresponded three distinct forms of nationalism: anticolonial nationalism, unification nationalism, and separatist nationalism.[44]

Revolutionary Independence

Most new states that entered the scene during the nineteenth century came into being in its first quarter, at the end of an Atlantic cycle of revolution.[45] This first wave of decolonization was part of a chain reaction that had begun in the 1760s with the roughly simultaneous (though causally unrelated) interventions by London and Madrid in their American colonies.[46] The reaction of the North Americans was prompt, that of the Spanish Americans a little delayed. When open revolt broke out in 1810 from the River Plate to Mexico, the wider context was different: not only was there the example of the United States, but the Spanish monarchy had collapsed in 1808 following Napoleon's invasion of the Iberian Peninsula (itself a sequel to the military expansionism that had marked the French Revolution almost from the beginning). The influence of 1789 made itself felt earlier and more directly on the island of Hispaniola, where an uprising of mulatto middle strata (*gens de couleur*) and black slaves got under way in 1792. Out of this genuine anticolonial and social revolution came the second republic in the Americas: Haiti.[47] It was recognized by France in 1825, and thereafter gradually by most other countries. On the mainland, a wave of revolutions gave birth to the independent polities that are still there today: Argentina, Chile, Uruguay, Paraguay, Peru, Bolivia, Colombia, Venezuela, and Mexico. But the larger entities envisaged by Simon Bolívar failed to materialize.[48] Later breakaways saw the emergence of Ecuador (1830), Honduras (1838), and Guatemala

(1839). Thus, after the interlude of a Mexican empire in 1822–24, a whole new archipelago of republics claimed and won external sovereignty, even if successes in internal nation building were often a long time in coming.

Developments were less revolutionary in Brazil, where Creole elites did not break with an unpopular imperial center. In 1807 the Portuguese dynasty managed to flee the French to its most important colony, and after the fall of Napoleon, the regent Dom João (later John VI) decided to remain in Brazil, raising it to the level of a kingdom and ruling it from 1816 as the King of Portugal, Brazil, and the Algarve. After his return to Europe, his son stayed on as prince regent and in 1822 had himself crowned as Emperor Pedro I of a Brazil now peacefully separated from the mother country. Only in 1889 did the most populous country of Latin America declare itself a republic.

In Europe, the only new state with origins in an empire was Greece. Here indigenous forces active both inside the country and in exile came together with vociferous philhellene movements in Britain and Germany to detach Hellas from the Ottoman Empire in 1827, eventually assisted by a naval intervention on the part of Britain, Russia, and France. For the time being, the borders encompassed only the south of present-day Greece plus the Aegean islands. If the period of Ottoman rule going back to the fifteenth century is baldly defined as "colonial," then liberated Greece was a postcolonial entity; it was the result, however, not of a wholly autonomous revolution but of a process supported by the Great Powers and lacking a broad social base. Greece then remained more dependent on the Great Powers than did the new states of Latin America. It won recognition, becoming a reality under international law, only in the London Protocol of February 1830. But the outer casing did not yet correspond to a social and cultural content: "A Greek state now existed, but a Greek nation still had to be made."[49]

Also in 1830–31 the Belgian state—traditionally the Southern Netherlands—came into being. Unlike the Greeks, the citizens of Brussels and its surrounding area could not complain of centuries of foreign rule. Their main grievance was what they saw as the autocratic policy of the Dutch king William I since the post-Napoleonic unification of the kingdom in 1815. But the conflict lacked an ideological dimension, such as the struggle of free Europeans against Oriental despotism that had won the Greeks so much publicity and support. More than Greece, Belgium was the progeny of a revolution. Amid the turmoil unleashed in many parts of Europe by the French revolution of July 1830, disturbances broke out in Brussels in August, during a performance of Auber's *La Muette de Portici* at the opera house. Uprisings ensued in other cities, and the Dutch sent in troops. Complete separation from the Netherlands, which in a few weeks became the goal of the fast-radicalizing movement, was here achieved without foreign military intervention, although the tsar and the king of Prussia had threatened to come to the aid of William, and for a time the related international crisis escalated dangerously. Like Greece, however, Belgium had its independence guaranteed by a great power treaty, in which Britain once again played the role of principal midwife.[50]

In 1804, much farther from the limelight in the pashalik of Belgrade—a border province of the Ottoman Empire, with a population of roughly 370,000—the Christians of Serbian origin rose up against the local Ottoman janissaries, who, barely under Istanbul control, had been exercising a reign of terror.[51] In 1830, after a long conflict, the sultan recognized the Principality of Serbia, nominally continuing as part of the Ottoman Empire. In 1867—at more or less the same time as similar developments in Canada—the Serbs reached a point where they no longer had to fear interference in their internal affairs by their remote suzerain; the last Turkish troops were withdrawn.[52] Finally, in 1878 the great-powers meeting at the Congress of Berlin recognized Serbia as an independent state in international law, as they also did Montenegro and Romania (long torn this way and that between Russian and Ottoman protection). Bulgaria profited from the Sultan's major defeat in the Russo-Turkish war of 1877–78, but it remained a tribute-paying principality of the Porte and achieved international recognition as a state with its own "tsar" only during the Young Turk Revolution of 1908–9 in the Ottoman Empire.[53]

Can it be said that all these new political structures were nation-states in an *internal* sense? There is reason to doubt it. After a hundred years of existence as a state, Haiti had to show for itself "a questionable past and a deplorable present"; neither its political institution building nor its social-economic development had made much progress.[54] In mainland South and Central America, the first half century after independence was not one of calm consolidation; most countries achieved political stability only in that crucial decade of the 1870s, which all over the world saw a centralization and reorganization of state power. Greece was at first subject to Bavarian tutelage; the Great Powers seconded Prince Otto, a son of Ludwig I of Bavaria, to reign as monarch. The country then experienced its first coups d'état (1843, 1862, 1909), and only after 1910, under the Liberal prime minister Eliftherios Venizelos, did it develop more stable institutions.[55] Even Belgium was no model nation-state. Its dominant nationalism, taking a clear distance from the Netherlands, rooted French in the constitution as the only official language, but from the 1840s it came under challenge from an ethnolinguistic Flemish nationalism. For this self-styled "Flemish movement," the issues were equal rights within the Belgian state and a cross-border unity with Dutch language and culture.[56]

Hegemonic Unification

State building through the voluntary union of allied peoples is a historically ancient model. When no single power is paramount, this involves the establishment of territorial statehood through a "multiheaded" federation of cities or cantons. The Netherlands and Switzerland are examples of such polycentric equilibrium, the basis for them having been laid long before the nineteenth century.[57] Even after 1800, facing much larger states in their vicinity, both retained a federative character that proved sufficiently flexible to cushion social and religious tensions. But whereas the Netherlands, marveled at as a curiosity in the early modern period, had become by 1900 more akin to a "normal" nation-state, Switzerland emphasized its special role by sticking to its loosely federal constitution and system of cantonal rights, with unusually direct forms of democracy.[58] The United States of America was typologically more complex, combining in its origins an independence revolution with a polycephalous federation; no such opportunity existed for the leaders of the Spanish American independence movement. The new United States aimed from the beginning to incorporate additional territories into the Union, and the Northwest Ordinance of 1787, a foundational document, laid down precise rules for this. There was nothing comparable in Europe to such a state with built-in mechanisms for further expansion.

Nation-state building in Europe at the time followed not a polycephalous but a *hegemonic* model, in which one regional power seized the initiative, brought its military strength into play, and put its stamp on the newly emergent state.[59] Such hegemonic unification "from above" was not a modern European invention. In 221 BC the Qin military state, on the geographical margins of the Chinese political world, founded the first imperial dynasty and went on to unify the Chinese empire. It displayed some affinities with eighteenth- and nineteenth-century Prussia: a crude military system (though in post–1815 Prussia less frightening than before) combined with access to the culture and technology of the neighboring civilization (eastern China and Western Europe, respectively). In much the same way as Prussia in Germany, the small border kingdom of Piedmont-Sardinia was the unifying hegemon in Italy, qualifying for this role as the only self-governing region of a land otherwise under the rule of Austria, Spain, or the Vatican. In both Prussia and Piedmont-Sardinia, there was in charge a strong-willed political realist with a wide constitutional scope for unhampered leadership—Bismarck or Cavour—who played on international differences to create the opportunities for his policy of national unification.[60] The Italians succeeded first, when a new all–Italian Parliament was established in February 1861. Austria's surrender of the Veneto in 1866, and the transfer of the capital in 1871 to a Rome wrested from Pope Pius IX in a rather symbolic conquest, completed the external building of the nation-state. The annexation of Rome was possible only after Napoleon III's defeat at the battle of Sedan robbed the pope of a reliable protector and forced the French garrison to pull out. Pio Nono grudgingly withdrew to the Vatican and threatened with excommunication any Catholic who became involved in national politics.

For all the similarities, the unification processes in Italy and Germany display a number of differences.[61]

First. Although the process was deeply rooted in the thinking of intellectuals in Italy, the practical preparations were more rudimentary there than in Germany.

There were no preliminary steps such as the Zollverein or the North German League, and in general the internal nation building, "understood as economic, social, and cultural integration of a space of communication,"[62] was less far advanced than in Germany. Mentally, too, there was almost nothing apart from the Catholic faith that linked together all Italians from Lombardy to Sicily—and from 1848 the church was on a collision course with Italian nationalism.

Second. The main reason for the lack of structural prerequisites of national unity was that external forces had been intervening in Italy for centuries. The country had to free itself of foreign occupation, whereas in Germany only the influence of the Habsburg emperor had to be driven back, albeit at the price of what has been called, with only slight exaggeration, a German civil war.[63] The military resolution was immediate, however: the Battle of Königgrätz (Sadová) on July 3, 1866, was the key date in the building of a "smaller German" nation-state. Prussia was an independent military power of a quite different caliber from little Piedmont-Sardinia. It was able to impose German unity by force in the international arena, whereas Piedmont had to rely on coalitions of powers in which it was always the weaker partner.

Third. In Italy, unification from above—as Cavour, allied to Napoleon III, pursued it mainly at the negotiating table, though certainly also on the battlefield—was supported by a stronger popular movement than in Germany and accompanied with greater public debate. Here too, of course, the state was not wholly reconstituted from below, and the national-revolutionary movement, headed by the charismatic Giuseppe Garibaldi, was not above manipulating "the masses." No constituent assembly was convened: the laws and bureaucratic order of Piedmont-Sardinia, largely resting on the prefecture system from the time of the Napoleonic occupation, were simply transposed to the new state. This Piedmontization met with considerable resistance. In Germany, constitutional issues (in the broad sense) had for centuries been at the forefront of politics. The early modern Holy Roman Empire, without any parallel in Italy or anywhere else in the world, had been less a union held together by force than an edifice of constantly honed compromise, and the same was true a fortiori of the Deutscher Bund, created at the Congress of Vienna and slowly evolving into the state framework of an emergent nation. The German constitutional tradition tended to be decentralized and federative, and even Prussia had to take account of this in its leadership of the North German Confederation (from 1866) and of the newly founded Reich (from 1871), as well as heeding for a long time anti-Prussian sentiments in the South. For the new Reich, federal statehood was "the central fact of its existence" (Thomas Nipperdey).[64] In Italy there was nothing comparable to the continuing dualism of Prussia and empire; Cavour's Piedmont-Sardinia was complete absorbed into the unitary Italian state. But social-economic differences remained (and remain today) a dominant problem within Italy. True unity between the prosperous North and the impoverished South was never achieved.

Fourth. In Italy the internal resistance was greater and lasted longer. The German princes accepted the material gifts on offer, and the population followed their lead. In Sicily and the southern mainland of Italy the rural underclasses, often allied with local notables, kept up a civil war all the way through the 1860s. This guerrilla-style struggle, officially known as "brigandage," typically involved horseback ambushes of anyone considered a collaborator with the North and the new order, and both the ferocity of the insurgents and the reprisals directed against them are less reminiscent of the "regular" unification wars of the age than of the no-holds-barred war in Spain from 1808 to 1813. Probably more people died in the *brigantaggio* wars than in all others fought on Italian soil between 1848 and 1861.[65]

Did anything similar happen in other parts of the world? Was there an Asian "founder of the empire," a Bismarck? There had been a distant parallel when Vietnam was unified in 1802 under the emperor Gia Long, but he had resided in the central city of Hué and been content to share power with the strong regional princes in the North (Hanoi) and the South (Saigon). In itself that was not necessary a disadvantage. More serious were the failure to build, or rebuild, a strong central bureaucracy (a Chinese influence with strong roots in the country), and Gia Long's neglect of his army. His successors did not correct these omissions, which contributed to Vietnam's weakness a few decades later when it came face to face with Imperial France.[66] The colonial intervention that began in 1859 with the conquest of Saigon held back the development of a Vietnamese nation-state for more than a century.

Evolution toward Autonomy

Apart from revolutionary secession from an empire—which in the nineteenth century occurred nowhere in Europe outside the Balkans and in the twentieth century was achieved in peacetime only by the Irish Free State in 1921—the other path involved gradual moves toward autonomy (or even peaceful separation) within a continuing imperial framework. Sweden and Norway ended their dynastic union in 1905, without internal convulsions or serious international tensions, after three decades of slow political estrangement and national identity formation on both sides. This amicable divorce took the form of a plebiscite on the independence of Norway, the junior partner, whereby the Swedish king lost the Norwegian throne formerly ceded to him by a Danish prince.[67]

By far the most important examples of evolutionary autonomy occurred within the British Empire. Apart from Canada, all the British settler colonies had come into being *after* the revolutionary independence of the United States (1783): Australia little by little after 1788, Cape Province after 1806, New Zealand after 1840. Thus, both the settlers and the imperial policymakers in London had time to digest the US experience, and until the secession of Southern Rhodesia (the future Zimbabwe) in 1965 there would be no further revolts by settlers of British origin. A critical point was reached in the second half of the 1830s in Canada (still called, more precisely, British North America). Until then local oligarchies had been firmly in the saddle in the various provinces; the elected assemblies did not even have control over finances, and the main conflicts were between the dominant merchant families and the respective governors. In the 1820s the assemblies increasingly became the forum for antioligarchic politicians seeking to bring about a gradual democratization of political life. They saw themselves as "independent cultivators of the soil" and defended positions similar to those of "Jacksonian democracy" (since 1829 in the United States). In 1837 several violent revolts broke out simultaneously, with the aim not of breaking away from the British Empire but of overthrowing the dominant political forces in individual colonies. These spontaneous uprisings did not come together in an organized rebellion and were brutally suppressed.

The government in London could have left things at that.[68] But instead, recognizing that the potential for conflict in Canada was more than a surface phenomenon, it sent out a commission of inquiry under Lord Durham. Although Durham did not stay long in Canada, his *Report on the Affairs in British North America*, issued in January 1839, was a profound analysis of the problems,[69] and his recommendations became a milestone in the constitutional history of the empire. Barely twenty years after the success of the Spanish American independence movements, and following the promulgation of the Monroe Doctrine in 1822, the Durham Report surmised that the days of imperial rule in America were numbered unless skillful political management was brought to bear. At the same time, Durham sought to apply recent experiences in India, where a period of ambitious reforms had begun in the late 1820s. The paths taken in India and Canada were quite different, but the basic idea—that imperial rule constantly needed reforms to be viable—would never again be entirely absent from the history of the British Empire. Lord Durham formulated the view that British political institutions, being the best suited in theory for overseas settler colonies, should be given the opportunity to serve the growing self-determination of colonial subjects. This radical proposal, only seven years after the Reform Bill of 1832 had opened up the political system, however timidly, in the mother country, involved establishing a Westminster-style lower house with powers to appoint and bring down the government.[70]

The Durham Report is one of the most important documents of global constitutional history. It argued that a balance needed to be found between the interests of settlers and the imperial center, within a framework of democratic institutions open to change; that the distribution of powers and responsibilities between the Whitehall-appointed governor and local representative bodies should be continually renegotiated. Many policy areas, especially foreign and military affairs, would remain under central control, and Canadian or Australian laws came into effect only when the Parliament in London had approved them. More important, however, the new constitutional framework meant that the dominions (as colonies with "responsible government" were now called) had the possibility of developing into fledgling nation-states.

This process took specific forms in Canada, Australia, and New Zealand. The confluence of several colonies into a federal state marked a key stage in Australian history. Only with the Statute of Westminster (1931) did the dominions—South Africa was a special case—become nominally self-governing states, linked only symbolically to the old colonial center by their recognition of the British monarch as head of state. But in the second half of the nineteenth century these countries passed through a series of stages of political democratization and social integration, which may be described as a combination of internal nation building and *delayed* external nation-state formation. This evolution toward autonomy within an empire that was more liberal than many others saw the emergence of some of the institutionally most stable, and socially and politically most progressive, states in the world, albeit ones burdened with disfranchisement and exclusion of the indigenous population.[71] The process was largely concluded before the First World War.[72]

Special Paths: Japan and the United States

Not all cases of nation-state formation in the nineteenth century fall under one of these three paths; some of the most spectacular were unique of their kind. Two Asian countries had never been part of a larger empire and were therefore, like Western Europe, capable of transforming themselves without the energy input of anti-imperial resistance: Japan and Siam/Thailand. Both had always (or more precisely, in the Siamese case, since the mid-eighteenth century) been independent in foreign policy and had never fallen under European colonial rule. Whether they should therefore be considered "new nation-states," in the external sense of the achievement of sovereignty, is questionable. For both countries reshaped themselves under considerable informal pressure from the Western powers—especially Britain, France, and the United States—the stimulus being a concern for communal and dynastic survival in a world where Western interference in the affairs of non-Western states seemed to be taken for granted.

In 1900 Japan was one of the most tightly integrated nation-states in the world, with a system of government approaching French levels of unification and centralization, regional authorities that did little more than follow instructions, a well-functioning internal market, and an exceptionally homogeneous culture (Japan had no ethnic or linguistic minorities, apart from the indigenous Ainu in the far North). This compact uniformity was the result of comprehensive reforms that began in 1868 and go by the name of the Meiji Renewal or the Meiji Restoration. It was one of the most striking instances of nation building anywhere in the nineteenth century, more dramatic in many respects than what happened in Germany.

This process was not associated with territorial aggrandizement. Japan did not expand beyond its archipelago until 1894—if one leaves aside the annexation, in 1879, of the formerly tribute-bearing Ryukyu Islands and an unimpressive naval expedition in 1874 to the Chinese island of Taiwan. Japan's closure from the outside world since the 1630s had entailed that until 1854 it scarcely had a foreign policy in the usual sense of the term. It maintained diplomatic relations with Korea but not China, and among European countries only with the Netherlands (which in the seventeenth century had had a high profile in Southeast and East Asia). This was not due to a sovereignty deficit, however: if Japan had wanted to "play the game" in the early modern world, it would undoubtedly have been recognized—like China—as a sovereign agent.

In the case of Japan, external nation-state formation means that after its "opening up" in the early 1850s the country gradually began to seek a role on the international stage. Internally, the order that survived until the Meiji Renewal was in essence the one created in 1600 by regional warrior-princes such as Hideyoshi Toyotomi or Tokugawa Ieyasu, which clever politics consolidated by the end of the seventeenth century into a political system with the greatest level of integration ever seen in the archipelago's history. The territorial aspect of this is not easy to grasp with Western categories. The country was split up into roughly 250 domains (*han*), with a prince (*daimyō*) at the head of each. These *daimyō* were not fully independent rulers. In principle they administered their territory autonomously, but they stood in a fief-like relationship to the most powerful princely house, the Tokugawa, presided over by the shogun. Legitimacy was vested in an imperial court in Kyoto that lacked all real power. The shogun in Edo (Tokyo), on the other hand, was a worldly figure with no sacred functions or royal aura: he could not base himself on any theory of divine right or celestial mandate. The *daimyō* were not organized as an estate; there was no parliament at which they could close ranks in opposition to the overlord. This at-first-sight highly fragmented system, reminiscent of the central European mosaic during the early modern period, was integrated through a rotation system that obliged princes to reside in turn at the shogun's court in Edo. This crucially assisted the flowering of cities and of an urban merchant class, especially in Edo itself. The development of a national market was far advanced by the eighteenth century. A functional equivalent of the German *Zollverein* was thus already a feature of early modern Japan.

In another similarity with (northern) Germany, politically influential circles in Japan understood that small-state particularism was no longer viable in a rapidly changing world. This did not lead all to agree voluntarily on a federative solution, which would have involved winding up the territorial principalities, and so the initiative had to come from a hegemon. The island empire under Tokugawa rule (the *bakufu* system) was already politically unified within the boundaries of Japanese settlement. The question was who would provide the impetus for centralization. In the end, the architects of change were not *bakufu* men but circles of samurai nobility in two peripheral principalities of southern Japan, Choshu and Satsuma, who made a grab for power in the capital, supported by officials of an emperor whose significance had long been merely ceremonial.

The Meiji "Restoration" of 1868 is so called because the authority of the imperial house was restored after centuries of retreat, and because the young emperor was thrust into the central position in the political system under the carefully chosen slogan "Meiji" (that is, "enlightened rule"). The rebellious samurai could not draw legitimacy either from traditional political thought or from democratic procedures. Behind the fiction or presumption of acting in the emperor's name lay an act of pure usurpation. In reality it revolutionized Japanese politics and society in the space of a few years; neither was it only a "revolution from above," in the sense of having a conservative social impact or of heading off a popular revolutionary movement. The samurai modernizers soon abolished the samurai status and all its privileges. This amounted to the most thoroughgoing revolution of the middle decades of the nineteenth century. It unfolded without terror or civil war; some *daimyō* put up resistance that had to be broken militarily, but there was nothing remotely like the drama and violence of the Austro-Prussian war of 1866, the Franco-Prussian war of 1870–71, or the war in northern Italy between Piedmont/France and Austria.[73] The *daimyō* were partly persuaded, partly bullied, and partly won over with financial compensation. In short, Japan needed relatively little force to achieve far-reaching changes: a peaceful convergence of internal and external nation-building in a protected international space outside the European system of states, without significant foreign military intervention and with no colonial subjugation.[74]

Isolation from European power politics linked Japan and the United States. At the same time, their political trajectories were quite different. In North America there were no "feudal" structures that had to be smashed. The rebellious colonies had won diplomatic recognition in 1778 from France, and in 1783 from the former imperial mother country, Great Britain. The United States, therefore, was from the outset an externally sovereign state. It was also remarkably well integrated at various levels, sustained by the unitary civic consciousness of its political elite, and appearing in every respect to be part of the modern world. The failure of these hopeful beginnings to translate into continuous and harmonious national development is one of the great paradoxes of the nineteenth century. A country that thought it had left behind the militarism and Machiavellianism of the Old World experienced the second-largest paroxysm of violence (after the Chinese Taiping Revolution of 1850–64) between the end of the Napoleonic Wars and the outbreak of the First World War. *Why* this was so cannot be explained here. Two processes interacted dynamically up to a point when secession of a large part of the territorial body politic became structurally almost inevitable: first, westward expansion proceeding without overall political guidance and generally in a highly haphazard fashion; and second, a broadening rift between the slave-based society in the eleven Southern states and the free-labor capitalism in the North.[75] The breaking point came in 1861, almost contemporaneously with Italian unification and the onset (in 1862) of a political-military dynamic that led to the founding of the German Reich in 1871. But there was something far more fatalistic about the prehistory of the American Civil War than about the Italian or German unification process, in which so much depended on the tactical skill and gambler's luck of men like Bismarck and Cavour. The break-away of the South became ever more unavoidable in the second half of the 1850s.

First of all, the secession broke up the United States as a unitary nation-state. The open-endedness of historical developments enters the picture only in the *aftermath* of great confrontations. On the eve of the Battle of Königgrätz in 1866 many people, if not most, expected that Austria would emerge the winner. With *hindsight* Prussia's victory is understandable: Moltke's mobile offensive strategy, together with the better weaponry and higher educational level of the Prussian conscript army, was the decisive factor. It was still a close call, however. If we allow ourselves a little thought experiment and imagine that the American Civil War ended in military stalemate, then the North would have had to accept the breakup of the republic. And if the Confederacy had been able to continue with its peaceful development, the slaveholder regime would probably have become a prosperous and internationally influential second great power in North America—a prospect to which even Britain's Liberal government began to warm in 1862, before the course of the war made it illusory.[76] Dwarfing the national risings in Poland (1830, 1867) and Hungary (1848–49), the secession of the Southern states was the most dramatic instance in the nineteenth century of a failed attempt to gain independent statehood.

After the end of the Civil War in 1865, the United States had to be refounded. In the years of the painful construction of a liberal Italy, the Meiji transformation in Japan, and the domestic consolidation of the German Reich, the United States—saved as a unitary state, but far from united internally—embarked on a new phase of nation building. The reincorporation of the South during the so-called Reconstruction period (1867–77) coincided with a further bout of westward expansion. The United States was unique in having to negotiate *simultaneously*, during its most intense period of internal nation building, three different processes of integration: (1) the annexation of the former slave states; (2) the incorporation of the Midwest behind the gradually advancing frontier; and (3) the social absorption of millions of European immigrants. The post-1865 refounding of the United States as a nation-state recalls most of all the model of hegemonic unification. In terms of pure power politics, Bismarck was the Lincoln of Germany, although the emancipator of no one. In the United States, the reintegration of a defeated civil-war adversary proceeded along traditional constitutional lines, without changes to the political system. This highlights the absolute symbolic centrality of constitutionalism in the political culture of the United States. The oldest of the world's great written constitutions has also been the most stable and the most integrative.

Abandoned Centers

Finally we consider a new situation for the nineteenth century: the abandoned imperial center. After 1945, several European countries woke up to the recognition that they were no longer in possession of an empire. Britain would have been more or less faced with this realization after the American War of Independence, had it not been able to compensate for the loss by building up its position in India and gaining new colonies and bases in the Indian Ocean. Spain did not have that chance: Cuba, Puerto Rico and the Philippines were all it had left after the liberation of the American colonies. Although Cuba in particular developed into a lucrative colony, Spain was from the 1820s confronted with the task of changing from the center of a world empire into an ordinary European nation-state—a special kind of nation building, involving contraction rather than expansion. For half a century it had relatively little success. Only in 1874 did political conditions stabilize. But in 1898 the shock of defeat in the war with the United States and the loss of Cuba and the Philippines threw everything into turmoil again. Spain, not the supposedly "sick men" on the Bosporus or the Yellow Sea, was the real imperial loser of the nineteenth century. Cuba, Puerto Rico, the Philippines, and the Pacific island of Guam were rich pickings for the United States; even the German Reich, which had played no part in the war, tried to help itself parasitically to a few morsels.[77] Spain was bitterly disappointed that the British did not support it against the United States—and resentfully felt targeted when Lord Salisbury, then the prime minister, made a speech in May 1898 about living and dying nations. The trauma of 1898 would weigh heavily for decades upon Spanish domestic politics.[78]

Brazil's independence similarly reduced the Portuguese empire to Angola, Mozambique, Goa, Macao, and Timor, but this was rather less dramatic than the shrinkage of Spain's position in the world. The total population of the empire fell from 7.3 million in 1820 to 1.65 million in 1850,[79] with only the African territories being of any real importance. It was a harsh blow when Britain demanded in 1890 that regions between Angola and Mozambique should be split off. Nevertheless, Portugal was not completely unsuccessful in building a "third," African empire: Angola and Mozambique, hitherto settled by Portuguese only in the coastal areas, were now subjected to "effective occupation" (as it is called in international law).[80] It was thus Spain, rather than Portugal, that became the first postcolonial country in Europe. In the looming "age of imperialism," the descendants of Cortés and Pizarro would have to learn with difficulty how to manage without an empire.

Which of today's nation-states came into being between 1800 and 1914? A first wave, lasting from 1804 to 1832, saw the creation of Haiti, the Empire of Brazil, the Latin American republics, Greece, and Belgium. Then a second wave, in the third quarter of the century, featured the hegemonic unification of the German Reich and the Kingdom of Italy. In 1878 the Great Powers decided at the Congress of Berlin that new states should be established in parts of the Balkans formerly under Ottoman rule. The Union of South Africa, formed in 1910, was in effect an independent state, more loosely connected than other dominions of Britain. The precise status of the other dominions, between reality and legal fiction, is hard to determine; in 1870 they ran their own internal affairs by means of representative institutions but were not yet sovereign under international law. The decades-long process of consensual transfer of powers was largely consummated in the First World War. The huge contribution in troops and economic assistance that Canada, Australia, and New Zealand made to the Allied victory, more voluntary than coerced, made it impossible for London after 1918 to continue treating them as quasi-colonies. On the eve of the First World War, the new nation-states on earth had not all come about through iron and blood—as Bismarck famously put it in 1862. Germany, Italy, and the United States did have such origins, but not Japan, Canada, or Australia.

3 What Holds Empires Together?

A Century of Empires[81]

Out of a world of empires, a small number of *new* nation-states struggled into existence in nineteenth-century Europe. When we turn to Asia and Africa, the picture is considerably more dramatic: here empires did triumph. Between 1757–64 (battles of Plassey and Baksar), when the East India Company appeared for the first time in India as a military great power, and 1910–12, when two medium-sized states, Korea and Morocco, were incorporated into colonial empires, the number of independent political entities on the two continents underwent an unparalleled decline. It is virtually impossible to say for sure how many such entities—kingdoms, principalities, sultanates, tribal federations, city-states, and so on—existed in eighteenth-century Africa or in fragmented regions of Asia such as the Indian subcontinent (after the fall of the Mogul empire), Java, and the Malay peninsula. A modern Western concept of the state is too angular and sharp-edged to do justice to the variety of such polycentric, hierarchically layered political worlds. What we can say for sure is that in Africa the several thousand political entities that probably existed in 1800 had given way a century later to roughly forty territories separately administered by French, British, Portuguese, German, or Belgian colonial authorities. The "partition" of Africa was, from an African point of view, the exact opposite: a ruthless amalgamation and concentration, a gigantic political consolidation. Whereas in 1879 some 90 percent of the continent was still ruled by Africans, no more than a tiny remnant was left by 1912,[82] and not a single political structure corresponded to the criteria of a nation-state. Only Ethiopia, though ethnically heterogeneous, administratively unintegrated, and (until his health broke down in 1909) ultimately held together by the towering figure of Emperor Menelik II, remained an autonomous player in foreign policy, signing treaties with several European powers and practicing with their forbearance "an independent African imperialism."[83]

In Asia the concentration of power was less drastic; this was, after all, the continent of ancient imperial formations. But here too, the big fish prevailed over the little. In the nineteenth century, for the first time in its history, India became subject to a central authority covering the whole subcontinent; even the Mogul Empire at its height in 1700 had not included the far South. On the Indonesian islands, following the great nobility-led Java rising of 1825–30, the Dutch gradually moved from a system of indirect rule that had left local princes a certain scope for cooperation to more direct forms of rule involving greater centralization and homogenization.[84] The Tsarist Empire after 1855 incorporated vast areas east of the Caspian Sea ("Turkestan") and north and east of the Amur River, and put an end to the independence of the Islamic emirates of Bukhara and Chiva. In 1897 the French finally merged Vietnam (the historic regions of Cochin China, Annam and Tonkin) with Cambodia and Laos into "L'Indochine," an assemblage without historical foundations. In 1900 Asia was solidly in the grip of empires.

China was and remained one such empire. In 1895 the new Japanese nation-state annexed the island of Taiwan at the expense of China, becoming a colonial power that followed Western methods, and soon gave itself up to grand geopolitical visions of pan-Asiatic leadership. Only Siam and Afghanistan retained a precarious independence. But Afghanistan was the utter opposite of a nation-state; it was—and remains today—a loose ethnic federation. Siam, thanks to the reforms of far-sighted monarchs since the middle of the century, had acquired many of the external and internal characteristics of a nation-state, but it was still a nation without nationalism. In official thinking and in the public mind, the "nation" consisted of those who behaved loyally toward the absolutist king. Only in the second decade of the twentieth century did conceptions of a Thai identity, or of the nation as a community of citizens, begin to take root.[85]

For Asia and Africa, the nineteenth century was even less than for Europe the age of nation-states. Previously independent polities, subject to no higher authority, found themselves absorbed into empires. Not one captive African or Asian country was capable of breaking free before the First World War. Egypt, governed since 1882 by the British, gained some amount of home rule in 1922 on the basis of a European-style constitution (though one more limited than Ireland's around the same time). But it remained an exception for decades to come. The decolonization of Africa began much later—in 1951 in Libya and 1956 in Sudan. The dissolution of the Ottoman Empire gave rise to "mandates" in the Middle East, which Britain and France, acting under the auspices of the League of Nations, treated as de facto protectorates. The first new Asian states subsequently developed out of these, beginning with Iraq in 1932, but they were all extremely weak structures subject to continuing "protection" and interference from outside.

The first genuine Asian nation-state might have been Korea, benefiting from a high level of integration inherited from its previous history. It suddenly lost its colonial master with the Japanese collapse in 1945. However, the division of the country at the onset of the Cold War blocked "normal" development. The real retreat of the European empires began in 1947—a year after the Philippines won sovereignty from the United States—with the proclamation of Indian independence. For Asia and Africa, only the twenty years after the end of the Second World War were the true era of the independent nation-state. The degree of preparedness for such independence had varied enormously in the late colonial era: intensive in the Philippines and India, almost nonexistent in Burma, Vietnam, and the Belgian Congo. Only in India, where the National Congress had since 1885 been the all-India rallying point for moderate nationalists, did the roots of emancipation as a nation-state lie in the nineteenth century.

All this points to the simple conclusion that the *twentieth* century was the great epoch of the nation-state. In the nineteenth-century world, empire remained the dominant territorial form of the organization of power.[86]

This finding casts doubt on the widespread image of "stable nation-states versus unstable empires"—a trope that goes back to the basic nationalist idea that the nation is natural and primal, whereas the empire it shakes off is an artificial imposition. Both Chinese and Western antiquity already thought of empires as subject to a cyclical fate, but this rests on an optical illusion. Since all empires decline sooner or later, it was believed that the seeds of their decline must be discoverable early on; and the availability of material from three millennia encouraged greater attention to this phenomenon than to the much younger nation-state. Nineteenth-century Europeans looked ahead contemptuously, triumphantly, or elegiacally to the decline of the Asiatic land empires, seeing them as unfit for survival amid the harsh international competition of the modern age. None of these prophecies held water. The Ottoman Empire dissolved only after the First World War. There was still a sultan when the last tsar lost his throne and his life and his Hohenzollern cousin was chopping wood for himself in exile. The whole field of Ottoman studies is nowadays agreed that the value-laden word "decline" should be erased from its vocabulary. In China the monarchy fell in 1911, but after four decades of confusion the Communist Party of China succeeded in restoring the empire at more or less the maximum extent it had achieved in 1760 under the Qing emperor Qianlong.

Much like the Habsburg Empire, which survived the existential threat of the Revolution of 1848–49 (especially strong in Hungary) as well as the defeat of 1866 at the hands of Prussia, the other nineteenth-century empires withstood major challenges. The Chinese empire eventually overcame the Taiping Revolution (1850–64) and the equally dangerous Muslim risings of 1855–73, while the Tsarist Empire recovered from its defeat in the Crimean War (1853–56). The Ottoman Empire suffered its worst blow in the devastating war with Russia in 1877–78, when it lost the greater part of the Balkans, hardly less valuable in geopolitical terms than the core Turkish region of Anatolia. No other empire had to absorb such a shock, after the loss of Latin America earlier in the century. Yet the rump empire soldiered on for several decades and, in its internal affairs, displayed trends that prepared the ground for the relatively stable Turkish nation-state that would be founded in 1923. If we add to this the fact that European colonialism survived two world wars, then the vulnerability of the empires appears less striking than their staying power and capacity for regeneration. They entered into the modern world as vastly modified "relics" from their formative centuries: the fifteenth (Ottoman), sixteenth (Portugal, Russia), or seventeenth (England, France, and Netherlands, or Qing China as the last chapter of an imperial history stretching back to the third century BC). In an early twentieth-century perspective, these empires appeared along with the Catholic Church and the Japanese monarchy as the oldest political institutions in the world.

Such survival would not have been possible without a considerable degree of cohesion and adaptability. The most successful survivors—above all, the British Empire in the nineteenth century—were even in a position to shape the circumstances in their particular space. They established conditions to which others had to respond by adjusting to them.

Types: Empire versus Nation-State

What differentiates an empire typologically from a nation-state? One possible criterion is how the elites that sustain or ideologically defend empire actually see the world—or in other words, which patterns of justification serve to legitimize the two political orders.[87]

1. The nation-state finds itself surrounded by other nation-states with a similar structure and clearly defined boundaries. An empire has its (less clearly defined) external boundaries where it encounters "wilderness" or "barbarians" or another empire. It likes to establish a buffer zone around itself. Direct borders between empires often have an unusually high level of military security (e.g., the Habsburg-Ottoman border in the Balkans, the borders between the Soviet and American empires in Germany and Korea).[88]

2. A nation-state, congruent in the ideal case with a single nation, proclaims its own homogeneity and indivisibility. An empire emphasizes all manner of heterogeneity and difference, seeking cultural integration only at the level of the top imperial elite. Core and periphery are clearly distinguishable in land as well as sea empires. Peripheries differ from one another according to their level of social-economic development and the degree to which they are ruled by the center (direct or indirect rule, dependence or sovereignty). Crises reaffirm the primacy of the core insofar as it is considered viable even without the periphery—an assumption widely confirmed in modern times.

3. Whether its constitution is democratic or authoritarian-acclamatory, the nation-state cultivates the idea that political rule is legitimated "from below"; government is just only if it serves the interests of the nation or the people. Empire, even in the twentieth century, had to make do with legitimation "from above"—for example, through loyalty symbols, the

establishment of domestic peace (*Pax*) and efficient administration, or the distribution of special benefits to clientele groups. Its form of integration was coercive, not voluntary: "intrinsically antidemocratic,"[89] "a sovereignty that lacks a community."[90] In almost every case where a colonial power allowed space for elections and political competition among its subjects, the gesture unleashed an irreversible dynamic toward emancipation. Empire and democracy are almost impossible to reconcile, whereas a nation-state depends on a general political awareness and involvement of the population, though not necessarily in the garb of democratic constitutionalism.

4. People as citizens directly belong to nation-states, with a general status based on equal rights and political inclusion. The nation is understood not as a conglomerate of subjects but as a society of citizens.[91] In an empire a hierarchy of entitlements takes the place of an equal citizenry. Insofar as there is such a thing as imperial citizenship that offers access to the metropolitan polity, it is restricted in the periphery to small sections of the population. Minorities must struggle to achieve special rights within the nation-state; empire rests from the beginning on the allocation of special rights and obligations by an unaccountable center.

5. Cultural affinities—language, religion, everyday practices—tend to be shared by the whole population of a nation-state. In an empire they are limited to the imperial elite in the core and its colonial offshoots. Moreover, differences between universal "great traditions" and local "little traditions" are generally preserved within an empire, whereas in a nation-state, mainly under the homogenizing influence of the mass media, they tend to be more blurred. Empires have a greater propensity than nation-states to religious and linguistic pluralism, that is, to conscious admission of plurality, which does not necessarily have to be based on universal moral principles of "tolerance."

6. By virtue of its supposedly higher civilization, the central elite of an empire feels that it has a kind of mission to create an educated social stratum at the periphery. The extremes of complete assimilation (France, at least in theory) and extermination (the Nazi empire in Eastern Europe) are rarely encountered. The civilizing task is normally understood in terms of a generous blessing. By contrast, analogous processes in nation-states—a universal school system, public order, a guarantee of basic subsistence, and so on—are not perceived as resulting from a *mission civilisatrice* but are defined as duties for the whole nation and as civil entitlements.

7. The nation-state traces its genesis back to the primal origins of its particular nation or even to a common biological ancestry, which may be a fabrication but is ultimately the object of genuine belief. In its clearest manifestations, what it constructs is a tribe-nation.[92] Empire, by

contrast, harks back to political founding acts of royal conquerors and legislators, often also utilizing the idea of an imperial *translatio* or continuation, for instance, when the East India Company, and later Queen Victoria, tried to derive legitimacy from their succession to the Mogul emperors. Empires have difficulty in (re)constructing their fragmented history other than in chronicles of supreme rulers. After the rise of national historicism, with its assumption of organic continuity, it became relatively easy to discover coherence in the past, not just in the political domain. While devising a social and cultural history of the nation—as in the nineteenth century Jules Michelet did for France—is facilitated by the focal role of an entity called "the [national] people," historizing an empire from the inside always has to grapple with the lack of a single historical subject.

8. The nation-state claims a special relationship to a particular territory, visible in places of remembrance that are often given the character of holy sites. The "inviolability" of a national geo-body is a "core belief of modern nationalism."[93] Empire has an extensive rather than an intensive relationship to the soil, which in its view is primarily an area of land available for it to rule. An exception to this premise is settler colonialism, because it tends toward an intensive relationship with the soil—a source of tension with the imperial administration as well as a major root of colonial nationalism.

Dimensions of Imperial Integration[94]

There are advantages in understanding nation-states and empires in terms of their different "logics" and of the meanings imputed to them. A complementary approach is to look for their distinctive modes of integration. What holds together a typical nation-state and a typical empire?

Empires are structures of rule on a large scale. They might be defined as the largest political entities possible under given geographical and technological conditions. They are composite structures. Imperial integration has a horizontal and a vertical dimension. Horizontally, territorial segments of the empire must be linked to the center; vertically, rule and influence must be secured in the colonized societies. First of all, horizontal integration requires coercive instruments and military potential. All empires rest on a latent threat of force beyond the imposition of a statutory legal system. Even if empires were not characterized by ongoing terror, even if the British Empire in the nineteenth and twentieth centuries bound itself to a basic rule of law (when it was not actually involved in cruelly suppressing revolts), an empire always stands in the shadow of a state of emergency. The nation-state has at worst—and rarely—to face revolution or secession, whereas empire must constantly be on the lookout for rebellion and treason on the part of disaffected subjects and allies. The ability to crush an uprising is a basic prerequisite for an imperial presence. The colonial state preserved this ability until very late in its existence. The British still had it in India during the Second World War, and in Malaya until the 1950s. The French, despite strenuous efforts, were unable to regain it in Vietnam after the Second World War and lost it in Algeria after 1954. Empires do not rely solely on local resources of violence; they retain the possibility of intervention from the center, symbolized in the punitive expeditionary force. One principle is to deploy special units from outside the area—Cossacks, Sikha, Gurkhas, Tirailleurs Sénégalais, Polish troops for the Habsburg wars in Italy—a kind of globalization of violence. This could sometimes bear strange fruit. The French intervention force in Mexico included 450 crack troops that Said Pasha, the ruler in Cairo, had lent at a price to his foreign protector, Napoleon III. These Egyptian troops remained until the end, providing cover for the French withdrawal and becoming among the most highly decorated troops of the Second Empire.[95]

Transportation and communications over long distances were constant necessities of empire.[96] Before regular telegraph services were introduced in the 1870s, news could not travel overseas faster than the ships and people who carried it. This alone is evidence that, even with the best organization of correspondence (the Spanish Empire in the sixteenth century, the East India Company), premodern empires were joined up very loosely by today's standards. Yet it is questionable whether modern communications technology made empires more stable. By no means did colonial authorities always have a monopoly over the transfer of information; their adversaries employed similar methods as well as countersystems, from the bush drum to the internet.

Whether an elaborate bureaucracy was created as an instrument of integration depended as much on the political system and style of the imperial center as on functional requirements. Although the Chinese Empire of the Han dynasty was much more tightly administered than the early Imperium Romanum during the same period, there was not a corresponding difference in the success of integration. Modern empires, too, have varied widely in their degree of bureaucratization, as well as in the mode and extent of links between the state personnel and the institutions of the core and the periphery. With the exception of China, there has seldom or never been a single administration throughout an empire. The British Empire, which managed to retain its cohesion over the centuries, had a confusing array of authorities held together at best by the general responsibilities of the cabinet in Westminster. As for the French, the startling multiplicity of their colonial institutions contradicts any idea of Cartesian clarity at the level of the state.

Unlike a nation-state, which has a more or less matching national society, an empire is a political but not a social association. There is no overarching imperial "society." The characteristic mode of imperial integration may be described as political integration *without* social integration. The social bonds were strongest among officials sent out for a limited term—that is, top cadres below the level of viceroy and governor. Until the introduction of competitive, efficiency-oriented examinations for the colonial service, family links and patronage played a major role everywhere in the filling of positions. Bureaucratization of imperial service led to a different, no longer kin-based kind of esprit de corps, but also to new kinds of career patterns and imperial circulation. A posting in the empire might result in either promotion or demotion.

The ties between social circles in Europe and settlers in the colonies were much weaker. Diverse processes of creolization, together with the formation of new settler identities, repeatedly made themselves felt. The strivings for autonomy were especially strong if they were directed, as in Spanish America, against newcomers with status in the home country, or if immigrants felt an especially great social distance from the metropolis, as they did in the (erstwhile) penal colony of Australia. Often the necessary demographic mass was lacking for self-reproducing settler societies. Things then remained at the level of insular, fragmented communities, such as one finds in urban trading bases and administrative centers or among a small settler population spread over a wide area (as in Kenya around 1890). Far looser still were relations across the barriers of ethnicity and skin color. Over time some empires permitted or facilitated the rise of colonial subjects within administrative, military, and ecclesiastical hierarchies; others persisted with an ethnic-racial exclusivism, which actually tended to grow in the course of the nineteenth century (and was absolute in the German and Belgian colonies in Africa, for example). A unique exception in modern times was the systematic recruitment of foreigners into the military elite of the Ottoman Empire and Mamluk Egypt. In general, it is questionable to equate political "collaboration" (structurally essential for the functioning of colonial state apparatuses) with social integration in such areas as marriage. Horizontal *social* relations were not the cement of empire.

Symbolic integration was another matter. The generation of identity through all manner of symbols is essential for nation-states, but it is at least as important for empires, which have draw upon them in compensation for the lack of other sources of coherence. Monarch and monarchy, as loci of symbolic condensation, had the dual advantage of rallying European colonials and impressing the natives. At least that is how it seemed. We cannot be sure whether many Indians felt excited by Queen Victoria's proclamation as Empress of India in 1876, but we do know that her grandfather, George III, served the North American revolutionaries as a useful *negative* symbol. Everywhere, monarchy was deployed as a focus of integration: in the Habsburg state, where on the occasion of the Imperial Jubilee in 1898, a *Reichspatriotismus* centered on the aged Franz Joseph was supposed to neutralize the newly rising nationalisms; in the Wilhelmine and Tsarist empires; very skilfully in the Qing Empire, with its Buddhist and Muslim minorities; heavy-handedly in the Japanese Empire, where Chinese (Taiwanese) and Korean subjects were forced to observe a cult of the tennō (emperor) that was culturally alien and repugnant to them.

Another popular symbol was the armed forces—in the British case, especially the ubiquitous Royal Navy. The bonding power of symbols, and perhaps of other kinds of affective (not primarily interest-related) solidarity, was particularly evident during the two world wars, when the dominions of Canada, Australia, and New Zealand (and sui generis South Africa) assisted Britain to an extent not explicable in terms of only the formal existence of the empire and the actual power relations in the world.

Finally, we need to mention four further elements of horizontal integration: (a) a shared religion or religious denomination; (b) the importance of a common legal system (e.g., Roman or British) for the unity of far-flung empires; (c) extensive market relations; and (d) the external relations of the empire. The last of these is by no means the least. Empires have always secured and defended their borders militarily: against neighboring empires, against pirates and other bandits, and against the constant threat of disturbances by "barbarians." But they have varied greatly in the extent to which they have protected themselves against the commercial activity of foreigners. Free trade, which Britain permitted in its own empire from the middle of the nineteenth century, while demanding the same of others, was a novel and extreme development. Most empires with sufficient organizational strength practiced some form of "mercantilist" control over their external economic relations. Some—for example, China from the early Ming period until the Opium War, or Spain for long stretches of its imperial rule—restricted third parties to activities within tightly supervised enclaves. Others, such as the Ottoman Empire, tolerated or even promoted the establishment of taxable commercial diasporas (Greeks, Armenians, Parsis, and so on). France awarded and guarded monopolies for colonial trade. In the nineteenth century Britain's free-trade policy helped to undermine the remaining systems of imperial protection, but in the twentieth it was unable to prevent the return of neo-mercantilism. In the 1930s and 1940s, the widespread practice of tariff preferences, trading blocs, and currency zones encouraged deeper integration of the British and French empires, as well as increased aggression on the part of the new fascist-militarist imperialisms.

One reason why it is essential to distinguish between horizontal and vertical integration is that empires, unlike hegemonic configurations or federations, are arranged in a radial structure.[97] Particular peripheries are only loosely in contact with one another; the metropolis seeks to direct all flows of information and decision making through the eye of the imperial needle; liberation movements are kept isolated from one another. This structural tendency to centralization stands in the way of broadly based horizontal solidarity and the formation of an empire-wide upper class. It is therefore also necessary to find local means of ensuring the loyalty of imperial subjects, the main purpose of vertical integration. In fact, most mechanisms of horizontal integration also have a vertical dimension: the recycling of violence through the recruitment of local *sepoy* troops and policemen provides a symbolic link with indigenous notions of political legitimacy;

the colonial government systematically observes and spies on the society in its charge; the controlled delegation of power to long-established notables or a wide range of new "collaborative elites" is tirelessly pursued.

The greater the perceived or "constructed" cultural and racial differences, the more plainly a tension develops between the need for political inclusion and the tendency to social-cultural exclusion. The white club remains closed to the politically useful local potentate, who takes umbrage at the slight. On the other hand, settlers are useful business partners even when they achieve political emancipation. This was the basis of the dominion model, which functioned well for both sides. Similarly, Britain and the United States maintained close economic links after the war they fought against each other in 1812 and went on gradually—and despite many upheavals—to build a wider "special relationship." At the other end of the typological spectrum are colonial systems with no vertical integration, most notably the slave societies of the eighteenth-century British and French Caribbean.

Theoretically, sources of *disintegration* may derive from the revaluation of integrative ties. But as it was already known in antiquity, most empires are prey not only to dissolution within but also to a combination of internal erosion and external aggression. Or, to put it more sharply, the greatest enemies of an empire are always other empires. It is striking that empires usually break up into smaller entities, realms, or nation-states; they seldom pass directly into hegemonic or federative structures. Plans for nations across the ocean, as mooted in the Bourbon reforms of Spanish America after 1760 or by the British colonial minister Joseph Chamberlain around 1900, inevitably fell short. The only success stories were a few (by no means all) federations under the umbrella of an overarching empire, such as Canada attempted in 1867 and Australia in 1901; similar projects for Malaya and British Central Africa during the decolonization period ended in failure.

Let us summarize what has been said so far in terms of an "ideal type." An empire is a spatially extensive multiethnic entity with an asymmetrical, and in practice authoritarian, core-periphery structure, which is held together by a coercive apparatus and political symbolism and by the universalist ideology of the imperial state and its elite bearers. Social and cultural integration does not take place beneath the level of the imperial elite; there is no homogeneous imperial society and no common imperial culture. Internationally, the center does not allow the periphery to develop external relations of its own.[98]

Relations within an empire involve constant contestation, bargaining, and compromise: it is not one huge barracks, and scope can be found on all sides for resistance and independent initiative. If conditions are favorable, people at all levels of society can live well and securely in an empire. But none of this should make us forget its essentially coercive character. An entity that many or all join of their own free will is not an empire but—as was the case of NATO before 1990—a hegemonic association with mainly autonomous partners and a primus inter pares at the center.

4 Empires: Typology and Comparisons

Empires differ from one another by their size on the world map, their total population, the number of their peripheries, and their economic performance. For the whole of the nineteenth century, the Netherlands had in Indonesia a colony that (after India) was economically the most successful of the age. Since it had no other colonies apart from Surinam and a few tiny islands in the West Indies, its "empire" was of quite a different caliber from that of the British. In a very different way, the same applies to the German colonial empire that came into being after 1884: a collection of thinly populated territories in Africa, China, and the South Seas that were expendable for the home country. Whereas the Netherlands was a small country with a large and wealthy colony, Germany was the opposite. In the nineteenth century, only the British and the French had what could be described as world empires. The Tsarist Empire was so extensive and so ethnically diverse that it constituted a world of its own; the Mongol "world empire" of the Middle Ages was not significantly larger.

Leviathan and Behemoth

It is not possible to translate the above ideal-typical empire into a neat and full typology; imperial phenomena are too diverse for that, both spatially and temporally, even in a single century. But a few points may help us to identify certain variants.

The distinction between land and sea empires is often considered the most important, not only academically but as a deep antagonism within the world of politics. Some geopoliticians and geophilosophers, from Halford Mackinder to Carl Schmitt, have even viewed the supposedly unavoidable conflict between continental and maritime powers as a fundamental trait of modern world history. The long-known problem with this is that the two types of empire are assumed, generally without proof, to be incomparable. Narrow conceptions of "overseas history" have prevented the historical experience of Russia and China or the Ottoman and Habsburg empires—not to speak of Napoleon or Hitler—from being used for a comparative analysis of empire. In reality, the distinction between land and sea empires is not always clear-cut or helpful. For England and Japan, everything was "overseas." The Imperium Romanum ruled both the Mediterranean and inland regions stretching all the way to Britain and the Arabian desert. A maritime empire in its pure form should be thought of as a transcontinental network of fortified ports, such as only the Portuguese, Dutch, and English constructed in the early modern period. Until the late eighteenth century, all of these contented themselves with controlling coastal footholds and their immediate hinterland. The sixteenth-century Spanish global empire already had a continental component insofar as it had to deploy techniques of territorial administration to consolidate its hold on the Americas. The East India Company had to develop similar techniques after it had gained control of Bengal in the 1760s.

Control problems appeared as soon as overseas bases expanded into, or were complemented with, territorial colonies. Geographical distance from the European imperial center was an important, though not the only, factor in their solution. Decentralization, one of the strengths of the British Empire, was a necessary result of the difficulties of communication in the days before the telegraph. Ever since the conquest of India, the British Empire was an amphibious structure, a Leviathan and Behemoth rolled into one. India and Canada were subordinate land empires of a special kind, gigantic countries that in the course of the nineteenth century were opened up, no less than the Tsarist Empire, by what geopoliticians considered the modern source of imperial land power: the railroad.[99] Logistics in the age of the steam engine, on wheels and at sea, did not unambiguously favor either of the two basic types. Both land and sea empires changed their character with the increase in transportation speed and volume. In preindustrial times, the same distance was easier and faster to cover on water than on land, but at the end of our period came a world war in which the resources of two vast land masses were pitted against each other. The Allies were victorious not because of a built-in superiority of maritime forces over land powers, but because their merchant naval capacity gave them access to the land-based industrial and agricultural potential of America, Australia, and India.[100] Meanwhile, the great battleship duel for which Germany and Britain had been steadily preparing failed to materialize.

Even so, a few differences between "pure" land and sea empires should not be overlooked. Foreign rule does not have the same meaning when it defines the relationship between old neighbors and when it comes about through the leap of an invasion; in the former it may be part of a long-term back-and-forth movement, such as that which occurred over centuries between Poland and Russia. In land empires, great efforts must be made to justify and assert an overarching claim to sovereignty. Examples include the dynastic unions that made Austria's emperor the king of Hungary, Russia's tsar the king of Poland, and China's Manchu emperor the great khan of the Mongols. The secession of part of a tightly knit contiguous empire tends to be more dangerous for the center than are Creole autonomy movements across the seas. They reduce the territory of the empire as a great power, possibly creating a new enemy or a satellite of a rival empire on its borders. The geopolitics of land empires is therefore different from that of sea empires. But it should not be forgotten that both Britain and Spain made huge military efforts to prevent the loss of their American possessions in the age of the Atlantic revolutions.

Colonialism and Imperialism

The artificial term "periphery," often used in this chapter, has a somewhat broader meaning than the more common "colony." In the nineteenth century, the power elites of the continental empires (Russian, Habsburg, Chinese, Ottoman) would have indignantly rejected any idea that they ruled over colonies, whereas others (e.g., the Germans) were proud of "possessing" some. In Britain, people insisted that India was not an ordinary colony but something unique; in France a sharp dividing line was drawn between Algeria (part of the French Republic) and the colonies proper. We should bear in mind that a *structural* definition of "colony" must be sufficiently tight to exclude other kinds of periphery.[101]

The late nineteenth-century term "colony" has a connotation of social-economic backwardness vis-à-vis the metropolis. But the Polish territories in the Tsarist Empire, Bohemia in the Habsburg Monarchy, and Macedonia in the Ottoman Empire were by no means underdeveloped—though they were certainly dependent peripheries whose political fates were decided in Saint Petersburg, Vienna, or Istanbul. Within the British Empire, there were few similarities in 1900 between Canada and Jamaica. Both were peripheries in relation to the imperial center, but one was a democratically self-governing proto-nation-state, the other a crown colony in which the governor exercised nearly unlimited power on behalf of the colonial minister in London. In many respects, the dominion of Canada was more akin to a European nation-state than to a Caribbean or African colony within the same empire. The same was true of peripheral lands in the Tsarist realm. For most of the nineteenth century Finland was a semi-autonomous grand duchy, occupied by Russian troops, in which a minority of originally German-speaking Swedish landowners and merchants set the social tone. Its dependence was thus scarcely of the same type as that of Turkestan, first conquered in the 1850s and (after the fall of Tashkent in 1865) treated more like an Asian colony of Britain or France than any other part of the Tsarist Empire.[102] Not all imperial peripheries were colonies, and colonial frontiers were not equally dynamic in all empires. Colonialism is but one aspect of nineteenth-century imperial history.

The rapid conquest and partition of the African continent, a new swashbuckling tone in international politics, and political support for European banks and resource-development corporations created a widespread impression around the end of the century that the world had entered a new "imperialist" phase. Many clever things were written to analyze this phenomenon. In particular, *Imperialism: A Study* (1902) by the British economist and journalist John A. Hobson can still be read today as a profound and partly prophetic diagnosis of the times.[103] This literature, including important contributions by Marxists such as Rosa Luxemburg, Rudolf Hilferding, and Nikolai Bukharin, sought above all to get to the bottom of Europe's (or even "the West's") new global expansionist dynamic.[104] For all their differences on points of detail, all were agreed that imperialism was an expression of tendencies characteristic of the modern age. Only the Austrian all-around social scientist Joseph A. Schumpeter raised the objection in 1919 that imperialism was in fact a political strategy of antiliberal pre-bourgeois elites, or of capitalist forces shying away from the world market.[105] In that, there was a lot of truth. Apart from the shock of the new that impressed people at the time, we can now see more clearly long-term continuities of European and other processes of expansion,[106] and the very different impetuses and motives that lay behind them.

A descriptive concept of imperialism therefore has the advantage that it does not bind one to a particular political, economic, or cultural explanation, since it refers to the sum of actions geared to the conquest and preservation of an empire. It would thus be possible to speak of Roman, Mongol, or Napoleonic imperialism. The phenomenon is characterized by a certain kind of politics that involves the crossing of borders, disregard for the status quo, interventionism, rapid military deployment at the risk of provoking war, and a determination to dictate the terms of peace. Imperialist politics bases itself on a hierarchy of peoples, always divided into the strong and the weak and usually graded by culture or race. Imperialists consider that their superior civilization entitles them to rule over others.

The theories that postulated an affinity between imperialism and capitalist modernity were referring to a special situation around the turn of the twentieth century, albeit one of exceptional significance. In the long sequence of empires and imperialisms, a "first age of global imperialism" began in 1760 with the Seven Years' War.[107] A second age got under way around 1880 and ended in 1918, while a third stretched from the Japanese invasion of Manchuria in 1931 down to the end of the Second World War. The second age of global imperialism, often known as High Imperialism, came about through the intertwining of four originally independent processes: (a) world economic integration in leaps and bounds (early "globalization"), (b) new technologies of intervention and domination, (c) the collapse of mechanisms to preserve the peace in the European system of states, and (d) the rise of social-Darwinist interpretations of international politics. Another novelty in comparison with the first age was that imperialist politics was no longer conducted only by Great Powers—or in other words that the Great Powers allowed weaker European powers a share of the imperial cake. King Leopold II, acting in an individual capacity, could even go over the head of Belgium's state institutions and get the Berlin Conference on Africa in 1884 to guarantee the giant Congo Free State as his private colony.[108]

It has often been claimed that High Imperialism was a direct result of industrialization, but things are not so simple. With the exception of Africa, the greatest territorial expansion took place *before* the industrialization of the imperial power in question: the Tsarist Empire in Siberia, the Black Sea, the steppes, and the Caucasus; the Qing expansion in Central Asia; the British conquest of India. India became an important market for British industry *after* it was conquered. Similarly, Malaya was not gradually brought under British control *in order to* open up access to rubber; its importance soon afterward as a supplier is another story. But it is true that there were indirect connections, for example, the American sales of the Lancashire cotton industry brought in Mexican silver that helped to finance Lord Wellesley's Indian conquests.[109] Industrialization does not necessarily push countries into an imperialist policy. If industrial capacity had translated directly into international strength, Belgium, Saxony, and Switzerland would have been aggressive great powers by 1860. The hunt for raw materials and "state-protected" markets—a hope repeatedly disappointed— was sometimes a not-insignificant motive; it played a certain role in France, for example. But not until the twentieth century did governments come to see control over foreign resources as a *national* objective of prime importance. Oil was the main spur for this strategic upgrading of raw materials, which began in the years before the First World War. Until then both resource extraction and direct capital investment had been a matter for private firms, although these could be sure of their government's support on an unprecedented scale. Imperialist politics in the second age of global imperialism was largely a matter of garnering favorable plantation, timber, mining, railroad, and canal concessions for private European business interests.[110] In the last third of the nineteenth century, an overall restructuring of the world economy was everywhere in evidence. Economic globalization was not a direct result of government policies but stood in a two-way relationship with it. Raw materials were no longer stolen but, rather, acquired through a mixture of extraction systems (e.g., plantations) and commercial incentives. The "mix of compliance mechanisms" changed, also depending on the type of colony.[111]

What direct effects did industrialization have on methods of imperial warfare? The conquest of India in 1800 was still accomplished with preindustrial military technology. Wellesley's chief adversaries, the Marathas, even had the better artillery (maintained by German mercenaries), but they were unable to deploy it to advantage.[112] Only steam-powered gunships brought industrial technology decisively into play, for the first time in the Anglo-Burmese war of 1823–24, and then in the Opium War against China in 1841.[113] A second phase of colonial conquest took place under the aegis of a (by European standards) relatively simple innovation: the Maxim gun. Invented in 1884, it was capable in the 1890s of turning clashes between European and indigenous troops into outright massacres.[114] The key factor was not the absolute level of industrial and technological development in the imperial heartland, but the capacity for coercion on the spot. Industrial strength had to be translated into *local* superiority case by case. Had this not been so, Britain would not have come off worse in the Second Afghan War (1878–90), or the United States in a whole series of twentieth-century interventions (Vietnam, Iran, Lebanon, Somalia, Afghanistan, etc.).

Not all imperialisms were equally active in the nineteenth century, and the differences between them did not follow the dividing line between land and sea powers. Three imperial powers in the European system of states were active all through the century: the United Kingdom, Russia, and France. Germany joined in as a colonial power in 1884, but under Bismarck it did not yet consciously pursue a *Weltpolitik*. This would be the Wilhelmine watchword around the turn of the century, once the modest colonial empire was felt to be too restrictive. Austria was a great power, though of second rank since the Prussian triumphs

of 1866–71, and it was also an empire, though it did not pursue a policy of imperialist expansion. The Netherlands, Portugal, and Spain, none of them a great
power, kept up old colonial possessions without adding anything major. The
Chinese and Ottoman empires, once highly bellicose and dynamic, were now on
the defensive in relation to Europe (though China less so than the Ottomans).
From 1895 on, Japan was a very active imperialist player. The empires of the nineteenth century differed in terms of their imperialist intensity. What might seem
at first sight, or in a very abstract theoretical perspective, to be a single closed imperialist system breaks down on closer inspection into imperialisms in the plural.

5   Central and Marginal Cases

The Habsburg Monarchy

The typical empire cannot be found in historical reality. And even a neat typology fails because of the multiplicity of possible criteria. Individual cases are
able to be defined, however, through a comparison of their specific characteristics.

An extreme case was the Habsburg Empire.[115] It was territorially overbur-
dened and penned in: an empire in the heart of Europe, the only one with problematic access to the sea (military ports of Trieste and Pula) and no navy worth
mentioning.[116] Metternich maintained at the Congress of Vienna that Austria
had reached its optimum extent, rejecting any further attempt at expansion.[117]
Yet he subsequently condoned the acquisition of Lombardy and the Veneto,
and Austria soon warmed to the idea of becoming a major power in Italy. It remained so until 1866. The occupation of Bosnia-Herzegovina in 1878, followed
by its annexation in 1908 that began the countdown to the First World War, was
less an act of calculated empire building than an anti-Serbian and anti-Russian
thrust by an irresponsible war party at the Viennese court.[118] No one wanted to
bring the two million South Slavs of Bosnia into the empire, upsetting the delicate balance of nationalities, and so Bosnia-Herzegovina was incorporated with
Reichsland status, which expressed the awkwardness of its position.

In no other empire was the term "colony" so out of place as in the Habsburg
Monarchy; there was not even a disadvantaged "internal colony," such as Ireland
represented in relation to England. Yet the imperial and royal (*kaiserlich und
königlich*, or k.u.k.) monarchy displayed many features of an empire.[119] It was a
weakly integrated multiethnic entity, a collection of territories with often ancient
historical identities of their own. Hungary, in particular, which in 1867 agreed to
a constitutional settlement as a semiautonomous kingdom (King Franz Joseph
being represented in Budapest by a Habsburg archduke), had its own government and two-chamber parliament within the newly created Dual Monarchy.
After the German-Austrians, no other ethnic group in the empire now had such
a strong position as the Magyars. In fact, Hungary was placed comparably to the
Dominion of Canada (formally created also in 1867) within the British Empire.

In both cases, the imperial framework was not experienced as coercive: Hungarians, like Canadians, could make a career for themselves within it; economic development was not seriously impeded by the imperial center, and much of the state expenditure was shared.[120] Like the British Empire, the Danube Monarchy did not develop into a federation; the whole state actually became more heterogeneous after 1867. The Slav nationalities rightly felt themselves to be the losers and, not seeing the emperor as an a neutral arbiter of interests, inwardly distanced themselves from the settlement. Right to the end, the various components of the Habsburg Monarchy were integrated in the imperial manner: a shared imperial culture and identity took shape to some extent, without being politically enforced, while horizontal *social* integration continued to be restricted. The empire was held together only at the top, through the symbols of monarchy and a multinational officer corps at least as mixed in composition as its counterparts in early modern Spain or British India. Yet it did not appear to most of its inhabitants as a military state. Only the Italians in Lombardy-Veneto had the sense of being under tyrannical alien rule. In a divided region such as Galicia, the Austrian part was typically far more liberal, as well as more enlightened than the Russian or Prussian zone, including toward its large Jewish population. The national groups that had been part of the Habsburg Empire for centuries were rather wary about their relations with one another. The notorious Habsburg "nationalities question" bore less on the links of peripheral regions with the center (as in the Tsarist Empire) than on their own conflictual relations with one another; Hungary, for example, had explosive minority problems of its own.[121]

The Habsburg Empire was unique in having no remnants of an open "barbarian frontier;" it no longer even had any settler colonialism. It was ethnically and culturally more uniform than the overseas empires of Western European powers, or than the Russian and Ottoman empires. Although the different languages, customs, and historical memories became ever more visible on the rising tide of national consciousness, all subjects of the emperor in Vienna had a white skin, and the great majority were Roman Catholics. Orthodox Serbs, the largest religious minority, made up just 3.8 percent of the population in 1910, and Muslims only 1.3 percent.[122] Compare this with the share of non-Muslims in the officially Muslim Ottoman Empire (roughly 40 percent before the major territorial losses in the Balkans after 1878) and of non-Orthodox in the officially Christian Orthodox Tsarist Empire (29 percent in 1897), or even with the situation in the British Empire, where all skin colors and all world religions were represented, and where Hinduism was numerically the preponderant religious orientation.[123] Even if people in Vienna, Budapest, or Prague looked on South Slavs or the Romanian minority as "barbarians," these peoples did not fit into the Western European, Russian, or Chinese discourse of noble and ignoble "savages." The Habsburg Empire was geographically and culturally a European/Western multinational structure. The equality of all citizens before the law made it in principle one of the most modern and "civic" of empires.[124] But this was not true in every respect. A sense of nationhood was more developed among Hungarians and Czechs, at least, than among German Austrians. In 1900 the latter did not yet constitute a nation, let alone a ruling one. Elsewhere in the world, a nation-state of the titular nation slumbered beneath the casing of the imperial metropolis, ready to stand on its own feet after the loss of peripheral regions; the Turkish Republic, for instance, emerged with astonishing speed out of the Ottoman Empire after the First World War. Not so in the Danube Monarchy. In this respect it was the most antiquated of all the empires, and therefore not by chance one of the first to disappear from the map.

The general secession that put an end to it has only one parallel: the decomposition of the Soviet Union in 1990–91. It followed military defeat in a world war that had strengthened rather than weakened in internal cohesion of the British Empire. Nevertheless, the most appropriate comparison is with that empire: Lombardy, Hungary, and the Czech lands had built up their respective nations within the Danube Monarchy so successfully that, like Australia, New Zealand, and Canada, they emerged from their imperial past without major convulsions as politically and economically viable nation-states. The same cannot be claimed of the Middle Eastern and Balkan successor states of the Ottoman Empire. At the other end of the spectrum lies the Chinese Empire, which suffered only one breakaway in modern times: Outer Mongolia in 1911. This state, after an early shaky autonomy and sixty years as the longest-lived satellite of the Soviet Union, regained only in 1991 the independence it had lost in 1690.[125]

France's Four Empires

For centuries the House of Habsburg competed with France for supremacy in continental Europe. In 1809, when Napoleon drove the Austrian monarchy to the brink of collapse and occupied Vienna, two nearly pure continental empires faced each other. The Napoleonic Empire, though so short-lived that most of the literature does not regard it as such, was indeed an empire of the first water. Despite the subordination of politics to military affairs throughout the sixteen-year period, evident particularly in the constant quest for money and recruits, it is possible to identify certain systemic contours[126] Two characteristics of empires in general were especially pronounced. First, Napoleon soon created a genuinely imperial ruling elite, which he allocated to, and rotated among, positions all over Europe; its core, the Bonaparte and Beauharnais families, supplied the most trusted marshals and a caste of professional administrators ready to serve anywhere.[127] The empire of Napoleon, the last and greatest ruler of Enlightenment absolutism, was an ultrastatist structure built similarly throughout, which professed to modernize in the general interest but allowed its subjects no institutionalized voice or scope for action. Like any empire, it relied on the collaboration of indigenous rulers and elites, without whom it would not have been able to mobilize the resources of subject societies. But they did not have even the modicum of formal representation granted under the British model.[128] No empire of the eighteenth or nineteenth century was more highly centralized. A law or decree issued in Paris had immediate validity in every nook and cranny.

Second, the whole Napoleonic project of expansion was forced through with a cultural arrogance rarely seen elsewhere, even between Europeans and non-Europeans, before the later age of fully fledged racism. This imperiousness, based on a conviction that postrevolutionary secular France represented the pinnacle of enlightenment and civilization, made itself felt least in the core regions identified by Michael Broers (eastern France, the Netherlands, northern Italy, and the German Rhine Confederation), and especially in the "outer empire" made up, above all, of Poland, Spain, and Italy south of Genoa.[129] Here the French conducted themselves as an occupying power, treating the "superstitious" and inefficient natives with contempt and engaging in outright colonial exploitation. The Napoleonic Empire exceeded all others in its objective of cultural uniformity. Influenced by Enlightenment utopias of a continent at perpetual peace with itself, Napoleon claimed in his memoirs to have dreamed of a united Europe "everywhere guided by the same principles, the same system."[130] First the non-French elites were to have been Gallicized, then a radical *mission civilisatrice* was to have freed the popular masses from the yoke of religion and localism. By 1808 this vision was already running into trouble in Spain.[131]

In October 1813 the Napoleonic Empire ended on the battlefields near Leipzig. France's nineteenth-century overseas empire, launched in 1830 by the conquest of Algiers (a typical opportunist diversion from internal political difficulties), was a completely new venture.[132] As there is often talk of a first and a second British Empire, separated from each other by American independence in 1783, so we might differentiate four French empires:

- a first, ancien régime empire, mainly covering the Caribbean, which ended at the latest with Haiti's independence in 1804; strongly mercantilist in its political outlook, weakly based on emigration, and built on slave labor;
- a second, Napoleonic empire, consisting of *France-Europe* conquered in a series of lightning wars;
- a third, colonial empire, built after 1830 on the slender foundation of the colonies returned to France in 1814–15 (e.g., Senegal) and dominated until the 1870s by Algeria; and
- a fourth empire, involving expansion of the third empire, which was now global in reach and, from the 1870s to the 1960s, had its geographical centers of gravity in North Africa, West Africa, and Indochina.

What remains today from this fourfold history are, of all things, remnants of the first empire: above all the overseas *départements* of Guadeloupe and Martinique, which are integral parts of the European Union. The post-Napoleonic empires were from beginning to end responses to the British Empire, never managing to extricate themselves from its shadow. The invasion of Algeria, easy to sell internationally as a punitive operation against a rogue state of Muslim pirates and kidnappers, was an attempt to intervene in a power vacuum that Britain had not yet picked out for itself. True, the British had controlled Gibraltar since 1713, confined Napoleon's navy to the Mediterranean, and held the island of Malta as a de facto possession since 1802 and as a crown colony and naval base since 1814. Nevertheless, until their occupation of Egypt in 1882 they had no other colonial interests in the region. Politicians and the public in France suffered for a long time from the trauma of their country's second-rank position in imperial geopolitics.

By other measures, however, France's colonial expansion was very successful. Its overseas empire, though far behind the British, was the second-largest in the nineteenth century. But territorial figures (9.7 million square kilometers in 1913 compared with the British 32.3 million[133]) are somewhat misleading on their own, since the latter figure includes the dominions and the former the uninhabited wastes claimed by Algeria. On the eve of the First World War, the British had important possessions on *all* continents, the French only in northern Africa (Algeria, Tunisia, Morocco), western and central Africa, Madagascar, Southeast Asia (Indochina, i.e., Vietnam and Cambodia from 1887, plus Laos from 1896), the Caribbean (Guadeloupe, Martinique), the South Seas (Tahiti, Bikini, etc.), and South America (French Guyana). France's colonial interests in Asia did not reach significantly beyond Indochina. In eastern and southern Africa it had no greater presence than in North America or Australia. And even in Africa, where French possessions were most numerous, Britain had the advantage of holding colonial positions on both the west and east coast all the way from Egypt to the Cape of Good Hope, together with the important Indian Ocean island of Mauritius.

Later conquests never dislodged Algeria from its number one place among French colonies. Chronologically, the Algerian story fits into a wider periodization. The original invasion met well-organized resistance under the leadership of Emir Abd al-Qadir (1808–83), who from 1837 to 1839 managed to maintain an Algerian counterstate with its own judicial and fiscal systems.[134] As was so often the case in the history of European imperialism (and of the North American frontier), the aggressors carried the day only because the indigenous forces were disunited. After four years of captivity following his capitulation in 1847, Abd al-Qadir was shown some respect as a "noble enemy" for the rest of his life—a fate similar to that of Shamil, the (in many respects) comparable leader of the anti-Russian resistance in the Caucasus.

While the conquest of Algeria was proceeding, the number of French and other (mainly Spanish and Italian) emigrants to the country shot up from 37,000 in 1841 to 131,000 ten years later.[135] Most of them did not become agrarian pioneers but settled down in the cities. Although the conquest of Algeria had begun at a time when the only other part of Africa with European settlers was the far south—it coincided with the Great Trek of the Boers—the 1880s were as much a watershed for the French colony in the north as for the rest of the continent. Napoleon III, an imperialist adventurer in Asia and Mexico, had never fully indulged the settlers' thirst for power and, at least on paper, had recognized the Algerian tribes as the owners of the land. But after the end of the Second Empire in 1870, this constraint ceased to apply. The French republic, unlike the British colonial power in the Cape, gave the *colons* a free hand in building their state, so that the 1870s and 1880s—after the brutal suppression of the last great Algerian rising in 1871–72—witnessed extensive land transfers through punitive expropriation, legislative measures, or judicial deception. The number of Europeans in Algeria climbed from 280,000 in 1872 to 531,000 twenty years later. Whereas the Second Empire had banked on private corporations to open up the country, the Third Republic propagated the model of farmers owning their own land. The aim was to produce a copy of rural France in the new colonial space.

There was no such thing as a typical European colony. Algeria was not one either, but it did play a major role in the emotional economy of the mother country and was at the origin of a new confrontation between Europe and the Islamic world; scarcely any other colony showed such disregard for the interests of indigenous people. Both logistically and historically, North Africa was not really "overseas" as far as Europe was concerned, and colonial apologists would exploit to the full the fact that it had been part of the Imperium Romanum. The sharpness of the clash with Islam in Algeria was paradoxical, because no other country than France has had closer and better contacts with the Islamic world in modern times.[136] In neighboring Morocco, moreover, the resident-general after 1912, Marshal Hubert Lyautey, conducted a conservative policy of minimal intervention in native society and knew how to curb the influence of the relatively small number of settlers.[137]

A second paradox is that despite their strong local position, the Algerian *colons* did not display the normal settler impulse of seeking political independence. Unlike their British counterparts in North America, Australia, or New Zealand, they did not try create a "dominion" type of state. Why not?

First, the settlers' weak demographic position meant that right until the end they were dependent on French military protection. Canada, Australia, and New Zealand, by contrast, could rely on their own security forces by 1870 or thereabouts. *Second*, from 1848 on Algeria was legally not a colony but a part of the French state, whose high degree of centralism allowed no scope for political autonomy or intermediate status of any kind. The result was more a tribal than a national consciousness among French Algerians, comparable to that of the Protestant British in Northern Ireland. On the other hand, Algeria was more marked by indigenous nationalism than almost any other European colony. After the humiliating French defeat in the war of 1870–71 with Prussia, it became an important arena of national regeneration through colonization.[138] *Third*, the Algerian colonial economy remained both dependent and precarious, being organized after 1870 mainly in small enterprises and with no reliable export other than wine—whereas the British dominions had large companies producing and exporting cereals, wool, and meat.

With the exception of Algeria, the French colonial empire got off to a late start. Only with the extensive conquests in western Africa, and eastward from there in what are now Mali, Niger, and Chad, did it create a territorial basis for competition with the British Empire. But in 1898, when colonial troops of the two powers clashed at Fashoda on the Upper Nile, the French retreat expressed the real relationship of forces. The African savannah belt offered little economically, whereas Vietnam proved from the beginning to be a productive colony ripe for exploitation. In the long process through which the three components of Vietnam (Cochin China, Annam, and Tonkin) lost their independence, the decisive year would be 1884. But even afterward resistance continued on a considerable scale, and it was only at the turn of the century that Vietnam and the other two parts of Indochina could be said to have been "pacified." In the next four decades Indochina became the main imperial turf for banks, mining companies, and agribusiness. Yet here too, there were limits to colonial economic influence: for example, it never became possible to replace the silver piastre and other local currencies with the French franc, so that Indochina, like China, remained on a silver standard that was exposed to major fluctuations.[139] For this reason—and also because of underdevelopment of the credit sector—the diversified activities of French banks were a symptom not only of aggressive finance imperialism but also of serious adjustment problems. Of all the French colonies, Indochina brought in the greatest yield for private businesses, both from exports and from the relatively large market in a densely populated region. Moreover, Vietnam had direct links with Marseille and functioned as a base for French economic interests in Hong Kong, China, Singapore, Siam, British Malaya, and the Dutch East Indies. A source of high profits for individual companies, Indochina also helped French capitalism in general to prosper.[140]

All in all, the French colonies were much less integrated than those of the British into the global system of the time. With the exception of Algeria, there was no significant movement of settlers from France; nor was Paris comparable to London as a center for the international movement of capital. The largest capital flows anyway went not to the colonial empire but to Russia, followed by Spain and Italy. France was also very active in lending to the Ottoman Empire, Egypt, and China, where much of the credit helped to develop outlets for French industry (especially weapons production) as well as to express an independent finance imperialism. Even less than in the British case did the geography of France's financial interests coincide with its formal empire; it did not have a tradition of overseas colonies comparable to those of England or the Netherlands. Until after the First World War, the French public showed relatively little interest in such matters; small lobbies—especially the colonial army and navy and geographers—were therefore a strong force in shaping colonial policy. On the other hand, there was less criticism of colonialism and imperialism in France than in Britain. In the 1890s a social consensus developed around the view that colonies were good for the nation, and that they provided an excellent opportunity to deploy its cultural prowess and *mission civilisatrice*.[141]

The political sterility of French imperialism is quite astonishing. The land of *citoyens* exported no democracy, most of its colonial regimes were exceptionally authoritarian, and later decolonization was relatively smooth only in West Africa. The early history of French expansion also involved far more frequent mistakes than those committed by the British. In 1882 Britain's success in snatching Egypt from under the noses of the French was an especially cruel blow. The main cultural effect of French expansion was the spread of the French language, with especially long-lasting results in western Africa. Otherwise, assimilation was left open for few members of the newly developing educated classes in the colonies, and the cultural change expected of them was extremely radical. Since this did not give rise to a genuinely integrative imperial culture, the French empire could not later develop into a looser structure along the lines of the British Commonwealth.

Colonies without Imperialism

There was also colonial possession without empire. An extreme case in point was the Belgian Congo (France had its own Congo-Brazzaville, created when the adventurer Pierre Savorgnan de Brazza raised the flag on its behalf in 1880[142]); it was only in 1908, after innumerable atrocities were uncovered, that the Belgian government took over responsibility for the territory from King Leopold II— or, in the language of international law, annexed it. Leopold was one of the most ruthless and ambitious imperialists of the age. The Congo under his rule was not even minimally developed: it was a pure object of exploitation. All kinds of violence and arbitrary action forced a defenseless population into hard labor to produce extremely high quotas of export goods such as rubber and ivory. The profits flowed into the pockets of the king and into public buildings that still adorn Belgian cities. The Welsh journalist and explorer Henry Morton Stanley, who in 1877 became the first European to cross Africa from east to west at the level of the Congo, later worked for Leopold II and organized armed expeditions that at first met with little resistance. From 1886 the Force Publique, an exceptionally brutal army of African mercenaries later supplemented by locally recruited warriors, was responsible for order in the Congo, while in the east of the country it fought Swahili slave dealers (often called "Arabs") in bloody operations that caused tens of thousands of deaths. The actual state apparatus, in the euphemistically named Congo Free State, was therefore extremely rudimentary, and Belgian settlers were few and far between; neither did the large concession companies that subsequently shared out the wealth of the Congo provide significant employment for Belgians. As for the Africans, they scarcely came into the field of vision of the whites, virtually none of them—unlike in the French or British empire—receiving higher education in the "mother country." Cultural transfers in either direction were close to zero.[143] Since Belgium's overseas interests were so slight, it played scarcely any role in high-level imperialist diplomacy, being a significant factor at most in the financing of the Chinese railroads.

The Netherlands did not have a colonial empire either, but it did have a vigorously governed colony. Between roughly 1590 and 1740 it had been the strongest single force in world trade, possessing a "seaborne empire" with bases from the Caribbean to Japan. By the nineteenth century, however, not much remained apart from the Dutch East Indies. In the 1880s the Netherlands was the only Western European country that did not participate in the division of Africa; it had even sold its last possessions on the Gold Coast (Ghana) to the British in 1872. The Dutch came to enjoy their position as a shrinking colonial power, with a self-image in which they appeared as a small neutral nation serving the cause of progress through a gentle colonialism quite different from that of the aggressive and rapacious Great Powers;[144] any expansion involved no more than a tightening of their control over the Indonesian islands, where they had first established themselves in the early seventeenth century (founding of Batavia in 1619) but had taken a long time to gain a firm hold. This centuries-long process concluded only with the Atjeh (or Aceh) war, which between 1873 and 1903 overcame fierce resistance to bring the northern tip of Sumatra under their rule. The military operations, which cost at least 100,000 people their lives, sparked considerable controversy in the Netherlands. The main factors were in fact international, since there were successive fears of an American or British, and then German or Japanese intervention.[145] As so often in the history of expansion, it was a case of aggressive defense, not of last-minute panic at the thought of being excluded from the spoils. If it gave the impression that the Netherlands was joining a fresh round of the imperialist game, this was not because any new impetus was driving it forward.[146] The large and wealthy Indonesian colony—in every respect second only to British India among European possessions in Asia and Africa—remained of interest to the Dutch for the same reasons as before 1870. The Netherlands was "a colonial giant but a political dwarf."[147]

Around 1900 there was a change in the methods of colonialism, not only on the part of the Dutch. The conquest of Africa was nearly complete, and in the new, more peaceful conditions the major colonial powers pursued a more systematic and less violent policy. The goal everywhere was what French colonial theory used to call "valorization" (*mise en valeur*). In Germany's African empire, especially in East Africa, the years after 1905 became known as the "Dernburg era," after the colonial secretary Bernhard Dernburg.[148] In British Malaya similar policies were observable at this time. But the most thorough *mise en valeur*, and the one most closely studied by other colonial powers, took place in Indonesia. Between 1891 and 1904, as many as twenty-five French delegations alone went out to study the Dutch East Indies, hoping to learn the secrets of how to use native labor most profitably.[149] Between the two wars, when colonialism entered its mature stage more or less worldwide, the Dutch East Indies could serve as a kind of model for good and for bad. India, though in many respects atypical, had played this role in the nineteenth century, but its liberation movement had raced ahead of most other colonies and was already on the road to a new future. The Dutch East Indies stood rather for continuity in colonial governance and ideology.

In the period from 1830 to 1870, the newly devised extractive institution of the so-called Cultivation System (*cultuurstelsel*), a kind of "planned economy" *avant la lettre*, allowed the Dutch to exploit Indonesia to a degree rarely paralleled in colonial history. One-fifth of the net revenue of the Netherlands treasury came directly from the colony. However, the system ushered in diminishing productivity and failed to provide the basis for sustainable economic growth.[150] In the three decades after 1870 there was a retreat from extreme forms of plunder and coercion, and in 1901, toward the end of the costly Atjeh war, the colonial power actually proclaimed a switch to an "ethical policy." This meant, above all, that the colonial state would invest in Indonesia for the first time, especially in infrastructure such as railroads, electricity generation, and irrigation (traditionally well developed, particularly in Java). The first moves were also made toward a colonial welfare state, such as never happened in India and only reemerged in post-1945 (western) Africa.[151] Scarcely any other colonial power in the long nineteenth century invested so much money in what would nowadays be called "development." Nor was it without successes: if the Indonesian economy had later grown as much as it did between 1900 and 1920, Indonesia would today be one of the richest countries in Asia.[152] This spurt, however, was mainly due not to the policies of the colonial state but to the hard work and entrepreneurship of the peoples of the Indonesian archipelago. Not enough was done in the post-1901 reform period to educate and train the local population of the colonies (to develop their "human capital"). This was perhaps the greatest sin of omission on the part of European colonialism.

Private Empires

Such forms of empire formation, though ultimately under the control of an autonomous metropolis and involving the projection of power from core to periphery, rarely had a grand strategy behind them. In this sense the historian Sir John Robert Seeley was not altogether wrong when he famously remarked in 1883, shortly after the highly planned occupation of Egypt, that the British Empire seemed to have been acquired "in a fit of absence of mind." It was an observation that also applied to other European empires.

But there were many deviations from the model: empires were not always propelled by military dynamics. In 1803 the Louisiana Purchase from France doubled the territory of the United States at a stroke, opening wide new spaces for settlement and the founding of new federal states. In 1867 the United States acquired Alaska from the Tsarist Empire. In 1878 Sweden sold its Caribbean island colony Saint Barthelemy to France, after the United States and Italy had both turned down the offer.[153] Such transactions were the modern counterpart of peaceful transfers of territory through dynastic marriages (Bombay, for example, was part of the dowry of the Portuguese princess Catherine when her marriage treaty with Charles II of England was agreed in 1661).

Another peaceful mode was for a land to place itself under higher protection, as the ruler of Bechuanaland (today's Botswana) did when he opted for British annexation over being ruled by Cecil Rhodes's private British South Africa Company.[154] "Voluntary" subjugation, whether in such a triangle or in direct recognition of vassal status, is one of the oldest and commonest mechanisms of imperial expansion. The system of US hegemony after the Second World War—which the contemporary Norwegian historian Geir Lundestad calls "empire by invitation"—bears traces of this variant.

Private empires also arose in the slipstream of Great Powers, Leopold II's in the Congo being only one such case. In Brunei and Sarawak (North Borneo) the Brooke family established itself as the ruling dynasty in an area of some 120,000 square kilometers. In 1839 the English adventurer James Brooke arrived on the island, in 1841 the sultanate (which had remained outside Dutch control) conferred on him the title of Rajah of Sarawak, and in the years until his death in 1868 he brought a large swathe of territory under his control. The second rajah, his nephew Charles Brooke, who ruled until 1917, expanded this still further. In 1941 the third rajah surrendered to the Japanese. The Brookes were not simply a band of robbers, but they did organize the extraction of considerable wealth, investing part of it in Britain and doing little for the long-term economic development of Sarawak. They regarded social change as detrimental to its indigenous people, yet allowed foreign corporations access to exploit the natural riches. Unlike King Leopold's Congo, however, Sarawak had at least the minimal trappings of statehood.[155]

Elsewhere, attempts were made to construct domains almost free of a state. Cecil Rhodes, who amassed a fortune from the South African diamond business, was relatively successful in building a private economic empire in southern Africa. For the British government, it was a cheap and easy option to cede the territory between Bechuanaland and the Zambezi river (Southern Rhodesia, today's Zimbabwe) to the British South African Company, which was endowed with a royal charter in 1889 and was largely funded by Rhodes and other South African mining magnates. The company undertook to "develop" the territory, and above all to meet all the necessary costs. In 1891 it was permitted to extend its operations north of the Zambezi, into what would become Northern Rhodesia (today's Zambia). For Rhodes and his company, the point was not to acquire and rule territory for its own sake but to exercise a monopoly over known and suspected mineral deposits, and to incorporate the mining areas into the South African economic space. For that, effective control was a necessity. "If we do not occupy, someone else will," he wrote in 1889, expressing as pithily as possible the logic of the scramble for Africa.[156] Rhodes made his plans even more palatable to

Whitehall by opening up the "Rhodesian territories" (the name came into use in 1895) for British settlers. "Company rule"—a method that had previously failed in German South West Africa—was vigorously criticized by missionaries, who in this instance complained of a colonial paternalism that was too indulgent toward the natives. But, in the eyes of other local whites, the semiprivate protectorate involved a successful symbiosis of big capital with the settler way of life.[157]

Large plantations and concessions, too, were often stateless zones in which the law of the land prevailed only indirectly, as on a Junker estate east of the Elbe.[158] Missionaries sometimes came to exert such influence that they built veritable protectorates of their own. Even with the end of the chartered companies in Asia, and finally of the East India Company in India (1858), new semiofficial colonization agencies came into being there. The most important of these was the South Manchurian Railroad Company (SMR), which after the Russo-Japanese War in 1905 took possession of the southern tip of Manchuria and the southern sections of the local Russian railroads. The SMR became a colonial power supported by the Japanese state, building the most lucrative railroad colony in history and a center of gravity for the whole economy of northeastern China. At the same time, Manchuria became the location of the largest heavy-industrial plants on the East Asian mainland.[159]

Secondary Empire Building

Japanese empire building was the only non-European instance after 1895 to be crowned with spectacular success—until 1945, that is—but we should not overlook a few others that for a while had a major regional impact. These cases of *secondary* empire building may be defined as military aggression plus territorial expansion, with the help of European military technology but not under the control of European governments. Africa of all places, which later became the main victim of European empire building, was an especially eventful arena in the first half of the century. At a time when Europeans were beginning to expand in three ways in Africa—fresh conquest beyond the South African frontier, military intervention in Algeria, and conversion of a trading frontier into a military frontier in Senegal[160]—the sub-Saharan savannah belt was witnessing several large and mutually independent processes of expansionist state building, with centralized and highly militarized structures, which correspond in many respects to our definition of empire. These formations, impelled by jihadi themes, derived their cohesion from two communicative elements that were lacking farther south: a script and cavalry animals.[161]

Other embryonic empires developed without Islam or cavalry: the Ganda (in Buganda), for example, built a fleet of war canoes in the 1840s and beyond, gaining a kind of imperial supremacy on and around Lake Victoria that exploited the labor of weaker peoples.[162] Often such operations used a far from modern, almost antiquated, technology. The military strength of the Boers in the early nineteenth century was based on horse-riding infantry equipped with muskets.

The Sokoto caliphate, built up roughly from 1804 to 1845, also supported itself on horses and muskets.[163] In all these cases there was no direct link with the Industrial Revolution in Europe. The technological gap was already smaller in the 1850s and 1860s, however, when the Muslim empire of Sheikh Umar Tal was taking shape in Upper Senegal.

The expansion of Egypt is a particularly good example of secondary empire building. It is one of the most remarkable facts of nineteenth-century imperial history that independent Egypt possessed an empire between 1813 and 1882— that is, an area under its military control that was more than just a sphere of influence. If we consider that the Japanese empire lasted a mere fifty years, then the Egyptian experience merits some consideration.[164] Pasha Muhammad Ali, a man of obscure Albanian origin and Egypt's de facto ruler from 1805 on, was never content with a realm along the Nile. It cannot be proved that he planned to supplant the sultan as universal caliph of Islam, but he set about building an empire that stood in a contradictory relationship with the Ottoman Empire (whose suzerainty over Egypt he never actually questioned). On the one hand, he openly defied the sultan as a rebellious satrap; on the other hand, the sultanate felt more threatened by the puritanical, fundamentalist, and antimodernist Wahhabi movement, founded in the Arabian Peninsula by Sheikh Muhammad ibn Abd al-Wahhab. The Wahhabis, who sought to return to the pure faith and ideal practices of the Prophet and the four rightful caliphs of the seventh century, branded all opponents as heretics and conducted a holy war against all other Muslims, including the Ottoman sultan. Indeed, until his death in 1792, the Wahhabi founder regarded the sultan as the greatest evil, calling upon Muslims to rise up and overthrow him. The movement displayed religious fervor and military skill in expelling the Ottomans from large parts of the peninsula. Its followers even occupied Mecca and Medina, in 1803 and 1805 respectively, and in 1807 denied Ottoman pilgrim caravans access to the holy sites. The sultan therefore welcomed Muhammad Ali's help in fighting the Wahhabis, while the pasha for his part cherished grand plans for the modernization of Egypt and had little time for a fundamentalist version of Islam. When the sultan assigned Muhammad Ali to put together an armed expedition against the Wahhabis, it was the starting signal for Egyptian empire building. In 1813 the Egyptian army recaptured the holy sites and the port of Jeddah, and a year later Wahhabi power crumbled, though not yet the movement and all resistance.

The geopolitical result was that Egypt's ruler established himself on the eastern shores of the Red Sea, entering a collision course with a great power, Britain, that had initially favored his operations against the unruly Wahhabis. In 1839 the British occupied the port of Aden in Yemen and put pressure on the pasha to withdraw from Arabia; this period is known in diplomatic history as the Second Muhammad Ali Crisis. In 1840 the pasha finally had to back down. His direct attack on the Ottoman Empire in Syria in 1831–32 confirmed his military strength (the Turkish army was crushed in December 1832 near Konya), but it also showed his political vulnerability. When the crunch came, Britain, Austria, and Russia all chose for reasons of their own to maintain the Ottoman Empire: only France backed Muhammad Ali. In September 1840 a British fleet bombarded Egyptian positions on the coasts of Syria and Lebanon, and shortly afterward Austrian and British troops landed in Syria as the Turkish army advanced from the north. Facing such pressure, Muhammad Ali agreed to a compromise whereby he was recognized as hereditary ruler of Egypt but gave up any claims within the Ottoman Empire.[165]

This settlement had no impact on Egypt's policies and positions in Africa. Under both Muhammad Ali and his successors, the power of the "Turkish-Egyptian" regime in Cairo was extended to the whole of the Sudan, in a campaign of conquest that uniquely combined European-trained military units with slaves bought in African markets and trained as soldiers. After a time, however, the pasha realized that conscripted Egyptian peasants fought better than African slaves. Under Egyptian rule, the mineral wealth of Sudan—especially its gold—was extracted on a large scale. The Sudanese became subject to unusual forms of high taxation. All Sudanese resistance was ruthlessly suppressed. And on the frontier, new warlords appeared in the violence markets and put an additional burden on local people.

Khedive Ismail cited the "politically correct" aim of the eradication of slavery as a pretext for further expansion, making use of the legendary general Charles Gordon (who had proved his worth in the 1860s against the Chinese Taiping) to drive the Egyptian administration into the far south of Sudan. Against these twin objectives, a messianic-revolutionary movement finally developed in 1881, with a leader, Muhammad Ahmed, that it saw as the longed-for "Mahdi," or redeemer. Its forces soon won control of most of Sudan and in 1883 annihilated a standing army under British command; Gordon, having exceeded his remit and hugely underestimated the enemy, now found himself completely isolated in Khartoum. Mahdi supporters caught up with him there in 1885. His killing drew a line under the Egyptian empire in Africa. The Mahdi's looser structure of rule rested on his charismatic authority and could scarcely survive his death. An extreme drought further weakened his authority so much that Lord Kitchener met little resistance when he moved to reconquer Sudan in 1898. The Mahdi movement arose in opposition to Egyptian-European incursions, with many typical features of an anti-imperial reaction. These included labeling the invaders as aliens—in this case, "Turks"—and as violators of religious norms.[166]

Conditions had been different in the similarly volatile world of late eighteenth-century Indian states. Most of those that succeeded the Mogul Empire, which had soon collapsed after the death of the Great Mogul Aurangzeb in 1707, were not what one would describe as empires. However, many did combine territorial expansion with rule over taxpaying farmers and elementary state-building measures often reminiscent of Muhammad Ali's in Egypt. The Sultanate of Mysore under Haidar Ali and his son Tipu Sultan, which might otherwise have followed an Egyptian path, took on the might of the East India Company and was destroyed in 1799. The tactically more cautious maharajah in the Punjab, Ranjit Singh, who like Tipu before him brought in European officers to reshape his army, managed to found a temporarily powerful Sikh state to which weaker polities—and this was its imperial aspect—had to pay tribute. Unlike in the jihadi empires of the African savannah, religious motives played no role in this Sikh expansion all the way to Peshawar at the foot of the Hindu Kush. Ranjit Singh created a typically imperial ("cosmopolitan") elite out of Sikhs, Muslims, and Hindus. But, in the age of Ranjit Singh, the British were already so strong that the new state could survive only while it remained useful as a buffer against the unpredictable Afghans. After the death in 1839 of the autocratic maharajah—who, unlike Muhammad Ali in Egypt, created no institutions capable of outliving him—the Sikh state was annexed in 1849 and turned into a province of British India.[167]

Internal Colonialism in the United States

The spread of the United States across the North American continent may be interpreted as a special kind of secondary empire building, and one of the most successful of all.[168] The United States of America began its existence in 1783 as one of the largest countries in the world, and over the next seventy years it further tripled in size. For Thomas Jefferson and many others with a keen sense of geopolitics, the advance to the Mississippi in the 1790s was an objective of prime importance. Beyond the river lay the vast land of Louisiana, stretching from the Great Lakes to the Gulf of Mexico, with New Orleans as its capital in the Deep South. In 1682 France had taken possession of it more in name than in reality, with no plans for intensive colonization. Indeed, the French king showed so little interest in it that he ceded to the Spanish king that part of Louisiana that he had kept after the Treaty of Paris in 1763. Charles III received the gift without enthusiasm, and it was a long time before the Spanish actually took possession of it.[169] By then American merchants had already reached the Mississippi from the north, so that considerable commercial interests were at stake. In 1801 Spain gave Louisiana back to France. Bonaparte, who once mooted a major military expedition to the Mississippi and fleetingly dreamed of Louisiana as an imperial jewel of the crown, performed a volte-face in April 1803. When President Jefferson instructed his ambassador in Paris to ask for talks about a cession of the mouth of the Mississippi, France's first consul—interested in good relations with the United States because of the prospect of a new war with Britain—surprisingly offered the whole of Louisiana (comprising all French territories in North America) at a bargain price. The American negotiators jumped at the opportunity. On December 20, La Nouvelle-Orléans was handed over to the US federal government.

Legally speaking, it was annexation. The 50,000 or so whites living in Louisiana, who had first been French, then Spanish, then French again, now found themselves subjects of the United States, without ever being asked for their views on the matter. At a stroke of the pen, and at very little cost, the largest republic in the world doubled in size. At the same time, it ended the potentially dangerous presence of another power (the militarily strongest of the age) on North American soil. Precisely twenty years after shaking off its colonial status, the United States swallowed up the first colony of its own—a case of secondary empire building without the use of force. Many characteristic problems of colonization then ensued: above all, a clash with the culturally foreign (French-speaking) population, which disliked the transfer of power and regarded as a hostile act the break with Spanish and French law and the introduction of the American system based on English common law. In Louisiana before 1803, free people of every color had enjoyed the same civil rights, whereas now they lost nearly everything as soon as an iota of "colored" blood was suspected.[170] In 1812, Congress in Washington made ex-French Louisiana the first of thirteen newly defined "federal states," but it took a long time to become Americanized. New immigrants came in dribs and drabs from France, and by the thousand from Cuba, where many planters, having fled the Haitian revolution, had found life unpleasant during the Spanish war of resistance against France. New Orleans, planned as a typical French colonial city, was divided into districts for English-speaking Americans and French-speaking Creoles even during the economic boom of the 1830s. Despite the harsh American race laws, however, the "color line" was less sharply drawn than elsewhere in the South. As Donald Meinig writes in his monumental geohistory of the United States, Louisiana was precisely what the country's self-image could not accept: an "imperial colony." That might perhaps have still been compatible with the ruling ideology if Louisianans had really been liberated from all forms of bondage. But they were "peoples of foreign culture who had not chosen to be Americans."[171] In this they did not differ from the original inhabitants of the continent, the Indians.

The question as to whether one should speak of "US imperialism," even in relation to the conquest of the Philippines after 1898, or to the numerous military interventions in Central America and the Caribbean during the early decades of the twentieth century, has long been the source of heated debate. Some regard the United States as an anti-imperialist power by definition; others see in it the acme of capitalist imperialism.[172] Donald Meinig frees the discussion from its ideological entanglements by convincingly pointing to structural similarities between the United States and other imperial formations. In the middle of the nineteenth century, he argues, the country was four things at once: a collection of regional societies, a federation, a nation, and an empire.[173] Why an empire?

The United States maintained a huge military apparatus, complete with forts, roadside checks, and so on, to repel and hold down the Indians. Special areas with even minimal autonomy were not tolerated. There were no protectorates for land belonging to the Indians, and no enclaves in the style of the princely states of India. During the years of the Indian wars, white America was in a position similar to that of the Tsarist Empire vis-à-vis the Kazakh steppe peoples. There, too, the imperial center asserted a general claim to sovereignty, costly military installations were created, and armed settlers were given encouragement at the frontier. The Kazakhs were more numerous and less divided among themselves, however, and they could not be subjected to totally arbitrary treatment. Their continuing cultural, and to some extent military, self-assertiveness underlined the multiethnic character of the Tsarist Empire. Today they have their own nation-state. The policy of military occupation and land acquisition makes it justifiable to speak of the imperial character of the United States. But it would be too simple to claim that the United States can be exhaustively described as an empire. It was an expanding nation with a federal type of organization, which could not derive a shared identity from a single national genealogy. All white and all black inhabitants of the United States were somehow "newcomers." The myth of the cultural melting pot, as remote as it was from reality, never corresponded to the nation's basic perception of itself. But neither did the "us" and "them" dichotomy of European nationalism enter the picture. It was never possible to say unequivocally who "we" were. Nineteenth-century Americans were obsessed with a fine hierarchy of differences, with the indispensability but also the instability of "race" as an category of imposing cognitive order.[174] This was a typically imperial mental grid that translated into manifold practices of segregation.

6 Pax Britannica

Imperial Nationalism and Global Vision

In the nineteenth century, the British Empire was by far the largest in both area and population,[175] but it also differed from others in its essential character. Britain was what one may call an imperial nation-state: that is, a nation-state that, by virtue of tendencies internal to it, became politically unified and territorially fixed in pre-imperial times, and whose politicians learned over time to define national interests as imperial and vice versa. Recent histories have pointed out that one should not exaggerate the national homogeneity of the United Kingdom; that Great Britain still contains four different nations (England, Scotland, Wales, and Northern Ireland). Much in its imperial history speaks in favor of this way of seeing things. Scots were disproportionately active within the British Empire—as businessmen, soldiers, and missionaries. The position of the Irish was ambivalent: the Catholic population of the island had every reason to feel itself disadvantaged in a quasi-colonial manner; yet many Irish—including Catholics—enthusiastically participated in the activities of the empire.[176] Nevertheless, the fact remains that Britain was perceived in the outside world as a closed imperial nation-state.

For a long time it was part of the self-image of the British upper classes and intellectuals that the country had been spared the virus of nationalism. Blinkered

Continentals might be nationalistic; Britons had a cosmopolitan way of thinking. Nowadays one would no longer put it quite like that. What was distinctive, rather, was the paradox of an *imperial* nationalism. This arose in the 1790s as a sense of nationhood that drew its energy mainly from the imperial victories of the day.[177] The (male) Briton thought his superiority to lie in the art of conquest, in commercial success, and in the benefits that British rule brought to all who came into contact with it. He was superior not only to colored peoples, who were in need of disciplined and civilizing leadership, but also to European peoples, none of which acted overseas with anything like the felicitous touch displayed by the British. This special imperialism lasted throughout the nineteenth century, its occasional jingoistic intensification being less important than its essential continuity over time. Imperial nationalism was associated with a Protestant sense of mission, in which values such as leadership and strength of character were of major importance. The idea that the British were a tool of Providence for the betterment of the world became a kind of ground bass among sections of the population whose gaze was directed beyond their own local sphere. Rather like the French after the revolution, the British felt themselves to be a kind of universal nation, both in their cultural achievements and in their resulting entitlement to spread them all around the world.

Throughout the nineteenth century, the British relationship to the rest of the world was based on a strong sense of a civilizing mission. This trope of a vocation to free other peoples from despotic rule and non-Christian superstitions rarely failed to produce its effect. Britain was the birthplace of humanitarian intervention, where the problem of human rights in relations between states was theorized (by John Stuart Mill, for example) in a way that is still topical today.[178] Whereas the first three wars against the Indian state of Mysore were interpreted in terms of pure power politics, the fourth one—which ended in 1799 with victory over Tipu Sultan—already appeared in British propaganda as a liberation struggle against a Muslim tyrant.

Much more important for the British self-image, however, was the open campaigning against the slave trade, which in 1807 led to victory for the abolitionists in Parliament. In the following decades, it became a primary task of the Royal Navy to force slave ships ashore in third countries and to release their captive cargo. That such pan-interventionism also furthered British strategic interests was a gratifying side effect. But what it involved was less global maritime supremacy than, as Schumpeter put it, a "global maritime police."[179] The civilizing mission was to be performed pragmatically, without fanatical dogmatism. At best, a mere glance at the British model would be enough to convince anyone of its unsurpassable wisdom.

Of course, the actual successes of the British Empire cannot be explained only by collective autosuggestion. Three factors lay behind the imperial rise of the small archipelago in the North Sea: (1) the decline of Dutch commercial hegemony and the successes of the East India Company; (2) an increase in global power during the Seven Years' War, reinforced by the Treaty of Paris (1763); and (3) the transition to territorial rule over wealthy regions of Asia capable of providing a handsome tribute. Moreover, since Britain's domestic finances were in better shape than those of any other state, and since its political elite had decided to make large and constant investments in a royal navy, the country was in a position to extinguish Napoleon's challenge at least at sea. As early as the 1760s the British elite had been the first in Europe to learn global thinking. Whereas it had previously been a question only of scattered possessions around the world, there was now a vision of a cohesive global empire; new approaches were devised in London and approved for general application.[180] They were ocean oriented, but with an eye to possible rule on land—unlike the earlier Habsburg version of the idea of a universal empire. At the end of the Seven Years' War, the conception burst forth of a country with seemingly unlimited horizons of influence, if not actual rule. The loss of the thirteen American colonies was a severe setback. But the continuity of empire could be saved, because the East India Company, even before 1783, had introduced energetic reforms and placed its rule in (not yet over) India on a new and solid foundation.[181]

The Navy, Free Trade, and the British Imperial System

Even during the Napoleonic Wars, not everything went as the British planned: defeats had to be swallowed at Buenos Aires (1806) and in the war with the United States (1812). When Napoleon was safely in Saint Helena and the threat from continental Europe had receded (only with Russia was there a kind of cold war in Asia, the so-called Great Game), the British Empire took on its mature form. What were its foundations?

First. Above-average population growth in the British Isles, together with an unusual propensity to emigrate (not to speak of deportation to Australia and elsewhere), produced demographic trends not seen in any other European country. Alongside the United States, first Canada and then the other dominions had a large British settlement that left a strong mark on their culture. Around 1900, smaller groups of British expatriates were to be found in India, Ceylon, and Malaya[182]; in Kenya and Rhodesia; and in port colonies such as Hong Kong, Singapore, and Shanghai. These formed quite a cohesive British world, in language, religion, and lifestyle, a global Anglo-Saxon community in a far-flung but never isolated diaspora.[183]

Second. Having gained a leading position at sea during the Seven Years' War, Britain could approach the showdown with Napoleonic France with the only navy capable of worldwide operations. This was the direct result of a unique mobilization of financial resources. Between 1688 and 1815 Britain's gross national product tripled in size, while tax revenue multiplied by a factor of fifteen. The British government could draw on a national income twice as high as that of the French. Since it raised most of its taxes indirectly from consumption, Britons *felt* their fiscal burden to be lighter than that of people across the Channel. In 1799

an income tax was introduced as an emergency measure, but this did last beyond
the end of the wars; it won broad public acceptance and became a cornerstone of
the British state. The foremost recipient of public funds was the Royal Navy.[184] It
could remain ready for action only because a global system of bases had already
been purposefully created. At the end of the nineteenth century, there was no
major waterway or strait in the world where the Royal Navy did not have a say.[185]

The navy rarely used its position to choke transport for strategic reasons (how
easy that would have been in Gibraltar, Suez, Singapore, or even Cape Town!)
or to hinder the trade of non-Britons. Its general objective, rather, was keep sea
routes open and to prevent others from blocking access to them. All through
the nineteenth century, Britain stood up for the principle of a *mare liberum*. Its
maritime superiority did not rest only upon its material edge; it also had political causes. Since the activities of the Royal Navy did not appear threatening to
European governments, they had no reason to engage in an arms race. In the second half of the nineteenth century, when France, Russia, the United States, Germany, and Japan strengthened their navies (while a country such as the Netherlands, which could have afforded a steamship fleet, kept out of the running),
Britain still managed to retain its place far out in front. Another factor in this
was the Royal Navy's superior logistics. Finally, British mastery on the world's
seas and oceans was underpinned by a large and efficient commercial fleet; in
1890 the country still had more merchant tonnage than the rest of the world
put together.[186] Ocean carriers and sea travel made a significant contribution to
Britain's balance of payments; some large fortunes were amassed in this domain.

Command of the seas made it unnecessary to maintain a large land army.
The principle of "No standing armies!" continued to apply. Home defense was
extremely skimpy, and on the eve of the First World War the largest section of
the UK land forces was still in India. Created after 1770 out of a developed mercenary market in the Subcontinent, the Indian Army was paradoxically among
the world's largest standing armies throughout the nineteenth century. It served
more purposes than one. Along with the bureaucracy, it was the second "steel
frame" (as Prime Minister David Lloyd George put it in 1922) that held the Indian giant together, but it also functioned as a colonial task force that could be
deployed elsewhere in Asia or Africa, or even for police operations in the International Settlement in Shanghai, where Sikh soldiers' brutal behavior triggered
Chinese mass protests as late as 1925.[187]

Third. Up to the last quarter of the nineteenth century, Britain had the most
efficient economy in the world. By 1830 it had become the "workshop of the
world," its light industry supplying markets on every continent. A majority of
iron ships, railroads, and textile machines were built in Britain; it offered goods
not available anywhere else, and with them came consumption models that took
root elsewhere and helped in turn to spread and stabilize the demand for such
goods. The high productivity of the British economy made it possible to sell
export products at a low price, undercutting all kinds of competitors. Those who

needed it also received cheap credit. The opportunities of empire were exploited by private companies, while the state itself, faithful to its liberal creed, practiced a hands-off approach. British businessmen could rely less than their French or (after 1871) German counterparts on local state action, even though UK diplomats and consuls all over the world looked to them as sources of information. Often the activities of businessmen contributed to the very instability that later offered politicians an excuse for intervention.[188] A kind of chain reaction generated a constant buildup of interests and openings. Thus, the imperial structures gave rise now and then to private economic empires that cared little about the limits of formal British sovereignty.

Unlike the empires of the eighteenth century, the British Empire in the high Victorian era was an enabling system for global capitalist operations. In this it also differed fundamentally from mercantilist formations, which sealed themselves off through external economic controls and monopolies, organizing themselves for economic warfare with neighboring empires. The dismantling—or, to use a more positive term, liberalizing—of its economic policy was the greatest contribution of the British state to an imperial system stretching far beyond the colonial territories under its formal rule. It was a twin-track process. In 1849 Westminster repealed the seventeenth-century Navigation Acts, under which all imports to England or Britain had to be carried in ships belonging either to British nationals or to citizens of the exporting country. Dutch middlemen were the first to feel the effects. By midcentury the economic freedom of the seas had been established.

The second track was the abolition of the Corn Law tariffs, a major theme of British domestic politics in the 1840s. In fact, the tariffs had only been introduced in 1815, to prevent the grain market from collapsing as a result of overproduction and rising imports. Purchases from abroad were prohibited unless and until the price of grain on the internal market reached a certain level. Corresponding to farmers' interests, this form of agricultural protection encountered growing opposition from manufacturers, who considered that artificially high food prices held back the demand for industrial goods. Furthermore, the system came under heavy fire as a symbol of aristocratic privilege. Sir Robert Peel, a leader of the mainly protectionist Tories, opposed powerful forces in his party and appealed to the interests of the country as a whole when as prime minister he pushed through the repeal of the Corn Laws in 1846 (it actually took effect three years later). A series of other measures to liberalize foreign trade followed in the 1850s, the breakthrough period for free trade, and the end of grain tariffs was soon viewed across party lines as a token of economic progress.[189]

It was unprecedented, indeed revolutionary, that Britain took these steps unilaterally, without expecting equivalent action from its trading partners. However, they unleashed a chain reaction—an appropriate image, since the United Kingdom never convened a major international conference to decide upon a new world economic order. The rapid spread of free trade meant that by the mid-1860s

tariffs had been largely dismantled between European states; the Continent became a free-trade area, from the Pyrenees to the Russian border. Free trade also
prevailed within the empire. In the surest sign of their growing strength, the dominions were able by the end of the century to carve out space for their own
independent tariff policies. But where the free world market (dominated by Britain because of its production superiority) ran up against trade barriers, energetic
measures were taken to remove them, with the whole British elite in support.[190]
Official doctrine saw national market protection—recommended by the US treasury secretary Alexander Hamilton in 1791, and the German economist Friedrich
List in 1831, to prevent a flood of British goods—as the expression of an unacceptable civilization deficit. The Latin American republics in the 1820s, the Ottoman
Empire in 1838, China in 1842, Siam in 1855, and Japan in 1858 were compelled
to relinquish virtually all market protection in a series of free-trade agreements,
mostly obtained through the threat or use of military force. This paradoxical phenomenon has been described as "imperialism of free trade."[191]

The global system of free trade offered extraordinary scope for British interests. But since it rested upon equal treatment for all and a strict antimonop-
olism, it was in principle equally open to members of other nations. The stronger
European and American economies became, the slimmer were the advantages
that British industry (finance was more robust) could derive from its waning
superiority. Although most European countries reverted to tariffs after 1878, and
although the United States seldom deviated from a basic protectionist mood
that often clashed with its demand for the opening of *other* markets, the United
Kingdom held firm to its free-trade policy. This enjoyed a broad consensus in
British society, stretching far beyond economic lobbies into the heart of the
working class, and by century's end it had become a pillar of the political culture and a basic emotional theme in the national self-consciousness.[192] The persistence of this unilateralism is as astonishing as its original appearance in the
middle of the century.

With its worldwide imperial system, Britain exercised a kind of benign—as
opposed to predatory—hegemony. It made public goods available free of charge:
law and order on the high seas (including the war on residual piracy), property
rights beyond national and cultural boundaries, voluntary migration flows,
an egalitarian and generally applicable system of customs duties, and a set of
free-trade agreements that included everyone by virtue of most-favored-nation
clauses. The latter provisions, the key legal mechanism of global liberalization,
implied that the most favorable terms of an agreement automatically applied to
all who participated in it.[193]

Costs and Benefits of the British Empire

In the second half of the 1980s, there was a dispute among historians over
whether the British Empire had been "worth it." A group of American researchers, with a major empirical input, came to the conclusion that it had ultimately

been a huge waste of money.[194] This was supposed to fatally undermine Marxist theses that British capitalism had expanded out of objective necessity, that the empire had been exploited on a massive scale, and so on. With the debate now over, it is possible to reach a more finely nuanced judgment. The first point to be made is that on longer time scales the empire was undoubtedly profitable for a large number of firms, and even for whole sectors of the economy. It allowed privatization of profits with a socialization of costs. Individual businesses could make a lot of money: one would have to look at their archives to ascertain how much. Since the British national economy was the only one in the world for which overseas trade had central importance, global commercial and financial relations played a greater role in defining its relative position than they did for any other European country. With the exception of India, however, such relations with the so-called dependent empire were far less important than economic links with continental Europe, the United States, and the dominions. In short, Britain made use of the empire without being dependent on it. A cross-check for this premise is that when decolonization began in 1947 with Indian independence, it had surprisingly few negative consequences for the British national economy.

If we narrow the question down to India, by far the largest colony, then the results are fairly unequivocal. By virtue of a well-organized colonial tax system, India in the long term covered the costs of the British administrative and military apparatus out of its own resources. Since political measures ensured that the Indian market remained open to certain British exports, and since India ran a long-term trade deficit that greatly contributed to the British balance of payments, the jewel in the imperial crown was anything but a loss-making enterprise during the half century before 1914.[195]

If we look a little beyond cost-benefit accounting, three further points appear more important.

1. Even if it is true that large sections of the British population gained little from the empire, millions were "proud of it" and consumed it as a status good. People reveled in the imperial pomp, even when the point of it was to impress them rather than the "natives."[196]

2. The empire created numerous job opportunities, especially in the armed forces. More important, however, was the scope that it opened up for emigration, which, economically speaking, afforded a more productive deployment of labor than in the home country, while politically it represented a safety valve for the outward channeling of social pressures. This effect was rarely a simple question of manipulation, however. Emigration was in most cases a personal decision: the empire created options.

3. The empire made it possible to conduct what (from the British point of view) was a highly rational foreign policy. It reinforced the advantage of an island position: namely, that one is not tied by nature to others that one would not choose to have as neighbors.

Britain had more leeway than any other great power when it came to policymaking: it could forge new ties if it wished, but it could also hold itself aloof. The United Kingdom had few friends in international politics, but it did not need to. It could therefore avoid being drawn into possibly fatal obligations. This low-commitment policy of managing all kinds of distance was practiced by all British governments in the nineteenth century, regardless of their party composition. But if a diplomatic understanding was reached with another power (the Anglo-Japanese Alliance in 1902, the Entente Cordiale with France in 1904, the Anglo-Russian Convention in 1907), it was never formulated in such a way as to entail automatic partnership in case of war. If the empire joined the First World War—it was declared on August 4, 1914, in the name of the *whole* empire—this was not because of an inescapable alliance mechanism but because Whitehall decided that it should be so. The possession of the empire meant that splendid isolation—which could function, however, only with a balance of power on the Continent—was one convenient policy option. The resources of empire were always available, and British policy was always pragmatic enough to keep open the possibility of a new orientation. At the beginning of the First World War, then, Britain was *not* isolated. The empire only really displayed its incomparable value in the years between 1914 and 1918.[197]

One does not have to be an apologist of imperialism to admit that the British Empire was a success by the standards of nineteenth- and twentieth-century imperial history. It survived the world crisis of the period between early modernity and the modern age (Koselleck's *Sattelzeit*), which witnessed the shipwreck of many another empire. It also pulled through a few dramatic setbacks. No major territory that came under British control was lost until the Second World War. (This is why the fall of Singapore to the Japanese army in February 1942 was such a devastating blow.) Retreats from unsustainable forward positions served to round out the contours of the empire. Thus, in 1904 an expeditionary force sent out from India under Sir Francis Younghusband advanced as far as Lhasa and, having failed to find suspected "Russian weapons," concluded an agreement for a protectorate in Tibet, a land over which China upheld vague suzerainty claims without being able to back them up at the level of power politics. The driving force behind this adventurist action was Lord Curzon, the ambitious viceroy of India. But London saw no reason to incur even minimal obligations to such an economically and strategically unimportant country, and so it disowned the local success achieved by Younghusband, this quintessential man on the spot.[198]

The British political class was also very successful in adapting to changed external conditions, when new Great Powers became active in the last third of the nineteenth century and Britain's comparative economic situation worsened as a result. It is true that Britain did not retain its global hegemony (i.e., a position whereby nothing really important happened against the wishes of the British Empire), but once again, with some difficulty, policymakers found a middle course between defense of the status quo and utilization of new economic and territorial opportunities.[199] In the course of the long nineteenth century, the British Empire displayed several different faces and passed through a number of metamorphoses. Yet it remained the most successful empire of the age and, after the First World War, even managed to extend its control over some League of Nations "mandates" (Iraq, Jordan, Palestine).

Factors of Stability

In addition to those already mentioned, a number of other factors explain this relative success.

First. As A. G. Hopkins and Peter Cain have shown, the main impetus for British expansion came not from industrialists but from a London-based financial sector closely linked to big agrarian interests looking to modernize their operations. The city was the home of the world's most influential banks and largest insurance companies. It financed the shipping and foreign trade of every nation. It was the focal point of the international business in private fixed incomes. Anyone who wished to invest in China, Argentina, or the Ottoman Empire used the financial services of the Square Mile. The pound sterling was the major world currency, and the mechanisms of the gold standard were kept going mainly from London. In comparison with industry, finance has the advantage of being less location dependent; it is therefore also less "national." Money from all over the world converged in the British capital, and so the city was not merely the economic center of the formal colonial empire, or even of the much larger sphere in which Britain exercised political influence. It was a global control center for flows of money and commodities, without rival until the rise of New York.[200]

Second. In the course of time—and having learned the lesson of disastrous blunders made during the American crisis of the 1770s—the managers of the British Empire developed and repeatedly put to the test a highly refined set of policy instruments. The basic principle of interventionism, in an age when the word "intervention" had fewer negative connotations than it does today,[201] was to use one's assets in the optimum manner. This is not self-evident in relation to empires, as we can see from the tendency of the United States in the twentieth century to deploy massive military force at an early stage. The British Empire always tried to keep this in reserve, developing an extraordinary virtuosity in the gradation of threats. British diplomats and army men were past masters in the art of persuasion and pressure, and so long as these could achieve the desired objective there was no need to resort to more expensive methods. One especially effective idea was to coordinate the application of pressure with a third power, preferably France; this was done in 1857 against Tunisia and in 1858–60 against China, while Siam was a touch more successful in playing off Europeans against one another.[202] British policy followed the principle that influence should be exerted for as long as possible and formal colonial rule be introduced only after the exhaustion of such informal options. A setup much favored by British imperialists involved the discreet presence of "residents" and other advisers to guide compliant local rulers. This could even result in an outright fiction. For example, Egypt after 1882 was for all intents and purposes a British colony, but the nominal suzerainty of the sultan in Istanbul was never actually disputed until 1914, and throughout the period in question an indigenous monarch sat on the throne and an indigenous prime minister remained in office. The all-powerful representative of Great Britain, who gave the government its instructions, bore the modest title of consul-general and had no formal attributes of sovereignty. In practice, this veiled protectorate allowed for measures no less drastic than in an autocratically governed crown colony.[203]

Third. The whole aristocratic stamp of British politics in the nineteenth century, so different from the bourgeois style prevalent in France, made it easy to practice elite solidarity across cultural boundaries. And, more than in the French case, the imperial apparatus incorporated subordinate local elites, albeit often only symbolically.[204]

Fourth. The British imperial class, especially toward the end of the nineteenth century, was no less racist in its attitudes than other European or North American colonial masters. It strongly emphasized social difference between people who did not have the same skin color. However, elite racism was virtually never taken to exterminist extremes; that was reserved for settlers—in Australia, for example—confirming James Belich's general observation that "Settlement colonies were usually more dangerous for indigenous peoples than subject colonies."[205] Uprisings such as the Indian "Mutiny" of 1857/58 might be brutally suppressed, and racism would then shed many inhibitions, but genocide or mass murder was never used as an instrument of rule in the British Empire, as it was in King Leopold's Congo or German South West Africa in 1904–8. A critical moment was the so-called Governor Eyre controversy. When Jamaicans in October 1865 resisted the colonial police during legal proceedings in the small town of Morant Bay, a protest action by small farmers led to the killing of a number of whites. Driven by paranoid fears of a "second Haiti," Governor Edward Eyre deployed a huge machinery of repressive "pacification," which in a few weeks left some 500 Jamaicans dead; many more were publicly whipped or tortured in other ways, and a thousand houses were burned to the ground. This reign of terror gave rise to a controversy in Britain that lasted nearly three years. The issue was whether Governor Eyre should be celebrated as a hero who had saved Jamaica for the Crown and prevented the massacre of whites on the island, or whether he was an incompetent murderer who had failed in his duties. Scarcely any other debate stirred and divided the Victorian public so deeply. The country's most prominent intellectuals took sides: Thomas Carlyle defended the governor with a racist diatribe; John Stuart Mill led the party of liberal opponents calling for a harsh punishment. Although the affair ended with a resounding victory for the liberals, Edward Eyre was not punished but merely dismissed from the colonial service; in the end he even received, however reluctantly, a pension awarded to him by Parliament.[206]

And yet 1865 was a milestone in the struggle against racism, comparable to the epochal decision of 1807 to abolish the slave trade. The vigilance of public opinion never flagged, and the foulest pages in the black book of colonialism were subsequently filled by nations other than the British.[207] When racism began to take extreme forms in Germany and Italy after the First World War (and especially in the 1930s), it had already ceased to be generally acceptable in British polite conversation. Race was not ignored, but discrimination in the colonies as well as in the British Isles did not result in state crimes.

So, what was Pax Britannica—from today's *analytical* vantage, not in the rhetoric of the time?[208] It is all too easy to say what it was not. Unlike the Imperium Romanum or the eighteenth-century Sino-Manchurian empire, the British Empire did not encompass a whole world civilization, an *orbis terrarum*. On no continent other than Australasia did Britain possess an undisputed imperial monopoly; everywhere and at every moment it was embroiled in rivalry with other powers. Its imperium was not a homogeneous territorial bloc but a complex network of global power, a structure with knotty bulges and uncontrolled spaces. Unlike the United States in the post-1945 Pax Americana, which had the technical means to reduce any corner of the planet to ruins, Britain in the nineteenth century did not have the military capacity to bring each and every land mass under its control. An intervention to save the Hungarian revolutionaries in 1849, though fervently demanded by sections of the British public, was scarcely feasible. Britain might appear in some measure as a gendarme of the seas, but not as a true global policeman.

Throughout the period from 1815 to 1914 (and despite the fact that after 1870 Britain found it somewhat, but not much, more difficult to have its way on the international stage) Pax Britannica mainly signified (a) an ability to defend the largest colonial empire in the world and even to expand it cautiously without a war with other powers; (b) an ability, beyond the limits of formal colonial empire, to utilize development disparities in such a way as to exercise strong or dominant informal influence in many countries outside the European system of states (China, Ottoman Empire, Latin America), backing this up with contractual privileges ("unequal treaties") and the Damoclean sword of military intervention ("gunboat diplomacy");[209] and (c) an ability to provide the international community with services (a free-trade regime, a currency system, rules of international law) that did not require the user to hold British citizenship. The British Empire was unique in that its territorial core (the "formal empire") had two concentric circles around it: the sphere without sharp contours in which Britain could informally exert decisive influence; and the space of a global economic and legal system that Britain had molded but did not control. Though exceedingly large, the empire did not contain the entirety or even the majority of British economic activity within its confines, not even in the midcentury decades when the United Kingdom was the only world power. Had it been otherwise, the transimperial, "cosmopolitan" free-trade policy would not have survived for long. This is another imperial paradox: for Britain during its period of industrialization and the classical Pax Britannica, the empire was economically less important than it had been before the loss of the United States or than it would be after the onset of the Great Depression in 1929.

7 Living in Empires

Ever since there have been empires, the verdict on them has oscillated between two extremes: on one side, the rhetoric of the imperialists, either triumphantly militarist or soothingly paternalist; on the other side, the rhetoric of resistance fighters (called nationalists in the nineteenth century) referring to oppression and liberation. These primal postures are repeated in today's controversies. Some see empires as violent machines of physical repression and cultural alienation—a view essentially developed in the age of decolonization[210]—while others conclude from the present world situation that empires did more than the chaos of immature nation-states to provide for peace and a modest degree of prosperity. Given the tensions built into this opposition, it is not easy to answer the question how "people" live in empires. Imperialist propaganda has drawn a veil over the realities, but this does not mean that every denunciation of an empire as a "prison of the peoples" is evidence of really unbearable suffering.

A second, related complication is that not all life in an empire or colony was shaped by imperial structures or a *situation coloniale*. It therefore makes little sense to treat the colonial world as a sphere closed in on itself, instead of attempting to understand it from the more general point of view of world history. Here it is difficult to find a middle way. Classical critics in the decolonization period were right to describe colonial relations as generally productive of deformations. By the measure of a fictitious normal condition, the ideal-typical colonizer and colonized both suffered damage to their personalities. However, we would be reinforcing the colonizer's fantasies of omnipotence if we were to see the whole of life in a colonial space as built upon heteronomy and coercion. Methodologically, it is also necessary to address the relationship between structure and experience, and here different approaches confront one another. A structural theory such as that associated with traditional Marxist interpretations often allows no room for the analysis of day-to-day realities and psychological situations within an empire. But, since the critical energies of Marxism have translated into postcolonialism, the opposite effect has made itself felt. An exclusive fixation on the microlevel of individuals, or at best small groups, has entirely blanked out wider contexts, making it difficult to grasp the forces that shape experiences, identities, and discourses in the first place.

Nevertheless, some general points can be made about typical and widespread experiences in nineteenth-century empires.

First. In most cases, an act of violence lies at the origin of a region's incorporation into an empire. This may be a lengthy war of conquest, but it may also be a local massacre—which seldom just happens and is often meant as an intimidating display of power.[211] If the operation is successful, the resulting "shock and awe" paralyzes the adversary, demonstrates the superiority of the conqueror's weapons, stakes out his claim to rule, and leads to the disarmament of the local population that is necessary for a monopoly of force. Unless it tiptoes in noiselessly through a trade agreement or has the way cleared for it by missionaries, an empire always begins with traumatic experiences of violence. To be sure, these often do not burst into a peaceful idyll: not infrequently, they encounter societies already weighed down by violent propensities, as in eighteenth-century India where many successor states of the Mogul Empire were locked in combat with one another, or in the large areas of Africa torn apart by the European or Arab slave trade. In reality, violent conquest frequently gives way to colonial peace.

Second. An imperial seizure of power does not necessary entail the sudden political decapitation of indigenous societies and their complete replacement by foreign authorities. Actually, this has rather seldom been the case. Dramatic examples are the Spanish conquest of America in the sixteenth century and the subjugation of Algeria after 1830. Imperial powers often look for members of the indigenous elite who are prepared to collaborate, some of whom, if only for cost reasons, can be assigned or reallocated to government functions. This strategy, which takes many forms, is called indirect rule. However, even in extreme cases where the practice of rule hardly seems to change under the new masters, the indigenous power holders end up damaged. The arrival of empire always leads to a devaluation of indigenous political authority. Even governments that have to make just a few territorial concessions under external pressure—as the Chinese did after the end of the Opium War in 1842—suffer a loss of legitimacy within their own polity. They become more vulnerable and have to reckon with resistance that at first, as in the Taiping movement after 1850, is by no means necessarily driven by anti-imperialist motives. As for the imperial aggressors, their legitimacy problem stems from the fact that colonial rule is always initially usurpation. Those who understand this soon make efforts to achieve at least rudimentary legitimacy, by gaining respect for their efficiency or by tapping local symbolic resources. But only in rare cases, and then almost always where cultural differences are not too great (as in the Habsburg Empire), does the usurpatory character of imperial rule become blurred over time. This is scarcely possible without mobilization of the symbolic capital of monarchy. If a society that came under an empire was not simply acephalous—as in parts of Siberia or Central Africa—but had a king or chief ruling over it, the colonial power tried either to drape itself in the mantle of imperial overlordship or to slip directly into the role of indigenous monarch. That this was not possible for republican France after 1870, proved to be a continual handicap at the level of symbolic politics.

Third. Incorporation into an empire involves linking up with a larger communicative space, where the flows typically radiate between the core and the periphery. Of course there is also communication among individual colonies and other peripheral areas of the empire, but it has rarely been dominant. The imperial metropolis often controlled the means of communication, viewing with particular suspicion any direct contacts between the subjects of various colonies. But, whenever it was technically possible and state repression did not prevent it, peripheral elites took advantage of the new opportunities.

One instructive field is the use of imperial languages.[212] Multilingualism used to be more or less the norm throughout history, until the nineteenth-century equation of a nation with a single language complicated matters. Thus, in the Muslim world it was very common for people to speak three tongues: Arabic, Persian, and Turkish. But there was a functional differentiation, since Arabic was the language of the (untranslatable) Koran, while Persian enjoyed especially high literary prestige and was the lingua franca in huge areas stretching from the eastern provinces of the Ottoman Empire to the Ganges. To see in the spread of imperial languages nothing but a diktat of European cultural imperialism is to oversimplify a complex reality. In early nineteenth-century India and Ceylon, it was the subject of extensive and sophisticated debates without a clear outcome.[213] Sometimes education in a foreign language was not imposed but freely accepted. Egypt, for instance, whose experiences of the French occupation between 1798 and 1802 were by no means uniformly pleasant, adopted French as the second language of the educated classes in the course of the nineteenth century. This was a voluntary measure on the part of the Egyptian elite, from a country considered to be the leading cultural nation in Europe. French maintained its status there even after the British occupation of 1882. In the Tsarist Empire too, as every reader of Tolstoy knows, French remained for a long time the prestige language of the aristocracy. Absorption by an empire did not automatically mean adoption of the new rulers' language.

Fourth. Many countries that were incorporated into an empire would previously have been part of an extensive economic circuit. Often, though not always, the imperial center broke these connections, by raising mercantilist tariff barriers, introducing a new currency, or closing down caravan or shipping routes. But it also created the possibility of linking up with a new economic context. In the nineteenth century that meant the "world market," which over the long run was growing in volume and density. By the eve of the First World War, few regions on the planet were completely unaffected by it. Insertion into the world market—or better, into particular world markets—took the most diverse forms. It always led to new kinds of dependence, and often also to new opportunities. Any empire is an economic space sui generis. Incorporation into it did not leave local relations unchanged either.

Fifth. Dichotomies between perpetrators and victims, colonizers and colonized are suitable at best for crude approximate models. They constituted a kind of founding contradiction in colonial societies. But only in extreme cases, such as Caribbean slavery in the eighteenth century, was this so dominant that it accurately described the social reality—and even then there were intermediate strata of "free persons of color," or *gens de couleur.* As a rule, societies incorporated into empires had a hierarchical structure that contact with the empire called into question. The empire differentiated between its friends and enemies. It divided indigenous elites and played their various factions off against one another; it sought collaborators, who had to be paid. The colonial state apparatus needed local personnel at every level—and on a large scale in the case of late-nineteenth-century telegraphy and railroads, and the customs service. Insertion into world markets created niches for upward social movement, in commerce or capitalist production, which minorities such as the Southeast Asian Chinese knew how to exploit. If European real estate law was introduced, it inevitably led to radical changes in property relations and rural stratification. In short, with the rare exception of low-key indirect rule in areas such as Northern Nigeria or Anglo-Egyptian Sudan, imperial absorption resulted in far-reaching transformation, sometimes approaching a social revolution within the space of a few years.

Sixth. Personal and collective identities change at the cultural frontier of an advancing empire. It would be too simple to see this as a transition from an equable self-image to "multiple" forms of personality and socialization. Even the emergence of what is sometimes called "hybridity" is not necessarily a distinctive feature of colonial and imperial constellations. The older sociological concept of "role" is more useful here. Any social situation becomes more complex if additional factors appear; the repertoire of roles grows larger, making it necessary for many people to master several at once. A typical colonial role, for example, is that of the middleman and interpreter. The position of women was also affected when new ideas about female conduct and labor were introduced, often by Christian missionaries. "Identity" is a dynamic category: it is recognized most clearly when it takes shape in acts of demarcation. This was not peculiar to colonial situations, of course, but perhaps we may say that in general it was important for imperial rulers to be able to sort their confusingly varied population into a number of clear-cut "peoples." Nation-states tend toward cultural and ethnic uniformity and seek to bolster it by political means. In empires, however, the emphasis is on difference. Postcolonial critics usually attack this as a grave offense to human equality, but it should not be evaluated in purely moral terms. Ethnic stereotyping undoubtedly intensified in the late nineteenth century under the influence of racial doctrines; it emanated, however, from various directions. Colonial systems tried to bring order into complexity by artificially creating "tribes" and other categories for the classification of their subject population. The aspiring science of anthropology/ethnology was influential here, and the census was useful in giving taxonomies some material weight. Certain social groups took shape in reality only once they had been defined in theory.[214] Colonial states first created difference, then went to great pains to order it. This happened in varying degrees of differentiation. The French presence in Algeria was constructed around a simple opposition between "good" Berbers and "degenerate" Arabs.[215] British India, on the other hand, elaborated a classificatory grid of pedantic sophistication.

The categorization and stereotyping of colonial subjects was not only a project of officialdom. To some extent the various peoples assumed the identities given them, but they also put up resistance and invested much energy in constructing an ethnicity of their own. Nationalism, an idea developed in Europe and imported from there, often reinforced formative processes already under way, constantly adapting to and changing them. The authorities thus faced a dilemma: the "divide and rule" principle tended to foster differences between ethnic groups, but they had to be prevented from escalating to a point where the groups became violent and hard to control. Collective identities were not always susceptible to manipulation, nor were they inevitably defined in *ethnic* terms. In fact, that was not much seen outside Europe in the nineteenth century. After the First World War, a wide range of options emerged for the creation of anti-imperial solidarity. The Indian freedom movement, in the phase that began in 1919 with Mohandas K. Gandhi's first campaign, was neither ethnically nor religiously based, and the idea that there should be a special Muslim state on Indian soil did not gradually mature over a long period but burst forth after 1940 in the tiny circle that went on to found Pakistan. From the middle of the nineteenth century on, empires were arenas for the formation of collective identities. These processes, already discussed as the "nationalities question" toward the end of many empires, were beyond anyone's capacity to channel them. Only in exceptional cases did a reasonably compact proto-nation become subject to an imperial power (Egypt in 1882, Vietnam in 1884, Korea in 1910) and then later, after the end of colonialism, successfully pick up the thread of its earlier quasi-national history. Elsewhere, empires generated willy-nilly the forces that would later turn against them.

Seventh. Of the political lessons that were learned in empires, the most widespread and important was that politics was possible only as resistance.[216] Empires know only subjects, not citizens, in their periphery. The dominions of the British Empire were the great exception in this respect. In 1867 the Hungarians managed to break the rule in the Habsburg Empire; and in 1910, with the founding of the Union of South Africa, the Afrikaners achieved a special variant of their own. Only in the French Empire after 1848 were a small number of nonwhites granted civil rights: in the *vieilles colonies* of Guadeloupe, Martinique, Guyane, and Réunion, and in the four coastal cities of Senegal.[217] Even when elite collaborators were integrated into the imperial state apparatus, they were barred from decision making at the top, remaining mere transmission belts from the real power center to the dependent society. Institutions that could articulate local interests were seldom created. For all the differences in detail, an empire is thus reducible to a one-way chain of command. Strong-willed men on the spot might make it looser, and smart imperial politicians kept their demands within limits and ensured that it was theoretically possible for their instructions to be carried out. The bow was not to be stretched too far; the empire must not appear to its subjects as no more than an apparatus of terror. Ever mindful of the cost-benefit relationship, imperial statecraft sought to establish firmly rooted interests, cultivating the perception that it was more advantageous to live inside the empire than outside.[218] This did not alter the general lack of indigenous political participation: the co-opting of a few elite figures into the "legislative council" of a British crown colony was window dressing designed to produce an illusion of representation; all nineteenth-century empires were autocratic systems from beginning to end. As in early modern variants of Western European "enlightened absolutism," this did not exclude a degree of legal security. Although it would be an exaggeration to describe the British Empire (where this was taken furthest) as a law-governed state, a kind of basic legality or "rule-based command" did generally prevail.[219] Indigenous people might still be denied some of the basic rights enjoyed by whites, and access to the justice system could be very difficult for them to obtain. But around the year 1900 it made some difference whether an African lived in King Leopold's Congo or British Uganda.

The nineteenth century was an age of empires, and it culminated in a world war in which empires fought one another. Each of the belligerents mobilized resources from its dependent peripheries. If it did not have any—Germany, for instance, could no longer profit from its colonies after 1914—then it became an major war aim to acquire additional quasi-colonial areas. After the end of the war, only a few empires were dissolved—and not the largest and most important. Germany lost its small, economically insignificant colonies; the Great Powers in the victorious coalition shared them out among themselves. The unique Habsburg Empire, a European multinational entity with no colonial possessions, broke up into its component parts. Of the Ottoman Empire there remained Turkey and the former Arab provinces (now mandated territories or semicolonies of Britain and France). Russia had to give up Poland and the Baltic, but under Bolshevik leadership it was able to reunify the great majority of non-Russian peoples of the Tsarist Empire within an imperial "union." The age of empires did not come to a close in 1919.

To be sure, generations of historians who have seen the rise of nationalism and the nation-state as key features of the nineteenth century are not wrong. But their judgment does need to be heavily qualified. Once all the new republics had emerged in Latin America by 1830, the formation of nation-states proceeded more slowly. The Balkans were the only (small) region where the pace was quicker. Elsewhere the opposite was the case. In Asia and Africa, independent political entities—one would not wish to describe them as "states"—disappeared in great number into the expanding empires, and no small nations freed themselves from coercive imperial relations. Not one of the numerous national movements in nineteenth-century Europe managed to help its national community to independence outside an empire; only Italy may in some sense be considered an exception. The partition of Poland continued, Ireland remained part of the United Kingdom, and Bohemia did not separate from the Habsburg Monarchy. Still less did any of the national movements destroy an empire.

Nationalism registered few palpable political successes in Europe, and fewer still in Asia and Africa. This must be distinguished from the fact that solidarity in the name of a nation was a twofold novelty of the century. On the one hand, nationalist intellectuals and their followers worked within imperial contexts to prepare the independent nation-states that many countries would become during the period from 1919 to around 1980. The great protest movements of 1919 in Egypt, India, China, Korea, and a few other countries of Asia and Africa were already nationalist in their motivation.[220] On the other hand, nationalism also became the mainstream rhetoric in fully consolidated states.[221] People began to understand themselves as a French or English/British or German or Japanese "nation"; they developed an appropriate cosmos of symbols, strove to differentiate themselves from other nations, talked themselves into competing with them, and lowered their tolerance threshold in relation to foreigners and foreign ideas. This happened in a world where exchange relations were multiplying and intensifying between members of different nations. Various kinds of nationalism were to be found in empires as well as nation-states. Pride in one's own empire, often fueled by official propaganda, became a widespread sentiment around the turn of the century, a constituent of the national self-image. Nationalism *within* empires was not always directed against the structures of imperial rule: it was thus not exclusively anticolonial. It might also—especially if reinforced by religious identities—fan the flames of conflict between subordinate groups. This would result in the breakup of the Habsburg Empire in 1918–19 and of unified India in 1947.

Nowadays, the word "empire" bears associations of unlimited power. Certain reservations are in order, however, even for the Age of Empire at its height. *Early modern* empires (with the exception of China) were loose political and economic networks rather than tightly integrated states or closed economic blocs. Even the sixteenth-century Spanish world empire, often cited as an early example of transoceanic territorial rule, rested to a large degree upon local autonomy, and mercantilist control over trade had to be constantly enforced in all empires against widespread smuggling. Empires were not the creatures of nations: their elites, and often the proletariat laboring on their ships or plantations, were composed of people from the most diverse countries. By 1900 most empires had become more "nationalized." Thanks to modern power techniques and media, they were more tightly integrated and therefore easier to control. Regions producing for export were closely tied into the world economy, often as small enclaves whose hinterlands became ever less interesting to imperial governments unless trouble was brewing there. Yet, in one way or another, every empire continued to rest upon compromises with local elites, upon an unstable equilibrium that could not be maintained only through the threat or use of force because military action was too expensive, difficult to justify, and productive of problems that were hard to calculate. In the club of imperialists, an empire counted as modern if it had a rationalized and centralized administration, made the exploitation of economic resources more effective and profitable, and took pains to spread "civilization."

Such activism, however, carried high risks. Reforms disturbed the existing equilibrium and always unleashed some kind of resistance whose strength it was never easy to predict;[222] North America in the 1760s was one cautionary example. But they also created new material, cultural, and sometimes political opportunities for particular groups, which in the long run, as bearers of a rival modernization, might develop into counterelites and social forces with a horizon beyond that of the empire. In the Ottoman and Chinese empires, notables in provincial cities strengthened centralizing initiatives;[223] this even contributed to the downfall of the Chinese monarchy in 1911. Restraint in the sensitive areas of law, finances, education, and religion was therefore a definite option for imperial centers. The British, for example, tended toward such conservatism in post-1857 India, and later wherever they practiced some form of indirect rule. "Empire light" did not disappear from the historical agenda. Indeed, in some circumstances the nation-state could weigh more heavily on its citizens, especially on members of an ethnic or religious minority, than many an empire did on its subjects.

CHAPTER IX

International Orders, Wars, Transnational Movements

Between Two World Wars

1 The Thorny Path to a Global System of States

Foreign policy players at the level of the globe or within one of its macro-regions—this chapter will refer to "spaces of power and hegemony"—together form a world of states, irrespective of the type and density of the relations among them. If these relations attain a certain threshold of structure and regularity, we should speak of a *system* of states or an "international system." Of all such systems in history, the best known is the modern European one that lasted, if we want attach precise dates to it, from 1763 to 1914—during a period between two world wars, the Seven Years' War and the Great War.[1] If an international system is held together by institutions *and* also by normative commitments to peace, without yet achieving the higher integration of a league or even federation of states, then the term "international community" is used.[2] In order to illustrate this distinction: the Second Hague Peace Conference in June 1907 brought together not only the European Great Powers (which had had their own "international system" for decades) but representatives of a total of forty-four states. It was the first time that nearly all of the world's states currently recognized as independent—the "*world* of states"—had gathered in a conference hall.[3] But this assembly failed to agree on institutions and conventions that would substantially further the cause of peace. An international *community* therefore did not take shape at The Hague.

The Two Phases of Peace in Europe

The European system of states was an action-guiding image in the heads of the foreign policy elites of individual countries. At least since the Congress of Vienna, it no longer produced fragile balances more or less automatically but required political management structured by a basic set of both manifest and unspoken rules. Statecraft, at least in theory, consisted in upholding national interests only so long as it did not threaten the functioning of the system as a whole. This worked for four decades—a long time in international politics.

469

But then came a period of eighteen years, from 1853 to 1871, in which five wars were fought with great-power participation: the Crimean War (1853–56), the war of 1859 in Italy that pitted France and Piedmont-Sardinia against Austria, the Danish-Prussian War (1864), the Austro-Prussian War (1866), and the Franco-Prussian War (1870–71). Austria was involved in four of these wars, Prussia in three, France in two, Britain and Russia in one. The Crimean War severely shook the cohesion among European nations, while the Italian and German wars of unification were accompanied with realpolitik that blatantly contradicted the spirit of the post-Napoleonic peace.

The Crimean War, the first in the series, differed from the others in two ways. On the one hand, its objectives were less clear. It came about "less through cool calculation or hostile intent than through a long chain of mistakes, wrong conclusions, misunderstandings, false suspicions, and irrational enemy-images."[4] It is remarkable that forces supporting war were at work in such diverse societies: in Russia a ruthless and ill-informed tsar, Nicholas I, who at the end of his reign was obsessed with his dilettante foreign policy; in France a political gambler, Napoleon III, who used risky maneuvers abroad to boost his prestige and popularity at home; and in Britain a Russophobic press capable of exerting pressure even on a supremely self-confident (though in the early 1850s by no means unanimous) political class. On the other hand, despite the chance events and short-term thinking that triggered its outbreak, the Crimean War involved a logic of geopolitical and economic interests that pointed beyond the European system of states. Its cause lay on the fringes of Europe, since the key issue was whether Ottoman-ruled lands would come under Russian control or would remain as a strategic buffer zone guaranteeing routes to India (the Suez Canal did not yet exist) and providing a new area for British economic penetration.

The Crimean War was essentially a conflict between the only two Great Powers of the day that had major interests in Asia. Its course and its outcome demonstrated the military weakness of both sides. The backwardness of the Tsarist Empire was plain to see, but serious doubts also became possible about Britain's ostensible position as the only world power; experienced veterans of France's colonial war in Algeria proved superior to the British units.[5] When in spring 1854 France and the United Kingdom entered the Russo-Ottoman war that had begun the previous year, this was a watershed in the international history of the nineteenth century. For the first time since 1815, war appeared an acceptable option—so much so that it actually happened.

The bellicose interlude in European history came to an end in 1871. If we think that by far the largest civil conflicts of the century—the American Civil War (1861–65) and the Taiping Revolution (1850–64)—as well as the Muslim unrest in China (1855–73) occurred in the third quarter, then clearly we are talking of a worldwide surge of violence with no common underlying causes.[6] The aftermath presents us with a major paradox. By 1871 there were no longer the simplest institutions or the most elementary values to preserve the peace, and yet peace did prevail in Europe for the next forty-three years—at least if we follow a convention among historians and disregard the Russo-Turkish war of 1877–78, fought mainly in what is now Bulgaria. The really astonishing thing about the First World War is not that it occurred at all but that it began so late. The "systemic" interpretation of European history developed by Paul W. Schroeder for the period between 1815 and 1848 may convincingly explain why peace reigned at that time: his argument, in a nutshell, is that the European system of states developed into an international community.[7] It is much harder to account for the stability of Europe in the age of industrialization, arms drives, and militant nationalism; each of the international crises (none of which led to war) would anyway have to be treated separately.[8] But a few general points may be made here.

First. For a long time no single power armed itself offensively for an intra-European war. A partial exception is the Anglo-French naval rivalry of the 1850s and 1860s, the first arms race in history that centered on a quest for the latest technology rather than quantitative accumulation of material.[9] The founding of a powerful German nation-state in the heart of Europe did not lead immediately to a new arms drive. Field Marshal von Moltke, the top strategist of the Reich, had concluded from the events of 1870–71 that Germany's interests would be best served by an armaments policy geared to deterrence. This changed only in 1897, when Admiral Alfred Tirpitz, Kaiser Wilhelm II, and "pro-navy" forces in the German public adopted a program of military shipbuilding that was not only part of an international trend to replace British hegemony at sea with a new balance of power but was from the outset offensively directed against Britain.[10] London took up the challenge, and in both countries—though in Germany without the basis in a dominant culture of seafaring—the navy was presented as the symbol of national unity, grandeur, and technological might. Of all people, it was an American naval officer, Alfred Thayer Mahan, who provided the historical and theoretical rationale on which the new worldwide (including German) enthusiasm for the navy based itself.[11] European politicians now had their first taste of an industrially accelerated arms race involving all the Great Powers.[12] The defensive goal of deterrence had an attack plan built into it. But, unlike after 1945, when Hiroshima and Nagasaki gave some inkling of what high-tech warfare would entail, the arms drive around the turn of the century pointed to a future whose gruesome shape lay outside the imagination of any but a handful of contemporaries. Nobody anticipated the horrors of Ypres and Verdun.

Second. For reasons that cannot be explained "systemically," no power vacuum appeared in Europe that could have led anyone to adopt an aggressive foreign policy. This was the paradoxical outcome of the successful building of nation-states in Germany and Italy, but also in France, which soon recovered from the military catastrophe of 1871. No state broke up. The Ottoman Empire was gradually driven from the Balkans in the years up to 1913, but it never collapsed in a way that gave its neighbors a chance to realize their fantasies of carving it up. In 1920, with the Treaty of Sèvres, these pipe dreams reached another climax in plans to confine Turkey to a rump state in Anatolia. But a great military effort under Mustafa Kemal (Atatürk) quickly put an end to such visions, in which the United States too had temporarily shared. In the Treaty of Lausanne (1923), the Great Powers accepted a Turkish nation-state as the strongest political force in the eastern Mediterranean. Still more important was the position of Austria-Hungary in the European world of states. Its internal evolution was contradictory: impressive economic development in several of its regions, combined with growing tensions among its nationalities. But this had little impact on the international position of Austria-Hungary. By any conceivable criteria, the Habsburg Monarchy remained the second-weakest great power throughout the century. During the four decades before the First World War, it was strong enough to retain its place in the European system, but too weak to behave aggressively against its two main rivals, Germany and Russia. This unintended optimization of Austria's power potential stabilized eastern-central Europe and left no room for any prospect of a "Central European" (*Mitteleuropa*) imperialism, such as many in Berlin as well as Vienna entertained in their dreams. The First World War was not a result of the collapse of the Habsburg Empire; the exact opposite was the case.

Third. Bismarck's policy after 1871 meant that a straightforward duel in Europe no longer made sense. Any conceivable war would have to involve rival coalitions. But the building of such alliances was far more difficult and tiresome, both politically and militarily. It was clear to all statesmen in Europe that the next war in the heart of Europe would not leave any of the Great Powers untouched.[13] The post-1871 "competitive alliance equilibrium"[14] suffered from a confidence and conciliation deficit, but it endured because all the alliances were defensively oriented: not a "balance of terror," as after 1945, but one of mistrust. Only after the turn of the century, as showdown fantasies ("Slavs versus Teutons," etc.) became virulent and developments in the Balkans enabled small countries to play on Europe's most dangerous fault line, that between Austria and Russia, did a fatal instability creep into the system.[15]

Fourth. The special relationship between Europe and overseas also helped to limit conflict. It was to be expected that the periphery would have various functions for the European system of states: as a safety valve for European tensions or conversely as a catalyst for conflicts that then impacted on Europe, but also as a field for trying out new weapons. Imperial powers could see that they were overstretched—the background to the Anglo-Russian Convention of 1907 on Asia—and decided to slow the dynamic of their expansion. Whenever and wherever this actually happened, the decisive point was that the uncoupling of the periphery for the purposes of security policy conflicted with its growing economic integration. The uncoupling proceeded all through the century, and attempts (such as Bismarck's at the Berlin Africa conference of 1884–85) to transpose the unwritten rules of the European system of states to the scramble for colonies were unsuccessful in the long run.[16] This, of course, is again a systemic argument. In the imaginative horizons of key players—especially Britain and Russia—there was by no means a sharp separation between Europe and the rest of the world. For example, a major reason why London continued to support the Ottoman Empire was that a course of action directed against the sultan (who also claimed the religious title of caliph) would have provoked unrest among millions of Indian Muslims.

Global Dualism

In contrast to various early modern peace agreements regulating colonial interests, the Congress of Vienna concerned itself only with the states of Europe.[17] The rest of the world was deliberately ignored, except insofar as slavery was taken up as a side issue. The very fact that the Ottoman Empire had no place at the conference table underlined this narrowly European focus, making it possible for the delegations to regard the Eastern Question as a special problem outside the framework of the settlement. All the mechanisms agreed upon at the Congress, whether counterrevolutionary interventions or diplomatic meetings to resolve conflicts in a timely manner, applied to Europe alone. It did not take long for this exclusion of the periphery to have practical consequences when the Great Powers, including the most reactionary of them, Russia, intervened under British leadership in the eastern Mediterranean. Flying in the face of all the moderate and conservative accords on Europe, it was a policy in favor of a revolutionary movement and against the oldest dynasty in sight: the Imperial House of Osman, in power since the fourteenth century. But this action on Greece had no repercussions for the relationship of the European powers with one another, and the potentially explosive creation of the Kingdom of Greece and almost simultaneously that of the new state of Belgium, showed the post-Vienna diplomatic mechanisms at their best.

In several respects, the insulation of Congress Europe from conflicts on its periphery was a brilliant peace-building idea.[18] It had an echo in 1823, when US President James Monroe proclaimed his famous doctrine that both North and South America "are henceforth not to be considered as subjects for future colonization by any European powers."[19] Thus, on both sides of the Atlantic, the years from 1814 to 1823 witnessed a conscious deglobalization of international politics. After the great world crisis of the preceding period, when revolutionary events in North America, France, and the Caribbean triggered effects as far away as South Africa, China, and Southeast Asia, international political relations became compartmentalized, while at the same time economic links continued to grow and intensify.

In a longer-term perspective, however, this also meant something else. During the early modern period, it had not been possible for Asian and European powers to construct a shared legal system; they had merely recognized that the other was in principle a legal subject of equal status, so that contracts or oaths had been valid across cultural boundaries. In the new order established in 1814–15, the

Europeans refrained from taking the initiative toward a new global legal order. The conditions were therefore lacking for the preservation of peace on a world scale. Even European international law, a major civilizing achievement, did not become part of a broader Western legal consciousness imposing certain obligations on Europeans overseas. Neither the *ius ad bellum*, which required a legal justification for war, nor the *ius in bello*, which regulated the conduct of war and was supposed to prevent excesses, found strict application outside Europe. In the age of rising global disparities and an ever sharper sense of cultural and ethnic differences, the globalization of law could consist only in the gradual imposition of European concepts, whose practical application, moreover, always tended to favor Europeans.[20]

The conceptual divide meant that overseas conquests and military interventions were not subject to the limits on warfare prevailing within Europe. Nor were there any normative rules in the European system of states that might have prevented or mitigated the most brazen forms of Western land grabbing, such as Russia's extortion of vast territories north of the Amur from China in 1860, the scramble for Africa, Italian operations in Tripolitania, or the US subjugation of the Philippines. The persistence of such dualism, even at the climax of imperialist aggression, served to maintain the insulation effect for Europe. From the 1870s on, the Great Powers grew used to the idea that their policies for equilibrium in Europe should also apply on the world stage—although this would really come into its own only during the Cold War after 1945–47. In the late nineteenth century, contradictory tendencies stood in opposition to each other: a growing certainty that all international relations should be seen as elements of a single global system, and a continuing conceptual separation of the periphery from the sphere of "true," European politics. The imperial powers tangled with one another in various places in the world: all parts of Africa, China, Southeast Asia, the South Seas, and in winter 1902–3 even in Venezuela. However, it was possible to solve all these conflicts or to limit their effects, not least because of unwritten rules of the game, such as the principle that "compensation" should be provided or tolerated elsewhere for the failed ambitions of an imperial power. Many of the imperial tensions fueled lasting mistrust between European governments, but not one impacted on European relations in a way that was directly productive of war.[21]

The European system of states in the decades before the First World War was not destabilized from without. Asia, Africa, and the Americas played an ever larger role in the overall political calculations of European governments, without leading them to suppose that a great war of the empires was unavoidable. It has even been suggested that, in the half century before 1914, the European interstate system of the five Great Powers established their collective "world supremacy."[22] Is this a reasonable proposition? It is certainly true that Britain, Russia, and France all had significant interests outside Europe, ruling or influencing large territories, as did the German Reich on a lesser scale after 1884. It is also true that the states of the pentarchy together had the world's greatest industrial and military potential and (with the exception of Austria-Hungary) were prepared to deploy it in interventions overseas. But this does not mean that Europe alone had attained the supreme cultural achievement of orderly "international relations," while the rest of the world remained mired in murderous anarchy.[23] The European system of states was never preponderant in the sense of acting as a single power, or even a coordinated collective, on the international stage. The main diplomatic congresses of the age were convened not by the system as such but by one individual power that saw it as being in its own interests to act as a "broker." The decisive alignments of overseas interests were invariably bilateral. Only once, in summer 1900, was there collective action outside Europe—when an eight-power expeditionary force broke the siege of the diplomatic missions in Beijing by the insurrectionary Yihetuan ("Boxer") movement. Japan and the United States played a major role in the relief operation, which was also the most ambitious action that Austria-Hungary carried out abroad in its entire history.[24]

Politically speaking, European imperialism was *less* than the sum of individual imperialisms. The mechanics of the system functioned, if at all, only among the five Great Powers *qua* European players, not among them as multicontinental empires. The system as such was not a supporting structure of "world politics."

2 Spaces of Power and Hegemony

The imperial expansion of Europe and North America did not occur in politically unstructured spaces; any simple opposition between Europe and "the rest" is misplaced. First of all, relations of quasi-colonial dependence were by no means absent within Europe itself. Traditional diplomatic history makes only marginal reference to what it calls the "lesser European powers," showing little interest in their scope for action in a world dominated by the Great Powers. Portugal, for instance, had an extreme economic dependence on Britain, keeping consumers there supplied with cork and port wine and sending 80 percent of its total exports in 1870 to the British Isles. Moreover, brutal exploitation was practiced in conditions that were no longer possible in Britain itself—for example, when British firms employed Portuguese children on piece rates to cut cork for bottles with cutthroat razors.[25] Such outsourcing of high-risk and low-paid jobs is always an important indication of asymmetry in the world system.

The Americas

One space with its own distinctive structures of hegemony was the Americas. In the 1820s, the separation of the Spanish colonies from the mother country, together with the slowly unfolding impact of the Monroe Doctrine, made the New World more detached from the Old World than it had been for centuries. For a brief historical moment, in 1806–7, Britain was tempted to take over the legacy of the *conquistadores* in the River Plate region and elsewhere, but in the end it never tried to intervene outside its already existing colonial domain. The United Kingdom would remain neutral in the conflict between Spain and its rebellious subjects in the Americas. Its trade with the region already increased during the independence wars, and by 1824 Latin America accounted for 15 percent of all British exports. London hastened to recognize the new republics, especially as US diplomats were by this time already seeking to extend the influence of their nation. Soon an international legal framework was in place that gave the protection of British laws to UK citizens in Latin America and, while not obliging Latin American countries to prefer British imports, required them to apply tariffs no higher than those imposed on most favored nations. Under this fairly light regime of "informal imperialism," Britain long remained the principal trading partner of many Latin American countries until the United States increasingly took over this role toward the end of the century.[26]

As early as the 1830s, having for twenty years been the most troubled continent in the world alongside Europe, and having in the same period aroused great interest abroad thanks to the travels of Alexander von Humboldt and others, Latin America slipped from the sight of international diplomacy.[27] Not a single country of the continent was drawn into intra-European power politics, nor did serious US-British rivalry break out in South America at any time in the century. Britain could not always successfully translate its economic weight into political influence. Its usual methods of diplomatic pressure failed to end slavery in Brazil (with which it otherwise had good relations). The Latin American countries themselves did not develop a distinctive interstate system; indeed, something closer to anarchy prevailed among the often arbitrarily defined splinters of the Spanish Empire. Simón Bolívar, the Liberator, ended his days despairing about the particularism of his compatriots. A genuine pan-Americanism, not instrumentalized by the United States, was never an important element in the situation. Many state borders were the object of dispute. Nothing was done for the external defense of the region; scarcely a single navy was capable of being deployed in battle.[28]

The grim War of the Triple Alliance, which in 1864–70 pitted Paraguay against Brazil, Argentina, and Uruguay, was not exactly characteristic, but its very possibility was eloquent testimony to the disunity of the continent. It was also the most costly conflict, in terms of human lives, in the whole history of South America. After 1814, under three successive dictatorships, little Paraguay developed into what has been described as an "enlightened Sparta": egalitarian, tightly disciplined, heavily armed, and comparatively literate.[29] A Brazilian border violation with Uruguay was the pretext for the dictator Francisco Solano López to march his well-trained army into battle with the second-rate troops of Brazil and Argentina. The first encounters ended in disaster for the Allies, who were soon joined by Uruguay. But in 1867 the war machine of Brazil, a country with a population twenty times larger than Paraguay's, went into top gear. By the end of the war, which Paraguay delayed in stubborn defensive actions, half

its population was dead—proportionally the highest military and civilian casualties of any war in modern times.[30] The conflict became the central event in the
history of Paraguay, the key datum in the collective memory, as well as a turning
point in the history of the continent. Argentina also suffered heavy military and
economic losses, and saw its previously unchallenged supremacy on the Plate
whittled away; Brazil's regional superiority was confirmed.[31]

The Pacific War or "Saltpeter War" (1879–83), which ended in victory for
Chile over Peru and Bolivia and gave it important reserves of nitrates, had a similar effect on the participants. The unparalleled mobilization of Chilean society
was its most intense collective experience since independence; and in Peru, where
guerrillas fought against the invaders, the violence ushered in a breakdown of the
state.[32] Given the lasting volatility, which might be characterized both internally
and externally with the formula "fragmentation plus weak stabilizing powers," it
is surprising that Latin America knew as much peace as it did.[33]

While the countries of South America did not develop a common security
system, those of Central America came increasingly under the influence of North
America. Here rivalry between Britain and the United States played a role, at
least indirectly, since the British, as Mexico's chief creditor, were able to bring a
certain political pressure to bear. Washington feared that London might thereby
lay hands on the Mexican province of California, but there is much stronger
evidence that the United States had been planning its annexation for some time.
President James K. Polk played the imperialist game long before all Europeans
had learned it. Having put the Mexicans under so much military pressure that
they finally withdrew, he went on to present Congress with evidence of a Mexican attack and convinced it to issue a declaration of war.[34] In the late summer
of 1847, a US expeditionary force reached Mexico City, and powerful political
actors in the United States, including the president himself, called for the whole
country to be annexed. The men on the spot—usually inclined to press on for
the maximum objectives of their paymasters—were in this case a moderating
influence. Yet the Treaty of Guadalupe Hidalgo (February 1848) was still a diktat. Paltry compensation was paid to Mexico for its forced surrender of territory
corresponding to the present states of Arizona, Nevada, California, and Utah, as
well as parts of New Mexico, Colorado, and Wyoming.

The United States and Britain found themselves on a collision course not
over Mexico and California but because of events farther south, in Central
America, where the British initially dominated a weak proto-system of states.
As the United States rapidly grew more interested in trade with Asia, attention
turned from the annexation of California and Oregon (1848) and the California
gold rush to the transit possibilities offered by Central America. In 1850 a British
envoy and the American foreign minister reached an agreement (the Clayton-
Bulwer Treaty), which stipulated that neither country would acquire new colonies in the region or build a canal across the isthmus without the other's consent. Symbolically this put Britain on an equal footing with the United States in

Central America. In the subsequent decades Washington continually expanded its influence, and in the 1870s and 1880s it landed combat forces several times in Panama (then a province of Colombia) "to restore order" and "to protect U.S. citizens."[35] The US-British equilibrium in the region disappeared over time, and in 1902 Congress unilaterally decided to build a canal through Panama. When Colombia balked at the purchase price offered for a canal zone, private interests arranged with US support for a new state of Panama to declare independence. The Canal Zone was then leased forthwith to the United States, and in 1906 several hundred workers were hired from Spain (soon to be joined by another 12,000 Spanish, Italians, and Greeks) to begin the construction work. The Panama Canal opened to shipping in August 1914.[36]

In South America the political map changed little after independence, with its mosaic of weakly articulated states all more or less in search of nationhood. Not even Brazil, on account of its Portuguese origins, was capable of soaring to hegemony in the continent; nor did Britain or (until the 1890s) the United States fill the gap. The Great Powers had clientelist relations with individual countries, but not a capacity to define a broad structural context that is implied in hegemony. No one any longer entertained the dreams of the bygone liberation period, when visions had been entertained of a great Hispanic American federation taking shape along the lines of the United States. Rather, the model was European diplomacy, in which secret treaties might be reached but no supranational forms of organization would come into being—not even a Latin American "concert." Compared with the climaxes of the military contest among Europeans, the countries of Latin America got on fairly peacefully with one another in the nineteenth century. The lack of genuine great powers was in *this* respect more a blessing than a disadvantage for the continent. On the other hand, South America was left without states and military forces capable of resisting the growing dominance of the United States.

President Monroe's message—"America for the Americans!"—became a "doctrine" and attained its maximum effect only in the decades after the French defeat in Mexico in 1867. In the Venezuela crisis of 1895–96, the United States for the first time used threats of war to assert its claims to leadership (against those of Britain) south of the Central American isthmus too. In 1904 President Theodore Roosevelt added a corollary to the Monroe Doctrine, whereby the United States claimed the right to "civilizing" intervention anywhere in South America. This effectively turned Monroe's idea on its head: he had protected Latin American revolutions, whereas Roosevelt now wanted to act against them; he had sought to keep South America free of armies, whereas Roosevelt looked to the supremacy of North American arms. The Roosevelt Corollary merely set the seal on current practice: US troops had already intervened twenty times in Latin America in the years between 1898 and 1902.[37]

What took shape in the 1890s was not a fully fledged American system of states but a rather a less-than-benevolent hegemony—"unilateralism" in today's parlance—practiced by the economically and militarily superior United States. This often remained only latent, however: Washington was not able to push through all its objectives. For example, the various regimes in Brazil kept up good relations with the United States without according the economic privileges that the latter desired. Nothing became of the idea of a pan-American free-trade zone.[38] It should also be recognized that, unlike Asia or Africa and partly as a result of the US umbrella, Latin America was spared involvement in two world wars. In the nineteenth century, the two countries of *North* America also did not constitute a "system" in keeping with the European model. Of greater importance was their agreement in 1817 to demilitarize the Great Lakes—an early example of bilateral disarmament. After the final resolution of all border issues in 1842, the United States and British Canada settled into a cool but peaceful relationship with each other, a zone of tranquility in the turbulent international history of the nineteenth century.

Asia

In other parts of the world, Europeans encountered older state formations that they were neither able nor willing to overturn. In South Asia in the eighteenth century, the French and British (and eventually only the latter) entered the power game along with the states that had succeeded the Mogul Empire. The British conquest of India can only be explained as a power grab from within the Indian world of states, supported by military and administrative forms of organization that the British brought with them or developed and tested on the spot. Under fully developed British rule—that is, after the annexation of the Punjab in 1849—there was only a semblance of pluralism in the Indian states. The five hundred or so remaining princedoms, where the East India Company and after 1858 the British Crown did not exercise direct rule, were in no position to pursue an independent external or military policy. A maharajah who took up with the Russians, for instance, would have been immediately removed from office. Any succession to the throne needed approval from the colonial authorities.[39] The British were also careful to ensure that horizontal links between princedoms were as weak as possible. The all-India princely assemblies (*durbars*), which from 1877 took place at wide intervals with great ceremonial pomp, had no political content and were in effect pseudo-feudal rituals of homage to the distant monarch and her (or later his) viceregal representatives.

In Malaya, the British operated for a long time within the plural world of local princedoms, which had never known an overarching imperial supremacy such as that of the Moguls in India.[40] In 1896 the four states on the east coast of the peninsula became the Federated Malay States, with their capital in Kuala Lumpur; in addition there were the Unfederated Malay States and the Straits Settlements. At no point before the Japanese invasion in 1941 did a single administrative structure exist for the whole of British Malaya. Annexation was a method that the British used more sparingly than in Africa, for instance, and their "residents" cultivated the art of diplomacy at the Malay sultans' courts for a long time yet. One reason for this was that in Southeast Asia, representatives of the British Crown had events largely under their control, and subimperialisms such as that of the early East India Company in India or Cecil Rhodes in southern Africa did not play a major role. To a great extent the various states were independent only on paper, but the precolonial pluralism of rule was not entirely swept away. Nevertheless, life was not breathed back into it after independence. From the patchwork quilt of the colonial period, all that remained after the 1960s was two sovereign states: Malaysia and Singapore.

Indochina, on the other hand, broke up again after the end of French rule into the three historical entities of Vietnam, Laos, and Cambodia. If we add Burma and Siam, both major powers toward the end of the eighteenth century,[41] then the surprising conclusion is that the age of colonialism did not fundamentally alter the precolonial configuration of states in mainland Southeast Asia. The pentarchy, which came into being more or less simultaneously with the arrival of Europeans, is still in existence today.

In China and Japan, Europeans and North Americans encountered highly complex political systems that could not be brought under colonial rule. Japan was never tightly integrated into any kind of international order. It was never part of a major empire, still less of a system of states roughly equal in strength, such as that which emerged in early modern Europe or in eighteenth-century India and Malaya. Even after Japan closed itself off in the 1630s, it maintained intensive commercial, artistic, and scholarly relations with China, and was thus an important component of the Chinese world order.[42] However, the later opening up of Japan was bound to lead to a particularly dramatic "clash of civilizations." Before the arrival of Commodore Perry in 1853, the Japanese knew a thing or two about international politics in Europe, though only at a theoretical level; they had scarcely any experience of diplomatic dealings with other nations.

The opening-up process involved *relatively* gentle methods: Japan was not overcome militarily or subjected to an occupation regime (as it would be after 1945). The United States as well as Britain, immediately behind it and soon taking the lead, enforced access to the island kingdom for their own citizens and the kind of trade concessions already familiar in the 1850s from other parts of the world (Harris Treaty of 1858).[43] They did not impose the full imperialist package of Western privileges pioneered in the China peace settlements of 1842 and 1858–60. Considering that Japanese negotiators had never had to face remotely similar problems before, they acquitted themselves surprisingly well. From the point of view of the Western powers, Japan did not have to be detached from a highly integrated world of states such as the traditional "Chinese world order." Few impediments had to be overcome for it to be tied into the modern state system under relatively favorable conditions. This process was effectively concluded in the 1870s, and in 1895 it received legal confirmation when the Great Powers agreed to revoke the "unequal treaties" negotiated with Japan between 1858 and

1871—something for which China would have to wait until 1942.[44] The Meiji government thereby achieved one of its main foreign policy objectives: to make Japan a sovereign subject with full rights under international law.

The situation in China was much more difficult.[45] Over many centuries, the Chinese Empire had built a world order of its own and maintained it politically as a fully developed, monocentric alternative to the polycentric state system of modern Europe. In many respects, it was the more "modern" of the two. For example, it had a more abstract concept of territorial inclusion: dynastic possessions or "crown land" (in the sense that nineteenth-century Luxemburg was a crown land of the Dutch House of Orange, etc.) were unknown, as was the feudal notion of overlapping claims to rule. In the seventeenth century, there had still been strong elements of polycentrism in East and Central Asia (the two should be seen as one geopolitical entity). A temporal cross section in 1620, for example, shows a number of formidable neighbors alongside the Ming Empire and not subject to it: Manchus in the north, Mongolians in the northwest, Tibetans in the south. After the completion of the Sino-Manchurian empire around 1760, the rulers in Beijing had to deal with a fast-strengthening Tsarist Empire but otherwise were surrounded by weaker tribute states in various kinds of symbolic vassalage toward them. This world order was a "system" in the broader sense of the term, consisting as it did of recognizable individual elements related to one another in accordance with explicit rules. But it differed fundamentally from the European state system in that the whole configuration radiated inward to the Chinese imperial court. The idea that each element was sovereign and enjoyed the same rights as the others played no role. Hierarchical thinking was deeply ingrained in the Chinese state, although historical experience had given it a repertoire much wider than the simple management of vassalage. Adaptation to the new international order of the nineteenth century was therefore bound to be much more difficult than it was for Japanese, Indians, or Malays.

The years between 1842 and 1895 were a striking period that used to be known euphemistically in the West as "China's entry into the family of nations." This involved a number of wars: in 1839–42, 1858–60, and 1884–85. Knowledge of the first of these, the Opium War of 1839–42, helped the Japanese in their negotiating tactics and underlined the risks of resistance. The Sino-Japanese Treaty of 1871, the first ever between those two states that respected the forms of international law, set the institutional seal on the opening up of China. China was opened to international commerce by means of "unequal" free-trade treaties. Foreigners were granted immunity from Chinese law and received the right to settle in a number of port cities. The old vassal belt of the Qing Empire was "decolonized" piece by piece, until Japan's annexation of Korea in 1910 and Mongolian independence in 1912 completed the dismantling of the old Chinese world order. The incorporation of China into the Western-dominated international arena was significantly more difficult and protracted than that of Japan; it involved a true clash of empires.

A further complication was that, in the eyes of Europeans and Americans, China stood at a lower "level of civilization" than Japan and deserved to be treated accordingly. Unlike Japan or India, it also became the locus of an international race for colonial bases and economic concessions. Yet, except for brief moments such as the defeat of the Boxer movement in 1900–1901 or the turbulent changeover from empire to monarchy in 1911–12, China never ceased to act as a sovereign state. In most cases, albeit from a position of weakness, it even played an active part in reshaping its external relations. Thus, the "unequal treaties" system was by no means a Western diktat alone. From the Chinese point of view, it continued the tradition of dealings with "barbarians," who could best be held at bay by giving them clearly defined areas of residence and negotiating only through their community leaders. The treaty ports and foreign consuls served this purpose. By the early 1890s, then, China had been quite stably integrated into the international hierarchy at a bottom-rung position.

The Sino-Japanese war of 1894–95 exposed in an instant the extreme military weakness of the Middle Kingdom, which until then no one, not even the Japanese, had really appreciated.[46] With this conflict, in which China lost almost all influence in Korea (traditionally its most important tributary state), the remnants of the old "Sinocentric" order in East Asia was fatally undermined—or at least that is how it appeared until Japanese historians traced deeper continuities beneath the surface of wars and treaties. In this new interpretation, the old Sinocentric order in East Asia passed much more imperceptibly into one dominated by the West and Japan in antagonistic cooperation with each other. In particular, China's trade within Asia—much more important for it than the trade with Europe and the United States—led to the development of hybrid forms of tribute and commerce. Seen from an Asian optic, the treaty ports were not so much bridgeheads for Western capitalist penetration of a passive, backward Chinese economy as relay points between different, but not incompatible, economic systems.[47] Similarly, the centuries-old thinking behind the "Chinese world order" did not vanish overnight amid the "onslaught of the West." Korea, for example, used the traditional pattern of relations with China for its handling of early foreign incursions, and strong forces in the country were careful until the last to avoid antagonizing the Qing court. In 1905, on the very eve of the Japanese declaration of a protectorate, Korea's ruling elite found it difficult to imagine any alternative to Chinese suzerainty, even though tributary practices had ended in 1895 and a modernizing current actually viewed China as a barbarian country on the fringes of the civilized world.[48] The Russo-Japanese War, which led to a completely new interstate structure and had a profound impact in the very heart of Europe,[49] finally drew a line under the Chinese world order. It was followed by four decades in which the Japanese attempted to construct their own hegemonic space in East Asia, known during the Second World War as the Greater East Asia Co-Prosperity Sphere. In this continuum, the First World War was not an event of prime importance. The international history of East Asia is framed by the years 1905 and 1945.

3 Peaceful Europe, Wartorn Asia and Africa[50]

Contemporary observers and more recently political theorists have given much thought to the question of what constitutes a "great power." Most of their considerations boil down to a simple core: a great power is a state that other great powers recognize in principle as coequal or (in the language of duels) "capable of giving satisfaction." This happens if it is prepared to defend its interests by military means, or if its neighbors believe that it could and will do so successfully. Although economic performance and territorial size were important criteria for the assessment of a given state's international stature, the fact is that the ranking order within the nineteenth-century international hierarchy was on several occasions established on the battlefield. Great-power status and military success were more closely correlated with each other than in the second half of the twentieth century. That an economic giant such as today's Japan carries virtually no military weight would have been inconceivable in 1900. However rapidly the United States grew after the end of the Civil War, and however much prestige it accumulated in foreign policy, only the victory over Spain in 1898 confirmed its claim to be a great power. Japan won respect as a regional power in East Asia through its triumph over China in 1895, but only its victory over the Tsarist Empire in 1905 gave it entry into the circle of the Great Powers. "Germany," hitherto mainly a cultural category, suddenly drew attention to itself as a great power in 1871.

Conversely, military disasters often exposed a mere pretense, as when China, the Ottoman Empire, and Spain suffered defeat and lost their claims to be taken seriously as "powers." Austria's prestige never really recovered from the debacle at Königgrätz in 1866; Russia's defeats in 1856 and 1905 precipitated major internal crises; and France's international position and self-esteem were so damaged by the traumatic events at Sedan in September 1870 that for decades they cast a shadow over its foreign policy and fueled a thirst for revenge. Even Great Britain, which between 1899 and 1902 had great difficulty in prevailing over the numerically and materially weaker Boers, fell into a self-critical mood at the height of the rivalry between the imperialist powers. If the period from 1815 to 1914 is taken as a whole, then only three states experienced uninterrupted ascents as political and military powers: Prussia/Germany, the United States, and Japan.

Behind this shift in the ranking of the leading states lay more general tendencies in the history of organized violence. These may be gauged most clearly within a long time span stretching from the French Revolution until the First World War.

Organization and Weapons Technology

First. The most general trend of the period was the systematic application of know-how, both organizational and technological, to problems of military effectiveness. Army organizers and battlefield commanders, and not only those in Europe, realized very early on that war is not just a matter of expressive combat rituals but requires careful planning with limited resources. The Chinese classic by Sun Zi (fifth century BC) formulated rules of strategy that were still being heeded in the twentieth century. The new element in the nineteenth century was the greater concentration of command structures, at once more flexible and more systematic. Prussia's rise among the European powers was thus based largely on the extensive reform of its army between 1807 and 1813, in response to the collapse of 1806. Prussia was the first state to raise the old leader-follower relationship between commanders and troops to a higher level of rationality. Beneath a royal supreme commander, expertise and authority were concentrated in a war ministry, and later also a general staff responsible for strategic planning, which guaranteed the continuity of military preparedness even in peacetime. The general staff, one of the most important military innovations of the nineteenth century, went decisively beyond the romantic heroism of the Napoleonic period, which could now act itself out only in colonial wars. Prussian officers were no longer principally fighting men and combat leaders but, in keeping with the times, highly trained professionals who practiced the "art of war" as a science. The Prussian army, especially from the 1860s on, gave its officers a completely new profile, which meant, among other things, that those in command at every level were carefully prepared for rational decision making on the field of battle. A dense network of communications was supposed to ensure that subordinate officers were aware of the overall plan and could, if necessary, react flexibly in the light of it. Even before Prussia had great industrial strength at its disposal, the rationalization of its army had enormously increased its military potential. A position in the aristocracy did not automatically translate into military rank; only the princes of the ruling family, and sometimes not even they, escaped the general demand for increased competence. Thus, especially after the victories of 1864, 1866, and 1870, Prussia set the world standard for a modern, professionally organized army.[51] The Japanese were the star pupils, whereas Britain and the United States adapted the Prussian model to their needs only around the turn of the century.

Second. In every civilization, technical know-how has manifested itself above all in relation to warfare, where "software" and "hardware" should always be seen as a single whole. Before the great innovations of the nineteenth century, the armies of Napoleon and Suvorov still fought essentially with an early modern weapons technology, and in general the armed forces of the Napoleonic age had many lines of continuity with the eighteenth century.[52] In military history too, it was a veritable bridge period. The superiority of these armies, especially the French forces, was due less to a technological lead over the enemy than to greater speed, smaller and more flexible units, and a new way of integrating artillery into the course of battle. The bayonet—that is, the firearm as spear—still played a highly important role, since infantry firepower was not effective at close quarters and was sensitive to weather conditions.

The great technological innovations made themselves felt only from midcentury on: the rifle, invented in 1848 by the French officer Claude-Etienne Minié, was adopted by all European armies in the 1850s, when it replaced the older musket as the standard issue for infantrymen.[53] Improving in accuracy and velocity as the century wore on, it also became easier to handle and generated less gun smoke. Meanwhile the average caliber of artillery increased, as did its power, mobility, and backfire safety. The naval corps of various powers profited from developments in ship-borne artillery, which permitted the deployment of calibers on board that were almost impossible to control ashore. Warships became larger with the advances in iron and steel, but also lighter and more maneuverable. In the course of the nineteenth century, "an industrial weapons complex developed out of the former semi-state arms depot economy."[54] This happened in a number of countries, which began to compete with one another in military clout and battle readiness. From midcentury on, *quantitative* differences in weaponry gained a determining effect on the course of wars. Arms races now became a permanent feature of international relations.[55]

Third. The fact that advanced weaponry could now be produced only at the cutting edge of industry, and therefore by only a few countries, did not prevent the global spread of newly developed infantry equipment. In some cases industrial potential translated directly into military superiority—for example, during the American Civil War, when the Confederacy often had tactical superiority over the North but could not keep pace with it industrially. Otherwise, the international arms trade was there to meet the requirements of any government in the world that could muster the financial wherewithal. Firms such as Krupp in Germany and Armstrong in England conducted a worldwide business. Already in the early modern period, Portuguese, German, and other producers had kept Indians, Chinese, Japanese, and many others supplied with muskets and cannons; the Ottoman Empire systematically went out of its way to acquire European-style weapons and related technology.[56] This global diffusion continued and grew in the nineteenth century.

A gap in military technology opened only slowly between the West and the rest of the world. The Chinese, seen by Europeans as having an "operetta army" since their defeat in the Opium War of 1842, proved able to construct port defenses that caused the British and French major problems in 1858. The French Lebel rifle, the first rapid-loading magazine rifle, went into mass production in 1886, soon followed by its German rival, the "Mauser." In the early 1890s, Emperor Menelik II of Ethiopia, an erstwhile Italian protégé and dedicated modernizer, purchased 100,000 units and two million rounds of ammunition. With the help of his long-serving adviser, the Swiss engineer Alfred Ilg, he also started up production in Ethiopia itself. Thus, when Italy set out to realize its dream of a colonial empire in East Africa, Menelik inflicted—at Adwa on March 1, 1896—the worst defeat that a European power ever suffered in a war of colonial conquest. In a single day, his artillery killed more Italian soldiers than were lost throughout the Italian independence wars of 1859 to 1861.[57]

In 1900, experienced Afrikaner troops were so well supplied with Mausers and machine guns that they caused the British unexpectedly high losses. In the Russo-Turkish war of 1877–78, the Ottoman side was by no means inferior technologically and proved outstanding at trench construction.[58] In the Manchurian theater of the Russo-Japanese war of 1904–5, the Russian Goliath was confronted with a Japanese David sporting the latest equipment and an army trained and organized along European lines. In many respects Japan was the more "Western" and more "civilized" country, and it was seen as such by international public opinion; it is therefore too simple to regard the Russo-Japanese War as a clash between Europe and Asia.[59] As for China, the darling of the Western public in the Sino-Japanese war of 1894–95, it paid a high price for many years of military neglect. The two main Chinese warships did not even have shells for their Krupp guns or powder for their Armstrong cannons. Nor did Beijing make any effort to develop an army medical corps, whereas the Japanese one was exemplary. The competence of Chinese officers was generally very poor, there was no unified command structure, and the wretched treatment of rank-and-file soldiers made low battlefield morale inevitable.[60] As early as the 1860s, leading Chinese statesmen had recognized the need for military modernization and even embarked on a national armaments program. But the acquisition of weapons was only part of the story; they also had to be handled properly.

Colonial Wars, Guerrilla Fighting

Fourth. Even at the height of the new conquests, imperialism did not mean that Europeans enjoyed a walkover against defenseless savages. Rather, as in the early modern period, they obtained *local* military advantages and proved adept at exploiting them. What the balance sheet shows beyond doubt, however, is that (except in relation to Japan and Ethiopia) Europeans were in the long run victorious. All in all, the age of colonial wars—a species practiced worldwide, including against the North American Indians—rained disaster on the heads of non-Europeans. The wars were also catastrophic for European soldiers who endured, and often fell victim to, hostile climes. Despite being, historically speaking, on the victor's side, in actual practice they had to contend with tropical diseases, appalling food, the misery of barracks life, prolonged tours of duty, and uncertainty about their return home.

Colonial war is not easy to define.[61] The boundary with other uses of force, such as police operations, was already unclear in the literature of the time; and it became even more blurred as colonial police forces gained in importance after the First World War. At first sight, colonial wars would seem to have the purpose of subjugating "foreign" territories. But was that not true of Napoleon's wars or the Franco-Prussian War, which ended in German control of Alsace-Lorraine? In their *result* many colonial wars led to the insertion of new areas into the world economy, but that was rarely the chief *motive* behind their conquest. Military force cannot simply pry open markets; one does not gain customers by killing them. Before 1914 wars were not often fought over industrially useful raw materials, and those that were mostly took place between sovereign nation-states, such as the Saltpeter War (1879–83) between Chile on one side and Peru and Bolivia on the other. Some of the largest territories in which colonial wars broke out—Afghanistan and Sudan, for instance—were of very little economic interest. There must be a further criterion: colonial wars were "extra-systemic;" they occurred outside the European system of states, with no reference to the "balance of power" and little or none to the sparse rules of international humanitarian law that existed at the time. In colonial wars no prisoners were taken—and the few exceptions could not look forward to a rosy future. That had already been the case in the *guerres sauvages* of early modernity (as well as in still earlier times): for example, the eighteenth-century Indian wars in North America, in which no distinction was made between combatants and noncombatants. As a repertoire of racist categories took shape in the nineteenth century, colonial wars were readily ideologized as wars against inferior races: that is, in practice, as wars that Europeans expected to win but were also ready to wage with greater cruelty than the "savages" employed.[62]

All the more traumatic was the occasional dashing of white troops' expectations: in 1879, when more British officers were killed at Isandlwana in Zululand than at the Battle of Waterloo; in 1876, when General Custer's cavalry was defeated by the Sioux at Little Big Horn; or in 1896, when the Italians at Adwa came under Ethiopian machine-gun fire and lost half their troops. Nor was racist ideology of much avail when the other side consisted of whites, so that the colonial war served not to conquer new territory but to avert or reverse a secession. This was the case not only in the Boer War but also, immediately before, on the island of Cuba, where the Creoles (locally born people of Spanish extraction) waged a revolutionary struggle for something like dominion status within the Spanish Empire. Since there was no provision for such an outcome in the Spanish Constitution, and since Madrid persisted with a hard line, a full-scale war broke out in 1895. At its height in 1897, a Spanish army 200,000 strong was engaged against a much smaller number of insurgents—a disproportion, incidentally, that proved ruinous for the Spanish budget.

The wars in South Africa and Cuba exhibit many parallels. First, the cruelty meted out against the (mostly white) adversaries was typical of colonial warfare. In 1896–97 the famous captain-general of Cuba, Valeriano Weyler y Nicolau, an admirer of General Sherman's campaign of devastation through Georgia in 1864 and a pioneer of anti-guerrilla warfare in the Philippines, herded the Cuban population of all races into *campos de concentración*, in which more than 100,000 died of undernourishment and neglect.[63] Shortly afterward, concentration camps holding 116,000 members of the Afrikaner nation and many of their black helpers, as well as the shooting of prisoners and hostages, were used by the British to break the morale of their South African adversaries.[64] A young journalist by the name of Winston Spencer Churchill, soon after his return from a trip to South Africa, counseled the Americans to use similar methods in the Philippines—which is what they did (not only because of Churchill's advice).[65] The Germans, for their part, followed suit after 1904 in the wars against the Herero and Nama peoples in South West Africa. The novelty in all this was the idea of the concentration camp, not the utter brutality of the operations. In the Zulu War of 1879, for instance, the British "man on the spot," the High Commissioner for South Africa Sir Bartle Frere, set out to free Zululand from the "tyrant" Ketchwayo, to disarm the Zulus, and to rule them indirectly through pliant chieftains under a British resident—in short, on the Indian model.[66] The military chances seemed fairly even, but this was no noble contest between warrior castes. When the British were staring defeat in the face, they answered Zulu atrocities by killing prisoners, burning down their kraals, confiscating livestock, and threatening the very basis of their existence.[67]

A "racial" interpretation alone cannot explain the brutality of the colonial wars. What happened between whites in the Balkan Wars of 1912–13 was no less horrific. Prisoners of war enjoyed no immunity, and terror was systematically employed for the purposes of ethnic homogenization. The conflicts around the turn of the century in Cuba, South Africa, Atjeh, and the Philippines, and the earlier ones in Algeria, Zululand, and the Caucasus, were not "small wars." Yet the notion persisted that every colonial war was in essence no more than a punitive expedition. Between 1869 and 1902, the British alone conducted a total of forty colonial wars and "punitive expeditions," most of them unprovoked attacks and a few operations to free hostages (as in Ethiopia in 1868).[68] Especially in Africa, the technical superiority of the invaders was overwhelming. It became dramatically evident on September 2, 1898, at the Battle of Omdurman, when Herbert Kitchener's Anglo-Egyptian forces suffered 49 dead and 382 wounded, whereas their heroic Mahdi enemy, unable to cope with their eight Krupp artillery pieces and numerous machine guns, ended up with losses of 11,000 to 16,000. (The British marched away from the battlefield, not bothering about the dying and wounded Sudanese.[69]) It was not always the case that the latest technology produced the greatest success. In the French conquest of large parts of West Africa, rapid cavalry movement and bayonet charges were decisive factors; the machine gun played no role, unlike in Britain's African wars or its invasion of Tibet in 1904 (the Younghusband Mission).[70] Colonial wars were embedded in a wider logistical context—steamship transport, railroads, telegraphic communications, tropical medicine—which made it easier to achieve results. Sometimes a railroad might be built purely for the sake of troop movements, as in Sudan or on the North-West Frontier. Two elements, in particular, favored the Europeans and North Americans in most of their colonial wars: better logistics and the use of local auxiliary troops (the *sepoy* principle).

Fifth. In many cases the weapon of the weaker side was guerrilla warfare. Here there were no major differences between Europe and elsewhere. As far back as

1592 the Koreans waged a guerrilla campaign against Hideyoshi's samurai.[71] The spiral of violence associated with such wars rarely ended in a stable civil order or in a lasting stalemate. In Spain in 1808–13, the prototypical guerrilla war, partisans also turned bandit-like on the civil population that by legend they would have been expected to protect.[72] Regular troops allied themselves only reluctantly with such forces. Military professionals distrusted freebooters on land or sea, even if—as with Spanish partisans and the British—they were fighting for the same cause. From the civilian point of view, there was not much to choose from between the two, since soldiers of any kind took whatever they wanted by force. Partisans are often hard to distinguish from what Eric Hobsbawm, in an influential book, called "primitive rebels."[73] The Robin Hood type of bandit is defined by his aims and his supporters, and the "small war" involving ambushes and other uncoordinated surprises is one of his characteristic ways of operating. Nearly all social rebels use such methods, but not all guerrilla fighters are social rebels or even social bandits. The two were closely associated in the Nian rebellion, which between 1851 and 1868 took control of several provinces of northern China away from the Qing government. The spear-wielding infantry and horseback swordsmen of the Nian made them one of the most effective guerrilla forces of the nineteenth century, and it took the Qing great effort, after the end of the Taiping Revolution of 1864, to crush this unrelated enemy too. Meeting the Nian cavalry with canals and ditches—a tactic that the Spanish would repeat on a large scale in Cuba in 1895–98—the Qing commanders also treated villagers well in an attempt to win them away from the rebels. As in many European wars in Africa, however, it was the technological gap between the two sides that decided the outcome in the end. The high-ranking official Li Hongzhang, who was eventually put in charge of the campaign and used it as a stepping stone for his career as China's premier statesman, deployed brand-new gunboats—not a weapon to be deployed by Europeans alone—from the West on the waterways of northern China and trained a well-paid elite corps that was superior in loyalty and motivation to the Qing conventional army.[74] A few years later, partisans appeared in Europe who were not social rebels but fighters for national defense: the *francs-tireurs* in the Franco-Prussian War, scarcely a formidable military force.[75]

Sixth. In 1793 the French Revolution invented the *levée en masse*, the mobilization of the entire (male) population in a spirit of patriotic enthusiasm. Some have seen in this the birth of "total war"—not false, but an exaggeration nonetheless.[76] Mass conscription grafted the energies of a new nationalism onto the dynamic earlier associated with social and religious movements. Yet the *levée en masse*, or rather its myth, might be interpreted in various ways: as a voluntary expression of spontaneity and enthusiasm, as a universal obligation to perform military service, or as a mobilization of all forces, including civilians, for war. If there was a *levée en masse* in the nineteenth century after 1815, then it was the short-lived mobilization during the Franco-Prussian War of 1870–71 (followed by the introduction of universal military service in France). The myth of the ubiquitous French *franc-tireur* later did the rounds in the German army, and in 1914 it was the pretext for preventive atrocities against the civilian population in Belgium and northern France. Genuine mass mobilization is found above all in civil wars: in the American War of Secession, and in the Chinese Taiping Revolution after 1850, in which a religiously motivated charismatic leader, Hong Xiuquan, amassed a huge following in the space of a few years.

In Europe, rulers ensured early on that the dangerous élan of military mass mobilization was diverted into disciplined institutional channels. Napoleon, too, was careful not to rely on enthusiasm: his armies were not carried to the ends of Europe on waves of patriotic excitement; their fighting core was made up of hardened veterans, more like military professionals than citizens in uniform.[77] Still, the extension of war required ever more manpower. A vast conscription apparatus held the whole Napoleonic empire in its grip, and for the Emperor's subjects of every nationality there was nothing more repugnant than the forced consignment of young men to the French war machine: a human harvest that reached its peak in 1811 with the recruitment of cannon fodder for the invasion of Russia.[78] For anyone who wished to hear, the wars of the Age of Revolution taught lessons about how to mobilize large populations. We see the new knowledge reflected in the work of a military theorist such as Carl von Clausewitz. But land armies, militias, partisans, and irregular troops of various kinds were a potential threat for any political and social order. Governments were therefore wary of letting them off the leash. The term "total war" is applicable not to people's war as such but to its bureaucratic organization within the framework of the state monopoly of force. And only the new communications technologies that emerged in the 1860s in the most advanced countries of the world made it possible for propaganda, coordination, and the planned use of productive resources to maintain its total character for a period of years.[79] The first total war was therefore the American Civil War. It remained the only one in the nineteenth century. The epoch prepared the ingredients of total war but did not suffer its consequences until 1914.

Seventh. This should not mislead us into thinking that the wars of the nineteenth century were less terrible than those of other eras. The statistics for those killed and wounded, especially in the civilian population, do not permit of any general assertion. But one thing is sure: the Napoleonic armies were larger than any in the early modern period, and the few *major* wars of the nineteenth century should be measured by that yardstick. In 1812 Napoleon led an army of 611,000 men into Russia; Tsar Alexander I could mobilize 450,000 troops against him. In March 1853 a Taiping army 750,000 strong appeared before the walls of Nanjing. On July 3, 1866, at Königgrätz, 250,000 fought on either side. Two weeks after the French declaration of war on July 16, 1870, Moltke had assembled 320,000 combat-ready troops on the frontier with France; a million reservists and home army members were waiting in the background. Also numbering 320,000 was the force that the British sent to South Africa by October 1899. In the winter of 1904–5, the Japanese fielded 375,000 men against the

Russians in Southern Manchuria (Port Arthur).[80] The Napoleonic format thus persisted until the First World War.

Only in autumn 1914 did the mass slaughter acquire a new dimension. In the greatest battle of the American Civil War, which took place near Gettysburg (Pennsylvania) between July 1 and July 3, 1863, the number of dead *and* wounded was 51,000, almost twice as much as in the big Austro-Prussian battle at König-grätz three years later. In the bloodiest conflict in Europe between 1815 and 1900, the Franco-Prussian War of 1870–71, a total of 57,000 soldiers lost their lives, while in the Crimean War of 1853–56 the figure was 53,000. In the fighting for the Russian fortress of Port Arthur, at the southern tip of Manchuria, nearly 81,000 men died between August 1904 and January 1905[81]—a bloodbath that was seen as shocking and unparalleled, although a few years later the killing on the fields of Flanders exceeded it by far. If any conflict between 1815 and 1913 gave a taste of things to come, it was the Russo-Japanese war.[82]

The horrors of war cannot be quantified and neatly inserted into a historical trend. They stretched from the Franco-Russian winter battle at Eylau (February 1807) and the excesses of the partisan struggle and its repression after 1808, so vividly depicted by Goya, to the massacres of numerous colonial wars to the pin-point artillery fire that rained down on Mukden and Port Arthur in 1904–5 and already foreshadowed Verdun. It is a striking feature of the nineteenth century that medical care kept pace less than ever with the capacity to kill and maim. The introduction of needle injections in 1851 was a great advance, making it possible to administer larger doses of opium as a painkiller. A young Genevan businessman called Henri Dunant, who had found himself on June 24, 1859, on the Solferino battlefield south of Lake Garda, had been so overwhelmed by the misery he saw around him that he provided the impulse for the founding of the International Committee of the Red Cross.[83] If "armies of cripples," such as those produced by the First World War, were not in evidence after 1871, it was not because few had been wounded but because their survival chances were extremely poor.[84] Despite all the horrors of war, as portrayed in literature from Erckman-Chatrian's *Histoire d'un conscrit de 1813* (1864) to Tolstoy's *War and Peace* (1868/69) to Stephen Crane's *The Red Badge of Courage* (1895), the hundred years from 1815 to 1914 in Europe was a period of relatively little violence among states, a peaceful interlude between the early modern age and the twentieth century. The few wars to be waged were neither protracted nor "total." The distinction between combatants and civilians was observed to a greater extent than in earlier or later European conflicts or in wars fought outside Europe. This was one of "the great, hitherto little recognized, cultural achievements of the century."[85]

Sea Power and Naval War

Eighth. Naval warfare requires equipment and skills that are harder to disseminate than the manual tools and dexterity of infantrymen. Two technical innovations came together. One was the quite leisurely replacement of wind power with coal-fired vessels: the Royal Navy's last great sailing ship was launched in 1848, although in the sixties the British flagships off Africa and in the Pacific still relied on wind in their sails. The other was the "ironclad revolution" in hull design, begun in 1858 and implemented at a faster pace. Soon ships were also being fitted with revolving gun turrets—a decisive advance on the wooden warship, with its more limited mobility. Ramming had no longer been practicable since the middle of the eighteenth century, and its use by Austrian and Italian ironclads in 1866 at the Battle of Lissa was based on a curious misunderstanding of the new technical possibilities. By that time little was left of the imposing timber structures and square rigging of Nelson's age: military specialists had replaced gentlemen officers, and crews were no longer press-ganged into service or terrorized with the cat-o'-nine-tails.[86]

In 1870 the only non-Western powers that had the new-style ships were the Ottoman Empire and Japan. China had begun in 1866 to develop a modern navy, through purchases abroad and the construction of its own shipyards; by 1891 these had turned out ninety-five modern ships, and a large number of naval cadets had been trained by foreign instructors.[87] This strengthened China's claim to be a regional great power, and indeed Western observers were highly impressed by its navy-centered military modernization.[88] However, the Chinese navy was a motley collection of ships divided into four separate fleets and placed under the governors of the respective coastal provinces. There was no overall strategic conception for their eventual deployment.[89] China's inferior performance in the Sino-Japanese War of 1895, and its lack of maritime ambitions over the following half century, should not obscure the amazing fact that, unlike the Ottoman Empire, it had no traditions as a sea power. The celebrated oceanic expeditions of Admiral Zheng He in the early fifteenth century could scarcely provide any bearings for the nineteenth. After the Opium War, for which it had been altogether unprepared, China therefore developed a new concept of sea-based defense (not prohibited under the "unequal treaties") and acquired the weaponry and know-how necessary for it. It was a huge challenge, which it seems to have almost mastered.

The situation in Japan was similar yet different. After an abortive invasion of Korea in 1592, which failed not out at sea (like the Spanish Armada of 1588) but only after bloody land battles, Japan refrained from building up its maritime armed forces. It had little reason to feel threatened, although after the Opium War numerous Western ships plied the waters around the archipelago. Commodore Perry's four heavily armed steamers were thus able to enter Tokyo Bay completely unopposed (and of course uninvited) on July 2, 1853. The largest of these was six to seven times more voluminous than anything the Japanese had in their fleet, or indeed than anything they had ever set eyes on before.[90] Just like the farsighted Chinese provincial governors, astute minds in the Japanese political elite, even before the Meiji Renewal, recognized the need for an efficient modern navy. After 1868, and especially from the mid-eighties on, this became a high national priority, together—and in rivalry—with expansion of the army. The naval program, not just the often-cited industrialization, was the secret of Japan's rise as a great power. Alongside a military fleet, the state supported the development of a private merchant navy that by 1910 was the third-largest in the world, after the British and the German.[91] The enormous war reparations imposed on China in 1895—a source of handsome profits for Western creditors—were comparable in their effect to France's burden after the war of 1871, helping to meet the costs of the Japanese armaments program.[92] Starting virtually from zero in 1860, Japan raced ahead to become the power which on May 27–28, 1905, near the island of Tsushima in the Korean Straits, fought and won the greatest naval battle since Trafalgar in 1805. Russia, the second maritime power in Europe, was defeated thanks to a combination of excellent ships, well-trained crews, masterly tactics, and a dose of good fortune—in a manner for which the cliché "annihilating" is for once appropriate.[93]

The age of ironclad battleship fleets with their coastal blockades and devastating victories was surprisingly brief. It began in the 1860s and ended in the Second World War. Aircraft carriers and nuclear submarines then became the central elements in sea warfare. During the battleship era, the final showdowns for which the European powers had planned for decades failed to materialize. In the First World War the only (indecisive) naval battle was fought off Jutland on May 31/June 1, 1916. The Second World War saw no classic sea battles in the Atlantic, and as early as 1942 Germany withdrew its surface ships from the high seas. The theater of the last sea battles in history was the Pacific Ocean, where in October 1944 the Americans and Japanese met in a gigantic confrontation in the Leyte Gulf. But the Battle of Midway, hinging entirely on aircraft carriers, had already shown in June 1942 that the classical age was well and truly over. Herein lies an irony of history. The era of surface warfare at sea, a European specialty ever since the Battle of Salamis, ended in a showdown between the two great powers that had emerged outside Europe at the turn of the century. Japan, bereft of maritime traditions, mastered the necessary technology and strategy to the limits of its industrial capacity, becoming a naval power second only to the United States in the first half of the twentieth century—until 1942.

4 Diplomacy as Political Instrument and Intercultural Art

Visions, Mechanisms, Norms

Of the opposing currents in the international relations theory of nineteenth-century Europe, one had older roots in the idea of a regulated world peace, the other in the principle of egoistic reasons of state. As we have seen, the Congress of Vienna in 1814–15 found an ingenious way of combining the two: the security of individual countries was to be guaranteed by mutually agreed conflict resolution within the system of states. However, with the turn to power politics in midcentury, the second of these tendencies came to the fore again. Cosmopolitan liberalism, whose chief representative was the British manufacturer and statesman Richard Cobden, had expected that the free movement of people, goods, and capital would lead to greater prosperity for all and lasting peace among the nations. Free trade, arms limits, and a degree of ethical principle—Cobden vigorously opposed British intervention in China in 1856—would finally extricate the planet from the bloody chaos of the premodern age.[94] In the political practice of Great Britain, the leading champion of free trade, this program was fraught with a contradiction: liberal statesmen such as Lord Palmerston had no misgivings about the illiberal *imposition* of worldwide freedom of movement. Until 1860 this was mostly a success: the last great act of "free-trade imperialism" was the opening up of Korea—a second-order phenomenon, as it were, since Japan appeared there as the trailblazer of the "civilized world" less than two decades after its own opening to it. The Kanghwa Treaty of 1876 between Japan and Korea was modeled on the "unequal treaties" that Japan itself had been forced to sign.[95] In the future, cosmopolitan liberalism would never disappear from thinking about international relations; it is today the dominant theory, or at least the dominant rhetoric, in international forums. But its influence sank to a low point in the final quarter of the nineteenth century, when imperialist thinking radicalized the return of continental Europe to realpolitik and (after 1878) protective tariffs.

Shared but seldom openly formulated by most of the political class in the powerful states of the day, including the United States, this bleak and fatalistic worldview consisted of the following elements:[96]

1. A struggle for existence marked not only society and nature (as the highly popular theory of social Darwinism now preached) but also the international stage. To stand still meant to be left behind. Only those who grew and expanded would have a chance of survival in a viciously competitive environment. Political systems had to be designed in such a way that they steeled the country for the battle of the giants. (Conversely, the ever sharper rhetoric of competition encouraged a reading of Darwin that emphasized the element of conflict in natural selection.)[97]
2. Success in these conflicts would depend on an ability to combine industrial strength and scientific-technological innovation with colonial possessions and a national fighting spirit.
3. The planet was becoming more and more "closed." The space in which new dynamic forces might seek an outlet was diminishing all the time. International conflicts would therefore increasingly result in struggles to divide up the world and to redivide what was already divided.
4. Weaker nations would not necessarily disappear altogether (the popular talk of "dying nations" should not always be taken literally), but their limited clout showed that they were not in a position to take control of their

destiny. Lacking the power to shape themselves politically and culturally, they should count themselves lucky to come under colonial tutelage.

5. International competition would demonstrate, somewhat tautologically, the superiority of the "white race." The uncommonly successful Anglo-Saxon race had a particular vocation to lead the rest of the world, whereas even southern Europeans or Slavs could not really be trusted to establish a viable order. The nonwhite races were not all equally capable of learning and being molded, but neither could they be categorized in terms of a static hierarchy. Special caution was needed with regard to the "yellow race." It was demographically stronger than the others, characterized by an aggressive business sense and, in the Japanese case, a feudal warrior ethic. If the West did not watch out, it would be threatened by a "yellow peril."[98]

6. The global sharpening of the struggle between the races meant that the militarized nation-state could not remain the only, all-embracing entity for the resolution of conflicts. The Anglo-Saxon nations of the world would have to strengthen their ties with one another; the Slavs to place themselves under Russian leadership; and the Germans to learn how to think "pan-Germanically," beyond the limits of the Bismarckian Reich.

Such thinking made the First World War possible, if not actually inevitable; the specter was conjured up and fantasized about—without any remotely realistic forecast of the carnages to come. Social Darwinism was not confined to "the West" (a term it increasingly used for itself), but reappeared across borders in different, though in many ways related, forms. It also resonated among the victims of imperialist aggression, although it did not then come with all the ideological baggage associated with it in the West. Japan, which at the latest by 1863 thought it enjoyed excellent relations with the Western Great Powers, suffered a mighty shock when France, Russia, and Germany—in what diplomatic history knows as the Triple Intervention—denied it some of the fruits of its 1895 military victory over Qing China. Among the Japanese public, this sowed distrust of visions of international harmony and replaced them with ideologies of heroic effort and readiness for war.[99] In China, then under quite different kinds of imperialist pressure (from Japan, among others), rising nationalism had tragic overtones, since the predatory world of the turn of the century threatened the old empire's very existence as a unified state and people. Internal reforms were thus mainly designed to strengthen China in the international struggle for survival. This was, for example, the view of the important scholar and journalist Liang Qichao, who was thoroughly modernist in other spheres and should not be regarded as "right-wing" in a European sense.[100] In a Muslim context, the no-less-complex and contradictory intellectual Sayyid Jamal al-Din al-Afghani also sought ways to overcome the lethargy of tradition and to awaken new political energies—for example, through the propagation of pan-Islamic unity.[101]

These visions of an international jungle took hold at the end of a century that had seen the diplomatic linking of the globe. Today even the smallest and poorest country maintains a worldwide network of missions; ministers are constantly meeting with one another, and heads of state attend regular summits. But this kind of diplomacy is only a product of the period after the First World War. The nineteenth century prepared the way for it, by spreading European theories and practices around the world; whether diplomacy was actually "invented" in Renaissance Italy or among ancient Indian princedoms is here irrelevant. For a long time the Ottoman Empire was the only non-Christian power involved in such relations: Venice, France, England, and the Viennese emperor all had missions in Istanbul. Practices were not uniform across cultural boundaries, however. In North Africa, French consuls of the eighteenth century conducted a flexible diplomacy in accordance with local conditions.[102] Japan allowed in only Dutch and Korean (not even Chinese) diplomats throughout the early modern period. China channeled its external contacts through the lavish ritual of tributary missions, and sometimes these were also sent out from Portugal, the Netherlands, and Russia. A less costly form of permanent contact, in some ways akin to diplomacy, took place between "supercargoes" (representing European East India companies in Canton) and the official Chinese "Hong merchants" resident there. This practice continued up to the Opium War.

In none of these early modern instances did either side insist on symbolic equality. Things changed only with the "new diplomacy" of the revolutionary age, sparser in protocol and resting upon symmetry and equal rights. One highly charged moment was the refusal of Lord Macartney, head of the first British mission to China in 1793, to perform the expected kowtow (*ketou*) or ninefold prostration before Emperor Qianlong, on the grounds that a freeborn Englishman did not indicate submission to an Oriental despot. The emperor remained surprisingly calm and saved the situation by acting *as if* the envoy had correctly observed the ritual.[103] At least Macartney bent his knee—a ceremonial gesture taken for granted even at European courts, although in those very years it was being discredited in the wake of the French Revolution.[104] In the Maghreb, the rituals of abasement that French consuls once reluctantly performed—for example, the unilateral kissing of a Muslim ruler's hand—were abandoned after the revolution. Whereas in principle European diplomats had previously accepted local customs, the rules of European diplomacy now came to be seen as generally binding. It was not possible to enforce these everywhere at once. State gifts of a tributary nature were replaced with "practical" tokens, such as the prosaic products of the English steel industry with which Lord Macartney disappointed the Chinese. In such small details too, there was a new attention to reciprocity. The spread of general norms also meant that diplomatic recognition was taken more seriously than in the past, with the result that it became possible to question the legitimacy of certain states whose sovereign existence had previously been tacitly accepted. The bey of Tunis was one case in point.

A set of rules for European diplomacy, partly written, partly unwritten, had taken shape by 1860. It was also expected of Oriental powers such as China and the Ottoman Empire that they would allow permanent missions in their capital cities and maintain their own missions in the capitals of the West. Ambassadors would have direct access to the head of state and top government circles—an unprecedented idea in China, for example, where no mortal had had the right to approach the emperor. Foreign ministries, hitherto known only in Europe, came into being and began to take charge of diplomatic contacts; but this, too, was far from a matter of course, so that even in such a centralized country as China governors of coastal provinces often meddled in foreign affairs until the very end of the empire in 1911, despite the creation in 1860 of the Zongli Yamen (a department with a lower rank than others in the bureaucracy, which gave way to a true foreign ministry only in 1901). Missions were also supposed to include a military attaché, who was not always above suspicion of espionage. Diplomatic immunity did have traditional roots in many parts of the world, but it was now strengthened and made explicit. An attack on a diplomat of any rank could even be a casus belli. In 1867 a British expeditionary corps was sent to Ethiopia to free the consul and several other imprisoned hostages; there could be no repetition of what happened in 1824 (in time of war), when the governor of Sierra Leone was overwhelmed by Ashanti warriors and his skull became a cult object in African rites.[105] The most dramatic of all conflicts directly involving diplomats was the siege during the Boxer Rebellion in summer 1900, which after the killing of a German and a Japanese diplomat escalated into an international war. Insurrectionary peasant militias, tolerated by the imperial court and eventually reinforced by regular Chinese troops, would probably then have massacred Western and Japanese representatives if the improvised fortifications had been breached before relief forces arrived on August 14. Thereafter, foreign troops were stationed for decades in Beijing and its surroundings to ensure the protection of diplomats. But "barbarians" overseas were not the only ones guilty of violating conventions. During the French Revolution, foreign diplomats were sometimes set upon by mobs, and envoys of Portugal and the Holy See were even temporarily held prisoner; a number of French diplomats actually met their deaths in Rome and Rastatt.[106] The new diplomacy of the revolution broke old rules, insofar as French emissaries openly interfered in the internal affairs of the country in which they were posted.

The new set of rules governing diplomacy and international action that took effect after 1815 was extolled as a normal product of advanced civilization. After the opening of non-European countries, treaties ensured that they would recognize civilized standards and observe them in practice.[107] Some elements in the package were therefore explosive, because they provided a basis for deviations from the general norm of noninterference in the internal affairs of another state.

Tricky situations could arise, for example, if European diplomats took the side of Christian groups in religious quarrels. From 1860 on, representatives of Western powers intervened everywhere in favor of European and North American missionaries. Sometimes they did so reluctantly, because many missionaries relied on such protection to engage in ill-considered provocations. Power politics then came into play, when European states proclaimed their role as defenders of Christian minorities. The French Second Empire did this in Ottoman Syria and Lebanon, and the interference of the Russian tsar in Levantine religious matters became the immediate cause of the Crimean War.[108]

A second source of interference was the protection of foreign property. Since the seventeenth century, the rights of foreign merchants in Europe had been formulated more and more clearly. But the problem became more acute as the development gap between countries grew wider and foreign investment more substantial. New legislation was passed to safeguard foreign-owned port installations, factories, mines (and later oil refineries), and valuable real estate. It is possible to interpret the early Chinese treaty system after 1842 not only as a bridgehead of imperialist aggression (as Chinese nationalists usually do) but also as a relatively successful attempt to contain foreign demands. It lost its effectiveness after 1895, when foreign investments were increasingly located outside the treaty ports and it became ever more difficult for Chinese authorities "up country" to ensure their protection. The Great Powers were tempted to take matters into their own hands; this is what happened wherever railroads had been built by foreign concessionaires or funded mainly by foreign investors.

A related problem arose if a debtor state failed to meet its financial obligations on time, or at all. Hardly a single country—Venezuela was an exception— did this with provocative intent, yet a new means of control was put in place. International supervisory bodies (often including representatives of private banks) insisted on prior approval of government financial measures and directly transferred large sums of revenue (from duties or a salt tax, for example) into the coffers of creditors. This is what happened, in various ways, between 1876 and 1881 in the Ottoman Empire, Egypt, and Tunisia. By 1907, forms of international public debt tutelage were also operating in China, Serbia, and Greece.[109] In the nineteenth century, government default took the place of the dynastic insolvency or former times, but under the conditions of financial imperialism it was a highly risky strategy that entailed various unpleasant consequences. No one yet dared to take the revolutionary step of expropriating foreign property, as occurred in the early Soviet Union, 1930s Mexico, and China after 1949. In the face of minor local infringements or nonservicing of private loans—the typical flashpoints in Latin America and China—Britain, as the leading investor nation, behaved with some restraint in comparison with the United States in the twentieth century. At first it was left up to private creditors to find ways of recovering their money, much as today's multinationals largely conduct their own diplomacy. The British state enforced legal claims to compensation, with the Royal

Navy as its most effective instrument of pressure, but it tried to avoid a situation where overzealous intervention would unleash a spiral of violence.[110]

For countries such as China, Japan, or Siam, it was a complete novelty to have to deal with foreign diplomats who insisted on symbolic equality, and often also came along with the affected peremptoriness of a great power. Diplomacy was changing in Europe as well during this period, but at a rather more leisurely pace. Foreign-policy apparatuses grew slowly: the diplomatic and consular services of the United Kingdom, for example, had a total staff of 414 on the eve of the First World War, fewer than 150 of whom were career diplomats. New consuls might be posted to the Americas and the noncolonial countries of Asia, where they often performed quasi-diplomatic duties more in the manner of imperial consuls than as mere representatives of their government, becoming typical "men on the spot" with enormous powers and vast freedom of action. British consuls in China had the right to call in a gunboat anytime on their own initiative.

Personnel

Since there were fewer states than today, diplomatic apparatuses remained manageable. The founding of the Latin American republics in the 1820s is said to have doubled the workload of the British Foreign Office at a stroke; it did not have to face such an experience again for a long time. Officials working in foreign capitals were not exactly busy. In 1870 the French finance ministry employed fifteen times as many civil servants as the foreign ministry. Foreign policy in Europe continued to be a domain of the aristocracy, and even in democratic systems of government it came under parliamentary control only in situations of acute crisis. The internal hierarchy within the diplomatic community reflected the changing importance of countries within the international system. The smaller ones carried less weight than previously. After 1815, countries such as the Netherlands, Denmark, Sweden, and the Swiss Confederation gradually developed a posture of neutrality, which made foreign policy in the usual sense more or less unnecessary. For representatives of the Great Powers, the most prestigious posts were for a long time in the capitals of the pentarchy. In midcentury the French government paid its representative (of envoy rank) in Washington a mere one-seventh of an ambassador's salary in London. Only in 1892 were the European legations in the United States upgraded to embassies. There were scarcely any diplomats at all in a political backwater like Tehran, which had a British embassy from 1809 but a French one only from 1855. After an early false start, the Ottoman Empire acquired a permanent network of missions in the 1830s; the exchange of envoys between Istanbul and Tehran in 1859 was the first example of modern diplomatic relations in the Muslim world. In 1860 China was forced to send diplomatic representatives to Europe, but they were conspicuously drawn from low down in the finely graded ranking order of Chinese officialdom. Only Japan, eager to be on a par with the West both practically and symbolically, threw itself with enthusiasm into the new diplomatic game. By 1873 it had nine legations in European capitals and Washington, and in 1905–6, in a clear sign of Japan's ascent in world politics, some Great Powers upgraded their missions in Tokyo from legations to embassies.[111] The telegraph created new scope for foreign-policy communications, although not from one day to the next. When France and Britain declared war on Russia in March 1854, the Sublime Porte in Istanbul learned of it more than a fortnight later, since the news traveled by wire only as far as Marseille and then had to be conveyed by ship.[112] Twenty years later, virtually the whole world was linked by cable. At first, the new medium also had the effect of shortening news reports and dispatches: it was too expensive to wire long documents around the globe.

One of the main tasks of diplomats in relation to the non-European world was to conclude all manner of treaties: trade agreements, protectorate treaties, border treaties, and so on. The idea of a treaty valid under international law was not entirely unfamiliar outside Europe (China had signed one in 1689 with Russia), but in many particular situations they led to cultural misunderstandings. The problems of translation alone could be very delicate and lead to serious complications at the implementation stage. A good example is the Treaty of Waitangi, which a British Crown representative signed on February 6 (now New Zealand's National Day), 1840 with a large number of local chiefs (as many as five hundred in the end), and which formed the basis for Britain's declaration of sovereignty. In actual fact, it was not a brutal imperialist diktat but bore the marks of the British humanitarian spirit of the age. Yet it became the most controversial element in New Zealand politics, since the English and the Maori wording were sharply at variance with each other. Given the military balance of forces at the time, the Crown could not simply have "taken possession" of the country without the agreement of the Maoris; Britain had not won a war against them (as it would against China two years later), and Captain William Hobson, the signatory to the treaty, had at his command no more than a handful of policemen. However, the interpretation of the text would bring many an unpleasant surprise for the Maoris.[113]

In societies without a written language, such as those in Africa or the South Seas, the conceptual gulf was by the nature of things especially wide. European notions concerning the validity and enforcement of contracts, and the sanctions to be applied in the event of a breach, were not everywhere immediately comprehensible. But even Asian cultures familiar with diplomatic correspondence in the region were not exempt from misunderstandings. Individual treaties piled up to form luxuriant sets involving a number of parties. The system of "unequal treaties" between various powers and the Chinese Empire had become so labyrinthine by the early twentieth century that virtually no one could master it in detail, except perhaps top Chinese legal experts employed to fend off Western claims. As early as 1868, amid the confusion of regime change, Japan's newly assembled imperial government raised objections under international law (with which it had only just become familiar) against the interventionist designs of the United States and various European powers.[114]

The impenetrability of the set of treaties was underlined by the fact that many of them had been kept secret. In Europe, the decades leading up to the First World War were the climax and terminus of secret diplomacy; opposition subsequently raged against such practices, in the name of a new diplomacy grounded on public legitimacy, which was championed above all by Woodrow Wilson. The new Bolshevik government in Russia published documents from the Tsarist archive, and in 1919 the charter of the League of Nations prohibited secret treaties.

A new element in international relations, or rather one revived in the second half of the nineteenth century, was the personal meeting between monarchs, often attended with great pomp and circumstance. Napoleon III, Wilhelm II, and Nicholas II wallowed in such occasions and staged them for a new mass public,[115] but the global radiance emanating from them was not very great. Monarchs did not even visit their own colonies—although Wilhelm II made it to Ottoman Palestine in 1898. In 1911–12 George V became the first British monarch to travel to India, in order to have himself crowned Emperor of India a year after his accession to the British throne. Meetings with non-European colleagues had rarity value. No European ruler ever saw Cixi, the Dowager Empress of China, or the Meiji Tennō, who in 1906 was ceremonially awarded the Order of the Garter as an almost routine follow-up to the signing of the Anglo-Japanese treaty of alliance in 1902.[116] Oriental rulers had to betake themselves to Europe. In 1867 Abdülaziz (r. 1861–76) set a precedent for an Ottoman sultan by traveling to Christian Europe for six weeks on the occasion of the Exposition Universelle in Paris—a trip whose main significance lay in the fact that his nephew, the future Sultan Abdülhamid II (an altogether weightier ruler), was a member of the party and received impressions that left a deep mark on him. The delegation was personally welcomed by Napoleon III at the Gare de Lyon, and it later met with Queen Victoria at Windsor Castle and visited the courts of Brussels, Berlin, and Vienna.[117] In 1873 Shah Nasir al-Din (r. 1848–96) became the first Iranian monarch to visit the lands of the infidel.[118] The Siamese king Chulalongkorn traveled to Europe in 1897 and 1907, meeting Queen Victoria and many other rulers. His policy aim was to raise the symbolic value of his country in European eyes, and to this end he awarded a number of honors to his hosts. He was rather put out, however, when the British did not reciprocate by decorating him with the Order of the Garter.

The cross-cultural relations were somewhat denser among other members of imperial and royal families. Queen Victoria may not have traveled to India, but Crown Prince Edward (the future Edward VII) did in her place. Empress Eugénie sailed aboard her luxury yacht to the opening of the Suez Canal in Egypt. Chulalongkorn sent two of his many sons to be trained at the Prussian military academy. The Qing dynasty was commanded to send a prince to European capitals in atonement for the Boxer Uprising. And in 1905 Wilhelm II succeeded in charming the Japanese crown prince and princess. The international of crowned heads remained a European affair: the New World lay outside its orbit, even more so after the Brazilians disposed of emperor Pedro II and turned their country into a republic in 1889. But at the latest Theodore Roosevelt signaled by the gravity of his conduct that American presidents were the equals of any monarchs on the planet. The Meiji Emperor, shrouded as few of his colleagues were in ceremony and mystique, is said to have been impressed in his forty-four-year reign by no one more than by the unassuming bourgeois Civil War hero and former president Ulysses S. Grant.[119] There would have been more cross-cultural encounters of this kind if it had been possible to expect some gain from them. Nineteenth-century Europeans cultivated the old image of the dumb and degenerate "Oriental potentate," fit only as operetta material. Gilbert and Sullivan's *The Mikado* (1885) evoked a fantasy-world Japan, with a fictitious ruler totally unrelated to the energetic and capable real-life Meiji emperor. In European cliché, the Ottoman sultans seemed to embody in person the "sick man of the Bosphorus." Clichés also obscured the achievements of competent and enlightened rulers such as Mongkut and Chulalongkorn of Siam or Mindon of Burma. What interested the public most about Mindon was a picturesque detail: the fact that for decades British ambassadors followed the prescribed custom of removing their shoes in his presence. When the government of British India in Calcutta put a stop to this in 1875, it was tantamount to the withholding of diplomatic recognition from Burma. The shoe question became one of the grounds for annexation of the Burmese rump state.[120]

Lack of deference to rulers reflected a lack of respect for their countries. The Law of Nations, which grew in importance after 1815 and, from the forties on, was mainly developed by British jurists and promoted by British politicians, did not afford protection to territories outside Europe. It also left large areas unregulated, especially at sea. Thus, whaler captains hunting in the same grounds made detailed agreements with one another to cover conflicts over the discovery and final ownership of prey.[121] Nineteenth-century European expansion tended to favor the English legal model of the protectorate. Originally all this meant was that a state transferred to a protector the task of looking after its external relations, but in colonial practice it often signified nothing less than a disguised form of annexation.[122] It was so popular as a legal form because it gave the protector country every opportunity for economic exploitation, without imposing the burden of responsible administration. So long as no third party, no other colonial power, opposed the creation of a protectorate relationship, there was nothing in international law to stand in its way. It often happened that contrary to legal doctrine, a protectorate was declared over a community that could not, with the best will in the world, be classified as a state.

At the other end of the spectrum, a state might be erased from the map after centuries of existence in which it had enjoyed a stable legitimacy at least as great as that of most European states. When Korea, with a continuous statehood stretching back to the fourteenth century, was declared a Japanese protectorate in 1905, it protested to the Second Peace Conference at The Hague (1907) against this degradation of its position. But the conference presidium did not even admit the Korean representatives, explaining that it considered Korea to be a nonexistent country. The enforcement of this view was then left up to power politics. The Japanese formally annexed Korea in 1910 and retained it as a colony until 1945. Nevertheless, decisions of this kind, so often made by ministers or a tiny group of great powers at an international conference, gradually became the object of public debates.

It is almost a commonplace that the years from 1815 to 1870 were the classical age of pure power play in foreign affairs, conducted by a narrow elite of aristocratic experts. *Previously*, dynastic considerations had often stood in the way of a "realistic" foreign policy, and the professionalization of diplomacy had still been in its infancy. *Subsequently*, the press and electoral moods would enter the picture as plebeian disruptive factors. The first Napoleon, no less than his great adversaries William Pitt the Younger and Metternich, kept the people well away from decisions about war and peace. The third Napoleon then played on the feelings of the masses, staging dramatic crises and organizing colonial conquests (as in Vietnam) to improve morale at home. Bismarck, who allowed no one a say in his external policy, did sometimes play the card of national mobilization, as in 1870 when Napoleon's declaration of war gave him the welcome pretext to weld the Germans together in the heat of patriotism. His long-term British opponent, Gladstone, who unlike Bismarck tended toward a moralistic-idealistic foreign policy, launched public campaigns in response to abuses and massacres in Italy and Bulgaria. Great waves of imperialist sentiment swept over Russia in 1877— when pan-Slav enthusiasts forced Tsar Alexander II into a declaration of war on the Ottoman Empire that he did not consider to be in the national interest[123]— and later over Japan in 1895 and the United States in 1898. The "jingoist" mood in America outstripped almost everything seen in Europe during the high tide of imperialism.[124] Everywhere two factors were involved: nationalism and the press.

Under such conditions, it was less and less possible to switch public reactions on and off as Bismarck had liked to do. A situation might arise in which politicians boosted nationalist expectations in the public that they were subsequently unable to deflate. A perfect illustration of this was the second Morocco crisis of 1911, when the German foreign secretary Alfred von Kiderlen-Wächter and his men in the media recklessly whipped up a war fever.[125] Classic arcane policies and secret diplomacy passed the peak of their effectiveness around the turn of the century. Thus, the Russo-Japanese peace talks sponsored by President Theodore Roosevelt after the war of 1904–5 took place in the limelight of a newly developing international public opinion. All parties to the negotiations felt the need to be skillful in their dealings with the press.

Resistance

This was also true of parts of the so-called periphery. In India, Iran, and China, anti-imperialist resistance went beyond hopeless military actions and turned to modern forms of agitation. In 1873 a number of Iranian notables and Koranic scholars attacked the extensive concession for railroad and other investment projects that the shah's government had awarded to Baron Julius de Reuter, owner of the press agency named after him. Later campaigns mobilized much larger numbers. In winter 1891–92, countrywide protests broke out against a monopoly for the production, domestic sales, and export of tobacco that the shah had conferred on a British businessman; even the shah's wives and non-Muslim minorities took part. Early in 1892 the concession was canceled outright, prompting a huge claim for damages that forced Iran to contract its first foreign loan. The success of this mass action, encompassing Muslim clerics, merchants, and large sections of the urban population, was unprecedented in the history of modern Iran. And the telegraph meant that it could be tactically coordinated over large distances.[126]

It was in 1905 that this kind of nationalist public made its presence felt right across Asia for the first time, with boycotts as its most important weapon. Large campaigns were organized against the British in India, while in China a near-nationwide boycott of American ships and goods, triggered by a tightening of US immigration policy, represented the country's first modern mass movement, just a few years after the archaic excesses of violence in the Boxers' rebellion and war. In 1906 the British envoy noted a "consciousness of national solidarity, which is an entirely new phenomenon in China."[127] In the Ottoman Empire, fired up by the recent Young Turk Revolution, large crowds gathered in Istanbul in October 1908 to protest against Austria's annexation of Bosnia and Herzegovina (two provinces that already had been under its de facto control since 1878), blocking access to Austrian businesses. The boycott soon spread to other cities and ended only the next year, after the Porte had recognized the annexation while Austria-Hungary had agreed to pay compensation.[128] All these movements, not related to one another in any obvious way, can be only superficially bracketed together as "nationalist." There were always specific local causes and driving forces. Nevertheless, not just spontaneous anger and direct material interests lay behind them; they were also bound together by a gradually mounting awareness of something like international injustice. To see the new demands and values as existing only in the mind of Woodrow Wilson, surfacing at the Paris Peace Conference of 1919, would be to overlook their earlier origins outside Europe in Asian as well as African reactions to European imperialism. In contrast to nearly all the primary resistance movements responding to the very first acts of European invasion, the overwhelmingly peaceful new mass protests were remarkably successful. Ad hoc alliances right across the urban social spectrum (the countryside was less involved) achieved more than government diplomacy could have done on its own. No Asian or African country of the pre-1914 period had the weight to protect its citizens living abroad in the West. Even Japan's influence was very limited on this count, as indicated by the failure to get US immigration laws revised. The Paris Peace Conference of 1919 would painfully demonstrate the limits of Japan's diplomatic clout when it failed to support Japan's push for a clause against racial discrimination in the Covenant of the League of Nations.[129]

5 Internationalism and the Emergence of Universal Norms

The compression and integration of the international community did not result only from the spread of European-style interstate relations and corresponding legal norms. Transnational private or nongovernmental networks developed in leaps and bounds in the second half of the century. Of course this was not a novelty of the nineteenth century. The Renaissance, the Reformation, and the Enlightenment had all been if not "transnational" then certainly intellectual movements that spread from country to country. Music and painting, science and technology had never allowed themselves to be contained by borders. From roughly the middle of the nineteenth century, transnational initiatives of a non-state character grew in both number and reach. International nongovernmental organizations, though few and far between until about 1890, subsequently multiplied to reach a peak in 1910 (not exceeded until 1945), before falling back again in the run-up to the First World War.[130] A separate history might be written about each of these initiatives; they varied greatly in their aims, their organization, and their support.

The Red Cross

Henri Dunant's Red Cross was the most successful of these organizations. It owed this to a brilliantly conceived division of labor: while the International Committee in Geneva concentrated on monitoring the world situation and verifying observance of the 1864 "Convention for the Amelioration of the Condition of the Wounded in Armies in the Field" and follow-up documents, national Red Cross Societies spread out from Württemberg and Baden (founded in 1863) until in 1870 they covered all the countries of western and northern Europe. A broad and highly diversified organization thus existed by the time of the First World War. What kept it going was the enthusiasm of hundreds and thousands of volunteers, with a structure sufficiently loose to draw on funding and individual commitment of every nature and magnitude. But again and again new solutions have had to be found for the relationship between national styles of work and a basic internationalist orientation. In its early phase, lack of symmetry posed a number of problems for the Red Cross: Prussia but not Austria implemented the Geneva Convention during the brief war they fought with each other in 1866; Japan but not China undertook to abide by its norms in their war of 1894–95.

In the 1870s a further issue was whether the Geneva Convention should apply to civil wars. This was answered in the affirmative with regard to the Balkans (then the scene of such conflicts), where it protected opponents of an Ottoman Empire deemed especially cruel by the West, but the Great Powers generally answered it in the negative in the period before the First World War. At the same time, the confrontation between the Muslim empire of the sultan and its Balkan enemies raised the question of whether the principles of the Geneva committee, originally understood to be Christian, could also claim validity outside the Christian West. The solution in the long run was to emphasize the transreligious humanitarian character of the Red Cross philosophy and the international laws of war. In the bloody tumult of the Balkan Wars after 1875, when Muslims acted aggressively against people wearing the Red Cross symbol, improvised talks were held to introduce the Red Crescent as an alternative.[131] The idea of the Red Cross also had an impact in more distant regions. In China, a long tradition of local philanthropic aid was revived by new social forces that saw it as a way of enhancing their reputation. The numerous civilian casualties and cases of homelessness during the Boxer uprising of 1900 led rich merchants in Jiangnan (the region on the Lower Yangtze) to send assistance to the North and to bring victims of the conflict down for care and treatment. This was the first time in China that aid had been made available across regions on a large scale. The Red Cross served as a model for this, and a Chinese Red Cross activism began to develop over the following decade.[132]

The humanitarianism of a number of Genevan citizens, and the Red Cross to which it gave rise, marked an important stage in the growth of an international social conscience.[133] The movement for the abolition of slavery and the slave trade had been an important forerunner. Humanitarianism represented a counterbalance to powerful trends of the age, a moral corrective to the normative minimalism of the anarchy among nations and states.

Political Internationalism

The numerous strands of nongovernmental political internationalism also regarded themselves as counterweights to pernicious tendencies of the age. Among these were the First International, personally founded by Karl Marx in 1864, and the much more stable and comprehensive Second International of the labor movement and its socialist parties, founded in Paris in 1889. Both remained confined to Europe; there was no broadly organized, politically influential socialism in the United States.[134] In Japan, the only non-Western country with fertile industrial soil, the first socialists—including Kōtoku Shūsui, also known as a theoretician of imperialism—suffered brutal persecution. A social-democratic party was founded in 1901, but both its organization and its press were immediately suppressed.[135] In China, both socialism and an initially strong anarchist movement spread beyond small intellectual circles during the First World War, becoming linked to the world revolution after 1921 by agents of the Third International (Comintern). In its many variants, socialism was from the beginning a transnational movement; the early Saint-Simonians had already traveled as far as Egypt. The extent to which socialist movements "nationalized" themselves in their respective political contexts has remained a major question for historians. In 1914 this process gained the upper hand over internationalism. Anarchism, the twin accompanying socialism in their formative period, sank deeper roots than ever before. It always centered on exile politics and conspiratorial action; the crossing of borders was part of its essence.

The women's movement—that is, above all, the struggle of women for civil and political rights—was in principle more mobile and capable of expansion than the socialist workers' movement, which could not exist without at least the rudiments of an industrial proletariat. Political women's movements arose not as a by-product of industrialization but, almost without exception, where "democratization was on the national agenda."[136] Since this was true from very early on in the United States, Canada, Australia, and New Zealand, suffrage movements developed in each of those countries. One finally appeared in Japan in 1919, at the very time when (as in China and Europe) the cultural image of the "new woman" was being discussed alongside the issue of voting rights.[137] In many respects, the women's movement was more internationalist than the labor movement. Its recruitment was at least potentially greater, and it was less likely to be suppressed as a threat to political stability. By 1914 no women's organizations existed anywhere in the colonies or the noncolonial Muslim world; the dominions were another matter, as was China (from 1913). In several countries, however, women were beginning before 1920 to occupy spaces outside the home, at first often through novel forms of charity work distinct from traditional religious care for the poor.[138]

As in the case of most other transnational networks, it would be too simple to analyze the history of the women's movement right from the beginning as a cross-border phenomenon. A more interesting question is the threshold beyond which particular institutional linkages hardened out. Where a movement is at issue, things are relatively straightforward for historians, since they can look for its organizational crystallization. The Second International Women's Conference, held in 1888 in Washington, DC, marked one such threshold, giving rise to the first transnational women's organization not fixed on a single objective: the International Council of Women (ICW). More than a suffragette union, the ICW came into being as an umbrella organization for national women's associations of every kind. By 1907 it could claim to speak on behalf of four to five million women, although outside Europe and North America it was represented only in Australia and New Zealand (South Africa would follow in 1908). The president of the council from 1893 to 1936 (with a few short breaks) was Lady Aberdeen, a Scottish aristocrat who at the time of her first appointment was living in Canada as the wife of the British governor-general. Of course, as with all overarching organizations of this kind, it was not long before fractures began to take place. The ICW was increasingly seen as conservative and likely to shy away from conflict, and many women considered it too close to the nobility and monarchy. Yet it performed the great service of bringing together women from different parts of the world and providing a stimulus for political work in their individual countries. The continuous history of feminist internationalism dates from 1888.[139]

It is surprising that this new beginning was necessary, for an international women's movement had already emerged in 1830 under the impact of discussions about the role of women in politics and society animated by writers such as Mary

Wollstonecraft and a few early socialists. George Sand, for example, had embodied a new type of emancipated and socially visible woman; Louise Otto-Peters had begun her many-sided career in journalism; the socialist Flora Tristan had written critical analyses of the new industrial society; and Harriett Taylor had formulated key ideas that her husband and widower, John Stuart Mill, would later take up in *On the Subjection of Women* (1869), the most emphatic defense of liberty in the work of the liberal philosopher. That first women's movement had culminated in continental Europe in the Revolution of 1848—and then had come to an end. The politics of reaction struck at public feminism in France, Germany, and Austria as new laws forbade women to attend political gatherings. The repression of socialist or independent religious associations in which women had participated was an additional blow to the infrastructure of civil society.

Paradoxically, however, this setback that often brought personal tragedy encouraged the development of an *international* movement, since some important representatives of that generation fled to freer countries, especially the United States, and continued their work there. Women's organizations already existing in America were themselves strengthened and given new life by this influx from Europe. But the upturn did not last long: activity had already peaked by the mid-fifties. Divisions then set in over the question of slavery (many feminists thought the struggle for women's rights should take a back seat for a while), while the increasingly national stamp of politics throughout Europe in the 1850s and 1860s precluded any new internationalist impetus. In the early sixties, the international ties of the women's movement decidedly slackened. The initiatives a quarter of a century later thus amounted to a fresh start—or at least that is how it looked in terms of organized movements.[140]

No less important than formal organizations were the informal personal networks that tied women to one another all through the century, as travelers, missionaries, and governesses, or as artists and entrepreneurs.[141] In the course of time, the British Empire became a space where female solidarity made itself felt at the levels of perception and action. Victorian feminists were active in seeking to improve the legal position of Indian women, and campaigns against the custom of compulsory foot binding found support among British and American women who came across it in China.[142]

Unlike the labor or women's movement, pacifists did not seek representation in national political systems.[143] They might fight from within against the militarization of individual nation-states (though seldom with noteworthy success), but they stood a chance of exerting only a minimum of influence at international level. Dread of war and a critique of violence constitute an old current in European (as well as Indian or Chinese) thought. In war-weary Europe after 1815, sometimes with older roots in Quakerism or Mennonism, they found a new lease on life, especially in Britain.[144] To have any public impact, pacifism needed to focus on a palpable experience of war or a strong and credible vision of the horror of future armed conflicts. This gave it strength in the 1860s, when, in an age of renewed warfare, it gained new supporters in Europe. In 1867 Geneva hosted the first "Congress of Peace and Freedom," which was followed by many similar gatherings on a smaller scale. In 1889 pacifism started to become a transnational lobby, and in the same year 310 activists attended the first Universal Peace Congress (in Paris). There would be a total of twenty-three congresses between then and 1913; the twenty-fourth was due to be held in Vienna in September 1914.

At the height of its significance, this international peace movement was sustained by approximately three thousand people.[145] It was a European project with North Atlantic extensions; otherwise there were peace societies only in Argentina and Australia. For the colonies, unable to be belligerent parties in their own right, pacifism was less relevant as an international attitude (Gandhi's later policy of nonviolence was a strategy for disobedience inside India); while in Meiji Japan, determined to build up the nation's military strength, it remained a cause taken up by only a few writers, with little or no impact outside their immediate circle. The earliest Japanese pacifist was Kitamoru Tōkoku (1868–94), who, like nearly all others who came after him, was inspired by Christianity and came close to being accused of treason. In 1902, the Chinese philosopher Kang Youwei composed in his Indian exile a grand utopian vision of world peace, *The Book of Great Unity*, which was published in full for the first time in 1935 and failed to have any political impact.[146] China and the Ottoman Empire were not a threat to other states, but they had to have a minimum of military strength to defend themselves. Pacifism therefore held no political attraction for them.

Since nineteenth-century pacifism had no natural social base or clientele and sprang above all from personal ethical convictions, it was more susceptible than the labor or women's movement to the charismatic power of individuals. This is why it was so important that Bertha von Suttner's rhetorically effective novel *Lay Down Your Arms!* (1889; Eng. trans. 1892) was an international success; that the Swedish explosives manufacturer Alfred Nobel created a prize for the furtherance of peace, which, like the other Nobel prizes, was awarded from 1901 on (the first going to Henri Dunant and the French politician Frédéric Passy, and the 1905 prize to Bertha von Suttner); and that in 1910 the American steel magnate Andrew Carnegie made part of his huge fortune available for the cause of peace and international understanding. The main currents of pacifism considered their objective to be not so much disarmament as a system of international arbitration. They had no great hopes in a reign of universal peace, but they realistically contented themselves with proposals for basic mechanisms of consultation, such as there had no longer been in the anarchic world of states since the Crimean War.

The activity of the international peace movement reached a peak in the 1890s, against the background of irresponsible war talk in Europe and a sharpening of imperialist aggression in Africa and Asia. Its greatest success was the convening of the First Hague Peace Conference in 1899, when the Great Powers had just descended upon China, the United States was waging a colonial war in the

Philippines, and the great struggle between Boers and British was getting under way in South Africa. Such a conference could not be a gathering of private individuals like the founding circle of the Red Cross; the formal initiative had to come from a government. Ironically, this was the government of the Tsarist Empire, the most authoritarian in Eurasia, whose motive was not a morally pure love of peace. The intensification of the arms race had put Russia in a financial squeeze, and it reacted by experimenting with new kinds of solution. A second conference followed at The Hague in 1907.

Both conferences led to important innovations in international law but failed to get any arbitration mechanisms up and running. They were not intended to reform the international state system, nor did they belong in the tradition of the great peace congresses. What reflected the real or perceived distribution of power in the international system was the fact that of the twenty-six countries represented in 1899, only six lay outside Europe: the United States, Mexico, Japan, China, Siam, and Iran. The Hague Peace Conferences grew out of closer cooperation less among states than among individual public figures—a kind of transnational peace milieu. The problem was that they achieved nothing at the level of great-power politics, and the "spirit of The Hague" changed nothing of note in the thinking of policymakers.[147]

If governments in the second half of the century gave any thought to international relations apart from military power games, then it was less to peace building than to the "mechanics" of internationalism.[148] Insofar as international law was an instrument and medium of such concretion below the level of grand politics, there was a transition "from coexistence law to cooperation law," the aim of which was "the joint achievement by states of transnational goals."[149] Strongly binding treaties, backed up by periodic conferences of experts, anticipated supranational legislation before any existed. The result was a historically unparalleled norm setting in countless areas of technology, communications, and cross-border trade. The unification of world time has already been discussed in chapter 2.[150] During the same period, weights and measures, international mail (Universal Postal Union of 1874, Universal Postal Convention of 1878), railroad gauges, train timetables, coinage, and much else besides were simplified and standardized for large areas of the world.[151] For large areas, but not really for the whole world: operational systems varied too much in complexity, and cultural and political resistance too tenacious. The international letter post could be homogenized more easily than the endless variety of currencies and means of payment. Not all the processes of adaptation and homogenization initiated in the nineteenth century had been completed by the First World War; many are still continuing today. The important point is that people in the nineteenth century saw the need for such regulations and took the first steps to bring them about. It is hardly surprising that much of the world was not yet integrated in this way. Once again the nineteenth century exhibits long-range continuity with the second half of the twentieth.

The continuities with the *past* were not very numerous. The early modern period in Europe knew many forms of philosophical and scientific universalism, but apart from transoceanic trade relations it created few trans-European systemic links. Its legacy lived on not so much in direct connections as in the revival of older programs. Thus, new proposals for a world language built on considerations that Leibniz had already presented. The best-known offering, alongside the Volapük invented by the Konstanz priest Johann Martin Schleyer, was the one that the Polish eye specialist Ludwik Lejzer Zamenhof submitted to the public in 1887 under the name "Esperanto." By 1912 there were more than 1500 Esperanto-speaking groups, some of them outside Europe and North America. A first world congress of the movement had been convened in 1905. This most effective kind of premeditated linguistic globalism created a truly planetary community of communication, but it never dislodged any of the national languages and did not gain widespread acceptance as a medium of scholarly exchange.[152]

Another initiative, which eventually proved much more successful, had roots far beyond the early modern period: the Olympic Games. Initially an obsession of a few (mostly English) philhellenes and sporting enthusiasts, joined by the Anglophile Frenchman Baron Pierre de Coubertin, the revival of this ancient idea led in 1896 to the first Olympiad of modern times and went on to become one of most inclusive, prestigious, and economically viable global movements. De Coubertin's original impetus had by no means derived from philosophical contemplation of a coming age of world peace. Rather, the young aristocrat formed a conviction that Germany had won the war of 1870–71 because of the superiority of its school gymnastics. In 1892 he put such nationalism behind him and argued instead for sportsmen of different countries to compete with one another.[153] The diffusion of other kinds of sport—especially the team games football (soccer) and cricket—also began in the last third of the nineteenth century.[154]

Like most dichotomies, the opposition between bellicose power politics and peaceful civil efforts of nongovernmental internationalists is too simple to be altogether convincing. In reality there were intermediate levels—above all, attempts by national governments to use internationalism for their own foreign policy ("internationalism to the advantage of nations," as the pacifist Alfred H. Fried put it in 1908[155]). Switzerland, and Belgium even more, pursued internationalization strategies in this way—for example, helping to create scientific and economic conferences with international participation, and letting no opportunity slip to put themselves forward as locations for international events and organizations. The key period for the founding of international governmental organizations (IGOs) was the 1860s—the same decade in which the Red Cross came into being as an international nongovernmental organization (INGO). Beginning in 1865 with the International Telegraph Union, more than thirty IGOs were established up to the outbreak of the First World War;[156] most of them saw the colonies as part of their sphere of activity. A large number of technical conferences also were held to coordinate new transportation and communications systems, such as the telegraph and regular steamship services, or to standardize legal norms in such matters as the cross-border movement of currencies. Especially important was the series of international public health conferences that began as early as 1851.[157]

As far as war, peace, and international politics are concerned, the nineteenth century began in 1815. It followed a long eighteenth century that for some parts of the world—Europe, India, Southeast Asia—had been an age of extraordinary military violence. In comparison with the periods before and after, the hundred years from 1815 to 1914 were unusually peaceful in continental Europe. Interstate wars had seldom been so limited in time and space, or casualties so low as a proportion of either troop strengths or the civilian population. The great civil wars took place in America and China, not in Europe. Weapons technology, railroads, general staffs, and compulsory military service revolutionized warfare. The built-up potential was discharged only in 1914, in a great war that lasted so long partly because the main belligerents had more or less the same means at their disposal. Lightning campaigns were still possible, but no longer those of the Napoleonic type that had crushed the enemy in a matter of days. Technological and organizational advances in Europe and the United States came into their own, especially after 1840, where no arms race could create a level playing field: that is, against preindustrial military cultures in Asia, Africa, New Zealand, and the North American interior. "Asymmetrical" colonial warfare became one of the forms of violence characteristic of the age. Another was the "opening-up war," a rather selective operation designed not for territorial conquest but to ensure that a country became politically amenable and geared its foreign policy to the West. Military strength was concentrated in the arsenals of an ever smaller number of great powers, which, with the exception of Japan after 1880, lay geographically in the North and culturally in the "West." For all the regional power differences, which made Egypt under Muhammad Ali, for example, appear a military factor deserving serious respect, this was the first time in centuries that not a single country in Africa, the Muslim world, or the Eurasian landmass east of Russia was in a position to defend its borders or to project its power beyond its own national or imperial limits. The Ottoman Empire definitively lost this capacity after its war with Russia in 1877–78. Brazil was a strong regional power as well, but no more than that.

In an age when migration, trade, currency coordination, and capital transfers were linking countries across the world, no global political order came into being. The most extensive of the European empires, though economically dominant for a time and normatively accepted as a model by many, was far from being a universal empire that created its own distinctive order. In 1814–15 the European Great Powers agreed among themselves on a surprisingly successful formula for peace. But something close to anarchy prevailed among the same powers qua empires with overseas interests, even if there were no great inter-imperial wars, and the opposition between France and Britain, having marked the eighteenth century down to the Battle of Waterloo, never again flared up into military conflict.

The old regional orders stretching back to time immemorial were dissolved and absorbed into something new. The Indian state order was transmogrified into the geopolitical patterns of British India. The ancient Chinese order, perfected by the Qing dynasty in the eighteenth century, receded and partly died away as its traditional tribute-paying periphery succumbed to foreign colonization. Japan did not yet have the will and the strength to shape a new order of its own; this would happen only after 1931, and all would be over within fourteen years at an untold cost in human lives. Thus, outside the Vienna Congress System, and even in Europe after the Crimean War, a kind of controlled anarchy prevailed. Its ruling ideology, around 1900, was an international liberalism inflected in a racist, social Darwinist direction. Regulation made strides in the pre-political sphere, emanating from private, or sometimes technical-administrative, initiatives aimed at international unity, solidarity, and harmony. All this was unable to prevent the Great War, and barely a decade after its conclusion hopes began to fade again that its lessons had been learned and that a viable peace was within reach.

CHAPTER X

Revolutions

From Philadelphia via Nanjing to Saint Petersburg

1 Revolutions—From Below, from Above, from Unexpected Directions

Philosophical and Structural Concepts of Revolution

More than in any other era, politics in the nineteenth century was revolutionary politics. It did not defend "age-old rights" but, looking ahead to the future, elevated particular interests such as those of a class or class coalition into the interests of a nation or even of humanity as a whole. "Revolution" became a central idea of political thought in Europe, serving as a yardstick that for the first time divided Left and Right. The entire *long* nineteenth century was an age of revolutions, as a look at the political map will make apparent. Between 1783, when the world's largest republic gained independence in North America, and the near worldwide crisis at the end of the First World War, some of the oldest and most powerful state organisms disappeared from the stage: the British and Spanish colonial states in the Americas (or at least south of Canada); the ancien régime of the Bourbon Dynasty in France; the monarchies in China, Iran, the Ottoman Empire, the Tsarist Empire, Austria-Hungary, and Germany. Upheavals of revolutionary dimensions occurred after 1865 in the Southern United States, after 1868 in Japan, and wherever a colonial power replaced indigenous groups with a form of direct rule. In each of these cases, what happened was more than a changeover of state personnel within an abiding institutional structure. New political orders came into being, with new bases of legitimacy. Any return to the world as it had been previously was barred; nowhere were prerevolutionary conditions restored.

The birth of the United States in 1783 was the first founding of a state of the new type. The revolutionary unrest that led to this event, and with it essentially the Age of Revolution, began in the middle of the 1760s. An Age of Revolution or Revolutions? A good case can be made for either. A view grounded in a philosophy of history prefers the singular noun; a structural approach, the plural. Those who initiated or lived through the revolutions in America and

France saw mainly the singularity of the new; the events in Philadelphia in 1776, when the thirteen colonies declared their independence of the British Crown, and the spontaneous emergence of a National Assembly in France in June 1789 appeared to be without parallel in any age. Whereas previous violent overthrows had merely led to external modifications of the status quo ante, the American and French revolutionaries expanded the whole horizon of the age, opening a path of linear progress, grounding social relations for the first time on the principle of formal equality, lifting the weight of tradition and royal charisma, and instituting a system of rules that made those in political authority accountable to a community of citizens. These two revolutions of the Age of Enlightenment, however different from each other in their aims, signaled the onset of political modernity. From then on, defenders of the existing order bore the mark of the old and obsolete, of reactionaries and counterrevolutionaries, or else they had to reinvent their posture as "conservative."

Both revolutions—though the French more than the American—polarized along new dividing lines: no longer between elite factions or religious groups, but between rival worldviews. At the same time, in a contradiction that would never be overcome, they raised the demand for human reconciliation, the "hope for the emancipation of all mankind through revolution."[1] Thomas Paine already set this new tone in 1776, combining a favorite theme of the European Enlightenment, the forward march of the human race, with the local protest of a British subject. "The cause of America," he wrote, "is in great measure the cause of all mankind."[2] Since then, what Hannah Arendt called the "pathos of an entirely new beginning"[3] and a claim to represent more than the self-interest of the protesters have been part of every self-styled revolution. In this sense, a revolution is a local event with a claim to universal validity. And every revolution has in a sense been imitative: it feeds off the potential of the ideas that first became a reality in 1776 and 1789.

Such a philosophical concept of revolution is admittedly very narrow, and it becomes still narrower if one insists that every authentic revolution must have happened under the banner of liberty and served the cause of progress. This would also be to generalize a claim to universality that was invented in the West and whose like is found nowhere else. A larger range of cases come under a concept that is not pitched in terms of aims or philosophical justifications, or the role of "great revolutions" in the philosophy of history, but bases itself on observable events and structural outcomes.[4] A revolution then denotes a collective protest of certain proportions: a systemic political change involving the participation of people who did not belong to the circle of the previous holders of power. In the language of a social scientist careful to keep his conceptual tools razor-sharp, it thus becomes "the successful overthrow of the prevailing elites . . . by new elites who, having taken power (usually with considerable violence and mass mobilization), fundamentally change the social structure and therewith also the structure of authority."[5]

Here nothing is said about a moment in the philosophy of history; the pathos of modernity vanishes. In this definition there have been revolutions almost everywhere and in almost every epoch. The whole of recorded history displays any number of radical breaks, including ones in which many people thought everything familiar was being turned upside down or torn up at the roots. If there were statistics for such things, they would probably show that really major watersheds have more often been a result of military conquest than of revolution. Conquerors do not only vanquish an army: they occupy a land, destroy or topple at least part of its elite, install their own men instead, and introduce foreign laws and sometimes also a foreign religion.

This also happened in the nineteenth century, all around the world. In terms of its effects colonial conquest was often "revolutionary" in a quite literal sense. In most cases the invaded and vanquished must have experienced it as a traumatic break with their previous way of life. Even where the old elite physically survived, it was degraded by the fact that a layer of new masters stood on top of it. The coming to power of alien colonial rulers through a military invasion, or less often through negotiations, was thus tantamount to a revolution for large numbers of Africans, Asians, or South Sea islanders. Furthermore, the *long-term* revolutionary character of colonialism lay in the fact that, after the original conquest, it created space for the rise of new groups in the indigenous society and thereby paved the way for a second wave of revolutions. In many countries, the true social and political revolution took place only during or after decolonization. Revolutionary discontinuity marked both the beginning and the end of the colonial period.

The idea of foreign conquest as "revolution" was more present to nineteenth-century Europeans than it is to us today. The Manchu takeover of China, for instance, which began with the fall of the Ming dynasty in 1644 and continued for several more decades, struck plenty of early modern European commentators as a dramatic example of a revolution. The older political language of Europe closely associated the term with the rise and fall of empires. Several factors came together here, in a way that Edward Gibbon synthesized between 1776 and 1788 (at the beginning of the Age of Revolution) in his great work on the decline and fall of the Roman Empire: namely, internal unrest and elite change, external military threat, secession on the imperial periphery, and spread of subversive ideas and values. The ingredients were no different in what we have called the "bridge period" (*Sattelzeit*). The old European conception of politics contained a complex picture of radical macrochange that led to an understanding of the novel events of the last third of the eighteenth century: they were simultaneously of unprecedented novelty and a repetition of familiar patterns. It would be too simple here to counterpose a new "linear" and an old "cyclical" view of history. What was the Battle of Waterloo if not the terminus of a *cycle* of French hegemony? Anyone looking for "premodern" patterns in their pure form could continue to discover them. For example, exactly at the same time as the revolutionary events in France, a drama unfolded in what is now Nigeria that could be copied straight out of Gibbon: the fall of the Oyo Empire as a result of elite infighting at the center and uprisings in the provinces.[6]

The chronological nineteenth century, from 1800 to 1900, does not have pride of place in the usual narrative histories of revolution; it witnessed the consequences of revolutions in North America and France, but it did not produce a "great" revolution of its own. The revolutionary dice, it would seem, were already cast by 1800, everything coming later being an imitation or lusterless rehearsal of the heroic beginning, farce after tragedy, petty disturbances aping to the great upheaval of 1789 to 1794. In this view it was only in Russia in 1917 that history once again threw out something unprecedented. The nineteenth century in Europe was less an age of revolutions than a rebellious century, an era of widespread protest that rarely achieved critical mass on the stage of national politics. In particular the period between 1849 and 1905 (the year of the first Russian revolution) was almost free of revolution in Europe, the one exception being the Paris Commune, which soon ended in failure. The statistics confirm this impression. Charles Tilly counts forty-nine "revolutionary situations" between 1842 and 1891, in comparison with ninety-eight in the period from 1792 to 1841.[7] And in most of those the potential did not translate into action with a lasting effect.

Variants and Borderline Cases

If, however, we use a structural concept that goes beyond the founding revolutions in America and France, the myth of their incomparability loses most of its dazzle, and various other kinds of system breakdown and violent collective action come into view. This raises two preliminary questions.

First. Should only successful revolutions be described as such? Or can the title also be conferred on power grabs which, though spectacular, did not achieve their goal? According to one of the best sociological surveys of theories of revolution, "revolutions are *attempts* by subordinate groups to transform the social foundations of political power."[8] This definition, then, includes major attempts with a radical intent. Anyway, can success and failure be clearly differentiated in every case? Does not victory sometimes come out of defeat, and might not triumphant revolutions destroy their own foundations by giving violence a momentum of its own? Such questions are often posed in too academic a manner. People in the nineteenth century saw things more dynamically: they were more inclined to use the adjectival form, looking for revolutionary tendencies, whether these were encouraged, welcomed, or feared. The historian can follow this lead, by employing the criterion of *actual* mobilization. One should speak of revolution if movements seeking to change the system—and they must always be popular movements—achieved such a position on the national political stage that they at least temporarily constituted a counterpower.

Let us take the two most important instances in the nineteenth century. Since a National Assembly did gather in the Paulskirche in Frankfurt, and since rebel governments with their own army did briefly hold power in Baden, Saxony, Budapest, Rome, Venice, and Florence, what happened in Europe in 1848–49 really was a revolution. Similarly, there was a Taiping Revolution in China between 1850 and 1864, not just (in the conventional Western terminology) a Taiping Rebellion. For a number of years the insurgents ran a complex counterstate, which in many respects was a variant of the existing order with reversed polarity.

Second. In order to be a revolution, must the serious shaking or successful elimination of the existing relations of authority always proceed "from below?" Must it stem from those in society whose interests are not regularly taken into account, and who resort to the collective use of force because the organized power of the state and elite groups has left them with no other course? Or should one also allow the possibility of a "revolution from above," that is, a systemic change going beyond merely cosmetic reforms, carried out by parts of the existing elite? This "revolution from above" is an equivocal figure, unless one casually treats it as just a façon de parler.[9] The revolution itself may lose its mass impetus as a result of inevitable "routinization," giving rise to a bureaucratic regime that puts into effect many of the revolution's proper goals with the instruments of state power, often without, against, or at the expense of the original revolutionaries. Napoleon and Stalin were "top-down revolutionaries" of this kind. A different possibility is a headlong conservative rush: modernization and strengthening of the state as a prophylactic defense against revolution. Anti-Jacobin statesmen like Otto von Bismarck (especially in his period as Prussian prime minister) or Camillo di Cavour in Italy were such "white revolutionaries." They saw that only those who kept abreast of the times could hope to maintain the initiative—an old insight of the British ruling class. However, "white" revolutions led not to a real change of elites but at best to the co-opting of new elite groups (e.g., bourgeois figures with a national-liberal coloration), and they saved the status quo more through its transposition into a different template than through reinvention. Bismarck preserved Prussia *within* Germany, and Cavour projected his Piedmont onto the larger canvas of Italy.

But there was one limiting case in which a subdominant elite reinvented a country's whole political and social system (and thereby also itself), in the most radical attempt at a revolution from above, but also one that eschewed the term "revolution" and sought legitimacy as the restoration of a previous state of affairs—the Meiji "Restoration" of 1868 and after. It lay outside the perceptual horizon of most European political commentators, and what knowledge there was of it had no influence on the European understanding of revolution and reform.

In Japan, a country whose elites felt threatened less by the specter of a "red" social revolution than by the incalculable consequences of a forced opening to the West, a radical system change disguised itself as a political "renewal" or "restoration" of legitimate imperial rule. For two and a half centuries, the powerless imperial court in Kyoto had led a shadowy existence, while the real authority in the country had lain with the supreme military commander, the shogun in Edo (Tokyo). In 1868 the shogunate was eliminated in the name of a newly active imperial power.[10] The driving forces were not members of the old dominant elite, the territorial princes, but small circles among their privileged vassals, the samurai. These constituted a lower, military nobility, which by the early nineteenth century performed little other than administrative duties.

This special kind of renewal for the sake of rapid increases in efficiency, neither motivated by counterrevolutionary aims nor propagating anything in the way of universal principles, was as momentous in Japan as the American or French revolution had been in its country of origin. But the historical context was not a revolt against injustice and participation deficits; the goal was to make an upcoming nation fit for global competition, using new international rules that it recognized from the start. The social content of the Meiji Renewal was thus incomparably more radical than Prussian-German nation-building was during the Bismarck period.

After a brief military conflict between the shogunate and imperial forces, a tiny oligarchy grasped hold of state power and introduced a reform policy which, though not sweeping away the existing social hierarchies, ran clearly counter to the interests of the samurai class from which the Meiji oligarchs themselves originated almost to a man. The European category of "revolution" is peculiarly unfocused in the Japanese case, and so too is the idea of revolution from above. The Meiji Renewal needs a different historical framing: as the most radical and successful self-empowerment operation of the nineteenth century, it belongs in the comparative context of similar state strategies of the time.[11] To describe it as a Japanese equivalent of "bourgeois revolution" would be formally correct insofar as it brought to an end the feudal ancien régime. The like cannot be said of any of the European "revolutions from above." It showed little respect for popular rights, and two decades would pass before middle and lower strata obtained some scope to articulate their interests within the Japanese political system. Implementation of the Meiji strategy did not even require a mobilization of the popular masses outside the increasingly disciplined world of labor. It was not the motives and methods of the Meiji Renewal but its consequences that were revolutionary: that is, an ideologically veiled radical break with the past, which suddenly opened up space for the future, and the return of a long-peripheral elite to the centers of power.

With regard to the mass experience of crisis, mention should finally be made of four other cases that do not fit unambiguously into the category of revolution. They are borderline or transitional phenomena, which bring out all the more graphically the peculiarities of real revolutions.

A revolution in the slipstream of history was the Tay Son uprising in Vietnam. In spring 1773 three brothers from the central Vietnamese village of Tay Son launched a protest movement that would become the largest revolt in the country's history before the twentieth century. They preached the equality of rich and poor, burned taxation registers, distributed the mobile property (but not the land) of affluent households to the poor, marched a 100,000-strong peasant army through the north of Vietnam (Tonkin), abolished the Lê Dynasty after more than three centuries in power, beat off Chinese and Siamese intervention in support of the Lê rulers, and attacked the neighboring kingdoms of Laos and Khmer. French, Portuguese, and Chinese mercenaries and "pirates" fought on both sides. Hundreds of thousands of people died in the fighting or from starvation. Once the Tay Son leaders had become masters of the whole of Vietnam, they established a tyrannical regime that brutally repressed the Chinese minority. Their support among the masses collapsed. Another warlord group put an end to their rule and in 1802 established the Nguyen dynasty in the city of Huê.[12]

Minor civil wars, often omitted in historical overviews, were also present in Europe and nearby, if by "civil war" we mean "armed combat within the boundaries of a recognized sovereign entity between parties subject to a common authority at the outset of the hostilities."[13] Following the death of Ferdinand VII, the last Spanish monarch with absolutist impulses, parts of Spain were turned into a battlefield during the First Carlist War (1833–40), which ranged parliamentary liberalism against a classical form of counterrevolution.[14] The Carlists, with their main stronghold in the Basque country, wanted to unify Spain along Catholic lines, to eradicate all liberal and "modern" tendencies, and to replace Queen Isabella II with her uncle Charles V, an absolutist pretender mentally stuck in the sixteenth century. In 1837 and 1838, whole armies were locked in a savage war reminiscent of the Napoleonic occupation. After their defeat, the Carlists did not surrender but continued a guerrilla campaign and made plans for a coup d'état; it was 1876 before the constitutional monarchy was firmly in the saddle, having seen off another attack by the Carlist "state within a state" in the Basque country, Navarre, and parts of Catalonia.[15] Comparable in brutality, though not in the scale of the fighting, were the civil war in Portugal (1832–34) and the chain of lesser revolts that followed it.[16]

In Ottoman Lebanon, a host of social conflicts, religious tensions, and capricious interventions by foreign powers after 1840 gave rise to "intercommunal" hostilities, which between 1858 and 1860 escalated into a civil war in which thousands were massacred and hundreds of thousands forced to become refugees. Here the outcome was not the fall of an ancien régime or the repulsing of a postrevolutionary counterrevolution but a kind of constitutional compromise reached through international negotiations; the history of an actual Lebanese state then began in 1861, albeit one that recognized French rights of protection and intervention.[17]

Peasant revolts disappeared in Europe (with the exception of the Balkans), after one last upsurge in 1848–49 from the east of the Habsburg Monarchy down to Sicily and up to southern and central Germany. These final outbursts of rural protest were quite in tune with the times and realistic in their goals and forms of action—by no means further instances of the blind, backward-looking outbreaks of violence that city dwellers, and even many historians, have tended to see in them. Outside the few European countries where its interests could find some representation in parliament, the peasantry resorted time and time again to violence or high-profile symbolic actions. Such protest movements were to be expected in every agrarian society, but they varied in size and scope. They took on larger dimensions in Mexico, for example, from 1820 to 1855, peaking in the years between 1842–46.[18] In Japan, where political life was more stable, they increased in frequency during the economically and ecologically harsh period of the thirties and then again, under different conditions that included alliances with urban forces, in the eighties.[19] Between 1858 and 1902, the Near and Middle East witnessed a number of peasant revolts, mostly in opposition to "modernizing" forces, a fiscally more demanding state, and absentee landowners seeking to boost their profits (and therefore the exploitation of labor) out of a structurally unreformed agriculture that was no more productive than before.[20]

Anticolonial resistance may acquire revolutionary forms and produce revolutionary effects.[21] The United States and the Latin American republics arose out of just such a situation. From the Greek revolt against Ottoman rule (1821–26), the great Java war of 1825–30, and the coeval resistance of the Kazakhs to Russian colonization to the Khoikhoi rebellion at the Cape of Good Hope (which did much to shape solidarity along the line of "black" and "white" racial stereotypes) to the Polish uprising of 1863, the Jamaican revolt of 1865, and the Cretan insurrection of 1866–69, a long chain of actions against foreign rule developed, up to the new great wave of anticolonial or anti-imperialist unrest in 1916–19 in Ireland, India, Egypt, China, Korea, and Central Asia. Anticolonial resistance is *revolutionary*, however, only when its aim is to establish a new and independent political order—such as a nation-state. This was relatively uncommon outside Europe before the First World War. One of the few instances is the Urabi movement in Egypt in 1881–82.[22]

Revolutions, as "accelerated processes" of a special kind,[23] are not distributed evenly along the temporal continuum. Often they appear clustered at critical junctures of historical change—which is why historians, especially since the French Revolution, have liked to use them as period markers. Even before the mid-eighteenth century, systemic crises or even breakdowns were plainly visible in several parts of the world: between 1550 and 1700, for example, in Japan, the Ottoman Empire, England, China, and Siam (to name only the most important cases). They occurred without having had a direct influence on, or encounter with, one another. The (temporary) fall of the Stuart dynasty in England in 1649 and the (definitive) removal of the Ming dynasty in China in 1644 had nothing causatively in common. Yet it has been argued that factors not recognized by people at the time—of which a similar demographic trend might be especially important—lay behind such conspicuous simultaneity.[24]

For us today the connections are much more apparent. Between roughly 1765 and 1830, the clusters of revolutionary events were so striking that it is possible to speak of a compact Age of Revolutions.[25] The imperial offshoots reached all continents, but the centers of interactive unrest were in the Americas and continental Europe. For this reason the "revolutionary Atlantic" is the most appropriate image. A *second* cluster of upheavals and revolutions may be found between 1847 and 1865: the European revolutions of 1848–49, the Taiping Revolution in China (1850–64), the so-called Mutiny or Great Rebellion in India (1857–58), and the special case of the American Civil War (1861–65). These events had a much weaker and less direct effect on one another than the comparable ones had had in the revolutionary Atlantic. They added up not so much to another compact age of revolutions as to a set of separate megacrises with rather weak "transnational" links. A *third* wave of revolution washed over Eurasia after the turn of the century: Russia in 1905, Iran in 1905, Turkey in 1908, China in 1911. The second Russian Revolution, born in February 1917 under the special conditions of the World War, also belongs in many respects to this context, as does the revolution in Mexico that began in 1910 and lasted a full decade. This time the mutual influences were more intense than in the mid-nineteenth century; the revolutionary events were expressions of a common background in the times.

2   The Revolutionary Atlantic

National Revolutions and the Atlantic Connection

Revolutions always have local roots, in the perceptions that individuals and groups have of injustices, alternatives, and opportunities for action. These particular perceptions give rise to acts of collective disobedience and to movements that grow, bring forth opponents, and take on a dynamic of their own. In rare cases the result is what the Marxist theory of revolution takes to be the norm: whole classes become historical actors. Since revolutions have often been seen in modern times as the founding acts of nations and nation-states, the history of revolution is essentially national history; the nation "invents" itself in the common endeavor. However, the dependence of revolutions on conditions that lie outside themselves, sometimes even on external midwifery, does not fit well in this narcissistic picture. The modern European concept of revolution is narrower than the old one that used to include war and conquest: it leaves out the external international dimension, disregarding nonlocal roots and emphasizing conflict within a particular society (hence the endogenous character of revolutions).[26] In extreme cases, the history of revolution was so nationally oriented that it was incapable of explaining central developments. Can one do justice to the Reign of Terror (1793–94) in the French Revolution if, like Hippolyte Taine (1828–93), one leaves out the key role of the external war danger in legitimating events?[27] It was at an amazingly late date that the French Revolution was first placed in its international (European) context: by the Prussian Heinrich von Sybel in his *Geschichte der Revolutionszeit* (1853–58), and in France only after 1885 by the historian Albert Sorel.[28] This never became the dominant perspective, however; it was more than once forgotten, then brought back into memory.

For a long time, historical work on the American Revolution also featured national navel-gazing, often known in the United States as celebration of American "exceptionalism."[29] The rebellious New Englanders, it was argued, turned their backs on the corrupt Old World and, in their undemanding isolation, created a polity of unique perfection. Since most revolutions are thought unique by their protagonists and by historians who come after them, comparison between revolutions—which always puts things into perspective and deflates the myth of singularity—did not play a major role, until philosophers of history and a number of sociologists finally began to take it seriously.[30]

The view that it is inadequate to regard the great revolutions of the *Sattelzeit* around 1800 in Europe and America as isolated from one another has two sources. From the 1940s on, a number of historians, especially in the United States and Mexico, began to treat the history of the New World as a single whole. In their view, elements of a common experience united its different histories of settlement and colonial rule. Then in the 1950s and 1960s the vision of an "Atlantic civilization" began to take shape, which at the height of the Cold War some authors gave a strongly anticommunist or even anti-Eurasian inflection: the "West" was supposed to have somehow expanded across the ocean. But it was not necessary to follow this descent into ideology in order to recognize that a transatlantic perspective made historiographical sense. The Frenchman Jacques Godechot and the American Robert R. Palmer simultaneously developed conceptions of an Atlantic Age of Revolution, which differed only in their finer details and took in both the American and the French Revolution.[31] Hannah Arendt approached the same theme from a philosophical point of view. Later historians added Haiti and the Spanish American revolutions to the picture.[32]

Only in the 1980s did historians begin to discover (or rediscover) a "black" Atlantic alongside the "white" and to study together a North shaped by Britain and a South molded by Spain and Portugal.[33] A further impetus to grasp the Age of Revolution as more than a pan-European phenomenon (at best) came from Leipzig, where Walter Markov, a Marxist specialist on the Left in the French Revolution, and his disciple Manfred Kossok developed a comparative approach combining the tradition of Karl Marx with that of the unconventional Leipzig historian Karl Lamprecht at the beginning of the twentieth century. Kossok's concept of "cycles of revolution" with a beginning and end made it possible to theorize interaction between the revolutionaries of different countries and regions and to arrive at a fairly well founded periodization of world history.[34]

North America, Britain, and Ireland

Which revolutions are at issue, what are their respective temporalities, and how do they relate to one another chronologically? It is not always equally obvious when a revolution (not just a *potential* revolutionary situation) began and when it ended; nor is the outcome unambiguous in every case. The American Revolution[35] reached a first peak with the Declaration of Independence on July 4, 1776, when all colonies except New York, representing the overwhelming majority of British subjects in North America, rejected once and for all the Crown's claim to sovereignty over them. Of course this did not come out of the blue. It was the culmination of resistance to British rule that had begun in March 1765 with protests against the Stamp Act, which, by imposing without consultation a new tax on newspapers and printed documents of any kind, had sharpened tensions between colonies and the mother country and triggered violent attacks on representatives of the colonial state.[36] The Stamp Act crisis mobilized North Americans (whose unaristocratic societies had long been receptive to republican ideas) on a scale that no previous political event had occasioned.[37] It created a new sense of unity among the elites of the various colonies, which differed quite considerably from one another in their forms of rule and social structure. The crisis between Britain and America escalated into economic warfare and finally, in April 1775, into open military confrontation, with General George Washington at the head of the rebellious colonies. The Continental Congress, which passed the Declaration of Independence drafted mainly by Thomas Jefferson, took place in the middle of the war. The public formulation of the reasons for independence was therefore above all a symbolic act.

The real watershed year was 1781, when two things happened: the colonies subscribed to the Articles of Confederation, a kind of constitution of the newly founded federation of states (not yet a federal state); and the British army surrendered at Yorktown, Virginia, on October 18. In the peace treaty, signed in Paris in 1783, Britain recognized the independence of the United States of America, largely on the terms laid down by the Americans, so that the United States then became a new entity in international law capable of acting in its own name. There is much to be said for the view that this marked the endpoint of the revolutionary process. But heated debates on the internal political system of the Union continued for a number of years. Only in June 1788 did the new Constitution come into effect, and spring 1789 saw the establishment of the main bodies of the state, including the presidency with George Washington as its first incumbent. In sum: the American Revolution lasted from 1765 until 1783; its chief outcome, the formation of a newly independent state, was concluded a few months before the storming of the Bastille in Paris.

The next act in the drama of the Atlantic Revolution took place not in France but in the British Isles. Between 1788 and 1791, independent revolts in Ireland, Yorkshire, and London challenged the existing order more profoundly than anything seen earlier in the century. Anyone in London who lived through the so-called Gordon Riots of June 1780, which were initially directed against fresh concessions to English Catholics, must have concluded that a great upheaval was brewing there rather than on the Continent. The disturbances caused enormous damage in the inner city; it took the army a great effort to restore order, and at the end fifty-nine rioters were condemned to death and twenty-six executed.[38] In Ireland too, the militia, including recruits from the Catholic population, had a hard time quelling unrest directly set in motion by events on the other side of the Atlantic, and after 1789, under the influence of the French Revolution, the island remained a hotbed of national rebellion. One leading historian in the field described the rising of 1798, which was supported by revolutionary France, as "the most concentrated period of violence in Irish history,"[39] in which probably as many as 30,000 people (on all sides) lost their lives. The merciless punishment of the rebels lasted until 1801. In 1798–99 alone, more than 570 death sentences were handed down.[40]

But we are running a little ahead. In England, as in many Continental countries, sympathizers with the French Revolution raised their heads and demanded a radical, or even republican, reform of the political system in accordance with the laws of reason. The agitation was mostly confined to a pamphlet war for and against the revolution and, unlike in 1780, did not lead to open revolt.[41] The conflicts became increasingly enmeshed in the threat (after February 1793, the reality) of war with France. And, as in France, criticism of the existing system might be represented as high treason. The radicalism of many intellectuals and artisans was compounded in the economically difficult war years by constant unrest in the countryside. The British state reacted with emergency laws and harsh repression (though by no means comparable to the *terreur*), so that by 1801 or thereabouts the last traces of a quasi-revolutionary challenge had disappeared and a new national consensus had formed around anti-French patriotism.[42] The great political overturn failed to materialize in Britain, but the country was nevertheless caught up in the tide of revolution. Some of the most important ideas of the revolutionary epoch came from its shores—whether from dead classics such as John Locke or living publicists and agitators such as Tom Paine, whose *Common Sense* (1776) gave a powerful impulse to the American Revolution at just the right moment. The political class stood in the other camp, waging wars with varying degrees of success against both the American and the French revolutionaries. During the decades of ferment, the British oligarchy understood what needed to be done to secure its rule.

The British near-revolution of the 1780s and 1790s gave way to thirty years of conservative buttressing of the system, then to a cautious reformism from above that set the tone for the rest of the century with the electoral Reform Bill of 1832. Things remained similarly (or more) peaceful in a few but not many countries of continental Europe. The revolutionary tendencies of the age recoiled from Russia in particular, leaving Tsarina Catherine II safely in power until her death in 1796. A great peasant rebellion led by Emilian Pugachev on the southeastern margins of the empire, in which several hundred nobles lost their lives, was crushed in 1775. That would be the last revolutionary challenge to the central government for more than a century. It is true that fear of a repetition lingered in the background as a policy factor. But Russia survived the onslaught of Napoleon's Grande Armée in 1812 without becoming infected with the ideas of Western liberalism. In 1825, in an attempt to profit from the unclear situation following the death of Alexander I, a group of noble conspirators staged a putsch to force liberalization, but the "Decembrists" were defeated within a few days and mostly vanished into Siberian exile.

France

The revolutionary turmoil on the Continent did not begin with the storming of the Bastille on July 14, 1789, but went back to the factional fighting of spring 1782 in the republican city-state of Geneva. In the eighteenth century Geneva had experienced several periods of unrest, but the rising of 1782 was the bloodiest of all and precipitated a joint intervention by France, Sardinia, and the Canton of Bern.[43] Of greater consequence, especially for the transnational concatenation of revolutions, were the events in the Netherlands where, as so often, revolution and war were closely interlinked. Once again Britain was one of the belligerents in late 1780, when, after a century of peaceful relations, it attacked the Netherlands on the grounds that Dutch ships had been supplying the rebellious North American colonies from the Caribbean. The short war resulted in military disaster for the Netherlands, unleashing the so-called Patriot Movement. This assertively nationalist initiative, influenced by the ideas of the American Revolution and the Enlightenment, sought to end the musty rule of the stadholder William V (a monarch in fact though not in name) and his clique. Anti-British and pro-French for internal more than external reasons, the Patriots triggered an onslaught from unexpected quarters: when one of their volunteer units arrested the Stadholder's spouse, a sister of the Prussian king Friedrich Wilhelm II, the Prussians sent in an army 25,000 strong with backing from London to free the good lady and to restore the incompetent Prince of Orange to power.[44] The Patriots then disappeared underground for the time being or fled abroad; after a period of harsh reaction, the Dutch ancien régime was swept away by the French invasion of 1795. The key point is that the French public, accustomed to confrontations with Britain and Prussia, regarded the inability of Louis XVI for financial reasons to come to the aid of the Dutch Patriots as a serious blow to the prestige of the French monarchy.

The chief causes of the French Revolution did not lie in foreign policy. Like all such phenomena in the history of revolutions, it was mainly "homemade."[45] But the dynamic of social conflict, the power of radical ideas, or the national will of an increasingly self-confident people cannot alone account for the king's dramatic loss of legitimacy from the mid-1780s on. The explanation of why a (potentially) revolutionary situation passed into the actuality of a revolutionary process must include both the strength of the rebellious dynamic and the weakness of its target of attack. Here begins a line of historical reasoning that considers, along with social tensions and ideological radicalization, the attempts made by the country in question to safeguard its place within the international hierarchy.[46] France had recently (in 1763) lost the struggle for global hegemony with Britain in the Seven Years' War, which, despite London's otherwise quite generous attitude at the peace negotiations in Paris, had driven it out of North America for good and greatly weakened its position in India.

The American Declaration of Independence offered the architects of French foreign policy a chance to take their revenge on the old rival. In 1778, on conditions seemingly favorable to the Americans, the French king and American antiroyalists agreed on a strategically motivated alliance against Britain that involved the first recognition of the rebels by a European power; Spain joined the alliance the following year. This European support helped the Americans out of difficulties at decisive moments in their struggle, especially when the French navy briefly gained control of the North Atlantic in 1781 and cut off British troops in America.

The Treaty of Paris in 1783, a major setback for Britain only twenty years after the triumph of 1763, strengthened the French position in the world. But it was a Pyrrhic victory, given that the price for the victory of the American allies and some rather token successes at sea against the world's premier navy was impending bankruptcy of the French state. Any other crisis—it happened to be France's lamentable failure to assist the Dutch against Prusso-British interference in 1787—would have cast a rude light on this desperate financial situation. It may not have been the deepest cause of the French Revolution, but in terms of *l'histoire événementielle* there was hardly a stronger impulse for the series of challenges that now faced the monarchy. Since the tax system offered no scope for rapid increases in revenue, and since the dynasty was too weak to cancel its debts out of hand, it found itself compelled to consult the notables of the kingdom. Instead of pragmatically clearing up the crisis, however, these now demanded that the consultation process should be formalized through the calling of the Estates-General, a representative body that had last met in 1614.

The result was a spiral of mounting demands on the Crown, which soon merged with other confrontational tendencies: clique struggles at court, unrest among the rural population in the provinces and the urban poor in the capital, conflicts between nobles and nonnobles within the upper classes. From the moment when the government, acting out of weakness, signaled its willingness to introduce reforms, new divisions appeared within an opposition that had originally been directed less against the system of rule as such than against its managerial deficiencies under Louis XVI. Faced with imminent changes and an uncertain future, groups and individuals made every effort to ensure that their own interests were respected, and in this jockeying for position it was not long before the monarchy's incapacity for reform was exposed.

Historians continue to argue about the precise role of foreign and colonial policy in the period from the outbreak of the revolution in 1789 to the beginning of the military contest with various European powers in April 1792.[47] One thing is clear: once the actual and symbolic weakening of France's external position in

1788–89 had crucially contributed to the collapse of the ancien régime, a correction of that position had to be an important objective for the new political forces that took the stage, especially as they expressed themselves in an ever sharper rhetoric of nationalism. The revolutionary wars and the later military expansion under Napoleon was therefore entirely in accordance with the logic of the long-standing global rivalry with Great Britain.

When did the French Revolution begin, and when did it end? There was not a single tumultuous process such as that which marked the American Revolution, stretching from the Stamp Act crisis of 1765 to the great revolutionary step of the Declaration of Independence in 1776. One might date the onset of the terminal crisis of the ancien régime to that same year, 1776, when the French foreign minister Vergennes, brushing aside the warnings of the great Turgot who had fallen from power in May, forced through his fateful policy of intervention in North America.[48] But one might also begin in 1783, when the consequences of that policy were already showing. Revolutionary violence, comparable to the American events of 1765, first broke out in 1789. The revolutionary point of no return was reached on June 17, when the Third Estate of the Estates-General constituted itself as the National Assembly, and when the king and his government lost what power remained in their hands. Already for contemporaries, it was the extraordinary acceleration of unprecedented events, most visible in Versailles and Paris, that gave the French Revolution its novel character. Such space-time compression had seldom happened before, even in North America after 1765.

It is not possible to describe here the further course of the revolution within France—the various stages at which options were closed (a parliamentary monarchy, for example, in summer 1792) and new horizons opened up.[49] When the revolution *ended* was and remains controversial. Its "hot" phase of heightened revolutionary violence began in August 1792 and lasted almost exactly two years until the fall of Robespierre at the end of July 1794. But political conditions became reasonably stable only when the Directorate took over the government in November 1795 and the new Constitution de l'An III was adopted in August of the same year. Did the revolution end with the seizure of power by General Bonaparte on November 9 (18 Brumaire), 1799, with the temporary external peace sealed by the Treaty of Amiens between Britain and France in March 1802, or with the downfall of Napoleon in April 1814? In a world-historical perspective, there is most to be said for the last of these dates. The impact of the French Revolution was slow to unfold, and it was the Napoleonic armies that first carried it out to the wider world, from Egypt to Poland and Spain.

Haiti

In 1804, the year when Bonaparte was crowned Emperor Napoleon, Jean-Jacques Dessalines proclaimed himself Emperor Jacques I in what had previously been France's richest colony. So ended, for once unambiguously, a revolutionary process that had been closely linked to the French and run almost in parallel to it. The revolution in the colony of Saint-Domingue, which occupied the western half of the Caribbean island of Hispaniola and almost coincided with the territory of what is now the Republic of Haiti, should be understood as a direct consequence of the revolution in France. Even before things came to a worldwide ideological civil war between revolutionaries and their enemies, as the Anglo-Irish politician and writer Edmund Burke had predicted, and indeed helped to bring about, through his *Reflections on the Revolution in France*, the events in Paris ignited a revolutionary process in the faraway Caribbean that would, in terms of violence between 1791 and 1804, overshadow anything seen in the American or French Revolution.[50] Since its history is less well known, a brief sketch will be in order here.

The social point of departure in the sugar-producing colony was quite different from that in North America or France. In the 1780s, Saint-Domingue had a prototypical slave society consisting of three classes: a large majority of black slaves (roughly 465,000 in 1789), very many of whom had been born in Africa; a white ruling elite of 31,000 plantation owners, bailiffs, and colonial functionaries; and between them about 28,000 *gens de couleur* with the status of free people, including some who were quite well off and even owned plantations with slaves.[51] Within this triangle three revolutions were acted out simultaneously: (1) a preemptive rebellion by conservative planters against the new antislavery regime in Paris; (2) a veritable uprising by the largest slave population outside the United States and Brazil; and (3) an attempt by the *gens de couleur* to break the dominance of whites in a society shot through with racial discrimination. No other land in the arc of Atlantic revolution had accumulated so much socially explosive material.

What was at stake in Saint-Domingue were not so much constitutional issues or the enforcement of legal principles as a sheer struggle for survival in an extremely brutalized society. Of all the great revolutions of the age, Haiti's was the one that can most clearly be described as social, in both its causes and its results. The American Revolution did not create a completely new type of society or eradicate any of the classes that had made up the colonial order; indeed, there are good reasons to claim that the social change during the period of the so-called "market revolution" (c. 1815 to 1848) went deeper than any that occurred after 1765.[52] The social effects of the French Revolution were more considerable: above all, the abolition of aristocratic privileges, the liberation of the peasantry from feudal constraints, the elimination of the church as a key social factor (with large landholdings, for example), and, mainly during the Napoleonic period, the creation of legal and administrative frameworks for bourgeois-capitalist economic forms. In neither of the two "great revolutions," however, was a whole social system destroyed along with a political order. That did happen in Haiti. The slaves emerged victorious from a long series of massacres and civil wars, and the colonial caste system gave way to an egalitarian society of free African American small farmers.

This drama unfolded in genuinely international context. In France, Enlight-enment champions of universal human rights pushed for liberation of the colo-nial slaves, while the question was posed from the beginning of the revolution as to how colonial Frenchmen and—most controversially—the *gens de couleur* should share in the democratization of French politics. A process of participa-tion got under way in Saint-Domingue as early as February 1790, with elections among the whites for a colonial representative assembly.[53] Earlier still, in Octo-ber 1789, a delegation of the *gens de couleur* had addressed the National Assem-bly in Paris. There was direct interaction between events in France and on the Caribbean island, although communication problems ruled out direct coordi-nation. When three commissioners representing the National Assembly arrived in Saint-Domingue in November 1791 to ensure orderly implementation of the new (though contradictory) policy decided in Paris, they did not yet know that a great slave uprising had broken out that August and been put down only with greatest difficulty.[54]

A symbolic watershed was reached in April 1792, when the National Assem-bly in Paris declared that white citizens, *gens de couleur,* and free blacks should enjoy political equality. This did not yet mean emancipation of the slaves, but it did establish for the first time the basic principle that civil rights did not de-pend on skin color. The various revolutionary groupings in Paris, however, had no intention of allowing their most valuable colony to go its own way. Under the leadership of a former slave, François Dominique Toussaint Louverture (or L'Ouverture, 1743–1803), who had risen in the service of the French government, revolutionary struggle was linked in a complicated way with cautious moves to-ward independence. France might possibly have tolerated an independent Haiti if it had received guarantees that the island would continue to play its role in the transatlantic French mercantile system. Toussaint Louverture, appointed gover-nor of Saint-Domingue in 1797, appears to have realized that a complete eco-nomic break was inadvisable. He also skillfully maneuvered between France and the two counterrevolutionary interventionist powers: Spain (which possessed the other half of Hispaniola) and Britain. In 1798 the British gave up a costly attempt to conquer the island.[55]

Then Napoleon wound up the experiment, rescinding the 1794 decree of the Convention Nationale that had abolished slavery throughout France's colonial possessions; once he had become first consul in April 1802 and signed a peace treaty with Britain, he sent a military expedition to the Caribbean to put an end to Toussaint Louverture's autonomy project. The governor was arrested and died soon afterward in captivity. Yet it proved impossible to reintroduce slavery in Saint-Domingue: the blacks defended themselves and, in an unusually destruc-tive guerrilla campaign, inflicted a crushing defeat on the French army in 1803. On January 1, 1804, the independent state of Haiti was proclaimed. Only in 1825 did France recognize it and effectively abandon any possibility of reconquering it through violence. In the teeth of opposition from the two strongest military powers of the age, Britain and France, the majority of the population on the island had abolished the institution of slavery dating back three hundred years. But revolution and war had also caused devastation on such a scale that it proved extremely difficult to build a liberated and prosperous new society.

Events in Haiti did not trigger a chain reaction. The spectacle of revolutionary self-emancipation would not be repeated in any other slave society in the nineteenth century. In France, the signal from the Caribbean came as a warning against too much pliability on the slavery issue; the country that had proclaimed total emancipation in 1794 did not free the rest of its slaves until 1848—fifteen years after Britain, which had resolutely fought against the revolution, did so. In all slave societies, nowhere more than the Southern states of the United States, paintings were commissioned to hang on walls as a reminder of the apocalyptic "Negro revolt" that would break out if the slaves were offered the least compromise. Until the Civil War, more than half a century after the denouement on Hispaniola, propagandists in the Southern states recalled that the French abolitionists (the Amis des Noirs) had opened the Pandora's box of slave rebellion. For their part, US abolitionists pointed out that only the ending of slavery could prevent the impending evil.[56]

Unlike the revolutions in North America and France, Haiti's did not occur in a society with a pronounced culture of writing and printing. Eyewitness accounts do exist, but not many, and there are only a few clearly formulated programmatic statements. Even some of the goals of Toussaint Louverture, a man with many sides, can only be deduced from his actions. Recent historians have shown great ingenuity in evaluating these sources and added an altogether new facet to the Age of Revolutions.[57] But for a long time the discursive paucity was one reason why Haiti was not taken seriously in histories of revolution; it seemed to emit no universalizable political message over and above a call for the liberation of slaves throughout the world. That is not untrue. But it must also be recognized that the revolution in the French Caribbean shared from the beginning in the Atlantic discourse of liberty. Both the British-American and the French critique of absolutism placed great emphasis on the image of shaking off the yoke of slavery.

Samuel Johnson, the English man of the Enlightenment, already expressed amazement that the loudest calls for freedom came precisely from slave owners.[58] Some of the founding fathers of the United States continued to own slaves (although George Washington set all of his free), and the Constitution of 1787 as well as subsequent amendments remained silent about the issue. Only in Haiti—and nowhere else after it—did a program of racial nondiscrimination, and then of slave emancipation, acquire a direct meaning for people who became active in the revolution. Blacks and coloreds suffering under a rigid system of oppression adopted the ideas, values, and symbols of the French Revolution, trying to find a place for themselves in the new world of "color-blind" citizenship that it had proclaimed in 1794.[59] Hence, the reintroduction of slavery unleashed an apocalyptic liberation war in 1802–3. And the persistence of colonialism everywhere

outside Haiti maintained for another century and a half the contradiction between legal norms of equality and its refusal in actual practice.

Latin America and North America Compared

The intellectual impact of the principles of 1776 and 1789 was unbounded in
time and space.[60] Almost everywhere (perhaps with the exception of Japan), people in all subsequent epochs have appealed to liberty, equality, self-determination,
and human and civil rights. A countercurrent in Western thought, from Edmund
Burke down to the French historian François Furet, reversed the polarity and
identified Jacobin radicalism as the wellspring of "totalitarian democracy" (to
quote Jacob L. Talmon, for whom Rousseau was the arch villain), and more
generally of any form of political fanaticism or fundamentalism. The *immediate*
global effects, in terms of actual interaction, were considerably more limited;
they ended, as we have seen, before the borders of Russia.[61] In China the French
Revolution had no real resonance until 1919, and even then, for good reason, the
anti-imperial struggle of the North American colonies aroused greater interest;
the revolutionary leader Sun Yat-sen (1866–1925) liked to think of himself as the
George Washington of China. In India, some opponents of the British hoped in
vain for French support, while the British for their part cleverly played on fears
of a French invasion as a pretext for the preventive conquest of large parts of the
Subcontinent under Richard Wellesley (brother of Arthur Wellesley, tested in
the Napoleonic wars and named Duke of Wellington in 1814).[62]

The greatest impact of the French Revolution outside or on the fringes of the
Atlantic area came about through Napoleonic military expansion in the Middle
East, beginning with the invasion of Egypt in 1798. The occupation there broke
the centuries-old power of the Mamluks and created space for new individuals and groups to seize power after the French withdrew in 1802. The Ottoman
Empire was a proven and once again important partner for the British, offering
security in the eastern Mediterranean. Sultan Selim III (r. 1789–1807), who had
coincidentally come to the throne in the epoch-making year of the French Revolution, failed in his attempt to curb the influence of the conservative military janissaries and to overcome their blanket opposition to reforms; this was achieved
only two reigns later, under Mahmud II, in 1826. Nevertheless, under the pressure of intense diplomatic and military activities, Selim embarked on modernization of the Ottoman army. Iran followed soon after with a similar program.
But nowhere in the Islamic world, or anywhere else in Asia or Africa, did the
French Revolution trigger independent revolutionary movements from below.[63]

How does Latin America fit into this picture?[64] It was the fourth of the regions bordering the Atlantic to be drawn into the Age of Revolution, but its
actual involvement varied from area to area, and only detailed studies of individual regions and cities seem to yield an adequate picture.[65] In North America the
colonies that would later form Canada stayed loyal to the British Crown. The
slave colonies of the Caribbean remained quieter than Saint-Domingue, and the

course of events varied there even among the French islands. In contrast, one of the most striking features of Hispanic America (Brazil went its own way under an offshoot of the Portuguese Crown) was the complete collapse of the Spanish colonial empire on the mainland. Within the space of a few years, a huge entity broke up into a mosaic of independent republics. The Spanish nation-state itself was in many respects an outcome of this disintegration—a process, best described as the "independence revolutions" (in the plural), which was the last of the great transformations in the Atlantic space. Its dating to the years between 1810 and 1826 is fairly uncontroversial.[66]

All three of the major revolutions may here serve as a point of reference. Haiti inspired fear wherever slavery played a large role, and especially where free colored people (paradoxically known in Hispanic America as *pardos* or "light browns") began to develop political goals of their own. Though less an example than a warning, Haiti did serve as a safe haven for rebels against Spain. As for the French Revolution, it was rather limited as a model since the leaders of the Hispanic American independence revolutions were mostly Creoles, that is, whites of Spanish descent born in the New World. Typically they belonged to the affluent upper strata of society, as landowners or members of the urban patriciate or both. However sympathetic they might be to the early liberal aims of the French Revolution, such people viewed Jacobin radicalism as a threat and were chary and suspicious about the (sometimes indispensable) arming of the popular masses.

The potential for large-scale protest action had already been demonstrated in the uprising of 1780–82 led by José Gabriél Condorcanqui, the self-styled Inca Túpac Amaru II. A few years after the Pugachev Rebellion in Russia, this in some ways similar event at the other end of the world rested on a broad but loose coalition of diverse forces and tapped the sources of a self-assertive popular culture. It, too, was directed against (and brutally suppressed by) the Spanish rulers, but its motives did not entirely coincide with the Creole oligarchic striving for autonomy. The size of the revolt, best gauged from the number of casualties, was certainly impressive: it probably claimed the lives of some 100,000 Indians and 10,000 Spanish.[67] So, for the "liberators" of Latin America, Jacobinism and the *levée en masse* held few attractions. Nor could they rely on revolutionary support from France, since the decisive years of the freedom struggle were during the Restoration period following the end of the Napoleonic Empire.

The link between the transformations in France and Latin America was at the level of power politics more than revolutionary substance. Moreover, we must go back to the 1760s, where the roots of both the North American and the Latin American revolutions lie. In that decade, for related but distinct reasons, the British and Spanish states simultaneously tried to tighten the leash on their American possessions by strengthening and reforming the apparatus of colonial rule, so that the colonies would be more economically useful to the mother country. Britain under its new king, George III, failed woefully in this ambition after just a few years. Spain under Carlos III (r. 1759–88) was at first more successful, or anyway encountered much less resistance from the colonists. The Spanish system of rule in the Americas had always been more homogeneous and centralist, and so it was easier for capable administrators to implement reforms; and the South American Creoles were less densely woven into the antiauthoritarian discourse of the Enlightenment, and less accustomed to expressing their will in representative bodies. For these and many other reasons, the Spanish colonial system did not break down in the same way that the British system did in the third quarter of the eighteenth century. In fact, it managed to hold on until Napoleon's invasion of Spain in 1808 brought down the Bourbon Monarchy itself.

Whereas the North American revolt targeted an imperial government that was increasingly seen as unjust and despotic, the critical junctures in Hispanic America arrived at a time when there was a vacuum at the center of the empire.[68] Two tendencies then came to the fore: on the one hand, local Creole patriotisms were much more prominent here than particular colonial identities in British North America; but on the other hand, there was a will to retain a looser political association with Spain, albeit within a new liberal-constitutional order. In a sense this was the mirror image of past development in North America. The "Creoles" (as they may be safely called) in the thirteen rebel colonies of North America still felt largely British at the beginning of the conflict, and it took quite a while for many of them to replace this solid identity with a still rather shaky American one.[69] Their resistance was therefore directed more against the real and symbolic figure of the king than against the boundless claims of the Parliament in London, which imposed arbitrary taxes on Americans without offering them more than a hollow pretense of representation.

In the Spanish case, the formation of a separate identity was more advanced. Yet, with the reactionary King Ferdinand VII a prisoner of Napoleon, the Hispanic American Creoles placed high hopes in the nonroyal government in the unoccupied part of Spain. The core of this was the Cortes that met at Cadiz in September 1810, Spain's first modern national assembly, which from the beginning was thought of as representing the whole Hispanic world, including the colonies.[70] The Cortes, which obviously had few Spanish Americans among its members, proved as stubborn on some issues (e.g., trade policy) as the British Parliament had a few decades before. An imperial federation, though perfectly conceivable in theory, could not be realized outside the framework of absolutism, and the Cortes also omitted to abolish slavery or the slave trade and, more generally, to take a stand on the problems of multiethnicity in the Americas. Nonetheless, Spain's early (and for the time, thoroughgoing) experiment with the rule of law habituated the Creoles both to the practice of a written constitution (the Spanish Constitution of 1812 became the formal model for proliferation in nineteenth-century Latin America) and to extensive male participation in politics free from such restrictions as a property-based suffrage.

Emancipation was much less of a linear process than in North America. The region was larger, the logistics more difficult, the town-country opposition sharper, and royalism stronger; and Creole divisions were often so great that they came close to civil war. Spatially, various armies and militias fought a number of independence wars that were only loosely related to one another. Temporally, two periods of war came in succession.[71] First, the whole new departure on both sides of the Atlantic was nullified overnight in May 1814 with the return of the neo-absolutist Ferdinand VII. It was only in the military resistance to a subsequent (and initially successful) attempt at reconquest that the liberation struggle led by men such as Simón Bolívar, José de San Martín, and Bernardo O'Higgins reached its heroic climax.[72] In 1816 it looked as if Spain had events under control, except in Argentina. The rebels were forced into the defensive in many parts of the continent; the imperial reaction was delivering its prisoners over to tribunals. But then the royal regime evinced its own inherent weaknesses and inconsistencies and squandered the last vestiges of loyalty and legitimacy it might have possessed.[73] A second phase gradually got under way, in which caudillos—warlords whose power rested on the war booty they made available to their armed bands and civilian followers, and who had little time for state institutions—already began to play an ominous role. All in all, the revolutionary process was socially far more multilayered than in North America, where it did not include peasant rebellions and popular uprisings within the elite revolution—uprisings which, as in rural Mexico, often served to defend a way of life under threat rather than to oppose the Spanish presence per se.[74] The last series of military victories in countries south of New Spain/Mexico had to do also with Spanish weakness, since there was little enthusiasm in the army for a *reconquista*, and without the army's presence in Europe the liberals would not have been able to force King Ferdinand to restore the Constitution in 1820. These new upheavals in Spain delayed the sending of fresh expeditionary troops. The recourse of the Spanish to French counterinsurgency methods, of which they had only recently been on the receiving end, shows once again a learning curve at work in the revolutionary Atlantic.

Finally, the international context. Unlike the North American insurgents after 1778, the Hispanic American freedom fighters lacked military support from outside, even from the United States. No other great power intervened directly in events, as had briefly happened in Haiti. The Royal Navy flung a protective cover over the Atlantic, but the decisive military confrontations took place entirely between Creoles and representatives of the restored Spanish monarchy. On the other hand, it should not be overlooked that at the beginning, in 1810, the fear that France might seize the Spanish colonies played a major role; no one in Latin America was eager to become a Napoleonic subject once the Spanish monarchy ceased to exist. In later phases, "private" backing was not unimportant. British and Irish soldiers and volunteers fought in various theaters (there were more than 5,300 of them in South America between 1817 and 1822),[75] US governments tolerated the action of American freebooters against Spanish ships, and British merchants provided some financial support, seeing it as a good long-term investment to open up new markets.

The independence revolutions throughout the Americas had—or at least tended to have—two fundamental consequences: subjects became citizens, and the structure of the old hierarchical societies started to totter.[76] However, colonial plurality gave way to different political landscapes: in Hispanic America sovereign nation-states brought with them an even greater diversity; in North America a federal state had a basic dynamic of territorial expansion to the west and south, at the expense of Mexico and Spanish civilization in general (and culminating with the Spanish-American war of 1898). In both hemispheres, there continued to be a large nonrevolutionary state: here the Empire (from 1889, the Republic) of Brazil, there the Dominion of Canada within the British Empire. Also in both, political revolution did not immediately result in stability, although the conditions for it were more favorable in the northern continent because the War of Independence was not at the same time a civil war, and because there was no equivalent to the *pardos*, that large stratum of free colored people wooed at times by both republicans and monarchists.[77] In North America the dividing line with Indians and blacks was clear cut: national politics remained white politics. In South America, where the colonial state had translated shades of skin color into legal status, the lines of conflict continued to be more complex. In the northern hemisphere a clearer balance persisted between town and country, whereas the period of wars in the southern hemisphere led to a "ruralization of power."[78] Over the following decades, the North American frontier promoted a certain democratization of landownership. In South America, by contrast, landowning oligarchies imprinted their stamp on the political system more powerfully than agrarian forces in the United States had been able to do at the height of their influence in the Southern states before the Civil War.

One of the great achievements of the early United States, not repeated farther south, was the avoidance of militarization and militarism. The nation in arms of the revolutionary period never became a military dictatorship; independent caudillos did not acquire any significance. Unlike South America and parts of Europe, North America did not evolve into a land of coups d'état.[79] Many countries of Hispanic America did not know internal peace until the 1860s or even 1870s, in the wake of their greater integration into the world economy.[80] If one were to define something like a peak period of political stability in Central and South America, then it would have to be the three decades between 1880 and the onset of the Mexican revolution in 1910.

As to the United States, its postrevolutionary stabilization really began with the election in 1800 of the third president, Thomas Jefferson, and was already well advanced by the time Latin America was embarking on its independence struggle.[81] Much of the consolidation was deceptive or provisional, however. Two questions in particular were unanswered: how slave society and the quite different Northern capitalism based on free wage labor could coexist within one and the same republic; and how new states could be integrated into the republic without upsetting the delicate constitutional balance. The outbreak of the Civil

War in 1861 did not come as a complete surprise, and in retrospect it seems much more "unavoidable" than the First World War, for example. A number of problems from the age of the revolution had been left unresolved. Only because the Founding Fathers had neglected to clarify the issue of slavery was it possible for anyone seriously to demand in the late 1850s that the African slave trade (banned since 1807) should be resumed, or for a level-headed politician like Abraham Lincoln to convince himself that the South was seeking to impose slavery in the free states of the North.[82] The Civil War was thus in a sense the last spinoff from the Revolutionary War. If one is not too afraid of belaboring the term, one might toy with the idea of a hundred-year cycle of revolutionary unrest in North America: from the Stamp Act crisis of 1765 to the defeat of the Confederacy in 1865.

The end of the Hispanic American independence revolutions was soon followed by the European revolutions of 1830–31, their Janus face turned to both past and future. They, too, should be classified as part of—and the closing of—the Age of Revolution. Triggered by unrest among Parisian artisans in late July 1830, revolutionary conditions prevailed in France, the southern Netherlands (which would emerge from these events as the autonomous state of Belgium), Italy, Poland, and some states of the German League (especially Kurhessen, Saxony, and Hanover). The results were rather modest. The restorationist tendency that had gained the upper hand in Europe after 1815 was weakened here and there, but politically defeated only in France—and even there the main social forces that increased their political room for maneuver, whether one calls them "notables" or "liberal bourgeoisie," had formed the core of the French elite even before the July Revolution.[83] What occurred in 1830 was more a political revolution than a social one. It did link up with 1789–91, inasmuch as it evoked the original revolutionary ideas of constitutionality and harked back strongly to the rhetoric and symbolism of the Great Revolution in its pre-Jacobin phase. But its heroic imagery of the urban barricade cannot hide the fact that some forms of rural protest—only loosely connected with events in the cities—were, to say the least, still distinctly "premodern."[84]

Transatlantic Integration

The Atlantic revolutions shared a new basic experience that debarred any return to prerevolutionary conditions: the ongoing politicization of broad sections of the population. Everywhere politics ceased to be merely elite politics. Some of this revolutionary legacy nearly always endured, even if the cooling-down period evolved in very different directions.[85] The most successful channeling into representative institutions took place in the United States, albeit with the exclusion of the nonwhite population. Where such an attempt at democratic reconstruction failed, as in France during the interlude of the Directorate (1795–99) and many countries of Latin America, new authoritarian systems could not dispense with a degree of popular legitimation, if only by acclamation. "Bonapartism" did not mean a return to the ancien régime. Even the Bourbon Restoration after 1814 accepted much from the post-1789 period, codifying some of its ideas in the Charte Constitutionnelle, for example, and taking on board the new aristocracy that Napoleon had created out of his generals and minions.[86] Nowhere outside Spain, Italy, and the German principality Hesse-Kassel did the forces of reaction completely erase the traces of revolution. Napoleon himself, a great institution builder, saw clearly that pure charisma was incapable of sustaining a postrevolutionary order. Bolívar too understood this and, despite a few dictatorial temptations in his years of triumph, he fought indefatigably for the rule of law and constraints on personal power. Yet he could not prevent his Venezuelan homeland and others like it from sliding into caudillismo over the decades.[87] Under such conditions, mass politics was reduced to keeping a narrow clientele happy.

The Atlantic revolutions arose out of a set of relations that had developed on both sides of the ocean since the time of Columbus. Five levels of integration overlapped:

1. *administrative* integration within the great empires of Spain, England/ Britain, and France, as well as the smaller ones of Portugal and the Netherlands
2. *demographic* integration through emigration to the New World, but also through reverse migration, especially of colonial personnel
3. *commercial* integration—from the fur trade in North America to the Angola-Brazil slave trade in South America—organized under the competitive rules of a national mercantilism that was ever harder to enforce and that was disturbed at first (up to around 1730) by endemic piracy; this gave rise to something like a pan-Atlantic consumer culture (the embryo of today's Western "consumerism"), whose interruption by politically motivated boycotts became a weapon in international relations[88]
4. *cultural* integration in many different shapes, from the transfer of West African lifestyles to the spread of performative practices right across the entire region to the modified reproduction of European architectural styles[89]
5. *normative* integration on the basis of common or similar normative foundations of "Atlantic civilization," borne and disseminated by growing numbers of books, pamphlets, and magazines (in 1828 the English essayist and literary critic William Hazlitt already described the French Revolution as a late effect of the invention of the printing press)[90]

This fifth point is of special importance for an understanding of the Atlantic revolutions, although it is not sufficient to explain political action as motivated by ideas alone and without reference to underlying interests. From the point of view of the history of ideas, all Atlantic revolutions were children of the Enlightenment. The Enlightenment was of European origin, and its effects on the other side of the ocean must be described first and foremost as a vast process of receptionand adoption. From the 1760s some American voices, heard also across the Atlantic, responded with indignation to European authors (such as the French naturalist Buffon or, at a later date, the German philosopher Hegel) who had spoken dismissively of nature and culture in the New World; among the most prominent were Benjamin Franklin, Thomas Jefferson, the authors of the *Federalist Papers* (1787–88), and the Mexican theologian Fray Servando Teresa de Mier.[91] Simón Bolívar too—Latin America's most important political thinker of the age, along with the versatile and for a long time London-based scholar Andrés Bello[92]—repeatedly insisted that the program of the Enlightenment should not be transferred without modification to the Americas. On this he could invoke Montesquieu, for whom the laws of a country always had to be adapted to its particular circumstances.

Various cores and peripheries took shape within the Atlantic Enlightenment as a whole. In comparison with France or Scotland, even Spain in the anticlerical reform period of Carlos III was an intellectual sideshow. It was a sign of the times, however, that people looked beyond the cultural boundaries within and around Europe. Britons and North American colonists, though often at odds religiously and otherwise, shared the same legal tradition and the same beliefs in individuality and personal safeguards.[93] Numerous pamphlets, and above all the Declaration of Independence, showed that John Locke's contractual theory of government, Algernon Sidney's doctrine of legitimate resistance, and the theories of Scottish moral philosophers such as Francis Hutcheson and Adam Ferguson were well known in North America.[94] Thomas Paine, a trained corset maker and self-taught philosopher, who first arrived in the New World in November 1774 and became one of the most influential journalists of all time, distilled British radical thought into his potent pamphlet of 1777, *Common Sense*; it was the product of an Atlantic radicalism that would find even more striking expression in his later work, *The Rights of Man* (1791–92).

Compared with the actual results of "enlightened absolutism" in Europe, the new United States embodied an advance of enlightenment in the real world. If there were any philosopher-kings at all in this age, then they were to be found— even more than in Frederick II's Prussia or Joseph II's Austria—in Napoleonic France or the America of George Washington's first three successors as president: John Adams, Thomas Jefferson, and James Madison. English-speaking America was also attentive to French authors, particularly Montesquieu, Rousseau, and the sharp critic of colonialism Abbé Raynal (whose name Denis Diderot sometimes used for his writing). Latin Americans also made the acquaintance of these *philosophes* early on. Simón Bolívar, a young man from a wealthy family in Caracas, read their works as well as those of Hobbes and Hume, Helvétius, and Holbach, and he was probably not altogether untypical.[95] In Mexico City in the 1790s, everyone from the viceroy on down studied what the critical minds of Europe had to say—without immediately putting it into practice.[96] More generally the zeitgeist, with its faith in progress, gripped not only intellectuals but also parts of the business world on both sides of the Atlantic.[97] For many Americans a visit to London, politically conservative but the world center of economic modernity, was therefore at least as exciting as a firsthand impression of the mood in revolutionary Paris.

Revolution is not a dinner party, wrote Mao Zedong—and he knew about such things—in 1927. So much is true of the Atlantic revolutions; none was as peaceful as that which took place in 1989–91 from the Elbe to the Gobi desert. The victims of the Terror of 1793–94 plus the civil war of 1793–96 in the Vendée, estimated at a minimum of 260,000 for the whole of France,[98] must be seen in the perspective of all those who died in the European wars between 1792 and 1815 (including the reign of terror on all sides in post-1808 Spain), the hundreds of thousands killed in Latin America from the Túpac Amaru rising of 1780 to the end of the liberation struggles and civil wars, sometimes waged as total wars of annihilation,[99] and all those who lost their lives in the worst revolutionary cauldron of the age, in Saint-Domingue/Haiti, including tens of thousands of ordinary French and British soldiers, most of whom died of tropical diseases. With justice the revolution of Thomas Jefferson and George Washington is favorably compared to that of Maximilien Robespierre; there was no American equivalent of the massacre of alleged traitors in France. But it should not be forgotten that the American war of independence, from 1775 to 1781, involved a mobilization on Britain's side greater than in any of its previous conflicts, making it in a sense the first modern war, and that it claimed some 25,000 lives among the rebel troops alone.[100] The war generated more refugees and emigrés than the whole of the French Revolution.[101] But it did not produce massacres of the civilian population—unlike the Russian-Ottoman war, for example, in which thousands of Turks were killed in one afternoon during the capture of the fortress of Ochakov (Özi) in 1789. By comparison, the second quarter of the nineteenth century was an innocuous period in world history, until the great Taiping bloodbath began in China in 1850–51.

Britain occupied a unique place in the Atlantic arena of revolution, being the strongest military power since at least 1763. The attempt to bring its willful colonials to heel was what originally unleashed the chain reaction of revolutions. Britain was involved everywhere: it waged war against all the revolutions of the age except in Latin America, and even there at least one early British military action, the occupation of Buenos Aires in June 1806, had far-reaching consequences. Yet the British political system survived throughout, unshaken by social protest and subversion in either the countryside or the new cities of the Industrial Revolution, achieving between 1775 and 1815 the largest military and economic mobilization before the First World War, and operating a leadership selection that brought to power uncommonly able politicians such as William Pitt the Younger (prime minister almost without a break from 1783 to 1806 and the most dangerous of all Napoleon's adversaries). Was Britain, for all its rapid social-economic change, therefore a pole of conservative quiescence in a world of upheavals?

The United Kingdom participated in the European revolutionary movement of 1830. Between summer 1830, when news of the July Revolution in France came hard on the heels of the death of King George IV, and June 1832, when Parliament finally passed the package of reform laws amid the most dramatic tension, the country lived through its greatest internal political crisis of the nineteenth century. Its vulnerability to revolution peaked neither in the 1790s nor in 1848, but at a point fifteen years after the end of a conflict that had lasted more than two decades. Uncontrolled aftereffects of the Napoleonic Wars came together with early industrialization to raise dissatisfaction with the prevailing order to extreme heights. Between 1830 and 1832 disturbances broke out in large parts of southern and eastern England and Wales; the port city of Bristol suffered considerable destruction; Nottingham Castle was burned down; both workers and the middle classes formed themselves into guards, militias, and unions. If the Duke of Wellington, the leading Conservative politician, had sought with the backing of a reactionary king to face down the public mood in spring 1832 (in the style of Prince Polignac two years earlier in France), then the House of Hanover might well have gone the way of the French Bourbons.[102] In the end, however, the duke helped the reform-minded Whig prime minister, Charles Grey (Second Earl Grey), to put together a majority in favor of the Reform Bill.

More important than the content of this legislation—which cautiously widened the male franchise and improved the parliamentary representation of the growing industrial cities—was the fact that it was adopted at all.[103] Reform from above preempted revolution from below. This added a new policy recipe for stability, while at the same time the conservative oligarchic regime for which Pitt had stood was replaced with a greater cross-party willingness to listen to the mood of the country, even of those still barred from elections. This was not enough for some. Disappointment with the limits of the Reform gave rise to the intellectually fertile Chartist movement. It failed politically in 1848 because it neither took the leap into violent revolution nor found sufficient reformist allies among the middle classes.

Another kind of British revolution had meanwhile had its first success in 1807, when a huge civil movement against the slave trade had secured the outlawing by Parliament of that monumental crime. In 1834 followed the suppression of slavery throughout the British Empire. This amounted to a revolution in morals and people's sense of justice, radically spurning an institution that had been taken for granted in Europe for centuries and regarded as conducive to its various national interests. The origins of this distinctively British revolution, which may be dated to 1787, lay with small numbers of religious activists (mostly Quakers) and humanitarian radicals. Its most tenacious and successful organizer was an Anglican priest, Thomas Clarkson, and its most prominent spokesman in Parliament the evangelical gentleman-politician William Wilberforce. At its height, abolitionism was a countrywide mass movement using a wide range of nonviolent techniques. It was the first broadly based protest movement in

Europe in which noble renegades played scarcely any role and the main leaders were businessmen (such as the pottery manufacturer Joseph Wedgwood).[104] Although abolitionism did not destroy the political system of a territorial state, it swept away a form of bondage and accompanying laws and ideology that had been part of the bedrock of the early modern Atlantic world.[105]

The revolutions did not impact on one another only through books and abstract discourse. Future revolutionaries learned on the spot. Between 1776 and 1785 Benjamin Franklin, the best-known American of the age by virtue of his scientific experiments and the incredibly broad range of his activities, embodied the new America as an envoy in Paris. The Marquis de Lafayette, the "hero of two worlds" who fought alongside many other European volunteers in the North American war of independence and was deeply marked by US constitutional principles, friendship with George Washington, and personal instruction from Thomas Jefferson, went on to become one of the leading moderate politicians in the early phase of the French Revolution. Soon he had to be told that France would not "slavishly" follow the American model. Fleeing abroad, he saw the inside of Prussian and Austrian dungeons as an allegedly dangerous radical and finally became for many—such as the young Heinrich Heine, who met him as a sprightly old man in Paris—the embodiment of the pure ideals of the revolution.[106]

Of course, individual revolutions went their own way. The French, for example, placed much less emphasis on checks and balances between different parts of the body politic, and much more on the articulation of an undivided national will in a Rousseauan sense. Here the North Americans were better disciples of Montesquieu, whose French compatriots lastingly embraced liberal democracy only in the 1870s. Yet the constitution adopted by the Directorate in 1795 was closer than its revolutionary predecessors to American political ideas, and General Bonaparte was celebrated by many as a second George Washington.[107] Moreover, the two revolutions continued to be mirrored in each other, until a wide mental gulf gradually opened between America and Europe in the nineteenth century.

The late Restoration period still showed signs of the experience of the revolution, many of its figures being directly linked to it by their biography. The age of the French Revolution, from the Tennis Court Oath to Waterloo, had lasted just twenty-six years. For someone like Talleyrand, who served each of the French regimes in a high capacity, they coincided with the active years of the middle of his life; others such as Goethe or Hegel followed the period as observers from beginning to end. Alexander von Humboldt heard Edmund Burke speak in London even before the revolution, held scientific discussions with Thomas Jefferson, was personally introduced to Napoleon, solicited sympathy in Europe for the Latin American independence struggle, and in March 1848 attended revolutionary gatherings in Berlin as an octogenarian.[108]

The Age of the Revolutions has presented itself as a great paradox since economic historians began to date industrialization to later in the nineteenth century. The plausible thesis of a dual revolution—political in France, industrial in

England—which was popularized by Eric Hobsbawm, is no longer sustainable. *Political* modernity begins with the great texts of the revolutionary age: above all, the American Declaration of Independence (1776), the US Constitution (1787), the Declaration of Human and Civil Rights (1789), the French decree on the abolition of slavery in the colonies (1794), and Bolívar's speech in Angostura (1819). These come from a time when even in Britain the Industrial Revolution was scarcely having a revolutionary impact. The dynamic of the Atlantic Revolution was not driven by the new social conflicts associated with industrialization. If there was in it anything "bourgeois," it had nothing to do with industry.

3 The Great Turbulence in Midcentury

There would be no second age of revolutions—if we leave aside the stormy years from 1917 to 1923 when revolutions and uprisings shook Russia, Germany, Ireland, Egypt, Spain, Korea, and China, and a number of new states came into being in Europe and the Middle East. In midcentury there were large-scale outbreaks of collective violence in various parts of the world; the most important were the revolutions of 1848–49 in Europe, the Taiping Revolution in China (1850–64), the Great Rebellion or the "Mutiny" in India (1857–58), and the Civil War in the United States (1861–65).[109] The fact that all these happened within a period of seventeen years suggests a revolutionary cluster; it looks as if the world as a whole was passing through a severe crisis. One might assume that, as global interconnections had been increasing since the age of the Atlantic revolutions, the revolutionary events in different parts of the world were more intertwined. Such was not the case. The midcentury cluster lacked the spatial unity of the revolutionary Atlantic. Each of these revolutions remained limited to part of a continent, although they were not national events: the 1848 Revolution immediately jumped across national boundaries; India and China were not nation-states at the time; and in the United States a precarious national unity was being openly called into question. The individual crises must therefore first be described separately from one another.

1848–49 in Europe

The course of the European revolutions of 1848–49 repeated the pattern of the French July Revolution of 1830: protests leaping between different political milieux with uncommon speed, only this time in many parts of Europe.[110] Historians used to like the naturalistic image of the wildfire, which of itself obviously explains nothing and cannot substitute for a more precise investigation of dispersion mechanisms. Anyway, this time the revolution did not spread abroad as in 1792 through the armies of the revolutionary state. Reports, often only rumors, summoned forth revolutionary action as a response to objective problems in each of the countries concerned. This happened so quickly because an outbreak of revolution had been widely expected since autumn 1847 and because a repertoire of rhetoric, dramaturgy, and actions—with barricades, for example, as the emblem of urban insurrectionary warfare—had been present since 1789, and reactivated by 1830, in the political culture of western and southern Europe. The forces of the established order also now thought they knew how a revolution "functioned," and they made their preparations accordingly. To be sure, the sensitivity of the different revolutionary centers to one another did not last very long; each one became localized, acquiring a distinctive power constellation and ideological coloration and following its course without any significant mutual assistance. Yet they remained parts of a synchronous epochal context within which they can be compared with one another.[111] In 1848–49 the individual revolutions did not flow into a single great European Revolution, but to a degree last seen in the Napoleonic Wars, Europe did become a "communications space," a "wide-ranging arena of action."[112] Particular theaters and events, though often only imbued with local meaning, became embedded in European contexts and horizons; political ideas, myths, and heroic images circulated all over the continent.[113]

The active participants in the events were Switzerland (which in 1847 went through a veritable civil war between Protestant and Catholic cantons), France, the German and Italian states, the whole of the multinational Habsburg Monarchy, and Balkan borderlands of the Ottoman Empire. The Netherlands, Belgium, and Scandinavia were affected insofar as ongoing reform processes were accelerated. All in all, this was the most violent and the most extensive (numerically and geographically) political movement in nineteenth-century Europe, often mobilizing large sections of the population. It is advisable to differentiate four components: peasant protests, civil rights movements, actions by the urban lower classes, and national-revolutionary movements sometimes involving a broad social alliance.[114] Farmers who illegally took wood from the forest did not always have a lot in common with urban notables who turned festive banquets into political forums.[115] The example of wood theft is not chosen at random: it shows that in 1848 nearly all latent conflicts became virulent. Access to forest resources was an especially heated issue: "Everywhere that there were forests, there were forest riots."[116] And Europe had plenty of forests.

If we look at events from the vantage of spring 1848, when power seemed to lie in the streets in a number of countries, we may find it surprising that all the revolutions ended in failure, in the sense that no group of actors lastingly imposed their objectives. But it is important not to rush to a blanket judgment, since in reality the "failures" varied in degree and form. A *socially* differentiated picture shows the peasantry winning the most: it shook off its servile status in the Habsburg Empire, where the earlier "emancipation" had failed to leave any mark, as well as in some of the German states. Where the legal situation of the peasantry had already improved, the process of emancipation accelerated toward completion; redemption payments, for instance, were reduced to realistic proportions.[117] Achievement of these aims meant that the peasantry generally lost interest in revolution; its discontent had anyway been directed only against the landowners, not the late absolutist monarchs whose power the civil rights movements sought to restrict. The lower levels of the peasantry, who received no concessions from the government, were among the losers from the revolution, like the urban poor who bore the full brunt of the repression. But it is still striking how many winners were to be found among those who failed on the surface. The nobility (if it was not, as in France, already emasculated) largely defended its position in society; the state bureaucracies learned much about how to handle a politicized population and the media; and the economic bourgeoisie found growing official understanding at least for its business interests.[118]

A *regionally* differentiated picture would show that the revolution failed less drastically in France than elsewhere. The last remnants of Legitimist monarchism were swept away, and a republic came into being for the first time since 1799. When Louis Bonaparte staged his coup d'état three years later, going on to claim his uncle's legacy as Emperor Napoleon III, this was in no sense the restoration of an earlier state of affairs. The Second Empire was a modernized version of the original Bonapartism, in many ways a synthesis of all the tendencies in French political culture since the end of the Terror in 1794.[119] The new regime began life with fierce repression of all its opponents, but over time it proved quite open to liberalization and established a framework within which France's bourgeois-capitalist society could peacefully flourish. In Hungary, by contrast, where national autonomy was at the center of all demands, the revolutionaries suffered a spectacular defeat. Since, alone in Europe, they armed themselves adequately, the conflict was bound to escalate into a war with the intractable imperial power, Austria, which lasted until the insurgents formally surrendered in August 1849. A tide of vengeance then swept over Hungary. All traces of the revolution would be erased, with the indulgent understanding of many Hungarian magnates. Army officers were tried before military courts, and the grim punishment of forced labor in chains (the Austrian equivalent of banishment to an island in the tropics) was meted out on a large scale. If losses on the Austrian side are included, approximately 100,000 soldiers alone lost their lives in Hungary in 1848–49—to which should be added thousands of peasants killed in rural conflicts among the nationalities of the Danube region.[120]

Finally, a longer time frame clearly places the "failure" of the 1848 Revolution in a different perspective. We can only speculate what its success would have produced: a recast republic in France, doubtless with unresolved contradictions; in the case of victory for the Italian and Hungarian rebels, most probably the breakup of the Habsburg Empire as a multinational state; a shortening of Germany's road to constitutional government and wider political participation. Tempting as such counterfactual speculation may be, the reality was this: The conservative oligarchies, having survived the storm, turned to neo-absolutist policies that left no doubt as to where the power lay (including the now stronger military power), but this does not mean they were set against any compromise. Napoleon III's quest for popular acclamation, Austria's conciliation of the Hungarian upper classes in the constitutional "settlement" of 1867 (unthinkable without its military defeat a year before at the hands of Prussia), and the granting of universal manhood suffrage in the 1871 Constitution of the German Reich were three examples, highly dissimilar no doubt, of a willingness to seek new solutions in a political middle ground. A second long-term effect, also irreversible, was that many social groups learned to mold the experience of politicization, which often came as a surprise even to themselves, into solid institutional forms. The years of the European revolutions therefore mark a turning point in the development "of traditional forms of collective violence into the organized assertion of interests."[121]

The revolutions of 1848 were not a global event. Here lies their supreme paradox: the greatest European revolutionary movement between 1789 and 1917 had an extremely limited impact around the world; it was not seen elsewhere as a beacon; unlike the French Revolution, it did not formulate any new universal principles. In 1848 continental Europe had fewer and less dense permanent contacts with the rest of the world than it had had fifty years earlier, or would have fifty years later. The paths of transmission were therefore rare and narrow, the most important being emigration across the Atlantic. The United States happily took in "forty-eighters" as refugees, seeing this as confirmation of its progressive superiority. Lajos Kossuth arrived there in late 1851 via the Ottoman Empire and was welcomed as a hero. In 1867 Emperor Franz Joseph granted him a pardon, but he remained until his death in exile in northern Italy. Carl Schurz, a participant in the Palatinate-Baden uprising of 1849, took the emigrant's road to America, becoming one of the most influential leaders of the newly founded Republican Party, a Civil War general, a senator from 1869 on, and secretary of the interior from 1877 to 1881. Gustav von Struve, somewhat less adaptable and successful than Schurz, was militarily active in both South Baden and the Shenandoah Valley, proud to have taken part in two great struggles for human freedom. A smaller revolutionary fish, the Saxon kapellmeister Richard Wagner, did not show himself in Germany again until 1862.[122] It is hard to judge the extent to which politics lay behind the growing midcentury emigration from central Europe, but there can be no doubt that revolution triggered a considerable brain drain to the more liberal countries of Europe and the New World, and that many emigrants took their political ideals with them.[123]

In 1848–49 Britain and Russia, the two powers at opposite ends of Europe and its most important links to other continents, were less caught up in the revolutionary events than in the earlier age when the heir and executor of the French Revolution had marched on Moscow with his huge army. The year passed peacefully in oppressive Russia, while in England the Chartist movement against arbitrary government and for the defense of ancestral rights, having peaked in 1842, flared up again in 1848 but again failed to produce results. The ideas and language of radicalism did not completely disappear, however, from the national culture. Nonradical currents in the public rejoiced over the superior performance of British institutions.

The most turbulent nations in the two empires—the Irish and the Poles—remained quiescent by the Western European standards of the time, but several hundred Irish rebels were deported as convicts to the colonies. This offers a first pointer to imperial interconnections.[124] As so often in the past, London used the soft option of "transportation" to clear away troublemakers. But many people in the colonies were tired of seeing their country used as a penal dumping ground; in 1848–49 Australia, New Zealand, and South Africa witnessed demonstrations thousands strong against the convict ships. Thus, although the British state was able to keep Chartists and Irish rebels at arm's length, it triggered unwelcome reactions in other parts of the Empire.

Finances were another link between world empire and the avoidance of revolution at home. Those in power in London saw that it was necessary at all costs not to raise the fiscal burden on the middle classes. A tax hike in the colonies (as in 1848 in Ceylon/Sri Lanka) risked the kind of protests familiar in Europe, which could be quelled only through repression. Where the colonial state reduced its personnel, as it did in Canada, settlers could obviously fill any gaps. And one of the reasons for the annexation of the Punjab in 1848–49 was that it would pacify a notoriously troubled frontier and enable defense costs to be lowered. Even if no sparks from the European revolutions flew as far as Britain's imperial periphery, opponents of the empire took heart when news eventually arrived from Europe (this was the age before telegraphic cabling), and French revolutionary rhetoric echoed in Ceylon, among French Canadians, and in radical circles in Sydney. Despite such links, imperial conflicts did not grow into political explosions in 1848–49. Yet there was something like a sharpening of political conflict in the wake of the revolutions. Colonial representative assemblies were given more leeway, while at the same time governors strengthened their control over the all-decisive area of finances. Symbolic concessions went hand in hand with a tighter grip on the levers of power.

The Taiping Revolution

There is nothing to suggest that the Taiping rebels in China heard a word about the '48 Revolution in Europe. Whereas in the mid-nineteenth century there were no Chinese observers in Europe who could report on political events there, European consuls, missionaries, and merchants residing in Hong Kong, as well as in the port cities opened up in 1842 under the Treaty of Nanjing, were relatively close to the events when the Taiping Revolution broke out in 1850. They did not learn much. The first reports, still based entirely on rumor, date from August 1850, when the movement was beginning to stir in the remote province of Guangxi. Only in 1854 did Western interest pick up, and then no more was heard about the Taiping for another four years; fleeting contact was made with them again only in 1858, during the Second Opium War. After 1860, when the movement was already on the wane and fighting to survive, contacts and reports finally began to multiply.[125] So, the leaders of the largest uprising by far in modern times knew nothing about the revolutions in Europe, while Europeans were largely in the dark about the scale of events in China; any direct interaction can be ruled out. Some Western mercenaries fought on the Taiping side, but it is not known whether there were any "Forty-Eighters" among them. In midcentury nothing connected the mental worlds of European and Chinese revolutionaries, and yet both must have a place in a global history of the nineteenth century.

What exactly was the Taiping uprising? With its construction of an alternative state and its virtual eradication of the old social elite in some provinces, it was at least as revolutionary as the 1848 Revolution, embroiling China in civil war for almost fifteen years.[126] The charismatic founder of the movement, Hong Xiuquan, was a farmer's son from the far south of China. Having entered a personal crisis after failing an exam in his home province, he experienced visions which, owing to his reading of Christian texts (in Chinese), led away from Chinese traditions. In 1847 he sought instruction from American revivalist preachers in Canton, and his conclusion from all he learned was that he was the younger brother of Jesus Christ, commanded by God to spread the true faith. Soon there came an additional mandate to liberate China from the Manchus. The Mormon sect in America arose in a similar way, and the idea of an apocalyptic clash between the powers of darkness and fighters for a new world order also existed on the margins of the French Revolution. Uniquely in China, however, the religious awakening of one individual led within a few years to a gigantic mass movement.

This would not have been possible if the potential for social revolution had not already existed in south central China, where the revolt had its origins, and in the other parts of the country that it soon overran. Alongside the political goal of driving out the ethnically alien Manchu, a program for the radical transformation of society took ever clearer shape. In the southern and central regions that fell under their control, the revolutionaries proceeded to expropriate land on a large scale, to hound officials and landowners, and to introduce new laws. The Heavenly Kingdom of Great Peace (*Taiping Tianguo*) proclaimed at the beginning of 1851, which two years later turned Nanjing, the old imperial city of the Ming dynasty, into its capital, implemented for several years a radical alternative to the old Confucian order, but it was not quite as egalitarian, or even proto-socialist, as official historians in the People's Republic later claimed.

The extraordinary military success of the Taiping is explicable by the initial weakness of the imperial armies and by the fact that some of the men who joined Hong Xiuquan were militarily and administratively more gifted than the rather confused prophet himself. In the course of time, however, these followers—who acquired regal titles in accordance with the old Chinese model of "rival kingdoms" (Northern King, Eastern King, etc.)—fell out with one another even more sharply than the European revolutionaries of 1848 had ever done. As a result of such dissension, the movement lost many a charismatic figure who had updated Hong Xiuquan's divine visions with new illuminations of his own. In 1853 Taiping troops were within sight of the walls of Beijing, from which the

Qing court had already fled, but their commander turned back and passed over this golden opportunity, supposedly because he lacked an order from heaven to capture the city. In 1856 the relative strength of forces began to tilt as the Qing rulers permitted some high regional officials to put together new armies and militias, which were far superior to the regular imperial troops and gradually got the better of the Taiping. The hiring of Western (mainly British and French) mercenaries without opposition from their governments further strengthened the imperial camp, although it cannot be said to have tipped the balance in the war. The Heavenly Capital, Nanjing, eventually fell to imperial forces in June 1864. The brutality that the Taiping showed toward their enemies, and the exterminatory impulses with which these responded, were without parallel in the history of the nineteenth century. To take just a couple of examples: when the Taiping captured Nanjing in March 1853, massacres and mass suicide claimed the lives of 50,000 Manchu soldiers and family members; and when Qing troops retook the city in 1864, it is estimated that 100,000 people died in a purge that lasted two long days, many preempting a grisly fate by taking their own lives.[127] In the three densely populated provinces of eastern China alone—Jiangsu (including Nanjing), Zhejiang, and Anhui—the population is thought to have declined by 43 percent between 1851 and 1864.[128] The overall loss of life resulting from the unrest in China, extremely difficult to estimate, has traditionally been put at 20 to 30 million. Historian Kent Deng has recently revised this number upward to arrive at 66 million.[129] The bitterness and violence were indeed symptomatic of a genuine civil war. Anyone identified as a Taiping leader was killed on the spot or executed after a trial. The Heavenly King, Hong Xiuquan, succumbed to disease or poisoning before the fall of Nanjing, but his fifteen-year-old son was among the victims of the repression. The wholesale elimination of the "bandits," as they were officially known, was the consequence not of innate Chinese cruelty but of political decisions. The Taiping, unlike the European revolutionaries, suffered a total defeat and left no legacy to be picked up later. After 1864, there would be no compromise and no reconciliation.

Some Westerners, especially missionaries, saw in the Taiping the founders of a new Christian China. Others regarded them as an unpredictable force of chaos and took sides with the tarnished Qing dynasty. In China itself the whole movement remained a taboo subject for decades: its survivors would not admit to having supported it, while the victors lived in the (justified) belief that it had been totally eradicated. There is surprisingly little evidence that the Taiping episode, with its radical programs and mass killing, constituted a lasting trauma for the country. The revolutionary leader Sun Yat-sen occasionally evoked the memory of the fallen Taiping, but only official Communist histories inserted the experience into a vision of antifeudal and anti-imperialist struggle. Now there is a retreat from such interpretations in China itself. Equally one-sided and outdated is the Cold War mirror image, in which the Taiping appeared as an early "totalitarian" movement.

In a world-historical perspective, four points are of particular interest.

First. The Taiping Revolution differed from all earlier Chinese popular movements by its Western inspiration. It is true that other elements also entered into its worldview, but it would not have taken the form that it did without the presence of European and American missionaries and their first Chinese converts in southern China. In the middle of the century, China's rulers and cultural elite knew next to nothing about Christianity. The ideas of the Taiping were therefore entirely alien to them and its peculiar amalgam of popular Chinese religion, Confucianism, and evangelical Protestantism quite incomprehensible. At the same time, the economic crisis in south central China, which brought the movement much of its support, was partly a result of the country's gradual opening to uncontrolled foreign trade in the years since 1842. Opium, together with the pressure of imports, led to social distortions that helped to bring about a revolutionary situation. The Taiping Revolution was among other things, but by no means exclusively, a phenomenon of globalization.

Second. Parallels between the Taiping and religious revivalist movements in other parts of the world are unmistakable. What was unique were the early militarization and military success of the movement and its far-from-otherworldly goal of overthrowing the existing political order. The Taiping was a charismatic movement but not a messianic sect awaiting salvation at the end of time. In keeping with Chinese tradition, it was much more concerned with life in this world.

Third. There were few programmatic elements in common between the Taiping and the European revolutions. The idea that an unsuccessful dynasty might forfeit its heavenly mandate did not originate in the West but came down from ancient Chinese political thought. At that time no one in China thought in terms of human and civil rights, the defense of private property, popular sovereignty, the separation of powers, or constitutions. But some elements within the Taiping movement—especially Hong Reng'an, a kind of chancellor of the Taiping Tianguo and a cousin of Hong Xiuquan—devised plans for the infrastructural and economic modernization of China that bore the clear mark of experiences in British-ruled Hong Kong and pointed far ahead into the future. Hong Reng'an could imagine a Christian China as an integrated part of the world community, and in this respect he was far in advance of most official representatives of the Chinese state, who still clung to the vision of an intrinsically superior Middle Kingdom. Hong already sought to introduce railroads, steamships, a postal system, patenting, and Western-style banking and insurance. And, not wanting the state alone to be responsible for these, he recommended the participation of private individuals ("prosperous people with an interest in public affairs").[130] This program was not unsuited to the needs of China. Like the European revolutions, it transgressed the mental boundaries of the ancien régime.

Fourth. The repression of the Taiping did not precipitate anything on the scale of the refugee flow from Europe after 1848–49. Where would so many Chinese have gone? But there were traces of some who made it to Southeast Asia, in particular, and among those who went abroad via the incipient coolie trade there must have been former revolutionaries who no longer felt safe in their homeland. Yet, the Taiping Revolution was not exported, and those who had fought for it did not take their goals with them into other milieux. In vain are we looking for a Chinese Carl Schurz.

The Great Rebellion in India

In spring 1857, Qing troops on several fronts were in full-scale retreat from the Taiping armies. On the other side of the Pacific, 1857 was the year of major agenda setting in the United States. A cumulative sharpening of conflicts had carried the North and the South to a point of no return beyond which a violent resolution seemed increasingly unavoidable. Some clear-sighted observers already suspected that a civil war lay around the corner, and four years later they would be proved right.[131] Not only was the world's oldest monarchy threatened with collapse; the largest of all republics, in many respects the world's most progressive polity, stood on the brink of an existential crisis. The largest state in Eurasia was also going through a period of special uncertainty. The rulers of Russia had just been plunged into deep self-doubt by the defeat in the Crimea. Tsar Alexander II and his advisers spent 1857 developing plans for the emancipation of the serfs, which by now seemed impossible to avoid.[132] No great peasant rising seemed imminent, but reforms were necessary to head one off.

Meanwhile, India showed what might befall an empire when its periphery rebelled. For precisely one century the British had been extending their power over the Subcontinent in one campaign after another, with no significant setbacks. Thinking their position assured, they came to believe not only that their Indian subjects accepted them as rulers but also that they were acting as benefactors by bringing them a superior civilization. The reality, and the British appreciation of it, changed within a few weeks. In July 1857 British rule collapsed in large parts of northern India, and pessimists thought it at least questionable whether the largest colony in the world could be kept within the empire.

The British spoke and speak of the Indian Mutiny. Horrific images such as the massacre at Kanpur (Cawnpore), when several hundred European and Anglo-Indian women and children were killed in July 1857, are still part of the mythology of imperial remembrance.[133] In India, where people are more likely to remember the atrocities committed against rebels—hundreds or thousands killed by cannon fire, Muslims sewn into pigskins before their execution—they speak rather of the Great Rebellion, an altogether preferable term.[134] Whether it may be seen as the beginning of the Indian independence movement has long been a politically controversial issue and does not need to be settled here. The important point is that it was a rebellion, not a revolution. The rebels had no other program than a return to pre-British conditions. Unlike the American and European revolutionaries, and also unlike their Taiping contemporaries, they outlined no vision of a new order adequate to the challenges of the day. In contrast to the Taiping, they never built a counterstate capable of lasting beyond a short-lived military occupation. Nevertheless, it is worth including the Indian Rebellion in a comparative series of the great midcentury convulsions.

The Great Rebellion differs from earlier and later Indian uprisings in that it was not a protest movement of the rural population but a soldiers' revolt—a constant danger in a military apparatus comprising (in 1857) 232,000 Indians alongside 45,000 British.[135] The unrest gathered momentum in the Army of Bengal, the largest of the three armies of the East India Company. Discontent had been growing for a century and a half among the Indian troops, or *sepoys*. Now rumors began to spread that they were to be forcibly converted to Christianity, and these were further fueled in 1856 when orders came for troops to be deployed overseas, where they might be required to violate religious prohibitions. For some time the upper castes of the Northwest Provinces (once the backbone of the army of British India) had been losing their privileges. Many members of this military elite came from the princely state of Awadh (or Oudh), which the British had annexed not long before in an action that was generally seen as arbitrary in the extreme. In Awadh a broad coalition of social forces joined the rebellious soldiers: peasants, large landowners (*taluqdar*), craftsmen, and so on. The beginning of the revolt can be precisely dated to May 10, 1857, the day on which three *sepoy* regiments mutinied in the city of Meerut in the vicinity of Delhi, after some of their comrades had been put in irons for refusing to use cartridges treated with animal fat (in violation of both Hindu and Muslim law). The soldiers killed their European officers and marched on Delhi; the revolt spread in no time at all. Attacks on European officers and their families were not only spontaneous expressions of rage but also a radicalizing tactic; there could be no return to normal conditions after that. From the British point of view, a nadir was reached when the rebels blocked the Great Trunk Road linking Bengal to the Khyber Pass. It was around this time that the British counteroffensive moved into top gear, with troops pulled out of Iran, China (where preparations for the Second Opium War were under way), and the Crimea. As in China (though there the scale was much larger), the investment and capture of rebel cities tipped the military balance. The fall of Lakhnau (or Lucknow) on March 1, 1858, signaled that the colonial power was winning through. The last battles were concentrated in central India, where the rani of Jhansi heroically fought the British at the head of her mounted troops. In July 1858 the governor-general announced that the rebellion was over.

The Indian rebels drove the colonial state closer to collapse than it had ever been before or would be again, much as the Taiping Revolution (strengthened by Nian rebels operating independently of it) did with the Qing dynasty. Despite a widespread hatred of the foreign rulers among the Indian population, the rebellion never had a broad social base outside Awadh comparable to that of the Taiping; it was much more of a regional phenomenon. The whole of southern India was unaffected, and the other two *sepoy* armies—those of Bombay and

Madras—played scarcely any role. In Bengal itself, the British had such a strong military presence that the region around Calcutta remained calm. In the Punjab, annexed only in 1848, it was an advantage for the colonial state that the local upper classes and indigenous Sikh fighters had been well treated. Indeed, Sikhs would form the core of the British army in India after the rebellion. With Scottish Highland troops and Gurkhas from the Nepalese Himalayas, they were among the most important of those heroic "martial races" which, in the eyes of the British public, ensured the security of the empire.

Thanks to a much greater media presence, exemplified by the superb *Times* correspondent William Howard Russell (who later also reported on the American Civil War), the international public was far better informed about the events in India than about the uprisings in the Chinese interior.[136] India was a step ahead of China in terms of communications technology. Telegraph links inside the country, unless cut by the rebels, served the British for both military and propaganda purposes. Moreover, a vast literature of memoirs subsequently appeared in Britain. The Great Rebellion lasted for a much shorter time than the Taiping Revolution: just twelve months, in comparison with fourteen years. Nor can it be said that it depopulated whole areas of the country or physically exterminated sections of the upper classes. Also the origins of the two movements were different: in India an army mutiny, in China a civilian religious movement that took up arms under the pressure of its enemies. Whereas Christianity gave the Taiping Revolution its initial impulse, the Indian rebellion sought to fend off the threat of Christianization. But in India too, millenarian religiosity played a certain role, more in the case of Muslims than Hindus. On the eve of the revolt, Muslim preachers had been predicting the end of British rule, and at its high point a call to jihad mobilized large sections of the population (leaving it open whether they should also vent their fury against Indian non-Muslims). Strategically acute leaders, however, tried to prevent hostilities between Muslims and Hindus from weakening the rebellion.[137] It was certainly not, as some British suspected at the time, the result of a great (even worldwide) Muslim conspiracy. Still, the religious dimension, which underlies some Indian myths of national resistance, should not be overlooked.

The revolts in India and China did have a patriotic side to them, and in this respect they were close to the Hungarian uprising of 1848–49. They might perhaps be described as proto-nationalist, although in India it is unclear how the traditional division of the Subcontinent would have been overcome if the rebellion had been successful. The movements in India and China failed more dramatically than the European Revolutions of 1848–49. In all cases, the existing social and political order initially came out of the challenge stronger than before. The East India Company dissolved, leaving the Crown to exercise direct rule over India right up to 1947; the Qing dynasty lasted only until 1911. In China, the Tongzhi Restoration (c. 1861–74) saw the Qing state embark on timid reforms of a military nature more than a political or social one. In India, British rule became conservative and conserving after 1858, based more than ever on traditional elites and marked by an increasingly racial sense of distance from the Indians. Only after the turn of the century would it be forced to respond to new political challenges from the Indian elite. In neither case can it be said that "reactionary forces" won out against the bearers of "progress." The Heavenly King Hong Xiuquan and Nana Sahib (actually Govind Dhondu Pant, the best-known leader of the Indian rebellion, at least abroad) were hardly people who could have led their country into the modern age. Here the analogies with Europe end.

Civil War in the United States

There can be nothing open in the verdict concerning the American Civil War.[138] Of all the great midcentury conflicts within a society, it was the one in which the forces of moral and political progress were most unambiguously victorious. Their victory was also associated with the "conservative" goal of preserving a nation-state that already existed. Such was not the case in India or China. The Indian *sepoy* rebels and the small number of princes who supported them would certainly not have been able to replace the great integrative framework that the British military and governmental apparatus constituted in its way; no Indian Prussia would have emerged from the successful rebellion to unify the Subcontinent. Instead, the outcome most likely would have been another string of statelets like that of the eighteenth century. A China ruled by the Taiping would definitely not have been a liberal democracy (whatever Hong Reng'an may have planned): perhaps an authoritarian theocracy or, with the passing of time, a refurbished variant of the Confucian order minus the Manchu component would have emerged. But the Taiping movement was so fissiparous that it is hard to imagine the preservation of a unified empire. If nineteenth-century China had been able to develop into a plurality of nation-states, would they have been economically viable? We have reason to doubt it.

Matters are clearer in North America. The victory of the North in 1865 prevented the lasting formation of a third independent state in the region, destroying the institution that went together with everything conservative or reactionary in the American context of the time. The political coordinates of the United States were quite different from those of Europe. Those who stood on the right in the United States in 1850 or 1860 were not partisans of authoritarian rule, neoabsolutist monarchy, or aristocratic privilege; they were defenders of slavery. Should we consider the American Civil War under the heading "revolution," as some contemporary observers did (e.g., Karl Marx or the young French journalist Georges Clemenceau)?[139] American historians have debated the issue more than once since the 1920s; a comparative treatment adds another dimension to it.[140] The same has been true of the Taiping movement. From a strictly Confucian viewpoint the participants were lawless bandits who deserved to be wiped out, whereas for later Sino-Marxists they were precursors of the revolution (not "bourgeois revolutionaries") that really began only with the founding of the

Communist Party of China in 1921.[141] But if we speak of the European events of 1848–49 as revolutionary, then the same ought to be said of the Taiping. Their revolution failed too. The social transformations they introduced were at least as radical as any that occurred in Europe in 1848–49. They did not build a lasting order, but they weakened key pillars of the old. The Chinese ancien régime collapsed in 1911, the central European régime not until 1918–19.

On a scale of violence and death, the American Civil War may be placed alongside only the much more violent Taiping Revolution; the events of 1848–49 in central Europe or 1857–58 in India pale in comparison. Even more than in other cases, a distinction must be drawn here between the revolutionary character of the causes and of the consequences. The immediate spur for the American Civil War was the development of two opposite interpretations of the US Constitution, the key symbolic bond that had held the Union together since 1787. In the preceding decades, tensions between the political elites in North and South had been mitigated by quite a robust two-party system that straddled regional (in America one would say sectoral) contradictions. This system polarized along regional lines in the 1850s: the Republicans stood for the North, the Democrats for the South. As soon as it became known that Abraham Lincoln, an opponent of slavery, had been elected president in late 1860, the champions of a new Southern nationalism began to put their program into practice. By the time of Lincoln's inauguration in January 1861, seven Southern states had already announced their exit, and in February a new Confederate States of America came into being and immediately proceeded to take over federal property on its territory. In his first inaugural address, delivered on March 4, Lincoln characterized the Southern action as secession and left no doubt that he would act to preserve the unity of the nation.[142] War broke out on April 14, after the South had attacked Fort Sumter, a federal garrison on an island off the coast of South Carolina.

The causes of the conflict, debated among historians ever since, were not the typical ones of European revolutions. There was no revolt by socially and economically underprivileged classes—slaves, peasants, or workers. Nor, of course, was freedom from autocracy an issue, although Barrington Moore is right to argue that "striking down slavery was . . . an act at least as important as the striking down of absolute monarchy in the English Civil War and the French Revolution."[143] Both sides spoke tirelessly of liberty: the North wanted freedom for the slaves; the South, freedom to keep them.[144] Whatever the background factors may have been (uneven economic development between North and South, clash of nationalist identities, inexperienced and overemotional handling of new political institutions, antagonism between an "aristocratic" South and a "bourgeois" North, etc.), the Civil War was not based on a European-style struggle for the rule of law. Rather, it was a postrevolutionary conflict, a follow-up to the earlier creation of a constitutional order. The fighting was not *for* a constitution but over the scope for different social models within an already existing constitution. The cohesive nation defined in the Constitution of 1787 had been undermined by the divergence of regional interests.[145] A readiness for war, eventually stretching well beyond the elites on either side, developed out of America's uncompleted eighteenth-century revolution, which had guaranteed white men their freedom but passed over the lack of freedom for Afro-American men.

As a series of events, the Civil War began when those who preferred this contradiction to be unresolved divided the unitary nation of the independence period and established a state of their own.[146] The years of the conflict break down into several histories. One that can be told concerns how the South, despite great material inferiority, acquitted itself surprisingly well and fell back before the stronger North only in the middle of 1863. A second would tell of the mobilization of ever greater areas of society on both sides; and a third, of Abraham Lincoln's massive feat of leadership, which, if such a superlative is in order, made him one among the few political figures of the nineteenth century who rose to a challenge with both strategic farsightedness and tactical mastery. The war ended in April 1865 with the surrender of the last Confederate troops.[147]

The failed revolt by large sections of the Southern white population had consequences that may be called revolutionary. The separate Southern state and its military apparatus were smashed; Lincoln's Thirteenth Amendment enshrined the freedom of the slaves in the constitution of the whole country. The conversion of four million people without rights into US citizens must count as one of the deepest possible inroads into society, even if discrimination would long restrict its practical effect. The liberation of the African Americans marked the South and the mentality of its people for decades. True, the old elite of slave owners was not physically eliminated, but it lost its slave property without compensation and, immediately after the end of hostilities, found itself excluded from decisions about the postwar order. The victors did not inflict bloody vengeance on their leaders, as the Qing generals did on the Taiping rebels, the British on the Indian mutineers, or the Habsburg military on the defeated Hungarian insurgents of 1849. Jefferson Davis, the president of the Confederacy, lost his citizenship, spent two years in prison, and died in poverty. Robert E. Lee, the military leader of the Southern states and perhaps the most brilliant of all the Civil War strategists, later became an advocate of reconciliation and ended up as the president of a university. Those were mild consequences for high treason. The South with its devastated landscapes and ruined cities—Atlanta, Charleston, and Richmond were hit especially hard—was at first placed under military occupation. But this soon gave way to a new civil order, which under the successor of Abraham Lincoln (who was assassinated on April 15, 1865) was buttressed by a general amnesty covering nearly all former officeholders in the Confederacy.[148] Everyone who lived in or, for whatever reason, moved to the South in the first few years after the war experienced it as a period of radical change. The erstwhile ruling class was dramatically weakened, war and abolition having stripped it of more than half of its assets. Before 1860 the plantation oligarchy of the South had been wealthier than the economic elite of the North. After 1870 four-fifths of superrich Americans lived in the former states of the North.[149]

That emancipated slaves lost no opportunity to take their fate into their hands was itself something new;[150] it had begun in the last two years of the Civil War. A total of 180,000 African Americans had served in the armies of the North, whereas unrest among the Southern slaves had grown with each military defeat of the Confederacy. In the situation opened up by the end of the war, various social groups fought for a position in the new order, now using legislative means rather than the force of arms: owners of the disintegrating large plantations, white farmers who had used little or no slave labor in the past, freedmen from the period before 1865, and former slaves. This happened in the framework of Northern-led "Reconstruction."

The federal drive for reforms in the South reached a climax in 1867–72: the age of radical reconstruction. It promoted greater political participation, whittling down the power of the old Southern oligarchs, but it left their social and economic position intact on the large plantations. By 1877 at the latest, the Republican Party gave up its attempts to impose a new distribution of power and came to an arrangement with the elite of the Southern states (one of the great placatory compromises of the age, along with the Austro-Hungarian "settlement" of 1867 and the welding together of the German Reich in 1871 out of the former princely states). But there was no going back to the conditions before 1865, and in this sense the turn may be described as revolutionary. The presence of African Americans in elected office at nearly every political level would have been inconceivable in 1860. On the other hand, the black population was not enabled to take real advantage of the new opportunities. Political emancipation did not go hand in hand with social and economic emancipation, and among most whites it did not lead to changed attitudes that cast aside racially motivated persecution and discrimination.[151]

The Civil War, then, remained an "unfinished revolution."[152] Hopes of a greater political role for women (of all colors) were also disappointed. However, many historians have described as revolutionary certain other impulses that flowed out of the 1860s and 1870s. After decades of far-reaching laissez-faire, the government—especially at the federal level—took on a more active role and greater responsibilities: construction of a state banking system that imposed unity on previously chaotic monetary conditions; a turn to protective tariffs (that is, an assertive foreign-trade policy, which the United States pursues to this day); greater government investment in infrastructure; and stricter regulation of westward expansion. This "American system," as it was called, was an important political prerequisite for the rise of the United States as the leading economic power. All this becomes apparent only if the Civil War and Reconstruction are taken as a single period stretching from 1861 to 1877,[153] just as the French Revolution and the Napoleonic era must be seen as a continuum from 1789 to 1815.

4 Eurasian Revolutions, Fin de Siècle

A Side Glance at Mexico

The third quarter of the nineteenth century was, with the relative exception of Africa, a period of great crises, many of them resolved through violence. Revolutionary challenges to the existing order began in 1847 in Western Europe and ended in 1873 with the crushing defeat of the last major Muslim revolts in southwestern China, due to both ethnic and religious tensions.[154] The largest European wars between 1815 and 1914, from Crimea to Sedan, fall within that time span. After the convulsions, many countries in the world entered almost simultaneously a phase of state consolidation, which in certain cases took the special form of the construction of a nation-state. Finally, in 1917, revolutionaries of a new kind triumphed in Russia, seeing revolution as a process that spreads across borders—as world revolution. With the founding of the Communist International in 1919, attempts began to help it along by sending out emissaries and providing military assistance. This was a new development in the history of revolutions. In the nineteenth century only anarchists had tried anything similar; the best-known of them, Mikhail Bakunin, seemed to appear at every crisis point in Europe, though neither he nor others achieved any results. The export of revolution, not borne as after 1792 by conquering armies, was a novelty of the twentieth century. Characteristically, the last great revolutionary event in Europe before 1917—the Paris Commune of 1871—had remained completely isolated, not conforming to the conflagrative pattern of 1830 and 1848. It was a local interlude, grown out of the Franco-Prussian War. What it did show was that more than eighty years after the Great Revolution, French society was not yet at peace.

In such a bird's-eye perspective, one may easily overlook some "minor" revolutions that seemed to lie on "the periphery," and of which one cannot really say whether they failed or succeeded by European standards. They all occurred in the period between 1905 and 1911, and were less spectacularly violent than the midcentury convulsions. The exception is the Mexican Revolution, which occupied the whole decade between 1910 and 1920; the whole of the twenties would be needed to contain its effects. Here revolution soon turned into civil war, passing through a number of phases and claiming the lives of one in eight Mexicans: a terrible toll in the history of revolutions, comparable only to the Taiping uprising in eastern China.[155] The Mexican Revolution was a "great revolution" in the French sense. It had a broad social base, being essentially a peasant uprising but also much more besides. It overwhelmed an ancien régime—not in this case an absolute monarchy but an ossified oligarchy—and replaced it with a "modern" one-party system that survived until the year 2000.

The Mexican Revolution was remarkable for the depth of its peasant mobilization, and also for the fact that it did not have to defend itself against a foreign enemy. The United States did intervene, it is true, but the importance of this should not be exaggerated. Unlike the Chinese or Vietnamese peasantry at a later date, the Mexicans were not fighting primarily against colonial masters and imperial invaders. Another peculiarity in comparison with the "great revolutions" in North America, France, Russia, and China (after the 1920s) was the lack of a worked-out revolutionary theory. A Mexican Jefferson, Sieyès, Lenin, or Mao never won international fame, and the Mexican revolutionaries never claimed that they wanted to make the rest of the world (or neighboring countries) happy. Thus, despite its great length and high levels of violence, the Mexican Revolution was a rather local or national event.

Eurasian Similarities and Learning Processes

The same may be said of the "minor" revolutions in Eurasia after the turn of the century. There were four series of events:

1. the 1905 Revolution in *Russia*, which actually unfolded between 1904 and 1907
2. what is usually called the Constitutional Revolution in *Iran*, which began in December 1905, produced country's first constitution a year later, and ended in 1911 with the breakdown of the transition to parliamentary rule
3. the Young Turk Revolution in the *Ottoman Empire*, which opened in 1908 when rebel army officers forced Sultan Abdülhamid II to revive the constitution suspended in 1878, and which did not really come to an end in a clearly discernible way but marked the beginning of a long transformation of the sultanate into a Turkish nation-state
4. the Xinhai Revolution in *China*, beginning in October 1911 as a military revolt in the provinces and leading promptly and without much bloodshed to the collapse of the Qing dynasty and, on January 1, 1912, the founding of a republic; it ended in 1913 when Yuan Shikai, a holder of high office in the old regime who had participated in the toppling of the Qing, but then turned against the revolutionaries, took power and ruled the republic as president-dictator until 1916

The societies and political systems within which these four revolutions took their course naturally varied in a number of respects. It would be irresponsible to speak of them as a single type. Nor did the revolutions directly ignite one another; in no case was the key spark a previous event in a nearby country. Thus, to construct an example, the Iranian Revolution was not the primary detonator of the Young Turk Revolution of 1908, but it is possible to play with the idea of certain causal chains. The Tsarist Empire would probably have remained more stable if it had not lost the war with Japan so shamefully in 1904–5 (much as Louis XVI disgraced himself with his incompetence during the Dutch crisis of 1787); and had the Tsarist Empire not been so weakened politically by the war and revolution of 1905, it probably would not have been willing in 1907 to divide Asian spheres of influence with the British. Furthermore, if the Russians and British had failed to reach that agreement, the panic among Turkish officers in Macedonia that the Great Powers were about to carve up the Ottoman Empire would have been less pronounced and not delivered the final signal for revolt.

Although there was no domino effect in the Eurasian revolutions, the various forces did act with an awareness of the range of revolutionary options available worldwide in their time. They were also aware of the recent history of their own countries. The constitution of 1876 that the Young Turks sought to reactivate had itself been wrested from the sultan of the day by "Young Ottomans" in the government and the civil service. From these predecessors the Young Turks inherited the idea that far-reaching change would have to originate from enlightened members of the elite. In China, the Taiping no longer served as a model in the run-up to 1911, but revolutionary activists bore in mind two more recent initiatives: the failed attempt by a section of officials in 1898 to win the Court over to an ambitious reform program (the Hundred Days' Reform movement); and the Boxer Rebellion of 1900–1901, which had proved unable to come up with any constructive perspectives. If the former was an example of a movement whose social base was too small to impose change, the latter stood for an intemperate outpouring of popular rage that held no promises for an enlightened nationalism.

In varying degrees, the Eurasian revolutionaries had some knowledge of European revolutions. The Young Ottomans of 1867–78, consisting of reform-oriented intellectuals and high state officials, had admired the French Revolution (though not the Terror), and in this the Young Turks later followed their lead.[156] Basic writings of the European Enlightenment, such as the works of Rousseau, had been translated into a number of Asian languages. In China the American Revolution was more popular than the French; historical literature about both had been published in Chinese. Most intellectuals around the turn of the century especially favored an energetic policy of modernization from above, such as that carried out by Peter the Great in Russia.[157] A still more important model, in both China and the Ottoman Empire, was Meiji Japan.[158] Here an enlightened elite had made the country rich and strong without undue bloodshed and presented a civilized face that impressed the West. Chinese revolutionaries saw their model partly in the political institutions of central Europe and North America, partly in the appropriation ("Asianization") of those institutions along lines similar to those adopted in Japan, though not necessarily in every detail.[159] The Young Turks warmed to Japan especially because it had just inflicted a heavy defeat on Russia, the archenemy of the Ottomans; they watched attentively the revolutionary turn of events "next door" in Russia and Iran, commenting on it in their press. In both cases, popular protests played a greater role than the Young Turks had expected in their scenarios. Developments in Russia in particular convinced them that the earlier Young Ottoman strategy of gaining the initiative inside the state apparatus was insufficient by itself.[160]

The revolutions in Iran, the Ottoman Empire, and China were not complete imitations of those of the West, nor were they mutual copies. But this does not mean they were unwilling to learn from one another. "Transfers," though never of decisive importance, occurred again and again. Iranian workers in the oil wells of Baku in Russian Azerbaijan brought revolutionary ideas home with them to Tabriz.[161] The Chinese Revolution of 1911 had much support among affluent overseas Chinese who, in the United States or European colonies in Southeast Asia, had come to know the advantages of a comparatively liberal economic policy. Such a learning process could be complicated at times. In March 1871 the Japanese Prince Saionji Kinmochi, from the noble Fujiwara clan, arrived in Paris to study French and the law. He observed the Commune at first hand, remained in the French capital for ten years, and went home convinced that Japan needed to establish basic civil liberties without exposing itself to the danger of unbridled people's power.[162] A friend of Georges Clemenceau, he served many times as minister and prime minister and became one of the key figures in Japan's liberal ruling elite and the longest-living elder statesman during the period of the country's rapid ascent.

In the quartet of fin-de-siècle revolutions, Russia's was in one respect a special case. Its economy was more developed than those of the other three countries, largely as a result of the modernization drive conducted under its finance minister Sergei Witte in the 1890s. Only in Russia was there already an industrial proletariat capable of giving political representation to its interests. In no Asian country would it have been possible to put together a demonstration like that of January 9, 1905, ("Bloody Sunday") in Saint Petersburg, when 100,000 workers marched peacefully to the Winter Peace to present a petition to the tsar. The massacre by Tsarist troops that ended the day unleashed an unprecedented strike wave throughout the empire from Riga to Baku in which more than 400,000 are estimated to have taken part.[163] Even larger was the general strike that, from October on, focused the unrest growing in many parts of the country. Where there was not yet sufficient industry and the railroad was still so rare that its paralysis could not cause major damage, another available form of struggle was the boycott: that is, the kind of shopkeepers' and consumers' strike that had already proved effective in Iran and China (where it would continue to be practiced until the 1930s). Whereas the Russian Revolution of 1905 was thus more modern in its social composition than the parallel movements in the three Asian countries, it was in other respects sufficiently like them for a comparison to be in order. In fact, the similarities among the four are at least as great as the differences, and even where there are divergences in their preconditions and national paths a comparative approach can help to bring these out more clearly.

Despotism and Constitutionalism

All four revolutions targeted old-style autocracies of a kind that had never existed in Western Europe. Traditional legal constraints on power were not entirely absent in Russia and Asia, but their force was altogether weaker there; the nobility and other landowning elite groups had never been strong enough to counter the absolute power of the ruler in something comparable to Western European (or Japanese) feudalism. The position of the monarch in the respective political systems was less open to challenge than that of Louis XVI and a fortiori George III of England. In theory they were despotisms, in which the ruler had the last word and did not have to pay heed to a parliament or an assembly of estates. But the practice was not always completely arbitrary. More than in other systems it depended on the personal qualities of the figure who sat on the throne. Sultan Abdülhamid II corresponded most closely to the Western cliché of a latter-day "Oriental despot." In February 1878 he brought a timid experiment with parliamentarianism to an end after just two years, dismissing the (hitherto rather ineffectual) assembly and suspending the Constitution of 1876.[164] From then on he ruled the Ottoman Empire as a remarkably active autocrat. Tsar Nicholas II (r. 1894–1917) was not far behind him on that score. His monarchical self-image made no concessions to liberal currents: he was perhaps a slightly less capable ruler than Abdülhamid, less in tune with the main tendencies of the age and (in his later years) increasingly prone to a bizarre obscurantism.[165] In Iran, Nasir al-Din Shah (r. 1848–96) was shot by an assassin after half a century on the throne; he introduced virtually no reforms, but he did bring the notoriously unruly tribes under control and thereby helped to hold the country together.[166] His son and successor Muzaffar al-Din Shah (r. 1896–1907) proved to be mild and irresolute, becoming little more than a plaything of other forces at the court; he was eventually replaced by his more vicious and tyrannical son Muhammad Ali Shah (r. 1907–9). Uniquely among the four revolutions, there was a change of ruler in Iran when the revolutionary process was already under way. The new shah's completely rigid attitude, closed to the slightest compromise (reminding us of Ferdinand VII of Spain), considerably exacerbated the situation in the country.

In China the age of powerful autocrats ended in 1850 at the latest, with the death of the Daoguang Emperor. The four emperors who followed him were all incompetent, or uninterested, in the affairs of state. After 1861 the Dowager Empress Cixi (1835–1908), an extremely energetic upstart woman, played the role of autocrat—and she knew how to defend artfully the interests of the dynasty. Being formally a kind of usurper, Cixi was never as safe from attack as the great Qing emperors of the eighteenth century had been. She ruled literally as "the power behind the throne"—the curtain behind which she used to sit is still on display in Beijing—for two feeble emperors. Her nephew, the Guangxu Emperor (r. 1875–1908), was kept under house arrest from 1898, when as a young man he dared to show sympathies with the liberal "Hundred Days" reformers; she probably had him poisoned shortly before her own death in 1908. After Cixi the Chinese throne was more or less vacant. A grandnephew of hers, the three-year-old Puyi, was placed on it in 1908, with his father, a half-brother of the deceased emperor, stepping in as regent. At the time of the 1911 Revolution, this Prince Qun effectively held monarchical powers. He was a narrow-minded man, and his aggressively pro-Manchu policy alienated him from the top Chinese bureaucrats.

So, on the eve of the respective revolutions, there were real autocrats in Russia and the Ottoman Empire, and to a more limited extent in Iran. Against these political systems, the revolutionaries—and this was the main thing they had in common—opposed the idea of the constitution.[167] Just as in Europe, with whose modern history they were familiar, it was the central plank of their political programs. In the Ottoman Empire and Iran, the Belgian constitution of 1831 (which provided for a parliamentary monarchy) enjoyed an especially high reputation.[168] Republican forces, unsatisfied with anything along the lines of the constitutional July Monarchy in France or the post-1871 German Reich, were a minority in the revolutionary camp. The one exception to this was China, where after more than two and a half centuries of Manchurian "alien" rule no clandestine survivors of an indigenous dynasty surfaced to offer an alternative to the Qing, and the absence of a high nobility excluded other routes of ascent to the emperorship. Each of the four revolutions produced a written constitution. Despite unavoidable borrowing from Western models, their authors consistently attempted to do justice to the peculiarities of their political culture. Constitutionalism was therefore a genuine political strategy, not at all a shiftless or opportunist imitation of Europe. A much-admired model was the Japanese constitution of 1889, largely the work of the worldly-wise statesman Itō Hirobumi, who appeared to have wrought a perfect blend of foreign and domestic elements. Japan also seemed to have demonstrated that the constitution could become a unifying political symbol in an emergent nation—not merely a plan for the organization of state organs but a cultural achievement of which people could be proud. The main difference with Western Europe was that the Japanese had not explicitly taken in its concept of popular sovereignty. Each of the new Asian constitutional traditions had to identify other sources of legitimacy, secular and religious, on which political rule should base itself.

Reforms as Triggers of Revolution

The French Revolution of 1789 was preceded not by a wave of repression and exclusion but by a cautious attempt, especially on the part of Minister Turgot, to open up and modernize the political system. This gave rise to the hypothesis— seemingly confirmed in the Soviet Union under Mikhail Gorbachev—that signs of liberalization smooth the path for revolutions by creating a spiral of rising expectations. On this point the Eastern revolutions we are considering differed from one another. Abdülhamid II was not altogether the tyrant depicted by hostile propaganda; he persevered with many of the reforms introduced before his time, such as the building of an education system and the modernization of the armed forces. But the sultan showed no willingness to compromise on the issue of political participation. In Iran there was little sign of reforms on the eve of the revolution; the shah, when faced with protests, had often revoked particular measures during the previous decades, but he had never countenanced actual changes to the system. In Russia, it was less out of perspicacity than because of external pressures that Nicholas II announced some minor reforms in summer 1904. But, instead of mitigating the climate of unrest, this long-awaited token of minimal concessions became the signal for the opposition to step up its activity against the autocracy.[169] Similarly, the decision of Louis XVI to convene the Estates-General had given a major impetus to public debate.

China was the big surprise, furthest of all from the cliché of Oriental despotism. The Empress Dowager had a reputation abroad of being perhaps the extreme hardliner among Asiatic monarchs, and it is true that nowhere else was life so dangerous for oppositionists. In 1898 Cixi had ruthlessly suppressed a moderate reform movement. But the catastrophic defeat of the Boxer Rebellion in 1900 convinced her of the need to reappraise the institutions of the Chinese state, to pursue modernization more actively, and to involve sections of the upper classes in policymaking. In the Tsarist Empire such participation had existed since 1864 in the form of the zemstvo: a rural administrative body at the governorate and district level intended to meet the needs of the local population in matters such as education, health care, and road construction. The zemstva were to some degree independent of the state bureaucracy, being constituted (after 1865) through elections in which only the nobility could vote. The peasantry was allowed to send representatives of its own, but from 1890 these were no longer directly elected. The creation of the zemstva served to politicize various sections of the population but also to divide them into mutually antagonistic tendencies. Where radical forces gained the upper hand, the zemstvo became an opposition forum in the early years of the twentieth century. Parliamentary-style local government was hard to reconcile, however, with an autocratic system subject to no constitutional safeguards and a bloated, increasingly self-assertive, bureaucratic apparatus. In the years before 1914, Russia was by no means on its way to ever broader self-government, let alone liberal democracy.[170]

In China it had traditionally been unthinkable to engage in politics outside the bureaucracy; the principle of representation was unknown. It therefore meant a radical break when the Dowager Empress, without facing significant resistance from the intrabureaucratic opposition, promised in November 1906 that work would begin on a constitution, and when the Court announced at the end of 1908 a transition to constitutional government within nine years. In October 1909 male members of the elite convened in provincial assemblies, much like similar bodies earlier in European history. China had not seen anything like it; officially sanctioned forums could for the first time freely discuss the affairs of their province and even of the country as a whole. At least as important was a whole series of reforms that high officials of the Qing Dynasty initiated in the first decade of the new century: creation of specialized ministries, suppression of opium growing, an accelerated construction of the railroad network, increased funding for universities and other educational establishments, and above all elimination of the examination system that had been used for more than eight hundred years to select top civil servants. This last measure changed overnight the character of the Chinese state and of the upper stratum in society. In none other of the four countries under comparison were such radical and farsighted reforms not only announced but implemented on the eve of the revolution. The oldest monarchy in the world seemed to prove itself particularly capable of learning, not least under the impact of events in Russia since 1905.[171] All the greater is the irony that none of the anciens régimes dropped out of the picture with such rapidity.

Intelligentsia

Behind each revolution stand special coalitions of social forces. The four countries under consideration traditionally had very different forms of society, yet they also displayed a number of common features. Everywhere the intelligentsia was a driving force. In Russia, where the term itself was coined, the *intelligentsiya* had come into being in the first third of the nineteenth century through Alexander I's modest, Enlightenment-inspired educational reforms. From then on, large sections of the nobility regarded a higher education as "an indispensable part of their life plan."[172] The original model was the European Enlightenment, and later heroic-romantic idealism, in the forms in which these reached the truly cosmopolitan culture of the Russian elite. Protected by the upper-class background of many of its members, the intelligentsia was able to develop even amid the censorship that gripped Russia from the 1860s on. The new liberal professions and growing elite education (limited though this was in comparison with Western Europe) expanded their recruitment base beyond the circles of the nobility. It began to define itself increasingly in opposition to the state: *intelligentsiya* versus officialdom. The lifestyles and values of the "nihilist" counterculture that arose in the sixties were also shaped by a protest symbolism that is harder to find in the other three countries where revolution was slowly ripening. After Tsar Alexander II was assassinated on March 1, 1881, by a terrorist group within one intellectual movement (the Narodniki or "Friends of the People"), the intelligentsia appeared more sharply as a force of radical *political* opposition.[173]

In the Ottoman Empire—comparable in a way to Prussia or the southern German states earlier on—reformist attitudes inspired by the Enlightenment spread in midcentury particularly among the upper reaches of the state bureaucracy. Here the critical intelligentsia was at first a special group close to the levers of power, until the authoritarian turn under Sultan Abdülhamid II made criticism of the status quo a dangerous business. Many independent minds then took the road of exile, to Western Europe and elsewhere, forming a diaspora that was by far the main source of revolutionary preparations.

This was paralleled in the Iranian case, though on a lesser scale. One peculiarity of Iran was that secularism, understood as a separation between religion and politics, made slower progress than in the Ottoman Empire. Shiite scholars of divinity and law, especially the high-ranking *mujtahids*, were able to preserve their general position in the culture more successfully than Sunni clerics. Indeed, their influence had grown in the eighteenth century, and under the Qajar Dynasty (from 1796) it became even stronger than it had been under the Safavids.[174] Equivalents of the European intelligentsia were therefore to be found not on the liberal wing of the state bureaucracy (as in the Ottoman Empire) but among more closed groups within the religious establishment.

In China, since time immemorial, there had been no room for a critical intelligentsia outside the elite of scholar-officials defined by success in competitive examinations. Any criticism had always been expressed from within the bureaucracy itself, often with considerable effect on imperial governance. After 1842, a modern press and the first scope for critical argument began to appear in the special areas to which the Chinese authorities had no direct access (above all the British colony of Hong Kong).[175] But so long as the examination system survived, mapping out life plans for young Chinese, a "free-floating" intelligentsia could develop on only a very limited scale. Its history therefore really dates from 1905, when thousands quickly took up the greater opportunities allowed for study abroad. A revolutionary intelligentsia did not develop in China out of a reform-minded state bureaucracy (the failed reformers of 1898 were outsiders) or, as in Russia, out of an elite culture oriented to the West; there was no clergy as in Iran. The concept of an intelligentsia (*zhishi fenzi*) can be applied at all only to circles of nationalist students educated mainly in Japan after 1905. Their chief organization was the Tongmenghui, a revolutionary association "bound by oath," which made an essential contribution to the program of the Chinese Revolution and eventually spawned the Guomindang, the National Party of Sun Yat-sen.

In scarcely any other country in the twentieth century did the intelligentsia shape history to the extent that it did in China, especially after 1915. Living mostly in exile, sometimes in Shanghai or Hong Kong, such intellectuals did not participate directly in the Revolution of 1911; they were a backstage influence rather, with only a weak presence in the spotlight of events. Their hour was to arrive *after* 1911. The role of the intelligentsia in the revolutions was most important in Russia and Iran. In the Ottoman Empire the initiative passed in spring 1908 from revolutionary exiles (including Armenians) to a group of officers in Ottoman Macedonia. It was from their ranks that the leadership of the Young Turk movement would be recruited after the success of the revolution.[176]

The Military and the International Setting

None of the four revolutions was a military coup d'état. In Russia and Iran the military was on the side of the old order. If the Russian armed forces had gone over to the striking workers, the rebellious peasantry, and the turbulent nationalities, the authoritarian regime would not have been able to survive. However, unrest in the Black Sea fleet did not escalate into a general mutiny; only on the battleship *Potemkin* did revolutionary sailors seize command and fraternize with radical groups in the port of Odessa. On June 16, 1905, the army put down the revolt with a brutality that went far beyond Bloody Sunday in Saint Petersburg; more than two thousand people were killed in the space of a few hours.[177] In Iran there was no army that could have been subordinated to the civil power. Following the early death of the resourceful Crown Prince Abbas Mirza in 1833 the regime had given up attempts to modernize the military. Then in 1879, after he had encountered Cossack troops on a visit to Russia, Nasir al-Din Shah established a Cossack brigade of his own under the command of Russian officers. It was a kind of Praetorian Guard, which, in addition to serving its own interests and those of the shah, represented the concerns of Russia. In June 1908 Muhammad Ali Shah deployed the Cossacks (now down to two thousand men, but a redoubtable force) in a coup d'état that disbanded the parliament, suspended the constitution and brought the first phase of the revolution to a sudden end.

The difference with the Ottoman Empire and China could hardly have been greater.[178] In those two countries, it was army officers who struck the decisive blows for the revolution. Both Sultan Abdülhamid and, beginning two or three decades later, the Qing rulers had founded military academies, recruited foreign advisers, and sought to raise some units to a European standard of training, equipment, and battle-readiness. Although this was quite successful, the central governments neglected to ensure the loyalty of their officers, who tended to be highly patriotic. The Young Turk movement, in whose emigré circles military men had initially played a very minor role, became the great danger for the sultanate once civilian organizations managed to win over numbers of officers.[179] Military pressure resulted in Abdülhamid's reactivation of the constitution on July 23, 1908, moving him away, at least in theory, from absolutist rule. In a similar way, ideas of a revolutionary conspiracy, first voiced among the exiled Tongmenghui radicals in Japan, found an echo among officers of modernized sections of the army in the Qing empire. Here regional militias had been formed in the wake of the anti-Taiping war. The new armies that had arisen in the 1890s were likewise concentrated not in the imperial capital (where the, by now, rather ineffectual Manchu troops were based) but in the various provincial capitals, where they were often in close contact with officials and other notables. This alliance boded ill for the fate of the dynasty.[180]

A chance uncovering of subversive activity among soldiers in Hankou (there was something similar in Ottoman Salonica in 1908) led to an improvised mutiny in several provinces. The Chinese Revolution of 1911 took the form of a defection of most provinces from the imperial house.[181] This set the power configuration for the next twenty years and more, in which the quest for autonomy on the part of military and civilian elites was the dominant tendency of Chinese politics. Things were much more centralized in the Ottoman system, where military leaders gradually ensconced themselves in positions of power after 1908. Turkey's major involvement in the First World War—in contrast to China's rather nominal participation—and the fact that not long afterward it had to fight another war against Greece, further strengthened the position of top officers. But one of the leading generals, Kemal Pasha (later Atatürk), managed to rein in and "civilize" the army in the 1920s by channeling its energies into the construction of a civilian republican nation-state, whereas the militarization of China—under warlords, Guomindang, and Communists—continued to proceed until the middle of the century.

"Military dictatorship" would not be the right word for what happened in either Turkey or China. Enver Pasha, the most influential military man in the Young Turk government (which was firmly in control by 1913), was certainly strong enough to take the Ottoman Empire into war in 1914 alongside the Central Powers, but he never had absolute authority and he remained primus inter pares in a mixed military-civilian ruling coterie. In China a powerful bureaucrat and military reformer from the Qing period, Yuan Shikai, took over the presidency within months of the revolution. Between 1913 and 1915 he ruled as de facto dictator, supported by the army but not by it alone. Yuan still shared some of the old Chinese mistrust of men-at-arms. Only after his death in 1916 did the country fragment into a mosaic of military regimes.[182]

The social coalitions on which the four revolutions based themselves varied considerably in breadth. The most intensive popular participation was in Russia, where the forces seeking to bring down the autocracy ranged from liberal nobles to starving peasants made destitute by the high redemption payments following the emancipation from serfdom. In China the revolution happened so quickly that its dynamic could not spread from the cities to the countryside. In the years before 1911 some parts of China had an above-average level of peasant protest, but the Qing were by no means driven from the throne by peasant uprisings like those that preceded the fall of the Ming Dynasty in 1644. "Bourgeois" forces were more important in Russia than in the other revolutionary processes, because it was more developed socially and economically. In Iran bazaar merchants played an active role by organizing boycotts. In China one cannot really speak of a bourgeoisie at all before 1911. The label "bourgeois revolution," as Lenin recognized, was hard to pin even on Russia, let alone on Iran, China, or the Ottoman Empire. None of the revolutions can be divorced from its international context. In all four, the existing regime was on the defensive, still reeling from military or political defeats in the wider world: the Tsarist Empire, from the war of 1904–5 with Japan; China, from the Boxer invasion of 1900; the Ottoman Empire, from fresh setbacks in the Balkans; and Iran, from the action of foreign concession hunters and the advance of the British and Russians into their respective spheres of influence. The revolutionaries did not just expect that a change or overthrow of the political system would solve their own material problems, guarantee civil liberties, and allow them a say in political life. They also hoped that a strong nation-state would be more assertive in standing up to the impositions of the Great Powers and foreign capitalists. This was less the case in relation to Russia, however, since it was itself an aggressive imperial power, and its costly and ultimately futile foreign policy was among the targets of protest action.

Outcomes of the Revolutionary Process

Where did the four different revolutions lead? In none of them would there be any going back to the old order. Russia's medium-term future was the Bolshevik Revolution. In Turkey and Iran, noncommunist, authoritarian regimes geared to economic development were established in the early 1920s. In China a regime of this type—the post-1927 Guomindang—managed to stabilize itself, though less comprehensively than the others, and after 1949 a second, Communist-led revolution finally halted and reversed the long-term process of political disintegration that the 1911 Revolution had speeded up.

But how did the results of the revolutions look in the short term, still on the horizon of the nineteenth century, as it were? In Russia the trend toward constitutional government, which for a short time had been more than what Max Weber, a keen observer of Russian politics, called "token constitutionalism," came to an end in June 1907 when prime minister Stolypin staged a coup d'état with the tsar's support, and the elected Second Duma—the successor to the First Duma conceded by the tsar during the 1905 Revolution—was disbanded. A Third Duma, elected on a highly skewed basis, proved to be accordingly tame and pliant, while the Fourth Duma (1912–17) was an almost total irrelevance. The all-decisive cutting short of Russia's parliamentarization had occurred in 1907.

In Iran there was a similar crushing of the blossoms of parliamentary democracy that had burst forth in the briefest space of time. The Majles (parliament) became a central institution of political life more than anywhere else in Asia and in Russia, carried along by a troika of bazaar merchants, liberal clerics, and secular intellectuals that would appear again in the Islamic Revolution of 1979.[183] The shah's putsch of June 1908 was a brutal affair. But whereas in Russia a general apathy set in after the dissolution of the Second Duma, resistance to the shah and his Cossacks led to a civil war that, in the north of the country, ended only with the intervention of Russian troops in winter 1911. Large numbers of constitutional politicians and revolutionary activists were removed from office and executed or deported.[184] Clearly there are parallels with Hungary in 1849, although there the Russians left it up to the Habsburgs to take revenge on the revolutionaries. Nevertheless, parliamentarianism had sunk deep roots in Iran, and since then, through all the regime changes, it has regarded itself as a constitutional country.

China was different. Before 1911 the demand for more efficient government, both internally and externally, had been much more important than the pressure for democratization. Since 1912 China has kept endowing itself with constitutions, but right up to today it has not been able to get a parliament up and running (except in Taiwan since the 1980s). The 1911 Revolution created no stable parliamentary institutions and, more important, no myth of parliamentary sovereignty that might be critically activated. Nowhere did an ancien régime crumble as quickly and noiselessly as in China. Nowhere did a republic arise so directly from its ruins. But nowhere either did the military, the only force theoretically capable of holding the country together, act with such little sense of responsibility. The revolution eliminated censorship and state-directed institutional conformism. In so doing, it opened at least the cities to a special kind of modernity, if only without constructing stable institutions.

In this respect the Ottoman-Turkish evolution was more successful; the transitions were smoother. The aged Sultan Abdülhamid even remained on the throne for another year, until his supporters foolishly attempted to remove the new rulers. His successor, Mehmed V. Reşad (r. 1909–18), was for the first time in Ottoman history a constitutional monarch without political ambitions. Neither the Romanov nor the Qing Dynasty would be granted such a mellow finale. It is true that the post-1908 period of freedom and pluralism ended in 1913 after the assassination of Mahmud Şevket Pasha, one of the leaders of the Young Turks, while at the same time the Balkan War plunged the empire into dire straits.[185] The outcome, however, was neither temporary restoration (as in Russia and Iran) nor territorial disintegration of the core country (as in China after the Yuan Shikai interlude in 1916) but a journey full of obstacles and detours toward one of Asia's less crisis-ridden and more humane polities of the interwar period: the Kemalist Republic.

Atatürk, certainly no democrat, was on balance an educator rather than a seducer of his people—not a warmonger, not a Turkish Mussolini. The Ottoman-Turkish revolutionary process therefore displays the clearest logic of the four Eurasian revolutions. It was more or less continuous and came to rest in the Kemalism of the 1920s. In 1925, when this goal was achieved, Russia (the Soviet Union) and China were entering new phases of their stormy history. Meanwhile in Iran the military strongman, Reza Khan, ended the (by then purely ornamental) Qajar Dynasty and elevated himself as the first shah of the new Pahlavi Dynasty. His autocratic rule, from his rise as war minister in 1921 to his eventual banishment in 1941, meant that Reza conformed more clearly than his contemporaries Atatürk and even Chiang Kai-shek (who alternated from 1926 in the roles of military and political leader) to the type of the violent military dictator with some modernizing ambitions. But, unlike Atatürk, he was neither an institution builder nor a man of political vision, and the weakness of Iran made him far more dependent on the Great Powers, until they finally brought him down because of his pro-German proclivities.[186] Twentieth-century Iranian history, following the Revolution of 1905–11, was more discontinuous than that of Turkey, and major goals of the revolutionaries remained unfulfilled. In 1979 a second revolution pursued a new, illiberal agenda for ten years before the order in Russia, too, was once more revolutionized. Only Turkey, through more or less the same time span as Mexico, experienced no further revolutions after the transitional years of 1908–13.

None of the four Eurasian revolutions suddenly erupted within contexts of utter stagnation and ossification. The popular European image of a "peace of the graveyard" supposedly created by bloodthirsty Oriental despots was a distortion of reality. Societies throughout Eurasia had been scarcely less in ferment than their European counterparts, exhibiting numerous forms of protest and collective violence.[187] In Iran, for example, sections of the population have rebelled on many occasions and, in much the same way as in early modern Europe, tried to assert their interests through pressure and theatrical actions: nomadic tribes, the urban poor, women, mercenaries, black slaves, and sometimes "the people" as a whole, especially directed against foreigners.[188] Other Asian countries were different only in degrees. In China the state had traditionally kept the population under tighter control than in the Muslim countries, using a system of collective liability and punishment (baojia) to hold whole families or villages accountable for individual rule breaking. But this was successful only so long as the bureaucracy remained reasonably efficient and people were not driven to desperation by conditions that threatened their survival. At the latest from the 1820s, when these premises began to crumble, China also became difficult to govern.[189] Thus, apart from anything else, the revolutions were responses to problems of governability. These problems were in turn partly the result of an intrinsic dynamic of social conflict and of changes in cultural values, partly of the kind of external destabilization that generally affected "peripheral" and socioeconomically "backward" countries.

It cannot be overstated how much Western European constitutional thought influenced Asia, from Russia to Japan, or how creatively it was adapted to the needs of individual countries.[190] The Ottoman constitutionalist movement of 1876 sent out a first important signal, and Japan after 1889 furnished evidence that a constitution was not merely a piece of paper but could also become, in an Asian context, a powerful symbol of national integration. The struggle for dominance and transformation in the state was invariably fueled by the rhetoric and practice of constitutionalism. The powers that be were no longer thought of as a fact of nature; power could be wrested in conflict and given a different institutional form. Time was running out for dynastic rule, now that its existence no longer seemed a matter of course. The age of ideologies and mass politics was beginning.

CHAPTER XI

The State

Minimal Government, Performances, and the Iron Cage

1 Order and Communication—The State and the Political

The variety of political forms was probably greater in the nineteenth century than at any previous time in history, ranging from the complete statelessness of hunting communities to the sophisticated systems of empires and nation-states. Already before the arrival of European colonialism, there was a considerable diversity of arrangements for exercizing power and regulating the affairs of the community, not all of them recognizable from a Western and modern point of view as a "state." The colonial state only gradually absorbed, or at least modified, these older forms on a case-by-case basis. It is possible to speak of the worldwide, though by no means uniform and complete, spread of the European state for the period shortly before the First World War, but certainly not in 1770, 1800, or 1830.

The nineteenth century began by inheriting the new states that had taken shape in the early modern period.[1] The generic term for those political orders used to be "absolutism." Today we know that Europe's "absolute" monarchies, with the possible exception of Petrine tsardom, did not enjoy such total freedom of action as contemporary apologists or later historians liked to imagine. Even "absolute" rulers were entangled in a web of reciprocal obligation. They had to pay heed to churches or landowning nobles, could not simply brush aside established legal conceptions, needed to keep their underlings in good spirits, and had to accept that even the most authoritarian practices could not be relied on to fill the state coffers. The European monarchies of the mid-eighteenth century were the result of an evolution that had not begun before the sixteenth. The same was true of most of Asia's monarchical systems, which, in their eighteenth-century form, were the creations not of a remote past but of quite recent military empire building. The hoary idea of an opposition between moderate monarchy in Western Europe and boundless despotism from the Tsarist Empire eastward—a contrast developed most sharply in Montesquieu's work of 1748 *De l'esprit des lois* (The spirit of the laws)—is not altogether aberrant. But the overriding impression we have of early modern Eurasia is of a spectrum of monarchical states that will not fit into such an East-West dichotomy.[2]

A further early modern innovation was the overseas colonial state, originally confined to the western hemisphere but from the 1760s on transplanted to India. Though it copied European state forms, it adapted to local circumstances and underwent many changes over the years. Its collapse in North America in the 1770s was soon followed by the momentous rise of the constitutional republican state. Political diversity reached its unprecedented peak in the middle of the nineteenth century. *Afterward* countries worldwide turned into territorially defined nation-states, a relatively uniform type that could be combined with various constitutional forms: democracy as well as dictatorship. The twentieth century was marked by further homogenization, so that in its second half an electorally legitimated constitutional state became the only recognized norm. Finally, the disappearance of communist party dictatorships masquerading as "people's democracies" left only one non-Western model consciously based on distinctive principles of its own: the theocratically oriented Islamic Republic.

Differentiation and Simplification

In the history of the organization of political power, the nineteenth century thus represents a transitional stage of differentiation and renewed simplification. It was also the starting point for four major trends that would come into their own globally in the course of the twentieth century: nation building, bureaucratization, democratization, and the rise of the welfare state. From the vantage point of post–World War I Europe, the nineteenth century must have appeared as the veritable golden age of the state. In that epoch, the American and French revolutions had associated the state with the principles of citizenship and the common good. Both moderate participation of the populace and the capacity to maintain the law in a just and equal way seemed to have found a fruitful balance. Moreover, until 1914 the European state even kept its dangerously growing military potential under control. In short: the state had steered clear of the twin political extremes of despotism and anarchy.

Let us take a closer look. The following were the main developmental tendencies of the state in the nineteenth century:

- construction of militarized industrial states with new capacities for empire building
- invention of the "modern" state bureaucracy based on principles of generality and rational efficiency
- systematic expansion of powers to extract taxes from society
- redefinition of the state as a provider of public goods
- development of the constitutional rule of law and a new idea of the citizen, involving a legitimate claim to the protection of private interests and a say in political life

· rise of the political type "dictatorship" as a formalization of clientelist relations and/or the exercise of technocratic rule by acclamation

Not all of these trends gradually spread out from Europe to the rest of the world through a process of conscious export or creeping diffusion. Some of them were of thoroughly *non*-European origin: the constitutional state arose in North America on the foundations of the English Glorious Revolution of 1688 and the political theory that had underpinned it; postmonarchic dictatorship flourished first of all in South America. It would be equally one-sided to see the main tendencies as unfolding "behind the backs of human subjects." The development of the state was not an automatic process independent of social change and political decisions. This becomes clear when we ask why one and the same main trend was more powerful here than there, and why its expressions were so different.

The problem is thrown into relief if the Western European state is no longer taken as the historical norm. Political systems in western Africa, for example, were by no means primitive or backward because they failed to correspond to the European model. The state in Africa did not involve military control over a precisely defined territory, in which a single power claimed sovereignty and expected people to obey it for that reason. Rather, Africa was organized as a patchwork quilt of overlapping and fast-changing obligations between subordinate and superordinate rulers. In the Arabian Peninsula too, there was no European-style "state organization" far into the nineteenth century, but there were complex relationships among a multiplicity of tribes under a form of Ottoman suzerainty that had for a long time been scarcely felt by those subject to it. The term "tribal quasi-states" has been used to refer to this.[3] The political landscape of Malaya, polycentric in a different way with its many princedoms (sultanates), represented a microcosm of the larger mosaic of Southeast Asia, in which only colonialism defined forms of rule on an unambiguously territorial basis.[4] To consider the European state normal would mean accepting that the history of that part of the world inevitably resulted in colonial conquest and imposed reorganization. In reality, colonialism was not the gentle telos of historical development but a foreign intervention that often appeared brutal to those exposed to it.

It is equally problematic to regard the state's "monopoly of the legitimate use of physical force in the enforcement of its order" (Max Weber)[5] not as a theoretical ideal but as an actual state of affairs. For some parts of the world it has never been a meaningful category—in Afghanistan, for example, as we can see today. In the major empires, independently armed minorities such as the Don Cossacks remained outside a central military command structure right up until the last quarter of the nineteenth century.[6] Piracy, thought to be a thing of the past, flared up again in the Caribbean in the wake of the Latin American independence struggles of the 1820s, and it was only with difficulty that the British and American navies stamped it out after 1830.[7] Monopoly of force, then, is not an intrinsic part of the definition of the modern state, but rather an exceptional historical circumstance that has been aimed at and achieved only provisionally. In revolutionary periods it has rapidly broken down. The Chinese state, for instance, tried with some success throughout the eighteenth century to disarm the population and keep it quiescent. After 1850, however, millions took up arms against the Qing state in the Taiping Revolution. It has seldom been a problem for revolutionaries to get their hands on weapons. A monopoly of force can be maintained only as long as a central state is able to tame bellicose elites and to convince a large part of the population that it can attend to law and order. If this is not the case, then markets in violence open up and the socialization of violence can quickly give way to its privatization. In one of the most stable democracies, the United States of America, the two tendencies were closely bound up with each other.

This leads to another general conclusion: that "strength" is not always just an independent variable in the development of the state. It would be a distorting idealization to see in the state an inexorable logic of growing depersonalization and rationality. The state does mold society, but it is also dependent on revolution and war, on the productive economy underlying its financial power, and on the loyalty of its "servants."

Types of Political Order

There are many possible typologies of political order; it all depends on the criterion of definition. One meaningful approach would be to ask where power is located and how intensely and in which ways it is exercised. A distinction might then be drawn between political orders in which power is deployed "extensively" to organize a large number of people over wide areas (a major empire, for example) and those in which "intensive power" within a smaller area induces people to engage in a high level of political activity (as in a Greek polis). Another useful distinction would be between "power of authority" (that is, the communication of commands within a hierarchy of subordination) and "dispersed power" that is less directly perceptible as a relation of obedience and works through more subtle constraints such as a legal system or ideological inputs. This second approach may be applied not only to whole political systems but also to particular organizations, such as an administrative office, a church, or a school.[8]

A criterion that suits the transitional age of the nineteenth century especially well is the extent to which there are checks on power. Liberalism, the most influential political theory of the time, saw the introduction of such controls as one of its chief objectives. And although in the period before the First World War, liberalism hardly anywhere fully realized the ideals of its leading champions, there was a visible tendency in many parts of the world to reduce individual arbitrariness in the exercise of power and to enforce the principle of accountability. From this point of view, the following basic types of political order may be identified around the year 1900.

Autocracy, where the will of a prince ruling with the help of advisers held ultimate sway (quite possibly within a system of codified law), was by that time uncommon. There was still absolutism of that kind in the Tsarist Empire, the Ottoman Empire (since 1878 once more), and in Siam. This did not necessarily mean that their systems were particularly backward—King Chulalongkorn of Siam was one of the least restricted rulers of the age, and yet, as an enlightened despot and far-sighted reformer, he made a host of decisions that served the overall interests of his country and led it into modernity.

Even in a monarchy, virtually unlimited executive powers may be given to a minister: Cardinal Richelieu, for example, or the Marquis of Pombal in Portugal in the 1760s. But they always remain dependent on the goodwill of the ruler, however weak he may be. *Dictatorships* are postrevolutionary or postrepublican systems in which a single individual, usually with a small group of helpers and subordinate rulers, enjoys a freedom of action comparable to that of a monarch. He does not, however, have the sanction of tradition, dynastic legitimacy, or religious consecration. The dictator, known in Europe since antiquity, keeps himself in power through the use or threat of violence, and by providing for a clientele of varying size. The army and police do well under his rule, and his control over them is an indispensable element. Having installed himself for life, he must ensure that the special conditions of his coming to power (whether a putsch or popular acclamation) are translated into durable institutions.

Examples of this type were few and far between in Europe after the fall of Napoleon I. The one that came closest was Field Marshal (later Count) Juan Carlos de Saldanha, who repeatedly intervened in Portuguese politics between 1823 and his death in 1876, but who in the long run served to establish less his own personal rule than an "oligarchic democracy."[9] Only with the Bolshevik Revolution of 1917 and its new-style party dictatorship, followed by the rise of rightist rulers in 1922 in Italy (Benito Mussolini) and 1923 in Spain (Primo de Rivera), did an age of dictatorship begin in Europe and, in the same decade, in noncolonial Asia (Iran, China).

In the nineteenth century, Hispanic America was the only stomping ground for dictators, most strikingly illustrated by Porfirio Díaz, who between 1876 and 1911 broke Mexico's vicious circle of political instability and economic stagnation, while reducing popular participation in decision making to a minimum and generally paralyzing public life. Don Porfirio was neither a warlord nor a murderous tyrant like Juan Manuel de Rosas—who ruled Argentina from 1829 to 1852 (and with particular brutality between 1839 and 1842) by means of a system of secret police, informers, and death squads[10]—nor a typical South or Central American caudillo, hostile to institutions, uninterested in economic development, using the business of violence to keep his direct supporters supplied with spoils and to afford "protection" to the property-owning classes. Díaz, rather, was obsessed with stability, although he did not manage to transform his well-oiled machine of personal patronage into a crisis-proof state apparatus.[11] Another strong president from the military, Julio Argentino Roca in Argentina, showed greater farsightedness in the 1880s and 1890s, using political parties and elections to improve the efficiency of the system and paving the way for an elitist "democracy."[12]

In *constitutional monarchies*, at least in the form they took around 1900, a written constitution contained some provision for parliamentary representation and authority, but it was not possible for the parliament to bring down the royal government. The executive was neither appointed by nor accountable to the elected national assembly. The monarch played a comparatively active role, and was usually required to arbitrate among the many informal groups into which the power elite was divided. Systems of this type existed, for example, in the German Reich, in Japan (whose constitution of 1889 borrowed heavily from the German one of 1871), and in Austria-Hungary from the 1860s on (where parliamentarianism was far less functional than in Germany, partly because of ethnic fragmentation among the imperial subjects).

Systems of parliamentary accountability could have a monarchical (as in Britain and the Netherlands) or a republican (as in the French Third Republic) head of state. This was of lesser importance than the fact that the executive was drawn from, and could be removed by, the elected parliament. A special variant was the US dualism of presidency and Congress, but although the president was not appointed by the people's representatives his election (direct or indirect) meant that his term of office was limited and even in wartime did not turn into a dictatorship. The American Revolution brought forth no Napoleons.

Material from all around the world has enabled political anthropologists to demonstrate the numerous ways in which power differences arise in society, and how political processes serving the goals of the collective, or some of its subgroups, get under way and remain in operation. More difficult to establish from the sources are the conceptual worlds or political "cosmologies" of societies with little or no written tradition. Highly complex conceptions of the political do not exist only in the great theoretical traditions of China, India, Christian Europe, and Islam. A static view of state-related institutions must accordingly be replaced with one more geared to the dynamic nature of events within political spaces and fields. The whole typological approach in which each state must be allocated to one particular form and a definite territory thus becomes open to question.[13] To complement the above fourfold characterization of the exercise and limits of power, we may posit a fifth category to cover the various possibilities of relatively weak institutionalization: that is, *systems of allegiance* or patron-client relations, in which—here lies the difference with dictatorship—an ancestrally defined prince, chieftain, or "big man" (women also can occasionally play this role) offers protection and serves as focus for the symbolic unity of the community. Here, too, there may be individual officials, but not a state hierarchy more or less independent of specific persons. The dynastic principle and the sacredness of the ruler are less pronounced than in more stable and complex forms of monarchy, and acts of usurpation are easier to pull off. The legitimacy of rulers rests in part on proven leadership qualities, and checks on the exercise of power mainly consist of deliberation and judgment concerning their performance record. For such an understanding of politics, hereditary kingship is more alien than a system involving election or acclamation of the First Man. Political systems of this kind existed at the beginning of the nineteenth century on every continent, including the world of the Pacific Isles, although the cultural prerequisites varied greatly. They made the "docking process" relatively easy for European colonialism, and after the period of actual conquest Europeans could try to situate themselves as top patrons in a chain of allegiances.

Vision and Communication

Such typologies make it possible to take orderly snapshots, as it were, but it must then be immediately asked which kinds of political process they capture. Political orders may here be distinguished from one another in two further respects. On the one hand, each has an underlying vision and imagery of the totality, which ideologues of the system, as well as a large number of people living within it, see not simply in terms of unequal power structures but as a framework of belonging. In the nineteenth century, the nation increasingly became the largest unit with which people could identify. But, under different conditions, other conceptions also persisted: for example, the idea of a paternalist bond between ruler and subjects, or, as in China, of the cultural unity of a great empire. Apart from a handful of anarchists, no one saw chaos as the optimum state of affairs; it was thought that integration of the ideal political order could come about in a number of ways. In the nineteenth century, too, religion defined the worldview of most and was a strong glue bonding people together.[14]

On the other hand, actual political orders exhibited different forms of communication, and it must be asked which of these were dominant and characteristic in each case. The traffic might take place inside ruling apparatuses—for example, between a monarch and his senior officials—or within cabinets or unofficial elite circles such as British clubs or the "patriotic societies" in the Tsarist Empire. But it could also—and this became increasingly important in the nineteenth century—involve two-way relations between politicians and their electorate or following. Kings and emperors had traditionally presented themselves before their people, usually at a ceremonial distance, unless they remained invisible or absent like the Chinese emperors after roughly 1820. Napoleon III, unlike his more solitary and despotic uncle, was a past master at appeals to the nation. And Wilhelm II, who under the constitution had to pay little heed to the views of his subjects, spoke more frequently at public gatherings than any other Hohenzollern monarch.[15]

It was a novelty of the nineteenth century that politicians directly addressed their voters and supporters, seeking to gain a mandate from them. This style of politics first established itself during the presidency of Thomas Jefferson

(1801–9), then grew apace with the so-called Jacksonian Revolution, named after President Andrew Jackson (1829–37), which replaced the elitism of the founding fathers with a more populist or "grassroots" conception of politics, and scorn for "factionalism" with an acceptance of competition among parties.[16] The number of elective offices increased by leaps and bounds, extending even to judges. In Europe, however (except for Switzerland), the practice of democracy remained much more oligarchical—even in Britain until 1867. The suffrage was more restrictive than in the United States.

To be sure, revolutions generated quite special surges of popular involvement. In periods without a revolution, the general election campaign—another "invention" of the nineteenth century—became a focus for direct communication between politicians and citizens. William Ewart Gladstone was a pioneer of this in his Scottish constituency of Midlothian, in 1879–80. Until then British election campaigns had been rather convivial affairs among a small circle of people, such as the one satirically described in Dickens's *Pickwick Papers* (1837). Gladstone was the first British politician (and the Conservative Disraeli followed in his tracks) who held mass rallies outside the context of protest actions, as a normal part of democratic life, in which the main tone was one of semireligious revival. The orator stirred up his audience, engaged in verbal jousting with hecklers, and ended proceedings by circulating among his supporters.[17] For Gladstone, if all this was responsibly managed, it helped to educate the ever widening electorate. The fine borderline with demagogy was crossed where—as in Argentina under Juan Manuel Rosas and his wife, a nineteenth-century Evita—inflammatory speeches against opponents were geared to the urban plebs: a primitive, personalized form of manipulation known since antiquity but little used outside revolutionary situations.[18] What was characteristic of the nineteenth century was the new taming of agitation for electoral purposes, within the regular functioning of the political system.

2   Reinventions of Monarchy

In the middle of the nineteenth century, long after the French Revolution, monarchy was still the most prevalent form of state. There were kings and emperors on every continent. In Europe the republics of the early modern period, as well as new ones summoned up in the Age of Revolution, had disappeared in a final wave of "monarchization."[19] If, as is sometimes claimed, the decapitation of Louis XVI removed the basis for monarchy as a political order and a focus of the collective imagination, its death agony nevertheless proved to be long and happy. In the years immediately after 1815, Switzerland was alone among the major European countries in not having a royalty. Monarchist sentiments were cultivated as far away as Australia—although no British monarch (as opposed to a succession of princes) put in an appearance there until 1954—and in 1901, when the Australian colonies established a federation, no one thought of proclaiming a republic.[20] There were rulers with a few thousand subjects and others with several hundred millions; autocrats in direct charge and princes who had to be content with a purely ceremonial function. A small kingdom in the Himalayas or the South Seas had two main things in common with the crowned heads of state in London and Saint Petersburg: a dynastic legitimacy that made the dignity of the monarch or emperor hereditary and an aura surrounding the throne that assured its occupant of a basic respect and veneration, irrespective of his or her personal qualities.

Monarchy in the Colonial Revolution

The labels "monarchy" and "kingdom" covered an extraordinary abundance of political forms. Even structurally similar instances varied widely in the cultural embeddedness of royalty: whereas the absolute Russian tsars cultivated a sacred aura right up to the end of the Romanov Dynasty, so that even Nicholas II cherished and celebrated a mutual bond of piety with the Russian people,[21] monarchs in France and Belgium after 1830 found themselves left with no more than the daily routine of bourgeois kings. The Russian Orthodox Church zealously preached the holiness of the tsar, while the church in Catholic countries exercised greater restraint, and Protestantism had only a rather abstract notion of a state church.

Southeast Asia offers a good example of the diversity of monarchy. At the beginning of the nineteenth century there were:[22]

- Buddhist kingdoms in Burma, Cambodia, and Siam, where the monarch lived in a closed palace world and could hardly take any political initiatives because of the power of his advisers or the burden of protocol;
- a Vietnamese imperial system modeled on China, in which the ruler stood at the apex of a pyramid of officials and customarily regarded neighboring peoples as barbarians in need of being civilized;
- Muslim sultanates in a polycentric Malayan world, whose positions were far less elevated than those of other rulers in the region, and who governed with less pomp over their mostly coastal or riverside capitals and their hinterlands; and
- colonial governors, especially in Manila and Batavia, who appeared as representatives of European monarchs and even sought to cultivate regal splendor as envoys of the rather frugal and republican-inclined Netherlands.

Along with revolution, European colonial rule was the great enemy of monarchies in the nineteenth century. Europeans destroyed indigenous royalty in many parts of the world, either totally eradicating it or weakening it beyond repair. Most often, the local monarch fell under their "protection" and was allowed to keep most of his revenue, together with a royal lifestyle and any religious role he had played until then. At the same time, his political powers were curtailed, and he lost both command of his army and traditional legal privileges such as the power of life and death over his subjects. The lengthy process of subjecting non-European kings (and chiefs) to indirect rule was completed shortly before the First World War. The Moroccan sultan was the last substantive monarch to be placed under a colonial resident (in 1912) while keeping his rank and dignity.[23] The decision as to whether a colonial power exercised direct or indirect rule was never made in accordance with general principles or an overarching strategic plan; the exact shape of colonial administrative despotism depended in each case on the local conditions.[24]

Sometimes a really serious misjudgment could be made. In Burma, King Mindon introduced a set of stabilizing reforms up to his death in 1878, intending them to remove the pretext that imperial control was necessary to end chaos and fill a power vacuum. But economic difficulties under the arbitrary rule of his successor, together with growing pressure from British economic interests, cleared the path for outside intervention. Fearing especially that the royal government in Mandalay was unable or unwilling to keep third parties out of what they regarded as their own sphere of influence, the British declared war in 1885 on the Kingdom of Upper Burma. Once the last resistance was overcome, they then annexed Upper Burma and in the following years added it to their long-standing administration in Lower Burma—and the government of British India. The Burmese monarchy was abolished. The British misjudgment lay in the fact that one of the traditional roles of the king had been to keep the large Buddhist clergy under control. The disappearance of the royal structures suddenly disempowered and devalued the whole world of the monasteries, so that, for example, there was no longer anyone to appoint a head of the hierarchy. Not surprisingly, therefore, the whole colonial period was marked by unrest among the Buddhist monks—an influential section of the population whose trust and support the colonial state never managed to obtain.[25]

No unified system was imposed in large areas of the colonies. The British illustrated this in India, where (a) some provinces were placed under the direct rule of the East India Company (after 1858, the Crown); (b) some five hundred other territories throughout India retained their maharajahs, nizams, and so on; and (c) a few border regions came under special military administration.[26] In the 1880s the French destroyed the Vietnamese kingdom, tapping it neither at the level of symbols nor through the incorporation of its administrative personnel. In other parts of the Indochinese federation they were more flexible: indigenous dynasties were left in place in Laos and Cambodia, but they had to accept that the French decided on the royal succession. There were—as in Africa—subtle nuances under a system of indirect rule; the colonial powers did not find it easy to manipulate the charisma of the rulers. Thus, King Norodom I (r. 1859–1904) and his ministers lost most of their powers after 1884, and the king, a strong character, was reduced to the role of lead player in court rituals, yet the colonial masters went in constant fear of royal resistance and were well aware that his removal might trigger uncontrollable reactions among the Cambodian population.[27] The

Cambodian monarchy was one of the few in Asia to survive the colonial period, and under King Norodom Sihanouk (r. 1941–2004, with interruptions) it played a not-insignificant role in the postwar history of Cambodia.

One of the strongest continuities anywhere in colonial history is to be found in Malaya, where no sultan was powerful enough to resist British influence effectively. The British wagered on close cooperation with the royal-aristocratic elite of Malaya, curtailing their privileges far less than those of the Indian princes. In a part of the world where, more than elsewhere in Asia, political rule did not come down to a simple matter of hierarchy, they strengthened the authority of the sultan in each of the states, simplified the rules of succession (but rarely intruded in it), laid ideological emphasis on the leading role of the Malay ruler in a multicultural society more and more dominated by Chinese economically, and eventually, on a much greater scale than in India, opened up the administration to princes from the sultan's family. Monarchy was therefore strengthened rather than weakened in Malaya during the colonial period—and yet, in the transition to independence in 1957, there was no centralized Malayan monarchy but only a set of nine thrones coexisting with one another.[28] The extreme Malayan example of indirect rule, fascinating though it is, was clearly an exception. Only in Morocco does one perhaps find a parallel, and there too, monarchy clung on more successfully than almost anywhere else in the Muslim world.

Where royal structures persisted outside Europe, they did not always remain on the paths of tradition. New contacts brought with them new models of rule and new scope for the appropriation of resources. If a king or chief managed to break into or even monopolize foreign trade, he might sometimes strengthen his position. This was the case in Hawaii, where in the 1820s and 1830s, long before the United States annexed the island in 1898, rulers were able to acquire luxury goods from abroad with the proceeds of the sandalwood trade, bedecking their persons and residences with costly, high-prestige objects in a hitherto unknown elevation of the monarchy.[29]

In short, few monarchies lasted through colonial times, and when they did it was under conditions of especially weak indirect rule. After independence, no country restored a fallen dynasty; a small number of monarchs appeared in the guise of republican presidents, such as the Kabaka of Buganda in the years from 1963 to 1966. The royal and imperial houses of Asia and Africa that lasted into the fourth quarter of the twentieth century—and sometimes up to this day— were essentially in countries that did not fall under colonial rule: above all Japan and Thailand, along with Afghanistan (until 1973) and Ethiopia (until 1974).

The Asian monarchies were not just fancy "theater states" in which inconsequential rituals were simply kept ticking over for the sake of aesthetics.[30] In the non-Muslim traditions of Asia, the ruler's task was to act as spiritual intermediary with higher powers, keeping up etiquette and ensuring that the correct forms of communication were used at court and in relations between the court and the population. Royal spectacles served to integrate the king's subjects symbolically; they were seldom merely ceremonial husks, as in the French Restoration monarchy between 1815 and 1830, which sought to cover up its legitimation deficit by means of nostalgic reenactments.[31] Asian monarchs, like their European counterparts, had to legitimize themselves performatively: the king had to be "just" and to order his country in such a way that a civilized life was possible. Fed by various sources, theories of worldly statecraft were of major significance for what people expected of their rulers, both in the great traditions of China and India and where they came together in Southeast Asia. The good king or emperor had to control resources, surround himself with dependable administrators, maintain a strong army, and wrestle with the forces of nature.[32] The monarchy itself stood above all criticism, but the individual who sat on the throne was obliged to prove his worth. These multiple tasks and expectations confronting the monarchy meant that its abolition by the colonial revolution created deep fissures in the social web of meaning. Transitions were especially difficult where a monarchical link to the symbolic repertoire of the past was totally lacking, and where, after the end of the colonial state, only the military or a communist party remained as a vehicle of national centralization.

By 1800 the age of unconstrained despots and arbitrary rulers was already over. Mass slaughter in the style of Ivan IV ("the Terrible," r. 1547–84), the Hongwu Emperor (founder of the Ming dynasty, r. 1368–98), or the Ottoman sultan Murad IV (r. 1623–40) was a thing of the past. The example of a "bloodthirsty monster" most widely publicized in Europe was the South African military despot Shaka. Europeans who visited him after 1824 unfailingly reported that, before their eyes, he had ordered an execution with a wave of his hand and dismissed the English penal system (which they described to him) as incomparably worse.[33] Shaka was a great exception. In Africa too, a simple opposition between total omnipotence and European monarchy bound by law and custom does not correspond to the true picture. Zulu kings and other rulers may or may not have had greater latitude than European monarchs in relation to local traditions. Their legitimacy did rest on arbitrary reserve powers, but clans and their main lineages always remained semiautonomous factors that the king had to take into account, and his control over the economic resources of his people (especially their livestock) was tightly circumscribed.[34]

In Southeast Asia, back in the precolonial years bridging the eighteenth and nineteenth centuries, monarchical systems had already moved away from extreme personalization toward greater institutionalization.[35] In China, with its strong bureaucratic proclivity, emperors repeatedly had to fight with officials to stamp their authority on the course of events. Those who ruled after the abdication of Qianlong in 1796 did so less and less successfully than their eighteenth-century predecessors had done. By the end of the nineteenth century, China's political system was in effect made up of an unstable equilibrium among Dowager Empress Cixi, the Manchu court princes, top officials resident in the capital, and some provincial governors with a semiautonomous power base. In addition, the general laws and statutes of the Qing state remained in force, as did the residual role models to which Cixi could do only limited justice. This, too, was a system of checks and balances, but not in Montesquieu's sense of a division of powers.

Constitutional Monarchy

Restricted monarchy, regulated to prevent excesses, was not a European invention, but constitutional monarchy was first conceived and tried out in Europe and then exported to other parts of the world. It is not an easy category to define unambiguously, since the mere presence of a written constitution is not a reliable guide to how things worked in practice. Cases where the royal will possessed ultimate force in every sphere of politics are relatively straightforward: one speaks then of "autocracy," with reference to Napoleonic France between 1810 and 1814 (although there were representative bodies even then) and above all to Russia until 1906 and the Ottoman Empire between 1878 and 1908. "Absolutism," on the other hand, signifies that the estates act as a force restricting the royal will, and that the monarch is usually less actively engaged in politics than an outright autocrat. Such conditions prevailed in Bavaria and Baden until 1818 and in Prussia until 1848. When they were reintroduced after an interlude of liberalization, the proper category would be "neo-absolutism"; an example is Austria between 1852 and 1861, essentially a form of bureaucratic reform-despotism with liberalizing tendencies. Within the group of law-governed states, historians like to differentiate between monarchical and parliamentary constitutionalism: the former involves a delicate balance between monarch and parliament that can tip either way, while the latter leaves no doubt in theory or practice that the parliament alone is sovereign. The monarch then rules as *king in parliament*, but he (or she) does not govern.[36]

This parliamentary sovereignty, so strong that it even excluded an independent role for a constitutional court patterned on the US Supreme Court (fully operative since 1801) was a British specialty that no one in the nineteenth century followed outside the empire: a special path resistant to export. Only Britain finally overcame the constitutional authoritarianism that still permeated the atmosphere in continental Europe as a late effect of absolutism. Only there, in a country without a written constitution, did it become clear at the latest by 1837 (the year of Victoria's accession to the throne) that the monarch had to respect the constitution even in times of crisis.[37] Victoria was one of the most diligent queens in history: she read mountains of legislation, kept informed about all possible matters, and allowed herself an opinion about nearly every political issue. But she refrained from meddling in politics beyond what was customary, or opposing the majority view in Parliament. Like her present-day descendants, she had a little leeway in appointing the government if the election result or the leadership situation in the political parties was unclear—but she was very reluctant to use it, and it never led to anything that might be described as a constitutional crisis. Queen Victoria had a trusting relationship with some of "her" ministers, especially Lord Melbourne and Benjamin Disraeli. The premier for four terms of office, however, was a man she did not like at all personally: William Ewart Gladstone. She had no way to avoid dealings with him.

The "absoluteness" of a monarchical system may be gauged from the extent to which the prime minister acts as arbitrator and initiator in political life. That never happened in the Tsarist Empire, for example. Bismarck, who as Prussian first minister complained that he had insufficient control over other ministers, wrote a stronger role for the chancellor into the Reich constitution of 1871. But it was only under the British solution of cabinet government, as it had gradually taken shape since the reign of William III and Mary II (1689–1702), that the prime minister's position became unassailable. In the nineteenth century—as still today—Parliament selected from its midst a head of government who, certain of a parliamentary majority behind him, could act confidently in his dealings with the monarch. At the same time, the cabinet as a whole was accountable to Parliament; the monarch could not go over its head by dismissing the prime minister or any other cabinet member. The cabinet was subject to the principle of collective responsibility, by virtue of which its majority decisions were binding on all. A minister who disagreed with his colleagues could express himself freely at cabinet meetings, but outside his hands were tied by cabinet discipline. This meant that the cabinet became in effect the strongest body of state—an ingenious way of getting around the dualism of parliament and monarch typical of constitutions on the Continent. Cabinet government was one of the most significant political innovations of the nineteenth century. Only in the twentieth century would it spread outside the ambit of British civilization.

In a parliamentary monarchy, especially one like Britain with a "first past the post" electoral system, the parliament can ideally serve as an efficient mechanism of leadership selection. In nineteenth-century Britain there was rarely a truly incompetent executive—yet another advantage in its competition with other nations. The strength of a parliament and government is also apparent in the fact that the individual personality of the monarch is relatively unimportant. Britain never had to face a stern test in this regard; Victoria, after sixty-four years on the throne, was replaced upon her death in 1901 by her (admittedly less well suited) son Edward VII (r. 1901–10). The German Reich was not so fortunate, since its constitution meant that the personality of the monarch was of much greater significance. Although one should not overstate, or indeed demonize, the role of Wilhelm II (r. 1888–1918), his numerous public appearances and political interventions seldom had constructive results.[38]

Contrary to a persistent legend, the succession problem was not necessarily solved in a more rational way in Europe than in Asia, where the practice of putting lesser brothers to death when a monarch ascended the throne belonged to the past. Europe's single advantage was that if a new dynasty had to be imported from one country to another, it could draw on a large reserve of ruling houses and an upper nobility capable of playing a role at court. Such interchange was unavoidable in the founding of monarchical states such as Belgium and Greece, and royal houses such as Saxe-Coburg and Gotha served as reliable suppliers of dynastic personnel. Mobility of this kind was lacking in Asia, where princes and princesses simply did not circulate around the continent. Ruling dynasties therefore had to find a way of regenerating themselves. In the nineteenth century, it was a point in favor of monarchy as a state form across the planet that the people occupying the throne in some of the most important countries in the world were not lacking in either energy or experience: Queen Victoria in Britain and the British Empire (r. 1837–1901), Franz Joseph I in Austria-[Hungary] (r. 1848–1916), Abdülhamid II in the Ottoman Empire (r. 1876–1909), Chulalongkorn in Siam (r. 1868–1910), the Meiji Emperor in Japan (r. 1868–1912). Where a formally powerful but personally incapable monarch chose weak ministers, as Victor Emmanuel II (r. 1861–78) did in Italy, the institution failed to fulfill its potential.

New Monarchical Fashions: Queen Victoria, the Meiji Emperor, Napoleon III

A certain revival of monarchy, associated with the prominent "Victorian" rulers, countered the worldwide decline mainly at the level of symbolic politics. This took a number of very different forms. Kaiser Wilhelm II (r. 1888–1918) used (and was in turn used by) the press, photography, and the newfangled motion picture, becoming Germany's first and last royal media star thanks to his frequent public appearances.[39] Ludwig II of Bavaria (r. 1864–86), who would probably have been good at such a role, belonged to a previous age in the development of the media, but he may also be understood as an early escapee from the obsolete hustle and bustle of the court.[40] Whereas Ludwig was an aficionado of the avant-garde music of Richard Wagner, Wilhelm II's passion was for the latest technology, especially if it had to do with war; he did not surround himself only with old Prussian nobles, but, as Walther Rathenau noted, felt best among "dazzling *grands bourgeois*, gracious Hanseatic citizens, and wealthy Americans."[41] The Russian tsars stuck to a more traditionalist image and, in the conflict with modern ideas of rationality, cultivated a political symbolism that emphasized the sacred aura of the ruler while in no way disdaining the new media. In three other high-profile cases—Queen Victoria, the Meiji Emperor, and Napoleon III—the monarchy was actually redesigned in accordance with nineteenth-century conditions.[42]

When Victoria was crowned in 1837, respect for the British monarchy was at an ebb. Supported by her capable husband Albert (named the first Prince Consort in 1857), she acquired over time the reputation of a conscientious mother of the nation with an exemplary family life. After Albert's death in 1861 at the age of 42, she withdrew for many years from all ceremonial functions and led a secluded life on her Scottish estates. This did not fail to have an impact on the public, some voices even calling the future of the monarchy into question, but it only underlines the importance of the role that the royal family had come to play in the emotional life of the nation. As the journalist Walter Bagehot put it in his influential book *The British Constitution* (1865), the monarchy was not a ruling apparatus of the British state machinery but a symbolic institution that ensured civic trust and community spirit.[43] Bagehot underestimated the momentary weakness of the British monarchy. Victoria reemerged from her solitary widowhood in 1872 and, thanks to her serious interest in public affairs, her ever more credible reputation of standing "above classes," and above all her carefully orchestrated political propaganda, she went on to become a genuinely popular queen. A number of her nine children and forty grandchildren ended up on European thrones, and when Disraeli had her elevated to Empress of India in 1876 she became a kind of global monarch, closely identified with and supportive of British imperialism. However, the youthful Victoria had already developed a strong sense of India's imperial belonging and of her own obligations to its peoples. Her Diamond Jubilee in 1897 aroused a degree of royalist enthusiasm right across British society that had never been seen before. When she died in 1901, most people in Britain could not remember a time without her. Critics of the monarchy fell almost completely silent.

Victoria, Albert, and their advisers adapted the institution to the modern age, both in its political functions and in its symbolic radiance.[44] As a woman at the head of the greatest world power, she stood more for matriarchal care than for a greater female role in politics and public life. Yet she embodied the political presence of women in politics as only one other imperial widow did: her slightly younger contemporary, the Dowager Empress Cixi of China. Originally close to the Liberals, Victoria ended her life supporting the conservative elements in British politics. But she held back from extreme forms of aggressive imperialism and left her family a legacy of solicitous care toward the poorer layers of British society.[45]

At first sight, Japan's imperial institution seems to move in a different orbit from that of the European monarchies. Documentary evidence allows it to be traced back to the end of the seventh century, when a centralized polity first took shape; this would make it approximately two hundred years older than the beginnings of the English (Anglo-Saxon) monarchy, in the time of Alfred the Great (r. 871–899). Despite the great model of the Chinese empire founded eight hundred years before, the tennō institution was from the outset rooted in the cultural and political specificities of Japan. In the nineteenth century, too, it developed outside the monarchical landscape of Europe, which incorporated the Meiji Emperor at best through symbolic acts.[46] He had no ties of kinship with the European monarchical class, whereas his only counterpart in the Americas, Emperor Pedro II of Brazil, was after all a cousin of the Austrian emperor. Asian sovereigns could appropriate European royal models only through literature, in the manner of Shah Nasir al-Din, who learned to admire Peter the Great, Louis XIV, and Frederick the Great by reading their biographies.[47] Monarchical solidarity across civilizations did not count for much. After a European trip that took him from capital to capital in 1867, Sultan Abdülaziz felt that only Franz Joseph had treated him in a brotherly spirit free of resentment.[48]

The Japanese emperor was a figure lost in reverie, not a European-style "bour-geois monarch," not the head of a courtly society open to the outside. Neverthe-less, there are a number of parallels with Europe. Unlike the imperial institution in China, which until its demise in 1911 clung to a self-image dating back to the seventeenth century, the Meiji Emperor was the product of a revolutionary age, a new beginning under the aegis of modernity. Japan, like Britain, underwent a huge revaluation of the monarchy in the course of the nineteenth century. In 1830, when immorality and abuse of authority had largely discredited British royalty, the imperial court in Kyoto was sunk in its customary powerlessness. The country's government revolved around the shogun in Edo. By the time of the Meiji Emperor's death in 1912, the imperial institution had become the high-est source of political legitimacy and the most important star in the firmament of national values. On paper as well as in reality, the tennō was more powerful within the Japanese political system than Queen Victoria was in Britain. In both cases, however, the monarchy performed a key integrative function in the na-tional culture. This was stronger in Japan than in Britain because of a conscious drive to breathe new life into the monarchy.

Here two things must be distinguished. On the one hand, the revolutionary edict of January 3, 1868, proclaiming the "restoration" of imperial rule made it the central element in the Japanese state—what Parliament was in Britain. Political power could henceforth have even minimal legitimacy only if it was exercised in the name of the young prince Mutsuhito, who had ascended to the throne at the age of sixteen under the governmental slogan of "Meiji." The architects of the Meiji Renewal used the emperor to lend authority to a regime that was in essence usurpatory; he more or less agreed with their aims, but he had a will of his own and never allowed himself to be simply instrumentalized. At the end of the century Japan thus became a constitutional state with an unusually strong impe-rial head—a form of sovereignty (in both senses of the word) that would not be similarly exercised by the Meiji Emperor's two successors. On the other hand, the symbolism of empire as a markedly *national* institution took some time to de-velop. Internally, it was supposed to cut across all social and regional boundaries, requiring discipline and obedience from the population, bearing a homogeneous national culture in contrast to the plurality of popular traditions, and imparting an outlook in which everyone could recognize themselves.

The emperor was not what the shogun of the House of Tokugawa had been from 1600 until 1868: the supreme feudal lord at the apex of a pyramid of priv-ilege and dependence. He was emperor of the whole Japanese people, an instru-ment and agency for educating it in a special kind of modernity. Outwardly the tennō was meant to embody this modern Japan, and he did so with a considerable degree of success. Courtly display became a mixture of old Japanese elements, either authentic or invented, with symbols and practices borrowed from European monarchies of the age. The emperor sometimes appeared in Japanese robes and sometimes in a European-style uniform or suit, his photographs presenting a dual official personality to his own people and the international public. His monogamous family life was in sharp contrast to the harems of his predecessors and other Asian monarchs. Successful symbolic strategies involving everything from imperial emblems to a national anthem had to be first developed and then conveyed to the population at large.

The Meiji Emperor's carefully prepared trips to various parts of his realm, the first undertaken by a Japanese monarch, served to carry the politically crafted national culture closer to his subjects.[49] In an age when mass media were not yet capable of forging a national consciousness, these direct encounters between emperor and people created a new sense of what it meant to be Japanese. To have seen the emperor meant to have participated in the awakening of national solidarity. In the 1880s the Japanese monarchy found a new respite: Tokyo was built up as the imperial metropolis, the symbolic and ritual core of the nation, whose displays were not a whit inferior to those of Western capitals. Here the spectacle of monarchy went hand in hand with normative disciplining and "civilizing" through institutions such as the school and the army.[50] This, too, was no different from trends in the monarchies and republics of the West, but Japan was especially skillful in its instrumentalization of the ruler, at first itinerant and later firmly established in his capital. As soon as the new political system was up and running, with all power concentrated in Tokyo, the emperor no longer needed to take to the road. In the more heterogeneous Russian Empire, despite the risks of assassination (the fate suffered by Alexander II in 1881 at the hands of revolutionaries), it was advisable for the tsar to seek personal contact from time to time with the provincial nobility. In the case of Abdülhamid II, such tensions led to a split between the ruler's self-image and the way he was seen by others. The sultan wanted to appear as a modern monarch, with a state more deeply rooted than ever before in the daily lives of the Ottoman population, but his obsession with personal security meant that he showed himself to his peoples less often than many of his predecessors had done and never traveled abroad. An extensive symbolic politics was therefore necessary to compensate for the visibility deficit.[51] It emphasized, for example, his religious role as caliph of all believers.

The caliph's supranational appeal was more serviceable for pan-Islamic aims than for the building of imperial or even national identity. In Japan, however, the monarchy became the most important integrative factor of the emerging nation-state. In the post-1871 German Reich, much more federal and less unitary than Meiji Japan, Kaiser Wilhelm I (r. 1871–88) cut a less dashing figure but played a roughly similar role, without inspiring a semireligious cult or making "loyalty to the emperor" the highest political criterion. In Britain, including

Scotland (for which Victoria had a special fondness), the integrative force of the renewed monarchy was also very strong. It was less powerful in the empire, although the persistence of the Commonwealth, adaptable across time and space, demonstrates even today the bonding power of the Crown. The second-largest European colonial empire, the one administered by the French Third Republic, did not bequeath such a lasting voluntary association of former colonies with the mother country.

The third new type of monarchy also fulfilled a primarily integrative function. Napoleon III's empire (1852–70) was the regime of an outsider and social climber who, while linking up with the myth of his uncle, could never make people forget that he did not come from one of Europe's great ruling houses. Unlike Yuan Shikai in a later age in China, he did manage in a postrevolutionary republic to convert himself from an elected president into an imperial dynast. Despite his putschist past, the parvenu gained respect among other European rulers; some monarchs in Asia even saw him as a paragon of enlightened autocracy.[52] Britain immediately recognized his regime, above all for foreign policy reasons, and Napoleon soon acquired the trappings of monarchy and learned to observe the correct etiquette. It was a triumph for him to receive Victoria and Albert in Paris as early as 1855— the first trip to the French capital since 1431 by reigning Britain monarchs. It was not a meeting between blue-blooded cousins but a modern-style state visit.[53] Like the Meiji Emperor, though in a very different way, Napoleon III was a revolutionary profiteer; he did not enter into an alliance with a revolutionary elite (the Japanese model) but created a power base of his own, first by being elected president of the republic in December 1848, then by staging a coup d'état in 1851 and founding a hereditary empire within the space of twelve months. Napoleon was a self-made man. Unlike Mutsuhito sixteen years later, he could not base himself on the institutional continuity of imperial office.

Historians are still debating the character of Napoleon III's rule, with the help of concepts such as Caesarism or Bonapartism.[54] But they generally agree with writers of the time such as Karl Marx or the Prussian journalist Constantin Frantz that it was a regime of a *modern* type. If we leave aside the question of its social foundation, this modernity is apparent in three features. *First*, the president and then emperor paid homage to the revolutionary rhetoric of popular sovereignty, grounding himself on the plebiscite of December 1851 that gave him a majority of 90 percent of the eight million French voters. The emperor considered himself accountable to the people and, in the Constitution of 1852, reserved the right to consult them again at any moment. He could be fairly sure that his rule would accord with the wishes of a large part of the French population, especially in the countryside. It was a monarchy that drew its legitimacy from popular consent, while taking greater care than any of its predecessors to humor the people by means of festivities, ceremonies, and gala events.[55] *Second*, by mid-nineteenth-century standards, it was modern that an initially bloody and repressive regime should seek to develop constitutional forms, hesitantly after 1861 but with greater energy from 1868 on. Louis Napoléon situated himself within the continuum of French constitutional history, and this enabled him, from the early sixties on, to conduct an orderly liberalization and gradually to endow other state bodies with considerable rights and scope for initiative. The monarch's position within the system, at first one of near omnipotence, could thus be reduced. *Third*, the emperor envisaged that the state would play an active role in producing the conditions for prosperity. His commitment to renovate the city of Paris was an expression of this attitude, as were a raft of economic measures. The regime was unprecedentedly interventionist in its economic policy.[56]

Certain parallels with Japan are undeniable. True, the Meiji project lacked the concept of popular sovereignty (never accepted by Emperor Franz Joseph, either),[57] but it also crowned the process of national integration with a carefully prepared constitution, and its economic interventionism after the early 1880s recalls the basic policy approach pioneered by Napoleon III. The Japanese monarchy also set itself the task of "civilizing" a nation that had fallen behind internationally, and did not shy from authoritarian measures to achieve that end. No one would have described the Meiji Emperor as a dictator. But then the use of that label for Napoleon III leads to misconceptions, at least if it suggests twentieth-century practices such as relentless mobilization of the population and long-term systematic repression or murder of political opponents. Napoleon III was normally not in a position to enforce his will by mere fiat. He had to take many different interests into account, including the aristocrats and *grands bourgeois* who had served the French state under the Restoration (1814–30) and the July Monarchy (1830–48). Genuine Bonapartists were rare even in the circle around the emperor.

The territorial pillars of the regime were the prefects responsible for manifold governmental and administrative tasks at the level of the *département*, who were subject to a range of local pressures and also had to deal with elected counselors. For although the head of state occupied a lifetime post (one important element of a monarchy), the common practice of elections in the *départements* amounted to what would today be called a system of "guided democracy." There were official candidates, and it was made difficult for others to win. But, as the opposition gathered strength and extracted compromises from the emperor, it acquired considerable leeway to express itself and to take independent initiatives.[58] A relatively free referendum in May 1870 made it clear how wide the support was for Napoleon III and his government, especially in the countryside and among the bourgeoisie. It showed how successful he had been in projecting himself as a bringer of prosperity and a bulwark against social revolution. When the Napoleonic system fell later in 1870, as a result of international politics and its own diplomatic incompetence, it had been heading less toward further internal "liberalization" (as many historians have claimed) than toward consolidation of an illiberal top-down democracy in monarchical garb.[59]

Courts

The convergence of monarchy and nation-state was a worldwide tendency in the nineteenth century. Some nations actually came into being through the founding act of a monarchy. In Egypt the foundations of a modern state were laid by a new dynasty that took de facto power in 1805, although it was only in 1841 that a *firman* of the sultan in Istanbul confirmed the hereditary rulership. A sumptuous court society, fusing elements of East and West, did not yet develop under the dynastic founder Muhammad Ali (r. 1805–48), a rather austere and modest general, but only under his successors from 1849 on.[60] Modern Siam/Thailand was also largely the creation of an enlightened despot, King Chulalongkorn (aka Rama the Fifth). Menelik II (r. 1889–1913) played a similar role in Ethiopia. In post-Napoleonic Europe, on the other hand, the initiative for major changes seldom came from crowned heads; no monarch after 1815, with perhaps the limited exceptions of Napoleon III and Alexander II of Russia (r. 1855–81), was a self-driven grand activator, reformer, or nation builder. But, once created, nation-states sought monarchical legitimation and tolerated even such bizarre figures as Leopold II of Belgium (r. 1865–1909), an unscrupulous imperial adventurer who was able to secure a position for himself above the internal strife between liberalism and political Catholicism. Rulers of multinational realms had a harder time of it, since they had to perform their integrative role over expanding (Russian) or shrinking (Habsburg, Ottoman) empires alive with centrifugal ethnic tendencies. There was no opportunity for the kind of national monarchic compromise that made the late century a powerful imperial moment in the special British case. The strongest identification between monarchy and nation, however, was not in Europe but in Japan, where a symbolic fusion under the Meiji Emperor's grandson, the Shōwa Emperor (Hirohito, r. 1926–89), contributed to the Asian cataclysm in the Second World War.

The survival of monarchy in large parts of the world gave a last lease on life to an ancient social form. There were court societies from Beijing, Istanbul, and the Vatican down to the small Thuringian city of Meiningen, whose orchestra in the 1880s was one of the finest in Europe (it gave the first performance of Johannes Brahms's Fourth Symphony in 1885). Germany was full of courts until 1918; they were the gravitational center of high society in numerous local capitals. Elsewhere too, disempowered potentates kept up the pomp and circumstance of court life as long as it remained within their means. India with its maharajahs was in this respect not dissimilar to the Germany of Bismarck's "wrens' nests."

None other than Bonaparte, the former general of the revolution, had revived court society in Europe from 1802 on, just a few years after the destruction of the Bourbon court; his brothers and legates would be the key players in Amsterdam, Kassel, and Naples. New liveries were tailored, new titles, offices, and noble rankings introduced, a military court established; and for her coronation on December 2, 1804, a proper empress wore a gold satin dress embroidered with bees symbolizing a busy and productive empire. Napoleon, though personally uninterested in pageantry, put on the whole show simply to keep his entourage, including the adventurous Joséphine, occupied and under control. He also thought that the French, like "savages," could be dazzled by displays of splendor.[61] The emperor cultivated an image of himself as a sober workaholic—in the style of Frederick the Great, previously perfected by the Yongzheng Emperor (r. 1712–35) in China. On the Potomac the second president of the United States, John Adams, already tried to produce something dimly resembling the Court of Saint James, but this was soon dismantled by his more easygoing successor, Thomas Jefferson, a widower bereft of a First Lady.[62] A key difference between the European and Oriental styles of court life was the public role of the royal couple. Japan's adoption of this Western symbol expressed more strongly than almost anything else its claim to be entering global modernity;[63] whereas nothing seemed to underline the antiquated decadence of Imperial China more than its exotic train of eunuchs and concubines instead of a settled bourgeois life at the head of the state.

3   Democracy

Whereas monarchy, as real concentration of power or mere ornament, was a ubiquitous feature of the nineteenth century, we have to look a little harder for the traces of democracy. It is not even altogether certain that more of the world's population than a century before had a direct influence on its political destiny. This was doubtless the case in Western Europe and America, but it has to be set against the unquantifiable barriers to participation associated with colonialism. Precolonial polities around the world were not liberal democracies in which all citizens theoretically enjoyed the same political rights and extensive protection against arbitrary action by the state. In many areas, however, at least among the elites, the scope for discussion and negotiation in public affairs was greater than under the authoritarian conditions of colonial rule. Democracy made advances during the nineteenth century, but it did not triumph everywhere and even in the best-functioning systems did not conform to the standards of stable mass democracy that are taken for granted today in most of Europe.

The American and French Revolutions formulated the concept of popular sovereignty and inscribed it in their constitutions. In France, in the wake of Jean-Jacques Rousseau, the ideal was taken as far as it has been anywhere in the world up to the present day. It is true that the authors of the US Constitution already provided for checks and balances to counter the tyranny of the majority, shielding themselves with an element of real fear from an unfiltered expression of the voters' will. The indirect choice of the president by an electoral college, which for a long time had some logistical foundation in view of the size of the country, still survives as a relic of this attitude. In Europe, memories of the Terror of 1793–94 ran deep: even property owners eager to avert princely absolutism of any description, including the Napoleonic variant, feared nothing more than "anarchy" or "mob rule" and took the corresponding precautions. Yet, once begotten, the twofold ideal of popular sovereignty—that the voters' will should find the truest possible expression and that the people should have the power to remove any kind of government—remained in principle a yardstick of all politics. This was the real novelty of the nineteenth century: a revolution in both expectations and anxieties. The struggle over political systems took on a new kind of dynamic. The main issue was no longer how just a ruler should be, or how one could best preserve the ancient rights of one's own status group, but rather who could and should participate, and to what extent, in decisions about the common good.

To specify how democratic a country is remains far from easy.[64] It may be difficult to distinguish between the reality and the democratic facade; the criteria may also be mixed up in an unclear way—for example, the legally prescribed opportunities for participation, along with the human rights record that is nowadays often the preferred measure of the morality of a political system. The oversized and diffuse question of democracy may be divided into several aspects for the nineteenth century. Here it is advisable to use a broadly defined concept. Female suffrage, for example, as a prerequisite of democracy would yield no democratic country in nineteenth-century Europe, and even an active suffrage of 45 percent of the male population—by no means an exacting criterion by today's standards—existed in only a minority of European countries around the year 1890.[65]

The Rule of Law and the Public Sphere

Both logically and historically, the image of the rule of law stands for all liberal restriction of power in the political system. A high value, whatever the cultural context, is placed on the protection of individuals from arbitrary official action: political power should be exercised in accordance with laws that are known to, and ideally valid for, everyone in society; some of them—especially those with a religious sanction—are binding even on the ruler and cannot be changed by an act of his will. This idea is not a European invention; it can be found long ago in China and the Islamic world, for example. But it took shape with special force and rigor in the political practice of England, where the rule of law came to be increasingly regarded as a matter of course. The core of this English conception, fully developed in the mid-eighteenth century, consisted of three points: (a) a professionally recruited and organized judiciary in charge of applying common law; (b) a real possibility of challenging government measures in the courts; and (c) a legislature and judiciary that respected the inviolability of the person and property and the freedom of the press.[66] On the Continent, it took longer for a similar legal culture to spread; basic rights became an issue much later than in the English-speaking countries. In the early nineteenth century, the rule of law chiefly referred to independence of the justice system, involving transparency and the safety of judges from dismissal, and to the legal propriety of all government action. The main focus was on the protection of property.

In actual practice, such forms of constitutional order could quite easily go together with "undemocratic," or even preconstitutional, conditions at the level of the political system. In the German states, for example, the rule of law was already widely respected before the principle of constitutional rule established itself. Indeed, according to some late-eighteenth-century theorists, such constitutionalism was a hallmark of "enlightened absolutism" that differentiated it from tyranny. The reforms of the 1860s in Russia also gradually promoted an awareness of "legality" (*zakonnost'*) in everyday life, which would coexist for half a century with the autocratic system.

In theory, European concepts of the rule of law were transferred to the colonial empires. Although special racist legislation increasingly applied to natives toward the end of the nineteenth century, nonwhite subjects of the British Crown did have chances of a fair trial that were not dramatically worse than those of lower-class people in the British Isles. The prominent role of lawyers in the Indian freedom struggle of the early twentieth century was due precisely to the significance of this nonpolitical legal sphere for the functioning of colonial society. As well as making them important intermediaries, it gave lawyers access to a realm of universal norms binding on the colonial rulers themselves. At least in the British Empire, then, the rule of law set some limits to colonial despotism. In emergencies such as the Great Indian Rebellion of 1857/58, or the Jamaican revolt of 1865, such legal guarantees were scandalously set aside. But the vehicle of empire nevertheless served to spread the British idea of the rule of law to all continents.

Despite the colonial nuances, the legal situation was not always less advantageous than that prevailing in neighboring territories under indigenous rule; a free Chinese press, for instance, developed not in the realm of the emperor but in colonial enclaves such as Hong Kong and the International Settlement in Shanghai, where British legal conceptions were in force. As to the French understanding of law, its development in the course of the nineteenth century attached less value to the legality of government action.[67] Legal checks on the administration were anyway less pronounced in France than in Britain, and in the colonies the law offered considerably weaker and less extensive protection for non-Europeans.

The most important legal peculiarity of the United States was the existence of a Supreme Court, empowered since 1803 to interpret the Constitution in a dynamic of long-term constitutional change not dictated by the politics of the day. No law-governed state in Europe had such an independent judicial body to which appeal could be made against the judgments of lower courts or the government. But some rulings of the Supreme Court polarized opinion and contributed to an exacerbation of political conflict. A direct result of the *Dred Scott* case of 1857, in which it was ruled that blacks never could be citizens of the United States, was the election of the antislavery candidate Abraham Lincoln to the presidency and the eventual outbreak of Civil War.[68] The fact that not even Supreme Court judgments commanded uncritical acceptance in the name of abstract principles of government became part of the political culture of the United States.

The new political and legal status of the citizen was a product of the American Revolution of the 1770s; former subjects of the British Crown were supposed to become citizens of an American republic. By 1900 notions of citizenship were widespread in Europe too.[69] In this respect the situation at that time differed from the rudimentary rule of law in late absolutist Prussia or Austria. A multiplicity of rights had given way to the idea of equality before the law—a status that presupposed the communicative compactness and homogenizing tendencies of a nation-state. Citizenship was one of the Western inventions that would prove to be universalizable in its cultural neutrality. Thus, in the Meiji reform era after 1868, all (male) Japanese were equal citizens subject to uniform national laws. Some rights were guaranteed by the state: a free choice of occupation, the right to alienate property, or freedom of movement from village to city. In other ways too, Japan in the year 1890 was not far behind European models of a law-governed state.[70]

Closely associated with political democratization was the rise of a public sphere of sociability and oral and written communication, located in a third space between the privacy of the home and the ceremony of organized state functions. The ongoing debates on the "public sphere," still often conducted in dialogue with Jürgen Habermas's work first published in German in 1962, cannot concern us here; they tend toward obscurity insofar as they take the public sphere to be an element of an even more broadly conceived "civil society," viewing it as the prerequisite, not the outcome, of democratic forms of politics. Even in an authoritarian state—to summarize a commonly used model—public spaces can emerge as the result of autonomous social development. Where they do not simply serve as an arena for acting out aesthetic visions of the political, they tend to take on certain functions of the state and to encourage the expression of antigovernment criticism. Habermas's book outlined a general model, loosely rooting it in time and space. For him the eighteenth century in Western Europe was both the formative period and the golden age of a "bourgeois" public sphere.[71] In the nineteenth century, its key principle of public criticism gradually waned. The public sphere lost its characteristic intermediate position insofar as its starting point, the sphere of private life, was weighed down by the manipulative force of the mass media. By the end of this process, the reasoning public *citoyen* had turned into a pacified cultural consumer.[72] The second, pessimistic part of Habermas's argument has seldom been addressed by historians; all the more eagerly has a new interest in communicative history led them to search for signs of the rise of the public sphere. Their findings, so extraordinarily rich in detail, can scarcely be reduced to a common denominator. But the following points seem to be clear.

First. There is a direct interplay between media technology and the intensity of communication. Wherever the technical and economic conditions exist for a culture of the printed word, the formation of a public sphere is not far off. Thus, we look in vain for such a sphere in the Muslim world before the spread of printing in the nineteenth century. But the development of technology was not always an independent driving force; sometimes it might be theoretically available yet fail to be matched by a demand for printed products.

Second. Public communication and its subversive content grow by leaps and bounds in a revolutionary period. It can be debated whether communication gives birth to revolution or vice versa; one is on safer ground if one sticks to an idea of them as simultaneous and interacting. Throughout the Atlantic space, the revolutionary age around the year 1800 witnessed a sharp rise both in book communication and in critical radicalization.[73] The same was observable during the Eurasian revolutionary surge after 1900.

Third. Where public spaces opened up outside Europe in the nineteenth century, they did not always reflect straightforward attempts to imitate the West. Within state bureaucracies (Chinese or Vietnamese, for instance), in churches, monasteries, and communities of clergy, or in feudal structures such as the Japanese prior to 1868 where spokesmen for particular regional interests competed with one another, there had for a long time also been institutionalized dialogue about matters of general concern. European rule suppressed some of these communicative structures, while others migrated to a subversive underground inaccessible to the colonial masters, and yet others—among the intelligentsia of Bengal, for instance—gained a new lease on life and became a factor in colonial politics. Relatively liberal colonial regimes, such as that of the British in Malaya, might give rise to lively debate among the indigenous public, in which a broad spectrum of political views, some sharply opposed to colonialism, found expression.[74]

Fourth. Public spheres could be built in the most varied of spaces. Microspheres, in which hearsay and rumor were often more important than the written word, sprang up alongside one another and to some extent overlapped, sometimes finding integration at a broader level. The public spheres of scholarship and religion crossed political boundaries quite easily; the Latin culture of medieval Christendom and the ecumenism of classical Chinese culture (which at least until the eighteenth century embraced Korea, Vietnam, and Japan) are two good examples. England and France in the second half of the eighteenth century had a *national* public sphere: anything of political or intellectual importance was acted out on the great stages of London and Paris. However, this was the exception rather than the rule. Where a single metropolis was less dominant, or where the means of state repression were also concentrated in such a center, the public sphere tended to emerge outside the royal court and the government: in Russian, Chinese, or Ottoman provincial capitals or in the newly founded states of the decentralized United States (where it was only later that New York came to be generally recognized as the cultural point of gravitation).[75] Often it was a major step forward when a communicative sphere first emerged across local boundaries, making it possible to address issues of power or status and matters of general concern on a basis that overcame political segmentation.[76] In the Hindu caste systems of India and other particularly inegalitarian societies, no one gave any thought to the European ideal of communication among equals. However, European-style institutions imparted a new meaning to status differences among individuals and groups and gradually introduced new rules governing competition. The word "public" was on everyone's lips in nineteenth-century India. In the early part of the century, the English-speaking elite (initially in Bengal) created many associations that defended its own interests and criticized the colonial state in writing. The colonial regime, by no means omnipotent, was sometimes helpless in the face of civil unrest and legal challenges. The courtroom became a new arena of status competition, and spectacular trials aroused great interest among the public.[77]

Fifth. In its early phases, the public sphere did not manifest itself always (or only) in explicit political criticism; the recent interest in "civil society" has also directed attention to prepolitical self-organization. In Europe or America this might take the form of religious communities or single-issue action groups; Alexis de Tocqueville registered the abundance of such associations in the United States in 1831–32.[78] In China after 1860, when the controlling power of the state further declined, it was common for charitable projects such as hospitals to bring together prosperous members of the nonbureaucratic elite. In Muslim countries religious institutions might play a similar role of integration and mobilization. It was then only a short step from such initially nonpolitical initiatives to other areas of individual concern and general significance. We must keep a sense of proportion, however. The degree of long-term politicization varied greatly among the urban populations. Only in a few European countries did it rise to the level of communal democracy practiced in the cities of the United States. Also the local public sphere often remained a very elitist affair in Europe, Asia, and elsewhere.

Constitution and Participation

What the great political scientist Samuel E. Finer called the constitutionalization of Europe began, following influential models of the past (United States in 1787, France in 1791, Spain in 1812), after the final downfall of Napoleon and essentially concluded with Germany's adoption of the Reich Constitution of 1871.[79] This process did not remain confined to Europe. In no part of the nineteenth-century world were as many constitutions written as in Latin America: eleven in Bolivia alone between 1826 and 1880, and ten in Peru between 1821 and 1867, although this cannot be taken as evidence that their political cultures were actually constitutionalist.[80] The Japanese Constitution of 1889 was the climax of the formation of the Meiji state as a Japanese-European hybrid. A new wave spread to the largest countries of eastern Eurasia around the turn of the century. The Morley-Minto reforms of 1909 pointed even British India, though still on a tight leash, onto a path of constitutional evolution that would lead through many stages to the Constitution of the Republic of India in 1950.[81]

There is no need here to describe Europe's progress in detail.[82] The key point is that on the eve of the First World War, after a full century of constitutionalization, only a few European countries had achieved a constitutional democracy with general elections and a system of majority government answerable to a parliament: Switzerland, France, Norway, Sweden, and Britain (as late as 1911, when the power of the unelected House of Lords was curtailed).[83] The main bastions of democracy were then, along with the United States, the newer European settler colonies of Canada, Newfoundland, New Zealand, the Australian Federation, and South Africa (where the black majority, however, was excluded from elections or hindered from using its right to vote).[84]

It is a great paradox that in a century when Europe, committed to the idea of progress, put its stamp on the world as never before, the most far-reaching political achievements were made in the colonial periphery. On the one hand, many of the world's peoples experienced the British Empire as an incapacitating apparatus of repression; on the other hand, it could operate as a stepping-stone to democracy. In the "white" dominions with a form of liberal government, settler societies were able to cover the road to modern democracy more swiftly than could the mother country, with its strongly oligarchic-aristocratic traditions. The nonwhite colonies were denied such a head start toward "responsible government," but India and Ceylon at least were drawn into a particular kind of a constitutional dynamic. Under the pressure of the nationalist freedom movement, the 1935 Government of India Act instituted a full written constitution that provided for Indian political participation at the regional level and was largely retained when the country later gained independence. In the case of its largest colony, an authoritarian empire thus created a framework for the independent evolution of a democratic constitutional order.

In nineteenth-century Europe, electoral democratization was not straightforwardly correlated with the consolidation of a parliamentary political system. To take a well-known example: all males over the age of twenty-five had active suffrage rights for Reichstag elections in post-1871 Germany, whereas a property qualification, a kind of census-based franchise, continued to operate in England and Wales. Even after the Reform Act of 1867, which extended the vote to a larger number of workers, electoral registers included only 24 percent of adult males in the (rural) counties and 45 percent in the (urban) boroughs.[85] Nevertheless, English voters decided on the composition of a parliament that, being the core of the political system, was far more powerful than the democratically elected Reichstag. In England parliamentarization preceded democratization; in Germany the opposite was the case, even though alongside Reichstag voting rights an extremely unequal "three-class franchise" persisted in elections to the Prussian regional parliament.

The history of the franchise is technically complicated for every country. It has a major territorial dimension, since even equal votes can lead to highly unequal results if the constituencies are divided up unevenly. It is also important whether constituencies send one or several representatives to the parliament, and whether special representatives of the estates continue to play a role—as delegates from the universities did for a long time in England. Proportional representation was uncommon in the nineteenth century: only Belgium, Finland, and Sweden had adopted it by 1914.[86] Secrecy of the ballot was a more flexible notion than it is today, since especially in the countryside pressure could easily be brought to bear on service personnel and other dependents. France (in 1820) was the first country to introduce secret votes; the process often took considerably longer elsewhere. Its pros and cons were debated until the end of the century and beyond. In Austria legislation to establish the practice was passed only in 1907.[87]

A step-by-step expansion of electoral citizenship was the norm, won partly through revolutionary struggles, partly through concessions from above. Fundamental considerations of a strategic nature were inevitably bound up with the various franchise reforms. In Britain, a country with no revolutions in modern times, the three reform acts (of 1832, 1867, and 1884) mark deep fissures in its political history, the last of the three having brought not only a major expansion of the franchise to roughly 60 percent of adult males and an end to de facto upper-class control over the composition of the two Houses of Parliament but also the removal of numerous exceptions and peculiarities. For the first time Britain had something like a rational electoral *system*. But it was not until 1918 that universal male suffrage was introduced in the United Kingdom.[88] As the electoral clientele expanded, the social composition and work style of Parliament changed. The mass electorate that appeared in France in 1848, the German Reich in 1871, and Great Britain after the (still not "universal") reform of 1884 required different kinds of party organization from those characteristic of an elitist democracy of notables. By 1900 programmatically defined parties had taken shape in most European countries with a constitutional government, many of them, as the sociologist Robert Michels noted in his *Political Parties* (1911), displaying a tendency to bureaucratic bloatedness and oligarchization. At the same time, a new type of professional emerged alongside the gentleman politician—although it did not become dominant so long as parliamentary deputies had no allowances on which they could live (as in Germany until as late as 1906). The way in which the "deputy" entered the public consciousness can be seen especially clearly in the France of the Third Republic;[89] his distinctive social figure, with his own independent weight, involved a rather detached relationship to direct representation. On the other hand, images of a direct expression of the popular will—if only in a Bonapartist plebiscite, not in new laws—had hung on firmly since the Great Revolution. In different political-cultural contexts, and in ways that changed over time, the election acquired a special symbolic significance. Voters may conceive of themselves as sovereign or as mere "voting fodder." This could be a topic for a comparative cultural history of political life.[90]

One great exception casts a shadow over the success story of advancing democratic participation. Although the United States was the largest and oldest of the modern democracies, it was difficult for its citizens to exercise their civil rights. The situation there is all the more obscure because the franchise was and is usually regulated at the level of individual states. The difficulties began (and still begin) with the electoral register, stretching all the way from property restrictions (whose significance receded over time) and residence qualifications to outright racist exclusion. Before the Civil War, blacks were virtually disenfranchised even in areas where slavery did not exist. After the Civil War this was hard to justify. Especially after the official end of Reconstruction in 1877 more and more inventive chicanery was tried out to prevent liberated Afro-Americans from exercising their right to vote.

New immigrants from supposedly uncivilized parts of Europe (e.g., Ireland) and Asia (China and Japan) also had major obstacles placed in their way.[91] The democratization of American citizenship thus encountered a severe setback. Comparatively, the United States remained one of the most democratic countries in the world, but it had great difficulty in harmonizing the universal principles of its republican order with the realities of a "multicultural" and racially divided society.

Local Democracy and Socialism

Outside England, with its old parliamentary tradition, the homogeneous representation of the nation in a central assembly was a new idea in the nineteenth century. Also without precedent was the idea that practices of representation might reflect existing hierarchies, and that social conditions might themselves be changed through electoral legislation. Of course, the importance of such issues should not distract attention from events at subnational level; for most people the political regulation of their everyday world is more important than high politics in a remote capital city. Local administration was even more multifarious than national political systems: it could lie in the hands of justices of the peace drawn from the local upper classes (the English model), be carried out by appointees of central government (the Napoleonic model), or be inserted into a form of grassroots democracy (the American model that was so admired by Tocqueville). In places where the central state refrained from direct intervention or lacked sufficient resources, space repeatedly opened up for limited consensus building of a deliberative, democratic nature. This might, as in Russia, occur within a peasant commune that had to agree on the allocation and use of common land. The same happened in local elite groups with little internal hierarchy, whether Hanseatic senates, consultation sessions (neither recognized as legitimate nor persecuted as illegitimate) among Syrian notables in the Ottoman Empire, or the city council in the Chinese part of Shanghai (which in 1905 became the first formally democratic working institution in the history of China).[92]

Politics in the early United States also had an elitist, patrician character, especially in the eastern cities, but a new conception of democracy broke through with the Jacksonian Revolution of the 1830s. The propertied classes, until then mostly large landowners, no longer made up the totality of politically responsible citizens as America abandoned the old idea, taken over from European republicanism, that only property ownership guaranteed independence and qualified people for rational political judgment. Now the autonomy of citizens rested upon ownership of their own person; property qualifications largely ceased to apply. Unusually high rates of electoral participation (often over 80 percent) were signs of the energy now invested in politics. As the young French lawyer Alexis de Tocqueville noted during his study trip to America, this kind of politics no longer had Washington, DC, as its principal stage; it gained its strength from local self-governing communities that elected their own officeholders (judges, sheriffs, etc.)—a radical alternative to the Western European model of authoritarian centralism going back to Napoleon. This kind of democracy involved far more than the right to vote. It meant a new kind of society in which the principle of equality, abstractly and negatively formulated by the French Revolution as the abolition of estate privileges, acquired the positive sense of self-empowerment of a citizenry enjoying equal personal rights. The tension between liberty and equality, which Tocqueville diagnosed with the eyes of a liberal European aristocrat, was not a problem for most (white) Americans of his time. What Europe would later call "mass democracy" arose in the United States as early as the 1820s and 1830s.[93] But its democratic efficiency was partly weakened by the characteristic federalism of the United States, the territorial side of its constitution. For how representative was Congress? Sectional interests stood opposed to one another: slave states against free states. And almost until the Civil War the slave states dominated national politics, to an extent that made the United States as a whole a slave-owners' republic. Their will prevailed again and again: from the "gag rule" that between 1836 and 1844 precluded any debate on slavery in the House of Representatives to the Kansas-Nebraska Act of 1854. The slave states enjoyed a structural majority by virtue of the Three-Fifths Clause: three-fifths of slaves were added to free persons for the purposes of direct taxation and the allocation of seats in the House of Representatives.[94]

With Jacksonian democracy, the United States struck out in a novel political direction for the second time since 1776. "Mass democracy" of this kind, overlaid with a competitive and sometimes violent rhetoric of freedom, did not exist before the last third of the nineteenth century anywhere in Europe—not even in France, where the local power of prefects remained unbroken through several regime changes and even after the introduction of universal manhood suffrage. Once again the British way and the American way parted company. In Britain, the supremacy of an elite group of gentlemen landowners, financiers, and industrialists reached its zenith in the period between the two reform acts of 1832 and 1867. Though tightly knit and homogeneous in its cultural perception of itself, this oligarchy did not operate as a caste: it was open at the margins to outsiders and developed a highly integrative understanding of politics. After 1832 it proved capable in principle of acting under the conditions of "modern" parliamentarianism, once the Crown was no longer in a position to keep a prime minister in office against the majority will of Parliament. From the 1830s on, Britain was not merely a constitutional monarchy but a parliamentary monarchy, in which the church, too, began to play a lesser political role than in many continental European countries. At the same time, politicians at Westminster scarcely had to take into account a socially and culturally remote mass electorate, since the Reform Act of 1832 expanded the electorate only from 14 percent to 18 percent of adult males. In Britain the middle decades of the century were thus a period of democratic procedures without broad democratic legitimation, but also of a widespread conviction that the middle classes would henceforth have to play an important role in politics.[95] Even the most progressive countries of Europe would close America's democratic lead, at both local and national levels, only with a delay of nearly half a century.

The vast majority of women stood outside active citizenship. Female suffrage made its American debut in Wyoming in 1869 (and more generally in the United States only after 1920). The first de facto sovereign country to adopt it, arousing worldwide attention and widespread celebrations, was New Zealand, initially (in 1893) as a right to vote but from 1919 also as a right to stand for office. Finland—then still part of the Tsarist Empire—led the way in Europe by introducing female suffrage in 1906, followed by Norway in 1913. In both cases, women were needed for their potential to enhance nationalist legitimacy.[96] Female suffrage movements grew strongly and at an early date in places where there had also been struggles for male voting rights. In Germany, where these were bestowed in 1867–71 as a "gift from above," the suffragette movement was weaker than in countries such as Britain.[97]

Democracy was in varying degrees built from the bottom up. The basic process of transforming customs into rights at the local level is not unique to postrevolutionary societies such as the United States; nor is it a Western peculiarity. In the late Tokugawa period, when hardly anyone in Japan could imagine the establishment of a national assembly, the scope for local participation gradually increased without being linked to a political revolution or a tradition of municipal self-government. Long-established families had to recognize the claims of rising "new families."[98] After the Meiji Renovation brought a degree of administrative decentralization from 1868 onwards, the boundaries between national and local government had to be redrawn. At first demands for village assemblies rang out on all sides, and these were established in many prefectures after 1880. At the same time, however, the central government began to sound a retreat by introducing controls on public activities, press freedoms, and new political parties, and in 1883 it banned the election of village and city mayors and insisted on their appointment from above. This called forth stormy protests. In 1888 legislation was introduced to regulate relations between the central state and the villages, so that mayors could be elected but only under close supervision by the relevant authorities.[99] What remained was a much greater scope for participation than under the pre-1868 ancien régime. In 1890 the first general election in Japanese history confirmed this by filling Parliament with representatives of the upper middle stratum, bringing a new class without a samurai background to the center stage of politics.[100] But it took another quarter of a century before Parliament, under constant threat of dissolution by the Emperor's government, could assert itself as a counter-weight to the executive. The first-ever elections held in China in the winter 1912–13, relatively free and honest, did not usher in stable democratic development. By 1933, all traces of democracy had been obliterated in China, Japan, and Korea.

It was not only in the United States and Britain that political movements and civic associations became schools of democracy, offering in their internal functioning a learning space not determined by status considerations. At first equality claims were often raised, and practiced at the level of social intercourse, among milieux, groups, and organizations made up of objective equals, capable of pursuing their interests all the more successfully in broader political arenas marked by intense conflict. This was the political kernel of socialism and related grassroots movements. There is much to be said, for example, for regarding early German Social Democracy less as a political party in today's sense than as an associative movement.[101] Socialism was a new language of solidarity among the nonprivileged layers of society, which came into being when corporate certain-ties had disappeared and a need was felt to move beyond the politically amor-phous existence of unorganized poverty. In institutional terms, before its bolshe-vization into a conspiratorial vanguard party, the socialist movement not only asserted collective interests in the struggle between classes but also involved the exercise of democracy. European socialism was a force for democratization. It combined the pre-Marxist or "utopian" early socialism represented by figures such as Robert Owen, Charles Fourier, or Pierre-Joseph Proudhon with the non-violent variant of anarchism (especially in the Russian prince and later émigré to Switzerland Pyotr Kropotkin)[102] and a majority of the parties (most of them explicitly Marxist) that came together in 1889 in the Second International.

The original ideals of economic decentralization, mutual aid, cooperative production, and sometimes even communal living outside the framework of bourgeois private property had become weaker by the turn of the century. But the memberships still aspired to express their individual wishes or ideas in par-ties and unions that represented their interests in the outside world yet were productive of mutual trust internally. Although no party of the labor move-ment came to power in Europe before the First World War, the formation of a democratic mentality in the numerous currents of European socialism played no small part in preparing the democratization process that would follow the war. Before its outbreak, Europe and the British dominions had already experi-enced a constant strengthening of social democracy, in which sizable tendencies had shaken off Marxist expectations of revolution. In Germany these operated as the "Revisionism" of Eduard Bernstein and his comrades, while in Britain they were close to the New Liberalism that, unlike the Old, no longer saw the social question as a necessary evil but placed it at the fore of politics. Social lib-eralism and democratic socialism converged in a reformist conception of poli-tics, but only in certain countries of central, western, and northern Europe, not under the conditions of Russian autocracy, which forced its opponents to take the revolutionary road, and not in the United States, where organized social-ism remained insignificant and where an intellectual rapprochement between liberal and moderate socialist thinking would have political consequences only in the New Deal of the 1930s.[103]

4 Bureaucracies

Even on the eve of the First World War, genuine democracy existed as a con-stitutional order in very few places in the world, and these did not include large republics such as China or Mexico. The state was much more widespread as an agency of rule than as an arena of participation.[104] "State" may be defined quite differently in either a broad or a narrow sense. Many small societies were "state-less," if this is understood to mean that they lacked even a staff in the ruler's household. In other cases, where the staff was unstable and poorly differenti-ated in institutional terms, the chances were often slight that something like "the functions of state" would be put on a regular footing. The state was weak not only in societies considered "primitive" in the normal parlance of the late nine-teenth century. In the United States too, an emphatically modern polity in many ways, people did not want to hear of the state in the European sense of an author-ity commanding obedience. In the eyes of American citizens, any authority not legitimated by the informed will of the electorate was very much a thing of the past. Government, unlike the state in the old European sense, had an obligation to give an account of itself. Around the turn of the century, only a few political theorists ventured to speak of a US "state" as an abstract category.[105] It is another matter that the prevailing ideology of statelessness, which in many ways harked back to old English conceptions of law, conflicted with reality on a number of points. At the US frontier, and especially in the newly incorporated territories in the West, the federal government and the local authorities (often with weak democratic legitimacy) fulfilled the classical political tasks of regional planning.

A narrower definition emphasizes the conceptual distinction between state and society. Breaking with older European political theory and similar con-ceptions elsewhere in the world, this moves away from the idea—or rather, im-age—of the state as a household or a body governed by its head. If state and soci-ety are taken as separate spheres, it is no longer true that the whole country can be regarded as one big family. The caring and punishing ruler worthy of respect: this view, violently attacked in John Locke's *First Treatise on Government* and eventually discredited, was already in retreat in eighteenth-century Europe, but it lingered on, for example, in the official rhetoric of late imperial China.

"Rational" Bureaucracy

Such a conception of the state as a structure outside society developed on several tracks in early modern Europe. It is by no means the case that a uniform absolutism was impelling all European societies, or even all the larger ones, in the same direction.[106] An inevitable part of this early modern state was a bureaucracy that had to tackle three main tasks: (a) to administer large states in a way that ensured their cohesion; (b) to keep the exchequer afloat (especially the war chest, given the importance of war for states in this period); and (c) to organize the administration of justice, in an age before an effective division of powers that gradually emerged in North America and Europe only in the late eighteenth century. Nowhere in Europe before 1800, however, was every level of the justice system in the hands of the state. Royal and imperial courts were never responsible for everything, even in the most centralist systems of absolute rule; there always remained special enclaves for cities, estates, corporations (e.g., the universities), or the landowning aristocracy (the so-called patrimonial courts in Prussia). Churches, monasteries, and other religious establishments often applied laws of their own to their members. In the Islamic world, secular and religious law was not sharply separated and had many overlapping elements. Imperial China in the eighteenth century, with no state-recognized churches and no equivalent of European canon law, was more marked than most parts of Europe by a state monopoly of justice. The lowest-ranking officials, of whom there were only a handful in each district (*xian*) throughout the Sino-Manchurian empire in the late eighteenth century, were generalists responsible for the dispensation of justice in all conceivable cases. Death sentences had to be personally upheld by the emperor. In terms of the degree of state involvement, the pre-1800 Qing justice system was therefore more modern than its European counterparts. It is hard to say whether the rule of law was equally pronounced, but from 1740 on there was a body of secular penal law altogether comparable with European codifications of the time.[107]

Since Max Weber, historical sociology has been agreed that in modern Europe patrimonial administrations turned into the rational bureaucracies we know today. This transformation took place in the nineteenth century and had its origins in the French Revolution, which paradoxically established a bureaucratic state dwarfing Bourbon absolutism in both scale and efficiency.[108] Napoleon spread this model beyond the borders of France, but the pace and degree of change varied from country to country.[109] The general political culture, together with infrastructural conditions and the nature of the political system, played a role in the development of state administrations as tightly integrated and smoothly functioning apparatuses of communication. Although the differences were not very great in these respects, no state bureaucracy was the same as another; the Bavarian bureaucracy in the mid-nineteenth century, for example, was plainly less hierarchical and authoritarian than the Prussian version.[110] Bourgeois or newly ennobled officials were characteristic of France and many parts of Germany, whereas in the lands of central and eastern Europe, from Austria to Russia, large state administrations offered employment mainly to declassed people from the lower ranks of the nobility. Since, with the limited exception of Hungary, this huge region had no representative institutions capable of exercising effective control over the executive, the second half of the nineteenth century there was the great age of bureaucratic domination within authoritarian-monarchical systems—more "Asiatic" than "European" in the modern sense.[111]

At the end of the nineteenth century, a "rational" state bureaucracy was not actually operational everywhere in Europe, but it had at least established itself as the ideal model. According to this, a modern state administration rested on an ethos of public service, and each ruler felt responsible for its adequate provision out of tax revenue. Corruption was neither desirable nor (with adequate salary levels) necessary. Civil servants were supposed to be above parties, bound by existing laws and subject to inspection. Hierarchies within the bureaucracy were transparent. Promotion followed open and familiar career paths, based on either seniority or performance. Officials were employed on the basis of expertise or special diplomas, not through nepotism or "connections." The buying of offices was ruled out. The work of administration took place by written communication. It stored up archives. It included special disciplinary proceedings where necessary, subject to the laws of the land.[112]

No rigid criteria can be applied to assess when an efficient state by modern standards was actually achieved. Pragmatically, a state may be considered modern if the following are true:

- Bands of robbers have ceased to terrorize the population and an effective police force (Max Weber's "monopoly of the legitimate use of physical force") is in operation.
- Judges are appointed and paid by the state, without being subject to dismissal or to outside control by other institutions within the system of government and administration.
- A treasury department regularly raises revenue through direct and indirect taxation, and the population recognizes the state's fiscal requirements as legitimate in principle (so that taxpayers are not in danger of being pummeled and tax evasion does not exist on a large scale).[113]
- Officials are appointed only on the basis of proven competence.
- Corruption in dealings between public and officials is not taken for granted but regarded as an evil deserving punishment.

This kind of state bureaucracy, increasingly copied in large private corporations since the last third of the nineteenth century, was a European invention with especially strong roots in Prussia and Napoleonic France. But this should not obscure the fact that there were imposing bureaucratic traditions outside Europe—for example, in China, Japan, and the Ottoman Empire—which should not be too hastily dismissed as "premodern" or "patrimonial." In the nineteenth century these traditions tended to converge with Western influences, producing highly varied results. Four examples will have to suffice: British India, China, the Ottoman Empire, and Japan.

Asian Bureaucracies: India and China

Nineteenth-century European colonies usually featured a low degree of bureaucratization in comparison with their mother countries. The colonial state had two aspects. On the one hand, it was often the only institution of any kind that—with the help of centralized powers such as the army, police, or customs and revenue—gave life to a territory as a single governed entity. The colonial state brought laws with it, along with judges who dispensed justice in accordance with those or special colonial laws. It recorded the population statistically, classifying it by ethnicity, religion, and other categories that had not previously been customary but now tended to shape the reality. "Tribes" or religious communities or (in India) whole castes were defined in such a way as to demarcate administrative districts or statistical objects, or to identify indigenous leaders with whom the colonial state wished to cooperate. In large parts of Africa, India, and Central Asia, such things became possible at all only through the establishment of European-style colonial state apparatuses. On the other hand, the colonial state was never an all-powerful monster. The forces it had in the field were so skimpy that it was seldom able to bring the vast colonial territories fully under its wing.

All this was true of the largest colony: India. Here the numerical relationship between European personnel and Indian subjects was especially unfavorable. Nevertheless, one of the full-scale bureaucracies of the era was built in India— the only such case in the colonial world of the nineteenth century. In 1880 India was more highly bureaucratized than the British Isles: not only in quantitative terms but, more decisively, because the bureaucracy did not perform the ancillary services of a purely administrative executive under political direction. Rather, it was the core of a system of rule that may best be described as bureaucratic autocracy. In this respect, the Indian colonial state had greater affinities with Imperial China than with any political system in Europe. Nor do the parallels end there. Both the Chinese state bureaucracy and the Indian Civil Service (ICS) revolved around a fairly small corps of highly qualified top officials who enjoyed great prestige in society. Outside the capital they were present at the lowest level of the hierarchy as district magistrates (*zhixian*) in China or "collectors" in India, the official duties of the two being very similar.[114] Specially educated for their posts in a system involving competitive examinations, both the Chinese and the British-Indian district officials were at once heads of local government, revenue collectors, and magistrates. There had been such exams in China for centuries. Europeans who knew of this practice often expressed admiration for it in the eighteenth century, and it would appear that the British had the model in mind when they introduced something similar not only for the Indian and colonial service but also—first proposed by experts in 1854, finally implemented after 1870—for the senior (ministry-level) bureaucracy at home.

The British colonial bureaucracy in India did not turn up one day in a political landscape previously free of the state. But the Mogul Empire and its successor states had not essentially been bureaucratic structures such as those of China or Vietnam; they had hierarchies of scribes and a developed chancellery but not a strictly or thoroughly organized civil service. The ICS could therefore build to only a limited extent on existing foundations. Its immediate predecessor was the administration of the East India Company, which, though one of the world's most complex formal organizations in the eighteenth century, was in many respects still premodern. The posts it had to fill were largely allocated through patronage, not by objective performance criteria. Such practices had been commonplace in the European state of the early modern period. In France, the Napoleonic rationalization of the state had replaced them early on with the advantages of an open career structure. In Britain, it was still possible until 1871 to purchase an officer's position in the army, and it was only around that time that it became the rule to recruit ministerial officials (with the exception of the strongly aristocratic Foreign Office) by means of aptitude tests. In India that had been the case since 1853, shortly before the East India Company was wound up in the wake of the Great Rebellion of 1857/58.[115]

The ICS was the second pillar, alongside the army, on which the British based their rule in India. If one judges an organization retrospectively by whether it achieved its own objectives, then the ICS was a very successful apparatus, at least until the First World War. Indian taxes flowed into the colonial coffers, and after the rebellion a high degree of internal peace was attained by means not confined to military force. Thanks to its high salaries and considerable prestige, the ICS became the civilian elite corps of the British Empire. The stresses and strains of life in the tropics were offset by the fact that one could accumulate quite a nest egg in colonial employ and enjoy early retirement back home as a gentleman of means. The Indian bureaucracy, as it still exists today, shows traces of its colonial origins. Since a slow Indianization of the service began after the First World War, the post-1947 Republic of India did not find itself in the position of having to repudiate the ICS as a symbol of colonialism. It therefore kept it going as the Indian Administrative Service.[116]

Though a European implant, the bureaucracy in India did not directly copy a European model but experimented with various forms under the special conditions existing in the country. China was not colonized. Colonial state apparatuses of a significant size arose only in marginal areas under Japanese rule: after 1895 on Taiwan; after 1905, and on a large scale after 1931, in Manchuria. China's ancient bureaucratic tradition therefore survived without direct colonial intervention until the end of the nineteenth century. Its old institutional forms ended when the Qing government abolished competitive state exams in 1905, but a kind of mental bureaucratism persisted under the new conditions of the republic and, after 1949, under the rule of the Communist Party of China. Still today the nationwide state and party hierarchies are the main braces holding the giant country together. At the modern peak of its efficiency, in the middle decades of the eighteenth century, the Chinese state bureaucracy was the most rationally organized in the world, the largest and most experienced, and the one responsible for the greatest number of tasks.[117]

For a late-nineteenth-century European, China had become the embodiment of a premodern bureaucratism out of tune with the requirements of the age. Observers from Western countries that had left behind the scourge of corruption perhaps only a few decades earlier now spoke contemptuously of the venality of the Chinese mandarinate.[118] Its inability to modernize the country economically also fueled doubts about the rationality of the Chinese state. Much in these contemporary judgments is justified. The Chinese bureaucracy did suffer from low pay that made its members dependent on "benefices" from office. It was also stifled by an overwhelmingly literary-philosophical education that, despite many attempted reforms,[119] fell far short of the requirements of modern technology. It was further impeded by the purchasing of office (bred by the dire condition of the state finances) that brought many unsuitable people into the apparatus, and from the fact that after the death of the Jiaqing Emperor in 1820 there was no strong monarch capable of imposing discipline and probity on officeholders. A more general problem, on top of all this, was that the Qing Dynasty failed to reform two central pillars of the state before 1895: the military and the exchequer. The army was just capable of defending the imperial borders in Central Asia, but it was in no position to stand up to the European Great Powers. The revenue system, based on a fixed land tax, was so antiquated that the imperial state was hopelessly impoverished by the time it neared its end.

It would have been even poorer if, of all things, the one example of a transfer of European administrative practices had not slowed the financial decline of the dynasty. After 1863 the Ulsterman Robert Hart (Sir Robert, from 1893) built up the Imperial Maritime Customs (IMC) in his capacity as inspector general, having been appointed to the post under Western pressure as the *homme de confiance* of the world trading powers. But he did have the rank of a high Chinese official, was formally a subordinate of the Chinese emperor, and interpreted his role as that of a dutiful intermediary between the two civilizations and economic systems. The IMC relied on rank-and-file Chinese assistants and even had a kind of Chinese shadow hierarchy, but it basically emulated the Indian Civil Service with its cadre force of highly paid European administrative experts. It was smaller in size than the ICS and less unambiguously under British control. The IMC ran an excellent customs service that made it possible for the Chinese state to profit from the growth in foreign trade. That would not have happened with the traditional techniques of Chinese district administration, which mainly involved exercising rule over the peasantry. Only after 1895 did the Great Powers gain a *direct* hold over the customs revenue, much to the displeasure of Sir Robert Hart. On the one hand, the IMC was an instrument of the Great Powers guaranteeing that China's customs sovereignty would remain limited under the unequal treaties. On the other hand, it was an agency of the Chinese state, operating in accordance with Western principles of administrative probity, formal regulations, transparent bookkeeping, and so on.[120]

Hart's IMC had a limited impact on the rest of the Chinese state administration. The Qing government began to introduce reforms only after the end of the century, and although these continued in the early republic, they had scant success. Nevertheless, it would be a mistake to take nineteenth-century European caricatures at face value; the Chinese bureaucracy (or the Vietnamese) cannot be simply dismissed as "premodern." One side of it, geared to impersonal rules transcending family or clientelist relations, attained a high degree of meritocracy in the selection of personnel. The case of Korea even showed that its principles were compatible with the continuing access of a hereditary aristocracy to the top posts in the state.[121] Administrative practices were in theory performance driven, objectively grounded, productive of individual accountability, and to some extent fashioned to comply with the law. All this is *modern* by sociological criteria. Another side of the bureaucracy, however, corresponded to a society permeated with ethical principles of substantive justice, which did not regard all citizens or subjects as equal (as a modern administration must), and whose Confucian understanding of family bonds, especially the subordination of sons to fathers, played a role in shaping action. This internal contradiction was the main problem with Chinese-style bureaucracies in the age of a near-global move toward rationalization of the state.[122] Finally, traditional formulas were incapable of handling political groups inspired by patriotic fervor. The bureaucratic apparatuses found themselves helpless in the face of the revolutionary movements that emerged in China around the turn of the century.

Asian Bureaucracies: The Ottoman Empire and Japan

China's bureaucratic tradition proved fairly resistant to Western influences in the nineteenth century. The structure and ethos of the state administration changed little. At least the bureaucracy could fulfill one of its main tasks, the territorial integration of the empire, until shortly before it came to an end. The path of change that the Ottoman Empire covered was rather longer. During the same period, the traditional Group of Scribes (*kalemiye*) turned into what became known after the 1830s as a civil service (*mülkiye*). It did not simply copy European examples—not even the French model, which resembled it in many respects. The need for change was felt most acutely in foreign ministry circles, where contact with the outside world was closest. But then the reform acquired a dynamic of its own, leading to the development of new norms, new role models, and new conceptions of administrative professionalism. In the Ottoman Empire, as in Europe and China, a centuries-old practice of patronage was not replaced overnight with a rational personnel policy based on objective criteria; the two orientations existed side by side in a relationship of mutual influence.[123] The post-1839 Tanzimat reforms made the new civilian officialdom the dominant elite in the empire—a professional corps that would number at least 35,000 by the year 1890. Whereas a century earlier, the thousands of scribes had been concentrated in the capital, Istanbul, only a minority of the new-style senior officials were employed there in 1890. The Ottoman bureaucracy thus spread out territorially at a very late date, following a course in the second half of the nineteenth century that had been characteristic of China for many hundreds of years.[124] Lacking in such experience, it could afford to be "more modern" than its Chinese counterpart, which, because of its strong path dependence, was less free to give up old ways and required exceptional energy to embark upon reforms.

In Japan a modern bureaucracy also took shape in the triangle formed by traditional foundations, Western models, and an indigenous modernization drive. Since the Tokugawa period there had been a large pool of administrative competence, but unlike in China or the Ottoman Empire, this had been concentrated at the level of lordly domains (*han*) more than of the central state. To an extent only really comparable with revolutionary France, the need to build a nationwide bureaucracy powerfully asserted itself after the Meiji Renewal of 1868; the administrative experience of the samurai, who had changed in the peaceful conditions of the Tokugawa period from warriors of the sword into masters of the pen, was now deployed in a growing number of fields. By 1878, just ten years into the Renewal, the state administration had been thoroughly rationalized along the lines of the professional system familiar from the Napoleonic Consulate, in which advisory bodies and any kind of self-government played only a subordinate role. A complete hierarchy of officials, such as had never existed on a nationwide level, stretched down from the chancelleries of state through the governors of newly created prefectures to local village heads.[125] In 1881, not much later than in Britain, examinations were introduced for the upper reaches of the civil service and soon replaced the traditional practice of patronage; only the most senior positions of all were now filled by government directive—a customary procedure in Europe as well. By the turn of the century, Japan's state administration had become a textbook example of Max Weber's "rational bureaucracy"; there were few so thoroughly modern elsewhere in the world. But since in Japan (as in Prussia, Austria, and Russia) the modernization of the bureaucracy preceded the rise of a critical public opinion and political parties, the danger of unchecked bureaucratism arose as soon as the political leadership of the Meiji oligarchs relaxed its vigilance. The consequences of such a trend would become apparent in the twentieth century.

The danger was still relatively slight during the first few decades of the Meiji period, partly because of the revolutionary origins of the new order. Since the political leadership had the legitimacy neither of tradition nor of representative or plebiscitary institutions (as in revolutionary France up to the time of Napoleon), it had to show by results that it had the ability and competence to rule. This included the creation of a public service ethos transcending patron-client relations, and of a bureaucracy dedicated to the goal of making Japan economically and militarily competitive among the Great Powers. A combination of samurai traditions of administration with loans from British, French, and German statecraft resulted, as in the Ottoman case, in something more than a simple import. Japan found its own form of bureaucratic modernity. But in fact it was a kind of semimodernity. Personal liberties and popular sovereignty remained alien ideas, and social contract theory was never understood. A monarchical patriarchalism was thus able to survive into the era of bureaucratic rationalization. The Constitution of 1889 diverged from its European models by declaring the person of the tennō to be "sacred and inviolable," with sweeping powers inherited from his imperial ancestors.[126]

To bolster this collectivist or organicist conception of the state, the late Meiji rulers took up the idea of a Japanese national essence (*kokutai*) first developed by the Confucian scholar Aizawa Seishisai in 1825;[127] the emperor was the head of a "family state" (*kazoku kokka*) that followed a single will, while his subjects owed loyalty and obedience to the political bodies he established.[128] The Japanese bureaucracy, though one of the world's most "rational" in form, therefore carried out its duties less as a service to citizens than as a fulfillment of national goals passed down from on high. A modernized authoritarian state—there are many parallels with the post-1871 German Reich—offered favorable terrain for the growth of a rational bureaucracy. The state administration was highly modern, but the same cannot be said in general of its ideology or of the political system in which it was inserted. In the end, it makes a difference whether or not bureaucratization develops in the context of a liberal political order and political culture.

An All-Pervasive State?

But this can be only one analytic approach to the phenomenon of the bureaucratic state. Another, equally important focus concerns how bureaucracy is experienced at the various levels of political life, including how "the state" is expressed in the village and how relations take shape in the triangle of peasant self-regulation, local upper-class hegemony, and intervention by ground-level organs of the state hierarchy.[129] Another important question for many countries in the second half of the nineteenth century was how the administrative integration of large territorial nation-states or empires was to be achieved. Old imperial federations, as in China or the Habsburg Empire (where military rather than civil administrators were the main lever), were successfully held together. Both Germany and Japan had to face huge challenges with regard to administrative convergence and standardization: the former after the foundation of the North German League in 1866 and, on a larger scale, after the establishment of the German Reich in 1871; the latter when the system of fiefdoms (*han*) was abolished, partly as a result of peasant revolts against their overlords, and prefectures were introduced along French lines.[130] A focus on the respective peripheries rather than the national centers makes more apparent the obstacles and limits of state-led centralization. It is therefore worth considering the founding of the German Reich from the point of view of a small component state, or Meiji unification

from that of a *han* turned into a prefecture, or the political history of late imperial China from that of an individual province.[131]

In Europe too, it was not the early modern period but the nineteenth century
that saw the transition from the traditional to the rational state.[132] Inevitably
this was bound up with the construction of bureaucracies and the expansion
of state activity—a process observable almost everywhere in the world. It was
not a by-product but often a premise of industrialization, where, as Alexander
Gerschenkron has pointed out, the state's structuring and initiating role increased in the case of countries that had fallen behind internationally. Russia
and Japan are good cases in point. The expansion of state institutions and activities proceeded in a number of different ways. Bureaucracies differed in efficiency (hence in their capacity to process information) and with regard to the
speed with which they made decisions and implemented them. The notoriously
cumbersome Habsburg bureaucracy needed a long time to reform itself. If self-
organization was strong in society at large, even in the construction of a capitalist market economy, then a lean state could be more effective than a bloated
bureaucracy obsessed with rules and regulations; the British example is evidence
of this. The pace of bureaucratization was rarely constant, and it even ran up
against countertendencies. In the United States, for example, the Northern
state apparatus grew strongly during the Civil War, and postwar Reconstruction
was an attempt to extend this experience to the South, but anticentralist forces,
sometimes hostile to the state as such, became stronger there as Reconstruction
shuddered to a halt. In the last quarter of the century, the spurt of liberal capitalism in the North led to a general decline in calls for state regulation.[133]

In Europe, by contrast, the politically driven conception of a "night watch-
man" state was dominant only in exceptional cases such as Britain. By 1914 at
least five characteristics of bureaucratization had established themselves in many
Continental countries: (1) regular salaries for work in the government service;
(2) employment and renewal of state personnel in accordance with an efficiency
criterion; (3) the grouping of individual authorities into official hierarchies with
a solid division of labor and chain of command; (4) the integration of all officials
into a national administration (harder to achieve and less complete in countries
with a federal system); and (5) a separation of powers between parliamentary
politics and bureaucratic executive, although the two were everywhere closely
linked at the very top.[134]

Even for Europe, however, it would be wrong to speak of an all-pervasive
state such as we know it today. Many spheres of life were not yet regulated by
laws and edicts; nor were there industrial standards, noise-control regulations,
construction licenses, or even compulsory general education. The bureaucratization of the state had been proceeding worldwide under scarcely altered technical
and media conditions. Communication in writing, already current in China at
a time when no one had thought of it in Europe, had become standard. Administration meant paperwork, and the telegraph, incapable of transmitting large

volumes of data, did not bring sensational advantages for the administration of distant regions. The omniscience and omnipotence of the state found their limits in logistics.

The growth of bureaucracies can be documented only approximately. The state "grows" if and when the number of public-service jobs increases at a faster rate than the population. By this criterion the state was shrinking both in China and in many European colonies. In a thoroughly administered country such as Germany the number of state employees began to soar only after 1871, but the threefold increase between 1875 and 1907 was disproportionately due to higher employment in transportation and the mail service, while jobs in the state administration proper and in public education actually declined.[135] The picture was the same in the colonies, especially those belonging to Britain and France. There, apart from the army and police, the largest share of both European and indigenous public employees worked in the railroads, the post office, and the customs service. The state intervened in society at the most diverse points. Its revenue department therefore required a reasonable monetary system, and in parts of Africa, for example, this first had to be created. State building and commercialization were mutually reinforcing.

The speed and scale of financial rationalization should not be exaggerated, however, even in the case of Europe. It took a long time for regular budgets to become the norm, for the state not only to record its income and expenditure but also to look ahead and more or less plan their future levels. In nineteenth-century Europe this was made easier by the fact that few wars needed funding, whereas in the previous century they had been the main burden on state finances—an area in which Britain, with its huge fiscal strength, outstripped all its rivals. Federal systems involved special complexities, since various taxes had to be raised at different levels and the problem of financial equalization also had to be addressed at some point.[136] When governments incurred debts in the nineteenth century, they too—like early modern princes—avoided becoming too dependent on individual financiers. Britain was the first country to introduce regular debt management, over and above ad hoc activity relating to particular business. Public borrowing to cover deficits became a normal instrument of financial policy, and one effect of this was to give investors a stake in the well-being of the state. The conflict between taxpayers and creditors who siphoned off revenue in the form of loan interest was not infrequently fought out in the open.

In the nineteenth century, the state was not yet thought of as a redistributive state; revenue was hardly ever used as a strategic instrument for shaping the stratification of society. In the conflict between cheap government and expensive public services, it was not just a liberal taxpaying public that opted for thrift. In the final decades of the century, as politics in Europe and Japan took an increasingly nationalist turn, a new dilemma between economical government and military spending came to the fore. Yet, on the eve of the First World War, state revenues reached 15 percent of GDP scarcely anywhere in Europe, and were well below 10 percent in the United States.[137] The rise to the levels of 50 percent or so that are taken for granted today would take place in the two postwar periods after 1918 and 1945.

One of the main fiscal innovations of the nineteenth century was a straightforwardly proportional tax on income. It operated continually in Britain after 1842 and allowed a cautious skimming of the growing affluence of middle and upper income groups. Between 1864 and 1900 many other European countries introduced such a tax.[138] In Britain, however, it was not a measure aimed at social reform and redistribution but was directly linked to the new policy emphasis on free trade. The new taxes offset loss of income due to tariff removal, while free trade promoted growth and increased prosperity.[139] Another modern feature of the new fiscal systems, above all in the West and Japan, was that at least in peacetime taxpayers did not have to face sudden or arbitrary impositions. Legislation set the level of taxation—a budget, too, is a kind of law—and clearly defined the region and time span for which it would apply. The tax-raising state and the rule of law went hand in hand.

5 Mobilization and Discipline

Conscription

Napoleon had first shown how a well-organized state could mobilize human as well as financial resources. General conscription of the young male population succeeded in exceptional circumstances, in polities that constituted themselves as belligerent formations and saw war as their principal raison d'être—for instance, the Zulus under King Shaka in the 1820s, or particular groups or tribes of mounted warriors in North America and Central Asia. Four types of military organization were especially common in the early modern period: (1) the mercenary army; (2) the warlord and his freebooting clientele; (3) feudal associations such as the Manchu bannermen of the Qing dynasty or the Rajputs in India; and (4) praetorian guards such as the Ottoman janissaries, active in political affairs especially in the capital city. Of these forms, two remained prominent in the nineteenth century: warlords (above all, in Latin America after independence or in comparable conditions following the breakup of an imperial order, as in China after 1916) and mercenary armies (especially in India's numerous markets for military labor and in parts of Africa).

In India, European rule was actually constructed on a military foundation, and the armed forces enjoyed priority funding. From the late eighteenth century on, the British spoiled their loyal mercenary troops and ensured they were adequately rewarded; the British and Indian military cultures merged into a martial *sepoy* world. Until 1895 there was a decentralized organization, so that different armies kept watch on one another. After the Great Rebellion the British relied more than before on Punjab Sikhs, who made up roughly half of the standing troop strength. At the end of their active service, they sank roots as army settlers and took on ancillary tasks such as horse breeding. In an age of increasing conscription, the Sikhs were perhaps the most highly decorated professional troops anywhere in the world.[140]

The nineteenth century also saw the peacetime debut of the standing army stationed in barracks.[141] This presupposed that all citizens stood on an equal footing, but at the same time it was an instrument whereby the state established such equality. General conscription, without which a people's army is unthinkable, thus stands in a complex interrelationship with the formation of nations and nation-states. In the revolutionary wars, soldiers on the French side fought as citizens for the fatherland, no longer as subjects for a king. The idea of the nation under arms was born. But it took universal peacetime conscription to bring about a new kind of relationship between state and society. The distinction between war and peace is important here, because the spontaneous self-mobilization of the popular masses under wartime conditions is something different from a routine annual levy of whole groups of young men. A conscript does not necessarily feel himself to be a *soldat-citoyen*. After its Jacobin origins, compulsory military service only gradually established itself in the face of major resistance. At the outbreak of the First World War, Britain was the only major power to rely on volunteers for the manning of its army.

Conscription did not necessarily imply democracy or fairness of the draft. In France, until 1872, it was nearly always possible for affluent citizens to buy themselves out of the army; there was a market for substitutes, with fluctuating prices. Until 1905, whole occupational groups (teachers, doctors, lawyers, and so on) were being spared. Well into the Third Republic, France had not so much a citizen army as an army of stand-ins. In Prussia, which introduced conscription early on as a matter of "national honor," the institution aroused less enthusiasm at the thought of service than ingenuity in the search for dodges. Only in the imperial period after 1871 did the army really become an important agency of socialization, a "school of nationhood" for nearly all layers of the population.[142] In Russia, conscription was part of a general duty of service to the tsar formalized in the early eighteenth century, and before the Crimean War any nonnoble who became caught up in the military machine had to remain there for twenty long years. Men were drafted from nearly all the peoples of the empire. But at first it was not possible to speak of universal conscription—that was officially introduced only in 1874.[143] The Tsarist army, like its Habsburg counterpart, was anything but a *national* army, rather comprising a mosaic of all possible ethnic and linguistic groups. The same was true of the force that Muhammad Ali began to put together in the 1820s in Egypt so that he could conduct his campaigns in Sudan and Arabia. Egypt turned into an aggressive military state, basing itself on the press-ganging of ordinary peasants, the fellahin. The officer corps, on the other hand, consisted not of Egyptians but of Turkish-speaking Turks, Albanians, Kurds, or Chechens, whose French instructors taught them the elements of modern warfare. Muhammad Ali did not yet think of involving the peasantry as active citizens in his authoritarian-dynastic project of nation building.[144]

Things were only slightly different in the Ottoman Empire during the second half of the century. The basis for military modernization was the suppression of the rogue janissaries (1826), a guard of originally non-Muslim (but later converted) groups based in Istanbul, who had degenerated into a self-perpetuating caste barely capable of performing its duties. In the 1840s, in the wake of the Tanzimat reforms, a new policy aimed to unify the status of male subjects and to close the gap between state and people by abolishing a range of intermediate bodies. The gradual introduction of universal military service after 1843 was part of this reorientation—here too a major intervention in society. As in many European countries, exceptions were made for certain groups such as nomads or residents of Istanbul. Non-Muslims were charged a special tax in lieu, becoming liable for conscription only much later, in 1909. Military service, which in practice could be extended far beyond the allotted term, was widely feared and detested, and the actual intake of recruits was comparatively small. After the turn of the century, the Ottoman army continued to rest on the sedentary Muslim farmers of the Anatolian core provinces. By then there was a competent officer corps, soon to prove itself the most active factor in Turkish politics, but a "school of nationhood" the Ottoman army would never become.[145]

Perhaps nowhere other than in Prussia-Germany did conscription acquire as much importance as in Japan. In stark contrast to the ethnic heterogeneity of the great continental armies, the post-1873 Japanese military was organized as a national force on the basis of universal conscription (three years in the field, four in the reserve), but as in France it was possible to buy exemption from service. Conscription had a directly revolutionary significance in Japan that was not present in any other country, since the Meiji military reformer Yamagata Aritomo opposed plans to convert the samurai of old into a neofeudal force of professional soldiers. A conscript army was supposed to avoid the formation of such an autonomous knighthood while providing an opportunity to tie the population to the new regime and to use its energies for national objectives. The prestige of the military grew enormously after the victories of 1895 and 1905. Japan's militarism in the early twentieth century was less a continuation of old martial traditions than the consequence of a new beginning that borrowed from the models of France and Prussia.[146] Above all, universal conscription made the military visible in civilian life during peacetime.

Police

In the army, mobilization converged with the disciplining of a certain population group. General order and discipline in peacetime was the responsibility of the police and the criminal justice system; the army had a hand only during periods of revolutionary turmoil, or else in rural contexts (as in Russia) where the police were too thin on the ground. The state withdrew earlier in

nineteenth-century Europe than elsewhere from spectacular acts of penal retribution. It no longer used ritual executions to stage its theater of horror. The
growing strength of humanitarianism gradually made such practices seem intolerable, and after the middle of the century they disappeared from Western
Europe: by 1863 in the German lands and by 1868 in Great Britain.[147] Something
like a global "premodernity" in the penal system ended wherever the state hang-
man vanished from the public eye as a skilled craftsman and entertainer. The
logic of the market also made such displays objectionable, since in many cities
the proximity of a place of execution interfered with the rising trend in real estate prices. Nonlethal state violence, of kinds also unthinkable in today's Europe,
persisted for rather longer. In 1845 Tsar Nicholas I forbade public floggings, but
the practice remained so widespread that it elicited protests until the end of the
century from humanitarian activists, as well as from nationalists who feared that
it threatened Russia's reputation as a civilized country.[148]

Greater penetration of society by the forces of law and order gave the state less
drastic means of exerting its power. The nineteenth century was the pioneering
age of the police. France had been the first European country—as early as 1700—
to have full-time police agents under central government control.[149] In Britain a
London-centered police *system* began to take shape in 1829, but the control of local
authorities remained greater there than on the Continent. It was 1848 before the
police in Berlin were provided with uniforms that made them clearly identifiable.
Meanwhile, the gendarmerie was responsible for government control in the countryside—a special force that had first acquired importance during the French Revolution, later serving as a model for the whole Napoleonic empire and beyond and
figuring as one of France's leading export articles throughout the nineteenth century.[150] The police and gendarmerie were in many countries the most lasting remnants of Napoleonic rule; the Restoration regimes took over few things so gladly.

The French police model also spread outside Europe. Whereas Japan (under
the influence of the Franco-Prussian War) had mainly imitated Germany in military matters, it looked to France for the building of its police force. In 1872 the
country's first justice minister sent a delegation of eight young officials to Europe
to study and compare its various police systems, and shortly after their return
Japan set to work (initially only in the capital) on organizing a modern one of
its own. The French system had rightly struck the visitors as the most clearly organized, and the ministry had already singled out France as the main model for
a new justice system. Over the next twenty years it would be the French police
system that the Japanese reproduced with a number of modifications. The Gendarmerie, for example, became the Kempetai.[151] After Japan began its imperial
expansion, it followed the French custom (unknown in the British Empire) of
placing its colonies under the control of its military police, and the Kempetai
took on this role in Taiwan and later Korea. Until 1945 it grew continually into
that brutal instrument of terror that kept the civilian population in fear and
trembling in all its conquered lands.

By 1881 Japan had completed the learning process in its police sector. What then followed were expanded adaptations of the imported system. Japan took the professionalization and training of its police more seriously than any country in Europe, covering the country with a dense network of stations. As the state's main agency for actual implementation of the many-sided Meiji reforms, the police nipped in the bud any resistance to the New Japan and ensured that social change would take place only from the top down.[152] Its greatest successes were in harassing undesirable political parties and organizations of the early workers' movement. Less effective were its operations against the spontaneous protests that became more frequent around the turn of the century. At the time of the Meiji Emperor's death in 1912, the typical Japanese policeman was not an Asiatic version of the friendly London bobby but a direct agent of the central government. Japan was then perhaps the society in the world with the most pervasive police presence.

Probably no colony in the nineteenth century was without the basic elements of a modern European-style police force, above all in the cities. In maintaining law and order in the countryside, the colonial rulers nearly always cooperated in one way or another with local elites, relying partly on patron-client relations, partly on mechanisms of collective responsibility. The revolts in Asia that repeatedly caught the colonial authorities by surprise show how little was known about what was happening in those large agricultural countries.[153] Whether a territory had been under European control for a long time (as had India and Indonesia), or whether it had been colonized only in the 1880s (tropical Africa or northern Vietnam), the colonial police began to tighten their grip in rural areas only in the 1920s, at a time when defiant workers were clashing with the authorities in the increasingly restless cities. Similar tendencies were apparent in noncolonial China, where halfhearted attempts at state building under the Guomindang government (1927–37) included the deployment of a rural police force such as had never existed before. Before 1920 it was only in exceptional cases, such as Cochin China, that colonial peoples experienced the kind of police control and village linkage to bureaucratic command chains familiar in continental Europe and Japan.

The worldwide evolution of police forces in the nineteenth and twentieth centuries offers good examples of all manner of transfers, not only from mother country to colonies or through export to independent countries (Siam or Japan) but also between parts of the same imperial system. Thus, after the British occupation of Egypt in 1882, the basic structures of the *Indian* police were introduced without any reference to local conditions. Other ways of establishing order in the colonies also impacted on Europe. The Indian penal code, for example, which Thomas Babington Macaulay, the famous historian, drafted in 1835 during his time as de facto justice minister of India, had a precision and consistency without precedent in the British Isles with its casuistic common law tradition; a comparably systematic English penal law followed in its wake only in the 1870s.[154] Just as the state in India adopted the drastic sovereign measures of a conqueror, legislator, and gendarme, many conservatives in the British mother country considered that the state's coercive power should be turned more strongly against the practice and rhetoric of democratization.[155] The forces opposing such an authoritarian backlash remained strong enough to stave off threats to the representative system at home. But critics of imperialism, such as the farsighted John Atkinson Hobson, expressed major concern that nine-tenths of the inhabitants of the empire lived under the yoke of "British political despotism," which threatened to poison the climate in the mother country.[156] Colonialism constantly spawned authoritarian challenges to metropolitan liberty—and regular demands for stronger police powers.

The police in the United States had its roots in England: first in the old tradition of community night watchmen transferred to the American colonies, then in the important modernization that gave rise in 1829 to the Metropolitan Police Force and its uniformed bobbies. This basic model was adopted with a delay of two or three decades by large cities in the United States,[157] and it was only in the 1850s that those in the East provided themselves with uniformed policemen on a permanent payroll. American peculiarities soon manifested themselves, however. A nationwide police force, such as existed in France and later in Britain, remained conspicuous by its absence, and it took many more decades before a further criterion of bureaucratic rationality—political independence—was fulfilled. Until then the police were often a tool of municipal party politics. Moreover, extreme decentralization contributed to wide variations in the intensity of policing, so that many areas (especially on the frontier) were virtually without a police presence and others were faced with a mosaic of overlapping jurisdictions. It was very difficult to bring to justice a criminal who managed to escape across such boundaries. This created a gap in the market that private detective agencies moved to fill, the best known being the one founded by Allan Pinkerton in 1850. Pinkerton's people initially guarded railroads and mail coaches, but in the 1890s they also became notorious for their attacks on striking workers. In no other country in the world did an incomplete state monopoly of physical force leave so much scope for private police forces; it was no easy matter to ensure that they were subject to judicial control. In the United States, the police force was regarded not as the organ of a hierarchical "state" but as a part of local government—the direct opposite of the situation in France or Japan, but also very different from England.

An English policeman in the late nineteenth century saw himself as acting under the authority of common law and the unwritten constitution, whereas his American counterpart thought of himself more as representing "justice" with the particular situation in which he operated. The "marshal" of the American West was the unmistakable embodiment of this type.[158] He was also often the only local representative of a distant state power. More typical of the nineteenth-century world was a division of labor between the police and the gendarmerie or army. The idea that the military should not be deployed to keep law and order inside a country was a new maxim of political culture in only a small number of countries. The police force was historically more recent than the army, emerged as the result of functional differentiation, and played a less prominent role in state-building processes. Its task was less to establish than to manage a state monopoly of legitimate force.

Discipline and Welfare

Although, organizationally speaking, state apparatuses had less scope for intervention in the nineteenth century, they sometimes took action in areas of daily life from which the (European) state of the early twenty-first century has long since retreated. This difference is directly linked to definitions of criminal behavior. History shows a wide variation according to whether the state attempts to impose religious conformity or, in one degree or another, considers itself the guardian of the private "morality" of its subjects and citizens. At least in Protestant Europe—and especially in Britain—there was a noticeable moralization of state functions, and hence of police activity, in the nineteenth century. In Victorian and Edwardian England, the police and courts became truly obsessed with the "prevention of vice," targeting prostitution, homosexuality (or "sodomy"), drunkenness, and a passion for gambling in particular—not only to protect the upright majority from such transgressions but also to carry forward the moral duty of lifting the moral condition of the population. The penal system was more than before an instrument of virtue, not without a nationalist ambition to make the country morally "fit."[159] In 1859 John Stuart Mill's essay *On Liberty* had warned against such an invasion of the private sphere, and soon after the turn of the century Karl Kraus exposed the contradictions of "morality and criminality" in an Austrian context.[160] That such polemics were necessary is an indication of the seriousness of the problem.

Criminalization also functioned in the colonies as a means of exclusion and control. The authorities in British India, for example, assigned people to tribes and castes that were graded on a scale of "hereditary criminality." By the end of the colonial era, in 1947, as many as 3.5 million individuals, or 1 percent of the total population, were classified as belonging to 128 mostly migrant "criminal tribes," which felt the full force of state persecution. Actual behavior, such as criminal practices handed down from generation to generation, interacted with official labeling to produce a stable definition of this minority, and in 1871 a Criminal Tribes Act fixed their position in relation to the colonial state. Among the methods of controlling them were police registration, compulsory residence in a certain village, and forced labor in land clearance. The analogies with Gypsies in Central Europe are evident. The "criminal tribes" were not pure inventions of a craze for taxonomy. It is now thought likely that these groups were descended from Central Asian nomadic tribes, which the collapse of the Mogul Empire in the eighteenth century condemned to a vicious circle of exclusion.[161]

It was not expected that the Indian "criminal tribes" would be "educated"; they stood outside the sphere in which "civilization" appeared possible and desirable. The same might happen where—almost simultaneously with the turn to greater compulsion—a policy of criminalization sought partly to reverse the consequences of emancipatory rhetoric. In Alabama, previously one of the largest slave states in the South, a mostly black convict population came into being after the Civil War and the Reconstruction period, especially from 1874 on. New crimes were introduced, and after a brief interlude of freedom prison posed a new threat to the black population. Under a new "convict lease system," profit-motivated penitentiaries began to offer cheap labor to the new industries and mines of the South.[162]

Japan's main borrowing from the European arsenal of discipline was the idea of prison as a place of surveillance *and* education. This entailed far-reaching changes in penal law. In the Tokugawa period, many jailed oppositionists had written about the appalling conditions in primitive dungeons similar to those in many other parts of the world. At the time there was no publicly recognized penal code; the first such codes, still little influenced by the West, would appear only in 1870 and 1873. Early Meiji rule books continued to specify details of corporal punishment, such as the number of blows in relation to the seriousness of the offense. In the 1870s support grew for the idea that useful labor should be introduced to improve the subjective state of the prisoner, and in 1880 the first penal code modeled on the West (in fact, drafted by a French legal expert) came into force.[163] The basic principle now was that any punishment must have the sanction of the law (*nulla poena sine lege*), and that it should not vary in accordance with social status. Also in the 1880s moves were begun to make education a systematic part of prison life.[164] In this respect, Japan soon took the lead over European countries. Penal reform became a major policy issue worldwide, a test of whether a country was part of "modern civilization" and had the capacity to take resolute action. Around the turn of the century, Chinese intellectuals who cared about China's future, for example, generally favored the creation of "model prisons" in the European or North American style.[165]

To what extent was the nineteenth-century state already a welfare state? Older policies aimed at "the poor and beggars" were dismantled in Europe over time. In France the revolution's plans to fund an equality-based system run by the state went unfulfilled. The hospitals, hospices, and other communal establishments characteristic of the ancien régime remained in existence, increasingly under the patronage of private benefactors. Governments in Western and Central Europe built many new complexes, often intentionally locating hospitals in the vicinity of mental asylums or workhouses. Poor relief and social disciplining were almost inextricably intertwined. Tight limits were set to independent workers' initiative, so long as freedom of association was denied to them. After 1848, in many countries of continental Europe, this became the basis for the formation of trade unions, consumer associations, and mutual insurance. In Britain "friendly societies" with similar aims had been in existence for an even longer time. The state sought to extend its control more than in the past, but its welfare spending had not risen appreciably by the end of the century. In some countries, such as England, it actually fell if it is measured by the share of poor relief in the national product.[166] Only after 1880 did governments begin to provide for welfare in general, not just for particular groups such as miners, by means of legislative and administrative measures and the incorporation of private or church institutions.[167] Poor relief was then gradually replaced with "welfare state transfers" and compulsory national insurance.[168]

A new definition of government tasks began to appear, with insurance covering the risks associated with paid employment. Sickness and accident insurance for workers in the German Reich, introduced in 1883–84 and supplemented in 1889 with disability and old-age insurance, opened the way internationally. At once, highly statist solutions placing the emergent welfare state in the hands of bureaucracies and interest groups overshadowed alternative ideas of social solidarity. Indeed, Bismarck's social insurance scheme went together with a ban on trade unions and Social Democratic endeavors (the *Sozialistengesetz* of 1878), one of his aims being to weaken the support funds autonomously managed by the labor movement.[169] The welfare state did not emerge from the very beginning as a complete package; Germany had to wait until 1927 for the unemployment insurance that was set up in 1907 in Denmark and in 1911 in Britain.[170]

The chronology of the transition to a state-funded and bureaucratically administered structure of legal entitlements appears very uneven if we look separately at the various kinds of insurance and support. Democracies did not consistently advance at a faster pace than authoritarian or semiauthoritarian political systems. In democratic France, for instance, the age of social insurance opened only in 1898, with the establishment of a scheme covering work accidents. Governments in various European countries, together with newly emerging small groups of "social experts," kept a close watch and learned from what others were doing, on the other side of the Atlantic too.[171]

This did not lead to the development of uniform systems. Rather, three different "worlds" took shape in the passage from the nineteenth to the twentieth century: a Scandinavian model that funded social security through redistribution; a British model, whose main aim was to avert poverty through tax-funded social provision; and a continental European model, financed by individual contributions and more strongly geared to social status (as in the privileged treatment of civil servants).[172] Nevertheless, it may be said that nowhere in the world other than in Europe and Australasia did the traditional municipal, philanthropic, religious-ecclesiastical, and official measures of poor relief evolve through their own dynamic into a new understanding of the tasks of the state. In the United States, where private charity enjoyed high esteem but the spending of tax revenue on the poor counted as waste, there were many local instances of borrowing from Europe, but comprehensive welfare programs were not rolled out until the

1930s. Japan too, in other ways so quick to follow Europe, took its time building a welfare state; only in 1947 did it become the last of the major industrial countries to introduce unemployment insurance. In many places, checks on the morals of welfare recipients lingered on as an ideological remnant of the nineteenth century.

From the point of view of global history, the welfare state belongs to the twentieth century. It was also then that, in an extraordinary development associated with state socialism, comprehensive (if low-grade) systems of social security were established in a number of economically backward countries. In China, which passed through this stage after 1949, the post-1978 liberalization has yet to put a new system of protection in place.

6 Self-Strengthening: The Politics of Peripheral Defensive[173]

Perceptions of Backwardness

The nineteenth-century state was a reforming state. It is true that in the dying years of the ancien régime, some rulers and ministers had seen the need to make government more efficient—hence to gain increased access to resources and, as far as possible, to widen the base of popular loyalty. Austria under Maria Theresa or Joseph II and, above all, during the reign of his brother Peter Leopold as grand duke of the model Enlightenment province of Tuscany (1765–90), as well as Prussia under Frederick the Great, were examples of such reforming states; Turgot wanted to make France follow suit; and after 1760 the Spain of Charles III undertook a general overhaul (by no means altogether unsuccessful in the medium term) of its huge overseas empire. In China too, it was a common idea that from time to time the state needed to be methodically regenerated, the last such repair of the bureaucratic machinery having been undertaken in 1730 by the Yongzheng Emperor. In the nineteenth century the impetus for reform came more than ever from outside, as international competition generated the necessary pressure. Of course, internal reform was also related to the threat of revolution. The events of 1789 had taught one or two things about the costs of delay, suggesting that reforms might serve to prevent something much more drastic. Here and there an *unsuccessful* revolution might also sow the idea of responding to some of its demands with timely reforms. The revolutions of 1848 did not remain totally without effect.

The reforms most typical of the nineteenth century, however, were triggered by a perception of national backwardness. Back in 1759, the Bourbon overhaul of the Spanish colonial empire had already been designed, inter alia, to dispel the notion that Spain was lagging behind other countries and to win the respect of enlightened public opinion in Europe. None of these perceptions was stronger than that resulting from failure in war. In 1806, the year of the great defeat at the hands of Napoleon, parts of the Prussian power elite concluded that the survival of the old order depended on a process of comprehensive renewal. The Crimean War had the same effect for the Tsarist Empire, as did the defeat of 1900 for the Qing Empire, when an international expeditionary force intervened against the Boxer Uprising. The actual reforms varied in each instance, but the underlying idea was that the operations of the state should be more rational, more subject to the equalizing influence of the law. Lost wars, then, did not generate only military reforms. It became a widespread view that military apparatuses could only be as good as the civilian structure of the state in which they were embedded. This was clear to the Prussian, Russian, and Chinese reformers (in the latter's case, too late), who all faced the task of transforming weakness into strength.

Behind this lay an even more general perception. Never before in history had so few societies been seen as the yardstick for so many others. To be sure, rather superficial attempts had been made to copy the outward forms of prestigious states and civilizations, as when France under the Sun King found imitators in large parts of continental Europe. The idea of political progress had also occurred to people in the early modern age. And before 1700 the Netherlands, England's great commercial and military rival, had seemed to offer a model in many areas of business, society, and politics. But these had been very limited perceptions of difference, which seldom crossed the boundaries between civilizations. In the seventeenth and eighteenth centuries, the enthusiasm of Jesuits and certain financial theorists for what they took to be the well-structured and wisely governed Chinese state of the great Qing Emperors had little transformative impact in Europe. Nor did the opening of the Ottoman Empire to Western European architectural and decorative styles during the so-called Tulip Age (1718–30) prove to have long-term consequences.[174]

The nineteenth century brought something new: Western European civilization became a model for large parts of the world. "Western Europe" meant first and foremost Great Britain, which by 1815 was being spoken of nearly everywhere as the richest and most powerful country in the world. Despite the fall of Napoleon and continuing political instability, France also counted as part of this Western European model. It was gradually joined by Prussia, although it took many decades to shake off its image as a Spartan military state on the eastern fringes of civilization, whose greatest king, disdainful of German literature and preferring to speak French, himself felt ill at ease there.

Outside this European core, nothing shaped the evolution of the state in the nineteenth century as much as the efforts of power elites to counter the dynamic of the West by preventively adopting elements of its culture. Around 1700 Tsar Peter the Great had already pursued such a policy of making Russia internally and externally strong, both with and against Western Europe. A century later the resistance to Napoleonic France triggered the first moves toward defensive modernization. The Ottoman Empire had already made a similar start under Sultan Selim III (r. 1789–1807), shocked by Russia's southward expansion under Catherine II and Bonaparte's invasion of Egypt in 1798. But his reforms ran up against strong internal opposition and made little progress. Less controversial, and therefore more successful, were the post-1806 reforms of the army, civil service, justice, and education in Prussia. The construction of a military state under Muhammad Ali in Egypt, begun in exactly the same period, is another facet of this moment in world history.

The success of Egyptian military expansion revealed the weakness of the Ottoman Empire. The fact that the Great Powers had to rush to its aid against its own vassal, Muhammad Ali, and that Greece came under their protection and was actually wrested from the empire, helped to push the sultan and leading statesmen in 1839 toward a bold policy of sweeping reforms, the so-called Tanzimat, which lasted for a quarter of a century.[175] The fruits were the creation of an educational system (with the suppression of some Islamic elements), reform of the state administration, legal changes tending toward a single citizenship, gradual alleviation of discrimination against non-Muslims, and a fiscal restructuring to replace one-off raids and tax farming. The figures leading this drive in the Sublime Porte knew the West from personal experience and formed ideas of their own about the goals, scale, and feasibility of partial Westernization under Ottoman conditions. Mustafa Reshid Pasha (1800–58), Ali Pasha (1814–71), and Fuad Pasha (1815–69), the key members of the reform generation, had at one time or another all been foreign minister or ambassador in London or Paris. The group of those able to combine Eastern and Western knowledge was very small, with the result that their initiatives had a strongly centralist and dirigiste character. A dynamic in civil society was not at the origin of the reform. But one could develop under favorable conditions, as soon as the impetus for reform in Istanbul had created the space for it. Cities such as Salonica and Beirut provided impressive evidence of this.[176]

Reforms

A sense of backwardness, for which causes were always to be found, also lay behind many reform drives in the second half of the century. Meanwhile the West, at once admired and feared, did not remain unchanged. Especially in the second half of the 1860s, the political systems of Britain, France, Prussia, and Austria-Hungary underwent remarkable, though not exactly revolutionary, changes. States everywhere were in the grip of reform.[177] On the edges of Europe and beyond, reluctant appreciation of the West's momentary superiority and genuine admiration for many of its civilizing achievements mingled in various ways with a lack of confidence in the reformability of the respective national institutions. Often there was also a hope that basic cultural values could somehow be rescued and preserved in the new age. Examples in this respect were the Russian reforms under Alexander II, centered on the abolition of serfdom in 1861 and the reform of the justice system in 1864;[178] the very cautious early reforms in China after the victory of the Qing Dynasty over the Taiping in 1864; and above all the radical "reformatting" of Japan after 1868 and its "little brother," the modernization of Siam/Thailand.[179] In each of these cases, major debates took place in ruling circles and a newly emerging public sphere. A comparative study of them has yet to be written. But the key issues were the scale and intensity of "Westernization," and the likelihood that it would be achieved. "Westernizers" clashed with "nativists," whether these were Russian Slavophiles or followers of orthodox Confucianism. Rulers who had previously had to bother little about such questions now found themselves facing risky political calculations. No amount of experience helped when it came to predicting the consequences of change. What was a reasonable price to pay? Who would be the winners and who the losers? Where might strong resistance be expected? What protection could be organized in the field of foreign policy? How should the reforms be financed? Where would the skilled personnel come from to implement them in different walks of life and geographical regions? The answers varied from case to case. But the similarity of the problems means that in principle the cases are susceptible to comparison.

All these reforms belong in a history of the state: that is, both in a history of how the European state spread through the world along several fault lines and with numerous modifications[180] and in a history of the mobilization of extra-European state resources in response to acute survival problems, at peripheral positions of international politics, global capitalism, and the dissemination of Western European civilization. The strategies differed considerably from one another, and varied enormously in their degree of success. Meiji Japan was in a category of its own, in terms of the pace and scale of system change—and it became a model much admired on all sides, though rarely copied successfully.[181] The defensive modernization of the Tsarist Empire, on the other hand, was a conservative holding operation. In the Ottoman Empire, the reform period issued in a new absolutism under Abdülhamid II, whose performance is still the subject of scholarly controversy. In China, several attempts at reform (1861–74, 1898, and 1904–11) failed to result in a viable renewal of the state and society. In Egypt, Westernization under the successors of Muhammad Ali ended in state bankruptcy and a colonial seizure of power (1882).

The "reform period" in Mexico, from the mid-fifties to the mid-seventies, is also part of this context, but like the Tanzimat it did not achieve a breakthrough to solid representative structures. Even the leading liberal statesman, Benito Juárez (in office 1860–72), sought refuge after 1867 in ad hoc authoritarian measures. And, like Abdülhamid II, Porfirio Díaz took sole power in the mid-seventies and continued to exercise it into the first decade of the new century. However, a flurry of reform legislation had been passed before the Díaz era, so that at least the influence of the church (a major adversary of Mexican liberals) was curtailed, and the principle of the equality of (white) citizens before the law was respected. The paternal supervision of life by secular and spiritual authorities went into decline.[182] A further example of post-reforming absolutism was the Russia of Alexander III (r. 1881–94). Many measures of his assassinated predecessor were rescinded, and although the successful justice reforms, at once expression and guarantee of a sophisticated legal culture in the late Tsarist period, were largely preserved, the powers of the police were significantly expanded. Again paralleling trends in the Ottoman Empire, the Russian authorities now viewed models from the West—especially its political liberalism—with much greater skepticism. Tsarist rule became more autocratic and internal repression more severe.[183]

New conceptions of the future were bound up with the reforms, but rarely from the very beginning. In the Ottoman case, it was only in the third Tanzimat decade that the original idea of reform as a timely restoration of precarious balances was replaced with a future-oriented vision of a definitive new order. The means changed with the end. Instead of a flexible combination of old and new techniques of rule, there came a stricter centralism and a new peremptoriness that cared less about compromises with local power holders than in earlier phases of the reform process.[184]

The deferred chronology of particular reform projects made it possible for them to learn from one another. The grand viziers and state philosophers of the Tanzimat era were still exposed to original Western European models; they had little more than France and Britain in mind. The Meiji leadership could already be influenced by the long-term consequences of the Prussian reforms, especially with regard to increased military strength. It saw itself in the role of a rational shopper, critically surveying a collection of models from the outside world. Hardly any of the smaller countries of Asia or Africa enjoyed such freedom of choice. Ahmad Bey (r. 1837–55), for example, the enthusiastic reformist ruler of Tunis, built his army—for lack of alternatives—with the help of the French, who were threateningly close just across the border in Algeria; British assistance would not have been viewed kindly in Paris.[185] As soon as the extent and success of the Japanese renewal became visible elsewhere, it set a new standard for other countries. The Chinese elite, for deeply rooted cultural reasons, did not find it easy to admit Japan's superiority, in the military field or anywhere else. But in the final years of the Qing period, Japan appeared to have caught up with—some would have said, overtaken—Europe and North America as the most attractive reference model. At the latest after its victory over Russia in 1905, Japan beckoned throughout Asia as the country that had broken the spell of European invincibility.

7 State and Nationalism

Strong State, Weak State

In the nineteenth century the strong state disappeared from political theory, at least in Europe. In the early modern period, leading theorists had concerned themselves with the greatest possible strengthening of the state, particularly of monarchies. A strong state was seen as something to strive for—a means of curbing anarchic private interests, breaking up small power enclaves, and purposively seeking the public good. Further justifications of absolute rule were added in the eighteenth century, with visions of enlightened princes and selfless officials; cameralism and a "science of public policy" (in German: *Polizeywissenschaft*) offered blueprints for state building. The picture was very similar at the time in China, where centralism and decentralization had clashed in the political culture for two thousand years. The old tradition of administrative theory was brought to a new peak in the eighteenth century. The three great Qing emperors who ruled successively between 1664 and 1796 were energetic and competent autocrats, not a whit inferior to Frederick II of Prussia or Joseph II of Austria. They defined their role very broadly, yet tirelessly sought to preserve and raise the efficiency of the bureaucratic apparatus. The state allowed some leeway: it was by no means the "totalitarian" Leviathan sometimes conjured up in the older sinology; it allowed niches of market economy, not as an institutional limitation to its power but as a generous favor from a ruler of unfathomable might.

Doctrines of the strong state were no longer publicly discussed in the nineteenth century. Even the Napoleonic regime, otherwise not averse to propaganda, did not present itself self-consciously as a modern command system. Liberal attempts to define the "limits of state action" (Wilhelm von Humboldt, 1792) were the norm at least until the second quarter. Conservatives did not openly champion "top down" neo-absolutist rule but embraced Romantic ideas of social estates, with special emphasis on the nobility's cultural leadership. Socialists and anarchists, between whom there was no fundamental difference on this point, developed few ideas about the state; the revolution would clear away the bourgeois capitalist system anyhow and institute a "realm of freedom."

While this distrust of an omnipotent state stretched far beyond the liberal parties of the time, developments in the real world were placing more and more means at the disposal of the state. Liberal thinkers as different as Herbert Spencer and Max Weber thought they had to warn against a new serfdom resulting from hypertrophy of the state, bureaucratization, and—in Weber's view—a tendency of capitalism to petrify. Paradoxically this accumulation of power, long undertheorized in discussions of the state, was sympathetically addressed in another field: in nationalist programs. Whereas the most reactionary monarch no longer dared to claim "L'État, c'est moi," the idea gained currency that the state was the nation: whatever served the state was useful to the nation. This displaced the basis for the legitimation of state power.

The nation-state had its own kind of reason: no longer the rightful claims of a princely dynasty rooted in the depths of history, or the organic harmony of a "body politic," but something called *national interests*. Who defined those interests and translated them into politics was a secondary matter. So long as politicians, at least in Europe, followed Giuseppe Mazzini's influential understanding of nationalism, the interests of a country—democratic order at home, peace with other nations—appeared to be simultaneously achievable. In the third quarter of the century, however, there was growing skepticism about such a utopian harmonization (it would be temporarily resurrected in 1919 with the founding of the League of Nations), and it became clear that the nation-state could go together with quite different political systems. Two things were decisive: internal homogeneity expressed at every possible level of integration, from language policy to religious uniformity to dense infrastructural projects such as a railroad network; and a capacity to take military action externally. Nationalism thus acquired huge importance for the theory of the state. "Pure" state theory would revive only when justifications began to be developed for the welfare state.

Divided Nationalism and State Legitimacy

The accumulation of state power in the course of the century, above all in its last quarter, was globally differentiated. The main reason for this was the extremely uneven distribution of industrialization. Whereas in the early modern period the states of Eurasia, in a great arc from Spain to Japan, grew stronger at the same time and on similar social foundations, the nineteenth-century accumulation of power was concentrated in three regions of the world housing the so-called Great Powers: Europe between the Pyrenees and the Urals; the United States of America; and, with a short delay, Japan. The strengthening of the state was thus by no means an advance in human evolution but a global redistribution of imbalances. Countries that weakened or fell behind became more vulnerable. Imperialism was the result of this power gap: weak states were in danger of being undermined or even subjugated. Europeans in the early modern period had imagined the "Oriental" state to be a crushing despotism, something it was decidedly not, even in China with its powerful bureaucracy. Ironically, nineteenth-century Asian rulers now tried to compensate for their weakness by assimilating the bureaucratic and centralist energy of the European nation-state.

Nationalism divided into two. One half became the doctrine of the strong, compact nation-states of the West, following a quite special agenda of their own; the other half appeared as a defensive program. States that had already lost their independence through conquest could do nothing other—on a larger scale after the First World War—than wage a defensive nationalist struggle within the framework of colonial rule. In other cases, defensive nationalism required a policy of self-strengthening in as many spheres as possible. Expansive and defensive nationalism thus stood in a dialectical relationship: each in its way was capable of extraordinary feats of mobilization in the name of solidarity among individuals not personally known to one another, and of drawing into politics social groups that previously had had no opportunity to participate.

More general still was the dialectic of nationalization and internationalization. Contrary to their self-image, nation-states by no means pursued their inner potential alone. Nationalism as an ideology and a program spread transnationally—across Europe, for example, through the ideas of Mazzini or the cult of a national freedom fighter such as the Hungarian Lajos Kossuth.[186]

During the second half of the nineteenth century, such direct transfer lost much of its importance as various nationalisms reacted to one another in antagonistic ways. However, the consolidation of national societies and the rise of a rhetoric of exclusion and superiority were closely bound up with the greater number and intensity of cross-border contacts at many levels.

Nation-states responded variously to this contradiction. Britain, for example, had long taken its empire for granted, so it was a possible strategy to simplify matters by rationalizing its variegated global presence and establishing closer links between individual colonies and the mother country. This is what the colonial secretary, Joseph Chamberlain, attempted around the turn of the century, though without success: to turn the loosely knit empire into a kind of super nation-state, a federation mainly of its "white" components. The German Reich was in quite a different situation. Founded at the very moment of a great worldwide advance of globalization, it immediately had to adapt its foreign economic policy to these conditions. It became primarily an industrial and military state because its politicians and entrepreneurs used the opportunities of internationalization to serve the national interest.[187]

Model Citizens and Intermediate Powers

The idea of democracy, whether direct in Rousseau's sense or within the indirect British tradition, envisaged a simplification of political mechanisms. Jeremy Bentham, the English Enlightenment thinker with a "utilitarian" leaning, expressed this perhaps more clearly than anyone else, but a basic point in all democratic programs was that accountable rule in the modern world required the elimination of intermediate powers. The people and those who governed them were to face each other as directly as possible. The link between them was to be one of representation: either democratic, through procedures of election and delegation, or a *unio mystica*, in which a monarch or dictator claimed to embody the nation, and the "people" endorsed this claim by acclamation or just supported it "virtually." In principle, therefore, the political system of nation-states rests on national homogeneity and simplicity of constitutional mechanisms.

Nation-states or modernizing empires strive for discursive simplification insofar as they establish, and seek to realize, norms for the "model citizen." In many civilizations, premodern debates on politics circled around the capacities, virtues, and devoutness of the model *ruler*. Modern debates center on the ideal *citizen*, defined in highly diverse ways but always expected to find a balance between the pursuit of private interests and service to the nation as a whole. Musings about national identity or "civilized behavior"—about how a Briton or Frenchman, Chinese or Egyptian, should comport himself (or herself), or what it meant to be British or French, Chinese or Egyptian—were a feature of public life in many countries around the turn of the century. They did not yet reach the collective excesses of the twentieth century, when "traitors to the fatherland," "class enemies," and "racial" minorities would be condemned to physical exclusion and persecution.

Nevertheless, the uniform simplicity of nations and "national organisms" remained an illusion. Empires could not conjure away their multinational character, and none took the radical step of introducing a single, "color-blind" citizenship. Whenever they attempted to create a common national foundation, they soon ran up against the contradictions that were part of their very essence. In colonial systems, political hierarchies could not fail to be complex; it was nearly always the case that many tasks relating to order and sovereignty had to be delegated. This also meant that colonial governments sometimes had to make others responsible for providing their funding. In a number of Southeast Asian colonies, compact Chinese minorities organized into *gongsi* (leagues or secret societies) helped out accordingly in their role as collective tax farmers and monopolists (e.g., in the opium trade).[188] The *gongsi* were not part of a formal system of rule, yet the state was not capable of functioning without them. Thus, even in a situation where democratic participation meant nothing, organized interests of recent creation might make themselves felt.

In the civil societies of the West, too, the ideal of simple government and a small state kept evaporating. New kinds of bodies proliferated between people and rulers: no longer the old estates but bureaucracies, political parties (increasingly compact organizations or, in the United States, local "machines"), syndicates, labor unions, all manner of lobbies and interest groups, desacralized churches representing special interests, and mass media under pressure to cut loose and play an independent role. The rational and simple political systems of classical liberalism became rather complicated affairs. By the First World War, the seeds had been sown in many places for those corporatist elements that would come to the fore in the 1920s, and not only in Europe.

PART THREE

THEMES

CHAPTER XII

Energy and Industry

Who Unbound Prometheus, When, and Where?

Few historiographical fields have recently been as innovative and exciting as (global) economic history. Cherished set pieces of historical lore such as the Industrial Revolution are undergoing critical reappraisal. Long-term developments across the centuries or even millennia find unprecedented attention. Culture and meaning are returning to economic analysis, "capitalism" has ceased to be vilified as a term of Marxist polemics, and a rehabilitation of materiality has prompted even committed students of discourse and imagination to turn to the world of objects and commodities. The towering issue of the "global rift," or "great divergence," is intriguing the most astute minds. Its extent, chronology, and causation remain hotly contested, and no consensus seems to be in sight.

Since almost anything is in flux, it may be appropriate to place an essay on industry and energy at the beginning of the third part of this book. No longer aimed at comprehensiveness the way the previous "panorama" chapters were, this essay and the ones that follow survey a given topic in a lighter, more playful, and more selective vein. Dealing with industrialization in such a more deliberative mode, with fewer pages and a smaller number of references, is intended to send two different messages to the reader. First, some of the issues at stake are too complex for the author—who is not a professional economic historian—to put forward his own considered solution with the necessary confidence; even a world historian is not obliged to have an opinion on everything. And second, the organization of production and the creation of wealth are absolutely crucial aspects of the nineteenth century. At the same time, it would be unduly reductionist to present them as independent variables and as the only sources of dynamism propelling the age as a whole. We are familiar with accounts of that kind where everything boils down to the "dual revolution" at the end of the eighteenth century. They retain some of their value. Nevertheless, it is time to decenter the Industrial Revolution.

1 Industrialization

If large parts of the world looked different around 1910 as compared with 1780, the main reason for the physical transformation was industry. The nineteenth century saw the spread of the industrial mode of production and of particular forms of society associated with it. But it was not an age of even, uniform industrialization. Whether industry sank root or whether it failed to catch on, whether it began late or was not even attempted: all this depended on complex combinations of multiple causes in specific local settings. Out of such combinations a new global geography of centers and peripheries, dynamic and stagnant regions, took shape. But what is industrialization? This concept, seemingly so straightforward, continues to arouse debate.

Controversies

Although the term "industrialization" was already in use in the 1830s, and "Industrial Revolution," first documented in 1799, had gained academic respectability in the English-speaking world by the mid-1880s, historians have been unable to agree on a precise usage.[1] The ramifications of the discussion among experts are difficult to penetrate. There is not just one question up for debate; rather, what is really at stake has to be repeatedly clarified anew. Another source of confusion is the fact that the historians involved each bring their own understanding of economic theory to bear. For example, some see industrialization as a process of measurable economic growth, driven mainly by technological innovations, whereas others attach greater importance to institutional change, seeing this as a contributory factor or even wishing to replace the term "Industrial Revolution" with "Institutional Revolution."[2] Scholars seem to be agreed mainly on two points:

1. that the economic and social changes associated with industry, visible on all continents by 1900, can be traced back to an innovatory impetus in England after 1760 (not even those who consider that impetus relatively undramatic and the term "Industrial Revolution" as exaggerated would deny this); and
2. that industrialization, at least in its beginnings, has always been a regional, not a national, phenomenon.

Even someone who underlines the significance of a legal-institutional regulatory framework, such as that which nation-states developed in the nineteenth century, will concede that industrialization is closely linked to resource supply at certain locations and that it does not necessarily mark *entire* national societies in the long term. By 1920 only a few countries in the world were "industrial societies," and even in parts of Europe such as Italy, Spain, or Russia the islands of industrial development by no means radiated out to mold society as a whole.[3]

The most interesting discussions today revolve around the following questions.

First. New and sophisticated evaluations of the fragmentary statistical material have demonstrated that in the last quarter of the eighteenth century and the first quarter of the nineteenth, the English economy grew more slowly and less regularly than champions of the big-bang theory have hitherto claimed. It has proved difficult to find data for a dramatic acceleration of growth, even in "leading sectors" such as the cotton industry. But if industrialization began gradually and proceeded at a gentle pace even in its "revolutionary" English period, then we have to ask in which older continuities it had its origins. Some historians actually go back to the Middle Ages, seeing the "Industrial Revolution" as one of several growth spurts since then.

Second. Even skeptics at pains to make an industrial revolution quantitatively invisible have to face the fact that numerous testimonies of the time saw the spread of industry and its social consequences as the dawning of a new age.[4] This was the case not only in England and in European countries that soon took a similar path of development but everywhere in the world where the advent of large-scale industry, new work regimes, and new social hierarchies had a perceptible impact. We must therefore always consider the relationship between quantitative and qualitative factors in describing and analyzing industrialization.[5] Representatives of so-called institutional economics, who see themselves as a (not too radical) alternative to the ruling neoclassical theory, have proposed a useful distinction between "formal" constraints on economic activity (above all, contracts, laws, etc.) and "nonformal" constraints within the respective culture (norms, values, conventions, etc.).[6] This richer, more variegated, picture of industrialization is certainly welcome. But there is a danger that too many aspects and factors will crowd in and make it necessary to give up the elegance of more "economical" explanatory models.

Third. Industrialization is generally thought of as the key to Europe's "special path" in history. The fact that unprecedented differences in prosperity and living standards were observable in various regions of the planet at the end of the nineteenth century can indeed be mainly attributed to the fact that many societies had embarked on the change to industrial society, while others had not.[7] But this can give rise to more than one problematic. In considering the reasons for this European "miracle" (Eric L. Jones), some conclude that England, Europe, and the West (or whichever counts as the main entity here) disposed of natural geographical, economic, and cultural prerequisites that were lacking in other civilizations—a familiar viewpoint going back to the studies of world economic history and the economic ethic of the world religions that Max Weber began after 1900. Others turn the tables and look for similar prerequisites in China, asking why it did *not* make a comparable breakthrough in productivity.[8] Should it turn out that the prerequisites for industrial development existed in China as well as Europe, it would have to be explained why they were not actually brought to fruition. All these debates hover delicately between informed numerical guesswork, anthropological assumptions about human behavior in different "cultural" framings, and counterfactual thought experiments.

Fourth. The older textbook accounts of industrialization used to assume, with Walt W. Rostow, that one national economy after another reached a "takeoff point" from whence it could progress along a stable, future-oriented path of self-sustaining growth. Determining the chronology of "takeoffs" provided a set of data marking the onset of economic modernity for various countries—an approximation still useful today. What is less convincing is the implied assumption that by some internal logic of its own, a standard model of industrialization was serially repeated from country to country. Against this, economic acceleration was in reality always fueled by both internal (endogenous) and external (exogenous) sources; the problem is to determine the proper proportions in each individual case. Since no catching-up industrialization took place without at least some transfer of technology, we may also say that "transnational" connections invariably played a role. No national or regional process of incipient industrialization has ever been entirely homemade and isolated from the larger world.

Early-nineteenth-century Britain was already swarming with technological spies from continental Europe and the United States, and there is much to be said for the view that (at least before 1914) extensive industrialization failed to occur in countries such as India, China, the Ottoman Empire, or Mexico largely because of a lack of political and cultural conditions for the successful import of technology. Only the adoption of new production and management know-how could have led to the modernization of their highly developed manufacturing traditions—as had already happened in France, the land of the artisan and the scientist.[9] Particular regional, and sometimes national, industrialization processes differ in their degree of autonomy. At one end of the spectrum, industrial forms of production take root almost entirely in small enclaves and as a result of the activity of foreign capital, without any noticeable, let alone beneficial, impact on the host country beyond the enclaves. At the other end, a whole national economy may be thoroughly industrialized under indigenous control, with very little "colonial" involvement. Most cases in historical reality were located somewhere between these polar opposites.

Classical Theories of Industrialization

Today's controversies among academic specialists have not entirely devaluated older or "classical" concepts of industrialization. Common to these is the idea that industrialization is part of a more comprehensive social-economic transformation.

Karl Marx and the Marxists (post-1867): industrialization as a transition from feudalism to capitalism by means of the accumulation and concentration of capital, factory organization, and the establishment of relations of production in which the owners of the means of production appropriate the surplus product created by nonservile wage labor—later supplemented by theories of the transformation of competitive capitalism into monopolistic (or organized) capitalism.[10]

Nikolai Kondratiev (1925) and Joseph A. Schumpeter (1922/1939): industrialization as a cyclically structured growth process of a capitalist *world* economy with changing leading sectors, joined on to older processes.[11]

Karl Polanyi (1944): industrialization as part of a wider Great Transformation, in which an autonomous market sphere detaches itself from exchange embedded in a regulatory economy focused on the satisfaction of needs rather than the realization of profit; more generally: the emergence of an autonomous economic logic.[12]

Walt W. Rostow (1960): industrialization as a temporally staggered but universal passage through five stages, of which the third and most important, "takeoff," ushers in durable, "exponential" growth—although this is not necessarily bound up with a qualitative remodeling of society.[13]

Alexander Gerschenkron (1962): industrialization as a process in which latecomers learn to overcome obstacles by using the advantages of imitation and state agency, thereby engendering special national forms and developmental paths within the framework of a single overall process.[14]

Paul Bairoch (1963): industrialization as the continuation of a previous agricultural revolution and the slow spread of industrial economic forms around the world, together with the marginalization of other, nonindustrializing economies.[15]

David S. Landes (1969): industrialization as a process of economic growth driven by the interplay of technological innovation and rising demand, which in the second half of the nineteenth century, through the imitation of England by Continental countries, led to a pan-European development model.[16]

Douglas C. North and Robert Paul Thomas (1973): industrialization as byproduct of Europe's centuries-long creation of an institutional framework guaranteeing individual property rights and hence an efficient use of resources.[17]

Not all these theories pose exactly the same questions, nor do they all use the term "Industrial Revolution."[18] What they share (with the exception of North and Thomas) is a rough chronology that situates the great transition between 1750 and 1850. Some emphasize the depth and dynamism of the break (Marx, Polanyi, Rostow, Landes)—we might call these the "hot" versions. Others are "colder," in identifying a long prehistory and a rather slow transition (Schumpeter, Bairoch, North and Thomas). The point of departure before the transformation is variously characterized as the feudal mode of production, agrarian society, traditional society, or premodernity. And the (provisional) endpoint is defined alternatively as capitalism in general, *industrial* capitalism, the scientific-industrial world, or (in Polanyi, less concerned with industry as such than with regulatory mechanisms in society) dominance of an unfettered market.

Last, the theories differ in the extent that their originators actually applied them to the whole world. Theoreticians are mostly a little more expansive than historians. Marx expected the homogenizing advance of capitalism as a revolutionary force destructive of feudalism in many parts of the world; only in his later years did he hint at the possibility of a special path in Asia (the "Asiatic mode of production"). Of the more recent writers, Rostow, Bairoch, and Gerschenkron were most inclined to express themselves on Asia, for example, although Rostow did so in a very schematic manner, taking little account of national peculiarities. By no means did all of the above theorists focus on the question of why the West developed dynamically and the East (ostensibly) remained static—that is, the "Why Europe?" question so much discussed since the late Enlightenment. Only North and Thomas (rather implicitly) and David Landes (especially in his later writings) considered it central.[19] Bairoch did not view civilizations as closed, monadic spaces but, like Fernand Braudel, studied in great detail the interaction between economies, applying the category "underdevelopment" to both the nineteenth and twentieth centuries. He did not, as Rostow did around the same time, assume that the whole world would eventually follow the same development path but placed the emphasis on divergences. Gerschenkron had no problem in applying to Japan his model of compensatory catching up from a position of backwardness; nonindustrialization interested him as little as it did Schumpeter (apart from the latter's interpretation of imperialism as driven by premodern impulses).[20]

The multiplicity of theories put forward since Adam Smith's pioneering work on the wealth of nations (1776) mirrors the complexity of the questions, but it also prompts the sobering conclusion that Patrick O'Brien drew in 1998: "Nearly three centuries of empirical investigation and reflection by the very best minds in history and the social sciences has not produced any kind of general theory of industrialization."[21] O'Brien naturally regretted this as an economist, but he was not too unhappy as a historian. What grand outline could do justice to the diversity of the phenomena yet retain the simplicity and elegance of good theory?

The British Industrial Revolution

Growth in GDP of 8 percent a year, such as China recorded around the year 2000 (against a paltry average of 3 percent in the industrial countries since 1950), was completely unimaginable in nineteenth-century Europe. Insofar as Chinese growth is driven by industrial expansion, and only secondarily by the "postindustrial" sector of services and telecommunications, the Industrial Revolution has been continuing with increased force. Industry has never been as revolutionary as it is today. To be sure, this is not the concept of Industrial Revolution used by historians[22]—that is, a complex process of economic construction that took place on the main island of the British Isles between 1750 and 1850. Anything else, they argue, should be called "industrialization," first of all in the formal sense of decades-long growth of more than 1.5 percent a year in real per capita output and, in the ideal-typical case, matched or exceeded by rising income levels among the population.[23] Such growth occurs on the basis of a new energy regime, which develops fossil fuels for material production and makes better use of traditional sources. Another characteristic feature is that in the organization of production in large mechanized enterprises, the factory does not radically displace all other forms but acquires a dominant position.

Industrialization mostly stands under the aegis of capitalism, but this is not necessarily so. In the twentieth century, a number of "socialist" countries carried out successful industrialization for a time. It would also be excessive to think that industrialization must permeate every sphere of a national economy; this may appear self-evident today, but it was almost never the case in the nineteenth century. Completely modernized "industrial societies" did not exist in any part of the world, and apart from the United States, Britain, and Germany few other countries came close to qualifying as "industrial" on the eve of World War I. On the other hand, large-scale factories and many pointers to industrially generated growth were to be found in mainly agrarian societies such as India, China, Russia, and Spain. We should therefore speak of industrialization even if the process was limited to a small number of sectors or regions.

Not all roads to the wealth of nations lead through industry. Successful economies such as those of the Netherlands, Denmark, Australia, Canada, and Argentina shared with highly industrialized countries the application of new technologies in all branches of production and transportation, and it is true that in the late nineteenth century roughly one-half of their economically active population was employed outside agriculture. But we would search in vain there for "industrial belts." Nor did every large military apparatus have an industrial foundation to sustain it in the long term. The key economic fact of modernity is not industrial growth per se but the general improvement in the conditions of human existence (shown by rising life expectancy, for example), along with increased polarization in terms of wealth and poverty among various regions of the planet.

The Industrial Revolution happened in England. Only there did the conditions permitting a new level of economic performance come together in a particular combination. The key factors that played a role in this can be easily enumerated (without regard for their intricate connectedness): a large national economic territory without tariff divisions; internal peace since the middle of the seventeenth century; favorable geographical conditions for transportation, especially along the coasts; "the cheapest energy in the world";[24] a highly developed tradition of precision engineering and toolmaking; extensive colonial trade bringing in raw materials and providing export markets; an unusually productive agricultural sector, making it possible to release manpower from the countryside; a high-wage economy of long standing that generated demand; an interest in improvement among large parts of the social elite; and a decidedly entrepreneurial spirit among small circles, especially of religious dissidents.[25]

From this long list, three points may be singled out by way of contrast with other countries.

First. In England, as a result of economic growth all through the eighteenth century, there was exceptionally high demand for "upmarket products," somewhere between the basic necessities and rare luxuries. The gradually developing middle classes became bearers of a consumption that was not, as in continental Europe, confined to the aristocracy and wealthy members of the mercantile elite. French observers, in particular, were repeatedly struck by the existence in Britain of something like a mass market for commercial products.[26]

Second. At the beginning of the eighteenth century, Britain was more involved in overseas trade than any other European country, more intensely even than the Netherlands. The North American colonies were increasingly important outlets for the British Isles, whose internal market alone could not absorb the growth in production. Conversely, Britain's international trade and sea links, whether colonial or not, provided access to key raw materials such as cotton, which at first came mainly from the West Indies and later was produced more cheaply by enslaved Africans working on newly developed land in the Southern states of the United States. Such trade was not the ultimate cause of the Industrial Revolution, but it was an important factor without which the technological innovations would not have had their full economic impact; the inputs of the Industrial Revolution would have been much more expensive to acquire. In the nineteenth century, Britain supplemented its role as "workshop of the world" with its function as chief organizer and distribution center of the trade in raw materials and semifinished products required for industrialization in continental Europe—an intermediate position that also had roots in the early modern period. These connections still await thorough investigation. But it is clear that the Industrial Revolution cannot be explained if we ignore the world economic context and especially the fact that Britain had already been highly successful in the Atlantic and later the global economy during the quarter millennium before 1760.[27]

Third. France and China, too, were countries with major scientific traditions and copious technological experience. In England and Scotland, however, the separate milieux of "theorists" and "practical men" were brought closer together than anywhere else. A common language of problem solving was gradually found, Newtonian physics was a way of thinking that could easily be translated into practice, and institutions such as patent law were created to consolidate the groundbreaking new processes. Britain thus developed for the first time what is another defining feature of industrialization: the normalization of technological innovation. Unlike in earlier epochs of history, waves of inventiveness did not suddenly break off or come to nothing. "Major" inventions did not come by themselves but rather in clouds or clusters. They were part of a process involving small steps and improvements and had spinoffs and follow-ups of their own. Techniques were acquired through ongoing practical effort. No really important knowledge was lost. This incremental stream of innovation, and its conversion into a technological culture, began in a country where an unusually high and widespread level of competence had already been achieved in the early eighteenth century before it was stabilized in the Industrial Revolution. But that country was not sealed off from the rest of the world. In the eighteenth century, scientific and technological knowledge circulated all over Europe and across the North Atlantic, and technological leadership, once attained, did not remain an English monopoly. In a number of spheres, French, German, Swiss, Belgian, or American scientists and engineers soon caught up and even over-took their British colleagues.[28]

If a utopian sketch of the coming Industrial Revolution had been drawn in 1720 for a seasoned observer, and if he had then been asked where it was most likely to occur, he would certainly have mentioned England and, in addition, the Netherlands and Flanders, northern France, central Japan, the Yangtze delta, and perhaps the areas around Boston and Philadelphia. All these regions displayed new forms of economic dynamism: a general and rapidly spreading emphasis on hard work and commercial endeavor; a high and still rising agri-cultural productivity; a developed market specialization among farmers, often bound up with sophisticated processing techniques; a considerable orientation to export markets; an efficient textile production, organized partly in peasant households and partly in large "manufactures." The institutional framework for all this was free (nonservile) labor, some property guarantees for productive capital, and a "bourgeois" business climate that included trust among market partners and faith in contracts. By 1720 England was ahead in many respects, but neither then nor later was it a unique case, an island humming with activity in a sea of agrarian stagnation.

This hypothesis has not yet been sufficiently verified for all of the regions just mentioned. Discussion today invokes the concept of an "industrious revolution," based on the observation that while output grew during the Industrial Revolu-tion, real incomes did not increase at the same pace. According to the theory, a similar trend had been operating in northwestern Europe, Japan, and colonial North America in the century before industrialization: households were raising their consumption demand and were prepared to work harder to fulfill it; people produced more in order to consume more. The Industrial Revolution was then able to link into this demand-driven dynamic. At the same time, the burden on manual workers was probably already increasing before the Industrial Revolu-tion and did not suddenly shoot up when happy peasants disappeared into dark satanic mills.[29]

Continuities

One aspect of the putative "industrious revolution" is the "proto-industrialization" that was invented as a concept in the early 1970s. Put very simply, this refers to the expanding production of goods in village households for translocal markets.[30] Typically outside the framework of the old municipal guilds, it was organized by urban entrepreneurs (in the "contracting out" sys-tem, for instance) and presupposed a manpower surplus as well as a readiness for self-exploitation in the village family. It was at its most thriving where the local power structure allowed peasants some scope for "entrepreneurial" decisions, but there were also cases where "feudal" landowners encouraged a degree of household industry and the collectivism of the village commune did not stand in its way.[31] Various forms of proto-industry have been detected in many countries, including Japan, China, and India, as well as in Russia, where the cotton and ironmongery trades have been especially well examined.

However, the assumption that this was a *necessary* transitional stage toward industrialization has not been confirmed, and the model does not appear to fit England itself very well. The Industrial Revolution did not grow in linear fashion out of a broad proto-industrialization.[32] Moreover, in England and southern Scotland, the first three quarters of the eighteenth century were a time of such lively enterprise that the installation of the first steam engines in large-scale production processes appeared less as a completely new beginning than as a consistent continuation of older trends. There was proto-industry, to be sure, but also a broad increase of output and productivity in the crafts and manufactures—for example, knife and scythe making in Sheffield.[33] In some cases, proto-industrialization made it easier for industry to be later organized on a factory basis. In others, proto-industrial arrangements settled in without setting up a dynamic that would eventually make them redundant.

As to longer-term continuities, the Industrial Revolution is seen as one of a series of economic upswings through which parts of western and southern Europe had passed since the Middle Ages—as had the Islamic Near and Middle East at the end of the first millennium, China under the Song dynasty in the eleventh and twelfth centuries or under the Qing emperors in the eighteenth, and maritime Southeast Asia between roughly 1400 and 1650. If the Industrial Revolution is compared with the upward phase of earlier cycles, its growth effect does not appear so out of the ordinary. What was new was that the Industrial Revolution, and the various national and regional industrialization processes, set up a *stable long-term* growth trend amid the cyclical fluctuations of "long waves" and conjunctures. This, together with other social changes associated with it, ended the epoch of stationary economies, in which productivity gains and rising prosperity were eventually canceled out by countervailing forces such as population growth. Along with demographic trends that had a largely independent dynamic, the Industrial Revolution and the ensuing industrializations brought a final escape from the "Malthusian trap" during the first half of the nineteenth century.[34]

Although the two opposing extremes of interpretation—quantifying growth skeptics and cultural theorists focused on an "institutional" revolution—have continued to raise objections, it remains to some extent justified to speak of a unique English Industrial Revolution. Yet the image, borrowed from aeronautics, of a powerful "takeoff" is an undue dramatization. On the one hand, economic dynamism did not break all of a sudden into stagnant conditions: the

British economy had already been experiencing long-term growth throughout the eighteenth century. On the other hand, growth in the first few decades of the nineteenth century was less spectacular than it was long assumed.[35] It took until midcentury for the new dynamic to free itself from various brakes on its unfolding. The early decades of the century were a time of sharp social conflicts, a period of transition or incubation rather than the actual breakthrough period of industrialization. Economic growth only just kept pace with population increase, yet for almost the first time in history demographic pressure did not hold down living standards. Some groups of workers plumbed the lower depths of poverty. New technologies, including the use of coal as an energy source, spread only slowly, and until 1815 war conditions imposed a heavy financial burden on the country. With an antiquated political system that had scarcely changed since 1688, governments had only limited capacity to build institutions in accord with the new requirements of the economy and society. Such initiatives became possible only with the Reform Act of 1832, which bridled the influence of uncontrolled "interests" (especially large landowners and commercial monopolists) on policymaking. Free trade and the gold standard (which automatically regulated the money supply) later enhanced the rationality of the system. But the transition from Industrial Revolution to genuine industrialization took place in Britain only after the year 1851, when the Great Exhibition at the Crystal Palace symbolically set the country on its way. It was now that per capita income grew appreciably; steam engines in factories, ships, and railroads became the chief means of energy transmission; and a declining trend in food prices shook the power basis of the landowning aristocracy.[36]

Britain's initial lead over continental Europe should not be exaggerated. Celebrated British inventions soon spread abroad, and by 1851 it was clear to everyone drifting through the marvels of the Crystal Palace that the United States had overtaken Britain in machine-building technology.[37] Despite early export prohibitions, British engineers and workers made the country's technology well known on the Continent and in North America.[38] In the time scales of economic history, a lag of three or four decades is by no means extraordinary; sometimes a particular invention needed so long to mature and to become economically relevant.

Repeated attempts have been made to date the starting point of national bursts of industrial activity. But this is largely a spurious problem. In some countries industrialization began with a bang, in others almost unnoticed; in some the economy immediately shot upward, in others several attempts were required before it got moving. Where the state actively promoted industrialization, as it did under Russia's finance minister Sergei Y. Witte from the nineties on, the break in continuity was greater than elsewhere. The sequence of European countries is reasonably clear even without exact dating: Belgium and Switzerland were early industrializers, France began after 1830, Germany after 1850, and other nations considerably later. More important, however, are the overall picture and the fundamental contradiction that it reveals. On the one hand, each European country took its own path of industrial development; there can be no suggestion that a British model was simply copied, if only because people elsewhere did not perceive one that was clear, unambiguous, and attractive. British peculiarities were so singular that such direct imitation would scarcely have been possible.[39]

On the other hand, from a greater distance we can discern amid the diversity of national paths a growing overlap that amounts to pan-European industrialization. Looking back from around 1900, countries like Britain, France, and Germany had arrived at similar performances through distinct trajectories. After the middle of the century, industrialization nearly everywhere received government support, while commercial links and international agreements (on free trade, among things) contributed to the integration of a European market, and the cultural homogeneity of the Continent made technological and scientific exchange ever easier to achieve.[40] By 1870 a few European economies had come so far that they were beginning to contend with British industry for markets. It was also generally apparent what was required, in addition to favorable natural conditions, for industrialization to be successful: that is, agrarian reform releasing the peasantry from extra-economic constraints, and investment in "human capital" ranging all the way from mass literacy campaigns to state research facilities. That well-educated manpower can make up for a shortage of land and natural resources—a conclusion still valid today—was first understood by certain European countries and Japan in the final third of the nineteenth century.[41]

An advantage of the industrial mode of production was that in at least one sense it was *not* revolutionary: it did not eradicate all earlier forms of value creation or bring about a radically new world. In other words, industry developed and develops in many different forms and can easily subordinate nonindustrial modes of production without necessarily having to destroy them. Large-scale industry, with thousands of employees in a single plant, was almost everywhere the exception rather than the rule. "Flexible production" maintained itself[42] even as mass production—probably an invention of the Chinese, who for centuries had been trying out modular series production based on a division of labor in ceramics and timber architecture[43]—advanced into one sector after another. Where flexibility bore greatest fruit, industrialization played itself out in a dialectic of centralization and decentralization.[44] Only Stalin's policy of industrialization under a central plan created a radical alternative from the late 1920s on, and the success of that was doubtful. The electric motor, which can be built in many different sizes, and wall-socket energy in general gave a new impetus to small-scale production at the end of the nineteenth century. The basic pattern was the same everywhere, taking in Japan, India, and China. Rings of small suppliers and competitors grew up around the conspicuous factories of large enterprises, and unless the state intervened, the conditions for workers in such sweat shops were much worse than in large-scale industry with its tightly regulated procedures, its premium on skilled labor, and its sometimes patriarchal social values.

The Second Economic Revolution

The term "second industrial revolution" has often been used to denote the period in the late nineteenth century when steel (Big Steel, on a far larger scale than before 1880), chemicals, and electricity replaced cotton and iron as the leading sectors. This was associated with a shift of industrial dynamism from Britain to Germany and the United States, which had both forged well ahead in the new technologies.[45] It seems more useful, however, to go beyond this narrowly technological focus and to speak of a second *economic* revolution.[46] It was this that shaped the modern "corporation," the dominant form of enterprise in the twentieth century. This key change, datable to the 1880s and 1890s, had an *immediate* global impact, whereas the effects of the first Industrial Revolution had only gradually made themselves felt outside its birthplace. During this watershed final quarter of the nineteenth century, a change in the leading technologies was not all that happened: complete mechanization in the most advanced economies swept away preindustrial "niches"; hired managers replaced owner-capitalists as the dominant agency of entrepreneurship; the limited liability company, funded through the stock exchange, rose to prominence; large-scale business spawned growing numbers of white-collar office workers; concentration and cartelization restricted the classical mechanism of competition; and multinational corporations, sporting brand names, took control of marketing their own products, founding global networks for this purpose together with numerous local partners.[47]

This last point gave particular global relevance to changes in the manner of industrial production. In China, for instance, American and European multinationals such as Standard Oil of New Jersey and the British-American Tobacco Corporation appeared on the scene in the 1890s and began to penetrate the consumer goods market with unprecedented directness. As vertically integrated companies, they controlled their own raw material sources as well as the processing and marketing side of their operations. Industry now became "business"—a new transnational complex in which industrial enterprises were more tightly interwoven with banks. It first developed into big business in the United States, where the first giant companies had earlier been confined to the railroad sector. Japan, which had begun industrializing in the mid-1880s, had a head start insofar as some of the great merchant houses of the Tokugawa period had changed with the times and reinvented themselves as *zaibatsu*: large, highly diversified, and often family-owned companies that took large parts of the economy under their common oligopolistic control. Their closest resemblance was not to the vertically integrated conglomerates that divided up whole sectors of American industry toward the end of the nineteenth century but rather to holding companies, with their set of loosely integrated commitments. After roughly 1910, the organization of major *zaibatsu* such as Mitsui, Mitsubishi, and Sumitomo became tighter and more centralized, with the result that Japan joined the United States and Germany—in this respect different from Britain or France—as a country of large corporations integrated horizontally as well as vertically.[48]

The Great Divergence

The discussion of industrialization in the last two or three decades, taking place mainly in journals or collective volumes and not yet condensed into a new synthesis, remained at a distance from the major theoretical work of an earlier period.[49] Research became modest and specific in focus, largely adhering to conventional definitions of growth. The most influential theorist of *global* history in the 1970s and 1980s, Immanuel Wallerstein, did not participate in the debate. Citing a long series of well-known objections, he considers the very concept of an Industrial Revolution "deeply misleading," on the grounds that it diverts attention from the key issue of the development of the world economy as a whole.[50] Paradoxically, a return of grand theory in the industrialization debate around the year 2000 was triggered by intensive historical research, though not in relation to Europe. Regional experts came to realize that in the seventeenth and eighteenth centuries both China and Japan, but also parts of India and the Muslim world, by no means corresponded to the stereotype of Asiatic impoverishment and stagnation that European social science had from its earliest days unquestioningly perpetuated on slender foundations of reliable knowledge. Certain prerequisites of the Industrial Revolution, according to the new consensus, had truly been present in those parts of the world. Meanwhile some authors, eager to dispense compensatory justice, went to the opposite extreme and painted premodern Asia in positively glowing colors, so that the "European miracle" appeared either as an optical illusion of Western image making or as the outcome of random concatenations with no inner necessity. In fact, it was argued, the Industrial Revolution *should have* taken place in China. That is certainly going too far. But the revaluation of "early modern" Asia has breathed new life into the "Why Europe?" debate, in which for a long time nearly everything seemed to have been said already.

It is no longer sufficient to present lists of Europe's advantages and achievements (from Roman law and Christianity to the printing press, exact sciences, rational attitudes to economics, a competitive system of states, and an "individualist picture of human beings") before moving on to the bald assertion that all this was lacking elsewhere.[51] The closer that Europe and Asia appear to each other in the premodern age, and the narrower the qualitative and quantitative differences between them, the more mysterious becomes the "great divergence" of the world into economic winners and losers after the middle of the nineteenth century.[52] Whereas Europe's success long seemed to have been programmed in the depths of its geographical-ecological setup (as in Eric L. Jones[53]) or in particular cultural dispositions (as self-proclaimed Weberian sociologists, David S. Landes, Niall Ferguson, and many other authors claim), detective work has now begun afresh on the question of what was Europe's real *differentia specifica*.

The point at which this difference became really telling keeps being pushed further into the nineteenth century as Asia's relative decline is set at a later and later date. The beginning of Europe's special path sometimes used to be placed as early as the Middle Ages (Eric L. Jones, and more recently Michael Mitterauer)—a time when other historians considered with good reason that China (especially in the eleventh century) and parts of the Muslim world were still ahead socioeconomically and culturally. More recently the point of bifurcation was shifted into the period commonly associated with the Industrial Revolution. The great divergence, then, first appeared in the nineteenth century. The issue has acquired a topicality and urgency that it did not have twenty years ago, because today's social and economic gap between Europe and Asia is beginning to close. The rise of China and India (Japan's has for some time been viewed with some equanimity) is currently perceived in Europe as little more than part of contemporary "globalization." But in reality it involves genuine industrial revolutions that, without precisely repeating the European experience, reenact much of what happened in the nineteenth-century West.

2 Energy Regimes: The Century of Coal

Energy as a Cultural Leitmotif

In 1909 Max Weber pulled out all the stops in polemicizing against "energy theories" of human culture, such as that which the chemist, philosopher, and Nobel Prize winner Wilhelm Ostwald had raised for discussion earlier in the year. According to Ostwald, as cited by Weber, "every turnaround in culture is determined by new energy circumstances," and "cultural work" is guided "by the endeavor to preserve free energy."[54] At the very time when the human sciences were struggling to emancipate themselves from the methodology of the natural sciences, their most distinctive area of study, human culture, was thus being incorporated into a monistic theoretical framework. We do not have to blunder into the trap identified by Weber, however, even if we regard energy as an important element in material history. In those days the discipline of environmental history did not yet exist, but since then—especially in light of today's energy problems—it has taught us the importance of this factor.

Energy theories of culture fit well into the nineteenth century. Hardly any other concept occupied scientists so intensively or cast such a spell over the public. Alessandro Volta's experiments with animal electricity in 1800, which had made possible the construction of the first source of electrical current, had led by midcentury to a whole new science of energy, and various cosmological systems—above all, that of Hermann Helmholtz in his epoch-making *Über die Erhaltung der Kraft* (1847)—had arisen on its foundations. The new cosmology left behind the speculations of the Romantic philosophy of nature; it had solid roots in experimental physics and formulated its laws in such a way that they stood up to empirical testing. The Scotsman James Clerk Maxwell discovered the basic principles and equations of electrodynamics and described the wealth of electromagnetic phenomena, after Michael Faraday had demonstrated electromagnetic induction in 1831 and built the first dynamo.[55] The new physics of energy, developed in tandem with optics, led to a great flow of technological transformation. A key figure of the times such as William Thompson (from 1892: Lord Kelvin, the first scientist to be raised to the peerage) shone both as science manager and imperial politician, groundbreaking researcher in physics and practical technologist.[56] Alongside the low-voltage technology needed for international news communication, with which the Siemens brothers made their first money, high-voltage technology appeared in 1866 when Werner Siemens discovered the principle behind the electrical dynamo.[57] From Siemens to the American Thomas Alva Edison to amateur enthusiasts, thousands of people with expertise in the field worked on the electrification of more and more parts of the world. From the eighties on, power stations came into operation and various municipalities introduced a regular electricity supply, and by the nineties it was possible to produce small three-phase current motors in large series.[58] But already in the first half of the century, the most important inventions for people's real lives had been those that generated and converted energy. The steam engine itself was nothing other than a device for the transformation of dead matter into technically useful power.[59]

Energy was a leitmotif of the whole century. What had previously been known only as an elemental force, especially in the shape of fire, now became an invisible but efficient power with unsuspected possibilities. The guiding scientific image of the century was no longer the mechanism, as in the early modern age, but the dynamic interrelationship of forces. Other sciences followed along the same path. In fact, political economy had already done this with much greater success than the energy theory of culture targeted by Max Weber. After 1870 neoclassical economics suffered from something like physics envy and began to make abundant use of energy images.[60] Ironically, it was just when the energy of animal bodies was losing its significance for economics that the significance of human corporeality became clear. Bodies were seen as necessarily participating in a universe where energy had no boundaries and—as Helmholtz had shown— did not vanish into thin air. Under the influence of thermodynamics, the still abstract philosophical "labor power" of classical political economy was replaced with the "human motor," which, as a combined muscular-nervous system, could fit into planned work processes, and whose ratio of energy output to input could be measured experimentally with precision. By midcentury Karl Marx's concept of labor power was reflecting the impact of Helmholtz's theories, and Max Weber too, at the beginning of his career, occupied himself in detail with the psychophysics of industrial work.[61]

It was no accident that nineteenth-century Europeans and North Americans found energy so fascinating. In one of its most important aspects, industrialization constituted a change of energy regime. All economic activity requires energy inputs, and poor access to cheap energy creates one of the most dangerous bottlenecks a country can face. Even when resources were otherwise quite plentiful, preindustrial societies everywhere were able to draw only on a handful of energy sources other than human labor: water, wind, firewood, peat, and work animals capable of converting fodder into muscle power. Given this limitation, energy supply could be assured only through more extensive farming and woodcutting, and more nutritious crops, but there was always a danger that the available energy would not keep pace with population growth. Societies differed in their proportional use of various kinds of energy. It has been estimated, for example, that in 1750 wood was the source of roughly a half of energy consumption in Europe, but of no more than 8 percent in China. Conversely, the use of human labor power was several times greater in China than in Europe.[62] Every society possesses its specific energy profile.

Fossil Fuels

With industrialization, one fossil fuel—coal—gradually came to dominate the energy scene, having been used increasingly since the sixteenth century, above all in England.[63] The speed of the change should not be overstated. In Europe as a whole, coal provided only a tiny fraction of energy use around the middle of the nineteenth century. Only subsequently did the share of traditional sources decline, while coal and later oil—as well as hydraulic power, now better harnessed by dams and new kinds of turbines—rose dramatically in importance.[64] The range of energy forms familiar to us today followed millennia dominated by wood, which in nineteenth-century Europe was still being used in quantities that now seem hard to believe.[65] Alongside the rise of coal and the decline of wood, wind continued to be used for transportation and mill power until the second half of the century. Combustible gas, initially obtained from coal, lighted the early public lamps in big-city streets; natural gas, which now covers one-quarter of world energy needs, was not yet available. World use of coal as a fuel reached its peak in the second decade of the twentieth century.

Whereas coal had long been known to humans, the history of petroleum can be precisely dated. The first successful drilling for commercial purposes took place on August 28, 1859, in Pennsylvania, immediately triggering an oil rush comparable to the Californian gold rush a decade before. From 1865 a young entrepreneur by the name of John D. Rockefeller made oil the foundation of big business. By 1880 his Standard Oil Company, founded ten years earlier, had near-monopoly control of the growing world market—a position that no individual supplier ever conquered in relation to coal. At first petroleum was mainly processed into lubricants and kerosene, a fuel for lamps and stoves. Only the spread of the automobile in the 1920s gave it major weight in the global energy balance.[66]

A demand still remained for animal energy: camels and donkeys (both unusually cost-effective) in transportation, oxen and water buffalos in agriculture, and (Indian) elephants in the rainforest. Part of the "agricultural revolution" was the growing substitution of horsepower for manpower: the number of horses doubled between 1700 and 1850 in England, and between 1800 and 1850 (at the height of the Industrial Revolution) the horse energy available per agricultural worker rose by 21 percent. In Britain as in France, the ratio of one horse to eight inhabitants remained fairly stable during the second half of the century.[67] The number of horses per hectare fell in Britain only after 1925—a process that had begun several decades earlier in the United States, the pioneer of this trend. Eventually the introduction of tractors expanded areas under cultivation without clearing new land, since it meant that less land was needed for the production of grass and oats to maintain workhorses.[68] Even in the United States, one-quarter of farmland in 1900 had been used to feed horses. The rice economies of Asia, where animal traction played scarcely any role and mechanization was more difficult to implement, lacked this important buffer for an efficiency-raising modernization of agriculture.

The industrial civilization of the nineteenth century rested upon fossil fuels and ever more efficient technical-mechanical conversion of the energy obtained from them.[69] The coal-guzzling steam engine set up a spiral of its own, since only steam-driven elevators and ventilators enabled the extraction of coal deposits deep below the earth's surface. In fact, the quest for better means of pumping water from mine shafts had been at the origin of the steam age; the earliest steam pumps, still primitive in their functioning, were built in 1697, and in 1712 the first of Thomas Newcomen's steam-driven vacuum pumps, indeed the first piston steam engine of any kind, was installed in a coal mine.[70] When the engineer James Watt and his business partner and capital provider Matthew Boulton launched their smaller and better steam engines from 1776 onward, the place they chose for the experiment was not a textile factory but a tin mine in Cornwall, a remote corner of England never of much importance industrially. The decisive technological breakthrough was then made by the tireless innovator James Watt, who in 1784 designed a much more efficient machine that could generate not only vertical but rotating movement.[71] The steam engine had come of age. Its coal consumption efficiency (that is, the proportion of freed energy usable for mechanical purposes) continued to increase throughout the nineteenth century just as, generally speaking, power-generating technology kept up with demand that was rising in quantity and changing in kind.[72]

Watt's machine made its debut in an English cotton-spinning mill in 1785, but it would be decades before the steam engine became the main energy source in light industry. In 1830 most textile factories in Saxony, one of the industrial heartlands of continental Europe, were still mainly using water power, and in many places it became profitable to switch to steam only after the railroad had facilitated access to cheaper coal.[73] To extract deposits with technologically advanced methods (themselves using steam power) and then to transport the coal at low cost by steam-powered trains and ships to the point of consumption became key conditions for successful industrialization.

Japan, with few coal reserves of its own, faced the greatest difficulties, and so it is not surprising that the age of the steam engine did not last long there. The first fixed machines (that is, not on board a ship) were imported from the Netherlands and installed at a state ironworks in Nagasaki in 1861. Until then most commercially used energy had come from water mills, which, as in England, had also driven the first cotton spinning mills. For some time the various kinds of energy existed alongside one another. But when Japan's industrialization got under way in the mid-1880s, it took only a few years for its factories to be equipped with steam engines, and their industrial use peaked by the mid-nineties. The Japanese economy was one of the first to employ electricity on a grand scale, obtaining it partly from water power, partly from the burning of coal, and this gave its industry major advantages. When the first steam engines began operating in Japan in the 1860s, the country was some eighty years behind Britain in energy technology. By 1900, advancing at breakneck speed, it had completely closed the gap.[74]

The statistics of rising coal production are an indication of the level of industrial development, but they also tell us something about the underlying causes. The figures should be taken with a pinch of salt, since no one has tried even to estimate the output of nonmechanized coal mines in China, for example. (Admittedly they almost never produced coal for industrial applications at that time.) The middle of the nineteenth century marked the turning point for pit coal production; it rose sixteenfold from a maximum of 80 million tons a year in 1850 to more than 1.3 billion tons in 1914. At the beginning of this period, Britain's 65 percent share made it by far the largest extractor, but on the eve of the First World War it had dropped to second place (25 percent), behind the United States (43 percent) and ahead of Germany (25 percent). All other countries were of secondary importance beside these giant producers. Russia, India, and Canada were climbing the ladder and would have a respectable coal industry within a few more years. But even the largest of these smaller producers—Russia—averaged only 2.6 percent of world output in the years between 1910 and 1914.[75] Many countries, such as France, Italy, or (southern) China, could not avoid making up their resource deficit by importing coal from regions with a surplus, such as those in Britain, the Ruhr, or Vietnam.

Whereas in the 1860s some commentators had gloomily forecast an impending exhaustion of global coal deposits, a half-century later the opening up of new fields had ensured an adequate supply and a geographical fragmentation of the coal market that meant Britain could no longer maintain its old dominance.[76] Some governments saw the need for an energy protection policy; others did not. Russia failed to develop an adequate coal base while Sergei Witte, the finance minister from 1892 and architect of late-Tsarist modernization, one-sidedly promoted high-tech projects in the steel industry and machine building.[77] In Japan, by contrast, the state encouraged coal mining to keep in step with industry; although the country certainly lacked the large reserves to be found in the United States or China, its output in the first post-1885 phase of industrialization was enough to cover its own needs. Only in a second phase, when the metalworking industry had undergone considerable expansion, was the quality of Japanese coal no longer sufficient. If Manchuria was of such interest to Japan as a colony, one reason was that its high-grade coal was better suited for carbonization, and deposits were being mined after 1905 in colonial territory controlled by the South Manchurian Railroad Company.[78] There are few clearer instances of "resource imperialism," that is, the subjugation of another country with the purpose of gaining control over raw materials necessary for one's own economic development.[79]

China offers the example of a colonial situation in reverse. Energy shortage was a chronic problem for that densely populated country, large stretches of which were almost completely denuded of forest. Northern and northwestern China sit on huge coal deposits, some of them even today not yet opened up for mining, and it cannot be said that they were unknown and unutilized. They were used early on for the production of iron on a grand scale; indeed, serious estimates suggest that around the year 1100, this may have been higher than the output of the whole of Europe (except Russia) in 1700.[80] It is difficult to tell why such levels were not maintained, but in any event China's coal production was sharply lower in the eighteenth and nineteenth centuries, especially because the fields in northwestern China were a long way from the commercial centers that sprang up after the opening of the treaty ports in 1842. The advantage of short distances and good waterways, which made it cost-effective to mine English coal so early on, was absent in China. When large enterprises began mechanical extraction there after 1895, the major mines were under foreign control, and those in Japanese possession either shipped their output straight to Japan or sent it to nearby Japanese-owned iron and steel factories. If, roughly after 1914, the newly emerging conurbations—above all, Shanghai—suffered from an energy shortage that presumably hindered their industrial development, this was due not only to insufficient output and colonial exploitation but also to the political chaos in the country, which meant, for example, that individual railroads were repeatedly out of service. China was potentially an energy giant, but in the first phase of its industrialization it was able to make only very limited use of its fossil fuels. Unlike in Japan, there was no central government that might have given priority to energy supply in its economic policies and its promotion of industrial growth.

A Global Energy Gulf

All in all, a deep energy gulf had opened up in the world by the early twentieth century. In 1780 all societies on the planet relied on the use of energy from biomass, differing from one another by the particular preferences they developed, or were forced to develop, under the pressure of their natural circumstances. In 1910 or 1920 the world was divided between a minority of countries that had gained access to fossil fuels and established the infrastructure necessary to use them, and a majority that had to cope with traditional energy sources under a growing threat of shortage. In terms of the distribution of world coal output, the gap between "the West" and the rest of the world was clear. In 1900 Asia accounted for just 2.82 percent of global production, Australia for 1.12 percent, Africa for 0.07 percent.[81] Country-by-country comparisons are another matter: Japan produced more coal on average than Austria-Hungary in the years from 1910 to 1914, with India only a short distance behind.[82]

Per capita consumption of commercially supplied energy in 1910 was probably a hundred times greater in the United States than in China. At the same time, new hydroelectric technologies made it possible for water-rich countries to raise the old water-mill principle to a new level. Whereas the steam engine at first generated energy more efficiently than the waterwheel, the water turbine had reversed the relationship by the second half of the nineteenth century.[83] For countries such as Switzerland, Norway, or Sweden and for some regions of France, dam and turbine technology offered a chance from the 1880s on to offset their dearth of coal. Outside the West, however, only Japan took advantage of these new possibilities. Under certain ecological conditions there were anyway no alternatives: huge areas of the Middle East and Africa had neither coal reserves nor water that could be used to produce energy. Egypt, for instance, which has little coal and can hardly use the weak Nile current for watermills, was at a strong disadvantage in comparison with Japan. During the first phase of industrialization, when processing factories were set up for the export economy and irrigation plants were partly mechanized, people still depended mainly on human and animal motive power.[84] And when oil extraction began in the Middle East in the early twentieth century—Iran, for example, virtually without an industry, first exported the fuel in 1912—it was destined entirely for abroad and had no other connection with the domestic economy.

The steam engine found many applications, not all of them in the production of industrial goods. In the Netherlands, it was introduced fairly late (around 1850) for drainage and polder installations, the higher costs being offset less by increased efficiency than by the greater control that steam engines permitted. By 1896 only 41 percent of reclaimed land was still being drained by windmills, and this aspect of the Dutch landscape, familiar from innumerable "Golden Age" paintings, gradually disappeared. More generally, there is much to be said for regarding the change in energy regime as one of the most important features of industrialization. But it did not happen overnight, in the form of a revolution, or as early as the British example might suggest. An energy economy with a broad mineral base developed worldwide only in the twentieth century, after oil came on stream in Russia, the United States, Mexico, Iran, Arabia, and elsewhere and began to be used alongside coal in the industrial countries.[85]

The energy-rich West confronted the rest of the world as more "energetic." The cultural heroes of the age were not contemplative idlers, religious ascetics, or tranquil scholars but practitioners of the *vita activa*: indefatigable conquerors, intrepid travelers, restless researchers, imperious captains of industry. Wherever they appeared on the scene, Westerners impressed, terrified, or bluffed people with a personal dynamism that was supposed to represent their society of origin. The actual strength of the West was projected as a force of nature and a mark of anthropological superiority. The racism of the age did not end with skin color: it classified the human "races" on a scale of potential physical and mental energy. At the latest by the end of the century, the West was typically characterized as "youthful" in the non-European world, while indigenous traditions and local rulers were seen as "old," passive, and lifeless. Patriots of the younger generation considered that their main task was to revitalize their own society, to kindle its slumbering energies, to give it a political direction. In the Ottoman Empire they were Young Turks; in China they called the standard-bearing journal championing political and cultural renewal "New Youth" (*Xin qingnian*). Nationalism, sometimes even socialist revolution, was discovered almost everywhere in Asia around this time as a vehicle of self-energization.

3 Paths of Economic Development and Nondevelopment

Although there was and is no unambiguous statistical measure for a country's degree of industrialization, it was fairly clear by the eve of the First World War who in Europe did and did not belong to the "industrial world." In absolute output figures there were two giants: Germany and the United Kingdom, followed at a considerable distance by Russia and France and in a third order of magnitude by Austria-Hungary and Italy. In terms of per capita industrial performance, the picture was a little different: Britain still had the lead over Germany; Belgium and Switzerland were on a par with Germany; and France and Sweden lay some distance behind. None of the other countries of Europe had achieved so much as a third of Britain's per capita industrial output; Russia was near the bottom of the league along with Spain and Finland.[86] Of course these figures, many of them estimates, tell us nothing about per capita income or average living standards. And a closer look shows that there can be no question of an "industrial Europe" as a whole, in contrast to an unmodernized rest of the world (except for the United States).

Export Orientation, Especially in Latin America

By roughly 1880, imperial geology—a science with eminently practical implications—was at work tracking down mineral deposits in every part of the world: manganese, the chief steel stabilizer, in India and Brazil; copper in Chile, Mexico, Canada, Japan, and the Congo; tin in Malaya and Indonesia. From the seventeenth century until 1914 Mexico was the world's largest silver producer—a position that South Africa had gained in relation to gold. Chile was the main source of saltpeter, then indispensable for the production of explosives, and in 1879–83 it even fought a war with Peru and Bolivia over deposits in their border areas. Many of these natural resources were also plentiful in North America, the best-endowed region of the world for industrial inputs. Outside Europe, mineral reserves rarely became the springboard for Western-style industrial development; they were often developed by foreign capital in export enclaves, without reshaping the respective national economy as a whole. The same was true of the production and export of agrarian inputs for the rubber, soap, chocolate, or other industries. In the two decades before the First World War, tin and rubber made British Malaya a particularly wealthy colony; production there was only partly in the hands of international corporations, and the Chinese minority played an important entrepreneurial role.

The new demand of European and American industry called export sectors into being in many countries around the world, whether formal colonies or not. In Latin America this put an end to the centuries-long dominance of precious metals in overseas trade. New products took over from silver and gold in a number of countries; Peru, the classical land of silver, became after 1890 an important supplier of copper for the electrical industry, so that by 1913 it represented one-fifth of all its export earnings. Silver also lost much of its significance in Bolivia, giving way especially to tin, which by 1905 made up 60 percent of exports. Chile first appeared on the world market as a copper producer, but the switch to saltpeter meant that by 1913 the mineral accounted for 70 percent of all exports.[87] Despite these changes, however, specialization in a small number of products remained a hallmark of many Latin American economies. Exports—which also included coffee, sugar, bananas, wool, and rubber—set up growth effects, but the narrower the range of products, the more vulnerable a country was to price fluctuations on the world market; Peru's guano boom ended in a crash, before the beginning of the great worldwide expansion in tropical raw materials.[88] Only Argentina managed to spread the risks sufficiently by diversifying before 1914. With less than 10 percent of the population in Latin America, it was then the region's most successful exporter and accounted for almost one-third of exports.[89] Other factors in the macroeconomic success of export orientations were whether (1) production took place in labor-intensive family enterprises, keeping the profits inside the country and sharing it relatively equally in society, or (2) the dominant form was plantations and mines, mostly worked by poorly paid wage laborers and owned by foreign companies that transferred a large part of their profits overseas. In general, type 2 was less advantageous than type 1 for the national economy and the overall development of society. If there was growth under type 2, it was often confined to isolated enclaves and did not have a stimulating effect on other branches of the economy. Only South Africa was a major exception to this rule.[90]

Not every industrializing country makes the best use of its opportunities. In the twentieth century there are several examples of failed industrialization strategies that did not take account of local specificities. With regard to export economies, the question constantly arises on every continent as to whether the profits were used to invest in industrial processing—in other words, whether productivity gains in export enclaves were transferred to nonexport sectors of the economy. One cannot speak of any kind of autonomous industrialization unless the industries in question mainly serve the internal market. This was rarely the case in Latin America before 1870. Later, at least in some countries, export earnings were distributed in society in such a way that domestic purchasing power increased. The spread of the railroad traditionally solved problems of clogging, and the adoption of electrical technologies removed energy bottlenecks. As in almost every other region of the world, the textile industry was present also in areas without a local supply of cotton or wool. Everyone needs clothing, and governments in the periphery that fought for protective tariffs did so primarily to keep out textile imports. Furthermore, the relatively high degree of urbanization in many parts of Latin America created a spatially concentrated market close to the location of textile factories.

In 1913, of all the Latin American republics, it was Argentina (where the textile industry played second fiddle) that had the highest level of industrialization, followed by Chile and Mexico. However, there was virtually no heavy industry in the region; the dominant sector was food and stimulants, followed by textiles. Although early industrialization reduced the level of imported consumer goods in comparison with machinery (including rail track and rolling stock), so that only the demand for luxuries had to be satisfied from Europe, a more complex industrial structure did not emerge anywhere. Even a large country like Brazil, which achieved high growth rates for a time, failed to escape the vicious circle of poverty and to stimulate industry by means of rising internal demand. And neither Brazil nor any other country progressed to industrial production capable of breaking into export markets. Nowhere did crafts or (widespread) proto-industry serve as a preparatory stage to autonomous industrialization, and in many of the smaller countries industrialization did not even begin.[91] Why did the countries of Latin America not succeed in linking up with the industrial dynamic in Western Europe, North America, and Japan *before* the experiments with state-sponsored import substitution in the period between the two world wars? This remains an unanswered question.

China's Impeded Start

Our aim here is not to tour the world looking systematically for evidence of newly emerging industry. A few remarkable cases may suffice. Just as interesting as the counterfactual problem of the "great divergence" debate—why did India and China not undertake their own industrial revolution before 1800?—is the fact that they *did* begin to industrialize a little more than a hundred years later. In China, with its major tradition of premechanical craft production and its widespread proto-industrialization, no direct path led from older forms of technology and organization to modern factory production. Until 1895 foreigners were not permitted to establish industrial enterprises on Chinese soil, even in the treaty ports; the handful that nevertheless managed to get off the ground were of little consequence. In this first phase of China's industrialization, the state took the levers of command. Beginning in 1862, several provincial governors—not the imperial court itself—embarked on a series of large-scale projects that all drew upon foreign technology and advisers: first arms factories and shipyards, then in 1878 a large coal mine in North China, a little later some cotton-spinning mills, and in 1889 the Hanyang ironworks in the province of Hubei. The chief motive for this policy was defensive; 70 percent of the capital was allocated to enterprises of military relevance. It would be too simplistic to write off all these early initiatives as failures. Most of them show that China was perfectly capable of adopting modern technology, and Hanyang, in the first few years after it started production in 1894, was actually the largest and most modern iron and steel plant in Asia. But it is true that the projects were uncoordinated, that none became a growth pole in even a regional industrialization strategy. Before the Sino-Japanese war of 1894–95, which ended in a resounding defeat, China had embarked on industrialization but not yet found its way to full-scale industrial transformation.[92]

After 1895, things became more complicated and more dynamic as companies from Britain, Japan, and elsewhere set up industrial enterprises in Shanghai, Tianjin, Hankou, and a few other large cities. With the state now largely inactive, Chinese entrepreneurs did not throw in the towel but began to compete with foreign interests in nearly every modern sector of the economy.[93] Steamship transportation had been introduced quite early on, in the 1860s, first by the Chinese state, then by private firms. The silk industry too, one of the country's leading export sectors since the eighteenth century, moved quickly to appropriate the new coal and steam technology. But since Japanese competitors did the same and worked more methodically to raise the quality and output of goods for the world market, Japan won the battle for international customers in the second decade of the twentieth century. The core industry in China too— apart from South Manchuria, Japanese-ruled after 1905 and a fast-growing coal and steel center—was cotton spinning. By 1913, of all spindles operating in factories on Chinese soil, 60 percent were Chinese-owned, while 27 percent and 13 percent were in the hands of European and Japanese corporations, respectively. On the eve of the First World War, however, China's cotton textile industry was still underdeveloped: it had installed 866,000 spindles, against Japan's 2.4 million and India's 6.8 million (roughly as many as in France). Only a wartime boom raised this total to 3.6 million. Between 1912 and 1920, modern Chinese industry actually notched up one of the highest growth rates in the world, so that by the end of the decade some of the foundations had been laid for industrialization—relatively weak, but capable of being built upon.[94] The internal chaos of the warlord period, the lack of vigorous development-oriented governments, and Japan's imperialist aggression were the main reasons why China had to wait more than another half century for a nationwide "take-off." The most characteristic feature of its industrial history before the great post-1980 upturn was not the cautious development that took place in the late imperial period with little or no state support but the braking in the 1920s of the process on which it had already embarked.

The argument that England's new mass production of cheap cotton material drove local spinners and weavers to the wall in China or India, seriously damaging the basis for autonomous industrialization, is not wrong but requires qualification. In China, despite a lack of tariff protection under the unequal treaty system, home weaving in the villages for local and regional demand stood up fairly well. And when cotton thread from new factories in the treaty ports (less so from abroad) increasingly supplanted hand milling in the early twentieth century, many weavers made the necessary switch and were able to continue functioning. In India the "flooding of Asian markets" thesis has long been discussed under the heading of "de-industrialization." Its starting point is the observation that, in the seventeenth and eighteenth century, Indian handicrafts had been capable of producing all grades of cotton goods in large quantities, that these goods entered distant commercial circuits channeling them to many parts of Asia, Africa, and the New World, and that their high quality ensured abundant demand in Europe.[95] The same was true, at lower levels of quantity and quality, for exports of Chinese cotton cloth. The fact that the printing was often done in Europe contributed to an interest there for cotton goods, and hence a demand for unprinted cloth that would later be met by the products of the mechanical mills at home. By 1840 or thereabouts, materials from Lancashire had driven Asian imports from the domestic market; English gentlemen no longer wore nankeens, the fine cloth trousers from the East. Such import substitution, economically viable because Britain enjoyed competitive advantages from its technology, thus marked the beginnings of Europe's industrialization.[96]

India's and China's loss of their export markets, similar to that experienced by the smaller Ottoman textile industry in the first half of the nineteenth century, had catastrophic effects in their cloth-producing regions. Qualitative evidence of mass destitution among Indian weavers is abundant, though its true extent remains unknown. As a recent survey of historiography concluded, "there has been very little serious historical investigation of the decline of cottons in India, especially in the major textile-manufacturing regions and in the first decades of the nineteenth century."[97] A clear regional differentiation seems to be helpful. Bengal was hit hard by the export crisis, whereas southern Indian weavers working for the home market were able to hold out much longer. Imported textiles never reached the standard of the finest Indian goods, so that luxury markets continued to be served by Indian producers. As in China, machine thread caught on in India to the extent that its lower price undercut even the most self-exploitative home spinning among rural families. At the same time, home weaving survived mainly because markets "segmented," as economists say; there was no general competition between imported and Indian-produced materials.[98]

India and the Relativity of "Backwardness"

Unlike in China, foreign capital had hardly any stake in the Indian cotton industry during the post-1856 period when it was being built up in Bombay and elsewhere. The early founders were Indian textile dealers who branched out into production.[99] The colonial state and British industry had no interest in such competition, nor were any insurmountable obstacles placed in their way. The fall in silver prices, which could not be checked by political means, entailed that the Indian rupee lost roughly a third of its value in the last quarter of the nineteenth century. This worked in favor of the Indian cotton mills,which were by no means technologically backward, even enabling them to beat back more expensive British thread in Asian markets. To look only at the trade between Europe and Asia is to miss the vitality of Asian producers on their home ground. Exports to China and Japan were the main factor in India's ninefold increase of its share of the world market for cotton thread—from 4 percent in 1877 to 36 percent in 1892.[100] The modern sectors of Indian industry were not primarily the result of capital and technology imports under colonial auspices; rather, the general commercialization that began in India in the eighteenth century expanded markets, stored up mercantile wealth, and—despite an abundance of cheap labor—created new incentives for technological improvement.[101] Historians are agreed that India's geographically concentrated industry played only a marginal role in the economy before the First World War. Nevertheless, it does not come out so badly from quantitative comparisons with Europe. Its total of 6.8 million spindles in 1913 was not worlds apart from the 8.9 million in the Tsarist Empire.[102] In purely quantitative terms, the Indian cotton industry looked more than respectable, and unlike its counterparts in China or Japan it developed without any state support.

Whereas China's early iron and steel industry (much of which fell under Japanese control after the First World War) grew entirely out of official initiatives, Indian steel was at first a one man show under Jamshedji Tata (1839–1904), one of the great entrepreneurial figures of the nineteenth century, a contemporary of such steel tycoons as the American Andrew Carnegie (1835–1919) or the German August Thyssen (1842–1926). Tata had made his money in the textile industry, but a visit to American steel plants prompted him to turn to metallurgy and to look for a location close to the coal and iron deposits of East India. Here, at Jamshedpur, the great steelworks of the Tata family came into being after his death. Advertised as a patriotic venture to be realized without recourse to the London capital market, it attracted investment from several thousand private individuals. The founder himself, realizing that India needed to become technologically independent, had contributed the startup capital for the Indian Institute of Science. And the Tata works, right from the beginning in 1911, strove to achieve product quality at the highest international level. Government orders played an important role, and the World War would set the firm on the road to

success. However, the efforts of the Tata Iron and Steel Company were not sufficient to create a heavy industrial sector before 1914, any more than the state-run
Hanyang Iron and Steel Works was able to do in China.

The case of India gives cause for reflection about general models in industrialization studies. "Backwardness" is a relative concept, and it is necessary to
specify the entities to which it refers. At a certain point, even at the end of the
nineteenth century, the socially and economically "backward" regions of Europe
were certainly not ahead of the more dynamic ones of India or China; the yardstick of economic success was the few large growth poles in Europe and North
America. In India, as we have seen, it was private entrepreneurs (not state officials) whose decisions led to the presence of large-scale factory production in a
number of sectors (the jute industry, dominated by British capital, is another
worth mentioning) and to the formation of an industrial proletariat that learned
to assert its interests. Industrialization, and many other processes included under
the heading of modernization, was under way in urban areas. The question of
whether India would have developed better without colonial rule—as nationalists and Marxists claim—will never be conclusively answered. Culturalist arguments, which see social structure (the "caste system"), mentalities, or religious
orientations ("profit-unfriendly Hinduism") as a basic obstacle to autonomous
development and even to successful learning from abroad, used to be popular
in Western sociology, but they have been under a cloud since India's high-tech
advances in the late twentieth century.

Similarly, Confucianism—a multifaceted term—and its ostensible hostility
to lucre have repeatedly been seen as the barrier to "normal" economic development in the nineteenth century and before. But since the spectacular economic successes of "Sinic" Taiwan, Singapore, and the People's Republic of
China (and of societies in Japan and South Korea inspired by Confucianism in
their own way), the old arguments have been quietly turned on their head and
Confucianism itself has come to be regarded as the cultural underpinning of
a distinctive East Asian capitalism. That such theories can explain success and
failure alike appears rather suspicious. Today many historians avoid asking why
countries such as India or China did *not* develop in accordance with a model
that they really "should" have followed. This leaves the task of carefully describing each special path.[103]

Japan: Industrialization as a National Project

Whereas it has been discussed since Max Weber's times why India and
China, despite many favorable conditions, did not take a "normal" path of economic development, the puzzle in the case of Japan has been why things worked
out so smoothly.[104] By the middle of the nineteenth century Japanese society
was highly urban and commercial; there were strong tendencies toward a unified national market, and the country's boundaries were clearly defined by its island position. Peace prevailed internally, and costly defenses against the outside

world were unnecessary. Administration was unusually good right down to the local level. People had experience with managing limited natural resources. The cultural level of the population, as seen in the estimated percentage of those able to read and write, was unusually high, not only by Asian standards. Japan thus had excellent conditions in which to adopt new technologies and new ways of organizing production.

Nevertheless, it would be superficial to see here only an objective logic of industrial progress serenely unfolding. It is not so clear that the conditions in Japan were decisively better than in certain parts of China or India. The key difference was the political project behind Japanese industrialization knitting together the state and private enterprise. The fall of the Tokugawa shogunate in November 1867 and the establishment of the Meiji regime two months later were less the result of changes in society and economy than a reaction to the sudden confrontation with the West. Japan's industrialization then got under way as part of a broader policy of national renewal, the most thorough and ambitious of its kind in the nineteenth century, though without a fully worked-out strategic plan. Close study of the Western powers had taught the Japanese elite that the development of industry would be central to the nation's future strength. As in China, therefore, but with central coordination and under much less foreign pressure, it was the government that launched the first industrial projects and supplied the foreign currency required for the purchase of industrial equipment.

Capital from outside the country played no significant role at this stage. At a time when Tsarist Russia was raising sizable loans in French and other European finance markets, and when the Ottoman Empire and China were being forced to borrow on unfavorable terms, Japan avoided any dependence on overseas creditors so long as it was economically vulnerable and had its sovereignty limited by the unequal treaties—that is, until well into the 1890s. Capital could be mobilized domestically, and there was a political will to invest it productively. Without any European influence (and, it would appear, uniquely in the non-European world), Tokugawa Japan had already introduced the practice of interbank lending, which would later greatly assist the funding of development projects. It did not take long after 1879 for a modern banking system to take shape, which was, like general financial and economic policy in the early industrialization period, largely the work of Matsukata Masayoshi, the son of a destitute samurai, who became a long-serving finance minister and one of the great economic wizards of the age.[105]

The fiscal policy of the Meiji state targeted an agriculture that was steadily increasing its yields. In fact, the agrarian sector was the most important source of capital in early Japanese industrialization; roughly 70 percent of state revenue after 1876 came from the land tax, and much of this was spent on industry and infrastructure. (In China, by contrast, agriculture was stagnating and a fiscally and administratively weak government profited little from any surplus.) Japan also had other advantages. Its population was large enough to generate internal demand, producers (especially of silk) methodically penetrated foreign markets, yet the development model, unlike in Latin America, was not one-sidedly geared to export growth. In several regions—around Osaka and Kobe, for example—an efficient proto-industry held up for quite a long time alongside steam-driven factory production, especially in the cotton goods sector. This was one of the chief differences between the English Manchester and the "Manchester of the East" that resembled it in many other respects.[106]

The Meiji state did not aim to construct a permanent state economy. Having provided an initial stimulus, the public sector gradually withdrew from most industrial projects, not least in order to ease the strain on the budget. Business pioneers also saw industrialization as a patriotic matter and, scorning American-style conspicuous consumption, cultivated a frugal ethos of service to the fatherland rather than individual profit maximization. One result of this was that firms generously shared with one another priceless knowledge about dealings with the world economy—knowledge that the Japanese had to acquire posthaste after the opening of the country. Bureaucrats and capitalists were successful in their efforts to achieve a diversified industrial structure so that Japan would be as independent as possible of imports.

Moreover, the Meiji oligarchs always kept the country's security policy in mind and were eager to prop up their own fragile legitimacy—after all, they had toppled the traditional political order—by the promise and reality of material progress. At the same time, there were sufficient private entrepreneurs willing to commit themselves through investment. At first, Japan could not avoid relying on Western technology, imported machinery, and foreign advisers. But the technology was often improved and adapted to Japanese conditions, and there could have been few other countries where the state took such an early systematic interest in this category of imports.[107] In many cases Japanese industry was not content to adopt simple technologies but attempted to acquire knowledge and to enter markets at the highest international level. All this was done in a relatively cost-effective way. It also involved an emerging legal framework of international patent law that, from the 1880s on, became yet another macrosystem knitting together economies in distant parts of the world.[108]

We shall not examine here Europe's most extraordinary industrial success story, that of post-1880 Sweden, or the great miracle that raised the United States in one generation (c. 1870–1900) to the position of the world's leading industrial power.[109] Two points should be noted, however. Even more than in Japan, industrialization in the non-slave-owning Northern States of the US took place on the basis of an "industrious revolution" and a palpable growth of per capita income during a period often identified as the "market revolution" (roughly 1815–50); international trade also played a greater role there than in Japan.[110] Consequently, rather than overdramatize the novelty of industrialization in the United States, we should recognize the long-term continuities. It is true that America's path primarily involved the free play of capitalist market forces, but they were not the only factor in operation. The federal government, controlled by the Republican Party between 1861 and 1913 with only two Democratic presidential interludes, pursued industrialization as a political project and set itself the task of ensuring national market integration, tariff protection, and a gold-backed currency.[111] Industrialization entirely without state assistance, which some liberal economists considered both possible and desirable, was historically a great exception. By no means did two grand models—one Western liberal, the other Eastern statist—stand facing each other in mutual opposition.

4 Capitalism

In the past twenty years, historians in many countries have fundamentally altered our view of global industrialization. For many parts of the world, the eighteenth century has come to be seen as a time of commercial expansion and dynamic enterprise. Markets grew larger and denser, and specialized production was encouraged for near and distant outlets, often for export to other countries or even continents. Officialdom, even the "Oriental despotism" that Europeans tended to paint in such lurid colors, rarely intervened to stifle this economic vivacity, which, after all, often helped to fill the state coffers. Demographic expansion, however, and the vulnerability of nearly every society to "Malthusian" counterforces, did not allow for genuine and stable growth of per capita income. It would therefore be more precise to say that although many economies were *moving* and even recording a slow rise in income, not one—with the exception of England from the last quarter of the eighteenth century—was *dynamically goal oriented*; none was growing in the modern sense of the term.

This new picture of the eighteenth century confounds the usual chronologies. The early modern "industrious revolution" sometimes extended far beyond the formal time threshold of 1800. When it came to changes, they rarely took place as sudden sprints, even though Alexander Gerschenkron seems to have been right in that *later* industrialization processes were more abrupt and temporally compressed than those of the first and second generations; Sweden, Russia, and Japan are good cases in point. But like the original Industrial Revolution in England, later industrializations did not begin from scratch; rather, they involved a change in the speed and type of advance within the general movement of the economy. Although industrialization got under way in a regional or increasingly national framework, the outcome was rarely a complete dominance of large-scale industry. What Marx called "petty commodity production" often stubbornly held its ground, sometimes in a symbiotic relationship with the world of the factory. Naturally, the early generation of factory workers had originated in the countryside, and many of them retained their connections with it for quite a long time. Factories and mines became magnets of industrialization, but also of innumerable migrant labor circuits between village and production site.

From midcentury on, the name for the new order was *capitalism*. Karl Marx, who seldom used the term as a noun but preferred to speak of "the capitalist mode of production," analyzed the new system in *Capital: A Critique of Political Economy* (1867–94) as a capital *relationship*, an antagonism between owners of labor power and owners of the means of material production. In a simplified form, interpreted by such loyal companions of the master as Friedrich Engels or Karl Kautsky or modified around the turn of the century by Rudolf Hilferding or Rosa Luxemburg, Marx's analysis of capitalism became the dominant theory in the European labor movement. Soon people less critical than Marx and his followers took up the term "capitalism," and early in the new century, especially in Germany, the research and debates of "bourgeois" economists, invariably under Marx's powerful spell, developed into a complex theory of capitalism represented by such figures as Werner Sombart and Max Weber.[112] These highly original thinkers, representing the "Historical School" of economics at its best, detached the concept of capitalism from its narrow association with nineteenth-century industry, seeing it as present not just in one particular stage of development, but in virtually all forms of economy, sometimes as far back as European antiquity. Various types were defined: agrarian capitalism, commercial capitalism, industrial capitalism, financial capitalism, and so on. The models of these German non-Marxists gave up Marx's central reliance on an "objective" labor theory of value, whereby all labor creates value in a way that is susceptible to measurement. At the same time, they did not embrace the new marginalist orthodoxy prevalent in British and Austrian economics since around 1870, for which the preferences of market participants are determined by their assessment of "subjective utility."

Capitalism theory around the turn of the century, as variously developed by Weber, Sombart, and other social theorists, did not neglect institutions. While by no means ignoring the contradiction between capital and labor, it placed greater emphasis than Marx had done on the workplace structure of production under capitalist conditions and the ways of thinking (the economic "attitudes" and worldviews) that kept the system going. Moreover, its main exponents had such a keen historical sense that they tended to regard analysis of the contemporary world as somewhat secondary. Even if Sombart often commented on economic life in his time, and Max Weber produced early empirical studies of the stock exchange, the press, and Prussian agricultural workers, their main research interests focused for many years on what would later be called "the early modern age." It was there that Weber found the origins of the "Protestant ethic" and Sombart a complex "commercial capitalism." From Karl Marx to Max Weber and Thorstein Veblen, capitalism was a central theme in social analyses of the age, and the radical-liberal and socialist theories of imperialism that were an offshoot of the debates about capitalism are among the most sophisticated accounts of the fin de siècle written at the time.[113] No uniform understanding of the term "capitalism" took shape, however, and by 1918, still in Max Weber's lifetime, one could find 111 definitions of it in the literature.[114]

Such indeterminacy did not mean that the concept of capitalism was abandoned: following the classical economists, it persisted not only in the Marxist tradition but also among open apologists of the system, although the new orthodoxy preferred to speak innocuously of "market economy." Events in the last two decades have now led to a certain revival of the term. Whereas its use was once associated with the rising power of industry, impoverishment of the early proletariat, and subjugation of the world by the spirit of business-oriented instrumental rationality, the most relevant tendencies today are the global presence of transnational corporations and the failure of all *non*capitalist alternatives, whether they ended with the hollowing out of socialism from within (as in China) or the straightforward collapse of any such order (as in the Soviet Union and its sphere of influence). Since the 1990s many attempts have been made to describe and explain "global capitalism," but a new synthesis is still lacking.[115] Today's typologies look different from those of a hundred years ago, with a special emphasis on regional capitalisms: European (itself differentiated into "Rhineland" and other forms), American, East Asian, and so on.[116] Many theorists have a strongly contemporary orientation, without the historical depth of the classics, and overlook what Fernand Braudel and some of his disciples, following in the tracks of Werner Sombart, have written about the early modern commerce centered on Europe (though not conducted by Europeans alone) as a first manifestation of something like "global capitalism."

One finds oneself agreeing with many observers of the world before 1920 who characterized the nineteenth century as a new, unprecedentedly dynamic stage of capitalism, and also with such interpreters as Sombart, Braudel, or Wallerstein who see the development of capitalism as a long-term process begun long before the nineteenth century. What general points, then, can be made about nineteenth-century capitalism?[117]

First. Capitalism cannot be purely a phenomenon of exchange and circulation. Long-distance trade in luxury goods may relocate and multiply wealth, but it does not institute a new economic order. That requires a special organization of production, as it came into being in the nineteenth century.

Second. Capitalism is such an economic order. It rests upon production for the market, involving a division of labor and organized by individual or corporate entrepreneurs who make a profit and mostly seek to reinvest it productively—in Marxist terms, "to accumulate."[118]

Third. Capitalism is bound up with general commodification, with a transformation of things and relations that makes perhaps not "everything" but every factor of production into a commodity exchangeable on the market. This is true of land as well as capital and knowledge, and above all of human labor power. Capitalism thus presupposes the presence of "free" (also in the sense of spatially mobile) "wage labor." It has often found ways of integrating unfree labor in the periphery of its systems but cannot tolerate it in the core. Slavery and other kinds of "extra-economic" bondage conflict with its logic of unlimited availability.

Fourth. Capitalism as an economic order has the flexibility always to use the most productive technologies and organizational forms (whose efficiency is tested by the market). In the nineteenth century, these included not only factory production but also large-scale, increasingly mechanized, agriculture, especially farm businesses of the North American type. Agrarian capitalism may be located upstream of industrial capitalism, in the sense of a preparatory agricultural revolution, but it equally exists alongside it in a symbiotic relationship.[119] Since the end of the nineteenth century, these forms have grown closer to each other within an internationally active agro-industry, which controls whole product chains from the original farming to processing stages to final marketing.[120]

Fifth. The famous Marxist question of the "transition from feudalism to capitalism" is rather academic and applies best to parts of Western Europe and Japan. In a number of other places where capitalism was particularly successful in the nineteenth century—the United States, Australia, and South African mining areas—there was never any "feudalism," any more than there was in China. The whole issue ought to be formulated more generally in terms of institutional frameworks for capitalism, which came about mainly through legislation and state action. But the state is not a product of the market. Although markets may also arise and grow spontaneously through the activity of private economic subjects, the free spaces for them to operate in are the result of political regulation or the lack of it, of state action or state inaction. Free trade in the nineteenth century was a creation of the British political elite. In the late twentieth century a one-party socialist dictatorship in China established a quasi-capitalist economic order. By means of detailed "bourgeois" legal systems—from the Napoleonic Code of 1804 to the German Civil Code of 1900 (still considered a model in many parts of the world)—state apparatuses everywhere safeguarded and made possible capitalist enterprise, first and foremost by providing the fundamental legal guarantee of private property. In East Asia and elsewhere, strong bonds of reciprocal trust among economic subjects in civil society fulfilled an analogous function. From German mining to Chinese industrialization, the state was also active as an entrepreneur in mixed public-private ventures.

Sixth. The links between capitalism and territory are especially controversial. Evidently, the global capitalism that spread after 1945 was less dependent than earlier forms on being anchored in a particular locality. Production is becoming ever more mobile, and with the Internet and advanced telecommunications many businesses can operate almost anywhere in the world. Early modern commercial capitalism too, featuring individual overseas merchants and chartered companies, wove its trading networks with often only a weak implantation in the Dutch or English mother country. In the nineteenth century, however, capitalism and the (national) territorial state stood in a close relationship with each other. *Before* capitalism could move beyond national boundaries, it profited from the state-backed integration of national markets—for example, in France, Germany (after the *Zollverein* of 1834), or post-1868 Japan. In the eyes of continental Europe and the United States, *extreme* free trade was an episode limited to the third quarter of the nineteenth century. Big business, emerging after roughly 1870 and taking shape, often with global reach, in the second economic revolution, displayed striking national styles beneath a general cosmopolitanism that was much more marked in finance than in industry.[121]

Seventh. Territorialization in the course of industrialization is bound up with the material character of industry. The overseas merchant of early modern times—think of Shakespeare's Antonio from *The Merchant of Venice*—whether alone or in a partnership, placed his productive assets in ships and transportable goods. The technological structures of the early industrial age opened up new opportunities for long-term material investment. Mines, factories, and rail networks were intended for a use cycle longer than the turnover time typical of the wholesale and overseas trade of early modernity; wealth was now tied up in machinery and infrastructure in a way that it had been earlier only in monumental buildings—which were unable to create further riches. This was linked to unprecedented intervention in the physical environment. No economic system has ever reshaped nature more drastically than the industrial capitalism of the nineteenth century.

Eighth. This materialization and crystallization of capital corresponded to its significantly greater mobility. In purely technical terms, this was at first the result of better-integrated money and finance markets; the transfer of monetary values from the colonies—still a major practical difficulty for the British in late-eighteenth-century India—became ever simpler as international means of payment were perfected in the nineteenth century. The rise of the City of London to become the center of the world capital market, together with the emergence of subordinate centers in Europe, North America, and (at the end of the century) Asia, made the network considerably denser. British, and increasingly other, banks and insurance companies offered financial services to the whole world. After 1870, capitalism discovered overseas investment as the way to export capital, although for a long time this remained a British specialty. Both the temporal dimension of amortization or debt repayment and the spatial horizon of planning grew wider; people planned not only further into the future but also across greater distances. Europe's textile industry had to make arrangements well ahead for its supply of raw materials from distant countries. The electrical industry came into existence only with the technical challenges of long-distance telegraphy, and right from the beginning it sold its products all around the world. Although the term "global capitalism" should be reserved for the period after 1945, or even 1970, many countries by 1913 had a national capitalism with a global radius of operations. Industrialization, defined as the development of mechanized factory production using local energy sources, was in each case a regionally specific process. Nineteenth-century capitalism, on the other hand, may be understood as an economic order that made it increasingly possible to insert local entrepreneurial activity into interactive circuits spanning large areas or even the globe as a whole.

CHAPTER XIII

Labor

The Physical Basis of Culture

At all times most people have worked.[1] Adults who did not do so—whether sick or disabled, fortunate in their circumstances, or belonging to an idle elite exempt even from military or priestly service—have been a minority in every society. Since work is performed in countless different ways and conditions, it is much more difficult to say anything general about it than about highly organized systems such as industry or capitalism. A history of work can be a history only of typical instances—or, where especially good data are available, of workloads and their gender distribution.[2] If work is regarded not as an abstract category but as an aspect of people's actual lives, then the worlds of work are legion. A butcher in Bombay in 1873, about whom we know from a record of court proceedings, lived in one such world. An opera singer in the Italy of Rossini's time, when patronage had all but given way to market employment, had to operate in quite another. And different again was the world of a Chinese coolie working as cheap labor down a South African mine, or of a ship's doctor as he accompanied every transoceanic voyage, under sail and under steam.[3]

Work produces something—and nothing more often than meals. Cooking must have been the most widespread, and generally most time-consuming, expenditure of labor throughout history. As this example shows, not all work is market oriented, and not all labor power is procured via the market. Work may take place at home, within a village community, or in a complex organization such as a factory, an army, or a municipal authority. The idea of a "regular job" appeared only in the nineteenth century; much work has been (and is) "irregular." Work usually follows a standard pattern, within the framework of "labor processes." These processes are social in nature. Most include direct cooperation with other people, and all are indirectly enmeshed in a social order. Certain kinds of worker and labor process typify a certain level of the social hierarchy. Relations of power and domination determine the extent to which work is autonomous or heteronomous. If standardized labor processes are combined with a consciousness defined primarily by work, the result is an "occupation." Workers who derive their identity from an occupation do not look only for an employer's approval but also set certain quality standards for their own work. But such standards are also corporately defined. In other words, practitioners of an occupation control, sometimes exclusively, the domain of their work: they regulate access to it "off market" and often receive state support to boot. This gives rise to niches in which the very membership of a profession (a trade, a guild, an occupational association, etc.) constitutes a form of income-generating capital.[4]

Given these multiple possibilities, it is hard to trace global tendencies for a whole century.[5] Yet, it is all the more important, though, because the nineteenth century had a special concern for questions of work. Where the culture held it in high esteem, as in Western Europe and Japan, capitalism created new scope for it to develop. In the West, "work" became both a high value and a favored category in the description that people gave of themselves; while idleness ceased to be a desirable norm even among the elites. Queens let themselves be seen in public with their knitting. In economic theory as in certain currents of anthropology, *Homo faber* became the mandatory model. Classical political economy explained creativity and physical effort as the source of value creation—a doctrine that also became the axiom of socialism and fueled demands for workers to be paid and treated well. Others went even farther, trying to conceive of work as the purification of humanity; alienated and exploited labor under capitalism was transformed into a utopia of emancipated labor. With the spread of machines, the superiority of manual labor became a distinctive theme: critics such as William Morris, the writer, early English socialist, influential designer, and founder of the Arts and Crafts movement, returned in theory and practice to the endangered ideals of premodern crafts. When the average workweek, having risen in the early period of industrialization, fell again in parts of Europe and the United States toward the end of the century, leisure time emerged as a new kind of time to be actively lived and not just idled away. This raised the issue of how to separate paid labor and nonwork—within each day and within a year or a lifetime. It has been argued that Europe drew a particularly sharp distinction between the two,[6] but even there different conceptions existed alongside one another, and the idea of a "typically European" understanding of work is not without its problems.[7]

Studies of the nineteenth-century work ethic in non-European civilizations are still lacking. What they would probably show is that attitudes to work differed not only, often not even primarily, along cultural lines of division; they were both class specific and gender specific, and external stimuli and a favorable institutional setting kindled work energies in the most diverse circumstances. A good illustration of this is the speedy and successful response of many West African farmers to new opportunities for export production. Efficient sectors—there were some in colonial times (e.g., cotton)—were adapted to the changed conditions, and new ones were created and built up.[8] Finally, in all or most civilizations, conceptions of work have been associated with different expectations regarding the "fair" treatment of workers.[9]

1   The Weight of Rural Labor

Dominance of the Countryside

In Europe, as everywhere else in the world, agriculture was the largest sector of employment throughout the nineteenth century.[10] Only in the years immediately after the Second World War did industrial society establish itself as the dominant type all over Europe, including the Soviet Union. Its supremacy was short-lived, however, since by 1970 the service sector accounted for a larger share of total employment in Europe. The classical industrial society was thus a fleeting moment in world history. There were only a few countries—Britain, Germany, Belgium, Switzerland—where industry was the leading sector of employment for more than half a century. It never reached that position in the Netherlands, Norway, Denmark, Greece, or even France, and it did so for only brief periods in Italy, Spain, Sweden, and Czechoslovakia. The brevity is even more striking if we look beyond Europe. Even in the two countries with the most productive industry, the United States and Japan, industrial work never overtook employment in farming and services. Of course, both there and elsewhere there were highly industrialized regions, but in 1900 industrial work had pride of place in only a few countries, such as Britain, Germany, and Switzerland.[11] In large parts of the world, agriculture grew in importance during the nineteenth century, since the advancing frontiers were mostly areas where new land was opened up for farming.[12] Sometimes the main type of pioneer was the planter or big rancher, but more frequently it was the small farmer: in the highlands of China, in Africa, in the Caucasian steppe, in Burma and Java. Some authors have spoken of a century of "peasantization" throughout Southeast Asia, and it is true that around 1900 its lowland areas were dotted with a myriad of tiny farms.[13] Peasants had not always "been there forever," since the Neolithic revolution. And they could still be "made" in the nineteenth century.

In 1900 or 1914 most people around the world were engaged in agriculture. They worked on and with the soil. They mainly toiled in the open air, where they were dependent on the elements. That an ever-increasing share of all work came to be performed indoors was a great novelty of the nineteenth century. For someone newly arrived from the country, the first impression of a factory must have been of a work*house*. At the same time, as a result of technical advances in mining, work penetrated deeper and deeper underground. Even the most widespread trends of the century—above all, urbanization—had little effect on the position of agriculture as some countertendencies, no less "modern," also grew stronger. The expansion of the world economy between 1870 and 1914 (especially after 1896) greatly stimulated agrarian production for export, and agrarian interests exerted huge political influence even in the most developed countries. Despite a relative decline in the weight of the upper nobility, large landowners put their stamp on the British political elite until the last quarter of the century, while in many Continental countries agrarian magnates continued to set the tone. Any regime in France, whether monarchy or republic, had to pay heed to a strong class of small farmers, and agricultural interests in the United States were consistently well represented in the political system.

Most people were tillers of the soil. What did this mean? A number of disciplines have long occupied themselves with this question: agrarian history, agrarian sociology, ethnology, and the largely related study of folklore. For premodern Europe and large parts of the nineteenth-century world, there was no need for a special "agrarian history"; farmers and rural society were anyway the central theme of economic and social history.[14] Of the numerous discussions since Alexander Chayanov's pathbreaking studies in the early 1920s, one of particular interest for global history is the debate of the 1970s between supporters of a "moral economy" approach and "rational choice" theorists.[15] For the former, the peasantry is subsistence oriented and hostile to the market, favoring communal over individual ownership, avoiding risks, and behaving defensively as a community toward the outside world; its ideal is justice within a traditional framework and relations of solidarity, also between landowners and tenants, patrons and dependents; the selling of land is seen only as a last resort. For the latter, peasants are at least potentially small entrepreneurs; they know how to use market opportunities when these present themselves, not necessarily to maximize their profits but to ensure their material existence by their own efforts, without completely abandoning group solidarity. Capitalist penetration leads to differentiation among such peasants, who may at first have been relatively homogeneous in social terms.

Each of these approaches refers to different examples, so that it is not possible to make a definite comparative judgment as to their empirical validity. In some historical situations one tends to find peasants with an individual business spirit; in others a community-centered traditionalism prevails. The important point here is that regionally or culturally specific classifications do not take us very far. There are no "typically Western European" or "typically Asian" farmers; very similar kinds of market-oriented entrepreneurialism may be found in the Rhineland, northern China, and West Africa. In the case of Japan, it is impossible already in the seventeenth century to find "traditional" peasants producing for no one but themselves in tiny isolated villages. Farmers who switched their crop mix according to market opportunity, using the best seeds or latest irrigation techniques and consciously striving to raise their productivity, were much more representative. They do not correspond to the image of primitive villagers imprisoned in narrow, unchanging life cycles.[16]

Villages

The actual work situation of people on the land differed in many respects. Nature smiled on many crops but ruled others out completely; it determined the number of harvests and the length of the harvest year. Irrigated agriculture, especially the intensive garden-style cultivation of rice in East and Southeast

Asia, where farmers stood directly in water, required an organization of work different from the hoeing of crops on dry soil. Household involvement was also highly varied, often with a sharp differentiation between the sexes and the generations. The main dividing line ran between two extreme situations: in one, the whole family, including children, would take part in rural labor and perhaps use any free time for domestic crafts; in the other, migrant workers lived apart from their families in all-male makeshift communities, with no insertion into village structures.

There were villages in most agrarian societies. The delineation of their function was of varying sharpness. In extreme cases, the village might be many things at once: "an economic community, a fiscal community, a mutual-assistance community, a religious community, the defender of peace and order within its boundaries, and the guardian of the public and private morals of its residents."[17] Village communities were especially strong where at least one of two factors played a role: (1) the village functioned as an administrative unit (e.g., a government tax-raising center), perhaps even being legally recognized as an independent corporation; (2) the village commune disposed of land for general use or even—as in the Russian *obshchina*—collectively decided on its distribution and redistribution. The latter was by no means a matter of course. In the intensive small-farmer economy of northern China, nearly all the land was privately owned; the state, whose agencies reached down only to district level, did not collect taxes from the village as a body but relied on an intermediary appointed by the community (the *xiangbao*) to work out the best method.[18] The village commune was thus less developed than in Europe. In southern China, on the other hand, extensive clan structures—which might be, but were not necessarily, identical with a self-contained settlement—undertook tasks of integration and coordination. It would be wrong to regard such clans as inherently backward or "primitive" in terms of the history of development; they might constitute the framework for highly efficient agriculture. A similar function was served (not only in China) by temple communities that held property in common.

The position of village communes in Eurasia therefore varied considerably. In Russia—at least until Prime Minister Stolypin's land reform of 1907—they played an important role in redistributing land in conditions where private ownership was little developed, whereas in Japan they were a multipurpose institution held together by an ideology of "community spirit" (*kyōdotai*), and in large parts of China (especially where clan ties were absent and the proportion of landless laborers was high) they exhibited a low degree of cohesion.[19] The Japanese case also raises the important question of the extent to which a stable village elite developed among the peasantry. In Japan, as in many parts of Western Europe, this happened through primogeniture: the eldest son inherited the farm. In China, and partly also in India, private landholdings were time and again parceled out among the male heirs, and it was difficult to maintain the continuity even of a modest family farm.[20]

An equally significant dimension of peasant life was access to the land. Who was the "owner"? Who held (possibly graduated) use rights? Was leaseholding (with its numerous variants) part of the picture,[21] and, if so, how much use was made of it? Did tenants have to pay a fixed sum of money or a share of the harvest? In what form was it handed over? In other words, to what degree was the rural economy monetized? Were noneconomic ("feudal") duties still expected of the peasantry—in particular, labor service for the landowner or the state (e.g., road or dike construction)? Were farmers free to sell their land? How was the land market organized?

A last major parameter is the extent to which production was geared to the market. Were the markets in question close or distant? Was there any network of local exchange relations, perhaps centered on a periodic market in a pivotal rural town? How much did farmers specialize, and how much was this at the expense of provision for themselves? Did they take their own produce to the market or rely on middlemen? Finally, what regular contact was there between farmers and nonfarmers? The latter might be city people, but they could also be nomads in the vicinity. The former would include absentee landlords who used local agents and had nothing in common culturally with the villagers. Large local landowners might nevertheless be seen in a church or temple, whereas urban magnates or landlords lived in an entirely separate world.

The Example of India

This diversity of agrarian forms of existence cannot be grouped only by continent or in the categories of East and West. Let us take a fully developed and at first sight typical peasant society, in the year 1863. Of its population of roughly a million, 93 percent live in village communities with two thousand inhabitants or less; nearly all are members either of a nuclear family (more than a half of cases) or of a family spanning several generations. Almost everyone owns some land, which is not generally in short supply; 15 percent of the total surface area consists of fields, pastures, and vegetable gardens. Anyone in need of land can obtain some from their village community. Large landholdings and tenancies are not a feature of the situation. Some peasants are richer than others, but there is no landlord class and no nobility. People work almost exclusively for their own subsistence, producing the food they eat as well as most of their clothing, footwear, household utensils, and furniture. Granaries help to guard against famines. There are few cities that require market relations for their supply. Cash for tax payments can be raised without difficulty through the selling of cattle. There are no railroads, virtually no roads that a horse and cart can use, scarcely any businesses or proto-industry, and no financial institutions. Ninety-eight percent of the rural population is illiterate. Although nominally belonging to a form of "high religion," people are guided by superstition in their everyday existence. They do not expect much from life and have little ambition to improve things or to work more than is necessary. Few and far between are those who plough more land than they need to feed their family. An abundance of natural resources means that the country does not strike one as poor. Its per capita national income is estimated to be roughly one-third of Germany's at that time.

This egalitarian idyll is not a "typically Asiatic" society made up of autarkic peasant settlements such as people in Europe imagined in the mid-nineteenth century: loosely controlled archipelagos of self-sufficient villages, with an immobile population consisting of self-sufficient households. Nor is it drawn from one of the fertile areas of tropical Africa. The land described above is European Serbia, at the time of its first more or less reliable census.[22]

This kind of peasant society, however, was not representative of either Europe or Asia. If we take India (the epitome of peasant archaism for nineteenth-century Europeans) as a second example from the wide range of Eurasian agrarian societies, then we obtain something like the following picture.[23] The basic unit of rural life was indeed the village. Its hierarchical structure nearly always included groups with high status, particularly members of upper castes or the army, but it is by no means the case that these were all big landowners. Unlike the typical landlord (*dizhu*) in the Chinese village, they rarely abstained completely from physical labor, but they did function as literate "managers" of village life and played a decisive role culturally. The deepest social rift was not, as in China, between a parasitic landlord class living off rents on the one hand and hardworking tenant farmers or peasant smallholders on the other, but rather between those (often a majority) with relatively stable land use rights and a landless underclass of wage laborers. The typical Indian village, then, was not governed by big landowners living in the city or on sumptuous rural estates, or by Chinese-style landlords under more modest material circumstances, but by a group of dominant peasants who controlled most of the resources (land, livestock, credit). While not deriving automatically from membership in an upper caste, their position did usually correspond to a superior caste ranking. As a rule they were active farmers themselves, working not only on land of their own but also on leaseholds. Colonial law regarded all farmers in principle as free subjects. Large-scale slavery was all but nonexistent in modern India, and remnants of household servitude disappeared with the abolition of slave status in 1848, fifteen years after it was prohibited by law elsewhere in the British Empire. Nevertheless, as in China, moneylending offered scope for reducing weaker members of the village hierarchy to a form of dependence.

The primary goal of Indian peasants was to ensure their family's subsistence. However, in a continuation of processes going back to precolonial times (which meant before c. 1760 in Bengal), commercial relations extended farther and farther beyond village boundaries. In some cases, the growing of cash crops—above all indigo and opium for the Chinese market—led to a concentration on exports. But this was less characteristic of India (or China) as a whole than it was of parts of the New World, Southeast Asia, or Africa, where export monoculture became widespread in the second half of the nineteenth century. The

Indian village usually had an open relationship with the city: it was inserted into trade networks. City-based middlemen would buy up surpluses and sell them in urban markets. Most peasant producers were anyway in no position to make "market decisions." Being restricted by property relations, environmental conditions (e.g., irrigation or the lack of it), and the power of dominant groups, they did not operate as the "independent entrepreneurs" of rational choice theory. In the first few decades after 1760, India's colonial status made itself felt in the material impact of devastating wars of conquest and in a higher average tax burden. The longer-term consequences were threefold: (1) stabilization of fiscal pressure at a high but predictable level; (2) gradual enforcement of contractual relations in private agriculture, under the supervision of colonial courts; and (3) the favoring of dominant village groups, rather than an acceptance of agrarian egalitarianism or the simple granting of privileges to existing or newly created aristocracies.

In the nineteenth century, India's precolonial social structures changed in various ways, which—as numerous protest movements illustrate—often involved a crisis of one kind or another. The colonial state was by no means the sole originator but interacted with autonomous trends in the economy and society. Agrarian India was flexible enough to adjust to new challenges, but the dynamic for a quite different, "capitalist" agriculture did not arise within it spontaneously. It would indeed be naive to expect that to have happened, since the liveliest counterfactual imagination can hardly conjure up a repetition in India of the northwest European agrarian revolution. In this respect, Indian, Chinese, and Javanese agriculture were similar: the easy availability of cheap labor, the limited scope for mechanical rationalization, and the lack of northwest Europe's distinctive combination of field and pasture stood in the way of radical change.

Types of Enterprise

Comparisons with China interspersed here and there have shown that the world's two largest agrarian societies resembled each other in many ways: farmers were in principle free agents; they produced partly for the market; and the main economic unit was the (often also proto-industrial) family household, which might be supplemented by a small number of servants and laborers. These three features indicate a certain affinity with Western Europe, or at least with France and Germany west of the Elbe, and underline the difference with parts of the nineteenth-century world where plantations, latifundia, and estates were run as large enterprises on the basis of servile or bonded labor. It would thus be wrong to think in terms of a split between free West and enslaved East. China's agrarian structure, with its numerous regional variants, was far freer than the rural order in eastern Europe.

It is hard to classify the variety of agrarian production and rural life, because several criteria that are not easy to line up with one another need to be taken into account. This is true even if we stick to the three most important ones:

(1) the biological-ecological foundations (which crops are grown?); (2) the form of enterprise and labor regime (who has how much leeway in deciding what to produce within a certain organizational framework?); and (3) property relations (who owns the land, who actually uses it, who profits from it and how?)—for example, whereas rice cultivation in irrigated fields (as opposed to wheat or cotton growing) poses great difficulties for a large enterprise, it may prosper under different property relations (individual small ownership, tenant farming, clan or temple ownership).

With regard to the first criterion, a distinction may be drawn between wet rice growing, mixed agriculture and livestock farming, horticulture, and so on.[24] Criteria 2 and 3 together yield another typology:

(a) a manorial system combining subsistence farming with unpaid labor on the domain of the landlord (who is at the same time politically dominant);
(b) family leasehold units (rentier landowner versus tenant farmer);
(c) family smallholdings with relatively secure property rights;
(d) plantations (capital-intensive export production of tropical crops using nonlocal, often ethnically foreign, manpower); or
(e) large-scale capitalist agriculture, for which the landowner employs wage labor.[25]

In reality, however, the transitions between (b) and (c) were fluid: someone who could invoke a hereditary tenancy relationship, whether in Java or the Rhineland, might be the titleholder of land without being its legal owner in the end.

All through the century, nearly everywhere in the world, farm labor remained manual labor; many parts of Europe were similar in this respect to Asia, Africa, and Latin America.[26] Class location, too, set up commonalities across cultures: agricultural wage labor on a Pomeranian or Polish estate was not fundamentally different from wage labor in India, although in each case it was embedded within particular hierarchies and cultural environments.[27] A material insecurity that made it necessary to move around in search of work meant that there was a basic affinity in people's living conditions and experiences; and, as in earlier ages, migration spread agricultural knowledge over large distances. Such parallels and linkages did not, however, result in transnational solidarity. Unlike in industry and transport, where an international orientation took root with the growth of the early workers' movement, there were no extensive links among agricultural laborers, no peasant international. A farm laborer or peasant in Bihar knew nothing about his counterparts in Mecklenburg or Mexico.

If work was changing, where and how did global processes influence the changes? In general, rising international demand for agrarian products, especially from the tropics, did not necessarily have a liberalizing impact on the conditions of rural labor. Liberal economic theory anticipated that international trade would dissolve "feudal" systems, free people from archaic constraints, and foster in them a spirit of hard work and enterprise. That was indeed a possibility, especially where small farmers took advantage of new overseas outlets for their produce without the interposition of foreign economic interests. Long-term success, however, required that the national government (as in Japan) should actually promote exports and create the appropriate legal and infrastructural framework, or that a colonial regime, often for the sake of political stability, should consciously side with local farmers against foreign plantation companies. In the absence of such conditions, foreign interests usually had the upper hand.

Plantations

The end of legalized slavery in the European colonies, the United States, and Brazil was by no means followed by the breakup of large plantations. These increasingly controlled production of such sought-after products as coffee, tea, or bananas, competing with (though not always eliminating) small producers in other parts of the world. After 1860, new plantation sectors began to appear: sugar in Natal, rubber in Malaya and Cochin China (South Vietnam), tobacco in Sumatra. The plantation was an innovative, "modern" form of enterprise, which Europeans had been introducing on a large scale in the New World since roughly 1600. Around 1900 it experienced another heyday, based not on the gradual continuity of local conditions but on active foreign intervention to found and organize new plantations. (Sometimes, though, as in Java and Ceylon, local businessmen also took advantage of the opportunities.)[28] A new plantation created a rift in the local society at least as deep as that brought by a new factory. Capital and management inevitably came from Europe or North America in the late nineteenth century, and planters tried to establish a rational, scientifically based form of cultivation with optimum yields. For this, apart from a few specialist jobs, they needed only an uneducated labor force. And since most plantations were in thinly populated areas, the workers often had to be brought in from far away. On the great tobacco plantations of East Sumatra, for example, the preference was to recruit Chinese and to house them in work camps. At best nominally free, and usually paid piece rates and further exploited by a contractor, the workers were often subject to such strict discipline and regimentation that conditions were not clearly distinguishable from those on a slave plantation. Abandoning the place of work was punished as a criminal offense.[29] In 1900 it was rare for a plantation to be family owned. Almost all belonged to capitalist companies, which invested considerable sums in a railroad and port facilities and kept a close watch on world market conditions. The colonial plantation was not a complete novelty but developed out of the old slave plantation. It was an instrument of global capitalism employed almost exclusively in tropical countries. In contrast to industry, it seldom formed part of a wider process of national economic development.[30]

Plantations had an industrial component if their produce was actually processed on the spot. The model of such an integrated business was the rubber plantation, since rubber trees can be tapped all through the year and allow a company to be independent of the seasons. This made the plantation even more factorylike. However, the new wave of plantations in Southeast Asia and Africa around the year 1900 did not mean that global agribusiness now swept everything before it; plantations and export-oriented small farmers would coexist and compete with each other for the whole of the twentieth century. The plantation economy was global also in the sense that both labor and capital came from several different countries. On the island of Sumatra, a core region for these new developments, only one-half of plantation investment in 1913 was Dutch owned; British, North American, Franco-Belgian, and Swiss interests accounted for the rest.[31] Plantations sprang up on land bought from indigenous princes, who then lost all influence on conditions within the huge areas in question. But the law of the Dutch colonial state also had limited validity there—and in some cases none at all. A special kind of plantation justice system might then come into force, displaying certain affinities with the patrimonial rule that was exercised independently of the state on large Prussian estates east of the Elbe.[32] Very similar conditions had begun to appear a few decades earlier in places such as southwestern India.

Haciendas

The plantation was not the only corporate response to export opportunities. In nineteenth-century Egypt, the application of political power led to the development of large estates, as the government handed over indebted villages to high officials in return for the guaranteed payment of taxes. Land thus became concentrated in the hands of a state class tied to the pasha, while the only way of halting the secular decay of Nile irrigation systems was the organization of renovation work by large enterprises with modern engineering expertise. The estates in question cultivated cash crops, particularly cotton and sugar beets, which promised their owners the high short-term income necessary to finance investments, as well as offering the state a secure source of fiscal revenue. Labor was often brought in from distant places to work alongside local fellahs. If Egypt became one of the world's main cotton exporters after the 1820s, this was due less to foreign initiative and forced incorporation of the country into the world economy than to the policies of Muhammad Ali and his successors. Egyptian estates were organized along plantation lines, even though foreign capital had not played a key role in their development.[33]

Not all large-scale farming in the nineteenth century should be seen primarily in world economic contexts. *Peones* working on Latin American haciendas were neither slaves nor wage laborers; rather, the model was that of a patriarchal family, with the *patrón* often appearing as a kind of godfather figure, and the ties of mutual obligation were of a noncontractual character resting on a "moral economy" outside the market. Often the physical location of the hacienda made it a closed world, the boss's precinct resembling a fortress surrounded by the *peones* in their villages. Unlike plantations, late-nineteenth-century haciendas were typically undercapitalized and technologically backward. The dependence of the *peones* rested not so much on overt compulsion as on a debt to the *haciendado* reminiscent of credit relations between ordinary peasants and the dominant elite in Chinese or Indian villages. Like the (slave) plantation, the hacienda was a relic from the early modern colonial period, and one need not consider it "feudal" (although many writers have) to identify it as a contrasting form to the plantation. The hacienda was geared more to self-sufficiency than to export production; labor relations had strong noneconomic overtones. Its characteristic social boding contributed to the fact that in the Latin American republics, the peasantry were unable to exercise their rights as free citizens. They had no opportunity to benefit from the freedom promised at the time of independence, and most of their protest movements failed to achieve results.[34]

In the case of Mexico, the years from 1820 to 1880 may be regarded as a transitional period for the hacienda.[35] With the demise of the colonial state, the *indios* lost a power that, however unreliably, had afforded them a degree of protection. Instead, the ruling Liberals and their ideology of progress viewed the *indios* as an obstacle preventing Mexico's development along European or (later) North American lines. They therefore showed no consideration for indigenous people. Whereas the colonial hacienda had still involved a certain balance between the interests of the landowner and the Indian commune, the policies of the new republic—and a fortiori of the Porfirio Díaz dictatorship after 1876—largely dispersed communal property and left the *indios* at the mercy of profit-hungry *haciendados*. This practice, however, did not make the hacienda a mainstay of the export economy, comparable to the plantations of Southeast Asia or Brazil. Nor was the hacienda in every case a historical dead end. After 1880, Mexico's industrialization slowly got under way with the laying of the railroads. Many haciendas took the opportunity to introduce less restrictive work contracts, a greater division of labor in production, more professional forms of management, and a move away from paternalist social relations.[36] Such modernized haciendas, often very large in size, existed alongside a multiplicity of smaller ones that continued to operate as they had in colonial times. On the whole, the nineteenth-century Latin American hacienda was a monadic structure in which the *patrón* largely did as he pleased. Although the body of laws was often highly progressive, the police and judiciary seldom intervened in favor of the *peones*, who no longer had the existential security provided by a functioning village community. The *peones* should not be thought of as a "landless proletariat" in the style of plantation workers or East Prussian, Chilean, or African migrant laborers; they remained in one place, geared to the life of "their" hacienda. But neither were they a peasantry structurally tied to a village, in the Russian, Western European, or Indian sense. This is not to deny that Latin America had a migrant proletariat without land of its own or (and this is decisive) the opportunity to acquire some. The phenomenon was widespread in Argentina,[37] where workers (and tenant farmers) tended to be Italian or Spanish in origin, typically single or with a wife and children living in the city.

2 Factory, Construction Site, and Office

Workshops

Work may be categorized according to where it takes place. Many workplaces in the nineteenth century had changed little in comparison with earlier times. Independent artisans in Europe—and a fortiori in Asia and Africa—worked essentially under "premodern" conditions, at least until the introduction of the electric motor late in the century and the spread of industrial mass production. In other civilizations too, older models continued to govern the organization of workshops. Knowledge transmission and market regulation through guilds or other collective institutions, which survived longer in the Ottoman Empire and China than in Europe, differentiated artisans from ordinary workers. The growth of industry devalued the productive activity of many craftsmen, but there were also plenty of cases where a workshop adapted successfully to changed market conditions. All in all, the crafts lost less of their economic importance in the nineteenth century than in the twentieth. In Europe too, good-quality clothing was still largely being made by a tailor, shoes by a shoemaker, and flour by a miller. A broader definition of crafts might take in hybrid forms of self-help, collaborative effort, and professional partnership. This is how a majority of private housing was built in most parts of the world—from simple western European half-timbered buildings to the wide range of African types.[38] House building was "preindustrial," and some of its work routines remain so to this day.

A number of crafts first developed in the nineteenth century, while others acquired a new significance. The steady or rising number of horses meant that smithies, for example, were used right through the century, and heavy industrial steelworks actually appeared as a reinvention of their art at a higher energy level, though without individual craftsmen. In many cultures the blacksmith enjoyed high esteem or even mythical status, as master of fire, physical strongman, toolmaker, and weapons producer—although in India it was seen as lower-caste work. In large areas of sub-Saharan Africa, it was not a craft with an ancient history but started up in the eighteenth century and reached its peak roughly between 1820 and 1920. Smithies produced things that were both useful and beautiful; jewelry, for example, was a prestige item for accumulation, as was coinage in places where there was not a state monopoly. They had a high degree of autonomy, being largely in control of their own production process. The conventional image of the village blacksmith is rather misleading, since he might well have worked also to satisfy demand outside his local area. In the Congo, for example, many had a far-flung clientele that was both ethnically and socially diverse.[39] The need to acquire raw materials tied them into wider trading circles and encouraged them to cultivate numerous social contacts.

The Shipyard

Any workplace might have assumed a new shape in the nineteenth century. One example is the shipyard, known for millennia in various civilizations as a hub of craft cooperation, and already in the early modern age it was one of the main sectors of large-scale enterprise in countries such as England, France, and the Netherlands. In those days it had been a domain of carpenters. Then shipbuilding became a leading branch in the industrialization process, and by 1900 it was one of Britain's most important industries, with a dominant position in the world market thanks to the productivity of the Scottish yards. This entailed radical changes in technology, which did not happen all of a sudden; only in 1868 did the total tonnage of *new* iron ships exceed for the first time that of wooden ones launched from British yards.[40]

Shipwrights and the new technicians or workers engaged in iron-hull construction had different types of social organization (the former trade still being rather closed) and for some time lived and worked alongside one another.[41] The switch from wood to iron did not occur everywhere, but it was by no mean confined to the West. The Indonesian shipbuilding industry was able to execute it at a time when the yards of the Dutch mother country were capitulating in the face of British competition.[42] Shipyard labor was overwhelmingly male and quite highly skilled, providing fertile ground for the early political organization of the working class, often in conjunction with other groups of workers in a port city. In some countries, such as China, shipyard and arsenal workers formed the oldest core of the industrial proletariat.

The Factory

The main novelty of the nineteenth century was the factory, in its dual nature as large production site and field of social activity.[43] Cooperative forms and power hierarchies first took shape here, before spreading to other parts of society. The factory was purely a locus of production, physically separate from the household; it required new habits and rhythms of work and a kind of discipline that left only limited meaning to the idea of "free" wage labor. Factory organization involved a division of labor adapted in sharply varying degrees to the capabilities of the workforce. Experiments with ways of making labor more efficient began very early on—long before 1911, when the American engineer and first high-profile management adviser Frederick Winslow Taylor developed a theory of psychophysical optimization, dubbed "Taylorism," to speed up labor processes and bring them under stricter "scientific" control.

The factory was also new in a more mundane sense in places where it appeared for the first time in history. Factories were not necessarily located in cities. Often the reverse was true: a city grew up around factories. Sometimes the factory remained as a freestanding complex in the "countryside," as in Russia, where in 1900 more than 60 percent stood in a nonurban setting.[44] In extreme cases, new factory sites became "total institutions," in which the owner provided food and board for the workers and largely sealed them off from the rest of society.[45] Such things did not happen in Russia alone. In the "closed compounds" introduced in 1885 into the South African diamond mines, black miners were housed in barracks or locked up under prisonlike conditions.[46] The concept of an autonomous factory world should not, however, be seen as entirely negative. On occasion a patriarchal-philanthropic entrepreneur, such as Robert Owen in Scotland, Ernst Abbe in Jena, or Zhang Jian in Nantong in southern China, attempted to create spaces for social improvement in the shape of model industrial communities.[47]

The first generation of factory workers were not always recruited from the surrounding area; those who went to work in the Ukrainian Donbas region, for example, often came from far away.[48] Entrepreneurs delegated the task of advertising for labor to local contractors, who might then spread their net far and wide to haul in workers from rural areas. Contract work existed almost everywhere that a culturally alien management confronted a mass of unskilled workers without being able to rely on an existing labor market. In return for a flat fee, the local contractor "procured" the necessary numbers to work for a fixed wage over a specified period of time; he was also responsible for their good behavior and therefore often served a disciplinary function, as well as lending money at an unfavorable rate to those dependent on him. No special skills, beyond a basic dexterity, were expected of the workforce during this first phase of light industrial development, and so the contractor did not have to select too carefully. Such a labor market substitute could be found in China, Japan, and India as well as in Russia and Egypt.[49] One of the earliest demands of the workers' movement in these countries was for a ban on the much-hated system of contract labor. In any case it proved to be a transitional phenomenon: management did not usually insist on keeping such indirect forms of control, since they prevented it from developing its own personnel policy. If the formation of the initial workforce became stalled in the circuit between village and factory, the workers' mentality might retain rural features for a long time to come.

Beyond all regional and cultural variants, the factory was everywhere associated with similar constraints for those who worked in it. In addition to better-known European or North American examples, a few from India and Japan might serve to illustrate what Jürgen Kocka called the "wretchedness of early factory work." In Japan the number of mechanical silk-spinning mills quadrupled between 1891 and 1899, most of them in the silkworm-producing regions in the center of the country. The workers, almost all of them women, typically came from families of impoverished tenant farmers; many were really still children, almost two-thirds being under the age of twenty.[50] Recruited by contractors who usually paid their wages straight to the family in their home village, most spent less than three years in the factory working in appalling conditions: accommodation in supervised dormitories, meager rice meals with nothing more than a few vegetables, a fifteen- to seventeen-hour workday with very short breaks, exposure to sexual violence. The work was monotonous but required constant attention; accidents were common at the cauldrons where the silkworm cocoons were boiled. Such factories were the worst breeding ground for tuberculosis.

Things were no better in the cotton industry, which experienced a boom around the same time and soon became an even larger employer of female labor. One of its distinctive features was exhausting night shifts, with a fourteen-hour day as the norm until 1916. Amid deafening noise and in air filled with toxic fumes, women slaved away on machines that claimed a continual toll of victims. Foremen ensured work discipline with the help of canes and whips, and it was only after 1905 that some positive incentives were gradually introduced. Here too, workers lived in prisonlike dormitories that were poorly ventilated and often without individual bedding. With virtually no provision for medical care, ill health resulting from the work conditions ensured that three-quarters of the women lasted less than three years in the factory.[51]

The variations were endless because of the very different surroundings in which the early factories sank root, but nearly everywhere the appearance of factories triggered a major shakeup of the labor market, redistributed people's life chances, and established new kinds of hierarchies.[52] The break with the past did not necessarily come with the very first factories, but rather with the first ones that managed to stabilize themselves and reach a sufficient size to make a wider impact on work organization. The decisive threshold to a new type of society was reached with the consolidation of a full-time labor force. A large number of people became industrial workers and were nothing other than that.[53]

For a long time the image of work in the nineteenth century was based on the "leading sector" iron and steel: that is, on heavy industry. Adolph von Menzel's painting *Iron Rolling Mill* (completed in 1875) strikes everyone who has seen it as a distillation of the age.[54] But in 1913, steel production was still fairly uncommon around the world, concentrated in a few countries and a few locations within them. The United States was by far the largest producer (31.8 million tons), followed by Germany (17.6 million) and then, well behind, Britain (7.7 million), Russia (4.8 million), France (4.7 million), and Austria-Hungary (2.6 million). Japan was still below 300,000 tons. In many countries there was one isolated steelworks (the Tata operation in India or the Hanyeping plant in China) but not an entire steel industry employing significant numbers of workers. No steel was produced anywhere in Africa, Southeast Asia, or the Near and Middle East (outside the Ottoman Empire), or in the Netherlands, Denmark, or Switzerland.[55] Only a tiny part of the working population worldwide was familiar with the most spectacular branch of production in the early industrial age.

Canal Construction

Less new, though hardly less emblematic of the age, and geographically more widespread than heavy industry, was a second kind of workplace: the large construction site. It had, of course, long existed, since the building of the pyramids, the Great Wall of China, and the medieval cathedrals, but in the nineteenth century major construction sites became more common and even larger in size. The main purpose of monumental building was no longer to glorify worldly and spiritual power, but rather to create basic infrastructure for the life of society.

Before the railroad came canals. Those of eighteenth-century England were built not by a fixed proletariat but by migrant workers, often of foreign origin and on a subcontractor's payroll. In the United States, the eight decades between 1780 and 1860, especially the 1820s and 1830s, were the great age of canal construction. A total of forty-four were built during this period in North America, and by 1860 there were approximately 6,800 kilometers of navigable canals in the United States.[56] At midcentury canal construction was among the most advanced industries, requiring the investment of large capital sums and cutting-edge technology, as well as the organization and disciplining of huge numbers of workers. It was *the* large-scale enterprise of the epoch—it opened new markets and called for business strategies of a new order. At the same time, it was an activity of great symbolic significance: no longer was the earth the sole preserve of farmers and miners; the arteries of capitalism were now being dug deep inside it. This was a new, often especially harsh experience in the world of work; the path from workshop to factory was not the only one taken in the nineteenth century. An army of mainly unskilled workers, drawn from the most diverse sources, came together for the construction of the American canals: job seekers from rural areas, new immigrants, slaves, free blacks, women, and children. They all lacked power and status, and control over their work conditions. The chances for solidarity were small, and organized workers' movements did not arise out of such kinds of work. Besides, canal workers were geographically marginal; their lifeworld was the construction site and the barracks camp.

In comparison with river regulation in the eighteenth century, the canal projects were truly gigantic. Whole areas had to be cleared, swamps drained, trenches dug, inclines secured, cliffs blown up, bricks produced and laid, and sluices, bridges, and aqueducts constructed. On an average day, lasting twelve to fourteen hours in summer and eight to ten in winter, ordinary digging cleared seventy wheelbarrows of material per worker.[57] It was pure drudgery, distinctly worse than the round of farm labor. And since the contractors were paid by results, they kept up a quasi-industrial pressure on the workers. At first horses were the only other source of energy; machines began to play a greater role only in the Suez Canal operations. Most demanding of all was the Erie Canal, a project of huge economic importance between 1817 and 1825 that involved digging a total of 363 miles from Albany to Buffalo.[58] Accidents were a frequent occurrence, as were outbreaks of mosquito-borne malaria, dysentery, typhoid fever, and cholera. Medical care was rudimentary, support for the bereaved or the incapacitated nonexistent. Canal construction was the grim material foundation of the rise of the United States.

Most spectacular of all was the Suez Canal, proposed as early as 1846.[59] In November 1854 the Egyptian authorities granted Ferdinand de Lesseps a first concession that enabled him to set up a fundraising company, and on April 25, 1859, after nearly two years of survey work, construction officially got under way at the Port Said lido. On August 12, 1865, the first convoy of coal ships completed the passage to the Red Sea. The Suez Canal Zone was marked out in February of the following year, and in July 1868, a regular train service began between Ismailia and Cairo. When the waters of the Red Sea were allowed into the Bitter Lakes on August 16, 1869, the ten-year-long construction program over a stretch of 101 miles was all but complete. The Suez Canal opened to shipping on November 20, 1869.

The canal was a French private venture, although the Egyptian government put up half the capital, incurring a huge debt that would be among the factors behind the British occupation in 1882. The construction site was one of the century's largest, intricately organized with a resident director-general at the top and a hierarchy of bureaucrats and engineers modeled on the French Ponts et Chaussées. The environment posed problems different from those encountered in America. Extreme temperatures meant that it was a challenge to ensure the workers' water supply: in April 1859 a Dutch firm installed a number of steam-powered desalination plants, but their high coal consumption made them unviable and water had to be brought in by barge and camel from Damietta. The pasha's very first *firman* stipulated that four-fifths of the labor had to be Egyptian. Unpaid corvée duty had normally been required of the fellah population since antiquity, but only on irrigation works in their local region. (This was not just a token of "Oriental" backwardness—witness the fact that until 1836 every peasant in France had been obliged to spend three days a year maintaining the roads in his local area, or that until the 1920s indigenous people in Guatemala were assigned to perform compulsory [paid] labor.)[60] In the case of the Suez Canal, fellahs had to be assembled from far and wide, and the Suez Canal Company, mindful of public opinion in France, initially advertised for free laborers in every Egyptian mosque, railroad station, and police station and handed out leaflets in villages as far away as Upper Egypt, Syria, and Jerusalem. This had little success, however, and most of those who took the offer soon melted away because of the terrible working conditions (digging in the mud of shallow lakes, etc.). European workers from places such as Malta were even harder to obtain. There were plans to hire as many as 20,000 Chinese—a remarkable idea at a time when "coolie exports" were just beginning.

Only when all else failed did de Lesseps and the khedive resort to the corvée, which was introduced in grand style in January 1862. But while the pasha supplied the promised labor, the company's French subcontractors did not keep their side of the bargain; the wages, if paid at all, were inadequate and often in the shape of useless French francs and centimes. A normal working day lasted seventeen hours, and no care was provided for workers who fell sick or suffered an accident. Discontent kept mounting, and more and more fellahs took to their heels. The forced labor in Egypt aroused revulsion among the British public, becoming an important weapon in Whitehall's attempts to sabotage the canal project. After all, the serfs had been emancipated in Russia and slaves freed in the United States. Under British pressure, the sultan—as the pasha's nominal overlord—banned the use of the corvée. In July 1864 Napoleon III declared an arbitration award that was accepted by both sides: French firms would cease to employ Egyptian forced laborers at the end of the year, but the Egyptian authorities agreed to sponsor their use for ancillary tasks. Figures are lacking for the total workforce, but it is estimated that 20,000 fellahs were taken on each month, and that altogether 400,000 Egyptians worked on the canal project.[61] The most important tasks, however, were performed by freely recruited workers. Fellahs working under the corvée system could be deployed only for short periods and in the vicinity of their home village. Any who came from Upper Egypt wasted half their requisition time on the journey.[62]

The construction of the canal drew on resources from many countries: coal from England for the steam dredgers and pumps (12,250 tons of coal a month were used for the final, technically most difficult, phase beginning in late 1867), wood from Croatia and Hungary for the construction of camp barracks, technical equipment and standard items of iron hardware from France. The accommodations improved over time, although the European camp for engineers remained strictly separate from the "Arab" tent city for the workers. Health considerations were an issue right from the start. Several hospitals in the new city of Ismailia and near the construction sites, as well as a fleet of ambulances, looked after Egyptians too. Preventive medicine and the quality of food also grew better—partly to take the wind out of the sails of British and other critics. All in all, a massive technological-administrative system took shape. The ceremonial opening of the canal took place on November 16–20, 1869, in the presence of Empress Eugénie and her itinerant court, Emperor Franz Joseph of Austria, and quite a number of European crown princes. Ismailia, usually a city of 5,000 souls at the most, received some 100,000 visitors. The khedive invited thousands of guests at his own expense; travel agencies organized tourist trips to the event of the century; public speakers and newspaper leaders compared Ferdinand de Lesseps, now sporting countless medals, to the greatest heroes in history.[63] Giuseppe Verdi, under contract to compose *Aida* for the canal's opening, had not exactly been idle, but he was able to deliver it only after the opening; the premiere took place on Christmas Eve 1871, before an international audience in Cairo.

Railroad Construction

While fellahs dug up the desert en masse, railroad tracks and stations were being built in many parts of the world. In principle, the work involved the same technical tasks in every continent: an accurate survey of the terrain, the installation of high-specification bridges and tunnels, and the development of civil engineering specialists in the field. The earth-moving requirements were greater than in road construction projects up to then; it was not uncommon for as many as 15,000 workers to be employed at the same time in coordination with one another. Construction work on the railroads, as on canals, involved basic manual labor with ax or shovel, but also highly advanced equipment such as steam cranes.[64] The Transcontinental Railroad from Chicago via Omaha (Nebraska) to Sacramento (California)—which was completed in 1869, the same year as the Suez Canal—deployed groups of workers as large as one of the smaller Civil War armies; indeed, it became a catchment basin for demobilized soldiers in the years immediately after the end of the conflict.

The Transcontinental also hired approximately 100,000 Chinese workers—although when this was first proposed, many had thought they would be physically too weak. "But they built the Great Wall," was the answer of the chief engineer, Charles Crocker.[65] Recruited via contractors, the Chinese were organized into gangs of twelve to twenty, each with its own cook and a responsible "headman." They proved to be capable workers, not only in implementing Western plans but also in finding solutions to difficult problems that cropped up along the way. They took charge of providing adequate food for themselves, and their custom of drinking tea and hot water reduced the number of accidents and protected them from many of the diseases that plagued Europeans. Unlike the Irish, who were also employed in large numbers, they did not have a problem with alcohol. They smoked opium only on Sundays, and violent quarrels and strikes were virtually unknown among them. However, though considered hardworking and conscientious, the Chinese were also treated with racist contempt.

Skillful teams of tracklayers could cover as much as three miles a day. Then came the hammer and screw men, performing a mechanical chorus like that of the hammering Nibelungs in Richard Wagner's *Rheingold* (composed in 1853–57, at a first climax of railroad construction in central Europe). Each mile required four thousand nails to be driven home with three hammer blows each. Once it was fully operational, in May 1869, the Transcontinental made it possible to travel from New York to San Francisco in seven days. It was the last major engineering project in the United States to be executed overwhelmingly by manual labor.

The great railroads of the world took shape along construction sites that were transnational in character.[66] British and French capital were dominant before 1860, but afterward national sources of finance made increasingly important contributions. The materials, craft labor, and technical know-how were seldom only local; European and North American planners and engineers everywhere monopolized the higher rungs of the jobs ladder. Skilled workers with experience were also in great demand. Only a few of the countries engaged in railroad construction had the heavy industry and machinery sector necessary to organize it by themselves. Even Witte, the Russian finance minister who wanted to construct the Trans-Siberian with national resources—and who was largely successful in his aim—could not dispense entirely with American steel. Local peasants helped with the construction in western Siberia, but as the work moved farther eastward the terrain became more difficult and the human resources harder to obtain. Laborers were recruited from European Russia, along with numerous Kazan Tatars and foreign workers from Italy and elsewhere. Soldiers were drafted in to work on the Ussuri section, alongside some eight thousand Chinese, Korean, and Japanese migrant workers. Prisoners were transported from Ukraine and other parts of the empire, and after a certain period they were even paid the full wage. It was a use of convict labor that prefigured Stalinist practices.[67]

In Rhodesia, where the peak years were from 1892 to 1910, the railroad labor force consisted of men from all over the world, including not a few Italians and Greeks. The highly skilled white workers were recruited in Britain, less-skilled ones in South Africa. In many places, Rhodesia and India among them, railroad companies went on to become the largest private employer outside agriculture. The Indian railroads were the largest construction project in Asia in the nineteenth century, as well as the greatest single capital investment in the British Empire. By 1901 the country's 25,000-mile network, fifth in the world behind the United States, the Tsarist Empire, Germany, and France, was longer than those of Britain (22,000 miles) and the Danubian Monarchy (23,000 miles) and not much shorter than the French (24,000 miles).[68] The construction work, which began in 1853, engaged more than ten million workers over the next five decades (the peak figure was 460,000 Indians in 1898). This unique labor density, roughly three times higher than on British sites,[69] was attributable in part to the large percentage of women and children; it was thought preferable to employ whole families, which could be had at rock bottom wages and often came from a landless underclass with no village ties. Many of these were, so to speak, unskilled "professionals," moving from site to site as and when they were needed. There are no precise statistics on the matter, but the number of human lives sacrificed in India's railroad construction must have been exceptionally high. It was more dangerous than the unhealthiest factory labor.[70]

On all continents, new labor markets with a translocal and often global reach took shape around the railroad construction sector. Many large sites tapped the vast pool of unskilled labor in Asiatic agrarian societies. But there was also a need for qualified train drivers, conductors, signalmen, and repair-shop technicians. This opened up new opportunities for local people below the color bar that, though always shifting, was never absent in the colonies. Such a move up the social hierarchy might be associated with demands of a nationalist inspiration. In Mexico, for example, before the revolution that began in 1910, local workers fought to gain access to highly skilled positions on the US-financed railroads. A new railwayman's habitus developed all over the world, most striking where the railroads were state owned and their officials came to represent public authority.

The Ocean Workplace

Ships were another typical workplace of the nineteenth century.[71] It is hard to imagine today, in the age of giant tankers and tiny crews, but sailing ships required a large, mostly unskilled, workforce. Long before industrialization, free wage labor was the norm on Europe's oceangoing merchant ships.[72] The early days of the steamship changed little in terms of crew size. Since passenger and freight volumes were increasing at the same time, including on rivers such as the Rhine, Yangtze, and Mississippi (Mark Twain closely described work routines from personal experience in his *Life on the Mississippi*, 1883), the ship reached its zenith as a workplace in the nineteenth century.[73] But it remained what it had been in the early modern period: a cosmopolitan space, with men recruited from all around the world. It was also, along with the army and the plantation, the workplace most heavily charged with violence: flogging was not banned on US ships until 1850; use of the cat-o'-nine-tails—an especially brutal form of chastisement—was permitted in the Royal Navy until the mid-1870s; and officers directly inflicted violence on sailors in the merchant navy too. Finally, the ship was an extremely hierarchical and segmented social space, with the quarterdeck an unmarked area reserved for the captain and the forecastle an inferno for the crew.

For all the *Moby-Dick* romanticism, whaling ships—together with premodern mines (still notoriously dangerous around 1900 even in an otherwise highly modern United States) and the Peruvian islands from which guano excrement was collected—were among the most unsavory workplaces imaginable, especially if, as in the Australian case, their main hunting grounds were far from home. In 1840 it was not unusual for a whaling expedition to last four years, with ports of call few and far between. The record was held by the *Nile*, which in April 1869 put into its home port in Connecticut after eleven years at sea. The food was usually appalling, the sleeping quarters cramped, and the medical care utterly minimal, while those who braved the dangers of the trip were subject to the discipline of an all-powerful captain. The dead whales, as much as ten tons in weight, were used for many purposes. No place was safe from the blubber-fueled fires that burned continually under the giant oil boilers; comparisons with hell came naturally to those who experienced the scenes on board. One of the reasons for the decline of whaling was that less-insalubrious job opportunities were available in other sectors.[74]

Office and Home

The office as such is not a nineteenth-century invention. As soon as there is a bureaucracy, its staff must have somewhere to sit—and so the office has existed in every civilization with a system of writing. In the Imperial Palace in Beijing, one can still see the austere workrooms used by high officials, and it is easy to imagine that they looked no different many centuries ago. The great East India companies ran administrative headquarters in London and Amsterdam, which required secretaries and administrative staff to handle the huge quantities of paper communication.[75]

What was new in the nineteenth century, especially after around 1870, was the bureaucratization of enterprises above a certain threshold in size; the male and female employee ensconced in their office workplace became an ever more important social category. However, "employee" is only a very broad and formal term covering the "white-collar" work situation of all those who did not have to get their hands dirty—from managers to humble bookkeepers to the female secretaries who proliferated from the mid-seventies on with the spread of the typewriter. The lower one's position on the ladder, the smaller was the scope for individual initiative and the greater the share of predefined executive tasks. Employees were also to be found in large industrial enterprises, especially the accounts and engineering departments, and they were predominant in sectors such as wholesale and overseas trade, banking, and insurance, where there were hardly any manual workers. The spread of white-collar activities created new functional and gender hierarchies. The market for female labor grew in this "tertiary" sphere—which included the retail trade, from small shops to department stores—more rapidly than it did in the "secondary" sector of commerce and industrial production. This cannot really be described as a "feminization" of given types of work, since women often found employment in *newly created* occupations. They had few opportunities to move into higher jobs. Women worked where the male management wished to put them.[76]

Outside Europe and North America, the first employees were in branches of foreign companies. As all these agencies operated in an unfamiliar business environment, they were almost always forced to employ some local people in managerial positions, naturally paying them a lower salary. So arose the figure of the "comprador," especially striking in China, but also important elsewhere— originally an indigenous merchant, in good repute and with some capital of his own, who was employed on a temporary basis by a European, North American, or even Japanese company. He took care of local business contacts, vouched for the trustworthiness of suppliers and customers, and had responsibility for a local workforce that he recruited and paid himself.[77] In the 1920s, some large Chinese enterprises, mainly banks at first, sowed the seeds of a local employee class by adopting a combination of indigenous and Western business principles. In Japan this process took place a few decades earlier, as the country took the lead in economic modernization and bureaucratization.

Whereas the white-collar job became a typical form in the West only with the expansion of the service sector and the bureaucratization of large-scale industry, domestic employment was one of the oldest economic activities in the world. Servants have existed always and everywhere in the homes of the rich and powerful, their number being an important indicator of status. In all civilizations, court life rested upon the service labor of thousands. Little changed in this respect so long as courts and "stately homes" remained in place, as they did worldwide through-out the nineteenth century. But in many countries there was also a fast-growing demand among the urban bourgeoisie for regular cooks, nannies, coachmen, and the like. At the beginning of the nineteenth century, many intellectuals from modest backgrounds were still able to—or had to—struggle along as private tutors; such was the case with the great German poet Friedrich Hölderlin, who never had any other position—in contrast to his friend Georg Wilhelm Friedrich Hegel, who went on to become a professor of philosophy.[78] By the end of the century, however, the spread of higher-quality public education meant that such individualized tutorships were more or less a thing of the past. At least outside the Tsarist Empire, where quite a few performers in distinguished string quartets or orchestras still had serf status in 1850, musicians also ceased to be in the ser-vice of the nobility, as Joseph Haydn had been for three decades with the Princes Esterházy—although it is true that King Ludwig II of Bavaria, Richard Wagner's patron, still allowed himself the luxury of an in-house quartet.[79]

Meanwhile, other forms of employment acquired a new salience. In Europe, the middle and upper bourgeoisie differentiated itself from the petite bourgeoi-sie in nothing as clearly as the employment of domestic servants, or at least of a maid. This was a token of luxury, however modest, and one of the most striking status symbols in the societies of the West.[80] Justified as complaints of exploita-tion often were, the position of maid offered young women from the country a chance to gain a foothold in the urban labor market under relative secure condi-tions. A life of cooking and washing was not necessarily an unacceptable alter-native to factory work or prostitution. In large Russian cities toward the end of the century, for example, most women newly arrived from the villages went into domestic employment rather than industry. In Moscow in 1882, more than 39 percent of households had one or more servants; the comparable figure in Berlin was around 20 percent.[81] The phenomenon continued to grow in importance, so that by 1911 it was the most extensive occupation recorded in the British cen-sus: 2.5 million domestics outside agriculture, as against 1.2 million employed in mines and quarries.[82] In the United States, even in the economically most developed Northeast, domestic service was by some way the largest female occu-pation in the third quarter of the nineteenth century. For black women, a small minority, there were scarcely any other options.[83]

Since the maid was often the only actual employee in less-affluent households, she differed in both function and gender from the hierarchical staff of a "stately home." There seems to have been a general trend to the feminization of domestic labor in the nineteenth century, but it was not everywhere as pronounced as it was in a few European countries. Female employment in private urban homes was especially attractive where agricultural work was on the decline and new opportunities were not yet sufficiently available in industry or the rest of the service sector.[84] Outside Europe and North America, households with a large staff of mostly male domestics remained somewhat longer. If market supply and demand was not the only regulator, it could sometimes become a burden for householders to maintain a large number of lackeys and dependents. In some societies—China, for instance, where concubines were a feature and adoption was common and easy to arrange—the boundary between family member and staff was more fluid than in Europe. In the colonies, the lowliest white representative of the state or a private company had a host of "boys" and other servants at his disposal; the availability of such cheap labor in Asia and Africa was one of the most prized benefits that came with the colonial way of life.

Domestic service was an important source of employment in areas where there was no shortage of willing and suitable people. The maid or butler had to be capable of communicating smoothly and correctly with Sir and Madam and their guests; some of the upper-class glitter had to rub off onto the staff. The solitary African servant was a rare exception in a European bourgeois household. The globalization of domestic service—Polish cleaners in Berlin, Filipino maids in the Gulf States of Arabia—came only in the late twentieth century. But migration in the opposite direction took place on a lesser scale in the nineteenth: European countries, especially Britain, exported governesses all around the world, and these became important agents of cultural transfer. The housekeeper or nanny-cum-teacher of young children was highly regarded not only among European expatriates but also in wealthy Oriental homes, where offspring were expected to speak English and/or French, to play the piano, and to master Western table manners. Sometimes, as in Istanbul, some uneasiness was felt in orthodox quarters about the corrupting influence of Christian staff in a Muslim household.[85] A governess in Europe and beyond occupied a high position in the service hierarchy, one of the most prestigious open to "honorable" women from the middle layers of society, so long as they had few opportunities to become a schoolmistress or college teacher.[86]

3 Toward Emancipation: Slaves, Serfs, and Peasants

Free Labor

In liberal economic theory, whose basic concepts mostly stem from the "long" nineteenth century, labor is free and obeys only the market laws of supply and demand. People are not compelled to work; they react to incentives. If this is meant as an actual description of reality, then some reservations are in order for the twentieth century. The Soviet gulag, its Chinese equivalent, and the Nazi camp system are the largest complexes of forced labor known to history. Only in the last few decades, has the world been largely free of such systems, although new forms of extreme heteronomy—sometimes referred to as "neoslavery"—have been gaining ground in the wake of globalization. In this respect too, the nineteenth century was an age of transition, when a historically new tendency toward free labor was set in motion. "Free" labor can be defined with reasonable clarity only in a formal legal sense: that is, as a contractual relationship agreed to without direct external constraint, in which employees hand over use of their labor power to employers in return for monetary compensation, usually for a specific period of time. In principle this relationship may be terminated by either side, and it does not give the employer any further rights over the person of the employee.

By 1900 such a concept of labor was taken for granted in most parts of the world, but in 1800 that was by no means the case. The same may be said if we use a definition that goes somewhat beyond wage labor: namely, free labor is what is performed without any restriction on the civil liberty or physical autonomy of the worker. In the early modern period, slavery (in its many different forms) was an important social institution in half the world: in North and South America including the Caribbean, in Africa, and in the whole Muslim area. China, Japan, and Europe were basically free of it, although the latter pursued it all the more vigorously in the New World.

If we use the more general concept of "servitude," then it encompasses at least four other forms in addition to slavery: (1) serfdom, (2) indentured service, (3) debt bondage, and (4) penal servitude.[87] These terms have universal applicability, but they are Western insofar as the dividing lines between them are less clear in other social contexts than in a Europe that had passed through centuries of Roman law and its training in disambiguation. In Southeast Asia, for instance, the passage from various shades of dependence to outright slavery was much less abrupt.

At least one of the basic categories, however, existed almost worldwide. In Europe in 1800 there was little debt bondage (on the other hand, people soon ended up in a debtors' prison), but serfdom was still widespread. In India the opposite was true. Australia was at first nothing other than a penal colony. So, in 1880—after the beacon of liberty from the French Revolution had faded—forms of legally sanctioned servitude were by no means discredited. Leading liberal nation-states with a constitutional political system, such as France and the Netherlands, abolished slavery in their colonies only in 1848 and 1863 respectively: France in the course of a revolution, the Netherlands because its plantations in Surinam were in danger of becoming unprofitable and reproduction of the slave population was causing difficulties. The tortuous imposition of "free" labor, with its pauses and setbacks (e.g., Napoleon's reintroduction of slavery in 1802 in the French colonial empire), was a complicated process that may be broken down into several strands in the European-Atlantic area.

Slavery[88]

In the early modern period, European slave production—which had largely disappeared from the Occident—was revived on a large scale in the American colonies and established within the framework of a highly productive plantation economy. The slave labor was imported from Africa, after the indigenous population in various parts of the Americas had been wiped out or proved itself unfit for heavy work duties, and after experiments with the European underclasses had ended in failure.[89] This tropical and subtropical plantation economy produced goods for European luxury consumption, such as sugar and tobacco, in addition to cotton, the main raw material in Europe's early industrialization. The first critique of slavery and the transatlantic slave trade, on which the human-devouring plantations depended, appeared in nonconformist Protestant (particularly Quaker) milieux; a broad abolitionist movement grew out of this on both sides of the Atlantic.[90] Its first two successes came in 1808, when Britain and the United States (independently of each other) declared the international slave trade to be illegal. From then on, the United States imported no more slaves, while Britain closed its colonies for new slave shipments, stopped the trade on all British ships, and assumed the right to deploy its naval power against the transportation of slaves by other countries.

Slavery was first destroyed in Saint-Domingue/Haiti during the revolution of 1791–1804. In all other cases its disappearance was brought about not by a slave revolution but under the pressure of liberal forces in the respective colonial metropolis. The illegalization of slavery in the European colonies began in 1834 in the British Empire and ended in 1886 in Cuba. The Latin American republics already prohibited it during the independence struggles, but slaves did not make up an appreciable part of the population in any of those countries. In Brazil the last slaves were emancipated in 1888, while in the United States the process dragged on for more than eighty years. Pennsylvania was the first of the North American colonies to come out against slavery, in 1780, and over the following decades all the Northern states in turn passed antislavery legislation. At the same time, the slaveholder system consolidated itself in the Southern states, reaching a climax amid the worldwide cotton boom. Its extension to new western parts of the federation became a central bone of contention and in 1861 eventually led to the secession of the Southern states and the beginning of the Civil War. At the end of the conflict, slaves were legally emancipated throughout the United States.

In countries where millions of African slaves worked in the most dynamic sector of the economy, slavery was anything but a disturbing relic of early modern times. In the Southern part of the United States and in Brazil, Cuba, and a number of other Caribbean islands, it was the fundamental social institution for as long as it lasted; the master-slave relationship expressed itself everywhere in daily life and shaped how people thought of society. Slavery, however one understands it, is an all-embracing form of existence: the slaves cannot define themselves in any other way, and the same is true of the slave owners, who have the institution to thank for their more or less luxurious lifestyle. In essence, however, slavery in the Atlantic world was a relationship geared to the exploitation of labor and should be discussed as such.

Atlantic slavery was no neo-archaic repeat performance of Ancient Roman slavery. The latter had not been based on racial hierarchization and had been closely bound up with a near-total authority of the paterfamilias over all members of his household, such as no longer existed anywhere in the lands of early modern Christendom.[91] Yet there were some legal similarities across the ages. In the Atlantic area, too, slaves were the property of their master: he had the right to use their labor power without limitation of time and to employ violence in extracting it; and he was under no obligation to reward his slaves or to provide for their keep. The general laws of the land did not apply to slaves, or they did so only in a highly restrictive sense. As a rule, slaves had no protection against the master's violence; they could be sold to a third party regardless of their family circumstances—a practice that was described to devastating effect in Harriet Beecher Stowe's bestseller *Uncle Tom's Cabin* (1852). Slave status usually lasted a lifetime and was often passed on to the children of female slaves. Resistance and flight were criminal acts and subject to the severest punishment.

Such was the basic model of Atlantic slavery—a very strong form of dependence, in which slaves were particularly lacking in rights. Historians have long debated whether the reality on the ground always corresponded to this extreme: whereas abolitionists, for tactical reasons and out of moral revulsion, presented slaves as mere objects, recent studies have uncovered a cultural richness in slave communities and shown that some scope still existed for individual personal action and life choices.[92] Yet the fact remains that in the West (and much more rarely in China and very seldom in Japan), in the first half of the nineteenth century and sometimes beyond, millions of people worked under conditions that could not have been further removed from the moral and economic ideal of "free labor" propagated by liberalism at that time. Nor did they do this in archaic or backward spheres of the national economy. There is some evidence—though the question "remains more or less open"[93]—that both in the Caribbean on the eve of British abolition and in the antebellum Southern United States, the slave plantation *could have been*, though was by no means necessarily, an efficient, profitable, and therefore economically "rational" form of production.

There was no short or direct path from slavery to the realm of freedom. Slaves freed by a stroke of the pen did not immediately enjoy new positive rights and an economic basis for existence; nowhere did they in fact become fully entitled citizens overnight. In the colonies they remained at first political minors, like the rest of the population, although in the British ones there was some possibility for their interests to be represented. In the United States universal male suffrage was introduced everywhere by 1870, without regard for skin color, but from the 1890s on the Southern states almost completely devalued this by way of special discriminatory rules (arduous registration procedures, property and education hurdles, and so on).[94] It took a whole century after the Civil War for blacks to achieve in practice their most important civil rights. In most processes of emancipation, former slaveholders were awarded some compensation: the state often paid out large sums, or ex-slaves might continue to perform labor duties during a transitional period.

Only in the United States was the end of slavery intertwined with the military defeat of slave-owning elites.[95] Here the outcome was most akin to a social revolution, with punitive confiscation of private property, but even so slaves did not change forthwith into free wage laborers in industry or small farmers on land of their own. Rather, the large plantations gave way to a sharecropping system, whereby the old planter remained in possession of the land and his ex-slaves had to share with him the fruits of their labor.[96] In general, the Southern plantocracy lost its ownership rights over human beings, but not over the land and other possessions. The plantation slaves thus became a landless underclass, subject to full-scale racist discrimination and the ever more precarious conditions of sharecroppers and wage laborers. Soon they were joined by poor whites, with a resulting intersection of class and race barriers.[97] In Haiti the plantations were physically destroyed during the revolution, and with them the whole sugar economy of a once highly profitable colony, so that to this day small farmers with tiny plots are characteristic of the country's agriculture. In the British and French Caribbean too, large-scale production did not continue without interruption. On some islands the planters managed to preserve the sugar plantations, using not so much ex-slaves as newly recruited contract workers ("coolies") to keep them going. As a rule, the slaves turned into allotment farmers, less discriminated against than their counterparts in the South of the United States, and enjoying at least minimal safeguards under British law.[98]

The lesson from the experiences of emancipation is that freedom was not an all-or-nothing matter; it came in various shapes and gradations. Whether someone was free or not was an academic question in comparison with the degree of freedom, what it could be used for, and which practices of exclusion, new or old, were in place.[99] It made a huge difference whether, as in Brazil, slaves were set free without any prior provision for their material existence, or whether someone showed a disinterested concern for them. Ex-slaves were weak and vulnerable people, without natural allies in society, and they needed some initial cushioning from the rigors of the struggle for survival in a market economy.

Serfs

In Christian Europe, especially north of the Alps, there had been no slaves since the Middle Ages. The characteristic type of servile labor was serfdom.[100] At the beginning of the nineteenth century, this institution survived mainly in Russia, where it had acquired a stricter form in the eighteenth century. To gain some idea of the scale of the phenomenon, we should recall that in 1860 just under 4 million people in the United States had the status of slaves: that is, 33 percent of the Southern and 13 percent of the Northern population. In Brazil, the number of slaves reached its peak in the 1850s (2.25 million, or 30 percent of the total population), a level close to that of the Southern United States.[101] The figures for serfs in 1858 in the Tsarist Empire (almost all of them in European Russia) were considerably higher: 11.3 million privately owned serfs and 12.7 million "state peasants," also not necessarily free. Serfs made up roughly 40 percent of the male population of Russia—or more than 80 percent if the state peasants are included.[102] The main demographic difference between Russia and the Southern United States at that time lay in the concentration of servile labor: in Russia it was not uncommon for several hundred serfs to be attached to a farm, whereas such an order of magnitude was rare on American plantations. Furthermore, the population of the Southern United States was far more urbanized, with a much larger share of whites who owned no slaves at all and many who had only a few domestic slaves. In 1860, only 2.7 percent of all Southern US slaveholders owned more than fifty slaves, while 22 percent of Russian noble landowners (*pomeshchiki*) had more than a hundred serfs at their disposal.[103]

Serfs were not slaves.[104] Those in Russia could mostly invoke certain land rights and engage in subsistence farming alongside their work on the landowner's estate. Usually a peasant from the locality, the serf had not—like the slave—been wrenched from the world of his origins and shipped across vast distances. Serfs remained embedded in the culture of the peasantry; they lived in their own village community. There was a sharper functional distinction between male and female labor under serfdom. Serfs had access to the landowner's patrimonial court system, whereas slaves were usually unable to invoke rights of any kind. In the realm of European law, serfs were accorded certain customary rights; slaves did not enjoy such rights. In short the serf was a peasant, the slave was not. No general statements can be made about the implications of the two systems; slavery tended to be harsher than serfdom, but it was not necessarily so in every case. Serfs in the narrower sense—that is, hereditary subjects according to Russian convention—could be sold, given away, or lost at the card table. They were not "tied to the soil" but theoretically mobile; they were therefore hardly less disposable than American slaves.

As the two systems were wound up at exactly the same time, they may be regarded as two strands in a wider process which, though not global, covered an area between the Urals and Texas. Russian serfdom, too, was profitable and economically viable. In neither country did capitalism yet appear as the most important solvent of traditional relations, although representatives of a new liberal-capitalist thinking expected that modes of production based on coerced labor would soon run up against their limits of expansion. In Western Europe, in the Northern states of the United States, and in the public opinion of the Tsarist Empire (mainly oriented to the Western model of civilization), there was agreement in the mid-nineteenth century that the permanent servitude of human beings innocent of any crime was an offensive relic of earlier times. The abolition of serfdom by decree in 1861was almost as revolutionary in a Tsarist context as Abraham Lincoln's Emancipation Proclamation on January 1, 1863—even if it was less of a frontal challenge to serfowners and gave them a say in implementing the new provisions.

At the time of emancipation, the freed black slaves in the United States seemed to have the better prospects for the future, since the Reconstruction policy of the victorious North was supposed to help ex-slaves acquire a respectable place in society. In comparison, the overcoming of serf status in Russia was quite a slow and gradual process, without the directness of America's new beginning or its grounding in principles held to be universally valid. Rather, it gave rise to a set of texts that were hard to understand and a confusing, spatially and temporally disparate series of rights and duties, which the justice system tended to interpret to the peasant's disadvantage.[105] Former owners of Russian "souls" received generous compensation, and the emancipated peasants were subject to a whole series of restrictions that continued to make their lives difficult. Only under the pressure of revolution in 1905 did the regime decree the abolition of all compensatory payments; and only in 1907 were the last debts remitted by law.

In 1900 or thereabouts, the two emancipation processes seemed to have worked out less differently than people had expected in the mid-sixties. They confirmed the rule that wherever slavery and serfdom were abolished, what first appeared in their place were not equality and prosperity but new, perhaps less oppressive, forms of dependence and poverty. In the United States, the well-intentioned Reconstruction broke down after a few years and gave way to renewed political dominance of the planters; a heavy price was paid for the failure to provide ex-slaves with land of their own. In Russia, the former serfs acquired legal title to roughly half of the land previously owned by the nobility—so that a new "peasant question" arose in place of the old. The twentieth century, from the time of Prime Minister Stolypin's agrarian reforms of 1907, then witnessed a series of experiments to solve the Russian peasant question, most of them not in the interests of the peasants themselves. Cautious attempts to foster a capitalist-style economy of middle-sized and large farmers were brutally cut short by the collectivization of 1928.[106] The emancipation of 1861 was no cultural revolution. It left untouched the raw customs of earlier times, doing little to raise educational levels or to reduce vodka consumption in a village life that was far from idyllic. "Emancipation," in the emphatic sense of the Western European Enlightenment, is therefore too strong a term to describe what happened.

Liberating the Peasantry

For liberal theorists, the Russian serf was unfree in two ways: he was the landowner's property, and he was trapped in the collectivism of the village commune. In 1861 the first of these ties was dissolved, and in 1907 the second. As for the rest of Europe, it is more difficult to say *from what* the peasantry was liberated. Attempts at defining particular types (such as "Russian-style" serfdom) should not mislead us into thinking that Europe can be straightforwardly divided into a free West and a servile East. There were many shades of servitude. Thus, the average situation of peasants in the mid-eighteenth century did not differ dramatically between Russia and German lands such as Holstein or Mecklenburg. In 1803 the German publicist Ernst Moritz Arndt still used the term "slavery" to characterize conditions in his home area of Rügen, an island in the Baltic Sea.[107]

"Freeing of the peasantry" commonly refers to the protracted process that by 1870, or at the latest 1900, had transformed the majority of European peasants into something they had not been a century earlier: citizens with the same rights as everyone else; legally competent economic subjects according to the respective national norms; tax and rent payers with no unwritten obligation to any "lord" to perform labor services. Such freedom was not necessarily bound up with land of their own; an English tenant farmer was better off than a northern Spanish smallholder. The key point was assured access to land under conditions favorable for the running of a business. This might come with a secure long-term tenancy, whereas the situation tended to be decidedly more unpleasant when landowners, profiting from a rural labor surplus, played off small farmers against one another in order to drive up rents. That kind of "rack renting" was a modern method with which completely "free" peasants might be confronted in Europe as well as in China. The old moral economy of rural life had involved a degree of paternalist care; its disappearance firmly yoked the peasant family to the vagaries of the market—unless governments, as they do in Europe or North America up to the present day, pursued an agrarian policy that gave the farmer some protection.

The freeing of the peasantry was a Europewide phenomenon, which by a legal definition was completed with the Romanian edict of 1864 but in actual fact lasted somewhat longer. It missed a few regions of Europe. In England—as Max Weber sarcastically remarked—"the peasantry are freed from the land and the land from the peasantry" by the enclosures of the eighteenth century,[108] so that the country's social structure at the dawn of the nineteenth century already presented the triad of large landowner, large tenant farmer, and agricultural laborer. The picture was similar in western Andalusia, where latifundia originating in the Middle Ages were mainly worked by day laborers (*jornaleros*) comprising as much as three-quarters of the farming population.[109]

Freeing of the peasantry meant an adaptation of rural society to general social and political roles that had only just taken shape. The peasant "estate" was stripped of its special character. It is fairly clear which forces lay behind this, but there is greater doubt as to their exact mix and the main causal factors. Jerome Blum, the great authority on the subject in a European perspective, saw the freeing of the peasantry (starting with the emancipation law of 1771 in the Duchy of Savoy) as the last triumph of Enlightenment absolutism.[110] He also noted that only in a few cases, most notably revolutionary France, was it carried through by a nonabsolutist regime. Yet it was precisely the French Revolution, and its diffusion by Napoleon, that in many cases was the impetus for a state-led initiative; often it was military disaster that pushed a monarchical regime to focus its attention on the peasant question. Prussia abolished serfdom in 1807 after its defeat at the hands of France. Failure in the Crimean War triggered the Russian reform package that included emancipation of the serfs; and the American Civil War was the cause of the liberation of the slaves.

Further factors were involved in the process of peasant emancipation: above all, a thirst for freedom going back to long before the French Revolution, when the peasantry had fought against restrictive "feudal" conditions and wrested some room for maneuver from various anciens régimes.[111] The specter of peasant revolt did not vanish from some parts of Europe in the nineteenth century. There is a direct parallel with the American planters' fear of slave uprisings following the bloody revolution in Haiti—a fear that took concrete shape in Jamaica in 1816 and 1823 and in Virginia in 1831 (Nat Turner's rebellion).

The freeing of the peasantry was almost always a reformist compromise; there were virtually no repetitions of the radical French solution of separating the aristocracy from the land. Landlord classes survived the freeing of the peasantry, and although their social and political position in most European countries was weaker in 1900 than it had been a century earlier, this was rarely due to a loss of their class privileges. For many landlords, the leeway was greater and the options clearer: either to go in for large-scale agribusiness or to withdraw into the passive existence of the rentier. Other goals and interests fed into the astonishingly convergent process of European peasant emancipation: the Austrian Crown, for example, even before the French Revolution, tried to increase its share of the agricultural surplus at the expense of aristocratic estate owners. Ironically, this very policy was often devised and supported by members of the *administrative* nobility drawing their main income from public service. However, even members of the landowning elite could veer toward a course of reform, especially if they sought political support from the peasantry, as they did in Poland to counter the effects of partition or in Hungary to resist Habsburg rule.

Finally, the overall development of society created a new general framework. Like New World plantation slavery, serfdom—above all, the "second serfdom" established in eastern Europe in the seventeenth century—had been a reaction to labor shortages. Rapid population growth in nineteenth-century Europe eliminated that problem, and at the same time urbanization and early industrialization opened up new job opportunities for people from the countryside. Labor markets thus became more flexible and needed much less coercive stabilization, increasingly difficult to impose, as it was, from an ideological point of view. In countries where something like "feudal" ties of dependence had still existed in the eighteenth century, large parts of the rural population were freed from extra-economic obligations to their landlord.

The results were varied. The peasant's lot improved most markedly in France, and things worked out quite well in Austria. Prussia and Russia made fewer concessions to the peasantry. At the other end of the spectrum were Pomerania, Mecklenburg, and Romania, where peasants were not substantially better off at the end of the century than they had been at the beginning. The biggest losers, apart from the prerevolutionary French nobility, were the millions of people in Europe who did not manage to shake off their status as landless laborers. The former landlords and former owners of human beings were far less seriously affected. The winners were the majority of peasants and, with unerring regularity, the state bureaucracies. By the end of the emancipation process the European peasantry had a more direct relationship with the state, without ending up as a state peasantry. Its old freedom had been acted out in the village, vis-à-vis the lord; the new freedom of the nineteenth century could not go beyond the framework set by the state. Over time even the most convinced liberals saw that agrarian markets cried out more than any other for political regulation; the last quarter of the century therefore gave birth to the agrarian policy on which European farmers have remained dependent ever since.

4 The Asymmetry of Wage Labor

A Long Transition

By the end of the emancipation process, two main roles had crystallized out in the countryside: the farming entrepreneur (large or small) and the wage laborer. These were two legally distinct kinds of "free" labor. But market freedom had little to do with the old freedom of peasant utopias. Such a genealogy does not explain the emergence of the concept of "free labor" outside what we may, with some simplification, describe as the transition from feudalism to capitalism. The legal historian Robert J. Steinfeld tells a different story for England and the United States, arguing that the decisive transition to "free" labor occurred when the workers had the power to withdraw it, when absence from the workplace was no longer prosecuted as a criminal offense. In this account, the starting point is not slavery or serfdom but a type of labor obligation that appeared with the colonization and settlement of the New World: indentured service.[112] By this is meant the pledging of the worker's labor power for a limited number of years against an advance payment to cover transportation costs: in other words, temporary bondage. In English legal culture, this kind of voluntary alienation of personal rights was always a contested area. The idea of the "freeborn Englishman," which originated in the seventeenth century and rapidly became the social norm, stood in contrast to such forms of bonded labor.

After roughly 1830, this opposition became more and more glaring in the United States as criticisms of slavery proliferated. There had been free labor in the American colonies as far back as the early eighteenth century, but for a long time it remained the exception rather than the rule across the spectrum of contractual labor services. The existence of a time-specific contract was the most important difference between indentured service and slavery or serfdom. Nor was indenture seen as a relic from archaic times; in fact, from the point of view of social and legal history, it was a thoroughly modern form of labor relation. All this made it easier to overcome. The critique of slavery called into question whether indentured service could really be considered a relationship that workers entered into of their own free will. This was the key issue, not the way in which indentured workers were actually *treated*. And since, unlike in the case of slavery, no one openly defended the practice, it was effectively wound up in the decade following its discursive delegitimation in the 1820s.

Legal thinking in the Anglo-American world henceforth considered free labor the self-evident norm. American courts first began in 1821 to rule that labor obligations must be freely entered into, and that such was not the case if a worker, having decided to leave his place of work, was physically prevented from doing so. This interpretation then fed back into the debate on slavery, and in the Northern states "free labor" became a rallying cry in the fight against Southern secessionists. At the same time, the use of physical violence against workers was defined as fundamentally illegitimate. The US legal practice moved ahead of England in the further sense that it no longer distinguished between a worker with a home of his own and a bonded laborer, maid, or servant kept as part of the master's household.[113]

The development of the concept of free labor in postrevolutionary France, as expressed in the Code Napoléon and echoed elsewhere in Europe, might be the theme of another story. And yet another could be told about domestic service law (*Gesinderecht*) in Germany, where, long into the high age of industrialization, "domestics" in Prussia and elsewhere remained subject to a number of extra-economic fetters on their freedom. It is true that the preamble to the Civil Code of 1896/1900, applicable throughout the Reich, ended the right of masters to chastise their servants. Yet in a weakened form ("indirect" powers of chastisement, etc.) it continued to haunt many areas of the law until the very end of the imperial period. Such chastisement must have continued to exist on a large scale.[114]

Robert Steinfeld's interpretation is especially interesting because it makes the nineteenth century the decisive period for the development of free labor. But decisive in what sense? For Steinfeld, free labor did not become dominant immediately after the end of indenture. As in the case of slavery, or Russian serfdom, there was a transitional phase. Even in English industry, nonmonetary coercion did not vanish from one day to the next. Statutory law and the actual administration of justice gave entrepreneurs and agrarian employers the means to enforce a continuation of the labor relationship. For many decades, relics of coerced labor lingered on within relations of free wage labor.

In a celebrated study published in 1974, Robert W. Fogel and Stanley L. Engerman claimed that, contrary to the view of classical economists, slave labor was hardly less efficient and rational than free labor on plantations and in crafts or industry.[115] Since then, the idea of a linear progress from the one to the other has become hard to credit. We should therefore give up the notion that free labor and coerced labor had nothing in common, belonged to distinct eras, and represented completely different social worlds. It makes more sense to think in terms of a continuum in which workers were subject to various forms and combinations of coercion.[116] This would push the great watershed further into the century, since even in England extra-economic coercion disappeared from the industrial wage relationship only after 1870 or thereabouts. Moreover, many of the functions of indenture did not immediately become redundant. Migrants to neo-European societies sought the help of fellow countrymen already living there: the Chinese soon had their Chinatowns, while southern Europeans, for example, had semilegal patrons who combined the roles of recruitment, protection, and exploitation, similar in this respect to the contractors who often organized the influx of first-generation workers into cities outside Europe.[117] This was not free labor in the sense of liberal theory. Besides, the kind of indirect labor relations mentioned above, with the contractor acting as a buffer, were not peculiar to areas outside Europe. The impresario, for example, who supplied singers to Italian opera-house owners until late in the nineteenth century, was nothing other than a contractor of this kind.[118]

Labor Market Imbalances

A new factor that appeared toward the end of the century was the rise of the organized labor movement. Little by little the growing capacity of collective demands to counter the power of the owners of capital corrected a fundamental imbalance in the labor market, the real breakthrough coming only when legislation created the scope for nationwide collective bargaining.[119] The difficult ascent of free labor led to a paradox: only the restriction of market freedom through monopolistic negotiations on the workers' side enabled individuals to free themselves from the instruments that the purchaser of labor had at his disposal—above all, the power to play workers competing for jobs against one another and to dismiss them at a moment's notice. Free labor, in a substantive sense of the term, arose out of the curbing of unlimited contractual freedom that came with the development of the welfare state. Contractualization of the labor relationship was not by itself capable of preventing or overcoming the "indignity of the workers" condition" (as the French sociologist Robert Castel put it). Endowed with nothing other than his physical labor power, the worker was a creature lacking rights or guarantees and was in this respect comparable to the slave; the pure freedom of the labor market was therefore inherently unstable.

After a few decades, the rudiments of the welfare state were laid through the interplay of workers' protests, elite moves to head off revolution, and a moral sense among small reform-minded groups. Philanthropic businessmen were the first to realize that a bare "freedom of labor" did little to further social integration, and the social welfare measures that began to be introduced in the 1880s systematized such concerns into a truly novel principle of compulsory insurance.[120] Behind this was a view of society as a tension-ridden plurality of collectives rather than an aggregation of individuals—a view shared in principle by conservatives and socialists. This alone made it possible to develop a conception of the welfare state that went beyond classical liberalism. On the other hand, not all the theoretical and political representatives of classical (essentially British and

French) liberalism had been extreme individualists adhering to the "Manchester School" of unfettered competition. The "new liberalism" could therefore attach itself to the general trend of the times toward state protection. In the last two or three decades before the First World War, definitions of the "social question" in the industrialized countries of Europe rested on a certain basic consensus. Social insurance—which in Germany was primarily a conservative project to stabilize the system—was taken up by a *liberal* government in Britain after 1906.[121]

Free wage labor, which appears to us today as such a natural relationship, did not appear desirable under all circumstances; "proletarianization," especially in agrarian societies, was seen with good reason as a move down the ladder. In Southeast Asia, for example—where work was held in high esteem, people had a strong attachment to their land, and traditional patron-client relations were not thought of as particularly exploitative—the idea that it was worthwhile voluntarily to "seek" work developed only gradually with the emergence of local urban labor markets. For a long time, employment in wealthy households and other nonmarket forms of dependence were considered preferable.[122] Only two basic survival strategies are open to weaker members of society: either reliance on the strong or solidarity with others who are weak. The first option generally offered greater security. It is true that colonial governments were often willing to abolish slavery, but they hesitated to allow the formation of a politically restive class of landless laborers—except in strictly controlled enclaves of the plantation economy. In the late nineteenth century, the sedentary farmer without political ambitions or pent-up grudges, working hard for subsistence or export and regularly paying his taxes, was the ideal subject for most colonial and other regimes around the world; "free wage labor" in the countryside was a suspect innovation. The picture was different in industry, although socialists were not alone in their doubts about completely individualist freedom in asymmetrical market conditions.

CHAPTER XIV

Networks

Extension, Density, Holes

"Network" is a metaphor, at once vivid and deceptive. Networks produce two-dimensional connections: they are flat, and they structure level spaces. A network has no relief. Network analysis in the social sciences, useful as it is, always risks overlooking or underestimating hierarchies, the third, vertical dimension. This is associated with the fact that networks are in a way democratic; all their nodes initially have the same value. Even so, a historian cannot do much with them unless the possibility is allowed that a network has strong centers and weak peripheries, that the nodes therefore vary in "thickness." Not every network has to be constructed like a spider's web, with a single center holding everything in place. The basic form of urban networks or trading networks is just as often polycentric as monocentric. The network metaphor is useful mainly because it permits the idea of multiple points of contact and intersection—and hence also because it draws attention to what is *not* networked. Each network possesses structural holes, and the current fascination with unfamiliar, previously unnoticed connections and relations, especially over long distances, should not make us forget the somber surfaces on the map indicating uninhabited nature or thinly populated countryside.

A network consists of relations that have attained a certain degree of regularity or permanence. Networks are traceable configurations of a repetitive relation or interaction. Hence they are structures with "medium" consistency: neither one-off chance relations nor organizationally entrenched institutions—although the latter may grow out of networklike relations. One of the outstanding features of the nineteenth century was the multiplication and acceleration of such repeated interactions, especially across national boundaries and often between regions and continents. Here we need to be more precise about dates: the six decades between midcentury and the First World War were a period of unprecedented network building. This is all the more striking because many of the networks were dismantled during the First World War, and particularist forces grew stronger in the decades following it. If the formation of worldwide networks can be described as "globalization" (a broad definition of this colorful term), then the period from roughly 1860 to 1914 witnessed a remarkable surge of globalization. We have already discussed two examples: intercontinental migration and the expansion of colonial empires.[1] This chapter will consider other global aspects that emerged here and there: transportation, communications, trade, and finance.

To think in terms of networks was a nineteenth-century development.[2] In the seventeenth century the English physician William Harvey discovered the body as a circulatory system, and in the eighteenth century the French doctor and "physiocratic" theorist François Quesnay applied this model to economy and society.[3] The next stage was the network. In 1838 the politician and scholar Friedrich List mapped out a railroad web—a "national transportation system"—for the whole of Germany: it was a bold vision of the future. Before 1850, however, it was not possible to speak of a railroad network in any European country. Friedrich List proposed the fundamental planning schema, and when the railroads were actually in place certain critics took up the web image and presented them as a dangerous spider stretched out over its victims. Later, the web came to stand for a way of visualizing a city, competing for a time with "labyrinth" or, especially in the United States, with "grid." The self-image of societies as networks thus has its roots in the nineteenth century, even if the full range of meanings—up to today's "social networks"—appeared only much later.

Perhaps the strongest everyday experience of a network, and also of dependence on functioning networks liable to break down, came with the linking of homes to centrally managed systems: water from a tap, gas from a pipe, electricity from a cable.[4] There was a difference as to the extent to which the private sphere was invaded: for instance, between the telegraph, an office machine that no one put in their living room, and the telephone, which after a slow start became a domestic fixture and an object of private use. At the beginning of the twentieth century, only a tiny minority of the world's population was linked to technical systems. "India" was said to be part of the international telegraph network, but the great mass of Indians had no direct experience of this—even if the influence of systems such as the railroad and telegraph on flows of products and information also made itself felt indirectly in daily life. Virtual opportunities must be distinguished from things that can actually be achieved. In the 1870s it was possible to circumnavigate the globe north of the Equator by steam-powered means of transport, without porters, horses, or camels, and without the effort of traveling on foot: London—Suez—Bombay—Calcutta—Hong Kong—Yokohama—San Francisco—New York—London. But who undertook this journey, aside from the gentleman Phileas Fogg in Jules Verne's novel *Around the World in Eighty Days* (1872; his model was the eccentric American businessman George Francis Train, who tried to set that record in 1870 and later cut it to sixty-seven days in 1890) and the American reporter Nellie Bly, who in 1889–90 needed no more than seventy-two days?[5]

1   Communications

Steamships

In the history of transportation, there is often no way around a mild form of technological determinism. New means of transportation do not appear because there is a cultural craving for them, but because someone comes up with the idea of creating them. It is another story whether they then catch on, fall flat, or are endowed with special meanings and functions. If we leave aside the towing of water vessels by sheer muscle power, ship travel—unlike land transportation—had always used nonorganic energy in the form of wind and current. Steam power added to these possibilities. In two parts of the world—England (with southern Scotland) and the northeastern United States, both pioneers of industrialization—prior modernization of the transportation landscape worked to the advantage of the steamship and railroad locomotive. Canal systems had already been laid out by commercially minded private landowners eager to increase the value of their land; in England the height of the enthusiasm for canals (also a highly popular investment) was reached between 1791 and 1794. They created part of the demand that the railroad would meet even better. Indeed, the "canal age" evoked by some historians stretched into the early part of the railroad age; the two forms of transportation partly competed with each other and partly linked into wider systems. By the middle of the nineteenth century, more than 25,000 cargo barges were operating on Britain's inland waterways, and a mobile, "amphibious" population of no fewer than 50,000 people, one-third of them employed by large companies, lived aboard them.[6] The boats were mainly drawn by horses, whereas in Asia human traction continued for a long time to perform the backbreaking work. Until the 1940s small ships were hauled upriver by "coolies" through the rapids of the Upper Yangtze that have now disappeared into the Three Gorges reservoir.

Steamships were too large for the canals of the eighteenth century. But since they could travel smoothly over still waters, they gave a major impetus to the construction of wider and longer canals. Many a city entered a new phase of development when it was connected to one: New York, for example, after the opening of the Erie Canal in 1825, or Amsterdam after the completion of the North Sea Canal in 1876. In the Netherlands, the personal interest of King William I helped in the creation of a closed canal network for the purposes of both transportation and water regulation. Its successful completion, thanks to a competent corps of engineers dating back to the time of the French occupation, meant that the country delayed the construction of a railroad system.[7] In the United States, the first railroads were seen as no more than feeders for canal transportation. In New York State, trains were prohibited until 1851 from carrying freight in competition with publicly owned canals.[8]

The steamship, which in long-distance traffic prevailed over the sailing vessel in the crucial decade of the 1860s, did not rely on an external energy supply.

It carried its own fuel on board: coal and then increasingly oil, after diesel engines—invented by the brilliant German engineer, Rudolf Diesel, during the 1890s—were introduced into shipping in the 1910s.[9] Being able to navigate independently, it was less at the mercy of the elements than the sailing ship had been and was therefore ideal for travel along coasts, against river currents, or on windless lakes and canals. This new freedom allowed shipping to keep to a schedule for the first time in history; the relations that made up a network became dependable and open to calculation. The early impact of the steamship was greatest within the technological and economic heartlands of Europe and North America: Glasgow saw one arrive every ten minutes in the 1830s,[10] while a regular service between Vienna and Budapest, inaugurated in 1826 and taken over in 1829 by the famous Donaudampfschifffahrtsgesellschaft (one of the longest words in the German language), had a fleet of seventy-one ships by 1850 for a trip lasting roughly fourteen hours.[11] The supply of transportation capacity interacted with new kinds of demand. Steamship expansion on the Mississippi and the Gulf of Mexico, for example, was closely bound up with the growth of cotton-producing slave plantations.

Not all steamships operated as part of a network. In some situations, where they spearheaded a drive to open up new regions for commercial activity, they were more like pioneering instruments of capitalist world trade. Nor were they necessarily under foreign control. From the 1860s on, the Chinese state took initiatives of its own (later supplemented by private companies) and successfully prevented the establishment of a foreign trading monopoly on the country's great rivers and coastal strips.[12] The competitive advantage of British (and later Japanese) shipping companies in China was less pronounced than in India, where indigenous shipowners were unable to secure a significant foothold in the market. One of the reasons for this was that British companies active in India were officially appointed to carry mail and received substantial subsidies for this service.

Moreover, in neither semicolonial China nor colonial India did indigenous forces (private or public) ever succeed in creating an overseas fleet. In this too, Japan was the great exception in the Afro-Asian world. The fact that, by 1918 at the latest, its military and mercantile shipbuilding industry had reached world level, making the country a leading force in commercial shipping as well as a top-class naval power, was both an expression of and a contributory factor in its national success.[13] Everywhere else in Asia (the same is true of Latin America) new relations of technological and economic dependence were visible in the control that foreign shipping lines had over overseas trade. It is characteristic that the Tata steel family in India, otherwise highly successful, failed in their attempt to open up a shipping route to Japan, largely because of British competition.[14] From 1828, when Lord William Bentinck arrived by steamship in Calcutta to assume his post as governor-general, the British attached great practical and symbolic significance to the vessels as heralds of a new era.

The first ocean steamship lines came into operation across the North Atlantic. The technological advances during the first half of the nineteenth century were so great that the journey time of fourteen days between Bristol and New York, already possible by midcentury, stood virtually unchallenged during the next few decades.[15] The beginning of the great migration to the New World then created a passenger demand of novel proportions. The same was not true of other parts of the globe, where, as in India, subsidized mail steamers became the driving force of maritime expansion. No imperial or colonial power thought it could afford to do without its own postal service between the mother country and its overseas possessions. The opening of the Suez Canal in 1869 triggered a further growth of passenger transportation between Europe and Asia, while the shipping lines also did a roaring trade in tropical exports. Although, thanks to its huge internal market, the United States rose after midcentury to become the largest shipping nation in the world, Britain hung on to its leading position in overseas transportation. In 1914 it still accounted for 45 percent of world commercial tonnage, followed by Germany (11 percent) and the United States (9 percent). Japan had reached 3.8 percent, just behind France (4.2 percent) but in front of the Netherlands (3.2 percent) that had dominated the seas in the seventeenth century.[16]

World maritime trade should not be thought of as an evenly connected network; it did not embrace vast regions such as northern Asia (which acquired an ice-free port only in 1860 with the founding of Vladivostok on the Pacific coast. By the criterion of seagoing tonnage, the world's four main ports in 1888 were London, New York, Liverpool, and Hamburg. Hong Kong—the gateway to the Chinese market and a major transhipment center for Southeast Asia—trailed behind in seventh place, but it was already far ahead of any other Asian port.[17] The major shipping routes were: (1) from Japan and Hong Kong to the Atlantic and North Sea ports, via the Strait of Malacca (Singapore), the northern Indian Ocean, Red Sea, Suez Canal, and Straits of Gibraltar; (2) from Australia to the Cape of Good Hope and then along the African West coast to Europe; (3) from New York to London and Liverpool (the widest shipping lane of all); (4) from Europe to Rio de Janeiro and the River Plate ports; and (5) across the Pacific from San Francisco and Seattle to Yokohama, the leading port of Japan.[18] Thus, although world shipping had a presence here and there in the remotest Pacific islands by 1900, it displayed a high degree of geographical concentration.

The sector itself was also highly concentrated. This was the great age of the private shipping companies (the state, despite a fin-de-siècle enthusiasm for "naval power," was much less involved than in the railroads), and some were among the best capitalized joint stock corporations in the world. Their hallmarks were regularity and punctuality, good service across a range of price categories, and safety standards which—despite some spectacular accidents, such as the sinking of the *Titanic* off Newfoundland on April 14, 1912—would have been scarcely imaginable in the age of the sailing ship, or even in the early decades of steamship travel. The major companies, such as the Holland-America Lijn, Norddeutscher Lloyd,

Hamburg-Amerika-Linie (or HAPAG), Cunard, Alfred Holt, or and Peninsula & Oriental Line, embodied at one and the same time a capitalism with global reach, a high level of technological perfection, and claims to superior civilization associated with sophisticated travel. The luxurious "swimming palaces" (a popular advertising cliché) became emblematic of the last three decades before the First World War.[19] From the 1860s on, national rivalry among the great shipping lines was repeatedly offset by the sharing out of markets and cartel-like "shipping conferences" that served to hold prices steady.

Although world shipping under northwestern European and North American control included all coastal regions between the 40th parallel south and the 50th parallel north in its global timetable, this was still not a truly global transportation network if measured against the airline yardstick of the last quarter of the twentieth century.[20] Only air travel would overcome the rift between land and sea, operating between air*ports* most of which are located inland. Virtually no large city in today's world lies outside the air network, and the frequency of contact is infinitely greater than it was in the heyday of passenger shipping. Moreover, the initial European-American monopoly was broken. From the 1970s on, even the smallest country set great store by having a national airline; only the collapse of Swissair in 2001 ushered in a new trend to privatization and the weakening of national transportation sovereignty. The largest globalization impetus in transportation history took place following the Second World War, especially in and after the sixties, when long-distance air travel ceased to be the preserve of politicians, managers, and wealthy individuals. The technological basis for this was jet propulsion. Since 1958, when the Boeing 707 came into service, and even 1970, when the Boeing 747 inaugurated the "jumbo" format, we have been living in a jet age beyond the dreams of the boldest visionaries of the nineteenth century.

The Railroad as Network Technology

The globalization effect of the railroad was not as great as that of the slightly older steamship. Railroads are systems with narrower spatial limits.[21] Technologically they were a complete novelty for which the world was unprepared, whereas the steamship had merged over a period of decades with an older infrastructure of water transportation. When the coal-based technologies arrived, there were already seaports but not yet railroad stations and iron tracks. Once built, however, the railroad was less dependent on climate and the environment; and this greater reliability meant that it could be better tied into production schedules. Only trains could guarantee the regular food supply to large cities and hence their future growth. The railroad was also less risky for the carriage of freight: a shipwreck might spell enormous financial losses, whereas a rail accident seldom destroyed wealth on a major scale, and insurance costs were accordingly lower. The techno-economic complex of the railroad gave rise to the first private companies of giant size that had ever existed: big business was a creature of the railroad.[22]

Even so, governments often had a great stake in railroad construction—not in Britain, but certainly in Belgium and several German states, and in China and Japan. There were also mixed forms, as in the Netherlands, where it became clear after several decades of experimentation that private initiative alone would not bring about an integrated network. Only special legislation in 1875 established a state railroad system, the organization of which—as well as its operational regulations—was brought over almost entirely from Germany.[23]

It is debatable what should be understood by a route *network*. Especially in the non-European world, there were various branch lines unconnected to anything else: for example, the Yunnan line built by the French between the Northern Vietnamese port of Haiphong and the terminus in the Chinese provincial capital Kunming. In Africa such branch terminal lines were the rule rather than the exception. Only in the south of the continent was there a two-dimensional network, which by the time it was completed in 1937 ran from the Cape up to the copper belt of Northern Rhodesia (Zambia).[24] The Trans-Siberian Railroad, despite a number of feeder lines, was and is a solitary arrow through the landscape. East of Omsk it served only strategic purposes, did not carry migrants on a large scale, and opened up no economic hinterland. There was a network in European Russia, but not in Siberia. In China, where railroad construction was continuous after 1897, a number of desirable and feasible stretches never left the drawing board for decades, so that the country had to make do with a fragmentary network with many loose ends, especially in the mountainous country south of the Yangtze.[25] Some parts of the interior were added only in the later twentieth century, Tibet only in 2006. In the case of Syria and Lebanon, where the railroads run by French companies had a different gauge from the Ottoman ones, two systems operated alongside each other with no points of contact.[26] Not everything that looks at first like a network holds up as one on closer inspection.

In first-generation countries, which could not yet simply import a readymade package—and even afterward most technologies retained a special national aspect[27]—the necessary experience had to be assembled from scratch.[28] The construction and running of a railroad required a large amount of know-how in iron and steel technology, machinery, geology, mining, telecommunications, site organization, finance, personnel relations, timetable coordination, and the design of bridges, tunnels, and stations. Much had to be improvised before all this was put on a "scientific" basis. While technical problems awaited a solution, legal matters such as land acquisition and related compensation also had to be addressed. Moreover, the railroad was often a political issue with a deeper military significance. In the United States, however, and to some extent in Britain, strategic considerations played a much smaller role than in continental Europe, so that the state—except during the interval of the Civil War—could safely forgo direct involvement.

The railroad network as we know it today (in some cases already reduced when compared with 1913 or 1930) was essentially complete by 1880 in Britain,

France, Germany, Italy, and Austria-Hungary, and by the end of the century in the rest of Europe. The spread of technology across borders meant that it was very difficult for a country to go its own way, the only partial exception being track gauges. George Stephenson, the "father" of the railroad, laid down a norm of 4 feet, 8.5 inches, which was also adopted elsewhere because of Britain's technological dominance in the field. The Netherlands, Baden, and Russia initially opted for a wider gauge, but in the end only Russia held out. By 1910, with only one short interruption to switch gauges, people could travel by train all the way from Lisbon to Beijing. In the same year, the transcontinental network also embraced Korea, where a railroad boom had started around 1900. This completed the unification of Eurasia in terms of railroad technology.

The Railroad and National Integration

The new "iron horse," initially competing with the fastest mail coaches ever put into operation,[29] offered a novel experience of the swiftly passing countryside and sparked debate about the desirability of the modernity that it seemed to epitomize.[30] It brought about the need and the chances of a new kind of spatial politics.[31] In France the "railroad question" became a central topic of elite discussions in the forties, and it was only in the face of great resistance, mainly from Catholic conservatives, that the new invention was held to serve the country's prosperity.[32] When the railroad later appeared in other parts of the world and unleashed similar reactions, people in Europe had long forgotten their early fears and held up backward, superstitious Orientals as figures of fun. The first project in China, the ten-mile Wusong railroad near Shanghai, was dismantled in 1877, just a year after its completion, because the local population feared it would destroy the harmony of natural forces (*feng shui*). This was ridiculed in the West as a primitive defense against the modern world. Yet it took only a few more years for the Chinese to understand the desirability of the railroad, and in the early years of the twentieth century patriotic members of the provincial upper classes collected large sums of money to buy back railroad concessions from foreigners. In 1911, an attempt by the imperial government to develop a centralized European-style railroad policy became the most important factor in the fall of the Qing dynasty. Regional and central forces fought for control over a modern technology that offered handsome profits to Chinese as well as foreign financiers and suppliers. In China, the railroad wrote history on a grand scale.

At that time, not long after its late entry into the railroad age, China was already capable of building and running its own subsystems of a national network. Until then most of the railroads, though under Chinese government ownership, had been funded by overseas capital and built by foreign engineers. An early major exception was the technically difficult stretch from Beijing to Kalgan (Zhangjiakou), whose completion in 1909, entirely devised and implemented by Chinese engineers, linked the state railroad system to the caravan trade from Mongolia. Foreign experts recognized it as an impressive feat, achieved at relatively low cost; the rolling stock, however, was not made in China. Thereafter, any railway built with Chinese capital made a point of dispensing with non-Chinese engineers.[33]

A similar symbol of resistance to European control and influence, motivated by geostrategic interests in the face of direct French and British penetration, was the Hijaz railroad from Damascus via Amman to Medina, with a branch line to Haifa. In the decade and a half before the First World War, the Ottoman Empire made the final bid for mobilizing its own resources in a great effort. Whereas other Ottoman railroads, including the famous one to Baghdad, had been established by Europeans, the Hijaz route was supposed to be funded, built, and managed by the Ottomans themselves. The plan was less successful in this respect than the Beijing-Kalgan railroad, since foreigners working under a German boss made up a much larger proportion of the construction engineers.[34] But the basic message was clear: a non-European state could best demonstrate its prowess by creating its own technostructures in line with European standards. This, of course, was the Japanese formula—much admired but not so easy to copy.

Unlike shipping and air transportation, the railroad was ultimately a vehicle of national integration. Back in 1828 Johann Wolfgang von Goethe, at almost eighty still a keen observer of his time, had assured Johann Peter Eckermann that he "was not uneasy about the unity of Germany," since "our good high roads and future railroads will of themselves do their part."[35] In particular, they integrated national markets or even created them where none had existed before. This is most visible in regional price differentials: today a loaf of bread costs more or less the same across a national economy. In 1870 wheat prices varied by as much as 69 percent between New York City and Iowa, but by 1910 this had fallen to 19 percent.[36]

The internationalism of the railroad leaps to the eye in Europe, where the confluence of national networks into a single (almost) continent-wide network was a major achievement.[37] It brought pan-European norms, such as a degree of timetable discipline and punctuality, and standardized many travel experiences. But since railroads could not cross the seas—even Napoleon's vision of a Channel tunnel was realized only in 1994—their *globalizing* effect was rather limited. The Trans-Siberian too, with its low passenger volumes, was no more than a modern Silk Route: a thin strip linking regions across huge distances, without joining them together into a quantitatively significant network. The Asian railroad systems remained unconnected with one another (the sole exception being the Siberia—Manchuria—Korea route). The Indian system, blocked to the north by the Himalayas, was never extended as far as Afghanistan, so that Russia would not have a gateway for an invasion of the Subcontinent; to this day, Afghanistan remains a country virtually without railroad facilities. Insofar as the railroads were instruments of "railroad imperialism," there was no need—outside India—to build them up into European-style national systems encompassing places of lesser strategic and economic importance.[38]

In Europe, governments ensured that railroad policy was conducted in the national interest. For a whole century the railroads were a focus of rivalry between France and Germany,[39] and their significance for troop mobilizations played a major role in conflict scenarios prior to the First World War. In large parts of the world, however—Latin America (except Argentina, which had a large network centered on Buenos Aires), Central Asia, and Africa—the train never had as great an impact on society as it did in western Europe, the United States, India, or Japan. Traditional forms of travel (walking, cart, or caravan) went unchallenged for a long time and had many advantages over the more expensive and inflexible railroad. Asian or African societies that had for good reason always been wheel-less remained so for the time being.[40] Indeed, it was not unusual for these regions to skip the railroad age, passing directly from human or animal motive power to the all-terrain vehicle and propeller-driven aircraft. Where railroads existed, their integrative effect sometimes remained weak because of the looseness of connections with rivers, canals, and highways. Tsarist transportation policy wagered everything on the train after the 1860s, but it neglected to construct paved feeder roads. The age-old impenetrability of the Russian and Siberian wastes therefore changed little, and huge regional variations in transport costs were a sure sign of the low level of integration.[41]

Cabling the World

The total length of submarine cables grew from 4,400 kilometers in 1865 to 406,300 kilometers in 1903.[42] Cable laying in the last four decades of the nineteenth century created a planetary network, ushering in a telegraph age that would last several decades until long-distance telephony became reasonably affordable.[43] For the first time in history, private correspondence involved a mixture of different media: handwritten or (from the 1870s onward) typed letters, interspersed with terse telegrams. Only in the last quarter of the twentieth century did the fax, e-mail, and mobile phone seal the fate of telegraphy.

The cabling of the world was an extraordinary feat, since it meant laying thousands of miles of thick, specially coated cable under the ocean waves, while the logistics on land was often not much more straightforward. It did not, unlike canal or railroad construction, require a huge deployment of manpower, and the technology was less invasive in urban environments. By the mid-eighties, the globe was, quite literally, wired up. In addition to the transoceanic cables, there were the much more numerous links over shorter distances: every medium-sized city, at least in Europe and North America, had its telegraph office, and the lonely operator in a godforsaken station in the Midwest became a stock-in-trade of later Hollywood movies. Rail track and telecommunication cables were often laid together, partly because a train was more or less essential to repair broken wires in remote areas. In Australia the first telegraph actually came into operation a few months before the first railroad line.[44]

The basic principle of telecommunication is that dematerialized information travels faster than people or objects.[45] This goal may be achieved in various ways. In the nineteenth century, the great new medium with a globalizing effect was the telegraph, not the telephone. The history of the latter began three or four decades later with the opening of exchanges in New York (1877–78) and Paris (1879), soon to be followed by interurban connections (United States in 1884, France in 1885). At first it did not mean the creation of an intercontinental network. The telephone, as it developed in the late seventies, still had a very short range and was limited to intra-urban communication—and a city like Shanghai, where it was introduced in the 1881, sported only a handful of devices. Its early history is overwhelmingly American.[46] In the 1880s and 1890s its potential increased not only within but also between cities; then technological progress speeded up after 1900, and once again after 1915. However, links between North America and the rest of the world were not possible until the 1920s, became reliable only in the 1950s, and could be afforded by ordinary individuals only from the late 1960s on. The original technology was developed almost entirely at Bell Laboratories, and subsequently AT&T enjoyed a kind of monopoly insofar as that was possible under antitrust legislation. Bell and AT&T held the key patents and marketed them internationally.

The national telephone networks that sprang up in the early twentieth century were nearly all state monopolies, and sometimes, as in Latin American countries, governments gave preference to state-run telegraphy.[47] Where it came into use early on, the telephone was a tool for people who had also rapidly adopted the telegraph. The first user groups were New York stock dealers, who soon learned to handle the inventions of Alexander Graham Bell.[48] Batch production of Thomas Alva Edison's later model began only in 1895. By 1900 one in 60 people in the United States owned a telephone; the figures for Sweden, France, and Russia were one in 115, one in 1216, and one in 7000, respectively. Such an important institution as the Bank of England had just connected up for the first time.[49] In 1900 the United States was on the way to becoming a telephone society, as the use of telegrams for private messages was on the decline; in Europe the new device made its mark only after the First World War.

It took an unusually long time for the technology to result in a fully operational network. National systems were generally in existence by the late 1920s, but for political rather than technical reasons several more decades passed before it was possible to have a reasonably comfortable international conversation. The fact that a public telephone company was established (in 1882 in India, 1899 in Ethiopia, and 1908 in Turkey) says little about the actual significance of the medium in a country's life.[50] It turned out to be unsuited for many of the purposes for which it was developed. In 1914, for example, the wired field telephony of the German army could not keep up with advances on the Western front, while the few radio telephones were not up to the task. Technology was therefore unable to provide the rapid and precise coordination of troop movements that the Schlieffen Plan required for the decisive breakthrough.[51]

Although the telegraph probably changed private lives less radically than the telephone and Internet did in later periods, its importance for commercial, military, and political activity cannot be underestimated. As far back as the Civil War, Abraham Lincoln directed his troops by means of what have been called his "T-mails."[52] Indeed, a cabled world had become imaginable already by 1800, long before the technology for its realization. Optical signals communication, such as Muhammad Ali Pasha introduced between Alexandria and Cairo in 1823, or the Russian government between Saint Petersburg and Warsaw in the 1830s, was a first practical step.[53] Other innovations, above all the gradual introduction of the steamship and the perfection of mail coach services, were concurrently diminishing global dispatch times which, on the eve of the telegraphic breakthrough, were already considerably shorter than they had been around 1820.[54] The electrical telegraph was tested in 1837, and Morse code was in commercial use by 1844. Underwater cables were laid all over the world in the third quarter of the century. Once it became possible to wire India (1870), China (1871), Japan (1871), Australia (1871), the Caribbean (1872), all large South American cities (by 1875), South and East Africa (1879), and West Africa (1886), an unprecedented density of information came on stream—even if it was only in October 1902 that a cable under the Pacific completed the global network.[55] In the 1880s, public business information from all around the world—such as stock exchange data and price quotations—could be obtained in London within just two or three days; private cable messages usually reached the recipient within one day. In 1798 the report of Bonaparte's invasion of Egypt took 62 days to arrive in London, hardly less than it would have taken 300 years earlier. In 1815 news of Napoleon's defeat at Waterloo reached Whitehall only two and a half days after it happened—although Nathan Mayer Rothschild had learned of it by private courier within 24 hours. On January 8, 1815, several hundred British and American soldiers met their end in the Battle of New Orleans because their commanders were unaware that the two sides had signed a peace agreement in Ghent on December 24. And just before the telegraph revolutionized the picture, letters to London were still taking 14 days from New York, 30 from Cape Town, 35 from Calcutta, 56 from Shanghai, and 60 from Sydney. A year before the transatlantic cable was laid, people in London learned only 13 days later of Lincoln's assassination in Washington, DC, on April 15, 1865. After the opening of the telegraph age, news of the assassination of Tsar Alexander II in Saint Petersburg on March 13, 1881, came through in just 12 hours.[56]

Individual markets now responded more quickly to one another, and price levels came into closer convergence. Since orders could be placed at short notice, it was no longer necessary in many business sectors to keep large stocks on the spot; this worked to the advantage of small firms. Telegraphy also smoothed the ascent of big business: large conglomerates could now operate with strewn-out locations, and communicative functions previously entrusted to agents could be brought in-house more easily. Middlemen and brokers became dispensable over time.

Nor was there a lack of political effects. The telegraph increased the pressure not only on diplomats serving abroad but also on cabinets and other decision-making bodies in capital cities. The response time in international crises grew shorter, and major conferences did not last so long. Encrypted messages could be wrongly decoded or give rise to a misunderstanding. Military headquarters and embassies were soon supplied with telegraphists, who went around with cumbersome codebooks vulnerable to espionage. The fear that someone might read confidential messages, or that the code might be cracked, was not always unfounded.[57] Such concerns cast a shadow over communications, and new opportunities—some hard to put into practice—opened up for censorship.

Hierarchy and Subversion in Telegraphy

The fact that the new medium was predominantly British—as telephony would later be American—had a certain influence on its military and political uses. By 1898 two-thirds of telegraph lines in the world were British owned, either by the Eastern Telegraphy Company and other state-licensed companies or directly by the Crown, US cables trailed behind in second place, while Germany accounted for just 2 percent of the total. Alongside the 156,000 kilometers belonging to British firms, a mere 7,800 were in the hands of the state—mainly in India. (Altogether, barely more than one-tenth of all lines in the world were directly controlled by governments.[58]) In other words, in terms of communication, the British Empire with its public and private representatives acted as a kind of hegemonic master empire with others partly dependent on it. However, fears that Britain would use its quasi-monopoly to spy on others or to establish a communicative stranglehold were not borne out. Even the British were not invariably successful in maintaining control. Shortly before the First World War, Americans owned more and more cables in the North Atlantic.

It soon became clear that access to the network would have to be carefully regulated. During the Crimean War, when the medium was deployed for the first time, British and French commanders found themselves bombarded with a welter of contradictory telegrams from civilian politicians.[59] In this respect, therefore, telegraphy tended to create new hierarchies rather than a level playing field. Only top officials permitted themselves access to it, and of course it became much easier to direct the course of negotiations abroad from headquarters in the mother country.[60] The age of the grand diplomacy of unencumbered plenipotentiaries was drawing to a close.

On the other hand, autonomy might assume a new awkwardness—if enforced in situations when the cable connection failed or, as often happened in wartime, was literally severed. In September 1898, when British and French troops met near Fashoda in the Sudan in one of the most famous "duels" in imperial history (the adversaries actually drank a bottle of champagne with each other), General Kitchener had access to the telegraph via Omdurman, while his French counterpart Major Marchand was denied it. The British used this advantage for a diplomatic stage performance that decisively sapped morale on the French side.[61]

In other circumstances, the telegraph could also be used for subversive effect. It enabled the coordination of political movements over large areas, as in India in 1908 (or the United States a year earlier), when the virtual community of telegraph operators organized a countrywide strike that crippled administrative and business life from Lahore to Madras and from Karachi to Mandalay. Cables were also an *object* of international (or even intra-imperial) politics. Canada fought for two decades for a Pacific cable that would give it greater freedom and draw it closer to its Western neighbors, while the government in London kept creating new obstacles in order to preserve the classical arrangement whereby peripheral regions communicated with one another via the imperial center only.[62] Little technology transfer to newcomers or imitators was involved in the twenty-year-long cabling of the world. Ownership of the hardware and control of know-how remained in the hands of a few inventors and investors.

As for the "old" means of communication, there was a major expansion of letter mail—up from 412 million to 6.8 billion units per annum inside Germany between 1871 and 1913, and by a proportionally similar amount for foreign mail.[63] Never had international correspondence been as common as it was in summer 1914. To be sure, it was still by no means the case that mail reached every inhabitant of the planet; services began to thin out even in peripheral areas of Europe. A large part of Russia's rural population had no access to mailboxes or post offices. But the vast expanses of the United States were already thoroughly covered by the eve of the Civil War; communications and literacy drove each other upward in a continual spiral.[64]

Another new network-creating technology was the electricity supply, which also appeared on the scene in the great watershed decade of the 1880s. Simple prototypes of central electricity stations were in operation from the beginning of the decade, and current transmission over a certain distance between cities became feasible in the early nineties. Networks had to combine the three functions of generation (mainly through waterpower), storage, and transmission of electricity. On the eve of the First World War, technology had matured to the point that systems of regional energy production and distribution were in place virtually all over the world. Urban households were connected to these systems, electricity entered the daily life of the affluent classes, and the electromotor found ever more practical applications in traffic and manufacturing.[65] Whereas Britain had pioneered the global telegraph and America the telephone, the world center of electricity was Germany—or rather, Berlin, "the electrical metropolis."[66] Major standardization comparable to that of the early railroads had to wait until after 1914; up to then there was a chaos of different voltages and frequencies, with few electrical networks transcending the confines of individual cities and regions and none crossing national borders. Only in the 1920s did the technical and political conditions exist for extensive energy linkups, and in 1924 the need for international regulation was made explicit at the first World Energy Conference.

2 Trade

World Market—Regional Markets—Niches[67]

For a long time the growth of the modern world economy was seen in the West as a spreading of links and contacts out from Europe; the memorable image of the phased development of an expansive "modern world-system" (Immanuel Wallerstein) also served to encourage this. Today, however, it is more plausible to suppose that the emerging world economy of the seventeenth and early eighteenth centuries was polycentric: several different commercial capitalisms flourished simultaneously in different parts of the world, each associated with the rise of production for distant markets.[68] European trade dominated the Atlantic and, from the middle of the eighteenth century, pushed back the Asian competition. But it would too simplistic to envisage the world economy, as it was restructured after the 1840s under the aegis of free trade, as a single network spanning the globe.[69] The world market is a rather abstract theoretical fiction. Depending on the commodity (which could also be human beings), many markets grew so large that one might describe them as global. But none of these can be separated from its specific geography; none covered the earth in a geometrically even manner.

Regional subsystems retained, or regained, a dynamic of their own. Between 1883 and 1928 trade within Asia grew significantly faster than between Asia and the West;[70] and for the most part the differentiation and distribution of economic roles within those regional systems was driven internally, not from Europe or by Europeans. Thus, after 1800—not earlier!—an international rice market developed in Asia: Burma, Siam, and Indochina exported the commodity, while Ceylon, Malaya, the Dutch East Indies, the Philippines, and China imported it.[71] Demand for rice was less an indicator of poverty than a result of regional specialization and, to some extent, higher consumption standards, given that it was seen in all Asian societies as a high-grade cereal comparable to wheat in Europe. The spread of modern technology did not necessarily mean that "premodern" forms of transportation and exchange disappeared from cross-border markets. The junk routes to Southeast Asia from the great port of Canton were by no means a legacy of "traditional" China but resulted directly from the trading monopoly that the Qianlong Emperor had awarded to the city in 1757. Junks did not become obsolete throughout the nineteenth century, any more than Arab dhows did in the Indian Ocean. European ships did not really drive out other vessels except in the transportation of cotton and opium. Trade links with Southeast Asia remained under Chinese control.[72]

From the point of view of commercial history, the nineteenth century was in many respects a continuation of the early modern period. In the seventeenth and eighteenth centuries, European merchants had already had great success in organizing intercontinental trade across cultural boundaries; the chartered company, especially in Asia, had been a highly efficient innovation, whose later decline was due not so much to shortcomings of its own as to the ideological reservations of ascending liberalism. From the 1870s on, it underwent a modest revival in the colonization of Africa, pioneered by the United African Company (1879). Apart from the great bureaucratic organizations of the East India Company and the Verenigde Oost-Indische Compagnie, individual merchants flourished on several continents, the core group in the eighteenth century being the "gentlemanly capitalists" of London and the southern English ports, along with a large contingent of Scots and growing numbers of North Americans. The trading networks into which such people were woven in the age of sailing ships and mercantilism largely anticipated the global commerce of the nineteenth century.[73] The internal means and flows of information at their disposal may be described (from the vantage of the cabled world of 1900) as harbingers of modernity: the large trading organizations and the business networks of individual entrepreneurs were held together and kept running by incessant correspondence. They were "empires in writing."[74]

To see the nineteenth century only as the age of the industrialism is to overlook the fact that merchants remained the most important force binding the world economy together. In adapting to circumstances and helping to mold them, they connected distant markets and various production regimes, accumulated capital that could flow into banking and industry, and created a need for transnational coordination and regulation leading to new practices of coordination and a body of border-crossing commercial law.

This capacity of merchants for organization was not a Western monopoly. Trading networks also contributed to China's prosperity in the eighteenth century, ensuring an optimum division of labor among the various provinces of the empire. They also presupposed a highly developed use of writing, and rested on solidarity among people from a shared locality rather than simply on kinship ties. Certain sectors all over China were in the hands of merchants from a particular city. Commercial techniques were often very similar in East and West, with partnership a major instrument joining together capital and skills both in Europe and in China or the Ottoman Empire;[75] one important difference, though, was that the state in Western Europe not only tolerated commercial capitalism but often explicitly promoted it. Outside Europe, merchant networks mostly survived into the nineteenth century and adapted to the new challenges, by no means disappearing overnight with the arrival of Western capitalism. One of their hallmarks was a close association with proto-industrial production. Many such networks— for example, the Chinese wholesale trade in cotton cloth during the second half of the nineteenth century—took care of the distribution of goods from quite different production contexts and "stages" of industrial development: household industry and early factories in China as well as imports from abroad.[76]

Another structural element to survive from the eighteenth century was the niches of religious and ethnic minorities spanning different countries and continents: Armenians, Greeks in the Ottoman Empire and Egypt, Parsis in India and Central Asia, Irish and Scots in the British world. Many of these groups, among which Jews were increasingly important, energetically took up the new opportunities that appeared with the nineteenth century. A growing European dominance in large parts of world trade did not prevent Hindu merchants from Sind province (in today's Pakistan) from continuing to build durable links in the Asian interior and establishing themselves as intermediaries among Chinese, British, and Russian interests. This was a specialty of the Shikapuri community. Another network, built and run by merchants from Hyderabad, resettled along the new tourist routes of the final decades of the nineteenth century and took advantage of the opportunities to market exotic textiles and Oriental handicrafts. Such groups were based mainly on kinship ties, although some of these were of a fictitious nature. They could not have been as successful as they were if they had not kept their eyes on the fast-changing markets and drawn the correct conclusions. Political borders meant little or nothing to them: they were "transnational" in their orientation.[77] Crossborder trading networks were closely linked to those existing inside South Asia or China. The rise of all-Indian networks and the expansion of activity in more distant areas were just as much two sides of the same coin as were the accelerated circulation between Chinese provinces and the expansion of Chinese business ties into Southeast Asia or the Americas.[78] In short, Asians and Africans were indispensable as a workforce for the new Europeanized world economy, but in many instances they also proved able to keep pace as merchants and to make the necessary adjustments. What was much more difficult for them was to break out of subordinate positions in industry and finance. By the beginning of the First World War only Japan had succeeded in these fields: its industry competed more and more with Europeans and Americans in Asian markets, and its trading and shipping companies had extended their operations far beyond their homeland archipelago.

Old Patterns, New Emphases

Between 1840 and 1913, statistics record an average expansion of world trade that had never been seen before and would be exceeded only in the "golden" postwar years of 1948 to 1971. Its value, at constant prices, increased tenfold between 1850 and 1913,[79] while its volume—which had crept up by an annual average of just under 1 percent between 1500 and 1820—jumped by 4.18 percent a year between 1820 and 1870, and by 3.4 percent between 1870 and 1913.[80] The bulk of international trade during this period took place among Europeans, or between Europeans and inhabitants of the neo-European settler colonies. From 1876 to 1880, Europe (including Russia) and North America together accounted on average for three-quarters of all international trade—a proportion that had changed only minimally by 1913.[81] Essentially this involved exchanges between economies with a relatively high level of income. European demand for tropical products declined slightly after the 1820s in comparison with the eighteenth century, when the attractions of sugar had been at their height. On the other hand, imports of foodstuffs and industrial raw materials from temperate regions became more important. Only in the mid-1890s did a new boom in tropical products from Asia, Africa, and Latin America get under way.

When Western import and export firms looked beyond the West, they almost invariably encountered trading structures with which they had to cooperate in order to open up Asian (and, to a more limited extent, African) markets. Before the end of the century, there could be no question of directly marketing Western products. Everywhere it was necessary to devise complicated mechanisms to mediate between different economic cultures. In Latin America too, where the cultural barriers were lower than in East Asia, European import-export houses seldom dominated a market completely and were forced to rely on the superior market knowledge and business ties of Spanish and Creole wholesale dealers. The telegraph weakened the position of the great trading houses, since less startup capital was necessary to enter a market and many smaller European or local firms (often run by recent immigrants from Italy, Spain, or elsewhere) jumped at the chance.[82] It was easier if the customer was a non-European *government*, since a business deal for weaponry or rolling stock could then be negotiated directly and the risks of default were smaller, though not entirely absent.

Imports from the future "third world" were another story. Western capital here managed to gain direct control over vital sources of production, such as plantations and mines, much earlier and more powerfully than over the marketing of its own products, so that its main dealings were not with self-confident local merchants but with a dependent labor force. The proliferation of export enclaves weakened indigenous businessmen unless they were able to gain a foothold there themselves—an eventuality that Latin America, in particular, shows to have been more common than was previously thought.[83] Export production in the periphery was always a patently new type of economy, for which integration with a hinterland was of lesser importance. In the case of such "dual" structures, the insertion into an overseas network was greater than into the local "national" economy. What has rightly been called a "Europeanization"[84] of the world economy in the first three quarters of the nineteenth century happened not through a uniform spread of European influence but because European firms (a) linked into trading networks already present in extra-European regions, (b) established bridgeheads for export production, or (c) restructured large frontier regions such as Australia, New Zealand, and Argentina in line with European requirements (so that the whole country became a kind of bridgehead).

In a broader perspective and going beyond the description of specific contacts, other processes were also at work in the nineteenth-century evolution of world trade: (a) the dismantling of customs barriers in Europe, the British Empire, China, the Ottoman Empire, and other parts of the world;[85] (b) the formation of new demand as industrialization and the opening of productive frontiers gradually raised income levels; (c) the creation of railroad access to new regions; and (d) the lowering of transportation costs for passengers and freight. The last of these factors is especially important. The opening of the Suez Canal in 1869 cut journey times from London to Bombay by 41 percent, while the North Atlantic passage fell from an average of thirty-five days in 1840 to twelve days in 1913. The improvement of sailing ships, followed by a smooth transition through the early steamships to ironclad vessels with efficient engines, steadily reduced the costs of freight and (to a lesser extent) passenger travel. In 1906 transportation costs per unit mass between Britain and India were a mere 2 percent of the 1793 level. At the same time, it cost only twice or three times as much to ship a ton of cotton goods from Liverpool to Bombay as it did to send them 45 kilometers by train from Manchester to Liverpool.[86] The effects of this revolution were similar all over the world.

The basic connections of the nineteenth-century system of trade were in place by the middle of the eighteenth century: the North and South Atlantic was crisscrossed by permanent shipping lanes, the fur trade integrated the northern latitudes of Eurasia and America, maritime commerce between Europe and Asia stretched from the Baltic to the South China Sea and the Bay of Nagasaki, continental Eurasia was covered with trade routes, deserts were traversed by caravans and the Pacific by Spanish Manila galleons.[87] Only Australasia and parts of southern Africa still resisted incorporation into global contexts. The organizational forms of trade scarcely underwent any revolutionary change until the appearance of multinational corporations in the late nineteenth century. As in the eighteenth century, individual and family businesses created extensive networks branching out into trade: for example, the Rothschild financial empire became a European player after about 1830, and Sir William Mackinnon's investment group, until its collapse in the 1890s, encompassed everything from Scottish shipbuilding to the Indian import-export trade to East African coastal shipping services.[88]

European and North American commercial capitalism did not sweep away existing networks, and since Western-generated commodities were not marketed exclusively by trading organizations, the exports of Western industrial economies gave a powerful boost to local commerce in many parts of the world. It even happened that Europeans were unable to gain a foothold in new and dynamic market sectors, such as the cotton or coal trade between China and Japan.[89] The "rise of Asia" becomes less of a mystery if we bear in mind that the business infrastructures of East Asia have been continually developing at least since the eighteenth century, damaged but not destroyed by imperialism and, later, Chinese communism. The general market expansion of the second half of the nineteenth century created opportunities that were not taken up only by people from the West.[90]

So, with all the continuities from the early modern period, what was new about the commercial networks that formed in the nineteenth century?

First. International trade in 1800 was by no means confined to light, high-value luxury products; raw cotton, sugar, and Indian textiles were already bulk goods. But only the transportation revolution, by dramatically lowering costs, made it possible to ship products such as wheat and rice, iron and coal, on such a scale that they dominated world trade in value terms too. High returns in early modern commerce were often due to a lack of competition in the destination country: tea in 1780 came only from China, sugar almost exclusively from the Caribbean. Eighty years later, long-distance shipping was worthwhile also for goods that could be produced in many different places. The great ports received products from literally all over the world; "natural" monopolies, not conferred by the state, were much fewer in number, and this meant that competition was considerably greater.[91]

Second. Without the factor of mass transportation, the quantitative soaring of intercontinental trade by both value and volume would remain inexplicable. Only with the record growth of freight in the 1850s, and again between 1896 and 1913, did external trade become crucially important for numerous societies and have an impact on living standards well beyond the rich. This expansion went hand in hand with market integration, apparent from the increasing convergence of international commodity prices. Before 1800 there was virtually no systematic relationship in price formation on opposite sides of an ocean. The picture changed enormously in the course of the nineteenth century as price levels began to match each other more and more closely.[92] Three-quarters of this was attributable to falling transportation costs, and one-quarter to the elimination of tariff barriers.[93] Market integration did not always conform to political boundaries: Bombay, Singapore, and Hong Kong, for instance, were integral parts of the *British* overseas economy. There, prices were more in line with London's than with those in their own Indian, Malayan, or Chinese hinterland.

Third. Since many of the shipments between continents—from raw cotton and iron to palm oil and rubber—ended up as industrial inputs, commodity chains now became more complicated. Additional processing stages inserted themselves between primary producer and end user.[94] The lack of mature industries in Asia, Africa, and Latin America also meant that value creation became more strongly concentrated in the leading industrial countries. Whereas early modern Europe had imported finished products from overseas (fine cotton towels from India, tea and silk materials from China, ready-to-use sugar from the Caribbean), processing now took place mainly in the metropolitan countries. It was there that cotton was machine spun, raw coffee roasted, and palm oil converted into margarine or soap. Indeed, some of these goods were then shipped back to be sold in the country that had produced the original input: cotton goods to India, for example.

3   Money and Finance

Standardization

The process of systematization and the creation of large-scale systems of exchange was even more dramatic in the realm of money and finance. Here Europeans had a bigger lead than in the field of commerce, as far as rationalization and efficiency were concerned, over economies that not so long before had been neck and neck with them. Complex currency relationships, multiple forms of money, and the difficulty of calculating the ratios between them always entail additional costs: this was true in early modern Europe, as it was in China until 1935. Despite several attempts, Imperial China did not manage to simplify its chaotic dual system of silver and copper money; trustworthy paper money gained ground only slowly, and the most diverse foreign means of payment remained in circulation—from the Spanish Carolus dollar that had been the standard currency in the Yangtze delta since the late eighteenth century to banknotes issued by foreign banks in the treaty ports. These were all major factors in the country's backwardness in the nineteenth and early twentieth centuries. Before 1914, when the Yuan Shikai dollar was introduced, there were not even the basic elements of a uniform national currency. A central bank finally came into being in 1928, but political turmoil kept it from functioning in more than a rudimentary fashion.[95]

Such conditions, characteristic of large parts of the world, contrasted with the creation of national monetary areas in nineteenth-century Europe. This certainly posed problems, especially for newly formed nation-states, but a combination of economic expertise, political will, and local interests proved decisive. The integration of national markets and economic growth, usually attributed to industrialization alone, would have been impossible without this far-from-marginal factor.[96] Only standardization and credible guarantees of stability enabled certain Western currencies (above all the pound sterling) to become sufficiently strong to operate internationally. Monetary and currency reforms were always a highly complex affair. It was necessary to borrow from successful models, and there had to be banks capable of issuing and managing a new currency. The obstacles to a uniform national system were often apparent also from the persistent fragmentation of credit markets. In Italy, for instance, a regional differentiation of interest rates persisted for several decades after the lira became the official currency in 1862.[97]

The logical next step—though one that had to be synchronized with national homogenization—was the international alignment of currencies. We should be wary, however, of assuming a general process of ever wider integration. In the eighteenth century, the Spanish Empire was the largest single currency and fiscal area in the world, and its dissolution between 1810 and 1826 eliminated the benefits it had brought and confronted each of the successor states with the problem of creating a monetary and financial system of its own. That this almost never succeeded upon the first attempt was one of the factors behind a vicious circle of political instability and economic inefficiency.[98]

For much of Europe, the Latin Monetary Union of 1866 finally created a de facto single currency that greatly facilitated business and travel. But this was not the main goal of the union. Rather, it reflected (a) France's emergence as a major exporter of capital; (b) France's political wish to make its bimetallic silver and gold currency hegemonic throughout continental Europe; and (c) the need to restore the balance between silver and gold prices upset by the discovery of new American and Australian gold deposits.[99] A further policy goal, addressed for the first time in such an international manner, was to create price stability. The currency of the Latin Monetary Union, embracing France, Belgium, Switzerland, and Italy (and later Spain, Serbia, and Romania), was really a silver currency, since each country defined its own money in relation to a fixed weight of silver. An unforeseen "extrasystemic" development caused this edifice to topple, when the discovery of new deposits lowered the price of silver and unleashed a flood of the metal in the countries of the union. There was then much to be said for an alternative gold-based currency, yet silver displayed an astonishing capacity to hang on.

Silver

The international monetary systems of the nineteenth century were the first coordinated attempts by a number of states to control the precious metal flows that had been circling the globe since the 1540s.[100] Even countries that aimed at strict regulation of their external economic (and other) relations—Japan, for example, and a fortiori China—were caught up in these flows and, often without understanding the causes, experienced the inflationary or deflationary effects of the global circulation of coins and metal. These effects could break through into politics. The Opium War between Britain and China (1839–42) had its main cause in the problems associated with silver. All through the eighteenth century, China had earned large quantities of silver in exchange for its export products (especially silk and tea), and these had breathed life into its domestic economy. But at the beginning of the nineteenth century this flow was reversed, until the British finally came up with something of interest to Chinese customers: opium produced in the Indian territories of the East India Company. This orientation also had consequences in far-flung parts of the world, since from the 1780s onward the need to sell something to the Chinese had been a major impetus for the exploitation of new Pacific resources, such as the sandalwood forests on the islands of Fiji and Hawaii. The more important opium became as an import good to China, the more the economic and ecological pressure on the Pacific subsided.

The beginning of the opium trade reversed China's insertion into the world economy, since it now paid for the opium with silver. The result was a serious deflation that affected South China down to the village level, as well as a threat to government tax revenue. In this situation the imperial court decided to put an end to opium imports (already regarded as illegal smuggling). The casus belli came when the emperor's special commissioner in Canton had stocks of British opium confiscated and destroyed, shortly after London's representative in the port had unceremoniously declared them to be Crown property. Since in this period opium revenues came second after the land tax among government income in India, there was a strong political interest in the continuation and expansion of the opium exports to China.[101] Of course, the silver-opium economy of China and India operated in wider global contexts, and the Chinese move was somewhat more complex than a simple measure against British corrupters of the Chinese people. A contraction of the country's export markets for silk and tea reduced its silver earnings after 1820, while at the same time lower output in South American mines raised the international price of silver and triggered further outflows of the metal from China. The aggressive and criminal British opium trade was therefore not the only reason for China's economic crisis of the 1830s.[102]

The economic fate of India was also strongly marked by silver. After 1820, large quantities of Chinese silver originating in the mines of Spanish America flowed into opium-producing India. In addition, silver from newly opened North American deposits was soon being used to pay for rising Indian exports of tea and indigo. When cotton supplies to European industry faltered during the American Civil War, Egypt and India leapt in to fill the breach. India's seemingly endless capacity to absorb silver—rather like that of China in the eighteenth century—was acceptable to the colonial power, since it facilitated the gradual monetization of the rural economy and the collection of land taxes on which British rule rested. From 1876 on, however, a steady decline in world silver prices dragged down the exchange rate of the Indian rupee, making exports cheaper. Because the dominant ideology of free trade ruled out tariff increases of any kind, the Indian government was unable to stem the resulting outward flow of agrarian products, and it faced growing difficulties as it struggled both to deliver promised pay raises to its officials and to cover the usual "home charges" to London.

Caught in an essentially silver-induced trap of inflexible revenue and rising expenditure, the government in Calcutta resorted in 1893 to a radical measure utterly at variance with market liberalism. It closed down the Indian mints, where until then anyone had been able to convert a small sum of silver into rupees. Now the country had a manipulated currency whose face value, decided by the Secretary of India in London, no longer corresponded to its metal value. This took India out of the global play of currency forces, tightening the British grip on the Indian economy.[103] This example shows how the free silver market—all in all, the chief globalizing factor from the early modern period down to the late nineteenth century (or in China even until 1931, when the

Great Depression hit the country)—could work itself out in practice. It also illustrates how, in the end, only the major Western states were capable of intervening in this interplay of forces.

Gold

Governments and investors fled from the risks of silver into the safety of gold. It was more by accident than design that the British economy, by far the strongest in the nineteenth century, had already used a de facto gold currency in the eighteenth. In medieval England, the pound sterling had still been fixed as a pound weight of "sterling" silver. But from 1774 on gold coins were the legal tender (the famous "guinea," so called after the main source region of the gold), soon displacing silver money in everyday use. After the Napoleonic Wars, the British government—alone in Europe at the time—committed itself to the gold standard. In 1821 a coherent monetary order was introduced by law: the Royal Mint had to trade gold in unlimited quantities at a fixed price; the Bank of England and any other British bank instructed by it had a legal obligation to exchange banknotes into gold; and the import and export of gold was subject to no restrictions. This meant that gold functioned as the reserve for the whole volume of money. Until the early 1870s Britain was the only country in the world with this kind of system. After the alternative model of the Latin Monetary Union collapsed within a short time of its introduction, the bimetallic solution fell by the wayside and one European government after another switched to the gold standard: Germany, Denmark, and Sweden in 1873, Norway two years later, France and other members of the Latin Monetary Union in the 1880s.

In each case there were major debates on the pros and cons of gold. It was not only in France that a gap opened between theory and practice. From 1879 the United States had in effect a (highly controversial) gold currency, although Congress did not officially admit to this until 1900. Russia—which entered the century with a silver standard and then printed quite large amounts of uncovered paper money—converted to the gold standard in 1897. Japan followed suit the following year, having used China's reparations from the war of 1895 to build up a gold reserve in its central bank. As so often in Japan at that time, this was associated with the aim of following the "civilized West"—unlike China, increasingly held in contempt by the Japanese, which did not manage to rid itself of its archaic silver currency.

But Japan was not alone in reacting as it did. Virtually all other countries, especially those outside Europe, were parvenus in comparison with Great Britain. Adherence to the gold standard signified international respectability and a will to respect the Western rules of the game. In some cases, there were also high hopes in foreign investment—a major factor for Russia, for instance, which by the end of the Tsarist period was the largest debtor country in the world.[104] Russia's switch to gold meant that all the major economies of Europe now had the same kind of currency; the integration of the continent was thus even greater than, at the level of commerce, under the free-trade system of the 1860s (which Saint Petersburg had never joined). A closer examination reveals some differences, however. Almost all countries but Britain, even financially strong creditor nations such as Germany and France, provided their monetary authorities with instruments to defend their gold reserves if they came under threat. In exceptional situations, the strict gold cover of paper money could be abandoned. None of the Continental countries (except France) was a net exporter of capital, and none had England's fully developed banking structure. The British model could therefore be imitated only in parts.

The Gold Standard as a Moral Order

The technical devices affording price and currency stability under the gold standard need not concern us here.[105] The key points in terms of network formation are the following.

First. Britain introduced the gold standard in the eighteenth century more or less by accident. In the next century too, the system did not display any clear-cut, intrinsic advantages over a bimetallic currency. An important element in the adoption of the gold standard by one European country after another was the fact that Britain—not mainly *because of* its gold currency—had become the world's leading industrial power and financial center. When Germany then caught up with it industrially, a chain reaction was unleashed. Anyone who wished to do commercial or financial business with Britain and Germany was well advised to adhere to their monetary system. Pragmatism was here mixed up with considerations of prestige. Gold counted as "modern," silver did not.

Second. It took a long time for a truly international gold-based monetary system to become operational—until the early years of the twentieth century, in fact. Soon afterward it was ripped apart by the First World War.

Third. The gold standard, as a regulatory mechanism effective across the world from North America to Japan, was not simply the abstract apparatus presented in textbooks. To quote the economic historian Barry Eichengreen, it was "a socially constructed institution whose viability hinged on the context in which it operated."[106] This institution required from participating governments an explicit or implicit willingness to do anything necessary to defend currency convertibility—hence a consonance at the level of economic policy. This meant, for example, that no one was supposed even to think of devaluation or revaluation, and that in a highly competitive international system, governments were ready to solve financial crises by mutual agreement and mutual assistance. This happened in the Baring crisis of 1890, for example, when a large British private bank declared itself insolvent and only prompt support from the French and Russian state banks maintained liquidity on the London market. The following years witnessed a number of similar cases in other countries. Such international coordination and fine-tuning, at a time when there was still no telephone and top officials did not hold regular meetings, was much more difficult to achieve than it is today. Yet the system proved its effectiveness thanks to the professional solidarity—"trust" would perhaps be too strong a word—among governments and central banks. In the world as it existed before 1914, there was a greater convergence of interests and spirit of cooperation in the field of monetary policy than in diplomacy and military affairs. This discrepancy between the different levels of international relations, involving an autonomy of prestige-centered power politics, was one of the main distinguishing features of globality during the quarter century before the outbreak of the First World War.

Fourth. The gold standard did not really function worldwide. Silver-currency nations such as China remained outside it, and colonial currencies, as the Indian example shows, operated independently of external interventions. The largest bloc of peripheral noncolonial countries that experimented with the gold standard were the states of Latin America. Until the 1920s these mostly lacked a central bank or private banking institutions that offered reasonable proof against crisis. No entity could intercept the inflow or outflow of metal money, and the public had little faith in government guarantees of gold cover for paper notes. South American as well as South European countries might be forced to suspend gold convertibility and to allow the value of their currency to decline—a not-unusual occurrence reflecting the influence of elite groups (landowners or exporters, often the same people) who had an interest in high inflation. Weak currencies and monetary chaos suited the oligarchies, which could make their will prevail with astonishing frequency against foreign capitalist allies and foreign creditors. Currency reforms were therefore usually halfhearted affairs that ended in failure; some countries never joined the gold standard, while others such as Argentina or Brazil hardly went beyond formal lip service. Despite British "hegemony" over Latin America, pressure from London never managed to force their compliance. The contrast with Japan is instructive. The archipelago was never a major exporter of raw materials, and special export interests carried little political weight there. The other way around, it had an interest in imports for its rapid modernization, and therefore in a stable currency. All the circumstances conspired to make Japan an ideal candidate for the gold standard.[107]

Fifth, and last but not least, the functioning of the gold standard presupposed free international trade such as the system created in the middle third of the nineteenth century had put in place. Paradoxically, it was the US economy—by then the largest in the world—that proved the greatest factor of instability after the turn of the century. Its huge agricultural sector, with no well-developed rural banking system behind it, had a periodic need for gold that put a great strain on European countries with sizable reserves. It is therefore not enough naively to hail the gold standard as a pure advance of globalization and network building. The risks inherent in this strongly Anglocentric system must also be appreciated. Above all, neither the colonial nor the noncolonial periphery of the world economy was integrated into it, however indirectly or lightly.

The gold standard was a kind of moral order. It universalized the values of classical liberalism: the autonomous individual pursuing his own interests, a reliable and predictable business environment, and a minimally active state. To function well, the system required its players to abide by these norms and to share the "philosophy" in which they were embedded. Conversely, a successful monetary system confirmed that the liberal worldview was fit for life's practical purposes.[108] The system was not invulnerable, however, being dependent on environmental and partly precapitalist conditions. It would not have attained its eventual form without the huge gold discoveries that happened to be made after 1848 on the frontiers of three continents. The mining of the new gold and silver deposits, though later put on a capitalist basis (especially in South Africa), initially owed everything to a primitive "grab and run" mentality in California, Nevada, and Australia.[109] A long chain of cause and effect stretched all the way from coarse gold panners to the refined gentlemen in the boardroom of the Bank of England.

There were wider repercussions of the system. For the dynamic stability of the pre-1914 Belle Époque, often glorified in retrospect, also rested on the fact that the working population was subject to a degree of discipline that later ceased to exist except in totalitarian systems. Since organized labor did not yet have the power to defend income levels or to fight successfully for higher wages, pay cuts could be used to head off a short-term crisis. It is true that workers in the literally "golden" age of capitalism were better off than in earlier times, and that in frontier areas where productivity gains could deliver more cash into their pockets or in those tropical export enclaves where farmers rather than plantation coolies were the force driving expansion, those who owned nothing but their labor power were able to make some headway. Yet the cost of adjustments could easily be loaded onto the backs of the weak. The gold standard was the mechanism symbolizing an order in which liberalism paradoxically went together with the subordination of both capital and labor to the "iron laws" of economics.[110]

The Export of Capital

If the nineteenth century was a time of network building in the world economy, then this applied not only to commerce (free-trade regime) and monetary matters (gold standard) but also to international finance markets.[111] Here too, as in currency relations, though not in international trade, the discontinuity with the early modern period is greater than the continuity. The "modern" European banking system gradually took shape from the sixteenth century on. Instruments such as long-term public debt and the financing of foreign governments were well developed, so that overseas investors subscribed on a considerable scale to Britain's national debt. The government of the newly independent United States of America tried to raise long-term loans on the Amsterdam money market, which was still in good shape despite the decline of Dutch commercial hegemony. The free transfer of capital characteristic of eighteenth-century Europe was severely limited by the wars that shook the continent between 1792 and 1815.

Subsequently, capital markets were rebuilt more as national institutions, with a greater level of government involvement, and only later gradually moved back to forms of international integration.[112]

The "cosmopolitanism" of the early modern period had been confined to Europe; no ruler and no private individual from Asia or Africa had thought of borrowing money in London or Paris, Amsterdam or Antwerp. This changed in the nineteenth century, especially during its second half. While tens of millions of Europeans and Asians migrated overseas, some nine to ten billion pounds sterling flowed out from just a few European countries (Britain being the front-runner by far) to nearly all parts of the world.[113] These sums took one of four forms: (a) credits to foreign governments; (b) loans to private individuals living abroad; (c) corporate stock and bonds held by foreigners; and (d) direct investment by European firms in other countries, often through branches and subsidiaries.

The export of capital was essentially an innovation of the second half of the nineteenth century. In 1820 there was very little foreign investment—all of it British, Dutch, or French[114]—but the period after 1850 saw the gradual emergence of the necessary prerequisites: special financial institutions in both lending and borrowing countries, accumulation of the savings of a new middle class, and a new awareness of the opportunities for foreign investment. Above all, liquid assets and the capacity to handle them came together in that unparalleled square mile called the City of London. The London capital market mobilized credit internationally and financed business far beyond the confines of the British Empire; it attracted funds from all over the world, and handled the issue of securities from many countries of origin as they became more and more important worldwide during the decades before 1914.[115]

By 1870 Britain, France, and Switzerland were the only countries in the world with significant foreign investments (the Netherlands no longer played any role). Germany, Belgium, and the United States joined the list during the great boom that followed, yet on the eve of the First World War, when Britain had long lost its industrial supremacy, its 50 percent share of capital invested overseas still made it the largest source of foreign investment, followed a long way behind by France and Germany. The United States, accounting for no more than 6 percent, was not yet a major factor in the equation. British capital was present everywhere in the nineteenth century. It financed the Erie Canal, the early railroads in Argentina and Japan, and conflicts such as the war of 1846–48 between the United States and Mexico. For a long time it enjoyed a front-ranking position such as that which the United States briefly held around 1960.

Though international finance developed in response to the needs of global trade and communications, it would be misleading to think of the basic structure of capital flows as a fully articulated network. They did not have the reciprocity of trade relations: capital was not exchanged but transferred from core to periphery. The reverse flow from countries in receipt of the credits and investments consisted not of loan capital but of profits, which disappeared into the pockets of the financiers. It was thus a typically imperial constellation, in which the asymmetry was plainly visible. The export of capital could be steered much better than trade flows, for there were only a few control centers.[116] And since, unlike trade, it presupposed the creation of modern institutions such as banks, insurance companies, and stock exchanges all around the world, it presented only weak analogies with the linkage between European merchants and preexisting local networks.

There were also considerable differences between capital flows and international monetary relations (although these were essential for the conduct of financial business). Before 1914 the circulation of investment capital was not regulated by any international agreements; there were no capital controls, no equivalent of the customs departments that had an effect on trade, and no limits on the sums that could be transferred. All that had to be paid to state treasuries was the capital gains tax, if there was one in the respective country. In Germany and France after 1871 the government had the right to block a public loan to another country (which seldom happened); in Britain and the United States even such instruments were lacking.

In contrast to the situation today, foreign loans were not generally issued by *governments*, and of course development aid was unknown. A foreign government in need of money turned to the free capital market. Large projects usually involved a consortium of banks, which either existed already or had to be assembled ad hoc. In many cases, such as that of Chinese government bonds after 1895, banks from different countries joined forces. All the major banks of the time had a branch in London, where most of the international loans were issued. The often colossal sums for war reparations, such as those incurred by China after its war with Japan in 1894–95, also had to be raised in the private money market.

Although European governments were not themselves active as creditors or donors, they did offer diplomatic and military support that made the banker's task easier. Many loans were foisted on reluctant parties, such as China or the Ottoman Empire, which found it more difficult to resist unfavorable terms if these had the backing of the British or French government. Diplomatic intervention was sometimes required to obtain securities from foreign borrowers. Whereas German and Russian banks worked closely with their governments from the 1890s on, the same was not true of their counterparts in Britain. The major British bankers of the age were never puppets of Whitehall, and the British state (or its avatar, the government of India) could sometimes stubbornly distance itself from private banking and business interests. High finance and international politics never overlapped completely. Otherwise, how would it have been possible for French bankers in 1887 to organize capital exports to Russia with great vigor, at a time when the Tsarist Empire was still in an alliance with Germany?[117] Yet the boundary between private interests and state strategies could become blurred, especially if foreign loans required official authorization or if the "good offices" of diplomats were brought to bear in the negotiation of concessions or contracts. In some striking cases—such as those involving China (1913) and the Ottoman Empire (1910), both at a time of weakness in the aftermath of revolution—a request for a loan was used to exert massive pressure. Those who felt its weight could not have been too concerned about the precise share of public and private elements in such financial imperialism.

The large-scale export of capital after 1870 was linked to expectations, especially among small private investors in Britain and France, that good and relatively secure profits were to be had overseas and in the Tsarist Empire. The ideal country for investors was one in the throes of modernization, politically stable, and with a high demand for Western railroads and other industrial inputs, yet sufficiently weak to accept and meet the conditions set by lenders. Such a scenario did not always correspond to the realities. Russia, Australia, and Argentina came close to it, but as for China, the Ottoman Empire, Egypt, or Morocco the average European "coupon clipper" (as Lenin put it) hoped that the Great Powers would prop up the government and that creditors would be indemnified in the event of a crisis. Were the general financial expectations fulfilled? In the period between 1850 and 1914, loans to the ten main borrowers did *not* secure an average return higher than that obtainable on domestic government bonds.[118]

Japan was anything but an unreliable debtor. It transformed itself into a model borrower, enjoying the highest trust on finance markets, but was forced to run up debts to cover its chronic balance-of-payments deficit. It also had to fund its costly wars with China and Russia, although, as we have seen, it managed to extract exorbitant reparations from China after its victory in 1895. By the end of the century the Bank of Japan was even strong enough to help out the Bank of England in case of need. Yet the Japanese government took care never to borrow under pressure or without due preparation, and avoided overreaching itself at all costs; even foreign business investments in Japan were made virtually impossible between 1881 and 1895. Japan was therefore not an easy customer to deal with, and over the years it proved capable of negotiating uncommonly favorable loan terms. Thanks to these farsighted policies, and to the mobilization of domestic capital through a reformed tax system and Asia's only network of savings banks, Japan offered no potential targets for European finance imperialism.[119] By contrast, one of the main obstacles to development in the Muslim world was that it lacked an efficient banking system under its own control—and failed to develop one after contacts with the West became more intensive.[120] The *domestic* impulse to run up foreign debts was therefore unusually strong, and little could be done in the face of Western attempts to gain financial supremacy.

Given the state of statistics in the nineteenth century, the export of capital was much more poorly recorded than international trade. In the end, the only source of accurate information remains the archives of participating banks. The extraordinarily high figures for "British" capital exports processed by the City of London include not just capital originating in the British Isles; investors from other countries that lacked financial institutions of their own usually had no other choice than to channel their funds via London. In 1850 roughly one-half of *British* foreign capital was invested in Europe and another quarter in the United States, followed by Latin America and finally the British Empire. The distribution pattern changed after 1865, however, and then remained essentially the same until 1914. During this period 34 percent of new issues went to North America (United States and Canada), 17 percent to South America, 14 percent to Asia, 13 percent to Europe, 11 percent to Australia and New Zealand, and 11 percent to Africa (mostly South Africa).[121] One is struck by the shrinking significance of Europe and the rise of the United States to become the number one destination for British capital. Almost exactly 40 percent went to the countries of the empire: India retained its importance throughout; Australia was the chief recipient of credit until 1890, after which the booming Canadian economy took the lead. Many small colonies in Africa or the Caribbean absorbed very little capital. Nevertheless, capital exports meant that major projects in the colonies no longer had to be accomplished with local resources alone. In the years around 1800, the development of Calcutta into the architectural pearl of the East had been funded entirely out of Indian taxes. That would have been altogether insufficient for the large-scale railroad construction that took place later in the century.

Within a few decades of the beginning of new-style capital exports, the "global South" was tightly enmeshed into the patterns of global cross-holdings. A comparison with the present day will make this apparent. In 1913–14, of all foreign investment around the world (not only British investment, as in the figures given in the last paragraph), no less than 42 percent was placed in Latin America, Asia, and Africa. In 2001 the corresponding total was only 18 percent. The share of Latin America had plummeted from 20 percent to 5 percent and that of Africa from 10 percent to 1 percent, while Asia had remained steady at the 1913–14 level of 12 percent.[122] In absolute terms, the figures are incomparably greater today than they were a century ago. But their geographical distribution, instead of widening out, is now concentrated to an extreme degree in western Europe and North America. The web of global capital did not become more even and dense, as did the networks of trade or (after 1950) air transportation. Latin America is today largely, and Africa almost completely, uncoupled from the great flows of finance. On the other hand, huge amounts of capital stream into North America or western Europe from regions that in 1913 were peripheral to the world financial system (Arabian oil states, China). The twentieth century witnessed a *de*globalization of international finance. Poor countries have worse access to external sources of capital than they did before the First World War. The good news is that political colonialism has been defeated; the bad is that economic development has become very difficult to achieve without the participation of foreign capital.

Whether as portfolio investment or foreign direct investment by companies using the capital on their own account, the largest share of British (and probably European) foreign investment before 1914 flowed not into the development of new industries but into infrastructural projects such as railroads, ports, and telegraph lines. The export of capital, itself only to a limited extent channeled through dispersed weblike structures, was thus a decisive element in the construction of *communication* networks around the world. Of course, a lot of this was used to fund the exports of Europe's machine-building industry (mainly railroads); many loans were directly linked to trade orders. The intertwining of *indigenous* financial systems with international flows of capital is a subject about which too little is known. The circuits of *agrarian* finance, supremely important for societies with a large farming sector, were little affected until around 1910, especially in business environments where efficient credit institutions survived from the period before contact with the West. Not all credit in Asia or Africa was—as Western clichés had it—"usury."[123]

Debts

The export of capital in the last five decades before the First World War projected the distinction between creditors and debtors onto the international level.[124] From now on there were creditor and debtor *countries*. Quite a few debtors actively sought to obtain capital. In the 1870s, large US banks sent representatives to London and various Continental financial centers to gather funds for investment in American infrastructure.[125] For those in search of capital, it was advisable to negotiate favorable rates of interest, maturity dates, and payment terms. Many overseas governments—not only Japan but post-1876 Mexico, for example, under Porfirio Díaz—went to great pains to shore up their reputation as financial partners who paid their debts on time. A country highly valued for its investment potential could hope to attract a continual inflow of foreign capital on tolerable conditions.[126] Elsewhere, a mixture of European predation and reckless non-European extravagance might lead to financial catastrophe in dependent countries. Egypt in the third quarter of the nineteenth century was a case in point. The government first sank £12 millions into the Suez Canal, out of which the country got nothing economically, and then had to sell its shares to the British government for £4 million. Ferdinand de Lesseps had palmed this huge commitment off onto Pasha Said, and in November 1875 Benjamin Disraeli used the impending financial collapse of the khedive to stage a coup that placed Britain alongside France at the center of political influence and promised rich pickings for the British state coffers. Since no session of Parliament capable of approving the financial outlay was imminent, the British prime minister borrowed it from the House of Rothschild, which charged a commission of £100,000. The Suez Canal business was highly complicated, and Disraeli soon had to recognize that Britain's stake of 44 percent did not give it a controlling interest. He could not foresee just how rich the pickings would be, or that the value of the shares would increase tenfold to £40 million.[127]

During the reign of Ismail, Egypt became seriously overstretched in other areas too. The khedive was unnecessarily generous in awarding concessions to foreigners and accepted loans at high real rates of interest and unusually low rates of issue. Between 1862 and 1872, Egypt took on loans with a nominal value of £68 million (for which interest then had to be paid), but received only £46 million in actual payments.[128] Ismail was not quite as irresponsible with money as foreign detractors have claimed to this day; some of the funds went into useful projects such as railroad construction or improvements to the port of Alexandria.[129] The real heart of the matter was a rigidly antiquated tax system, which did not allow the government to profit from the expansion of dynamic sectors of the economy, and the sharp decline in revenue from cotton exports after the end of the American Civil War in 1865. By 1876 the Egyptian state had to declare bankruptcy, and in the following years its financial affairs were placed under near-total Anglo-French control. The Egyptian Commission de la Dette grew into a major department of the central government, almost exclusively staffed by foreigners.[130] From this it was only a small step before the British took charge alone in 1882, in a quasi-colonial setup. Egypt's fate as a debtor country was thus even harsher than that of the Ottoman Empire, which, already insolvent by 1875, was subjected to a somewhat less invasive debt-management regime.

Failure to keep up repayments to foreign creditors was not an "Oriental" speciality. Every country of Latin America found itself in this situation at one point or another, as did the Southern US states before the Civil War, Austria (five times), the Netherlands, Spain (seven times), Greece (twice), Portugal (four times), Serbia, and Russia.[131] On the other hand, there were highly in-debted countries outside Europe that scrupulously paid off their debt—above all China, whose railroad bonds plunged into crisis only amid the political tur-moil of the 1920s. From the 1860s on, the country's customs department was run by an authority—the Imperial Maritime Customs—which, though not a direct instrument of the imperial powers, was under strong European influence; in the late nineties it was even authorized to bypass the Chinese Finance Ministry and to pay revenue straight into the accounts of foreign creditor banks.

In the 1870s at the latest, a new kind of crisis, already common in Latin Amer-ica since 1825, became a characteristic phenomenon at the interstate level: the international debt crisis. Mostly this involved a conflict between extra-European governments and European private creditors, but it seldom remained without political or diplomatic consequences. The lenders wanted their money back, but that was possible, if at all, only if governments became involved on either side. An ever-present tendency to financial imperialism therefore lurked within the international bond market. Debt was as unavoidable as it was risky for everyone involved.[132] But for almost a whole century—from 1820 to 1914—no tears in the web of international loans were so radical that they could not be repaired through intervention. Such breakdowns would become a feature of the twentieth century: in 1914 the state coffers in Mexico lay empty in the aftermath of revolution, in 1918 the new Soviet regime in Russia repudiated the Tsar's foreign obligations, and after 1949, in an exact replay, the People's Republic of China unilaterally canceled all debts to "imperialist" creditors. Such financial radicalism was unthinkable in the nineteenth century.

CHAPTER XV

Hierarchies

The Vertical Dimension of Social Space

1 Is a Global Social History Possible?

"Society" has many dimensions. One of the most important is hierarchy.[1] The majority of societies have an *objectively* unequal structure: some of their members dispose of more resources and life chances than others, perform less hard physical labor, enjoy greater respect, and command obedience for their wishes and orders. As a rule, people also perceive these *subjectively* as a set of relations of superiority and subordination. The utopian dream of a society of equals has existed at various times in many civilizations—utopian, because it contradicted the reality of life as a hierarchy in which the individual sought to find his or her place. In the Victorian era, even in a distinctly modern society such as Britain, the image of society as a kind of step ladder was widespread even among the working population.[2]

"Hierarchy" is only one of several approaches to social history. Historians have variously focused on classes and social strata; groups and milieux; family types and gender relations; lifestyles, roles, and identities; conflict and violence; communicative relations; and collective symbolic universes. Many of these aspects lend themselves to comparisons between societies at a geographical distance from one another. Often it is worth pursuing the hypothesis that there were influences and transfers across civilizations—more plausible and easier to demonstrate in the case of economic networks, cultural orientations, and political institutions than in the formation of social structures. Society grows out of everyday practice at specific places and times. It is also dependent on local ecological conditions: collective human life inevitably varies according to whether the location is tropical rainforest, desert, or Mediterranean coast. Beijing and Rome are at approximately the same latitude, yet for long periods of time they have had very different forms of society. The ecological framework defines possibilities, but it does not explain why some of these rather than others actually become reality.

There is a further difficulty. In the course of the nineteenth century, it came to be taken for granted that a distinctive national society must correspond to the nation-state within its political boundaries. To some extent this was indeed the case. Nation-states often developed out of older social ties; societies began by thinking of themselves in terms of national solidarity and then looked for an appropriate political form. Conversely, the political framework and the constant influence exerted by the state strongly marked the forms of society. The law is the original expression of this, insofar as it is validated by the authority of the state. "National" societies may thus be usefully characterized by their particular legal institutions. Alexis de Tocqueville underlined this point in 1835 with reference to inheritance law: provisions for the distribution of a dead person's effects "do belong, true enough, to the civil code but they ought to take their place at the head of every political institution since they have an unbelievable effect upon the social conditions of the people, while political laws only mirror what the state actually is. They have, moreover, a reliable and consistent method of operating on society since they take a hold to some degree on all future generations yet unborn."[3] So it was that quite different types of agrarian society crystallized around distinctive legal institutions regulating the passing on of property. Much depended on whether land and agricultural enterprises were kept together by primogeniture (England) or split up through the partition and distribution of real estate (China).

Despite the framing and shaping of social life by political authorities in bounded territories, it is not easy, and often quite pointless, to make statements about Chinese, German, or Mexican society *in general*. One may well doubt, for instance, whether it is possible to speak of just one "society" in Germany across a multitude of sovereign territories around 1800,[4] and in the case of China no fewer than ten different "regional societies" have been identified within the overarching framework of the Qing empire.[5] The British colonies that formed themselves into the United States of America were in essence thirteen different countries, with characteristic forms of society and regional identities. Little changed in this respect during the subsequent decades; many differences actually increased. Around 1850, extraordinary diversity persisted between the Northeast (New England), the Southern slave states, the Pacific coast (California), and the frontier in the interior. A similar kind of heterogeneity is discernible in the vertical dimension: Egyptian society had been so strictly stratified over centuries that it cannot be described as an even minimally coherent totality. A Turkish-speaking Ottoman Egyptian elite ruled over an Arabic-speaking majority to which it was bound by little more than the tax nexus.[6] To an extent scarcely imaginable today, older or even archaic social forms survived in ecological, technological, or institutional niches around the world, long after they had ceased to be progressive or dominant.[7]

Even more questionable are sociological generalizations at the higher, supranational level of "civilizations." Historians trained in subtle distinctions and the study of change over long periods are reluctant to operate with static macro-constructs such as "European," "Indian," or "Islamic" society. Numerous attempts to define the cultural or social peculiarities of Europe suffer from the juxtaposition of such phantoms and from the untested claim that salient European virtues are absent in other parts of the world. In the worst cases, the clichés about Europe itself are no less crude than those about Indian or Chinese society.[8]

Grand Narratives

There are still no synthetic accounts of all-European or North American social history in the nineteenth century—not for want of research, but because of the difficulties of organizing and conceptually working through the vast amount that is known about it. All the harder must it be to outline such syntheses for other parts of the world, where many empirical questions are still unresolved and sociological or social-historical concepts of Western origin cannot be simply applied without further ado. To embark on a social history of the *world* for a whole century would be the height of presumption. It would have no identifiable object, since no uniform "world society" can be uncovered for 1770 or 1850, or indeed for 1900 or 1920.

Historians in the nineteenth century itself were less cautious. Building on Enlightenment ideas of progress, some of the leading minds of the age elaborated theories of social development and in many cases held them to be universally valid. Eighteenth-century Scottish moral philosophers, economists, and philosophers of history—such as Adam Ferguson and Adam Smith—postulated a material progression of the human species through stages of hunting and gathering, pastoralism, and agriculture to modern life in the "commercial society" of emergent capitalism. The German Historical School adopted such conceptions, while in France Auguste Comte constructed a stages model that placed the emphasis on the intellectual development of mankind. Karl Marx and his disciples thought they could discern a necessary succession from primitive society to slavery and feudalism to bourgeois or capitalist society, though Marx himself, in his later years, hinted that there might have been a deviation from this normal path: the so-called Asiatic mode of production.

Other authors thought less in terms of sequential stages of development than of great transitions. In the 1870s, the English philosopher Herbert Spencer suggested a progression from "military" to "industrial" society—an idea rooted in a complex theory of social growth through phases of differentiation and reintegration. The legal historian Sir Henry Maine, who was familiar with India, observed how in many societies contractual relations made status relations obsolete. Ferdinand Tönnies, one of the founders of sociology in Germany, perceived a trend away from "community" to "society;" Max Weber analyzed the "rationalization" of many areas of life, from the economy to the state to music; and Émile Durkheim thought that societies based on "mechanical" solidarity were being superseded by others based on "organic solidarity." Although at least Maine, Durkheim, and Weber were interested in societies outside Europe, it is hardly surprising that all these theories were "Eurocentric" in the spirit of the age. But this was mostly true in an inclusive rather than an exclusive sense: those lagging behind in non-European civilizations, regardless of skin color or religion, could in principle be fitted into general models of social progress. Only toward the end of the century—and then only seldom among the truly important authors—did modernization theory take on racist hues, in the sense that scriptless "primitives" and sometimes even "Orientals" were denied any capacity to rise to higher cultural achievements.[9]

From Status to Class?

To this day, the schemas and terminology of (late-) nineteenth-century sociology have not disappeared from discussion, but they have remained too general for use in descriptions of actual change. Historians prefer to cultivate their own grand narratives—of industrialization, urbanization, or democratization. One such model is of a transition from a "society of estates" or "corporate-feudal society" to "class society" or "bourgeois society." The opposition between the two was already carried to a high pitch of intensity in Enlightenment polemics against the feudal-monarchical order, and in the nineteenth century it became a key theme in the account that European societies gave of themselves. Toward the end of the early modern period, it was argued, the basic organizational principle of society changed: an immobile stratification into clearly defined status groups, each with its particular rights, duties, and symbolic markers, gave way to a structure in which property ownership and market position determined the life chances of individuals and their place in the occupational and class hierarchy. Upward and downward mobility, with formal legal equality as its prerequisite, was much more likely to occur under such circumstances than in the rigid status system.[10]

This model, originating in Western Europe, was by no means equally applicable to other parts of the continent—or unconditionally even to Britain, the "modern" pioneer. England in 1750 was rather a "commercial society," in Adam Smith's sense, than a status society of the Continental type. In the Scottish highlands, however, which had no transitional stage of estates, the old Gaelic clan structures—not incomparable to those in Africa—passed directly in the last quarter of the century into the social relations characteristic of agrarian capitalism.[11]

Eighteenth-century Russia also lacked estates in a French or German sense: that is, no corporate groups with a separately defined juridical status and territorial basis, rooted in local legal traditions and opportunities for political participation. The division of society (and, more narrowly, of the elite servicing the state) into ranking classes, and the allocation of collective privileges, proceeded outward from the state. Thus, no group rights were safe from retraction by the monarch.[12] Russia was a relatively open society, in which it was possible to climb the ladder by serving the state, and nonpeasant city dwellers could not be precisely or stably demarcated from other segments of the population. Persistent attempts by the Tsarist authorities to impose a system of legally defined ranks came into constant conflict with the plasticity of actual status ascriptions. This has led some scholars to speak of a general "lack of structure," or an absence of universally recognized concepts of social order, in the late Tsarist Empire.[13]

Since the initial situation varied from region to region, the "from status to class" model only imperfectly describes social change in a more comprehensive sense.[14] Not everywhere in Europe around 1800 was "estate" or "status group" the main principle of social classification; and elsewhere in the world, status societies were rarely to be found. The term may best be applied to Tokugawa Japan, with its deep social and symbolic cleavage between nobles (samurai) and commoners—although status groups there did not exercise representative political functions such as we know them from France or the Holy Roman Empire.[15]

Status group criteria of social hierarchy were less pronounced in Asia than in the center of Europe. Siam was an extreme example of an Asian country where a deep gulf separated the nobility (*nai*) from ordinary people (*phrai*), although both groups were jointly subject to the limitless power of the king.[16] Elsewhere, as in China, state rhetoric had propagated since ancient times a fourfold division of society into scholars, peasants, artisans, and merchants. But these vague distinctions did not crystallize into clear-cut legal categories or systems of privilege, and in the historical reality of the eighteenth century they were overlaid with more sophisticated hierarchies. Any part of the world living mainly under tribal conditions, whether in Africa, Central Asia, or Australasia, exhibited an organizational principle quite different from that of status society. Hindu societies had yet another form of differentiation, with hierarchies based on endogamy and purity taboos. The concept of caste may now be under a cloud, suspected of having been a phantasm of the colonial state and Western ethnology, but it is clear that important forms of society in premodern India differed in their classificatory rules from European status society. Those rules were, however, given additional force for traditionalist ends. When the British extended their rule to Ceylon after 1796, they perceived social relations there through an Indian lens and introduced a kind of caste system that had not previously existed on the island.[17]

The old European status society was transferred to overseas colonies only in a disaggregated form. In British North America, the fine distinctions that marked society in the British Isles were preponderant from the beginning. Hereditary aristocracies with the privileges of an estate never gained a foothold there, and the prevailing image of society was one of Protestant egalitarianism with only small internal gradations. In all the settler societies of the Americas, ethnic inclusion and exclusion played a role it could never have had in Europe. In North America, the principle of equality was from the beginning valid only for whites, while in Hispanic America—as one of its most acute observers, Alexander von Humboldt, already showed near the end of the colonial era—skin color operated on top of everything else as a criterion of stratification.[18] Estate elements, migrating across the Atlantic in the sixteenth century and contributing to the formation of a conquistador nobility, soon had this new principle of hierarchy superimposed onto them. As late as the second half of the nineteenth century, Mexicans still defined their place in society primarily in terms of color or "blood mix," and only secondarily by occupation or class.[19]

For vast areas, the *global* social history of the nineteenth century is identical with the history of migration and closely bound up with the history of diaspora formation and the new frontiers resulting from it.[20] After 1780, neo-European settler societies either were newly founded against indigenous resistance that might be weak (as in Australia) or strong (as in New Zealand), or they received large new waves of immigration that built them up from thinly populated peripheries into sizable countries in their own right (United States, Canada, Argentina). In not one of these cases, however, were European social structures exported en bloc. Noble strata capable of reproducing themselves as such never sank roots in the British settler colonies, while the *very poor* underclass at the other end of the spectrum was not disproportionately represented except for those driven from home in conditions of extreme misery, as during and after the Great Famine in Ireland. Australia was a special case, because settlement began there (in New South Wales) with convict transportations.[21] But an underclass removed from the context of its original classification is not automatically an underclass in the open situation of a settlement frontier. Other groups crossing the Atlantic consisted of millions of people from the middle layers of European society, as well as declassed nobles and less privileged members of noble families. Worldviews and patterns of social differentiation had to be invented and negotiated anew in the colonies.[22] Opportunities to climb the social ladder were greater than in Europe. The process whereby European migrants built new societies transcending the status orders of the Old World is one of the most striking developments in the global social history of the nineteenth century.

In the nineteenth century, societies around the world practiced a multiplicity of hierarchical rules alongside one another, differing in their property relations and the dominant ideals of social ascent. A clear classification covering most possibilities is scarcely feasible. In addition to market-regulated societies of property owners ("bourgeois" society), which in a western or central European or North American perspective were the characteristic type in the nineteenth century, there were residual status societies (e.g., Japan until about 1870), tribal societies, theocratic societies in which clerics were the dominant stratum (e.g. Tibet), societies with a meritocratic elite selection (China, precolonial Vietnam), slave societies (Southern US states until 1863–65, Brazil until 1889, remnants in Korea),[23] "plural societies" where various ethnic groups coexisted in the framework of colonial rule, and mobile "frontier societies." The transitions were fluid, and hybrids more or less the rule. Comparison becomes easier if we focus not on the whole profile of hierarchy but on individual positions within it. Let us take two examples, initially from a European vantage point: the nobility and the bourgeoisie.[24]

2 Aristocracies in (Moderate) Decline

International Scope and National Profiles

The nineteenth century was the last in which the nobility, one of the most ancient social groups, played an important role. In eighteenth-century Europe it still "had no social competitor,"[25] but by 1920 it was no longer possible to say anything of the kind. In no European country had the nobility survived as a primary political force, or as the main one setting the cultural agenda. This decline was a result partly of revolutions in the late eighteenth and early twentieth centuries, partly of the reduced value of land as a source of wealth and prestige. Where the revolutions overthrew monarchies, the nobility was deprived of its imperial or royal protector. But even in Great Britain where the old order did not collapse and nobles were able to preserve greater influence than elsewhere, the section of the population endowed with a knighthood or peerage lost its virtual monopoly over top positions in the political executive. From 1908 on, all but three British prime ministers have had a bourgeois family background and only one was heir to a noble title. The fall of the age-old European institution of the nobility took place in the relatively short space of time between 1789 and 1920. Those two years are not, of course, joined by an ever dropping curve. East of the Rhine, the political situation of the nobility did not become critical until the final period of the First World War. On the whole, "the nineteenth century was a good time to be an aristocrat."[26]

Nobilities have existed almost everywhere in the world, except in "segmentary" societies. A small minority of the population concentrates in its hands the means to exercise violence, enjoys favorable access to economic resources (land, manpower), belittles manual labor (except for war and hunting), cultivates a high-profile lifestyle with an emphasis on honor and refinement, and passes on its privileges from generation to generation. Nobilities often consolidate themselves into aristocracies. Over and over again in history, such aristocracies have been decimated or even perished as a result of war. In modern times, colonial conquest hit them especially hard: they suffered destruction or drastic political-economic degradation, beginning with the Aztec nobility in sixteenth-century Mexico and continuing around the globe. But it sometimes happened that an aristocracy was incorporated into an empire in a subordinate position and managed to retain its symbolic distinction. Thus, after 1680 the Manchurian Qing Dynasty, already commanding the allegiance of its own nobility, disempowered the Mongol aristocracy and bound it through a set of vassal relations. Indirect rule in the European colonial empires involved a similar technique. Other empires, however, did not allow local aristocracies to survive. The Ottoman Empire suppressed forms of Christian feudal rule in the Balkans and did not give leeway for a new landowning elite to take shape. At the beginning of the nineteenth century, Serbia and Bulgaria were without an aristocracy but had a relatively free peasantry by eastern European standards.[27] Where the nobility remained in existence under foreign rule, it was often denied a say in politics—such was the case in pre-unification Italy, for example—and nobles failed to develop a substantial experience of public service.

In eighteenth-century Europe—unlike in the Arab world, for example—the days of knightly chivalry were over. But even without this primal function it was clear in 1800, as it still was in 1900, who did and did not belong to the European nobility. Only in England, with its elastic social attribution, did many people on their way up have to ask themselves whether they had crossed the critical threshold.[28] In regions where certain legal privileges survived until the end of the First World War—above all, in the eastern half of the Continent—there could anyway be no mistaking the scale of the phenomenon and its subtle internal hierarchies. Elsewhere the boundaries were defined by titles, additions to names, and other symbolic markers. No other social grouping attached so much value to distinction. The fact that one belonged to the nobility had to be visible and unambiguous.

Apart from tiny transnational elites such as the top of the Catholic hierarchy or Jewish financiers, the nobility was the segment of European societies with the strongest international orientation. Its members knew of one another, could gauge positions in the ranking order, shared a series of behavioral norms and cultural ideals, spoke French when they needed to, and participated in a cross-border marriage market. The higher their rank and wealth, the more they were integrated into such wider networks. On the other hand, close associations with landownership, agriculture, and country life meant that nobles often had strong local roots and were less mobile than certain other sections of society. Between the internationalist and local levels of orientation lay a *national* arena for noble life, where solidarity and a sense of identity were strengthened in the nineteenth century. While the nobility became more international thanks to new communications technologies, it was also increasingly "nationalized."[29] A new conservative nationalism thus appeared on the scene alongside an older liberal nationalism—above all in Prussia and later Germany.

Three Paths in the History of European Nobility: France, Russia, England

In France, the nobility was stripped of all its titles and privileges during the revolution. Its special rights were generally not restored in later years, especially in the case of émigrés, so that "empty" titles were all that remained behind. Although the importance of landownership should not be underestimated, the French nobility played only a secondary role in a society that was exceptionally "bourgeois." In addition to those who had survived from the ancien régime, a new nobility emerged under Napoleon (himself a scion of the lower Corsican nobility), which the old aristocracy often regarded with a mixture of disparagement and admiration as a breed of parvenus: mostly military dignitaries endowed with rights of succession and forming the core of a new hereditary elite.[30] The rise of a miller's son to become Duc de Danzig (in 1807), in recognition of his services as an army marshal, would have been unthinkable under the ancien régime. Ennoblement along similar lines, serving as an instrument of state patronage, was then liberally applied almost everywhere in nineteenth-century Europe. Napoleon also created a merit-based Légion d'Honneur, a kind of postfeudal elite corporation without hereditary rights, which was later converted unproblematically into republican forms.

After 1830 in France there was no strong central institution, like the House of Lords in Britain or the royal court in most other countries, around which the nobility could gather. Neither the "bourgeois monarch" Louis Philippe nor the imperial dictator Napoleon III built up extensive court structures or supported the grandeur of their rule on a strong upper nobility. The vestiges of court life then disappeared along with the emperor in 1870. Insofar as there remained an identifiable French nobility during the first two-thirds of the nineteenth century, it was less of a self-conscious class than its counterparts farther east in Europe. And the impoverished nobleman as a type was encountered far more often in France (and Poland) than elsewhere. Wealthy property owners of diverse origin—local notables, as they were known early on—became the opinion leaders who set the tone for the wider society.[31] In the Third Republic this aristocratic-bourgeois composite stratum, typically residing in provincial cities, became more and more marginalized. In no other major European country did the nobility have such a small superiority in power and landownership at the decisive local level.[32]

At the other end of the spectrum stood the heterogeneous Russian nobility,[33] more dependent on the crown than its counterparts in other major European countries and empires. Catherine the Great's "Charter" of 1785 relaxed the state's chokehold, giving nobles full property rights and putting them more or less on a legal par with their fellows in Western Europe. The state and the imperial house remained by far the largest owners of land, however, and tsars since the time of Peter the Great gave out land and "souls" (i.e., serfs) as gifts. Ennoblement, a simple procedure, was widely practiced at the end of the nineteenth century. Some of the largest landowning magnates could trace their wealth and privileges back only a few decades or even years. There was also a large "minor gentry," consisting of people who in England would not have been considered part of the nobility. The diffuse picture of an upper class based on landownership had greater affinities with an old European conception of *nobilitas*. Moreover, since the abolition of serfdom in 1861 did not dramatically affect the financial status or social position of large landowners, it was not comparable in its effects to the simultaneous ending of slavery in the Southern US states. The reform was incomplete, and the political dominance of the former masters remained intact, so there was limited incentive for landowners to turn themselves into agrocapitalist businessmen.

The English nobility was unique: the richest class of its kind in Europe, enjoying relatively few legal privileges but present in the political and social control centers. Primogeniture could be relied on to keep large assets in one piece, so that younger sons and their families drifted to the periphery of noble society. But there were few of the trappings of a caste; the only thing to be precisely defined was the right to sit as a peer of the realm in the upper house of Parliament. In 1830 there were 300 family heads belonging to the upper nobility, and in 1900 more than 500.[34] As far back as the 1780s, under William Pitt the Younger, the government had increased the rate of ennoblement and made ascent into the lower nobility a fairly simple process. What remains unclear, even today, is the extent to which Victorian nouveaux riches bought land in order to boost their image.[35] In any event, a country house was indispensable as a stage for social intercourse. Inversely, even the largest landowners were not averse to participation in "bourgeois" society.

The ideal of the gentleman cultivated by the English nobility had an extraordinary integrative force, generating a lifestyle and culture at home and in the Empire that often lacked the razor-sharp distinctions of continental European elites.[36] The gentleman increasingly became an ideal without fixed social moorings; "blue blood" played scarcely any role. Even if the prerequisites were considered innate, male offspring still had to be socialized as gentlemen at elite public schools and at Oxford or Cambridge. A gentleman could be anyone who, whatever the basis of his prosperity, appropriated and practiced the lifestyle, values, conduct, and bodily practices associated with the ideal. An education at Eton, Harrow, or Winchester was not narrow status training in the manner of early modern "knight academies" (*Ritterakademien*) on the Continent, nor did it focus primarily on intellectual development; its main purpose was aristocratic-bourgeois character formation of an integrative kind—with a growing tendency to imperial militarism as the century progressed.[37] The principle of achievement was in command. Noblemen might have things easy in English society, but they had to face up to competition. There was a permanent need to forge and sustain alliances beyond their own social stratum. The English aristocrat was not dependent on the Crown; there was no longer a court nobility under Queen Victoria. The nobility assigned themselves leadership tasks and expected gratitude and deference in return. This was not the same as passive obedience but rather an attitude capable of being channeled through the institutions of a political life that was being slowly democratized.[38] More plainly than anywhere else, nobility was not so much a precise legal status as a mental disposition: a self-assurance in setting the tone for others.

Survival Strategies

If the European nobility went downhill, it was not for want of trying out survival strategies.[39] The most promising of these (at the very time, roughly after 1880, when agricultural yields were declining in large parts of Europe) was to replace the traditional rentier mentality with an opening to the business world, new kinds of investment portfolios, social fusion with the upper bourgeoisie (on its part strongly inclined toward land acquisition and gentry lifestyles), a marriage policy to protect against the splitting and dispersal of estates, and the adoption of national leadership roles, especially when there were not enough other candidates for them.

Although such a reorientation, practiced in various combinations right across Europe, might achieve its immediate goal in particular cases, the European nobility had lost its leading cultural position by the turn of the century. A market-oriented cultural industry had appeared in place of the aristocratic patronage that had still sustained the European fine arts and music in the age of Haydn and Mozart. Musicians obtained funding from performances in the opera house and the concert hall, painters from public exhibits and the nascent art trade. Noble subjects became less common in literature, still lingering, for example, in Anton Chekhov's melancholy stories and plays about the twilight of the Russian nobility. Only a few prominent—and bourgeois—thinkers, such as Friedrich Nietzsche and Thomas Carlyle, still preached aristocratic ideals, though these were detached from any clear social foundation and referred to nobilities of the spirit rather than the blood.

Were empires a stomping ground for European aristocrats? This can safely be said only of the British Empire. The colonialism of Napoleon III and the Third Republic, by contrast, had a decidedly bourgeois veneer. Top positions in the army and civil service of the British Empire continued to be staffed by members of the nobility, who seemed best suited to bridge the different civilizations and political cultures of colonial society by cultivating a supposed affinity with Asian or African nobilities in the service of higher imperial ends.[40] India was the most promising field of application in this respect, while bourgeois specialists were on the advance in Africa and elsewhere. A certain romanticism of decline ensured a minimum of cross-cultural sympathy for non-European subjects of the empire.[41] A special kind of aristocratic consciousness was to be found in the Southern US states before the Civil War, where the numerically small planter elite harbored fantasies of itself as a "natural" ruling class in charge of huge slave plantations— veritable lords of the manor in a rerun of the Middle Ages. Personal detachment from manual labor, revulsion from the "materialistic" vulgarity of the industrial North, unfettered lordship over those subjugated to them: all this seemed to permit a flowering of anachronistic chivalry.[42]

In comparison with the post-1917 "aristocide,"[43] the nineteenth century was rather like a golden October of the European nobility, especially for its upper ranks. The embourgeoisement of the world ground on relentlessly, though at a moderate pace. Precipitate declines occurred elsewhere—for instance, in Mexico during the revolution of 1910–20,[44] or in the three main societies of Asia.

India: A Neo-British Landed Aristocracy?

In India, the princes and their feudal retinues were initially stripped of their functions in one region under British rule after another. The British abandoned this policy after the Great Rebellion of 1857/58, however, as the utopia of a middle-class India, dreamed up by influential English utilitarians in the 1820s and 1830s, lost its attractiveness. Henceforth the main effort went into feudalizing at least the external appearance of British rule. So long as the maharajahs and nizams remained loyal, disarmed, and financially spoon-fed, serving as picturesque disguises for the bureaucratic character of the colonial state, they had nothing to fear.[45] A new, specifically Indian nobility was invented, with Queen Victoria as its distant empress from 1876 on. The romantic chivalry of Victorianism, which in the British Isles expressed itself in neo-Gothic architecture and the odd tournament, had a much larger stage in India and was accompanied with much more colorful pomp.

The details of nobility in India are a complicated matter. As in other parts of the world, the British—or anyway the aristocrats who found employment early on in the "bourgeois" East India Company—looked for an Indian counterpart (a "landed nobility") but had considerable trouble finding one because of the different legal frameworks. Early modern European observers had recognized the problem when they pointed out that private individuals did not really own land in Asia; everything was subject to the monarch's overlordship. In some theories of "Oriental despotism," Montesquieu's being the most notable, this was blown up into the idea that private property in general (not only landownership) was completely insecure—but the Montesquieu school was not altogether on the wrong track. However much Asian countries differed in their legal relations, the link between a particular piece of land and a noble family was seldom as safe from monarchical infringement as it was in most parts of Europe. In Asia, the status and income of the upper classes often derived less from direct landownership than from a (perhaps fleeting) enfeoffment or from tax-farming privileges assigned by the ruler to individuals or groups. On the eve of the East India Company's power grab, the zamindars of Bengal, for example, were not an entrenched landed nobility in the English sense but a rural elite with rights to a sinecure—though admittedly they kept up a grand lifestyle and held the real power in villages. For the British, they were a quasi-aristocracy that promised to guarantee social stability in the countryside, both then and in the future. For a time every effort was made to transform them into a genuine aristocracy more suited to a "civilized" country, except that they were not left with their old police and judicial powers.[46]

The promotion of the Bengal zamindars, including their provision with enforceable land deeds, was only a prelude to their fall. Some of them were no match for the market forces that the colonial order now unleashed; others lived to see the British impose crippling financial demands that could and did end in expropriation. Old established families faced ruin, while new ones arose out of the merchant class. The consolidation of the zamindars into a European-style hereditary aristocracy was a failure, and the hope that they would become "improving" landlords able to invest and develop scientific farming methods ended in disappointment. In the early twentieth century, it would not be zamindars but middle peasants who became the dominant rural stratum in Bengal and many other regions of India, as well as the social base for the independence movement. By 1920, lofty mind-sets and lifestyles were no less marginal in India than in Europe.

Japan: The Self-Transformation of the Samurai

Japan took a sui generis path.[47] In no other major country did a privileged status group undergo such a transformation. The Japanese equivalent of the European nobility was the samurai, originally warriors bound to a lord by strong ties of loyalty and mutual advantage. After the pacification of Japan in the decades around 1600, most samurai remained in the service either of the shogun or of the 260 or so feudal princes (daimyō) among whom the Japanese archipelago was divided. Integrated into the elaborate hierarchy underpinning the shogunate, they were endowed with a number of symbolic markers that identified them as members of a special warrior aristocracy at the very time when no more wars needed to be fought. Many samurai swapped the sword for the paintbrush and took on bureaucratic tasks, making Japan one of the most densely (though not in every respect most efficiently) administered countries in the world. Yet for many samurai and their families there was literally nothing to do. Some worked as teachers, others as foresters or doormen, while others still were secretly active in the despised world of commerce. All the more stubbornly did they cling to their hereditary privileges: the right to bear a family name, to carry a sword and wear special clothing, to ride on horseback and force others to make way for them in the street. All this meant that they closely resembled the nobles of Europe. But their 5 to 6 percent share of the population in the early nineteenth century was comparable only to that of exceptional European countries (Poland and Spain) and much larger than the European norm of well below one percent.[48] The lack of meaningful functions was therefore a major problem, even in quantitative terms, and exacted a high toll from society in general. The main difference with Europe lay in the isolation of the samurai from the countryside: they generally owned no land, let alone any with legally enforceable deeds. Instead, they were paid stipends measured in rice and usually dispensed in kind. The typical samurai, then, did not control any of the three factors of production: land, labor, and least of all capital. He was a particularly vulnerable element of Japanese society.

When the post-1853 confrontation with the West brought Japan's chronic problems to crisis point, it was primarily samurai from princedoms remote from the House of Tokugawa who supplied the initiative for change at national level. This small group, which overthrew the shogunate in 1867–68 and set about building the new Meiji order, recognized that samurai could survive as a distinctive section of society only if they lost their antiquated status. The most important props of their existence were removed with the disempowerment of the princes and the sweeping transformation of the daimyates, and from 1869 on, samurai status was gradually dismantled. The harshest economic blow was the abolition of stipends (cushioned at first by their replacement with government bonds); the worst symbolic humiliation, in 1876, was the revocation of the privilege of the sword. Individual samurai now had to fend for themselves; an important step in 1871 had been the recognition of the freedom to choose an occupation (which the revolution had decreed in France back in 1790). After the final samurai revolts of 1877, there was no longer any resistance to this policy turn.[49] It brought great hardship for many samurai and their families, and the social policies of the government offered only partial alleviation.

The samurai lingered on as an ethos and a myth, but in the 1880s they evaporated as a recognizable element of Japanese society. A new upper nobility, created by the Meiji state in imitation of the British peerage, is reminiscent of a Napoleonic artefact; it was embraced by remnants of the *daimyō* families and the old court aristocracy in Kyoto, while the oligarchs—mostly men under forty at the time of the regime change in 1867–68—bestowed it on themselves as a reward. In the new political system, which from 1890 included a second chamber along the lines of the House of Lords, this nobility would play a significant role as a buffer between the revered and remote tennō and the "common people."

China: Decline and Transformation of the Mandarins

China came closer to European conditions; indeed, it was in many ways ahead in its modernity. It had already had a largely unrestricted market for land in the eighteenth century, and feudal burdens and obligations to private lords had almost disappeared. There was no way of legally enshrining in perpetuity a family's control of a particular piece of land, yet—as in Europe—the acquisition of title deeds made it largely secure from state intervention. But can China's scholar-officials, often called "mandarins" or "literati" by European observers, really be seen as the equivalent of a European nobility? In many respects they certainly can. They had effective control over the bulk of land used for agriculture, and they were the dominant force culturally, far less challenged in this than the European nobility of the early modern period. The most important difference was that, although ownership of land was passed down within the family, status could not be inherited; the two were almost entirely separate from each other. The stratum known in Chinese as *shenshi* and often translated in English as "gentry" represented approximately 1.5 percent of the population—between the percentage share of the nobility in Europe and that in Japan. Entry to it was achieved through state examinations held at regular intervals.[50] Only those who scored at the least the lowest of nine passing grades could enjoy the reputation and palpable benefits of a *shenshi*, including such things as tax exemption and immunity from corporal punishment.

A *shenshi* could consider himself and his family to be part of the local upper class, entrusted with a range of leadership tasks. Where clan organizations existed, he belonged to their inner elite. He shared in the cultural and social world of the Confucian *junzi*, whose basic normative structure corresponded in many ways to that of the English gentleman. Imperial officials, on the other hand, were appointed only from among those who had achieved the top passing grades, usually in an examination in the capital conducted by the emperor himself. To place one of its sons as a court official or a member of the provincial administration was the highest ambition of a family in the hierarchy-conscious society of Imperial China.

Historians have repeatedly contrasted Japan's success with China's failure: the one converted the shock of being "opened up" into a major program of modernization and nation building, whereas the other misread the signs of the times and missed the opportunity to strengthen itself through renewal. China's immobility had various causes. At least as important as a "culturally" determined lack of interest in the outside world were the lack of a strong monarchy after 1820 and the delicate balance in the state apparatus between Manchu dignitaries and Han Chinese officials; any strong impetus to reform threatened this unstable equilibrium. That is one way of reading Chinese history, but we might also experiment with posing the key question in a different way. Why was it that in Japan a much lesser impulse from abroad—Commodore Perry's theatrical intrusion was by no means comparable to the Opium War of 1839–42—triggered a much sturdier response than in China?

Two answers are possible. The first is that the Chinese official elite, having previously concerned itself with border issues, had infinitely greater experience of dealing with aggressive foreigners of every kind; the Japanese samurai, disoriented by the arrival of red-haired barbarians from across the ocean, had no schemas of conduct to fall back upon and were forced into a radical reorientation. So long as the external threat did not reach China's real power center in Beijing (it came near to that only in 1860, with the plunder and destruction of the Summer Palace), the old methods of keeping foreigners at bay still seemed reasonably effective and prevented a loss of bearings that would have made a completely new approach to the problems unavoidable. Only the humiliation of the dynasty by the eight powers that invaded the northern provinces during the Boxer War (1900) marked a point of no return.

The second possible answer is that China's state apparatus and the *shenshi* class on which it rested were less weakened than the samurai in Japan. After all, at exactly the same time as the dramatic developments in Japan, China's dominant class had managed to survive physically and politically (albeit with numerous casualties) the shattering social revolution of the Taiping. Around 1860, something like a modus vivendi was found with the aggressive Great Powers (Britain, France, and Russia), and this reduced the political and military pressure on the country for more than three decades. At the moment when the old order collapsed in Japan, it seemed to have recovered in China without the need for too many destabilizing reforms.

In 1900, however, when the fate not only of the dynasty but of the whole empire hung in the balance, sizable forces at the head of the Chinese state, both Han and Manchu, were prepared to undertake radical reforms.[51] Abolition of the centuries-old practice of state examinations, hitherto the only mechanism for elite recruitment, was a fairly precise equivalent of the abolition of samurai status in Japan three decades earlier. In both cases, active elements in the elite undermined the basis of their own social formation. The Chinese reform lacked both the systematic character of Meiji politics and the foreign-policy breathing space that had allowed it to be implemented. When the dynasty collapsed in 1911, the not-very-large Manchu nobility lost its privileges from one day to the next.[52] From then on, however, hundreds of thousands of Han Chinese gentry families were cut off both from the old fountains of honor and prestige and from employment opportunities in the central civil service.

The educated, competent, and (in theory if not always in practice) public-spirited scholar-officials of the high imperial period soon became in reality, as well as in the perception of society, a seedy, parasitic landlord class, while at the same time (or, to be more precise, after the beginning of the New Culture Movement in 1915) the newly emerging intelligentsia in the big cities vehemently opposed the whole worldview that the mandarinate had embodied and represented. Deserted by the state, combated and treated with contempt by politicized intellectuals, locked in a structural conflict with the peasantry, the old upper stratum of Imperial China became one of the most vulnerable elements in Chinese society. The samurai path of salvation through self-effacement was no longer available to it. Those whom Chinese Marxists vilified from the 1920s on as the "landlord class" had neither the material means to defend themselves nor the vision of a national future for which allies might have been found. Further debilitated after 1937 by the Second Sino-Japanese War, the old rural upper class of China no longer had any way of resisting the Communist peasant revolution of the late 1940s.

The Chinese *shenshi* were not a warrior aristocracy in the European or Japanese sense. Merit, not birth, was the criterion by which they were recruited. Nor did elite positions last such a long time: the cycle of rise and fall for individual families often spanned only a few generations. The continuity of the elite was secured not through genealogy but through the strength of government-related institutions constantly drawing on fresh talent. Nevertheless, the *shenshi* resembled a classical aristocracy in their proximity to the ruler, their role in supporting the state, and an agonistic conception of the world geared not to physical competition in war and hunting but to intellectual rivalry in mastering the inherited educational canon. Two further common elements were control of land and a detachment from physical labor. All in all, the similarities outweigh the differences. The *shenshi* were in many ways a functional equivalent of a European nobility, and they too got off fairly lightly during the chronological nineteenth century. After the Taiping threat

subsided in 1864, the direct competition they faced in society was comparatively weak; the challenge of a nascent "bourgeoisie" to *shenshi* hegemony was
on a much lesser scale than anything seen in analogous situations in Europe.
In China, the threat came mainly from peasant revolts and *foreign* capitalism.
The terminal point was 1905, which was for the *shenshi* what the 1790s represented for the French aristocracy, 1873 for the samurai, or 1919 for the nobility
in Germany. The *shenshi*, too, were a land-based elite in decline, the largest
anywhere in the world.

The fate of aristocratic and quasi-aristocratic elites was partly homespun,
partly influenced by wider developments. Here there were two opposing
trends. On the one hand, it turned out that the radiance and attractiveness
of aristocratic ideals was wearing thin. Societies took shape in America and
Australia that were, in a historically novel sense, immune from and toward
nobility, and even the colonial empires managed to stabilize things only on a
makeshift basis. In the early modern period, Europe's colonial outthrust had
hugely extended the geographical sphere of operations of the European nobility, but although there was a degree of solidarity across cultures, non-European
nobles seldom adopted European worldviews or role conceptions. In comparison, the cultural package offered by the European bourgeoisie was a much
more attractive export item. The new colonies of the late nineteenth century
did not bear an aristocratic stamp. In Africa and Southeast Asia after 1875 the
European powers together spawned a new type of bourgeois functionary, and
even in India feudal mummery could not disguise the bureaucratic character
of the colonial state.

On the other hand, a number of general changes made themselves felt. The
beginning of the end was in sight for the aristocratic "international" when the
foreign offices and diplomatic services of the Great Powers ceased to recruit
exclusively princes, counts, and lords. Before 1914, the foreign policy of the
United States and the French Republic was already being shaped almost entirely by bourgeois politicians. State building in the nineteenth century led
nearly everywhere to a greater distance between the central institutions of government and a nobility struggling to control its own local power resources. If
the state employed aristocrats, these were no more than its "servants" either. At
the same time, the nobility had less access to its old agrarian sources of income,
power, and prestige: all manner of peasant emancipation, together with the
whittling down of local privileges and the decline of agricultural income in an
age of industrial development and world economic expansion, restricted the
traditional opportunities that had enabled the aristocratic classes to flourish.
The nobility kept control of its destiny, even in the early twentieth century,
mainly where it saw itself as part of a broader elite no more than weakly defined by inherited status, where it reined in its habitual conceit and where it
pragmatically forged new social and political alliances.

3 Bourgeois and Quasi-bourgeois

Phenomenology of the Bourgeois

The nineteenth century was the century of the bourgeoisie, at least in Europe.[53] A social space marked by its distinctive values and lifestyles opened up in the cities—between a declining nobility that made offers of class compromise among the prosperous layers of society, on the one side, and a class of wage laborers that, by the last third of the nineteenth century, had evolved from a plebs into a proletariat and achieved a degree of political self-organization and cultural independence, on the other. The mansion suburbs that sprang up in many European cities during the last two decades before the First World War are visible relics of this bygone world of a bourgeoisie eager to put its hallmark features on display. Who was a bourgeois and what it meant to be one cannot be reliably defined by objective criteria of family origin, income level, and profession.[54] People were bourgeois—such is the near-tautological conclusion of extensive research and discussion—if they *considered* themselves bourgeois and gave this belief practical expression in the way they led their life. Radical skeptics have called into question the whole construct of "the bourgeoisie." We can doubtless identify individual bourgeois and whole generational chains of bourgeois families, both in literary fiction (Thomas Mann's *Buddenbrooks*, 1901) and in historical reality.[55] But, it is argued, the bourgeoisie as a social stratum or class escapes definition. Was it not simply a myth?[56]

It is easier to define a bourgeois negatively: he is not a feudal lord deriving his conception of himself from landownership plus genealogy, and not a manual worker in dependent employment. Otherwise, the category seems broader than any other classificatory social construct. If we think of the period around 1900, for example, it encompasses some of the richest people in the world—industrialists, bankers, shipowners, railroad magnates—and also professors and judges earning an adequate but not lavish salary, members of the liberal professions with an academic qualification (e.g., doctors and lawyers),[57] as well as storekeepers, master artisans, and policemen. Around 1900 the new "white-collar" employee was also becoming more visible: a subordinate figure on the margin of bourgeois life, but one who attached great value to the fact that by working at a cash desk in a bank or the accounts department of an industrial enterprise, he did not have to get his hands dirty. Now that a growing number of large firms were run by managers rather than their owners, there was even a layer of "executive" employees who looked upper middle class and had wide scope for independent initiative, apparently on a par with the most zealous guardians of bourgeois values.

So, one reason why the concept of a bourgeoisie is so misleading is that it breaks up so quickly into individual life paths. The bourgeois strives to rise in society and is afraid of nothing as much as the opposite: a fall into the ranks of the poor and despised. A ruined aristocrat is always still an aristocrat; a ruined bourgeois no more than a déclassé.[58] The successful bourgeois owes his position to self-reliance and achievement; nothing inborn seems to him dependable. Society in his eyes is a ladder: he is somewhere in the middle, constantly under pressure to move upward. Ambition is not just a matter of personal ascent, family prosperity, and a perception of direct class interest. The bourgeois wants to shape and organize things; he has a lofty conception of his responsibility and, by making his own life, wishes to play a role in giving a direction to society.[59] In the most rapacious bourgeois there is still a spark of the public-spirited *citoyen*. Bourgeois culture, more than any other nonreligious system of values, raises a claim to universality and thus contains an urge to move beyond its original social bearers. The bourgeois always has many beneath him toward whom he cultivates an attitude of superiority, and as a rule he has at least a few above him. So long as there are nonbourgeois elites—a nobility or a prestigious clergy (such as the Muslim *ulama*)—even the wealthiest bourgeois does not stand at the top of the social hierarchy. Only in a few societies were things otherwise in the nineteenth century: for example, Switzerland, the Netherlands, post-1870 France, or the East Coast of the United States. The most "bourgeois" society is one in which bourgeois players in *every* sphere of life themselves set the rules for their competition with one another. This tended to become the norm in the twentieth century; it was the exception worldwide in the nineteenth.

But the twentieth century also witnessed the long fall of the bourgeoisie as a class, a radical de-bourgeoisification and de-feudalization of whole societies. The drama began to be acted out in 1917 in Russia and was soon repeated in central Europe and (after 1949) in China. The twentieth-century revolutions lumped the bourgeoisie and residual aristocracy together. In nineteenth-century Europe, however, it was often difficult—though never really life threatening—to be a bourgeois. Before 1917, the European bourgeoisie as a social group never suffered the fate that befell sections of the French aristocracy after 1789. The Bolshevik Revolution destroyed ways of life opposed to it much more radically than any previous revolution had been able to do. The world of the Russian economic bourgeoisie, which came into being only after 1861 and had had only five decades to develop, looked like a sunken civilization in the optic of the late 1920s.[60] And until the great postwar inflation in Germany and Austria (the harshest blow yet to the classical bourgeoisie in Europe) and the subsequent onset of the Great Depression in 1929, large parts of the bourgeoisie had never been collectively deprived of the supports for its claim to a "refined" standard of living. The nineteenth century was quite a good time to be a bourgeois too.

Petit Bourgeois

How large was the bourgeoisie? The terminological proximity between the bourgeoisie proper and the petite bourgeoisie of storekeepers and independent artisans still causes confusion. What did a steel magnate and a chimney sweeper have in common? The differences were much more obvious. The social characteristics of "large" and "small" bourgeois are at first sight easy to distinguish; the two groups evolved along different tracks. Thus, in many European countries in the second half of the nineteenth century, the mentality and politics of the educated property-owning bourgeoisie differed considerably from those of a petite bourgeoisie anxious to distance itself from industrial workers. France actually became a nation of petit bourgeois, while in Russia, quite short of small and medium-sized cities, the new stratum of capitalist pioneers and educated dignitaries could support itself on only a thin cushion of petit bourgeois.

The petite bourgeoisie is conceptually hard to grasp. The term "middleclass," preferred in Britain and the United States (although its first appearance in an American dictionary was only in 1889[61]), does not satisfy everyone as a solution to the problem, since its unity and homogeneity are not easy to demonstrate even for the United States, where the bourgeois consensus was from the beginning broader than in Europe. Theorists have made a more persistent effort (though without generalizable results) to identify the social membrane between lower middle class and upper middle class, and they have rarely been able to avoid drawing internal dividing lines: in the English case, for example, between a capitalist middle class and a noncapitalist or professional middle class, roughly (but only roughly) corresponding to the German *Wirtschaftsbürgertum* and *Bildungsbürgertum*.[62] "Middle class" or "middle stratum" is poorer in cultural content than "bourgeoisie," and so it can be used in a larger number of contexts and is better suited for a global social history. Not every member of a middle stratum carries around a complete bourgeois value system.

A particularly useful distinction is the one between different milieux, each with its sphere of sociability and shared beliefs. Thus, Hartmut Kaelble proposes to distinguish between a bourgeois milieu in the narrow sense (the "upper middle class") and a petit bourgeois milieu.[63] These milieux are not precisely circumscribed groups but social fields with fuzzy boundaries, which may overlap and influence one another. Milieux may also be thought of more specifically as arenas of local life. The first to take shape are based on friendship, marriage, and clubs or associations, their composition and subculture varying from place to place; *then* perhaps come translocal strata and classes.

"Petite bourgeoisie" has yet to be developed as a theme in *global* social history. This is unsurprising in the case of the nineteenth century: the lives in question were local to a quite exceptional degree,[64] and their economic radius of action seldom stretched beyond a neighborhood of people in constant contact with one another. Shopkeepers knew their customers by name. After a youthful period of travel and companionship, the subject of so much Romantic verse, the typical petit bourgeois rarely went outside the boundaries of their locality. The culture, too, was limited in reach. The petite bourgeoisie, in particular, was not an international stratum (although there was a first world congress of *petits bourgeois* in 1899!): it was less mobile than migrant underclasses; and it had few cross-border links in comparison with the far-flung families of the aristocracy or the business connections of the upper bourgeoisie. For this reason, the very term "petit bourgeois" is hard to transfer from one context to another. What is to be gained from using it to describe a silversmith in Isfahan or the owner of a teahouse in Hankou? Similarly, some of its pejorative connotations have little purchase outside the cultural or political circles in which they originally developed.

Under the blanket term "petit bourgeois," never altogether free of disdain, lies many a local artisan with his own ethos and the pride that comes from self-confident mastery of a trade.[65] Such cultures, sometimes (as in parts of India) involving a caste-like exclusiveness, existed all around the world and often enjoyed higher esteem than the sphere of commerce: fixed and stable spheres of the social middle, supported by monopolies of know-how that no upper class could contest or replace. Traditional knowledge is more able than property or legal privilege to escape devaluation through political revolution; there is always a need for artisans and basic service providers. Only machine production presents a challenge, without necessarily rendering time-tested skills superfluous. This staying power counterbalances an ubiquitous fear of proletarianization. Thus, the petit bourgeois (in a broad sense) does not always look up obsequiously to the higher ranks of the social hierarchy. Not aspiring to be the originator or bearer of a superior culture, he does not invest much of his cultural capital in education (as distinct from vocational training); he has a pragmatic attitude to it, weighing up how useful it might be for his offspring.

Petit bourgeois are certainly capable of collective political action. If they control major channels of social communication, they may exercise greater power than many a captain of industry. Strikes by merchants in Middle Eastern bazaars or Chinese port cities have repeatedly generated significant political pressure, and when directed against foreign interests they became early expressions of nationalist politics. The key international experience for the petit bourgeois was war. Along with peasants and workers, they formed the main bulk of armies nearly everywhere. Noncommissioned ranks (corporals and sergeants) were petit bourgeois in both origin and habitus. In general, military hierarchies often accurately mirrored grading systems in civilian life. In scarcely any other domain can the nationally variegated rise of the European bourgeoisie be observed more clearly than in the struggle for officers' commissions and for acceptance by aristocratic general staffs.[66]

Respectability

The true bourgeoisie, corresponding to the "upper middle class," consisted of people who had a wider mental horizon than the petite bourgeoisie, operating with capital (also the cultural capital of academic knowledge) and managing not to get their hands dirty. The "bourgeois," remarked Edmond Goblot smugly in an unsurpassed essay from the 1920s, "wears kid gloves."[67] This was a key element of a specifically bourgeois habitus. Another was concern for one's reputation. Instead of honor, as in the case of noblemen, the typical bourgeois was preoccupied with respectability—even if he occasionally subjected himself to the aristocratic code of the duel. Individual bourgeois sought to appear respectable above all in the eyes of other bourgeois, but also in those of the upper classes (who must not be offered a chance to treat one with condescension) and in those of people lower down the social ladder (who were expected to behave with deference and to recognize one as an opinion leader). This middle-class striving for respectability is also found outside Europe. Its economic expression is creditworthiness: the bourgeois has a reasonably secure income, and if he needs money the lender can expect to be repaid. A respectable bourgeois obeys the law and observes moral prohibitions, knowing what is expected and behaving accordingly. If female, she avoids idleness but also physical labor outside the home. The wife and daughters of a bourgeois man do not need to work in the service of others, while a high member of bourgeois society is in a position to employ domestic servants of his own.

"Respectability," like the character model of the English gentleman, was a mobile cultural ideal capable of being learned and transmitted. Europeans and non-Europeans alike could aspire to it—white and black middle layers in nineteenth-century urban South Africa, for example, until racism put more and more obstacles in the way of such convergence.[68] Arab, Chinese, and Indian merchants, too, cultivated an aloofness from manual labor, set a high value on domestic virtues (no less achievable in a polygamous setup), emphasized the degree of foresight required for their activity, planned according to the rules of rational business calculation, and took pains to demonstrate their high standing. Something like a bourgeois habitus is not necessarily tied to Western cultural presuppositions. The huge middle classes numbering overall hundreds of millions that emerged in countries such as Japan, India, China, and Turkey in the last third of the twentieth century cannot therefore be adequately explained as a mere import of Western social forms. They would have been unimaginable without indigenous foundations.

The educated property-owning bourgeoisie was everywhere a minority in the nineteenth century, rarely exceeding the 5 percent of the population (or 15 percent including the urban petite bourgeoisie) that has been estimated for Germany.[69] In the United States, however, there is an influential tradition still alive today that sees the country as made up of nothing other than "middle classes." The American people, wrote the historian Louis Hartz in 1955, are "a kind of national embodiment of the concept of the bourgeoisie."[70] Social historians have deconstructed this myth of classlessness, a twin of the "melting pot" legend, and exhaustively described the *differentia specifica* of bourgeois situations and worldviews in the United States. For the American grande bourgeoisie did not demarcate itself any less sharply than its European counterparts from lower strata of society.[71]

If the bourgeoisie in 1900 was still thinly spread even in most of the "West"— the main exceptions being Britain, the Netherlands, Belgium, Switzerland, northern France, Catalonia, western Germany, and the northeastern states of the United States—how much more was this true of the global scene. In the "bourgeois age," the educated property-owning bourgeoisie comprised a tiny minority of the world population, distributed very unevenly around the planet. However, its distribution did not follow the simple schema of "the West and the rest." Europe *as a whole* was not at all living in a bourgeois age, and sprouts of bourgeois or quasi-bourgeois development were by no means completely lacking outside Europe and North America.

The Universality of Middle Ranks in Society

At this point a global social history starts to become interesting. To be sure, the bourgeoisie and bourgeois values were products of Western European urban culture and early modern long-distance trade, which were then reshaped in the nineteenth century under the conditions of industrial capitalism and revolutionary theories of equality. Moreover, the idea and to some extent the actual practice of "bourgeois societies" were among the most striking aspects of the (Western) European special path in modern history. Nowhere except in Europe and the neo-European settler societies does the belief seem to have existed that the middle orders could stamp their lifestyle ideals on society as a whole. Nevertheless, it is worth asking whether and how in the nineteenth century, outside the North Atlantic, social milieux arose that could be described as similar or even equivalent in their roles to the Western middle classes.

The following remarks do not amount to a panorama of bourgeois existence outside Europe;[72] they aim to throw light on a number of analogies and relations and to illustrate them with examples taken mainly from Asia. That was the continent, in the early modern period, that gave rise to merchant cultures by no means inferior to those of Europe in complexity and efficiency.[73] It was there too, that by 1920 at the latest, embryonic bourgeoisies emerged in many regions in a tense interplay between capitalism and higher education: societal strata who— and this was new—thought in categories of *national* politics. Similar processes began to occur in many parts of Africa too, but the social discontinuities were typically sharper in sub-Saharan Africa than in Asia. There were two reasons for this. First, European control over new modern sectors of the economy (mining, plantations) was even more comprehensive, with Africans serving only as wage laborers or small agricultural suppliers. Second, the appearance of Christian missionaries in Africa led to a much deeper social-cultural rift than almost anywhere in Asia. The mission and its educational facilities alone led to the formation of a Western-oriented elite, whereas in East or South Asia indigenous cultures of knowledge were converted in a series of complex processes.[74]

Broadly speaking, the relative weight of middle ranks in society increased in many parts of the world especially after the mid-nineteenth century. This had to do with the greater social differentiation fueled by population growth and with the general expansion of regional and supraregional trade and business activity— processes that left no continent untouched, even involving sub-Saharan Africa long before the colonial conquests began there.[75] Merchants and bankers—experts in exchange and circulation—were the main impetus and beneficiaries in many cultural contexts. A third factor was the establishment of state administrations and associated job opportunities at middle levels of the hierarchy—hence for nonnoble functionaries with at least some formal schooling if not a full liberal education. In the nineteenth century, social groups were bourgeois if they occupied a "third" position on the margins, or in the vertical middle, of social hierarchies.

An image of society constructed in this way was not self-evident. Societies could be visualized from within as egalitarian-fraternal, as dichotomous (top/bottom, insiders/outsiders), or as finely graded into ranks and status groups. The idea of an intermediate level between the elite and the peasant or plebeian masses—that is, a middle position filled with significance—became characteristic of the nineteenth century only after the eighteenth century had seen the strengthening of a capitalist bourgeoisie in many European and Asian countries. Not only tolerated and secretly respected, the merchant or banker now also gained "theoretical" acceptance in the dominant value structure of society. This revaluation did not necessarily involve the immediate "rise of the bourgeoisie." Sometimes, the shift in favor of large merchants and notables could be seen only in the finer shades of social interaction. But the trend was global in reach: activities, lifestyles, and mentalities that had more to do with commerce and noncanonical knowledge than with agriculture, country life, and cultural orthodoxies, and whose horizons surpassed the "view from the church tower," acquired growing importance in comparison with earlier epochs.

The subjects of such activities, lifestyles, and mentalities defined their social identity more in terms of achievement and competition than of adaptation to existing status hierarchies. They strove to accumulate and protect movable wealth, even if they invested some of their money in real estate for reasons of security and prestige. Quasi-bourgeois groups were nowhere "in power" in Asia, but despite their small size they were often influential and had a considerable impact on the modernization of their society. In many cases, this happened in the absence of a thought-out program of bourgeois activism and without a self-conscious expression of bourgeois norms and values. Advanced techniques of production and commercial organization were brought into use, investment flowed into sectors such as export agriculture or mechanized mining, and methods for the mobilization of capital were implemented that were beyond traditional indigenous capacities. In their objective effect, these bourgeois were economic pioneers with the calculating mentality of entrepreneurs. But they seldom came forward as self-confident representatives of economic or even political liberalism. This obscured their visibility in the eyes of European contemporaries, and of historians looking first of all for liberal rhetoric and only then for the people behind it.

The quasi-bourgeois of Asia would anyway not have been able to indulge in antistatist liberalism, since they themselves stood in an ambiguous relationship to the state. As with the economic bourgeoisie elsewhere, their twin objectives were to clear all possible obstacles to self-organization and to exercise control over the operation of the market. One such kind of market economy had existed in eighteenth-century China, and the next period when the Chinese bourgeoisie won space for initiative was not accidentally between 1911 and 1927, when the state was as weak as it had ever been or would be in the future.[76] In many other Asian countries, however, the commercial middle classes entered into a symbiotic relationship with the state, financing it as taxpayers or bankers and relying on its support in return. The state, whether indigenous or colonial, often had to protect them in an unfavorable environment and to guarantee a minimum of legal security. There was a wide range of scenarios—from monopolistic advantages for commercial minorities in some European colonies in Southeast Asia (opium monopoly for Chinese dealers, for instance[77]) to a laissez-faire colonial state, as in British Hong Kong, that provided the freedom to operate abroad. In most cases, the relationship with the state was closer than in Western Europe. True, the Asian bourgeoisies that developed toward the end of the nineteenth century were not primarily classes in the service of the state and were seldom directly created by it; they had their own histories of mercantile success behind them. Yet, from the Ottoman Empire to Japan, they were at first state-protected niche groups. In the nineteenth century, the institutional requirements for autonomous systems of private market regulation were lacking in the major part of the world.

Fully developed bourgeois societies, especially ones with a "bourgeois" political system, were therefore few and far between. More characteristic, not only in the colonies but also in independent countries of Asia and the southern or eastern periphery of Europe, was what the Hungarian-born historian Ivan T. Berend (with eastern Europe in mind) has called a "dual society."[78] In this asymmetrical formation, the economic importance of the bourgeoisie was growing but older elites retained their political preponderance and, to some extent, their cultural authority—even if the industrious, education-oriented, and self-disciplined middle of society often regarded them as decadent and ineffectual.

Commercial Minorities in the Growing World Economy

Not all quasi-bourgeois outside the West had an orientation to the world economy, but their network-building functions were undoubtedly one of their most striking features. Whole societies of traders, such as the Swahili in East Africa, could hold their ground for a long time through adaptation to changing external conditions.[79] Quasi-bourgeois were for the most part active in trade and finance, two fields in which many families had acquired great wealth as far back as the eighteenth century. This was true of the *bania* in India, for example, on whom the British remained partly dependent long after they were able to dispense with Indo-Islamic administrative officials, or the Hong merchants who had conducted Chinese trade with Europeans before the Opium War. Such groups suffered in various ways from the expansion of European, especially British, commerce after 1780, losing much of their prosperity and prestige: Indian merchants because of the East India Company's trading monopoly; their Chinese colleagues because the imperial foreign trade monopoly was undermined and eventually abolished, and China was opened up to a limited regime of free trade in which old commercial dynasties accustomed to parasitic bureaucratism and immobile monopolism found no new role for themselves.[80] There was no straightforward path from these "early modern" merchant classes to a modern bourgeoisie, any more than merchant princes in Europe regularly mutated into industrial entrepreneurs. Everywhere except in Japan and the west of India (where Parsi merchants in the Bombay region got a cotton industry on its feet), little scope existed even in 1900 for entrepreneurial involvement in industry. The railroads, which gave such an impetus to private entrepreneurship in Europe and the United States, were mostly in foreign hands. At best, plantations offered a favorable, low-tech opportunity to break into capitalist production. The Singhalese bourgeoisie of colonial Ceylon, one of the oldest and most durable in Asia (some of its pioneering families still dominate Sri Lankan politics), owed its rise in the nineteenth century to such an early involvement in the plantation economy. Arab merchant dynasties in Malaya and Indonesia also invested in this sector.[81]

From the beginning of trade contacts with Europe, non-European quasi-bourgeois often exercised "comprador" functions as middlemen.[82] In this way they were able both to widen their experience with indigenous trade networks and to link them into the world economy. First of all they facilitated exchange between different business cultures—for example, between those of India or China (the word *comprador* stems from an early modern Portuguese-Chinese context) and the West. They tapped sources of finance and used their contacts with business partners in the interior. In China alone there were roughly 700 compradors in 1870, and as many as 20,000 in 1900.[83] Often religious or ethnic minorities (Jews, Armenians, Parsis in India, Greeks in the Levant) played such a role.[84] (Nor was this an extra-European peculiarity: in Hungary, for instance, where a strong nobility had little interest in modern economic life, Jewish and German entrepreneurs occupied a central position in the emergent business community.[85]) In China intermediary functions remained in the hands of special groups of Chinese merchants in the treaty ports; émigré Chinese were active in commerce, and to some extent mining (Malayan tin) and plantations, in every Southeast Asian country. They also formed internal hierarchies of wealth and prestige, stretching from family shopkeepers in a village in the interior to immensely rich, multifunctional capitalists in Kuala Lumpur, Singapore, or Batavia.[86] In the Dutch colony of Java, virtually the whole of internal trade was in Chinese hands at the beginning of the nineteenth century. For its exploitation of the island, the colonial power depended almost entirely on a minority that had dominated business life in the capital, Batavia, since its founding in 1619. Although European interests later intruded more actively in Java, the Chinese (comprising less than 1.5 percent of the population) remained indispensable to the colonial system and profited handsomely from it, acting as intermediaries between foreign firms and local Javanese until the end of Dutch rule in 1949.[87] Sometimes commercial minorities conducted business over very large distances. Russian wheat exports via Odessa to the United States in the early nineteenth century were in the hands of Greek merchant families, most of whom originally came from the island of Chios.[88]

The position of such minorities was rarely protection against crisis, and there is little to suggest they enjoyed a self-confident bourgeois existence. After the Ottoman Empire adopted free trade in 1838, the proud Greeks from Chios were demoted to agents of Western firms and often acquired British or French citizenship. The ethnic Chinese compradors, for their part, were gradually replaced with Chinese employees working for large Japanese or Western import-export businesses along the coast of China.

State protection could not prevent repeated attacks and acts of expropriation, which became more virulent as nationalism grew among the majority population and reached dramatic proportions in the twentieth century. In the nineteenth century, there were not yet events on the scale of the expulsion of European minorities from Egypt after the Suez crisis of 1956 or the massacre of Chinese in Indonesia in 1965.[89] European colonial governments often protected minorities, on whom they relied for tax revenue. The weakness of quasi-bourgeoisies outside Europe, vis-à-vis both their indigenous society and world market forces, did not prevent them from deploying their own business policy and expanding their room for maneuver. But they were on their guard against one-sided dependence and often sought the security of property accumulation within their close or extended family—a way of minimizing risk that features in many variants of Asian capitalism. Another business strategy was to diversify as widely as possible, into trade, manufacturing, moneylending, agriculture, and urban real estate. If the main characteristic of bourgeois economic culture is self-reliant operations in high-risk environments, without much of an institutional safety net, then this was present to a high degree among self-made men on the "periphery" of the world economy.[90]

Modernity and Politics

Outside Europe, groups that may be regarded as quasi-bourgeois seldom exhibited an offensive political self-confidence; they had little influence in politics and tended to be socially isolated. Where they formed a conspicuous minority, as did Greeks in the Ottoman Empire or Chinese in Southeast Asia, their ability, and sometimes willingness, to adapt to the social environment was often limited. All the more did they cultivate a niche culture of their own, though in many cases it clashed with their striving to link up with global trends and conceptions of normality. A similar contradiction was present in the Jewish bourgeoisie of Western Europe: an interplay among assimilation to the social surroundings, belief-driven adoption of universal cultural values, and a wish to preserve the tradition-based solidarity of a religious community.

If we look for an orientation common to various parts of the world, then it was an aspiration not so much for political power or independent cultural hegemony as for *civilization*. A bourgeois existence in Asia and Africa from the late nineteenth century on (as for Western European Jews since the time of Moses Mendelssohn) meant linking into the development of "civilized" morals and lifestyles, not necessarily seen as an emanation from Europe and by no means perceived by those involved only as a process of slavish imitation. Unmistakable as the civilizing trends were in metropolises such as Paris, London, or Vienna, quasi-bourgeois forces outside Europe were sufficiently self-aware to see them as a general feature of the times in which they could have an active share. Istanbul, Beirut, Shanghai, and Tokyo were being modernized, and in writing about them indigenous intellectuals created the city as "text."[91]

All around the world, middle classes recognized one another by their wish to be modern, any limiting epithet being of secondary importance. Modernity should and did acquire an English, Russian, Ottoman, or Japanese flavor, but what mattered more was its indivisibility. Only thus was it possible to avoid the fatal distinction between the genuine article and imitations. The program of multiple modernities, already outlined in the late nineteenth century before being assigned such a major role in present-day sociology, was therefore a double-edged gift for Asia's newly emerging quasi-bourgeois elites. Modernity had to have a culturally neutral, transnational appeal if it was to command acceptance and be generally comprehensible. It should be a single symbolic language with local dialects.[92]

If middle classes were to found on different sides of the colonial divide—as they were first in India and by 1920 in Indonesia and Vietnam—the relationship between them was ambivalent. Partners could turn into economic and cultural rivals. However useful Europeanized Asians or Africans might be as cultural intermediaries, they disturbed the value system of modern Europeans. Indigenous claims to modernity were sharply rejected, and the insults were felt with special bitterness. Failure to be recognized as equals—also in the sense of citizenship— converted some of the most "Western" Asians into implacable opponents of colonialism. Middle classes in Asia and Africa took to a nationalist politics of their own only after about 1900—or, to be more precise, after the First World War, when waves of protest shook the imperial world from Ireland to Russia, Egypt, Syria, and India to Vietnam, China, and Korea. Even in Japan, the country with Asia's most progressive constitution, it was only around this time that representatives of bourgeois values were able to gain a hearing in a political system that until then had been dominated by Meiji figures with a samurai background. In general, the impulse from the revolutions of the twentieth century (including post-1945 decolonization) was a precondition rather than a result of opening up spaces of "civil society" to be filled with the political life of freshly emerging citizens.

Elements of civil society had, of course, been widely present earlier in pre-political spheres. The European culture of clubs and associations, which extended eastward as far as Russian provincial cities, found equivalents in other parts of the world. Prosperous merchants in China, the Middle East, or India, often pooling their efforts across regions, involved themselves in disaster relief, founded hospitals, collected money to build temples or mosques, and supported preachers, scholars, and libraries.[93] In many cases, organized philanthropy was the innocuous starting point for a wider preoccupation with public affairs, as well as an arena in which private individuals from the "middle" of society rubbed shoulders with aristocrats and representatives of the state. Another element of civil society was the municipal guilds, which in the central Chinese metropolis of Hankou, for example, took over more and more functions from the 1860s on and played an important role in crystallizing a community that encompassed a broad cross-section of the urban elite."[94]

"Bildungsbürgertum": Education, Cultural Hegemony, and the Middling Ranks

Some "bourgeois" social types were more universal than others. The Protestant high school teacher in Imperial Germany (carrying the title and prestige of a "professor") or the coupon-clipping rentier in the French Third Republic who derived his income from Chinese government bonds was a special local product, less exportable than the industrial or financial entrepreneur to be found almost everywhere around the year 1920. Middle-class traders were anyway common enough, but the *Bildungsbürger* was a specifically central European, indeed German, phenomenon.[95] What was so distinctive was not only the content of his education (its linguistic form, its expression in aesthetic and philosophical idioms incomprehensible elsewhere) but also the value attached to it in society. On the ground of the educational reforms of 1810 and the subsequent years, and often with an input from the distinct cultural world of the Protestant parsonage, the educated middle classes in Germany spread their wings in opposition to the less-than-intellectual priorities of the average nobleman and the forms and themes of aristocratic culture. The bourgeois could assert his aspirations and superiority only because the values of premodern elites lay in other domains—which did not exclude extraordinary connoisseurship and practical competence on the part of aristocrats, for example, in Viennese musical life in the age of Haydn and Beethoven. It was possible only under particular historical conditions for people without roots in genealogy and tradition to become creators and guardians of the national culture and enthusiasts for an ideal of individual fulfillment through self-education. The most important of these conditions was the state promotion of the educated classes, carried to its highest pitch in the German lands. It lastingly associated professions with a comprehensive education and created public-service opportunities for social ascent that did not obey the laws of the free labor market.[96] We need look no farther than Switzerland or England (not to speak of the United States) to find market regulation of the "liberal professions" without the heavy state intervention typical of Prussia or Bavaria. On the other hand, such a system did not yet guarantee homogeneity in the development of an educated middle class. In the Tsarist Empire, the self-assurance of senior officials—most of whom had a legal background—was based not on higher education but on their place in the formal hierarchy of ranks.[97]

The *Bildungsbürger* was such a rare breed that it is unnecessary to explain why the type did not flourish elsewhere; the very word *Bildung* is notoriously untranslatable. Evidently, however, ideals of a literary-philosophical education and of intellectual and spiritual maturing and perfectibility are to be found in a number of civilizations with a system of written communication. Self-perfecting of the inner world through traditionalist character formation, sometimes understood in Asia as a task of the individual and, for example, actively pursued even by nonmandarin merchants in late imperial China, was not so far removed from the European or German ideal of *Bildung*. In Japan too, the late Tokugawa period saw a similar rapprochement of values and tastes between the cultures of the samurai and commercially active city dwellers (*chōnin*).[98] But why were there no *Bildungsbürger* in China, the most plausible candidate for them given the profound admiration of nonreligious learning in that bookish civilization?

Such a social group could not appear where the established elite already defined itself in terms of *Bildung* and held a monopoly over its institutions and forms of expression. That was indeed the case in late imperial China, where no superior conception could challenge the canonical idea of education until the end of state examinations in 1905 and of the dynasty itself in 1911. The Confucian tradition did not allow itself to be outtrumped; it could only be overthrown by a cultural revolution. After reform movements among the literati ended in failure around the turn of the century, a general offensive against China's ancient worldview began in 1915. It was conducted not by the capitalist bourgeoisie or civil servants but by iconoclastic intellectuals, including many from the fallen mandarinate, who lived off the emerging literary market or worked in one of the new educational institutions.[99] What developed in China, then, was not a politically indifferent or quietist layer of *Bildungsbürger* but a highly politicized intelligentsia concentrated in the big cities, which later produced many leaders of the Communist Revolution. Certain affinities with the European *bohème* and its antibourgeois subculture are unmistakable.[100] Yet, since the selective intellectual Westernization of China was limited by the political conditions during a time of chaos and violence, no new posttraditional world was allowed to emerge on a broad social basis. The infatuation of parts of the new Chinese middle strata with European classical music—today China is the most rapidly growing market for pianos in the world—is a recent phenomenon, unknown before the 1980s.

The second prerequisite of a *Bildungsbürgertum* was the freeing of the mind from the all-pervasiveness of religion—which in Europe was the work of the Enlightenment and its critique of religion. Only then could secular knowledge be held in high esteem, not to speak of the glorification of education or even the elevation of art and science to the status of substitute cults and creeds in their own right. That kind of differentiation between the godly and worldly realms did not go so far in Islamic or Buddhist cultures, for example, where challenges to religious authority in matters of value orientation tended to get stuck, as did the downplaying of religious obligations in everday life in the name of "educated" lifestyles. The *Bildungsbürger*, understood as the serene exponent of a consensus on high-cultural values and taste, was something of a rarity even in the heart of Europe. In many other cultural and political contexts, there was a sharp antagonism between upholders of orthodoxy and radical intellectuals influenced by Western dissident traditions such as anarchism or socialism.

Colonial and Cosmopolitan Bourgeoisies

Western colonial bourgeoisies were surprisingly relatively weak in the nineteenth century. On the whole, colonialism contributed little to the export of European bourgeois culture, and European societies were reproduced in only fractured and fragmentary fashion in the colonies, with few exceptions such as Canada, New Zealand, and, in a special way, Australia. Distortions in the process of transfer were unavoidable because all Europeans automatically fell into the role of masters. In terms of social rank and often income, the humblest white civil servant or employee of a private company stood above the whole of the colonized population except for its princely apex, if there was one. Colonial bourgeoisies were thus distorted mirror images of bourgeois groups in the European metropolises and remained to a large extent dependent on them culturally. In only a few nonsettler colonies was there sufficient mass for a local *society* to come into being. Of course, the social profiles of particular colonies differed considerably from one another. In India, where Britons were relatively little involved in the private economic sector, bourgeois lifestyles were taken up mainly in the colonial state apparatus, only the upper levels of which were dominated by the aristocracy. Here a distinction was made between "official British" and "unofficial British," which together constituted local mixed societies of civil servants, officers, and businessmen. After the Great Rebellion, these became increasingly compartmentalized along color lines. Family members circulated between India and Britain, and as a rule they did not become "Indianized" even over several generations, rarely shifting the main focus of their family life to India.[101] Europeans were not so much settlers as temporary "sojourners." A microcosm of all this was Malaya, where the settler element was more strongly represented than elsewhere in British Asia.[102]

South Africa was a rather special case, because the discovery of gold and diamonds soon paved the way for a tiny, ultrarich plutocracy—an isolated capitalist bourgeoisie of "Randlords" such as Cecil Rhodes, Barney Barnato, and Alfred Beit—to emerge in the mining districts. Such men were not embedded in a multifarious bourgeoisie and had only weak relations with long-established bourgeois families in the Cape. For the most part, white hierarchies in settler colonies were only indirectly linked into the mechanisms of social reproduction in the mother country; they were not mere copies of social relations back home. Normally their members sank permanent family roots in the colony, often developing a colonial spirit tinged with local chauvinism. In France's largest settler colony, Algeria, farms growing cereals and wine were quite widely spread around the end of the nineteenth century, and the resulting society of farmers and petit bourgeois *colons* of French descent felt rather distant from bourgeois strata in the large French cities. Algeria was the model of a *petit* bourgeois colony, in which, despite many forms of discrimination, a small but growing *indigenous* middle class of merchants, landowners, and state functionaries also found a place for itself.[103]

Another hallmark of bourgeois life is domesticity. It is not necessarily associated with particular forms such as the central European monogamous, two-generation family. But the basic features are plain: the domestic sphere, clearly separate from the public, is a refuge to which strangers are denied entry. For upper layers living in the lap of luxury, the dividing line between private and semipublic space runs through the house or apartment: guests are received in the lounge or dining room but have no access to the inner sanctum. It was a code practiced as much in Western European bourgeois families as in the Ottoman home. Even the functional allocation of spaces in the home is common to nineteenth-century Europe and the cities of the Ottoman Empire.[104] Where emergent bourgeois groups looked to Europe, they filled their homes with Western features: tables, chairs, metal cutlery, even open fireplaces in the English style—but selectively. Japan resisted the chair, China the knife and fork. The colorless, close-fit clothing of the European bourgeoisie became the public costume of the whole "civilized" and would-be civilized world, but in their own homes people stuck to older indigenous forms. Global bourgeois culture manifested itself in sartorial uniformity, assisted by missionary notions of decent clothing in lands remote indeed from the homeground of the bourgeoisie. If there was an insistence on local touches, that itself could have a "bourgeois" sense. For example, headgear of widely varying form and material quality had always symbolized rank in the Ottoman Empire, until Sultan Mahmud II declared in 1829 that the fez should be obligatory for all state officials and subjects;[105] the Oriental object, in its very sameness, acquired the significance of bourgeois *égalité*. Thus, the Tanzimat decree of 1839, making all Ottoman subjects equal regardless of the group they belonged to, had been anticipated a decade earlier on the heads of the male population.

A final aspect spanned East and West. The Atlantic had already been commercially integrated in early modern times by European and American traders, as had the Indian Ocean by Arab seafarers and merchants; the great Dutch and English trading companies, run by bourgeois patricians, had also commercially linked continents. What was new in the nineteenth century was the emergence of a *cosmopolitan* bourgeoisie. Two things may be understood by this. On the one hand, a rentier public living off faraway earnings took shape over time in the wealthier countries of the West. The global capital market that developed after the mid-nineteenth century made it possible for bourgeois investors (and others, of course) in Europe to profit from business in other continents—whether Egyptian or Chinese government bonds, Argentine railroads, or South African gold mines.[106] Cosmopolitanism in this sense lay not so much in the variety and reach of entrepreneurial activity as in their consequences: the consumption of profits, though drawn from all parts of the world, took place in the metropolises, since the beneficiaries resided in Parisian apartments and English suburban mansions. On the other hand, there was what might be called the failed utopia of bourgeois cosmopolitanism.[107] An idealized vision of liberalism at its midcentury peak consisted of free trade in goods between countries and continents, unconstrained by government action or national boundaries, impelled by enterprising individuals of every religion and color. Nationalism, colonialism, and racism would put a brutal end to this vision in the last third of the century.

The cosmopolitan bourgeoisie never developed into an actual social formation with a shared consciousness. Nationalization of the different bourgeoisies prevented this, and uneven economic development around the world took away its material foundation. What remained were nationally based entrepreneurs, many of whom became true "international operators," part adventurers, part corporate strategists (the boundaries between the two were fluid). On all continents, raw materials were exploited, mines operated under license, loans granted, and transportation connections put in place. In 1900 the British, German, North American, and even Belgian and Swiss, capitalist bourgeoisies operated on a scale that would have been unimaginable to any earlier elite. No one from a non-Western country was yet in a position to make a breakthrough at this level of early global capitalism. Even Japanese corporations (save a few shipping companies) limited their expansion before the First World War to a politically secure colonial territory and sphere of influence on the Chinese mainland.[108]

At various times in the nineteenth and twentieth centuries, many societies—at a regional or even national level—arrived at a hard-to-define threshold at which a multitude of "middling sorts" (to use an eighteenth-century Anglo-American term) turned into a social formation displaying solidarity beyond one's town or part of town, congregating around institutions such as the humanities-centered "gymnasium" in Germany, reflecting on a shared universe of values, and developing a politically articulated consciousness of itself as separate from the top and bottom in society. In France this threshold was reached in the 1820s, in the Northeast of the United States or urban Germany around the middle of the century (although the German bourgeoisie, for example, remained significantly more heterogeneous than the French).[109]

As an age of transition, the nineteenth century witnessed the rise, but not necessarily the triumph, of the bourgeois conception of the world and human existence. In Europe this came under challenge from the growing ranks of labor. The partial embourgeoisement of the working population did not inevitably strengthen the bourgeoisie, and at the end of the century upwardly mobile groups of employees in some parts of Europe and the United States came dangerously close to it, though seldom gaining political independence in the manner of the workers' movement. Bourgeois culture itself acquired mass aspects, even before an entertainment industry took widespread hold after the First World War. Alongside classical high culture and the new mass culture, a third position associated with the avant-garde appeared around the turn of the century. Small circles of creators, such as the Viennese composers around Arnold Schönberg who claimed to be emancipating musical dissonance, retreated from the bourgeois public sphere and chose to launch their work at private events. Visual artists in Munich, Vienna, and Berlin proclaimed "secessions" from the aesthetic mainstream. This was an almost unavoidable reaction to the museumization and historicization of bourgeois culture, from which the artistic production of the time increasingly distanced itself. Finally, the suburbanization process fueled by railroads and the automobile undermined bourgeois sociability in the early twentieth century. The classical bourgeois is a "man about town," not a suburbanite. As the housing sprawl robbed cities of their shape, the intensity of bourgeois communication began to slacken.

So, it was not only the shock of the First World War that ended a *belle époque* for the nobility and the upper middle classes. Tendencies to disintegration were already building up before 1914. The crisis of the European bourgeoisie in the first half of the twentieth century passed into the huge post-1950 expansion of middle-class societies, which substituted consumerism for the ideals of virtue and respectability of the "classical" bourgeoisie. This was a worldwide process, although it made itself felt unevenly. Even where the bourgeoisie had been weak in the nineteenth century, middle strata now grew markedly in size and influence. Communist rule acted as a brake, but "goulash communism" was perfectly consistent with petit bourgeois ways, and the nomenklatura parodied high bourgeois or even aristocratic precursors in such things as its passion for hunting. In Eastern Europe and China, the history of the bourgeoisie could recommence only after 1990. Some continuities then led back into the nineteenth century.

A global social history of the nineteenth century can set itself many tasks other than those outlined here. For example, it can ask which positions were occupied by custodians of knowledge and "knowledge workers" in various social spaces, such as the type of intellectual developed in the West that others modified and adopted elsewhere in the world—a process that appears to have accelerated soon after 1900.[110] It can take an interest in the development of gender roles and family types, whose great variety makes generalizations especially difficult. Whether there was and is a typically European model of the family and kinship relations, and what specific changes it underwent in the nineteenth century, is a controversial question that only extensive comparisons can help to elucidate.[111] We can be sure that European family ideals did not spread around the world through simple diffusion and force of example. The merits of European technology or military methods could be easily appreciated and copied, but not those of modes of biological and social reproduction. Such basic elements of sociality did not travel well. Colonial governments showed much greater caution here than in other domains, and reform initiatives on the part of public and private bodies began on a large scale only after the turn of the century.[112] Even the war on polygamy and concubinage—the most visible and, for Christians, most objectionable deviations from European standards—was in most cases only halfhearted, being left to missionaries and seldom producing the success expected of it.

CHAPTER XVI

Knowledge

Growth, Concentration, Distribution

"Knowledge" is a particularly ephemeral substance. As a social quantity, distinct from its various philosophical concepts, it is the invention of a discipline scarcely a hundred years old: the sociology of knowledge. It took what German idealism had called *Geist* ("spirit") and placed it at the heart of society, relating it to existential practices and social locations. "Knowledge" is somewhat narrower than the all-embracing concept of "culture." It does not for our purposes include religion and the arts;[1] it will refer here to cognitive resources for the solution of problems and the mastering of life situations in the real world. This is a preliminary decision in conformity with the nineteenth century itself, when, at least in Europe and North America, a rationalist, instrumental understanding of knowledge came to the fore: knowing served a purpose. It was supposed to enlarge the mastery of nature, increase the wealth of whole societies through its technical application, liberate worldviews from "superstition," and be generally "useful" in as many respects as possible. Nothing was a more conspicuous measure of progress—the hallmark of the age for European elites—than the expansion and improvement of knowledge.

From the "Res Publica Litteraria" to the Modern System of the Sciences

The formation of "modern knowledge society" has been situated in a long early modern period that lasted until approximately 1820.[2] The next hundred years then witnessed its constant enlargement, institutionalization, and routinization, and even the beginnings of its globalization. Such a continuity should not be exaggerated, however. Only in the nineteenth century was the old concept of "science" enriched with aspects that we now firmly associate with it. The subject classification still in use today goes back no further. Modern institutional forms for the acquisition and dissemination of knowledge were created at that time: the research university, the laboratory, the humanities seminar. The relations between science and its applications in technology and medicine grew closer; the scientific challenge to religious conceptions of the world became weightier. Many terms for disciplines such as "biology"—first used in 1800—or "physics" only now established themselves. The "scientist" (another neologism, coined in 1834) developed into a social type who, despite much overlapping, differed from the "scholar" or "intellectual" (one more nineteenth-century creation). Science as a whole was demarcated more sharply than ever from philosophy, theology, and other traditional branches of learning.

In the middle of the nineteenth century, a new concept of science prevalent among scientists gave up the old claim to strict universality, unconditional necessity, and absolute truth and emphasized the reflexive character of knowledge— its conditional validity, intersubjectivity, and autonomy—within the social system of science.[3] The old imaginative community of scholars, the *res publica litteraria* that cultural historian Peter Burke, following Coleridge, described as a "clerisy," broke open and yielded a special scientific community with narrower membership criteria.[4] The scientist saw himself as a "professional," a specialist in a clearly defined area, having little in common with literary "intellectuals," who addressed a wider public and were politically committed. This was a big step on the way to "two cultures," and only a small number of natural scientists, such as Alexander von Humboldt, Rudolf Virchow, or Thomas H. Huxley, sought and found a hearing for their views on nonscientific matters. Toward the end of the nineteenth century, governments began to take a greater interest in science; science policy became a new branch of systematic statecraft. Big industry (e.g., the chemicals sector), too, increasingly regarded scientific research as one of its tasks. The links between science and war or imperial expansion became closer than ever before.

The Cultural Authority of Science

By the eve of the First World War, the modern system of science had come of age institutionally in a number of countries. Science was a force in the work of interpreting the world and a cultural presence enjoying extraordinary prestige. Anyone who did not observe its standards of argument and justification was thrown into defensive mode, so that even Christians had to make concessions to scientific thinking. It became a compulsory part of the school syllabus, as well as a profession for large numbers of (overwhelmingly male) individuals. Whereas in the seventeenth and eighteenth centuries—up to the time of Alexander von Humboldt, who spent his inheritance on his research interests—many heroes of the "scientific revolution" had lived *for* science on other sources of income, their successors in 1910 lived *on* it. The amateur was retreating on a wide front before the expert. No one could gain recognition as a scientific dilettante, as Goethe had still been able to do in the theory of colors, morphology, and anatomy.

All this holds true only for parts of Europe and for the United States. A global historical approach would not radically alter the picture, however. Modern industry, based on the use of fossil energy, came into being in Europe, and so did the science that has now swept everything before it. Yet a global perspective can place these developments in a comparative context and draw attention to the worldwide impact of the Western explosion of knowledge. A first requirement for this is to expand our concept of knowledge beyond science. Insofar as science itself is understood as a communicative enterprise and its results passed on through channels of communication to a wider public, it relies on a system of symbols that makes scientific contents intellectually transmittable in the first place. Mathematics—an important element also in economics from about 1875—and some natural languages with transcontinental reach guaranteed the mobility of scientific meaning. But, of course, languages are also the most important vehicle for many kinds of knowledge other than organized science. It is therefore impossible to speak of the history of knowledge in the nineteenth century without taking a closer look at language and languages. Their spread and use is a good indicator of the ever-changing geography of political and cultural dominance.

1 World Languages

In the nineteenth century, some language areas became larger than they had been in the early modern period. By 1910 the "world languages" (a term now justified for the first time) had been distributed around the globe in a pattern that is still largely with us today. Here two aspects must be distinguished from each other—although often in practice no clear dividing line can be drawn between them. It makes a difference whether a majority of the population adopts a foreign language as its chief means of everyday communication, a kind of second-order mother tongue, or whether the language remains "foreign" while being used for functional purposes such as trade, scholarship, religious worship, administration, or contact across cultures. The expansion of a language is made easier by political and military empire building, without being an inevitable outcome of it. For example, in the early modern period in Asia, Persian and Portuguese became more widely spoken without being carried into new territories by the colonial rule of Portugal or Iran. On the other hand, relatively short-lived formations such as the Mongol Empire of the Middle Ages or the Japanese Empire in the first half of the twentieth century left behind hardly any lasting linguistic traces. In Indonesia too, despite three hundred years of colonial rule, Dutch did not maintain itself alongside indigenous languages, since unlike the British in India, the Netherlanders never took pains to develop a culturally Europeanized layer of the population.

Portuguese survived around the Indian Ocean into the 1830s as a lingua franca of multicultural merchant milieux. The flowering of Persian between the thirteenth and seventeenth centuries in western, southern, and west-central Asia was followed by the collapse of its literary ecumene in the eighteenth.[5] But until the 1830s it continued to play its old role as an administrative and commercial language beyond the borders of Iran. Both Portuguese and Persian were then replaced by English, which in 1837 became the only recognized language of administration in India and, at the latest with the opening of China in 1842, the dominant non-Chinese language in the Eastern seas. By the end of the century, the Portuguese-speaking world had been whittled down to Portugal, Brazil, Goa, and a few possessions in southern Africa. Spanish was a legacy of colonial settlement in South and Central America, its geographical extent remaining more or less unchanged in the nineteenth century. Chinese spread slightly as a result of coolie emigration from China, but it never moved outside the overseas Chinese communities to become a language of education reaching into the environment around them. The fact that most of the overseas Chinese originated in Fujian or Guangdong province and used dialects barely intelligible to Mandarin speakers contributed to this isolation of the Chinese language.

Winners of Linguistic Globalization

The German language spread to only a very limited extent in the wake of colonization and had no real lasting effect in Africa. But its position strengthened in east-central Europe with the founding of the German Reich in 1871 and the literary and scientific esteem it enjoyed from the eighteenth century on. It continued to be the administrative language of the Habsburg Empire and, until the end of the Tsarist period, it remained with French and Latin a major language of communication among scholars in Russia; the papers of the Saint Petersburg Academy of Sciences, for example, were largely composed in German. Wherever the Reich pursued a policy of Germanization in its border areas, compulsory use of the German language became more common.

Russian expanded to an even greater degree, as a direct result of Tsarist empire building and the cultural Russification associated with it after midcentury. Russian was imposed as the only official language in the Tsarist Empire, meeting resistance from Poles and subject populations in the Caucasus. Apart from being a symbol of Tsardom, it was also the main cultural cement of the empire. In contrast to the great ethnic diversity of the Habsburg armies, the Tsarist military consisted overwhelmingly of Russian-speaking soldiers.[6] This was also the time when Russian developed as the language of a world-class literature. Nevertheless, it may be doubted whether the Tsarist Empire really did become an integrated linguistic community. Especially in the Baltic provinces in the Northwest and the Muslim lands in the South, the Russian language did not penetrate beyond circles of immigrants from Russia and a stratum of administrative officials.

At a time when the use of French was gradually declining among scholars and educated people in Europe, the number of French speakers in the colonial empire was on the rise. Moreover, the French Canadians in Quebec (since 1763 no longer part of the French empire) were maintaining themselves as a separate linguistic group. It was the only territory ever ruled by France where the language remained in everyday use beyond elite circles in the late nineteenth century (even today it is the mother tongue for roughly 80 percent of the population). Things were different in the African and Asian colonies. Almost half a century after the end of colonial rule, the number of Algerians who speak or understand French is estimated at up to a quarter of the population.[7] In countries that used to belong to France's West African empire, French is still the official language (alongside English in Cameroon), although it is probably used by just 8 percent of people in daily life.[8] Haiti sticks to French two hundred years after its revolutionary separation from France. If a traveler in 1913 could get by with French better than with any other language except English, this was due to France's military-colonial expansion after 1870 and the high cultural prestige it enjoyed among Middle Eastern elites in particular. From 1834, French was part of the training program for Ottoman elite officers, and in Egypt it held its ground among the upper classes even after the British occupied the country in 1882.[9] At the end of the nineteenth century, a kind of *francophonie* reached far down into the Pacific, where political control had weakened other culturally autonomous forces and broken up their coherence.

The biggest winner from nineteenth-century globalization was English. In 1800, although already respected throughout Europe as a language of business, poetry, and science, it had by no means been the undisputed number one. But by 1920 at the latest, it had become geographically the most widespread language in the world and culturally the most influential. At a rough estimate, for the period between 1750 and 1900, one-half of the "weightiest" publications on natural science and technology appeared in English.[10] As early as 1851 Jacob Grimm, the leading linguist of his age, noted that no other language carried so much force.[11] In North America (where, contrary to legend, German never had a chance of becoming the national language of the United States), English was as firmly rooted as in Australia, New Zealand, or Cape Province. In all these cases, it was the language of settlers and invaders little open to the influence of indigenous languages (which were never of any importance officially).

In India, by contrast, English became the standard language in the higher law courts only in the 1830s, while the lower courts continued to operate in local languages, often with the help of interpreters. Here and in Ceylon, English did not spread through European settlement, or a fortiori as a result of ruthless Anglicization policies on the part of the colonial rulers, but because a combination of cultural prestige and mundane career advantages made it advisable to master the language.[12] New educated strata first emerged in Bengal and around the colonial metropolises of Bombay and Madras, then in other parts of the Subcontinent. In the 1830s there was a heated debate between "Anglicists" and "Orientalists" about the pros and cons of an education in English versus one of the indigenous Indian languages.[13] The Anglicists won out in 1835 at the level of countrywide politics, but in practice there was scope for pragmatic compromises. The British language export to India was at the same time a voluntary import by Indian citizens and intellectuals who hoped to link up with more extensive circles of communication. During the second half of the nineteenth century, English spread along with British colonial administrators and missionaries to Southeast Asia and Africa. In the Pacific (Philippines, Hawaii) the US influence was decisive.[14] But the global fortunes of the English language in the nineteenth century were driven by Britain more than America. The triumph of English in education, business, mass media, pop music, science, and international politics got under way only after 1950, this time spurred on by the dynamism of the United States.

Language Transfer as a One-Way Street

Outside the colonies too, there was growing pressure and incentive to learn European foreign languages. The Chinese state, which in the Qing period was officially trilingual (Chinese, Manchu, Mongolian), had never felt it necessary to promote the study of European languages. Paradoxically, this was one of the reasons for the high linguistic competence of Jesuit missionaries during the early modern period, so high that many served as interpreters for the Qing Emperor in contacts with emissaries from Russia, Portugal, the Netherlands, or Britain. But since the ex-Jesuits who remained behind in China after the abolition of their order in Europe had no knowledge of English, the British envoys who established the first diplomatic contacts in 1793 could in some cases communicate only through a prior translation into Latin for the Jesuits' benefit. When much more serious negotiations had to be conducted, after 1840, such go-betweens were no longer available. China initially lacked any personnel trained in languages—another disadvantage in the general asymmetry between China and the West—and the emperor long adhered to the old Qing policy of making it as difficult as possible for foreigners to study Chinese.

In the Ottoman Empire too, no encouragement for the study of European languages was given until well into the nineteenth century. But after 1834 (the comparable Chinese date was 1877), when the Sublime Porte began to establish permanent diplomatic representations in the main European capitals, some of the leading Tanzimat reformers got to know foreign languages and foreign countries while serving as diplomats abroad. The new power elite of the Tanzimat period was recruited less from the army and the *ulama* (clergy trained in law) than from the State Translation Bureau and embassy chancelleries.[15] In China, meanwhile, the Qing government changed course only after the Second Opium War ended in defeat in 1860. Two years later the Tongwenguan translation school—the first Western-style educational institution of any kind—was founded in Beijing; its dual task was to train English speakers and to translate technical literature from the West (no mean feat, given that, as in Turkey a few decades earlier, much of the vocabulary first had to be created in the destination language).[16] Even some of the large state arsenals and shipyards that sprang up in this period had language departments attached to them. The most important channel of linguistic transfer, however, was the mission schools and universities. At the Paris peace conference of 1919, China fielded a young guard of capable diplomats who impressed others with their proficiency in foreign languages.

In Japan, where classical Chinese remained the most prestigious language of education down to the end of the Tokugawa period, specialist hierarchies of translators were responsible for contacts with the Dutch in Nagasaki; the world of true scholarship had little to do with them. It was through this needle's eye of Dutch trade, alone in having official approval, that European knowledge found its way into the sealed-off archipelago. Only after 1800 did it gradually became clear to the Japanese government that Dutch was not the most important European language, and greater efforts now went into translation from Russian and English.[17] Since the seventeenth century, Japanese had also been familiar with translations of Western scientific and medical texts into classical Chinese, made by Jesuits in China with the help of indigenous scholars;[18] "Holland studies" (*rangaku*), in which scientific material had featured prominently since the 1770s, were not the only transmission route of Western knowledge into Japan. But in the end, the more intensive introduction of that knowledge in the Meiji period was possible only because in addition to the hiring of Western experts, there was a more systematic drive to develop translation skills among the Japanese themselves.

European languages were included only late and sporadically in the official educational syllabus of non-European countries, even though these often had a multilingual dimension in that scholars were required to show proficiency in Turkish, Arabic, and Persian. Knowledge of Europe was for a long time the reserve of indispensable, but not very highly regarded, specialists modeled on the dragomans in the Ottoman Empire—a small group of state-appointed interpreters and translators dominated by Christian Greeks until 1821.[19] Conversely, it never occurred to anyone in Europe to honor a non-Western language by including it in the school curriculum. Among European linguists, Persian and Sanskrit (first known in Europe in the late eighteenth century) were considered the height of perfection. But if they could ever have seriously competed with Greek and Latin (perhaps in 1810 or 1820), that brief opportunity was missed.[20] The humanism of the *Gymnasien*, *lycées*, and public schools remained purely Greco-Roman; European intellectual formation centered on the West. Only in recent times has Chinese made a breakthrough into the syllabus of a growing number of high schools in Australia or a few European countries.

Linguistic Hybridity: Pidgin

World languages—that is, ones in which people could make themselves understood outside their land of origin—were for the most part loosely superimposed on a multiplicity of local languages and dialects. Even in postcolonial India, a maximum of 3 percent of the population could understand English (the figure in today's Republic of India is around 30 percent).[21] In many cases, simplified hybrids made communication easier. These seldom replaced the original languages, however, and demonstrated by their very existence how strongly local languages resisted the colonial ones they encountered. Not a few pidgin languages were older than colonialism. And when, following the Peace of Utrecht in 1713, French replaced Latin as the usual language of negotiation and treaty among the representatives of European states, diplomats in the eastern Mediterranean and Algeria were still using the old lingua franca (i.e., language of the Franks), a kind of pidgin Italian.[22] In other parts of the world—for example, the Caribbean and West Africa—Creole tongues developed into independent language systems.[23]

Pidgin English, originally known as "Canton jargon," took shape in a long process after the 1720s as the second language on the South Chinese coast. After the opening of China it served throughout the treaty ports as a means of communication between Chinese and European traders. It was later forgotten that it had originated in a reluctance or inability on the Western side to learn Chinese; the risibility of pidgin, with its reduced and inflected forms ("likee soupee?"), became a key element in the racist cliché of "primitive" Chinese. Conversely, a striving to overcome this humiliation was a major reason why nationalistic Chinese intellectuals, in particular, learned foreign languages in the early twentieth century. This went hand in hand with drastic "depidginization." On closer examination, however, the mature China Coast English that pidgin became around the turn of the century proved to be a communicative medium well suited to the situation. Blending many other sources into the mix, from Malay to Portuguese to Persian, it offered a rich vocabulary for the realities of life on the Chinese coast.[24]

As in India, sophisticated communication in a European language did not mean subjugation to linguistic imperialism so much as an important step to cultural acceptance and equality. Pidgin remained a language of the business world; Western-oriented intellectuals learned proper English. Pidgin did not persist in twentieth-century China, leaving only scattered lexical remnants even in Hong Kong. Chinese as a language of education easily survived contacts with the West, while in Japan there was not even an embryonic pidgin. Classical Chinese also continued to fulfill practical objectives in the region where Chinese culture has always radiated outward. When in 1905 Phan Boi Chau, the most famous Vietnamese patriot of his time, visited the great Chinese intellectual Liang Qichao in his Tokyo exile, the two men found they had no spoken language in common. But since Phan had mastered classical Chinese writing, for centuries the medium of communication used by Vietnamese mandarins, they were able to engage in what Phan in his memoirs calls "brush conversation."[25]

Knowledge travels in the baggage of languages. Not only did the expansion of major language areas in the nineteenth century strengthen local linguistic diversity and the practical necessity of multilingualism at a time when an extra language required close attention; it also opened up new spaces of horizontal communication and increased the mobility of knowledge. Colonialism and globalization created cosmopolitan language systems. In Chinese civilization, which had never lost its linguistic unity and capacity for resistance, this spelled a less dramatic change than in regions such as South Asia, where in the preceding centuries local vernaculars had gained ground at the expense of a single overarching language, Sanskrit, and where new semantic ranges were now developing at the level of the elite. After its linguistic fragmentation, India was reunified communicatively through the appropriation of English.[26]

Limits of Linguistic Integration

We should not, however, exaggerate the integrative effects outside the ranks of small elites. In Europe too, linguistic homogeneity within nation-states often emerged only in the course of the nineteenth century. The national language, rising above a multiplicity of regional idioms, did become the ideal norm for communication and the measure of correctness, but it was rather a slow process putting such an ideal into practice.[27] This was true even of France, with its strong centralist traditions. In 1790 an official investigation established that a majority of people in France spoke and read a language other than French: Celtic, German, Occitan, Catalan, Italian, or Flemish. Even in 1893 every eighth schoolchild between ages seven and fourteen knew no French.[28] The situation was even more discrepant in Italy, where in the 1860s less than 10 percent of the population could understand effortlessly the Tuscan Italian that had been declared the official language in the process of nation building.[29] Nor were things necessarily different in the successor states of the Spanish colonial empire. The Porfirio Díaz regime in Mexico did not think of creating schools for the Indian or mestizo population, so that in 1910 as many as two million Indios—14 percent of the total population—spoke no Spanish.[30]

As scholars all over Europe collected languages (and added neologisms) in dictionaries, described them in grammar books, and laid down rules for spelling, pronunciation, and style, whole nations were conceived and promoted as speech communities, and a cultivated language began to be considered a key achievement of every nation. Yet the language that ordinary people spoke in many regions remained stubbornly tied to the locality of their birth. If scientists and intellectuals in Asian countries—around 1862 (and even more after the turn of the century) in the Ottoman Empire, or after 1915 in China—created simpler forms of language, writing, and literature to bridge the gulf between elite and popular culture, they were doing only what had been done in European countries a few decades earlier, or was even then being done, without engaging in anything that might be described as direct imitation. In Europe too, the linguistic divide in the nineteenth century between elite and people, between written and spoken language, was more extensive than we can easily imagine today. For mature nation-states, however, this became intolerable a few decades later, and great efforts were made to impose a uniform national language or at least to preserve the external appearance of one. After the Second World War, European regional and national movements—from Catalonia via Wales to the Balkans—set a countertrend in motion.

2 Literacy and Schooling

One of the most important cultural processes of the nineteenth century was the spread of mass literacy. Having begun centuries earlier in many societies, and developing now at a highly uneven regional or local pace, it should not be too hastily attributed to other basic processes such as state building, the growth of confessionalism or a science society, or even industrialization.[31] One can argue at length about the precise meaning of "literacy," the spectrum of which runs from the ability to sign a marriage certificate to regular reading of religious texts to active involvement in public literary life. The crux of the matter is clear, however: literacy is a cultural technique of reading (and secondarily writing) that makes it possible to participate in communicative circles wider than those of face-to-face speech and hearing. Someone who is able to read becomes a member of a translocal public. This also opens up new opportunity for manipulating and being manipulated. By 1914 the male population of Europe had attained such a degree of literacy that soldiers on all sides could read weapon instruction manuals, absorb the propaganda that warmongers wrote for them, and keep their family posted with news from the front. The scope and scale of the Great War is hardly imaginable without comprehensive literacy.

The Trends in Europe

The nineteenth-century spread of mass literacy was first of all a process of *European* cultural history. On that continent—only in China do we find anything comparable, with no influences on each other—roots existed here and there in an older tradition of book reading that went back to the age of the Reformation or the "popular enlightenment" and its emphasis on practical pedagogy. The nineteenth century continued these trends and gave them a certain finality. It was the rise of mass education that, in conjunction with the "scientific revolution" of the early modern period, laid the key foundations of our age. Beyond the functional aspect of increased competence, literacy gained new symbolic significance as the expression of progress, civilization, and national cohesion by creating an imagined community of people capable of communicating with one another but also of being steered toward common goals.[32] By 1920 the male population of the major European countries, as well as part of the female population, was in possession of reading and writings skills.

Lest we create the impression of an educated continent facing a world sunk in ignorance, some distinctions need to be drawn within Europe itself. Only Britain, the Netherlands, and Germany were 100 percent literate in 1910; the rate in France was 87 percent, while in Belgium, the least literate of the "developed" European countries, it was 85 percent. Then, a long way behind, came southern Europe: 62 percent were literate in Italy, 50 percent in Spain, only 25 percent in Portugal;[33] the picture was certainly no better on the eastern and southeastern periphery of Europe. Nevertheless, there were certain continent-wide tendencies:

the proportion of literate males and females was rising constantly and in no case stagnating. Some countries—Sweden, for example—was advancing rapidly from a high initial level.

The period around 1860 was a watershed for the whole of Europe. Before, only Prussia had come close to the goal of completely eradicating illiteracy, but a quickening of the pace after 1860 is apparent not only from the statistical data but also from the general climate in society. By the turn of the century, widespread illiteracy was no longer taken for granted even in Russia or the Balkans; an ability to read and write was seen more or less everywhere as a normal state of affairs and a political objective worth striving for. It was achieved not only in the nobility and urban middle classes but also among artisan strata in town and country, skilled workers, and ever larger numbers of the peasantry.[34] Regional differences did not completely disappear. In the 1900 census, the Vorarlberg region of Austria recorded just 1 percent illiteracy, while the figure in Habsburg Dalmatia was 73 percent.[35] It would be a while longer before reading and writing skills permeated the last village in Russia or Serbia, Sicily or the Peloponnese.

Full literacy did not come overnight: it was a long process that did not embrace whole countries all at once. It began in small groups. Some family members, mostly the younger generation, learned to read, others did not. This had consequences for parental authority. Villages, neighborhoods, or parishes gradually changed their mix of cultural techniques. It would be too simple to assume that there was a wholesale transition from orality to literacy; competence in writing continued to impart cultural authority, and oral communication persisted in many of its old forms. The fact that from about 1780, urban intellectuals in Europe were transcribing fairy tales, legends, and folk songs, giving them a tone of highly artificial naturalness, was a sign that oral traditions were losing their spontaneous impact. Examples in Germany included Johann Gottfried Herder (who published several sets of folktales from 1778 on), Achim von Arnim and Clemens Brentano (*Des Knaben Wunderhorn*, 1805–8), and the Brothers Grimm, whose first collection, *Children's and Household Tales* (1812), would become the hardiest perennial of German literature.[36] Only that which is, or is becoming, "alien" can be rediscovered. Mass literacy first developed in the cities and often percolated very slowly into village society, so that during a transitional period it actually widened the cultural gap between town and country. It also changed the parameters of *Bildung*. Only those who read much and without difficulty could participate in the semantic universe of high culture. But the spread of reading also increased the demand for popular material—from the farmer's almanac to pulp fiction. Historians have closely studied these fine shades of democratization between the two poles of "high culture" and "popular culture."[37]

Elites reacted to mass literacy in contradictory ways. On the one hand, the enlightenment of "simple people," dispelling superstition with rational literature and generally standardizing cultural practices, appeared as a prime instance of "civilizing from above" that spread modernity and promoted national integration. On the other hand, mistrust lingered on (though everywhere in a downward curve), since the cultural emancipation of the masses—as the workers' associations soon showed—was bound up with demands for social and political betterment. This attitude on the part of the powerful and well educated was not without a basis in reality. More democratic access to literary forms of communication did usually lead to restructuring of the hierarchies of prestige and power, opening up new possibilities for an attack on the existing order. The cultural worries of the elite also reverberated in gender politics. The idea that immoderate reading could lead to fanciful illusions and (especially among women) to an overheated erotic imagination—a satirical theme in literature up to Gustave Flaubert's *Madame Bovary* (1856) and beyond—was a source of concern for male guardians of morality.[38]

Mass literacy campaigns were mostly initiated by the government of the day. Elementary schools were their chief instrument, although for a time many European governments were content to leave them in the hands of the church. The weaker a state was, the stronger the educational role of religious institutions, if only in the modest form of Sunday schools, remained. Or to put it in another way: the state, churches, and private providers competed with one another to serve a burgeoning education market. Nor was this in essence a purely European phenomenon. The English education system, for example, had many similarities to that which existed around the same time in Muslim countries: for example, the primary level was largely controlled by religious institutions, whose main aims were to teach reading and writing, to inculcate moral values, and to protect children from "bad influences" in their everyday environment. The differences were a matter of degree more than principle. In England there was less learning by rote, less recitation of sacred texts, a slightly greater practical orientation, and a moderately better provision of material aids and furnishings for schools.[39]

Popular education could not be simply forced down people's throats. It could be successful only if they associated their own desires and interests with it. The difficulties that every country faced in actually enforcing compulsory education (at various moments in the nineteenth or twentieth century) point to the extraordinary importance of parental cooperation. Economic requirements had to be fulfilled if mass literacy was to be achieved. Of course, it would be wrong to underestimate the genuine thirst for education in many societies: the motivation to learn reading and writing, both for oneself and for one's children, was not only a question of material gain and utility. Nevertheless, only above a certain income threshold were families able to release their children from production and to cover the costs of regular schooling. Mass education with fixed hours of attendance and set tasks that had to be done regardless of the rhythm of the local economy was possible only where children did not have to work to keep the home in one piece. On average, it was in the last quarter of the nineteenth century that European families became prepared to send their seven- to twelve-year-old children to the special world of the school, where professional teachers (whose professionalism was often debatable) had an authority that could hardly be challenged from outside.[40] The actual figures should not be exaggerated, however. In Britain in 1895, only 82 percent of children registered to attend primary school were regularly present in the classrooms.[41] In many other countries of Europe, the proportion was far smaller.

An Age of Reading in the United States

Were there similar developments outside Europe? The school uptake in countries such as Mexico, Argentina, or the Philippines was not dramatically lower than in southern Europe or the Balkans.[42] As far as literacy is concerned, comparative research is still in its infancy, and in many parts of the world, statistics are lacking for the whole of the nineteenth century. Of course, this is not the case for North America, where the early colonies already had high levels of literacy comparable with those in the most advanced European countries. Increased immigration in the nineteenth century meant that an ability to read and write in English was often equated with "Americanization." Many new arrivals, especially Catholics, accepted this imperative, but created educational institutions of their own where learning was closely associated with religion and ethnic identity. From the 1840s on, there was a growing sense in the United States that an "age of reading" had dawned. Rapid expansion of the press and book production contributed to this, as the Northeast in particular became the locus of a vigorous print culture.

By 1860 the male literacy rate in New England was already 95 percent, and uniquely in the world, women there had reached a similar level. The fact that the national average (an especially unhelpful term in the United States) was considerably lower had to do less with a certain backwardness of the white population in the West and South than with the low literacy rate among blacks and Native Americans. Some slaves learned to read the Bible from their mistress, but normally they were kept well away from such things: a literate slave could become a fomenter of rebellion and was treated with constant suspicion. As for the Northern states, despite much discrimination, freed slaves showed a great interest in written forms of communication—as several hundred autobiographies from the two decades before the Civil War eloquently testify. The nationwide literacy rate among African Americans rose from 39 percent in 1890 to 89 percent in 1910, but then fell back to 82 percent in 1930;[43] it was thus higher than in any population group of comparable size in black Africa or much of rural eastern and southern Europe. After the restoration of white hegemony in the Southern states in the 1870s, however, African Americans had to fight for an education through common efforts against a hostile white environment and an (at best) indifferent government.[44] The same was true for other disadvantaged ethnic segments of US society. Some Indian peoples, though facing great resistance, used literacy as an instrument of cultural affirmation; the most notable case was the Cherokee Nation, which had had a written language since 1809 and was able to use this as the basis for a simultaneous acquisition of reading and writing skills in both Chero-kee and English. Similarly, in many other parts of the world, languages first had to be given an alphabet and lexically recorded (often, though not always, by missionaries); then parts of the Bible were translated and used as exercise material, providing the basis for the enrichment of communication through writing.

Asia's Old Literate Cultures

The picture was different again in civilizations that had treasured writing and learning since time immemorial: the Islamic countries, with their strong focus on the Koran and legal-theological commentaries, and the regions influenced by Chinese culture. In Egypt less than 1 percent of the population was able to read in 1800; this rose to 3–4 percent by 1880 as a result of modernization policies, and the 1897 census, the first in modern Egypt, recorded 400,000 literate people, or roughly 6 percent of the population over the age of seven (excluding nomads and foreigners).[45]

In 1800, even by strict European standards, Japan was already a society permeated with writing. A literary mass market had emerged as early as the seventeenth century in the cities; all samurai and the numerous village headmen had to be literate and to read Chinese characters in order to carry out their administrative tasks. On the whole, the authorities did not fear educated subjects, and some princely houses saw it as their duty to raise the moral and technical level of the population at large. In the early decades of the nineteenth century, elementary education already went beyond the circle of rural notables, and by the end of the Tokugawa period in 1867 as many as 45 percent of boys and 15 percent of girls (some estimates are even higher) had regular instruction outside the home in reading and writing.[46] All this happened without the slightest European influence, missionaries having been banned from the country since the 1630s. In 1871 a national education ministry was created, and the Meiji government made it a high priority to develop every level, from the village school to the university, under close central supervision. Many schools and teachers from the Tokugawa period were incorporated into the new system, which provided for a compulsory four-year course. Pedagogues now began to study Western models and brought over some elements from it, but isolated premodern Japan had already set its sights on state-run education, and an independent direction was much more in evidence than in the army reforms introduced during the same period. By 1909, near the end of the Meiji period, the number of illiterates among twenty-year-old recruits was below 10 percent almost throughout the archipelago—a success without parallel elsewhere in Asia.[47]

In 1912 Japan was one of the world leaders in literacy. In China, where the standard textbook went back to 500 AD, the literacy rate seems to have stag-nated in the nineteenth century, though at a comparatively high level for a pre-modern society. For many centuries, China had shown great reverence for the written word and refined calligraphy that permitted the dissemination of all manner of books, and the flourishing of a varied landscape of private education as well as community, welfare, clan, and temple schools little regulated (and by no means systematically shaped) by the government. During the greater part of the nineteenth century, most of these were one-teacher schools rooted in a local initiative; their organizers could draw on a huge pool of some five million people with a training in high culture, who, having failed at some stage in the state examination system, were excluded from the status group of title bearers and often worked as home tutors for upper-class families.[48] For want of statistics, we have to rely on good-quality anecdotal evidence, and this does permit the conclusion that 30 to 45 percent of the male and 2 to 10 percent of the female population had at least basic reading and writing skills.[49] This did not mean, of course, that they met the high standards of elite communication, but they understood a basic repertory of written characters and therefore edicts and proclamations of a hortatory, admonitory, or interdictory nature that the government issued to its subjects, and often also simplified versions of classical texts. The imperial state made some commitment to education and the funding of schools, but without asserting the kind of general authority in the matter that slowly developed in Europe during the nineteenth century. For centuries the legitimacy of the political and social order had rested on the fact that access to education, and hence to status and prosperity, was not reserved only for the offspring of upper-class families. Possibilities of upward movement therefore had to be kept open, such as those offered at least by the church in early modern Europe. Practices on the ground were quite flexible: for example, elementary education for peasant children was concentrated in months when there was no work to be done in the fields.

Why Did China's Culture of Education Fall Behind?

The Chinese elementary school system, like the institutional arrangements for education in general, did not keep abreast of international competitors in the nineteenth century. The traditional system, efficient though it was in many respects, contained no potential for modernization (unlike the Tokugawa system in Japan). The imperial government itself recognized this after a long period of hesitation. In 1904 it issued a national schools ordinance and declared its intention to build a countrywide, three-tier educational system modeled on those of the West and above all Japan (which in turn had used Europe as its template). One year later, the old system of status assignment and civil service recruitment through state examinations was abruptly discontinued, with little or no provision for transitional measures.[50] Korea—the third Asian country after China and Vietnam with an old tradition of state exams—had executed a similar radical step in 1894, an astonishingly early date.[51] The collapse of central state power in China, beginning with the 1911 revolution and unstoppable throughout the period of the Republic (until 1949), frustrated the plans that had been worked out at the turn of the century. If China's educational system today is highly differentiated and efficiency oriented, having successfully blended assistance from abroad with the country's own resources to rise up the international rankings, this is mainly a result of state policy after 1978. The gap that appeared around 1800 has now been corrected, two hundred years later. But how did that gap come about? Three reasons suggest themselves:

First. The traditional education system was shaped entirely "from above" and geared to state examinations. Even if the great majority of peasant school-children were not expected to undergo one day the full rigors of the examination procedures, they did have to memorize the simpler writings of the Confucian canon as soon as they had learned a basic stock of characters. This unitary conception of education left no room for the particular skills required by various layers of the population. It is true that—in contrast to the modern European notion (now highly developed in China too) of school as a special space removed from ordinary life—a dense web of connections integrated schools into everyday existence. But the subject matter was frozen into a curriculum increasingly divorced from practical concerns—an obvious definite loss of creativity in comparison with earlier times, when the curriculum had repeatedly been a hotly debated bone of contention.

Second. The failure of China's educational system to keep up with its international rivals first became evident when the previously uncontested empire began to suffer military defeats after 1842. But it took decades before an analysis was made of the reasons for China's military weakness and economic stagnation. For the scholar-officials who governed and administered the empire, nothing was more difficult than to admit that the education to which they owed their social rank and personal identity could be somehow to blame, or that adjustments were required to meet the new challenges. The superiority of Western knowledge (*xixue*) in some domains was soon recognized, but there was an unwillingness to grant equal value to Western culture as such. The fact that aggressors and invaders were the bearers of the new knowledge, and that Christian missionaries in the forefront often behaved without the necessary tact, contributed to the general sense of mistrust. After 1860, small circles of Chinese opened up intellectually to the West, and the state established a number of translation bureaus. But a sterile counterposition of Chinese to Western knowledge became a dogma among the majority of literati in the second half of the nineteenth century.[52] When after the turn of the century the mood shifted into one of acute national crisis, Chinese tradition came to be seen as deeply problematic. Elements of Western knowledge were imported as a matter of urgency (mainly via a grudgingly admired Japan); the Japanese educational system (or anyway some of its elements) was hastily adopted in a spirit of panic. Throughout the period of the Republic (1912–49), Chinese intellectuals and educational reformers wrestled with the problem of how to assimilate and integrate knowledge from diverse sources. Some tried to salvage valuable parts of the tradition by scrutinizing and cleansing them with the methods of source criticism, while others looked for salvation either in Bolshevik-inspired anti-Western Marxism or in full-scale Westernization. Given the weakness of the Chinese state, however, no solutions of any kind could be converted into policies applicable to all parts of the country. The basic intellectual and educational problems of the nineteenth century would have to be tackled anew in the People's Republic after 1949.

Third. The late imperial state would have had neither the administrative nor the financial resources to take charge of education. The size of the country, the traditional underdevelopment of religious/church education as a third way between the private home and state institutions, the weak presence of the bureaucracy at village level, and the deficient fiscal base of the central government together conspired to rule out resolute policies along the lines of Meiji Japan.[53]

School State and State Schooling

A discussion that starts with literacy as a knowledge indicator soon broadens out into a comparative account of institutional education as a whole. Here we may draw two general conclusions. On the one hand, it was only in the nineteenth century that the many forms of practical learning and moral instruction in society came to be thought of, and actually organized, as an educational *system*. The idea that schools should have a standard form and be connected by a common syllabus, that pupils should pass through classes grouped by age, that teachers should receive a professional training and have the appropriate qualifications, that special ministries should direct and monitor changes to the system: all this acquired practical importance in Europe and elsewhere only in the nineteenth century. On the other hand, the state—in competition with private bodies, including religious communities—began to aim for a monopoly in the education of children and young people of compulsory school age. In many countries, such as the Netherlands, a deep political gulf developed over whether the state or the church should control education. A state monopoly took a long time to come into effect even in centralist France, while in some leading Western societies such as the United States or Britain it never came close to being achieved. Today it is being increasingly undermined by private schools in mainland Europe too and is certainly not a distinctive feature of "the West" as a whole. It was taken furthest in the socialist party dictatorships of the twentieth century—one among few achievements brightening up their historical record. Since the state relaxed its grip in the 1990s, even the People's Republic of China has experienced a dramatic rise in the number of illiterates (those unable to read at least 1,500 characters).[54]

The state's claim to sovereign control over the formal education of young people was a revolutionary innovation of the nineteenth century. Children from the lower and middle strata of society entered state schools for the first time, while those from rich families were more often educated together in special institutions rather than by private tutors at home. The state became a "school state," society a "school society"—as historian Thomas Nipperdey put it with reference to the German lands.[55] The trend was most evident there, but it made itself felt worldwide; Germany—especially Prussia—became the closely observed model to be copied elsewhere. It was Prussia's organizational and bureaucratic measures that counted most here, rather than the idealistic ambition of its early reform period to reinvent Prussia as a *Bildungsstaat*. Such noble policy objectives were a thing of the past by midcentury.[56] Governments around the world had various aims and priorities in their development of public education: to discipline the population, to shape "model citizens" for a "model state,"[57] to improve military effectiveness, to create a homogeneous national culture, to integrate empires culturally, to promote economic development by raising the skill levels of "human capital." To be sure, such a top-down perspective needs to be set alongside the view from below. Whatever the intentions of the state elite, people in many societies around the world saw in education the promise of upward mobility and a better life. This translated into a demand for opportunities that could be satisfied by the state, the church, or private philanthropy—or else by self-help.

Colonial governments were the least ambitious and forthcoming. At the minimalist end of the spectrum, they showed no concern at all for education and left the initiative entirely to missionaries. This was the case in the Congo Free State (after 1908, Belgian Congo), where at the onset of decolonization in 1960, after some eighty years of colonial rule, there was virtually no European-educated elite and only patches of literacy in a few local languages. The situation looked better in colonies such as Nigeria (British since 1851/62) or Senegal (French since 1817), but secondary schools were very thin on the ground. In Algeria a state education system competed with Koranic schools that the colonial authorities found very difficult to control: an educational dualism, in fact.[58] The other extreme was represented by the Philippines, under US control from 1898 on, which by 1919 already boasted 50 percent literacy. The main European colonies in Asia had much lower rates: 8 percent in Indonesia, 10 percent in French Indochina, and 12 percent in British India.[59]

India was in some ways exceptional: the colonial regime promoted middle and higher education even in the period before the First World War, although the number of schoolchildren and students who benefited from it was fairly small in comparison with the huge population. The Hindu College in Calcutta opened its doors as early as 1817; universities followed in 1857 in Calcutta, Bombay, and Madras; 1882 in Lahore; and 1887 in Allahabad. They were not fully fledged teaching and research universities, however, but essentially institutes that awarded grades and diplomas to students scattered among all manner of colleges in the region; teaching took place only at Lahore University. The colleges taught little else than the "liberal arts," since the British were interested mainly in developing a culturally Anglicized Indian stratum that could be involved in administering the country. Science and technology occupied a much humbler place. Only after Lord Curzon, then the viceroy of India, pushed through the Indian Universities Act in 1904 did some Indian universities create research departments—including in princedoms such as Baroda and Hyderabad that were not subject to the Raj bureaucracy and sometimes had ambitious modernization plans of their own. Insofar as research in India took place under the aegis of British rule, it was strongly oriented to practical applications; theory and pure research had a harder time. Sciences such as botany (which had uses in agriculture) received the greatest encouragement.[60]

Independent Asian governments saw things differently and sought to develop the sciences on a broad base. In Japan the importance of technical skills was understood early on, while in China a few reformers fought unsuccessfully for decades against the pride in "humanist culture" of a majority of officials. Science and technology were given major importance only in a number of American missionary schools and universities founded after 1911 in Beijing and Shanghai. In the Ottoman Empire, where many architecturally imposing new schools had been built, similar trends came into conflict with one another. The question was whether higher education should serve mainly to give civil servants a training based in Islam or to cultivate practical, "productive" individuals versed in technology and economics? Until the turn of the century it was the former that prevailed.[61] As in China (much less in Japan), foreign educational institutions in the Ottoman Empire, often run by missionaries, competed heavily with government initiatives. They offered foreign languages and in many cases had a better reputation than public schools. The presence of foreign schools and universities was less a sign of imperialist cultural aggression than an inducement for the indigenous state to widen and improve its own educational opportunities.[62] It would be wrong, however, to draw conclusions regarding "the Muslim world" as a whole. Until the first decade of the twentieth century, the kind of educational reforms that had already visibly changed Egypt and the Ottoman Empire were almost completely lacking in Iran. There, in the second largest noncolonial Muslim country in the world, the state did not interfere with the near-total control that the *ulama* retained over schooling.[63]

Schooling the World

The schooling of society was a European/North American program of the early nineteenth century that gradually became the goal of official policy worldwide. The school became a major tool for the state penetration of society and also a focus of civic commitment. The key issue was and is whether the state, local communities, or parents themselves should finance the running of schools. In the view of international organizations, school attendance and literacy rates are still today important indices of social development—hence of what, in the nineteenth century, used to be called a country's "level of civilization." Three aspects came together in the school: the socialization aspect, or the shaping of personality and particular human types; the political aspect, essentially concerning the relationship between secular government and religious educational institutions; and the instructional aspect, or the securing and dissemination of knowledge. The insight that science, as a cognitive and productive power and a vital social force, required well-run schools to train its future practitioners took the nineteenth century beyond the earlier threshold period of the scientific revolution. But the leading scientific countries of the age—Britain, France, Prussia/Germany, and the United States—differed considerably in the educational strategies they adopted. Nowhere did so much weight and government attention center on the secondary stage of education as in Germany (especially the pioneering lands of Prussia and Bavaria). This was the birthplace of the "humanistic gymnasium" with its enormous emphasis on Greek and Latin, which in the middle of the century was joined by a different type of high school catering more to the needs of technology and business. Standardized since the 1830s, the gymnasium provided the foundation for the rise of German science in the Kaiserreich from 1871 onward. In Britain, to take an example at the opposite extreme, various private schools certainly produced excellent results, but before the 1902 Education Act there was nothing that could be described as a secondary school *system*.[64] Only in the military field was Germany at that time as much of an inspiration to the world as it was in education. This was also true of its universities.

3 The University as a Cultural Export from Europe

The Break with the Early Modern Period

The nineteenth century witnessed the emergence of the modern university in its three dimensions: (a) a training center that structures, preserves, and transmits knowledge; (b) a place for research or the generation of new knowledge; and (c) an agency of socialization, character formation, and self-discovery for young people after they complete their compulsory schooling. In most European countries, the reorganization of university training and scientific research preceded the reshaping of high schools. Educational systems were dynamized from the top down.

The university as an autonomous corporation of scholars was a time-honored institution characteristic of Latin Europe. Other civilizations such as the Chinese or Islamic had no less effective means of establishing and transmitting knowledge: monasteries, religious high schools, or academies (e.g., the Chinese *shuyuan*), where scholars would gather together informally. "Forums for rigorous intellectual debate" were not peculiar to Europe in premodern ages.[65] In this diversity of scholarly cultures, the European university shaped in the Middle Ages stood out because it was relatively independent of external powers and constituted a space with its own laws. The Chinese state—to take an extreme counterexample—did not allow for a semiautonomous res publica of knowledge bearers. Either scholars were firmly integrated into the state apparatus (many as "compilers" at the Imperial Hanlin Academy in Beijing) or they congregated in semiprivate circles that the emperor viewed with suspicion. In China there were no legally protected corporations of scholars—still less ones comparable to the English universities, which had their own political representatives in Parliament.

Such "premodern" conditions disappeared at various points in the nineteenth century—in China and Japan between 1870 and 1910, although for the time being private academies in Japan held their ground alongside the state school system, with a teaching program less strongly geared to the West. Only in the Islamic world did some of the old institutions—above all, the religious schools (madrasas) independent of the state—survive in a modified form; al-Azhar ("the Luminous") in Cairo, a place of theological and legal learning dating back to the tenth century, is the oldest university in the world.[66] The European university, by contrast, having undergone *fundamental* reform in the nineteenth century, spread all around the world. The modern university, as a place where secular knowledge is *produced*, arose after 1800 in close association with the emergence of nation-states in Europe, becoming in the last third of the century one of the basic institutions of the modern world. Its inventors and the place and time of the invention can be identified with precision: namely, a handful of aristocratic reformers (Freiherr vom Stein, Hardenberg) and idealist philosophers (Fichte, Hegel, Schleiermacher), in Berlin in the years after 1803 and especially 1806—when the near collapse of the Prussian state had left a power vacuum, suddenly opening up a space in which new unorthodox approaches were on offer to save the state and the nation. Although the modern university that came into being in those years, with Berlin University (founded in 1810) as its flagship, preserved many rituals and symbols from its medieval past, it was in essence a revolutionary invention in the Age of Revolutions.[67]

The new university brought with it a number of distinctive social types: for example, the Oxbridge "don" or the German *Ordinarius*, ruling in authoritarian fashion over institutes and flocks of assistants.[68] New above all was the youthful "student," who in Europe replaced an older type of the more or less ageless "scholar"; the consequences are still visible today. In some countries, the nonacademic observer becomes aware of the university's existence only when students call attention to themselves through political activity. The chain of association "students—young people—rebellion" was forged in the early nineteenth century. In Germany it was the student fraternities (*Burschenschaften*), first appearing in public in 1815, which made student protest a factor in politics. In the case of France, "the birth of students as a social group" has been dated to the three decades after 1814;[69] they played a significant role in all the revolutions of the nineteenth century. Later, students and graduates of modern educational institutions became active in radical, and increasingly also nationalist, politics. A Russian student movement developed in the years after the Crimean War at the five universities of the time, although in its early stages it was tightly controlled; the first disturbances associated with it broke out in 1861.[70] In India, students played a leading part in the mass actions of 1905 against the partition of Bengal—key events for the founding of Indian nationalism—and in the

Japanese colony of Korea they led the nationwide movement of March 1919 that mobilized more than two million people in anti-Japanese protests.[71] In China, only two months later, student unrest linked to the Fourth of May Movement provided an anti-imperialist and cultural spark that ignited the next stage of the revolutionary process. In each of these cases, national universities had borrowed from Western models in which free space existed for the development of political consciousness.

Colonial Universities

Before 1800, universities of the European type had been founded elsewhere only in the New World. In Spanish America they were inserted into a system of church control over cultural life. Conditions were freer in those that sprang up in North America, already conspicuous by their number alone; the United States today has thirteen universities founded before 1800, compared with a mere two in England. In Canada there was clearly less interest. As for the non-Spanish Antilles, no effort was made to found independent universities; the sons of the Creole elite went to Europe for their higher education. In Portuguese America, there had been no high schools at all. The first university was established in Brazil only in 1922.

The founding of a college near Boston in 1636, named three years later after an ecclesiastical patron, John Harvard, set the English colonies across the Atlantic on their way to becoming the third growth center for universities alongside Europe and Spanish America. Yale, Princeton, and Columbia Universities, the University of Pennsylvania, and Rutgers University already existed before the American Revolution. Each had a character and organizational forms peculiar to itself, enjoying considerable independence from the political authorities; none of them adopted the Oxbridge model unaltered, and the influence of Scottish universities and Presbyterian/nonconformist academies was hardly less important. Common to them all was a relative impoverishment: John Harvard's generous legacy had been a great exception. The land donations that most of them received were in a part of the world where land was available in abundance and did not yet have much value. The early colleges had to raise their funds from a wide variety of sources, the main one being student fees. Teaching was on a very modest scale: probably no more than 210 professors were active in 1800 in all the North American colleges combined. Their main goal was the training of clerics, and preparation for other professions developed only slowly.[72]

The idea and practice of the university spread worldwide only after the middle of the nineteenth century. In the semiautonomous settler colonies within the British Empire, it became a matter of honor for the colonial authorities and municipal dignitaries to lay the foundations for a local university, even if for a long time there was no chance of departing from the great British models. Australia's first university came into being in 1850 in Sydney; New Zealand followed in 1869. As for Europe's "nonwhite" colonies, universities were created if they seemed to fit the purpose of training indigenous personnel. The sons of colonial functionaries and settlers were sent to the mother country to complete their education. Not only were colonial universities starved of funds, they were unable to confer doctorates; Europeans always stood at the top of the academic hierarchy, irrespective of their individual talents. Even in Algeria, a comparatively old colony close to the metropolis, there was no full university until 1909, and the later renowned University of Hanoi, the most original French creation in the sphere of colonial education, had its launch only in 1919. Where a prestigious, high-quality university stood out amid the varied landscape of secondary and tertiary education, it was founded after the turn of the century, and in most cases after the First World War. In Egypt, a number of institutes of learning fused together in 1908 to form a (private) Egyptian University. In West Africa, the ideas that led to the founding of universities in the twentieth century were already being formulated by Africans after 1865; but it was only in the 1940s that capable universities were created in the British colonies of tropical Africa. The widest tertiary education in the colonies was offered by the American Philippines, where a state university along the lines of US agricultural and engineering colleges opened its doors in Manila in 1908; there were also a number of private universities, many of them run by missionaries.

A German-style system of higher education did not develop in a single colony; nor was the English model of democratically constituted, self-governing colleges in the loose overall framework of a university exported to Asia and Africa. Colonial universities had an authoritarian structure, and their curriculum largely depended on the metropolis and the special objectives of the colonial authorities. Sometimes tertiary education was dispensed with altogether. Dutch universities, especially the old "Rijksuniversiteit" of Leiden, contained important centers for Asian studies; very little research was conducted in Indonesia itself (in contrast to British India or French Indochina), and before the Second World War the Dutch did not think of satisfying the educational needs of an Indonesian elite. The fleeting vision of an "imperial science" in which all the talents of the empire would participate—an idea propagated under Lord Curzon's viceroyalty—had absolutely no counterpart in the Dutch colonies. Only in 1946, three years before independence, was a "Provisional University of Indonesia" launched with faculties of law, medicine, and philosophy—the germ of the later Universitas Indonesia.[73]

Scholarly Traditions and New Approaches in Noncolonial Asia

In the politically independent countries of Asia and Africa too, the adoption of European university models did not begin until the turn of the century. South Africa, even as a British colony, had had a larger number of educational institutions than any other African country, but the foundations of the university system that we see today were not laid there until after 1916. In the Middle East, Lebanon was a special case: higher education developed there earlier than anywhere else in the region, though not on the initiative of the central Ottoman state but as missionary implants. In 1910 the Protestant American University of Beirut took shape out of a series of precursors, while the Université Saint-Joseph, run by French Jesuits, opened in the same year on the foundations of what had originally been a theological institute, later supplemented by a medical college whose degrees were recognized even by the secular state of the French Third Republic.[74] The most important new creation in the Turkish part of the Ottoman Empire was the University of Istanbul (1906), successful at the fourth attempt, which was explicitly modeled on American and European universities and had a total of five faculties. In contrast to the Lebanese universities, the natural sciences occupied an important place in Istanbul right from the beginning.[75] It marked a clear break with older Islamic institutions centered on law and religion; its precursors, rather, were the (often ephemeral) semiprivate circles in which individuals had grappled with Western knowledge and its relationship to the indigenous heritage.

The development of higher education in China was parallel in time and similar in substance. The first universities appeared there after 1895, the Imperial University (embryo of the future Beijing University) in 1898. Traditional institutions of learning had all but disappeared by the time of the 1911 Revolution, but—again as in the Ottoman Empire—many of the values and attitudes associated with classical scholarship had survived. There was great resistance to subject specialization, for instance, and until the abolition of the state examinations system in 1905 the Confucian scholar had to demonstrate his competence in nearly every branch of knowledge. It must be said that a critical spirit was not absent from Imperial China: philological methods fostered doubts about the written tradition, and there was a right to criticize the highest dignitaries, including the emperor himself, if their policies were thought to be deviating from the principles of the classical teachings. However, the cultural authority of the top bureaucracy, which set the tasks for the state examinations until the system was wound up, was considered unassailable. The frank criticism voiced outside its ranks—for example, in local private academies—first had to gain entry to the public space of the newly emerging universities.[76]

Chinese universities drew on a variety of sources. The Imperial University of 1898 was founded with an eye to Tokyo University, itself shaped by French and German examples. When Japan intensified its aggressive policy against China during the First World War, parts of the new Chinese academic intelligentsia turned more toward European and North American models; mission universities—some considered excellent, even for the sciences, after the First World War—had the same horizons already. Only in the 1920s did the landscape become more diversified and give birth to a real academic community. The main impetus for reform came from the important scholar-administrator Cai Yuanpei, who from 1917 built Beijing University into a fully fledged research institution along German lines, while also observing the principle of the unity of research and teaching (scarcely a feature in colonial universities). Under extremely difficult external circumstances, China in the Republican period developed an academic life (including the Academia Sinica, founded in 1928) that was capable of top-class achievements. Despite ancient traditions of scholarship, it was only the early Republic that laid the foundations for China's present-day status as a major player in the world of international science.

Japan was the only country in Asia that evolved differently. Its premodern conditions were not necessarily more favorable, but the reception of European knowledge was not broken off as dramatically as it was in late-eighteenth-century China, when the flow of information via the Jesuits came to an abrupt end. In the early nineteenth century, "Holland studies" became a wider opening to European science, and from the 1840s it was possible to study Western surgery and medicine in Edo (Tokyo). After 1868, the Meiji leadership set out to make systematic use of Western knowledge: Tokyo University, founded in 1877, was completely oriented to Western sciences and refrained from giving courses in Japanese and Chinese literature. Although private initiatives should not be overlooked, the state stood more solidly than anywhere else in Asia behind the building of universities. A decree of March 1886 explicitly stated that the planned new crop of imperial universities should "teach those arts and sciences essential in the nation."[77] After the First World War, with a group of well-developed universities at its core, Japan's diversified system of higher education was surpassed only by the United States and a few European countries. Despite the unusually strong role of the state, university professors in the late Meiji period (from roughly 1880 on) were by no means spokes in a wheel happy to take orders from above. Along with French and German forms of organization had come an ethos of the university as a free space for research and debate. The academic elite of the Meiji period linked up with two different mandarin traditions and their related role models: on the one hand, it could identify with the self-confidence and autonomous tendencies of classical Chinese scholars; on the other, it took up the authoritarian habits, but also the pride, of German academic "mandarins," as Fritz K. Ringer memorably called them.[78] However, they were paid more like Chinese than German mandarins: badly.[79]

Ideal and Model of the Research University

The ideal of reliably funded research, free of immediate utility pressures and provided with the necessary material trappings (laboratories, libraries, external research stations, etc.), was essential to the nineteenth-century European conception of a university, though much more difficult to export or import than its general framework as an *educational* institution. A few premodern universities—most notably Leiden in the Netherlands—had already thought of themselves as research universities. But today's conception of it as a "total package" first emerged during the Age of Revolution, or, to be more precise, between the 1770s and 1830s in Protestant Germany: in Göttingen, Leipzig, and eventually the Berlin of Wilhelm von Humboldt and Friedrich Daniel Schleiermacher.[80] By no means were all German universities research universities. However, they were examples of the few high performers that echoed around the world. The research university model had as its core a centralization of tasks that until then were scattered around in the "republic of scholars." Even if other places of research continued to exist in Germany, and even if new ones were added toward the end of the nineteenth century (the Physikalisch-Technische Reichsanstalt, the Kaiser-Wilhelm-Gesellschaft, etc.), a basic idea of German reformers was to move research out of the academies into the universities and to bring various "schools" under their roof as institutes and seminars.

The university thus acquired much wider objectives than before. Initially existing alongside academies and learned societies (such as the Royal Society in Britain), museums, and botanical gardens, it became the dominant scientific institution and the decisive social space in which academic communities developed.[81] It also offered opportunities to conduct research without an eye on how it could be turned to account. Only in this way was it possible to separate theoretical physics (a new field whose great age began at the turn of the century) from the hold of experimental physics.[82] Together with classical and Romantic music (in which Austria, too, was involved, of course), the research university model became Germany's most important cultural export since the Reformation—a complex with a global, though highly varied, impact. Nor should its disadvantages be overlooked: since school qualifications such as the *Abitur* exam guaranteed access to higher education, a danger of overloading was built into the university system. In Imperial Germany, the fact that the educated middle classes and technical specialists were products of an educational system completely run by the state (albeit decentralized at *Länder* level) contributed to an illiberal fixation on state authority among large sections of the German elite. The nonvocational "liberal education" that in Britain or America is still seen as a task of the tertiary phase of the education system ended in Germany when students graduated from the gymnasium. The Germany university trained people in a particular subject and did not care for character formation. Nowhere was specialization taken so far in both research and teaching.[83]

Delayed Adoption of the German Model in Europe

The German formulas did not at once find enthusiastic imitators elsewhere in Europe. In 1800, with individual exceptions, the advance of science was concentrated in Britain, France, and the German lands. Italy and the Netherlands had failed to keep up. Breakthroughs in linguistics and archaeology came from Scandinavia, and Russia later contributed major achievements in the natural sciences (e.g., Mendeleev's periodic table of the elements, in 1869). It seemed to many observers that the relative weight of the countries in the Big Three shifted in the course of the nineteenth century. Important scientific discoveries continued to be made in France and Britain too, but to a much greater degree than in Germany this happened outside university structures. Under Napoleon, the Grandes Écoles had developed into sophisticated, authoritarian training centers for the state bureaucracy and civil engineering, with inadequate emphasis on the "pure" natural sciences and the humanities. In England, Oxford and Cambridge—traditionally geared to training the priesthood—long steered clear of the sciences and showed no interest in building laboratories. As in China, it seemed self-evident that higher education should proceed through the study of texts, in sharp contrast to practical education in hospitals, law courts, or museums. Appropriately enough, the first science to take up residence in the universities was geology: the science of reading the stone "book of nature."

Gentlemen scholars such as Charles Darwin, the son of a wealthy doctor and speculator (and grandson of Josiah Wedgwood, one of the great pioneers of industrialization), continued to play a role in English science that was no longer possible in Germany after the death of Alexander von Humboldt in 1859. (A special case was Gregor Mendel, whose brilliant discoveries in genetics, made at the secluded Augustinian abbey in Brünn [Brno, Czech Republic], had no impact on the scientific public for more than three decades.) Scientific societies, many newly founded in the nineteenth century, retained special importance for a long time in France and Britain. As in the early modern period, London was a much more important center for the sciences than Oxford or Cambridge and the location of all the learned societies active on a national level. Modern developments in higher education emanated mainly from particular institutions within the University of London or from later foundations in cities such as Manchester (1851).

There were not yet any Nobel prizes; the first were awarded in 1901. Nor did quantified rankings form part of academic life. Reputations had to be built up through individual work within webs of exchanges with other scholars, which from the beginning had an international as well as national dimension. Decades before the unification of Germany as a nation-state, its scientists formed a community which, thanks to its own performance and the diplomatic efforts of Alexander von Humboldt, was well integrated with the rest of Europe. From roughly midcentury on, academic communities in different countries kept a close watch on one another's activities. Science became a public arena of international competition—for example, between the microbiologists Louis Pasteur and Robert Koch. When Wilhelm Röntgen's recent discovery of X-rays became known in 1896, Emperor Wilhelm II sent a telegram to the later Nobel laureate, in which he thanked God for this triumph of the German fatherland.[84] At the same time, the links between science, technology, industry, and national power became more apparent. In Britain, an impression spread among the public that the country had come off badly at the International Exhibition in Paris in 1867. In France, the military defeat of 1871 at the hands of the new German Reich was put down to a backwardness in education and science. But demands that the state should build large "German-style" universities yielded results only after the political consolidation of the Third Republic in 1880, the legal foundations for a new system finally being laid in 1896.

Even then, however, the research imperative had less force than in Germany.[85] A modern system of higher education developed in France no earlier than in Japan, while in Britain the decentralized structures of academic life made it difficult to speak of a university system at all until far into the twentieth century. Oxford and Cambridge, which after the turn of the century modernized their teaching methods, stopped giving grades without written tests, and ended the requirement that fellows remain single. They converted themselves only after the First World War into research universities with a strong scientific component, following the lead of Imperial College in London, established in 1907 and soon acknowledged as one of the top research institutions in the world. The high costs of modern laboratory work required central financial planning beyond the budgets of traditional colleges and individual faculties. Specialized technical colleges have continued to play a lesser role in Britain than in Germany, France, Switzerland (where the prototype of such an institution, the Eidgenössische Technische Hochschule in Zurich dates back to 1858), or Japan. The PhD, initially awarded for science subjects too, was not introduced in Cambridge until 1919, by which time it had long been customary in Germany and the United States.[86] It also took many years before restrictions on internal appointments of teaching staff in Oxbridge allowed fresh ideas to penetrate from outside.

The Rise of Universities in the United States

The German research university was thus adopted in modified form by other European nations with an important scientific life, though only after an extraordinary delay of at least half a century. Its influence was felt earlier outside Europe. However, the performance of American universities should not be exaggerated, either in colonial times or during the period up to the Civil War. One of their principal historians speaks of the years from 1780 to 1860 as a "false dawn" and dates the real hegemony of the American research university to the period after 1945.[87] Only in the two decades after the Civil War did academic communities take shape in the main scientific disciplines, whereas similar trends had been operating in Britain, France, and Germany since the 1830s. The German model of the research university was then comprehensively studied in the United States, and in 1876 the founding of Johns Hopkins University in Baltimore signaled the emergence of the full university on the other side of the Atlantic. It is true, though, that it spread only slowly elsewhere; in many cases, research was seen as a prestigious luxury, not as the very essence of a university.[88]

The spectacular rise of certain American universities would have taken much longer if they had not been able to profit from the economic boom of the last quarter of the century. Ever since the days of John Harvard and Elihu Yale they had been dependent on private donations and foundations, but around 1850, wealthy individuals began to show an increased willingness to support the academic world philanthropically. After 1880, as the great American fortunes were being made, sponsors sought to perpetuate their memory in the title of universities: whereas John D. Rockefeller, for example, had contributed anonymously to Columbia University, many institutions now bore the names of railroad, tobacco, or steel barons. Often religious motives also lay behind this. New university buildings were built in a uniform neo-Gothic style—sometimes, as at Palo Alto at the Stanford family's request, in accordance with Mediterranean taste. The old American colleges had been small and plain, and in their architecture too. Now large spaces were required to accommodate new libraries, laboratories, and sports facilities. More than in Europe, affluent civic pride found expression in splendid university buildings that were the architectural highlight even of a city as large as Chicago. German influence was evident in the ambitious orientation to research and the allocation of subjects and faculties, but state planning, direction, and funding, essential to the German model, were confined to a minority of universities in the public sector. The fast-growing top universities built up their own internal bureaucracies; professors, though held in ever-higher social esteem, were regarded as employees subject to management. University presidents saw themselves increasingly as entrepreneurs. Among administrators and those involved in teaching and learning, pride in the institution was combined with a cool, market-oriented vision of education and science. All this made late nineteenth-century American research universities an unmistakably original development on their side of the Atlantic.[89]

Japan: A Semi-Import of the German Model

In comparison with the United States, Japanese universities were still weakly developed on the eve of the First World War. All sciences considered at all modern had a place in Tokyo or one of the other imperial universities, but the lavish funding received by American and some German universities was not forthcoming. The two faculties enjoying the most generous support were medicine and engineering, where Japan's early successes had attracted attention abroad. In other spheres, the dependence on the West was still so strong that teaching did not progress beyond the repetition of textbook wisdom. Meanwhile hundreds and thousands of Japanese went to study in Europe and the United States, and those who returned to take up a responsible academic post imitated their Western teachers in every detail for the time being. Western advisers and lecturers had formerly played a major role in building certain departments, but this gradually declined in the late Meiji period. Altogether some eight thousand such experts were employed,[90] giving a crucial impetus not only in natural science or medicine but also in law or history. Since it was not possible to recruit abroad systematically, and since a career in Japan, despite quite high pay, was not everyone's dream in life, much depended on luck and chance. The example of modern historiography, introduced by the Berlin-trained Ludwig Rieß shows the limits of the transfer.[91] Academics in Japan adopted the positivist source criticism of the German historical school (which fit in well with national traditions of textual criticism originating in China), but not its philosophical program and literary techniques. Nor could they claim to have the same public appeal that Rieß's German masters enjoyed. Historiography remained narrowly specialist and did not dare to tackle the new national myths of the Meiji regime, such as its fictitious imperial genealogy. Unlike in the admired German example, history did not become the leading discipline in the humanities or among the educated middle-class public.

Another weak point of the early Japanese university system was the extreme hierarchy that made Tokyo the unchallenged top dog. This prevented the kind of competition to be found among American universities as well as in the strongly decentralized federal German system, where the job market encompassed not only the German Reich but also Austria, Bohemia (mainly Prague), and German-speaking Switzerland. Nevertheless, by the 1920s at the latest, it was clear to the international scientific public that a start had been made in East Asia on the development of a research-oriented academic system—not only the organizational forms of the European university but also its research imperative. This was one of the differences between Japan and China on the one hand and the Ottoman Empire on the other. In the view of the Turkish historian Ekmeleddin İhsanoğlu, the considerable efforts of the Ottoman reform elite (decades before similar initiatives in China) to translate or "buy" Western knowledge from European experts stopped at the threshold of an experimental spirit and a research culture capable of learning from results.[92]

4 Mobility and Translation

Patterns of Perception

The science that blossomed in these new organizational forms was European in origin; only a few other elements entered into the edifice of what by 1900 was universally valid science. The study of nature in the medieval Arab world might have been superior to that in the Latin West, and the ancient Indians might have been supreme mathematicians and linguists: yet nineteenth-century European science was less in debt to non-Europeans than the early modern collectors, classifiers, and cartographers in Asia, whose work could be carried out only with the help of local experts. In the eighteenth century, Europeans had still believed they could learn from Asian textile technologies or agrarian practices such as fertilizer use or crop rotation.[93] In the nineteenth century, such trust in the practical knowledge of others was on the wane. "Scientific" colonialism, much vaunted at the end of the century, often arrived at agronomic insights that had long been known to peasants living in the area, or made mistakes against which they could easily have been warned. At the height of colonial narrow-mindedness, local topographical expertise and the skills of indigenous craftsmen were used at best in the construction of roads and houses, but otherwise no serious notice was taken of other people's knowledge. It would, however, be naive to romanticize "local knowledge" in non-European cultures, and unjust to charge an expanding Europe with its wholesale suppression—a sin more grievous than that of simply ignoring it.

Asian and African elites recognized the significance of the scientific and technological knowledge coming out of Europe, and increasingly the United States. They tried to acquire it, to put it to the test, to translate it into non-Western languages and conceptual frameworks, and to relate it to their own traditions and experience. The mobility of individual complexes of knowledge proved to be quite varied: some "traveled" easier and faster than others. The old idea that the worldwide "diffusion" of European sciences, by virtue of their innate superiority, was a more or less natural process is not altogether misguided, but it is simplistic insofar as it overlooks the particular cultural and political conditions under which contact was made and knowledge transferred.[94]

Nakayama Shigeru, a historian of science who has studied various patterns of transfer in East Asia, argues that since Japanese mathematics was self-enclosed and incompatible with European mathematics in its structure and notation system, it dropped out of the picture soon after the Meiji Renewal. This did not happen because it was more primitive, but because it was more practical and economical for Japanese mathematicians to adopt the new system en bloc than to tinker with the old one. In medicine, by contrast, Chinese or Japanese systems survived intact alongside others imported from the West; the two were never fused into one. The combination was (and is) effected at the level of practice rather than theory. In Japan, however, where all transfer decisions reflected the drive to shake off China's long-term tutelage and to become the star pupils of Western modernity, indigenous medicine lost its scientific status during the Meiji period; either it was not taught at all at the new universities or it was demoted to a popular (but widely used) art. Nakayama finds yet another pattern in astronomy. Jesuit missionaries introduced the European science into China as early as early as the seventeenth century, but their data and calculation methods could be incorporated without too many problems into Chinese calendar astronomy. The Jesuits thus helped to reinforce the traditional role of court astronomy as a support for the emperor's legitimacy. For two and a half centuries no one ever thought of regarding Western astronomy as "modern" or superior. The main reason why its indigenous equivalent disappeared was not that it was defeated in a battle of ideas, but that it lost its function in society. When the offices of court astronomer and state custodian of the calendar were eventually abolished—not before the late nineteenth century!—the game was up; young astronomers trained in Europe and America soon built up a new discipline in the universities. Until then, however, the imported science had actually served to strengthen indigenous traditions.[95]

The dissemination routes of Western knowledge were tortuous and unpredictable. An international community of researchers, such as we take for granted today, came into being only in the late twentieth century. In the nineteenth century, non-European cultures had to acquire not simply existing stocks of knowledge but complete scientific worldviews. Thus, although the Jesuits acquainted Chinese scholars with Euclidian geometry and Newtonian physics back in the seventeenth and eighteenth centuries, full translations of the *Elements of Geometry* and the *Principia Mathematica* were not completed until the 1860s.[96] At that time, when Protestant missionaries and Chinese scholars were beginning to work closely together on translation projects, there was a preference for compact information in Western textbooks, which were themselves popular digests of previous research. By the early twentieth century, Chinese scientists were nearly always capable of understanding specialist literature in English or German. Their efforts tended to be derided in the West, both then and later, as attempts to catch up that often took them down a blind alley. But a different way of looking at things is also possible. Given the inertia of traditional scholarly cultures, it was a respectable performance to absorb Western knowledge within just a few decades in countries such as Japan, China, or the Ottoman Empire. Only in Japan did the state give systematic financial support. Where missionaries were the decisive agency of transfer, as they were in China, many initiatives remained private.

The challenges were huge, starting with formidable problems of terminology. The adaption of scientific Latin had begun here and there in the early modern period, but by no means always did this result in a stable nomenclature; the terms chosen by the Jesuits were frequently criticized and corrected in nineteenth-century China. As in Japan, several translators might work alongside one another in a single discipline, so that long and ramified discussions were often necessary to reach lexical agreement. In philosophy and theology, in law and the humanities, the difficulties were especially great. Concepts such as "freedom," "right," or "civilization," each with complex semantics of Western origin, could not be represented directly and unambiguously in Japanese, Chinese, Arabic, or Turkish. These cultures had their own no-less-intricate worlds of meaning, so that a new Western concept had to be interpreted within the reception context, where it would nearly always pick up nuances alien to it in the original language. For example, by 1870, Japanese lexicographers and translators were conveying the English word "liberty" by means of four different terms in Chinese characters, each of which added a special sense of its own. Only gradually did one of these, *jiyū* ("following one's intentions without restriction"), became accepted as the standard translation.[97]

"Science" was another concept over which translators wrestled. The classical vocabulary in China had more than one expression that came close, without corresponding to it precisely: the traditional *zhizhi* signified "extending knowledge to the full," while *gezhi* meant rather "investigating and developing knowledge." Any Chinese scholar in the nineteenth century knew that these verbal expressions, both containing the character *zhi* (knowledge), should be seen against the background of twelfth-century neo-Confucian philosophy. From the 1860s, *gezhi* gradually stabilized as the translation of "science," but also of "natural philosophy." But then the term *kexue*, imported via Japan, appeared on the scene and after 1920 or thereabouts became the standard translation that it still is today. *Kexue* places the emphasis less on the process of knowledge acquisition than on the categorization of knowledge, especially its curricular organization. When the leading minds of the post-1915 New Culture Movement began to feel that the narrow, static quality of this term did not reflect the novelty of the modern concept of science, they actually turned for a while to the rough phonetic imitation *saiyinsi*. This post-Confucian neologism, devoid of the semantic baggage of centuries past, was supposed to convey the idea of a moral awakening from the slumber of sterile tradition, a renewal of Chinese civilization and nationhood through enlightenment and critical thinking.[98]

Science in Exchange for Art and Irrationalism?

More than ever before, the flow of knowledge around the world in the long nineteenth century was a one-way street. Western natural science devalued the stock of knowledge about nature in other regions, with the result that there was little or no interest in even Chinese or Indian medicine and pharmacology—which has since been rediscovered in the West and is becoming increasingly influential over the last half century or so. All that traveled in an east-west direction was aesthetic and religious impulses. The knowledge involved here did not have transcultural validity underpinned by verifiable research procedures and scientific criticism. Rather, it offered Asian, and later African, responses to the Western quest for spirituality and new sources of artistic inspiration. Indians, Chinese, Japanese, and inhabitants of Benin in West Africa (where a British "punitive" expedition in 1897 hauled off a fortune in ivory and bronze objects highly valued in Europe) did not propagate their culture in the West. Western artists and philosophers themselves went in quest of the unfamiliar and adjusted what they found to their requirements. Romantic poets and thinkers, such as Friedrich Wilhelm Joseph Schelling or Friedrich Creuzer, became excited about Eastern mysteries, and for a few decades the ancient Sanskrit literature, translated into European languages since the 1780s, aroused much interest among intellectuals in the West. Recent translations of the classical books of Hinduism fascinated Arthur Schopenhauer, while Ralph Waldo Emerson, the leading North American philosopher of his time, delved deeply into Indian religious thought, criticized the absolute claims of Christianity and Enlightenment rationalism, and advocated a spiritual rapprochement between East and West.[99]

In 1857, Japanese artists, most notably Takahashi Yuichi, began to practice European techniques of oil painting and triggered a new wave of interest in Western art. In the same decade, the first Japanese woodcuts reached Europe in the baggage of travelers and diplomats. Some were put on display for the first time at a public exhibition in London in 1862, but this and later collections by no means gave a representative overview of ancient and modern Japanese art.

Nevertheless, individual prints by masters such as Hokusai or Hiroshige were a source of lasting excitement to artists and critics. The so-called Japonism that grew out of these encounters was something new: art from outside Europe was no longer used only for decoration or costumes, in the way that Chinese and Turkish material had been in various Oriental fashions of the eighteenth century, or that North Africa had featured as an exotic setting for desert or harem motifs in French painting between 1830 and 1870 (Eugène Delacroix, Jean-Auguste-Dominique Ingres, Eugène Fromentin, and others). Japanese art gave answers to problems with which artists in the forefront of European modernism were then wrestling; they observed its independent achievements and realized the close affinities with their own efforts. Thus, the European enthusiasm for Japanese art and the Japanese enthusiasm for European art peaked at exactly the same time, but for different reasons. The fascination of the Western aesthetic for Japanese people began to wear off after Ernest Fennelosa—an influential figure in both East and West—alerted them to the wealth of their own artistic heritage and placed himself at the head of a movement that, with the support of official cultural policy, advocated the patriotic renewal of genuinely Japanese painting. An American Japanophile thus became the founder of Japanese neotraditionalism. Fennelosa's writings elicited a strong response in Europe too, raising the interest in things Japanese to a new level of art criticism.[100]

The musical influence of East Asia was also important, though rather less epochal. The old prejudice that Chinese music was intolerable to Western ears remained alive for a long time, based only on the impressions of individual travelers and their incomplete attempts to transcribe exotic tunes into European notation. In the 1880s, the invention and rapid proliferation of the phonograph finally created the conditions for non-Western music to become better known in Europe. Giacomo Puccini and Gustav Mahler, for example, studied phonogram recordings of East Asian music, the former turning them to account in *Madame Butterfly* (1904) and *Turandot* (1924–25), the latter in *Das Lied von der Erde* (1908) and his *Ninth Symphony* (1909); Puccini, it has been alleged, ultimately relied on a musical clock imported from China. Composers of light music were content simply to evoke Oriental moods by means of instrumentalization and tone color. Musical inspirations that often sounded like clichés could lead to fresh inventions in the hands of such masters as Giuseppe Verdi (*Aida*, 1871), Camille Saint-Saëns (*Suite algérienne*, 1881), or Nikolai Rimsky-Korsakov (*Sheherazade*, 1888). The Asian influence ran deeper where the Western tone system was allowed to be destabilized by alien elements. Claude Debussy led the way in this respect, after he had heard authentic gamelan music at the Exposition Universelle in Paris in 1889.[101]

After its heyday in the period between 1860 and 1920, the European fascination with Asia gradually subsided. Postwar Europe was more preoccupied with itself, while "Oriental" Asia seemed to lose its magic as urban modernization got under way, revolutions and anti-imperialist movements flared up, and harbingers of military rule appeared here and there. The small minority of fin-de-siècle European intellectuals who looked east to Asia did so with little concern for its contemporary reality, in a spirit of *Kulturkritik* or with hopes of salvation. The attraction was the inexhaustible depths of various "Eastern wisdoms," amid a crisis that seemed to many to be affecting Christianity as much as the rational worldview of natural science. In Germany the publishing house of Eugen Diederichs, a lawyer who espoused conservative lifestyle reforms, brought out the *Analects of Confucius*, the *Book of Laozi*, and other texts of the ancient Chinese canon, in a series of translations by the missionary-sinologist Richard Wilhelm that were of a high philological and literary quality. From 1875 the system of so-called theosophy, preached with bizarre appurtenances by Helena Petrovna Blavatsky, had a particular impact, even in India and Ceylon. It was a syncretic version of conventional occultism combined with the most diverse Middle Eastern and Asiatic traditions, from the Kabbalah to the Hindu Vedas, with a sprinkling of Aryan racism.[102] Rudolf Steiner, a master to a huge number of devoted followers in Germany, Switzerland, the Netherlands, and the United States, came out of this mystical milieu; in 1912 he created a doctrinally more temperate Anthroposophical Society of his own.

An undifferentiated "Asia," *fons et origo* of salvationist doctrines, thus became the symbol of an irrationalism polemically counterposed to the Western faith in reason that seemed to reach even into the well-tempered culture of orthodox Protestantism. Such impulses were not expected to come from Islam. There was an aesthetic appreciation of Muslim poetry and architecture, but its main currents were quite rationalist and did not seem to offer an alternative religious worldview. A paradoxical situation therefore developed in the last third of the nineteenth century. Painfully aware of the gap that had opened up, elites in the non-Occidental world strove to appropriate advanced science and technology from the West, often regarding it as a universal achievement of the modern age that would forearm them against the supremacy of the major Western powers,[103] while also—especially in India and, a few decades later, in China—sharply criticizing elements of irrationalism and "superstition" in their own traditions.[104] At the same time, minorities of intellectuals in Europe and North America instrumentalized "Eastern wisdom" in their struggle against the faith in reason that characterized Western scientific culture. The ironical counterpoint that Max Weber presented in his late studies of the economic ethos of world religions escaped public notice in this regard. In his view, the tension between worldliness and otherworldliness was a source of the economic dynamism of the Occident, whereas India was too strongly, and China too weakly, oriented to spiritual hopes of salvation.

Around the turn of the century, Asia thus acquired greater importance than ever in certain fields of Western thought, but it also became a projection screen for European irrationalism that seemed to leave it with no opportunities for development of its own. Revered for its "spirituality," Asia was stuck in limbo, with no present and no future. Only Mohandas K. Gandhi, the later "Mahatma" who first attracted Western attention after his return, in 1915, from a long sojourn in South Africa, managed (at least in European eyes) to combine the air of an Asian prophet and holy man with a cunning politics to empower the powerless.

5 Humanities and the Study of the Other

By 1900 the sciences had acquired unprecedented cultural authority in Europe, the United States, and some Asian countries like Japan and India.[105] At first small, then rapidly growing communities of scholars had taken shape in newly formed disciplines. The great majority of the world's scientists were no longer educated amateurs but salaried professionals working in universities, industry, or government research institutions. The system of education in the most advanced countries now included both "pure" and "applied" science—a distinction that had only just appeared on the scene. A foundation in mathematics *and* (ancient) languages, universally applicable, meant that the sciences could be extended into further domains through the training of new generations. Admittedly the total volume of creativity did not keep pace with the number of scientists, since there was a disproportionate growth of mediocrity and routinism. The production of geniuses can be socially managed to only a very limited extent.[106]

The Human and Social Sciences

Institutionalized expansion took in not only natural science and medicine, which by the early twentieth century was no longer understood as a proto-scientific craft and an art, but also the human and social sciences (*Geistes- und Sozialwissenschaften*)—two terms that were, if not coined, then first popularized among the scientific public toward the end of the nineteenth century. The "humanities" was another neologism of this kind. "Social science" went back a few decades earlier, used from the beginning not as an umbrella term for older discourses such as "statistics" (= the description of states) or "political economy," but as an indication that the rigor of modern natural science was being claimed for the study of society, with practical purposes, chiefly social reform, in view. If we leave aside early theorists with a background in philosophy, such as Auguste Comte or Herbert Spencer, the discipline was at first closer to empirical investigation than to theory (in Lorenz von Stein or the early representatives of the German Verein für Sozialpolitik founded in 1873). Karl Marx, not just a speculating theorist but a tireless student of social reality, was one of the few who transcended this opposition in their work.

No attempt was made before 1890 to define a common identity that differentiated the social sciences from other fields of learning; only then did professorships in "sociology" start to become common in Europe and the United States.[107] For the time being, sociology and economics remained closely intertwined, especially in the two German traditions of Marxism and the Historical

School of *Nationalökonomie* (up to and including Max Weber). After 1870, economic science in most countries moved away from the older tradition of political economy—which focused on production and labor in their social interrelationship—and turned to theories of marginal utility and equilibrium primarily concerned with the market and the structure of subjective needs. This separation of economic behavior from its social preconditions was part of a general differentiation within the social sciences during the last four decades before the First World War.[108] By 1930, at least outside Germany, where remnants of the Historical School stood their ground, there was an almost unbridgeable gulf between economics and sociology—as well as a split between the social conformism of economic science and the sociological interest in the dark sides of capitalist development and the chances for reforming society. In Japan, the Western social sciences met with greater interest than anywhere else. But they were received selectively. *Gemeinschaft* was more important than *Gesellschaft*, the collective rated higher that the individual, for early Japanese sociologists and political scientists. Since their work involved them in the grand national project of neo-traditionalist integration through a strong state, they were wary of subjecting the new myths of the Meiji period—above all, the emperor cult and the fiction of Japan as "one big family"—to rigorous criticism.[109]

Humanities faculties began to take shape in European universities, especially in France and Germany, in the middle of the nineteenth century; the individualist gentleman-scholar held sway for a little longer in the British Isles. The academization of the human "sciences" was something new. Historians, for example, had existed for more than two thousand years in Europe and China, but never before had history been taught in educational institutions as a methodical science. The first history *professors* still worth mentioning in a history of science were to be found after 1760 in Göttingen, then the most highly regarded university in the German-speaking world, but they also taught politics or topical matters relevant to the life of the state ("statistics," *Polizeywissenschaft*, etc.). At the same time, the greatest European historian of the age, Edward Gibbon, was writing his monumental *Decline and Fall of the Roman Empire* (1776–88) in the comfortable circumstances of a prosperous private scholar on the shores of Lake Geneva. In Britain the first significant historian to occupy a university chair was William Stubbs, in 1886. After Germany had once again taken the lead (Leopold Ranke's professorship in Berlin began in 1834 and lasted until 1871), it took several decades for history faculties to become established in all European countries. This happened quite early in Russia, where Sergei Mikhailovich Solovev helped to create a school in Moscow in the 1850s. In France, it was only in 1868 that the founding of the École Pratique des Hautes Études initiated a similar process of "scientific" historical research in the Ranke tradition. Even Jules Michelet, both then and now the most famous French historian of the nineteenth century, was noted more as an orator and writer than as an educator. After Louis Napoléon removed him in 1851, for political reasons, from his positions at the National

Archives and the Collège de France, Michelet lived off the royalties from his numerous publications.

In Europe and the United States, the professionalization of historical science was a phenomenon of the period after 1860.[110] It took a little longer to develop in the aesthetic disciplines. Intellectually rigorous criticism had existed in Europe since at least the middle of the eighteenth century,[111] but it was only shortly before 1900 that university departments of art, music, and various national literatures came into being alongside (not in place of) the freer public discourse of literati, journalists, private scholars, clerics, artists, and professional musicians. There was a less clear-cut separation between public criticism and academic science than in the case of history; the distinction between amateur and professional remained more permeable than in other fields of knowledge. Scholarship differed from aesthetic argument by virtue of its strict philological methods and its careful attention to ancient or medieval sources. As nations increasingly defined themselves in terms of a shared and distinctive cultural legacy, literary critics acquired a prominent new role as literary *historians*. The history of the nation's great poets, dramatists, and prose writers joined its political history as a second prop of national identity and pride. Not infrequently, as in the German case, language and literature were a more important element in mental nation building than the memories of a rather unglamorous record of political togetherness. The *Geschichte der poetischen National-Literatur der Deutschen* (1835–42), by the historian and liberal politician Georg Gottfried Gervinus, became a fundamental work of the age.

Orientalism and Ethnology

The study of other civilizations developed on the margins of the human sciences, never coming to play a central role in European universities.[112] More important to this day has been the reaffirmation of Europe's own roots, partly in Greco-Roman antiquity, partly in the early medieval social formations that are seen as the origins of nationhood. It is true, though, that contacts with foreign civilizations have always aroused curiosity about the Other. Accompanying the ideological glosses on European expansion and aggression, a huge literature developed in the early modern period in which Europeans—often travelers not directly associated with imperial operations—reported on their overseas experiences and adventures and tried to understand the customs, religions, and social institutions of the peoples they encountered. The study of language was a special concern. The interest in Arabic language and literature, particularly the Koran, had been constant since the twelfth century, while the Chinese language became known after 1600 via Jesuit missionaries. In places that had regular contact with the Ottoman Empire—Venice or Vienna, for example—experts in the field developed early on. As to the New World, missionaries began systemic study of indigenous languages soon after the Conquest. In close collaboration with Indian savants, European scholars based in Calcutta and Paris discovered, or rather rediscovered, the old language of high culture, Sanskrit, in the 1780s.[113] Thanks to the decoding of hieroglyphs by the French linguist and traveler Jean-François Champollion in 1822, Pharaonic Egypt became legible at last. And in 1802 Georg Friedrich Grotefend, a young teacher at a secondary school in Göttingen, discovered the key to unlock the ancient Persian cuneiform script.

Over several centuries, a varied literature of travelogues, country studies, botanical encyclopedias, dictionaries, grammars, and translations accumulated as a result of countless individual efforts, often outside the major centers of learning. Only the study of Arabic and Middle Eastern languages (important for biblical theology) had roots in early modern university chairs in places such as Leiden and Oxford. Nevertheless, the overall perception of the non-European world since the Middle Ages was saturated with scholarly seriousness. Even travel reports were not usually naive accounts of exciting adventures and strange fables, but were penned by observers who carried the most advanced knowledge in their baggage. This intellectual curiosity about the outside world was specific to Europeans in the early modern period. Other civilizations did not establish colonies overseas and, apart from rare diplomatic emissaries, sent no travelers to distant lands. Although a few Ottomans reported on their journeys, Muslims generally had little interest in "infidel" lands. The Japanese state forbade its subjects to leave the archipelago, on pain of severe punishment. Chinese scholars, to be sure, studied any "barbarians" who showed up at the imperial court, but only in the nineteenth century did they compose firsthand works on the non-Chinese periphery of the Qing empire. Before 1800, and even as late as 1900, the huge European literature on foreign civilizations was matched by very few texts giving an external view of Europe.[114] Whereas "Oriental studies" got off the ground in Europe, it would be the late twentieth century before one could speak of the beginnings of "Occidental studies" in Asia and Africa.

The character of European Orientalism changed in the early nineteenth century. As it divided more sharply than before by region (Chinese, Arabic, Persian, etc.), it also defined itself more narrowly as the study of ancient texts and sought the same kind of scientific detachment that its model, Greek and Latin philology, had already achieved. This entailed a lack of interest in the contemporary Orient; everything that seemed of value in Asia lay deep in the past, accessible only in a dubious inheritance of written texts and material relics over which Asian or Egyptian archaeology claimed an interpretive monopoly. Ancient Egypt was rediscovered by the scientists who accompanied Bonaparte on his Nile expedition of 1798. This initiated a continuous history of Egyptology, in which French, British, Germans, and Italians long played a greater role than Egyptians themselves. In Mesopotamia, archaeological excavations began during the second decade of the century, encouraged (as later in Anatolia and Iran) by British consular officials.[115] These men were well educated and, often having little else to do, could turn their hand to Middle Eastern research, much as army officers played a major role in uncovering the Indian past.[116]

In 1801 Thomas Bruce, the Seventh Earl of Elgin and then British ambassador to the Sublime Porte, obtained permission from the Ottoman government to take large parts of the Parthenon friezes (already badly damaged by Venetians and Turks) back to London; the famous Elgin Marbles. A hundred years later, with archaeology turned increasingly professional since midcentury, public museums and private collectors in major European countries had accumulated huge quantities of Oriental "antiquities" alongside treasures from ancient Greece and Rome. Manuscripts from all cultures found their way into special sections of the great Western libraries. In regions such as East Asia, where colonial control was more elusive, the market stepped in to assist the acquisition of art objects (stone testimony being less common than in Europe, because of local traditions of timber construction). But there was also theft on a massive scale, as in China during the Second Opium War (1858–60), which reached a climax with the plunder and burning of the Summer Palace in Beijing by British and French troops, and again during the foreign occupation of the imperial capital after the defeat of the Boxer Rebellion in summer 1900. Shortly after the turn of the century, a huge number of documents from the fourth to the eleventh century were "acquired" for a token price and carted off from caves near Dunhuang (in today's northwestern province of Gansu) to European libraries and museums. Archaeology was not simply a colonial pursuit, however; it could and does also serve to build a sense of nationhood, by uncovering cultural roots long before the historical invasions recorded in written documents.

Beginning in the nineteenth century, the material appropriation of Asia, North Africa, and Central America by Europeans (and North Americans) snatched numerous relics of the past from sandy or tropical oblivion, probably saved others from destruction, and laid the foundations for scientific knowledge about Egyptian tombs and Chinese ceramics, Mayan sculpture and Cambodian temples, Persian inscriptions and Babylonian reliefs. Doubts about the propriety of Western actions were seldom voiced at the time, and indigenous governments sometimes gave their approval for excavations and the shipping abroad of cultural treasures. Only since the end of the colonial era has the public become aware of the legal and ethical problems with such pillage.

In 1780 only a few specialists in Europe had linguistic access to religions, philosophies, literature, or historical documents from other parts of the world, and Oriental objects were lost amid the colorful diversity of princely "wonder chambers." By 1910, however, a highly sophisticated academic study of the Orient in France, Germany, Russia, Britain, and the United States was in charge of, and kept adding to, a colossal store of knowledge about foreign civilizations. Archaeology, Oriental studies, and comparative religion (a newly emerging discipline pioneered in Oxford in the 1870s by the Saxon scholar Friedrich Max Müller) contributed some of the titanic feats of the nineteenth-century human sciences. Yet, contemporary non-European societies that had no system of writing and little or no urban life could not be studied with the methods of Oriental philology. The science of ethnology that came into being from the 1860s onward developed a professional interest in these "primitive" peoples or (in German) *Naturvölker*, as they were called at the time. Strongly aligned in its first few decades with evolutionist theories of a general progression of humanity, this new science looked for social conditions in other parts of the world that, for Westerners, represented an earlier stage of development they had left behind long ago. Many of the early ethnologists did not travel themselves. Some classified and interpreted the tools, weapons, clothing, and cult objects that had been collected by scientific expeditions and colonial armies; others tried to identify basic patterns hidden in popular myths. The Enlightenment ambition to develop a general "science of man," a comprehensive "anthropology," gave way over time to detailed research into particular ethnic groups.

Bronisław Malinowski (a Pole) and Franz Boas (an American immigrant from Westphalia), working independently of each other, transformed ethnology (or anthropology, as Boas called it) from a series of speculations based on discrete anecdotal material into a science with empirical procedures centrally involving long-term participant observation. By 1920 the paradigm shift was complete, so that it was now possible and normal to describe the distinctive logic underlying a given non-Western society. This had a paradoxical effect. On the one hand, despite its many links to colonialism, the discourse of ethnology was relatively nonracist. Franz Boas's theory of "cultural relativism," in particular, was a weighty counter to the racist zeitgeist. On the other hand, the transition from full-scale evolutionism to the new emphasis on specialized modes of inquiry detached nonliterate societies from a comprehensive history of the human species, placing them in a space of their own outside the parameters of history and sociology. This also bred a certain isolation of ethnology/anthropology among the sciences, least marked in relation to the kind of sociology practiced by Émile Durkheim in France. Only in the 1970s—when its heroic period of description and classification of ethnic groups around the world was essentially over—did anthropology begin to have a major influence on other human and social sciences.

There has been much argument as to whether Oriental studies, archaeology, and ethnology should be regarded as handmaidens of colonialism.[117] It is clear enough that the simple existence of an empire offered fertile opportunities for many sciences such as botany, zoology, or tropical medicine.[118] But otherwise the balance sheet has to be mixed.[119] On the one hand, from the vantage point of the early twenty-first century, the arrogant conviction of European scientists about the all-around superiority of their own civilization is truly astounding. The assumption seemed, however, to be borne out by major successes in the study of other cultures—successes that were not without an eminently practical side, since anyone with good maps, linguistic competence, and knowledge of the morals and customs of others finds it easier to conquer, govern, and exploit them. To this extent it may be said that Oriental studies and ethnology (sometimes against the intentions of their representatives) produced knowledge for the sake of colonial domination. On the other hand, it is doubtful how useful this knowledge actually was, and how much it served practical purposes. Attempts to place colonial rule on a scientific foundation became a policy objective only after the First World War, and then the key experts were economists, not ethnologists. Before 1914 ethnologists—and, more important, colonial administrators for many of whom ethnology was a hobby—played a role above all where attempts were made to classify imperial subjects according to their degree of ability and cultural achievement.

But there were very few ethnologists around in those days, and when their numbers increased after the First World War they often proved troublesome critics of colonial practices.[120] Philological studies of ancient India or Vietnam, for their part, offered little knowledge that was directly serviceable to colonial rulers. Some have argued that precisely because of this apolitical conception of itself, Oriental studies "objectively" played into the hands of Western world domination—a charge that would be serious indeed if the supremacy of Western knowledge had demonstrably incapacitated Asians and Africans or reduced them to silence. However, it is not easy to find evidence that colonialism suppressed the knowledge of indigenous peoples about their own civilization. The academic revival of Indian traditions was in principle a joint European-Indian project, and it continued without interruption after independence came in 1947. In noncolonial countries such as Japan, China, and Turkey—to take the example of historiography—the encounter with Rankean critical methods led to a pluralist approach to the past and a more discriminating attitude toward the cultural heritage. In the nineteenth century, therefore, Western academic study of other cultures, in spite of all the annoying arrogance that came within it, was not just a destructive intrusion into vibrant non-European cultures of scholarship but also a founding impetus for the globalized human sciences of the contemporary world.

Geography as an Imperial Science

If any discipline was complicit in European expansion, it was geography.[121] In the first three decades of the nineteenth century, it developed from the descriptive collection of data about countries into a complex discourse about natural and social contexts on the earth's surface, within clearly definable spaces and landscapes. Its chief founders were far removed from European colonialism: Alexander von Humboldt, who had studied conditions in late colonial Spanish America in greater detail than anyone else, was one of its sharpest contemporary critics. Carl Ritter, the great encyclopedist at Berlin University, espoused—long before Franz Boas explicitly formulated the approach—a cultural relativism that recognized the equal value of social and cultural forms around the world. This detachment from politics was not a matter of course. House geographers already accompanied Napoleon, a zealous promoter of the subject, in his building of the empire, and geographical elements were present in many other imperial ventures throughout the century. Official cartographers mapped newly occupied territories. Geo-experts helped to draw boundaries, gave advice on the location of naval bases, and always had things to say about mineral wealth, transportation, or agriculture. These functions were sustained by a broad public interest in geography. School courses included classes about other continents, and imperial expansion found lively approval among lay members of geographical societies. From 1880 on, a special colonial geography emerged in the European metropolises, the conditions for truly global visions of exploration and "valorization" being particularly favorable in the British Empire. With a characteristically British interpenetration of private and public initiative, the founding of the Royal Geographical Society in 1830 created a kind of headquarters for the organization of research trips and the collection of geographical knowledge from all around the world. Imperial uses, though not always foregrounded, were never overlooked. Of all the branches of learning, geography had the greatest affinity with the imperial expansion of the West.[122]

It does not follow, however, that geography as such should be blamed for collaborating with the suppression of foreign peoples. It found a place in the university at only a very late date—not before 1900 in Britain and in the last third of the century in Germany, France, and Russia. For a long time it trailed behind the more respected discipline of history, although in the nineteenth century, under the philosophical aegis of "historicism" (*Historismus*), historiography distanced itself from anything that looked like a natural determination of human freedom. The physical and cultural aspects of geography, still united in Humboldt, later moved apart from each other, without abandoning the common academic umbrella; it was a necessary separation, but it created an insoluble identity problem and caused geography to fall somewhere between natural science (strictly geared to physics) and the "true" human sciences. Furthermore, with the exception of specialist colonial geographers, few representatives of the discipline were directly serving the imperial project. Many saw their main task as being to describe the territory of their own nation.

The close link between expansion and exploration went back a long time. Ever since the days of Columbus, overseas voyages and the urge to occupy and colonize new lands had been two sides of the same coin. Discoverers and conquerors came from the same cultural backgrounds in Europe; their education and goals in life were similar, as was their conception of the global position and mission of their own country, Christendom, or Europe as a whole. In the eighteenth century, it was taken for granted that major powers should use the resources of the state to help in unveiling the world. Britain and France sent out lavishly equipped scientific expeditions to circumnavigate the world. Tsarist Russia staked its claim to equal imperial and scientific status by following on the same path (the Kruzenstren mission of 1803–6). The first crossing of North America from east to west during the same years, initiated by President Thomas Jefferson and led by Meriwether Lewis and William Clark, may be seen as the US equivalent of these maritime operations. Even the details of its scientific tasks were similar to those of the great sea voyages since the time of Captain James Cook.

The "discoverer" type was compromised from the beginning. Columbus and Vasco da Gama already made use of violence. But over the next four hundred years there were at least as many examples of peaceful research trips; the most important were those of Alexander von Humboldt, Heinrich Barth, and David Livingstone. The age of high imperialism did, however, witness a final blossoming of the conquistador traveler. Bismarck, King Leopold II of Belgium, and the French Republic used the services of research explorers (widely varying in scientific competence) to register ownership claims to territory in Africa or Southeast Asia. Henry Morton Stanley, a reporter by training whom Leopold chose as his man in Africa, embodied this type in the eyes of the media of several continents (three Africa expeditions between 1870 and 1889). In the subsequent generation, Sven Hedin, having started his long career in 1894 with a research trip to Central Asia, became the most famous Swede of his age, with unfettered access to monarchs and heads of government in both West and East and adorned with countless decorations, gold medals, and honorary doctorates. Hedin's life encapsulates the contradictions of Europe's relationship with Asia. Convinced of the general superiority of the West over the East, Hedin was an excellent linguist and scholar and at the same time a Swedish (and, from personal choice, German) nationalist and militarist, a man of the political Right, who enjoyed taking part in geopolitical fantasizing about a "power vacuum" in the heart of Asia. But he was also one of the first Westerners to take contemporary Chinese science seriously and to cooperate with Chinese experts. He is held in high esteem in China today: a not atypical posthumous reputation, since quite a few European explorers, despite their activity in the service of empire, have been integrated into the collective memory of postimperial countries.[123]

Folklore and the Discovery of Country Life

Last but not least of the "alien" groups that became the object of scientific study in the nineteenth century were those living in the same country as the learned professors. Rationalist elites during the Age of Revolution had regarded the lifestyle and thinking of the peasantry, urban lower classes, and vagabonds as an obstacle to social modernization and relics of a superstitious mind-set. Military and civilian administrators in the Napoleonic Empire had as little time for Catholic popular beliefs in Italy or Spain as supporters of the utilitarian philosopher Jeremy Bentham within the East India Company had for the Hindu or Muslim traditions of India. Attitudes and procedures toward Europe's "internal savages" did not differ essentially from the situation in the colonies. In both cases, the authorities preached and practiced "education for work."[124] The reliance on government action or naked coercion varied, but the aim was much the same: to make human capital more effective, in association with a genuine, often Christian-inspired effort to raise the "level of civilization" among the lower orders. The Salvation Army, founded in London in 1865 and gradually spreading internationally, was an expression of such a charitable vision, and the overseas "mission to the pagans" was paralleled in Protestant Europe by an "internal mission" to assist the weaker members of society. Apart from such early social policies, whether philanthropic or bureaucratic in inspiration, there was sometimes a reverence for popular ways of life that bordered on glorification. Johann Gottfried Herder had been the original intellectual force behind such attitudes. Linguists, legal historians, and collectors of "popular verse" strengthened them in the early nineteenth century.

Social romanticism was linked to very different points in the political spectrum. In the great French historian Jules Michelet, it signified a radical admiration for the creators of the nation and the revolution, whereas in Wilhelm Heinrich Riehl—who published a four-volume social history of the German people (1851–69)—there was an underlying mistrust of the socially destructive consequences of urbanization and industry. Both men, writing at almost the same time but with quite different premises, described the life of poor and simple folk, including women, both past and present with a sympathy and accuracy rarely seen before. Riehl became a founder of what was called *Volkskunde*, a study of the "spirit" and customs of peoples rooted in conservative Romanticism.[125] He found admirers above all in Russia, who saw his work as confirmation of their own (politically opposite) leanings. The newly emancipated peasantry and its age-old communes were glorified by upper-class urban intellectuals as the natural agents of an impending revolution. These "friends of the people," the *narodniki*, opened a new chapter in the history of Russian radicalism.[126]

Folk elements also attracted fresh attention in the arts, as the internal exoticism of folklore traditions within Europe ran almost exactly parallel to the external exoticism of Orientalist persuasion. A search for inspiration in the anonymous music of ordinary people and for characteristic national styles soon resulted in a versatile melodic idiom. A kind of musical exoticism developed within Europe itself. French composers conjured up Spanish color (Georges Bizet: *Carmen*, 1875; Edouard Lalo: *Symphonie espagnole*, 1874); and "typically Hungarian" Gypsy touches, which the cosmopolitan Franz Liszt (born in Austria's hilly Burgenland) turned into a national trademark in 1851 with his *Hungarian Rhapsodies* for piano, slipped easily into the tone language of the native Hamburger and Viennese resident Johannes Brahms. In 1904, dissatisfied with the kitsch of national Romantic clichés, the young Hungarian Béla Bartók and his compatriot Zoltán Kodály went in search of authentic music among the Hungarian rural population, as well as non-Magyar minorities in the Habsburg kingdom of Hungary. The new methods of ethnomusicology were then applied in the same way to musical production outside Europe. Bartók, a composer who had moved beyond Romanticism,[127] proved it was possible to engage in top-class research on ethnic subjects without succumbing to the ideology of *völkisch* nationalism.

In the nineteenth century, writing gave many people in the world greater scope for extensive communication, thanks to the spread of literacy and the growing availability of print media. Distribution of the ability to read and write was extremely uneven, depending on levels of prosperity, political objectives, missionary goals, and the educational ambition of individuals and groups. Usually it required a local impetus, which then had to be translated into some kind of sustainable institutional form, with compulsory schooling as the logical *terminus ad quem*. The spread of world languages further widened communicative spaces, at least for those who took the opportunity to learn one or several of them. As a rule, Europe's expanding languages did not obliterate and replace existing linguistic worlds but were superimposed on them.

Access to knowledge became easier. But it had to be acquired—or rather, worked for—with considerable effort. Reading is a cultural technique that demands a lot from individuals: an illiterate person can much more easily install a radio or television set and follow the readymade programs. In this respect, twentieth-century technologies reduced the level of cultural effort, but also the threshold for at least passive participation in communication. But what kind of knowledge became more accessible? Little can be said that applies worldwide. Structured knowledge outside the realm of everyday life—what people were generally beginning to call "science"—certainly increased in the nineteenth century on an unprecedented scale; and there were more and more scientists who produced it. This happened within institutions, universities above all, which not only created a loose framework for the scholarly activity of individuals (as academies had in the early modern period) but systematically endeavored to acquire new knowledge and provided means to that end. Science expanded also because whole areas of social discourse were put on a scientific footing: the literary and textual criticism that had been blossoming in Europe became the discipline of literary studies (at the end of the century), while the collection of words and grammatical elements became a methodical, historically based search for laws and eventually, in Ferdinand de Saussure (*Cours de linguistique générale*, 1916), a science that postulated deep structures of language. Before 1800 the human and social "sciences," in the sense of established disciplines, had not existed in Europe. By 1910 the matrix of disciplines and the range of academic institutions that we know today had taken shape—first in several European countries, then a little later in the United States, but in a process that was increasingly internationalized, not locally disparate.

By 1910 a number of cross-border scientific communities had come into being, where information circulated at great speed, academics competed to take the lead, and procedures were in place for quality assessment and the allocation of prestige. These circles were entirely male dominated; non-Westerners gradually gained entry to them—first a number of Japanese scientists, joined after the First World War by colleagues from India and China. Transnational standards operated in the natural sciences. This made interwar attempts to establish a special "German," "Japanese," or (in the Soviet Union) "socialist" science seem regressive and ridiculous. It was another matter that scientists often felt an urge to ensure that their work was of benefit to the nation. However transnational the communicative infrastructures and scientific standards, scientists everywhere felt under an obligation to their national institutions (never more than during the First World War), and arts scholars—the inheritors of ancient rhetoric—operated first and foremost in the public arena of their own country. As far as science was concerned, internationalization and nationalization stood in a tense and contradictory relationship to each other.

CHAPTER XVII

Civilization and Exclusion

1 The "Civilized World" and Its "Mission"

For thousands of years, some human groups have considered themselves superior to their neighbors.[1] City dwellers looked down on villagers, settled populations on nomads, literate on illiterate, pastoralists on hunters, rich on poor, practitioners of complex religions on "pagans" and animists. The idea of different degrees of refined living and thought is widespread across regions and epochs. In many languages it is expressed in words that roughly correspond to "civilization" in European usage—a term that has meaning only in a relationship of tension with its negative twin. Civilization prevails where "barbarism" or "savagery" lie defeated; it needs its opposite to remain knowable as such. Were barbarism to disappear altogether from the world, there would no longer be a foil for "civilized" people to measure themselves against, either taking the offensive in a spirit of self-satisfaction, or bemoaning the fate of superior humanity amid crudeness and decline. The less civilized are a necessary audience for this grand theater, for the civilized need the recognition of others, preferably in the form of admiration, reverence, and peaceful gratitude. They can live with envy and resentment if they have to; any civilization must arm itself against the hatred and aggressiveness of barbarians. The sense of worth felt by civilized people arises from an interplay between self-observation and attention to the various ways in which others react to them, with an awareness that their own attainments are constantly at risk. A barbarian attack or a revolt by plebeian "internal barbarians" might bring ruin at any moment, but an even greater danger, and one harder to discern, is the slackening of moral endeavor, cultural ambition, and realistic tough-mindedness. In China, Europe, and elsewhere this has traditionally been denoted by "corruption" in the broad sense of the term; *fortuna* enters a downward spiral when the power to stick to high ideals begins to wane.

Civilization, in the normative sense of socially determined refinement, is thus a universal concept not limited to the modern age. Frequently it is associated with the idea that civilized people have a task, or even a duty, to propagate their cultural values and way of life: whether in order to pacify barbarians living in the surrounding world, to spread the one true doctrine, or simply to do what is good. Such varied motives fuel all kinds of "civilizing mission" that cover more than just the dissemination of a religious faith. It involves a self-given assignment to transmit one's norms and institutions to others, sometimes by exerting pressure of varying degrees of intensity. This presupposes a firm belief in the superiority of one's own way of life.

Contradictions of the Civilizing Mission

Civilizing missions may be found in the relationship of ancient Chinese high culture to various barbarians living nearby, as well as in European antiquity and in all expansive religions. Never was the idea as powerful as it was in the nineteenth century. In the case of early modern Europe, the Protestant Reformation may be interpreted as a huge movement to civilize a corrupt culture, and the Counter-Reformation, its mirror image, as a defensive impulse designed to regain the initiative for civilizing work in the reverse. Cultural monuments such as the Luther Bible or the great Baroque churches may be understood as instruments of a civilizing mission. But the missionary dynamic of the early modern period should not be overestimated, especially in the context of European overseas expansion. The early modern empires were seldom driven by the idea of a mission, and outside the Spanish monarchy no one dreamed of fostering a homogeneous imperial culture.[2] For the Dutch and English, *imperium* meant a commercial undertaking that required little moral regulation; missionary zeal was not supposed to get in the way of business or to disturb the unstable fiction of imperial harmony. Protestant governments therefore did not usually allow missionaries to operate in their colonial territories until the end of the eighteenth century, and the Catholic mission in the Iberian empires lost much of its support from the state during the second half of the eighteenth century. The idea that European law should apply to the "natives" was rarely entertained and almost never practiced.

The early modern period still lacked a conviction that European civilization was the sole standard for the rest of the world. Normative globalization, as it followed in the long nineteenth century, presupposed an end to older military, economic, and cultural equilibria between Europe and other continents (especially Asia). We should note a paradox. On the one hand, the civilizing mission of Europeans was an ideological instrument of imperial world conquest; on the other, it could not easily be spread by means of gunboats and expeditionary corps. The success of the civilizing mission in the nineteenth century rested on two further premises: (a) a conviction among European power elites and the most diverse private agencies of globalization that the world would be a better place if as many non-Europeans as possible took in the achievements of an allegedly higher civilization; and (b) the emergence, in numerous "peripheries," of social groups that shared this point of view. The original ideal of the civilizing mission was strictly Eurocentric, and in its claim to absoluteness it was directed against any kind of cultural relativism. It was therefore inclusive: Europeans did not want to keep their higher civilization for themselves; others should also have a share in it. It was also politically polymorphous, in that the civilizing work was supposed to unfold both inside and outside colonial systems. It could precede European territorial conquest, be independent of it, or serve as its a posteriori justification. The rhetoric of civilization could also accompany processes of state building and political consolidation in which Europe played no part. Founded on an optimistic vision of progress and growing rapprochement among the cultures of the world, it also justified and provided propagandistic cover for all manner of "projects" that claimed to be serving the cause of progress. Thus, it was perfectly possible to "civilize" not only barbarians and different faith groups but also flora, fauna, and landscapes. The land-clearing settler, big-game hunter, and river tamer were emblematic figures of this drive to civilize the whole planet. The great opponents that had to be defeated were chaos, nature, tradition, and the ghosts and phantoms of any kind of superstition.

The theory and practice of the civilizing mission have a history. It began in the late eighteenth century, shortly after the term "civilization" became a central category used by European societies to describe themselves, first of all in France and Britain. The prestige of European civilization reached its peak outside Europe in the middle of the nineteenth century, before the emphasis on a civilizing mission, came to be seen as increasingly hypocritical in the decades around 1900, in view of the massive use of force in pursuit of imperialist aims. The First World War then severely damaged the white man's aura, although it by no means buried his civilizing urge.[3] After 1918, all the colonial powers sooner or later switched to a "developmental" style and rhetoric of colonial rule, more in tune with the times. It would be continued in the postcolonial policies that national governments and international organizations adopted after the winning of sovereignty.

The transformative period around 1800 was when civilizing missions began to be practiced in grand style. Two developments in the history of ideas lay behind this: (a) the confidence of late Enlightenment thinkers in pedagogy, that is, a belief that truths, once recognized as such, had only to be learned and applied; and (b) the formulation of universal models of progress in which humanity passed through various stages from humble beginnings to the full blossoming of a civil society based on legality and diligence. Various options now opened up. Those who trusted in the automatic working of the evolutionary process were less inclined to intervene actively than those who felt an urge to combat barbarism in the world.

The "civilizing" concept was also applied closer to home in the nineteenth century. Influential intellectuals—such as the future Argentine president Domingo Faustino Sarmiento, in his groundbreaking *Facundo: Civilización y barbarie* (1845)—constructed whole national histories out of the opposition between civilization and barbarism.[4] Internal peripheries in old as well as newly emerging nation-states were regarded as culturally remote areas left behind from earlier stages of development. Remnants of archaic clan structures in the Scottish Highlands, for example, turned into folklore for tourists from the South, so that a region discovered as a kind of Africa of the North in the 1770s became an open air museum of social history in the age of the Crystal Palace Exhibition (1851). Harsher and less forgiving than the English gaze toward the North was the attitude of northern Italians, when it came to Sardinia, Sicily, or the Mezzogiorno. The more that national unification led to disappointment over the difficulties of integrating peripheral regions, the more the language in the North came to resemble the racist rhetoric used about Africa.[5] The underclasses in the big industrial cities also appeared as alien tribes, in whom state and market, private charity, and religious persuasion had to inculcate a minimum of civilized, in other words, bourgeois behavior.

National Variants: Bavarian, French, British

Civilizing missions also had national peculiarities. Until 1884 the Germans had no overseas colonial empire in which they could carry out "cultural work" (as it was called in those days). The German idea of education in the Classical and Romantic periods was a program of personal self-cultivation, not without a strong dose of political utopianism. For lack of barbarians in the flesh, the civilizing process turned reflexively inward at an individual level. But once Germans had the chance to take part in a grand civilizing project, they did so with particular relish. In 1832 the Great Powers placed the newly founded country of Greece under Bavarian custodianship: it acquired a Bavarian prince as king, a Bavarian bureaucracy, and a Bavarian ideology of "elevating" reforms. There was a contradiction, however: every German high school student dreamed of reviving classical Hellas after the end of Turkish "despotism," but it seemed beyond doubt that the Greeks actually living there were completely useless for that sublime task.

The Bavarian regency council later withdrew, and the Greeks eventually deposed their unloved King Otto and left him to withdraw into exile in Franconia.[6] It is an irony of history that soon afterward they hit upon their own variant of the civilizing mission—the "big idea" (*megali idea*) that they directed against the Turks with the aim of wresting from them as much as possible of the ancient Hellenic and Byzantine lands. In 1919 they suffered a crushing defeat when, spurred on by Britain and their own inflated estimate of their strength, they made the error of attacking the Turkish army. The collapse of Greek expansionist ambitions after the First World War was one of the most spectacular reverses for the civilizing mission.

Previously, Napoleon's project of spreading civilization on horseback, beginning with early campaigns in Italy and Egypt that his propaganda presented as one great liberation, had had mixed results. In Egypt, as later in Spain and the French Caribbean, the mission had ended in failure. Slaves who were already emancipated in the West Indies found themselves reduced to slavery again in

1802. On the other hand, the French regime in the German Confederation of the Rhine did generally have a civilizing and modernizing impact; it introduced French laws and institutions of a bourgeois hue and swept away traditions that had lingered beyond their time. Indirect French influence worked in the same direction in Prussia and, less powerfully, in the Ottoman Empire. The French civilizing style was distinctive. In occupied areas of Europe, especially where popular traditions had been shaped by Catholicism, French officers and functionaries behaved with extreme arrogance and condescension toward local people they considered to be backward. The occupation regimes were highly efficient and rational, but also utterly remote. In Italy, for instance, French rule seldom managed to create any link with the indigenous population beyond a small circle of trusted collaborators.[7]

Napoleonic France was the first specimen of an authoritarian civilizing state in Europe. The state became the instrument for a planned transformation of elements of the ancien régime, both inside the country and farther afield. The reformers' aim was no longer, as in the early modern period (or anyway, before Joseph II's energetic initiatives in the Habsburg lands), to remove particular grievances but rather to bring a completely new order into being. This technocratic reshaping of society from above was also found in various forms in the colonial world. Lord Cromer, for example, who concentrated nearly all power in his hands after the British occupation of Egypt in 1882, cut a "Napoleonic" figure with his fondness for cold administrative rationality—though with the difference that the idea of "liberating" the indigenous population never entered his thoughts. In 1798 Bonaparte had wanted to take the torch of Enlightenment to Egypt, whereas after 1882 Lord Cromer's only aim was to ensure that all remained quiet (and fiscally sound) at a major bridge between Asia and Africa—an improved application of the techniques used to rule post-Mutiny India. Detached from the mass of the Egyptian population, the "civilizing" work served only the interests of the occupying power and made no claim whatsoever to be transforming the society.[8] At the same time, it has to be said that later French colonial policy had nothing Napoleonic about it either. At most, in western Africa, it came a step closer to the establishment of a rational state that concerns itself with the education of the population. But even there the state had to strike compromises that made the idea of perfect direct rule illusory—compromises not with settlers (as in Algeria) but with indigenous power holders.

In contrast to the interventionism of the Napoleonic state, which was hostile to organized religion, the early British civilizing mission was driven by strong religious impulses. Its first weighty advocate, Charles Grant—a high-ranking official in the East India Company and author of the influential *Observations on the State of Society among the Asiatic Subjects of Great Britain* (1792)—was representative of evangelical revivalism in the age of the French Revolution. The Protestant call of duty to "better" the Indians was a distinctively British type of colonial romanticism, compounded by an English form of late Enlightenment thinking (Jeremy Bentham's utilitarianism) which, in its pursuit of rationaliza-tion and its authoritarian tendencies, was not all that far from the Napoleonic conception of the state.[9] In India, this singular alliance of pious evangelicals with utilitarians who were mostly indifferent to religion succeeded in eradicat-ing such practices as the burning alive of widows (*sati*) in 1829—after seventy years in which the British authorities in Bengal had tolerated this cruel custom and its annual toll of hundreds of victims.[10] Attempts to civilize India in a West-ern sense reached their peak in the 1830s and ended in 1857 with the shock of the Great Rebellion.

Many other spheres for missionary activity were discovered around this time. After the middle of the century, the British-style civilizing mission developed less as an unconditional blueprint than as a set of attitudes, strongly shaped by a Protestant ethical sense that the famous explorer, missionary, and martyr David Livingstone expressed most clearly. The spread of secular cultural values was also largely the work of missionaries. The state and Christian missionary socie-ties were here much less close to each other than in the French colonial empire, where Napoleon III used Catholic missions directly as a policy instrument, and even the Third Republic did not shrink from collaborating with them. In the British Empire, missionaries aimed at fundamental changes in the everyday life of their charges and converts even if the colonial state was much more reticent.[11] Not all of the numerous Protestant missionary societies thought it their task to change anything other than religious beliefs, but most of them did not draw a radical distinction between religion and other spheres of life.

The typical British missionary in the late nineteenth century had much to offer: bibles and primers, soap and monogamy. A second aspect of the program for the education of humanity was the growing confidence among missionaries, at least by midcentury, that their work among particular peoples was assisting the breakthrough of civilization as such. Despite its rival claims to universalism, the French *mission civilisatrice* had a stronger patriotic foundation. The universal character of British civilization mirrored the global reach of its empire, but it also reflected a greater identification with two normative practices whose opera-tional radius was theoretically unlimited: international law and the free market.

Law and the Standard of Civilization

Toward the middle of the century, the old *ius gentium* was refashioned into a legally binding "standard of civilization." Law became the most important me-dium of cross-cultural civilizing processes, more effective than religion because its importers could adapt it to local needs even when indigenous values and norms proved immune from foreign faiths. Japan—where Christian missionar-ies of every denomination never really gained a foothold, even after they were allowed back in 1873—adopted major elements of European legal systems. The resistance to Christian proselytism was at least as strong in the Islamic world, with its dense interweaving of faith and law, but central pillars of European law were introduced into noncolonial countries here too, such as the Ottoman Empire and Egypt (even before the British occupation of 1882). The prestige and effectiveness of the law have to do with its twofold nature: both a political instrument in the hands of legislative authorities and a product of autonomous or—as German Romantic theorists would have it—"anonymous" developments in the moral concepts of society. This duality of construction and evolution was apparent also in colonial contexts, where the body of law, together with its enforcement by judges and policemen, was often a sharp weapon of cultural aggression. Bans on the use of indigenous languages, for example, were among the most hated measures in the whole history of colonialism, "own goals" that never had the intended "civilizing" effect. But one of the strengths of the British Empire, in comparison with others, was the versatility of the English legal tradition, whose pragmatic application in many colonies left some latitude for compromise and coexistence with local forms. An awareness of law, unmatched elsewhere in Europe, meant that the accountability of officeholders not only to their superiors but to a morally and legally vigilant public was seen as a central pillar of civilized existence. An internationally applicable standard of civilization was therefore the counterpart of the rule of law in the internal governance of society.[12]

For the Victorian mind, the standard of civilization had its source less in the constructed aspect than in the evolutionary aspect of the law.[13] An ethnocentric precursor of today's human rights, it had come to be understood as a universally valid bedrock of norms defining what it meant to belong to the "civilized world." Such norms existed in several areas of the law—from the prohibition of cruel forms of corporal punishment through the inviolability of property and civil contracts to the exchange of ambassadors and (at least symbolic) equality in dealings between states. The evolutionary side consisted in the idea that the standard of civilization was the outcome of a long civilizing process in Europe, and its so-called leading nations—often denoting only Britain and France until 1870—were called upon to guard this state of legal perfection.[14] Europe based its claim to moral authority on the success it had had in educating itself. Had it not, in the eighteenth century, left behind the open brutality of the wars of religion, stripped criminal law of its archaic features at least with regard to white people, and developed practical rules for social interaction among its citizens?

Up to the 1870s, European legal theorists used the standard of civilization to criticize barbaric practices elsewhere in the world, but there was not yet any thought of large-scale intervention to enforce it directly. Even the opening up of China, Japan, Siam, and Korea, through war or gunboat diplomacy, was seen less in terms of a general civilizing mission than as a necessary measure to facilitate intergovernmental relations. The treaty-port system established in 1842, for example, was not so much a Western triumph as a compromise. China was forced to grant special extraterritorial rights for foreigners, but no pressure was put on it to change its *whole* legal system; the Westernization of Chinese law would be a long-drawn-out process that began only after 1900 and is still not completed today. In the nineteenth century, the next step by the West that followed any "opening up" was to demand reforms in a few particular areas of the law: property and inheritance, but also matrimonial affairs, were brought more into line with "civilized" customs in countries such as Brazil and Morocco.

The Market and Violence

The second major vehicle of the Victorian civilizing mission was the market. In the liberal utopia of passions tamed by interests, markets made nations peaceful, warrior classes superfluous, and individuals industrious and ambitious. But in the nineteenth century a new idea emerged: namely, that the market was a natural mechanism for the generation of wealth and the distribution of life chances. All that was needed for the maximum development of human capacities in any culture was to clear away obstructive traditions and to give up interference in self-regulating systems. Classical liberalism assumed that anyone would respond enthusiastically to market incentives; steam transportation and telegraphic communication would weave markets into ever wider spheres of activity, and the Victorian trade revolution would make itself felt on a planetary scale. Not all economists in the mid-nineteenth century shared this naive optimism. Sharp-eyed observers of society soon saw that the market economy would not necessarily serve to perfect human beings or to raise the general level of morality. The market civilized some but left others untouched, and in a third group it brought out the ugliest side of human nature. As John Stuart Mill and some of his contemporaries suspected, *Homo oeconomicus* required a degree of education and maturity too. Politically this was a double-edged argument: on the one hand, it sought to prevent the consequences that would follow if premodern economic cultures suddenly had to face untrammeled competition; on the other hand, it could imply that a tutelary colonialism was needed to open up a cautious path to economic modernity for non-Europeans. In colonial reality the slogan "educating for work" often meant a great deal of work and very little education.

Market economy, law, and religion were the three pillars supporting the British civilizing mission, the most effective of its kind. In the French case—though not so emphatically anywhere else—assimilation to the high culture of the colonial power added a further dimension.[15] Particular civilizing initiatives differed not only from country to country but also according to their time frame, principal agencies, local conditions, and the degree of the perceived cultural gap. If this gap was deemed unbridgeable, candidates for civilization appeared incapable of meeting the demands of the "superior" culture, and therefore soon came to count as useless and dispensable. Repression, marginalization, and even physical annihilation were possible consequences, but they were exceptional even in the age of high imperialism. No colonial power had a rational interest in systematic genocide in peacetime. However, King Leopold II of Belgium allowed large-scale atrocities to be perpetrated in his euphemistically named Congo Free State from the 1880s on, and German troops deliberately committed them in 1904–5 against the Herero and Nama peoples of South-West Africa. Some colonial wars of the epoch—for example, the US war of conquest in the Philippines—were waged with such single-minded brutality that historians have used the term "genocide" to describe them.[16]

Care and Self-Consciousness

The civilizing mission as a project to reshape whole ways of life lay midway between two extremes of nonintervention. At one end, coexisting with the humanitarianism of a morally solicitous Europe, was a calm and arrogant acceptance that "primitive peoples" were doomed to extinction. Hard-boiled economists had already interpreted the Irish famine of 1846–50 as a necessary adjustment crisis.[17] Around the turn of the century, there was much talk of "dying races" on the periphery of empires—that is, of peoples whose demise should not be halted. At the other end, all European colonial powers opted in special circumstances for a policy of indirect rule, avoiding any deep intrusion into indigenous social structures. Local people were left to their own devices so long as they kept the peace, paid their taxes, listened to the "advice" of colonial agents, and could be relied on to deliver goods for export. Indigenous law, including "barbaric" forms of punishment, then often remained untouched. The colonial authorities reined in overzealous missionaries and sometimes cultivated relations of mutual respect with the local upper classes, reluctant to allow the uniformity of Westernization to dampen their colorful exoticism. Such was the relationship of the British to Indian princes or Malayan sultans, or of the French to the elite in their post-1912 Moroccan protectorate.[18]

Under such circumstances, the social and lifestyle engineering involved in a civilizing mission would only have disturbed longstanding balances of power and cultural compromise. Civilizing missions, taken seriously, aimed at a wholesale refashioning of any society on which its guns came to be trained. They usually were the program and the work of activist minorities. Even in European societies, high-minded bourgeois reformers found themselves in the midst of "uncivilized" majorities of peasants, urban plebeians, and mobile vagrants. The growing metropolises were a magnet for large migratory movements, which called forth an ambivalent mix of rejection and philanthropic eagerness to change the newcomers. Observers such as Friedrich Engels and Henry Mayhew saw only slight differences between English slum dwellers and the impoverished masses in the colonies. Mayhew, indeed, thought of the destitute "urban nomads" at home as closely analogous to the true nomads far away in the desert. For reform-minded middle-class minorities, the "internal barbarians" were scarcely less alien and frightening than exotic savages. Nor was this a European peculiarity. In Mexico the liberal *científicos*, a bureaucratic elite that modeled itself on Europe's municipal oligarchies and efficient state administrators, waged a lengthy campaign against rural *indios* and their supposedly backward ideal of common land-ownership. Racist representatives of this elite, however, considered them to be biologically inferior and therefore impervious to improvement and education.[19] In Tokyo, Istanbul, and Cairo, urban intellectuals and bureaucrats similarly regarded their country regions as remote, primitive, and menacing worlds.

The most spectacular outbreak of "savagery" in a society proud of its civilized refinement took place during the Paris Commune rebellion of 1871. After its suppression—no less violent than British operations against the Indian Rebellion of 1857/58—four thousand surviving Communards were deported to New Caledonia, a recently colonized archipelago in the South Pacific, where they were subjected to a harsh "civilizing" program not unlike that inflicted on the native Kanak people.[20] From the point of view of nineteenth-century civilized elites, barbarism was lurking everywhere in the most diverse guises and called for vigorous countermeasures in every corner of the world. Only where the demographic preponderance of the white population was unmistakable could the work of civilization proceed from a position of unchallenged superiority—above all, in North America after the end of the Indian Wars and in the Philippines (where the United States introduced systematic reform programs even before the First World War).

The language of civilization and civilizing was the dominant idiom of the nineteenth century. In the decades around the turn of the century, it was briefly undermined or called into question by extreme forms of racism that doubted whether certain peoples were capable of being educated. After the First World War, when racist rhetoric generally became more muted (though not in Germany or east-central Europe), the idea of civilizing others underwent a revival. But in the 1930s Italians, Japanese, and Germans began to argue that they were superior human beings who, by virtue of the law of the strongest, were justified in ruling colonial peoples without the minimum of fellow feeling essential to the transformative relationship of the civilizing mission. Three different paths therefore led out of the Victorian civilizing mission: one ended in violent collapse, when the civilizer's denial of the humanity of others exposed the fictitiousness of his own civilized character; one led via embryonic "colonial development" in the period of late colonialism to the national and international development aid of the second half of the twentieth century; and one ended in indifference after major material and moral investment had borne no fruit.

The optimistic civilizer is constantly at risk of seeing his efforts fail. The British lived through such a moment in 1857, when "India" (perceived as a uniform whole) shocked them after decades of reforms by giving dramatic evidence of its ingratitude and "unteachability." Missionaries repeatedly had similar experiences: their implanted Christianity failed to take root, or else proved so successful that the new converts went their own way. All kinds of movements for political autonomy were often seen as unintended side effects of the spread of Western thought. Using the law they had learned from Europeans, Asians and Africans turned the universalism of its lofty principles against the culpability of colonial practice. European languages taught with great zeal became instruments of anti-imperialist rhetoric.

The nineteenth century stands out from the sequences of ages by the fact that never before, and never again after the First World War, were the political and educational elites of Europe so sure of marching at the head of progress and embodying a global standard of civilization. Or, to put it the other way around: Europe's success in creating material wealth, in mastering nature through science and technology, and in spreading its rule and influence by military and economic means, brought about a sense of superiority that found symbolic expression in talk of Europe's "universal" civilization. Toward the end of the century, a new term for this made its appearance: modernity. The word had no plural; only in the final years of the twentieth century would scholars begin to speak of "multiple modernities." The concept of modernity has to this day remained enigmatic: there has never been agreement as to what it means and when the corresponding phenomenon emerged in historical reality. Its geographical compass has also varied over time. It often applied to Western European civilization *as a whole*, distinct from all other cultures, but then two levels of contradiction *within* Europe itself were built into the picture. First, "modernity" and "modernism" referred to avant-garde attitudes among small circles opposed to the traditionalism and philistinism of the majority—a narrow sense covering various movements of renewal in the arts that went beyond accepted aesthetic norms. Second, in many parts of Europe around 1900, it was at most the lifestyles, consciousness, and taste of urban elites that counted as modern; the rest of the country vegetated in a rural torpor. From the viewpoint of London, Paris, Amsterdam, Vienna, Berlin, and Budapest, but also from Boston and Buenos Aires, it was questionable whether large areas at the respective periphery were dispensers or needy recipients of civilization. Did the Balkans, Galicia, and Sicily; Ireland and Portugal; or the rustic frontier societies of the American continent belong at all to the "civilized world?" In what sense were they part of "the West?"

Arrogant pride in one's own civilized status, and a belief that one was entitled or even duty-bound to spread it throughout the world, were in one respect pure ideology. In numerous cases this was used to justify aggression, violence, and plunder. Civilizational imperialism lurked within every kind of civilizing mission.[21] On the other hand, the relative dynamism and ingenuity of Western European and neo-European societies should not be ignored. The asymmetry at the level of historical initiative was temporarily in favor of "the West," so that others appeared to see no future for themselves except in imitating it and trying hard to catch it up. For those who were convinced of the West's lead in civilization, the rest of the world was trapped in a primeval condition with no history to speak of or had been left wrestling with the dead weight of tradition.

By 1920 the material differences between the rich Western countries and the poorest societies elsewhere had grown much larger than they had been a century before when such theories first came to be proposed. And yet the first forces challenging the West's claim to universality, though very weak at the end of the First World War, were beginning to stir. The League of Nations, newly founded in 1919, did not yet offer them the forum that the United Nations would become after 1945. The promise of 1919, when US president Woodrow Wilson awakened hopes of emancipation with his vague talk of "self-determination," soon lost the wind in its sails.[22] The colonial empires of the victorious powers remained intact. For the time being, the disenchantment of Europe, whose self-butchery others had watched with stupefaction, produced few tangible consequences. Though prone to doubt internally (see Oswald Spengler's German bestseller *The Decline of the West*, 1918–22) and faced with challenges externally, above all from the rise of the Japanese empire, the pride of Europe and North America in the superiority of their civilization was not yet seriously endangered. Mahatma Gandhi, the greatest Asian adversary in the interwar period, put it in a nutshell when a journalist asked him what he thought of Western civilization. "I think it would be a good idea," he replied, tongue in cheek.[23] Yet many nationalist leaders of the Indian freedom struggle did not hesitate to side with their British oppressors in the late 1930s, when a rift opened in the West and British arrogance began to pale alongside the Nazi's murderous race hatred.

2 Slave Emancipation and White Supremacy

More Slavery in the West than in East Asia

In 1800 barbarism still nestled at the heart of civilization. The countries that thought of themselves as the world's most civilized still tolerated slavery in their areas of jurisdiction, which included their overseas empires. By 1888, a hundred years after the first small abolitionist groups were founded in Philadelphia, London, Manchester, and New York, slavery had been declared unlawful throughout the New World and in many countries elsewhere.[24] It was then but a small step to the present legal situation, where slavery is considered a crime against humanity. The traces of an institution that for centuries had underpinned large parts of the Americas, including the Caribbean, did not disappear overnight. The mental and social consequences of slavery persisted for decades, and many are still discernible today. In Africa, which supplied the slaves for American plantations, remnants of the slave trade and slavery itself survived until well into the twentieth century. Only in the 1960s, a full century after the abolition of slavery in the United States, did the Islamic world reach a broad consensus against its juridical legitimacy and social acceptability. In 1981, Muslim Mauritania became the last country in the world to outlaw the practice.[25]

Nevertheless, 1888 marked a watershed in the history of humanity. The institution that, more than any other, contradicted the liberal spirit of the age was largely delegitimized and spurned outside the Muslim cultural area. If there were still societies *with* slaves, there were no longer any outright *slave societies*. A last relic of the seventeenth century, when slavery had enjoyed its first great blossoming in postmedieval times, withered away once every region in the European and neo-European sphere ceased to allow the treatment of human beings as property to be bought and sold and inherited.

Although people in the West congratulated themselves on this great advance of civilization, which was supposed to have finally established a truly Christian society, it would be only fair to point out that the barbarism of slavery and the slave trade had not marked every region in the world in the eighteenth and early nineteenth century. The two previous centuries had seen a reversal of the European tendency to freer labor relations, both in the overseas colonies and—in the form of a "second serfdom"—east of the Elbe. During the same period, in China a long-term trend toward negotiated labor relations softened the harsher forms of social subordination and led to the retreat of the most degrading forms of bondage. Things there had become more complicated after the victorious Manchus had grafted military slavery and other Inner Asian concepts of slavery onto Chinese notions and practices of servitude and dependency. By the end of the eighteenth century, bondage still affected millions of people in the Qing Empire. But in contrast to the mature systems of slavery and serfdom in the Americas and Russia, the state, its laws, and its courts did not explicitly uphold coercive relationships. Qing policy, with some success, sought to move against the penchant of landlords to debase the status of agricultural workers and, to some extent, also against the sexual exploitation of women. Where it continued to exist, slavery was not seen as the core institution of society, but as an aberration from the norm of the legally free commoner tilling the soil as the owner or tenant of his land. In this sense, China around 1800 was a decidedly "freer" country than Russia, Brazil, Jamaica, Cuba, or the southern United States.[26] Slavery was even rarer in Japan where both the external and the domestic slave trade had been banned since 1587 and agriculture came to rely on unencumbered labor. "Historians are generally agreed that slavery, as a significant form of labor relationship, had more or less ceased to exist in Japan by the end of the seventeenth century."[27] The story was different in Korea, where slavery was abolished as late as 1894 under Japanese influence.[28] In Vietnam, shaped by Confucianism, servile relations gradually declined in the course of the eighteenth century; nor were they reintroduced in the early modern period, the time of their great Western renaissance.

In the late eighteenth century, then, China and, a fortiori, Japan—but not the West—were civilizations where slavery was absent or on the wane. Buddhism, whose influence was greatest in Southeast Asia, dissociated itself from slavery more strongly than either Islam or mainstream Christianity, although formal abolition was decided upon only in the nineteenth century. When Siam, after decades of rolling back servile labor and extreme forms of social stigmatization, passed a first abolitionist decree in 1874 and lifted the few remaining exemptions in 1908, it did so less in response to direct Western pressure than as the result of a Buddhist revival centered on the exemplary life of the Buddha. This was supported by a determination on the part of the monarchy to strengthen its newly emerging image of modernity. In the early twentieth century, the enlightened absolutist monarchy that lasted until 1932 ensured that the old land of slavery acquired a new identity. The essence of modern Thailand was shaped precisely by its lack of extreme forms of coerced servility.[29]

Chain Reactions

Such comparative observations were rarely made in the West around the turn of the century. People generally had such a low opinion of Eastern societies, with the exception of Japan, that they were unwilling to perceive the major historical leap that had been achieved there. Another point that got lost amid the self-congratulation over the end of slavery was that it had not been brought about automatically through the march of progress, that it would not have advanced as far as it did if a sizable number of individuals had not been prepared to convert moral sensitivities into political action. There was a real struggle against slavery. Its opponents in Europe and America had to swallow many a setback, and the powerful interests supporting slavery meant that many of the victories were meager and precarious. It did not "die out" in the course of time, did not disappear because it became outdated. Its fate was bound up with the great convulsions of the age. Slavery suffered its main defeats not in peacetime but in the context of revolutions, civil wars, and sharp international rivalry.

In the late nineteenth century, the end of slavery at home provided Europeans and North Americans with fresh grounds to assert their civilizing missions. The "civilized world" appeared once more to have demonstrated its right to global leadership; it was possible—not without reason—to adopt an attitude of serene moral superiority, particularly in relation to the Islamic world, where slavery was not yet considered wrongful. In Africa, the European war on slavery even became a primary motive and justification for military intervention, enabling colonialism to present itself as being on the side of progress. Progressive imperialists, white abolitionists, and African American opponents of slavery joined forces to carry the battle to the African side of the Atlantic,[30] pushing into the interior to stamp out the slave trade and to destroy the political power of slave owners.

Slavery did not return to the lands colonized at the height of the imperialist age. Harsh forms of compulsory labor were certainly the rule, but none of the European overseas empires accepted the slave trade or inscribed slave status into colonial law. Whereas Europeans in the early modern period had sharply separated their legal systems at home from those in their foreign possessions, high imperialism brought about a unified jurisdiction at least in this special regard. Nowhere in the British, Dutch, or French empires was it permissible to sell, buy, or give away other human beings, or to subject them to serious physical cruelties without the sanction of the penal code.

The suppression of slavery and the slave trade developed as a transatlantic chain reaction, in which each local incidence acquired additional meaning from a broader context. British abolitionists saw themselves from the beginning as activists working for a global cause. After victory was achieved in their own territories, they sent delegations to various slave states and organized international congresses.[31] Opponents and supporters alike kept a close eye on what was happening around the world and tried to assess the changing balance of forces. The chain reaction was not without interruptions: the emancipation process was punctuated by long periods of stagnation, or even by revivals of slavery.

The historical location of the Haitian revolution was thus thoroughly double edged. On the one hand, a slave system was overthrown by revolution in the 1790s in the French colony of Saint-Domingue, which became the independent state of Haiti in 1804. Wherever slaves in the Atlantic area heard of it, the event operated as a signal for liberation. On the other hand, the outcome in the former sugar colony strengthened slavery elsewhere. French planters flocked from there to British Jamaica and the Spanish island of Cuba, contributing in each case to the consolidation of a slave economy. It was this inflow of capital and migrant energies that changed Cuba from a forgotten corner of the colonial world into a country with an export-oriented agribusiness.[32] Anyone there or in the southern United States who wanted arguments for avoiding any concession to restless slaves could find them in the fact that the looser grip during the years of the French Revolution had opened the way for militant protests among the slave population.

The pattern of making slave systems harsher as a reaction to emancipatory advances was repeated in the 1830s and 1840s. After a brief transitional period of dwindling quasi-slavery, emancipation became legally binding in 1838 in the British Caribbean colonies and South Africa, bringing freedom to 800,000 men, women, and children. This was state emancipation, not the result of a Haitian-style revolutionary war, but the economic and social consequences were similar in the British West Indies. With the dissolution of large plantation enterprises on islands such as Jamaica, Barbados, Trinidad, and Antigua, agriculture reverted to a pattern of small-scale subsistence farming and largely ceased to generate imperial wealth through exports. Monetary compensation flowed from the public purse into the hands of the planters, who often lived in England as absentee owners and failed to invest the money in the Caribbean. (In South Africa, similar compensation was to a large degree injected into the local economy, with vitalizing results.) For apologists of slavery, especially in the southern United States, all this confirmed that the supposedly moral progress of emancipation was more than outweighed by an economic regression that was harmful to everyone concerned. The experience of the British Caribbean hardened the resolve of plantation owners to prevent the same from happening elsewhere.[33]

Antislavery: A British Answer to the French Revolution

In the "Age of Reason," few Europeans took exception to the slave trade, which acquired growing importance in the eighteenth-century Atlantic area. Individual critical voices such as those of Montesquieu, the Abbé Raynal, or Condorcet could not disguise the fact that slavery seldom clashed with the moral sensitivities or even the natural law theories of the Enlightenment. Since it was almost exclusively a question of enslaving black Africans, a traditional European revulsion at all things black also came into play. Although Enlightenment thinkers still cherished the unity of the human race—and did not, like many theorists of the nineteenth century, seek to divide humanity into separate species defined by race—it was nevertheless a common view in early modern Europe that people with black skin were outsiders, more alien than Arabs or Jews.[34]

The humanitarianism motivating the founders of antislavery societies in the 1780s stemmed less from the high theory of the age than from two other sources: (a) a renewal of Christian ideas of brotherhood on the margins of established religion, and (b) a new patriotism that saw the superiority of a nation not only in its economic achievement or military strength but also in its ability to show the way for the rest of the world in law and morality. This combination was peculiar to Britain. More an attitude than an articulated theory, it initially fired only a small number of activists, including some former black slaves such as Olaudah Equiano (1745–97).[35] But it soon found a strong resonance in the British public, which indeed entered a new phase of its development as a result of the antislavery movement. Antislavery became a watchword that at its height rallied hundreds of thousands in nonviolent extraparliamentary action. In a political system in which the sovereignty of Parliament still lay with a tiny oligarchy, they donated money to support runaway slaves, attended mass events that reported on the horrors aboard Atlantic slave ships and on Caribbean plantations, and signed petitions to the lawmakers in Westminster. Consumer boycotts of Caribbean sugar kept up the pressure on slaveholder interests. Against this background, and following a series of detailed hearings, members of both Houses of Parliament voted in March 1807 to prohibit the slave trade on ships flying the British flag as of January 1, 1807. A similar decision had been thwarted in 1792, but at this second attempt it actually went through. The poet Samuel Taylor Coleridge voiced in 1808 what was in the minds of many: the conquests of Alexander and Napoleon looked "mean" in comparison with the triumph over the slave trade.[36]

Historians are agreed that this spectacular demise of a core imperial institution cannot be explained by economic factors alone.[37] The slave-based plantation economy had reached a peak of efficiency and profitability toward the end of the eighteenth century, some owners had amassed huge fortunes, and nothing in the national economy required change in the existing practices. Adam Smith's argument that free labor was more productive than forced labor was by no means the majority view among British economists. What tipped the scales were motives at the level of ideas, capable of inspiring sufficient members of the political elite who had no direct stake in the West Indies. Taken together, these may be seen as Britain's ideological response to the French Revolution and Napoleon.

Especially in its initial phase, before the Reign of Terror, the revolution had inscribed on its banner a universalist conception of humanity to which the mere affirmation of particular national interests was not a convincing response. There was little that conservative ideologues could marshal against the powerful Declaration of Human and Civil Rights, unless one defined an alternative field of transnational universalism. One such field was slavery. The revolutionary National Assembly in Paris, in which plantation interests carried considerable weight (as they did in the British Parliament), had engaged in petty delaying maneuvers. It is true that in 1794 the Convention finally prohibited slavery in all French possessions and extended citizenship to all male inhabitants of France and the colonies regardless of skin color. But in 1802 Bonaparte as first consul made both slavery and the slave trade legal again. Within a few years, therefore, France lost its position as opinion leader on this issue and reverted to the self-seeking habits of the ancien régime. In Britain, locked in struggle with Napoleon during the years before the Act of 1807, the patriotic public took the ideological initiative, relying on the fact that no other country in the world had *institutional* guarantees against (monarchical or revolutionary) arbitrary rule. These guarantees had only to be applied to the colonies.

Such political motives might easily be combined with individual reasons for action. Active support for the abolitionist cause made it possible for a huge number of male and female citizens to display their commitment ahead of the still pending democratization of the British political system, and to find relief from a burden that was increasingly experienced as collective guilt. The rhetoric of leading abolitionists was designed precisely to convey an identification with the victims, for which the way had been paved by sentimental novels of the eighteenth century and popular themes of liberation from tyranny (Beethoven's *Fidelio* dates from 1805).[38]

The major abolitionist literature mixed humanitarian-ethical appeals with arguments relating to the military and imperial interests of the nation;[39] the great global contest with France inevitably affected all areas of British politics. But this backdrop changed in 1815. The slave trade was sharply reduced as a result of Britain's withdrawal, and the Royal Navy, ruler of the waves, assumed the right to seize ships of other countries and to free any slaves they found in them without regard to formal ownership. This could not eliminate the trade altogether (there are grave doubts as to the effectiveness of naval police actions), but it did prevent others from filling the gap left by the British, albeit at the price of a number of diplomatic incidents (e.g., with France). Also in 1807 the US Congress forbade the involvement of its citizens in the African slave trade, effectively making it illegal to import any more slaves.

The moral impetus of abolitionism was strong enough to ensure that, even in the later times of intensified imperialism, a horror of slavery remained alive in the British public. Antislavery continued to be a watchword capable of mobilizing people. Thus, when it was discovered in 1901 that the chocolate firm Cadbury's—to the disgust of its Quaker founders –had been using cocoa beans produced with slave labor on the Portuguese Atlantic island of São Tomé, humanitarian groups started a fierce campaign against both Cadbury and the Portuguese government, eventually forcing the Foreign Office to take the matter up diplomatically.[40]

India: Abolition in a Caste Society

Slavery in the British Empire unraveled in several different ways. In the Caribbean, abolition weakened the plantation economy, but British planters received compensation for their losses. In South Africa, the whites (Afrikaners, but also British) whose agriculture rested on the exploitation of slaves—notably in wheat and wine production—experienced the new law as a direct assault. The Great Boer Trek to the interior, which began in the mid-1830s, was not least a response to the new humanitarian rhetoric of the twenties, to the undermining of patriarchal authority by egalitarian legislation, and to the liberalization of labor relations in the Cape of Good Hope.[41] In the multifarious social landscape of India, abolition was gradually enforced only from the early 1840s on. Here, unlike in the Caribbean, there was not a single, clearly structured system of slavery; the boundary between chattel slavery and other forms was hard to define, and fine gradations of servitude existed in the legal codes and customary laws of various communities. There was domestic slavery and agricultural forced labor; women were sold for sexual services and children given away to strangers in times of famine; the bondage of insolvent debtors often bordered on slavery, especially if parents passed their debt on to the next generation.

In such conditions, British and Indian reformers had to proceed cautiously and with regionally differentiated strategies. In Muslim parts of the country, where slavery had deep roots, the ruling elites were not challenged in too provocative a manner, while Hindu areas raised the difficult problem of establishing the point at which the subjugation of lower castes could be described as a form of slavery. The situation was not always as clear-cut as in Kerala, where members of the lower castes in the early nineteenth century could be legally bought and sold, pledged as security, or even killed by their master with impunity. (Indian society today is still quite receptive to debt bondage and forms of child labor akin to slavery.) On the whole, however, a long process of emancipation got under way there before the middle of the century. The year 1843 was a legal turning point, since from then on the courts in India refused to enforce claims based on a debtor's ostensible slave status.[42] Many who left India in later decades to work as contract laborers did so to escape the even harsher conditions of a slavery that was in only slow retreat.

French and Dutch Abolition

Despite strong British pressure, France took its time over abolition. Until 1848 governments were content to placate London by paying it lip service, and a humanitarian abolitionist movement found little support among the French public. During the Restoration period (1815–30), the colonial administration acted in close concert with planters' interests. In the Caribbean, the slavery system of the ancien régime was revived in a weaker form, and on the Indian Ocean sugar island of Réunion (at exactly the same time as in Spanish Cuba) a plantation economy was actively built up. Before the free-trade era, moreover, it remained possible to reserve the French sugar market for colonial produce. Paradoxically, it was the government of Charles X, a particularly reactionary regime even by the standards of Restoration Europe, which signed a bilateral trade agreement with Haiti in 1825 and set a European precedent by recognizing the breakaway black republic in return for exorbitant compensation to dispossessed French landowners.[43] The July Monarchy, which replaced Bourbon rule in 1830, ended the secret slave trade in the French colonies that had so antagonized Britain, keeping the planters on a shorter leash and looking more to the political model of contemporary Britain than to the past of the ancien régime.

Yet it was only during the 1848 Revolution that a small group led by Victor Schoelcher (a businessman's son who had seen the poverty of Caribbean slaves at first hand in 1829–30) successfully campaigned for the legal suppression of slavery. This breakthrough was due to the fact that aside from narrow interest groups, the institution of slavery had fewer and fewer supporters. Many prominent intellectuals, from Tocqueville to Lamartine to Victor Hugo, had championed the abolitionist cause in the 1840s, and the new republican regime could bring the situation in the colonies under control only by subduing the planter elites. The republic and the subsequent monarchy of Napoleon III cast themselves in the role of well-meaning patriarchal overlords ruling dark-skinned colonial subjects.[44] France never had a broadly based mass movement against slavery that remained active over a long period.

In 1863 the Netherlands became the last Western European country to abolish slavery in its American possessions, above all in Surinam. Here too, there was a transitional period of quasi-slavery, which lasted until 1873. And, as in the cases of Britain and France (but unlike in the United States or Brazil), slave owners were compensated out of the public purse; the funds were covered directly by income from the Netherlands East Indies, which in the middle decades of the century sharply increased under the so-called cultivation system. Indonesian forced laborers thus paid for the liberation of Caribbean slaves.[45]

The abolition of slavery in the colonies was a delayed domino effect, in a way already triggered by the Haitian Revolution that sent shock waves through the world of Western slavery. After Britain's opening move, no Western European country that wanted to be seen as civilized could long afford to remain outside the ongoing dynamic. Russia's elimination of serf status in 1861 should also be seen as part of this European trend; it was largely a state-driven project, in which neither peasant revolts nor a public movement in favor of free labor conditions played much of a role. In the eyes of Tsar Alexander II and his advisers, serfdom was a blot on the international reputation of the Empire, and one that stood in the way of social modernization.

The End of Slavery in the United States

Things ran a very different course in the United States.[46] Nowhere were the foundations of slaveholder society more stable than in the Southern states, whose very population, despite the lack of new imports from Africa, continued to grow by leaps and bounds. On the eve of the Civil War, four million slaves were living in the United States; twenty years earlier, in 1840, there had been no more than 2.5 million.[47] These people lived in the same society with whites, whereas the British and Continental abolitionists who took up the cause of slaves in their own country or across the ocean rarely had any actual contact with Africans. The abolitionist movement campaigned in the slave-free North but had no chance of catching on in the South. What developed in the South in the years before the Civil War was increasingly a laager mentality that brooked no opposition to the system. Even the nonslaveholding majority of whites identified as voters with a propagandistic image of the South and helped to sustain social relations from which they did not themselves profit directly; after all, the life of a big plantation owner often corresponded to their vision of an ideal existence. Northern abolitionists—who launched a militant crusade, with strong female and Afro-American involvement, only in the 1830s—succeeded in mobilizing a smaller base than British opponents of slavery had had at various peaks before 1833. They also fought in a more difficult situation, since the North and South of the United States had much closer ties with each other than Britain had with its distant sugar colonies. Besides, racism was much more pervasive in Northern society than in early-nineteenth-century Britain and reflected the influence of the North's own past history of slavery.

Confronted with an elaborate apologia for slavery, North American abolitionists drew upon more radical religious sources than their British counterparts, and appeared more fanatical within the ideological spectrum of the United States. Many believed, with an almost obsessional conviction, that the practice of slavery, or indeed its craven tolerance, was a sin worthy of punishment. Since their primary aim was often more to eradicate the evil of slavery than to integrate blacks into American society, proposals to solve the problem by repatriating emancipated slaves to Africa found approval well beyond the limited abolitionist circles; it was quite possible for criticism of slavery to go hand in hand with rejection of a dark-skinned presence in America.[48]

However, radicals grouped around the publicist William Lloyd Garrison would have nothing to do with such plans.[49] In the North too, abolitionists encountered stronger resistance than ever existed in Britain; it might involve physical attacks on their persons and stocks of literature, but also a conspiracy of silence. After the Missouri crisis of 1819 to 1821 there was a secret agreement among leading political forces to make slavery a taboo subject, and between 1836 and 1844 an actual "gag rule" prohibited any treatment of the subject in Congress. For a long time, then, the United States lacked the political will that in

Britain associated the slave question with changes in voting law and resulted in the great reform package of 1832–33.

The struggle of white and black abolitionists would not by itself have brought about the Civil War, and without the Civil War the "peculiar institution" would probably have held on a while longer. The North did not wage war *directly* to end slavery. Lincoln's Emancipation Proclamation of January 1, 1863—the most important turning point in nineteenth-century African American history—also had the pragmatic intent of mobilizing black resistance for the Northern military effort. Lincoln himself had not originally been an abolitionist, believing that blacks would not be able to live with equal rights in a majority white society, but he publicly changed his position in the mid-fifties. On the other hand, he had long held the conviction that everyone should harvest the fruits of their labor and be free to develop as individuals. When he became president, Lincoln first acted cautiously with regard to the goal of full emancipation, but he then took the leap with characteristic determination.[50]

Thus, although abolition in the United States came as a by-product of war rather than as the culmination of public campaigns or a steadily growing openness among top political leaders toward decisive reforms, the problem of slavery was at the origin of the Civil War. It was the main reason why the weak institutions of central government—presidency, Congress, Supreme Court—could no longer hold in check the centrifugal tendencies asserting themselves in the regions. The pros and cons of extending slavery into the new territories in the West was the fulcrum of US domestic politics between 1820 and 1860. The revolutionary process in the Americas—which included the revolution in Haiti, the decree of 1794 (later repealed in 1802) abolishing slavery in all French colonies, and the British Slave Trade Act of 1807—had weakened the foundations of slavery, while at almost the same time the Deep South of the United States became the new center of gravity of the plantation economy.

Notwithstanding the principle of the indissolubility of the Federal Constitution, this might well have led to the coexistence of two differently constituted sovereign countries on the territory of the (former) United States, if slave-owning Southerners had not exercised a dominant influence in national politics all the way from George Washington to the election of Abraham Lincoln. The laager mentality of the antebellum South therefore went together with attempts to force the North to abandon its own moral principles. In 1850 Congress passed a Fugitive Slave Law permitting the federal authorities to pursue runaway slaves into free states and to return them by force to their owners in the South. There were several other provocations of a similar kind, which made it clear that the normative unity of the country—an emotive myth cultivated since the earliest days—was breaking down. The mounting tensions finally resulted in the outbreak of the Civil War, each side declaring itself blameless.

Both made a strong pitch to the outside world: Britain, despite many conflicts, remained the chief political and cultural reference for the industrializing

North, while the Southern elites had greater affinities with slave owners in Brazil and Cuba. Lincoln's Emancipation Proclamation also had an impact overseas, by aligning British public opinion more clearly with the North. Lord Palmerston's government, partly under pressure from Napoleon III, had previously been toying with the idea of intervening in the Civil War, but the proclamation rekindled the moral energies of the abolitionist movement thirty years on and led to mass demonstrations that prevented Whitehall from siding with the Confederacy.[51]

Brazil, Compared with the United States

In Haiti and the United States, the ending of slavery was attended with violence on a large scale. The process was more peaceful in other Caribbean colonies and the new republics of Hispanic America (where abolition laws were mostly enacted in the 1850s), as was the emancipation of the serfs in the Tsarist Empire in 1861. Cuba and Brazil, too, saw relatively low levels of violence. It is still debated among historians whether slavery was less brutal for those concerned in Latin America than in the Caribbean and the United States. The fact that toward the end of the system, the mortality rate of slaves in Brazil was distinctly higher than in the Southern US states would seem to suggest the opposite.

Strong forces supported slavery both in politically independent Brazil and in the Spanish colony of Cuba—otherwise it would not have lasted until 1886 and 1888, respectively. In Brazil, slave owners still defended their property with gun in hand at the beginning of the 1880s; slave resistance, overt or covert, never slackened, but the repression was strong enough to head off major revolts. In Cuba, which was still receiving shipments of slaves in the mid-sixties, the question of emancipation became caught up in the wider program of the independence war of 1868 to 1878. The war ended in failure, but both Creole rebels and Spanish rulers courted slaves to fight on their side and offered them the prospect of freedom.[52]

In contrast to Cuba and elsewhere in Hispanic America, Brazil had a slave system that was one of the two largest in the Western world (together with the Southern US states). Its abolition did not come suddenly, as in the United States, but through a long decline in significance involving numerous acts of manumission and culminating in the "Golden Law" of 1888, the last spectacular official action of the monarchy under the princess regent, Isabel. In 1831, under British pressure, new legislation had promised freedom to all human beings recently imported as slaves, but smugglers went unpunished and largely circumvented its provisions; in the forties there was even a rise in the import trade. More effective measures were introduced in 1850, partly out of fear that slave ships would bring cholera with them. The price of slaves then increased, while the numbers set free declined.

Manumission had always been much simpler in Brazil than in the United States, and the chances of gaining freedom therefore higher. So long as the slave trade and smuggling ensured fresh supplies from Africa, it made economic sense for Brazilian slave owners to allow the freeing of slaves and to use the proceeds of their sale to fill the resulting gaps. In the United States this was no longer an option after 1808, because of the ending of imports and the sharp rise in slave prices—and the same now happened in Brazil. Around the middle of the century, albeit with wide regional variations, slavery began to lose its importance for the Brazilian economy, first in the cities, then on the sugar plantations (increasingly equipped with modern steam mills and run by British investors using free labor), and finally in the coffee-producing areas. Unlike in the Southern United States, slavery was no longer an economically vigorous institution at the time when it was abolished. It had been hanging on mainly in less productive, technologically backward sectors and in regions with poor transportation links.

The relatively peaceful road to abolition in Brazil may be attributed to three factors. First, the main economic center of gravity shifted from the sugar-producing Northeast to the expanding coffee zones in the South, where immigrants from southern Europe played a growing role. Slavery held its ground there, amid an insatiable labor market, but the advantages of free immigrant labor were ever more apparent to entrepreneurs. Second, Brazil was the only part of the Americas south of California where abolitionism (from the 1860s on) spread beyond circles of intellectuals, liberal politicians, and capitalist employers to gain a mass following among the urban middle classes and free workers (immigrants as well as locally born men and women). Third, despite the formation of centers of gravity, the slave population was distributed across the regions of Brazil. The slave question was therefore not a source of polarization between free and unfree zones, as it had been in the US drift toward secession and civil war.[53]

To a greater extent than in the Caribbean and the Southern United States, slavery ceased to be a suitable form of labor in its final decades. Nevertheless, it did not simply evaporate. In a constitutional monarchy, its abolition without compensation (as in the United States, though not in the British Caribbean) became the object of political power struggles, in which the victors were those who saw a slave-free republic as the prerequisite for a modern nation-state, especially one oriented to the Anglo-Saxon countries.[54] In the end, the decision was made under the impact of a kind of mass strike, as slaves deserted the plantations by the thousands.

The long history of the rise and fall of slavery and the "second serfdom" in the civilization of "the West" therefore came to a final conclusion in the 1880s. Later it would be Eurasia, but not the Western hemisphere, that witnessed the camps and exterminatory worlds of the National Socialists and Soviet or Chinese Communists (Hitler's armaments minister, Albert Speer, spoke after 1945 of an SS "slave state")—a phenomenon ultimately worse than classical African slavery, as it rested not upon *trade* in human beings but upon a system in which labor was more a by-product of organized repression than a reason for its exercise.[55] In the liberal-capitalist West itself, however, not even extreme reactionaries or antihumanists thought of making slavery once more a "normal" social institution.

Emancipation in the Muslim world?

Developments were different in the Muslim world. Slavery, traditionally approved of, underwent a decline in the eighteenth century. In the middle third of the nineteenth, however, rising demand for exports led again in many places to a sharp upward trend in the employment of slaves, typically in small-scale production rather than on large plantations. In the Egyptian cotton boom of the 1860s and 1870s, resulting from the interruption of American supplies to European factories during the Civil War, even ordinary farmers reached the point of employing black African slaves. At the same time, the state's need for forced labor and slave troops kept growing, while servile concubines from the Black Sea area became a status symbol among wide circles in Egypt. Slaves were also a common sight in the homes and fields of Anatolia, Iraq, and Muslim parts of India.

Production slavery probably declined in the Islamic countries from the 1880s on.[56] But the views of intellectual opinion leaders, especially of clerics schooled in law, and the value system of society as a whole did not display the same decisive turn against slavery that had taken place in the West. Proposals for its sudden abolition almost never became politically influential. One early exception, though, was Ahmad al-Husain, the bey of Tunis, who in 1846 (two years before France) became the first ruler in Muslim history to lay the basis for the end of slavery, combining personal convictions with an attempt to gain British respect and to deny France any pretext to intervene from across the border in Algeria. On a visit to France in late 1846, he enjoyed being feted by French liberals as a civilized champion of liberty.[57]

But Ahmad remained an exception from the rule. In the Ottoman Empire, such liberals as Grand Vizier Midhat Pasha were unable to prevail for long. Sultan Abdülhamid II moved only very reluctantly against the slave trade from Africa and the Caucasus, and he did not end the old practice of harem bondage; in 1903 there were still 194 eunuchs and nearly 500 women in his own seraglio.[58] Only after the Young Turk Revolution of 1908 did slavery start to decline sharply, although after 1915 part of the surviving Armenian community knew a fate akin to that of slaves. In Egypt, the khedive Ismail, so open to the West in other respects, was the country's largest slave owner, and it was only the British occupation after 1882 that put an end to all forms of slavery. Iran, like other Middle Eastern countries, signed the Brussels Convention against the slave trade as early as 1890, but only its radical secular modernization *à la turque* under Reza Shah led to the prohibition of slavery itself in 1928–29.[59]

Hampered by dissension among the many local schools of law, abolition in Muslim parts of the world was a more gradual, less dramatic process than in the West. Not all moves to end slavery there should be attributed to Western pressure. There was an indigenous basis for its rejection, including in certain readings of the Koran, but before the First World War this seldom led to vigorous action on the part of nation-states.

Passages from Slavery

What came after slavery? Ideally the moment of liberation, symbolized by the springing of chains, should have been carried over into legal-political systems and social structures that safeguarded the new freedom. Such systems and structures could be created with the help of former slaves, but not by them alone; the framework of the national or colonial state also had to be transformed. Mentalities had to change—from contempt, or at best condescending sympathy, to a real preparedness to recognize ex-slaves not only as abstract human beings but as neighbors, citizens and useful members of society. Such a liberal utopia was virtually never realized in the nineteenth century. Suspecting that this would be the case, a number of early abolitionists played down local successes and set themselves more demanding goals based on the idea of a global civilizing mission. The world, it seemed, would be safe from a relapse into barbarism only if slavery was everywhere torn up by the roots. Particularly active in this regard was the African Civilization Society, founded in London in 1840 and supported by a large part of the Victorian establishment, including Prince Albert and several dozen members of Parliament. One of its first actions, in 1841–42, was to send an antislavery mission to the Niger region in West Africa. This ambitious nonimperialist venture, which encountered numerous difficulties and proved unable to achieve its lofty objectives, was a remarkable expression of the sense of mission that sometimes drove opponents of slavery in the early nineteenth century.[60]

The Niger mission—like the later trips to Africa by the missionary David Livingstone—was rooted in Christian, humanitarian, and patriotic impulses. But these played little role in the construction of systems after the ending of slavery, when problem solving was consistently local in character and involved few international transfers. Besides, the multiplicity of development paths makes comparison especially difficult here.[61] Microhistorical investigations have focused on life destinies that can be reliably documented, on the conversion of plantations into mosaics of more or less independent small farms, or on the processes (barely discernible to those involved) whereby bonds of slavery passed into relations of compulsion bearing different names and having a different status in the eyes of the law. The general term used today is "postemancipation societies"[62]—which differ from one another in respect of the number and percentage of former slaves in the total population, the type and intensity of racism in society, the employment and promotion opportunities, the prevalence of violence, and gender disparities in life: in short, "degrees of freedom."[63]

The plantation economy was not destroyed everywhere. In Haiti it disappeared along with export production. There was a similar, though less dramatic, development in Jamaica (which remained a British colony). In Trinidad, plantation output was restored after a few decades, although the workforce now mainly consisted not of local ex-slaves but of indentured laborers from Asia; much the same happened on the Indian Ocean island of Mauritius, also under British rule. Cuba, which ended slavery eighty years after Haiti, took yet another path: changes in sugar technology and white immigration from Spain smoothed the transition, so that output declined only slightly after emancipation and was back above previous levels within a few years.[64] These changes operated within an agrarian framework. For the time being, even in the Southern US states, large-scale industrialization did not occur in the aftermath of emancipation.

The outcomes were interpreted by affected groups on a case-by-case basis. The interests of former slaves were different from those of former slaveholders; colonial governments and abolitionists each nursed expectations of their own. Slave emancipation—one of the most ambitious reform projects of the nineteenth century—was associated with unusually high and widespread disappointment. Sometimes this was hypocritical: the same colonial regimes that complained of the difficulty of eradicating indigenous slavery in Africa had few scruples about establishing new forms of bondage, whether compulsory labor in all its guises (the corvée was prohibited in the French Empire only in 1946), fiscal exactions, or direct intervention in agriculture. Only rarely, however, did these crystallize into stable structures of extreme subjugation. Under pressure from unrest abroad and public criticism at home, the European colonial systems were capable of substantial self-correction. Ultracoercive labor regimes and excesses of violence were therefore much less common after the First World War than they had been before. It would be wrong to underestimate the deep moral and political break that the abolition of legalized slavery represented wherever it came about. By 1910, with minor exceptions, the eradication of slavery had been achieved throughout sub-Saharan Africa.[65]

Postemancipation Society in the American South

In no other country did the abolition of slavery expand the scope for action as dramatically as it did in the United States. During the Civil War hundreds of thousands of African Americans took their fate in their hands, fighting on the Union side or otherwise assisting it as free blacks from the North or runaway slaves from the South, and taking possession of land in the South that had been left without an owner. At the time of the Emancipation Proclamation, a great uprising of black Americans was already under way.[66] In the transition to freedom, former slaves gave themselves new names, moved into new homes, brought their scattered families together again, and looked for ways of becoming economically independent. Those whom a master had previously denied free speech could now openly express themselves in public; black community institutions that had been operating underground—from churches and schools to burial societies—found their way to the surface. As slaves, black women and men had been their master's property and therefore not legal subjects in their own right. Now they could step out into the world, give testimony in court, conclude mutually binding contracts, sit on juries, cast their vote at elections, and stand for office.[67]

But then this great new start turned into its opposite: into sharp racial discrimination. By the end of the 1870s the gains of the emancipation period had been largely obliterated; and in the 1880s, race relations in the former slave states of the South took a dramatic turn for the worse. True, African Americans were not slaves again after 1890, but they were subject to an extremely restrictive racial system that went hand in hand with white terror and lynch law. For blacks, there could no longer be any talk of exercising their civil rights. There would be only three instances of such a harsh racial order outside the context of slavery: in the American South between the 1890s and the 1920s; in South Africa after 1948; and in Germany after 1933 and German-occupied Europe during the Second World War. If we leave aside the case of Germany, there are rough similarities between the United States and South Africa, whose apartheid system had roots stretching far back into the nineteenth century.[68] In 1903 W.E.B. Du Bois, the leading, and universally respected, African American intellectual of his time, opened his "electrifying manifesto"[69] *The Souls of Black Folk* with the prediction that the "color line" would be *the* problem of the twentieth century—not just in the United States, but on a worldwide scale.[70] In places where this prognosis proved most accurate, slavery was replaced by white supremacy, and state or nonstate violence enforced privileges for groups defined by nothing other than their skin color.

Whereas hierarchical relations in slave societies had rested on the evident fact that manual work was performed almost exclusively by slaves and freedmen, and that neither of these two groups had much chance of social advancement, the situation in postemancipation society was that ex-slaves competed directly with poor whites on the labor market. Under conditions of political freedom, blacks defended their own interests and did not allow themselves to be downgraded into acolytes of white leaders. Parts of white society responded to this dual challenge by means of discrimination and violent hostility. Racism was a premise of such thinking and structures, and they in turn gave it additional force. Thus, a racist ostracism built on white supremacy appeared in place of the repressive racism of slave society. It had already been a common attitude in the Northern states of the United States that gave up legalized slavery during the revolutionary period; now it became more widespread and more radical in the New South of the late nineteenth century, undermining the Fourteenth Amendment of the Constitution that declared anyone "born or naturalized in the United States" to be a citizen with equal rights before the law. The laws in individual states did incorporate this formulation, and the departure of the last federal troops from the South removed the protection that the less racist central government might have given to the black minority there. The new racial order in the South, symbolized by the growing activity of the Ku Klux Klan from 1869 on, reached its height of virulence around the turn of the century, then grew milder in the 1920s and was finally toppled by the civil rights movement of the sixties.[71]

South Africa, United States, Brazil: Racial Orders

Although the differences with South Africa are too great to permit a comprehensive comparison, some instructive cross-references may be found here and there. Developments in the two countries, including a few influential transfers, did not occur synchronously: South Africa emancipated its slaves nearly three decades before the United States, but by 1914 the ideologies and instruments of racist hierarchy and exclusion were present in both. Then, in the 1920s, South Africa once more took the lead by making apartheid a basic principle of national legislation, so that the race system could be removed there only through regime change at the center (as happened in 1994), not through the kind of "gradualist" changes to the law and the justice system that occurred in the United States after the Second World War. In both countries, black civil rights movements supported by white liberals played a very important role. In both too, the racial order of the early twentieth century had old historical roots. Free-labor ideologies—represented in one case by the industrialized North, in the other by the British presence at the Cape—came into conflict with the preference of Afrikaners or Southern plantation oligarchs for race-based subordination and the political power monopoly of a purely white master people. The secessionist wars of 1861–65 in the United States and 1899–1902 in South Africa went in favor of liberal-capitalist forces: militarily overwhelming in the United States, only just so in South Africa. But within a decade and a half in the Southern United States and roughly half that time in South Africa, the two white camps reached a compromise with each other at the expense of the black population.

In 1910 the British Empire granted autonomy to the white settlers in South Africa. In a process of "national" reunification actually limited to whites, the black majority was denied rights that it had previously enjoyed or been promised. In the United States, after the end of Reconstruction in 1877, the North failed to prevent the Southern states from depriving blacks of their rights and putting a color bar in place. In the North, for all the discrimination it faced in everyday life, the black population never ceased in principle to have access to the ballot box. Legalized discrimination thus remained a local or regional peculiarity, not a national norm.[72] The humanitarianism that in both countries drove the process leading to the abolition of slavery—at first in South Africa via impulses imported from Britain—had disappeared from their politics by the early twentieth century. The struggle against white supremacy dragged on for many decades. Colorblind democracy, having asserted itself as a political program in the middle decades of the nineteenth century, had suffered a setback that could be reversed with only the greatest difficulty in both South Africa and the American South.

In the other major nineteenth-century instance of slavery on a mass scale, white supremacy was not the sequel to its abolition. There are several reasons why slavery persisted longer in Brazil than anywhere else in Latin America: not the least important is the fact that Brazilians did not fight a war of independence against the colonial power, so that, unlike neighboring countries in the struggle against the Spanish, they did not have to recruit any black soldiers. There were repeated cases of black resistance to slavery, but nothing comparable to the black African armed with British approval who could demand something in return for his services. One of the key sources of "modern" politicization was therefore lacking in Brazil. But why did no formal racial order develop there after 1888? After the end of slavery, which coincided with a peaceful transition from monarchy to republic, a long debate began over the country's national and racial identity and its opportunities for modernization. Since manumission had been easier in Brazil than in the United States, and miscegenation had been less severely dealt with, there was not such a strict overlap between skin color and social status, and people generally were less inclined to sharp dichotomies. Freed slaves thus found a place earlier in the conceptions of modernization entertained by sectors of the white elite.

Even more important was the strategy of replacing slaves in dynamic sectors of the economy with newly recruited immigrants from Europe. The ex-slaves found themselves economically marginalized, inhabiting a different labor market from that of the new immigrants, so that the fierce competition that typically fueled racism elsewhere in the world was a factor of minor significance. Nor did race ever become a contentious issue in regional politics in Brazil; no special areas defined themselves in terms of a racial identity that suggested secession as a solution to their problems, as the South did in the United States. Indeed, the elite took pains to preach an inclusive nationalism and a myth that the older slave system had been exceptionally humanitarian. This made it possible to construct the national history as a continuum stretching from colonial times to monarchy to the republic.

The material position of blacks in postemancipation Brazil was in no way better than in Alabama or South Africa; the state simply did not concern itself with them. There was no equivalent of Reconstruction, but also no backlash in the shape of official apartheid; the authorities did not think it was their job to uphold racial barriers. If much racist violence went unpunished, this was not because it directly emanated from the state but because the state itself was too weak. The abolitionists were incapable of influencing the social order after emancipation.[73] Meanwhile in Cuba, whites and blacks fought together against the Spanish in the war of independence, and the workforce in the sugar economy had a wider mix of skin colors. After the end of slavery, politics was therefore more "colorblind" than in other postemancipation societies, especially the United States. White racial supremacy did not assert itself on the island.[74]

All processes that led to the abolition of slavery in the West had one thing in common apart from Christian and humanitarian aspects: namely, a liberal hope that under free market conditions ex-slaves would respond to positive incentives and work as productively as before in export agriculture. Economists and politicians saw emancipation as a great experiment. Former slaves would have a chance to prove their "rationality" (their human worth by the standards of an enlightened age) by behaving in the manner of the *Homo oeconomicus* of liberal theory, oriented to hard work, profit, and accumulation. An organized transition from slavery to freedom (often, as in the British Empire, conceived as an apprenticeship) was supposed to facilitate this for them. The granting of full civil and political rights would then crown this development of a "moral personality."[75]

The reality often looked different. Freed slaves tended to behave in unexpected ways, preferring the security of their own small plot of land to wage labor in a large enterprise, or opting for some combination of the two. The result was a reduced market orientation in comparison with the age of plantations producing for export. The reformers experienced another disappointment when many ex-slaves failed to aspire to bourgeois ideals of family life. The two together seemed to demonstrate that, because of anthropological peculiarities, black Africans were unable to cope either with market rationality or with civilized norms of personal conduct. Although this was not a *cause* of racism, it did strengthen racist tendencies. The great experiment of emancipation left largely unfulfilled the illusory, self-serving hopes of its liberal protagonists.[76]

3   Antiforeignism and "Race War"

The Rise and Fall of Virulent Racism

In 1900 the word "race" was in common usage in many languages around the world. The global climate of opinion was saturated with racism.[77] At least in the global "West," to be found in every continent in the age of imperialism, few doubted that mankind was divided into races with different biologically determined capacities, and that therefore they did not all have the same right to shape their own existence. Around 1800, although practices in the colonies and the transatlantic slave trade were based on differences in skin color, such ideas were mainly being developed in European academic circles. By 1880 they were a basic part of the collective imaginary in Western societies. Fifty years on, racism was already a touch less acceptable around the world.

In the "white" West, prosperous African Americans with a bourgeois appearance still found it difficult to find a hotel room, but academics at least tended to be less uncritical in dealing with the concept of "race." Japan's attempt at the Paris Peace Conference in 1919 to have a clause against racial discrimination written into the charter of the newly founded League of Nations failed mainly because of resistance by the British dominions and the United States, but it showed the extent to which racist discourse and practices were by then subject to challenge.[78] After 1933, the racist rhetoric and actions of the German National Socialists caused greater consternation in international public opinion than they would have done around the turn of the century, though they were often negligently played down abroad as a German "quirk." By 2014 racism is discredited throughout the world, its propagation is a punishable offense in many countries, and any claim to scientific credentials are laughed out of court. The rise and fall of racism as a force capable of shaping history occupied the relatively short period between 1860 and 1945. Its macabre cycle spans the nineteenth and twentieth centuries.

Race was a central issue in 1900 not only in countries where the "white man" (as he was now known) formed the majority of the population. Ruling white minorities in the colonies worried over the threat to their supremacy from subject "inferior" races, and in Japan or China groups of intellectuals were appropriating the vocabulary of European racial theories. "Race" was taken seriously as a scientific concept. Biologists and ethnologists especially liked to talk about it. But in neighboring disciplines too, a *Volk* or "people" referred increasingly to the common biological descent of an *ethos*, and less than in earlier decades to the political community of a *demos*. Such discourse did not leave sections of the political Left untouched; there was even a socialist variant of eugenics, a theory for planning healthy heredity, which claimed to serve the advent of an ideal society of equals.

But racial thinking was essentially situated on the political Right. It contradicted Enlightenment ideas such as the natural equality of human beings, their inborn rights, and their striving for freedom, peace, and happiness. Racial thinking tended to be collectivist rather than individualist; terms such as the German *Volk* or *völkisch* became its most important semantic bearers, even if there was not a complete identity between theories based on race and *Volk*. "Social Darwinist" conceptions of race war and the inevitable subjugation of the weakest were part of the picture. In fact, whites could end up the losers as well: some of the early racial theorists had pessimistic inclinations, and many colonial practitioners waited in a mood of imperial melancholy for the white man to be ground down by the rigors of tropical life.[79]

Racial thinking cultivated certain aversions and hate objects: Jews and coloreds, democrats, socialists, and feminists. Heads of state, scholars, and street mobs, who otherwise had nothing in common, were united in their racist prejudices. The main imagery was of bodies and physicality: people spoke of threats to "the national body" from enemies and pests. The old physiognomy of the eighteenth century reappeared in theories suggesting that the body expressed racial "inferiority" or a criminal disposition. Racial thinking caused, made possible, or facilitated genocide in the Congo Free State, German South-West Africa, and Amazonia; anti-Jewish pogroms in the Tsarist Empire; and sadistic lynchings or attacks on ethnically alien immigrants in the Southern United States. Aggression and fear were usually closely associated with each other; simple race hatred was never the only, and seldom the principal, source of such acts of violence. Homicidal masses and college professors who would never harm a fly found themselves spontaneous accomplices in the business of fabricating "purity" of the race and the nation. So it was that a brief period of virulent racism began around 1870. It paved the way for the German mass murder of European Jewry—without making it inevitable, since further elements of extremism had yet to appear after the First World War.

Race Theories: Prerevolutionary and Postrevolutionary

Racism is extremely difficult to break down into types and to classify. Looking only at the strategies proposed and the practices implemented we may distinguish between four variants along with their different consequences:

1. *repressive* racism leading to the formation of politically and economically deprived underclasses
2. *segregatory* racism culminating in the establishment of formal or informal ghettos
3. *exclusionary* racism fostering suspicion of the outside world and aiming at closely patrolled borders of the nation-state
4. *exterminatory* racism stigmatizing specific groups as "racial enemies" and persecuting them to the point of systematic annihilation

The arguments and narratives associated with race were of different kinds. The picture would also have to include a whole series of transnational connections. Just as, in the decades around 1900, race was the Western intellectual's favorite category in building macropictures of the relations among states and nations, national racisms reacted to one another and thinkers who believed in the "breedability" of man were especially inclined to join forces across frontiers.[80]

As an extreme form of ethnocentrism, which sees the chief distinction among human groups not in changeable modes of cultural behavior but in immutable, biologically inherited physical properties, racism came into being during the early modern period, when contacts between societies became more intense across the globe. But it was not the dominant worldview among Europeans, not even among seafarers and colonial conquerors, until well into the nineteenth century. Any quotation from an early modern travel report that may be read as a disparaging remark on non-European human groups is more than outweighed by expressions of respect and admiration; travelers were more interested in the morals and customs of other peoples than in any phenotype.

Racist attitudes and stereotypes, but not yet elaborate racial theories, developed in the various milieux of the Atlantic slave trade, the American plantations, and the immigrant societies of the Western hemisphere where perceived color differentials served to construct social hierarchies. The first extensive apologia in racist language for the institution of slavery, based on references to the anthropology of the age, was *The History of Jamaica* (1774) by the planter Edward Long. Racism was not the cause of slavery, but in the late eighteenth century and especially the first half of the nineteenth, it increasingly served to justify it.[81] At many frontiers of European expansion, differences between settlers and indigenous people were still being given a cultural rather than a biological interpretation.

In general, the relationship between slavery and imputed racial characteristics is flexible. Numerous slave systems in history did not rest ultimately upon physical differences. Slavery in Greco-Roman antiquity and military servitude in the Ottoman Empire (where recruits were supplied from the Balkans or the Black Sea area) are two good examples of this. Even in North and South America, there were slaves lighter in color than many of their European owners and guards.[82]

In the last quarter of the eighteenth century, classification and comparison became fashionable scientific methods among European intellectuals. Proposals were made to divide mankind into "types," and comparative anatomy and phrenology (cranial measurement as a pointer to intelligence) gave such approaches a veneer of credibility according to the standards of the time. Some authors, consciously spurning the Christian doctrine of the Creation, went so far as to postulate the separate origins of various races (polygenesis) and hence to question the basic affinity between whites and blacks emphasized by abolitionist movements. Until the mid-twentieth century, racial classification remained a pet activity for many anatomists and anthropologists, while colonial administrators tried using it to bring order into the motley variety of their subjects.

Like phrenology, this diversity was a popular theme throughout the nineteenth century, regularly presented in visual displays at world's fairs and special exhibitions. Some of the categories developed before 1800 clung on stubbornly: "the yellow race," "Negro," or "Caucasian" (the latter going back to the Göttingen scientist Johann Friedrich Blumenbach and still employed today in the United States as a euphemism for "white"). Classificatory systems led to endless confusion, especially since the English word "race" was also used to refer to nations, as in "the Spanish race," and so on. By the late 1880s, the number of races distinguished in the US literature alone varied between 2 and 63.[83] There is no straight line leading from Blumenbach or Kant to the exterminatory racism of the past century. At worst, late Enlightenment taxonomies and early attempts to rank racial types or subspecies of humanity could serve to justify a repressive, exploitative racism, but not one with murderous intent. Nor could they legitimate a demand for the segregationist color bars that were characteristic of racism after 1900, but much less significant in colonial practice before the 1850s or thereabouts. Late-nineteenth-century racism was not an uninterrupted continuation of eighteenth-century developments.

The racial theories of the nineteenth century were postrevolutionary. They presupposed a loosening of the ties of Christendom but, above all, a world in which hierarchies were no longer seen as part of a divine or natural order. They emerged less in the largest colonial power (Britain) than in France or the United States. British political thought has never been emphatically egalitarian, so that the tension between the theoretical promise of equality and the unequal reality on the ground was never felt as strongly as in the lands of the Declaration of Independence and the Déclaration des Droits de l'Homme et du Citoyen. After roughly 1815, racial theories of a new type became possible. The first premise for this was a farewell to the idea that environmental influences could lastingly shape human nature, even in its phenotypical variations.[84] The idea of "betterment" dropped out of racial thinking, to stage a comeback only in the last third of the century as eugenic biotechnology. Concepts of race thus began to clash with the idea of a civilizing mission. The second premise was a claim—much more sweeping than any advanced by late Enlightenment naturalists—that race was a central category in the philosophy of history, a universal key to understanding both past and present, in direct competition with such terms as "class," "state," "religion," or "national spirit."

A striking feature of such racial thinking—as Alexis de Tocqueville was one of the first to recognize—was its strong propensity to determinism and hence to the marginalization of politics and any active shaping of history.[85] Only after 1815, and particularly after the revolutions of 1848 caused intense disquiet among conservatives, did race-based universal theories or—to put it more critically—closed systems of delusion come to the fore. Two authors played a leading role in this. In 1850, the Scottish doctor Robert Knox published a collection of lectures, *The Races of Men*, with the aim of alerting readers to the racial backdrop of political conflicts in Europe at that time.[86] His influence, certainly sizable, was surpassed by the impact of the *Essai sur l'inégalité des races humaines* (1853–55), whose author, the French count Arthur de Gobineau, was obsessed with the dangers of racial mixing. The two men were only early and prominent representatives of a Euro-American racial discourse that quickly gathered momentum after the middle of the century. Natural scientists had never abandoned the theme, although one of their greatest figures, Alexander von Humboldt, remained an uncompromising opponent of all racial thinking. Later, the revolutionizing of biology and anthropology by Charles Darwin and his followers again changed the parameters of the debate.[87]

German scholars and writers figured rather little among the international champions of racist thought after the Age of Revolution. In a new situation, where the principal dynamic was no longer one of revolution and counterrevolution but of national self-assertion in a Europe changed by the upheavals between 1789 and 1815, some followed the philosopher Johann Gottlieb Fichte (*Speeches to the German Nation*, 1807–8) in seeking an ethnic unity for the German nation, which for the time being could not be constituted through political action. Inspired by a new historical interest in origins (those of the Roman state, for example, in the emerging field of ancient history), they pursued fantasies about "Teutonic" roots of the elusive German nation.[88] In fact, *germanisch* was an enigmatic cultural-biological hybrid category, later capable of being interpreted in a number of different ways. In the hands of Romantic nationalists, it served to prove the superiority of their own nation over its eastern (Slav), western, and southern neighbors, and ultimately also over the cultural models of ancient Greece and Rome. Even in England, never fertile ground for extreme racist ideas, writers sought to derive the present day not from medieval Norman principles of community and law but from germs among the pagan Anglo-Saxons. In the age of slowly spreading industrialization, it was not only the "Germanic" European countries that began to study and imaginatively re-create the pre-Christian beginnings of their nationhood. New national epics came into being, such as the Finnish *Kalevala* (final version 1849) that the doctor and song collector Elias Lönnrot put together as a verse mosaic from original sources.

Almost the whole of Europe (though not Finland) became fascinated with a theory of its "Indo-Germanic" or "Aryan" origins, which initially had more to do with common linguistic roots than with biological links, and whose success was based on a deceptively simple opposition between Aryan and Semitic. This conceptual antinomy, dignified by scientific credentials, was taken over later in the century by anti-Semites, who used it to exclude non-Aryan Jews from the European cultural community. But the myth of Aryanhood provoked others to contradict it. The British, for example, were far from enthusiastic about the view that they were related to Indians, especially after the Great Rebellion made them inclined to see India as completely "other."[89] Not all racial thinking was antinomic or binary. There were people who racked their brains over shades of skin color and "mixed blood" percentages, or drew up gradations between noble (for the British: manly or martial) and nonnoble "savages."[90] In any event, racism meant thinking in terms of differences, both coarse and fine.

Dominant Racism and Its Opponents

From the 1850s on, it is possible to speak of a *dominant* racism. Though very unevenly distributed through the Western world and its colonies, it was never absent there and underlay a picture of the world that was one of the most influential of the age. From a penchant of outsiders and minorities it became a classificatory schema that marked the perception of cultural and political elites; the emergent mass electorate could be won over to it in special cases. It seemed natural to look down on "inferior races" with at best well-meaning condescension. Extreme expressions of racism, such as had been unthinkable in 1820 and would have caused a scandal in 1960, could be voiced with impunity. The production of racially skewed worldviews reached a peak in Richard Wagner's son-in-law, the British writer Houston Stewart Chamberlain, whose German-language work *The Foundations of the Nineteenth Century* (1899) was an instant bestseller in Europe and a major source for Nazi racial ideology.[91] Austrian racist circles in particular, following the lead of Gobineau, became increasingly puffed up with talk of race and blood. International politics, too, could be explained in terms of a "race war"—a fateful conflict between "Germanic" and "Slavic" peoples for the influential Pan-German League. A "yellow peril" seemed to threaten from Asia in the shape of cheap Chinese laborers and Japanese marching columns.[92]

There were certainly individuals who escaped what David Brion Davis calls the "official racism in Western culture."[93] In a dramatic intervention on the Jamaican Morant Bay scandal in 1865, John Stuart Mill spoke out against the racist polemic of his fellow intellectual Thomas Carlyle.[94] Others registered doubts about the idea that modern civilization stemmed from Germanic or "Aryan" roots. W.E.B. Du Bois and the German-born Franz Boas (one of the founders of ethnology and cultural anthropology) waged decades-long campaigns against pseudoscientific racism,[95] while Rudolf Virchow combated it with the authority of a great natural scientist. The new discipline of sociology, represented by Émile Durkheim, Max Weber, Georg Simmel, and Vilfredo Pareto, also stood from the beginning in opposition to the zeitgeist, refusing to accept any biological or genetic factors in its explanations. Some sociologists in this pioneering generation did invoke race—for example, the Austrian Ludwig Gumplowicz—but their work led down an academic blind alley. After the First World War, racial classifications began to lose their scientific respectability, at first in Britain and the United States.[96]

The State, Immigration Policy, and Racism

Another feature of the dominant racism after the 1860s was the leading role of the state. Older racisms had had the character of personal attitudes, but now there was a built-in tendency to seek the *realization* of a racial order. This required the help of the state: or, in other words, racists struggled to capture state power. They succeeded mainly in the Southern United States, in Nazi Germany (although Fascist Italy and Japan between 1931 and 1945 showed similar trends, they cannot be described as fully fledged racial states), and in the former settler colony of South Africa. The European colonies were not really racial states: they did not make *official* racism a guiding ideological and practical principle; the general rule was that colonial subjects (most of whom paid taxes) might not be worth as much as whites but should nevertheless be treated "decently."

What was new in the last third of the nineteenth century was that national governments and, in a weaker form, empires saw it as their task to safeguard cultural homogeneity and ethnic purity within their borders. This happened in various ways and with varying degrees of intensity. Free movement across borders had become more widespread in the first two-thirds of the century, except for members of the lower classes. Many requirements to carry identity documents disappeared.[97] But this trend went into reverse toward the end of the century, as passports and passport controls erected a paper wall of differing heights around nation-states.

Britain remained a liberal exception. Until the First World War, citizens of the United Kingdom had no identity documents; they could leave their country without a passport or official approval and convert their money straightforwardly into foreign currencies. Conversely, foreigners were not prevented from entering Britain; they could spend their life there without having to register with the police. Nor were passport formalities usually necessary for travel between colonies of a single empire. In continental Europe, sharper dividing lines were drawn between citizens and aliens toward the end of the century. Entry, residence, citizenship, and naturalization became subject to legal regulation and administrative processing—an expression not so much of growing racism as of the widening scope of state activity and increased migration flows.[98] The internal consolidation of nation-states meant that the question of membership in the majority "state nation" had to be posed more energetically. The reintroduction of protective tariffs on the Continent in the late 1870s showed how governments were capable of regulating cross-border flows in the case of material goods. As for persons, the issue was who should be kept out as undesirable, and who should be placed where on a scale of "naturalization worthiness."

In many parts of Europe toward the end of the century, there was a growing tendency to regard aliens with mistrust or even animosity. However, nation-states by no means shut the door completely, and racial criteria for inclusion did not gain the upper hand. This was true not only of Britain; the French Third Republic, permeated by high patriotic sentiment, placed few obstacles in the way of immigration, partly because its unusually low demographic growth engendered a certain mood of crisis. Waves of foreign workers came into the country from midcentury on, gradually developing into ethnic communities with a high propensity to assimilation. Xenophobic campaigns were never able to have a significant impact on national legislation. France had great faith in the integrative power of its language, its educational system, and its armed forces.[99] In the German Reich too, where much stronger forces on the Right were agitating for a racial concept of the nation and, in the years before the First World War, stirred panic over the influx of Poles and Jews from the East, the nation-state did not become a "racial state" in its immigration policy. A major overhaul of citizenship legislation in 1913 did not evince a Reichstag majority in favor of biological conceptions of race. Nor was it agreed to incorporate into the law of the land such colonial administrative practices as the obstruction of "racial intermarriage."[100]

Racist Protectionism

It was not in Europe but in the democratic societies of North America and Australasia that a political majority was secured for racial protectionism.[101] This was directed mainly against Asians. Chinese had migrated for various reasons to the United States: as gold prospectors to California, as railroad workers, and as plantation coolies to Hawaii. Many of them later drifted into the cities, working as cooks or launderers and living together in their own communities. Although they were initially welcomed as hard workers, white Americans later turned against them and demanded a halt to immigration from Asia. In a language that had much in common with the attacks on postemancipation African Americans, the Chinese were increasingly branded as "half-civilized" people incapable of fitting into their American surroundings. Leaders of labor unions feared their presence would depress wages. Disgust over prostitution became the pretext for limiting the influx of Chinese women and thereby curbing the growth of the Asian population in the United States. California, in particular, witnessed pogrom-like incidents that resulted in deaths and injuries. Finally, in 1882, supporters of a federal ban had their way; Congress passed a Chinese Exclusion Act that virtually banned immigration from China for an initial ten-year period. This proved to be the first in a long series of measures that followed until the ending of the exclusion policy in 1943.[102] Even more bitter were the attacks on Japanese, who in many cases had come to the United States not as coolies but in response to the emigration policies of their own government. They were also more active than Chinese in sectors of the economy where they competed directly with whites and encountered especially strong resistance.

As in the American West, Asian emigration to Australia from the 1880s on became the trigger for labor union mobilizations and a burning issue in election campaigns. A "swamping" hysteria took on such proportions there that the book market produced a special genre peddling fantasies of an imminent invasion.[103] Asians already living in the country were better treated than in the United States, enjoying a degree of state protection and many civil rights. But official support for a white Australia was even stronger than comparable tendencies in America. For a whole century—from the 1860s to the 1960s—the Australian colonies and then the federation pursued a policy of hindering immigration of nonwhites. Its rational kernel was a wish to prevent the formation of a nonwhite underclass, but the justifications acquired an ever shriller racist tone, so that any further immigration became extremely difficult from 1901 on.[104] In 1910, Canada switched to a white Canada policy. In 1903 Paraguay adopted a highly restrictive law on immigration, and in 1897 the colony of Natal in South Africa tried to prohibit the influx of Indians, ostensibly to the advantage of the African population.

This Pacific exclusionism, concentrated on the West Coast in the case of the United States, was the most drastic concretization of global racism (along with racial discrimination in the American South and various colonial practices) around the turn of the century. Behind it lay ideas of white superiority and of a need to protect its valuable substance from alien hordes. A further problem in the United States was that the Anglo-Irish-German majority among the population was challenged by new arrivals from southern and eastern Europe whom established citizens regarded with suspicion. This gave rise to endless debate on gradations of skin color and cultural competence.[105] A contradiction in America's perception of itself, still visible today, became apparent for the first time. The United States—which sees itself as superior in every respect and therefore as a savior for the peoples of the world—also has a pervasive fear of being infected and ruined by those same peoples.[106]

Non-Western Racism: China

Of course, according to the conceptions of the time, every sovereign nation-state had the right to decide who lived within its borders. If thousands took to the streets in China to protest against strict US immigration policies, one reason why they did so was that China had no way of giving tit for tat. In 1860 the country had been forced to allow free entry to nonnationals. It therefore had abundant grounds for a restrictive attitude to foreigners, but not for racist protectionism. There was no ethnic minority that had previously been enslaved. Small numbers of Jews had for centuries been well-integrated subjects of the emperor, and no Chinese Judeophobia existed to fuel anti-Semitism. And yet there too, it is not difficult to find a "discourse of race."[107] China thus exemplifies the fact that racism was not limited to the West in the nineteenth century. Racial prejudices, which in a world marked by postcolonial guilt feelings are seen as a special defect of white Westerners, can certainly be identified in non-Western civilizations. The traditional weakness of racial prejudice in China makes the nineteenth-century experience there all the more interesting.

Imperial China knew all manner of "barbarian" stereotypes and recorded physical peculiarities of the most diverse peoples at its frontiers. Without exception, however, the barbarian was considered a culturally deficient being through no fault of his own, and therefore as a candidate for benevolent civilizing. The path from a culturally to a biologically alien status was blocked in traditional Chinese thinking. This changed in the late nineteenth century, as a result of new contacts with the West. The greater physical and cultural foreignness of Europeans and North Americans (in comparison with neighboring Asiatic peoples with which the Chinese had had dealings over the millennia), as well as their unusually aggressive behavior, were the reasons why elements of an ancient religious demonology were now grafted onto older images of barbarians. There was talk of foreign devils (*yang guizi*) and red-haired barbarians (*hongmaofan*), for example. This negative stereotyping applied indirectly to Africans too, although scarcely any Chinese had an opportunity to meet a visitor from Africa. It was comforting to some to think that other victims of European imperialism stood even lower in the eyes of the colonial masters.

China's growing acquaintance with Western racial theories toward the end of the nineteenth century was one condition for the development of Chinese racism; the other was the catastrophic military defeat at the hands of Japan in 1895, the last nail in the coffin for a Sinocentric view of the world. In their question for an alternative conception of China's place in the international order, a number of leading intellectuals were attracted by the vision of a struggle between the races (*zong*) and eagerly began to assemble the kind of ranking tables that had existed in Europe for hundreds of years. Africans inevitably found themselves in bottom place, reproducing the worst "white" prejudices toward them. The "yellow race"—a term that did the rounds until the end of a temporary Sino-Japanese rapprochement in 1915—was by no means permanently inferior to the white; rather, the two were locked in a struggle for world supremacy. Such notions, found in Europe at the Rightist end of the political spectrum, were characteristic of reform currents in turn-of-the-century China. Political liberalization and social modernization were supposed to serve the purpose of steeling China for the coming battle between races—an objective that would require overthrowing the

Qing dynasty. The fact that the imperial house was formed by a non-Han ethnic group had not featured prominently in earlier criticisms of the Qing political order, but new racial theories made the Manchu appear as an inferior alien race against which all means were justified. During the revolution of 1911, threats on the part of literary pamphletists gave way to massacres not only of defeated Manchu troops but also of their families—although not everywhere in the country, and not as a strategic aim of the revolutionaries.[108]

A further racial theme was the conversion of the ancestral figure of the Yellow Emperor from a mythical cult hero into a biological precursor of the "Chinese race"—although this never acquired the same significance as in Japan, where parallel genealogical moves created one of the main pillars of the emperor cult from the Meiji period on. The Chinese example shows that European racial thinking could not be easily introduced into societies that had not developed something similar of their own, and that it did not spontaneously find its way there. Particular groups outside Europe, mostly small circles of intellectuals, first had to become familiar with such theories and then recast them for their own ends. Discourses of race became internationally mobile only when they were formulated in the universalist idiom of (natural) science, acquiring an aura of robust objectivity. Such mobility presupposed, in turn, the special climate of opinion that existed at the turn of the century, when even black Americans campaigning for civil rights and incipient pan-Africanists automatically thought in categories of racial difference and invoked the unity of the "negro race" in support of their political projects.

4 Anti-Semitism

Jewish Emancipation

The prototypical outsiders in European societies of the early modern period had been the Jews. Their history in the nineteenth century can be narrated and explained in various ways, with the necessary distinctions of time and place. One possible perspective is that of civilization and exclusion. The nineteenth century was a time of successes without precedent in the history of the Jewish religious community. Between roughly 1770 and 1870, as the great historian Jacob Katz showed, the Jewish communities of Western Europe experienced deeper changes in their whole way of life than any other population group of comparable size: it was a transmutation of "the very nature of their entire social existence," in short, a "social revolution."[109] In this period an Enlightenment reform movement among the Jews, beginning with Moses Mendelssohn and some of his younger contemporaries in the 1770s, radically transformed the Jewish understanding of religion, community practices, cultural relations with the non-Jewish world, and attitudes toward social changes in Europe. This self-reform, seen by many of its protagonists as a self-civilizing process, brought measured adjustment to the surrounding world, while also preserving a core Jewish identity.

It led to emancipation, to an improved or even equal position of Jews in the eyes of the law, since enlightened liberal forces in West European governments supported such aspirations of their own accord. Especially in Germany and France, emancipation was seen as a state-led process for "civilizing" and integrating Jews. This congruence of internal and external impulses placed growing numbers of people of the Jewish faith in a position where they could profit from the new economic opportunities in a modernizing Europe.[110] The ghetto walls behind which Jews had lived until then came down everywhere to the west of the Tsarist Empire. Career paths opened up in business and the liberal professions, although access to the civil service remained much more difficult for a long time. An active, successful minority in the rising European bourgeoisies belonged to the Jewish faith. Benjamin Disraeli, a Jew baptized in childhood, went on to become prime minister of the foremost world power and the Earl of Beaconsfield. His older contemporary, the financier and philanthropist Sir Moses Montefiore, has been described as "one of the first truly global celebrities."[111] Men of Jewish origin, some of them baptized Christians, rose to leading positions in the cultural life of the continent: Felix Mendelssohn-Bartholdy was a composer, pianist, and conductor of European reknown; Giacomo Meyerbeer dominated opera stages between Rossini's falling silent and Verdi's rise to preeminence; Jacques Offenbach created the art form of the satirical operetta and brought it to its highest point.

An old hostility, mainly based on religion, did not disappear overnight. Even prominent artists ran up against aversion and rejection. Poor Jews in the country were the most vulnerable. There continued to be attacks on Jews. But in Germany, for example, these died down after the first third of the century. Never before had Jews in Western Europe felt as safe as they did in the middle decades. They were no longer, like early modern "court Jews," under the personal protection of whimsical princes but under the protection of the law.

The Rise of Anti-Semitism

After 1870, anti-Jewish polemics began to regain momentum almost everywhere in Europe. Enemies of the Jews went onto the offensive.[112] In France and Germany, the old theological image of Jews was not discarded but supplemented with secular-rationalist arguments. Accusations that Jews were both protagonists and profiteers of a disconcerting modernity escalated into full-blown conspiracy theories; nationalist reproaches of disloyalty compounded prejudices concerning the supposed moral inferiority of Jews. Under the impact of new biological thinking, Jews were increasingly constructed as a "race" apart. Those who thought and wrote along such lines implied that Jewish assimilation was no more than a maneuver, that individual conversion to Christianity had no significance, that Jews would never change. Before the First World War, however, the racist aspect was not dominant among the numerous facets of European anti-Semitism.

It was not just a question of books and pamphlets by intellectuals such as Richard Wagner (whose *Jewishness in Music*, first published in 1850, really made an impact only in a second and more vociferous edition of 1869). Anti-Semitic associations and political parties also came into existence. Accusations of ritual murder gathered fresh momentum, especially in rural areas, having been on the wane for decades. In France, Britain, Italy, and Germany, Jews still did not have to fear for their lives or property; more typical were the insults and rejection that one routinely came across in certain German spas, for example, which advertised themselves as *judenfrei*. But anti-Semitism also met with social and political resistance. In Germany it was more virulent in the late 1870s than a decade later, while in France it suffered a major setback at the end of the century in the Dreyfus Affair, when the Left and the bourgeois Center successfully exposed a military plot driven by hatred of the Jews.[113]

Anti-Semitic agitation also intensified in Austria and Hungary, where it followed the German example but mainly reflected local circumstances. It was more violent in the Tsarist Empire than anywhere else. A majority of European Jewry lived in its Polish part and faced a particularly contradictory situation there. On the one hand, a large number of eastern European Jews had not been affected by Reform Judaism and—except in Austrian Galicia—received no help from an emancipation-minded government. The tsars had even practiced a discrimination bordering on apartheid, and the material position of the *Ostjuden* was in most cases quite desperate. On the other hand, the Tsarist Empire housed some very successful Jewish entrepreneurs who corresponded to the clichéd figure of the "plutocrat," and Jews were also prominent in the leadership of the newly emerging revolutionary groups. This made eastern Europe fertile ground for a rabid anti-Semitism more social and antimodernist than biological-racist in its foundations. In several waves of pogroms, especially those of 1881–84 and 1903–6, a considerable number of Jews lost their lives (more than three thousand in the disastrous year 1905 alone) or were injured or deprived of their property. These mainly urban riots had a spontaneous form, but they were usually covered up, or at least not punished, by the authorities. They triggered hasty emigration and a belief that (eastern) European Jews had to create a homeland of their own in Palestine: that is, Zionism. The key text of that movement, *Der Judenstaat* (1896), was written by the Austrian journalist and foreign correspondent Theodor Herzl, though mainly under the impact of the Dreyfus Affair and anti-Semitic disturbances in France.

At the end of the nineteenth century, the West of the Tsarist Empire was the most dangerous area in the world for Jews. Anti-Semitism there was not simply copied from Germany or Austria but had a real ideological autonomy. The years 1902–3 saw the appearance of an ominous document, *The Protocols of the Elders of Zion*, which conjured up plans for Jewish world domination. They were later shown to be a forgery, but especially after the First World War this product of Russia's highly paranoid anti-Semitism aroused discussion all around the world.[114] Two readers who helped it on its way were the Austro-German rabble-rouser Adolf Hitler and the American car tycoon Henry Ford—by no means the only anti-Semite in a country where social discrimination against Jews was widespread and physical violence not uncommon.[115]

There is no simple explanation for the simultaneous, but far from uniform, rise of anti-Semitism in the countries of Europe. Anyone who ventured to predict in 1910 where mass murder of the Jews would begin thirty years later would probably have named Russia, Romania, or even France, putting well-ordered Germany much further down the list.[116] The various anti-Semitisms were primarily shaped by their national contexts. Judeophobic discourse initially surfaced in a series of national public spaces, resonating differently according to the economic, social, and political circumstances of the country in question. But there was also a supranational level: older conceptions of race had developed in what counted as the international public of the time; individual experiences of "race relations" during trips abroad or in émigré communities were sometimes transferred to other settings; and academic eugenicists or "racial hygienists" organized internationally. To be sure, such "transnationality" had its limits. Sometimes anti-Semitism was more of a subnational, local phenomenon. In 1900, for example, it played a major role in Vienna (which had had an anti-Semitic mayor, Karl Lueger, since 1897) but not necessarily in other Austrian cities.

Continental Europe as a Special Case

Anti-Semitism was found where there were Jews. But a Jewish presence did not lead automatically to anti-Semitic reactions—in the late Ottoman Empire, for instance. Nowhere in the Muslim Orient did Jews face anything comparable to Europe's rising tide of religiously motivated anti-Semitism. Indeed, until the First World War, they enjoyed the protection of the Ottoman state, which for its part regarded them as pillars of support. The real danger for Jews was Christian anti-Semitism, which in the nineteenth century nearly always made itself felt as soon as Ottoman rule was rolled back: in Serbia, Greece, Bulgaria, and Romania. There anti-Jewish and anti-Muslim violence ratcheted up in close parallel. Jews in the new Balkan states were exposed to persecution by Christian neighbors, the authorities, and the church (above all the Orthodox Church). In many cases they had been integrated into the financial and commercial networks of the Ottoman ecumene, so that, if a region broke away and constituted itself as a separate peasant state, that section of the Jewish population might be threatened in its economic existence. Many Jews from the Balkans found refuge in the territories of the sultan—when they did not emigrate to France, Palestine, or the United States.[117] The attacks on Jews in the post-Ottoman Balkans did not remain hidden from the international public. At the Congress of Berlin in 1878, the Great Powers dictated to the Balkan states a number of clauses offering protection to non-Christian minorities. Since no major power was prepared to use force to defend Jews in a faraway land, such threats never went beyond declarations on paper. Yet, for the first time, new international legal instruments for the protection of minorities made it possible to conceive of limiting national sovereignty in the name of human rights—an anticipation of the future.[118]

Anti-Semitism in the form it took between 1870 and 1945 was peculiar to continental Europe, where in 1900 four-fifths of the world's 10.6 million Jews lived.[119] In Britain, which had few Jews (36,000 in 1858; 60,000 in 1880) as compared with 462,000 in the German lands in 1852 (and 587,000 in 1900), members of the Jewish faith who were unable to swear a Christian oath nevertheless enjoyed full civil rights from 1846 on—a few decades later than in France, the pioneer of Jewish emancipation. In 1858 they also won the right to stand for Parliament—later than in France but earlier than in Germany, where full legal emancipation came only in 1871 with the founding of the German Empire.

In Britain there had never been a "Jewish question" in the Continental sense. English law in the early modern period did not discriminate against Jews as aliens or force them to live in ghettos, and it imposed on them only such restrictions as applied also to Christian nonmembers of the Church of England, mainly Catholics and Protestant Nonconformists. At the beginning of the nineteenth century, Jews were British citizens, albeit with unequal rights in some respects. Emancipation was therefore not, as in Germany, a long process involving state-led integration of a distinctive minority into civil society, but rather a constitutional act at the level of the central state that extended to Jews the equality of rights that had earlier been granted to Catholics.[120] Against this background, no articulated or organized anti-Semitism along German or French lines developed in the British Isles before 1914, and the same was true in principle of the British settler regions and overseas offshoots.

Did anti-Semitism also have more remote effects? In Japan, with its tendency to follow European crazes, there was an imitative anti-Semitism without a physical presence of Jews. The *Protocols of the Elders of Zion*, translated in 1924, strengthened conspiracy fears and fueled a xenophobic nationalism that small circles in the country had long been cultivating. Jews appeared as accomplices of a West that was supposedly challenging Japan's right to existence.[121] In China there was an opposite reaction. The translation of Shakespeare's *The Merchant of Venice* in 1904 first acquainted people there with a European stereotype, but the Jewish Shylock was sympathetically regarded as a suffering victim inviting global solidarity among the oppressed. Phantom anti-Semitism *à la japonaise* failed to emerge in China.

Anti-Semitism and Racial Orders

It would be too shallow to interpret post-1870 European anti-Semitism as a direct application of race doctrines. Some of the early racial theorists had already lined up Jews in their sights: Robert Knox in 1850, for example, had described them as culturally sterile parasites.[122] Other founders of racist discourse, such as Gobineau, could not be described as anti-Semites. The basic ideas of biological racism were applied to American blacks much earlier than to Jews in Germany or France.[123] Before the First World War, the arguments used in support of anti-Semitism were not mainly racist—and, insofar as they were, they represented a consequence rather than a form of racial theory.[124] For anti-Semitism to sink roots in society, there first had to be a crisis potential and a political fallout from democratization and the quest for national identities.[125]

The anti-Semitism of the long nineteenth century did not take material shape in racial orders—on the contrary. The premodern segregation of Jews in ghettos was abolished, and no new formal apartheid appeared in its place. The Jews of Europe, at least outside the Tsarist Empire, no longer lived beneath the Damoclean sword of expulsion. The concentration of their communities into a huge "Pale of Settlement" between the Baltic and the Black Sea—a measure decreed by Catherine the Great in 1791 and reinforced in the Jewish Statute of 1804—was the most important fetter on Jewish mobility during this period.[126] Jews in the Tsarist Empire, like other non-Orthodox groups, were denied equal civil rights. Many discriminatory regulations remained in force; others were relaxed or revoked in the reform period under Alexander II. After the assassination of the Reform Tsar in 1881, Jews saw their legal position deteriorate once more and were unable to achieve civil emancipation along French or German lines until the Revolution of 1917. In 1880 Romania was the only other country in Europe where, despite pressure at the Congress of Berlin in 1878, Jews continued to live under degrading special laws.[127] The last major Jewish ghettos were wound up after the middle of the century: in 1852 in Prague, in 1870 in Rome.

To the west of Poland, anti-Semitism was a postemancipation phenomenon, much like the aggression against blacks in the postbellum American South. It forms part of the context of intensified demarcation between those who "belong" and those do not, national majorities and migrant or cosmopolitan minorities. By 1900 a unified racist vocabulary could be mobilized to justify these highly disparate cases of exclusion in program and praxis. This by no means necessarily led to imperial outcomes. It was in the logic of radical racism (and can already be found in Robert Knox) to avoid imperial rule as it necessarily involved close contact with ethnic Others. Before the German war of extermination in eastern Europe after 1941, there had been no case in history of an imperialism or colonialism that had sought to rule over other peoples *in order to* suppress or annihilate them on racist grounds; colonialist programs had always, in one way or another, had some constructive tones. The civilizing mission was a stronger impetus for colonial expansion in the nineteenth century. And conversely, it was extreme racists who advocated sending black Americans back to Africa or, at a later date, deporting European Jews to Madagascar. In 1848, plans to annex even larger parts of Mexico failed because of fears of ethnic swamping, and until the late 1890s the threat of contamination by "inferior races" meant that white-supremacist ideology curbed rather than encouraged further possible territorial expansion.[128] Also the early promise of independence to the Philippines—a unique case in colonial history—was not given only for philanthropic reasons. Some of its advocates were mainly concerned to separate the United States as quickly as possible from its "racially alien" colony.[129]

Two Emancipations in Peril

Comparison between the emancipations in North America and Central Europe may be taken a little further, drawing on the work of George W. Fredrickson.[130] The abolition of slavery and the liberation of the great majority of European Jews from a ghettoized underdog existence required help from outside: in the first case from abolitionists, in the second from enlightened representatives of the upper state bureaucracy. Common to both was a conception of reform as a civilizing mission: Afro-Americans were to be "raised up," Jews to be "improved" in their cultural level, while maintaining a proper social distance from the dominant majority.

In the United States, the end of the Civil War finally offered an opportunity to implement this program under the auspices of "Radical Reconstruction." Integration of the Jewish minority into American society took place under unevenly favorable conditions. During the interval when the old hatred of Jews had abated and modern anti-Semitism had not yet emerged, ideological hostility remained at a relatively low level. It was certainly far from comparable to the racism that affected all blacks, including "free" African Americans in the North, and grew more intense after the end of Reconstruction in 1877, coinciding almost to the year with the new rise of anti-Semitic discourse in France or Germany and anti-Semitic pogroms in the Tsarist Empire. On both sides of the Atlantic, the international economic crisis after 1873 and the decline of liberal forces in the domestic politics of at least the United States, Germany (following Bismarck's break with the Liberals), and the Tsarist Empire were aggravating factors. Jews, just as blacks, were robbed of important allies.

Jewish minorities in the nation-states of Europe found themselves in a more vulnerable position than African Americans in the United States. It is true that many had come to occupy respectable and respected positions in the business world and in public intellectual life, but this very success made them objects of greater resentment among the majority population than African Americans experienced in their almost invariably lower place in the social hierarchy. In the view of white supremacists, "Negroes" had only to be deprived of rights and subjected to intimidation; open struggle was not necessary to contain them. It was easier to establish whether someone belonged to the group of African Americans, especially because of the taboo on cross-color sexuality and the persecution inflicted on those who violated it.

This strict insistence on the purity of the white race found its way into European anti-Semitism, with a delay of a few decades. Since Jews could not be identified by their appearance, pseudoscientific elements of "racial biology" came into play, much more elaborate than the criterion of skin color routinely applied in the United States. Finally, contacts between the African American diaspora and colonized Africa did not seem sufficiently threatening to make whites fear some harm to the national interest, whereas the multifarious international links among Jewish communities provided fuel for national-populist conspiracy fantasies about Jewish capital and the Jewish world revolution. In both Germany and the United States, a majority of the population directed its antipathy against those who contradicted common visions of the national character. African Americans were *not sufficiently* modern in a society obsessed with modernity, while Jews appeared *too* modern in the eyes of mainstream German society.[131] When the immigration of "caftan Jews" from eastern Europe, with premodern "Oriental" habits, increased around the turn of the century, the two stereotypes merged into one.

African Americans in the South saw a turn for the worse in their situation barely a decade after Abraham Lincoln's Emancipation Proclamation, whereas on the whole the newly unified Kaiserreich offered to German Jews physical security and *relatively* good opportunities for advancement. Omens of a new era in the history of European Jewry appeared immediately after the end of the First World War. In 1919–20, during the Civil War in Russia and Ukraine, counter-revolutionary "white" troops and militias engaged in the mass murder of Jews, often regarding them en bloc as Bolshevik sympathizers. These killings were not simply a fresh wave of pogroms but, in their scale and their sadism, went far beyond what was familiar from the period before 1914. The unleashing of a destructive soldiery on whole Jewish communities had been a rare exception in the nineteenth century.[132] In the 1920s, when the position of African Americans had begun slowly to improve, an exterminatory anti-Semitism was brewing also in Germany and parts of east-central Europe (especially Romania), having until then been limited to isolated rhetorical threats without any support from the state. There was a line from pre-1914 anti-Semitism to the post-1933 *Judenpolitik* of the Nazis, but not a direct and untwisted one.[133]

CHAPTER XVIII

Religion

There are strong reasons why religions and religiosity should occupy center stage in a global history of the nineteenth century.[1] Only for a few Western European countries at most would it be justified to treat religion as one more subdivision of "culture" and to limit oneself to its organizational constitution as a church or churches. Religion was a force in people's lives throughout the nineteenth-century world, giving them bearings and serving to crystallize the formation of communities and collective identities. It was an organizing principle of social hierarchies, a driving force of political struggles, a field of demanding intellectual debates. In the nineteenth century, religion was still the most important provider of meaning for everyday life, and hence the center of all culture associated with the mind. It took in the whole spectrum from universal churches to local cults with few participants. It encompassed in a single cultural form, and often constituted the main link between, both literate elites and those illiterate masses who could communicate only through the spoken word and religious images. Only very exceptionally in the nineteenth century did religion become what sociological theory calls a functionally differentiated subsystem, alongside other systems such as law, politics, or the economy, and hence a reasonably distinct sphere with identifiable patterns of reproduction, renewal, and growth. The huge diversity of religious phenomena, and the great abundance of literature in disciplines from the history of religion to anthropology to Oriental philology, places any kind of comprehensive account beyond reach in this book. What follows is a rough sketch of a number of selected topics.

1 Concepts of Religion and the Religious

Vagueness and Disambiguation

Globally speaking, religious phenomena do not fit together into a single overarching history like those covering the macroprocesses of urbanization, industrialization, or the spread of literacy. The claim that the nineteenth century overall was an age beyond religion cannot be sustained, and a grand narrative other than the well-known one of "secularization" is nowhere in sight.[2] Another way of connecting things up also turns out to be a great oversimplification: No doubt the conquering and colonizing, traveling and proselytizing expansion of Europeans around the globe from the sixteenth century on created better conditions for the spread of the principal European religion, yet it seemed to keen observers in 1900 or 1914 that the influence of Christianity in the world was far slighter than Europe's political-military strength or that of the West as a whole. In many non-Western societies that were in regular contact with Europe during the nineteenth century, and in which a Westernization of lifestyles has persisted to this day, Christianity was unable to gain a real foothold. It became a global religion but was not globally dominant; the Christian offensive encountered resistance and renewal movements in its path. Religious change, however, must be seen not only as a conflictual process of expansion and reaction but also, under different circumstances, as a result of interrelations and a shared history, or as "analogous transformation" in the West and in other parts of the world, fueled by local sources and linked up only loosely or not at all.[3] Processes such as nation-state formation or mass distribution of printed matter stood in a mutual relationship with changes in the religious field that was in principle similar worldwide.

The concept of religion is notoriously hazy, and Max Weber, one of the pioneers of the comparative sociology of religion, never allowed himself to be drawn into defining it. Some old problems in this field have never been solved unambiguously, beginning with the distinction between "true" religion, "superstition," and inner-worldly (or "philosophical") belief systems. For instance, is Confucianism the "religion" that Western textbooks often claim it to be, even though it has no church, no conception of salvation or an afterlife, and no elaborate ritual obligations? And what of Freemasonry, an equally worldly organization? Should any cult and any religious movement be called a religion, or should the term apply only to worldviews, organizations, and ritual practices beyond a certain threshold of complexity? How important is the way in which its adepts and others see it? As conventional faiths lose support, under what conditions is it justified to speak of art or certain forms of ritualized politics as an ersatz religion? We should hesitate to follow those theorists who are interested only in discourses about religion and maintain that religious phenomena are not discernible in the reality of history. Such radical skepticism, reflecting the "linguistic turn" in the study of history, goes too far. Insight into the constructed character of concepts may then easily lead into a denial of their practical effect in people's lives. What does it mean for someone who cultivates a Hindu identity to say that "Hinduism" is a European invention? It would be problematic to conclude that because the concept of "religion" was developed in nineteenth-century Europe, the term is merely a hegemonic imposition on the part of an arrogant West.[4]

Even so, an abstract, universal concept of religion is a product of nineteenth-century European intellectuals, most of them with Protestant leanings.[5] It included the idea of a plurality of religions beyond the monotheistic trio of Christianity, Judaism, and Islam, but often rested on an unspoken assumption that Christianity, seen as the most advanced in terms of cultural evolution and spiritual authenticity, was the only truly universal religion. The concept combined at least four elements:[6]

1. the existence of a pivotal holy text (such as the Bible or Koran) or a clearly defined canon of sacred writings;
2. exclusivity, that is, an unambiguous religious loyalty and identification with a religion that people consider as their own spiritual possession;
3. separateness from other spheres of life; and
4. a certain detachment from charismatic leader-figures and from excessive personalization—even if such a detachment does not always lead to the founding of a hierarchically organized church.

Toward the end of the nineteenth century, this concept of religion percolated into non-Occidental cultural worlds, not only via colonial channels. It was by no means always unattractive on its own grounds. There was much to be said for re-interpreting, concentrating, and systematizing religious programs and practices, following the models of Christianity and, in a different way, Islam.

In China, for example, people had for centuries spoken only of *jiao*—roughly translatable as "doctrines" or "orientations," mostly with a plural sense. In the late nineteenth century, a wider concept was imported via Japan from the West and incorporated into the Chinese lexicon as *zongjiao* (the sign prefix *zong* denotes ancestor or clan, but also model or great master).[7] This neologism shifted the emphasis from a plural simultaneity of teachings to the historical depth of a convergent tradition. At the same time—and here lies the special interest of the Chinese case—a limit of adaptation was reached. For the Chinese elite refused to go along with the attempt by a number of late imperial scholars (and ultimately, in 1907, by the Qing Dynasty itself) to turn the prestigious Confucian worldview (*ru*) into a Confucian religious faith (*kongjiao*).[8] "Confucius"—the iconic sage whom the Jesuits created around 1700 out of a complex legacy handed down over the centuries—was presented by Kang Youwei and his comrades with some success as the symbol of "Chinesehood" and then of the Chinese nation.[9] The revolution later dethroned this figure in the name of Marx and Mao, but he underwent an amazing rebirth in the late twentieth century and, with the founding of the first Confucius Institute in 2004 (in Seoul), became the patron of the foreign cultural policy of the People's Republic of China. Under imperial China (until 1911) and the Republic of China (1912–49), all endeavors to impose a state Confucianism by analogy with Japan's state Shintoism ended in failure. The European concept of religion here reached the limits of its exportability, and around the turn of the century China's opinion leaders (without always being aware of it) paradoxically inclined toward an *older* construct in which Europeans had had a hand:

the "philosophical" Confucius, whom the Jesuits had rehabilitated against the "neo-Confucianism" prevailing at the time.

Elsewhere, this concept of religion imported from Europe had a strong social, and sometimes also political, impact. In Islam, Buddhism, and Hinduism, there were efforts to combine tradition and fresh imagination in carving out a more distinctive religious profile. This led in Islam, for example, to an emphasis on the sharia as binding religious law and in Hinduism to a stronger canonization of the Vedic scriptures as against other writings in the classical heritage.[10] Moreover, the plethora of new nation-states that emerged in the twentieth century established the idea of one official religion in place of the premodern hierarchy of different faiths. This made a new type of *religious* minority possible, in a situation where all citizens were formally equal, and at the same time bred religious conflicts that special laws for each group could resolve only with difficulty. Tendencies to religious uniformity and a more clear-cut identity mostly developed with other religions in mind, and often in direct confrontation with them. This complex reordering of the global religious landscape through emulation and demarcation was a major new development in the nineteenth century.

World Religions

One legacy of the nineteenth century that still marks public language is the idea of "world religions" towering like mountains above the topography of human faith. In the new discourse of religious studies (*Religionswissenschaft*), a wide range of orientations was condensed into macrocategories such as Buddhism or Hinduism, and these "world religions," together with Christianity, Islam, Judaism, and not uncommonly Confucianism, allowed for a mapping of religions that allocated them to major "civilizations." Experts used the crude grid of "world religions" as the basis for elaborate classifications of faith systems or sociological types of religion, with the underlying assumption that all non-Europeans were firmly in the grip of religion, and "Oriental" or "primitive" societies could best be described and understood in terms of religion; only enlightened Europeans were credited with the achievement of breaking the intellectual constraints of religion and even to relativize their own belief system, Christianity, by looking at it from the outside in.[11]

This approach, shallow as it might seem to us, made some sense in the nineteenth century. On the one hand, these societies—with the partial exception of China, because of its rich historiographical tradition—revealed themselves to Western scholars mainly through texts of a religious character (Max Müller's famous fifty-volume collection of translations, *Sacred Books of the East*, appeared between 1879 and 1910). On the other hand, it seemed to Europeans that the most threatening resistance to colonial conquest came from religious dignitaries and religiously inspired movements.

The thesis of the primacy of the religious in non-Western societies contributed to a lasting dematerialization, dehistoricization, and depoliticization of the way in which those societies were understood in the West. Clichéd equations ("Hindu India," "Confucian China") still imply that religious modernization is confined to the West, that it is the only civilization in the world to have declared religion a private affair and grounded its image of itself on secular "modernity." Talk of "world religions" is not entirely misguided. But it should not mislead us into considering particular religions as self-enclosed spheres in which any development is autonomous and barely subject to outside influences. Such an approach brings with it a level of political drama: visions of a clash of civilizations presuppose powerful blocs defined in terms of religion.

Revolution and Atheism

The nineteenth century opened in Europe with a general assault on religion. To be sure, elites had been overthrown and rulers executed in previous revolutions too. But the attacks of the French Revolution on the church and religion as such, prepared by theoretical critiques and anti-ecclesiastical polemic among radical Enlightenment authors, had no historical precedent and were one of the most extreme aspects of the whole upheaval. Church property was nationalized as early as the end of 1789. And although clerical deputies representing the first estate had made possible the conversion of the three-tier Estates-General into a revolutionary national assembly in June of the same year, the church was quickly excluded as a factor in the French power game. Catholicism lost its status as the *religion d'État*, and the clergy forfeited a large part of its traditional income. All monasteries were dissolved—a process that Emperor Joseph II had already initiated in the Habsburg Empire. The break with the pope, now regarded as one foreign monarch among others, came in 1790 over the Civil Constitution of the Clergy. Priests, or anyway a section of them, had already joined the state payroll without offering much resistance. Now the revolutionary legislators went a step farther, declaring them to be civil servants and incorporating them into the new administrative hierarchy, so that they were now chosen by secular bodies and had to swear loyalty to the state. This led to a deep split between those who agreed to take the oath and those who refused, between the French (Constitutional) Church and the Roman Church. It would be the basis for the persecutions that hit parts of the French clergy over the following years—although the conflicts seem rather innocuous in comparison with the religious civil wars in early modern France.

This radical assault on organized religion was a French peculiarity, whose long-term consequence was the ending of the Catholic monopoly. Earlier, the North American revolutionaries had freed themselves from the supremacy of the Anglican state church, but had not initiated anything like the French "dechristianization" policy or the violent iconoclasm associated with Robespierre's sponsorship in 1793 of the Cult of the Supreme Being. Church representatives were not subject to physical repression in the United States; antichurch sentiment or state-supported atheism was not a legacy of the Atlantic revolution as a whole. Anyway, during his period as first consul, Bonaparte already showed himself willing to neutralize a potentially dangerous enemy by striking a deal with the Holy See (Concordat of 1801) and recognizing it as a power in European diplomacy. After 1815, under the Restoration monarchy, the church regained much of its former influence, and Napoleon III, whose most loyal supporters were in the Catholic countryside, treated it with respect. Only under the Third Republic did a thoroughly secular separation between church and state become a basic feature of French politics, although it was a far cry from any state-imposed atheism. The radical character of the French handling of organized religion in the 1790s looked ahead rather to the twentieth century, where it reappeared in more violent forms in the Soviet Union, in revolutionary Mexico (refigured there in the vehemently anticlerical 1870s), and in the later Communist dictatorships. No other part of the nineteenth-century world saw a comparable offensive against organized religion. No state declared itself to be atheist.

Tolerance

The Atlantic revolution left behind a less spectacular but continuing legacy in the shape of religious tolerance.[12] The basic idea had originated in Europe during the religious wars of the sixteenth and seventeenth centuries; since Pierre Bayle and John Locke, it had been one of the pillars of Enlightenment thought, soon coming to define not only relations among religions in Europe but also the equal rights of others outside the West.[13] In 1791 the principle that the state should not dictate the private beliefs of its citizens or favor one religion over others was simultaneously established in France (Constitution of September 3) and the United States (First Amendment to the Constitution). The United States therefore guaranteed religious freedom from its earliest days, even if being a Protestant long remained advantageous for a career in politics.[14] In Britain it took several more decades before Catholics (1829) and Jews (1846/58) won full civil equality, while on the Continent freedom of religion and freedom of the press were main planks in liberal programs. For Jews in Germany, the first key dates were 1862 in Baden and 1869 in the North German League. In 1905 the Tsarist Empire became the last major country in Europe to accept religious toleration, issuing an edict that promised "freedom of conscience." Those who profited most from this were not the Jews but Muslims and sectarian offshoots of the Russian Orthodox Church. In fact, Catherine II had granted legal security to Islam back in 1773—the first step in a retreat from state persecution.

The fact that religious tolerance was first codified in the countries of "applied Enlightenment" (the United States and France), and that this set in motion a process that culminated in the UN Declaration of Human Rights in 1948, does not mean that it was an unknown practice in other parts of the world. In the early modern period, Europe's bitter religious wars and antagonisms were rather an exception to the rule of peaceful religious pluralism. In the multinational empires ruled by Muslim dynasties, aggressive Islamicization would not have been practicable; it would also have contradicted old political customs. The Prophet Muhammad himself reached various agreements with "People of the Book" in the Arabian Peninsula, and the Ottomans granted "protection" to non-Muslim *millets* (chiefly Christians, Jews, and Parsis, whose economic activities were beneficial to the state) in return for tribute-like payments; Christian peasants in the Balkans were an exception, however. In the Indian Mogul Empire, a Muslim conquering dynasty ruled over a non-Muslim majority with many different religious orientations. Here raison d'état demanded a policy of toleration, such as that which was pursued with impressive effect especially in the sixteenth century. When the dynasty under Emperor Aurangzeb (r. 1658–1707), the only jihadist Mogul ruler, changed course and tried to impose sharia throughout the empire, it contributed to the tensions that resulted in the collapse of Mogul rule in the early eighteenth century. In principle, however, Islam ruled out the equality of other religions with the one Truth revealed to Muhammad, the "Seal of the Prophets." We should not idealize the religious pluralism that existed in the Islamicate empires; non-Muslims were tolerated and largely protected from persecution, but only as second-class subjects. Nevertheless, there is a striking contrast with the ruthless exclusion of religious aliens in early modern Western Europe. Around 1800, religious minorities still had an easier time in the Muslim Orient than in the Christian Occident.

In China the Manchu conquerors, whose religious background lay in North Asian shamanism, operated a finely calculated system of balances among the various schools of thought and religious currents. They showed special care in cultivating Lama Buddhism, in view of its important political role for Mongols and Tibetans. But there were major structural tensions between the Qing State and its Muslim subjects, whose position in the hierarchy of minorities deteriorated in comparison with the Ming period (1368–1644). As far as "traditional" African societies are concerned, their characteristic hospitality was recognizable also in an openness to outside religious influences, which greatly facilitated missionary work for Islam and Christianity in the nineteenth century.[15] Since the idea of religious toleration is linked to the modern constitutional state, it cannot be applied sensu stricto to all these cases. But religious coercion was not the normal practice in non-Western societies before they were exposed to the influence of European liberalism. In the early modern period, which in terms of religious policies began with the compulsory baptism or expulsion of Jews (1492) and Muslims (1502) by the Crown of Castile, Europe's record in accepting religious diversity shows a deficit in comparison with the rest of the world. And once liberalism got into power, established churches might be in for a hard time—and not just in Europe. "Power," says John Lynch in view of the period 1870 to 1930, "could change Latin American liberals into monsters of illiberalism."[16]

2 Secularization

Dechristianization in Europe?

The nineteenth century has often been viewed as the age of "secularization."[17] Until the middle of the nineteenth century, this word was understood to refer to the transfer of church lands to lay owners. Then it acquired a new meaning: the decline of religious influence over human thought, the organization of society, and government policies. To simplify somewhat, the issue in the case of Europe has been to plot the graph of dechristianization that began with the Enlightenment and the French Revolution and has continued to this day. Here, historians have come to very different conclusions, irrespective of what they understand by "religion." Hugh McLeod, a British specialist in comparative religion, identifies six distinct areas of secularization: (1) personal faith, (2) participation in religious practices, (3) the role of religion in public institutions, (4) the significance of religion in public opinion and the media, (5) the contribution of religion to individual and collective identity-formation, and (6) the link between religion and popular beliefs and mass culture. For Western Europe between 1848 and 1914, his conclusions are as follows. In the first two respects, secularization was most evident in France, Germany, and England. The share of the population who regularly attended religious services and took part in communion showed a considerable decline. This cannot be quantified, but a jigsaw of discrete observations yields that overall impression. At the same time, there was a clear rise in the share of the total population (not only small intellectual circles) who expressed personal indifference, aversion, or hostility to the Christian faith. This trend was essentially the same in all three countries.

The differences were greater with regard to the significance of religion in public life. State and church were most clearly separated in France, especially from the 1880s on, and it was there, too, that Catholics had great success in building a "counterworld" out of their own organizations. Victorian England witnessed what might be called a creeping secularization, but no explicit ideology corresponded to it. Officially the country claimed to be devout and churchgoing. The much noted piety of William Ewart Gladstone (1809–98), who now and then felt divine inspiration for his political decisions, stood in sharp contrast to the religious indifference of another prime minister, Lord Palmerston (1784–1865), from a previous generation. In Germany, amid continuing opposition between Protestants and Catholics, the churches were well funded and could secure for themselves an unusually large role in education and social welfare.[18] Everywhere, religious orientations had by far their deepest roots in popular culture. Even those who did not go to church regularly or consider themselves part of the faithful clung to elements of a religious worldview, recognized and used religious symbols, observed the calendar of feast days, and sought help from religion in times of crisis.

Nationalism and socialism also offered all-embracing worldviews, but they were never able to supplant Christianity. Denominational subcultures proved more elastic than ever before in the three countries—even more so in the Netherlands—and had political parties attached to them (though not in Britain). The great majority of people in Europe (including the Jewish communities) held on to at least some outward religious forms.[19] The absorptive capacity of official Christianity was so great that even an Enlightenment agnostic like Charles Darwin was buried in a state funeral at Westminster Abbey. It is true, though, that the Archbishop of Canterbury sent his apologies.[20]

Symbolism and Law

Did this restrained secularization of Western Europe reflect a general trend? Little is known about the evolution of individual belief in many parts of the world. Where religious law and informal controls made participation in religious community life more or less obligatory, and where religiosity was expressed less in conventional acts of worship than in relations between individual masters and pupils, attendance at services is no longer a significant measure. On the other hand, we have estimates for the size of the monastic population. In 1750 Catholic Europe, from Portugal to Poland, had the highest figures since the Reformation: 200,000 monks and 150,000 nuns, or just under 0.3 percent of the total population of Europe west of Russia.[21] The dimensions were very different in Buddhist countries, the second great area of monastic culture. In Burma the number of monks seems to have remained constant throughout the century, or even to have grown: it represented 2.5 percent of the male population in 1901.[22] Tens of thousands of men in saffron robes, recruited from every section of the population and by no means divorced from worldly life, formed an important cement of Burmese society. In Tibet around 1800, there are said to have been 760,000 monastery residents—a quite staggering figure, twice as high as in the whole of Europe before the French Revolution.[23] In 1900 too, the country at the roof of the world was a monastery-dominated theocracy with the Dalai Lama as its spiritual and political leader—not at all peaceful, though, but in a constant state of unrest as various sects and monasteries fought it out with one another. Monastic rule was not altogether peculiar to the Orient, for at the same time, almost until the end of the colonial period, Spanish monks constituted the strongest political force in the Philippines; the independence revolution of 1896–98 was directed mainly against their unpopular ascendancy. Even in the case of Tibet, however, it is possible to speak of a kind of secularization. The Thirteenth Dalai Lama (called the "Great Thirteenth" in Tibet, r. 1894–1935), far from being an unworldly dreamer, was a priest-king who saw early on the opportunity for Tibet to develop into a nation-state and, with Britain's support (but without its direct colonial input), devised plans to lead his country out of the Chinese sphere of influence into an independent modernity.[24]

Insofar as secularization means the withdrawal of religious symbols from public space, the gap between Europe and Asia remained small. So long as there were monarchies that invoked at least a minimal degree of religious sanction, state rituals continued to have a religious character. Sultan Abdülhamid II (who also bore the title of caliph) played this role with at least as much calculation as the last two tsars or Emperor Franz Joseph in Vienna. Wherever revolutions swept monarchy aside, the secularization of power also came to a conclusion. From 1912 on there was no longer an emperor in China who might have performed the rites at the Temple of Heaven, and after the end of the sultanate-caliphate, secular symbols of the Kemalist republic appeared in place of the religious account that the bygone dynasty used to give of itself.

The secularization issue was (and still is) posed especially where a clear separation did not exist between secular and religious law. In such conditions—Egypt is a good case in point—secularists were those who sought to wrest space for European-style legislation away from the authority of religious law (e.g., the sharia). Legal reform, pursued by indigenous intellectuals with support from the protectorate power, became the first stage in secularization of the state as a whole. It was seen as part of a comprehensive process that would transform the premodern jumble of laws and jurisdictions into an orderly modern system.[25] Secularization of the state, first launched in reality with some Ottoman reforms after 1826, became a central theme in the Islamic world.[26] The postimperial countries, beginning with the Turkish Republic under Kemal Atatürk, transformed themselves in the twentieth century overwhelmingly into secular orders—a process whose reversibility would be dramatically demonstrated in 1979 with Khomeini's revolution in Iran.

Religious Fervor in the United States

Although major doubts have been raised whether, by most of Hugh McLeod's criteria, secular tendencies actually asserted themselves before 1910 in the non-Western world, a glance at the United States shows that the West, too, followed a number of different paths. In Western Europe, the cautious secularization after the turn of the century was by no means a linear continuation from the decline of religion around 1800. The Age of Revolution, when the greatest minds from Kant to Jefferson and Goethe serenely distanced themselves from belief in supernatural powers, gave way in the name of Romanticism to a rediscovery of the religious among large sections of the European intelligentsia. "Godlessness" was with some justification imputed to the underclasses living from hand to mouth in the heartlands of early industrialization, but a middle-class way of life, at least in the Protestant countries, included a new culture of piety and Christian moralizing. As we saw in chapter 17, one of its by-products was the successful antislavery movement. The religious dynamic in England, a pioneer of the new tendencies, was at first concentrated in revivalist groups outside the state church (which was seen as spiritually sterile and morally degenerate) and later in an opposition inside the Church of England. Wherever it took root, this evangelism emphasized the ubiquity of spiritual conflicts, the active intervention of Satan in the workings of the world, the personal sinfulness of the individual, the certainty of a coming Last Judgment, the possibility of salvation through belief in Jesus Christ, and the unrestricted authority of the text of the Bible. At the individual level, the experience of awakening and conversion to true "living" Christianity was fundamental; then came the obligation to prove oneself in the world.[27]

This evangelical revival got under way in the 1790s, and after a few decades it began to abut against reform initiatives within the Anglican establishment itself. In the second half of the nineteenth century, however, the rapturous zeal cooled down and passed into the secular tendencies described above, which in England were only slightly more hidden than elsewhere in Europe. In the United States, a similar revival occurred among Protestants, continuing a chain of energizing movements that had punctuated the eighteenth century; it ran parallel to a prophetic mobilization among Indians in the Northwest, led by the Shawnee warrior prince Tecumseh (1768–1813) and his inspired brother Tenskwatawa. The Great Awakening (as historians later called it) of the early nineteenth century grew into a vast self-Christianizing movement among North Americans, which, unlike in Europe, was never reined in by ecclesiastical establishments but preserved its dynamism in a fluid landscape of churches and sects. Between 1780 and 1860, when the population of the United States increased eightfold, the number of Christian communities rose by a factor of 21, from 2,500 to 52,000.[28] This permanent revival, lasting in essence to the present day, made the United States an intensely Christian nation that sees itself as morally and materially "civilized," and in which the greatest religious pluralism has prevailed.[29]

Immigrants from all around the world sought to stabilize their identity through religion. Migration in general not only spreads religious forms spatially but often modifies them and deepens the practices associated with them. Irish carried their Catholicism wherever they went, and the church sent priests out with them from Ireland. Thanks to Irish and southern European immigrants, the share of Catholics in the total population of the United States increased from 5 percent in 1850 to 17 percent in 1906.[30] The trend reversal toward secularism that became unmistakable in Europe toward the end of the century did not happen among either Protestants or Catholics in the United States. The American case also shows that religious vitalization—or what Enlightenment critics referred to as *Schwärmerei* (raptured enthusiasm)—did not inevitably lead back into theocracy, fanatical social controls, and irrationalism in other areas of life. The consequences of religious excitement can be contained if the distinction between private and public space has already been solidly established at an earlier stage.

Religion, State, and Nation

Western Europe trod a separate path in the nineteenth century, in the sense that church influence on the internal politics of nation-states became only here a central conflict of the age. What was at issue was not essentially the secular character of the modern state; that had already been secured after protracted struggles at the end of the revolutionary period. Europe's last theocracy disappeared in 1870, when the Italian Republic annexed the Papal States. Only in Russia did the Orthodox Church and tsarism form a symbiotic relationship, but this only alienated the emerging liberal public from the church and ultimately failed to prop up imperial rule. The conflicts in continental western Europe—Britain was affected only by the problem of Home Rule in Catholic Ireland—resulted from a combination of three factors: (1) the aversion that liberalism, at the height of its influence in midcentury, felt toward the Catholic Church; (2) the strengthening of the papacy, especially under Pius XI (r. 1846–78), which set itself openly against the national and liberal tendencies of the age and tightened the leash on national churches; and (3) the homogenizing tendencies involved in building nation-states, which made external, "ultramontane" masterminding of any section of the population unacceptable even in the eyes of nonliberal politicians. Catholics in the United States, for example, found themselves in a long-lasting conflict of loyalties. As citizens, especially if they were of Italian origin, many could not hold back their sympathy with the founding of a liberal Italian nation-state; but as members of the Roman Church, they were sworn to support the papacy in its battle against that nation-state and its founding principles.[31]

Three issues kept flaring up in Europe: the right to appoint bishops, the recognition of civil marriage, and influence in the education system. In the 1860s and 1870s, this tangled conflict escalated into a struggle between church and state of almost pan-European dimensions. In countries such as Belgium and the Netherlands, education was for decades at the top of the domestic political agenda.[32] We can see today that it was all a matter of rearguard actions. The years between 1850 and 1859 were, in the words of the great church historian Owen Chadwick, "the last years of Catholic power in Europe."[33] The political power of the papacy collapsed in 1859, when its two protectors—Austria and France (then under the far-from-devout Napoleon III)—ended their alliance with each other. In individual countries, a compromise solution was found over time. The battles over church and culture had fizzled out by 1880. However, even after the passing of the mulish Pius IX ("Pio Nono"), the Catholic Church had difficulty adjusting to the modern world: small wonder in an institution that could still afford the hoary authority of the Inquisition and even had a "Grand Inquisitor" post until 1929.

Defense against the "transnational" disloyalty (real or imaginary) of the Catholic Church was mirrored in the most diverse rapprochements between religion and nationalism. When there was a reasonably unified vision of the nation's future, a religious legitimation of it was not slow in coming; otherwise, rival blueprints expressed themselves in denominational forms. Little can be found elsewhere in the world that corresponds to this peculiarly European development. Some nationalisms were neutral as to religion and could be effective only by remaining so—for example, the All India movement that appeared in the 1880s, whose always-shaky foundation was unity across the boundaries between religious communities (above all, Hindus and Muslims). Chinese nationalism too, from its beginnings around the turn of the century down to the present day, has had no religious connotations. The United States was a Christian country through and through, but one in which church and state were strictly separate, churches never had deeply rooted privileges or large landholdings, and the state did not subsidize religion. The multiplicity of Protestant sects and denominations, alongside Catholicism and Judaism, prevented the correlation of any specific religion with the nation. American nationalism had a strongly Christian charge, but this remained supradenominational, unlike the Protestant nationalism that had marked the German Empire even after 1879 and the end of the *Kulturkampf* against Catholicism. Its core was a vague sense that white America had been chosen to play a key role in the plan of salvation. And it had to be equally congenial to Methodists and Mormons, Baptists and Catholics.

In no other major country in the nineteenth century was religion such a potent religious force as it was in Japan. Even during the Meiji period, the country's elite remained deeply suspicious of Christianity, which had almost disappeared after it was torn up at the roots in the early seventeenth century. It came as a complete surprise in 1865 when communities totaling some 60,000 "native Christians" were discovered to have kept the faith underground for more than two hundred years in the Nagasaki region. But this was more a curiosity than the prelude to a new growth of Christianity in Japan. After the ban on Christian proselytism was lifted in 1873, Catholic, Protestant, and Russian Orthodox missions had little or no success, and the stigmatization of Christianity as "un-Japanese" in the rising nationalist tide of the 1890s reduced its public presence still farther. The Japanese elite mobilized resources of its own to endow the newly created imperial state with religious and nationalist legitimacy, placing the indigenous Shinto tradition at the center of national religious life.

Before 1868, Shinto shrines and Buddhist temples existed alongside each other on an approximately equal footing, and thousands upon thousands of local shrines serving to honor divine spirits (*kami*) were integrated into people's everyday lives. The new Meiji oligarchy decided to create an orderly national hierarchy out of the chaos and to establish State Shinto as the basis for a new cult of the emperor. Right at the top was the Ise shrine dedicated to the goddess Amaterasu, the mythical ancestor of the emperor's family and protector of the whole nation. The imperial and national shrines were lavishly funded by the central government, their priests acquired the status of civil servants, and every household was officially allocated to a shrine. New sites such as the Yasukuni shrine in Tokyo would later be used for war-remembrance ceremonies. The old religious landscape of Japan, locally fragmented and remote from politics, was reshaped from above and pressed into a national mold. Buddhism was humiliated, its monasteries and temples reined in amid a kind of religious *Kulturkampf*. Within a few years, one-fifth of Buddhist temples were closed down, many thousands of monks and nuns were forced out into the world, and large numbers of cult objects and artistic treasures were destroyed. If US museums today house the largest stocks of Japanese Buddhist art outside Asia, it is because American collectors seized the opportunity for a bargain and saved numerous objects from destruction. New charismatic religions that had emerged in the early nineteenth century also had to yield to State Shinto.

The Japanese state intervened in religious life more than any other state in the nineteenth century. State Shinto standardized the practice of religion by means of a new ritual calendar and a nationwide liturgy, while the Shinto clergy became an important pillar of the political order. The state founded new religious traditions, and the sacralization of rule went far beyond any alliance of throne and altar imaginable in the most conservative parts of Europe. This laid the basis for the nationalism that saw the wars of aggression between 1931 and 1945 as the fulfillment of a divine mandate to a chosen master people.[34] State Shinto was not the result of a transfer from abroad. The young leaders of the Meiji Renewal understood that their goal of national integration could scarcely be achieved without ideological centralization under state control. The idea of a nation-state was vaguely known to them from contemporary Europe, but their ideological blueprint drew more on the traditional concept of *kokutai*, which the scholar Aizawa Seishisai had revived in 1820s. Since its golden age of antiquity, the theory suggested, Japan had stood out by virtue of its harmonious fusion of state and religion.[35] *Kokutai*, with its myth of a "national essence," gave a religious gloss to the elevation of the Meiji Emperor as the key bonding figure; nothing then really stood in the way of a racist-imperial interpretation of this concept of unity. Japan's new integral nationalism did not lag behind Western precedents. It was ahead of its time.

Shinto, as the national integration project of the Meiji period, stands in a paradoxical relationship to other tendencies of the age. It was a state-prescribed cult, demanding little in the way of faith or "piety" from those who observed it—more an orthopraxis than a theologically developed orthodoxy. In this sense it fit well into the cooling of religious sentiment, being the very opposite of a revivalist movement. On the other hand, since State Shinto was not one religion among others (or a "world religion") but the national religion of Japan, it conflicted with the tendency toward pluralism in modern conceptions of religion. Completely subordinate to state objectives, it was the antithesis of a view of religion as a matter of private religiosity and one sphere of the social among others. The contrast with China, where both the late imperial state and the Republic (1912–49) invested little in religion, could not be greater—unless, that is, we wish to regard the official Marxism (or "Maoism") of the three decades after 1949 as a functional equivalent of State Shinto.

3   Religion and Empire

Religious Pluralism

Conquest brings with it subjects whose religions are different from those of the imperial power. Jews came under the Roman Empire, Coptic Christians under Arab invaders in seventh-century Egypt, Orthodox Christians under Muslim rulers in the Balkans, Aztec polytheists under Catholics, and Irish Catholics under Protestants. Outside the Ottoman Empire—which itself lost much of its Christian population through territorial shrinkage and therefore became more Islamic for demographic reasons alone—there were no longer any Muslim empires in the nineteenth century. On the other hand, Muslims formed large population groups in the empires of Britain, Russia, France, the Netherlands, and China. At the latest after the British incorporation of Egypt and large parts of sub-Saharan Africa shaped by Islam, no monarch in the world ruled over as many Muslims as Queen Victoria did; she was also empress of the great majority of Hindus. The British had to govern Buddhist majority populations in Ceylon and Burma, as did the French in Cambodia and Laos. In Africa, parts of Southeast Asia, and the South Sea Islands, Europeans only slowly discovered and described the welter of religious forms of expressions with which they were confronted. Their first impression was that the peoples in question had no religion the speak of and were therefore, according to one's point of view, either wide open to Christian missionary work or immune from any "civilizing" mission.[36] In the 1860s Edward Burnett Tylor, one of the founders of ethnology, coined the blanket term "animism," which soon caught on as a neutral replacement for the early modern "idol worship" or "idolatry," once viewed with horror as the opposite of all forms of monotheism.[37]

Beneath the regulated surface of the organized "world religions," every region in the world, including Christian Europe, harbored all manner of superstitions with which the guardians of orthodoxies usually came to some arrangement, even if Enlightenment secularists and religious missionaries disapproved of compromises in principle. In the colonies, there were often complex religious structures by no means subject to clear authority relations. The more that Europeans were accustomed to transparent church hierarchies with vertical chains of command, the harder it was for them to decide where to begin implementing their religious policy amid the "chaos" of orders and fraternities, temples and shrines. The early modern Ottoman state was more successful in this regard. The sultan-caliph insisted on channeling all contact with his non-Muslim subjects through their religious leaders, who enjoyed considerable autonomy within niches of the system assigned to them by agreement. This in turn promoted the grouping of religious minorities into churches.[38] The religious leaders were in some cases brutally held to account. In 1821, when news of the Greek uprising reached Istanbul, the Ottoman government ordered the summary execution of Patriarch Grigorios V, even though he was not implicated in the revolt.

What was the significance of the fact that a larger number of non-Christians came under Christian rule than in any previous century? Although the self-assigned civilizing mission, the main ideological justification of imperial rule, could be easily formulated as a religious duty, the colonial powers almost never pursued an active policy of converting their subjects to Christianity. Provision was made for the spiritual care of European colonizers, and the ritual facade of colonial rule invariably included Christian symbols, but otherwise it made sense to keep the peace by avoiding provocations to the various religious groups in the land. In the late nineteenth century, empires therefore still tended to be structurally neutral in religious matters. After the Great Rebellion, Queen Victoria's Parliament confirmed to the princes and peoples of India that from the following November the Raj would observe the principle of not interfering in the affairs of the country's religious communities.[39] Clauses with similar effect were also written into treaties signed after 1870 with the sultans of Malaya. The promises were not always kept, but both the British and the Dutch maneuvered very cautiously in relation to Islam. Of course, the creation of hierarchies and bureaucracies was designed to make it easier in the long run to monitor what was happening in the realm of religion.[40]

That is how empires have always liked to operate. After the first partition of Poland in 1772, Maria Theresa introduced in Galicia the new function of state-appointed chief rabbi, with the aim that he would keep his coreligionists under reliable supervision.[41] In its Tibetan protectorate in the eighteenth century, the Qing government restructured the Lamaist hierarchy and attempted to reshape it into a docile instrument of control. One of the many other methods used to manipulate religious powers without disabling them was intervention in the filling of offices—in the same way that European governments valued having a say in the choice of Catholic bishops. Muslim subjects were particularly difficult to handle, partly because many of them had contacts as businessmen or pilgrims beyond the frontiers of the colony. Colonial powers therefore thought it advisable to isolate "their" Muslims from the rest of the community of believers and to limit their opportunities for a pilgrimage to Mecca.[42]

In seeking to maintain contact with "reliable" religious leaders, imperial administrators could sometimes land themselves in a paradoxical situation. In the Islamic world, for example, mystical Sufi orders were rather suspect as partners in cooperation; functionaries preferred to deal with sedentary local authorities that behaved in a reasonably "rational" manner. But in Senegal the French gradually learned before 1914 that, in the interests of internal order, it made more sense to collaborate not with "chiefs" but with marabouts, the somewhat intractable spiritual leaders of the Sufi brotherhoods, who were less corrupt, more respected by the population, and therefore more likely to get things done.[43] Whether in the British, Russian, or French Empire, religious policy was a constant and unavoidable concern of the colonial state; any mistakes could trigger unrest that was very difficult to subdue. The whole of nineteenth-century imperial history, including that of Qing China, is shot through with fears of a Muslim revolt. In the Western perception, the "revolt of Islam"—the memorable title of a long poem by Percy Bysshe Shelley (1818), which actually deals more with the French Revolution—began not with the triumph of Ayatollah Khomeini in 1979 or the events of September 11, 2001, but with the militant Muslim movements around the year 1800.

Empires have always intervened in one way or another in the religious topography and hierarchy of their colonies, but they have seldom altered them fundamentally. Forced conversions or baptisms happened here and there but were generally considered undesirable and prohibited. Outside its own colonies, however, a major European power might deliberately stir things up by intervening to protect a Christian minority within an Oriental empire. Russia did this with the Greeks in the Ottoman Empire, and France with the Christians in the mountain areas of Lebanon—in both cases triggering complications that led to war—and Sultan Abdülhamid II, for his part, declared himself the protector of all Muslims living under Christian rule. It was German strategy to incite religious, ethnic, or protonational minorities against the British Empire during the First World War, and the British did the same against the Ottoman Empire, culminating in the memorable exploits of T. H. Lawrence "of Arabia."[44] It had already been tried out in the Anglo-Russian Great Game of the nineteenth century.

Missionaries: Motives and Driving Forces

One of the main lines of global religious history in the nineteenth century is the rise and fall of Christian missions.[45] In the early modern period, although European missions had huge cultural consequences—notably the role of Jesuits as a bridge between Europe and China in the seventeenth and eighteenth centuries—they remained quantitatively modest. Mass conversions in Asia were neither desired nor tolerated by European colonial powers or indigenous rulers; Africa was still outside the sphere of missionary operations. Of the million or so people who went to Asia in the seventeenth and eighteenth centuries with the Dutch East India Company, only a thousand were men of the cloth—and their main task was to fight off the competition from Catholicism.[46] In contrast, the nineteenth century saw a major development of missions to Christianize large populations or even whole peoples. This was a Protestant phenomenon, which—with an antecedent around 1700, in the mission of German Pietists from Halle to the Danish colony of Tranquebar in Southeast India—developed first in Britain, and a little later in the United States, out of the surplus energies of the evangelical revival. In contrast to early modern attempts to win foreign rulers to the Christian faith, it involved a mission to the "pagan" masses. If we

were to name a starting date, it would be not so much the year when a particular organization was founded (Baptist Missionary Society in 1792, the originally nondenominational London Missionary Society in 1795, or the Anglican
Church Missionary Society in 1799) as the opening of British India to missionaries in the new East India Company charter of 1813. From that point on, merchants and missionaries appeared in growing numbers in the Subcontinent; the
markets for goods and for souls now mirrored each other. As always, however,
selling beliefs was more difficult than peddling material goods. Tiny groups of
initial converts were instrumental in igniting "explosions of spiritual energy . . .
that brought whole communities into the new faith."[47] The second major missionary region in Asia, the Chinese Empire, was opened up in 1858–60 by several
"unequal" treaties, after a period since 1807 when missionaries had been working
in restrictive and dangerous conditions out of the Canton trading post and the
Portuguese enclave of Macau.[48] In 1900 there were roughly two thousand missionaries in the whole of China.

In Africa the missionary presence took longer to establish itself and was more
decentralized, beginning around 1800 in the south and west of the continent.
Here, of course, there was no central government to regulate access, so that by
midcentury the whole spectrum of Protestant orientations and churches was
represented. In the 1870s, on the eve of the great European invasion, missionary
activity increased again, and a little later it became caught up in the wake of
military conquests that were advantageous but also created new problems for
it.[49] The Catholic mission—which, like the Roman Church in general, took a
long time to recover from the Age of Revolution—followed a few decades later,
sustained mainly by the ambitions of Napoleon III in international and colonial
politics. By 1870 it was active worldwide, and the much more numerous Protestant missionaries looked upon it as a dangerous rival.

Much was new in the nineteenth-century Protestant mission. Its basic purpose was to save thousands—or in China, as its propaganda tirelessly proclaimed, millions—of souls from eternal damnation. It mobilized tens of thousands of men and women, who were often ill prepared for hazardous and often
materially unrewarding service in remote tropical areas. Martyrdom also was still
a possibility; more than two hundred missionaries and family members lost their
lives during the anti-Christian Boxer Rebellion in China. Missionary work was a
huge achievement on the part of a quite special "civil society" organization resting on voluntary initiative. Most of the Protestant societies in question relied on
donations and set great store by their independence of the state and church hierarchies. Indeed, they were the very first organizations to elevate fundraising to a
fine art. Sponsors in the home country had to be continually humored, remoti-
vated, and persuaded of the spiritual benefits of their mundane investment. Missionary activity also involved a combination of business and logistical planning.

Mission history is today a huge research field, which easily merges with the
history of Christianity outside Europe. What went on between missionaries and

natives is increasingly seen as a symmetrical interaction and is elucidated from more than one viewpoint.[50] One particularly controversial question, to which no general answer can be given, is whether and how missionaries were "accomplices" of imperial expansion and colonial rule. The extraordinary spread of missionary activity is, of course, unthinkable outside the wider context of European global conquest, and there are many cases where missionary penetration into a well-known region followed its political appropriation. Missionaries were often direct beneficiaries of imperial protection. They belonged to "white society" in the colonies—but at a low level of prestige, at least in the British case, since their typically petit bourgeois habits made them appear out of place in elite circles. On the other hand, missionaries pursued objectives of their own that did not always overlap with those of a colonial state to which they definitely did not belong. Often they were at odds with the aims of private settlers too. From the point of view of the colonial state, missionaries were welcome if they built schools and provided as much as possible of the funding for them. The enthusiasm of governors or (in a noncolonial country like China) consuls was far less profuse if missionaries "irresponsibly" sowed unrest among the indigenous population and then expected a European government representative to bail them out. Where nationalist aspirations appeared in the open, individual missionaries were invariably suspected of backing them.

The numerous missionary societies varied in their theological beliefs and in their objectives, methods, and willingness to take risks. It made a difference whether one wore Chinese dress (as members of the fundamentalist China Inland Mission did) and tried to spread the word of God in a provincial village backwater, or whether one stuck to European sartorial markers and concentrated one's efforts in higher education and the provision of health care in the cities. Nineteenth-century missionaries were scarcely less cosmopolitan than their distant precursors in the Jesuit order of the early modern period. English-speaking evangelism had from the outset been a transatlantic project, and missionary work in faraway places often bridged doctrinal conflicts and strengthened ecumenism. Missionaries from continental Europe had their own societies but were also to be found working in the Anglo-Saxon organizations. It was rare for a missionary society to be composed only of one country's nationals, and, at least during the first three quarters of the nineteenth century, national identity did not play a primary role for missionaries. At the same time, many had no reason to commit themselves to the imperial ambitions of a foreign government. In its early days, the Church Missionary Society employed more Germans and Swiss than British.[51] In 1914, when national tendencies had become stronger, more than a tenth of the 5,400 Protestant missionaries active in India still came from continental Europe.[52]

Adventurous migration across cultural boundaries was not uncommon. An amazing life such as that of Samuel Isaac Joseph Schereschewsky, though far from usual, was still a possibility. Early in the nineteenth century, Anglicans launched a mission among the Jews in the Prussian, Russian, and Austrian parts of Poland—a "transnational" project in itself. One of the converted Jews was Samuel Schereschewsky, who had received a rabbinical education in Lithuania and been strongly influenced by the Jewish enlightenment (Haskalah). The young man then studied theology in Breslau (Wrocław) and made his way to the United States, where Baptists only then actually baptized him. After a further seminary program in theology, he put himself forward for missionary service in China with the Episcopalian Church. Having arrived in Shanghai in 1859, he spent the years from 1862 to 1874 in Beijing, and was consecrated the first Anglican bishop of Shanghai in 1877. Schereschewsky became one of the great Sinologists of his age. The first Chinese translation of the Hebrew Old Testament, still in use today, comes largely from his pen. He always kept a great distance from imperial politics and did not share the proselytizing ardor of his prophet-like contemporary J. Hudson Taylor, who in 1865 had founded the China Inland Mission. There was space for very different characters under the broad roof of the mission.[53]

The Christian Mission: A Balance Sheet

It is hardly possible to establish an overall balance sheet of the Christian mission. Conversion statistics should always be treated with suspicion. The utopian goal of drawing whole peoples into the global flock of Christians was achieved in only exceptional cases. Nor was conversion necessarily definitive. When the British relaxed the requirements of the law in Ceylon after 1796, many indigenous Protestants reverted to Buddhism or Hinduism.[54] Missionary success often occurred where links with the colonial state were especially weak; there is good evidence of this in India.[55] Marginal and underprivileged groups, as well as many women, were especially likely to let themselves be approached. Yet, after several centuries of zealous missionary work, only 2 percent of Indians had been converted to Christianity. In China, the mismatch between huge investment and modest results is perhaps even more striking. The greatest breakthroughs were in West and Southern Africa. The indigenous churches that sprang up there—at the same time as among the Maoris in New Zealand—often had missionary backing, but they soon developed a communal and theological life of their own. Undoubtedly missions made a decisive contribution to the globalization of Christianity, and the churches existing today are by no means dependent on mother institutions in Europe. Global Anglicanism, for example, is a product of imperial expansion, but it has long since left behind its past in empire.[56]

Things look slightly different when we turn to the objects of missionary zeal. Asian governments and a fortiori local authorities feared little more than the arrival of a Christian mission. Its workers did not think like the diplomats or soldiers with whom they were used to dealing; theirs was not the familiar transcultural logic of power politics but a program for the overthrow of existing relations. Missionaries often appeared to be creatures from a different planet, challenging the authority of local rulers and (especially if they knew an imperial gunboat was available) setting themselves up as local counterpowers. Even if they did not explicitly intend it, missionaries always called the existing social hierarchy into question. They freed slaves, gathered marginal elements of the local society around themselves, raised the position of women, and—as the archmissionary, Saint Boniface, had done eleven hundred years earlier—undermined the prestige of priests, medicine men, or shamans. Missionaries were guests who invited themselves, not wise men called in like the Buddhist monks of early Tang China. Although they might be given a hospitable welcome at first, they soon broke with convention by staying on and trying to change the rules of the social game.

One thinks of missionaries as operating mainly under colonial or "stateless" conditions—in Africa or the South Seas, for example. But quite well established states such as the Ottoman Empire also felt the challenge of this new breed of holy warrior, who let no opportunity slip to project the image of someone representing a "higher civilization." Such ideological militancy, especially among American Protestants, reached its high point around the turn of the century, when 15,000 men and women from various US churches and missionary societies were active in foreign lands. The Ottoman state found itself in a comparatively favorable situation, since the Treaty of Berlin in 1878 had acknowledged its right to oppose the conversion of Muslims to another religion; China had no longer had such an option since 1860. Nevertheless, some circumspection was to be recommended. Since missionaries controlled their own media and had good contacts in the Western press, they were capable of doing serious harm to the empire's image abroad. Catholics were reasonably familiar as envoys of the pope in Rome, who counted as a kind of colleague of the sultan in his religious capacity as caliph. But American Protestants, in particular, caused great confusion with their brisk self-assurance, appearing not only as religious rivals but as apostles of earthly objectives similar to those of the late Ottoman state: they, too, promoted the emergence of an educated middle class.[57]

Missionaries differed in their effect from representatives of international capitalism, who in the space of a few years could revamp whole countries and integrate them into the international division of labor. Missionaries worked under particular local conditions, building a church here or a schoolhouse there, and in the process reshaping the space in which others lived. They intervened directly in the course of people's lives, not in a roundabout way through abstract powers such as the world market or the colonial state. Local individuals acquired new opportunities and might even receive an education in the metropolis; others gained a new purpose in life by trying to repel the missionary invasion. The effects of missionary work therefore went beyond the circles of proselytes and sympathizers. Local societies did not automatically become more modern through exposure to missionary activities, since missionaries—above all, the "faith missions" geared around a fundamentalist reading of the Bible—brought in their baggage a West that was not the one of liberalism, reforms, and technological mastery of nature. In any event, local societies faced an unprecedented challenge to their traditional certitudes.

In some countries, the main historical contribution of Christian missions was to assist in the appropriation of Western sciences, including medicine. This was especially true of the work of Protestant missionaries in China from midcentury on. Only a small percentage of their translations were of texts with a Christian content; most related to science, technology, and practical issues facing society. Beginning in the 1920s, the sciences in China reached a level where they were independent of the initial missionary impetus. The Protestant mission played a similar role in Latin America (above all Brazil) and Korea, where it got under way only in the 1880s. One reason why the (mainly US) mission was more successful in Korea than in China, and a fortiori Japan, was that it offered a moderately oppositional alternative to official Confucianism and the dead weight of Chinese cultural hegemony without suffering from the burden of complicity with imperialism (Korea was annexed by Japan in 1910). The fact that the mission in Korea used both English and Korean (a language scorned by the old elite), thereby offering cultural space for the articulation of a rising tide of nationalism, was another of its attractions. A tortuous process stretching from 1884 to the present day has made South Korea's Christian third of the population one of the highest proportions anywhere in Asia.

4 Reform and Renewal

Charisma and State Building

Even more than the eighteenth century, the nineteenth was an age of reforms and new departures in religion. Many of these, though not all, can be explained with the cliché of "the challenge of modernity." Many, though not all, were responses to the global hegemony of Europeans. Much as basic themes of Pietism—here understood in a broad sense going beyond Germany—took on new shape in the various evangelical movements of the nineteenth century, the eighteenth century in the Islamic world had been an age in which movements of renewal, also seeking authentic roots of piety, appeared outside the established clerical hierarchies.[58] These outbreaks of fervor originated less in the centers of Islamic learning than in peripheries such as Southeast Asia, Central Asia, or the Arabian desert (which in the eighteenth century was an Ottoman frontier territory but also the oldest Muslim region). The best known of these movements is Wahhabism, so called after the fiery preacher Muhammad ibn Abd al-Wahhab (1703–91), who condemned nearly all existing variants of Islam as heretical and demanded a radical purification. The fury of the Wahhabis was so great that between 1803 and 1813 they caused serious damage even to some of the holy sites in Mecca and Medina, arousing revulsion in large parts of the Muslim world.

The significance of this movement lies not so much in its (modest) theological originality as a "grim and narrow theory of unbelief," and it should not
be seen as representative of eighteenth-century reformist Islamic thought.[59] It
is mainly interesting for its temporarily successful state building. The founder
joined forces with a local ruler, giving rise to a militant state based on Islamic
renewal. In 1818 Muhammad Ali, the pasha of Egypt, conquered the first Wahhabi desert state with the approval of the Ottoman sultan and put an end to the
experiment. But in 1902 the Wahhabi ruling house of Saud began a period of renewed ascent, which directly preceded the step-by-step formation of the Kingdom of Saudi Arabia in the early twenties. In 1925 the holiest places of Islam
again came under Wahhabi control.[60] Unlike later forms of militant Islam in
India, North and East Africa, or the Caucasus, original Wahhabism cannot be
seen as a movement of resistance to the West (which did not have the slightest
influence in Arabia in the late eighteenth century). The road from a heterodox
breakaway to a newly founded state was an exception rather than the rule. In
the nineteenth-century Islamic world, religious energy often streamed precisely
out of the tension between state structures—whether colonial as in India, Indonesia, and Algeria, or indigenous as in Iran and the Ottoman Empire—and
vibrant, less institutionally hardened orders and brotherhoods.[61]

Other examples of attempts to build a state on religious charisma were the
Taiping movement in China and Mormonism in the United States. First appearing on the scene in 1850, the Taiping under their prophet Hong Xiuquan
were a social revolutionary movement that constructed a complex worldview
out of Protestant missionary propaganda and the traditions of Chinese sects.[62]
Their state-building efforts finally came to naught, but the Mormons had greater
political success. The Church of Jesus Christ of Latter-day Saints, as they also
call it, was founded in 1830 by an American prophet, Joseph Smith, who, like
Hong Xiuquan seven years later, experienced visions as a young man and interpreted them in the sense of a prophetic mission. After Smith was murdered by
a hostile mob in 1844, his successor Brigham Young led an adventurous exodus
in 1847–48 to the uninhabited Great Salt Lake region, taking several thousand
followers with him. Other converts joined them, some from Britain and Scandinavia, and by 1860 some 40,000 Mormons were living in the state of Utah.
The Latter-day Saints were not permitted to establish a theocratic republic
where the people deferred to their inspired leaders.[63] Utah was founded as a territory under the direct control of the US presidency, and from 1857 to 1861 (as
it happened, the high point of the Qing government's war against the Taiping)
the Mormon zone was actually under military occupation. If the Taiping doctrine may be understood as an indigenized Christianity, evidently remote from
biblical sources, Mormonism was also a version of Christian doctrine adapted
to local circumstances, complete with a holy book of its own from its founder.[64]
Its characterization as "Christian" is still disputed today. To many people living
at the time of its founding, its adoption of polygamous practices made it as

alien as an American Islam. But Mormonism answers the question of why the Bible is silent about America. With its bold speculations about westward migration in the age of the Old Testament, it includes the American landscape in the biblical plan of salvation and is thus the most American of all the religions in the United States.

Prophetic movements, some believing a messianic end of time to be imminent, also appeared in other parts of the world. The Mahdi movement in Sudan (1881–98) had such a character, as did the Spirit Dance movement among American Indians in the northern Midwest (1889–90), the movement of Sayyid Ali Muhammad Shirazi (aka "the Bab") that appeared in Iran in 1844, or the movement that inspired the Maji-Maji uprising against colonialism in German East Africa between 1905 and 1907. Resistance to imperial invasions or a tightening of colonial rule was very often led by prophetic figures and accompanied with millenarian expectations.[65] All these movements promised radical change: their goal was not adaptation to the modern world but its downfall and a return to conditions they thought of as self-determined. Messianism was not a necessary requirement of radical politics. Compact religious communities of any kind could, out of religious motives, respond to outside pressure with active resistance. In parts of Southeast Asia, therefore, well-organized Buddhist monks—for whom messianism was an alien phenomenon—mounted effective opposition to the colonial powers.[66]

In the sphere of religion, it is problematic to advance a dichotomy between revolt and reform, between messianic movements, often visualizing the future as a return to a mythical golden age of the past, and religious doctrines and practices that propose a rational, cautious adaptation to the changing times. Such a distinction becomes more plausible, however, if one and the same movement can be shown to shift from the first pole to the second. This was the case with the Bab movement, a Shi'ite heresy in Iran, in which Sayyid Ali Muhammad Shirazi advocated the divine rule of God's chosen representatives on earth and finally claimed for himself a prophetic authority, based on direct communication with the Almighty, that superseded the teaching of the Qur'an.[67] After his execution by firing squad as an apostate and political rebel in 1850, the movement did not collapse but passed on the original charisma as it implemented a series of reforms. A comrade of the Bab, Mirza Husain Ali Nuri (or Bahaullah), undertook this task during his decades of exile in the Ottoman Empire. While referring to himself as the universal messiah—or the reborn Christ, Mahdi, and Zoroaster rolled into in one—he took great pains to recast the teachings of the movement in cosmopolitan terms attuned to the modern world. Following his death in 1892, the Shi'ite messianism of the founder became the modern Bahai religion, which after 1910 spread to Europe and America and today has its spiritual and organizational center in Haifa (Israel). Along with Mormonism and Indian Sikhism, it is one of the few new religious creations to have survived from the nineteenth century. Bahaullah, together with the exiled Chinese philosopher Kang Youwei (the creator of "Great Community" utopianism), was one of the major thinkers of the late nineteenth century who cut across cultural boundaries. The modern elements in the Bahai faith were its advocacy of a constitutional state and parliamentary democracy, its support for an expansion of women's rights, its rejection of religious nationalism, its renunciation of the doctrine of holy war, its concern for world peace, and its openness to science.[68]

Modernity and Modernism

This last point was and is—precisely in the light of early twenty-first century debates between Darwinists and biblical "creationists"—perhaps the most important criterion of all for religious modernity. Not all facets of scientific knowledge were equally well known or accessible to lay people. According to this yardstick, religious modernization meant not discarding the latest science per se as a source of truth. Through the astronomical demonstration of plural worlds, the discovery of deep time by geology and paleontology, and above all the evolutionary theory of Charles Darwin (with its more radical, militant expressions in the work of men such as Thomas H. Huxley in England or Ernst Haeckel in Germany), the natural sciences confronted all religions and denominations with major challenges.[69] The relationship between faith and religion therefore became, at least in Europe, a central theme in philosophies offering a world orientation; and hopes for a harmonization of religion and science, nurtured in Biedermeier Germany and early Victorian England, have been under a cloud ever since. Explicitly rationalist post-theistic quasi-religions have not been able to bridge the gulf for long: neither the elevation of science into a creed (the "religion of science"), nor various secret doctrines based on Freemasonry, nor the "social religion" sketched around 1820 by the French socialist Claude-Henri de Saint-Simon and cultivated for decades after his death in the form of a sect, in which scientists and artists were supposed to give the new industrial age an ethical foundation and thereby make possible its fully productive blossoming. As to the positivism of Auguste Comte (which already the master's late works elevated to a "religion of humanity"), it was principally in Mexico, Brazil, and Bengal that it was understood as a secular salvationism, a message of scientifically guided progress, in which political liberalism and economic laissez-faire retreated behind visions of technocratic order. Comte had expected the triumph of his doctrine in Western Europe, but he also sought, in vain, to win the support of Pasha Muhammad Ali in Egypt, where Saint-Simonists tried for a time to realize their utopian ideas of communes. Positivism did not catch on there, but rather in countries where it was regarded as a comprehensive worldview allowing them to catch up in the modernization race.[70]

Similar problems to those resulting from scientific corrections to the biblical story of the Creation emerged through new historical approaches in the humanities. The arts, philosophy, and sciences were now studied as they took shape over the course of time; historical accounts of the slow evolution of national literatures took their place alongside literary criticism. Kant had still sketched only in broad strokes the development of philosophical thought up to his own time, but just a few years after his death Hegel was giving richly detailed lectures on the subject. Nor did religion escape historicization. The clash between conventional beliefs and the new sense of the historical became an issue for many communities and churches, causing am even greater stir in Judaism, for which the idea of reform was traditionally alien, than in the Christian denominations. A relationship to history and temporality became the core of the modernization of the Jewish faith.[71] In Christianity, Bible criticism had similar dramatic effects. Insofar as it involved the investigation of Old Testament sources and textual tradition, it also directly affected the way in which Jews perceived their history. In the long run, patient philological work did more than a sharp polemic attack such as David Friedrich Strauss's *Life of Jesus* (1835) to advance an historical approach to Christianity, also among sections of the public with a broad education. Its methods yielded ever more accurate knowledge of the historical facts, as well as a wide range of interpretations that each took a critical distance from the biblical narrative. Nineteenth-century liberal Protestant theology and church history tended to portray Jesus as an ethical teacher of transcendent values. When the comparative history of religions became a rival focus of interest, this led to a different image of Jesus as an Oriental prophet, who urged the world to change direction in the face of impending doom.[72] The fact that European scholars also subjected the founding of other religions (e.g., Islam or Buddhism) to historicist critique or critical historicization came to be seen in the eyes of their devotees as a challenge and a desacralizing affront—one source of present-day accusations of "Orientalism."

It would be wrong to construct a stark opposition between the West (religion safely channeled along the lines of bourgeois rationality) and the non-Christian remainder of the world (religious dynamic expended in militant fervor, charismatic leader cults, and holy wars). Diehard traditionalism, charismatic challenges to it, subsequent development through reform: these existed in both East and West in the nineteenth century. Under Pius IX the Catholic Church explicitly came out against the legacy of the Enlightenment, and in truth Pius was almost as reactionary as he was painted in the polemics of liberal Europe at the time. To his successor Leo XIII (r. 1878–1903) he bequeathed a Catholic laager mentality. Yet Leo, also a man with deeply conservative inclinations, risked a cautious opening when he turned to the social issues of the age and tried to find a third way between laissez-faire capitalism and socialism. On the whole, the church remained a force of inertia.

At the time of these two long pontificates, modernist renewal was becoming a factor of major significance in Islam. The beginning of its rise may be dated to the 1840s. This multistrand reform movement, encompassing the whole Muslim world from North Africa to Central Asia to Malaya and Indonesia, was an important element in the nineteenth-century history of ideas. Borne along by legal and religious scholars, at times also by political leaders with very different backgrounds, it displayed a common concern that faults of Islam's own making would throw it into the intellectual defensive in an age of European world hegemony and exacerbate its political weakness. The modernizers—the best known internationally were the charismatic Sayyid Jamal al-Din al-Afghani (1838–97), a restless wanderer through Muslim lands; Muhammad Abduh in Egypt, a high clerical dignitary, politician, and (in comparison with al-Afghani) systematic theoretician; the philosopher and educationist Sir Sayyid Ahmad Khan in northern India; and the Crimean Tatar intellectual Ismail Gasprinskii—searched for compromises between the stock of Islamic tradition with its ubiquitous defenders and the challenges and opportunities presented by the modern world. They fought for free critical spaces where a reinterpretation of Islamic textual sources would be possible. Their debates, given novel resonance by the emergence of a Muslim press after about 1870, addressed the requirements and opportunities of modernization, forms of constitutional government, the rapprochement with modern science, the content and methods of education, and the rights that should be accorded to women.[73] Of the dozens who spoke their mind, some women among them, there were virtually no freethinkers who sought to question Islam at a fundamental level. The intention was to refute, with demonstrations to the contrary, the widespread view among Europeans that Islam was a rigid and tyrannical dogma. New things had to become capable of expression in the language of Islam.

The Islamic modernists did not stop at theory. Many of them were especially active in the realm of education—for example, the scholarly polymath (originally judge) Sayyid Ahmad Khan, who, at the Cambridge-style Muhammadan Anglo-Oriental College (today the Aligarh Muslim University in Uttar Pradesh) that he founded in 1875 with the primary aim of training senior officials for the colonial state, tried to link Muslim identity building with the role model of the English gentleman.[74] Even more successful in terms of recruitment and growth was a network of reform-minded theological colleges, designed mainly to educate *ulama* to serve the spiritual needs of the community. The first to get off the ground, in 1867, was the Koranic school (madrasa) in Deoband in northern India—hence the title of the Deoband Movement, whose affiliates subsequently spread to many other areas of the Subcontinent. The traditional jumble of religious institutions in Muslim India thus became subject to a degree of streamlining and bureaucratization, but it kept its distance from the colonial state, which, not needed as a source of funding, looked with some suspicion on such "civil society" initiatives from its Muslim subjects.[75] In 1900 it was not unrealistic to foresee a great future for the highly diversified modernist tendencies within Islam. Their decline in the midst of secular (e.g., Kemalist) nationalism, fascism, and Bolshevik socialism belongs in a different epoch—and in a history of missed opportunities.

Reform movements also appeared in the variegated religious worlds of non-Muslim India, often aiming at broad cultural renewal rather than simply a purification of religion.[76] Ram Mohan Roy, Ramakrishna Paramahamsa, and his disciple Svami Vivekananda (who in the early 1890s developed a monist conception of the Absolute that gave Hinduism a more universalist dimension) are just three thinkers who became known far beyond the confines of India.[77] The superiority that Christians of all denominations increasingly claimed over educated Asian elites also forced the other side into a stronger identification with religion that helped to build up "Hinduism" (the term first appeared in the early nineteenth century) as a uniform doctrine and social institution. Drawing on cultural resources of their own, reform movements reacted in various ways to new impulses: to the European discipline of Oriental studies (often taught in India itself), to the Christianity on offer from missionaries, and also to one another, since, as in Christianity and Islam, modernist forces in turn triggered neo-orthodox responses. In exceptional cases the impetus came from outside: in the 1880s in Ceylon, for instance, US theosophists and local disciples reinvented Buddhism by writing a catechism, restoring monuments, and popularizing Buddhist symbols.[78] The basic options regarding the problems of the age were everywhere the same, in a spectrum that went from militant rejection of the new and alien to large-scale adaptation to what were considered the dominant forces of the contemporary world. More interesting than the extremes are the many intermediate solutions, which cannot be grasped through a simple counterposition of "tradition" and "modernity."

Religious Communication

Along with science, religion belongs among the great creators of extensive communication networks. It would be a banality to call such networks transnational. Many of them are today more comprehensive than modern nation-states, and nearly all are older. Not necessarily having to rely on state structures, they operate across existing borders and also create new ones. By no means do they survive only in the form of official church organizations. Over many centuries, mystical orders within Islam have developed huge networks stretching from China to Central Asia to the Mediterranean.[79]

Apart from the frontiers of Christian or Islamic conversion and such one-off events as the World's Parliament of Religions that convened in Chicago during the international "Columbian Exhibition" of 1893, religious communication in the nineteenth century occurred mostly within the framework of a single religion.[80] Some of these spheres were quite large, though, and new means of transportation were developing them better than ever before. Muslims from many parts of Asia and Africa traveled by steamship to the holy sites in Arabia or to centers of learning such as Cairo, Damascus, or Istanbul. In Malaya, it was journeys to Mecca that actually brought into being something that could be described as a tourist industry.[81] The railroad made it affordable to visit Islamic shrines in the Tsarist Empire or sacred places in Catholic Europe (in 1858, apparitions of the Virgin in Lourdes helped this little town in the French Pyrenees to become a major center of pilgrimage). The new logistics also bolstered Rome's image as the Eternal City, since believers could now travel there en masse even outside holy years.[82] Pius IX, who scarcely ever left Rome and declared war on the modern zeitgeist, paradoxically became the creator of the worldwide papal church. Less a bureaucrat than a pastoral worker, he actively sought contact with the faithful, encouraged their financial contributions to the papal coffers, and was the first pontiff to call bishops from all over the world to Rome. An unprecedented total of 255 bishops gathered there in 1862, several years before the First Vatican Council (1869–70), on an occasion that was itself "global" in character: the canonization of twenty-six individuals who had suffered religious martyrdom in Japan more than a quarter of a millennium before.[83] This ceremony happened to come at the end of a year when many new martyrs had been created among missionaries and converts in Vietnam, in the last major "old style" persecution of Christians in Asia, before the rise of atheistic state apparatuses in the twentieth century.

The new media did the rest to speed up the circulation of religion. One factor in Rome's role as world capital of Catholicism was that the foreign press began to post correspondents there; the papacy had become newsworthy. The Mormon leader Brigham Young, at once theocratic ruler, sect boss, and businessman, realized which way the wind was blowing and soon had telegraph cables laid all the way to Utah. Railroad links to Salt Lake City made it harder to keep temptations at bay and easier for the federal army to send in troops, but the farsighted leader of the sect also saw that they would steer Mormons away from extreme navel-gazing tendencies.[84] In the second half of the nineteenth century, cheaper and simpler publishing techniques made it possible for the first time to print the Bible by the million and to favor exotic peoples with the holy book in their own language. The numerous translations done for this purpose belong among the greatest achievements of intercultural transfer in the nineteenth century. Catholic milieux that were less focused on the Bible itself now began to consume huge quantities of cheap tracts, pamphlets, and almanacs, giving a boost to new forms of popular religiosity on the margins of the official church. Popular religions blossomed wherever the respective orthodoxy lost some of its capacity for control. A major prerequisite for this was the decline of illiteracy and the growing potential to provide a mass public with printed matter. Both in Europe and in missionary regions, the possibility of feeding the Bible to new readers became an important motive for the religious (especially Protestant) commitment to education. Where people felt defensive toward the torrent of words coming from an expanding Christianity, the printing press offered itself as a weapon of resistance. This was one reason why in the last third of the century, after ages of skepticism, the Islamic clergy (*ulama*) enthusiastically embraced it for its own ends.[85]

CONCLUSION

The Nineteenth Century in History

"A general history of the world is necessary but not possible in the present state of research. . . . But we need not despair: particular research is always instructive when it produces results, and nowhere more so than in history, where even in deep recesses it always encounters a living element with universal significance."[1] These words of Leopold von Ranke, written in 1869, still hold true today. This book has attempted a piece of impossible, though perhaps not "general," global history. In the end, both reader and author should return to particular concerns, not soar upward into even more ambitious generalizations. The panoramic view from a summit is an impressive experience. But—as the great German medievalist Arno Borst asks—how long can a historian remain on a summit?[2] The following remarks do not offer the distilled essence of an epoch or a speculation about the spirit of the age. They are meant as a final comment, not as a summation.

1 Self-Diagnostics

The opening chapter presented the nineteenth century as an age of increased self-reflection. From Adam Smith in the 1770s until Max Weber in the early decades of the twentieth century, grandiose attempts were made to grasp the whole of the contemporary world and to place it within the historical *longue durée*. Diagnoses of the age did not appear only in Europe. They are found wherever societies developed the type of the scholar or intellectual, wherever ideas were written down and discussed, wherever observation and criticism gave an impetus to reflections on one's own lifeworld and its broader spatial and temporal preconditions. Such reflections did not always take a form that can be easily identified from today's retrospect as "diagnosis of the times" or "theory of the contemporary age."[3] They could be clad in the most diverse genres: as contemporary history in the Egyptian Abd al-Rahman al-Jabarti, for example, who experienced the Napoleonic occupation of his country and gave a detailed account of it,[4] or in the famous historian of antiquity

Barthold Georg Niebuhr, who also lectured on his own times, the "Age of Revolution"; as taking a position on political events of the day, as in Hegel's 1831 essay on the English Reform Bill or Marx's stirring polemic against Louis Napoleon and his shift from president by election to dictator by acclamation (*The Eighteenth Brumaire of Louis Napoleon*, 1852); as philosophical criticism of contemporary culture in Madame de Staël (*De l'Allemagne*, 1813), Alexis de Tocqueville (*Democracy in America*, 1835–40), or the Egyptian educational reformer and translator Rifaa al-Tahtawi (*A Paris Profile*, reporting on his stay in the French capital in 1826–31, first published in 1834);[5] as a regular journal in Edmond and Jules de Goncourt (covering the years 1851–96) or the Japanese army doctor and poet Mori Ōgai (for his stay in Europe between 1884 and 1888); as autobiography in the black ex-slave, intellectual, and civil rights activist Frederick Douglass (the most important of his three books of memoirs: *My Bondage and My Freedom*, 1855) or the American historian Henry Adams (*The Education of Henry Adams*, 1907 privately, published in 1918); or, finally, as disparate journalism in John Stuart Mill (whose diagnosis of the age is found more in short *pièces d'occasion* than in his principal works) or Liang Qichao (who for three decades commented on and helped to shape political events in China).

Sociology, as it emerged around 1830 on older foundations, was an endeavor to interpret the contemporary world. Initially associated with political economy and the newly rising science of ethnology, it developed basic models for an understanding of the age that are still discussed today: for example, the transition from status to contract as the organizing principle of society (in the legal historian Sir Henry Maine, *Ancient Law*, 1861) or the related opposition between community and society (*Gemeinschaft* and *Gesellschaft*) in the eponymous book by Ferdinand Tönnies (1887). Karl Marx analyzed capitalism as a historically determinate social formation—and Friedrich Engels added many insightful points relating to the diagnosis of his time. John Stuart Mill had earlier produced a great synthesis of classical political economy (*Principles of Political Economy*, 1848). Herbert Spencer tried to show how a peaceable industrialism had evolved out of a military barbarism into which it might one day relapse (*Principles of Sociology*, vol. 1, 1876). Fukuzawa Yukichi inserted Japan into the general development of civilization (*Bummeiron no gairyaku* [Sketch of a theory of civilization], 1875);[6] the Armenian Iranian Malkom Khan interpreted European modernity in the light of Islamic values (*Daftar-i Tanzimat* [Book of reform], 1858).[7] Philosophers and literary critics such as Friedrich Schlegel and Heinrich Heine (especially in his *History of Religion and Philosophy in Germany*, 1835), Ralph Waldo Emerson and Matthew Arnold, Friedrich Nietzsche, and at the end of our period, Karl Kraus and Rabindranath Tagore registered the cultural sensibilities and contradictions of their age.[8] The rich self-diagnoses of the nineteenth century must be the starting point for any attempt to grasp its specific signature.

2 Modernity

On top of these come the interpretations offered by present-day sociology, which revolve around the concept of modernity.[9] Mostly they also have something to say about the past, therefore referring explicitly or between the lines to the nineteenth century, but often the net is cast more widely to take in the whole of the European modern age. A category such as "individualization" can hardly be pinned down to a particular period. By tradition and custom, virtually the entire modern discourse of sociology limits itself to Western Europe and the United States. Since about 2000, however, the research agenda of "multiple modernities," championed by the great sociologist S. N. Eisenstadt, has brought an important advance. What Eisenstadt sees in the nineteenth century is above all a divergence between European and North American paths, so that modernity for him has by no means shaped a homogenous West, while in the non-Western world the characteristic features of modernity are recognizable only in Japan, if only with many special twists.[10] It is indeed difficult, for the period roughly between 1800 and 1900, to find distinctive Indian, Chinese, Middle Eastern/Islamic, or African paths to modernity independent of the West European model. Such differentiation became noticeable only after the turn of the century, at first less structurally than in the history of ideas.

If historians today want to operate meaningfully with the category "modernity," they must guide themselves by theories at the highest level that sociology has to offer. At the same time, they should bear in mind how the nineteenth century interpreted itself, and they ought to strive for greater spatial and temporal precision than is usually to be found in social science literature. Sweeping conceptions of "the bourgeois subject," "functional differentiation," or "civil society" become serviceable only if it is possible to specify their reference in historical reality. Any attempts to postulate the spontaneous emergence of modernity in the course of the nineteenth century only remain contentious. The intellectual foundations of modernity were laid during the "early modern" age in Europe, between Montaigne and Bacon at the beginning and Rousseau and Kant toward the end of the period.

What is the *primary* understanding of modernity? Is it an incipient long-term rise in national income; the conduct of life involving rational calculation; a transition from status to class society; the growth of political participation; a legal basis for relations of political rule and social intercourse; destructive capacities of a quite new dimension; or a shift in the arts away from imitation of tradition to the creative destruction of aesthetic norms? There is no concept that would hold all these aspects (and others) in neutral equilibrium, and a mere listing of characteristics would remain unsatisfactory. Concepts of modernity always pose priorities and—even if they are not monothematic—place the various aspects in a ranking order. As a rule, they do not disregard the fact that these aspects were in harmony with one another in only a few historical cases. It is enough to look closely at a country like France, a pioneer of modernity, to encounter discrepancies and obstructions. The Enlightenment philosophes were in their century the most "modern" group of thinkers anywhere in the world, and the French Revolution, especially the phase before the execution of Louis XVI and the onset of the Terror, appears to many historians and theoreticians even today as a highly important source of political modernity. On the other hand, France was a country where, outside Paris and a few other large cities, archaic social forms persisted well into the nineteenth century, at a time when they were much rarer in England, the Netherlands, or southwestern Germany.[11] Moreover, it took a full ninety years after the beginning of the Great Revolution for the French political system to stabilize as a parliamentary democracy. Lengthy processes were necessary to translate the "birth of modernity" at the level of ideas into institutions and mentalities that came close to the definitions of modernity used in today's social theory. Also the experience of the nineteenth, and even more the twentieth, century shows that economic modernity can go together with politically authoritarian conditions. It is also true that aesthetic innovation is improbable under extreme repression (Dmitri Shostakovich or Anna Akhmatova were exceptions that proved the rule in the Stalinist period), but it does not necessarily flourish where the most modern political conditions prevail. Thus, around 1910 the capital of the Habsburg monarchy was in no way inferior as a cultural center to London and New York, the metropolises of democracy and liberal capitalism.[12]

There is a further problem with "modernity." Are we interested mainly in its "birth," which by definition could happen only once at a particular time and place? Is it enough that modern principles came into the world somewhere and sometime? Or are we more concerned with how it spread and took effect, and with the point at which whole societies could be described as modern or thoroughly modernized? How can such gradations of modernity be determined? When fully developed, "high" modernity is no longer an insular tendency but has become the dominant way of life; it is no longer norm-breaking and revolutionary, as in the period of its "birth," but an everyday routine productive in turn of antimodern or postmodern tendencies. Since the concept of modernization receded in the late twentieth century before the concept of modernity, such questions about the breadth or systematic character of modernity are seldom raised. One would not wish to describe many countries in the world around 1900 as predominantly modern; the list would include Britain, the Netherlands, Belgium, Denmark, Sweden, France, Switzerland, the United States, the British dominions (Canada, Australia, and New Zealand), and with some reservations Japan and Germany. In relation to Europe east of the Elbe or Spain and Italy, there would be doubts about whether they were ripe for modernity. But what is to be gained by such evaluations?

3   Again: The Beginning or End of a Century

Historians today need not allow political rhetoric to drive them into making essentialist statements about Europe. Their discipline is in the fortunate position of being able to leave behind old political-ideological struggles over the conception of Europe. The issue is now seldom any more whether it should be Catholic or Protestant, Latin or Germanic (or Slav), socialist or liberal-capitalist, although older cleavages along a north-south axis have reemerged during the financial crisis of the 2000s. Also there is broad agreement in the literature about Europe's most important characteristics and tendencies in the long nineteenth century.[13] For the most part, however, it cannot clarify the extent to which such features and processes constituted a special European role in history, because it still rarely uses the possibilities of a comparison with regions outside Europe. We should note with the German historian Jost Dülffer: "Europe cannot be presented or understood from within itself";[14] only comparison with Japan or China, Australia or Egypt, can bring out its distinctive profile. This is especially productive if it is undertaken by non-Europeans, since they are struck by many cultural peculiarities that Europeans take for granted. Of course, a global historical perspective must do without such opportunities for an external or ex-centric viewpoint. The world as a whole cannot be contrasted with anything else.

What other picture of the nineteenth century results if the vantage is not purely European? The first point to make is that a long nineteenth century, from the 1780s to the First World War, remains a useful assumption or auxiliary construction, but it should not be taken as a natural or globally valid form of the past. Even if one does not stick pettily to the European outer dates of 1789 and 1914, whole national and regional histories elude this framework. It may be applicable elsewhere, but sometimes for reasons that have little to do with Europe. The fact that Australia's recorded history begins in 1788, with the first penal convoy, is not related in any way to the French Revolution. And if the years between the abdication of the Qianlong Emperor in 1796 and the Revolution of 1911 have a certain unity within the political history of China, this has reasons internal to the dynasty and cannot be attributed to European activities in East Asia. There are numerous instances in which a different periodization should be preferred. In Japan, the years between the opening of 1853 and the collapse of the empire in 1945 constitute a complete historical cycle. Latin America's nineteenth century stretches from the independence revolutions of the 1820s (whose causes go back to the 1760s) to the eve of the Great Depression of 1929. As far as the United States is concerned, the Civil War of the 1860s ended a first era that had begun with the transatlantic crisis of the 1760s, and the new epoch of political and social history certainly did not end in 1914 or 1917–18 but rather in 1941 or 1945 or, from the important point of view of race relations, as late as the 1960s. For the whole of Africa—with the exception of Egypt and South Africa—neither the years between 1800 and 1900 nor the "long" nineteenth century seem a relevant time frame. The colonial invasion of the 1880s opened an age that lasted beyond the First World War to the peak of decolonization in the 1960s. It follows that a global historical periodization cannot work with precise cutoff points like those that mark particular national histories or the history of Europe. The beginning and end of the nineteenth century must remain open.

Yet the various narrative threads of this book do yield a pragmatic solution. A new era gradually began in the 1760s with a multiple political crisis throughout the Atlantic space, Britain's colonial implantation in India, and the development of new production techniques. It ended in the course of the 1920s, as the manifold consequences of the First World War (some of them positive in East Asia and Latin America) became visible, and movements for national autonomy arose all over the colonial world (except for tropical Africa) and other regions held down by the West. Another process with far-reaching implications was the transformation of the Soviet regime from a center of world revolution to a neo-imperialist power. Over a vast territory, the most important nineteenth-century current of dissident ideas—socialism—crystallized into a state with no precedent in history, introducing new polarities into world politics and, in the initial period, a new kind of revolutionary ferment.

The First World War had disenchanted the West and placed a question mark over its claim to rule over, or at least to act as a civilizing guardian for, the rest of humanity. Many global interrelations of the prewar period had thinned out.[15] The new order that emerged from the peace conferences of 1919–1920 was not totally misconceived, but it was not capable of fulfilling many expectations; Wilson had not brought about perpetual peace. The forces of capitalist regeneration seemed to be stretched beyond the limit, at least in Europe. Liberalism in all its four aspects—moral/individual-ethical, constitutional-political, international, and economic—was under strong legitimation pressure and losing influence worldwide.[16] The 1920s marked the decisive passage from the nineteenth century to another age.

4 Five Characteristics of the Century

How should this long nineteenth century, open at either end, be characterized from the point of view of global history? We cannot try to summarize the content of this book in a few sentences, nor will it advance our knowledge to repeat the headings conventionally, and accurately, used to describe the main trends of the age: industrialization, urbanization, state building, colonialism, globalization, and a few more besides. Instead, let us propose five less common angles of vision.

(1) The nineteenth century was an age of *asymmetrical efficiency growth*. An overall gain in efficiency manifested itself in three spheres. First, the productivity of human labor increased in a degree that outstripped growth processes in earlier

epochs. Even if statistics cannot meet the challenge of quantification, no one disputes the fact that in 1900, material value creation per capita of the world population was considerably higher than a century before. Per capita income had risen, humanity had become materially richer, and long-term growth, with conjunctural fluctuations oscillating around a steady upward trend, had been achieved for the first time in history. One of the two factors underlying this was the spread of the industrial mode of production, marked by an extensive division of labor, factory organization, and coal-powered machinery—a process with a very uneven regional distribution, even in the most developed industrial heartlands of northwestern Europe and the northern United States. It rested to some extent on scientific principles known for some time. Innovative routines, together with new market structures and legal conditions capable of making them worthwhile, developed in a few countries in Europe and North America and, as the century wore on, gave rise to self-reproducing systems of knowledge production and "human capital" formation, both in public or private higher education and within industry itself. "The greatest invention of the nineteenth century," as the philosopher Alfred North Whitehead trenchantly remarked in his Lowell Lectures of 1925, was "the invention of the method of invention."[17]

The other source of increased wealth was the opening up of new frontiers in every continent: from the American Midwest to Argentina, from Kazakhstan to Burma. This, too, was bound up with particular visions of modernity; not every kind of nineteenth-century modernity was placed in an industrial frame. A kind of agrarian revolution preceded the Industrial Revolution, above all in England. Later, accompanying the uneven and often less-than-revolutionary spread of industrialization, there was a much wider extension of land use, resulting in higher productivity for individual producers in some frontier areas. The typical products of these frontiers were geared not to local consumption but to intercontinental trade, which was no longer simply trade in luxury goods. The application of industrial technology in the form of steamships and railroads rapidly lowered transport costs, thereby boosting the export of classical frontier items such as wheat, rice, cotton, and coffee. The opening up of agrarian frontiers was linked to industrialization insofar as demand grew for raw materials, and food had to be found for the industrial workforce newly released from the land. But only in the twentieth century do we see an industrialization of agriculture itself and the global rise of agroindustry.

A third domain clearly displaying efficiency growth was the armed forces. The killing capacity of an individual soldier increased, not as a direct result of industrialization but in close parallel to it. Along with innovation in weapons technology, advances in organization and strategy were an independent factor in efficiency gains—another precondition being the political will to divert government resources to the military. International discrepancies in these respects became noticeable in the German wars of unification, the numerous colonial wars of the time, and the Russo-Japanese War. In 1914, military apparatuses scarcely susceptible to political control entered into open conflict with one another. These apparatuses with their real or imagined inherent dynamism—one famous example of such a clockwork-like autonomous logic was the war plan of Alfred von Schlieffen, the chief of the Imperial German General Staff—made an incompetent or irresponsible foreign policy even more dangerous than in the past. The potential destructiveness of the instruments multiplied the risks of political folly.

The World War itself created the occasion for further efficiency gains at several levels, including the organization of a war economy in Germany, Britain, and the United States. At the end of the century, the unevenness of the distribution of military power around the world was without precedent. It had become identical with industrial might, in a way that had not been the case at all in 1850. There were no longer any nonindustrial great powers. Although Afghans, Ethiopians, and Boers scored some momentous victories in passing, no non-European military player—bar Japan—could withstand the armored powers of "the West." This military "great divergence" gradually receded again only in the early 1950s, when China resisted the United States in the Korean War and the Vietnamese defeated the French at Dien Bien Phu in 1954.

A fourth field of increased efficiency was the greater control of state apparatuses over their own population. Administrative regulations became denser; local authorities took on additional powers; official agencies registered and classified the population, as well as its ownership of land and fiscal potential; taxes were skimmed off more fairly and with greater regularity from a growing number of sources; police forces were strengthened in both depth and breadth. On the other hand, there was no straightforward correlation between the political system and the intensity with which government steered people's lives. Up to today, a democracy may be densely administered, while a despotic regime may have only a weak presence at the base of society. The nineteenth century saw the emergence of new technologies of local governance—prerequisites also for universal conscription and state education and welfare systems. The state began to develop into a new Leviathan, but one that did not necessarily have to be a monster.

This increase in the effective reach of the state was also very unevenly distributed: Japan was more thoroughly penetrated by the state than China; Germany more than Spain. Almost everywhere, the colonial state had the will to register and regulate its subjects, but often it lacked the financial resources and the personnel to carry it out. The idea of the nation-state that emerged in the nineteenth century, involving a coincidence of state form, territory, and culture (language), stood in a mutually determining relationship with state intervention. Members of a nation wanted not to be subjects but rather free citizens treated equally within a homogeneous collective; they strove for their country to be recognized internationally and to be held in high esteem. Yet, in the name of national unity, national honor, and the national interest, people endured a regulation frenzy that they would have opposed in earlier times.

Partial efficiency increases occurred in many places around the world. In no way was industrialization an independent variable or a demiurge unleashing all other kinds of dynamic: agrarian frontiers were more widely spread than industrial heartlands; Washington and Suvorov, Napoleon and Wellington conducted preindustrial wars. Nor did the three spheres of growing efficiency—the economy, the armed forces, and the state—reinforce one another in a predictable manner. In the Ottoman Empire, a "modern" state bureaucracy began to develop without a significant industrial backdrop. The United States in the decades after the Civil War was an economic giant but a military dwarf. Russia industrialized and had a huge army, but it is questionable how deeply its state penetrated society before 1917, especially in the countryside. In fact, only Germany, Japan, and France remain as models of a modern nation-state in every conceivable dimension. Britain, with its modest territorial army and relatively nonbureaucratic local government, was as much a case on its own as the United States.

Nevertheless, the rise of Europe, the United States, and Japan in comparison with the rest of the world was more than ever before or since an incontrovertible fact. There were a whole series of reasons for this. At least until the First World War, their success story was self-sustaining. The dominant countries profited from a liberal world economic order of their own creation, which in turn supported economic growth that could be profitably steered in such a way as to finance a position of power in the international arena. Imperialism could also be a good investment. Although colonial expansion may not in every case have directly yielded monetary gains to the national economy, military superiority meant that it was relatively cost-effective to conquer and administer a colony. Imperialism was politically worthwhile so long as it cost the state little or nothing; and it called forth vested interests prepared to lend it political support.

(2) Less need be said about the epochal marker of increased *mobility*, since the relevant chapters above speak for themselves. The whole of recorded history is rife with movement: travels, mass migration, crusades, long-distance trade, spread of religions, languages, and art styles. Three things were new in the nineteenth century.

First, the scale of human mobility sharply increased. Earlier history knows no examples on a par with the emigration to North and South America, Siberia or Manchuria, nor has the magnitude of permanent relocation during the years between 1870 and 1930 been repeated since. It is a striking global characteristic of that period. The circulation of goods reached a new level too, when the luxury businesses of early modern merchants trading in silk, spices, tea, sugar, and tobacco were overshadowed by mass transfers of food staples and industrial raw materials. Aggregate figures for the expansion in world trade, far exceeding increases in output, clearly demonstrate this point. Capital in general was mobilized on a large scale only during this period. Before the middle of the century, wealthy individuals had lent money to princes and certain others who needed it. The early modern chartered companies had been, by the standards of their time, complex financial operations. But it was only after 1860 or thereabouts that something like a capital market came into being. Driven even more by railroad construction than by the industrial factory economy, paper capital "flowed" for the first time around the globe—no longer (or not only) as actual bullion in ships' bellies. The age of liquidity was dawning. The steamship and the railroad increased the mobility of people and goods, while the telegraph, and later the telephone, facilitated the communication of information.

Second, these technical innovations speeded up all forms of circulation. Things moved faster even within cities, as the pedestrian gave way to the streetcar. To see this acceleration as a hallmark of the age has become almost a banality, but it is difficult to exaggerate the historic impact on human experience represented by the ability to move faster and more reliably than a horse or to travel on water without being at the mercy of the wind. By 1910 the railroad was established on every continent, even where there was little industry to speak of. For ordinary people in India, the chance of working on the railroad or one day traveling by train was considerably greater than that of seeing the inside of a factory.

Third, mobility was only now underpinned by infrastructure. Although we should not underestimate the complexity of communications in the Inca world, in the thirteenth-century Mongol Empire, or the mail coach network of Regency England, the fact remains that the laying of railroads, the initiation of global shipping lines, and the cabling of the planet brought a quite different level of technological application and organizational permanence. Mobility was no longer just a way of life for nomadic peoples, an emergency for refugees and exiles, or a way for seamen to keep body and soul together. It had become a dimension of organized social existence whose rhythms differed from those of local everyday routines. These trends continued without interruption into the twentieth century. The keyword "globalization" finds its place here, if we define it roughly—without exhausting the potential scope of the term—as accelerated and spatially extended mobilization of resources across the boundaries of states and civilizations.

(3) A further striking feature of the nineteenth century may be described, somewhat technically, as its tendency to *asymmetrical reference density*. "Increased perception and transfer across cultures" would be a less cumbersome, but also less precise, formulation for the same phenomenon. What is meant is that ideas and cultural content in general—more than the pieces of information transmissible by telegraph—became more mobile in the course of the nineteenth century. Again, we should not underestimate what happened in earlier epochs. The diffusion of Buddhism from India to many regions of Central, East, and Southeast Asia was an immense, multifaceted process of cultural migration often quite literally carried by the feet of itinerant monks. The novelty of the nineteenth century was the spread of media that allowed people to send news over great

distances and across cultural boundaries and to make themselves familiar with the ideas and artifacts produced in distant lands. There were more translations than in previous times: not only within Europe (where the eighteenth century had already been a great age of translation) but also in the more difficult interchange between European languages and others more remote from them. In 1900, the major libraries of the West had available in translation the basic texts of the Asiatic tradition, while European textbooks in many branches of knowledge, as well as a selection of writings in political philosophy and legal or economic theory, were accessible to readers of Japanese, Chinese, and Turkish. The Bible, of course, was translated into a great number of languages some of which had lacked a script before the advent of Christian missionaries. Some grasp of foreign languages, especially English and French, made it easier for educated elites in the East to become familiar with Western ideas at first hand.

"Greater reference density" means more, however, than a mutual widening of horizons. The American sociologist Reinhard Bendix has underlined the power of the "demonstration effect" in history: that is, the existence of "reference societies" serving as a model for imitation but also as a focus for the formation of identities through rejection and discriminating critique.[18] In the eighteenth century, France with its tension between court and salon was such a reference for large parts of Europe; and long before, Vietnam, Korea, and Japan had taken their bearings from China. Two things happened in the nineteenth century. On the one hand, such external orientations grew in number: while a great majority of the world population continued to know nothing of life in foreign countries, or else associated it with only the haziest imaginings, the educated elites began to observe the outside world more closely than ever before. On the other hand, the reference became asymmetrical or unipolar. Instead of a multiplicity of cultural models, the West now appeared as the global standard. But "the West" certainly did not mean the whole of Europe, nor did it always include the United States (which acquired importance as a distinct civilizational model only around the end of the century). For China, Japan, Mexico, or Egypt in 1870 or 1880, "the West" was first Britain, then France. Where the elite was impressed by the military and scientific achievements of the Bismarckian state—which it was in Meiji Japan, for example—Germany came to feature as an additional model.

Peripheries whose "Western" credentials were not entirely beyond doubt could also be found within the geographical confines of Europe. Russia, with its long experience as an outpost of Christianity, continued to see itself as a periphery in relation to the French, British, and German West. Debates there between "Westernizers" and "Slavophiles" bore more than a passing resemblance to those in the Ottoman Empire, Japan, or China. The spectrum of possible attitudes ranged from genuine enthusiasm for Western civilization—associated with a critical, indeed iconoclastic, relationship to one's own tradition—to contemptuous dismissal of Western materialism, superficiality, and arrogance. The convictions of most "peripheral" intellectuals and statesmen hovered in an ambivalent middle. In many places across the globe, debates were raging about whether or how it might be possible to appropriate the technological, military, and economic achievements of the West without capitulating to it culturally. In China this was expressed in the pithy *ti-yong* formula: Western knowledge for application (*yong*), Chinese knowledge as cultural substance (*ti*). The same challenging paradox was familiar in a wide variety of contexts.

A perception that the Western model of civilization, with all its unconcealed internal differences, made it essential to find some political response resulted in various strategies of defensive modernization, from the Tanzimat reforms in the Ottoman Empire to technocratic rule in the Mexico of Porfirio Díaz. In general, these were motivated by a sense that something useful could be learned from the West, but usually they also involved strengthening the country to forestall military conquest or colonization. Sometimes this was successful, but in many other cases it was not.

Liberal patriots, spread widely outside Europe if only in tiny circles, found themselves in a particularly difficult position. As liberals they enthusiastically read Montesquieu, Rousseau or François Guizot, John Stuart Mill or Johann Kaspar Bluntschli, and demanded freedom of the press and association, religious tolerance, a written constitution, and representative government. As patriots or nationalists they had to oppose the very West from which all these ideas stemmed. How was it possible in practice to separate the good West from the bad? How could controlled imports of culture or even finance be achieved *without* imperialism? This was the great dilemma of politics in the nineteenth-century peripheries. But once imperialism had struck, it was too late to oppose it for the time being. The room for maneuver shrank dramatically, the range of options was hugely reduced.

Greater reference density was neither something as innocuous as a simple gain in knowledge and education nor so free of contradictions that it could be summed up in the crude term "cultural imperialism." In most cases it was a question of politics, but not always with one clear way forward. Almost never was the power of European colonial masters great enough to force on unwilling subjects the most prestigious of all Western cultural exports: the Christian religion. Reference density was asymmetrical not only within the (always unbalanced) colonial relationship but for two other reasons besides. First, the major European powers repeatedly abandoned their fragile alliances with Western-oriented reformers in the East and the South, if this seemed to be advisable in pursuit of national or imperial interests. By the turn of the century, scarcely anyone in Asia or Africa believed that the West, committed to hard-nosed *realpolitik*, was interested in the genuine modernization of colonies and of those independent peripheral states that thought of themselves as promising aspirants to modernity. The utopia of a benevolent West-East partnership in modernization, having peaked in the 1860s, 1870s, and 1880s with the late Tanzimat reforms, the khedive Ismail in Egypt, and the Rokumeikan period in Meiji Japan,[19] had given way to a deep mistrust of Europe.

Second, knowledge of the non-European world increased appreciably in the West, thanks to the rise of Oriental philology, ethnology, and comparative religious studies, but it yielded no practical consequences. Whereas the East borrowed all it could from the West—from legal systems to architecture—no one in Europe or North America thought that Asia or Africa offered a model in anything. Japanese woodcuts or West African bronzes found admirers among Western aesthetes, but no one suggested, for example, taking China as a model for the organization of the state in Western Europe (as some had done in the eighteenth century when the Chinese bureaucracy won a number of admirers in the West). To some degree reciprocal in theory, cultural transfer was in practice a one-way street.

(4) Another feature of the century was the *tension between equality and hierarchy*. In a major textbook, the Swiss historian Jörg Fisch has rightly described "the successive realization of legal equality through the removal of particular areas of discrimination and the emancipation of groups affected by them" as one of the central processes in Europe during the second half of the nineteenth century.[20] This tendency toward legal equality was associated with rules and patterns of societal stratification that reduced the importance of family origin, making the market more important than ever before in determining social position and possibilities of advancement up the ladder. With the abolition of slavery, the transatlantic part of the West, already less marked by status hierarchies than the Old World, joined the trend toward *general* equality.

Europeans were thoroughly convinced of the perfection and general validity of their conceptions of social order. As soon as elites in non-European civilizations became familiar with European legal thinking, they realized that it was both specific to Europe and capable of universalization; it contained a threat and an opportunity, according to circumstances and political belief. This applied especially to the postulate of equality. If Europeans denounced slavery, the inferior position of women, or the repression of religious minorities in non-European countries, this was liable to present an explosive challenge to the established order. The outcome had to be radical changes in power relations: a limitation of patriarchy, the toppling of slave-owning classes, or the ending of religious and ecclesiastical monopolies. Social equality was not just a European idea: utopian visions of leveling, fraternity, and a world without rulers were widespread in many different cultural contexts, In its modern European guise, however, whether based on Christian humanitarianism, natural law, utilitarianism, or socialism, the idea of equality became a matchless weapon in internal politics. Conservative reactions were inevitable, cultural battles between modernists and traditionalists became the rule.

The commitment of Westerners to their own principle of equality, however, proved to be limited. New hierarchies formed in international relations, for example. The Peace of Westphalia (1648) had substituted a simpler ranking system

for the older plethora of finely shaded relations of subordination and privilege—
even if it is much too simple to imagine that the diplomats at the peace congress
instantly created a "Westphalian system" that would last until 1914 or even 1945.[21]
Only in the nineteenth century, and above all after the geopolitical upheavals of
the 1860s, do we see the disappearance of small and medium-sized international
actors from the European political scene (temporarily, as developments in the
late twentieth century were to show). Only then did the famous "pentarchy" of
great powers have things all to themselves. Any country that could not keep up
in the arms race ceased to count in world politics. The Netherlands, Belgium,
and Portugal, for example, were demoted to the status of low-ranking owners
of colonies without political clout. The extent to which the weaker countries
of Europe became irrelevant was demonstrated in 1914 when Germany violated
Belgian neutrality without any scruples.

Non-European countries, including the Ottoman Empire (a sixteenth-century
superpower) but obviously not the United States, were assigned places at the
bottom of the hierarchy. Only Japan, relying on unprecedented national exertions, an astute foreign policy, and a little luck, managed to break into the exclusive circle of the major powers. But it did so at the expense of China and Korea,
after one of the bloodiest wars of the age, and not without some galling snubs
from the "white" protagonists of world politics. The decisive turning point came
at the Washington Conference of 1921–22, which finally recognized Japan's
position as a front-ranking naval power in the Pacific and hence its great-power
status.

What might be called the "secondary" hierarchies, newly established in the last
third of the century, further sabotaged the postulate of equality. The achievement of equal civil rights by the Jewish population of Western Europe was
followed in short order by their subjection to social discrimination. And the
abolition of slavery in the United States soon led to novel practices of segregation. The new social distinctions were formulated at first in the language of
fully attained versus deficient "civilization," and later in a racist idiom scarcely
ever called into question in the West. The racist cancellation of the principle of
equality pervaded the international order for an entire century, from about the
1860s through decolonization. Only a quiet revolution in international human
rights norms, also involving antiracism, more robust principles of territorial
sovereignty, and a strengthening of the right to national self-determination, has
finally led since the 1960s to a turning away from the nineteenth century.

(5) The nineteenth century was also a century of *emancipation*. This will hardly
sound surprising. We read again and again about an Age of Revolution, stretching either from 1789 to 1849 or covering the whole period down to the Russian
revolutions of 1905 and 1917, and also about "emancipation and participation"
as basic tendencies of the epoch.[22] This always refers to Europe alone. The word
"emancipation," derived from Roman law and emphatically European, is far less

likely to be applied to the world as a whole. Emancipation means, in the words of a political scientist, "the self-liberation or release of groups in society from intellectual, legal, social, or political tutelage or discrimination, or from forms of rule that are perceived as unjust."[23] The term also often refers to national liberation from the rule of an empire or neighboring state. Should we then extend to the rest of the world Benedetto Croce's idealist view of 1932, in which the drive for liberty was a major motivating force of nineteenth-century Europe?[24] To some extent, yes.

A number of emancipation processes were successful. They led to greater freedom and equal rights, more rarely to actual equality. Slavery disappeared as a legal institution from the countries and colonies of the West. European Jews to the west of the Tsarist Empire achieved the best legal and social position they had ever had. The European peasantry was released from feudal burdens. The working classes fought for and won the freedom of association and, in many European countries, the right to vote. The balance sheet is harder to draw in the case of women's emancipation, which first became a theme of public debate only in the nineteenth century. Here the British dominions and the United States led the way in terms of political rights and opportunities. But it is not possible to say in general, even for Europe, whether the position of women in relationships and the family also improved. The bourgeois family brought constraints of its own into play.

If we assume that the revolutions of the age were also about emancipation, the successes are more conspicuous than the failures—perhaps an illusion, given that history prefers to remember the victors. There were ambiguous cases, such as the French Revolution: its early goals of representative democracy were finally achieved in the Third Republic after many system changes, whereas the direct democracy model of the Jacobin dictatorship foundered and sank, making only one brief reappearance in the Paris Commune of 1871. Nor were the revolutions of 1848–49 unequivocal in their effect; complete failures they were undoubtedly not, if compared with such abortive and ultimately inconsequential experiences as the Tupac Amaru uprising in Peru or the Taiping Revolution in China.[25] In the interplay between revolution and reformist prophylaxis or postrevolutionary absorption of revolutionary impulses, Europe did in the end—at least west of the Tsarist Empire—achieve a gradual broadening of constitutional provisions for civic involvement. The fact that representative government had deeper roots here than in other parts of the world made this evolution easier. But on the eve of the First World War there were not so many democracies in the late-twentieth-century sense of the term. Not every state that had given itself a republican form, as most Latin American countries and China recently (1912) had done too, thereby provided the substance for democratic politics. The vast colonial sphere was divided between the very democratic British dominions (by now essentially independent nation-states) and the invariably autocratic colonial systems of what was then known as the "colored world."

All in all, the picture is ambiguous and contradictory even for Europe. In 1913, with regard to the trends of recent decades, it was possible to speak of the spread of democracy but not of its irresistible triumph, while political liberalism already had its best years behind it. Nevertheless, it was a century of emancipation or, more plainly put, a century of revolt against coercion and humiliation. Traditional forms of domination were less routinely perpetuated than in previous ages. The development of a huge federal polity in North America showed that contrary to all theoretical prognoses, a major country was capable of surviving on the basis of citizenship and participation. Monarchical absolutism was in crisis far beyond the borders of Europe—seemingly least in the Tsarist Empire, but all the more dramatically there as things turned out in 1917–18. Where the legitimation model of divine right persisted (as it did in Russia), major propaganda efforts were required to make it palatable to the population. Strong monarchies, such as Japan's system of imperial rule, did not rely on an uninterrupted continuum with the past but were self-consciously neo-traditionalist. European constitutionalist theory found serious and enthusiastic advocates in large parts of noncolonial Asia and Africa. The British Empire, by far the largest, sported constitutional rule in its dominions and, shortly before the First World War, indicated a willingness to consider timid constitutional concessions in India.

Emancipation pressure kept mounting "from below," from a "people" that, by virtue of the great revolutions at the beginning of the period, had become a real player as well as a legend that was often evoked. Slaves put up resistance, therefore making modest but incremental contributions to their own liberation. The Jewish population of Western Europe did not wait for effusions of grace from enlightened rulers but set in motion a great project of self-reform. Social interests organized themselves on a permanent basis; never before had there been anything like labor unions or mass socialist parties.

Even at the height of colonialism and imperialism, the concept of emancipation was not entirely out of place. Despite the fact that things quieted down in many colonies after the wars of conquest, perhaps even bordering on something like internal peace, foreign colonial rule could base itself on scant legitimacy. There was a thoroughly pragmatic reason for this, since the most popular justification—the "civilizing mission"—could easily be measured by its results. The colonized peoples might accept the self-serving rhetoric of the colonizers if the intervention actually brought the much-heralded benefits: security, justice, a little more prosperity, slightly better health care, and new educational opportunities not offered in exchange for complete cultural estrangement. Alien rule is an age-old phenomenon in history. So, in the eyes of many of its subjects, European colonialism was not more objectionable per se than any other kind of foreign rule: that of the Moguls in India, the Ottomans in Arabia, the Manchus in China, and so on. But if the promised advances failed to materialize or if living conditions became worse, the colonial reserves of legitimation soon ran out. This was the case in many places even before the First World War. The liberation movements of the later Third World—whether or not we call them "nationalist" for the early twentieth century—emerged in response to this credibility deficit. It was not difficult for critical intellectuals in the colonies or in exile to uncover the contradictions between the West's universal principles and its often deplorable behavior on the spot. After the Age of Revolution, colonialism was therefore ideologically unstable (and controversial also among the public of the colonial powers);[26] and even before any nationalist program entered the equation, pressure for emancipation was part and parcel of a colonial system resting on inequality, injustice, and hypocrisy—on "the unblushing selfishness of the greatest civilized nations" (as the outspoken naturalist and explorer Alfred Russel Wallace put it in 1898 in his review of the period).[27]

The nineteenth century did not end abruptly in August 1914, before Verdun in 1916, or with Lenin's arrival at the Finland Station in Petrograd in April 1917. History is not a theater where the curtain suddenly falls. In autumn 1918, however, it was widely noted that the "world of yesterday" (the title of the Austrian writer Stefan Zweig's important memoirs, posthumously published in 1942)[28] had gone up in smoke. In Europe some felt nostalgic for it, while others glimpsed the opportunity for a new beginning beyond the now disenchanted "belle époque." The US president Woodrow Wilson and his supporters around the world hoped to have finally overcome the discredited past. The twenties became the decade of global reorientation, a hinge period between the centuries, at least in a political sense.[29] Economically, they turned out to be the prelude to the Great Depression, a crisis more global still than the World War. Culturally, they prolonged in Europe the artistic avant-garde of the prewar period, while elsewhere they marked the start of something new in aesthetic terms. Whether it serves historical understanding to apostrophize the years between 1914 and 1945 as a "Second Thirty Years' War" must remain undecided. In any event, the analogy could apply only to Europe.

Let us try a different tack. Between 1918 and 1945, the world came up with unusually few constructive and durable solutions. The First World War had revealed many problems of the nineteenth century, while the interwar period offered not enough responses to those that still persisted. Many questions that had arisen in the nineteenth century retained their virulence even after 1945. Tendencies carried over from the late nineteenth to the late twentieth century. The *second* postwar period attempted a reset—not always successfully, but on the whole more so than the first. Some of the older men and women looking for new directions after 1945 had been born and socialized in the nineteenth century. Many had already been politically influential, or at least gained political experience, in 1919 or the years immediately after: for example, Winston Churchill, Konrad Adenauer, John Foster Dulles, Joseph Stalin, Yoshida Shigeru, and Mao Zedong. Others, such as John Maynard Keynes and Jean Monnet, had been active as advisers. Great philosophers, scientists, engineers, writers, composers, painters, and architects who had left their mark on the times before 1914 continued their labors. The nineteenth century had paved the way for the disasters since 1914; the philosopher Hannah Arendt and others held it responsible for them.[30] But other traditions in readiness after 1945 (liberalism, pacifism, trade unionism, or democratic socialism, for example) were not completely tainted or decrepit. From the retrospect of 1950, the year 1910—when, as Virginia Woolf once quipped, human character changed—appeared to be infinitely remote. In many respects, however, it was closer than the horrors of the most recent war.

ABBREVIATIONS

AER	American Economic Review
AES	Archives européennes de sociologie
AHR	American Historical Review
AJS	American Journal of Sociology
CSSH	Comparative Studies in Society and History
EcHR	Economic History Review
EHR	English Historical Review
EREH	European Review of Economic History
GG	Geschichte und Gesellschaft
GWU	Geschichte in Wissenschaft und Unterricht
HAHR	Hispanic American Historical Review
HEI	History of European Ideas
HJ	Historical Journal
HJAS	Harvard Journal of Asiatic Studies
HT	History and Theory
HZ	Historische Zeitschrift
IHR	International History Review
IJMES	International Journal of Middle Eastern Studies
IRSH	International Review of Social History
JAfH	Journal of African History
JAH	Journal of American History
JAS	Journal of Asian Studies
JbLA	Jahrbuch für Geschichte Lateinamerikas
JBS	Journal of British Studies
JEEcH	Journal of European Economic History
JEH	Journal of Economic History
JESHO	Journal of the Economic and Social History of the Orient
JGH	Journal of Global History
JGO	Jahrbücher für Geschichte Osteuropas
JHG	Journal of Historical Geography

JICH	Journal of Imperial and Commonwealth History
JIH	Journal of Interdisciplinary History
JLAS	Journal of Latin American Studies
JMEH	Journal of Modern European History
JMH	Journal of Modern History
JPH	Journal of Pacific History
JPS	Journal of Peasant Studies
JSEAS	Journal of Southeast Asian Studies
JTS	Journal of Turkish Studies
JWH	Journal of World History
LARR	Latin American Research Review
LIC	Late Imperial China
MAS	Modern Asian Studies
NPL	Neue Politische Literatur
P&P	Past and Present
PHR	Pacific Historical Review
VSWG	Vierteljahresschrift für Sozial- und Wirtschaftsgeschichte
WP	World Politics
ZHF	Zeitschrift für Historische Forschung

NOTES

INTRODUCTION

1. J. R. McNeill and McNeill, *Human Web*.

2. On the present situation see S. Conrad, *Globalgeschichte*; Osterhammel, *Global-geschichte* (2007).

3. Acham and Schulze, *Einleitung*, p. 19.

4. The title of an essay by Tony Judt, in *New York Review of Books*, September 21, 2000.

5. Bayly, *Birth of the Modern World*. See my review article: Osterhammel, *Baylys Moderne*.

6. An earlier work of mine, *Geschichtswissenschaft*, referred in its subtitle to "the history of connections" and "comparison between civilizations."

7. J. M. Roberts, *Twentieth Century*, p. xvii.

8. Hobsbawm, *Age of Revolution*; idem, *Age of Capital*; idem, *Age of Empire*.

9. Bayly, *Birth of the Modern World*, pp. 202ff.

10. Ibid., p. 4.

11. The dialectic of integration and differentiation is a commonplace of functionalist sociology, but for historians it is no more than a plausible-sounding phrase that has to be filled with content.

12. Bayly, *Birth of the Modern World*, pp. 451–87.

13. Fernand Braudel, "On a Concept of Social History," in: idem, *On History*, p. 131.

14. Acham, *Einleitung*, p. 16.

15. See the summary of many arguments in P.H.H. Vries, *Via Peking*.

CHAPTER 1: Memory and Self-Observation

1. *Süddeutsche Zeitung*, 24 June 2006. The giant tortoise Adwaita that died in March 2006 in Calcutta is said to have reached the age of 250, having been, in his youth, a pet of Robert Clive, the British conqueror of Bengal. http://news.bbc.co.uk/2/hi/south_asia/4837988.stm.

2. *New York Times*, 31 May 2009; *Süddeutsche Zeitung*, 28 May 2008.

3. Gluck, *Past in the Present*, p. 80.

4. Quoted in Blight, *Race and Reunion*, p. 1.

5. Peterson, *Lincoln*, pp. 320f.

6. Schreiber, *Kunst der Oper*, vol. 1, pp. 28–36; Mackerras, *Peking Opera*, p. 11.

7. J. H. Johnson, *Listening in Paris*, p. 239.

8. M. Walter, *Oper*, p. 37.

9. Scherer speaks of the "magnet city" in his splendid *Quarter Notes*, p. 128.

10. Burns, *Brazil*, p. 335.

11. McClellan, *Performing Empire*, p. 154.

12. Bereson, *Operatic State*, pp. 132f.

13. Roger Parker, "The Opera Industry," in: Samson, *Cambridge History of Nineteenth-Century Music*, pp. 87–117, at 88.

14. Takenaka, *Wagner-Boom*, pp. 15, 20.

15. Cf. Rutherford, *Prima Donna*.

16. See also chapter 6.

17. Pomian, *Sur l'histoire*, p. 347; Fohrmann et al., *Gelehrte Kommunikation*, pp. 326f.

18. Esherick and Ye, *Chinese Archives*, pp. 7, 10.

19. See the account of Turkish archiving in Faroqhi, *Approaching Ottoman History*, pp. 49–61.

20. D. M. Wilson, *British Museum*, p. 118 (Fig. 19: photo of the cast iron construction).

21. The Japanese Diet Library became the country's largest only in 1948, when it incorporated the holdings of the former Imperial Library.

22. MacDermott, *Chinese Book*, p. 166.

23. Kornicki, *Book in Japan*, pp. 364, 382, 384, 407f., 410, 412.

24. On the periods of book history in the Arab world, see George N. Atiyeh, "The Book in the Modern Arab World: The Cases of Lebanon and Egypt," in: idem, *Book in the Arab World*, pp. 233–53.

25. Sheehan, *Museums*, pp. 9ff.

26. Samuel J.M.M. Alberti: "The Status of Museums: Authority, Identity, and Material Culture," in: Livingstone and Withers, *Geographies*, pp. 51–72, at 52.

27. Plato, *Präsentierte Geschichte*, pp. 35ff.

28. Tseng, *Imperial Museums*, p. 4; a useful survey is Knell et al., *National Museums*.

29. Hochreiter, *Musentempel*, p. 64.

30. On the collectors of antiquities in India and Egypt between 1750 and 1850, see Jasanoff, *Edge of Empire*.

31. D. M. Reid, *Whose Pharaohs?* pp. 104–6.

32. In keeping with the German tradition of Ottoman studies, the more familiar name "Istanbul" (first made official in 1930) is here used throughout for the city. In the nineteenth century, it featured in the everyday speech of ordinary Turks, and one often finds the forms "Stambul" or "Stamboul" in Western sources. Few historians of diplomacy still prefer to speak of Constantinople.

33. See Anja Laukötter, "Das Völkerkundemuseum," in: Geisthövel and Knoch, *Orte der Moderne*, pp. 218–27; and, for a first-rate account of the politics and culture of collections and exhibitions, with examples especially from Scotland and New Zealand, Henare, *Museums*, esp. chs. 7 and 8.

34. Penny, *Objects of Culture*, p. 2.

35. A. Zimmerman, *Anthropology*, pp. 173f.

36. See the account of the plundering of Benin City in West Africa by a British "punitive expedition" in 1897, after which the famous "Benin bronzes" were shipped off to the British Museum, in: Coombes, *Reinventing Africa*, pp. 9–28.

37. P. Conrad, *Modern Times*, p. 347.

38. See the moving story of a group of Australian aborigines who were put on display in the 1880s in, among other places, Brussels, Paris, Gothenburg, Moscow, Wuppertal, and Istanbul in: Poignant, *Professional Savages*.

39. There is a vast literature on this subject. Especially worthy of note are Greenhalgh, *Ephemeral Vistas*; Hoffenberg, *An Empire on Display*; Tenorio Trillo, *Mexico*; Geppert, *Fleeting Cities*.

40. Auerbach, *Great Exhibition*; P. Young, *Globalization and the Great Exhibition*.

41. See the overview in Headrick, *Information*, pp. 142ff.

42. Sayer, *Bohemia*, p. 96.

43. Rétif, *Pierre Larousse*, pp. 165ff.

44. Çikar, *Fortschritt durch Wissen*, pp. 35f., 74–76.

45. Kaderas, *Leishu*, esp. pp. 257–80.

46. See details in Ogilvie, *Words of the World*.

47. Schumpeter, *Economic Analysis*, pp. 519ff.

48. Stierle, *Paris*, p. 108.

49. "Essai politique sur l'île de Cuba," in: Humboldt, *Relation historique*, vol. 3, pp. 345–501; volume 3, with the imprint "1825" was actually published only in 1831. English translation: Humboldt, *Political Essay*.

50. Buchanan, *Journey from Madras*.

51. Marx and Engels, *Collected Works*, vol. 4, p. 302.

52. Mayhew, *London Labour*, vol. 1, p. v.

53. Among his many works, *Les ouvriers européens* (Paris 1855), stands out in particular.

54. This kind of realism may also be found in other genres, such as painting or the operas of Verdi.

55. Lepenies, *Between Literature and Science*, pp. 4–5.

56. See Moretti's *Atlas of the European Novel* and volume 3 of his five-volume collection *Il romanzo*.

57. Schmidt-Glintzer, *Geschichte der chinesischen Literatur*, pp. 490–93.

58. Kato, *Japanese Literature*, ch. 9.

59. Although most German historians of Eastern Europe prefer the term "Russia's empire" (on the correct grounds of its multiethnicity), I shall use "Russian Empire" or "Tsarist Empire" throughout in the interests of readability; this also concurs with the usage of such an authority as Kappeler, *Russian Empire*.

60. On most of these authors, see the internationally unrivaled entries in D. Henze's *Enzyklopädie*.

61. See Robertson, *Raja Rammohan Ray*. Li Gui's diary has been translated in: Desnoyers, *A Journey to the East*.

62. Wang Xiaoqiu, "A Masterful Chinese Study of Japan from the Late-Qing Period: Fu Yunlong and His *Youli Riben tujing*," in: J. A. Fogel, *Sagacious Monks*, pp. 200–217. The term "volume" here refers to a *juan*, a fascicle in Chinese bookbinding.

63. Cf. Das, *History of Indian Literature*, pp. 83ff., 100ff., 132ff.

64. Keene, *Japanese Discovery of Europe*. See also chapter 3, below.

65. Godlewska and Smith, *Geography and Empire*.

66. On the irrational side of travel and exploration, see Fabian, *Out of Our Minds*; Driver, *Geography Militant*.

67. On the modern history of cartography, see Headrick, *Information*, pp. 96–141; Akerman and Karrow, *Maps*; Akerman, *The Imperial Map*. See also chapter 3, below.

68. Yonemoto, *Mapping Early Modern Japan*, pp. 173f.

69. On the rise of the *kaozheng* school, see Elman, *From Philosophy to Philology*, pp. 39–85.

70. These were at their grandest in the Napoleonic period: see Godlewska, *Geography Unbound*, pp. 149–90.

71. Dabringhaus, *Territorialer Nationalismus*, pp. 57ff.

72. Dahrendorf, *LSE*, pp. 3ff., 94ff.; D. Ross, *Origins*, p. 123.

73. Schwentker, *Max Weber in Japan*, pp. 62–64.

74. Gransow, *Geschichte der chinesischen Soziologie*, esp. pp. 51f.

75. Lai, *Adam Smith across Nations*.

76. Ho Ping-ti, *Studies*, p. 97.

77. Hanley and Yamamura, *Preindustrial Japan*, p. 41; Hayami, *Historical Demography*, pp. 21–38; Hayami, *Population*, p. 167.

78. Karpat, *Ottoman Population*, p. 22.

79. Livi-Bacci, *World Population*, p. 30.

80. For a detailed account of the rise of statistical bureaux in Europe, see Dupâquier and Dupâquier, *Histoire de la démographie*, pp. 256ff.

81. P. C. Cohen, *A Calculating People*, p. 176.

82. What follows draws on Cohn, *An Anthropologist*, pp. 231–50.

83. Maheshwari, *Census Administration*, pp. 62ff.

84. Christopher, *Census of the British Empire*.

85. Joshua Cole, *Power of Large Numbers*, pp. 80–84.

86. Bourguet, *Déchiffrer la France*, pp. 68f., 97f.

87. Cullen, *Statistical Movement*, pp. 45ff.

88. Patriarca, *Numbers*, p. 4.

89. See R. D. Brown, *The Strength of a People*, a study focused mainly on the history of ideas.

90. Stöber, *Deutsche Pressegeschichte*, p. 164.

91. Ibid., pp. 136f.

92. Robin Lenman, "Germany," in: R. J. Goldstein, *War for the Public Mind*, pp. 35–79.

93. On "seditious libel," see L. W. Levy, *Emergence of a Free Press*, esp. ch. 1.

94. Bumsted, *Peoples of Canada*, pp. 1f.

95. Macintyre, *Australia*, p. 118; Rickard, *Australia*, p. 93.

96. Carr, *Spain*, p. 287.

97. R. J. Goldstein, *Political Censorship*, pp. 34–43 (and Tab. 2.1, p. 35).

98. R. Price, *French Second Empire*, pp. 171–87; Charle, *Le siècle de la presse*, p. 111.

99. Robert Justin Goldstein, "France," in: idem, *War for the Public Mind*, p. 156; Livois, *Histoire de la presse française*, vol. 2, p. 393.

100. Lothar Höbelt, "The Austrian Empire," in: R. J. Goldstein, *War for the Public Mind*, pp. 226f.

101. Hildermeier, *Geschichte Russlands*, pp. 1261–69.

102. Bayly, *Empire and Information*, p. 239.

103. Abeyasekere, *Jakarta*, pp. 59f.

104. See Janku, *Nur leere Reden*, e.g., p. 179.

105. Huffman, *Creating a Public*, p. 222.

106. Judge, *Print and Politics*, p. 33.

107. Vittinghoff, *Journalismus in China*, esp. pp. 73ff.

108. Ayalon, *Press in the Arab Middle East*, p. 30.

109. Christoph Herzog, "Die Entwicklung der türkisch-osmanischen Presse im Osmanischen Reich bis ca. 1875," in: Rothermund, *Aneignung*, pp. 15–44, at 31, 34.

110. Ayalon, *Press in the Arab Middle East*, p. 41.

111. Ayalon, *Political Journalism*, pp. 103, 108.

112. See the case study of Aleppo in Watenpaugh, *Being Modern*, pp. 70 ff.

113. W. König and Weber, *Netzwerke*, pp. 522–25; Smil, *Creating the Twentieth Century*, pp. 204–6.

114. On the beginnings of investigative journalism and its later turn to "muckraking," see T. C. Leonard, *Power of the Press*, pp. 137 ff.

115. T. C. Leonard, *News for All*, p. 47.

116. A fine description of the paper is given in Emery, *Press and America*, pp. 225–35.

117. Livois, *Histoire de la presse française*, vol. 1, p. 274.

118. Juergens, *Joseph Pulitzer*, p. vii.

119. Cranfield, *Press and Society*, pp. 160, 220.

120. Emery, *Press and America*, p. 345.

121. For the United States, see Baldasty, *Commercialization of News*, pp. 59ff.; and for Britain, L. Brown, *Victorian News*, p. 16f.

122. Huffman, *Creating a Public*.

123. A large amount of Russell's journalism is available in themed collections, some of them recently reprinted by Cambridge University Press. See also Daniel, *Augenzeugen*.

124. Headrick, *Tools*, p. 158. See also chapter 14, below.

125. S. J. Potter, *Communication*, p. 196.

126. Read, *Power of News*, pp. 7, 32, 40 (quotation); cf. S. J. Potter, *News*, pp. 16–35, 87–105; on the history of cable corporations and news agencies, see Winseck and Pike, *Communication and Empire*, esp. chs. 6 and 7.

127. On foreign correspondents, see L. Brown, *Victorian News*, ch. 10.

128. Briggs and Burke, *Media*, pp. 155–63.

129. Pre-photographic techniques of observation cannot be considered here: see Crary, *Techniques of the Observer*. Nor will I broach the difficult question of the relationship between photography and "realist" art: see, inter alia, the remarks in Fried, *Menzel's Realism*, pp. 247–52.

130. Hörisch, *Sinn*, pp. 227–29.

131. T. C. Leonard, *Power of the Press*, p. 100.

132. Gernsheim, *History of Photography*, p. 159.

133. Newhall, *History of Photography*, p. 89.

134. M. Davis, *Late Victorian Holocausts*, pp. 147f.

135. Jäger, *Photographie*, pp. 48, 51.

136. Stiegler, *Philologie des Auges*, pp. 136–41.

137. A key book for the history of ethnographic photography is the exhibition catalog: Theye, *Schatten*, esp. pp. 61ff.

138. Gernsheim, *History of Photography*, p. 447.

139. J. R. Ryan, *Picturing Empire*, pp. 73ff.

140. See the examples in Gernsheim, *History of Photography*, p. 116.

141. See the wonderful material in Majluf et al., *La recuperación de la memoria*.

142. Faroqhi, *Subjects of the Sultan*, pp. 258f.

143. A particularly successful example is Ayshe Erdogdu, "Picturing Alterity: Represen-tational Strategies in Victorian-Type Photographs of Ottoman Men," in: Hight and Sampson, *Colonialist Photography*, pp. 107–25.

144. On the early history of the cinema, see also Hörisch, *Sinn*, pp. 284–92.

145. Rittaud-Hutinet, *Le cinéma*, pp. 32, 228–39.

146. Leyda, *Dianying*, p. 2.

147. Harding and Popple, *Kingdom of Shadows*, p. 20.

148. Leyda, *Dianying*, p. 4.

149. Toeplitz, *Geschichte des Films*, p. 25.

150. There is a growing literature on the history of sound recording. A classic in the field remains Gelatt, *Fabulous Phonograph*.

151. See also chapter 16, below.

CHAPTER II: Time

1. J. M. Roberts, *Twentieth Century*, p. 3.

2. Wills, *1688: A Global History*; Bernier, *The World in 1800*.

3. Pot, *Sinndeutung*, p. 52, referring to such authorities as Jan Romein, Lucien Febvre, and R. G. Collingwood.

4. Tanaka, *New Times*, p. 112.

5. On the theories of time underpinning this, see Kwong, *Linear Perspective*.

6. Kirch, *On the Road*, pp. 293f.

7. Today's standard work sets this out in detail; see Strachan, *First World War*.

8. See Manela, *Wilsonian Moment*.

9. Eichhorn, *Geschichtswissenschaft*, pp. 145–52.

10. Evidence for the varied use of "modernity" is collected in Corfield, *Time*, pp. 134–38.

11. Wolfgang Reinhard suspects that this is the case; see his "The Idea of Early Modern History," in Bentley, *Companion*, pp. 281–92, at 290.

12. Cf. P. Nolte, *Einheit*.

13. Hobsbawm, *Revolution, Capital*, and *Empire*.

14. E. Wilkinson, *Chinese History*, p. 509.

15. On the pragmatic reasons why the Gregorian calendar was preferred, see Watkins *Time Counts*, p. 47. Along with Watkins's classic, the best modern history of the calendar is E. G. Richards, *Mapping Time*.

16. E. G. Richards, *Mapping Time*, p. 114.

17. See Gardet et al., *Cultures and Time*, pp. 201, 208.

18. E. G. Richards, *Mapping Time*, p. 236.

19. Wilcox, *Measure of Times Past*, p. 8.

20. Tanaka, *New Times*, p. 11.

21. Brownlee, *Japanese Historians*, p. 209.

22. Coulmas, *Japanische Zeiten*, p 127; Zöllner, *Japanische Zeitrechnung*, p. 9; Tanaka, *New Times*, pp. 5f., 9.

23. Zerubavel, *Time Maps*, pp. 89ff., speaks of "firstism."

24. Keirstead, *Inventing Medieval Japan*.

25. Pot, *Sinndeutung*, p. 63.

26. Troeltsch, *Historismus*, pp. 756, 765.

27. However, some historians have made bold suggestions for carving up world history into fairly thin temporal slices of three to four decades. See Wills, *The World from 1450 to 1700*.

28. Cited in Raulff, *Der unsichtbare Augenblick*, p. 19.

29. Barry, *Influenza*.

30. See Wigen, *Japanese Periphery*, p. 19. The author had in mind 1868, the central date in nineteenth-century Japanese history.

31. Hans-Heinrich Nolte has even postulated a major epoch in world history stretching from the fifteenth to the end of the nineteenth century; see his *Weltgeschichte*.

32. Cf. Green, *Periodization*, pp. 36, 46, 50, 52f.

33. Schilling, *Die neue Zeit*, pp. 10–15.

34. Gerhard, *Old Europe*. A similar approach had been taken previously by the famous historians Otto Brunner and Otto Hintze.

35. Braudel, *Civilization and Capitalism*.

36. Macfarlane, *Savage Wars of Peace*; A. Reid, *An Age of Commerce*, pp. 5f.; idem, *Charting the Shape*, pp. 1–14, esp. 7.

37. The concept of a "long" eighteenth century (c. 1680–1830) has been argued for in Osterhammel, *Entzauberung Asiens*, pp. 31–37. On the enlarged meaning of the eighteenth century, cf. Blussé and Gaastra, *Eighteenth Century*; Nussbaum, *Global Eighteenth Century*.

38. Quataert, *Ottoman Empire*, p. 54; Kreiser, *Der osmanische Staat*, pp. 36ff.

39. See the authoritative account in Totman, *Early Modern Japan*; cf. J. W. Hall, *Cambridge History of Japan*, vol. 4.

40. R. Oliver and Atmore, *Medieval Africa*.

41. Quoted in Jordheim, *Against Periodization*, p. 156.

42. For a first impression of the period turn to Blom, *Vertigo Years*.

43. Nitschke et al., *Jahrhundertwende*.

44. E.g., Dejung and Petersson, *Foundations of Worldwide Economic Integration*.

45. This is a periodization suggested in the six-volume *History of the World* edited by Akira Iriye and Jürgen Osterhammel (Cambridge, MA 2012ff.). See E. S. Rosenberg, *A World Connecting*.

46. The term should not be used naively, without an awareness of the rich history behind it. On post-Victorian (British) perceptions of the Victorians, see Gardiner, *The Victorians*.

47. G. M. Young, *Portrait*, p. 151.

48. For example, Searle, *A New England?*

49. Rudolf Vierhaus, "Vom Nutzen und Nachteil des Begriffs 'Frühe Neuzeit': Fragen und Thesen," in: Vierhaus et al., *Frühe Neuzeit*, p. 21.

50. One might also turn the question around and focus on the new beginning in the 1840s, as the great social historian Jerome Blum does convincingly in his last work, *In the Beginning*.

51. Bayly, *Birth of the Modern World*, pp. 110ff. The argument is more striking in Bayly's earlier writings, when it was not yet mixed in with a particular interpretation of globalization: see esp. *First Age*.

52. This emerges clearly in F. Anderson, *Crucible*; McLynn, *1759*; and above all, in a masterly work, Marshall, *Making*, pp. 86–157.

53. Palmer, *Democratic Revolution*; Godechot, *France*. For the background, cf. Bailyn, *Atlantic History*, pp. 15–15, 24–30.

54. See Bayly, *Imperial Meridian*, p. 164; Förster, *Weltkrieg*, especially on the global military context; and Michael Duffy, "World-Wide War and British Expansion, 1793–1815," in Louis, *Oxford History of the British Empire*, vol. 2, pp. 184–207.

55. Here the founding of the United States, the Haitian revolution, and the independence of South and Central America should be seen as a single interlinked process, as they are, for example, in Langley, *The Americas*.

56. See Meinig, *Shaping of America*, vol. 2, pp. 81–96.

57. C. A. Bayly, "The British and Indigenous Peoples, 1760–1860: Power, Perception and Identity," in: Daunton and Halpern, *Empire and Others*, pp. 29–31. See also chapter 7, below.

58. Again, it is Bayly, *Imperial Meridian*, where this point is underscored forcefully.

59. A model for such research may be found in Dipper, *Übergangsgesellschaft*.

60. Maddison, *World Economy*, p. 27; and *Contours*, pp. 73f.

61. Wrigley, *People*, p. 3.

62. Ibid., pp. 10f.; J. R. McNeill, *Something New under the Sun*, p. xxiii, 298; Smil, *Energy*, pp. 156ff.

63. Foucault, *Order of Things*, pp. 248ff.

64. C. Rosen, *Classical Style*; idem, *Romantic Generation*.

65. P. Nolte, *1900*, p. 300.

66. J. R. McNeill, *Something New under the Sun*, p. 14; and see Fig. 6.5 in Smil, *Energy*, p. 233.

67. Stearns, *Industrial Revolution*, pp. 87ff.

68. Smil, *Creating the Twentieth Century*, pp. 33–97 ("The Age of Electricity").

69. A. D. Chandler, *Visible Hand*, ch. 5 and passim; Zunz, *Making America Corporate*, pp. 40f.

70. Woodruff, *Impact*, p. 150 (Tab. IV/1).

71. Nugent, *Crossings*, p. 12.

72. Or, to put it differently, the 1880s ushered in the "fourth wave of globalization": Therborn, *Globalizations*, p. 161.

73. In his great history of Western music, Richard Taruskin diverges from this view by arguing in detail that nineteenth-century music ended only with the First World War. The "long" fin de siècle, he maintains, was as an age of "maximalist" intensification of the Romantic striving for expression (Mahler, Debussy, Scriabin, Richard Strauss's early operas, the Schönberg of the *Gurrelieder* and the Stravinsky of the Ballets Russes). The musical twentieth century began only with the advent of a greater artistic stringency, with the emphasis on irony, pastiche, and constructivism under the aegis of neoclassicism, New Objectivity and twelve-tone technique. See Taruskin, *Western Music*, vol. 4, pp. 448, 471.

74. The comparison between India and Italy is drawn in: Antony Copley, "Congress and Risorgimento: A Comparative Study of Nationalism," in: Low, *Indian National Congress*, pp. 1–21; see also Marr, *Vietnamese Anticolonialism*, p. 47.

75. A. Black, *Islamic Political Thought*, pp. 295–99, 301–4.

76. On the complicated dating of Kang Youwei's intellectual evolution, see Hsiao Kung-chuan, *A Modern China*, p. 56.

77. See R. W. Bowen's major study, *Rebellion*.

78. Though never fully elaborated, these ideas appear most clearly in Braudel, *History and the Social Sciences*.

79. J. Goldstone, *Problem*, p. 269.

80. Koselleck speaks of "structures of repetition," in *Zeitschichten*, p. 21. Charles Tilly has developed similar ideas in a number of writings.

81. An original way of differentiating these modes of historical change may be found in Laslett, *Social Structural Time*.

82. See Koselleck, *Futures Past*, p. 96.

83. Schumpeter, *Economic Analysis*, pp. 738–50.

84. These approaches are synthesized in Rasler and Thompson, *Great Powers*. Three major representatives are George Modelski, Joshua S. Goldstein, and Ulrich Menzel.

85. For a concise discussion, see Schmied, *Soziale Zeit*, pp. 144–63.

86. Gardet et al., *Cultures and Time*, p. 212.

87. Aung-Thwin, *Spirals*, pp. 584, 590, 592, 595.

88. See Osterhammel, *Entzauberung Asiens*, pp. 390–93. Around 1900, Japanese intellectuals saw the rest of Asia, especially Korea, in a similar light.

89. At its most effective in Fabian, *Time and the Other*. He speaks there of a "denial of coevalness."

90. A brief introduction to the question is given in Östör, *Vessels of Time*, pp. 12–25.

91. Surveys are Wendorff, *Zeit und Kultur*; J. T. Fraser, *Voices of Time*; and a classical text, Needham, *Grand Titration*, pp. 218–98.

92. This is one of the main themes in Galison, *Einstein's Clocks*.

93. See Blaise, *Time Lord*.

94. Dohrn-van Rossum, *History of the Hour*, p. 348.

95. Bartky, *Selling the True Time*, pp. 93, 114.

96. Whitrow, *Time*, p. 164.

97. Bartky, *Selling the True Time*, pp. 139f., 146.

98. I am borrowing Vanessa Ogle's argument about the interplay of nationalizing and internationalizing time. See her forthcoming book on the topic: *Contesting Time: The Global Struggle for Uniformity and Its Unintended Consequences* (Harvard University Press).

99. Galinson, *Einstein's Clocks*, pp. 153, 162 ff.

100. Landes, *Revolution in Time*, pp. 97, 287.

101. Mumford, *Technics*, p. 14.

102. Coulmas, *Japanische Zeiten*, pp. 142, 233.

103. Kreiser, *Istanbul*, p. 181.

104. E. P. Thompson, *Time*.

105. Gay, *Clock Synchrony*, pp. 112, 136.

106. Voth, *Time and Work*, p. 257 and passim. This also contains a summary and evaluation of older studies.

107. Ibid., pp. 47–58.

108. David Landes gave an unambiguous answer in his great work on the history of clocks: "The clock did not create an interest in time measurement, the interest in time measurement led to the invention of the clock." *Revolution in Time*, p. 58.

109. On the (not very precisely developed) concept of metronomization, see Young, *Metronomic Society*. And on the mechanization of classical labor, see the work by the Swiss architectural historian and theorist Sigfried Giedion, *Mechanization*.

110. For the example of a seminomadic tribe in Morocco, see Eickelman, *Time*, esp. pp. 45f., and for present-day Bali, Henk Schulte Nordholt, "Plotting Time in Bali: Articulating Plurality," in: Schendel and Schulte Nordholt, *Time Matters*, pp. 57–76.

111. The founder of ethnological functionalism, Bronisław Malinowski, already noted this in the early twentieth century. See Munn, *Cultural Anthropology*, pp. 96, 102–5.

112. T. C. Smith, *Peasant Time*, pp. 180f., 184–89, 194f.

113. M. M. Smith, *Mastered by the Clock*, pp. 5–7.

114. A lot of relevant material, especially from Western Europe, has been collected and discussed in: Borscheid, *Tempo-Virus*, esp. chs. 5–7, and Kaschuba, *Überwindung*; also still valuable is Kern, *Culture*, pp. 109–30.

115. See the historical phenomenology of rail travel in Schivelbusch, *Railway Journey*; Freeman, *Railways*.

116. Cvetkovski, *Modernisierung*, pp. 192, 222, 236f., 242f.

117. Berlioz, *Memoirs*, pp. 456f.

118. Koselleck, *Zeitschichten*, p. 153.

119. See, in addition to Koselleck: E. W. Becker, *Zeit der Revolution*, pp. 14–16; and numerous works by Lucian Hölscher.

120. Litwack, *Been in the Storm so Long*, p. 172 and passim.

121. Shih, *Taiping Ideology*, p. 75.

CHAPTER III: Space

1. Koselleck, *Zeitschichten*, p. 9.

2. Harvey, *Postmodernity*, p. 240.

3. The key text here is Livingstone, *Geographical Tradition*; see also Marie-Claire Robic, "Geography," in: T. M. Porter and Ross, *Modern Social Sciences*, pp. 379–90.

4. On the example of the Dufour Map of Switzerland, see Gugerli and Speich, *Topografien*, p. 76.

5. See, e.g., Dabringhaus, *Territorialer Nationalismus*, pp.57ff.

6. This story has become widely known from Sten Nadolny's novel *Discovery of Slowness*.

7. Excellent on the Royal Navy and Exploration: Angster, *Erdbeeren und Piraten*.

8. Japan is the only non-European country for which there is a near-complete edition of available European reports up to approximately 1830: Kapitza, *Japan in Europa*, which includes extracts from Kaempfer.

9. On the organization of research journeys, see the exemplary study of Murchison's travels: Stafford, *Scientist of Empire*.

10. A. v. Humboldt, *Relation historique*.

11. The results of Wilhelm von Humboldt's trip are contained in "Die Vasken, oder Bemerkungen auf einer Reise durch Biscaya und das französische Basquenland im Frühling des Jahres 1801" (*Werke*, vol. 2, pp. 418–627).

12. Isabella Bird's travel books were republished in a twelve-volume edition in 1997: *Collected Travel Writings*.

13. Cosgrove, *Apollo's Eye*, p. 209 (and Fig. 210).

14. See Carter, *Botany Bay*, esp. pp. 4–33.

15. Barrow, *Making History*, pp. 101, 103.

16. M. W. Lewis and Wigen, *Myth of Continents*, p. ix.

17. Ibid., p. 181; Foucher, *Fronts et frontières*, p. 156.

18. M. W. Lewis and Wigen, *Myth of Continents*, p. 172.

19. J. D. Legge, "The Writing of Southeast Asian History," in: Tarling, *Cambridge History of Southeast Asia*, p 1–50, at 1.

20. Sinor, *Introduction*, esp. p. 18.

21. Such a concept works best for premodern periods: Beckwith, *Empires of the Silk Road*.

22. Mackinder's lecture was published in: *Geographical Journal* 23, pp. 421–37.

23. See the useful collection Bonine et al., *Is There a Middle East?*, esp. Huseyin Yilmaz, "The Eastern Question and the Ottoman Empire: The Genesis of the Near and Middle East in the Nineteenth Century" (pp. 1–35); see also Scheffler, *"Fertile Crescent."* Said, *Orientalism* effectively deconstructed (and rightly criticized) the concept of "the Orient" as a typical example of European "Othering."

24. J. A. Fogel, *Articulating the Sinosphere.*

25. See the extract from Fukuzawa's text in: Lu, *Japan*, vol. 2, pp. 351–53. Cf. Tanaka, *Japan's Orient*, whose evidence is taken mainly from the period *after* 1890.

26. See, in particular: Sven Saaler, "Pan-Asianismus im Japan der Meiji- und der Taishō-Zeit: Wurzeln, Entstehung und Anwendung einer Ideologie," in: Amelung et al., *Selbstbehauptungsdiskurse*, pp. 127–57.

27. C. Ritter, *Erdkunde*, vol. 1: Der Norden und Nord-Osten von Hoch-Asien, p. xv.

28. C. Ritter, *Einleitung*, p. 161.

29. On the development of descriptive geographical terminology, see Godlewska, *Geography Unbound*, pp. 41–45.

30. See, most generally: C. Ritter, *Erdkunde*, vol. 1: Der Norden und Nord-Osten von Hoch-Asien, p. 63f.

31. C. Ritter, *Erdkunde*, vols. 1–3.

32. Ratzel, *Politische Geographie*, chs. 11–28; on islands ch. 24. These parts of the book are more valuable than Ratzel's notorious "fundamental law of the spatial growth of states" (chs. 8–10).

33. Reclus, *L'Homme et la terre*, vol. 1, p. 123.

34. Ibid., pp. 348–53.

35. W. D. Smith, *Sciences of Culture*, pp. 154–61; Petermann, *Geschichte der Ethnologie*, pp. 583ff.

36. Bonnett, *Idea of the West*, pp. 14ff.

37. Bulliet, *Islamo-Christian Civilization*, pp. 5f. A giant "History of the West" by the eminent German historian Heinrich-August Winkler sees monotheism as a distinctive cultural trait of the West and traces the origins of a "Western project" back to the Egyptian pharaoh Echnaton: Winkler, *Geschichte des Westens*, vol. 1, pp. 25, 27f.

38. Carmagnani, *The Other West*,

39. See chapter 17, below.

40. Asbach, *Erfindung*; Woolf, *European World View*.

41. Boer, *Europa*, pp. 99–110; for a marvelous history of iconic representation see Wintle, *Image of Europe*.

42. Boer, *Europa*, pp. 181ff. (esp. the map on p. 182).

43. Schroeder, *Transformation*, esp. pp. 575–82.

44. Isabella, *Risorgimento in Exile.*

45. Gollwitzer, *Geschichte des weltpolitischen Denkens*, vol. 2, pp. 83ff.

46. See the map in Lichtenberger, *Europa*, p. 43.

47. Malia, *Russia*, p. 92.

48. Bassin, *Imperial Visions*, pp. 37ff.

49. Hauner, *What Is Asia to Us?* chs. 2–4.

50. See Malia, *Russia*, p. 165.

51. Kreiser and Neumann, *Türkei*, p. 283. An alternative view sees the turning point as the Treaty of Karlowitz with Austria.

52. Nouzille, *Histoire de frontières*, p. 254. Cf. Hösch, *Balkanländer*, p. 91.

53. See the fine illustration in Ruthven and Nanji, *Historical Atlas of Islam*, p. 89.

54. On its conference activities after 1856, see Baumgart, *Europäisches Konzert*, pp. 155f.

55. One of the few recent exceptions is J. Fisch, *Europa*, pp. 228–35.

56. In perceptions of the Balkan see Mazower, *The Balkans*; Todorova, *Imagining the Balkans*.

57. This has been done for Scotland in Withers, *Geography*, pp. 142ff.; for Thailand in Thongchai, *Siam Mapped*; and for Mexico in Craib, *Cartographic Mexico*.

58. There is now an extensive literature on the phenomenology and psychology of spatial perception. One of the pioneers, Tuan Yi-fu, still stands out in the field, especially for his *Space and Place*.

59. Richter, *Facing East*, p. 11 and passim.

60. Rowe, *Saving the World*, p. 356.

61. Eggert, *Chinesische Reiseschriften*, p. 283.

62. J. K. Leonard, *Wei Yuan*, pp. 121ff.

63. Drake, *Hsu Chi-yü*, general appreciation on pp. 67f.

64. See Karl, *Staging the World*.

65. Toby, *State and Diplomacy*, pp. 161–67.

66. See Beasley, *Japan Encounters the Barbarian*; the vivid account of these efforts in Pantzer, *Iwakura-Mission*; and Duus, *Japanese Discovery of America*.

67. Konvitz, *Urban Millenium*, pp. 82–85.

68. Types (a) to (d) follow A.R.H. Baker, *Geography and History*, chs. 2–5, a fundamental work in this field.

69. This is the approach in traditional accounts of modern historical geography: e.g., Pounds, *Historical Geography*.

70. There is a noteworthy "island discourse" in historical geography: see the references to themes and literature in Dodds and Royle, *Rethinking Islands*. Cf. Pocock, *Discovery*, who defines the British Isles from the Channel Islands to the Shetlands as "the Atlantic archipelago" (p. 78).

71. Fernández-Armesto, *Civilizations*, passim. In the background here is Braudel's conception of "civilizations" as spaces: *History of Civilizations*. Another important author uses a regional rather than systematic approach: Richards, *Unending Frontier*.

72. See François Walter's major work *Les figures paysagères de la nation*.

73. A.R.H. Baker, *Geography and History*, p. 112.

74. This is one of the main theses in Elvin, *Elephants*.

75. Fundamental here is: Dunlap, *Nature*.

76. Cain and Hopkins, *British Imperialism*.

77. This is one of the key methodological ideas in the new comparative studies: see, e.g., Pomeranz, *Great Divergence*, pp. 10 and passim. On the possibilities of regional history (with special reference to identity), see Applegate, *Europe of Regions*

78. Werdt, *Halyč-Wolhynien*, p. 98.

79. An example of the rapidly growing literature is: B. Klein and Mackenthun, *Sea Changes*.

80. Braudel, *Mediterranean*.

81. Horden and Purcell, *Corrupting Sea*, p. 25—a work that shares the order of magnitude of Braudel's classic.

82. C. King, *Black Sea*; Herlihy, *Odessa*; Farnie, *East and West of Suez*.

83. Horden and Purcell (*Corrupting Sea*, pp. 461 ff. and passim) are the foremost representatives of this "Mediterraneanism," which has been opposed most notably in numerous writings by the American anthropologist Michael Herzfeld.

84. K. N. Chaudhuri, *Trade*; idem, *Asia*.

85. Cf. Wong, "Between Nation and World," p. 7.

86. See J. de Vries, "Connecting Europe and Asia: A Quantitative Analysis of the Cape-Route Trade, 1497–1795," in: Flynn et al., *Global Connections*, pp. 35–106, at 69 (Tab. 2.5). A masterly essay.

87. On the many different forms of mobility, see: Bose, *A Hundred Horizons*. On Indian Ocean history in general, see Pearson, *Indian Ocean*.

88. Kirch, *On the Road*, pp. 293, 300, 302.

89. The classic on the Pacific is Spate, *Pacific*—once again the work of a historical geographer, though unfortunately stopping around 1800.

90. See the bleak balance sheet in Scarr, *Pacific Islands*, pp. 134–44.

91. See Flynn et al., *Pacific Centuries*; E. L. Jones et al., *Coming Full Circle*.

92. Korhonen, *Pacific Age*, p. 44.

93. See Heffer, *United States*, pp. 249ff.—which, despite its title, is actually a general history of the modern Pacific.

94. Crucial here is Brading, *First America*, pp. 447ff., which also helps to clarify the term "patriotism."

95. See, for example, Zeuske, *Schwarze Karibik*.

96. For syntheses and surveys of the literature, see: Bailyn, *Atlantic History*; Pietschmann, *Atlantic History*; Armitage and Braddick, *British Atlantic World*; Benjamin, *Atlantic World*; Falola and Roberts, *Atlantic World*; Greene and Morgan, *Atlantic History*; Canny and Morgan, *Oxford Handbook of the Atlantic World*. The most exacting attempt at a theoretical synthesis has come from a sociologist: Jeremy Smith, *Europe and the Americas*.

97. See, above all, the work of Nicholas Canny and his circle.

98. By now we do have some "partial integrations," though. Attempts have been made to define the "limits of the early modern world" in the northern and southern Atlantic by the reach of certain legal concepts: Gould, *A World Transformed?*; Benton, *Legal Regime*.

99. Figure from Bade et al., *Encyclopedia of European Migration*, p. 210.

100. P. D. Curtin, *Slave Trade*, p. 266 (Fig. 266), 268 (Tab. 77).

101. Berlin, *Many Thousands Gone*, pp. 95ff.

102. See here one of the great modern masters of Japanese historiography: Amino, *Les Japonais et la mer*, p. 235. Amino also stresses, however, that, despite the geographical similarities between the two countries, the Japanese never developed a maritime identity like that of the British.

103. Lemberg, "*Zur Entstehung*," esp. pp. 77ff.; Kirby, *Baltic World*, p. 5; Mead, *Scandinavia*, pp. 9–13, 210–12; J. Fisch, *Europa*, p. 148.

104. On the diverse conceptions of Central Europe, see H.-D. Schultz, *Deutschlands "natürliche Grenzen"*; and idem, *Raumkonstrukte*. A map showing their various boundaries may be found in Dingsdale, *Mapping Modernities*, p. 18.

105. For example, L. R. Johnson, *Central Europe*.

106. Berend, *History Derailed*, p. xiv.

107. Halecki, *Limits and Divisions*.

108. Szücs, *Les trois Europe*. On the binary pattern of "two Europes," see Valerie Bunce, "The Historical Origins of the East-West Divide: Civil Society, Political Science, and Democracy in Europe," in: Bermeo and Nord, *Civil Society*, pp. 209–36.

109. See Laruelle, *Russian Eurasianism*, pp. 16–49. Three further concepts of "Eurasia" are discussed in Schmidt-Glintzer, *Eurasien*, pp. 189–92.

110. Hawes, *Poor Relations*, pp. 10f., 39, 152f., 168 (on Skinner); Stoler, *Sexual Affronts*.

111. Buettner, *Empire Families*.
112. Fletcher, *Integrative History*; Lieberman, *Binary Histories*; Lieberman, *Strange Parallels*, vol. 1, pp. 77–80; vol. 2, pp. 1–11 and passim.
113. See Osterhammel, *Entzauberung Asiens*.
114. See esp. Findley, *Turks*, chs. 2–3; and Wong, *Entre monde et nation*, pp. 18ff.
115. See also Beckwith, *Empires of the Silk Road*, chs. 10, 11.
116. Guy, *Qing Governors*, p. 31.
117. The various contributions by the editor are fundamental in Skinner, *City*. On macro-regions as social-historical entities, see Naquin and Rawski, *Chinese Society*, pp. 138–216.
118. Meinig, *Shaping*, vol. 2, p. 3.
119. Stefan Kaufmann, "Landschaft beschriften. Zur Logik des 'American Grid System,'" in: idem, *Ordnungen der Landschaft*, pp. 73–94, quotation 78.
120. Reardon-Anderson, *Reluctant Pioneers*, p. 72.
121. Edney, *Mapping an Empire*, p. 200; Bayly, *Empire and Information*, pp. 303ff.; O'Cadhla, *Civilizing Ireland*.
122. Lappo and Hönsch, *Urbanisierung Russlands*, p. 34.
123. See Planhol, *Les fondéments*.
124. See J. C. Scott, *Seeing Like a State*, pp. 37–47.
125. See Maier, *Consigning the Twentieth Century*, pp. 808, 814, 816.
126. Charles Tilly, "Reflections on the History of European State-Making," in: idem, *Formation of National States*, pp. 3–83, at 15.
127. Lieberman, *Strange Parallels*, vol. 1, p. 455, speaks of "three post-1750 consolidations of the mainland."
128. Ratzel, *Politische Geographie*, pp. 193ff.
129. Kashani-Sabet, *Frontier Fictions*, p. 23.
130. R. Cohen, *Global Diasporas*, pp. 26 and 177ff.
131. Ibid., chs. 2–6.
132. Takaki, *Mirror*, p. 247.
133. Nadel, *Little Germany*, p. 10.
134. The fashionable claim that a diaspora is inherently "deterritorialized" goes too far.
135. Unfortunately we must pass over here contributions from the new field of "borderland studies."
136. On the concept, see Böckler, *Grenze*.
137. Nordman, *Frontières*, pp. 486ff.
138. T. M. Wilson and Donnan, *Introduction*, p. 25, also 9; Windler, "Grenzen vor Ort," p. 143.
139. Nordman, *Frontières*, p. 40.
140. Bitsch, *Belgique*, p. 83.
141. It is hard to find details about the state of the frontiers in Africa. But precise information about West Africa is contained in John D. Hargreaves, "The Berlin Conference, West African Boundaries, and the Eventual Partition," in: Förster et al., *Bismarck*, pp. 313–20, at 314–17.
142. Foucher, *Fronts et frontières*, pp. 114, 135ff.
143. Ibid., p. 122.
144. The "natural" border is a special case of a content-defined frontier. See the fine general considerations in Burnett, *Masters*, pp. 208ff.
145. Kashani-Sabet, *Frontier Fictions*, pp. 24–28.
146. S.C.M. Paine, *Imperial Rivals*, pp. 90f.

147. Thongchai, *Siam Mapped*, pp. 68–80: one of the most original works on the "construction" of spaces.

148. Excellent (with many hints for further reading) are: Windler, "Grenzen vor Ort," pp. 138–45; and Baud and Schendel, *Comparative History*, esp. 216ff.

CHAPTER IV: Mobilities

1. Khater, *Inventing Home*, pp. 52–63.

2. Rallu, *Les populations océaniennes*, p. 222.

3. Schmid, *Korea*, p. 101; Etemad, *Possessing the World*, p. 225.

4. Lavely and Wong, *Malthusian Narrative*, p. 719; and on the sources cf. J. Z. Lee and Wang, *One Quarter of Humanity*, pp. 149–57.

5. Calculated from Livi-Bacci, *World Population*, p. 31 (Tabs. 1–3).

6. See also the graphic illustration in McEvedy and Jones, *Atlas*, p. 349.

7. Livi-Bacci, *World Population*, p. 31 (Tabs. 1–3).

8. Bähr, *Bevölkerungsgeographie*, p. 217 (Tab. 23).

9. Bardet and Dupâquier, *Histoire des populations de l'Europe*, p. 469 (Tab. 84); Marvin McInnis, "The Population of Canada in the Nineteenth Century," in: Haines and Steckel, *Population History*, pp. 371–432, at 373 (Tab. 9.1); M. Reinhard et al., *Histoire générale*, pp. 391, 423, 426; R. V. Jackson, *Population History*, p. 27 (Tab. 6); *Meyers Großes Konversations-Lexikon*, vol. 12 (6th ed. 1905), p. 695; vol. 18 (6th ed. 1907), p. 185.

10. Karpat, *Ottoman Population 1985*, p. 117 (Tab. I.6). With Egypt, 40.5 million.

11. Maddison, *Chinese Economic Performance*, p. 47.

12. See, for example, Rudolf G. Wagner, "Taiping-Aufstand," in: Staiger et al., *China-Lexikon*, pp. 735–39, at 736.

13. Deng, *China's Political Economy*, p. 38,

14. Cf. J. Z. Lee and Wang, *One Quarter of Humanity*, pp. 14–23.

15. J. Z. Lee and Campbell, *Fate and Fortune*, p. 70.

16. Hanley and Yamamura, *Preindustrial Japan*, p. 320.

17. Totman, *History of Japan*, pp. 326f.

18. Wolfram Fischer, "Wirtschaft und Gesellschaft Europas 1850–1914," in: Fischer, *Handbuch*, vol. 5, pp. 1–207, at 14 (table 3).

19. For a survey of growth rates in Europe, see Tortella, *Modern Spain*, p. 33 (Tab. 2.2.).

20. Saunders, *Russia*, p. 270.

21. The basis for my data here is Maddison, *World Economy*, p. 241 (Tab. B-10).

22. Dupâquier, *Histoire de la population française*, p. 293.

23. On the current state of research, see Bardet and Dupâquier, *Histoire des populations de l'Europe*, pp. 287–325.

24. O'Gráda, *Ireland's Great Famine*, p. 16.

25. O'Rourke and Williamson, *Globalization*, pp. 150–52.

26. Figure from McPherson, *Battle Cry*, p. 854.

27. Ricklefs, *Modern Indonesia*, p. 142.

28. Figures from J. Levy, *War*, p. 90; Rasler and Thompson, *War*, p. 13 (Tab. 1.2).

29. Schroeder, *International System*, p. 11; secondary analysis of the data in Eckhardt, *Civilizations*.

30. Rallu, *Les populations océaniennes*, p. 6.

31. See Thornton, *American Indian Holocaust*, pp. 107–9; and the even bleaker estimates in Nugent, *Into the West*, p. 35.

32. For a good summary and discussion of the present state of knowledge, see Broome, *Aboriginal Victorians*, pp. 79–93.

33. R. V. Jackson, *Population History*, p. 5 (Tab. 1).

34. Many data, of highly varied quality, may be found in Ferro, *Livre noir*.

35. Etemad, *Possessing the World*, p. 70. Since Etemad includes the Spanish war of the 1920s in Morocco, the total figure for this period should probably be closer to 280,000. Etemad does not take Russian or Japanese expansion into account.

36. Ibid., pp. 93, 94 (Tab. 5.1).

37. Coquery-Vidrovitch, *Africa*, p. 10; Vanthemsche, *La Belgique et le Congo*, pp. 40–42.

38. Ruedy, *Modern Algeria*, p. 93.

39. S. Doyle, *Population Decline*, p. 438.

40. See the carefully considered account and evaluation in C. Marx, *Geschichte Afrikas*, pp. 143–47.

41. The following draws on Bähr, *Bevölkerungsgeographie*, pp. 219–29; Dyson, *Population and Development*.

42. Bähr, *Bevölkerungsgeographie*, p. 222.

43. H. S. Klein, *Population History*, pp. 77–79.

44. Bardet and Dupâquier, *Histoire des populations de l'Europe*, p. 149 (Tab. 9).

45. Livi-Bacci, *World Population*, p. 113.

46. See chapter 7, below.

47. Gelder, *Het Oost-Indisch avontuur*, pp. 14, 41, 64.

48. Liauzu, *Histoire des migrations*, pp. 66–73; Nicholas Canny, "In Search of a Better Home? European Overseas Migration, 1500–1800," in idem, *Europeans*, pp. 263–83.

49. Canny, *Europeans*, p. 279.

50. For a survey, see the historical-sociological investigations in Ribeiro, *Americas*.

51. Robert W. Slenes, "Brazil," in: Paquette and Smith, *Oxford Handbook of Slavery*, pp. 111–33, at 114f. A useful tabulation of the data is Stephen D. Behrendt, "The Transatlantic Slave Trade," in ibid., pp. 251–74, at 263 (Tab. 11.1); detailed numbers for all destinations in Eltis and Richardson, *Atlas*, pp. 200–203.

52. Slenes, "Brazil," pp. 119–21.

53. H. S. Klein, *Slave Trade*, p. 45.

54. H. S. Klein, *Population History*, p. 83.

55. Gudmestad, *Troublesome Commerce*, pp. 3f., 8 (figure).

56. P. D. Curtin, *Slave Trade*, p. 27 (n. 16).

57. Meyer and Sherman, *Mexican History*, p. 218.

58. Stanley L. Engerman and Barry W. Higman, "The Demographic Structure of the Caribbean Slave Societies in the Eighteenth and Nineteenth Centuries," in Knight, *Slave Societies*, pp. 45–104, at 50 (Tab. 2-1).

59. Higman, *Concise History of the Caribbean*, p. 159.

60. Kaczyńska, *Gefängnis*, pp. 24f., 44, 53f. (total figure).

61. Jonathan W. Daly, "Russian Punishments in the European Mirror," in: McCaffray and Melancon, *Russia*, pp. 161–88, at 167, 176.

62. Waley-Cohen, *Exile*; Lary, *Chinese Migrations*, pp. 79, 83.

63. Bullard, *Exile*, p. 17.

64. Pérennès, *Déportés*, p. 483.

65. Bouche, *Colonisation française*, pp. 185f., and for the greatest detail Pérennès, *Déportés*.

66. Rickard *Australia*, pp. 21–25; Marjory Harper, "British Migration and the Peopling of the Empire," in: Louis, *Oxford History of the British Empire*, vol. 2, pp. 75–87, at 78.

67. See the case study in C. Anderson, *Convicts*, which emphasizes the cultural autonomy of the Indian convicts and the fact they were not slaves.

68. Isabella, *Risorgimento in Exile*, ch. 1.

69. Marrus, *The Unwanted*, p. 17.

70. Reiter, *Asyl*, pp. 28–33.

71. Alexander, *Geschichte Polens*, pp. 203f.

72. N. Davies, *God's Playground*, pp. 276, 287–89.

73. Reiter, *Asyl*, p. 38.

74. See Hanioğlu, *Young Turks*, pp. 71–78.

75. Suny, *Looking toward Ararat*, pp. 67ff.

76. Bergère, *Sun Yat-sen*, is the best biography.

77. Amrith, *Migration and Diaspora*, pp. 59–62.

78. Hsiao Kung-chuan, *A Modern China*, pp. 409ff.

79. M. C. Meyer and Sherman, *Mexican History*, pp. 498–500.

80. Marrus, *The Unwanted*, p. 18.

81. K. Schultz, *Tropical Versailles*, pp. 4, 76.

82. Todorov, *Balkan City*, p. 328.

83. B. G. Williams, *Crimean Tatars*, pp. 106–8, 119, 138, 148; Kirimli, *National Movements*, pp. 6–11.

84. J. H. Meyer, *Immigration*, pp. 16, 27f.

85. Jersild, *Orientalism*, pp. 25f.

86. Utley, *Sitting Bull*, pp. 182, 191, 231.

87. Marrus, *The Unwanted*, p. 23.

88. Neubach, *Ausweisungen*, p. 129 (totals) and passim.

89. Shannon, *Gladstone*, vol. 2, pp. 166f., 171.

90. Karpat, *Ottoman Population*, p. 49. These are high figures: one must put one's faith here in Karpat's great authority.

91. McCarthy, *Death and Exile*, p. 90 (Tab. 90).

92. Malcolm, *Bosnia*, pp. 139f.

93. Mazower, *Salonica*, pp. 298–304, 349.

94. Boeckh, *Von den Balkankriegen*, pp. 257–75; total figure calculated from data on pp. 271f. The book does not deal with Romania and Albania.

95. The following draws on Marrus, *The Unwanted*, pp. 2739; Kappeler, *Russian Empire*, pp. 267–73; Haumann, *History of East European Jews*, pp. 84ff.

96. See Klier and Lambroza, *Pogroms*.

97. Marrus, *The Unwanted*, p. 32.

98. Ibid., p. 34; Fink, *Defending the Rights of Others*, pp. 22–24, 27–30.

99. Volkov, *Juden*, p. 58.

100. Bade, *Migration in European History*, pp. 40, 129ff.

101. See Hoerder, *Cultures*, pp. 288–94 on the regional systems of labor migration before the middle of the nineteenth century. Hoerder's book remains the standard text on migration history. It should now be complemented by Dirk Hoerder, "Migrations and Belongings", in: Rosenberg, *A World Connecting*, pp. 435–589, and Gabaccia and Hoerder, *Connecting Seas*.

102. Bade, *Migration in European History*, pp. 46ff.

103. N. G. Owen, *Paradox*, p. 48.

104. Naquin and Rawski, *Chinese Society*, p. 130.

105. Stephan, *Russian Far East*, pp. 71–73, 79f.

106. Gottschang and Lary, *Swallows*, pp. 2, 38: an exemplary case study.

107. Adas, *Burma Delta*, pp. 42–44, 85ff.

108. Brocheux and Hémery, *Indochina*, pp. 121f.

109. Woerkens, *The Strangled Traveler*, pp. 43 ff.; A. J. Major, *State and Criminal Tribes*.

110. Hoerder, *Cultures*, pp. 381f.; Macfarlane and Macfarlane, *Green Gold*, pp. 141ff.

111. The following is based on Hoerder, *Cultures*, pp. 306–21; Harzig and Hoerder, *What Is Migration History?* pp. 35–42; Kappeler, *Russian Empire*, pp. 50f., 168–212. See also Dirk Hoerder, "Migrations," in: Bentley, *World History*, pp. 269–87.

112. James Forsyth uses this image in his excellent general account: *Peoples of Siberia*, p. 216.

113. A consistently interesting study, which uses a very broad concept of nomadism leaning on French research, is: Ilja Mieck, "Wirtschaft und Gesellschaft Europas von 1650 bis 1850," in: Fischer, *Handbuch*, vol. 4, pp. 1–233, at 72–74. Cf. W. Reinhard, *Lebensformen*, pp. 325–30. Barfield, *Nomadic Alternative*, is a good ethnic-historical introduction.

114. Paul, *Far West*, p. 195.

115. A.K.S. Lambton, "Land Tenure and Revenue Administration in the Nineteenth Century," in: Avery et al., *Cambridge History of Iran*, vol. 7, pp. 459–505, at 470f.

116. Abrahamian, *Iran*, pp. 141f.

117. Donald Quataert, "The Age of Reforms," in: İnalcık and Quataert, *Ottoman Empire*, vol. 2, pp. 759–943, at 768f., 873f.

118. Kasaba, *A Moveable Empire*, p. 86.

119. The standard ethno-archaeological and ecohistorical work is A. B. Smith, *Pastoralism in Africa*, esp. chs. 6–9.

120. On "le grand nomadisme," see Planhol, *Les nations*, pp. 313f.

121. J. Fisch, *Geschichte Südafrikas*, p. 92; Robert Ross, "Khoesan and Immigrants: The Emergence of Colonial Society in the Cape, 1500–1800," in: C. Hamilton, et al., *Cambridge History of South Africa*, vol. 1, pp. 168–210, at 203.

122. Here see especially Zeleza, *Economic History of Africa*, pp. 72, 117ff.

123. Austen, *African Economic History*, p. 162.

124. Eltis and Richardson, *Atlas*, p. 89 (Tab. 4).

125. Ibid., p. 203 (Tab. 6).

126. Lovejoy, *Transformations*, p. 154; very important is Ewald, *Soldiers*, pp. 53–56, 163–66.

127. Lovejoy, *Transformations*, p. 155.

128. Clarence-Smith discusses all the present-day estimates in: *Islam*, pp. 11–13; the figure is based on Lovejoy's latest revision.

129. Manning, *Slavery*, p. 83 (Fig. 4.20).

130. Lovejoy, *Transformations*, p. 142, who estimates 3.46 million slaves for the nineteenth-century Atlantic trade. Eltis (*Volume*, 2001, p. 43, Tab. 1) confirms this with his figure of 3.44 million, drawn partly from other sources.

131. Newitt, *Mozambique*, pp. 268–72; on Mauritius as a slave market, see Vaughan, *Creole Island*, pp. 103–8.

132. H. S. Klein, *Slave Trade*, p. 210f. (Tab. A.1).

133. See the illuminating study Law, *Ouidah*, pp. 189–203. See also Eltis and Richardson, *Atlas*, 287.

134. J. Fisch, *Geschichte Südafrikas*, p. 103.

135. For Senegal, see Searing, *West African Slavery*, p. 166.

136. Manning, *Slavery*, p. 84.

137. H. S. Klein, *Slavery and Colonial Rule*, p. 55.

138. Lovejoy, *Transformations*, pp. 65 ff.; Law, *Ouidah*, p. 77.

139. See Isichei, *History*, pp. 290–312.

140. The following draws on a number of sources: Zeleza, *Economic History of Africa*, pp. 73–75; Etemad, *Possession*, pp. 264f. (Tab. 26); J. Fisch, *Geschichte Südafrikas*, p. 405; M. Daly, *Cambridge History of Egypt*, vol. 2, p. 7.

141. Zeleza, *Economic History of Africa*, pp. 74f.

142. Iliffe, *Tanganyika*, pp. 138–40.

143. This is the estimate in Zeleza, *Economic History of Africa*, p. 75.

144. Amsden, *Rise of "the Rest,"* p. 21 (Tab. 1.11).

145. Bade, *Migration in European History*, p. 124.

146. For a more detailed account, see Grabbe, *Flut*, pp. 333–64.

147. Hoerder, *Cultures in Contact*, p. 331.

148. Grabbe, *Flut*, p. 94 (Tab. 13).

149. See the graph in Michael R. Haines, "The White Population of the United States, 1790–1920," in: M. R. Haines and Steckel, *Population History*, pp. 305–69, at 345 (Fig. 8.1.).

150. Ibid., p. 346 (Tab. 8.5.).

151. Nugent, *Crossings*, p. 43 (Tab. 9).

152. Ibid., pp. 29f.

153. Ibid., p. 30 (Tab. 8).

154. Marvin McInnis, "The Population of Canada in the Nineteenth Century," in: M. R. Haines and Steckel, *Population History*, pp. 17, 422.

155. Nugent, *Crossings*, pp. 137f., 112.

156. See a study that considers both sides of the Atlantic: Moya, *Cousins*.

157. Rosselli, *Opera Business*.

158. Rock, *Argentina*, pp. 133–43.

159. Bernand, *Buenos Aires*, pp. 194f.

160. Galloway, *Sugar Cane Industry*, p. 132.

161. Kale, *Fragments of Empire*, p. 1.

162. On passenger densities, see Northrup, *Indentured Labour*, p. 85.

163. Ibid., p. 9.

164. Ibid., p. 149 (Tab. 6.1); David Northrup, "Migration from Africa, Asia, and the South Pacific," in Louis, *Oxford History of the British Empire*, vol. 3, pp. 88–100, at 96.

165. Calculated from Northrup, *Indentured Labour*, pp. 156f. (Tab. A.1).

166. The following draws on: Tinker, *New System of Slavery*; Northrup, *Indentured Labour*, pp. 59–70; A.J.H. Latham, "Southeast Asia: A Preliminary Survey, 1800–1914," in Glazier and Rosa, *Migration*, pp. 11–29.

167. P. C. Emmer, "The Meek Hindu: The Recruitment of Indian Indentured Labourers for Service Overseas, 1870–1916," in idem, *Colonialism and Migration*, pp. 187–207.

168. On the early criticisms see Kale, *Fragments of Empire*, pp. 28–37.

169. Philip D. Curtin, "Africa and Global Patterns of Migration," in Wang Gungwu, *Global History*, pp. 63–94, at 83.

170. Tinker, *New System of Slavery*, p. 334.

171. Richardson, *Chinese Mine Labour*, pp. 177f. and passim.

172. G. William Skinner, "Creolized Chinese Societies in Southeast Asia," in: Reid, *Sojourners*, pp. 51–93, at 52.

173. Good introductions to the history of the emigration are Wang Gungwu, *The Chinese Overseas*, and Kuhn, *Chinese among Others*.

174. Skinner, *Chinese Society*, pp. 30f., 73.

175. Wang Sing-wu, *Chinese Emigration*, pp. 50–53, quotation on p. 62.

176. Irick, *Coolie Trade*, p. 183.

177. On coolie protection policies in the late Qing period, see Yen Ching-hwang, *Coolies*.

178. David Northrup, "Migration from Africa, Asia, and the South Pacific," in: Louis, *Oxford History of the British Empire*, vol. 3, pp. 88–100, at 94 (table 5.3).

179. For China, see Hunt, *Special Relationship*, p. 64; for Europe: Baines, *Migration*, p. 126.

180. Gyory, *Closing the Gate*, p. 67.

181. See in general McKeown, *Chinese Migrant Networks*.

182. Amrith, *Migration and Diaspora*, p. 32. Many of the often-cited data for South Asia are from Kingley Davis, *The Population of India and Pakistan*, Princeton, NJ 1952.

183. McKeown, *Global Migration*, p. 157; so also McKeown, *Melancholy Order*, pp. 43–65.

184. Susan Naquin and Yü Chün-fang, "Introduction: Pilgrimage in China," in idem, *Pilgrims*, pp. 19f.

185. Peters, *The Hajj*, mainly a rich collection of translated sources; Faroqhi, *Herrscher über Mekka*, pp. 223 ff., 252 (table 7); Mary Byrne McDonnell, "Patterns of Muslim Pilgrimage from Malaysia, 1885–1985," in: Eickelman, *Muslim Travellers*, pp. 111–30, at 115.

186. Umar Al-Naqar, *Pilgrimage Tradition*, pp. 82 ff.

187. Hayami, *Population*, p. 37.

188. For the US: Bodnar, *The Transplanted*, pp. 117–43.

189. Hochstadt, *Mobility*, p. 218.

190. This was not only relevant for relatively poor countries like Japan and China, but also for Britain. See Magee and Thompson: *Empire and Globalisation*, pp. 97–105.

191. T. M. Devine, *To the Ends of the Earth*, p. 31.

192. E. Richards, *Poor People*, pp. 251–53; R. F. Haines, *Emigration*.

CHAPTER V: Living Standards

1. W. Reinhard, *Lebensformen*, p. 453. Cf. Michael Argyle, "Subjective Well-Being," in Offer, *In Pursuit*, pp. 18–45, an attempt to develop parameters of happiness.

2. On various debates surrounding the standard of living, cf. Carole Shammas, "Standard of Living, Consumption, and Political Economy over the Past 500 Years," in Trentmann, *Oxford Handbook of the History of Consumption*, pp. 211–26.

3. Bengtsson et al., *Life under Pressure*, p. 33.

4. For a summary see Van Zanden, *Wages*, pp. 191–93.

5. The estimates are slightly different in G. Clark, *Farewell to Alms*, pp. 319f., 324f.

6. Bourguignon and Morrison, *Inequality*, pp. 731, 743.

7. Cf. the independent estimates in ibid., p. 728.

8. See, on the basis of Maddison: Easterlin, *Worldwide Standard of Living*, p. 10.

9. Lavely and Wong. *Malthusian Narrative*, p. 723.

10. Maddison, *World Economy*, p. 30 (tab. 1–5a); cf. R. W. Fogel, *Escape*, p. 2 (tab. 1–1).

11. See Hanley, *Everyday Things*, and the comparison between Japan and England in Macfarlane, *Savage Wars of Peace*.

12. Imhof, *Lebenszeit*, p. 63. Imhof's wide-ranging researches are fundamental for this topic.

13. G. Clark, *Farewell to Alms*, pp. 45, 95f.

14. R. W. Fogel, *Escape*, pp. 2f., 8.

15. See, for example, Szreter and Mooney, *Urbanization*, pp. 108f.

16. Hans-Joachim Voth, "Living Standards and the Urban Environment," in: Floud and Johnson, *Cambridge Economic History of Britain*, vol. 1, pp. 268–94, at 293.

17. Wehler, *Gesellschaftsgeschichte*, vol. 2, pp. 281–96.

18. R. W. Fogel, *Escape*, pp. 11, 18, 35f., 38, 40.

19. Riley, *Rising Life Expectancy*, p. 34; Imhof, *Lebenszeit*, p. 84.

20. Cameron Campbell, "Mortality Change and the Epidemiological Transition in Beijing, 1644–1990," in: Liu Ts'ui-jung et al., *Asian Population History*, pp. 221–47, at 222f., 243. Today, life expectancy in the People's Republic is about 5 years lower than in the richest countries of the West.

21. Riley, *Rising Life Expectancy*, p. 39.

22. There is a good description in C. King, *Black Sea*, pp. 168–72.

23. D. Fraser, *Evolution*, pp. 66–78. Though focused on the United States, the best *general* introduction is Melosi, *The Sanitary City*.

24. Burrows and Wallace, *Gotham*, pp. 625–27.

25. R. Porter, *London*, pp. 265f.

26. Münch, *Stadthygiene*, pp. 128f., 132–36, 191.

27. Weintraub, *Uncrowned King*, pp. 430, 435. The real cause of death was probably stomach cancer; the rumors in question were the result of the official diagnosis.

28. R. Porter, *London*, pp. 263f.; Inwood, *London*, pp. 433f.; Halliday, *Great Stink*, pp. 84, 91–99. A history of the city as a history of smells, taking Cairo as its main example, is Fahmy, *Olfactory Tale*.

29. Verena Winiwarter emphasizes the important of cultural perceptions of dirt, in "Where Did All the Waters Go? The Introduction of Sewage Systems in Urban Settlements," in: Bernhardt, *Environmental Problems*, pp. 106–19.

30. Halliday, *Great Stink*, p. 103.

31. Wedewer, *Reise nach dem Orient*, p. 216; on Istanbul's water supply before the beginning of modernization, see Kreiser, *Istanbul*, pp. 58–64, and on other Middle Eastern cities, Raymond, *Grandes villes arabes*, pp. 155–67.

32. Dossal, *Imperial Designs*, p. 116.

33. Arnold, *Colonizing the Body*, p. 167.

34. MacPherson, *Wilderness*, pp. 116 f., 120; Dikötter, *Exotic Commodities*, p. 145.

35. A. Hardy, *Health and Medicine in Britain*, pp. 12f.

36. Daunton, *Progress*, p. 439.

37. D. C. North, *Understanding*, p. 97, fig. 7.10.; Richard H. Steckel and Roderick Floud, "Conclusions," in: idem, *Health*, pp. 423–49, at 430f., figures p. 424 (Tab. 11.1). The retrospective construction of a Human Development Index for the United States points to stagnation between 1830 und 1860, followed by a continuous rise.

38. Steckel and Roderick Floud, *Health*, p. 436.

39. Vögele, *Sozialgeschichte*, pp. 84, 87 ff.; Arnold, *Colonizing the Body*, p. 167; Harrison, *Public Health*, pp. 99ff.

40. Vögele, *Urban Mortality Change*, p. 213.

41. Labisch, *Homo Hygienicus*, p. 134.

42. G. Rosen, *History of Public Health*, pp. 147–51.

43. Huerkamp, *Aufstieg der Ärzte*, pp. 177ff.

44. Witzler, *Großstadt und Hygiene*, pp. 131–38.

45. Higman, *Slave Populations*, pp. 262–64, 271f., 328, 341.

46. Riley, *Rising Life Expectancy*, pp. 21–24.

47. See Dormandy, *White Death*, a rather anecdotal treatment of the subject; and, above all, D. S. Barnes, *Making*.

48. A sparkling account may be found in Hays, *Burdens of Disease*, pp. 168–71.

49. Kiple, *Human Disease*, p. 403

50. Johnston, *Modern Epidemic*, pp. 70f., 73, 90, 135ff., 305–8 (statistics).

51. S. Watts, *Epidemics*, p. 25.

52. Barry, *Influenza*, pp. 398, 450.

53. Kiple, *Human Disease*, p. 1012.

54. Kuhnke, *Lives at Risk*, pp. 113–15.

55. R. Porter, *Greatest Benefit*, p. 420.

56. A good overview is Glynn and Glynn, *Smallpox*, pp. 115–29.

57. Jannetta, *Vaccinators*, pp. 71, 145.

58. Winkle, *Geißeln*, pp. 893f.; Smallman-Raynor and Cliff, *War Epidemics*, pp. 452–69.

59. Higman, *Slave Populations*, pp. 278f.

60. D. R. Hopkins, *Princes*, pp. 149–54.

61. John R. Shepherd, "Smallpox and the Patterns of Mortality in Late Nineteenth-Century Taiwan," in: Liu Ts'ui-jung et al., *Asian Population History*, pp. 270–91.

62. Kiple, *Human Disease*, pp. 403f.

63. D. R. Hopkins, *Princes*, pp. 194, 303.

64. Huerkamp, *Smallpox Vaccination*, pp. 622f.

65. D. R. Hopkins, *Princes*, pp. 186. 189.

66. Terwiel, *Acceptance*.

67. Much the same was observable in "internal peripheries" such as Hokkaido in northern Japan, where the central state presented itself as a "civilizer" of the indigenous Ainu population. See B. L. Walker, *Early Modern Japanese State*, esp. pp. 156f.

68. On the early history of medical bacteriology, see Gradmann, *Krankheit im Labor*, pp. 31ff.

69. Riley, *Rising Life Expectancy*, p. 113.

70. Iliffe, *East African Doctors*, p. 11.

71. Rosner, *Medizingeschichte Japans*, pp. 113–17; Nakayama, *Traditions*, pp. 197–200.

72. For the period up to the Opium War see L. L. Barnes, *Needles*.

73. S. Watts, *Epidemics*, p. 24.

74. On the practice of quarantine in the eighteenth century, see Panzac, *Quarantaines*, pp. 31–56 (ports), 61 (*cordon* in France), 67–78 (Balkan *cordon*).

75. Winkle, *Geißeln*, pp. 498f.

76. For further details, see Panzac, *La peste*, pp. 134–73.

77. Even this figure paled in comparison with the losses from disease (three-quarters of all deaths) in the French expedition to Haiti in 1802. See Laurens, *L'Expédition d'Égypte*, p. 468.

78. Moltke, *Briefe*, pp. 146–51.

79. On the end of the plague in the Ottoman Empire, see Panzac, *La peste*, pp. 446ff., 509.

80. Panzac, *Quarantaines*, p. 79.

81. Bickford-Smith, *Cape Town*, p. 19.

82. The standard work is still Hirst, *Conquest of Plague*, esp. pp. 254ff., 378ff. On Australia cf. Christabel M. Young, "Epidemics and Infectious Diseases in Australia prior to 1914," in: Charbonneau and Larose, *Mortalities*, pp. 207–27, at 216.

83. Terence H. Hull, "Plague in Java," in: N. G. Owen, *Death*, pp. 210–34, at 210f.

84. Mollaret and Brossolet, *Alexandre Yersin*.

85. Papin, *Hanoi*, p. 252.

86. Jannetta, *Epidemics*, p. 194.

87. Benedict, *Bubonic Plague*, pp. 25f. The rest of this section draws on Benedict's outstanding study.

88. The following is based on Arnold, *Colonizing the Body*, pp. 200–239.

89. Ibid., p. 203.

90. Huber, *Unification of the Globe by Disease?*

91. Echenberg, *Pestis Redux*, pp. 432, 444f.

92. C. E. Rosenberg, *Cholera Years*, p. 38.

93. Kerrie L. MacPherson, "Cholera in China: An Aspect of the Internationalization of Infectious Disease," in: Elvin and Liu Ts'ui-jung, *Sediments of Time*, pp. 487–519, at 498, 511. Cf. Harrison, *Climates and Constitutions*, pp. 190f. on the resistance in India to the water theory, which was viewed as simplistically monocausal.

94. Hamlin, *Cholera*, p. 3.

95. Koch, *Disease Maps*, chs. 6–11.

96. The approach of cholera to Western Europe from 1823 is impressively described in Dettke, *Die asiatische Hydra*, pp. 26ff.

97. Winkle, *Geißeln*, p. 191.

98. Bourdelais and Raulot, *Peur bleue*, p. 85.

99. C. E. Rosenberg, *Cholera Years*, p. 226.

100. This was already argued in Strachey, *Eminent Victorians*, pp. 132–36.

101. Smallman-Raynor and Cliff, *War Epidemics*, p. 417; Gruzinski, *Mexico*, p. 413; Echenberg, *Africa in the Time of Cholera*, pp. 56–59.

102. Münch, *Stadthygiene*, pp. 134f.

103. R. J. Evans, *Death in Hamburg*, pp. 285ff.

104. Rodney Sullivan, "Cholera and Colonialism in the Philippines, 1899–1903," in: MacLeod and Lewis, *Disease*, pp. 284–300, at 284.

105. Snowden, *Naples*, pp. 247ff.

106. Arnold, *Colonizing the Body*, p. 161.

107. R. J. Evans, *Death in Hamburg*, pp. 293f.

108. See Delaporte, *Disease*, pp. 10–18, 47ff., 97ff. (on the crisis of the sense of civilizational superiority), and Briese's monumental *Angst*.

109. Echenberg, *Africa in the Time of Cholera*, p. 75.

110. Vigier, *Paris*, pp. 76, 80, 85.

111. Kudlick, *Cholera*, pp. 81ff.

112. Arnold, *Colonizing the Body*, p. 178.

113. Baldwin, *Contagion*, p. 140.

114. Ibid., pp. 43–45.

115. Ibid., p. 190.

116. Kassir, *Beirut*, p. 112.

117. On the example of Canada, see Bilson, *Darkened House*, pp. 8ff.

118. An interesting work here is Igler, *Diseased Goods*, which also emphasizes the function of sailors as carriers. On medical and political views on invasion and quarantine see Harrison, *Medicine*, p 254–85.

119. Virchow, *Sämtliche Werke*, vol. 4, pp. 357–482, quotation 374, on hunger and epidemic disease: 420ff.

120. Smallman-Raynor and Cliff, *War Epidemics*, pp. 370ff.

121. P. D. Curtin, *Disease*, p. 177.

122. W. H. McNeill, *Plagues*, p. 261; P. D. Curtin, *Death by Migration*, p. 13.

123. P. D. Curtin, *Death by Migration*, pp. 62–68.

124. Bowler and Morus, *Making Modern Science*, p. 450.

125. A superb characterization of this "hygienic moment" in nineteenth-century history, centered on the example of France, is: La Berge, *Mission and Method*.

126. Tables in Easterlin, *Growth Triumphant*, pp. 161f.

127. For a brilliant argument about the imperial consequences of this fact see J. R. McNeill, *Mosquito Empires*.

128. S. W. Miller, *Environmental History*, p. 110; Sachs, *Tropical Underdevelopment*, pp. 15–18.

129. D'Arcy, *People of the Sea*, p. 128.

130. Winchester, *Crack*, pp. 259f., 271.

131. J. A. Lockwood, *Locust*, pp. 83f.

132. Winchester, *Krakatoa*.

133. DeJong Boers, *Tambora 1815*, pp. 375–77, 382–85.

134. Kaiwar, *Nature*, p. 25.

135. Ali, *Punjab*, pp. 8–61; Beinart and Hughes, *Environment and Empire*, pp. 130–47. On the link between irrigation and malaria, see especially: Radkau, *Nature and Power*, pp. 127–31.

136. See the Chinese case study: Schoppa, *Xiang Lake*.

137. Amelung, *Der Gelbe Fluß*, pp. 1f., 28–37, 43f., 55; Esherick, *Boxer Uprising*, pp. 7ff.

138. Amelung, *Der Gelbe Fluß*, pp. 379–81; cf. the somewhat different considerations in Elvin, *Elephants*, pp. 115–24.

139. T. N. Srinivasan, "Undernutrition: Concepts, Measurements, and Policy Implications," in: Osmani, *Nutrition*, pp. 97–120, at 97. On the emergence of "starvation" as a concept of "nutritional sciences" in the nineteenth century see Vernon, *Hunger*, ch. 4.

140. Wilhelm Abel, "Landwirtschaft 1648–1800," in: Aubin and Zorn, *Handbuch*, pp. 524f. Woolf, *Italy*, p. 279. Even the wealthy Netherlands had 60,000 deaths from starvation in the 1840s, with another 50,000 in Flanders.

141. Wells, *Wretched Faces*.

142. Tortella, *Modern Spain*, pp. 33f.; Yrjö Kaukiainen, "Finnland 1860–1913," in: Fischer, *Handbuch*, vol. 5, p. 274.

143. Nelson, *Bitter Bread*, pp. 117ff. on the relief measures.

144. Devine, *Great Highland Famine*, pp. 33ff.

145. The following is based on Daly, *Famine in Ireland*; Kinealy, *Death-Dealing Famine*; Kinealy, *Great Irish Famine*; O'Gráda, *Ireland*, pp. 173–209, and 85, 97; O'Gráda, *Ireland's Great Famine*; Clarkson and Crawford, *Feast*.

146. Floud, et al.: *Changing Body*, p. 116.

147. Robbins, *Famine in Russia*, pp. 3, 10, 176f.

148. Robert McCaa, "The Peopling of Mexico from Origins to Revolution," in: M. R. Haines and Steckel, *Population History*, pp. 241–304, at 288; Livi-Bacci, *Population*, pp. 68f.

149. A.K.S. Lambton, "Land Tenure and Revenue Administration in the Nineteenth Century," in: Avery et al., *Cambridge History of Iran*, vol. 7, pp. 459–505, at 469.

150. Gado, *Sahel*, pp. 67–88, 104.

151. J. C. Miller, *Significance*, pp. 21, 23, 25–31.

152. Zeleza, *Economic History of Africa*, pp. 35–40; Coquery-Vidrovitch, *Africa*, p. 32.

153. Hayami, *Population*, pp. 142f.

154. Harold Bolitho, "The Tempo Crisis," in: J. W. Hall et al., *Cambridge History of Japan*, vol. 5, pp. 116–67, at 117–20; Totman, *Early Modern Japan*, pp. 236–45, 511-18.

155. Davis, *Holocausts*, p. 7 (figures).

156. Wallace, *Wonderful Century*, p. 375.

157. For data on individual famines (including minor ones), see also Bhatia, *Famines in India*.

158. M. Davis, *Holocausts*, p. 50.

159. Bhatia, *Famines in India*, pp. 241f.

160. Ibid., p. 9.

161. Ludden, *Agrarian History*, pp. 199–201. Especially important on the role of money-lenders (and the lack of government interference in their activity) is Hardiman, *Feeding the Baniya*, pp. 57–61, 272ff. More generally, cf. Seavoy, *Famine*, pp. 241–85. A plausible new interpretation in the light of Amartya Sen's "entitlement" approach is Chakrabarti, *Famine of 1896–1897*.

162. The following is based on L. M. Li, *Fighting Famine in North China*, pp. 272–77. For the context, see ibid., chs. 8–10.

163. Bohr, *Famine in China*, pp. 13–26.

164. See the study Rankin, *Managed by the People*.

165. See Will, *Bureaucracy*.

166. Will and Wong, *Nourish the People*, pp. 75–92.

167. Robert Tombs, "The Wars against Paris," in: Förster and Nagler, *On the Road to Total War*, pp. 541–64, at 550.

168. Crossley, *Orphan Warriors*, pp. 132f.

169. Only a few brief remarks are possible here on this inexhaustible theme, which has been little discussed by historians. A still pioneering work is Bairoch, "Les trois révolutions agricoles".

170. Grigg, *Transformation*, p. 19 (tab. 2.2).

171. Federico, *Feeding the World*, pp. 33f. (tab. 4.1).

172. Ibid., pp. 18–19 (tab. 3.1, 3.2).

173. Bairoch, *Victoires*, vol. 1, p. 278.

174. Bray, *Rice Economies*, p. 95.

175. Data collated from Pohl, *Aufbruch*, pp. 99ff.

176. Wolfram Fischer, "Wirtschaft und Gesellschaft Europas 1850–1914," in: Fischer, *Handbuch*, vol. 5, pp. 1–207, at 137f.; Grigg, *Transformation*, p. 19 (diag. 3.1.)

177. On the concept, see Overton, *Agricultural Revolution*, ch. 1. There is a very broad term that denotes the overall transformation of rural societies in the wake of industrialization (as in Marx, Tawney, and the Hammonds), but that is not what is meant here.

178. Tracy Dennison and James Simpson, "Agriculture," in: Broadberry and O'Rourke, *Cambridge Economic History of Modern Europe*, vol. 1, pp. 148–63, at 162.

179. Overton, *Agricultural Revolution*, pp. 8, 206; see the broad panorama in Grigg, *Transformation*.

180. There is still something to be said for the older thesis that the greatest productivity increases in English agriculture occurred only after 1800. This would mean that the "agricultural revolution" was not a in a neat way preliminary to the Industrial Revolution but a synchronic part of a comprehensive process of transformation. See M. E. Turner et al., *Farm Production*. A new general discussion is Mokyr, *Enlightened Economy*, pp. 170–84.

181. Bairoch, *Victoires*, vol. 1, pp. 273f.; Daunton, *Progress*, p. 44; Robert C. Allen, "Agriculture during the Industrial Revolution," in: Floud and Johnson, *Cambridge Economic History of Britain*, vol. 1, pp. 96–116, at 96.

182. Overton, *Agricultural Revolution*, pp. 121f., 124.

183. Grigg, *Transformation*, pp. 48–50.

184. There is a good survey of eighteenth-century European agriculture in: Cameron, *Economic History*, pp. 109–14.

185. Braudel, *Civilization and Capitalism*, vol. 1, p. 155; see the brief account in Chaudhuri, *Asia*, pp. 233–38.

186. Huang, *Peasant Family*, pp. 77ff.; Pomeranz, *Great Divergence*, pp. 215f.

187. Bray, *Rice Economies*, pp. 55, 205.

188. Achilles, *Deutsche Agrargeschichte*, p. 206.

189. Wolfram Fischer, "Wirtschaft und Gesellschaft Europas 1850–1914," in: Fischer, *Handbuch*, vol. 5, pp. 1–207, at 140 (tab. 38).

190. Overton, *Agricultural Revolution*, p. 131.

191. For 1700–1850: Robert C. Allen, "Agriculture during the Industrial Revolution," in: Floud and Johnson, *Cambridge Economic History of Britain*, vol. 1, pp. 103f.

192. See the broad panorama, mainly focused on environmental history, in Dunlap, *Nature*; for India, Markovits, *Modern India*, pp. 306–8.

193. Offer, *First World War*, pp. 404 and passim.

194. This, slightly modified, is the interpretation in Koning, *Failure*, esp. pp. 71 ff.

195. See Stedman Jones, *End to Poverty*.

196. See the resumé in Kaelble, *Industrialisierung*, p. 55; Colin Heywood, "Society," in: Blanning, *Nineteenth Century*, pp. 47–77, at 57f.; and the quantitative case put forward in Hoffman et al., *Real Inequality*, pp. 348, 351.

197. D. Lieven, *The Aristocracy*, ch. 2.

198. A splendid description of this world may be found in the social and architectural study: J. M. Crook, *Rise of the Nouveaux Riches*, esp. pp. 37ff.; cf. Mandler, *Fall and Rise*.

199. See the studies of Britain, France, Italy, and the United States in: Rubinstein, *Wealth*; and, for the United States, the overviews in Lee Soltow, "Wealth and Income Distribution," in: Cayton, *Encyclopedia*, vol. 2, pp. 1517–31, and Ronald Story, "The Aristocracy of Inherited Wealth," in: ibid., pp. 1533–39 (percentage on p. 1536).

200. Williamson and Lindert, *American Inequality*, pp. 75–77; Huston, *Securing the Fruits*, pp. 339 f.

201. Homberger, *Mrs. Astor's New York*, pp. 1ff. and passim; Bushman, *Refinement*, p. 413; Sarasin, *Stadt der Bürger*, ch. 4.

202. G. Clark, *Farewell to Alms*, pp. 236, 298f.

203. Carosso, *The Morgans*, p. 644. In today's values that would have been equivalent $800 million.

204. W. D. Rubinstein, "Introduction," to idem, *Wealth*, pp. 9–45, at 18–21; Cannadine, *Decline and Fall*, pp. 90f.; and Beckert, *Monied Metropolis*, p. 28.

205. Naquin, *Peking*, pp. 392–94 (sketch of such an estate: p. 393).

206. Abeyasekere, *Jakarta*, p. 62.

207. There is a fine description of the mood among the samurai in McClain, *Japan*, pp. 120–24.

208. Ravina, *Land and Lordship*, pp. 68f.

209. A conceptually clear analysis of *waqfs*, especially in the eighteenth century, is Leeuwen, *Waqfs*; see esp. the overview of their integrative functions on p. 207.

210. Iliffe, *African Poor*, pp. 14, 29, 114, 124, 143, 148, 164ff. and passim.

211. D. Lieven, *Aristocracy*, p. 39; Freyre, *Mansions*, p. 22; Abeyasekere, *Jakarta*, p. 37.

212. Iliffe, *African Poor*, pp. 65–81. On the structural conditions of the nomadic way of life, see Khazanov, *Nomads*.

213. George R. Boyer, "Living Standards, 1860–1939," in: Floud and Johnson, *Cambridge Economic History*, vol. 2, pp. 280–313, at 298f.

214. Özmucur and Pamuk, *Real Wages*, pp. 316f.; G. Clark, *Farewell to Alms*, p. 49 (tab. 3.5); see also Şevket Pamuk and Jan-Luiten van Zanden, "Standards of Living," in: Broadberry and O'Rourke, *Cambridge Economic History of Modern Europe*, vol. 1, pp. 217–34; Malanima, *Pre-Modern European Economy*, p. 271.

215. Bishnupriya Gupta and Debin Ma, "Europe in an Asian Mirror: The Great Divergence," in: Broadberry and O'Rourke, *Cambridge Economic History of Modern Europe*, vol. 1, pp. 263–85, at 273.

216. See Lindenmeyer, *Poverty*, pp. 142–44. A survey of poor relief in "Northern Europe," including France, Germany, and Russia, may be found in Grell et al., *Health Care*.

217. Ener, *Managing Egypt's Poor*, pp. 19–23.

218. Braudel, *Civilization and Capitalism*, vol. 1, pp. 187ff.

219. See the overview in Wendt, *Kolonialismus*, pp. 83–85, 184–90, 372f.; and the product by product account in Kiple, *Movable Feast*.

220. E. N. Anderson, *Food of China*, pp. 97f.

221. Yves Péhaut, "The Invasion of Foreign Foods," in: Flandrin and Montanari, *Food*, pp. 457–70, at 457–61.

222. Peter W. Williams, "Foodways," in: Cayton, *Encyclopedia*, vol. 2., pp. 1331–44, at 1337.

223. G. G. Hamilton, *Commerce*, pp. 76f.; Dikötter, *Exotic Commodities*, pp. 222–24, 228f., 231.

224. J.A.G. Roberts, *China to Chinatown*, chs. 6–7. Cf. Goody, *Food*, pp. 161–71.

225. Walvin, *Fruits of Empire*, p. 168–73.

226. Ibid., p. 30.

227. Pohl, *Aufbruch*, p. 111.

228. Mintz, *Sweetness*, pp. 78, 114–20, 133f., 148f., 180f., Mintz considers this to have been specific to Britain.

229. Galloway, *Sugar Cane Industry*, p. 239.

230. Vigier, *Paris*, p. 316.

231. D. J. Oddy, "Food, Drink and Nutrition," in: Thomson, *Cambridge Social History of Britain*, vol. 2, pp. 251–78, at 270f.

232. Hanley, *Everyday Things*, p. 162.

233. Mokyr, *Lever of Riches*, p. 141.

234. Pohl, *Aufbruch*, pp. 106f.

235. Rock, *Argentina*, pp. 171f.

236. Cronon, *Nature's Metropolis*, pp. 207–12, 225–47. On the comparative strength of the American meat culture, see Horowitz et al., *Meat*.

237. Peter W. Williams, "Foodways," in: Cayton, *Encyclopedia*, vol. 2, p. 1336; and D. J. Oddy, "Food, Drink and Nutrition," in: Thomson, *Cambridge Social History of Britain*, vol. 2, pp. 274f.

238. Ellerbrock, *Geschichte der deutschen Nahrungs- und Genußmittelindustrie*, p. 235.

239. Pounds, *Hearth and Home*, pp. 394f..

240. Benjamin, *Arcades Project*, e.g., pp. 40ff.; see also Crossick and Jaumain, *Cathedrals of Consumption*.

241. Higonnet, *Paris*, pp. 194–200.

242. R. Porter, *London*, p. 201.

243. Burrows and Wallace, *Gotham*, pp. 667f.

244. Bled, *Wien*, p. 216.

245. Seidensticker, *Low City*, pp. 110–14.

246. For the prosaic explanation: Jean-Robert Pitte, "The Rise of the Restaurant," in: Flandrin and Montanari, *Food*, pp. 471–80; and for the complex perspective of cultural studies: Spang, *Restaurant*, p. 150 (quotation) and passim.

247. Walton, *Fish and Chips*, pp. 5, 8, 25.

248. Hanley, *Everyday Things*, p. 164; Nishiyama, *Edo Culture*, pp. 164–78.

249. W. König, *Konsumgesellschaft*, p. 94.

250. Tedlow, *New and Improved*, pp. 14f., on Coca Cola pp. 23–111 (Tab. 2-2, p. 29).

251. W. König, *Konsumgesellschaft*, pp. 94f.

252. McKendrick et al., *Birth of a Consumer Society*; Brewer and Porter, *Consumption*.

253. See the chapter "The Consuming City" in: Boyar and Fleet, *Ottoman Istanbul*, pp. 137–204.

254. Brook, *Confusions of Pleasure*, pp. 190–237; on fashion: pp. 218ff.

255. Hannes Siegrist, "Konsum, Kultur und Gesellschaft im modernen Europa," in: Siegrist et al., *Europäische Konsumgeschichte*, pp. 3–48, at pp. 18f.

256. Freyre, *Mansions*, pp. 206ff. On the similar symbolism of the watch, see chapter 2, above.

257. Bernand, *Buenos Aires*, pp. 187–89, 98.

258. Cohn, *Colonialism*, p. 112 (also pp. 123f. on the later reorientalization of military uniforms in British India); Mukherjee, *Calcutta*, p. 90.

259. Purdy, *Tyranny of Elegance*, pp. 215.19.

260. Ross, *Clothing*, p. 87.

261. C. J. Baker and Phongpaichit, *Thailand*, p. 100.

262. Zachernuk, *Colonial Subjects*, p. 30.

263. On the problem that the British and many Indians had with "nakedness" in India, see Cohn, *Colonialism*, pp. 129ff.

264. A. J. Bauer, *Goods*, pp. 130, 138–64; Needell, *Tropical "belle époque,"* pp. 156ff. and passim. Needell speaks of "consumer fetishism" (p. 156). On Egypt, see the rather scanty collection of material in Luthi, *La vie quotidienne en Égypte*.

265. Charlotte Jirousek, "The Transition to Mass Fashion Dress in the Later Ottoman Empire," in: Quataert, *Consumption Studies*, pp. 201–41, at 208, 210, 223f., 229; on earlier Ottoman dressing habits see Boyar and Fleet, *Ottoman Istanbul*, pp. 175–82.

266. Abu-Lughod, *Rabat*, p. 107.

267. Seidensticker, *Low City*, pp. 97, 10; Hanley, *Everyday Things*, pp. 173–75, 196; above all Esenbel, *Anguish*, esp. pp. 157–65.

268. On Lodz: Pietrow-Ennker, *Wirtschaftsbürger*, p. 200.

269. Esenbel, *Anguish*, pp. 168f.

270. Finnane, *Changing Clothes in China*, p. 77: the standard work on the subject.

271. Nuckolls, *Durbar Incident*; Cohn, *Colonialism*, pp. 127–29.

272. See the sparkling and entertaining account in: Dalrymple, *White Mughals*.

273. J. G. Taylor, *Social World of Batavia*, pp. 112f.; Abeyasekere, *Jakarta*, p. 75.

274. Papin, *Hanoi*, pp. 197, 200.

275. G. Wright, *Politics of Design*, pp. 236–43.

276. See Radkau, *Nervosität*, pp. 17–23.

277. Klaus Tenfelde, "Klassenspezifische Konsummuster im Deutschen Kaiserreich," in: Siegrist et al., *Konsumgeschichte*, pp. 245–66, at 256–59.

278. Montanari, *Hunger*, pp. 155ff., 189.

279. A comprehensive, in part formal-mathematical, conceptualization of all possible aspects of the standard of living is: Dasgupta, *Inquiry*.

CHAPTER VI: Cities

1. The contextual opposition of city and country is therefore too narrow. In nineteenth-century Brazil, for example, the relevant contrast was between the city and the plantation: see Freyre, *Mansions*, p. 26 and passim.

2. See H. S. Jansen, *Wrestling with the Angel*.

3. Bairoch, *Cities and Economic Development*, pp. 19 ff., 93ff. applies the concept of urbanization as far back as the Ancient East.

4. E. Jones, *Metropolis*, p. 76.

5. Coquery-Vidrovitch, *History of African Cities*, pp. 263–79; P. B. Henze, *Layers of Time*, p. 154.

6. Geertz, *Local Knowledge*, p. 137.

7. Kanwar, *Imperial Simla*; D. Kennedy, *Magic Mountains*.

8. Kent, *Soul of the North*, p. 320.

9. Morse, *Japanese Homes*, pp. 12f.

10. Seidensticker, *Low City*, p. 263.

11. A masterful analysis of this process in Europe, considering all its aspects, is Lenger, *European Cities* in the Modern Era; for a concise quantitative overview of the major European regions see P. Clark, *European Cities and Towns*, pp. 221–35.

12. Lepetit, *Les villes*, p. 94.

13. Martin Daunton, "Introduction," in: P. Clark, *Cambridge Urban History*, vol. 3, pp. 1–56, at 6ff.

14. Girouard, *English Town*, p. 190.

15. See Pike, *Subterranean Cities*. On the obsession with the Paris catacombs since Victor Hugo's time, see the literary study: Prendergast, *Paris*, pp. 74–101.

16. This often neglected point is emphasized in Dodgshon, *Society*, p. 159; a key work on the theory and history of infrastructural development is Grübler, *Infrastructures*, whose main focus is on intercity transportation. Cf. Laak, *Infra-Strukturgeschichte*.

17. An interesting argument in de Soto, *Mystery of Capital* is that the chronic undervaluation of urban land has been one reason for the "poverty" of the "third world."

18. Chudacoff, *American Urban Society*, p. 37.

19. Chartier et al., *La ville des temps modernes*, p. 567.

20. These important issues have rarely been given serious attention in economic and social history. But see the exemplary work: Day, *Urban Castles*.

21. Important here are: P. Clark, *British Clubs*; and Hardtwig, *Genossenschaft*. For China see Rankin, *Elite Activism* and the debate it triggered about the beginnings of a "public sphere" in China.

22. Lees, *Cities Perceived*, p. 79. In Walter Benjamin, who made this expression famous, it was of course more than a mere cliché.

23. David Ward and Olivier Zunz, "Between Rationalism and Pluralism: Creating the Modern City," in: idem, *Landscape of Modernity*, pp. 3–15; Harvey, *Postmodernity*; Berman, *All That is Solid*.

24. J. de Vries, "Problems in the Measurement, Description, and Analysis of Historical Urbanization," in: Woude et al., *Urbanization*, pp. 43–60, at 44. This article is an excellent introduction to the theory of urbanization.

25. Hohenberg and Lees, *Making of Urban Europe*, pp. 200–205.

26. Reulecke, *Urbanisierung in Deutschland*, pp. 11f.

27. Hohenberg and Lees, *Making of Urban Europe*, p. 244.

28. Bairoch, *Cities and Economic Development*, pp. 258ff.

29. E. A. Wrigley, "A Simple Model of London's Importance in Changing English Society and Economy, 1650–1759," in: idem, *People*, pp. 133–56, quotation on p. 146 (first published in 1967).

30. Martin Daunton, "Introduction," in: P. Clark, *Cambridge Urban History*, vol. 3, pp. 1–56, at 42.

31. Gerhard Melinz and Susan Zimmermann, "Großstadtgeschichte und Modernisierung in der Habsburgermonarchie," in idem, *Wien–Prag–Budapest*, pp. 15–33, at 23.

32. Daniel R. Brower, "Urban Revolution in the Late Russian Empire," in: Hamm, *City in Late Imperial Russia*, pp. 319–53, at 325.

33. Adler, *Yankee Merchants*, pp. 1, 4 and passim.

34. Olsen, *City*, p. 4.

35. Still the classic text on the world history of the city is: Mumford, *The City in History*, even though—or because—many of its judgments invite contradiction. An equally ambitious work is: P. Hall, *Cities in Civilization*.

36. Inwood, *London*, pp. 270, 411; P. Clark, *Cambridge Urban History*, vol. 2, p. 650 (Tab. 19.1). Bairoch, *Cities and Economic Development*, p. 81, thinks that Rome may have reached a total of 1.3 million—as much as the largest European city in 1823.

37. A. F. Weber, *Growth of Cities*, p. 122.

38. T. Chandler and Fox, *3000 Years*, p. 313.

39. Ibid., p. 321.

40. Ibid., p. 323; P. Clark, *European Cities and Towns*, p. 131 (tab. 7.2).

41. On the 1790s as the decade of New York's great surge, see Burrows and Wallace, *Gotham*, pp. 333–38.

42. A. F. Weber, *Growth of Cities*, p. 139.

43. Kumar, *Java*, p. 180.

44. Maddison, *Chinese Economic Performance*, p. 35, which for China supports itself on Gilbert Rozman.

45. Gilbert Rozman, "East Asian Urbanization in the Nineteenth Century: Comparisons with Europe," in: Woude et al., *Urbanization*, p. 65, Tab. 4.2b.

46. Ibid., p. 64, tab. 4.1a/4.1b.

47. Bairoch in: Bardet and Dupâquier, *Histoire des populations de l'Europe*, pp. 212 (tab. 21).

48. J. de Vries, *European Urbanization*, pp. 28, 39, 258f.

49. Ibid., p. 84.

50. Jan de Vries, "Problems in the Measurement, Description, and Analysis of Historical Urbanization," in: Woude et al., *Urbanization*, pp. 43–60, at 58f.; H. S. Klein, *Population History*, pp. 142f.

51. Lappo and Hönsch, *Urbanisierung Russlands*, p. 38; Goehrke, *Russischer Alltag*, p. 290. See also Hildermeier, *Bürgertum*, pp. 603f.

52. Paul Bairoch, "Une nouvelle distribution des populations: Villes et campagnes," in: Bardet and Dupâquier *Histoire des populations de l'Europe*, pp. 193–229, at 204f.

53. Palairet, *Balkan Economies*, pp. 28f.

54. Skinner, *Chinese Society*, pp. 68ff.

55. Anthony Reid, "South-East Asian Population History and the Colonial Impact," in: Liu Ts'ui-jung et al., *Asian Population History*, pp. 45–62, at 55.

56. In 1910 Bangkok was twelve times larger than Siam's second city: C. J. Baker and Phongpaichit, *History of Thailand*, p. 99.

57. Doeppers, *Philippine Cities*, pp. 783f., 791f.

58. Narayani Gupta, "Urbanism in South India: Eighteenth–Nineteenth Centuries," in: Banga, *City in Indian History*, pp. 121–47, at 137f., 142; Mishra, *Economic History*, p. 23. Ramachandran, *Urbanization*, pp. 61f.

59. The following figures draw mostly on Chandler and Fox, *3000 Years*, passim.

60. M. Reinhard et al., *Histoire générale*, p. 426.

61. Hofmeister, *Australia*, pp. 54, 64–67.

62. Monkkonen, *America Becomes Urban*, p. 70.

63. Ibid., p. 81.

64. A. F. Weber, *Growth of Cities*, p. 450.

65. Chudacoff, *American Urban Society*, p. 36.

66. Monkkonen, *America Becomes Urban*, p. 85.

67. Abu-Lughod, *New York*, p. 134.

68. Boyer and Davis, *Urbanization*, p. 7 (Tab. 2).

69. A. F. Weber, *Growth of Cities*, p. 450.

70. Bairoch, *Cities and Economic Development*, p. 217.

71. Bardet and Dupâquier *Histoire des populations de l'Europe*, pp. 193–229, at 227 (Tab. 24); Karpat, *Ottoman Population*, p. 103 (Tab. 5.3).

72. Ruble, *Second Metropolis*, pp. 15f., 25.

73. T. O. Wilkinson, *Urbanization of Japanese Labor*, pp. 63–65.

74. Hohenberg and Lees, *Making of Urban Europe*, p. 42; Meinig, *Shaping of America*, vol. 2, pp. 318–21.

75. Kassir, *Beirut*, pp. 110ff.

76. This point follows C. Tilly, *Coercion*, p. 51; it is more sharply expressed by Tilly in C. Tilly and Blockmans, *Cities*, p. 6; cf. Hohenberg and Lees, *Making of Urban Europe*, pp. 169ff.

77. Lemon, *Dreams*, p. 78; G. B. Nash, *First City*, pp. 45ff.

78. Hohenberg and Lees, *Making of Urban Europe*, p. 241.

79. Coquery-Vidrovitch, *History of African Cities*, pp. 226–27.

80. Kaffir, *Beirut*, pp. 28, 110f., 122f.; Hanssen, *Beirut*, pp. 84ff.

81. Cf. Lepetit, *Les villes*, p. 51.

82. George Modelski, "World Cities in History," in: W. H. McNeill, *Berkshire Encyclopedia*, vol. 5, pp. 2066–73, at 2066.

83. Braudel, *Civilization and Capitalism*, vol. 3, pp. 21ff.

84. See Sassen, *Global City*.

85. Paul Knox, "World Cities in a World System," in: Knox and Taylor, *World Cities*, pp. 3–20, at 12.

86. Snouck Hurgronje, *Mekka*, p. 7.

87. Coquery-Vidrovitch, *History of African Cities*, p. 236.

88. See the exemplary analysis: David D. Buck, "Railway City and National Capital: Two Faces of the Modern in Changchun," in: Esherick, *Remaking the Chinese City*, pp. 65–89. "National capital" refers to the fact that from 1932 to 1945 Changchun was the capital of the Japanese puppet state of Manchukuo. On Nairobi: Karl Vorlaufer, "Kolonialstädte in Ostafrika. Genese, Funktion, Struktur, Typologie," in: Gründer and Johanek, *Kolonialstädte*, pp. 145–201, at 164f.

89. Mommsen, *Das Ringen um den nationalen Staat*, p. 230.

90. Shannon, *Gladstone*, vol. 2, p. 572.

91. Girouard, *English Town*, pp. 289–91.

92. Walton, *English Seaside Resort*, pp. 5ff. and passim.

93. Collier and Sater, *Chile*, pp. 76–80, 161.

94. See the exemplary history of the city: Rohrbough, *Aspen*, esp. pp. 13, 288 ff.

95. J. M. Price, *Economic Function*.

96. Mantran, *Istanbul*, p. 258.

97. Bled, *Wien*, pp. 183f.

98. Coquery-Vidrovitch, *History of African Cities*, pp. 291–300.

99. K. Schultz, *Tropical Versailles*, pp. 101ff.

100. Raymond, *Cairo*, pp. 300f.

101. Perkins, *Modern Tunisia*, p. 14.

102. Letter from London, October 15, 1826, in: Pückler-Muskau, *Briefe eines Verstorbenen*, p.432.

103. Kuban, *Istanbul*, p. 379.

104. Naquin, *Peking*, p. 684. A good new history of the city is: L. M. Li et al., *Beijing*.

105. Dong, *Republican Beijing*, pp. 90–100 on "tourist Beijing."

106. Berelowitch and Medvedkova, *Saint-Pétersbourg*, pp. 317f.

107. See Reps, *Making of Urban America*, pp. 240–62.

108. Dickens, *American Notes*, p. 129.

109. Gerhard Brunn, "Metropolis Berlin. Europäische Hauptstädte im Vergleich," in: Brunn and Reulecke, *Metropolis*, pp. 1–39, at 13f.; P. Hall, *Cities in Civilization*, pp. 377, 386.

110. Kenneth T. Jackson, "The Capital of Capitalism: The New York Metropolitan Region, 1890–1940," in: Sutcliffe, *Metropolis*, pp. 319–53, at 347.

111. Ball and Sunderland, *Economic History of London*, p. 313.

112. A recent general account rejects the idea of London's decline as an industrial city in the nineteenth century: ibid., pp. 55–66, esp. 65; cf. Martin Daunton, "Introduction," in: P. Clark, *Cambridge Urban History*, vol. 3, pp. 1–56, at 45.

113. A first-rate piece is R. J. Morris, "The Industrial Town," in: Waller, *English Urban Landscape*, pp. 175–208; for data across Europe see P. Clark, *European Cities and Towns*, pp. 246f.

114. Briggs, *Victorian Cities*, p. 96. As Briggs shows, after 1851 scarcely anyone got worked up about Manchester any more (p. 112).

115. Girouard, *English Town*, pp. 249f., 253f.

116. On the perception of Manchester, see Lees, *Cities Perceived*, pp. 63–68; cf. 49–51 (praise for Manchester).

117. Bairoch, *Cities and Economic Development*, p. 254 (Tab. 15.1).

118. See Konvitz, *Urban Millennium*, pp. 98f.

119. Lichtenberger, *Die Stadt*, pp. 41, 43.

120. See Dennis, *English Industrial Cities*, pp. 17f.

121. Jürgen Reulecke, "The Ruhr: Centralization versus Decentralization in a Region of Cities," in: Sutcliffe, *Metropolis*, pp. 381–401, at 386.

122. Hohenberg and Lees, *Urban Europe*, pp. 188ff, 213, 234.

123. Barrie Trinder, "Industrialising Towns 1700–1840," in: P. Clark, *Cambridge Urban History*, vol. 2, pp. 805–829; David Reeder and Richard Rodger, "Industrialisation and the City Economy," in: ibid., vol. 3, pp. 553–592, at 585ff. On Lancashire as an innovative milieu, see the eloquent discussion in P. Hall, *Cities in Civilization*, pp. 314f., 334ff.

124. Goehrke, *Russischer Alltag*, pp. 292ff. Positive side: on the industrialist Carl Scheibler in Lodz, see Pietrow-Ennker, *Wirtschaftsbürger*, 2005, p. 187; Shao Qin, *Culturing Modernity*.

125. Lepetit, *Les villes*, p. 123,

126. Lis, *Social Change*, pp. 27ff. The conversion was preceded by a false start in the machine-based textile industry.

127. See F. W. Knight and Liss, *Atlantic Port Cities*, especially Barry Higman's contribution on Jamaica (pp. 117–48).

128. L. Ray Gunn, "Antebellum Society and Politics (1825–1860)," in: M. M. Klein, *Empire State*, pp. 307–415, at 319.

129. Fernández-Armesto, *Civilizations*, pp. 381–84.

130. Konvitz, *Cities and the Sea*, p. 36.

131. Corbin, *Lure of the Sea*; Girouard, *English Town*, p. 152.

132. Kreiser, *Istanbul*, pp. 218–25.

133. Amino, *Les Japonais et la mer*, p. 235.

134. Recent exceptions are two valuable collections on Asian port cities edited by Frank Boeze: *Brides* and *Gateways*.

135. Friel, *Maritime History*, p. 198.

136. Hugill, *World Trade*, p. 137.

137. Borruey, *Marseille*, pp. 5, 10, passim.

138. Dyos and Aldcroft, *British Transport*, p. 247.

139. Konvitz, *Urban Millennium*, p. 65; R. Porter, *London*, pp. 188f. There is a superbly detailed account of the old London docklands in Bird, *Major Seaports*, pp. 366–90.

140. Grüttner, *Arbeitswelt*, p. 19.

141. Dossal, *Imperial Designs*, p. 172; Ruble, *Second Metropolis*, pp. 222–26, esp. 222; Abeyasekere, *Jakarta*, pp. 48, 82; Chiu, *Port of Hong Kong*, p. 425.

142. Bourdé, *Urbanisation*, pp. 56–60.

143. Worden et al., *Cape Town*, p. 166; Bickford-Smith et al., *Cape Town*, p. 26.

144. Bergère, *Shanghai*, pp. 52–54.

145. John Butt, "The Industries of Glasgow," in: W. H. Fraser and Maver, *Glasgow*, vol. 2, pp. 96–140, at 112ff.

146. As in a famous essay: J. M. Price, *Economic Function*.

147. Robert Lee and Richard Lawton, "Port Development and the Demographic Dynamics of European Urbanization," in: idem, *Population and Society*, pp. 1–36, at 17. On the important concept of "casual labour," see Phillips and Whiteside, *Casual Labour*.

148. Excellent on this is: Linda Cooke Johnson, "Dock Labour at Shanghai," in: S. Davies, *Dock Workers*, pp. 269–89.

149. Marina Cattaruzza, "Population Dynamics and Economic Change in Trieste and Its Hinterland, 1850–1914," in Lawton and Lee, *Population and Society*, pp. 176–211, at 176–78; Herlihy, *Odessa*, pp. 24ff., 248ff.

150. Panzac, *Barbary Corsairs*, p. 270.

151. Auslin, *Negotiating with Imperialism*, p. 97: 1,500 Japanese fatalities against 18 British. Four years previously the French had burned Saigon to the ground—an unprovoked act of vandalism. And earlier still, Napoleon's troops had committed similar depredations in Spanish cities (although Madrid itself was spared).

152. Robert Lee and Richard Lawton, "Port Development and Demographic Dynamics of European Urbanization," in: idem, *Population and Society*, pp. 1–36, at 3.

153. Josef W. Konvitz, "Port Functions, Innovation and Making of the Megalopolis," in: T. Barker and Sutcliffe, *Megalopolis*, pp. 61–72, at 64f.

154. On the following, see also Lees and Lees, *Cities*, pp. 244–80; a different take on the same subject: Thomas R. Metcalf, "Colonial Cities," in: P. Clark, *Oxford Handbook of Cities*, pp. 753–69.

155. Doeppers, *Philippine Cities*, pp. 778, 785.

156. See the illustrated volume: Losty, *Calcutta*. On the background: P. J. Marshall, "Eighteenth-Century Calcutta," in: R. Ross and Telkamp, *Colonial Cities*, pp. 87–104.

157. See Raymond F. Betts, "Dakar: Ville impériale (1857–1960)," in: ibid., pp. 193–206.

158. Whelan, *Reinventing Modern Dublin*, pp. 38, 53, 92f.

159. Irving, *Indian Summer*, p. 42; on hybrid architecture in India see also Chopra, *A Joint Enterprise*, pp. 31–72.

160. Papin, *Hanoi*, pp. 233–46; Logan, *Hanoi*, pp. 72, 76f. 81, 89; G. Wright, *Politics of Design*, pp. 83, 162, 179.

161. Papin, *Hanoi*, p. 251.

162. Of a whole number of attempted definitions, the most useful is still that in A. D. King, *Colonial Urban Development*, pp. 18, 23–26, 33f., and—despite its scholastic overcomplexity— in idem, *Global Cities*, pp. 39–49. Cf. the skeptical view of generalizations in Franz-Joseph Post, "Europäische Kolonialstädte in vergleichender Perspektive," in: Gründer and Johanek, *Kolonialstädte*, pp. 1–25. A good overall account in unexpected areas: Beinart and Hughes, *Environment and Empire*, pp. 148–66.

163. E. Jones, *Metropolis*, pp. 17f.

164. Hamm, *City in Late Imperial Russia*, p. 135; Bled, *Wien*, p. 178. Other capital cities under military occupation: Mexico City 1847–48, Budapest 1849–52, Beijing 1900–1902.

165. Häfner, *Gesellschaft*, pp. 75f.

166. Mantran, *Istanbul*, p. 302.

167. See the exemplary analysis of Salonica in Anastassiadou, *Salonique*, pp. 58–75; and the fundamental work: Raymond, *Grandes villes arabes*, pp. 101ff., 133 ff., 175 ff., 295 ff.

168. Dalrymple, *Last Mughal*, pp. 454–64, on the destruction.

169. N. Gupta, *Delhi*, pp. 15, 17, 58–60.

170. Kosambi, *Bombay*, pp. 38, 43, 44.

171. Lichtenberger, *Die Stadt*, pp. 240ff.

172. On the difficulty of distinguishing ethnic from social segregation, with reference to the Irish in Victorian cities, see: Dennis, *English Industrial Cities*, pp. 221–33.

173. Jacobson, *Whiteness of a Different Color*.

174. Bronger, *Metropolen*, p. 174 (Tab. 19: Megacities), p. 191 (Tab. 55: Global cities).

175. W. J. Gardner, "A Colonial Economy," in: Oliver, *Oxford History of New Zealand*, pp. 57–86, at 67.

176. On the significance of this "agency system," see, e.g., Davison, *Marvellous Melbourne*, p. 22.

177. For greater detail on the treaty ports, see the articles Jürgen Osterhammel, "Konzessionen und Niederlassungen," "Pachtgebiete," and "Vertragshäfen," in: Staiger et al., *China-Lexikon*, pp. 394–97, 551–53, 804–8. Similar regulations applied with Siam, Morocco, and the Ottoman Empire.

178. For a case study of a modern variant of such a port of trade in Morocco, see Schroeter, *Merchants of Essaouira*. However, Essaouira is reminiscent not so much of the post-1842 treaty ports in China (which Schroeter has in mind) as of the Old China Trade (that is, late-eighteenth-century Canton).

179. Osterhammel, *China*: e.g., pp. 167, 176f.

180. See Hoare, *Japan's Treaty Ports*; and Henning, *Outposts of Civilization*.

181. The standard work: Bergère, *Shanghai*. There is no Western monograph on Tianjin, but see the detailed study in Chinese: Shang Keqiang and Liu Haiyan, *Tianjin*.

182. Schinz, *Cities in China*, p. 171.

183. Several examples in Esherick, *Remaking the Chinese City*, and a detailed case study: Zhang Hailin, *Suzhou*.

184. Raymond, *Cairo*, pp. 299–308; cf. Fahmy, *Olfactory Tale*.

185. Raymond, *Cairo*, p. 309.

186. Abu-Lughod, *Cairo*, pp. 98, 104–6.

187. Fahmy, *Olfactory Tale*, pp. 166–69.

188. Raymond, *Cairo*, pp. 309–17. T. Mitchell, *Colonising Egypt* puts forward the interesting, if somewhat overdrawn, thesis of an Egyptian self-colonization.

189. Kassir, *Beirut*, pp. 129ff.; Çelik, *Remaking of Istanbul*, esp. chs. 3 and 5; Eldem et al., *Ottoman City*, pp. 196ff.; Seidensticker, *Low City*. "Chicago/Melbourne" was an observation of the English globetrotter Isabella Bird, quoted on p. 60; M. E. Robinson, *Korea's Twentieth-Century Odyssey*, p. 8.

190. For Morocco see Abu-Lughod, *Rabat*, pp. 32, 98f.

191. Coquery-Vidrovitch, *History of African Cities*, pp. 242–44. On developments in East Africa, see ibid., pp. 213–14.

192. The following is based on Rowe's monumental *Hankow*, one of the milestones of the social history of China.

193. On the economic rise of Hong Kong, see the excellent work: D. R. Meyer, *Hong Kong*, chs. 4–5. On the political and social aspects of its colonial status: Tsang, *Hong Kong*, chs. 2, 4 and 5.

194. Rowe, *Hankow*, vol. 1, pp. 19, 23.

195. Gruzinski, *Mexico*, pp. 326, 329, 332.

196. See David Atkinson et al., "Empire in Modern Rome: Shaping and Remembering an Imperial City," in: Driver and Gilbert, *Imperial Cities*, pp. 40–63.

197. Port, *Imperial London*, pp. 7, 14f., 17, 19, 23.

198. Schneer, *London 1900*, esp. ch. 3.

199. Abu-Lughod, *Cairo*, p. 85.

200. Abu-Lughod, *Rabat*, p. 117.

201. Lichtenberger, *Die Stadt*, p. 153.

202. The following draws on ibid., pp. 154f.

203. Chartier et al., *La ville des temps modernes*, p. 563.

204. Michel, *Prague*, p. 202.

205. Woud, *Het lege land*, pp. 324–28.

206. Sarasin, *Stadt der Bürger*, pp. 247f.

207. Lichtenberger, *Die Stadt*, p. 154.

208. Olsen, *City*, p. 69.

209. Lavedan, *Histoire de l'urbanisme à Paris*, pp. 376, 494; and, with an even sharper sense for spatial design, Rouleau, *Paris*, pp. 316ff.

210. Catherine B. Asher, "Delhi Walled: Changing Boundaries," in: Tracy, *City Walls*, pp. 247–81, at 279f.; N. Gupta, *Delhi*, p. 79.

211. Steinhardt, *Chinese Imperial City Planning*, pp. 178f.; Naquin, *Peking*, pp. 4–11.

212. L. C. Johnson, *Shanghai*, pp. 81, 320.

213. If a city decided to skip the railroad age, it was relatively easy to open new gates in the city wall for automobile traffic, as in the western Chinese city of Lanzhou in the 1930s. See Gaubatz, *Beyond the Great Wall*, p. 53.

214. Carla Giovannini, "Italy," in: R. Rodger, *European Urban History*, pp. 19–35, at 32.

215. An excellent account is: Pounds, *Historical Geography*, pp. 449–61. On the prerequisites of a "system," see F. Caron, *Histoire des chemins de fer en France*, vol. 1, p. 281; and on the technical aspects, W. König and Weber, *Netzwerke*, pp. 171–201.

216. Kellett, *Impact of Railways*, p. 290.

217. Dennis, *English Industrial Cities*, pp. 128f.; Brower, *Russian City*, p. 53, and see the remarkable sociology of urban immigration in ibid., pp. 85ff.

218. Kellett, *Impact of Railways*, p. 18.

219. Mak, *Amsterdam*, pp. 206–10.

220. Sutcliffe, *Paris*, pp. 97f.

221. Brower, *Russian City*, p. 52—remarks that railroad engineers and station architects took decisions about the shape of the city out of the hands of the authorities.

222. See the abundant material in Parissien, *Station to Station*.

223. Pinol, *Le monde des villes*, pp. 73ff.

224. Kreiser, *Istanbul*, p. 53; Kuban, *Istanbul*, p. 369.

225. Frédéric, *La vie quotidienne au Japon*, p. 336.

226. Vance, *Continuing City*, p. 366.

227. Merchant, *Columbia Guide*, p. 109.

228. McShane and Tarr, *The Horse in the City*, p. 124f.

229. In his essay on noise ("Über Lärm and Geräusch"), published in 1851.

230. K. T. Jackson, *Crabgrass Frontier*, p. 41.

231. Çelik, *Remaking of Istanbul*, pp. 90–95, 102; on dog life in Istanbul, see Boyar and Fleet, *Ottoman Istanbul*, pp. 273–75.

232. Dennis, *English Industrial Cities*, p. 125.

233. John Armstrong, "From Shillibeer to Buchanan: Transport and the Urban Environment," in: P. Clark, *Cambridge Urban History*, vol. 3, pp. 229–57, at 237.

234. Data from Roche, *Le cheval moteur*, p. 65.

235. Ball and Sunderland, *Economic History of London*, p. 229.

236. Bouchet, *Le cheval à Paris*, pp. 40, 45, 83f., 123, 170–76, 215, 254–56—a richly detailed work. See also the social and organizational study: Papayanis, *Coachmen*; and now a wealth of material and insight in Roche, *Le cheval moteur*, pp. 57–120, the first volume of an intended trilogy on the horse in modern European civilization (vol. 2: 2011).

237. Dyos and Aldcroft, *British Transport*, pp. 74f.; Ball and Sunderland, *Economic History of London*, pp. 204f.; Ransom, *Archaeology of the Transport Revolution*, pp. 95–116; Grossman, *Charles Dickens's Networks*, ch. 1.

238. Bartlett, *New Country*, pp. 293, 298f.

239. Kassir, *Beirut*, pp. 115–21.

240. Bouchet, *Le cheval à Paris*, p. 214.

241. The pedicab, a cross between rickshaw and bicycle, was invented in the 1940s.

242. Frédéric, *La vie quotidienne au Japon*, p. 349.

243. Bairoch, *Cities and Economic Development*, p. 314; Merki, *Siegeszug des Automobils*, pp. 39–40 (also Tab. 1), 88f., 95; Hugill, *World Trade*, pp. 217–20.

244. Wolmar, *Subterranean Railway*, chs. 1–7; on the Paris Metro see Pike, *Subterranean Cities*, pp. 47–68.

245. Gruzinski, *Mexico*, pp. 321, 323.

246. Bradley, *Muzhik and Muscovite*, pp. 55, 59.

247. K. T. Jackson, *Crabgrass Frontier*, pp. 13f.; Vance, *Continuing City*, p. 369. The literature on suburbanization is especially abundant in the fields of urban sociology and geography.

248. Fogelson, *Fragmented Metropolis*, p. 2.

249. Girouard, *Cities and People*, pp. 275–79, 282 (Taine); Girouard, *English Town*, p. 270. The great account of *villa suburbia* is Olsen, *City*, pp. 158–77.

250. Escher and Wirth, *Medina von Fes*, p. 19.

251. H. J. Dyos and David A. Reeder, "Slums and Suburbs," in: Dyos and Wolff, *Victorian City*, pp. 359–86. Not all slums were products of industrialization; the notorious ones in Dublin resulted from economic decline. For a broad survey of housing conditions in Europe see Lenger, *European Cities*, pp. 97–112.

252. Pooley, *Housing Strategies*, pp. 6, 328–32.

253. See the discussion with reference to Moscow in Brower, *Russian City*, p. 79.

254. D. Ward, *Poverty*, pp. 13, 15, 52.

255. Yelling, *Slums*, pp. 153f. On the "discovery" of the English slums, see also Koven, *Slumming*.

256. On the urban abodes of the nobility, see Olsen, *City*, pp. 114–31; Lichtenberger, *Stadt*, pp. 208–16.

257. Plunz, *Housing in New York City*, pp. 60–66, 78–80.

258. See the description in Vigier, *Paris*, p. 314. Although it had some model housing for workers, the Ruhr was generally no exception to the grim picture in Europe. See Reulecke, *Urbanisierung in Deutschland*, p. 46, 98.

259. Frost, *New Urban Frontier*, pp. 21f., 34f., 92f., 100, 128f.; cf. Davison, *Marvellous Melbourne*, pp. 137ff.

260. Inwood, *London*, p. 372.

261. This aspect is explored in Schivelbusch, *Disenchanted Night*; Schlör, *Nights in the Big City*; P. C. Baldwin, *In the Watches of the Night*.

262. Pounds, *Hearth*, p. 388.

263. Frédéric, *La vie quotidienne au Japon*, pp. 341–44.

264. Daniel, *Hoftheater*, p. 370.

265. Schlör, *Nights in the Big City*, p. 68; P. C. Baldwin, *In the Watches of the Night*. pp. 157–61, on the introduction of electric lighting.

266. Two masters of such description are the sociologist Richard Sennett and the historian Karl Schlögel.

267. Oldenburg, *Colonial Lucknow*, pp. 24, 36f., 96ff.

268. Ruble, *Second Metropolis*, pp. 221f.; Frédéric, *La vie quotidienne au Japon*, p. 340.

269. C. J. Baker and Phongpaichit, *History of Thailand*, p. 72.

270. Conner, *Oriental Architecture*, pp. 131–53.

271. Sweetman, *Oriental Obsession*, pp. 218ff.

272. A good overview: MacKenzie, *Orientalism*, pp. 71–104.

273. See T. Mitchell, "World as Exhibition."

274. Girouard, *Cities and People*, pp. 291–93; Girouard, *English Town*, pp. 229f.

275. The history is recounted in detail in: Solé, *Le grand voyage de l'obélisque*.

276. Girouard, *Cities and People*, pp. 301–3.

277. Kassir, *Beirut*, p. 114.

278. See, for Europe: Lenger, *European Cities*, pp. 165–72.

279. Briggs, *Victorian Cities*, p. 115; C. Zimmermann, *Metropolen*, p. 66.

280. Konvitz, *Urban Millennium*, pp. 132f.

281. For a fine account of Urbana, IL, in 1869: Monkkonen, *America Becomes Urban*, p. 133.

282. Reps, *Making of Urban America*, p. 380; also pp. 349ff.

283. Ruble, *Second Metropolis*, p. 216.

284. Bessière, *Madrid*, pp. 135f.

285. H. M. Mayer and Wade, *Chicago*, pp. 117f., 124.

286. Brower, *Russian City*, p. 14.

287. Gruzinski, *Mexico*, pp. 57, 59, 339f.

288. C. Zimmermann, *Metropolen*, p. 162: "the largest urban redevelopment in nineteenth-century Europe."

289. Bernand, *Buenos Aires*, pp. 209f., 213. A final imitation of England, in 1878, was the building of a prison in the style of a medieval fortress: ibid., p. 191.

290. Horel, *Budapest*, p. 183.

291. Ibid., pp. 93, 155, 174.

292. P. Hall, *Cities in Civilization*, pp. 707–45 (quotation p. 737): the best introduction to Paris under Haussmann. See also Sutcliffe, *Planned City*, pp. 132–34, and the detailed studies D. P. Jordan, *Transforming Paris*, and Van Zanten, *Building Paris*.

293. Sutcliffe, *Paris*, pp. 83–104, esp. 86–88.

294. Sutcliffe, *Planned City*, pp. 9ff.

295. There is an exemplary analysis in S. Fisch, *Stadtplanung*. On a striking parallel in Japan: Hanes, *City as Subject*, esp. pp. 210ff.

296. Irving, *Indian Summer*; Volwahsen, *Imperial Delhi*; Ridley, *Edwin Lutyens*, esp. pp. 209ff.

297. This moment in architectural history is extensively documented in H. M. Mayer and Wade, *Chicago*, pp. 124ff.

298. Bessière, *Madrid*, p. 205.

299. E. Jones, *Metropolis*, p. 76; Vance, *Continuing City*, pp. 374–76; Girouard, *Cities and People*, pp. 319–22.

300. On the persistent identity of the European city, see the major contributions by H. Häußermann and H. Kaelble in *Leviathan* 29 (2001).

301. Frost, *New Urban Frontier*, p. 14.

302. There are fine case studies in Esherick, *Remaking the Chinese City*.

303. See Leeuwen, *Waqfs*, esp. pp. 206f. On other distinctive features, see the detailed research report in Haneda and Miura, *Islamic Urban Studies*.

CHAPTER VII: Frontiers

1. K. L. Klein, *Frontiers*, pp. 145f.

2. For a brief account of this evolution in American historiography, see Walsh, *American West*, pp. 1–18.

3. R. White, *Middle Ground*.

4. A major work integrating the family perspective is Hyde, *Empires, Nations, and Families*; see also T. Jordan, *Cowgirls*.

5. See chapter 12 (pp. 432–50) in Bayly, *Birth of the Modern World*.

6. In F. J. Turner, *Frontier*, pp. 1–38.

7. Key texts are Waechter, *Erfindung*, esp. pp. 100–120; Jacobs, *On Turner's Trail*; Wrobel, *End of American Exceptionalism*.

8. An important author in this context, with many works to his name, is Richard Slotkin.

9. Billington, *Westward Expansion*, pp. 3–7.

10. W. P. Webb, *Great Frontier*, first published in 1952.

11. Hennessy, *Frontier in Latin American History*, pp. 22, 144, and Toennes's related *Die "Frontier"*. Another fine elaboration of this approach is Cronon, *Changes in the Land*.

12. W. H. McNeill, *Europe's Steppe Frontier*.

13. Important suggestions have been taken here from Howard Lamar and Leonard Thompson, "Comparative Frontier History," in idem, *Frontier*, pp. 3–13, esp. 7f.; C. Marx, *Grenzfälle*; Walter Nugent, "Comparing Wests and Frontiers," in: Milner et al., *American West*, pp. 803–33; Hennessy, *Frontier in Latin American History*; Careless, *Frontier and Metropolis*, p. 40.

14. The viewpoint of "shared history" is impressively developed in E. West, *Contested Plains*: "The frontier never separated things. It brought things together" (p. 13). A vision of the "global frontier land" as an irredeemable no-man's-land is set forth in Bauman, *Society under Siege*, pp. 90–94.

15. Cf. the considerations in Maier, *Among Empires*, pp. 78–111, especially the typology of frontiers on pp. 99f. The most substantial contribution to the debate is now Belich, *Replenishing the Earth*, with its grand analogy between the westward advance in North America and British empire building.

16. Moreman, *Army in India*, pp. 24–31, and passim.

17. Mehra, *An "Agreed" Frontier*.

18. Quoted in Adelman and Aron, *From Borderlands to Borders*, p. 816. The term is used somewhat differently in Baud and Schendel, *Comparative History*, p. 216. A major review of the American debate is Hämäläinen and Truett, *On Borderlands*.

19. See the final version of his theory of imperialism: Ronald Robinson, "The Excentric Idea of Imperialism, with or without Empire," in: Mommsen and Osterhammel, *Imperialism and After*, pp. 267–89, at 273–76.

20. Lattimore, *Inner Asian Frontiers*.

21. J. F. Richards, *Unending Frontier*, pp. 5f.

22. P. D. Curtin, *Location*, pp. 49ff.; on Australia's frontier history see Rowley, *Destruction*.

23. Adelman, *Frontier Development*, pp. 21, 96.

24. Rohrbough, *Days of Gold*, p. 1; E. West, *Contested Plains*, p. xv. On the social history of gold digging, see Finzsch, *Goldgräber*; and for the general context, cf. Nugent, *Into the West*, pp. 54–65.

25. Hine and Farragher, *American West*, pp. 36–38, 71–73, 79; and the detailed D. J. Weber, *Spanish Frontier*.

26. Prucha, *Great Father*, pp. 181ff.; Banner, *How the Indians*, pp. 228–56.

27. This is a hallmark of the school of historians around William Appleman Williams, summarized in Waechter, *Erfindung*, pp. 318–28. For a good development of this approach by a French historian, see Heffer, *The United States and the Pacific*.

28. A recent attempt at stock taking (with an introductory bibliography) is Stephen Aron, "Frontiers, Borderlands, Wests," in: Foner and McGirr, *American History Now*, pp. 261–84.

29. Jennings, *Founders of America*, p. 366.

30. Hurtado, *Indian Survivals*, p. 1.

31. See Dowd, *A Spirited Resistance*.

32. Hämäläinen, *Plains Indian Horse Cultures*, summarizes the conventional story and adds his own interpretation.

33. E. West, *Contested Plains*, p. 78.

34. Hurt, *Indian Agriculture*, p. 63.

35. Isenberg, *Destruction of the Bison*, pp. 25f.

36. On this energy argument, see E. West, *Contested Plains*, p. 51.

37. Hämäläinen, *Comanche Empire*, pp. 240f., quotation 241.

38. Utley, *Indian Frontier*, p. 29.

39. On the mobility at the heart of the Indian way of life, see Cronon, *Changes in the Land*, pp. 37f. and passim.

40. Kavanagh, *Comanche Political History*; Hämäläinen, *Comanche Empire*.

41. Krech, *Ecological Indian*, esp. pp. 123–49 on the tension between conservation and dissipation in the Indians' relationship to the bison.

42. Isenberg, *Destruction of the Bison*, p. 83.

43. Hämäläinen, *Plains Indian Horse Cultures*, p. 844. The Lakota-Sioux in the North were more successful in finding an equilibrium and were therefore able to resist Euro-American encroachment for a few more decades (p. 859).

44. Isenberg, *Destruction of the Bison*, pp. 121, 129, 137, 139f.

45. Farragher, *Sugar Creek*, pp. 22f.

46. Nugent, *Into the West*, p. 24.

47. Walsh, *American West*, p. 46 (Tab. 3.1).

48. Unruh, *The Plains Across*.

49. Limerick, *Legacy of Conquest*, p. 94.

50. Faragher, *Sugar Creek*, p. 51.

51. Danbom, *Born in the Country*, pp. 87, 93.

52. Nugent, *Into the West*, pp. 83–85.

53. Gutiérrez, *Walls and Mirrors*, p. 14; Walsh, *American West*, p. 62. In 1900 there were as many as 500,000 people of Mexican origin in the Southwest.

54. Walsh, *American West*, pp. 58ff., esp. 68; Limerick, *Legacy of Conquest*, p. 260.

55. For greater detail: Walsh, *American West*, p. 27.

56. Paul, *Far West*, pp. 189, 199f.; Hennessy, *Frontier in Latin American History*, p. 146.

57. See the illustrated volume: Axelrod, *Chronicle*.

58. Unruh, *The Plains Across*, pp. 189, 195–98.

59. Clodfelter, *Dakota War*, pp. 2, 66f.

60. Ibid., p. 16.

61. For a map of the military frontier, see Howard R. Lamar and Sam Truett, "The Greater Southwest and California from the Beginning of European Settlement to the 1880s," in: Trigger and Washburn, *Cambridge History of the Native Peoples of the Americas* vol. 1, pt. 2, pp. 57–115, at 88f.

62. Vandervort, *Indian Wars*. The Indian experience in the generation of the Indian wars is vividly presented in the biography: Utley, *Sitting Bull*.

63. See Peter Way, "The Cutting Edge of Culture: British Soldiers Encounter Native Americans in the French and Indian War," in: Daunton and Halpern, *Empire and Others*, pp. 123–48.

64. Richard Maxwell Brown, "Violence," in: Milner et al., *American West*, pp. 293–425, at 396, 399, 412f., 416; cf. R. M. Brown, *No Duty to Retreat*, pp. 41, 44, 48 and passim. Others have countered Brown's grim view, arguing that everyday life on the frontier was much less violent than in American inner cities today.

65. Richter, *Facing East*, p. 67.

66. On the early treaties see Prucha, *Great Father*, pp. 7, 19ff., also 140f., 165ff.

67. Ibid., p. 44.

68. Quoted in Hine and Faragher, *American West*, p. 176.

69. Rogin, *Fathers and Children*.

70. Richter, *Facing East*, pp. 201–8, 235f.

71. J. L. Wright, *Creeks*, p. 282.

72. Hine and Faragher, *American West*, pp. 179f. Today the Seminole tribe of Florida is very active in business. In 2006 they bought the worldwide Hard Rock Café chain.

73. Utley, *Indian Frontier*, pp. 59f.; Prucha, *Great Father*, p. 97.

74. Prucha, *Great Father*, p. 83, and Prucha's account of the Indian removal, pp. 64ff.

75. Hine and Faragher, *American West*, p. 231.

76. Michael D. Green, "The Expansion of European Colonization to the Mississippi Valley, 1780–1880," in: Trigger and Washburn, *Cambridge History of the Native Peoples of the Americas*, vol. 1, pt. 1, pp. 461–538, at 533.

77. On the continuing Apache resistance, see Vandervort, *Indian Wars*, pp. 192–210.

78. Limerick, *Something in the Soil*, pp. 36–64. See the depressing description of the Comanche's defeat and decline in Hämäläinen, *Comanche Empire*, chs. 7–8.

79. On the history of barbed wire, see Krell, *The Devil's Rope*, at p. 12.

80. Nugent, *Into the West*, p. 100; Hine and Farraghar, *American West*, pp. 324ff.

81. Meinig, *Shaping of America*, vol. 2, p. 100.

82. Utley, *Indian Frontier*, p. 60.

83. Perdue, *China Marches West*, pp. 292–99.

84. Prucha, *Great Father*, p. 186; an excellent case study is Monnett, *Tell Them We Are Going Home*.

85. Careless, *Frontier and Metropolis*, p. 41. On relations between whites and Indians in Canada, see J. R. Miller, *Skyscrapers*.

86. Cronon, *Changes in the Land*, pp. 65f., 69.

87. The most comprehensive debate on communal ownership in this period took place in Russia. See Kingston-Mann, *In Search of the True West*.

88. Hurt, *Indian Agriculture*, p. 68.

89. Jennings, *Founders of America*, pp. 304f.

90. Hurt, *Indian Agriculture*, pp. 78f., 84f., 90–92.

91. This has been splendidly done in Parker, *Native American Estate*.

92. M. D. Spence, *Dispossessing the Wilderness*.

93. That there were many frontiers is clear from the case studies in Guy and Sheridan, *Contested Ground*. In Argentina, long before F. J. Turner, Domingo Fausto Sarmiento developed his own theory of the frontier: see Sarmiento, *Civilization and Barbarism*, and Navarro Floria, *Sarmiento*.

94. See K. L. Jones, *Warfare*.

95. Hennessy, *Frontier in Latin American History*, p. 84.

96. Garavaglia, *Les hommes de la pampa*, p. 396.

97. Amaral, *Rise of Capitalism*, pp. 286f.

98. Hennessy, *Frontier in Latin American History*, pp. 19, 92; Hoerder, *Cultures in Contact*, p. 359.

99. An especially acute social-historical account of the gaucho, focused on southern Brazil rather than Argentina, may be found in Ribeiro and Rabassa, *Brazilian People*, pp. 293–303.

100. Another genealogical root is the North American "pathfinder," whose heyday was between 1820 and 1840: see Bartlett, *New Country*, p. 88.

101. Slatta, *Gauchos*, pp. 2, 5, 9, 22, 35, 180ff. A comparative history of cowboys in the Americas is Slatta, *Cowboys*.

102. Lombardi, *Frontier*, is a useful introduction.

103. Amado et al, *Frontier in Comparative Perspective*, p. 18.

104. Bernecker et al., *Geschichte Brasiliens*, p. 181.

105. Walter Nugent, "Comparing Wests and Frontiers," in: C. A. Milner et al., *American West*, pp. 828f.

106. See esp. vol. 2 of the trilogy: Hemming, *Amazon Frontier*. Langfur, *Forbidden Lands*, has now broken new ground methodologically for the period before 1830.

107. Norman Etherington et al., "From Colonial Hegemony to Imperial Conquest," in: Hamilton et al., *Cambridge History of South Africa*, vol. 1, pp. 319–91, at 384.

108. See Gump, *Dust*.

109. Fundamental for the 1820s and 1830s in South Africa is Etherington, *Great Treks*, esp. chs. 5–9.

110. J. Fisch, *Geschichte Südafrikas*, pp. 138f.

111. Giliomee, *Afrikaners*, pp. 186–90.

112. Leonard Thompson and Howard Lamar, "The North American and Southern African Frontiers," in idem, *Frontier in History*, pp. 14–40, at 29.

113. This is a central theme in Feinstein, *Economic History of South Africa*.

114. Allister Sparks, *The Mind of South Africa* (London 1991), quoted in Maylam, *South Africa's Racial Past*, p. 55.

115. Ibid., pp. 51–66.

116. P. D. Curtin, *Location*, p. 67.

117. Ibid., pp. 74–76, 87–90; for a clear discussion of Boer evaluations and objectives, see Nasson, *South African War*, pp. 47–49.

118. See Fredrickson, *White Supremacy*, pp. 179–98.

119. On the concept of Eurasia, see chapter 3, above, and von Hagen, *Empires*, esp. pp. 454ff.

120. See, for example, Markovits et al., *Society and Circulation*.

121. Barfield, *Nomadic Alternative*, pp. 7–9, passim.

122. Khazanov, *Nomads*, pp. 198–227.

123. See the synthetic discussion in Perdue, *China Marches West*, pp. 524–32.

124. Findley, *Turks*, p. 93.

125. See the overview in Osterhammel, *China*, pp. 86–105.

126. J. A. Millward, *Eurasian Crossroads*, esp. chs. 4–5.

127. See S.C.M. Paine, *Imperial Rivals*, chs. 4–6.

128. Rogan, *Frontiers of the State*, pp. 9–12; Kieser, *Der verpasste Friede*, pp. 24, 43–44.

129. There are good overviews in the collective volume Brower and Lazzerini, *Russia's Orient*, and the concise Moshe Gammer, "Russia and the Eurasian Steppe Nomads: An Overview," in: Amitai and Biran, *Mongols*, pp. 483–502.

130. Seely, *Russian-Chechen Conflict*, p. 32. The standard work on the Caucasus is still Gammer, *Muslim Resistance*.

131. LeDonne, *Russian Empire*: a geopolitical study of the Tsarist Empire, a little schematic in its division of the western, southern, and eastern frontiers.

132. Khodarkovsky, *Russia's Steppe Frontier*, pp. 137–38.

133. From 1819 the war against the Chechens has been described as "mass terrorism bordering on genocide": Seely, *Russian-Chechen Conflict*, p. 34.

134. Overviews: LeDonne, *Russian Empire*; D. Lieven, *Empire*, pp. 208–13; Kappeler, *Russian Empire*, pp. 114ff..

135. Kappeler, *Russian Empire*, p. 159.

136. Barrett, *Edge of Empire*; O'Rourke, *Cossacks*, chs. 2–3; Alfred J. Rieber, "The Comparative Ecology of Complex Frontiers," in: Miller and Rieber, *Imperial Rule*, pp. 177–207, at 188f.

137. Kappeler, *Russian Empire*, p. 193.

138. Forsyth, *Peoples of Siberia*, p. 130; Rossabi, *China and Inner Asia*, pp. 167–79; Jersild, *Orientalism*, p. 36.

139. The standard account is now J. F. Richards, *Unending Frontier*, pp. 463–546.

140. Forsyth, *Peoples of Siberia*, pp. 123, 190f.

141. Slezkine, *Arctic Mirrors*, pp. 97–99.

142. Forsyth, *Peoples of Siberia*, pp. 159f., 163, 177–79, 181, 216–18.

143. On the following, see Kappeler, *Russian Empire*, pp. 185–90.

144. Virginia Martin, *Law and Custom*, pp. 34ff. (also pp. 17–24 for a precise description of Kazakh nomadism and its political organization).

145. Sunderland, *Taming the Wild Field*, p. 223.

146. Cf. the case study of the Molochna River plains northeast of the Crimea: Staples, *Cross-Cultural Encounters*.

147. On the civilizing mission see chapter 17, below.

148. Bassin, *Turner*.

149. Jersild, *Orientalism*, pp. 56, 87, 97.

150. Breyfogel, *Heretics*, p. 2.

151. For Russia: Layton, *Russian Literature*; and on the perception of Siberians: Slezkine, *Arctic Mirrors*, pp. 113–29.

152. Forsyth, *Peoples of Siberia*, pp. 118, 120, 164–66, 176. For an exhaustive account of the Buryats: Schorkowitz, *Staat und Nationalitäten*.

153. Excellent on this is Blackbourn, *Conquest of Nature*, pp. 280ff.

154. J. C. Scott, *Seeing Like a State*, pp. 181ff.

155. The image of congealment is drawn from the important work Weaver, *Great Land Rush*: "frontiers congealed into settler societies" (p. 69).

156. The following partly follows Osterhammel, *Colonialism*, pp. 4–10.

157. See the theoretical definition in McCusker and Menard, *Economy of British America*, p. 21.

158. See Marks, *Road to Power*, pp. 196ff.

159. See Mosley, *Settler Economies*, pp. 5–8, 237 (note 1).

160. R. W. Fogel, *Without Consent or Contract*, pp. 30f.

161. See Mark Thomas, "Frontier Societies and the Diffusion of Growth," in James and Thomas, *Capitalism in Context*, pp. 29–49, at 31.

162. Adelman, *Frontier Development*, p. 1.

163. See Stefan Kaufmann, "Der Siedler," in Horn et al., *Grenzverletzer*, pp. 176–201, esp. 180–86.

164. There are numerous historical case studies from every continent. Especially systematic is Janssen, *Übertragung von Rechtsvorstellungen*, pp. 86–134. On Africa see, e.g., various works by Martin Chanock.

165. Fundamental on colonial land policy is Weaver, *Great Land Rush*, pp. 216ff.

166. Dunlap, *Nature and the English Diaspora*, p. 19.

167. Crosby, *Ecological Imperialism*, pp. 217–69; M. King, *Penguin History of New Zealand*, pp. 196f.

168. Tyrrell, *Peripheral Visions*, pp. 280f.

169. Ibid., pp. 286f. See also idem, *True Gardens*, esp. chs. 2–4.

170. Nor are there are any synopses of the literature comparable to J. F. Richards, *Unending Frontier* or J. R. McNeill, *Something New Under the Sun*; for the time being Krech et al, *Encyclopedia* remains the most important source of information.

171. Naquin and Rawski, *Chinese Society*, pp. 130–33.

172. This shift is briefly discussed in Coates, *Nature*, pp. 129–34. Interest in, and fear of, the mountains persisted during and after the period of Alpine Romanticism.

173. J. R. McNeill, *Something New under the Sun*, p. 229.

174. Chew, *Ecological Degradation*, p. 133; percentages from John F. Richards, "Land Transformation," in B. L. Turner et al., *The Earth*, pp. 163–78, at 173 (Tab. 10–2).

175. J. R. McNeill, *Something New Under the Sun*, p. 232; Delort and Walter, *Histoire de l'environnement européen*, p. 267.

176. The standard work on the question (in which many more histories are narrated) is M. Williams, *Deforesting the Earth*.

177. Elvin, *Elephants*, p. 85.

178. Guha, *Environment and Ethnicity*, pp. 62ff.

179. Elvin, *Elephants*, p. 470. This also contains much material on regional differences in the Chinese cultural attitude to wood, trees, and forest.

180. Totman, *Early Modern Japan*, pp. 226f., 268f.

181. A. Reid, *Humans and Forests*, p. 102.

182. The following is based on Boomgaard, *Forest Management*.

183. Radkau, *Nature and Power*, p. 152.

184. See the standard accounts in R. H. Grove, *Green Imperialism*, esp. chs. 6–8; Rangarajan, *Fencing the Forest*; Beinart and Hughes, *Environment and Empire*.

185. See the synthesis of copious research in M. Williams, *Deforesting the Earth*, pp. 354–69.

186. An example from the Himalayas: Singh *Natural Premises*, pp. 147f., 153.

187. Guha, *Environment and Ethnicity*, p. 167.

188. For the United States see Jacoby, *Crimes*; for France: Whited, *Forests*, esp. ch. 3.

189. See the global survey in R. H. Grove, *Ecology*, pp. 179–223.

190. M. Williams, *Deforesting the Earth*, pp. 368f.

191. Ibid., pp. 371–79; cf. the extensive account in Dean, *Broadax*, esp. ch. 9.

192. Simmons, *Environmental History*, p. 153.

193. John F. Richards, "Land Transformation," in: B. L. Turner et al., *The Earth*, pp. 163–78, at 169.

194. M. Williams, *Americans and Their Forests*, pp. 332f.

195. M. Williams, *Deforesting the Earth*, p. 360.

196. J.L.A. Webb, *Desert Frontier*, pp. 5, 11, 15f., 22.

197. Few things are as elusive as this world for European historians of the twenty-first century. See Brody, *Other Side of Eden*.

198. Mumford, *The City*, pp. 269–73.

199. Boomgaard, *Frontiers of Fear*, pp. 56, 111.

200. Ibid., pp. 121, 125, 127.

201. Mackenzie, *Empire of Nature*, p. 182.

202. Rothfels, *Savages and Beasts*, pp. 44–80, esp. 51f., 57f., 76–80.

203. Planhol, *Le paysage animal*, p. 689.

204. Ibid., pp. 70f.

205. Beinart and Coates, *Environment and History*, pp. 20–27.

206. On the history of whaling until around 1800, see J. F. Richards, *Unending Frontier*, pp. 574–607.

207. Ray Hilborn, "Marine Biota," in B. L. Turner et al., *The Earth*, pp. 371–85, at: 377 (Fig. 21.7).

208. Mawer, *Ahab's Trade*, pp. 23, 179, 213.

209. Ellis, *Men and Whales*, pp. 101–13.

210. Ibid., p. 166; Bockstoce, *Whales, Ice, and Men*, pp. 24, 159.

211. Bockstoce, *Whales, Ice, and Men*, p. 208.

212. Pasquier, *Les baleiniers français*, pp. 28f., 32f., 194.

213. Mawer, *Ahab's Trade*, pp. 319–21.

214. Bockstoce, *Whales, Ice, and Men*, p. 324.

215. Ellis, *Men and Whales*, p. 166.

216. Kalland and Moeran, *Japanese Whaling*, p. 74.

217. President Millard Fillmore to the Emperor of Japan, 1852/11/13, in: Beasley, *Select Documents*, pp. 99–101 (p. 100 on whaling).

218. Kalland and Moeran, *Japanese Whaling*, p. 78.

219. Nickerson and Chase, *Loss of the Ship "Essex"*.

220. Blackbourn, *Conquest of Nature*, pp. 71–111.

221. *Faust. Der Tragödie zweiter Teil*, 11091–11094. See the commentary in Johann Wolfgang von Goethe, *Sämtliche Werke, Briefe, Tagebücher und Gespräche*, vol. 7/2, ed. Albrecht Schöne, Frankfurt am Main 1994, pp. 716f. English translation: http://www.poetryintranslation.com/PITBR/German/Fausthome.htm.

222. J. R. McNeill, *Something New under the Sun*, pp. 188f.

223. J. de Vries and Woude, *First Modern Economy*, pp. 28f., 31.

224. Ven et al., *Leefbar laagland*, pp. 152f.

225. Woud, *Het lege land*, pp. 83f.

226. Jeurgens, *De Haarlemmermeer*, pp. 97, 99, 167.

227. Ven et al., *Leefbar laagland* 1993, p. 192.

228. The classical text on this is Nash, *Wilderness and the American Mind*.

CHAPTER VIII: Imperial Systems and Nation-States

1. For this, see Burbank and Cooper, *Empires in World History*.

2. Darwin, *After Tamerlane*, p. 254.

3. In this respect, my main inspiration is a classic work of political science: Finer, *History of Government*.

4. See the world-historical survey in Hansen's monumental *City-State Cultures*.

5. With the exception of the Russian war in the Caucasus, which followed old imperial lines of conflict.

6. Blanning, *French Revolutionary Wars*, pp. 100f. For an English translation of the decree, see F. M. Anderson (ed.), *The Constitutions and Other Select Documents Illustrative of the History of France*, 1789–1907, 2nd ed., Minneapolis 1908, pp. 184–85.

7. Duroselle, *Tout empire périra*, pp. 67f.

8. See the fine sketch in Girault, *Diplomatie européenne*, pp. 13–19.

9. D. Geyer, *Der russische Imperialismus*, pp. 47ff.

10. Joseph Smith, *Spanish-American War*, pp. 32f., 198.

11. On the changes in military communications, see Kaufmann, *Kommunikationstechnik*; and in warfare more generally, Hew Strachan, "Military Modernization, 1789–1918," in Blanning, *Oxford Illustrated History*, pp. 69–93.

12. Figures from P. M. Kennedy, *Great Powers*, p. 203 (Tab.19).

13. See the brilliant sketch in Paul W. Schroeder, "International Politics, Peace, and War, 1815–1914," in Blanning, *Nineteenth Century*, pp. 158–209; and compare this with Doering-Manteuffel, *Internationale Geschichte*, pp. 94–105. Schroeder and Doering-Manteuffel put forward strikingly distinctive theses. The best "neutral" textbook is Rich, *Great Power Diplomacy*; very succint is Bridge and Bullen, *Great Powers*; and excellent for the period until 1815 is Scott, *Birth*.

14. For a detailed discussion of the movements up and down, see Duchhardt, *Balance of Power*, pp. 95–234.

15. M. S. Anderson, *Eastern Question*, still valuable.

16. Rusconi, *Cavour e Bismarck*; Gall, *Bismarck*.

17. On German foreign policy in Europe, see Mommsen, *Großmachtstellung*, and Hildebrand, *Das vergangene Reich*.

18. Mommsen, *Großmachtstellung*, p. 107.

19. Girault, *Diplomatie européenne*, pp. 151–69.

20. For a long-term perspective: Gillard, *Struggle for Asia*. After 1907, however, the tensions between Russia and Britain continued in different forms.

21. See Mulligan, *Origins*, as a judicious synthesis of a huge literature; still important: Joll, *Origins*.

22. The significance of developments in East Asia between roughly 1895 and 1907 for the international system cannot be overestimated. See Nish, *Origins*.

23. What that meant for world politics is superbly analyzed in Marshall, *Remaking the British Atlantic*.

24. Yapp, *Strategies*, pp. 419–60; M. C. Meyer and Sherman, *Mexican History*, pp. 385–401. On colonial wars, see chapter 9, below.

25. Labanca, *Oltremare*, pp. 108–22.

26. Wesseling, *Divide and Rule* is the standard narrative account. For more recent approaches, see Pétré-Grenouilleau, *From Slave Trade to Empire*.

27. Even the best of them all: Gildea, *Barricades*, pp. 326ff., and Sperber, *Europe 1850-1914*.

28. This is especially marked in J.-C. Caron and Vernus, *L'Europe au XIXᵉ siècle*.

29. Koebner and Schmidt, *Imperialism*, p. 50.

30. An important theme in Winkler, *Long Road West*, broached already in vol. 1, p. 5.

31. Otto Dann, *Zur Theorie*, p. 69. Such a definition was still lacking from the "basic concepts" in Dann, *Nation*, pp. 11–21.

32. Voigt, *Geschichte Australiens*, p. 114; M. King, *Penguin History of New Zealand*, pp. 266f.

33. The literature on nationalism is no longer manageable. For Europe, the main focus of research, recent works include: v. Hirschhausen and Leonhard, *Nationalismen*, esp. the editors' introduction (pp. 11–45); Leerssen, *National Thought*; Baycroft and Hewitson, *What is a Nation?*.

34. H. Schulze, *States, Nations, and Nationalism*.

35. W. Reinhard, *Staatsgewalt*, p. 443.

36. On the "modernity" of nation building and the epochal change around 1800, see the argument summarizing the discontinuity thesis in Langewiesche, *Nation*, pp. 14–34.

37. For an attempt to grasp this ideal-typically as an opposition between perennialism (the Romantic idea of the nation as a primal entity) and modernism (the nation as a construct), see Smith, *Nationalism and Modernism*, pp. 22f.

38. See Guibernau, *Nationalisms*, p. 48. Although I have many points in common with Connor (*Ethnonationalism*, 1994), I differ from him in this stress on internal "nation building" and, more generally, on objective, nonascriptive factors.

39. W. Reinhard, *Staatsgewalt*, p. 443.

40. See the original map in Buzan and Little, *International Systems*, p. 261.

41. Schölch, *Egypt for the Egyptians!*; J. R. Cole, *Colonialism*; Marr, *Vietnamese Anticolonialism*, pp. 166f.

42. E. Weber, *Peasants into Frenchmen*.

43. Schieder proposes a similar but slightly different typology in *Nationalism*, pp. 110f. These should not be confused with typologies of nation *building*: see Hroch, *Europa der Nationen*, pp. 41–45.

44. Breuilly, *Nationalism*, chs. 4–7.

45. The concept of the cycle of revolution was first introduced by the Leipzig historian Manfred Kossok. See chapter 10, below.

46. For an initial orientation, see Wood, *American Revolution*, pp. 17–30; Rodríguez, *Independence of Spanish America*, pp. 19–35; and, in a broader comparative framework, Elliott, *Empires*.

47. Dubois, *Avengers*. See also chapter 10, below.

48. The classic account is J. Lynch, *Spanish American Revolutions*, a masterpiece of narrative history.

49. Seton-Watson, *Nations and States*, p. 114.

50. Bitsch, *Histoire de la Belgique*, pp. 79–86; Rich, *Great Power Diplomacy*, pp. 59–61.

51. Jelavich, *Balkans*, vol. 1, pp. 196f.

52. Sundhaussen, *Geschichte Serbiens*, p. 130.

53. Jelavich and Jelavich, *Establishment*, p. 195.

54. Bernecker, *Geschichte Haitis*, p. 106.

55. Clogg, *Greece*, p. 73. See also chapter 17, below.

56. Bitsch, *Histoire de la Belgique*, pp. 119ff.

57. On the concept of the "polycephalic federation," see Rokkan, *State Formation*, pp. 111, 220.

58. Blom and Lamberts, *Low Countries*, p. 404; J. Fisch, *Europa*, p. 171.

59. Another approach that does not use the concept of hegemony is Ronald Speirs and John Breuilly, "The Concept of National Unification," in idem, *Germany's Two Unifications*, pp. 1–25.

60. On the two different styles of constitutional-authoritarian rule see Rusconi, *Cavour e Bismarck*, esp. 169ff.

61. Summary discussions for Italy: Beales and Biagini, *Risorgimento*; Banti, *Il Risorgimento italiano*; a synopsis of recent research is Banti and Ginsborg, *Il Risorgimento*. Of the numerous studies of Germany, a particularly good one is Lenger, *Industrielle Revolution*, pp. 315–81.

62. Lenger, *Industrielle Revolution*, p. 348.

63. Blackbourn, *History of Germany,*, p. 184.

64. Nipperdey, *Deutsche Geschichte 1866–1918*, vol. 2, p. 85.

65. Francesco Leoni, "Il brigantaccio postunitario," in Viglione, *La Rivoluzione Italiana*, pp. 365–85.

66. N. G. Owen et al., *Emergence*, p. 115.

67. Kirby, *Baltic World*, pp. 185–89.

68. Bumsted, *History*, pp. 132–42.

69. Extracts are published in Keith, *Selected Speeches*, vol. 1, pp. 113–72.

70. Mansergh, *Commonwealth Experience*, vol. 1, pp. 34–46.

71. See chapters 7 and 17.

72. See the exemplary analysis in Voigt, *Geschichte Australiens*, esp. pp. 170–84.

73. The dramatic story of the resistance is recounted in Ravina, *Last Samurai*, esp. chs. 5–6.

74. M. B. Jansen, *Modern Japan*, pp. 343–47.

75. I have found most convincing the analysis in Potter, *Impending Crisis*.

76. H. Jones, *Union in Peril*; cf. the speculations on the consequences a possible Confederate victory in R. W. Fogel, *Without Consent or Contract*, pp. 411–17.

77. See Dülffer et al., *Vermiedene Kriege*, pp. 513–25.

78. Carr, *Spain*, p. 347ff.; Balfour, *End of the Spanish Empire*, pp. 44–46; A. Roberts, *Salisbury*, p. 692.

79. Engerman and Neves, *Bricks*, p. 479.

80. Clarence-Smith, *Third Portuguese Empire*.

81. There are few general books covering the history of empires in the nineteenth century in their entirety. For overseas empires see Wesseling, *European Colonial Empires*; and for the later part of our period: Butlin, *Geographies of Empire*; also two excellent French textbooks: Surun, *Les sociétés coloniales*; Barjot and Frémeaux, *Les sociétés coloniales*. A pioneering attempt to consider continental and overseas empires within one framework is v. Hirschhausen and Leonhard, *Comparing Empires*.

82. R. Oliver and Atmore, *Africa since 1800*, p. 118.

83. C. Marx, *Geschichte Afrikas*, p. 70.

84. Ricklefs, *Modern Indonesia*, pp. 144–60.

85. C. J. Baker and Phongpaichit, *Thailand*, p. 105.

86. See the more detailed argument in Osterhammel, *Geschichtswissenschaft*, pp. 322–41.

87. The following draws on suggestions in classics of nationalism theory, such as the work of Benedict Anderson and Ernest Gellner, and Calhoun, *Nationalism*, pp. 4f. It also expands upon the argument in Osterhammel, *Expansion*.

88. On borders see Münkler, *Empires*, pp. 5ff.; also Osterhammel, *Geschichtswissenschaft* pp. 210–13, and chapter 3, above.

89. Charles Tilly, "How Empires End," in Barkey and von Hagen, *After Empire*, p. 7.

90. M. W. Doyle, *Empires*, p. 36.

91. Langewiesche, *Nation*, p. 23.

92. Thom, *Republics*.

93. Langewiesche, *Nation*, p. 23.

94. Integration is an aspect not often discussed in the literature on empire. But see the empirically rich study Magee and Thompson: *Empire and Globalisation*.

95. Dunn, *Africa*, pp. 29, 33.

96. A new survey, which places great emphasis on the role of private firms, is Winseck and Pike, *Communication and Empire*.

97. This has often been noted before—most recently in Motyl, *Revolutions*, pp. 120–22. However, the same may be found in nation-states such as Spain and even France.

98. This structural definition builds on and modifies suggestions in Motyl, *Imperial Ends*, pp. 4, 15–27, and M. W. Doyle, *Empires*, pp. 19, 36, 45, 81. Cf. the excellent little book: S. Howe, *Empire*, esp. pp. 13–22.

99. In 1900 railroad mileage was of the same order of magnitude in these three countries. See Woodruff, *Impact of Western Man*, p. 253, Tab. VI/1.

100. See Offer, *First World War*.

101. See Osterhammel, *Colonialism*, pp. 10–18; cf. von Trotha, *Kolonialismus*.

102. Kirby, *Baltic World*, pp. 52, 79f; Brower, *Turkestan*, pp. 26ff.

103. See Cain, *Hobson*.

104. Still invaluable is Mommsen, *Theories of Imperialism*. On the "classical" theories up to 1919, see Semmel, *Liberal Ideal*; and for a good survey of recent historical interpretations, Porter, *European Imperialism*, chs. 1–5.

105. Schumpeter's classic essay on imperialisms (in the plural) is translated in his *Economics and Sociology*, pp. 141–219, esp. 190–213. The central concept here is "export monopolism."

106. See W. Reinhard, *Expansion*; idem, *Colonialism*; Adas, *Islamic and European Expansion*.

107. Bayly, *First Age*. See also chapter 2, above.

108. Wesseling, *Divide and Rule*, pp. 119ff.

109. J. R. Ward, *Industrial Revolution*, p. 62.

110. Many detailed examples are given in Brötel, *Frankreich im Fernen Osten*.

111. Abernethy, *Global Dominance*, p. 101.

112. J. Black, *War and the World*, p. 152.

113. Headrick, *Tools*, pp. 20f., 43–54.

114. Ibid., p. 117.

115. On the following, a good synthesis is Okey, *Habsburg Monarchy*.

116. For a brief sketch of the Habsburg position in Europe, see P. M. Kennedy, *Rise and Fall*, pp. 215–19.

117. Bérenger, *History of the Habsburg Empire*, p. 134.

118. On the disastrous sequel, see Bridge, *Habsburg Monarchy*, pp. 288ff.

119. There is a tendency in the recent literature to distinguish between a pre-1867 "empire" and a looser post-1867 "monarchy." See, e.g., Ingrao, *Habsburg Monarchy*; and Okey, *Habsburg Monarchy*.

120. Cf. the evaluation in Hoensch, *Modern Hungary*, pp. 20ff.

121. See the sophisticated discussion of nationalism in Okey, *Habsburg Monarchy*, pp. 283–309.

122. Bérenger, *History of the Habsburg Empire*, p. 214.

123. İnalcık and Quataert, *Ottoman Empire*, vol. 2, p. 782; Kappeler, *Russian Empire*, pp. 285f.

124. Cf. D. Lieven, *Empire*, pp. 184f.

125. Bawden, *Mongolia*, pp. 187ff.

126. For a brief sketch of the Napoleonic Empire, see Boudon, *Histoire du consulat et de l'Empire*, pp. 283–303, see also Dwyer and Forrest, *Napoleon and His Empire*.

127. There is a brilliant portrait of this new ruling class in Woloch, *Napoleon and His Collaborators*, esp. pp. 156f.

128. Broers, *Europe*, esp. pp. 125–38, 202–30.

129. See the map in ibid., p. 181.

130. Quoted from Jourdan, *L'Empire de Napoléon*, p. 120.

131. On the aspect of economic integration, which I will pass over here, see Woolf, *Napoleon's Integration of Europe*, pp. 133–56.

132. On the French colonial empire in general, see Bouche, *Histoire de la colonisation française*; J. Meyer et al., *Histoire de la France coloniale*; Aldrich, *Greater France*; Liauzu et al., *Colonisation*; Wesseling, *European Colonial Empires*, which displays the author's excellent knowledge of France.

133. Etemad, *Possessing the World*, pp. 220–25 (Appendices C and D). See also chapter 4, above.

134. Ruedy, *Modern Algeria*, pp. 60, 62, 66; Danziger, *Abd al-Qadir*, pp. 180–205. Abd al-Qadir was not a Westernizer, however: ibid., p. 200.

135. Ruedy, *Modern Algeria*, p. 69 (table 3.1).

136. The standard account of policy toward the Muslims in Algeria is Ageron, *History of Modern Algeria*, pp. 47–81.

137. See Rivet, *Le Maroc*.

138. The comparison is very interesting in Lustick, *State-Building Failure*.

139. Brocheux and Hémery, *Indochina*, pp. 138f.

140. Ibid., pp. 165–73.

141. Wesseling, *European Colonial Empires*, p. 127. Good on French colonial ideology is Aldrich, *Greater France*, pp. 89–111.

142. Wesseling, *Divide and Rule*, pp. 92ff.

143. The standard history of the Congo is now Vanthemsche, *La Belgique et le Congo*. On the crimes in the Congo, see also Ewans, *European Atrocity*.

144. H. L. Wesseling, "The Strange Case of Dutch Imperialism," in idem, *Imperialism and Colonialism*, pp. 73–86, here 77.

145. Ricklefs, *Modern Indonesia*, pp. 176–79.

146. Everything in the Netherlands that can be called imperialist appears in Kuitenbrouwer, *The Netherlands*.

147. Wesseling, *European Colonial Empires*, p. 141.

148. See Gründer, *Geschichte der deutschen Kolonien*, pp. 163–66.

149. Gouda, *Dutch Culture Overseas*, p. 45.

150. Van Zanden and Marks, *Economic History of Indonesia*, pp. 46–72.

151. Booth, *Indonesian Economy*, pp. 149–54, 160; v.d. Doel, *Het Rijk van Insulinde*, pp. 157–66.

152. Booth, *Indonesian Economy*, p. 328.

153. Kent, *Soul of the North*, pp. 368f.

154. See Parsons, *King Khama*, pp. 201ff.; Rotberg, *The Founder*, pp. 486f.

155. Tarling, *Imperialism*, pp. 55–62; Kaur, *Economic Change*.

156. Quoted from Rotberg, *The Founder*, p. 290.

157. See Shula Marks, "Southern and Central Africa, 1886–1910," in: Fage and Oliver, *Cambridge History of Africa*, vol. 6, pp. 422–92, here 444–54; Rotberg, *The Founder*, chs. 12–13.

158. See Breman, *Taming the Coolie Beast*.

159. Matsusaka, *Japanese Manchuria*, pp. 126–39.

160. C. Marx, *Geschichte Afrikas*, p. 60.

161. Ibid., pp. 72f.

162. R. Reid, *Ganda*, p. 362.

163. M. Last, "The Sokoto Caliphate and Borno," in Ajayi, *General History of Africa*, pp. 555–99, here 568f.

164. The following draws on Hassan Amed Ibrahim, "The Egyptian Empire, 1805–1885," in Daly, *Cambridge History of Egypt*, vol. 2, pp. 198–216; and Fahmy, *All the Pasha's Men*, pp. 38–75.

165. Rich, *Great Power Diplomacy*, pp. 69–74.

166. F. Robinson, *Muslim Societies*, p. 170; see also the good brief account of the Mahdi state on pp. 169–81.

167. Grewal, *Sikhs*, pp. 99–128, who speaks of a "Sikh empire."

168. The following draws on Meinig, *Shaping of America*, vol. 2, pp. 4–23.

169. J. Meyer et al., *Histoire de la France coloniale*, pp. 209–13.

170. Meinig, *Shaping of America*, vol. 2, p. 17.

171. Ibid., p. 23.

172. See the contrary arguments of Klaus Schwabe and Tony Smith, often repeated by other authors until today, in Mommsen and Osterhammel, *Imperialism and After*.

173. Meinig, *Shaping of America*, vol. 2, p. 170.

174. See, e.g., Jacobson, *Whiteness of a Different Color*.

175. The standard work, already slightly dated, is Louis, *Oxford History of the British Empire*, vol. 3, vol. 5 (on the state of research); good overviews are B. Porter, *Lion's Share*; Hyam, *Britain's Imperial Century*; Darwin, *Empire Project*; Darwin, *Unfinished Empire*.

176. See Fry, *Scottish Empire*; S. Howe, *Ireland*, which also discusses the effects down to the present day.

177. A thesis first put forward in Colley, *Britons*.

178. John Stuart Mill, "A Few Words on Non-Intervention [1859]," in Mill, *Collected Works*, vol. 21, pp. 109–24.

179. Schumpeter, *Economics and Sociology of Capitalism*, esp. p. 196.

180. H. V. Bowen, *British Conceptions*, p. 1.

181. Marshall, *Making*, p. 228; also H. V. Bowen, *Business of Empire*.

182. A fine case study of the smaller "expatriate communities," which by 1911 added up to some 3,500 persons, is Butcher, *The British in Malaya*, figure from p. 30. On Kenya, see D. Kennedy, *Islands of White*.

183. A broad perspective from a New Zealand point of view: Pocock, *Discovery*, esp. pp. 181–98.

184. N.A.M. Rodger, *Command of the Sea*, p. 579; Daunton, *Progress*, pp. 518–20.

185. See the maps in P. M. Kennedy, *British Naval Mastery*, p. 207, and A. Porter, *Atlas*, pp. 146f. (with coaling stations).

186. P. M. Kennedy, *British Naval Mastery*, p. 151.

187. Kolff, *Naukar* on the formation of the Indian Army; Metcalf, *Imperial Connections*, pp. 68–101 on its deployment outside India.

188. On the whole range of possibilities in Southeast Asia, see Webster, *Gentlemen Capitalists*.

189. On the domestic political background, see Hilton, *A Mad, Bad, and Dangerous People?*, pp. 543–58; important is A. Howe, *Free Trade*.

190. Darwin, *Imperialism*, pp. 627f. The climax came with Palmerston's speech of June 25, 1850.

191. Gallagher and Robinson [1953], in Louis, *Imperialism*, pp. 53–72.

192. Frank Trentmann, "Civil Society, Commerce, and the 'Citizen-Consumer': Popular Meanings of Free Trade in Modern Britain," in idem, *Paradoxes*, pp. 306–31; and, in greater detail, idem, *Free Trade Nation*.

193. Patrick K. O'Brien, "The Pax Britannica and American Hegemony: Precedent, Antecedent or Just Another History?" in O'Brien and Clesse, *Two Hegemonies*, pp. 3–64, esp. 13f., 16f., 21.

194. L. E. Davis and Huttenback, *Mammon*.

195. Offer, *British Empire*, p. 228. Cf. P. M. Kennedy, *Costs and Benefits*.

196. Cannadine, *Orientalism*.

197. Offer, *First World War*, pp. 368ff.

198. Gilmour, *Curzon*, pp. 274–76, 287–90; Verrier, *Younghusband*, pp. 179ff.

199. Friedberg, *Weary Titan*.

200. Cain and Hopkins, *British Imperialism*, chs. 3–4; on the City and its personnel, see Kynaston, *City*.

201. Neff, *War*, p. 217. This work develops a whole typology of international intervention in the nineteenth century (pp. 215–49).

202. Perkins, *Modern Tunisia*, p. 19; Hsü, *Modern China*, pp. 205–12; Wyatt, *Thailand*, pp. 184f.

203. Key texts are Fisher, *Indirect Rule in India* and, on Egypt, Owen, *Lord Cromer*, chs. 10–16.

204. This is argued in Cannadine, *Orientalism*.

205. Belich, *A Cultural History of Economics?*, p. 119.

206. The standard work remains Semmel, *Jamaican Blood*; see also C. Hall, *Civilising Subjects*, pp. 23–27 and passim.

207. See the collection of material (varying greatly in quality from region to region): Ferro, *Le livre noir*.

208. See also the considerations in Hildebrand, *No Intervention*, pp. 27f.

209. Following Ronald Robinson and John Gallagher, this has also been termed "informal empire." On the definition, see Jürgen Osterhammel, "Britain and China 1842–1914," in Louis, *Oxford History of the British Empire*, vol. 3, pp. 146–69, esp. 148f.

210. Georges Balandier, Albert Memmi, and others. On these classical interpretations, see Young, *Postcolonialism*.

211. Trotha, *Koloniale Herrschaft*, pp. 37ff.

212. See also chapter 16, below.

213. Zastoupil and Moir, *Great Indian Education Debate*.

214. T. R. Metcalf, *Ideologies of the Raj*, pp. 66ff.; Forsyth, *Peoples of Siberia*, pp. 156f.

215. Lorcin, *Imperial Identities*.

216. There is a good brief account of the history of resistance in Abernethy, *Global Dominance*, pp. 254ff.

217. Aldrich, *Greater France*, p. 212.
218. Maurice Duverger, "Le concept d'empire," in idem, *Le concept d'empire*, pp. 5–24, at 11.
219. See the study of legal history: Chanock, *Law*, p. 219.
220. See Manela, *Wilsonian Moment*.
221. See the detailed discussion in Bayly, *Birth of the Modern World*, ch. 6.
222. This is another of the "dilemmas of empire" uncovered in D. Lieven, *Empire*.
223. There are fine case studies for the Ottoman Empire in Hanssen et al., *Empire in the City*.

CHAPTER IX: International Orders, Wars, Transnational Movements

1. H. M. Scott, *Birth* convincingly situates the beginnings of the European system in the 1760s (pp. 121, 143ff.). His book should be read in comparison with Bois, *De la paix des rois*. On the Seven Years' War as a world war see Füssel, *Der Siebenjährige Krieg*. In a sense, already the Nine Years' War (1688–97) might be called a "world war."
2. The classical distinctions derive from Bull, *Anarchical Society*, pp. 8ff. More complex and historically more specific are the considerations in Buzan and Little, *International Systems*, pp. 9off. and passim.
3. Dülffer, *Regeln*, p. 300.
4. Baumgart, *Europäisches Konzert*, p. 343.
5. Wawro, *Warfare*, pp. 55–57; Figes, *Crimea*, pp. 117–19, 178–80.
6. See the wide panorama in Geyer and Bright, *Global Violence*.
7. Schroeder, *Transformation*, and a collection of his essays: Schroeder, *Systems, Stability and Statecraft*. Another key text is Dülffer et al., *Vermiedene Kriege*, one of the most important works on international relations in the nineteenth century.
8. This is done persuasively in F. R. Bridge, "Transformations of the European States-System, 1856–1914," in: Krüger and Schroeder, *Transformation*, pp. 255–72.
9. C. I. Hamilton, *Anglo-French Naval Rivalry*, pp. 273–74.
10. Mommsen, *Bürgerstolz*, p. 305.
11. Mahan's famous book is titled *The Influence of Seapower upon History 1660–1783*.
12. For a comparative overview of armaments in the 1840–1914 period, wider than its title would suggest, see R. Hobson, *Imperialism at Sea*.
13. Eberhard Kolb, "Stabilisierung ohne Konsolidierung? Zur Konfiguration des europäischen Mächtesystems 1871–1914," in: Krüger, *Das europäische Staatensystem*, pp. 188–95, at 192.
14. I. Clark, *Hierarchy of States*, p. 133.
15. D. Lieven, *Russia*, pp. 40f.
16. Mommsen, *Großmachtstellung*, p. 69.
17. Zamoyski, *Rites of Peace* is rich on descriptive detail concerning the Congress.
18. Schroeder, *International System*, pp. 12–14.
19. James Monroe, Message to Congress, December 2, 1823, quoted from D. B. Davis and Mintz, *Boisterous Sea*, p. 350.
20. Heinhard Steiger, "Peace Treaties from Paris to Versailles," in: Lesaffer, *Peace Treaties*, pp. 59–99, at 66f.
21. See the case studies in Dülffer et al., *Vermiedene Kriege*.
22. Schieder, *Staatensystem*.
23. As Baumgart still argued in *Europäisches Konzert*.
24. Xiang Lanxin, *Origins of the Boxer War*.
25. Birmingham, *Portugal*, pp. 133, 135.

26. D.A.G. Waddell, "International Politics and Latin American Independence," in: Bethell, *Cambridge History of Latin America*, vol. 3, pp. 197–228, at 99, 216–18; Alan Knight, "Britain and Latin America," in: Louis, *Oxford History of the British Empire*, vol. 3, pp. 122–45; Cain and Hopkins, *British Imperialism*, pp. 243–74.

27. D. Gregory, *Brute New World*.

28. Sondhaus, *Naval Warfare*, p. 15.

29. Landes, *Wealth and Poverty*, p. 331.

30. Kraay and Whigham: *I Die with My Country*, p. 1. German data from Wehler, *Gesellschaftsgeschichte*, vol. 4, p. 944.

31. Hans Vogel, "Argentinien, Uruguay, Paraguay, 1830/1852–1904/1910," in: Bernecker et al., *Handbuch*, vol. 2, pp. 694–98.

32. Collier and Sater, *Chile*, p. 139; Klarén, *Peru*, pp. 183–91; Riekenberg, *Ethnische Kriege*, pp. 101–9.

33. H.-J. König, *Geschichte Lateinamerikas*, p. 392.

34. LaFeber, *American Age*, p. 110.

35. Ibid., p. 164.

36. J. Major, *Prize Possession*, pp. 34ff., 78ff. (figures from p. 83).

37. LaFeber, *American Age*, p. 234; the other major history of US foreign relations is Herring, *From Colony to Superpower*.

38. Topik, *Trade*, p. 209.

39. Fisher, *Indirect Rule in India*, pp. 255–57.

40. The standard regional history is Andaya and Andaya, *Malaysia*.

41. Lieberman, *Strange Parallels*, vol. 1, p. 302.

42. See chapter 1 of M. B. Jansen, *China in the Tokugawa World*, a most important work for the international history of the early modern period.

43. Text in Lu, *Japan*, vol. 2, pp. 288–92.

44. Auslin, *Negotiating with Imperialism*, which lists all sixteen treaties of friendship and trade. The most important Western work on Japan's "limited sovereignty" is Hoare, *Japan's Treaty Ports*, esp. chs. 4 and 8.

45. The following draws on Osterhammel, *China*, chs. 9–10; Dabringhaus, *Geschichte Chinas*, pp. 56–59, 145–57; and, of the older literature, especially Kim, *Last Phase*. Westad, *Restless Empire*, is the most up-to-date long-term survey; Suzuki, *Civilization and Empire*, is a well-informed comparative interpretation in the light of "English School" international relations theory.

46. S.C.M. Paine, *Sino-Japanese War*.

47. Hamashita Takeshi, "Tribute and Treaties: Maritime Asia and Treaty Port Networks in the Era of Negotiations, 1800–1900," in: Arrighi et al., *Resurgence*, pp. 17–50, and Hamashita's other pathbreaking essays: *China, East Asia and the Global Economy*.

48. Schmid, *Korea*, pp. 56f.

49. Klaus Hildebrand, "Eine neue Ära der Weltgeschichte." Der historische Ort des Russisch-Japanischen Krieges," in: Kreiner, *Der Russisch-Japanische Krieg*, pp. 27–51, at 43.

50. We have to skip the precolonial military history of these continents. See, e.g., the major study R. Reid, *War in Pre-colonial Eastern Africa*.

51. Howard, *War in European History*, pp. 100f.

52. Connelly, *Wars of the French Revolution*, p. 115.

53. Wawro, *Warfare*, p. 33.

54. Pröve, *Militär*, p. 4.

55. Dieter Storz, "Modernes Infanteriegewehr und taktische Reform in Deutschland in der Mitte des 19. Jahrhunderts," in Epkenhans and Groß, *Militär*, pp. 209–30, at 217.

56. Agoston, *Guns for the Sultan*, mostly on the seventeenth and eighteenth centuries. On the nineteenth century see Ralston, *Importing the European Army*, pp. 43–78, and especially Grant, *Rulers*.

57. Jonas, *Battle of Adwa*; Pankhurst, *Ethiopians*, pp. 188–93.

58. Wawro, *Warfare*, p. 127.

59. On the international impact of the Russo-Japanese War, see Aydin, *Politics of Anti-Westernism*, pp. 71–92; Kowner, *Russo-Japanese War*.

60. S.C.M. Paine, *Sino-Japanese War*, p. 182; Sondhaus, *Naval Warfare*, pp. 133f., 152.

61. See the considerations in Dierk Walter, "Warum Kolonialkrieg?" in: Klein and Schumacher, *Kolonialkriege*, pp. 14–43, esp. 17–26, and the case studies in ibid., as well as in Moor and Wesseling, *Imperialism and War*. Important here is Wesseling, *Imperialism and Colonialism*, esp. the first essay: "Colonial Wars and Armed Peace, 1871–1914—A Reconnaissance," pp. 12–26.

62. Belich, *New Zealand Wars*, pp. 323f.

63. Tone, *War and Genocide*, p. 193 (with a portrait of Weyler on pp.153–77); cf. Everdell, *The First Moderns*, pp. 116–26; Gott, *Cuba*, pp. 93–97.

64. Nasson, *South African War*, pp. 220–24.

65. S. C. Miller, *"Benevolent Assimilation"*, pp. 164, 208–10.

66. Laband, *Kingdom in Crisis*, p. 14.

67. M. Lieven, *Butchering*, p. 616.

68. Spiers, *Late Victorian Army*, p. 335, and see the good analysis of this kind of warfare in ibid., 272–300. Wesseling puts the number of colonial wars between 1871 and 1914 at 23 for the British, 40 for the French, and 32 for the Dutch (*Imperialism and Colonialism* [1997], pp. 13f.). See also C. Marx, *Geschichte Afrikas*, pp. 133f.

69. Vandervort, *Wars of Imperial Conquest*, pp. 174–77.

70. Ibid., p. 49.

71. Lee Ki-baik, *Korea*, p. 212.

72. Esdaile, *Fighting Napoleon*, p. 176, and for a wider range of references: Broers, *Napoleon's Other War*.

73. Hobsbawm, *Primitive Rebels*, esp. ch. 2.

74. Teng Ssu-yü, *Nien Army*.

75. Showalter, *Wars of German Unification*, pp. 315–27.

76. Blanning, *French Revolutionary Wars*, p. 101; and, more dramatically, Bell, *First Total War*.

77. See Forrest, *Napoleon's Men*.

78. Broers, *Europe*, pp. 70–77; and for another view Connelly, *Wars of the French Revolution*, p. 117; a good overall chronicle is Esdaile, *Napoleon's Wars*; on Napoleon's disastrous invasion of Russia see the authoritative study D. Lieven, *Russia against Napoleon*.

79. Stig Förster and Jörg Nagler, "Introduction," in idem, *On the Road*, pp. 1–25, at 6f.

80. Wawro, *Warfare*, pp. 19. 89, 155f.; idem, *Franco-Prussian War*, pp. 75, 84; Nasson, *South African War*, p. 75; Elleman, *Modern Chinese Warfare*, p. 41.

81. McPherson, *Battle Cry*, p. 664; Wawro, *Warfare*, p. 155; Urlanis, *Bilanz*, p. 99, 122—still the best source for data on military casualties in modern Europe (new Russian edition in 1994).

82. See the overview in J. W. Steinberg et al., *Russo-Japanese War*.

83. Moorehead, *Dunant's Dream*, pp. 1–7.

84. Steinbach, *Abgrund Metz*, p. 45.

85. Langewiesche, *Kriegsgewalt*, p. 27.

86. E. Grove, *Royal Navy*, pp. 39–68.

87. Deng Gang, *Maritime Sector*, p. 195 (Tab. 4.3).

88. See the evidence in M. C. Wright, *Last Stand*, p. 220.

89. Sondhaus, *Naval Warfare*, pp. 3, 52, 73, 103, 133f., 150–52.

90. M. B. Jansen, *Modern Japan*, p. 277.

91. Simon Ville, "Shipping Industry Technologies," in: Jeremy, *International Technology Transfer*, pp. 74–94, at 83 (tab. 5.2). Essential on the naval program is D. C. Evans and Peattie, *Kaigun*, pp. 1–31.

92. Josef Kreiner, "Der Ort des Russisch-Japanischen Krieges in der japanischen Geschichte," in: idem, *Der Russisch-Japanische Krieg*, pp. 53–76, at 57.

93. D. C. Evans and Peattie, *Kaigun*, p. 124.

94. The best biography of one of the most influential nineteenth-century politicians never to have held public office is Edsall, *Richard Cobden*.

95. Lee Ki-baik, *Korea*, pp. 268f.; W. G. Beasley, "The Foreign Threat and the Opening of the Ports," in: J. W. Hall, *Cambridge History of Japan*, vol. 5, pp. 259–307, at 307.

96. See the masterful summary in Gollwitzer, *Geschichte des weltpolitischen Denkens*, vol. 2, pp. 23–82; cf. the depiction of the prewar mood in Joll, *Origins*, ch. 8; Cassels, *Ideology*, chs. 3–6. The most comprehensive association of these themes may be found in the work of the little known Swedish geopolitican Rudolf Kjellen (1864–1922).

97. D. P. Crook, *Darwinism*, p. 63.

98. Gollwitzer, *Die gelbe Gefahr*.

99. Gluck, *Japan's Modern Myths*, p. 206.

100. Chang Hao, "Intellectual Change and the Reform Movement, 1890–8," in: Fairbank and Twitchett, *Cambridge History of China*, vol. 11, pp. 274–338, at 296–98; cf. Pusey, *China and Charles Darwin*, pp. 236–316 (which is critical of Liang).

101. A. Black, *Islamic Political Thought*, p. 304.

102. For a survey, see Windler, *La diplomatie*, and the grand view of early modern "global" diplomacy in Bély, *L'art de la paix en Europe*, pp. 345–73.

103. On the Macartney mission: Hevia, *Cherishing Men from Afar*.

104. H. M. Scott, *Birth*, p. 278.

105. Vandervort, *Wars of Imperial Conquest*, p. 85.

106. H. M. Scott, *Birth*, pp. 275f.

107. Gong, *Standard of "Civilization"*; Frey and Frey, *Diplomatic Immunity*, pp. 384–421; Jörg Fisch, "Internationalizing Civilization by Dissolving International Society: The Status of Non-European Territories in Nineteenth-Century International Law," in: M. H. Geyer and Paulmann, *Mechanics*, pp. 235–57; for Japan: Henning, *Outposts of Civilization*.

108. Farah, *Politics of Interventionism*, is rich in details.

109. R. Owen, *Middle East*, pp. 122–35; Osterhammel, *China*, pp. 211–18. On financial imperialism, there is still nothing superior to Mommsen, *Der europäische Imperialismus*, pp. 85–148. Exemplary on Germany is B. Barth, *Die deutsche Hochfinanz*.

110. Excellent on this is Lipson, *Standing Guard*, pp. 37–57.

111. M. S. Anderson, *Rise of Modern Diplomacy*, pp. 103–11; Girault, *Diplomatie européenne*, pp. 13–19.

112. Headrick, *Invisible Weapon*, p. 17.

113. For greater detail see M. King, *Penguin History of New Zealand*, pp. 156–67; Belich, *Making Peoples*, pp. 193–97.

114. Kinji Akashi, "Japanese 'Acceptance' of the European Law of Nations: A Brief History of International Law in Japan, c. 1853–1900," in: Stolleis and Yanagihara, *Perspectives*, pp. 1–21, at 9.

115. Paulmann, *Pomp und Politik*, pp. 295ff.

116. Keene, *Emperor of Japan*, p. 632.

117. Georgeon, *Abdulhamid II*, pp. 31–35.

118. D. Wright, *The Persians amongst the English* is a fine study of intercultural diplomacy: see pp. 121–40 on the shah's two visits, in 1873 and 1889.

119. Keene, *Emperor of Japan*, p. 308.

120. D.G.E. Hall, *South-East Asia*, pp. 629f.

121. Mawer, *Ahab's Trade*, pp. 97f.

122. Grewe, *Epochen*, p. 554.

123. Jelavich, *Russia's Balkan Entanglements*, p. 172.

124. This mood is well captured in a journalistic book: Traxler, *1898*.

125. Dülffer et al., *Vermiedene Kriege*, pp. 615–39; Mommsen, *Großmachtstellung*, pp. 213–27.

126. Nikki R. Keddie, "Iran under the Later Qajars, 1848–1922," in Avery et al., *Cambridge History of Iran*, vol. 7, pp. 174–212, at 195f.; Keddie, *Qajar Iran*, pp. 37–39.

127. Quoted from Osterhammel, *China*, p. 222. On the boycott see Wang Guanhua, *In Search of Justice*.

128. Quataert, *Social Disintegration*, pp. 121–45.

129. Lauren, *Power and Prejudice*, pp. 57, 76–101; cf. the excellent study: Shimazu, *Japan, Race and Equality*.

130. John Boli and George M. Thomas, "INGOs and the Organization of World Culture," in idem, *Constructing World Culture*, pp. 13–49, at 23 (Fig. 1.1).

131. Moorehead, *Dunant's Dream* p. 125; cf. Riesenberger, *Für Humanität*, pp. 35f.

132. Chi Zihua, *Hongshizi yu jindai Zhongguo*, pp. 52ff.

133. F.S.L. Lyons, *Internationalism*, p. 263—an unsurpassed standard work.

134. See above all the older literature: Braunthal, *History of the International*, vol. 1; Joll, *Second International*.

135. M. B. Jansen, *Making*, p. 491.

136. Bock, *Women*, p. 118.

137. McClain, *Japan*, pp. 381f.

138. See the case study for Egypt: Badran, *Feminists*, pp. 47–51.

139. Rupp, *Worlds of Women*, pp. 15–21. Karen Offen sees the years from 1878 to 1890 as a first great period of the internationalization of feminism: *European Feminisms*, pp. 150ff.

140. B. S. Anderson, *Joyous Greetings*, pp. 24f., 204f.

141. See McFadden, *Golden Cables*, esp. apps. A–F.

142. On India: Burton, *Burdens of History*, chs. 4–5.

143. They could not, however, avoid sometimes taking a "national" position against aggressors: see Grossi, *Le pacifisme européen*, pp. 219ff. (the best overall account).

144. Ceadel, *Origins of War Prevention*.

145. S. E. Cooper, *Patriotic Pacifism*, pp. 219f.

146. Translation: K'ang Yu-wei, *One-world Philosophy*; an authoritative reference is Hsiao Kung-chuan, *A Modern China*, pp. 456ff.

147. Unsurpassed on the Hague Conferences is Dülffer, *Regeln*. For a more optimistic interpretation of them as a symbolic step forward, see I. Clark, *International Legitimacy*, pp. 61–82.

148. See the important collection M. H. Geyer and Paulmann, *Mechanics*.

149. Vec, *Recht und Normierung*, p. 379.

150. See chapter 2, above.

151. Some of these processes are discussed in Martin H. Geyer, "One Language for the World," in M. H. Geyer and Paulmann, *Mechanics*, pp. 55–92. But the best overview is Murphy, *International Organization*, pp. 46–118.

152. Forster, *Esperanto Movement*, p. 22 (Tab. 3).

153. D. C. Young, *Modern Olympics*, pp. 68–70, 85.

154. On football: Goldblatt, *The Ball Is Round*, pp. 85–170.

155. Quoted from Herren, *Hintertüren zur Macht*, p. 1.

156. See the list in Murphy, *International Organization*, pp. 47f.

157. Ibid., pp. 57–59.

CHAPTER X: Revolutions

1. Arendt, *On Revolution*, p. 11.

2. T. Paine, *Common Sense*, p. 63.

3. Arendt, *On Revolution*, p. 37.

4. "Great revolutions" are those that (a) led to the consolidation of a revolutionary state power *and* (b) had a program that at least for a time commanded worldwide attention.

5. E. Zimmermann, *Political Violence*, p. 298, slightly simplified.

6. Law, *Oyo Empire*, pp. 245ff.

7. C. Tilly, *European Revolutions*, p. 243 (Tab. 7.1, curiously lacking Germany).

8. Kimmel, *Revolution*, p. 6.

9. See the striking theoretical elaboration in Moore, *Social Origins*, pp. 433–52, and on the Japanese case pp. 228ff.

10. Beasley, *Meiji Restoration*; Jansen, *Modern Japan*, pp. 333–70, and the primary sources in Tsunoda et al., *Sources of Japanese Tradition*, vol. 2. See also the interpretation of the Meiji Renewal as a revolution or "revolutionary restoration" in Eisenstadt, *Japanese Civilization*, pp. 264–77.

11. See chapter 11, below.

12. Dutton, *Tây Son Uprising*.

13. Kalyvas, *Logic of Violence*, p. 5.

14. Later there were a second (1846–49) and a third (1870–75) Carlist war.

15. Carr, *Spain*, pp. 184–95.

16. Labourdette, *Portugal*, pp. 522–27.

17. Farah, *Interventionism*, pp. 695f.

18. Tutino, *Revolution in Mexican Independence*.

19. McClain, *Japan*, pp. 123f., 193f.

20. Edmund Burke III, "Changing Patterns of Peasant Protest in the Middle East, 1750–1950," in Kazemi and Waterbury, *Peasants and Politics*, pp. 24–37, at 30.

21. See chapter 11, below.

22. Schölch, *Egypt for the Egyptians!*; J. A. Cole, *Colonialism*.

23. Jacob Burckhardt, "Die geschichtlichen Krisen," in idem, *Werke*, vol. 10, p. 463.

24. See, e.g., Goldstone, *Revolution and Rebellion*.

25. See chapter 2, above.

26. Early opposition to the internally focused view came from within sociology: see Skocpol, *States*, one of the classics of the comparative history of revolutions.

27. Schulin, *Französische Revolution*, p. 37.

28. On Sorel, see Pelzer, *Revolution und Klio*, pp. 120–41.

29. Mainly programmatic: Bender, *Rethinking American History*.

30. Pioniers in this respect were Eugen Rosenstock-Huessy, *Die europäischen Revolutionen* (Jena 1931) and Crane Brinton, *The Anatomy of Revolution* (New York 1938).

31. Godechot, *France*; Palmer, *Age*.

32. Very ably so Klooster, *Revolutions in the Atlantic World*, and, assembling a galaxy of first-rate authors, Armitage and Subrahmanyam, *Age of Revolutions*. Very important is also Belaubre et al., *Napoleon's Atlantic*.

33. Bailyn, *Atlantic History*, pp. 21–40.

34. Kossok, *Ausgewählte Schriften*, esp. vol. 2.

35. A good introduction is Countryman, *American Revolution*, and the work by one of the most influential interpreters: Wood, *American Revolution*. On historians' controversies surrounding the revolution see A. F. Young and Nobles, *Whose American Revolution Was It?*; and Woody Holton, "American Revolution and Early Republic," in: Foner and McGirr, *American History Now*, pp. 24–51.

36. That this was not only a tactical error on London's part but reflected a different conception of empire from that of the colonists, is well brought out in Gould, *Persistence of Empire*, pp. 110–36.

37. Wood, *Radicalism*, p. 109.

38. Langford, *A Polite and Commercial People*, pp. 550f.

39. Foster, *Modern Ireland*, p. 280.

40. Ibid., p. 281.

41. Texts in Hampsher-Monk, *Impact*.

42. Mark Philp, "Revolution," in McCalman, *Romantic Age*, pp. 17–26.

43. Godechot, *France*, pp. 54f.

44. Schama, *Patriots*, pp. 120–31.

45. A good discussion of the explanatory power of the interpretations current today is Spang, *Paradigms*; see also P. R. Campbell, *Origins*.

46. A fundamental theoretical contribution is Skocpol, *States*.

47. See esp. Whiteman, *Reform*; and the earlier overview, less well "focused" and resting on secondary literature, in B. Stone, *Reinterpreting the French Revolution*; on transatlantic interaction during the 1780s see also Andress, *1789*.

48. W. Doyle, *French Revolution*, p. 66; and on detailed points Whiteman, *Reform*, pp. 43ff.

49. My favourite history of the French Revolution remains, against huge competition, W. Doyle, *French Revolution*. The unity of the 1789–1815 period is brought out well in Sutherland, *French Revolution and Empire*. A good approach to the Revolution is through its exciting historiography, see the collection on major historians: Pelzer, *Revolution und Klio*.

50. For an analytical sketch cf. F. W. Knight, *Haitian Revolution*; the standard work is Dubois, *Avengers*, another major study Popkin, *You Are All Free*.

51. Figures from Dubois, *Avengers*, p. 30.

52. On the "market revolution," see Sean Wilentz, "Society, Politics and the Market Revolution, 1815–1848," in: Foner, *New American History*, pp. 61–64 and esp. 62–70.

53. Dubois, *Avengers*, p. 78.

54. Ibid., p. 125.

55. See Geggus, *Slavery* for an extensive account of this important episode in the Atlantic war.

56. Fox-Genovese and Genovese, *Mind of the Masterclass*, p. 38.

57. This is especially impressive in Dubois, *Colony of Citizens*, a study mainly of Guadeloupe.

58. Davis, *Problem of Slavery*, p. 3.

59. Dubois, *Colony of Citizens*, pp. 7, 171ff.

60. See the case studies in Klaits and Haltzel, *Global Ramifications*.

61. On the French Revolution as a "catalyst of political cultures in Europe," see Reichardt, *Blut der Freiheit*, pp. 257–334.

62. Förster, *Die mächtigen Diener*.

63. See Keddie, *Iran*, pp. 233–49; Shaw, *Between Old and New*; Laurens, *L'Expédition d'Égypte*, esp. pp. 467–73.

64. For an overview of the most important interpretations, see Uribe, *Enigma*.

65. A model of its kind is C. F. Walker, *Smoldering Ashes*, esp. chs. 4–5.

66. The standard account remains J. Lynch, *Spanish-American Revolutions*, now to be supplemented with Rinke, *Las revoluciones en América Latina*. Of new interpretations especially notable: Adelman, *Sovereignty*, esp. chs. 5, 7; on broad historiographical trends see idem, "Independence in Latin America," in: Moya, *Oxford Handbook of Latin American History*, pp. 153–80. Anna, *Spain and the Loss of America*, focuses on the losing Spanish side.

67. Elliott, *Empires*, p. 360.

68. Ibid., p. 374.

69. Wood, *Benjamin Franklin*.

70. Rodríguez, *Independence of Spanish America*, p. 82.

71. Graham, *Independence*, pp. 107ff.

72. J. Lynch, *Simón Bolívar* is a masterly study.

73. For Mexico: Anna, *Fall of the Royal Government*, pp. 225f.

74. This is shown in the monumental work Van Young, *Other Rebellion*.

75. J. Lynch, *Simón Bolívar*, p. 122.

76. Graham, *Independence*, pp. 142f.

77. J. Lynch, *Simón Bolívar*, p. 105.

78. Ibid., p. 147.

79. On the (post-) revolutionary militarization of Latin America, see Halperin-Donghi, *Aftermath*, pp. 17–24.

80. Colin Lewis, "The Economics of the Latin American State: Ideology, Policy and Performance, c. 1820–1945," in A. A. Smith et al., *States*, pp. 99–119, at 106.

81. Finzsch, *Konsolidierung*, pp. 25ff.

82. Ibid., pp. 596f.; D. B. Davis, *Inhuman Bondage*, p. 262.

83. Pilbeam, *1830 Revolution*, esp. p. 149.

84. On one such case from the French Pyrenees in 1829–31: P. Sahlins, *Forest Rites*.

85. A good overview of this "exit" process in France is Jourdan, *La révolution*, pp. 71–83.

86. Woloch, *New Regime*, pp. 380–426.

87. Dominguez, *Insurrection*, pp. 227f.

88. Breen, *Marketplace*, esp. pp. 235ff.

89. See R. G. Kennedy, *Orders from France*; also Roach, *Cities of the Dead*.

90. "Life of Napoleon Buonaparte," in: *The Complete Works of William Hazlitt*, vol. 13, ed. by P. P. Howe, London 1931, p. 38.

91. Brading, *First America*, pp. 583–602.

92. See the anthology: Bello, *Selected Writings*.

93. See Gould, *A World Transformed?*, and several contributions in Gould and Onuf, *Empire and Nation*.

94. On the reception of the European Enlightenment, see May, *Enlightenment in America*, although its periodization is rather overschematic.

95. J. Lynch, *Simón Bolívar*, p. 28.

96. J. Lynch, *Spanish American Revolutions*, p. 27.

97. This is the main theme in Liss, *Atlantic Empires*, which despite its title is a study of the ideas of American political economy.

98. Gough, *Terror*, p.77.

99. John Lynch speaks of the ten year war in Venezuela as "a total war of uncontrolled violence": *Spanish American Revolutions*, p. 220.

100. Conway, *British Isles*, pp. 43f.

101. Langley, *The Americas*, p. 61.

102. Royle, *Revolutionary Britannia?*, pp. 67f.

103. Hilton, *A Mad, Bad, and Dangerous People?*, p. 421.

104. For a full account, see Hochschild, *Bury the Chains*.

105. This aspect is discussed in greater detail in chapter 17, below.

106. L. S. Kramer, *Lafayette*, pp. 113f.

107. Jourdan, *La révolution*, p. 357.

108. Beck, *Alexander von Humboldt*, vol. 1, pp. 223f.; ibid., vol. 2, pp. 2f., 194–200 (Humboldt 1848).

109. Other protest movements requiring detailed analysis would include the Mahdi uprising of Bu Ziyan in the Algerian Atlas in 1849: see Clancy-Smith, *Rebel and Saint*, pp. 92–124.

110. For an all-European perspective, see Hachtmann, *Epochenschwelle*; Mommsen, *1848*; Sperber, *European Revolutions*—the best overall account; and many individual contributions in Dowe et al., *Europe in 1848*.

111. John Breuilly, "1848: Connected or Comparable Revolutions?" in Körner, *1848*, pp. 31–49, at 34f.

112. Dieter Langewiesche, "Kommunikationsraum Europa. Revolution und Gegenrevolution," in idem, *Demokratiebewegung*, pp. 11–35, at 32.

113. This is brilliantly demonstrated in Ginsborg, *Daniele Manin*—one of the classic works on 1848–49.

114. Mommsen, *1848*, p. 300.

115. Sperber, *European Revolutions*, p. 62.

116. Ibid., p. 124.

117. Blum, *End of the Old Order*, p. 371.

118. See the balance sheet in Hachtmann, *Epochenschwelle*, pp. 178–81.

119. Tombs, *France*, p. 395.

120. Deák, *Lawful Revolution*, pp. 321–37 (figure from p. 329).

121. Langewiesche, *Europa*, p. 112.

122. A paradigmatic life: from revolt to exile to rehabilitation. See Gregor-Dellin, *Richard Wagner*, pp. 288ff.

123. Hachtmann, *Epochenschwelle*, pp. 181–85; Brancaforte, *German Forty-Eighters*; Levine, *Spirit of 1848*; Wolfram Siemann, "Asyl, Exil und Emigration," in Langewiesche, *Demokratiebewegung*, pp. 70–91.

124. The following builds on M. Taylor, *1848 Revolution*.

125. Clarke and Gregory, *Western Reports* contains excellent documentation.

126. The best account and analysis of the movement, especially of its beginnings, is J. Spence, *God's Chinese Son*. Also still important are Michael, *Taiping Rebellion*, vol. 1; Jen Yu-wen, *Taiping*; and Shih, *Taiping Ideology*.

127. Spence, *God's Chinese Son*, p. 171; Michael, *Taiping Rebellion*, vol. 1, p. 174.

128. Cao Shuji, *Zhongguo yimin shi*, p. 469.

129. Deng, *China's Political* Economy, p. 38.

130. Michael, *Taiping Rebellion*, vol. 1, pp. 135–68; sources in ibid., vol. 3, pp. 729–1378, esp. 754ff.

131. Stampp, *America in 1857*, p. viii.

132. W. B. Lincoln, *Great Reforms*, pp. 68f.

133. P. J. O. Taylor, *Companion*, p. 75.

134. The events are described in every up-to-date history: e.g., Markovits et al., *Modern India*, pp. 283–93. A good introduction is Llewellyn-Jones, *Great Uprising in India*. (More) sources may be found in Harlow and Carter, *Archives of Empire*, vol. 1, pp. 391–551. Herbert, *War of No Pity*, is an interesting study from the point of view of the history of mentalities and human psychology. Wagner, *Marginal Mutiny*, discusses recent research; Pati, *Great Rebellion*, samples this work.

135. Omissi, *Sepoy*, p. 133. After the "Mutiny," the ratio of 5:1 was lowered to 2:1.

136. Russell's reporting was astonishingly impartial, not hostile to the Indians. See his *My Diary in India* (reprint 2010).

137. Cook, *Understanding Jihad*, pp. 80f.; Bose and Jalal, *Modern South Asia*, p. 74.

138. On the course of events, see any textbook, or Guelzo, *Fateful Lightning*; for additional context Ford, *Companion to the Civil War*; readers of German will profit from a non-American perspective: Finzsch, *Konsolidierung*, pp. 561–741.

139. McPherson, *Abraham Lincoln*, esp. pp. 6f.

140. In his global comparative study of paths to modernity, Barrington Moore spoke of the American Civil War as the "last capitalist revolution": *Social Origins* pp. 111ff., esp. pp. 151–55. The main proponent of the revolution thesis is the great Civil War historian McPherson, *Abraham Lincoln*, esp. pp. 3–22.

141. Even one great non-Marxist authority has used the term revolution: Jen Yu-wen, *Taiping*.

142. A. Lincoln, *Speeches and Writings*, vol. 2, p. 218.

143. Moore, *Social Origins*, p. 153.

144. Foner, *American Freedom*, p. 58.

145. See the brief discussion in John Ashworth, "The Sectionalization of Politics, 1845–1860," in Barney, *Companion*, pp. 33–46. The standard work is Freehling, *Road to Disunion*.

146. Potter, *Impending Crisis*; Levine, *Half Slave*.

147. On the course of the war, see McPherson, *Battlecry*.

148. W. J. Cooper and Terrill, *American South*, vol. 2, p. 373.

149. R. W. Fogel, *Slavery Debates*, p. 63.

150. Still relevant is Litwack, *Been in the Storm so Long*.

151. Boles, *Companion*, chs. 16–18.

152. The expression is borrowed from Eric Foner, the author of the most comprehensive history of this episode: *Reconstruction*.

153. Or even from 1848 to 1877, as in the standard account, Barney, *Battleground*, which uses "1848" as a formal starting date ("mid-century").

154. Atwill, *Chinese Sultanate*, p. 185: not a purely religious conflict, and one largely provoked by Han Chinese.

155. Figure from M. C. Meyer and Sherman, *Mexican History*, p. 552; the standard accounts are an analytical work by a Swiss expert: Tobler, *La revolución Mexicana*, and, more narrative than analysis, A. Knight, *Mexican Revolution*.

156. Mardin, *Genesis of Young Ottoman Thought*, pp. 169–71; he stresses, however, that it was several decades before the French Revolution showed any impact in the Ottoman Empire.

157. D. C. Price, *Russia*.

158. Fundamental for China is Reynolds, *China*.

159. Gasster, *Chinese Intellectuals*, esp. pp. 106ff.

160. Sohrabi, *Global Waves*, p. 58.

161. Gelvin, *Modern Middle East*, p. 145.

162. Yoshitake, *Five Political Leaders*, pp. 180, 193, 222.

163. Ascher, *Revolution of 1905*, p. 28. This is the abridgement of a work that originally appeared in two volumes.

164. Kreiser and Neumann, *Türkei*, pp. 341f.; Georgeon, *Abdulhamid II*, pp. 87–89.

165. See D. Lieven, *Nicholas II*.

166. The standard biography, focused particularly on the early period of his rule, is Amanat, *Pivot of the Universe*.

167. Fundamental here are Arjomand, *Constitutions*, pp. 49–57; and Sohrabi, *Historicizing Revolutions*.

168. For the text see Gosewinkel and Masing, *Verfassungen*, pp. 1307–22.

169. Ascher, *Revolution of 1905*, pp. 16f.

170. Janet M. Hartley, "Provincial and Local Government," in D. Lieven, *Cambridge History of Russia*, vol. 2, pp. 449–67, at 461–65; Philippot, *Les zemstvos*, pp. 76–80.

171. For a detailed account of the late Qing reforms: Chuzo Ichiko, "Political and Institutional Reform, 1901–11," in Fairbank and Twitchett, *Cambridge History of China*, vol. 11, pp. 375–415; also Reynolds, *China*.

172. Sdvižkov, *Zeitalter der Intelligenz*, p. 150.

173. A classical account is Venturi, *Roots of Revolution*, chs. 21–22. For a more succinct portrait, see Sdvižkov, *Zeitalter der Intelligenz*, pp. 139–83.

174. See Vanessa Martin, *Islam and Modernism*, pp. 18f.

175. See the case study: P. A. Cohen, *Between Tradition and Modernity*.

176. On the politicization of the military within the anti-Hamid movement, see the (somewhat confused) analysis in Turfan, *Rise of the Young Turks*. The standard work in English on the Young Turk Revolution is Hanioğlu, *Preparation*.

177. Ascher, *Revolution of 1905*, pp. 57f.

178. Keddie, *Qajar Iran*, p. 59.

179. Zürcher, *Turkey*, pp. 93f.

180. Fung, *Military Dimension*; a key work is McCord, *Power of the Gun*, pp. 46–79.

181. For a history of the events, see J. Spence, *Modern China*, pp. 249–68.

182. On Yuan, a fascinating figure in the transition from the nineteenth to the twentieth century, see E. P. Young, *Presidency of Yuan Shih-k'ai*.

183. On this early parliamentarianism, see Abrahamian, *Iran*, pp. 81–92.

184. Afary, *Iranian Constitutional Revolution*, pp. 337–40.

185. Kreiser and Neumann, *Türkei*, p. 361.

186. On the Pahlavi regime down to 1941, see Gavin R. G. Hambly, "The Pahlavi Autocracy: Riżā Shāh, 1921–1941," in: Avery et al., *Cambridge History of Iran*, vol. 7, pp. 213–43.

187. There is a trend in recent research to identify revolution as an extreme form or macro-variant of collective violence: see, e.g., C. Tilly, *Collective Violence*.

188. Excellent on this is Vanessa Martin, *Qajar Pact*, which gives an especially impressive account of the "agency" of women (pp. 95–112).

189. Hsiao Kung-chuan, *Rural China*, pp. 502f.

190. See Gelvin, *Modern Middle East*, pp. 139–46.

CHAPTER XI: The State

1. For a long-term perspective see Charles S. Maier, "Leviathan 2.0: Inventing Modern Statehood," in: E. S. Rosenberg, *A World Connecting*, pp. 29–282. Any global history of the state has to engage with this brilliant analysis. Since this is impossible here, I leave the chapter unrevised as it was written in 2006 and published in 2009.

2. On the novelty of the Asian "gunpowder empires," see Finer, *History of Government*, vol. 3, chs. 1–4; also Lieberman, *Beyond Binary Histories*.

3. Ernest Gellner, "Tribalism and the State in the Middle East," in: Khoury and Kostiner, *Tribes*, pp. 109–26, at 109.

4. Carl A. Trocki, "Political Structures in the Nineteenth and Early Twentieth Centuries," in: Tarling, *Cambridge History of Southeast Asia*, vol. 2, pp. 79–130, at 81.

5. M. Weber, *Economy and Society*, vol. 1, p. 54.

6. O'Rourke, *Warriors*, p. 43.

7. Earle, *Pirate Wars*, pp. 231ff.

8. M. Mann, *Sources of Social Power*, vol. 2, p. 6.

9. Birmingham, *Portugal*, p. 125.

10. See J. Lynch, *Argentine Dictator*; pp. 201–46 on Rosas's reign of terror.

11. There is a fine character portrait of Díaz in A. Knight, *Peculiarities*, pp. 102f.; also M. C. Meyer and Sherman, *Course of Mexican History*, pp. 453–57. On classical *caudillismo* before ca. 1850, see J. Lynch, *Caudillos*, esp. pp. 183–237, 402–37. Excellent on violence and the rudimentary state organization in nineteenth-century Hispanic America is Riekenberg, *Gewaltsegmente*, pp. 35–79, incl. 59–63 on caudillos. For Central America, in particular, see Holden, *Armies*, esp. pp. 25–50.

12. Unlike in neighboring Uruguay, the Argentine caudillos who came after Rosas were "tamed" by the landowning oligarchy: R. M. Schneider, *Latin American Political History*, p. 139.

13. For an overview, see Herzfeld, *Anthropology*, pp. 118–32.

14. A key text is Newbury, *Patrons*, esp. the survey on pp. 256–84.

15. C. M. Clark, *Kaiser Wilhelm II*, p. 162. The standard biography is J.C.G. Röhl, *Wilhelm II*.

16. Schudson, *Good Citizen*, p. 132.

17. H.C.G. Matthew, *Gladstone*, pp. 293–312, esp. 310f.

18. J. Lynch, *Argentine Dictator*, p. 112; also Bernand, *Buenos Aires*, pp. 149f., 155–57.

19. P. Brandt et al., *Handbuch*, p. 42.

20. Rickard, *Australia*, p. 113.

21. Wortman, *Scenarios of Power*, p. 347.

22. The panorama of monarchical forms is described in D. J. Steinberg et al., *Southeast Asia*, pp. 57–91.

23. Pennell, *Morocco*, pp. 158–63.

24. For India, see Fisher, *Indirect Rule in India*.

25. Thant, *Modern Burma*, pp. 209f.; Kershaw, *Monarchy in South-East Asia*, p. 25.

26. Among numerous studies of princely states in India, an outstanding work (on Southeast India) is P. G. Price, *Kingship*.

27. Kershaw, *Monarchy in South-East Asia*, p. 26.

28. Ibid., pp. 28f.

29. M. D. Sahlins, *Anahulu*, pp. 76f.

30. As Geertz, for example, argues in *Negara*; the one-sidedness of this influential account has been corrected by Schulte Nordholt, *Spell of Power*, which above all places Geertz's static view in its own historical context (esp. pp. 5–11).

31. Kroen, *Politics and Theater*.

32. There is a fine discussion of this in relation to Burma in Koenig, *Burmese Polity*, pp. 16–84.

33. Morris, *Washing of the Spears*, pp. 79f., 91, 98f. On Shaka and his demonization, see C. Hamilton, *Terrific Majesty*.

34. Laband, *Kingdom in Crisis*, pp. 22 f.; Cope, *Characters of Blood*, pp. 266f. A classic on forms of monarchy in Africa is Fortes and Evans-Pritchard, *African Political Systems*; see esp. Fortes on the Zulus (pp. 25–55).

35. David K. Wyatt, "The Eighteenth Century in Southeast Asia," in: Blussé and Gaastra, *Eighteenth Century*, pp. 39–55, here 47.

36. On the taxonomy of constitutional forms, see Kirsch, *Monarch*, pp. 412f. (chart); cf. P. Brandt et al., *Handbuch*, pp. 41–51.

37. E. N. Anderson and Anderson, *Political Institutions*, pp. 35—a classic.

38. C. M. Clark, *Kaiser Wilhelm II*, pp. 259f.

39. Kohlrausch, *Monarch im Skandal*, pp. 45ff. On Wilhelm's interest in technology in general, see W. König, *Wilhelm II.*, esp. pp. 195–33 on the "traveling emperor."

40. Daniel, *Hoftheater*, p. 369.

41. Rathenau, *Der Kaiser*, p. 34.

42. Another case in point would be Emperor Pedro II of Brazil (1825–91, r. 1840–89); for his biography, see Barman, *Citizen Emperor*.

43. Bagehot, *English Constitution*, pp. 61, 82ff.

44. Outstanding among the plethora of recent literature are D. Thompson, *Queen Victoria* and Homans, *Royal Representations*.

45. D. Thompson, *Queen Victoria*, pp. 144f.

46. Keene, *Emperor of Japan*, pp. 632–35.

47. Amanat, *Pivot*, p. 431.

48. According to the biographer of his nephew and successor: Georgeon, *Abdulhamid II*, p. 33.

49. Fujitani, *Splendid Monarchy*, p. 49.

50. Ibid., p. 229: one of the best books of any kind on monarchy in the nineteenth century.

51. Deringil, *Well-protected Domains*, p. 18.

52. This was the view of the young shah of Persia: see Amanat, *Pivot*, p. 352.

53. Paulmann, *Pomp und Politik*, p. 325; there is an impressive account of the visit on pp. 301–31.

54. On the debate see Kirsch, *Monarch*, pp. 210ff. There is an original discussion in Rosanvallon, *La démocratie inachevée*, pp. 199ff., which sees Louis Napoléon as a theorist of "Caesarism."

55. R. Price, *French Second Empire*, p. 95; on the permanent *fête imperiale* see Baguley, *Napoleon III*.

56. R. Price, *French Second Empire*, p. 211.

57. Beller, *Franz Joseph*, p. 52.

58. Price, *People*, pp. 67–120, on the main currents in the opposition.

59. Here I am following Rosanvallon, *La démocratie inachevée*, pp. 199ff., esp. 237f.

60. Toledano, *State and Society*, pp. 50f.

61. Bernier, *The World in 1800*, pp. 76, 78.

62. Ibid., p. 150.

63. Fujitani, *Splendid Monarchy*, pp. 182–85.

64. There is a good and historically well founded discussion of the methodological problems in C. Tilly, *Democracy*, pp. 59–66.

65. Caramani, *Elections in Western Europe*, p. 53, Tab. 2.3.; Fenske, *Verfassungsstaat*, p. 516.

66. Somewhat shortened from Raphael, *Recht und Ordnung*, p. 28.

67. Rosanvallon, *L'État en France*, p. 99.

68. Fehrenbacher, *Slavery*.

69. National conceptions of political participation varied greatly, however. See the magisterial work Fahrmeir, *Citizenship,* esp. chs. 2–4.

70. Ikegami, *Citizenship*—with an emphasis on the role of opposition and protest movements.

71. Thorough recent research has dated the rise of a "Habermasian" public sphere to an earlier period—to the 1640s in the case of England (see McKeon, *Secret History* [2005], pp. 56 and passim).

72. Habermas, *Structural Transformation*, pp. 159 ff.

73. H. Barker and Borrows, *Press*; Uribe-Uran, *Birth of a Public Sphere*.

74. A key work here is A. Milner, *Invention of Politics*.

75. Apart from numerous studies of particular cities, see (for the United States) M. P. Ryan, *Civic Wars* and (for China) Ranking, *Elite Activism*.

76. P. G. Price, *Acting in Public*, pp. 92f.

77. Ibid., p. 113. On the diversity of voices in (Southern) India, which it was impossible for the colonial regime to control, see Irschick's exemplary *Dialogue and History*.

78. See the chapter "Political Associations in the United States" (I.2.iv) in: Tocqueville, *Democracy*, pp. 219–27.

79. Finer, *History of Government*, vol. 3, pp. 1567ff.; the main constitutional texts have been consulted in Gosewinkel and Masing, *Verfassungen*.

80. See the list in Navarro García, *Historia de las Américas*, vol. 4, pp. 164–73.

81. On the revolutions around the turn of the century, see chapter 10, above.

82. See Fenske, *Verfassungsstaat*. There is a succinct account in W. Reinhard, *Staatsgewalt*, pp. 410–26; Kirsch and Schiera, *Verfassungswandel* gives a general European overview at the highpoint of the interest in constitutional history.

83. Fenske, *Verfassungsstaat*, pp. 525f. Curiously, though, Fenske does not include Britain among the demopcracies, on the grounds that its political class was still marked to an unusual degree by the aristocracy.

84. J. Fisch, *Geschichte Südafrikas*, pp. 203f.

85. Hoppen, *Mid-Victorian Generation*, p. 253.

86. Caramani, *Elections in Western Europe*, p. 60.

87. Ibid., p. 65.

88. Ibid., p. 952; Searle, *A New England?*, p. 133.

89. Rosanvallon, *La démocracie inachevée*, pp. 299–302.

90. A key reference here is Rosanvallon, *Le sacre du citoyen*.

91. The story of this is related in Keyssar, *Right to Vote*, pp. 105ff.

92. On the latter see Mark Elvin, "The Gentry Democracy in Chinese Shanghai, 1905–1914," in idem, *Another History*, pp. 140–65.

93. The standard account is Wilentz, *Rise of American Democracy*, chs. 9–14; see also D. W. Howe, *What Hath God Wrought*.

94. Fehrenbacher, *Slaveholding Republic*, pp. 24f., 76f., 236f.

95. Wahrman, *Imagining the Middle Class*, chs. 9–11.

96. Dates from Bock, *Women*, p. 143.

97. On the history of female suffrage, see ibid., ch. 4.

98. T. C. Smith, *Agrarian Origins*, p. 197.

99. M. W. Steele, "From Custom to Right: The Politicization of the Village in Early Meiji Japan," in: Kornicki, *Meiji Japan*, vol. 2, pp. 11–27, here 24f.

100. Mason, *Japan's First General Election*, p. 197.

101. See, e.g., Welskopp, *Banner der Brüderlichkeit*.

102. On early socialism and anarchism, see Stedman Jones and Claeys (eds.), *Cambridge History of Nineteenth-Century Political Thought*, chs. 14, 16.

103. Still worth discussing as a diagnosis of the times is Sombart, *Why Is There No Socialism in the United States?* (first published in German in 1906).

104. See also the four "figures" of the state in Rosanvallon, *L'État en France*, p. 14.

105. Rodgers, *Contested Truths*, pp. 146, 169.

106. Of the various interpretations of this path, see, for example, P. Anderson, *Lineages*, which also includes the Ottoman Empire, and Ertman, *Birth of the Leviathan*.

107. W. C. Jones, *Great Qing Code*.

108. H. G. Brown, *War*, p. 9 and passim.

109. An excellent survey is still G. E. Aylmer, "Bureaucracy," in Burke, *Companion Volume*, pp. 164–200.

110. Krauss, *Herrschaftspraxis*, p. 240 and passim.

111. Berend, *History Derailed*, pp. 188f., 259. Not by chance was the Habsburg Monarchy known, at least until 1859, as the "China of Europe": Langewiesche, *Liberalism in Germany*, p. 63.

112. European "rationalization" paths are well characterized in Breuer, *Der Staat*, pp. 175–89.

113. The legitimacy of taxation is an important but frequently overlooked element in the efficiency of the state. See Daunton, *Trusting Leviathan*.

114. China: Watt, *District Magistrate*; India: Gilmour, *Ruling Caste*, pp. 89–104.

115. Gilmour, *Ruling Caste*, p. 43.

116. Misra, *Bureaucracy in India*, pp. 299–308.

117. Guy, *Qing Governors*; R. J. Smith, *China's Cultural Heritage*, pp. 55–67; and Hucker, *Dictionary*, pp. 83–96. On the Chinese state's limited scope for action in the early nineteenth century (with particular reference to the opium bans), see Bello, *Opium*.

118. For a critique of many clichés concerning Chinese corruption, see Reed, *Talons and Teeth*, pp. 18–25.

119. Elman, *Civil Examinations*, pp. 569ff.

120. Osterhammel, *China*, pp. 163f.; account is taken of more recent literature in Eberhard-Bréard, *Robert Hart*.

121. Hwang, *Beyond Birth*, p. 334.

122. Woodside, *Lost Modernities*, p. 3: a highly stimulating interpretation.

123. Findley, *Ottoman Civil Officialdom*, p. 292 and passim.

124. Findley, *Turks*, p. 161.

125. Silberman, *Cages of Reason*, p. 180.

126. Constitution of the Japanese Empire, Preamble and Paragraph 1 (Clause 3).

127. Wakabayashi, *Anti-Foreignism*; see the translation there of Aizawa's Nine Theses (pp. 147–277).

128. Wolfgang Schwentker, "Staatliche Ordnungen und Staatstheorien im neuzeitlichen Japan," in W. Reinhard, *Verstaatlichung der Welt?*, pp. 113–31, at 126f.

129. A key article here is Lutz Raphael, "L'État dans les villages: Administration et politique dans les sociétés rurales allemandes, françaises et italiennes de l'époque napoléonienne à la Seconde Guerre Mondiale," in Mayaud and Raphael, *Histoire de l'Europe rurale contemporaine*, pp. 249–81.

130. Baxter, *Meiji Unification* is a fine study of Japan in the 1870s. The pressures stemming from grassroots protest and a precarious external policy differentiate the Japanese from the German case. See Yoda, *Foundations of Japan's Modernization*, pp. 72f.

131. Baxter, *Meiji Unification*, pp. 53–92.

132. See Breuer, *Der Staat*. Cf. M. Mann, *Sources of Social Power*, vol. 2, pp. 444–75.

133. Bensel, *Yankee Leviathan*, p. 367.

134. M. Mann, *Sources of Social Power*, vol. 2, p. 472.

135. Wunder, *Bürokratie*, pp. 72f.

136. Ullmann, *Steuerstaat*, pp. 56ff.

137. M. Mann, *Sources of Social Power*, vol. 2, p. 366 (Tab. 11.3).

138. Raphael, *Recht und Ordnung*, p. 123.

139. Daunton, *Progress*, p. 519.

140. Ali, *Punjab*, pp.109ff.; Heathcote, *Military in British India*, pp. 126f. On the formation of the "garrison state" in India, see Peers, *Mars*.

141. See the studies in Frevert, *Militär und Gesellschaft*; and Foerster, *Wehrpflicht*.

142. Frevert, *A Nation in Barracks*, pp. 149ff.

143. Dietrich Beyrau, "Das Russische Imperium und seine Armee," in: Frevert, *Militär und Gesellschaft*, pp. 119–42, at 130–33.

144. Fahmy, *All the Pasha's Men*, esp. pp. 76ff.

145. Eric J. Zürcher, "The Ottoman Conscription System in Theory and Practice," in: idem, *Arming the State*, pp. 79–94, esp. 86, 91.

146. McClain, *Japan*, p. 161.

147. R. J. Evans, *Rituals of Retribution*, pp. 305–21. In France, though, there were occasional public executions until 1939.

148. Schrader, *Languages of the Lash*, pp. 49, 144ff.

149. See the overview: David Bayley, "The Police and Political Development in Europe," in: C. Tilly, *Formation of National States*, pp. 328–79, esp. 340–60; and, for a comparative study in terms of political sociology, Knöbl, *Polizei*.

150. Emsley, *Gendarmes and the State*.

151. Westney, *Imitation*, pp. 40–44, 72f.

152. Ibid., pp. 94f.

153. For the example of South India, see Arnold, *Police Power*, pp. 99, 147.

154. Clive, *Macaulay*, pp. 435–66.

155. Townshend, *Making the Peace*, pp. 23–29.

156. J. A. Hobson, *Imperialism*, p. 124.

157. Monkkonen, *Police*, pp. 42, 46.

158. See the sketch in Eric H. Monkkonen, "Police Forces," in: Foner and Garraty, *Reader's Companion*, pp. 847–50.

159. This is the main theme in Petrow, *Policing Morals*.

160. Kraus's writings on the subject, which appeared between 1902 and 1907 in *Die Fackel*, are now collected in *Schriften*, vol. 1, Frankfurt a.M., 1987.

161. A. J. Major, *State and Criminal Tribes*, pp. 657f., 663; see also T. R. Metcalf, *Ideologies of the Raj*, pp. 122–25, and chs. 3–4 on ethnic classification in general.

162. See, e.g., M. E. Curtin, *Black Prisoners*, pp. 1ff.

163. Karl-Friedrich Lenz, "Penal Law," in: W. Röhl, *History of Law in Japan*, pp. 607–26, at 609ff.

164. Umemori Naoyuki, "Spatial Configuration and Subject Formation: The Establishment of the Modern Penitentiary System in Meiji Japan," in: Hardacre and Kern, *New Directions*, pp. 734–67, esp. 744–46, 754, 759f.

165. Dikötter, *Crime*, pp. 56–58; the plans were implemented on a large scale, however, only under the Republic.

166. Lindert, *Growing Public*, pp. 46f.

167. Rosanvallon, *L'État en France*, p. 175; Raphael, *Recht und Ordnung*, p. 102; Lindert, *Poor Relief*.

168. The standard work here is Lindert, *Growing Public*, pp. 171ff.; see also W. Reinhard, *Staatsgewalt*, pp. 460–67.

169. Eichenhofer, *Geschichte des Sozialstaats*, p. 54.

170. Comparative data in M. G. Schmidt, *Sozialpolitik*, p. 180 (Tab. 5).

171. See Rodgers, *Atlantic Crossings*, esp. pp. 209ff. (on social insurance).

172. Esping-Andersen, *Three Worlds*.

173. On this whole section, see P. D. Curtin, *World*, pp. 128–91.

174. See Faroqhi, *Subjects of the Sultan*, p. 19.

175. There are summaries in all histories of the Ottoman Empire, for instance Findley, *Turkey*, pp. 88–106; Hanioglu, *Brief History*, pp. 72–108. Reforms on a more modest scale took place in Iran, influenced by the Ottoman example: see Bakhash, *Iran*.

176. See Anastassiadou, *Salonique*; Hanssen, *Beirut*.

177. Rich, *Age of Nationalism*, pp. 145 ff. draws interesting parallels between the more or less "liberal" reforms occurring at the same time in Britain and Russia.

178. W. B. Lincoln, *Great Reforms*; Eklof et al., *Russia's Great Reforms*; Beyrau et al. *Reformen*.

179. Even Korea, the last East Asian country by far to remain sealed off from the West, embarked on a policy of self-strengthening reform. See Palais, *Politics and Policy*.

180. W. Reinhard, *Verstaatlichung*.

181. See the study by Roussillon, *Identité et modernité*.

182. Paul Wanderwood, "Betterment for Whom? The Reform Period, 1855–1875," in: M. C. Meyer and Beezley, *Oxford History of Mexico*, pp. 371–96.

183. Polunov, *Russia*, pp. 123f., 174–89.

184. Maurus Reinkowski, "The State's Security and the Subjects' Prosperity: Notions of Order in Ottoman Bureaucratic Correspondence," in: Karateke and Reinkowski, *Legitimizing the Order*, pp. 195–212, at 206; Reinkowski, *Dinge der Ordnung*, pp. 284, 287.

185. Perkins, *Modern Tunisia*, pp. 14f.

186. Isabella, *Risorgimento in Exile*.

187. See Torp, *Herausforderung*.

188. Trocki, *Opium and Empire*.

CHAPTER XII: Energy and Industry

1. W. Hardy, *Idea of the Industrial Revolution*, p. 3.

2. Wischermann and Nieberding, *Die institutionelle Revolution*, pp. 17–29.

3. See Pollard, *Peaceful Conquest*.

4. See Riello and O'Brien, *Future*.

5. A model for this kind of analysis is Mokyr, *Enlightened Economy*.

6. D. C. North, *Institutions*, pp. 36f.

7. See chapter 5, above.

8. Pomeranz, *Great Divergence*; P.H.H. Vries, *Via Peking*. The question was first posed in the 1950s by Chinese historians.

9. See the illuminating thesis in Amsden, *Rise of "the Rest"*, pp. 51ff.

10. A famous undogmatic interpretation along these lines is Hobsbawm, *Industry and Empire*, pp. 34–78.

11. Schumpeter, *Business Cycles*. Kondratiev was shot in September 1938 in a Moscow prison. His collected works were first published in full in 1998, in the West.

12. Polanyi, *Great Transformation*. Influential developments of this approach may be found in the anthropology of peasant communities, and in the theory of a "moral economy" in E. P. Thompson and James C. Scott.

13. The final version of the theory is: Rostow, *World Economy*.

14. Gerschenkron, *Economic Backwardness*; see the good discussion in Verley, *La Révolution industrielle*, pp. 111–14, 324, 26.

15. Bairoch, *Révolution industrielle*; for a later version: Bairoch, *Victoires*, vol. 1.

16. Landes, *Unbound Prometheus*—a masterpiece of historical synthesis and still a basic work on the subject.

17. D. C. North and Thomas, *Rise*.

18. Schumpeter explicitly rejects it, Max Weber mentions it only in passing when he takes issue with technological determinism: see Swedberg, *Max Weber*, pp. 149f.

19. Landes, *Wealth and Poverty*.

20. Sylla and Toniolo, *Patterns*, tests Gerschenkron's theory country by country, but has little to say about Japan.

21. Patrick K. O'Brien: "Introduction," in idem, *Industrialisation*, vol. 1, p. xliii.

22. An exception is Stearns, *Industrial Revolution*.

23. According to Easterlin, *Growth Triumphant*, p. 31; E. L. Jones, *Growth Recurring*, p. 13, speaks of "intensive" growth.

24. R. C. Allen, *British Industrial Revolution*, p. 80.

25. Still a good example of this kind of plurifactorial analysis is Mathias, *First Industrial Nation*.

26. Verley, *La Révolution industrielle*, pp. 34–36.

27. Allen, *British Industrial Revolution*, p. 15. Findlay and O'Rourke come to similar conclusions in *Power and Plenty*, pp. 330–52, esp. 339–42. A thorough study is Inikori, *Africans*.

28. Mokyr, *Gifts of Athena*, in the tracks of the pathbreaking Jacob, *Scientific Culture*; for a later period, see Smil, *Creating the Twentieth Century*; see also Inkster, *Science*, which ranges into world history and is especially interesting on the theme of transfers.

29. J. de Vries, *Industrial Revolution*, esp. pp. 255f.; in much greater detail idem, *Industrious Revolution*; the concept was taken up by Bayly, *Birth of the Modern World*, pp. 51–59. Earlier, the Japanese economic historians Akira Hayami und Osamu Saito had made similar points regarding Japan: see the overview in Hayami et al., *Economic History of Japan*, esp. chs. 1, 9–11; and Austin and Sugihara, *Labour-Intensive Industrialization*.

30. Ogilvie and Cerman, *Proto-Industrialization*; Mager, "Protoindustrialisierung".

31. This was the case in the Tsarist Empire, on which see Gestwa, "Proto-Industrialisierung," pp. 345ff.

32. This is the prudent judgment in Daunton, *Progress*, p. 169.

33. Especially rich in insights is M. Berg, *Age of Manufactures*.

34. Komlos, *Industrial Revolution*.

35. Findlay and O'Rourke, *Power and Plenty*, p. 313: the formulation of a new consensus in economic history.

36. Martin Daunton, "Society and Economic Life," in C. Matthew, *Nineteenth Century*, pp. 41–82, at 51–55.

37. Verley, *La Révolution industrielle*, p. 107.

38. A key work here is Jeremy, *Transatlantic Industrial Revolution*.

39. Cameron, *New View*.

40. Craig and Fisher, *European Macroeconomy*, pp. 257ff., 280, 309; Pollard, *Peaceful Conquest*; Teich and Porter, *Industrial Revolution*.

41. A new macrohistorical theory even sees this as the ultimate cause of economic growth: "Growth is generated overwhelmingly by investments in expanding the stock of production knowledge in societies." G. Clark, *Farewell to Alms*, pp. 197, 204–7.

42. Sabel and Zeitlin, *World of Possibilities*.

43. See Ledderose, *Ten Thousand Things*, esp. pp. 2–4; a key work on mass production in the West is Hounshell, *From the American System*.

44. For Germany see Herrigel, *Industrial Constructions*.

45. There is a good textbook account of this in Matis, *Industriesystem*, pp. 248–65. For a long time the most influential analyst was Alfred D. Chandler: see his *Visible Hand* and *Scale and Scope*. An exemplary study of a national transformation process is M. S. Smith, *Emergence*, pp. 325ff. More recently, Peter Temin and others have proposed an alternative paradigm.

46. I follow Werner Abelshauser, "Von der Industriellen Revolution zur Neuen Wirtschaft. Der Paradigmenwechsel im wirtschaftlichen Weltbild der Gegenwart," in Osterhammel et al., *Wege*, pp. 201–18.

47. Important here is G. Jones, *Multinationals*, a book with plenty of material on the nineteenth century, dispersed over topical chapters; see also Geoffrey Jones, "Globalization," in: G. Jones and Zeitlin, *Oxford Handbook of Business History*, pp. 141–68, esp. 143–47.

48. Blackford, *Rise of Modern Business*, pp. 103ff.; Boyce and Ville, *Modern Business*, pp. 9f.

49. Outstanding collections are O'Brien, *Industrialisation*; Church and Wrigley, *Industrial Revolutions*; J. Horn et al., *Reconceptualizing the Industrial Revolution*; Austin and Sugihara, *Labour-Intensive Industrialization*.

50. Wallerstein, *Modern World-System*, vol. 3, p. 33.

51. Recently in this manner: Ferguson, *Civilization*.

52. The fundamental work here is Pomeranz, *Great Divergence*. Less spectacular in its theses, but empirically groundbreaking, is Blussé and Gaastra, *Eighteenth Century*.

53. E. L. Jones, *European Miracle*, p. 160.

54. M. Weber, *Wissenschaftslehre*, p. 407; translation: "'Energetic' Theories of Culture," in: *Mid-American Review of Sociology* 9:2, pp. 33–58, at 36.

55. C. Smith, *Science of Energy*, esp. pp. 126–69.

56. There is a monumental portrait of Lord Kelvin and his times in C. Smith and Wise, *Energy and Empire*. Here too—as in the case of Siemens in Germany—it is interesting to note the huge significance of the telegraph as a scientific challenge (pp. 445 ff.).

57. Feldenkirchen, *Siemens*, pp. 55ff.

58. W. König and Weber, *Netzwerke*, pp. 329–40; Smil, *Creating the Twentieth Century*, ch. 2.

59. The central importance of this machine for the history of technology in the nineteenth century is unmistakable in Wagenbreth et al., *Dampfmaschine*.

60. Mirowski, *More Heat than Light*.

61. Rabinbach, *Human Motor*.

62. See Malanima, *Economia preindustriale*, p 98. A briefer and updated version of this fundamental book is Malanima, *Pre-Modern European Economy*, here ch. 2.

63. Malanima, *Uomini*, p. 49; Wrigley, *Energy*, pp. 91–101. See also two world histories of mining: C. E. Gregory, *Mining* and M. Lynch, *Mining*, both of which deal only with coal. On the energy problem and industrialization in general, see Sieferle et al., *Ende der Fläche*, esp. chs. 4–5.

64. Paolo Malanima, "The Energy Basis for Early Modern Growth, 1650–1820," in Prak, *Early Modern Capitalism*, pp. 51–68, at 67. A basic text for the history of technology is Hunter, *Industrial Power*.

65. Malanima, *Uomini*, p. 45 estimates that, in the early modern period, per capita use in Europe was 2 kilograms—a minimal quantity, perhaps obtained within a southern Italian perspective.

66. Grübler, *Technology*, p. 250. On the history of oil before 1914, see Yergin, *The Prize*, chs. 1–8.

67. Roche, *Le Cheval moteur*, p. 38.

68. Overton, *Agricultural Revolution*, p. 126; Grübler, *Technology*, p. 149 (Fig. 5.8).

69. Wrigley, *People*, p. 10; idem, *Energy*, with plenty of evidence from England.

70. M. Lynch, *Mining*, pp. 73f. On the technological and international spread of Newcomen's machines, see Wagenbreth et al., *Dampfmaschine*, pp. 18–23.

71. Marsden, *Watt's Perfect Engine*, pp. 118f.

72. Grübler, *Technology*, p. 209 (Fig. 6.3).

73. Wagenbreth et al., *Dampfmaschine*, p. 240.

74. Minami, *Power Revolution*, pp. 53f., 58, 331–33.

75. Percentages calculated from Pohl, *Aufbruch*, p. 127 (Tab. VI. 4).

76. See general histories of power such as Debeir et al., *In the Servitude of Power*; Smil, *Energy*.

77. Trebilcock, *Industrialization*, p. 237.

78. W. W. Lockwood, *Economic Development of Japan*, p. 91.

79. The term appears in Sugihara, *Japanese Imperialism*, p. 13.

80. Pomeranz, *Great Divergence*, p. 62.

81. R. Reinhard, *Erdkunde*, p. 119.

82. Pohl, *Aufbruch*, p. 127, (Tab. VI.4).

83. Smil, *Energy*, p. 228.

84. Alleaume, *Industrial Revolution*, p. 341.

85. Verley, *La Révolution industrielle*, pp. 492f.

86. Wolfram Fischer, "Wirtschaft und Gesellschaft Europas, 1850–1914," in Fischer, *Handbuch*, vol. 5, p. 149 (Tab. 42).

87. Bulmer-Thomas, *Economic History*, pp. 58f. For a survey of the historiography, see Haber, *How Latin America Fell Behind*; a country-by-country analysis of the Latin American export experience since 1880 is provided in Cárdenas, *Economic History*.

88. Klarén, *Peru*, pp. 180f.

89. Bulmer-Thomas, *Economic History*, p. 61.

90. Feinstein, *Economic History of South Africa*, pp. 90–99.

91. Bulmer-Thomas, *Economic History*, pp. 130–39.

92. There is a more detailed account in Osterhammel, *China*, pp. 188–94; on the comparison between Japan and China, see Yoda, *Foundations of Japan's Modernization*, pp. 119–25.

93. Köll, *From Cotton Mill*; Cochran, *Encountering Chinese Networks*; Bergère, *Capitalisme et capitalistes*, pp. 86ff.

94. Osterhammel, *China*, pp. 263f.; Bergère, *Capitalisme et capitalistes*, pp. 96–104.

95. Plenty of evidence in Riello and Roy, *How India Clothed the World*.

96. Inikori, *Africans*, p. 428; Parthasarathi, *Why Europe Grew Rich*, pp. 89–114.

97. Prasannan Parthasarathi and Ian Wendt, "Decline in Three Keys: Indian Cotton Manufacturing from the Late Eighteenth Century," in: Riello and Parthasarathi, *Spinning World*, pp. 397–407, at 407.

98. Dietmar Rothermund, "The Industrialization of India: Technology and Production," in B. B. Chaudhuri, *Economic History of India*, pp. 437–523, at 441f.; Roy, *Economic History*, pp. 123–31.

99. Farnie and Jeremy, *Fibre*, p. 401, see 400–413 on the early history of the Indian cotton industry.

100. Ibid., p. 418.

101. Roy, *Economic History*, pp. 131–33.

102. Arcadius Kahan, "Rußland und Kongreßpolen 1860–1914," in W. Fischer, *Handbuch*, vol. 5, pp. 512–600, at 538 (Tab. 11).

103. Stimulating on this is Chandavarkar, *Imperial Power*, pp. 30–73.

104. Good introductions are McClain, *Japan*, pp. 207–45; and Janet E. Hunter, "The Japanese Experience of Economic Development," in O'Brien, *Industrialisation*, vol. 4, pp. 71–141. The specialist debates are documented in Church and Wrigley, *Industrial Revolutions*, vol. 7.

105. Tamaki, *Japanese Banking*, pp. 51ff.

106. Mosk, *Japanese Industrial History*, p. 97. Fundamental on the rise of the Japanese cotton industry in its international context is Howe, *Origins*, pp. 176–200.

107. Morris-Suzuki, *Technological Transformation*, p. 73.

108. See Pierre-Yves Donzé, "The International Patent System and the Global Flow of Technologies: The Case of Japan, 1880–1930," in: Dejung and Petersson, *Foundations of Worldwide Economic Integration*, pp. 179–201.

109. On both cases see the discussions of recent research in Horn et al., *Reconceptualizing the Industrial Revolution*, chs. 7 and 9; see also Schön, *Modern Sweden*, pp. 117–26.

110. See the summary in Sean Wilentz, "Society, Politics, and the Market Revolution, 1815–1848," in Foner, *New American History*, pp. 61–84; and chs. 9–10 of Barney, *Companion*.

111. This is the central thesis in Bensel, *Political Economy*.

112. Takebayashi, *Kapitalismustheorie*, pp. 155ff. For a more general overview see Muller, *The Mind and the Market*, ch. 9 and passim.

113. See Mommsen, *Theories of Imperialism*; Semmel, *Liberal Ideal*.

114. Grassby, *Idea of Capitalism*, p. 1.

115. Appleby, *Relentless Revolution*, is surprisingly reticent on global aspects; but see Frieden, *Global Capitalism*; for a theoretical perspective, Leslie Sklair, "Capitalism: Global," in Smelser and Baltes, *International Encyclopedia*, vol. 3, pp. 1459–63.

116. The reader will find an enormous literature on "varieties of capitalism," little of which has a satisfactory historical background. But see Kocka, *Writing the History of Capitalism*; on the historiography of the US variant, see Sven Beckert, "History of American Capitalism," in: Foner and McGirr, *American History Now*, pp. 314–35.

117. Two books published at the beginning of the recent debate on capitalism, and sharply opposed in their judgements, are still stimulating: P. L. Berger, *Capitalist Revolution,* and Heilbroner, *Nature and Logic*. The *longue durée* is (in critical opposition to Braudel) carefully formulated in Arrighi, *Long Twentieth Century*. An excellent starting point for further theoretical discussion is Richard Swedberg, "The Economic Sociology of Capitalism: An Introduction and Agenda," in: Nee and Swedberg, *Economic Sociology*, pp. 3–40.

118. A modification of P. L. Berger, *Capitalist Revolution*, p. 19. For a motivational rather than a systemic definition of "capitalism," see Appleby, *Relentless Revolution*, pp. 25f. In this work, the story of capitalism is that of profit seeking within different cultural contexts.

119. A fundamental Marxist work here is Byres, *Capitalism from Above*.

120. A paradigmatic case is the creation of Lever Brothers/Unilever and the development of its overseas activities since 1895.

121. See the national profiles in A. D. Chandler et al., *Big Business*. There is an interesting contrary view in Arrighi, *Long Twentieth Century*, pp. 33f., which sees a sharp opposition between "capitalism," and "territorialism."

CHAPTER XIII: Labor

1. Many ideas in this chapter draw on Kocka and Offe, *Arbeit*, esp. pp. 121ff.

2. See chapter 5, above.

3. Siddiqi, *Ayesha's World*; Rosselli, *Singers*, chs. 3–4; Richardson, *Chinese Mine Labour*; Druett, *Rough Medicine*. Such studies, reconstructing particular worlds of work on the basis of firsthand documents, are of enormous value.

4. Chris Tilly and Tilly, *Work*, p. 29.

5. For a list of research reports and literature on this chapter, see Lucassen, *Global Labour History*.

6. Kaelble, *Erwerbsarbeit*, pp. 22–25.

7. This is shown in Biernacki, *Fabrication of Labor*.

8. See Lynn, *Commerce*, a study of the palm oil trade, esp. pp. 34–59.

9. For Africa see Atkins, *The Moon Is Dead*, p. 128.

10. There are scarcely attempts at a global history of agriculture, but on the Atlantic there is Richard Herr, "The Nature of Rural History," in idem, *Themes*, pp. 3–44. See the masterly panorama of rural Europe (with various side glances) in Hobsbawm, *Age of Capital*, ch. 10.

11. Kaelble, *Erwerbsstruktur*, pp. 8, 10.

12. See chapter 7, above.

13. Elson, *End of the Peasantry*, pp. 23f.

14. For a brief overview of the research, see M. Kearney, "Peasants and Rural Societies in History," in Smelser and Baltes, *International Encyclopedia*, vol. 16, pp. 11163–71. There is also a good survey of theories of "peasant society" in Wimmer, *Die komplexe Gesellschaft*. Much of the theoretical elaboration is based upon Russian and Southeast Asian examples.

15. There is a good summary in Little, *Understanding Peasant China*, pp. 29–67.

16. Hanley and Yamamura, *Preindustrial Japan*, p. 332.

17. Blum, *Internal Structure*, p. 542.

18. Huang, *Peasant Economy*, pp. 225–28.

19. On the village commune in Europe, see (in addition to Blum) Rösener, *Peasantry*; and on Russia, Ascher, *Stolypin*, pp. 153–64. There are still few comparative works on Asia. Fukutake, *Asian Rural Society*, is an ethnological study (Japan, China, India) with little historical depth of focus. See also Gilbert Rozman, "Social Change," in J. W. Hall, *Cambridge History of Japan*, vol. 5, pp. 499–568, at 526f. It goes without saying that there is no such thing as "the" European or Japanese village.

20. Fukutake, *Asian Rural Society*, p. 4.

21. In Japan alone, a study in 1885 discovered more than twenty different forms of leasehold. See Waswo, *Japanese Landlords*, p. 23.

22. Palairet, *Rural Serbia*, pp. 41–43, 69ff., 78, 85–90.

23. The main source here is Robb, *Peasants' Choices?*; a synthesis of recent research may be found in Jacques Pouchepadass's chapters in Markovits et al., *Modern India*, pp. 294–315, 410–31. See also Ludden, *Agrarian History*.

24. See, e.g., Grigg, *Agricultural Systems*.

25. Stinchcombe, *Stratification*, pp. 33–51.

26. See chapter 5, above.

27. For India see Prakash, *World of the Rural Labourer*.

28. Peebles, *Sri Lanka*, p. 58.

29. For a detailed account of working conditions, see Breman, *Taming the Coolie Beast*, pp. 131f.

30. There is a brief description of the type in Grigg, *Agricultural Systems*, pp. 213–15.

31. Stoler, *Capitalism*, p. 20.

32. Ibid., pp. 25–36.

33. Alleaume, *Industrial Revolution*, pp. 331, 335, 338, 342f.; R. Owen, *Middle East*, pp. 66–68.

34. For Mexico and Peru, see Mallon, *Peasant and Nation*.

35. Nickel, *Soziale Morphologie*, pp. 73–83.

36. Ibid., pp. 110–16. See also the overview of the hacienda in Wasserman, *Everyday Life*, pp. 23–29, 70–72, 150–54.

37. Adelman, *Frontier Development*, p. 130.

38. There is a fine passage on African housing in Zeleza, *Economic History of Africa*, pp. 213–16.

39. Kriger, *Pride of Men*, p. 119.

40. Friel, *Maritime History*, p. 228.

41. At least this was so in Hamburg until the end of the century; the guild element was weaker in England and Scotland. See Cattaruzza, *Arbeiter*, pp. 118f.

42. Peter Boomgaard, "The Non-Agricultural Side of an Agricultural Economy: Java 1500–1900," in Alexander et al., *Shadow*, pp. 14–40, at 30.

43. Labor historians think mainly in terms of urban factory work; see the balance sheet in Heerma van Voss and Linden, *Class.*

44. Bradley, *Muzhik and Muscovite,* p. 16.

45. See R. E. Johnson, *Peasant and Proletarian,* p. 26.

46. Turrell, *Capital and Labour,* pp. 146–73.

47. On the less-well-known Chinese case, see Shao Qin, *Culturing Modernity.*

48. Friedgut, *Iuzovka and Revolution,* vol. 1, pp. 193ff.

49. See, e.g., Beinin and Lockman, *Workers on the Nile,* p. 25; Tsurumi, *Factory Girls,* pp. 59–67.

50. The important theme of child labor, for which there is a lack of studies outside Europe, will have to be passed over here. Ten European countries feature in Rahikainen, *Centuries of Child Labour.* The general conclusion is probably that children always worked everywhere until the 1880s, when a number of European countries—above all, Britain and Germany— introduced protective legislation, though only for industrial work (ibid., pp. 150–57). See also Cunningham, *Children and Childhood;* for a rich and precise model study on Britain see Humphries, *Childhood and Child Labour.*

51. Johnston, *Modern Epidemic,* pp. 74–80; Tsurumi, *Factory Girls,* esp. pp. 59 ff. For Germany see, e.g., Kocka, *Arbeitsverhältnisse,* pp. 448–61.

52. For a full-scale discussion, see G. A. Ritter and Tenfelde, *Arbeiter,* pp. 265ff.

53. See, e.g., the study of early industry in New England: Prude, *Industrial Order,* pp. 76ff.

54. Important on work conditions in the iron and steel industry in Germany and beyond is Kocka, *Arbeitsverhältnisse,* pp. 413–36.

55. P. M. Kennedy, *Great Powers,* p. 200 (Tab. 15); B. R. Mitchell, *Europe,* pp. 456f.

56. Way, *Common Labor,* p. 8.

57. The realities of canal work are described in ibid. pp. 133–43.

58. Meinig, *Shaping of America,* vol. 2, pp. 318–21.

59. The following borrows from the excellent study based on the archives of the Suez Canal Company: Montel, *Le chantier.* For the context, see Karabell, *Parting the Desert,* and on the later significance of the canal, Farnie, *Suez Canal;* Huber, *Channelling Mobilities.*

60. McCreery, *Sweat,* pp. 117f.

61. Montel, *Le chantier,* p. 64.

62. Diesbach, *Ferdinand de Lesseps,* p. 194.

63. For a detailed description of the festivities, see ibid., pp. 261–72.

64. On railroad construction workers in Germany, see Kocka, *Arbeitsverhältnisse,* pp. 361–66.

65. Ambrose, *Nothing Like It,* p. 150. A more systematic study is still Licht, *Railroad.*

66. Shelton Stromquist, "Railroad Labor and the Global Economy," in Lucassen, *Global Labour History,* pp. 623–47, esp. 632–35.

67. Marks, *Road to Power,* pp. 183–85.

68. *Meyers Großes Konversations-Lexikon,* 6th ed. (Leipzig 1903), vol. 5, p. 505.

69. Kerr, *Building,* pp. 200, 214 (Tab. 2).

70. Ibid., pp. 88–91, 157f.

71. This was closely associated with dock labor, already discussed in chapter 6, above. A key work is S. Davies et al., *Dock Workers.*

72. Greater detail in an unexpected place: Stinchcombe, *Sugar Island Slavery,* pp. 57–88.

73. This has often been overlooked—though not in a still useful older work: Fohlen and Bédarida, *Histoire générale du travail,* pp. 166–73.

74. Some of the voices are quoted in Mawer, *Ahab's Trade*, pp. xiv, 73–75, 230. See also chapter 7, above.

75. H. V. Bowen, *Business of Empire*, ch. 6.

76. Simonton, *European Women's Work*, p. 235.

77. Osterhammel, *China*, pp. 185–88.

78. Except for a very brief stint as court librarian in Hesse-Homburg.

79. Stites, *Serfdom*, pp. 71–82; Finscher, *Streicherkammermusik*, p. 84.

80. Gunilla-Friederike Budde, "Das Dienstmädchen," in Frevert and Haupt, *DerMensch*, pp. 148–75; and the wide-ranging Simonton, *European Women's Work*, pp. 96–111, 200–206.

81. Rustemeyer, *Dienstboten*, p. 88.

82. MacRaild and Martin, *Labour in British Society*, p. 21 (Tab. 1.1).

83. Dublin, *Transforming Women's Work*, pp. 157–62.

84. L. A. Tilly and Scott, *Women*, p. 69.

85. Boyar and Fleet, *Ottoman Istanbul*, p. 297.

86. See Hardach-Pinke, *Gouvernante*, pp. 206–40, which is concerned mainly with female German teachers abroad, also K. Hughes, *The Victorian Governess*.

87. The classification follows Bush, *Servitude*.

88. This theme will be taken up in a different perspective in chapter 17.

89. Eltis, *Rise of African Slavery*, esp. pp. 137ff.

90. On the controversy surrounding this theme, see the work by two eminent scholars of slavery: Davis, *Inhuman Bondage*, chs. 12–13; Drescher, *Abolition*, esp. ch. 5. There is a lively but somewhat naïve account in Hochschild, *Bury the Chains*.

91. I am grateful to Norbert Finzsch for helping me clarify this point.

92. A pioneering and now classical example of this approach is Genovese, *Roll, Jordan, Roll*.

93. Peter Coclanis, "The Economics of Slavery," in: Paquette and Smith, *Oxford Handbook of Slavery*, pp. 489–512, at 498.

94. Cooper and Terrill, *American South*, vol. 2, pp. 517–19.

95. See also chapter 10, above.

96. For a systematic analysis, see Byres, *Capitalism from Above*, pp. 282–336.

97. The fate of these "rural dispossessed," both black and white, is movingly portrayed in Jones, *The Dispossessed*.

98. Ward, *Poverty*, pp. 31ff.

99. See the excellent comparative study in Scott, *Degrees of Freedom*.

100. We shall pass over the difficult issue of the terminological relationship with the manorial system (*Gutsherrschaft*) that from 1570 on gradually became the dominant form east of the Elbe.

101. Berlin, *Generations of Captivity*, Tab. 1 (appendix); Drescher and Engerman, *World Slavery*, pp. 69f.

102. Kolchin, *Unfree Labour*, p. 52 (Tab. 3).

103. Ibid., p. 54 (Tabs. 5, 6).

104. See Bush, *Servitude*, pp. 19–27; Stanley L. Engerman, "Slavery, Serfdom and Other Forms of Coerced Labour: Similarities and Differences," in Bush, *Serfdom*, pp. 18–41, at 21–26.

105. Kolchin, *Sphinx*, pp. 98f.

106. Kolchin, *Unfree Labour*, pp. 359–75; idem, "After Serfdom: Russian Emancipation in Comparative Perspective," in Engerman, *Terms of Labour*, pp. 87–115.

107. Blickle, *Leibeigenschaft*, p. 119.

108. M. Weber, *General Economic History*, p. 108.

109. Teófilo F. Ruiz, "The Peasantries of Iberia, 1400–1800," in Scott, *Peasantries*, pp. 49–73, at 64.

110. Blum, *End of the Old Order*, p. 373.

111. This is the basic argument in Blickle, *Leibeigenschaft*.

112. See chapter 4, above, and Northrup, *Indentured Labour*.

113. Steinfeld, *Invention of Free Labor*, pp. 4–7, 147f., 155–57.

114. Vormbaum, *Politik und Gesinderecht*, pp. 305, 356–59.

115. Fogel and Engerman, *Time on the Cross*.

116. Steinfeld, *Coercion*, p. 8.

117. Peck, *Reinventing Free Labor*, esp. 84ff.

118. Rosselli, *Singers*, p. 5.

119. Details on individual European countries may be found in van der Linden and Rojahn, *Formation*.

120. Castel, *Metamorphosen*, pp. 189, 254f. There is an English translation of this important work: *Manual Workers to Wage Laborers: Transformation of the Social Question* (New Brunswick, NJ 2003).

121. Hennock, *Origin of the Welfare State*, p. 338.

122. Elson, *End of the Peasantry*, pp. 23f.

CHAPTER XIV: Networks

1. See chapters 5 and 8, above.

2. There is now a sizable theoretical and historical literature on networks. Beyrer and Andritzky, *Das Netz*, is an especially instructive exhibition catalogue focusing on visual images.

3. What this meant for the conception of cities is well brought out in Sennett, *Flesh and Stone*, pp. 256–81.

4. For a European overview, see R. Millward, *Enterprise*.

5. Dehs, *Jules Verne*, pp. 211, 368.

6. Bagwell, *Transport Revolution*, pp. 17, 33.

7. Woud, *Het lege land*, pp. 115–32.

8. L. Ray Gunn, "Antebellum Society and Politics (1825–1865)," in M. M. Klein, *Empire State* pp. 307–415, at 312.

9. Smil, *Two Prime Movers*, p. 381.

10. P. Clark, *Cambridge Urban History of Britain*, vol. 2, p. 718.

11. Bled, *Wien*, p. 199.

12. Rawlinson, *China's Struggle*.

13. C. Howe, *Origins*, p. 268.

14. Broeze, *Underdevelopment*, p. 445.

15. Hugill, *World Trade*, p. 127.

16. R. Reinhard, *Erdkunde*, p. 194.

17. Sartorius von Waltershausen, *Weltwirtschaft*, p. 269. On the extraordinary rise of Hong Kong from "fishing village" to Asia's main transshipment center, see D. R. Meyer, *Hong Kong*, pp. 52ff.

18. Maps in Hugill, *World Trade*, p. 136 (Fig. 3–3); R. Reinhard, *Erdkunde*, p. 201.

19. Rieger, *Technology*, pp. 158–92.

20. Hugill, *World Trade*, pp. 249ff.

21. This is reflected in the national framework typical even of the best survey literature, e.g., Roth, *Jahrhundert der Eisenbahn*; Wolmar, *Fire and Steam*.

22. Youssef Cassis, "Big Business," in: G. Jones and Zeitlin, *Oxford Handbook of Business History*, pp. 171–93, at 175f.

23. Veenendaal, *Railways*, pp. 29, 50.

24. Map in Fage and Oliver, *Cambridge History of Africa*, vol. 7, p. 82.

25. Huenemann, *Dragon*, gives details of the construction progress (pp. 252–57).

26. R. Owen, *Middle East*, p. 246.

27. On national "technology styles," in early American and German railroad construction, see Dunlavy, *Politics*, pp. 202–34.

28. François Caron, "The Birth of a Network Technology: The First French Railway System," in M. Berg and Bruland, *Technological Revolutions*, pp. 275–91.

29. A charming discussion of this rivalry is Grossman, *Charles Dickens's Networks*, ch. 1.

30. Schivelbusch, *Railway Journey*; Freeman, *Railways*; Desportes, *Paysages en mouvement*.

31. See the general reflections in R. White, *Railroaded*, pp. 140–78.

32. F. Caron, *Histoire des chemins de fer en France*, vol. 1, pp. 84, 113, 169.

33. Chang Jui-te, "Technology Transfer in Modern China: The Case of Railway Enterprise in Central China and Manchuria," in: Elleman and Kotkin, *Manchurian Railways*, pp. 105–22, at 111.

34. Ochsenwald, *Hijaz Railway*, pp. 30ff., 152.

35. October 23, 1828: Eckermann, *Conversations*, p. 279.

36. Ronald Findlay and Kevin H. O'Rourke, "Commodity Market Integration, 1500–2000," in Bordo et al., *Globalization*, pp. 13–62, at 36.

37. See the first points made in Florian Cebulla, "Grenzüberschreitender Schienenverkehr. Problemstellungen, Methoden, Forschungsüberblick," in Burri et al., *Internationalität*, pp. 21–35.

38. There are good country studies in C. B. Davis, *Railway Imperialism*.

39. A. Mitchell, *Train Race*.

40. Bulliet, *Camel*, pp. 216ff.

41. Cvetkovski, *Modernisierung*, pp. 79, 167f., 189.

42. Wenzlhuemer, *Connecting the Nineteenth-Century World*, p. 119 (Tab. 5.1)

43. See also chapters 1 and 9, above.

44. Briggs and Burke, *Media*, p. 134. More on railroads and the telegraph in Wenzlhuemer, *Connecting the Nineteenth-Century World*, pp. 31–34.

45. Dematerialization as the crucial feature of telecommunication has been highlighted by Wenzlhuemer, *Connecting the Nineteenth-Century World*, pp. 30, 62.

46. C. S. Fischer, *America Calling*.

47. Hugill, *Global Communications*, pp. 53f.; Hills, *Struggle for Control*, p. 168.

48. Winston, *Media Technology*, p. 53.

49. Ibid., pp. 254f.

50. Horst A. Wessel, "Die Rolle des Telephons in der Kommunikationsrevolution des 19. Jahrhunderts," in: North, *Kommunikationsrevolutionen*, pp. 101–27, at 104f.

51. Strachan, *First World War*, pp. 233f.

52. Wheeler, *Mr. Lincoln's T-Mails*.

53. Wobring, *Globalisierung*, pp. 39ff., 80ff.

54. Kaukiainen, *Shrinking the World*—an article of major importance.

55. On the construction of the network, see Headrick, *Invisible Weapon*, pp. 28–49.

56. Jorma Ahvenainen, "The Role of Telegraphs in the 19th-Century Revolution of Communications," in: North, *Kommunikationsrevolutionen*, pp. 73–80, at 75f.; G. Clark, *Farewell to Alms*, pp. 306f.; Ferguson, *Rothschild*, vol. 1, p. 98.

57. There is good illustrative material in Roderic H. Davison, "Effect of the Electric Telegraph on the Conduct of Ottoman Foreign Relations," in: Farah, *Decision Making*, pp. 53–66.

58. Headrick, *Invisible Weapon*, pp. 38f. (Tabs. 3.2, 3.3); Jürgen Wilke, "The Telegraph and Transatlantic Communications Relations," in: Finzsch and Lehmkuhl, *Atlantic Communications*, pp. 107–34, at 116.

59. Nickels, *Under the Wire*, p. 33.

60. Ibid., pp. 44–46.

61. Headrick, *Invisible Weapon*, pp. 84f.

62. R.W.D. Boyce, *Imperial Dreams*, p. 40.

63. Cornelius Neutsch, "Briefverkehr als Medium internationaler Kommunikation im ausgehenden 19. und beginnenden 20. Jahrhundert," in: M. North, *Kommunikationsrevolutionen*, pp. 129–55, at 131f.

64. Cvetkovski, *Modernisierung*, pp. 135f., 149; Henkin, *Postal Age*, chs. 1–2.

65. Hausman et al., *Global Electrification*, pp. 18f.

66. Hughes, *Networks of Power*, pp. 232 and 175–200—one of the most important books on the technological history of the long "turn of the century."

67. On world trade since c. 1850 see in exhaustive detail: Steven C. Topik and Allen Wells, "Commodity Chains in a Global Economy," in: E. S. Rosenberg, *A World Connecting*, pp. 593–812.

68. A pioneer of this interpretation was Frank Perlin. See the collection of his influential essays: *Invisible City*.

69. The copious literature on free trade is strongly geared to Britain: see above all A. Howe, *Free Trade*. Trentmann, *Free Trade Nation*, esp. chs. 1–3, breaks new ground on free trade as a central element of Britain's political culture. The classical all-European perspective is Kindleberger, *Rise of Free Trade*.

70. Sugihara, *Japan as an Engine*.

71. Latham, *Rice*.

72. Cushman, *Fields from the Sea*, p. 66.

73. Hancock, *Citizens of the World*, esp. pp. 279ff., on the integrative lifestyle of the gentleman.

74. See H. V. Bowen, *Business of Empire*, pp. 151ff. for the East India Company.

75. Çizaka, *Business Partnerships*.

76. Gary G. Hamilton and Chang Wei-an, "The Importance of Commerce in the Organization of China's Late Imperial Economy," in: Arrighi et al., *Resurgence of East Asia*, pp. 173–213.

77. Markovits, *Global World*, esp. ch. 5.

78. Claude Markovits, "Merchant Circulation in South Asia (18th to 20th Centuries): The Rise of Pan-Indian Merchant Networks," in Markovits et al., *Society and Circulation*, pp. 131–62, and see a recent collection of Markovits's seminal papers: *Merchants*. On Chinese networks, often centered on Hong Kong, see D. R. Meyer, *Hong Kong*, pp. 91–98.

79. Torp, *Herausforderung*, p. 41; and more data in Rostow, *World Economy*, p. 67 (Tab. II-7). There is a brilliant interpretation of the consequences in Rogowski, *Commerce and Coalitions*, pp. 21–60.

80. Maddison, *Contours*, p. 81 (Tab. 2.6).

81. Kenwood and Loughed, *Growth*, p. 80.

82. R. Miller, *Britain and Latin America*, pp. 79, 83f., 98.

83. For a good example, see Topik, *Coffee Anyone?*, esp. pp. 242ff.

84. Sydney Pollard, "The Europeanization of the International Economy, 1800–1870," in: Aldcroft and Sutcliffe, *Europe*, pp. 50–101.

85. For data on import tariffs, see Amsden, *Rise of "the Rest,"* pp. 44f. (Tab. 2.3).

86. G. Clark, *Farewell to Alms*, p. 309.

87. More work remains to be done on caravan traffic and trade. But see the excellent study Lydon, *On Trans-Saharan Trails*, esp. pp. 206–73 on the organization of caravans and the importance of trust among their participants.

88. Ferguson, *Rothschild*, vol. 1; Munro, *Maritime Enterprise*.

89. Sugihara, *Japan*, chs. 2–4; several examples, drawing also on Korea, are given in Sugiyama and Grove, *Commercial Networks*, esp. chs. 1, 3, 5, 6.

90. The pioneer of this interpretation has been Hamashita Takeshi, see *China, East Asia and the Global Economy*—a collection of his papers.

91. Findlay and O'Rourke, *Power and Plenty*, pp. 307f.

92. Torp, *Herausforderung*, pp. 34–36.

93. Peter H. Lindert and Jeffrey G.Williamson, "Does Globalization Make the World More Unequal?" in: Bordo et al., *Globalization*, pp. 227–71, at 233.

94. See the exemplary case studies in Topik et al., *From Silver to Cocaine*.

95. Akinobu Kuroda, "The Collapse of the Chinese Imperial Monetary System," in: Sugihara, *Japan*, pp. 103–26, esp. 106–13.

96. See the meticulous analysis in Otto, *Entstehung eines nationalen Geldes*.

97. Toniolo, *Economic History*, p. 59.

98. Irigoin, *Gresham on Horseback*.

99. See Flandreau *Monetary Unions*.

100. See the research conclusions in Flynn and Giráldez, *Cycles of Silver*.

101. Roy, *India in the World Economy*, p. 127.

102. Lin Man-houng, *China Upside Down*, p. 114. For a more nuanced analysis than can be attempted here, see Hamashita, *China, East Asia and the Global Economy*, pp. 39–56. A classic on the origins of the Opium War is Chang Hsin-pao, *Commissioner Lin*; and a recent exhaustive treatment of opium in modern world history is Derks, *Opium Problem*.

103. Rothermund, *Economic History of India*, pp. 43f.

104. P. R. Gregory, *Before Command*, p. 67.

105. Eichengreen, *Globalizing Capital*, pp. 24–29; there is also a good account in Frieden, *Global Capitalism*, pp. 6f., 14–21, 48f.

106. Eichengreen, *Globalizing Capital*, p. 29.

107. Cecco, *Money and Empire*, p. 59; R. Miller, *Britain and Latin America*, pp. 168, 174f. (doubts about the stabilizing influence from abroad in the Chilean case); Richard Salvucci, "Export-Led Industrialization," in: Bulmer-Thomas et al., *Cambridge Economic History of Latin America*, vol. 2, pp. 249–92, at 256–60.

108. See the discriminating analysis in Gallarotti, *Anatomy*, pp. 207–17.

109. On the gold supply as an independent variable, see Eichengreen and McLean, *Supply of Gold*, esp. p. 288, which shows that output was to only a limited extent triggered by demand.

110. Here I am following Frieden, *Global Capitalism*, p. 121.

111. There is a somewhat incoherent introduction to the subject in Allen, *Global Financial System*, pp. 8–9, 12.

112. Neal, *Financial Capitalism*, p. 229. For reasons of space, this brief chapter cannot even attempt to sketch the global development of financial institutions. While there is no comprehensive history of banking or stock exchanges, the history of insurance is now superbly covered in Borscheid and Haueter, *World Insurance*.

113. Kenwood and Loughed, *Growth*, p. 6.

114. See the data for 1825–1995 in Maurice Obstfeld and Alan M. Taylor, "Globalization and Capital Markets," in Bordo et al., *Globalization*, pp. 121–83, at 141f. (Tab. 3.2).

115. See Kynaston, *City*; Michie, *London Stock Exchange*, ch. 3; idem, *Global Securities Market*, chs. 4 & 5.

116. See Cassis, *Capitals of Capital*.

117. Girault, *Diplomatie européenne*, p. 39.

118. Peter H. Lindert and Peter J. Morton, "How Sovereign Debt Has Worked," in: Sachs, *Developing Country Debt*, pp. 225–35, at 230.

119. Excellent on this is Suzuki, *Japanese Government Loan Issues*, which also has a good account of the London capital market (pp. 23ff.). See also Tamaki, *Japanese Banking*, pp. 87ff.

120. Kuran, *Islam and Mammon*, pp. 13f.

121. I. Stone, *Global Export*, pp. 381, 409 (rounded up or down).

122. Schularick, *Finanzielle Globalisierung*, p. 44 (Tab. 1.10, rounded up or down).

123. G. Austin and Sugihara, *Local Suppliers of Credit*, pp. 5, 13.

124. See chapter 9, above.

125. Kindleberger, *Financial History*, p. 222.

126. Topik, *When Mexico Had the Blues*.

127. Blake, *Disraeli*, pp. 581–87.

128. R. Owen, *Middle East*, p. 127, Tab. 19.

129. On Ismail's costly embellishment of Cairo, see chapter 6, above.

130. R. Owen, *Middle East*, pp. 130–35.

131. On the problem of state bankruptcy before 1914, see Petersson, *Anarchie*, ch. 2.

132. See Marichal, *Debt Crises*, on Latin America; similar overviews are still lacking for Asia.

CHAPTER XV: Hierarchies

1. I use this term, which sociological theory considers imprecise, as a rough synonym for the somewhat narrower and technical-sounding "stratification." What interests me here are only certain positions (especially "above," "in the middle," and "outside") in social structures that participants perceive or "imagine" to be unequal. To speak in general of "hierarchy" in the nineteenth century is not to take "stratificatory differentiation" as typical of the epoch worldwide or to deny that processes of transition to "functional differentiation" (Niklas Luhmann) may be empirically observable.

2. Cannadine, *Rise and Fall*, pp. 88f., 91, 99.

3. Tocqueville, *Democracy*, p. 60 (pt. 1, ch. 3).

4. Kocka, *19. Jahrhundert*, p. 100.

5. Naquin and Rawski, *Chinese Society*, pp. 138ff.

6. Toledano, *State and Society*, pp. 157f.

7. Stinchcombe, *Economic Sociology*, p. 245—an unusually stimulating book for social history. There are fine examples relating to France in G. Robb, *Discovery of France*.

8. This is the theme of Goody, *Theft of History*.

9. See the overview in Burrow, *Crisis of Reason*, ch. 2.

10. See Gall, *Bürgertum*, pp. 81f. A precise discussion of the concepts may be found in Kocka, *Weder Stand noch Klasse*, pp. 33–35.

11. Devine, *Scottish Nation*, pp. 172–83.

12. Wirtschafter, *Structures of Society*, p. 148; Elise Kimerling Wirtschafter, "The Groups Between: *Raznochintsy*, Intelligentsia, Professionals," in: D. Lieven, *Cambridge History of Russia*, pp. 245–63, at 245.

13. Hartley, *Social History*, p. 51.

14. On West European society in the late eighteenth century, see Christof Dipper, "Orders and Classes. Eighteenth-Century Society under Pressure," in Blanning, *Eighteenth Century*, pp. 52–90.

15. See the chapter "Status Groups," in M. B. Jansen, *Modern Japan*, pp. 96–126.

16. This is an extreme simplification. For an example of the extraordinary complexity of social hierarchies in early nineteenth century Asia and of the terminology used to describe them, see Rabibhadana, *Thai Society*, pp. 97–170.

17. V. Das, "Caste," in Smelser and Baltes, *International Encyclopedia*, vol. 3, pp. 1529–32; Peebles, *Sri Lanka*, p. 48.

18. A. von Humboldt, *Studienausgabe*, vol. 4, pp. 162ff.

19. Wasserman, *Everyday Life*, p. 12.

20. See chapters 4 and 7, above.

21. See chapter 4, above.

22. Rickard, *Australia*, p. 37.

23. Korea war the only society to have slavery in modern East Asia, with remnants lasting into the nineteenth century. See Palais, *Korean Uniqueness*, p. 415.

24. On workers and farmers, see chapter 13, above.

25. Walter Demel, "Der europäische Adel vor der Revolution: Sieben Thesen," in: Asch, *Adel*, pp. 409–33, at 409. See also Lukowski, *European Nobility*.

26. D. Lieven, *Aristocracy*, p. 1. But for the roots of the slow decline of the nobility, see Demel, *Der europäische Adel*, pp. 87–90.

27. Maria Todorova, "The Ottoman Legacy in the Balkans," in: L. C. Brown, *Imperial Legacy*, pp. 46–77, at 60.

28. Beckett, *Aristocracy*, p. 40.

29. Demel, *Der europäische Adel*, p. 17.

30. Woloch, *Napoleon and His Collaborators*, pp. 169–73.

31. This term is also used in the history of the Near and Middle East, though with stronger reference to a political role of mediation between ruler and people (somewhat similar to that of the *gentry* or the *shenshi* in China). See Albert Hourani, "Ottoman Reform and the Politics of Notables," in: Hourani et al., *Modern Middle East*, pp. 83–109.

32. Charle, *Histoire sociale de la France*, pp. 229ff.

33. There is a good characterization in D. Lieven, *Empire*, pp. 241–44.

34. Beckett, *Aristocracy*, p. 31.

35. Two chief adversaries in the debate have been F.M.L. Thompson and W. D. Rubinstein.

36. Asch, *Europäischer Adel*, p. 298.

37. Searle, *A New England*, pp. 37f.

38. See the overview in Beckett, *Aristocracy*, pp. 16–42.

39. Maria Malatesta, "The Landed Aristocracy during the Nineteenth and Early Twentieth Centuries," in: Kaelble, *European Way*, pp. 44–67.

40. Cannadine, *Ornamentalism*, pp. 85ff.

41. Liebersohn, *Aristocratic Encounters* develops this theme with reference to North America and remarks by European aristocratic travellers on elements of nobility among the Indian population.

42. Fox-Genovese and Genovese, *Mind of the Master Class*, pp. 304–82.

43. This is the dramatic, but not altogether inappropriate, term used in Wasson, *Aristocracy*, p. 156.

44. Nutini, *Wages of Conquest*, p. 322 argues that in Mexico an aristocracy going back to early colonial times and ultimately resting on hacienda ownership was able to maintain itself unchallenged.

45. The classical analysis is Cohn, *Anthropologist*, pp. 632–82.

46. Panda, *Bengal Zamindars*, p. 2.

47. This section follows Schwentker, *Samurai*, pp. 95–116. Another very interesting account (essentially of the Tokugawa period), by a sociologist working in the field of history, is Ikegami, *Taming of the Samurai*. For a vivid evocation of the life of a low-ranking samurai, see Katsu Kokichi, *Musui's Story*.

48. Demel, *Der europäische Adel*, p. 88.

49. Ravina, *Last Samurai*, pp. 191ff. The leader of the revolt was not a direct victim of the Meiji Restoration but one of its main protagonists.

50. Elman, *Civil Examinations*; and a classic of social history, Chang Chung-li, *Chinese Gentry*. See also R. J. Smith, *China's Cultural Heritage*, pp. 55–64, 71–75; and Joseph W. Esherick and Mary Backus Rankin, "Introduction," in idem, *Chinese Local Elites*, pp. 1–24.

51. Reynolds, *China*, offers an overview.

52. Crossley, *Orphan Warriors*.

53. For Germany (with a few doubts), see Kocka, *19. Jahrhundert*, pp. 98–137. Bank and Buuren, *1900*, offers a comprehensive snapshot of a prototypical bourgeois society in Europe; Tanner, *Arbeitsame Patrioten*, is an empirically dense portrait of the most "bourgeois" country in the world.

54. Europe-wide studies of the middle classes kept many German historians busy in the 1980s and 1990s. Summaries of their work are Lundgreen, *Sozial- und Kulturgeschichte*; Kocka and Frevert, *Bürgertum*; Gall, *Stadt und Bürgertum*; a critical comparison of the various schools is Sperber, *Bürger*.

55. The Bassermann family serves as the example in Gall, *Bürgertum in Deutschland*.

56. Maza, *Myth of the French Bourgeoisie*.

57. See the overview in Pilbeam, *Middle Classes*, pp. 74–106.

58. Goblot, *Barrière et niveau*, p. 7—one of the most intellectually stimulating books ever written about the bourgeoisie.

59. Daumard, *Les bourgeois*, p. 261.

60. See J. L. West and Petrov, *Merchant Moscow*—including the photographs.

61. Cindy S. Aron, "The Evolution of the Middle Class," in Barney, *Companion*, pp. 178–91, at 179.

62. For England, see Perkin, *Origins*, pp. 252f.

63. Hartmut Kaelble, "Social Particularities of Nineteenth- and Twentieth-Century Europe," in idem, *European Way*, pp. 276–317, at 282–84.

64. For Europe (Germany, England, France, Belgium), see Crossick and Haupt, *Petite Bourgeoisie*.

65. Farr, *Artisans*, pp. 10ff.

66. Pilbeam, *Middle Classes*, p. 172.

67. Goblot, *Barrière et niveau*, p. 40.

68. R. Ross, *Status*. Other examples of "Victorian values" among the educated African elite (in this case in Lagos) may be found in K. Mann, *Marrying Well*.

69. Jürgen Kocka, "Bürgertum und bürgerliche Gesellschaft im 19. Jahrhundert. Europäische Entwicklungen und deutscher Eigensinn," in: Kocka and Frevert, *Bürgertum*, vol. 1, pp. 11–76, at 12; Jürgen Kocka, "The Middle Classes in Europe," in: Kaelble, *European Way*, pp. 15–43, at 16. These are two fundamental texts on the subject.

70. Quoted in Blumin, *Emergence of the Middle Class*, p. 2. This contemporary idealization recalls, across a large space of time, the wishful image of a "classless civil society" that Lothar Gall has traced in detail for early nineteenth-century Germany.

71. Beckert, *Monied Metropolis*.

72. This would require a number of case studies as high in quality as Pernau, *Bürger mit Turban*.

73. Braudel, *Civilization and Capitalism*, vol. 2, still offers a first overview of this field.

74. A. Adu Boahen, "New Trends and Processes in Africa in the Nineteenth Century," in Ajayi, *General History of Africa*, pp. 40–63, at 48–52.

75. Long known in the case of West Africa, this is now apparent also for a less noticed emporium; see G. Campbell, *Imperial Madagascar*, pp. 161–212.

76. Bergère, *Golden Age*.

77. Trocki, *Opium and Empire*.

78. Berend, *History Derailed*, p. 196.

79. See the splendid monograph: Horton and Middleton, *The Swahili*.

80. Markovits et al., *Modern India*, pp. 320, 325f.; Cheong, *Hong Merchants*, pp. 303f.

81. Jayawardena, *Nobodies*, pp. 68ff.; Freitag, *Arabische Buddenbrooks*, pp. 214f.

82. See also chapter 14, above.

83. Bergère, *Golden Age*, p. 40; Hao, *Commercial Revolution*.

84. Dobbin, *Asian Entrepreneurial Minorities*.

85. Györgi Ránki, "Die Entwicklung des ungarischen Bürgertums vom späten 18. zum frühen 20. Jahrhundert," in: Kocka and Frevert, *Bürgertum*, vol. 1, pp. 247–65, at 249, 253, 256.

86. Robert E. Elson, "International Commerce, the State and Society: Economic and Social Change," in Tarling, *Cambridge History of Southeast Asia*, vol. 2, pp. 131–95, at 174.

87. Dobbin, *Asian Entrepreneurial Minorities*, pp. 47, 69, 171.

88. Frangakis-Syrett, *Greek Merchant Community*, p. 399.

89. As early as 1740, though, a massacre of Chinese in Java had had similar causes.

90. See the remarks on the business policy of North Indian merchant families in Bayly, *Rulers, Townsmen and Bazaars*, pp. 394–426.

91. For a fine example, see Hanssen, *Beirut*, pp. 213–35. See also chapter 6, above.

92. See the fundamental considerations in Watenpaugh, *Being Modern*, pp. 14f.

93. Rankin, *Elite Activism*, pp. 136ff.; Kwan, *Salt Merchants*, pp. 89–103; Freitag, *Indian Ocean Migrants*, pp. 9, 238–42.

94. Rowe, *Hankow*, vol. 1, pp. 289ff.

95. On the semantics, the key work is still Engelhardt, "*Bildungsbürgertum*." Although it had many precursors, the actual concept goes back only to the 1920s. See also Conze et al., *Bildungsbürgertum*, especially the various cross-European comparisons in volume 1.

96. Peter Lundgreen, "Bildung und Bürgertum," in idem, *Sozial- und Kulturgeschichte*, pp. 173–94, at 173.

97. Dietrich Geyer, "Zwischen Bildungsbürgertum und Intelligenzija: Staatsdienst und akademische Professionalisierung im vorrevolutionären Russland," in: Conze et al., *Bildungsbürgertum*, vol. 1, pp. 207–30, at 229.

98. Lufrano, *Honorable Merchants*, pp. 177ff. The term "self-cultivation," which Lufrano uses for Chinese merchants, reminds one of a classic book on the idea of *Bildung*: Bruford, *German Tradition of Self-Cultivation*. For Japan, see Gilbert Rozman, "Social Change," in: J. W. Hall et al., *Cambridge History of Japan*, vol. 5, pp. 499–568, at 513. The question is whether eighteenth-century merchant culture was more autonomous, or less "embedded," in Japan than in China.

99. See, e.g., Schwarcz, *Chinese Enlightenment*.

100. Kreuzer, *Bohème*, is an important work for both social and cultural history.

101. Buettner, *Empire Families*; anecdotally Yalland, *Boxwallahs*; and, for a comparison with the completely nonaristocratic Chinese treaty ports, Bickers, *Britain in China*.

102. This is well researched in Butcher, *British in Malaya*.

103. Ruedy, *Modern Algeria*, pp. 99f.

104. Quataert, *Ottoman Empire*, p. 153.

105. Ibid., p. 146. See also chapter 5, above.

106. The most recent general account of this international financial world is Cassis, *Capitals of Capital*, pp. 74ff.

107. I refer to an interesting thesis in C. A. Jones, *International Business*.

108. Wray, *Mitsubishi*, p. 513.

109. Model analyses of this process are (for France) Garrioch, *Formation of the Parisian Bourgeoisie*; and (for the United States) Blumin, *Emergence of the Middle Class*; and Bushman, *Refinement*.

110. See the few hints in this direction in chapter 16, below.

111. For many areas there is still no synthesis of research such as we have for Europe in Gestrich et al., *Geschichte der Familie*.

112. See the fine case studies in Clancy-Smith and Gouda, *Domesticating the Empire*.

CHAPTER XVI: Knowledge

1. On religion, see chapter 18, below.

2. Dülmen and Rauschenbach, *Macht des Wissens*.

3. H. Pulte, "Wissenschaft (III)," in: *Historisches Wörterbuch der Philosophie*, vol. 12, Darmstadt 2004, col. 921.

4. Burke, *Social History*, pp. 19f.

5. Fragner, *"Persophonie,"* p. 100.

6. Ostler, *Empires of the Word*, pp. 438f.

7. Ibid., pp. 411f.

8. Mendo Ze et al., *Le Français*, p. 32.

9. B. Lewis, *Emergence*, p. 84.

10. Crystal, *English*, p. 73.

11. Ibid., p. 66.

12. The degree to which the speaking of English was "ordered" from above is discussed at length in Phillipson, *Linguistic Imperialism*.

13. Zastoupil and Moir, *Great Indian Education Debate*, esp. the introduction (pp. 1–72).

14. Crystal, *English*, pp. 24ff. gives a (rather superficial) region-by-region overview.

15. B. Lewis, *Emergence*, pp. 88, 118.

16. Adamson, *China's English*, pp. 25f.

17. Keene, *Japanese Discovery of Europe*, pp. 78f.

18. Elman, *Modern Science*, pp. 86f.

19. B. Lewis, *Emergence*, p. 87.

20. This moment in intellectual history was identified in Schwab, *Oriental Renaissance*.

21. Ostler, *Empires of the Word*, p. 503.

22. H. M. Scott, *Birth*, pp. 122f.; Haarmann, *Weltgeschichte der Sprachen*, p. 314.

23. See the overview in Haarmann, *Weltgeschichte der Sprachen*, pp. 309–34.

24. Bolton, *Chinese Englishes*, pp. 146–96.

25. Marr, *Reflections from Captivity*, pp. 30, 35.

26. Pollock, *Cosmopolitan Vernacular*.

27. Sassoon, *Culture*, pp. 21–40, on the rise of national languages in Europe.

28. Vincent, *Mass Literacy*, pp. 138f., 140.

29. Janich and Greule, *Sprachkulturen*, p. 110.

30. M. C. Meyer and Sherman, *Course of Mexican History*, p. 457.

31. For an introduction to the problem, see Ernst Hinrichs, "Alphabetisierung. Lesen und Schreiben," in Dülmen and Rauschenbach, *Macht des Wissens*, pp. 539–61, esp. 539–42. The theoretical complexity of the theme is shown in Barton, *Literacy*.

32. See Graff, *Legacies*, p. 262—the unsurpassed standard work on the subject.

33. Tortella, *Patterns of Economic Retardation*, p. 11 (Tab. 6).

34. Vincent, *Mass Literacy*, p. 11. There are still not many national studies comparable in quality to Brooks, *When Russia Learned to Read*.

35. Graff, *Legacies*, p. 295 (Tab. 7–2).

36. With reference to Europe in general: Sassoon, *Culture*, pp. 93–105.

37. For Germany, cf. Engelsing, *Analphabetentum*. A number of works by Roger Chartier and Martyn Lyons cover the field in France.

38. M. Lyons, *Readers*, pp. 87–91.

39. Starrett, *Putting Islam to Work*, p. 36.

40. Vincent, *Mass Literacy*, p. 56.

41. Gillian Sutherland, "Education," in F. M. L. Thompson, *Cambridge Social History of Britain*, vol. 3, pp. 119–69, at 145.

42. See the estimates for 1882 in Easterlin, *Growth Triumphant*, p. 61 (Tab. 5.1.).

43. William J. Gilmore-Lehne, "Literacy," in Cayton, *Encyclopedia*, vol. 3, pp. 2413–26, at 2419f., 2422.

44. Graff, *Legacies*, p. 365.

45. Ayalon, *Political Journalism*, p. 105.

46. Gilbert Rozman, "Social Change," in J. W. Hall et al., *Cambridge History of Japan*, vol. 5, pp. 499–568, at 560f.

47. Rubinger, *Popular Literacy*, p. 184.

48. Pepper, *Radicalism*, p. 52.

49. Rawski, *Education*, p. 23.

50. P. Bailey, *Reform the People*, pp. 31–40.

51. M. E. Robinson, *Korea's Twentieth-Century Odyssey*, p. 11.

52. This remained the case until the end of the system; see Elman, *Civil Examinations*, pp. 597–600.

53. Alexander Woodside, "The Divorce between the Political Center and Educational Creativity in Late Imperial China," in: Elman and Woodside, *Education and Society*, pp. 458–92, at 461.

54. *BBC News Service,* April 2, 2007.

55. Nipperdey, *Napoleon to Bismarck*, p. 398.

56. Karl-Ernst Jeismann, "Schulpolitik, Schulverwaltung, Schulgesetzgebung," in: C. Berg et al., *Handbuch*, vol. 3, pp. 105–22, at 119.

57. An influential Foucault-oriented analysis of Egypt along these lines is T. Mitchell, *Colonising Egypt*. But see the criticisms of such an approach in Starrett, *Putting Islam to Work*, pp. 57–61.

58. Bouche, *Histoire de la colonisation française*, pp. 257–59.

59. Wesseling, *European Colonial Empires*, p. 60.

60. D. Kumar, *Science*, pp. 151–79; Ghosh, *History of Education*, pp. 86, 121f.; Arnold, *Science*, p. 160; Bhagavan, *Sovereign Spheres*.

61. Somel, *Modernization of Public Education*, pp. 173–79. Somel sums this up as the "duality of technological modernism and Islamism" (p. 3). On architecture, see also Fortna, *Imperial Classroom*, pp. 139–45.

62. Somel, *Modernization of Public Education*, p. 204.

63. Szyliowicz, *Education and Modernization*, pp. 170–78; Keddie, *Modern Iran*, p. 29; Amin et al., *Modern Middle East*, pp. 43f.

64. Ringer, *Education and Society*, p. 206.

65. Goonatilake, *Toward a Global Science*, p. 62, drawing on Benares as an example. Cf. Burke, *A Social History*, pp. 50–52.

66. For an (implicitly comparative) account of Islamic institutions of learning, see Huff, *Early Modern Science*, pp. 147–79.

67. Björn Wittrock, "The Modern University: The Three Transformations," in: Rothblatt and Wittrock, *European and American University*, pp. 303–62, at 304f., 310ff.

68. There is a wonderful sociological character sketch in Rothblatt, *Revolution of the Dons*, pp. 181–208.

69. J.-C. Caron, *Générations romantiques*, p. 167.

70. Brim, *Universitäten*, p. 154.

71. Lee Ki-baik, *Korea*, p. 342; Lee Chong-sik, *Korean Nationalism*, pp. 89–126.

72. John Roberts et al., "Exporting Models," in: Rüegg, *History of the University*, vol. 2, pp. 256–83.

73. Edward Shils and John Roberts, "The Diffusion of European Models outside Europe," in Rüegg, *History of the University*, vol. 3, pp. 163–231. Interesting on Africa is Nwauwa, *Imperialism*.

74. Rüegg, *History of the University*, vol. 3, pp. 187ff.

75. İsanoğlu, *Science*, Text III, pp. 38f.

76. Hayhoe, *China's Universities*, p. 13. Cf. Lu Yongling and Ruth Hayhoe, "Chinese Higher Learning: The Transition Process from Classical Knowledge Patterns to Modern Disciplines, 1860–1910," in Charle et al., *Transnational Intellectual Networks*, pp. 269–306.

77. Quoted in Shils and Roberts, "The Diffusion of European Models outside Europe," in Rüegg, *History of the University*, vol. 3, p. 225.

78. Ringer, *German Mandarins*; B. K. Marshall, "Professors and Politics: The Meiji Academic Elite," in Kornicki, *Meiji Japan*, vol. 4, pp. 296–318.

79. Bartholomew, *Science in Japan*, pp. 84f.

80. W. Clark, *Academic Charisma*; cf. Schalenberg, *Humboldt auf Reisen?* pp. 53–75. That Humboldt's university was not a radically new departure but part of a wider European conception of "enlightened absolutism," is shown in R. D. Anderson, *European Universities*, ch. 2, see also ch. 4 (on Humboldt).

81. David Cahan, "Institutions and Communities," in idem, *From Natural Philosophy*, pp. 291–328, at 313–17.

82. Jungnickel and McCormmach, *Intellectual Mastery*, vol. 2, pp. 166ff.

83. Konrad H. Jarausch, "Universität und Hochschule," in C. Berg et al., *Handbuch*, vol. 4, pp. 313–39, at 38f.

84. R. D. Anderson, *European Universities*, p. 292.

85. W. Clark, *Academic Charisma*, p. 461.

86. Leedham-Green, *Concise History*, p. 195.

87. John R. Thelin, "The Research University," in Cayton, *Encyclopedia*, vol. 3, pp. 2037–45, at 2037.

88. Veysey, *Emergence*, p. 171.

89. Thelin, *American Higher Education*, pp. 114, 116, 122–31, 153f. Still useful for the period around the turn of the century is Veysey, *Emergence*.

90. Bartholomew, *Science in Japan*, pp. 64, 68ff., 123.

91. On Rieß see Mehl, *History and the State*, pp. 94–102.

92. İsanoğlu, *Science*, Text X, p. 53.

93. Goonatilake, *Toward a Global Science*, pp. 53–55.

94. See the fundamental considerations in Raina, *Images and Contexts*, pp. 176–91; and the superb collections (of reprints) Habib and Raina, *Social History of Science*.

95. Nakayama, *Traditions*, pp. 195–202.

96. Elman, *On Their Own Terms*, p. 298.

97. Howland, *Translating the West*, p. 97.

98. Wang Hui, "The Fate of 'Mr. Science' in China: The Concept of Science and Its Application in Modern Chinese Thought," in Barlow, *Formations*, pp. 21–81, at 22f., 30f., 33, 56. There are many excellent case studies of terminological transfer into Chinese in Lackner et al., *New Terms*, and Vittinghoff and Lackner, *Mapping Meanings*.

99. C. T. Jackson, *Oriental Religions*, p. 57.

100. See Sullivan, *Meeting of Eastern and Western Art*, pp. 120–39, 209–29; cf. K. Berger, *Japonisme*.

101. Fauser, *Musical Encounters*; Locke, *Musical Exoticism*.

102. There is a brief characterization of theosophy in Burrow, *Crisis of Reason*, pp. 226–29; see also Aravamudan, *Guru English*, pp. 105–41.

103. On India see Arnold, *Science*, p. 124; also important is Yamada Keiji, *Transfer of Science*.

104. Prakash, *Another Reason*, pp. 6, 53.

105. Especially good on India is the analysis in ibid., pp. 52ff.

106. Bowler and Morus, *Making Modern Science*, p. 338.

107. Theodore M. Porter, "The Social Sciences," in Cahan, *From Natural Philosophy*, pp. 254–90, at 254. See also chapter 1, above.

108. Dorothy Ross, "Changing Contours of the Social Science Disciplines," in D. Porter and Ross, *Modern Social Sciences*, pp. 205–37, at 208–14.

109. Barshay, *Social Sciences*, pp. 40–42.

110. There are brief overviews in Iggers and Wang, *Modern Historiography*, pp. 117–33; and D. R. Woolf, *A Global History of History*, pp. 364–97; and on a grander scale, idem, *Oxford History of Historical Writing*, vol. 4.

111. René Wellek, the standard authority, traces the beginnings of literary criticism to 1750. Art criticism went back earlier in Europe, to the time of Giorgio Vasari (1511–74).

112. There is more on this in Osterhammel, *Entzauberung*.

113. Still unsurpassed, after numerous more recent studies, is Schwab, *Oriental Renaissance*.

114. A few classics are Tahtawi, *An Imam*; Kume Kunitake, *Iwakura Embassy*; Parsons, *King Khama*. More in Osterhammel, *Ex-zentrische Geschichte*.

115. Gran-Aymerich, *Naissance de l'archéologie moderne*, pp. 83–86.

116. Peers, *Colonial Knowledge*.

117. Said, *Orientalism* launched this debate and is still one of its most important texts. On the discussion in English and Arabic, see Varisco, *Reading Orientalism*; a model of a sober empirical study on orientalist scholarship is Marchand, *German Orientalism*.

118. See the case studies in Stuchtey, *Science*.

119. This ambiguity is well brought out from the French example in Singaravélou, *L'École Française d'Extrême-Orient*, pp. 183ff.

120. Stocking, *Victorian Anthropology*; idem, *After Tylor*.

121. See also chapters 1 and 3, above.

122. Stafford, *Scientist of Empire*; Robert A. Stafford, "Scientific Exploration and Empire," in Louis, *Oxford History of the British Empire*, vol. 3, pp. 224–319; Driver, *Geography Militant*.

123. Brennecke, *Sven Hedin*. There seems to be no adequate biography in English.

124. S. Conrad, *Globalisation*, ch. 2.

125. Schleier, *Kulturgeschichtsschreibung*, vol. 2, pp. 813–41.

126. Venturi, *Roots of Revolution*, pp. 633ff.

127. As a young man, however, Bartók had learned the habits of high Romantic virtuosos from his teacher István Thomán, one of Liszt's most gifted disciples.

CHAPTER XVII: Civilization and Exclusion

1. On the following, see B. Barth and Osterhammel, *Zivilisierungsmissionen*; Mazlish, *Civilization*; and, with special reference to South Asia, Fischer-Tiné and Mann, *Colonialism*. There is a good succinct overview in an unexpected place: Costa, *Civitas*, vol. 3, pp. 457–99.

2. Pagden, *Lords*, pp. 79f.

3. Adas, *Contested Hegemony*.

4. Sarmiento, *Civilization and Barbarism*. The centrality of the barbarism/civilization opposition, with a wider reference than Argentina, is shown in Brading, *First America*, pp. 621–47 and Manrique, *De la conquista a la globalización*, pp. 147–66.

5. Nani, *Ai confini della nazione*, pp. 97ff.; Moe, *View from Vesuvius*.

6. Seidl, *Bayern in Griechenland*.

7. Broers, *Napoleonic Empire*, pp. 245f. and passim.

8. R. Owen, *Lord Cromer*, esp. pp. 304ff.

9. The classic text on the impact of utilitarians in India is Stokes, *English Utilitarians*.

10. J. Fisch, *Immolating Women*, pp. 376ff., 232f. In noncolonial Nepal, widow burning remained legal until 1920!

11. See the distinction between a "state model" of colonization and a missionary-borne "civilizing colonialism," in Comaroff and Comaroff, *Ethnography*, pp. 198–205.

12. On the example of the Jamaica (Morant Bay) affair of 1865, see Kostal, *Jurisprudence of Power*, pp. 463 and passim.

13. Gong, *Standard of "Civilization."*

14. Koskenniemi, *Gentle Civilizer*, pp. 49, 73.

15. The standard work is still Betts, *Assimilation*.

16. Data on colonial crimes may be found in Ferro, *Le livre noir*. German operations in South-West Africa have recently attracted particular attention.

17. Brantlinger, *Dark Vanishings*, pp. 94ff.

18. Rivet, *Le Maroc*, pp. 36–77.

19. M. C. Meyer and Sherman, *Course of Mexican History*, p. 457.

20. Bullard, *Exile*, pp. 17, 121f.

21. The term is borrowed from Stephanson, *Manifest Destiny*, p. 80—a good introduction to American ideas about "civilizing." For different approaches see Ninkovich, *Global Dawn*; Tyrrell, *Reforming the World*.

22. Manela, *Wilsonian Moment*.

23. The quote was doing the rounds in 1930, but it is hard to track down the precise source.

24. The authoritative account of antislavery is Drescher, *Abolition*.

25. Clarence-Smith, *Islam*, p. 146.

26. This is my reading of the circumspect discussion in Pamela Kyle Crossley, "Slavery in Early Modern China," in: Eltis and Engerman, *Cambridge World History of Slavery*, vol. 3, pp. 186–213, esp. 206f.

27. Botsman, *Freedom without Slavery*, p. 1327.

28. Palais, *Korean Uniqueness*, p. 418.

29. Thanet Aphornsuvan, "Slavery and Modernity: Freedom in the Making of Modern Siam," in: Kelly and Reid, *Asian Freedoms*, pp. 161–86, esp. 177.

30. Sanneh, *Abolitionists Abroad*.

31. Temperley, *British Antislavery*, gives a clear account of this kind of internationalism.

32. Gott, *Cuba*, pp. 45f.

33. Green, *British Slave Emancipation*, is still a fundamental work on the subject.

34. D. B. Davis, *Inhuman Bondage*, p. 79.

35. For a profound analysis of the conceptual world of British abolitionists see D. B. Davis, *Slavery and Human Progress*, pp. 107–68. On the "egoism" of such thinking, see C. L. Brown, *Moral Capital*; and on the general "culture" of the movement, Turley, *English Anti-Slavery*.

36. Quoted in C. L. Brown, *Moral Capital*, p. 8.

37. The prolific work of Seymour Drescher has been especially influential in the formation of this consensus.

38. Carey, *British Abolitionism*.

39. D. B. Davis, *Inhuman Bondage*, p. 236.

40. Satre, *Chocolate on Trial*, pp. 77ff.

41. Keegan, *Colonial South Africa*, pp. 35f.

42. Dharma Kumar, "India," in: Drescher and Engerman, *Historical Guide*, pp. 5–7.

43. Blackburn, *Overthrow*, p. 480; Bernecker, *Geschichte Haitis*, p. 69.

44. N. Schmidt, *L'Abolition de l'esclavage*, pp. 22ff.

45. Emmer, *Nederlandse slavenhandel*, pp. 205f.

46. See also, in a different perspective, chapter 10, above.

47. The precise figures are given in Berlin, *Generations of Captivity*, Appendix, Tab. 1.

48. Drescher, *From Slavery to Freedom*, pp. 276f.

49. See the overview in Stewart, *Holy Warriors*. On the most famous (though perhaps not most influential) white abolitionist, see H. Mayer, *All on Fire*.

50. As a way into the huge literature on Lincoln and slavery, see Oakes, *The Radical and the Republican*, esp. pp. 43ff.; Foner, *Fiery Trial*.

51. D. B. Davis, *Inhuman Bondage*, pp. 317f.

52. Zeuske, *Geschichte Kubas*, pp. 124 ff.; Schmidt-Nowara, *Empire and Antislavery*.

53. Viotti da Costa, *Brazilian Empire*, pp. 125–71; A. W. Marx, *Making Race*, p. 64.

54. Bernecker et al., *Geschichte Brasiliens*, p. 210.

55. On slavery and Holocaust, see Drescher, *From Slavery to Freedom*, pp. 312–38.

56. Clarence-Smith, *Islam*, pp. 10f.

57. Ibid., pp. 100f.

58. Ibid., pp. 107f.

59. Ibid., p. 116.

60. Temperley, *White Dreams*.

61. Fundamental (though rather skeptical) on the possibilities of comparative slavery research is Zeuske, *Sklaven*, pp. 331–60. However, some authors such as Seymour Drescher have made very successful use of comparative methods.

62. See, e.g., F. Cooper, *Beyond Slavery*; and the important regional analyses in Temperley, *After Slavery*.

63. A key case study is R. J. Scott, *Degrees of Freedom*.

64. Stanley Engerman, "Comparative Approaches to the Ending of Slavery," in: Temperley, *After Slavery*, pp. 281–300, at 288–90.

65. On the many strands of slavery in Africa, see the collective volume Miers and Roberts, *End of Slavery*, as well as F. Cooper et al., *Beyond Slavery*, pp. 106–49 (on the significance of the year 1910 see p. 119).

66. This is emphasized in Berlin, *Generations of Captivity*, pp. 248–59.

67. Ibid., pp. 266f.

68. Keegan, *Colonial South Africa* finds them already before 1850, not only after the "mineral revolution."

69. D. L. Lewis, *W.E.B. Du Bois*, p. 277.

70. Du Bois, *Writings*, p. 359. Thanks to Scaff, *Max Weber in America*, pp. 98–116, we now know that Max Weber, perhaps the greatest European observer of his time, was extraordinarily receptive to Du Bois's diagnosis.

71. The reasons for this development are still hotly debated. For a report on the controversies, see James Beeby and Donald G. Nieman, "The Rise of Jim Crow, 1880–1920," in Bowles, *Companion*, pp. 336–47.

72. Fredrickson, *White Supremacy*, p. 197.

73. Winant, *The World Is a Ghetto*, pp. 103–5; A. W. Marx, *Making Race*, pp. 79, 178–90; Drescher, *From Slavery to Freedom*, pp. 146f.

74. R. J. Scott, *Degrees of Freedom*, pp. 253ff.

75. F. Cooper et al., *Beyond Slavery*, p. 18.

76. Drescher, *Mighty Experiment*, pp. 158ff.; see also Holt, *Problem of Freedom*.

77. There is no adequate account of racism in the history of ideas, the closest to one being Mosse, *Final Solution*. A very brief introduction is Geulen, *Rassismus*.

78. Shimazu, *Japan, Race and Equality*; on the interpretation, see Lake and Reynolds, *Global Colour Line*, pp. 285–309.

79. Frank Becker, "Einleitung: Kolonialherrschaft und Rassenpolitik," in idem, *Rassenmischehen*, pp. 11–26, at 13.

80. Christian Geulen, "The Common Grounds of Conflict: Racial Visions of World Order 1880–1940," in: S. Conrad and Sachsenmaier, *Competing Visions*, pp. 69–96.

81. This is argued in detail in one of the great classics on the history of racist ideas: W. D. Jordan, *White over Black*. As so often, the influence of authors remains an open question. Did Long really represent "planters" or even "the British public"? Drescher, *From Slavery to Freedom*, p. 285, casts doubt on the latter.

82. Patterson, *Slavery and Social Death*, p. 61.

83. Roediger, *Working toward Whiteness*, p. 11; on the history of racial classification, see esp. Banton, *Racial Theories* and, for a very general survey, Fluehr-Lobban, *Race*, pp. 74–103.

84. Augstein, *Race*, p. xviii.

85. The dispute between Gobineau and Tocqueville in the 1850s elucidated the alternatives with unparalleled clarity. See Ceaser, *Reconstructing America*, ch. 6.

86. Banton, *Racial Theories*, pp. 54–59.

87. There is a good survey of nineteenth-century biological theories of race in Graves, *Emperor's New Clothes*, pp. 37–127.

88. Hannaford, *Race*, pp. 226f., 232f., 241.

89. Ballentyne, *Orientalism*, p. 44. Still fundamental are Poliakov, *Aryan Myth*; Olender, *Languages of Paradise*; Trautmann, *Aryans*.

90. Lorcin, *Imperial Identities*; Streets, *Martial Races*.

91. Hannaford, *Race*, pp. 348ff.

92. Lauren, *Power and Prejudice*, pp. 44ff.; Gollwitzer, *Die gelbe Gefahr*; Mehnert, *Deutschland*; Geulen, *Wahlverwandte*, pt. 2.

93. D. B. Davis, *Inhuman Bondage*, p. 76.

94. See chapter 8, above.

95. L. D. Baker, *From Savage to Negro*, pp. 99ff.

96. Barkan, *Retreat of Scientific Racism*.

97. Torpey, *Invention of the Passport*, pp. 91f.

98. For a thorough account mainly focused on Europe, see Caplan and Torpey, *Documenting Individual Identity*.

99. Noiriel, *Immigration*, pp. 135ff.

100. Gosewinkel, *Einbürgern*, pp. 325–27.

101. For a broad overview, see Lake and Reynolds, *Global Colour Line*.

102. See Reimers, *Other Immigrants*, pp. 44–70. Also Takaki, *Strangers*; Gyory, *Closing the Gate*; and, on the Chinese experience, E. Lee, *At America's Gates*.

103. D. R. Walker, *Anxious Nation*, p. 98.

104. A good account is Markus, *Australian Race Relations*.

105. Jacobson, *Whiteness*.

106. Jacobson, *Barbarian Virtues*, pp. 261f.

107. Dikötter, *Discourse of Race*.

108. Rhoads, *Manchus and Han*, p. 204.

109. Katz, *Out of the Ghetto*, p. 1. A more recent account, focused especially on political emancipation, is Vital, *A People Apart*.

110. For a general history of the Jewish reform movement, see M. A. Meyer, *Response to Modernity*.

111. A. Green, *Moses Montefiore*, p. 2.

112. Katz, *Prejudice*, pp. 245–72.

113. There is a summary of the extensive literature in Noiriel, *Immigration*, pp. 207–86.

114. Research on Russian anti-Semitism is summarized in Marks, *How Russia Shaped the Modern World*, pp. 140–75.

115. Sorin, *A Time for Building*, p. 55; Dinnerstein, *Antisemitism in America*, pp. 35ff.

116. See also Mosse, *Final Solution*, p. 168.

117. Shaw, *Jews of the Ottoman Empire*, pp. 187–206.

118. Fink, *Defending the Rights of Others*, pp. 5–38.

119. On the geographical distribution of the world's Jewish population, see Karady, *Jews of Europe*, pp. 44f.

120. Reinhard Rürup, "Jewish Emancipation in Britain and Germany," in: Brenner et al., *Two Nations*, pp. 49–61.

121. Goodman and Miyazawa, *Jews in the Japanese Mind*, p. 81.

122. Poliakov, *Aryan Myth*, p. 232.

123. Fredrickson, *Racism*, p. 72.

124. Geulen, *Wahlverwandte*, p. 197.

125. A cross-European perspective is offered in Brustein, *Roots of Hate*.

126. Haumann, *East European Jews*, pp. 78f.

127. Ibid., pp. 171f.; Weeks, *From Assimilation*, pp. 71ff.

128. Love, *Race over Empire*, pp. 1–5, 25f.

129. P. A. Kramer, *Blood of Government*, pp. 356f.

130. Fredrickson, *Racism*, pp. 75–95.

131. Ibid., p. 95.

132. Vital, *A People Apart*, pp. 717f., 725.

133. This is also the conclusion in Volkov, *Germans, Jews, and Anti-Semites*, pp. 67f.

CHAPTER XVIII: Religion

1. This chapter owes some important suggestions to an excellent sociological study: Beyer, *Religions*.

2. On "master narratives" in the modern history of religion, see D. Martin, *On Secularization*, pp. 123–40.

3. "Analogous transformation": Beyer, *Religions*, p. 56; as a model of "entangled history" in Britain and India, see Veer, *Imperial Encounters*.

4. See the persuasive critique in Graf, *Wiederkehr*, pp. 233–38; and Beyer, *Religions*, pp. 62ff.

5. On the various concepts of religion in the "world religions," see Haußig, *Religionsbegriff*.

6. J. R. Bowen, *Religions in Practice*, pp. 26f. (expanded).

7. On emerging Japanese notions of "religion," see the brillant study: Josephson, *Invention of Religion in Japan*.

8. Jensen, *Manufacturing Confucianism*, p. 186.

9. Hsiao Kung-chuan, *A Modern China*, pp. 41–136.

10. Beyer, *Religions*, pp. 83f.

11. Masuzawa, *Invention*, pp. 17–20.

12. For nineteenth-century Europe and America, see Helmstadter, *Freedom and Religion*; and for a comparison between emancipatory processes, Liedtke and Wendehorst, *Emancipation*.

13. Cassirer, *Philosophy of the Enlightenment*, pp. 160ff.

14. Zagorin, *Toleration*, p. 306.

15. Sanneh, *Crown*, p. 9. Islamic and Christian proselytism is a central theme in Coquery-Vidrovitch, *Africa and the Africans*.

16. Lynch, *New Worlds*, p. 228.

17. The argument in this section focuses mainly on social history. There is a fine discussion of the history of ideas (taking France as its example) in Lepenies, *Sainte-Beuve*, pp. 317–62.

18. McLeod, *Secularisation*, p. 285.

19. Ibid., pp. 224, 262.

20. Browne, *Darwin*, vol. 2, p. 496.

21. Beales and Dawson, *Prosperity and Plunder*, pp. 291f.

22. Spiro, *Buddhism*, p. 284.

23. Joseph Fletcher, "Ch'ing Inner Asia," in: Fairbank and Twitchett, *Cambridge History of China*, vol. 10, pp. 35–106, at 99.

24. M. C. Goldstein, *Modern Tibet*, pp. 41ff.

25. Asad, *Formations*, pp. 210–12, 255.

26. Berkes, *Secularism in Turkey*, pp. 89ff.

27. Hilton, *A Mad, Bad and Dangerous People?* p. 176.

28. Butler, *Sea of Faith*, p. 270.

29. See the analysis of the overall process in Casanova, *Public Religions*, pp. 134ff.

30. Finke and Stark, *Churching of America*, p. 114 (Tab. 4.1).

31. D'Agostino, *Rome in America*, p. 52.

32. See the European overview in C. M. Clark and Kaiser, *Culture Wars*.

33. Chadwick, *History of the Popes*, p. 95.

34. Hardacre, *Shinto*, pp. 27ff.; McClain, *Japan*, pp. 267–72.

35. See Wakabayashi, *Anti-Foreignism*, a translation and interpretation of the principal sources.

36. Chidester, *Savage Systems*, pp. 11–16.

37. Petermann, *Geschichte der Ethnologie*, pp. 475f.

38. Hösch, *Balkanländer*, p. 97.

39. Keith, *Speeches*, pp. 382–86.

40. Tarling, *Southeast Asia*, pp. 320f.; Gullick, *Malay Society*, pp. 285ff.

41. Bartal, *Jews of Eastern Europe*, p. 73.

42. Federspiel, *Sultans*, pp. 99f.

43. F. Robinson, *Muslim Societies*, p. 187. On the attitudes of Sufi brotherhoods to colonial rule, see Abun-Nasr, *Muslim Communities of Grace*, pp. 200–235.

44. Strachan points out that the Germans first developed this kind of imperial subversion as a strategy in a real world war: Strachan, *First World War*, p. 694, and, in greater detail, ch. 9.

45. There is no up-to-date and comprehensive history if the Christian mission. Excellent on the British point of view is Andrew Porter, "An Overview, 1700–1914," in: Etherington, *Missions and Empire*, pp. 40–63. For original sources, see Harlow and Carter, *Archives of Empire*, vol. 2, pp. 241–364 (with an emphasis on missionary activism). Islamic missions and expansion, above all in Africa, must be left out of account here; see Hiskett, *Islam in Africa*. Missionary tendencies also developed in Buddhism (in Ceylon, for example), partly as a reaction to Christian penetration.

46. Tarling, *Southeast Asia*, p. 316.

47. Frykenberg, *Christianity in India*, p. 206.

48. For a succinct overview, see R. G. Tiedemann, "China and Its Neighbours," in: Hastings, *World History of Christianity*, pp. 369–415, at 390–402.

49. Surveys: Kevin Ward, "Africa," in: Hastings, *World History of Christianity*, pp. 192–237, at 203ff.; C. Marx, *Geschichte Afrikas*, pp. 90–100; Coquery-Vidrovitch, *Africa and the Africans*, pp. 207–31.

50. The work that set the standard in this field is a major anthropological study of South Africa: Comaroff and Comaroff, *Of Revelation and Revolution*. A. Porter, *Religion versus Empire?* is an outstanding recent account of missionary strategies, while a different view is offered in C. Hall, *Civilising Subjects*; see also Veer, *Conversion*.

51. Brian Stanley, "Christian Missions, Antislavery, and the Claims of Humanity, c. 1813–1873," in: Gilley and Stanley, *Cambridge History of Christianity*, pp. 443–57, at 445.

52. Andrew Porter, "Missions and Empire, c. 1873–1914," in: ibid., pp. 560–75, at 568.

53. See the biographies of the two mission leaders in Eber, *Jewish Bishop;* and A. J. Austin, *China's Millions.*

54. Peebles, *Sri Lanka*, p. 53.

55. Robert Eric Frykenberg, "Christian Missions and the Raj," in: Etherington, *Missions and Empire*, pp. 107–31, at 107, 112.

56. K. Ward, *Global Anglicanism.*

57. Deringil, *Well-Protected Domains*, pp. 113, 132.

58. For an account of the varied religious landscape, see Voll, *Islam*, ch. 3.

59. A point made strongly by Ahmad S. Dallal, "The Origins and Early Development of Islamic Reform," in: M. Cook, *New Cambridge History of Islam,* vol. 6, pp. 107–47, at 108, 111, quote at 115.

60. D. Cook, *Understanding Jihad*, pp. 74f.

61. John Obert Voll, "Foundations for Renewal and Reform: Islamic Movements in the Eighteenth and Nineteenth Centuries," in: Esposito, *Oxford History of Islam*, pp. 509–47, at 523, 525.

62. See chapter 10, above.

63. Bigler and Bagley, *Mormon Rebellion*, p. 263.

64. Shipps, *Mormonism.*

65. See the broad overview in J. R. Bowen, *Religions in Practice*, pp. 216–28.

66. Reynaldo Ileto, "Religion and Anti-colonial Movements," in: Tarling, *Cambridge History of Southeast Asia*, vol. 2, pp. 198–248, at 199ff.

67. Dabashi, *Shi'ism*, p. 182.

68. See the fascinating study of Bahaullah in J. R. Cole, *Modernity.*

69. Nikolaas A. Rupke, "Christianity and the Sciences," in: Gilley and Stanley, *Cambridge History of Christianity*, pp. 164–80.

70. Wernick, *Auguste Comte*, is very critical. There is a concise account of the Comte reception outside Europe in Forbes, *Positivism*, esp. pp. 147–58.

71. Funkenstein, *Perceptions of Jewish History*, pp. 186–96.

72. John Rogerson, "History and the Bible," in: Gilley and Stanley, *Cambridge History of Christianity*, pp. 181–96, at 195.

73. Fundamental here are Kurzman, *Modernist Islam*, esp. the editor's excellent introduction (pp. 3–27); A. Black, *Islamic Political Thought*, pp. 279–308; Hourani, *Arabic Thought*, still a classic in its field.

74. R. Guha, *Makers of Modern India*, pp. 53–70 (with sources).

75. B. D. Metcalf, *Islamic Revival*; F. Robinson, *Islam*, pp. 254–64; Pernau, *Bürger mit Turban*, pp. 219–24.

76. See the overview in Stietencron, *Hinduismus*, pp. 83–88; and the detailed regional accounts in K. W. Jones, *Reform Movements*, which also considers Muslim movements. On the nineteenth-century roots of today's Hindu nationalism, see Bhatt, *Hindu Nationalism*, chs. 2–3; and on the origins of the idea of "Indian spirituality," Aravamudan, *Guru English*.

77. See Sharma, *Modern Hindu Thought*.

78. Peebles, *Sri Lanka*, pp. 74f.

79. On the best example of this, see Weismann, *Naqshbandiyya*.

80. Lüddeckens, *Weltparlament*.

81. Gullick, *Malay Society*, p. 299.

82. Boudon et al., *Religion et culture*, pp. 39ff., 134; Chadwick, *History of the Popes*, p. 113

83. Chadwick, *History of the Popes*, pp. 159, 181f.

84. Arrington, *Brigham Young*, pp. 321ff.

85. This is the argument in F. Robinson, *Islam*, pp. 76f.

CONCLUSION: The Nineteenth Century in History

1. Ranke, *Aus Werk und Nachlaß*, vol. 4, p. 463.

2. Borst, *Medieval Worlds*, p. 71.

3. The latter is the title of an influential book by the German sociologist Hans Freyer (*Theorie des gegenwärtigen Zeitalters*, Stuttgart 1955); Freyer also published a *Weltgeschichte Europas* (1948).

4. There is a full translation of this important text in Philipp and Schwald, *Abd-al-Rahman al-Jabarti's History of Egypt*.

5. There is a recent English translation: Tahtawi, *An Imam in Paris*.

6. Blacker, *Japanese Enlightenment*, pp. 90–100. There are English translations of some of Fukuzawa's major writings, especially Fukuzawa, *Autobiography*.

7. See A. Black, *Islamic Political Thought*, pp. 288–91; and Abrahamian, *Iran*, pp. 65–69. Source excerpt in Kurzman, *Modernist Islam*, pp. 111–15.

8. Karl Kraus (1874–1936), one of the greatest minds of his age, deserves to be better known outside the German-speaking world; see a two-volume biography by Edward Timms (1986/2005). There is a memorable portrait of Tagore in Sen, *Argumentative Indian*, pp. 89–120. Hay, *Asian Ideas*, is still a key work on his influence; see also P. Mishra, *Ruins of Empire*, pp. 216–41. For the full richness of Indian (political) thought in our period see Bayly, *Recovering Liberties*.

9. There are now dozens of theories of modernity: see the anthology Waters, *Modernity*. Particularly fruitful approaches for historians are the (otherwise very different) ones proposed by S. N. Eisenstadt, Anthony Giddens, Richard Münch, Alain Touraine, Johann P. Arnason, Stephen Toulmin, and Peter Wagner.

10. Eisenstadt's still-powerful initial statement was *Multiple Modernities*; it was later spelled out in a number of papers and lectures, see Eisenstadt, *Comparative Civilizations*; see also Eisenstadt (ed.), *Multiple Modernities*.

11. Numerous examples are given in G. Robb, *Discovery of France*.

12. Janik and Toulmin, *Wittgenstein's Vienna*; Schorske, *Fin-de-Siècle Vienna*; but see also Hamann, *Hitler's Vienna*.

13. Two brief syntheses are: F. J. Bauer, *Das "lange" 19. Jahrhundert*, and Langewiesche, *Neuzeit*.

14. Dülffer, *Im Zeichen der Gewalt*, p. 245.

15. Adas, *Contested Hegemony*.

16. A classical diagnosis for Europe, first published in Italy in 1925, is Ruggiero, *Liberalism*.

17. Whitehead, *Science*, p. 96.

18. Bendix, *Kings or People*, vol. 1, p. 12.

19. Rokumeikan: so called after a Tokyo government building built in 1881–83 in the Italian style by the English architect Josiah Conder. It consisted of a billiards room, a reading room, and a number of guest suites. See Seidensticker, *Low City*, pp. 68f., 97–100.

20. J. Fisch, *Europa*, p. 29.

21. This notion is often met in the literature on political theory.

22. F. J. Bauer, *Das "lange" 19. Jahrhundert*, p. 41–50.

23. Martin Greiffenhagen, "Emanzipation," in *Historisches Wörterbuch der Philosophie*, Basle 1972, vol. 2, col. 447.

24. Croce, *History of Europe*, esp. ch. 1.

25. But see a pioneering study on the enormous *human* consequences of the Taiping Revolution: Meyer-Fong, *What Remains*.

26. On European anticolonialism, see the broad survey in Stuchtey, *Die europäische Expansion*.

27. Wallace, *Wonderful Century*, p. 379.

28. Zweig, *World of Yesterday*—still in print. The German original was published in Stockholm as *Die Welt von gestern: Erinnerungen eines Europäers*.

29. This emerges forcefully from a comparative study of postliberal politics in the 1920s: Plaggenborg, *Ordnung und Gewalt*; see also Nicholas Doumanis, "Europe and the Wider World," in Gerwarth, *Twisted Paths*, pp. 355–80.

30. Arendt, *Totalitarianism*.

BIBLIOGRAPHY

Only the first name is given for works with more than two authors or editors, and only the first place of publication where there are several. Multiple entries for the same author are listed in chronological order. In Chinese and Japanese names the family name is put first. Italics signify shortened titles used in the notes.

Abelshauser, Werner: *Umbruch* und Persistenz. Das deutsche Produktionsregime in historischer Perspektive, in: GG 27 (2001), pp. 503–23.

Abernethy, David B.: The Dynamics of *Global Dominance*. European Overseas Empires, 1415–1980, New Haven, CT 2000.

Abeyasekere, Susan: *Jakarta*. A History, 2nd ed., Singapore 1989.

Abrahamian, Ervand: *Iran*. Between Two Revolutions, Princeton, NJ 1982.

Abu-Lughod, Janet L.: *Cairo*. 1001 Years of the City Victorious, Princeton, NJ 1971.

———: *Rabat*. Urban Apartheid in Morocco, Princeton, NJ 1981.

———: *New York*, Chicago, Los Angeles: America's Global Cities, Minneapolis, MN 1999.

Abun-Nasr, Jamil M.: A History of the *Maghrib* in the Islamic Period, Cambridge 1987.

———: *Muslim Communities of Grace*. The Sufi Brotherhoods in Islamic Religious Life, London 2007.

Acham, Karl: *Einleitung*, in: idem (ed.), Geschichte der österreichischen Humanwissenschaften, vol. 4, Vienna 2002, pp. 5–64.

Acham, Karl, and Winfried Schulze: *Einleitung*, in: idem (eds.), Theorie der Geschichte, vol. 6: Teil und Ganzes, Munich 1990, pp. 9–29.

Achilles, Walter: *Deutsche Agrargeschichte* im Zeitalter der Reformen und der Industrialisierung, Stuttgart 1993.

Adamson, Bob: *China's English*. A History of English in Chinese Education, Hong Kong 2004.

Adas, Michael: The *Burma Delta*. Economic Development and Social Change on an Asian Rice Frontier, Madison, WI 1974.

———: *Machines* as the Measure of Men. Science, Technology, and Ideologies of Western Dominance, Ithaca, NY 1989.

——— (ed.): *Islamic and European Expansion*. The Forging of a Global Order, Philadelphia 1993.

———: *Contested Hegemony*. The Great War and the Afro-Asian Assault on the Civilizing Mission Ideology, in: JWH 15 (2004), pp. 31–63.

Adelman, Jeremy: *Frontier Development*. Land, Labour, and Capital on the Wheatlands of Argentina and Canada, 1890–1914, Oxford 1994.

———: *Republic of Capital*. Buenos Aires and the Legal Transformation of the Atlantic World, Stanford, CA 1999.

———: *Sovereignty* and Revolution in the Iberian Atlantic, Princeton, NJ 2006.

Adelman, Jeremy, and Stephen Aron: *From Borderlands* to Borders. Empires, Nation-States, and the Peoples in Between in North American History, in: AHR 104 (1999), pp. 814–41.

Adler, Jeffrey S.: *Yankee Merchants* and the Making of the Urban West. The Rise and Fall of Antebellum St. Louis, Cambridge 1991.

Afary, Janet: The *Iranian Constitutional Revolution*, 1906–1911. Grassroots Democracy, Social Democracy, and the Origins of Feminism, New York 1996.

Ageron, Charles-Robert: A *History of Modern Algeria*. From 1830 to the Present, London 1991.

Agoston, Gabor: *Guns for the Sultan*. Military Power and the Weapons Industry in the Ottoman Empire, Cambridge 2005.

Ajayi, J. F. Ade (ed.): *General History of Africa*, vol. 6: Africa in the Nineteenth Century until the 1880s, Paris 1989.

Akerman, James R. (ed.): *The Imperial Map*. Cartography and the Mastery of Empire, Chicago 2009.

Akerman, James R., and Robert W. Karrow, Jr. (eds.): *Maps*. Finding Our Place in the World, Chicago 2007.

Aldcroft, Derek H., and Anthony Sutcliffe (eds.): *Europe* in the International Economy 1500–2000, Cheltenham 1999.

Aldcroft, Derek H., and Simon P. Ville (ed.): The *European Economy*, 1750–1914: A Thematic Approach, Manchester 1994.

Aldrich, Robert: *Greater France*. A History of French Overseas Expansion, Basingstoke 1996.

Alexander, Manfred: Kleine *Geschichte Polens*, Stuttgart 2003.

Alexander, Paul, et al. (eds.): In the *Shadow* of Agriculture. Non-Farm Activities in the Javanese Economy, Past and Present, Amsterdam 1991.

Ali, Imran: The *Punjab* under Imperialism, 1885–1947, Princeton, NJ 1988.

Alleaume, Ghislaine: An *Industrial Revolution* in Agriculture? Some Observations on the Evolution of Rural Egypt in the Nineteenth Century, in: Proceedings of the British Academy 96 (1999), pp. 331–45.

Allen, Larry: The *Global Financial System* 1750–2000, London 2001.

Allen, Robert C.: The *British Industrial Revolution* in Global Perspective, Cambridge 2009.

———: *Global Economic History*. A Very Short Introduction, Oxford 2011.

Amado, Janaina, et al.: *Frontier in Comparative Perspective*. The United States and Brazil, Washington, DC 1990.

Amanat, Abbas: *Pivot of the Universe*. Nasir al-Din Shah Qajar and the Iranian Monarchy, 1831–1896, Berkeley, CA 1997.

Amaral, Samuel: The *Rise of Capitalism* on the Pampas. The Estancias of Buenos Aires, 1785–1870, Cambridge 1998.

Ambrose, Stephen E.: *Nothing Like It* in the World. The Men Who Built the Transcontinental Railroad, 1863–1869, New York 2000.

Amelung, Iwo: *Der Gelbe Fluß* in Shandong (1851–1911). Überschwemmungskatastrophen und ihre Bewältigung im China der späten Qing-Zeit, Wiesbaden 2000.

Amelung, Iwo, et al. (eds.): *Selbstbehauptungsdiskurse* in Asien. China—Japan—Korea, Munich 2003.

Amin, Camron Michael, et al. (eds.): The *Modern Middle East*. A Sourcebook for History. Oxford 2006.

Amino Yoshihiko: *Les Japonais et la mer*, in: Annales HSS 50 (1995), pp. 235–58.

Amitai, Reuven, and Michal Biran (eds.): *Mongols*, Turks, and Others. Eurasian Nomads and the Sedentary World, Leiden 2005.

Amrith, Sunil S.: *Migration and Diaspora* in Modern Asia, Cambridge 2011.

Amsden, Alice H.: The *Rise of "the Rest."* Challenges to the West from Late-Industrializing Economies, Oxford 2001.

Anastassiadou, Meropi: *Salonique*, 1830–1912. Une ville ottomane à l'âge des réformes, Leiden 1997.

Andaya, Barbara Watson, and Leonard Y. Andaya: A History of *Malaysia*, 2nd ed., Basingstoke 2001.

Anderson, Bonnie S.: *Joyous Greetings*. The First International Women's Movement, 1830–1860, New York 2000.

Anderson, Clare: *Convicts* in the Indian Ocean. Transportation from South Asia to Mauritius, 1815–53, Basingstoke 2000.

Anderson, Eugene N.: The *Food of China*, New Haven, CT 1990.

Anderson, Eugene N., and Pauline R. Anderson: *Political Institutions* and Social Change in Continental Europe in the Nineteenth Century, Berkeley, CA 1967.

Anderson, Fred: *Crucible* of War. The Seven Years' War and the Fate of Empire in British North America, 1754–1766, New York 2000.

Anderson, Matthew S.: The *Eastern Question*, 1774–1923, Basingstoke 1966.

———: The *Rise of Modern Diplomacy*, 1450–1919, London 1993.

Anderson, Perry: *Lineages* of the Absolutist State, London 1974.

Anderson, Robert D.: *European Universities* from the Enlightenment to 1914, Oxford 2004.

Andress, David: *1789*: The Threshold of the Modern Age, London 2008.

Angster, Julia: *Erdbeeren und Piraten*. Die Royal Navy und die Ordnung der Welt 1770–1860, Göttingen 2012.

Anna, Timothy E.: The *Fall of the Royal Government* in Mexico City, Lincoln NE 1978.

———: *Spain and the Loss of America*, Lincoln, NE 1983.

Appleby, Joyce: *Inheriting the Revolution*. The First Generation of Americans, Cambridge, MA 2000.

———: The *Relentless Revolution*. A History of Capitalism, New York 2010.

Applegate, Celia: A *Europe of Regions*. Reflections on the Historiography of Sub-national Places in Modern Times, in: AHR 104 (1999), pp. 1157–82.

Aravamudan, Srinivas: *Guru English*. South Asian Religion in a Cosmopolitan Language, Princeton, NJ 2006.

Arendt, Hannah: The Origins of *Totalitarianism*, New York 1951.

Arendt, Hannah: *On Revolution*, rev. ed., New York 1965.

Ariès, Philippe, and Georges Duby (eds.): *A History of Private Life*, 5 vols., Cambridge, MA 1987–91.

Arjomand, Saïd Amir: *Constitutions* and the Struggle for Political Order. A Study in the Modernization of Political Traditions, in: AES 33 (1992), pp. 39–82.

Armitage, David, and Michael J. Braddick (eds.): The *British Atlantic World*, 1500–1800, Basingstoke 2002.

Armitage, David, and Sanjay Subrahmanyam (eds.): The *Age of Revolutions* in Global Context, c. 1760–1840, New York 2010.

Arnold, David: *Police Power* and Colonial Rule. Madras 1859–1947, Delhi 1986.

———: *Colonizing the Body*. State Medicine and Epidemic Disease in Nineteenth-Century India, Berkeley, CA 1993.

———: *Science*, Technology and Medicine in Colonial India, Cambridge 2000.

Arrighi, Giovanni: The *Long Twentieth Century*. Money, Power, and the Origins of Our Times, London 1994.

Arrighi, Giovanni, et al. (eds.): The *Resurgence of East Asia*. 500, 150 and 50 Year Perspectives, London 2003.

Arrington, Leonard J.: *Brigham Young*. American Moses, New York 1985.

Asad, Talal: *Formations* of the Secular. Christianity, Islam, Modernity, Stanford, CA 2003.

Asbach, Olaf: Die *Erfindung* des modernen Europa in der französischen Aufklärung, in: Francia 31/2 (2005), pp. 55–94.

Asch, Ronald G. (ed.): Der europäische *Adel* im Ancien Régime. Von der Krise der ständischen Monarchien bis zur Revolution (ca. 1600–1789), Cologne 2001.

———: *Europäischer Adel* in der Frühen Neuzeit. Eine Einführung, Cologne 2008.

Ascher, Abraham: *P. A. Stolypin*. The Search for Stability in Late Imperial Russia, Stanford, CA 2001.

———: The *Revolution of 1905*. A Short History, Stanford, CA 2004.

Atiyeh, George N. (ed.): The *Book in the Arab World*. The Written Word and Communication in the Middle East, Albany, NY 1995.

Atkins, Keletso E.: *The Moon Is Dead*! Give Us Our Money! The Cultural Origins of an African Work Ethic, Natal, South Africa, 1843–1900, Portsmouth, NH 1993.

Atwill, David G.: The *Chinese Sultanate*. Islam, Ethnicity, and the Panthay Rebellion in Southwest China, 1856–1873, Stanford, CA 2005.

Aubin, Hermann, and Wolfgang Zorn (eds.): *Handbuch* der deutschen Wirtschafts- und Sozialgeschichte, vol. 1, Stuttgart 1971.

Auerbach, Jeffrey A.: The *Great Exhibition* of 1851. A Nation on Display, New Haven, CT 1999.

Augstein, Hannah Franziska: *Race*. The Origins of an Idea, 1760–1850, Bristol 1996.

Aung-Thwin, Michael: *Spirals* in Early Southeast Asian und Burmese History, in: JInterdH 21 (1991), pp. 575–602.

Auslin, Michael: *Negotiating with Imperialism*. The Unequal Treaties and the Culture of Japanese Diplomacy, Cambridge 2004.

Austen, Ralph A.: *African Economic History*. Internal Development and External Dependency, London 1987.

Austin, Alvyn J.: *China's Millions*. The China Inland Mission and Late Qing Society, 1832–1905, Grand Rapids, MI 2007.

Austin, Gareth, and Kaoru Sugihara (eds.): *Local Suppliers of Credit* in the Third World, 1750–1960, Basingstoke 1993.

——— (eds.): *Labour-Intensive Industrialization* in Global History, London 2013.

Avery, Peter, et al. (eds.): The *Cambridge History of Iran*, vol. 7: From Nadir Shah to the Islamic Republic, Cambridge 1991.

Ayalon, Ami: The *Press in the Arab Middle East*. A History, New York 1995.

———: *Political Journalism* and Its Audience in Egypt, 1875–1914, in: Culture & History 16 (1997), pp. 100–121.

Aydin, Cemil: The *Politics of Anti-Westernism* in Asia. Visions of World Order in Pan-Islamic and Pan-Asian Thought, New York 2007.

Baczko, Bronislaw: *Ending the Terror*. The French Revolution after Robespierre, Cambridge 1994.

Bade, Klaus J.: *Migration in European History*, Oxford 2003.

Bade, Klaus J., et al. (eds.): *Encyclopedia of European Migration* and Minorities. From the Seventeenth Century to the Present, New York 2011.

Badran, Margot: *Feminists*, Islam and Nation. Gender and the Making of Modern Egypt, Princeton, NJ 1995.

Bagehot, Walter: The *English Constitution* [1867], ed. R.H.S. Crossman, London 1964.

Baguley, David: *Napoleon III* and His Regime. An Extravaganza, Baton Rouge, LA 2000.

Bagwell, Philip S.: The *Transport Revolution* from 1770, London 1974.

Bähr, Jürgen: *Bevölkerungsgeographie*, 4th ed., Stuttgart 2004.

Bailey, Paul: *Reform the People*. Changing Attitudes towards Popular Education in Twentieth-Century China, Edinburgh 1990.

Bailyn, Bernard: *Atlantic History*. Concept and Contours, Cambridge, MA 2005.

Baines, Dudley: *Migration* in a Mature Economy. Emigration and Internal Migration in England and Wales, 1861–1900, Cambridge 1985.

Bairoch, Paul: *Révolution industrielle* et sous-développement, Paris 1963.

———: *De Jéricho à Mexico*. Villes et économie dans l'histoire, 2nd ed., Paris 1985.

———: *Les trois révolutions agricoles* du monde développé. Rendements et productivité de 1800 à 1985, in: Annales ESC 44 (1989), pp. 317–53.

———: *Victoires* et déboires, 3 vols., Paris 1997.

Baker, Alan R. H.: *Geography and History*. Bridging the Divide, Cambridge 2003.

Baker, Christopher J., and Pasuk Phongpaichit: A History of *Thailand*, Cambridge 2005.

Baker, Lee D.: *From Savage to Negro*. Anthropology and the Construction of Race, 1896–1954, Berkeley, CA 1998.

Bakhash, Shaul: *Iran*. Monarchy, Bureaucracy and Reform under the Qajars, 1858–1896, London 1978.

Baldasty, Gerald J.: The *Commercialization of News* in the Nineteenth Century, Madison, WI 1992.

Baldwin, Peter: *Contagion* and the State in Europe, 1830–1930, Cambridge 1999.

Baldwin, Peter C.: *In the Watches of the Night*. Life in the Nocturnal City, 1820–1930, Chicago 2012.

Balfour, Sebastian: The *End of the Spanish Empire*, 1898–1923, Oxford 1997.

Ball, Michael, and David Sunderland: An *Economic History of London*, 1800–1914, London 2001.

Ballantyne, Tony: *Orientalism* and Race. Aryanism in the British Empire, Basingstoke 2002.

Banga, Indu (ed.): The *City in Indian History*. Urban Demography, Society, and Politics, New Delhi 1991.

Bank, Jan, and Maarten van Buuren: *1900*. The Age of Bourgeois Culture, Assen 2004.

Banner, Stuart: *How the Indians* Lost Their Land. Law and Power on the Frontier, Cambridge, MA 2005.

Banti, Alberto M.: *Il Risorgimento italiano*, Rome 2004.

Banti, Alberto M., and Paul Ginsborg (eds.): *Il Risorgimento*, Turin 2007.

Banton, Michael: *Racial Theories*, Cambridge 1987.

Bardet, Jean-Pierre, and Jacques Dupâquier (eds.): *Histoire des populations de l'Europe*. Vol. 2: La révolution démographique, 1750–1914, Paris 1998.

Barfield, Thomas J.: The *Nomadic Alternative*, Upper Saddle River, NJ 1993.

Barjot, Dominique, and Jacques Frémeaux (eds.): *Les sociétés coloniales* à l'âge des empires: des années 1850 aux années 1950, Paris 2012.

Barkan, Elazar: The *Retreat of Scientific Racism*. Changing Concepts of Race in Britain and the United States between the World Wars, Cambridge 1991.

Barker, Hannah, and Simon Burrows (eds.): *Press*, Politics and the Public Sphere in Europe and North America, 1760–1820, Cambridge 2002.

Barker, Theo, and Anthony Sutcliffe (eds.): *Megalopolis*. The Giant City in History, Basingstoke 1993.

Barlow, Tani E. (ed.): *Formations* of Colonial Modernity in East Asia, Durham, NC 1997.

Barman, Roderick J.: *Citizen Emperor*. Pedro II and the Making of Brazil, 1825–1891, Stanford, CA 1999.

Barnes, David S.: The *Making* of a Social Disease. Tuberculosis in Nineteenth-Century France, Berkeley, CA 1995.

Barnes, Linda L.: *Needles*, Herbs, Gods, and Ghosts. China, Healing, and the West to 1848, Cambridge, MA 2007.

Barney, William L.: *Battleground* for the Union. The Era of the Civil War and Reconstruction, 1848–1877, Englewood Cliffs, NJ 1990.

—— (ed.): A *Companion* to Nineteenth-Century America, Malden, MA 2001.

Barrett, Thomas M.: At the *Edge of Empire*. The Terek Cossacks and the North Caucasus Frontier, 1700–1860, Boulder, CO 1999.

Barrow, Ian J.: *Making History*, Drawing Territory. British Mapping in India, c. 1756–1905, New Delhi 2003.

Barry, John M.: The Great *Influenza*: The Epic Story of the Deadliest Plague in History, New York 2004.

Barshay, Andrew E.: The *Social Sciences* in Modern Japan. The Marxian and Modernist Traditions, Berkeley, CA 2004.

Bartal, Israel: The *Jews of Eastern Europe*, 1772–1881, Philadelphia 2005.

Barth, Boris: *Die deutsche Hochfinanz* und die Imperialismen. Banken und Außenpolitik von 1914, Stuttgart 1995.

Barth, Boris, and Jürgen Osterhammel (eds.): *Zivilisierungsmissionen*. Imperiale Weltverbesserung seit dem 18. Jahrhundert, Konstanz 2005.

Barth, Volker: *Mensch versus Welt*. Die Pariser Weltausstellung von 1867, Darmstadt 2007.

Bärthel, Hilmar: *Wasser für Berlin*, Berlin 1997.

Bartholomew, James R.: The *Formation of Science* in Japan. Building a Research Tradition, New Haven, CT 1989.

Bartky, Ian R.: *Selling the True Time*. Nineteenth-Century Timekeeping in America, Stanford, CA, 2000.

Bartlett, Richard A.: The *New Country*. A Social History of the American Frontier, 1776–1890, New York 1976.

Barton, David: *Literacy*. An Introduction to the Ecology of Written Language, 2nd ed., Malden, MA 2007.

Bary, William T. de, et al. (eds.): *Sources of Chinese Tradition*, 2 vols., New York 1960.

Bassin, Mark: *Turner*, Solovev, and the "Frontier Hypothesis." The Nationalist Significance of Open Spaces, in: JMH 65 (1993), pp. 473–511.

————: *Imperial Visions*. Nationalist Imagination and Geographical Expansion in the Russian Far East, 1840–1865, Cambridge 1999.

Baud, Michiel, and Willem van Schendel: Toward a *Comparative History* of Borderlands, in: JWH 8 (1997), pp. 211–42.

Bauer, Arnold J.: *Goods*, Power, History. Latin America's Material Culture, Cambridge 2001.

Bauer, Franz J.: *Das "lange" 19. Jahrhundert*. Profil einer Epoche, Stuttgart 2004.

Bauman, Zygmunt: *Society under Siege*, Cambridge 2002.

Baumgart, Winfried: *Europäisches Konzert* und nationale Bewegung. Internationale Beziehungen 1830–1878, Paderborn 1999.

Bawden, C. R.: The Modern History of *Mongolia*, rev. ed., New York 1989.

Baxter, James C.: The *Meiji Unification* through the Lens of Ishikawa Prefecture, Cambridge, MA 1994.

Baycroft, Timothy, and Mark Hewitson (eds.): *What Is a Nation?* Europe 1789–1914, Oxford 2006.

Bayly, C. A.: *Rulers, Townsmen and Bazaars*. North Indian Society in the Age of British Expansion, 1770–1870, Cambridge 1983.

————: *Indian Society* and the Making of the British Empire, Cambridge 1988.

————: *Imperial Meridian*. The British Empire and the World 1780–1830, London 1989.

————: *Empire and Information*. Intelligence Gathering and Social Communication in India, 1780–1870, Cambridge 1996.

————: The *First Age* of Global Imperialism, c. 1760–1830, in: JICH 26 (1998), pp. 28–47.

————: The *Birth of the Modern World*, 1780–1914, Oxford 2004.

————: *Recovering Liberties*. Indian Thought in the Age of Liberalism and Empire, Cambridge 2012.

Beachey, R. W.: A History of *East Africa*, 1592–1902, London 1996.

Beales, Derek, and Eugenio F. Biagini: The *Risorgimento* and the Unification of Italy, 2nd ed., London 2002.

Beales, Derek, and Edward Dawson: *Prosperity and Plunder*. European Catholic Monasteries in the Age of Revolution, 1650–1815, Cambridge 2003.

Beasley, W. G. (ed.): *Select Documents* on Japanese Foreign Policy, 1853–1868, London 1955.

————: The *Meiji Restoration*, Stanford, CA 1973.

————: *Japanese Imperialism, 1894–1945*, Oxford 1989.

————: *Japan Encounters the Barbarian*. Japanese Travellers in America and Europe, New Haven, CT 1995.

Beck, Hanno: *Alexander von Humboldt*, 2 vols., Wiesbaden 1959–61.

Becker, Ernst Wolfgang: *Zeit der Revolution*! Revolution der Zeit? Zeiterfahrungen in Deutschland in der Ära der Revolutionen, 1789–1848/49, Göttingen 1999.

Becker, Frank (ed.): *Rassenmischehen—Mischlinge—Rassentrennung*. Zur Politik der Rasse im deutschen Kolonialreich, Stuttgart 2004.

Beckert, Sven: The *Monied Metropolis*. New York City and the Consolidation of the American Bourgeoisie, 1850–1896, Cambridge 2001.

————: *Emancipation and Empire*. Reconstructing the Worldwide Web of Cotton Production in the Age of the American Civil War, in: AHR 109 (2004), pp. 1405–38.

Beckett, J. V.: The *Aristocracy* in England, 1660–1914, Oxford 1986.

Beckwith, Christopher I.: *Empires of the Silk Road*. A History of Central Eurasia from the Bronze Age to the Present, Princeton, NJ 2009.

Beinart, William, and Peter Coates: *Environment and History*. The Taming of Nature in the USA and South Africa, London 1995.

Beinart, William, and Lotte Hughes: *Environment and Empire*, Oxford 2007.

Beinin, Joel, and Zachary Lockman: *Workers on the Nile*. Nationalism, Communism, Islam and the Egyptian Working Class, 1882–1954, Princeton, NJ 1987.

Belaubre, Christophe et al. (eds.): *Napoleon's Atlantic*. The Impact of Napoleonic Empire in the Atlantic World, Leiden 2010.

Belich, James: The *New Zealand Wars* and the Victorian Interpretation of Racial Conflict, Montreal 1986.

———: *Making Peoples*. A History of the New Zealanders from Polynesian Settlement to the End of the 19th Century, Honolulu 1996.

———: *Replenishing the Earth*. The Settler Revolution and the Rise of the Anglo-World, 1780–1930, Oxford 2009.

———: *A Cultural History of Economics?* in: Victorian Studies 53 (2010), pp. 116–21.

Bell, David A.: The *First Total War*. Napoleon's Europe and the Birth of Modern Warfare, London 2007.

Beller, Steven: *Franz Joseph*. Eine Biographie, Vienna 1997.

Bello, Andrés: *Selected Writings*, Oxford 1997.

Bello, David Anthony: *Opium* and the Limits of Empire. Drug Prohibition in the Chinese Interior, 1729–1850, Cambridge, MA 2005.

Bély, Lucien: *L'art de la paix en Europe*. Naissance de la diplomatie moderne XVIᵉ–XVIIIᵉ siècle, Paris 2007.

Bender, Thomas (ed.): *Rethinking American History* in a Global Age, Berkeley, CA 2002.

———: *A Nation among Nations*. America's Place in World History, New York 2006.

Bendix, Reinhard: *Kings or People*. Power and the Mandate to Rule, 2 vols., Berkeley, CA 1978.

Benedict, Carol: *Bubonic Plague* in Nineteenth-Century China, Stanford, CA 1996.

Bengtsson, Tommy, et al.: *Life under Pressure*. Mortality and Living Standards in Europe and Asia, 1700–1900, Cambridge, MA 2004.

Bengtsson, Tommy, and Osamu Saito (eds.): *Population* and Economy. From Hunger to Modern Economic Growth, Oxford 2000.

Benjamin, Thomas: The *Atlantic World*. Europeans, Africans, Indians and Their Shared History, 1400–1900, Cambridge 2009.

Benjamin, Walter: The *Arcades Project*, Cambridge, MA 1999.

Bensel, Richard Franklin: *Yankee Leviathan*. The Origins of Central State Authority in America, 1859–1877, Cambridge 1990.

———: The *Political Economy* of American Industrialization, 1877–1900, Cambridge 2000.

Bentley, Jerry H. (ed.): The *Oxford Handbook of World History*, Oxford 2011.

Bentley, Michael (ed.): *Companion* to Historiography, London 1997.

Benton, Lauren: The *Legal Regime* of the South Atlantic World, 1400–1750. Jurisdictional Complexity as Institutional Order, in: JWH 11 (2000), pp. 27–56.

———: *Law and Colonial Cultures*. Legal Regimes in World History, 1400–1900, Cambridge 2002.

Berelowitch, Wladimir, and Olga Medvedkova: Histoire de *Saint-Pétersbourg*, Paris 1996.

Berend, Iván T.: *History Derailed*. Central and Eastern Europe in the Long Nineteenth Century, Berkeley, CA 2003.

Bérenger, Jean: A History of the *Habsburg Empire*, 1700–1918, New York 1994.

Bereson, Ruth: The *Operatic State*. Cultural Policy and the Opera House, London 2002.

Berg, Christa, et al. (eds.): *Handbuch* der deutschen Bildungsgeschichte, vols. 3 and 4, Munich 1987, 1991.

Berg, Maxine: The *Age of Manufactures*, 1700–1820, London 1985.

Berg, Maxine, and Kristine Bruland (eds.): *Technological Revolutions* in Europe. Historical Perspectives, Cheltenham 1998.

Berger, Klaus: *Japonisme* in Western Painting from Whistler to Matisse, New York 1992.

Berger, Peter L.: The *Capitalist Revolution*. Fifty Propositions about Prosperity, Equality, and Liberty, New York 1986.

Berger, Stefan (ed.): A *Companion* to Nineteenth-Century Europe, 1789–1914, Malden, MA 2006.

Bergère, Marie-Claire: The *Golden Age* of the Chinese Bourgeoisie, 1911–1937, Cambridge 1989.

———: *Sun Yat-sen*, Stanford, CA 1998.

———: *Capitalisme et capitalistes* en Chine, XIXe–XXe siècle, Paris 2007.

———: *Shanghai*. China's Gateway to Modernity, Stanford, CA 2009.

Berkes, Niyazi: The Development of *Secularism in Turkey*, 2nd ed., New York 1998.

Berlin, Ira: *Many Thousands Gone*. The First Two Centuries of Slavery in North America, Cambridge, MA 1998.

———: *Generations of Captivity*. A History of African-American Slaves, Cambridge, MA 2003.

Berlioz, Hector: The *Memoirs* of Hector Berlioz [1870], ed. David Cairns, New York 2002.

Berman, Marshall: *All That Is Solid* Melts into Air. The Experience of Modernity, New York 1982.

Bermeo, Nancy, and Philip Nord (eds.): *Civil Society* before Democracy. Lessons from Nineteenth-Century Europe, Lanham, MD 2000.

Bernand, Carmen: Histoire de *Buenos Aires*, Paris 1997.

Bernecker, Walther L.: Kleine *Geschichte Haitis*, Frankfurt a.M. 1996.

Bernecker, Walther L., et al. (eds.): *Handbuch* der Geschichte Lateinamerikas, 3 vols., Stuttgart 1992–96.

———: Kleine *Geschichte Brasiliens*, Frankfurt a.M. 2000.

Bernhardt, Christoph (ed.): *Environmental Problems* in European Cities in the 19th and 20th Century, Münster 2001.

Bernier, Olivier: *The World in 1800*, New York 2000.

Bessière, Bernard: Histoire de *Madrid*, Paris 1996.

Best, Geoffrey: *War and Society* in Revolutionary Europe, 1770–1870, London 1982.

Bethell, Leslie (ed.): The *Cambridge History of Latin America*, 11 vols., Cambridge 1984–95.

Betts, Raymond F.: *Assimilation* and Association in French Colonial Theory, 1890–1914, New York 1970.

Beyer, Peter: *Religions in Global Society*, London 2006.

Beyrau, Dietrich, et al. (eds.): *Reformen* im Rußland des 19. und 20. Jahrhunderts. Westliche Modelle und russische Erfahrungen, Frankfurt a.M. 1996.

Beyrer, Klaus: Die *Postkutschenreise*, Tübingen 1985.

Beyrer, Klaus, and Michael Andritzky (eds.): *Das Netz*. Sinn und Sinnlichkeit vernetzter Systeme, Heidelberg 2002.

Bhagavan, Manu: *Sovereign Spheres*. Princes, Education, and Empire in Colonial India, New Delhi 2003.

Bhatia, Bal Mokand: *Famines in India*, 3rd ed., Delhi 1991.

Bhatt, Chetan: *Hindu Nationalism*. Origins, Ideologies and Modern Myths, Oxford 2001.

Bickers, Robert: *Britain in China*. Community, Culture and Colonialism, 1900–1949, Manchester 1999.

Bickford-Smith, Vivian, et al.: *Cape Town* in the Twentieth Century, Claremont 1999.

Biernacki, Richard: The *Fabrication of Labor*. Germany and Britain, 1640–1914, Berkeley, CA 1995.

Bigler, David L., and Will Bagley: The *Mormon Rebellion*. America's First Civil War, 1857–1858, Norman, OK 2011.

Billingsley, Philip: *Bakunin* in Yokohama. The Dawning of the Pacific Era, in: IHR 10 (1998), pp. 532–70.

Billington, Ray Allen: *Westward Expansion*. A History of the American Frontier, New York 1949.

Bilson, Geoffrey: A *Darkened House*. Cholera in Nineteenth-Century Canada, Toronto 1980.

Bird, Isabella: *Collected Travel Writings*, 12 vols., Bristol 1997.

Bird, James: The *Major Seaports* of the United Kingdom. London 1963.

Birmingham, David: A Concise History of *Portugal*, Cambridge 1993.

Bitsch, Marie-Thérèse: *Histoire de la Belgique*. De l'Antiquité à nos jours, Brussels 2004.

Black, Antony: The History of *Islamic Political Thought*. From the Prophet to the Present, Edinburgh 2001.

Black, Jeremy: *War and the World*. Military Power and the Fate of Continents, 1450–2000, New Haven, CT 1998.

Blackbourn, David: *History of Germany*, 1780–1918. The Long Nineteenth Century, 3rd ed., Malden, MA 2003.

――――: The *Conquest of Nature*. Water, Landscape and the Making of Modern Germany, London 2006.

Blackburn, Robin: The *Overthrow* of Colonial Slavery, 1776–1848, London 1988.

Blacker, Carmen: The *Japanese Enlightenment*. A Study of the Writings of Fukuzawa Yukichi, Cambridge 1964.

Blackford, Mansel G.: The *Rise of Modern Business* in Great Britain, the United States, and Japan, 2nd ed., Chapel Hill, NC 1998.

Blaise, Clark: *Time Lord*. Sir Sandford Fleming and the Creation of Standard Time, New York 2000.

Blake; Robert: *Disraeli*, London 1966.

Blanning, Timothy C. W.: The *French Revolutionary Wars*, 1787–1802, London 1996.

―――― (ed.): The *Oxford Illustrated History* of Modern Europe, Oxford 1996.

―――― (ed.): The *Eighteenth Century*. Europe 1688–1815, Oxford 2000.

―――― (ed.): The *Nineteenth Century*. Europe 1789–1914, Oxford 2000.

Bled, Jean-Paul: *Wien*. Residenz, Metropole, Hauptstadt, Vienna 2002.

Blickle, Peter: Von der *Leibeigenschaft* zu den Menschenrechten. Eine Geschichte der Freiheit in Deutschland, Munich 2003.

Blight, David W.: *Race and Reunion*. The Civil War in American Memory, Cambridge, MA 2001.

Blom, J.C.H., and E. Lamberts (eds.): History of the *Low Countries*, New York 1999.

Blom, Philipp: The *Vertigo Years*. Change and Culture in the West, 1900–1914, London 2008.

Blum, Jerome: The *Internal Structure* and Polity of the European Village Community from the Fifteenth to the Nineteenth Century, in: JMH 43 (1971), pp. 541–76.

――――: The *End of the Old Order* in Rural Europe, Princeton, NJ 1978.

———: *In the Beginning. The Advent of the Modern Age; Europe in the 1840s*, New York 1994.

Blumin, Stuart M.: The *Emergence of the Middle Class*. Social Experience in the American City, 1760–1900, Cambridge 1989.

Blussé, Leonard, and Femme Gaastra (eds.): On the *Eighteenth Century* as a Category of Asian History, Aldershot 1998.

Boahen, A. Adu (ed.): *General History of Africa*, vol. 7: Africa under Colonial Domination, 1880–1935, Paris 1985.

Bock, Gisela: *Women* in European History, Oxford 2002.

Böckler, Stefan: *Grenze*. Allerweltswort oder Grundbegriff der Moderne? in: Archiv für Begriffsgeschichte 45 (2003), pp. 167–220.

Bockstoce, John R.: *Whales, Ice, and Men*. The History of Whaling in the Western Arctic, Seattle, WA 1986.

Bodnar, John: *The Transplanted*. A History of Immigrants in Urban America, Bloomington, IN 1985.

Boeckh, Katrin: *Von den Balkankriegen* zum Ersten Weltkrieg. Kleinstaatenpolitik und ethnische Selbstbestimmung auf dem Balkan, Munich 1996.

Boemeke, Manfred F., et al. (eds.): *Anticipating Total War*. The German and American Experiences, 1871–1914, Cambridge 1999.

Boer, Pim den: *Europa*. De Geschiedenis van een idee, Amsterdam 1999.

Bohr, Paul Richard: *Famine in China* and the Missionary. Timothy Richard as Relief Administrator and Advocate of National Reform, 1876–1884, Cambridge, MA 1972.

Bois, Jean-Pierre: *De la paix des rois à l'ordre des empereurs*, 1714–1815, Paris 2003.

Boles, John B.: *The South* through Time. A History of an American Region, Englewood Cliffs, NJ 1995.

——— (ed.): A *Companion* to the American South, Malden, MA 2002.

Boli, John, and George M. Thomas (eds.): *Constructing World Culture*. International Nongovernmental Organizations since 1875, Stanford, CA 1999.

Bolton, Kingsley: *Chinese Englishes*. A Sociolinguistic History, Cambridge 2003.

Bonine, Michael E., et al. (eds): *Is There a Middle East?* The Evolution of a Geopolitical Concept, Stanford, CA 2012.

Bonnett, Alastair: The *Idea of the West*. Culture, Politics and History, Basingstoke 2004.

Boomgaard, Peter: *Children* of the Colonial State. Population Growth and Economic Development in Java, 1795–1880, Amsterdam 1989.

———: *Forest Management* and Exploitation in Colonial Java, 1677–1897, in: Forest and Conservation History 36 (1992), pp. 4–21.

———: *Frontiers of Fear*. Tigers and People in the Malay World, 1600–1950, New Haven, CT 2001.

———: *Southeast Asia*. An Environmental History, Santa Barbara, CA 2007.

Booth, Anne: The *Indonesian Economy* in the Nineteenth and Twentieth Centuries. A History of Missed Opportunities, London 1998.

Bordo, Michael D., and Roberto Cortés-Conde (eds.): *Transferring Wealth* and Power from the Old to the New World. Monetary and Fiscal Institutions in the 17th through the 19th Centuries, Cambridge 2001.

Bordo, Michael D., et al. (eds.): *Globalization* in Historical Perspective, Chicago 2003.

Borruey, René: Le port moderne de *Marseille*. Du dock au conteneur 1844–1974, Marseille 1994.

Borscheid, Peter: *Das Tempo-Virus*. Eine Kulturgeschichte der Beschleunigung, Frankfurt a.M. 2004.

Borscheid, Peter, and Niels Viggo Haueter (eds.): *World Insurance*. The Evolution of a Global Risk Network, Oxford 2012.

Borst, Arno: *Medieval Worlds*. Barbarians, Heretics, and Artists in the Middle Ages, Cambridge 1991.

Bose, Sugata: *A Hundred Horizons*. The Indian Ocean in the Age of Global Empire, Cambridge, MA 2006.

Bose, Sugata, and Ayesha Jalal: *Modern South Asia*. History, Culture, Political Economy, 2nd ed., New York 2004.

Bossenbroek, Martin: The *Living Tools* of Empire. The Recruitment of European Soldiers for the Dutch Colonial Army, 1814–1909, in: JICH 23 (1995), pp. 26–53.

Botsman, Daniel V.: *Freedom without Slavery*? "Coolies," Prostitutes, and Outcastes in Meiji Japan's "Emancipation Moment," in: AHR 116 (2011), pp. 1323–47.

Bouche, Denise: *Histoire de la colonisation française*, vol. 2: Flux et reflux (1815–1962), Paris 1991.

Bouchet, Ghislaine: *Le cheval à Paris* de 1850 à 1914, Geneva 1993.

Boudon, Jacques-Olivier: *Histoire du consulat et de l'Empire* (1799–1815), Paris 2000.

Boudon, Jacques-Olivier, et al.: *Religion et culture* en Europe au 19e siècle (1800–1914), Paris 2001.

Bourdé, Guy: *Urbanisation* et immigration en Amérique Latine. Buenos Aires (XIXe et XXe siècles), Paris 1974.

Bourdelais, Patrice, and Jean-Yves Raulot: Une *Peur Bleue*. Histoire du choléra en France, 1832–1854, Paris 1987.

Bourguet, Marie-Noëlle: *Déchiffrer la France*. La statistique départementale à l'époque napoléonienne, Paris 1988.

Bourguignon, François, and Christian Morrison: *Inequality* among World Citizens, 1820–1992, in: AER 92 (2002), pp. 727–44.

Bowen, H. V.: *British Conceptions* of Global Empire, 1756–83, in: JICH 26 (1998), pp. 1–27.

——: The *Business of Empire*. The East India Company and Imperial Britain, 1756–1833, Cambridge 2006.

Bowen, John R.: *Religions in Practice*. An Approach to the Anthropology of Religion, 3rd ed., Boston 2006.

Bowen, Roger W.: *Rebellion* and Democracy in Meiji Japan. A Study of Commoners in the Popular Rights Movement, Berkeley, CA 1980.

Bowler, Peter J., and Iwan Rhys Morus: *Making Modern Science*. A Historical Survey, Chicago 2005.

Boyar, Ebru, and Kate Fleet: A Social History of *Ottoman Istanbul*, Cambridge 2010.

Boyce, Gordon, and Simon P. Ville: The Development of *Modern Business*, Basingstoke 2002.

Boyce, Robert W. D.: *Imperial Dreams* and National Realities. Britain, Canada and the Struggle for a Pacific Telegraph Cable, 1879–1902, in: EHR 115 (2000), pp. 39–70.

Boyer, Richard E., and Keith A. Davis: *Urbanization* in 19th-Century Latin America, Los Angeles 1973.

Brading, D. A.: The *First America*. The Spanish Monarchy, Creole Patriots and the Liberal State, 1492–1867, Cambridge 1991.

Bradley, Joseph: *Muzhik and Muscovite*. Urbanization in Late Imperial Russia, Berkeley, CA 1985.

Brancaforte, Charlotte L. (ed.): The *German Forty-Eighters* in the United States, New York 1989.

Brandt, Peter, et al. (eds.): *Handbuch* der europäischen Verfassungsgeschichte im 19. Jahrhundert. Institutionen und Rechtspraxis im gesellschaftlichen Wandel, vol. 1: Um 1800, Bonn 2006.

Brantlinger, Patrick: *Dark Vanishings*. Discourse on the Extinction of Primitive Races, 1800–1930, Ithaca, NY 2003.

Braudel, Fernand: *History and the Social Sciences*, in: American Behavioral Scientist 4 (1960), pp. 3–13.

———: The *Mediterranean* and the Mediterranean World in the Age of Philip II, 2 vols., London 1972.

———: *On History*, Chicago 1980.

———: *Civilization and Capitalism*, 15th to 18th Century, 3 vols., London 1981–84.

———: The *History of Civilizations* [1963], London 1995.

Braunthal, Julius: *History of the International*, vol. 1: 1864–1914, London, 1966.

Bray, Francesca: The *Rice Economies*, Oxford 1986.

Breen, T. H.: The *Marketplace* of Revolution. How Consumer Politics Shaped American Independence, Oxford 2004.

Breman, Jan: *Taming the Coolie Beast*. Plantation Society and the Colonial Order in Southeast Asia, Delhi 1989.

Brennecke, Detlef: *Sven Hedin*, Reinbek 1986.

Brenner, Michael: *Zionism*. A Brief History, Princeton, NJ 2003.

Brenner, Michael, et al. (eds.): *Two Nations*. British and German Jews in Comparative Perspective, Tübingen 1999.

Breuer, Stefan: *Der Staat*. Entstehung, Typen, Organisationsstadien, Reinbek 1998.

Breuilly, John: *Nationalism* and the State, new ed., Manchester 1993.

Brewer, John, and Roy Porter (eds.): *Consumption* and the World of Goods, London 1993.

Breyfogle, Nicholas B.: *Heretics* and Colonizers. Forging Russia's Empire in the South Caucasus, Ithaca, NY 2005.

Bridge, Francis R.: The *Habsburg Monarchy* among the Great Powers, 1815–1918, New York 1990.

Bridge, Francis R., and Roger Bullen: The *Great Powers* and the European States System, 1815–1914, Harlow 1980.

Briese, Olaf: *Angst* in den Zeiten der Cholera. Seuchen-Cordon, 4 vols., Berlin 2003.

Briggs, Asa: *Victorian Cities*, Harmondsworth 1968.

Briggs, Asa, and Peter Burke: A Social History of the *Media*. From Gutenberg to the Internet, Cambridge 2002.

Brim, Sadek: *Universitäten* und Studentenbewegung in Russland im Zeitalter der großen Reformen, 1855–1881, Frankfurt a.M. 1985.

Brocheux, Pierre, and Daniel Hémery: *Indochine*. La colonisation ambiguë (1858–1954), Paris 1995.

Broadberry, Stephen, and Kevin H. O'Rourke (eds.), The *Cambridge Economic History of Modern Europe*, 2 vols., Cambridge 2010.

Brody, Hugh: The *Other Side of Eden*. Hunter-Gatherers, Farmers and the Shaping of the World, London 2001.

Broers, Michael: *Europe* under Napoleon 1799–1815, London 1996.

Broers, Michael: The *Napoleonic Empire* in Italy, 1796–1814. Cultural Imperialism in a European Context? Basingstoke 2005.

———: *Napoleon's Other War*. Bandits, Rebels and Their Pursuers in the Age of Revolutions, Oxford 2010.

Broeze, Frank: *Underdevelopment* and Dependency. Maritime India during the Raj, in: MAS 18 (1984), pp. 429–57.

——— (ed.): *Brides* of the Sea. Port Cities of Asia from the 16th–20th Centuries, Honolulu 1989.

——— (ed.): *Gateways* of Asia. Port Cities of Asia in the 13th–20th Centuries, London 1997.

Bronger, Dirk: *Metropolen*, Megastädte, Global Cities. Die Metropolisierung der Erde, Darmstadt 2004.

Brook, Timothy: The *Confusions of Pleasure*. Commerce and Culture in Ming China, Berkeley, CA 1999.

Brook, Timothy, and Bob Tadashi Wakabayashi (eds.): *Opium Regimes*. China, Britain, and Japan, 1839–1952, Berkeley, CA 2000.

Brooks, Jeffrey: *When Russia Learned to Read*. Literacy and Popular Literature, 1861–1917, Princeton, NJ 1985.

Broome, Richard: *Aboriginal Victorians*. A History since 1800, Crows Nest, NSW 2005.

Brötel, Dieter: *Frankreich im Fernen Osten*. Imperialistische Expansion und Aspiration in Siam und Malaya, Laos und China, 1880–1904, Stuttgart 1996.

Brower, Daniel R.: The *Russian City* between Tradition and Modernity, 1850–1900, Berkeley, CA 1990.

———: *Turkestan* and the Fate of the Russian Empire, London 2003.

Brower, Daniel R., and Edward J. Lazzerini (eds.): *Russia's Orient*. Imperial Borderlands and Peoples, 1700–1917, Bloomington, IN 1997.

Brown, Christopher Leslie: *Moral Capital*. Foundations of British Abolitionism, Chapel Hill, NC 2006.

Brown, Howard G.: *War*, Revolution and the Bureaucratic State. Politics and Army Administration in France, 1791–1799, Oxford 1995.

Brown, L. Carl (ed.): *Imperial Legacy*. The Ottoman Imprint on the Balkans and the Middle East, New York 1996.

Brown, Lucy: *Victorian News* and Newspapers, Oxford 1985.

Brown, Richard D.: *The Strength of a People*. The Idea of an Informed Citizenry in America, Chapel Hill, NC 1996.

Brown, Richard Maxwell: *No Duty to Retreat*. Violence and Values in American History and Society, New York 1991.

Browne, Janet E.: Charles *Darwin*. A Biography, 2 vols., New York 1995–2002.

Brownlee, John S.: *Japanese Historians* and the National Myths, 1600–1945, Vancouver 1997.

Bruford, Walter H.: The *German Tradition of Self-Cultivation*. "Bildung" from Humboldt to Thomas Mann, London 1975.

Bruhns, Hinnerk, and Wilfried Nippel (eds.): *Max Weber* und die Stadt im Kulturvergleich, Göttingen 2000.

Brunn, Gerhard, and Jürgen Reulecke (eds.): *Metropolis Berlin*. Berlin als deutsche Hauptstadt im Vergleich europäischer Hauptstädte, 1870–1939, Bonn 1992.

Brunner, Otto, et al. (eds.): *Geschichtliche Grundbegriffe*. Historisches Lexikon zur politisch-sozialen Sprache in Deutschland, 8 vols., Stuttgart 1972–97.

Brustein, William I.: *Roots of Hate*. Anti-Semitism in Europe before the Holocaust, Cambridge 2003.

Bryant, G. J.: *Indigenous Mercenaries* in the Service of European Imperialists. The Cause of the Sepoys in the Early British Indian Army, 1750–1800, in: War in History 7 (2000), pp. 2–28.

Buchanan, Francis: A *Journey from Madras* through the Countries of Mysore, Canara, and Malabar, 3 vols., London 1807.

Buettner, Elizabeth: *Empire Families*. Britons and Late Imperial India, Oxford 2004.

Bull, Hedley: The *Anarchical Society*. A Study of Order in World Politics, London 1977.

Bullard, Alice: *Exile* to Paradise. Savagery and Civilization in Paris and the South Pacific, 1790–1900, Stanford, CA 2000.

Bulliet, Richard W.: The *Camel* and the Wheel, New York 1975.

———: The Case for *Islamo-Christian Civilization*, New York 2004.

Bulmer-Thomas, Victor: The *Economic History of Latin America* since Independence, Cambridge 1994.

Bulmer-Thomas, Victor, et al. (eds.): The *Cambridge Economic History of Latin America*, 2 vols., Cambridge 2006.

Bumsted, J. M.: *The Peoples of Canada*. A Post-confederation History, Toronto 1992.

———: A *History of Canadian Peoples*, Toronto 1998.

Burbank, Jane, and Frederick Cooper: *Empires in World History*. Power and the Politics of Difference, Princeton, NJ 2010.

Burbank, Jane, and David L. Ransel (eds.): *Imperial Russia*. New Histories for the Empire, Bloomington, IN 1998.

Burckhardt, Jacob: *Werke*. Kritische Gesamtausgabe, Munich 2000ff.

Burke, Peter (ed.): The Cambridge Modern History, vol. 13: *Companion Volume*, Cambridge 1979.

———: A *Social History of Knowledge*. From Gutenberg to Diderot, Cambridge 2000.

Burnett, D. Graham: *Masters* of All They Surveyed. Exploration, Geography, and a British El Dorado, Chicago 2000.

Burns, E. Bradford: A History of *Brazil*, 2nd ed., New York 1980.

Burri, Monika, et al. (eds.): Die *Internationalität der Eisenbahn*, 1850–1970, Zurich 2003.

Burrow, John W.: The *Crisis of Reason*. European Thought, 1848–1914, New Haven, CT 2000.

Burrows, Edwin G., and Mike Wallace: *Gotham*. A History of New York City to 1898, Oxford 1999.

Burton, Antoinette: *Burdens of History*. British Feminists, Indian Women, and Imperial Culture, 1865–1915, Chapel Hill, NC 1994.

——— (ed.): *After the Imperial Turn*. Thinking with and through the Nation, Durham, NC 2003.

Bush, Michael L. (ed.): *Serfdom* and Slavery. Studies in Legal Bondage, London 1996.

———: *Servitude* in Modern Times, Cambridge 2000.

Bushman, Richard L.: The *Refinement* of America. Persons, Houses, Cities, New York 1992.

Bushnell, David, and Neill Macaulay: The *Emergence of Latin America* in the 19th Century, 2nd ed., New York 1994.

Butcher, John G.: The *British in Malaya* 1880–1941. The Social History of a European Community in Colonial South-East Asia, Kuala Lumpur 1979.

Butler, Jon: Awash in a *Sea of Faith*. Christianizing the American People, Cambridge, MA 1990.

Butlin, Robin A.: *Geographies of Empire*. European Empires and Colonies c. 1880–1960, Cambridge 2009.

Buzan, Barry, and Richard Little: *International Systems* in World History. Remaking the Study of International Relations, Oxford 2000.

Byres, Terence J.: Historical Perspectives on *Sharecropping*, in: JPS 10 (1983) pp. 7–40.

———: *Capitalism from Above* and Capitalism from Below. An Essay in Comparative Political Economy, Basingstoke 1996.

Cahan, David (ed.): *From Natural Philosophy* to the Sciences. Writing the History of Nineteenth-Century Science, Chicago 2003.

Cain, Peter J.: *Hobson* and Imperialism. Radicalism, New Liberalism, and Finance, 1887–1938, Oxford 2002.

Cain, Peter J., and A. G. Hopkins: *British Imperialism*, 2 vols., 2nd ed., London 2001.

Calhoun, Craig: *Nationalism*, Minneapolis 1997.

Cameron, Rondo: A Concise *Economic History* of the World. From Paleolithic Times to the Present, 3rd ed., New York 1997.

Campbell, Gwyn: An Economic History of *Imperial Madagascar*, 1750–1895, Cambridge 2005.

Campbell, Judy: *Smallpox* in Aboriginal Australia, 1829–1831, in: Australian Historical Studies 20 (1983), pp. 536–56.

Campbell, Peter R. (ed.): The *Origins* of the French Revolution, Basingstoke 2006.

Cannadine, David: The *Decline and Fall* of the British Aristocracy, New Haven, CT 1990.

———: The *Rise and Fall* of Class in Britain, New York 1999.

———: *Ornamentalism*. How the British Saw Their Empire, London 2001.

Canny, Nicholas (ed.): *Europeans* on the Move. Studies on European Migration, 1500–1800, Oxford 1994.

Canny, Nicholas, and Philip Morgan (eds.): The *Oxford Handbook of the Atlantic World*, 1450–1850, Oxford 2011.

Cao Shuji, *Zhongguo yimin shi* [History of migrants in China], vol. 6: Qing-Minguo shiqi [1644–1949], Fuzhou 1997.

Caplan, Jane, and John Torpey (eds.): *Documenting Individual Identity*. The Development of State Practices in the Modern World, Princeton, NJ 2001.

Caramani, Daniele: *Elections in Western Europe* since 1815. Electoral Results by Constituencies, London 2004.

Cárdenas, Enrique, et al. (eds.): An *Economic History* of Twentieth-Century Latin America, vol. 1, Basingstoke 2000.

Careless, James M. S.: *Frontier and Metropolis*. Regions, Cities, and Identities in Canada before 1914, Toronto 1989.

Carey, Brycchan: *British Abolitionism* and the Rhetoric of Sensibility. Writing, Sentiment, and Slavery, 1760–1807, Basingstoke 2005.

Carmagnani, Marcello: *The Other West*. Latin America from Invasion to Globalization, Berkeley, CA 2011.

Caron, François: *Histoire des chemins de fer en France*, 2 vols., Paris 1997–2005.

Caron, Jean-Claude: *Générations romantiques*. Les étudiants de Paris et le Quartier Latin (1814–1851), Paris 1991.

Caron, Jean-Claude, and Michel Vernus: *L'Europe au XIXe siècle*. Des nations aux nationalismes, 1815–1914, Paris 1996.

Carosso, Vincent P.: *The Morgans*. Private International Bankers, 1854–1913, Cambridge, MA 1987.

Carr, Raymond: *Spain, 1808–1975*, Oxford 1982.

Carter, Paul: The Road to *Botany Bay*. An Essay in Spatial History, London 1987.

Casanova, José: *Public Religions* in the Modern World. Chicago 1994.

Cassels, Alan: *Ideology* and International Relations in the Modern World, London 1996.

Cassirer, Ernst: The Philosophy of the *Enlightenment* [1932], Princeton 1951.

Cassis, Youssef: *Capitals of Capital*. A History of International Financial Centres, 1780–2005. Cambridge 2005.

Castel, Robert: From *Manual Workers* to Wage Laborers. Transformation of the Social Question, New Brunswick, NJ 2003.

Cattaruzza, Marina: *Arbeiter* und Unternehmer auf den Werften des Kaiserreichs, Wiesbaden 1988.

Cayton, Mary Kupiec, et al. (eds.): *Encyclopedia* of American Social History, 3 vols., New York 1993.

Ceadel, Martin: The *Origins of War Prevention*. The British Peace Movement and International Relations, 1730–1854, Oxford 1996.

Ceaser, James W.: *Reconstructing America*. The Symbol of America in Modern Thought, New Haven, CT 1997.

Cecco, Marcello de: *Money and Empire*. The International Gold Standard, 1890–1914, Oxford 1974.

Çelik, Zeynep: The *Remaking of Istanbul*. Portrait of an Ottoman City in the Nineteenth Century, Seattle, WA 1986.

Chadwick, Owen: A *History of the Popes*, 1830–1914, Oxford 1998.

Chakrabarti, Malabika: The *Famine of 1896–1897* in Bengal. Availability or Entitlement Crisis? New Delhi 2004.

Chandavarkar, Rajnarayan: *Imperial Power* and Popular Politics. Class, Resistance and the State in India, c. 1850–1950, Cambridge 1998.

Chandler, Alfred D., Jr.: The *Visible Hand*. The Managerial Revolution in American Business, Cambridge, MA 1977.

——: *Scale and Scope*. The Dynamics of Industrial Capitalism, Cambridge, MA 1990.

Chandler, Alfred D., Jr., et al. (eds.): *Big Business* and the Wealth of Nations, Cambridge 1997.

Chandler, Tertius, and Gerald Fox: *3000 Years* of Urban Growth, New York 1974.

Chang Chung-li: The *Chinese Gentry*. Studies on Their Role in Nineteenth-Century Chinese Society, Seattle, WA 1955.

Chang Hao: *Liang Ch'i-ch'ao* and Intellectual Transition in China, 1890–1907, Cambridge, MA 1971.

Chang Hsin-pao: *Commissioner Lin* and the Opium War, Cambridge, MA 1964.

Chang Kwang-chih (ed.): *Food in Chinese Culture*. Anthropological and Historical Perspectives, New Haven, CT 1977.

Chanock, Martin: *Law*, Custom and Social Order. The Colonial Experience in Malawi and Zambia, Cambridge 1985.

——: A *Peculiar Sharpness*. An Essay on Property in the History of Customary Law in Colonial Africa, in: JAfH 32 (1991), pp. 65–88.

Charbonneau, Hubert, and André Larose (eds.): The Great *Mortalities*. Methodological Studies of Demographic Crises in the Past, Liège 1979.

Charle, Christophe: *Histoire sociale de la France* au XIXᵉ siècle, Paris 1991.

Charle, Christophe: *Le siècle de la presse* (1830–1939), Paris 2004.

Charle, Christophe, et al. (eds.): *Transnational Intellectual Networks*. Forms of Academic Knowledge and the Search for Cultural Identities, Frankfurt a.M. 2004.

Chartier, Roger, et al.: *La ville des temps modernes*. de la Renaissance aux révolutions, Paris 1980.

Chaudhuri, Binay Bhushan (ed.): *Economic History of India* from Eighteenth to Twentieth Century, New Delhi 2005.

Chaudhuri, K. N.: *Trade* and Civilisation in the Indian Ocean. An Economic History from the Rise of Islam to 1750, Cambridge 1985.

———: *Asia* before Europe. Economy and Civilization of the Indian Ocean from the Rise of Islam to 1750, Cambridge 1990.

Cheong, Weng Eang: The *Hong Merchants* of Canton. Chinese Merchants in Sino-Western Trade, Richmond 1997.

Chew, Sing C.: *Ecological Degradation*. Accumulation, Urbanization, and Deforestation, 3000 B.C.–A.D. 2000, Walnut Creek 2001.

Chi Zihua: *Hongshizi yu jindai Zhongguo* [The Red Cross and Modern China], Hefei 2004.

Chidester, David: *Savage Systems*. Colonialism and Comparative Religion in Southern Africa, Charlottesville, VA 1996.

Chiu, T. N.: The *Port of Hong Kong*. A Survey of Its Development, Hong Kong 1973.

Ch'oe, Yŏng-ho, et al. (eds.): *Sources of Korean Tradition*, vol. 2: From the Sixteenth to the Twentieth Centuries, New York 2000.

Chopra, Preeti: *A Joint Enterprise*. Indian Elites and the Making of British Bombay, Minneapolis MN 2011.

Christian, David: *Maps of Time*. An Introduction to Big History, Berkeley, CA 2004.

Christopher, A. J.: The Quest for a *Census of the British Empire* c. 1840–1940, in: JHG 34 (2008), pp. 268–85.

Chudacoff, Howard P.: The Evolution of *American Urban Society*, 2nd ed., Englewood Cliffs, NJ 1981.

Church, Roy A., and E. A. Wrigley (eds.): The *Industrial Revolutions*, 11 vols., Oxford 1994.

Çikar, Jutta R. M.: *Fortschritt durch Wissen*. Osmanisch-türkische Enzyklopädien der Jahre 1870–1936, Wiesbaden 2004.

Çizakça, Murat: A Comparative Evolution of *Business Partnerships*. The Islamic World and Europe, Leiden 1996.

Clancy-Smith, Julia A.: *Rebel and Saint*. Muslim Notables, Populist Protest, Colonial Encounters (Algeria and Tunisia, 1800–1904), Berkeley, CA 1994.

———: *Mediterraneans*. North Africa and Europe in an Age of Migration, c. 1800–1900, Berkeley 2011.

Clancy-Smith, Julia A., and Frances Gouda (eds.): *Domesticating the Empire*. Race, Gender, and Family Life in French and Dutch Colonialism, Charlottesville, VA 1998.

Clarence-Smith, William Gervase: The *Third Portuguese Empire*, 1825–1975. A Study in Economic Imperialism, Manchester 1985.

———: *Islam* and the Abolition of Slavery, London 2006.

Clark, Christopher M.: *Kaiser Wilhelm II*, Harlow 2000.

———: *Iron Kingdom*. The Rise and Downfall of Prussia, 1600–1947, Camridge, MA 2006.

Clark, Christopher M., and Wolfram Kaiser (eds.): *Culture Wars*. Secular-Catholic Conflict in Nineteenth-Century Europe, Cambridge 2003.

Clark, Gregory: A *Farewell to Alms*. A Brief Economic History of the World, Princeton, NJ 2007.

Clark, Ian: The *Hierarchy of States*. Reform and Resistance in the International Order, Cambridge 1989.

——: *International Legitimacy* and World Society, Oxford 2007.

Clark, Peter: *British Clubs* and Societies, 1580–1800. The Origins of an Associational World, Oxford 2000.

—— (ed.): The *Cambridge Urban History* of Britain, 3 vols., Cambridge 2000–2001.

——: *European Cities and Towns*, 400–2000, Oxford 2009.

—— (ed.): The *Oxford Handbook of Cities* in World History, Oxford 2013.

Clark, William: *Academic Charisma* and the Origins of the Research University, Chicago 2006.

Clarke, Prescott, and J. S. Gregory: *Western Reports* on the Taiping. A Selection of Documents, London 1982.

Clarkson, L. A., and E. Margaret Crawford: *Feast* and Famine. Food and Nutrition in Ireland, 1500–1920, Oxford 2001.

Clive, John: *Macaulay*. The Shaping of the Historian, New York 1973.

Clodfelter, Michael: The *Dakota War*. The United States Army versus the Sioux, 1862–1865, Jefferson, NC 1998.

Clogg, Richard: A Concise History of *Greece*, Cambridge 1992.

Coates, Peter: *Nature*. Western Attitudes since Ancient Times, Berkeley, CA 1998.

Cochran, Sherman G.: *Encountering Chinese Networks*. Western, Japanese, and Chinese Corporations in China, 1880–1937, Berkeley, CA 2000.

Cohen, Patricia Cline: A *Calculating People*. The Spread of Numeracy in Early America, Chicago 1982.

Cohen, Paul A: *Between Tradition and Modernity*. Wang T'ao and Reform in Late Ch'ing China, Cambridge, MA 1974.

——: *History in Three Keys*. The Boxers as Event, Experience, and Myth, New York 1997.

Cohen, Robin (ed.): The Cambridge Survey of *World Migration*, Cambridge 1995.

——: *Global Diasporas*. An Introduction, London 1997.

Cohen, Robin, and Paul Kennedy: *Global Sociology*, 2nd ed. Basingstoke 2007.

Cohn, Bernard S.: *An Anthropologist* among the Historians and Other Essays, Delhi 1987.

——: *Colonialism* and Its Form of Knowledge. The British in India, Princeton, NJ 1996.

Cole, Joshua: The *Power of Large Numbers*. Population, Politics, and Gender in Nineteenth-Century France, Ithaca, NY 2000.

Cole, Juan R.: *Colonialism* and Revolution in the Middle East. Social and Cultural Origins of Egypt's Urabi Movement, Princeton, NJ 1993.

——: *Modernity* and the Millenium. The Genesis of the Baha'i Faith in the Nineteenth Century Middle East, New York 1998.

Colley, Linda: *Britons*. Forging the Nation 1707–1837, New Haven, CT 1992.

Collier, Simon, and William F. Sater: A History of *Chile*, 1808–1994, Cambridge 1996.

Comaroff, Jean, and John L. Comaroff: *Of Revelation and Revolution,* 2 vols., Chicago 1991–97.

——: *Ethnography* and the Historical Imagination, Boulder, CO 1992.

Connelly, Owen: The *Wars of the French Revolution* and Napoleon, 1792–1815, London 2006.

Conner, Patrick: *Oriental Architecture* in the West, London 1979.

Connor, Walker: *Ethnonationalism*. The Quest for Understanding, Princeton, NJ 1994.

Conrad, Peter: *Modern Times*, Modern Places, London 1998.

Conrad, Sebastian: *Globalisation* and the Nation in Imperial Germany, Cambridge 2010.

———: *German Colonialism*. A Short History, Cambridge 2011.

———: *Globalgeschichte*. Eine Einführung, Munich 2013.

Conrad, Sebastian, and Dominic Sachsenmaier (eds.): *Competing Visions* of World Order. Global Moments and Movements, 1880s–1930s, New York 2007.

Conway, Stephen: The *British Isles* and the War of American Independence, Oxford 2000.

Conze, Werner, et al. (eds.): *Bildungsbürgertum* im neunzehnten Jahrhundert, 4 vols., Stuttgart 1985–92.

Cook, David: *Understanding Jihad*, Berkeley, CA 2005.

Cook, Michael (ed.): The *New Cambridge History of Islam*, 6 vols., Cambridge 2010.

Coombes, Annie E.: *Reinventing Africa*. Museums, Material Culture and Popular Imagination in Late Victorian and Edwardian England, London 1994.

Cooper, Frederick, et al.: *Beyond Slavery*. Explorations of Race, Labor, and Citizenship in Postemancipation Societies, Chapel Hill, NC 2000.

Cooper, Sandi E.: *Patriotic Pacifism*. Waging War on War in Europe, 1815–1914, New York 1991.

Cooper, William J., and Thomas E. Terrill: The *American South*. A History, 2nd ed., 2 vols., New York 1996.

Cope, R. L.: Written in *Characters of Blood*? The Reign of King Cetshwayo Ka Mpande 1872–9, in: JAfH 36 (1995), pp. 247–69.

Coquery-Vidrovitch, Catherine: *Africa*. Endurance and Change South of the Sahara, Berkeley, CA 1988.

———: *L'Afrique* et les Africains au XIXᵉ siècle. Mutations, révolutions, crises, Paris 1999.

———: The History of *African Cities* South of the Sahara. From the Origins to Colonization, Princeton, NJ 2005.

Corbin, Alain: The *Lure of the Sea*. The Discovery of the Seaside in the Western World, 1750–1840, Berkeley, CA 1994.

——— (ed.): *L'Invention du XIXᵉ siècle*. Le XIXᵉ siècle par lui-même (littérature, histoire, société), Paris 1999.

Corfield, Penelope J.: *Time* and the Shape of History, New Haven, CT 2007.

Corvol, Andrée: *L'Homme aux bois*. Histoire des relations de l'homme et de la forêt (XVIIᵉ–XXᵉ siècle), Paris 1987.

Cosgrove, Denis: *Apollo's Eye*. A Cartographic Genealogy of the Earth in the Western Imagination, Baltimore, MD 2001.

Costa, Pietro: *Civitas*. Storia della cittadinanza in Europa, 4 vols., Rome 1999–2001.

Coulmas, Florian: *Japanische Zeiten*. Eine Ethnographie der Vergänglichkeit, Reinbek 2000.

Countryman, Edward: The *American Revolution*, 2nd ed., New York 2003.

Craib, Raymond B.: *Cartographic Mexico*. A History of State Fixations and Fugitive Landscapes, Durham, NC 2004.

Craig, Lee A., and Douglas Fisher: The *European Macroeconomy*. Growth, Integration and Cycles, 1500–1913, Cheltenham 2000.

Cranfield, Geoffrey A.: The *Press and Society*. From Caxton to Northcliffe, London 1978.

Crary, Jonathan: *Techniques of the Observer*. On Vision and Modernity in the Nineteenth Century, Cambridge, MA 1990.

Croce, Benedetto: *History of Europe* in the Nineteenth Century, New York 1933.

Cronon, William: *Changes in the Land*. Indians, Colonists, and the Ecology of New England, New York 1983.

———: *Nature's Metropolis*. Chicago and the Great West, New York 1991.

Crook, David P.: *Darwinism*, War and History. The Debate over the Biology of War from the "Origin of Species" to the First World War, Cambridge 1994.

Crook, J. Mordaunt: The *Rise of the Nouveaux Riches*. Style and Status in Victorian and Edwardian Architecture, London 1999.

Crosby, Alfred W.: *Ecological Imperialism*. The Biological Expansion of Europe, Cambridge 1986.

Crossick, Geoffrey, and Heinz-Gerhard Haupt: The *Petite Bourgeoisie* in Europe 1780–1914, New York 1998.

Crossick, Geoffrey, and Serge Jaumain (eds.): *Cathedrals of Consumption*. The European Department Store, 1850–1939, Aldershot 1999.

Crossley, Pamela Kyle: *Orphan Warriors*. Three Manchu Generations and the End of the Qing World, Princeton, NJ 1990.

———: A *Translucent Mirror*. History and Identity in Qing Imperial Ideology, Berkeley, CA 1999.

Crouzet, François: A History of the *European Economy*, 1000–2000, Charlottesville, VA 2001.

Crystal, David: *English* as a Global Language, Cambridge 1997.

Cullen, Michael J.: The *Statistical Movement* in Early Victorian Britain. The Foundations of Empirical Social Research, New York 1975.

Cunningham, Hugh: *Children and Childhood* in Western Society since 1500, 2nd ed., London 2005.

Curtin, Mary Ellen: *Black Prisoners* and Their World. Alabama, 1865–1900, Charlottesville, VA 2000.

Curtin, Philip D.: The Atlantic *Slave Trade*. A Census, Madison, WI 1969.

———: *Cross-Cultural Trade* in World History, Cambridge 1984.

———: *Death by Migration*. Europe's Encounter with the Tropical World in the Nineteenth Century, Cambridge 1989.

———: *Disease* and Empire. The Health of European Troops in the Conquest of Africa, Cambridge 1998.

———: *Location* in History. Argentina and South Africa in the Nineteenth Century, in: JWH 10 (1999), pp. 41–92.

———: The *World* and the West, Cambridge 2000.

Curwen, Charles A.: *Taiping Rebel*. The Deposition of Li Hsiu-ch'eng, Cambridge 1977.

Cushman, Jennifer Wayne: *Fields from the Sea*. Chinese Junk Trade with Siam during the Late Eighteenth and Early Nineteenth Centuries, Ithaca, NY 1993.

Cvetkovski, Roland: *Modernisierung* durch Beschleunigung. Raum und Mobilität im Zarenreich. Frankfurt a.M. 2006.

Dabashi, Hamid: *Shi'ism*. A Religion of Protest, Cambridge, MA 2011.

Dabringhaus, Sabine: *Territorialer Nationalismus*. Historisch-geographisches Denken in China 1900–1949, Cologne 2006.

———: *Geschichte Chinas* 1279–1949, 2nd ed., Munich 2009.

D'Agostino, Peter R.: *Rome in America*. Transnational Catholic Ideology from the Risorgimento to Fascism, Chapel Hill, NC 2004.

Dahlhaus, Carl: *Nineteenth-Century Music*, Berkeley, CA 2009.

Dahrendorf, Ralf: *LSE*. A History of the London School of Economics and Political Science, 1895–1995, Oxford 1995.

Dalrymple, William: *White Mughals*. Love and Betrayal in Eighteenth-Century India, London 2002.

———: The *Last Mughal*. The Fall of a Dynasty: Delhi 1857, London 2006.

Daly, Jonathan W.: *Autocracy* under Siege. Security Police and Opposition in Russia, 1866–1905, DeKalb, IL 1998.

Daly, Martin (ed.): The *Cambridge History of Egypt*, vol. 2: Modern Egypt from 1517 to the End of the Twentieth Century, Cambridge 1998.

Daly, Mary E.: The *Famine in Ireland*, Dundalk 1986.

Danbom, David B.: *Born in the Country*. A History of Rural America, Baltimore, MD 1995.

Daniel, Ute: *Hoftheater*. Zur Geschichte des Theaters und der Höfe im 18. und 19. Jahrhundert, Stuttgart 1995.

——— (ed.): *Augenzeugen*. Kriegsberichterstattung vom 18. zum 21. Jahrhundert, Göttingen 2006.

Dann, Otto: *Nation* und Nationalismus in Deutschland 1770–1990, 2nd ed., Munich 1994.

———: *Zur Theorie* des Nationalstaates, in: Bericht über das 8. deutsch-norwegische Historikertreffen in München, Mai 1995, Oslo 1996, pp. 59–70.

Danziger, Raphael: *Abd al-Qadir* and the Algerians. Resistance to the French and Internal Consolidation, New York 1977.

D'Arcy, Paul: The *People of the Sea*. Environment, Identity, and History in Oceania, Honolulu 2006.

Darwin, John: *Imperialism* and the Victorians. The Dynamics of Territorial Expansion, in: EHR 112 (1997), pp. 614–42.

———: *After Tamerlane*. The Global History of Empire since 1405, London 2007.

———: The *Empire Project*. The Rise and Fall of the British World-System, 1830–1970, Cambridge 2009.

———: *Unfinished Empire*. The Global Expansion of Britain, London 2012.

Das, Sisir Kumar: A *History of Indian Literature*, vol. 8: 1800–1910. Western Impact—Indian Response, New Delhi 1991.

Dasgupta, Partha: An *Inquiry* into Well-Being and Destitution, Oxford 1993.

Daumard, Adeline: *Les bourgeois* et la bourgeoisie en France depuis 1815, Paris 1991.

Daunton, Martin J.: *Progress* and Poverty. An Economic and Social History of Britain, 1700–1850, Oxford 1995.

———: *Trusting Leviathan*. The Politics of Taxation in Britain, 1799–1914, Cambridge 2001.

———: *Wealth* and Welfare. An Economic and Social History of Britain, 1851–1951, Oxford 2007.

Daunton, Martin J., and Rick Halpern (eds.): *Empire and Others*. British Encounters with Indigenous Peoples, 1600–1850, Philadephia 1999.

Davies, Norman: *God's Playground*. A History of Poland, vol. 2: 1795 to the Present, Oxford 1981.

Davies, Sam, et al. (eds.): *Dock Workers*. International Explorations in Comparative Labour History, 1790–1970, Aldershot 2000.

Davis, Clarence B., and Kenneth E. Wilburn, Jr. (eds.): *Railway Imperialism*, New York 1991.

Davis, David Brion: *Slavery and Human Progress*, New York 1984.

———: *Inhuman Bondage*. The Rise and Fall of Slavery in the New World, Oxford 2006.

Davis, David Brion, and Steven Mintz (eds.): The *Boisterous Sea* of Liberty. A Documentary History of America from Discovery through the Civil War, Oxford 1998.

Davis, John A. (ed.): *Italy* in the Nineteenth Century, 1796–1900, Oxford 2000.

Davis, Lance E., and Robert A. Huttenback: *Mammon* and the Pursuit of Empire. The Economics of British Imperialism, Cambridge 1986.

Davis, Mike: *Late Victorian Holocausts*. El Niño Famines and the Making of the Third World, New York 2001.

Davison, Graeme: The Rise and Fall of *Marvellous Melbourne*, Melbourne 1979.

Day, Jared N.: *Urban Castles*. Tenement Housing and Landlord Acitivism in New York City, 1890–1943, New York 1999.

Deák, István: The *Lawful Revolution*; Louis Kossuth and the Hungarians, 1848–1849, New York 1979.

Dean, Warren: With *Broadax* and Firebrand. The Destruction of the Brazilian Atlantic Forest, Berkeley, CA 1995.

Debeir, Jean-Claude, et al.: *In the Servitude of Power*. Energy and Civilisation through the Ages, London 1991.

Dehio, Ludwig: *The Precarious Balance*. Four Centuries of the European Power Struggle [1948], New York 1962.

Dehs, Volker: *Jules Verne*. Eine kritische Biographie, Düsseldorf 2005.

DeJong Boers, Bernice: *Tambora 1815*. De geschiedenis van een vulkaanuitbarsting in Indonesië, in: Tijdschrift voor Geschiedenis 107 (1994), pp. 371–92.

Dejung, Christof, and Niels P. Petersson (eds.): The *Foundations of Worldwide Economic Integration*. Power, Institutions, and Global Markets, 1850–1930, Cambridge 2013.

Delaporte, François: *Disease* and Civilization. The Cholera in Paris, 1832, Cambridge, MA 1986.

Deloria, Philip J., and Neal Salisbury (eds.): A *Companion to American Indian History*, Malden, MA 2004.

Delort, Robert, and François Walter: *Histoire de l'environnement européen*, Paris 2001.

Demel, Walter: *Der europäische Adel*. Vom Mittelalter bis zur Gegenwart, Munich 2005.

Deng Gang: *Maritime Sector*, Institutions, and Sea Power of Premodern China, Westport, CT 1999.

Deng, Kent: *China's Political Economy* in Modern Times. Changes and Economic Consequences, 1800–2000, London 2012.

Dennis, Richard: *English Industrial Cities* of the Nineteenth Century. A Social Geography, Cambridge 1984.

Denoon, Donald: *Settler Capitalism*, Oxford 1983.

——— (ed.): The *Cambridge History of the Pacific Islanders*, Cambridge 1997.

Denoon, Donald, and Philippa Mein-Smith: A *History of Australia*, New Zealand and the Pacific, Oxford 2000.

Deringil, Selim: The *Well-Protected Domains*. Ideology and the Legitimation of Power in the Ottoman Empire, London 1998.

Derks, Hans: History of the *Opium Problem*. The Assault on the East, ca. 1600–1950, Leiden 2012.

Desnoyers, Charles: *A Journey to the East*, Ann Arbor, MI 2004.

Desportes, Marc: *Paysages en mouvement*. Transports et perception de l'espace XVIIIᵉ–XXᵉ siècle, Paris 2005.

Dettke, Barbara: *Die asiatische Hydra*. Die Cholera von 1830/31 in Berlin und den preußischen Provinzen Posen, Preußen und Schlesien, Berlin 1995.

Devine, T. M.: The *Great Highland Famine*. Hunger, Emigration and the Scottish Highlands in the Nineteenth Century, Edinburgh 1988.

———: The *Scottish Nation*. 1700–2000, New York 1999.

———: *To the Ends of the Earth*: Scotland's Global Diaspora, 1750–2010, London 2011.

Dickens, Charles: *American Notes* for General Circulation [1842], ed. Patricia Ingham, London 2000.

Diesbach, Gislain de: *Ferdinand de Lesseps*, Paris 1998.

Dikötter, Frank: The *Discourse of Race* in Modern China, London 1992.

———: *Crime*, Punishment and the Prison in Modern China, New York 2002.

———: *Exotic Commodities*: Modern Objects and Everyday Life in China, London 2006.

Dingsdale, Alan: *Mapping Modernities*. Geographies of Central and Eastern Europe, 1920–2000, London 2002.

Dinnerstein, Leon: *Antisemitism in America*, New York 1994.

Dipper, Christof: *Übergangsgesellschaft*. Die ländliche Sozialordnung in Mitteleuropa um 1800, in: ZHF 23 (1996), pp. 57–87.

Disney, Anthony R.: A *History of Portugal* and the Portuguese Empire, 2 vols., Cambridge 2009.

Dobbin, Christine: *Asian Entrepreneurial Minorities*. Conjoint Communities in the Making of the World Economy, 1570–1940, Richmond 1996.

Dodds, Klaus, and Stephen A. Royle: *Rethinking Islands*, in: Journal of Historical Geography 29 (2003), pp. 487–98.

Dodgshon, Robert A.: *Society* in Time and Space. A Geographical Perspective on Change, Cambridge 1998.

Doel, H.W. van den: *Het Rijk van Insulinde*. Opkomst en ondergang van een Nederlandse kolonie, Amsterdam 1996.

Doeppers, Daniel F.: The Development of *Philippine Cities* before 1900, in: JAS 31 (1972), pp. 769–92.

Doering-Manteuffel, Anselm: *Die deutsche Frage* und das europäische Staatensystem 1815–1871, Munich 1993.

———: *Internationale Geschichte* als Systemgeschichte. Strukturen und Handungsmuster im europäischen Staatensystem des 19. und 20. Jahrhunderts, in: Wilfried Loth and Jürgen Osterhammel (eds.): Internationale Geschichte, Munich 2000, pp. 93–115.

Dohrn-van Rossum, Gerhard: *History of the Hour*. Clocks and Modern Temporal Orders, Chicago 1996.

Dominguez, Jorge I.: *Insurrection* or Loyality. The Breakdown of the Spanish American Empire, Cambridge, MA 1980.

Donald, David Herbert: *Lincoln*, New York 1995.

Dong, Madeleine Yue: *Republican Beijing*. The City and Its Histories, Berkeley, CA 2004.

Dormandy, Thomas: The *White Death*. A History of Tuberculosis, London 1999.

Dossal, Mariam: *Imperial Designs* and Indian Realities. The Planning of Bombay City, 1845–1875, Bombay 1996.

Doumani, Beshara: *Rediscovering Palestine*. Merchants and Peasants in Jabal Nablus, 1700–1900, Berkeley, CA 1995.

Dowd, Gregory Evans: A *Spirited Resistance*. The North American Indian Struggle for Unity, 1745–1815. Baltimore, MD 1992.

Dowe, Dieter, et al. (eds.): *Europe in 1848*. Revolution and Reform, New York 2001.

Doyle, Michael W.: *Empires*, Ithaca, NY 1986.

Doyle, Shane: *Population Decline* and Delayed Recovery in Bunyoro, 1860–1960, in: JAfH 41 (2000), pp. 429–58.

Doyle, William: The *Oxford History of the French Revolution*, 2nd ed., Oxford 2002.

Drake, Fred W.: China Charts the World. *Hsu Chi-yü* and His Geography of 1848, Cambridge, MA 1975.

Drescher, Seymour: *From Slavery to Freedom*. Comparative Studies in the Rise and Fall of Atlantic Slavery, Basingstoke 1999.

———: The *Mighty Experiment*. Free Labor versus Slavery in British Emancipation, Oxford 2002.

———: *Abolition*. A History of Slavery and Antislavery, Cambridge 2009.

Drescher, Seymour, and Stanley L. Engerman (eds.): A Historical Guide to *World Slavery*, New York 1998.

Driver, Felix: *Geography Militant*. Cultures of Exploration and Empire, Oxford 2001.

Driver, Felix, and David Gilbert (eds.): *Imperial Cities*. Landscape, Display and Identity, Manchester 1999.

Druett, Joan: *Rough Medicine*. Surgeons at Sea in the Age of Sail, New York 2000.

Dublin, Thomas: *Transforming Women's Work*. New England Lives in the Industrial Revolution, Ithaca, NY 1994.

Dubois, Laurent: *Avengers* of the New World. The Story of the Haitian Revolution, Cambridge, MA 2004.

———: A *Colony of Citizens*. Revolution and Slave Emancipation in the French Caribbean, 1787–1804, Chapel Hill, NC 2004.

Du Bois, W.E.B.: *Writings*, New York 1996.

Duchhardt, Heinz: *Balance of Power* und Pentarchie. Internationale Beziehungen 1700–1785, Paderborn 1997.

Dudden, Alexis: *Japan's Colonization of Korea*. Discourse and Power, Honolulu 2005.

Duggan, Christopher: The *Force of Destiny*. A History of Italy since 1796, London 2007.

Dülffer, Jost: *Regeln* gegen den Krieg. Die Haager Friedenskonferenzen 1899 und 1907 in der internationalen Politik, Frankfurt a.M. 1981.

———: *Im Zeichen der Gewalt*. Frieden und Krieg im 19. und 20. Jahrhundert, Cologne 2003.

Dülffer, Jost, et al.: *Vermiedene Kriege*. Deeskalation von Konflikten der Großmächte zwischen Krimkrieg und Ersten Weltkrieg, Munich 1997.

Dülmen, Richard van, and Sina Rauschenbach (eds.): *Macht des Wissens*. Die Entstehung der modernen Wissensgesellschaft, Cologne 2004.

Dumoulin, Michel, et al.: *Nouvelle histoire de Belgique*, 2 vols., Brussels 2006.

Dunlap, Thomas R.: *Nature and the English Diaspora*. Environment and History in the United States, Canada, Australia and New Zealand, Cambridge 1999.

Dunlavy, Colleen A.: *Politics* and Industrialization. Early Railroads in the United States and Prussia, Princeton, NJ 1994.

Dunn, John: *Africa* Invades the New World. Egypt's Mexican Adventure, 1863–1867, in: War in History 4 (1997), pp. 27–34.

Dupâquier, Jacques: *Histoire de la population française*, vol. 3: De 1789 à 1914, Paris 1988.

Dupâquier, Jacques, and Michel Dupâquier: *Histoire de la démographie*. La statistique de la population des origines à 1914, Paris 1985.

Duroselle, Jean-Baptiste: *Tout empire périra*. Une vision théorique des relations internationales, Paris 1992.

Dutton, George: The *Tây Son Uprising*. Society and Rebellion in Eighteenth-Century Vietnam, Honolulu 2006.

Duus, Peter: The *Abacus* and the Sword. The Japanese Penetration of Korea, 1895–1910, Berkeley, CA 1995.

—— (ed.): The *Japanese Discovery of America*. A Brief History with Documents, Boston 1997.

Duverger, Maurice (ed.): *Le concept d'empire*, Paris 1980.

Dyos, H. J., and D. H. Aldcroft: *British Transport*. An Economic Survey from the Seventeenth Century to the Twentieth, Leicester 1969.

Dyos, H. J., and Michael Wolff (eds.): The *Victorian City*. Images and Realities, 2 vols., London 1973.

Dyson, Tim: *Population and Development*. The Demographic Transition, London 2010.

Dwyer, Philip G., and Alan Forrest (eds.): *Napoleon and His Empire*. Europe, 1804–1814, Basingstoke 2007.

Earle, Peter: The *Pirate Wars*, London 2003.

Easterlin, Richard A.: *Growth Triumphant*. The Twenty-First Century in Historical Perspective, Ann Arbor, MI 1997.

——: *How Beneficent Is the Market?* A Look at the Modern History of Mortality, in: EREH 3 (1999), pp. 257–94.

——: The *Worldwide Standard of Living* since 1800, in: Journal of Economic Perspectives 14 (2000), pp. 7–26.

Eastman, Lloyd E.: *Family*, Fields, and Ancestors. Constancy and Change in China's Social and Economic History, 1550–1949, New York 1988.

Eber, Irene: The *Jewish Bishop* and the Chinese Bible. S.I.J. Schereschewsky (1831–1906), Leiden 1999.

Eberhard-Bréard, Andea: *Robert Hart* and China's Statistical Revolution, in: MAS 40 (2006), pp. 605–29.

Echenberg, Myron: *Pestis Redux*. The Initial Years of the Third Bubonic Plague Pandemic, 1894–1901, in: JWH 13 (2002), pp. 429–49.

——: *Africa in the Time of Cholera*. A History of Pandemics from 1817 to the Present, Cambridge 2011.

Eckermann, Johann Peter: *Conversations of Goethe* [1836–48], ed. J. K. Moorhead, London 1930.

Eckhardt, William: *Civilizations, Empires and Wars*. A Quantitative History of War, Jefferson, NC 1992.

Edney, Matthew H.: *Mapping an Empire*. The Geographical Construction of British India, 1765–1843, Chicago 1997.

Edsall, Nicholas C.: *Richard Cobden*. Independent Radical. Cambridge, MA 1986.

Eggert, Marion: Vom Sinn des Reisens. *Chinesische Reiseschriften* vom 16. bis zum frühen 19. Jahrhundert, Wiesbaden 2004.

Eichengreen, Barry: *Globalizing Capital*. A History of the International Monetary System, 2nd ed., Princeton, NJ 2008.

Eichengreen, Barry, and Ian W. McLean: The *Supply of Gold* under the Pre-1914 Gold Standard, in: EcHR 47 (1994), pp. 288–309.

Eichenhofer, Eberhard: *Geschichte des Sozialstaats* in Europa. Von der "sozialen Frage" bis zur Globalisierung, Munich 2007.

Eichhorn, Jaana: *Geschichtswissenschaft* zwischen Tradition und Innovation, Göttingen 2006.

Eickelman, Dale F.: *Time* in a Complex Society. The Moroccan Example, in: Ethnology 16 (1974), pp. 39–55.

———(ed.): *Muslim Travellers*. Pilgrimage, Migration, and the Religious Imagination. Berkeley, CA 1990.

Eisenstadt, S. N.: *Japanese Civilization*. A Comparative View, Chicago 1996.

———: *Multiple Modernities*, in: Daedalus 129 (2000), pp. 1–30.

———(ed.): *Multiple Modernities*, New Brunswick, NJ 2002.

———: *Comparative Civilizations* and Multiple Modernities, 2 vols., Leiden 2003.

Eklof, Ben, et al. (eds.): *Russia's Great Reforms*, 1855–1881, Bloomington, IN 1994.

Eldem, Edhem, et al.: The *Ottoman City* between East and West. Aleppo, Izmir, and Istanbul, Cambridge 1999.

Elleman, Bruce A.: *Modern Chinese Warfare*, 1795–1989, London 2001.

Elleman, Bruce A., and Stephen Kotkin (eds.): *Manchurian Railways* and the Opening of China. An International History, Armonk, NY 2010.

Ellerbrock, Karl-Peter: *Geschichte der deutschen Nahrungs- und Genußmittelindustrie* 1750–1914, Stuttgart 1993.

Elliott, John H.: *Empires* of the Atlantic World. Britain and Spain in America 1492–1830, New Haven, CT 2006.

Ellis, Richard: *Men and Whales*, New York 1991.

Elman, Benjamin A.: *From Philosophy to Philology*. Intellectual and Social Aspects of Change in Late Imperial China, Cambridge, MA 1984.

———: A Cultural History of *Civil Examinations* in Late Imperial China, Berkeley, CA 2000.

———: *On Their Own Terms*. Science in China, 1550–1900, Cambridge, MA 2005.

———: A Cultural History of *Modern Science* in China, Cambridge, MA 2006.

Elman, Benjamin A., and Alexander B. Woodside (eds.): *Education and Society* in Late Imperial China, 1600–1900, Berkeley, CA 1994.

Elson, Robert E.: The *End of the Peasantry* in Southeast Asia. A Social and Economic History of Peasant Livelihood, 1800–1990s, Basingstoke 1997.

Eltis, David: The *Rise of African Slavery* in the Americas, Cambridge 2000.

Eltis, David, and Stanley L. Engerman (eds.): The *Cambridge World History of Slavery,* vol. 3: AD 1420–AD 1804, Cambridge 2011.

Eltis, David, and David Richardson: *Atlas* of the Transatlantic Slave Trade, New Haven, CT 2010.

Elvin, Mark: *Another History*. Essays on China from a European Perspective, Broadway (New South Wales) 1996.

———: The Retreat of the *Elephants*. An Environmental History of China, New Haven, CT 2004.

Elvin, Mark, and Liu Ts'ui-jung (eds.): *Sediments of Time*. Environment and Society in Chinese History, Cambridge 1998.

Elwin, Verrier: *Myths of the North-East Frontier* of India, Itanagar 1993.

Emery, Edwin: The *Press and America*. An Interpretative History of Journalism, 2nd ed., Englewood Cliffs, NJ 1954.

Emmer, Pieter C. (ed.): *Colonialism and Migration*. Indentured Labour before and after Slavery, Dordrecht 1986.

Emmer, Pieter C.: De *Nederlandse slavenhandel*, 1500–1850. 2nd ed., Amsterdam 2003.

Emsley, Clive: *Gendarmes and the State* in Nineteenth-Century Europe, Oxford 1999.

Ener, Mine: *Managing Egypt's Poor* and Politics of Benevolence, 1800–1952, Princeton, NJ 2003.

Engelhardt, Ulrich: *"Bildungsbürgertum."* Begriffs- und Dogmengeschichte eines Etiketts, Stuttgart, 1986.

Engelsing, Rolf: *Analphabetentum* und Lektüre. Zur Sozialgeschichte des Lesens in Deutschland zwischen feudaler und industrieller Gesellschaft, Stuttgart 1973.

Engerman, Stanley L. (ed.): The *Terms of Labor*. Slavery, Serfdom, and Free Labor, Stanford, CA 1999.

Engerman, Stanley L., and João César das Neves: The *Bricks* of an Empire, 1415–1999. 585 Years of Portuguese Emigration, in: JEEcH 26 (1997), pp. 471–509.

Engerman, Stanley L., and Robert E. Gallman (eds.): The *Cambridge Economic History of the United States*, vol. 2: The Long Nineteenth Century, Cambridge 2000.

Epkenhans, Michael, and Gerhard P. Groß (eds.): *Das Militär* und der Aufbruch in die Moderne 1860 bis 1890, Munich 2003.

Erbe, Michael: *Revolutionäre Erschütterung* und erneuertes Gleichgewicht. Internationale Beziehungen 1785–1830, Paderborn 2004.

Erdem, Y. Hakan: *Slavery* in the Ottoman Empire and Its Demise, 1800–1909, Basingstoke 1996.

Ertman, Thomas: *Birth of the Leviathan*. Building States and Regimes in Medieval and Early Modern Europe, Cambridge 1997.

Escher, Anton, and Eugen Wirth: Die *Medina von Fes*, Erlangen 1992.

Esdaile, Charles J.: *Fighting Napoleon*. Guerrillas, Bandits, and Adventurers in Spain, 1808–1814, New Haven, CT 2004.

———: *Napoleon's Wars*. An International History, 1803–1815, London 2007.

Esenbel, Selçuk: The *Anguish* of Civilized Behavior. The Use of Western Cultural Forms in the Everyday Lives of the Meiji Japanese and the Ottoman Turks during the Nineteenth Century, in: Japan Review 5 (1994), pp. 145–85.

Esherick, Joseph W.: The Origins of the *Boxer Uprising*, Berkeley, CA 1987.

——— (ed.): *Remaking the Chinese City*. Modernity and National Identity, 1900–1950, Honolulu 1999.

Esherick, Joseph W., and Mary Backus Rankin (eds.): *Chinese Local Elites* and Patterns of Dominance, Berkeley, CA 1990.

Esherick, Joseph W., and Ye Wa: *Chinese Archives*. An Introductory Guide, Berkeley, CA 1996.

Esping-Andersen, Gøsta: The *Three Worlds* of Welfare Capitalism, Cambridge 1990.

Esposito, John L. (ed.): The *Oxford History of Islam*, Oxford 1999.

Etemad, Bouda: *Possessing the World*. Taking the Measurements of Colonization from the Eighteenth to the Twentieth Century, New York 2007.

Etherington, Norman: The *Great Treks*. The Transformation of Southern Africa, 1815–1854, Harlow 2001.

——— (ed.): *Missions and Empire*, Oxford 2005.

Evans, David C., and Mark R. Peattie: *Kaigun*. Strategy, Tactics, and Technology in the Imperial Japanese Navy, 1887–1941, Annapolis, MD 1997.

Evans, Eric J.: The *Forging* of the Modern State. Early Industrial Britain, 1783–1870, 2nd ed., London 1996.

Evans, Richard J.: *Death in Hamburg*. Society and Politics in the Cholera Years, 1830–1910, Oxford 1987.

———: *Rituals of Retribution*. Capital Punishment in Germany, 1600–1987, Oxford 1996.

Everdell, William R.: *The First Moderns*. Profiles in the Origins of Twentieth-Century Thought, Chicago 1997.

Ewald, Janet J.: *Soldiers*, Traders, and Slaves. State Formation and Economic Transformation of the Greater Nile Valley, 1700–1885, Madison, WI 1990.

———: *Crossers of the Sea*. Slaves, Freedmen, and Other Migrants in the Northwestern Indian Ocean, c. 1750–1914, in: AHR 105 (2000), pp. 69–91.

Evans, Martin: *European Atrocity*, African Catastrophe. Leopold II, the Congo Free State and Its Aftermath, London 2002.

Fabian, Johannes: *Time and the Other*. How Anthropology Makes Its Objects, New York 1983.

———: *Out of Our Minds*. Reason and Madness in the Exploration of Central Africa, Berkeley, CA 2000.

Fage, J. D., and Roland Oliver (eds.): The *Cambridge History of Africa*, 8 vols., Cambridge 1975–86.

Fahmi, Halid: *All the Pasha's Men*. Mehmed Ali, His Army and the Making of Modern Egypt, Cambridge 1997.

Fahmy, Khaled: An *Olfactory Tale* of Two Cities. Cairo in the Nineteenth Century, in: Jill Edwards (ed.), Historians in Cairo. Essays in Honor of George Scanlon, Cairo 2002, pp. 155–87.

Fahrmeir, Andreas: *Citizenship*. The Rise and Fall of a Modern Concept, New Haven, CT 2007.

———: *Europa* zwischen Restauration, Reform und Revolution 1815–1850, Munich 2012.

Fairbank, John K., and Denis Twitchett (eds.): The *Cambridge History of China*, Cambridge 1978 ff.

Falola, Toyin, and Kevin D. Roberts (eds.): *The Atlantic World. 1450–2000*, Bloomington, IN 2008.

Faragher, John Mack: *Sugar Creek*. Life on the Illinois Prairie, New Haven, CT 1986.

Farah, Caesar E. (ed.): *Decision Making* and Change in the Ottoman Empire, Kirksville, MO 1993.

———: The Politics of *Interventionism* in Ottoman Lebanon, 1830–1861, Oxford 2000.

Farnie, Douglas A.: *East and West of Suez*. The Suez Canal in History, 1854–1956, Oxford 1969.

———: The *English Cotton Industry* and the World Market, 1815–1896, Oxford 1979.

Farnie, Douglas A., and David J. Jeremy (eds.): The *Fibre* that Changed the World. The Cotton Industry in International Perspective, 1600–1990s, Oxford 2004.

Faroqhi, Suraiya: *Herrscher über Mekka*. Die Geschichte der Pilgerfahrt, Munich 1990.

———: *Approaching Ottoman History*. An Introduction to the Sources, Cambridge 1999.

———: The *Ottoman Empire* and the World around It, London 2004.

———: *Subjects of the Sultan*. Culture and Daily Life in the Ottoman Empire, London 2005.

——— (ed.): The *Cambridge History of Turkey*, vol. 3: The Later Ottoman Empire, 1603-1839, Cambridge 2006.

Farr, James R.: *Artisans* in Europe, 1300–1914, Cambridge 2000.

Fauser, Annegret: *Musical Encounters* at the 1889 Paris World's Fair, Rochester 2005.

Federico, Giovanni: *Feeding the World*. An Economic History of Agriculture, 1800–2000, Princeton, NJ 2005.

Federspiel, Howard M.: *Sultans, Shamans, and Saints*. Islam and Muslims in Southeast Asia, Honolulu 2007.

Fehrenbacher, Don E.: *Slavery*, Law, and Politics. The Dred Scott Case in Historial Perspective, abridged ed., Oxford 1981.

———: The *Slaveholding Republic*. An Account of the United States Government's Relations to Slavery, New York 2001.

Feinstein, Charles H.: An *Economic History of South Africa*. Conquest, Discrimination and Development, Cambridge 2005.

Feldbauer, Peter, et al. (eds.): *Die vormoderne Stadt*. Asien und Europa im Vergleich, Munich 2002.

Feldenkirchen, Wilfried: *Siemens*. Von der Werkstatt zum Weltunternehmen, 2nd ed., Munich 2003.

Fenske, Hans: Der moderne *Verfassungsstaat*. Eine vergleichende Geschichte von der Entstehung bis zum 20. Jahrhundert, Paderborn 2001.

Ferguson, Niall: The House of *Rothschild*, 2 vols., New York 1998–99.

———: *Civilization*. The West and the Rest, London 2011.

Fernández-Armesto, Felipe: *Civilizations*, London 2000.

Ferro, Marc (ed.): *Le livre noir* du colonialisme, XVIe–XXIe siècle, Paris 2003.

Figes, Orlando: *Crimea*. The Last Crusade, London 2010.

Findlay, Ronald, and Kevin H. O'Rourke: *Power and Plenty*. Trade, War, and the World Economy in the Second Millenium, Princeton, NJ 2007.

Findley, Carter V.: *Ottoman Civil Officaldom*. A Social History, Princeton, NJ 1989.

———: The *Turks* in World History, Oxford 2005.

———: *Turkey*, Islam, Nationalism, and Modernity. A History, 1789–2007, New Haven, CT 2010.

Finer, Samuel E.: The *History of Government* from the Earliest Times, 3 vols., Oxford 1997.

Fink, Carole: *Defending the Rights of Others*. The Great Powers, the Jews, and International Minority Protection, 1878–1938, Cambridge 2004.

Fink, Leon: *Sweatshops at Sea*. Merchant Seamen in the World's First Globalized Industry, from 1812 to the Present, Chapel Hill, NC 2011.

Finke, Roger, and Rodney Stark: The *Churching of America*, 1776–1990. Winners and Losers in Our Religious Economy, 5th ed., New Brunswick, NJ 2002.

Finnane, Antonia: *Changing Clothes in China*. Fashion, History, Nation, New York 2008.

Finscher, Ludwig: *Streicherkammermusik*, Kassel 2001.

Finzsch, Norbert: Die *Goldgräber* Kaliforniens. Arbeitsbedingungen, Lebensstandard und politisches System um die Mitte des 19. Jahrhunderts, Göttingen 1982.

———: *Konsolidierung* und Dissens. Nordamerika von 1800 bis 1865, Münster 2005.

Finzsch, Norbert, and Ursula Lehmkuhl (eds.): *Atlantic Communications*. The Media in American and German History from the Seventeenth to the Twentieth Century, Oxford 2004.

Fisch, Jörg: *Die europäische Expansion* und das Völkerrecht. Die Auseinandersetzungen um den Status der überseeischen Gebiete vom 15. Jahrhundert bis zur Gegenwart, Stuttgart 1984.

———: *Geschichte Südafrikas*, Munich 1990.

———: *Europa* zwischen Wachstum und Gleichheit 1850–1914, Stuttgart 2002.

———: *Immolating Women*. A Global History of Widow Burning from Ancient Times to the Present, Delhi 2005.

Fisch, Stefan: *Stadtplanung* im 19. Jahrhundert. Das Beispiel München bis zur Ära Theodor Fischer, Munich 1988.

Fischer, Claude S.: *America Calling.* A Social History of the Telephone to 1940, Berkeley, CA 1992.

Fischer, Wolfram (ed.): *Handbuch* der europäischen Wirtschafts- und Sozialgeschichte, 5 vols., Stuttgart 1980–93.

———: *Expansion,* Integration, Globalisierung. Studien zur Geschichte der Weltwirtschaft, Göttingen 1998.

Fischer, Wolfram, et al. (eds.): The *Emergence* of a World Economy, vol. 2: 1850–1914, Wiesbaden 1986.

Fischer-Tiné, Harald, and Michael Mann (eds.): *Colonialism* as Civilizing Mission. Cultural Ideology in British India, London 2004.

Fisher, Michael H.: *Indirect Rule in India.* Residents and the Residency System, 1764–1858, Delhi 1991.

———: *Counterflows to Colonialism.* Indian Travellers and Settlers in Britain, 1600–1857, Delhi 2004.

Flandreau, Marc: The Economics and Politics of *Monetary Unions.* A Reassessment of the Latin Monetary Union, 1865–71, in: Financial History Review 7 (2000), pp. 25–44.

Flandrin, Jean-Louis, and Massimo Montanari (eds.): *Food.* A Culinary History from Antiquity to the Present, New York 1999.

Fletcher, Joseph: *Integrative History.* Parallels and Interconnections in the Early Modern Period, 1500–1800, in: JTS 9 (1985), pp. 37–57.

Flores, Dan: *Bison Ecology* and Bison Diplomacy. The Southern Plains from 1800–1850, in: JAH 78 (1991), pp. 465–485.

Floud, Roderick, et al.: *Height,* Health and History. Nutritional Status in the United Kingdom, 1750–1980, Cambridge 1990.

———: The *Changing Body.* Health, Nutrition, and Human Development in the Western World since 1700, Cambridge 2011.

Floud, Roderick, and Paul Johnson (eds.): The *Cambridge Economic History of Britain,* 3 vols., Cambridge 2004.

Fluehr-Lobban, Carolyn: *Race and Racism.* An Introduction, Lanham, MD 2006.

Flynn, Dennis O., et al. (eds.): *Pacific Centuries.* Pacific and Pacific Rim History Since the Sixteenth Century, London 1999.

——— (eds.): *Global Connections* and Monetary History, 1470–1800, Aldershot 2003.

Flynn, Dennis O., and Arturo Giráldez: *Cycles of Silver.* Global Economic Unity through the Mid-Eighteenth Century, in: JWH 13 (2002), pp. 391–427.

Foerster, Roland G. (ed.): Die *Wehrpflicht.* Entstehung, Erscheinungsformen und politisch-militärische Wirkung, Munich 1994.

Fogel, Joshua A. (ed.): *Sagacious Monks* and Bloodthirsty Warriors. Chinese Views of *Japan in the Ming-Qing Period,* Norwalk, CT 2002.

———: *Articulating the Sinosphere.* Sino-Japanese Relations in Space and Time. Cambridge, MA 2009.

Fogel, Robert W.: *Without Consent or Contract.* The Rise and Fall of American Slavery, New York 1989.

———: The *Slavery Debates,* 1952–1990. A Retrospective, Baton Rouge, LA 2003.

———: The *Escape* from Hunger and Premature Death, 1700–2100. Europe, America, and the Third World, Cambridge 2004.

Fogel, Robert W., and Stanley L. Engerman: *Time on the Cross*. The Economics of American Negro Slavery, Boston 1974.

Fogelson, Robert M.: The *Fragmented Metropolis*. Los Angeles, 1850–1930, Cambridge, MA 1967.

Fohlen, Claude, and François Bédarida: *Histoire générale du travail*, vol. 3: L'Ère des révolutions (1765–1914), Paris 1960.

Fohrmann, Jürgen, et al.: *Gelehrte Kommunikation*. Wissenschaft und Medium zwischen dem 16. und 20. Jahrhundert, Vienna 2005.

Foner, Eric: *Reconstruction*. America's Unfinished Revolution, 1863–1877, New York 1988.

——— (ed.): The *New American History*, rev. ed., Philadelphia 1997.

———: The Story of *American Freedom*, New York 1998.

———: The *Fiery Trial*. Abraham Lincoln and American Slavery, New York 2010.

Foner, Eric, and John A. Garraty (eds.): The *Reader's Companion* to American History, Boston 1991.

Foner, Eric, and Lisa McGirr (eds.): *American History Now*, Philadelphia 2011.

Forbes, Geraldine Hancock: *Positivism* in Bengal. A Case Study in the Transmission and Assimilation of an Ideology, Calcutta 1975.

Ford, Lacy K. (ed.): A *Companion to the Civil War* and Reconstruction, Malden, MA 2005.

Foreman-Peck, James: A History of the *World Economy*. International Economic Relations since 1850, Brighton 1983.

Forrest, Alan I.: *Napoleon's Men*. The Soldiers of the Revolution and Empire, London 2002.

Forster, Peter G.: The *Esperanto Movement*, The Hague 1982.

Förster, Stig: *Die mächtigen Diener* der East India Company. Ursachen und Hintergründe der britischen Expansionspolitik in Südasien, 1793–1819, Stuttgart 1992.

———: Der *Weltkrieg* 1792–1815. Bewaffnete Konflikte und Revolutionen in der Weltgesellschaft, in: Jost Dülffer (ed.), Kriegsbereitschaft und Friedensordnung in Deutschland 1800–1914, Münster 1995, pp. 17–38.

Förster, Stig, et al. (eds.): *Bismarck*, Europe and Africa. The Berlin Africa Conference 1884–1885 and the Onset of Partition, Oxford 1988.

Förster, Stig, and Jörg Nagler (eds.): *On the Road to Total War*. The American Civil War and the German Wars of Unification, 1861–1871, Cambridge 1997.

Forsyth, James: A History of the *Peoples of Siberia*. Russia's North Asian Colony, 1581–1990, Cambridge 1992.

Fortes, Meyer, and E. E. Evans-Pritchard (eds.): *African Political Systems*, London 1967.

Fortna, Benjamin C.: *Imperial Classroom*. Islam, the State and Education in the Late Ottoman Empire, Oxford 2003.

Foster, Roy F.: *Modern Ireland*, 1600–1972, London 1988.

Foucault, Michel: *The Order of Things*. Archaeology of the Human Sciences, London 1989.

Foucher, Michel: *Fronts et frontières*. Un tour du monde géopolitique, Paris 1991.

Fox-Genovese, Elizabeth, and Eugen D. Genovese: The *Mind of the Master Class*. History and Faith in the Southern Slaveholders' Worldview, Cambridge 2005.

Fragner, Bert G.: Die *"Persophonie."* Regionalität, Identität und Sprachkontakt in der Geschichte Asiens, Berlin 1999.

Frangakis-Syrett, Elena: The *Greek Merchant Community* of Izmir in the First Half of the Nineteenth Century, in: Daniel Panzac (ed.), Les villes dans l'Empire ottoman, vol. 1. Marseille 1991, pp. 391–416.

Fraser, Derek: The *Evolution* of the British Welfare State. A History of Social Policy since the Industrial Revolution, 3rd ed., Basingstoke 2003.

Fraser, J. T. (ed.): The *Voices of Time*, 2nd ed., Amherst, MA. 1981.

Fraser, W. Hamish, and Irene Maver (eds.): *Glasgow*, vol. 2: 1830 to 1912, Manchester 1996.

Frédéric, Louis: *La vie quotidienne au Japon* au début de l'ère moderne (1868–1912), Paris 1984.

Fredrickson, George M.: *White Supremacy*. A Comparative Study in American and South African History, New York 1981.

————: *Racism*. A Short History, Princeton, NJ 2002.

Freehling, William W.: The *Road to Disunion*, 2 vols., Oxford 1990–2007.

Freeman, Christopher, and Francisco Louçã: *As Time Goes By*. From the Industrial Revolutions to the Information Revolution, Oxford 2001.

Freeman, Michael J.: *Railways* and the Victorian Imagination, New Haven, CT 1999.

Freitag, Ulrike: *Arabische Buddenbrooks* in Singapur, in: Historische Anthropologie 11 (2003), pp. 208–23.

————: *Indian Ocean Migrants* and State Formation in Hadhramaut. Reforming the Homeland, Leiden 2003.

Freund, Bill: The *African City*. A History, Cambridge 2007.

Frevert, Ute (ed.): *Militär und Gesellschaft* im 19. und 20. Jahrhundert, Stuttgart 1997.

————: *A Nation in Barracks*. Modern Germany, Military Conscription, and Civil Society, New York 2004.

Frevert, Ute, and Heinz-Gerhard Haupt (eds.): Der *Mensch* des 19. Jahrhunderts, Frankfurt a.M. 1999.

Frey, Linda, and Marsha Frey: The History of *Diplomatic Immunity*, Columbus, OH 1999.

Freyre, Gilberto: *The Mansions and the Shanties*. The Making of Modern Brazil, New York 1966.

Fried, Michael: *Menzel's Realism*. Art and Embodiment in Nineteenth-Century, Berlin, New Haven CT 2002.

Friedberg, Aaron L.: The *Weary Titan*. Britain and the Experience of Relative Decline, 1895–1905, Princeton, NJ 1988.

Frieden, Jeffry A.: *Global Capitalism*. Its Fall and Rise in the Twentieth Century, New York 2006.

Friedgut, Theodore H.: *Iuzovka and Revolution*, 2 vols., Princeton, NJ 1989–94.

Friedrichs, Christopher R.: The *Early Modern City*, 1450–1750, London 1995.

Friel, Ian: *Maritime History* of Britain and Ireland, c. 400–2001, London 2003.

Frost, Lionel: The *New Urban Frontier*. Urbanisation and City-Building in Australasia and the American West, Kensington, New South Wales 1991.

Fry, Michael: The *Scottish Empire*, Phantassie 2001.

Frykenberg, Robert E.: *Christianity in India*. From Beginnings to the Present, Oxford 2008.

Fueter, Eduard: Die *Schweiz* seit 1848. Geschichte–Politik–Wirtschaft, Zurich 1928.

Fujitani Takashi: *Splendid Monarchy*. Power and Pageantry in Modern Japan, Berkeley, CA 1996.

Fukutake Tadashi: *Asian Rural Society*. China, India, Japan, Seattle, WA 1967.

Fukuzawa Yukichi: The *Autobiography* of Fukuzawa Yukichi [1899], rev. transl. by Eiichi Kiyooka, New York 1966.

Fung, Edmund S.K.: The *Military Dimension* of the Chinese Revolution. The New Army and Its Role in the Revolution of 1911, Vancouver 1980.

Funkenstein, Amos: *Perceptions of Jewish History*, Berkeley, CA 1993.

Füssel, Marian: *Der Siebenjährige Krieg*. Ein Weltkrieg im 18. Jahrhundert, Munich 2010.

Gabaccia, Donna R., and Dirk Hoerder (eds.): *Connecting Seas* and Connected Ocean Rims. Indian, Atlantic, and Pacific Oceans and China Seas Migrations from the 1830s to the 1930s, Leiden 2011.

Gado, Boureima Alpha: Une histoire des famines au *Sahel*. Étude des grandes crises alimentaires (XIXᵉ–XXᵉ siècles), Paris 1993.

Galison, Peter: *Einstein's Clocks* and Poincaré's Maps. Empires of Time, New York 2003.

Gall, Lothar: *Bismarck*. The White Revolutionary, London 1986.

———: *Bürgertum in Deutschland*, Berlin 1989.

——— (ed.): *Stadt und Bürgertum* im Übergang von der traditionalen zur modernen Gesellschaft, Munich 1993.

———: *Bürgertum*, liberale Bewegung und Nation. Ausgewählte Aufsätze, Munich 1996.

———: *Krupp*. Der Aufstieg eines Industrieimperiums, Berlin 2000.

Gallarotti, Giulio M.: The *Anatomy* of an International Monetary Regime. The Classical Gold Standard, 1880–1914, New York 1995.

Galloway, J. H.: The *Sugar Cane Industry*. An Historical Geography from the Origins to 1914, Cambridge 1989.

Gammer; Moshe: *Muslim Resistance* to the Tsar. Shamil and the Conquest of Chechnia and Daghestan, London 1994.

Garavaglia, Juan Carlos: *Les hommes de la pampa*. Une histoire agraire de la campagne de Buenos Aires (1700–1830), Paris 2000.

Gardet, Louis, et al.: *Cultures and Time*, Paris 1976.

Gardiner, John: *The Victorians*. An Age in Retrospect, London 2002.

Garrioch, David: The *Formation of the Parisian Bourgeoisie, 1690–1830*, Cambridge, MA 1996.

Gasster, Michael: *Chinese Intellectuals* and the Revolution of 1911. The Birth of Modern Chinese Radicalism, Seattle, WA 1969.

Gaubatz, Piper Rae: *Beyond the Great Wall*. Urban Form and Transformation on the Chinese Frontiers, Stanford, CA 1996.

Gay, Hannah: *Clock Synchrony*, Time Distribution and Electrical Timekeeping in Britain 1880–1925, in: P&P 181 (2003), pp. 107–40.

Geertz, Clifford: *Agricultural Involution*. The Processes of Ecological Change in Indonesia, Berkeley, CA 1963.

———: *Negara*. The Theatre State in Nineteenth-Century Bali, Princeton, NJ 1980.

———: *Local Knowledge*. Further Essays in Interpretive Anthropology, New York 1983.

Geggus, David: *Slavery*, War and Revolution. The British Occupation of Saint-Domingue, 1793–1798, Oxford 1982.

Geisthövel, Alexa, and Habbo Knoch (ed.): *Orte der Moderne*. Erfahrungswelten des 19. und 20. Jahrhunderts, Frankfurt a.M. 2005.

Gelatt, Roland: The *Fabulous Phonograph*, 1877–1977, 2ⁿᵈ rev. ed. London 1977.

Gelder, Roelof van: *Het Oost-Indisch avontuur*. Duitsers in dienst van de VOC (1600–1800), Nijmegen 1997.

Gelvin, James L.: The *Modern Middle East*. A History, New York 2005.

Genovese, Eugene D.: *Roll, Jordan, Roll*. The World the Slaves Made, New York 1972.

Georgeon, François: *Abdulhamid II*. Le sultan calife, Paris 2003.

Geppert, Alexander C. T.: *Fleeting Cities*. Imperial Expositions in "Fin-de-Siècle" Europe, Basingstoke 2010.

Geraci, Robert P., and Michael Khodarkovsky (eds.): *Of Religion and Empire*. Missions, Conversion, and Tolerance in Tsarist Russia, Ithaca, NY 2001.

Gerhard, Dietrich: *Old Europe*. A Study of Continuity, 1000–1800, New York 1981.

Gernsheim, Helmut: The *History of Photography* from the Camera Obscura to the Beginning of the Modern Era, rev. ed., London 1969.

Gerschenkron, Alexander: *Economic Backwardness* in Historical Perspective, Cambridge, MA 1962.

Gerwarth, Robert (ed.): *Twisted Paths*. Europe 1914–1945, Oxford 2007.

Gestrich, Andreas, et al.: *Geschichte der Familie*, Stuttgart 2003.

Gestwa, Klaus: *Proto-Industrialisierung* in Rußland. Wirtschaft, Herrschaft und Kultur in Ivanovo und Pavlovo, 1741–1932, Göttingen 1999.

Geulen, Christian: *Wahlverwandte*. Rassendiskurs und Nationalismus im späten 19. Jahrhundert, Hamburg 2004.

———: Geschichte des *Rassismus*, Munich 2007.

Geyer, Dietrich: *Der russische Imperialismus*. Studien über den Zusammenhang zwischen innerer und auswärtiger Politik 1860–1914, Göttingen 1977.

Geyer, Martin H., and Johannes Paulmann (eds.): The *Mechanics* of Internationalism. Culture, Society, and Politics from the 1840s to the First World War, Oxford 2001.

Geyer, Michael, and Charles Bright: *World History* in a Global Age, in: AHR 100 (1995), pp. 1034–60.

———: *Global Violence* and Nationalizing Wars in Eurasia and America. The Geopolitics of War in the Mid-Nineteenth Century, in: CSSH 38 (1996), pp. 619–57.

Ghosh, Suresh Chandra: The *History of Education* in Modern India, 1757–1998, 2nd ed., Hyderabad 2000.

Giedion, Siegfried: *Mechanization* Takes Command. A Contribution to Anonymous History, New York 1948.

Gildea, Robert: *Barricades* and Borders. Europe 1800–1914, Oxford 1996.

Giliomee, Hermann: The *Afrikaners*. Biography of a People, London 2003.

Gillard, David: The *Struggle for Asia*, 1828–1914. A Study in British and Russian Imperialism, London 1977.

Gilley, Sheridan, and Brian Stanley (eds.): The *Cambridge History of Christianity*, vol. 8: World Christianities, c. 1815–c. 1914, Cambridge 2006.

Gilmour, David: *Curzon*, London 1995.

———: The *Ruling Caste*. Imperial Lives in the Victorian Raj, London 2005.

Ginsborg, Paul: *Daniele Manin* and the Venetian Revolution of 1848–49, Cambridge 1979.

Girault, René: *Diplomatie européenne* et impérialismes. Histoire des relations internationales contemporaines, vol. 1: 1871–1914, Paris 1979.

Girouard, Mark: *Cities and People*. A Social and Architectural History, New Haven CT 1985.

———: The *English Town*, New Haven, CT 1990.

Glazier, Ira A., and Luigi de Rosa (eds.): *Migration* across Time and Nations. Population Mobility in Historical Contexts, New York 1986.

Gluck, Carol: *Japan's Modern Myths*. Ideology in the Late Meiji Period, Princeton, NJ 1985.

———: The *Past in the Present*, in: Andrew Gordon (ed.), Postwar Japan as History, Berkeley, CA 1993, pp. 64–95.

Glynn, Ian, and Jenifer Glynn: The Life and Death of *Smallpox*, Cambridge 2004.

Goblot, Edmond: *La barrière et le niveau*. Étude sociologique sur la bourgeoisie française moderne, Paris 1925.

Göçek, Fatma Müge: *Rise of the Bourgeoisie*, Demise of Empire. Ottoman Westernization and Social Change, New York 1996.

Godechot, Jacques: *France* and the Atlantic Revolution of the Eighteenth Century, 1770–1799, New York 1965.

Godlewska, Anne, and Neil Smith (eds.): *Geography and Empire*, Oxford 1994.

Goehrke, Carsten: *Russischer Alltag*, vol. 2: Auf dem Weg in die Moderne, Zurich 2003.

Goldblatt, David: *The Ball Is Round*. A Global History of Football, London 2006.

Goldstein, Melvyn C.: A History of *Modern Tibet*, 1913–1951. The Demise of the Lamaist State, Berkeley, CA 1989.

Goldstein, Robert Justin: *Political Censorship* of the Arts and the Press in 19th Century Europe, Basingstoke 1989.

———— (ed.): The *War for the Public Mind*. Political Censorship in 19th Century Europe, Westport, CT 2000.

Goldstone, Jack A.: *Revolution and Rebellion* in the Early Modern World, Berkeley, CA 1991.

————: The *Problem* of the "Early Modern" World, in: JESHO 41 (1998), pp. 249–84.

Gollwitzer, Heinz: *Die gelbe Gefahr*. Geschichte eines Schlagworts. Studien zum imperialistischen Denken, Göttingen 1962.

————: *Geschichte des weltpolitischen Denkens*, 2 vols., Göttingen 1972–82.

Gong, Gerrit W.: The *Standard of "Civilization"* in International Society, Oxford 1984.

Goodman, Bryna: *Native Place*, City, and Nation. Regional Networks and Identities in Shanghai, 1853–1937, Berkeley, CA 1995.

Goodman, David G., and Masanori Miyazawa: *Jews in the Japanese Mind*. The History and Uses of a Cultural Stereotype, Lanham, MD 2000.

Goody, Jack: The *East in the West*, Cambridge 1996.

————: *Food* and Love. A Cultural History of East and West, London 1998.

————: The *Theft of History*, Cambridge 2006.

Goonatilake, Susantha: *Toward a Global Science*. Mining Civilizational Knowledge. Bloomington, IN 1998.

Goor, Jurrien van: *De Nederlandse koloniën*. Geschiedenis van de Nederlandse expansie 1600–1975, Den Haag 1994.

Gosewinkel, Dieter: *Einbürgern* und Ausschließen. Die Nationalisierung der Staatsangehörigkeit vom Deutschen Bund bis zur Bundesrepublik Deutschland, Göttingen 2001.

Gosewinkel, Dieter, and Johannes Masing (eds.): Die *Verfassungen* in Europa 1789–1949, Munich 2006.

Gott, Richard: *Cuba*. A New History, New Haven, CT 2004.

Gottschang, Thomas, and Diana Lary: *Swallows* and Settlers. The Great Migration from North China to Manchina, Ann Arbor, MI 2000.

Gouda, Frances: *Dutch Culture Overseas*. Colonial Practice in the Netherlands Indies, 1900–1942, Amsterdam 1995.

Gough, Hugh: The *Terror* in the French Revolution, Basingstoke 1998.

Gould, Eliga H.: The *Persistence of Empire*. British Political Culture in the Age of the American Revolution, Chapel Hill, NC 2000.

————: *A World Transformed?* Mapping the Legal Geography of the English-Speaking Atlantic, 1660–1825, in: Wiener Zeitschrift zur Geschichte der Neuzeit 3 (2003), pp. 24–37.

Gould, Eliga H., and Peter S. Onuf (eds.): *Empire and Nation*. The American Revolution in the Atlantic World, Baltimore, MD 2005.

Grabbe, Hans-Jürgen: Vor der großen *Flut*. Die europäische Migration in die Vereinigten Staaten von Amerika 1783–1820, Stuttgart 2001.

Gradmann, Christoph: *Krankheit im Labor*. Robert Koch und die medizinische Bakteriologie, Göttingen 2005.

Graf, Friedrich Wilhelm: Die *Wiederkehr* der Götter. Religion in der modernen Kultur, Munich 2004.

Graff, Harvey J.: The *Legacies* of Literacy. Continuities and Contradictions in Western Culture and Society, Bloomington, IN 1987.

Graham, Richard: *Independence* in Latin America. A Comparative Approach, 2nd ed., New York 1994.

Gran-Aymerich, Ève: *Naissance de l'archéologie moderne, 1798–1945*. Paris 1998.

Gransow, Bettina: *Geschichte der chinesischen Soziologie*, Frankfurt a.M. 1992.

Grant, Jonathan A.: *Rulers*, Guns, and Money. The Global Arms Trade in the Age of Imperialism, Cambridge, MA 2007.

Grassby, Richard: The *Idea of Capitalism* before the Industrial Revolution, Lanham 1999.

Graves, Joseph L.: The *Emperor's New Clothes*. Biological Theories of Race at the Millennium. New Brunswick, NJ 2001.

Green, Abigail: *Moses Montefiore*. Jewish Liberator, Imperial Hero, Cambridge, MA 2010.

Green, William A.: *British Slave Emancipation*. The Sugar Colonies and the Great Experiment, 1830–1865, Oxford 1976.

———: *Periodization* in European and World History, in: JWH 3 (1992), pp. 13–53.

Greene, Jack P., and Philip D. Morgan (eds.): *Atlantic History*. A Critical Appraisal, Oxford 2009.

Greene, Jack P., and Jack R. Pole (eds.): A *Companion* to the American Revolution, Malden, MA 2000.

Greenhalgh, Paul: *Ephemeral Vistas*. The "Expositions universelles," Great Exhibitions and World's Fairs, 1851–1939, Manchester 1988.

Gregor-Dellin, Martin: *Richard Wagner*. His Life, His Work, His Century, London 1983.

Gregory, Cedric E.: A Concise History of *Mining*, Lisse, Netherlands, 2001.

Gregory, Desmond: *Brute New World*. The Rediscovery of Latin America in the Early Nineteenth Century. London 1992.

Gregory, Paul R.: *Before Command*. An Economic History of Russia from Emancipation to the First Five-Year Plan. Princeton, NJ 1994.

Grell, Ole Peter, et al. (eds.): *Health Care* and Poor Relief in 18th and 19th Century Northern Europe. Aldershot 2002.

Grewal, J. S.: The *Sikhs* of the Punjab, Cambridge 1990.

Grewe, Wilhelm G.: *Epochen* der Völkerrechtsgeschichte, 2nd ed., Baden-Baden 1988.

——— (ed.): *Fontes historiae iuris gentium*, 3 vols. (in 5 parts), Berlin 1988–1995.

Grigg, David: The *Agricultural Systems* of the World. An Evolutionary Approach, Cambridge 1974.

———: The *Transformation* of Agriculture in the West, Oxford 1992.

Grossi, Verdiana: *Le pacifisme européen* 1889–1914, Brussels 1994.

Grove, Eric: The *Royal Navy* since 1815. A New Short History, Basingstoke 2005.

Grove, Richard H.: *Ecology*, Climate and Empire. Colonialism and Global Environmental History, 1400–1940, Cambridge 1995.

———: *Green Imperialism*. Colonial Scientists, Ecological Crises and the History of Environmental Concern, 1600–1800, Cambridge 1995.

Grübler, Arnulf: The Rise and Fall of *Infrastructures*. Dynamics of Evolution and Technological Change in Transport, Heidelberg 1990.

———: *Technology* and Global Change, Cambridge 1998.

Gründer, Horst: *Geschichte der deutschen Kolonien*, 5th ed., Paderborn 2004.

Gründer, Horst, and Peter Johanek (eds.): *Kolonialstädte*. Europäische Enklaven oder Schmelztiegel der Kulturen? Münster 2002.

Grüttner, Michael: *Arbeitswelt* an der Wasserkante. Sozialgeschichte der Hamburger Hafenarbeiter 1886–1914, Göttingen 1984.

Gruzinski, Serge: Histoire de *Mexico*, Paris 1996.

Guelzo, Allen C: *Fateful Lightning*. A New History of the Civil War and Reconstruction, Oxford 2012.

Gugerli, David (ed.): *Vermessene Landschaften*. Kulturgeschichte und technische Praxis im 19. und 20. Jahrhundert, Zurich 1999.

Gugerli, David, and Daniel Speich: *Topografien* der Nation. Politik, kartografische Ordnung und Landschaft im 19. Jahrhundert, Zurich 2002.

Guha, Ramachandra (ed.): *Makers of Modern India*, Cambridge, MA 2011.

Guha, Sumit: *Environment and Ethnicity* in India, 1200–1991, Cambridge 1999.

———: *Health and Population* in South Asia. From Earliest Times to the Present, London 2001.

Guibernau, Montserrat: *Nationalisms*. The Nation-State and Nationalism in the Twentieth Century, Cambridge 1996.

Gullick, J. M.: *Malay Society* in the Late Nineteenth Century. The Beginnings of Change, Kuala Lumpur 1987.

Gump, James O.: The *Dust* Rise Like Smoke. The Subjugation of the Zulu and the Sioux, Lincoln, NE 1994.

Gupta, Narayani: *Delhi* between Two Empires 1803–1931. Society, Government and Urban Growth, Delhi 1981.

Gupta, Partha Sarathi, and Anirudh Deshpande (eds.): The *British Raj* and Its Indian Armed Forces, 1857–1939, Delhi 2002.

Gutiérrez, David G.: *Walls and Mirrors*. Mexican Americans, Mexican Immigrants and the Politics of Ethnicity, Berkeley, CA 1995.

Guy, Donna J., and Thomas E. Sheridan (eds.): *Contested Ground*. Comparative Frontiers on the Northern and Southern Edges of the Spanish Empire, Tucson, AZ 1998.

Guy, R. Kent: *Qing Governors* and Their Provinces. The Evolution of Territorial Administration in China, 1644–1796, Seattle, WA 2010.

Gyory, Andrew: *Closing the Gate*. Race, Politics, and the Chinese Exclusion Act, Chapel Hill, NC 1998.

Haarmann, Harald: *Weltgeschichte der Sprachen*. Von der Frühzeit des Menschen bis zur Gegenwart, Munich 2006.

Haber, Stephen (ed.): *How Latin America Fell Behind*. Essays on the Economic Histories of Brazil and Mexico, 1800–1914, Stanford, CA 1997.

Habermas, Jürgen: The *Structural Transformation* of the Public Sphere. An Inquiry into a Category of Bourgeois Society, Cambridge, MA 1989.

Habib, S. Irfan, and Dhruv Raina (ed.): *Social History of Science* in Colonial India, Oxford 2007.

Hachtmann, Rüdiger: *Epochenschwelle* zur Moderne. Einführung in die Revolution von 1848/49, Tübingen 2002.

Häfner, Lutz: *Gesellschaft* als lokale Veranstaltung. Die Wolgastädte Kazan' und Saratov (1870–1914), Cologne 2004.

Hagen, Mark von: *Empires*, Borderlands, and Diasporas. Eurasia as Anti-Paradigm for the Post-Soviet Era, in: AHR 109 (2004), pp. 445–68.

Haines, Michael R., and Richard H. Steckel (eds.): A *Population History* of North America, Cambridge 2000.

Haines, Robin F.: *Emigration* and the Labouring Poor. Australian Recruitment in Britain and Ireland (1831–60), London 1997.

Haj, Samira: *Land*, Power and Commercialization in Lower Iraq, 1850–1958. A Case of "Blocked Transition," in: JPS 2 (1994), pp. 126–63.

Halecki, Oskar: The *Limits and Divisions* of European History, New York 1950.

Hall, Catherine: *Civilising Subjects*. Metropole and Colony in the English Imagination, 1830–1867, Cambridge 2002.

Hall, D.G.E.: A History of *South-East Asia*, 4th ed., Basingstoke 1981.

Hall, John Whitney, et al. (eds.): The *Cambridge History of Japan*, 6 vols., Cambridge 1989–1999.

Hall, Peter: *Cities in Civilization*. Culture, Innovation, and Urban Order, London 1998.

Halliday, Stephen: The *Great Stink* of London. Sir Joseph Bazalgette and the Cleansing of the Victorian Capital, Thrupp 1999.

Halperin-Donghi, Tulio: The *Aftermath* of Revolution in Latin America, New York 1973.

Hamann, Brigitte: *Hitler's Vienna*, London 2010.

Hämäläinen, Pekka: The Rise and Fall of *Plains Indian Horse Cultures,* in: JAH 90 (2003), pp. 833–62.

———: The *Comanche Empire*, New Haven CT 2008.

Hämäläinen, Pekka, and Samuel Truett: *On Borderlands,* in: JAH 98 (2011), pp. 338–61.

Hamashita Takeshi: *China, East Asia and the Global Economy*. Regional and Historical Perspectives, London 2008.

Hamilton, C. I.: *Anglo-French Naval Rivalry*, 1840–1870, Oxford 1993.

Hamilton, Carolyn: *Terrific Majesty*. The Powers of Shaka Zulu and the Limits of Historical Invention, Cambridge, MA 1998.

Hamilton, Carolyn, et al. (eds.): The *Cambridge History of South Africa*, vol. 1: *From Early Times to 1885*, Cambridge 2010.

Hamilton, Gary G.: *Commerce* and Capitalism in Chinese Societies, London 2006.

Hamlin, Christopher: *Cholera*. The Biography, Oxford 2009.

Hamm, Michael F. (ed.): The *City in Late Imperial Russia*, Bloomington, IN 1986.

Hamnett, Brian: *Juárez*, Harlow 1994.

Hampsher-Monk, Iain (ed.): The *Impact* of the French Revolution. Texts from Britain in the 1790s, Cambridge 2005.

Hancock, David: *Citizens of the World*. London Merchants and the Integration of the British Atlantic Community, 1735–1785, Cambridge 1996.

Haneda Masashi, and Miura Toru (eds.): *Islamic Urban Studies*. Historical Review and Perspectives, London 1994.

Hanes, Jeffrey E.: The *City as Subject*. Seki Hajime and the Reinvention of Modern Osaka, Berkeley, CA 2002.

Hanioğlu, M. Şükrü: The *Young Turks* in Opposition, New York 1995.

———: *Preparation* for a Revolution. The Young Turks, 1902–1908, Oxford 2001.

———: A *Brief History* of the Late Ottoman Empire, Princeton, NJ 2008.

Hanley, Susan B.: *Everyday Things* in Premodern Japan. The Hidden Legacy of Material Culture, Berkeley, CA 1997.

Hanley, Susan B., and Yamamura Kozo: Economic and Demographic Change in *Preindustrial Japan*, 1600–1868, Princeton, NJ 1977.

Hannaford, Ivan: *Race*. The History of an Idea in the West, Washington, DC 1996.

Hansen, Mogens Herman (ed.): A Comparative Study of Thirty *City-State Cultures*, Kopenhagen 2000.

Hanssen, Jens: Fin-de-siècle *Beirut*. The Making of an Ottoman Provincial Capital, Oxford 2006.

Hanssen, Jens, et al. (eds.): The *Empire in the City*. Arab Provincial Capitals in the Late Ottoman Empire, Würzburg 2002.

Hao Yen-p'ing; The *Commercial Revolution* in Nineteenth-Century China. The Rise of Sino-Western Mercantile Capitalism, Berkeley, CA 1986.

Hardach-Pinke, Irene: Die *Gouvernante*. Geschichte eines Frauenberufs, Frankfurt a.M. 1993.

Hardacre, Helen: *Shinto and the State. 1868–1988*, Princeton, NJ 1989.

Hardacre, Helen, and Adam L. Kern (eds.): *New Directions* in the Study of Meiji Japan, Leiden 1997.

Hardiman, David: *Feeding the Baniya*. Peasants and Usurers in Western India, Delhi 1996.

———: *Usury*, Dearth and Famine in Western India, in: P&P 152 (1996), pp. 113–56.

Harding, Colin, and Simon Popple: In the *Kingdom of Shadows*. A Companion to Early Cinema, London 1996.

Hardtwig, Wolfgang: *Genossenschaft*, Sekte, Verein in Deutschland, vol. 1, Munich 1997.

Hardtwig, Wolfgang, and Klaus Tenfelde (eds.): *Soziale Räume* in der Urbanisierung. Studien zur Geschichte der Stadt Munich im Vergleich 1850 bis 1933, Munich 1990.

Hardy, Anne: *Health and Medicine in Britain* since 1860, Basingstoke 2001.

Hardy, William: The Origins of the *Idea of the Industrial Revolution*, Victoria, British Columbia, 2006.

Harlow, Barbara, and Mia Carter (eds.): *Archives of Empire*, 2 vols., Durham, NC 2003.

Harrison, Mark: *Public Health* in British India. Anglo-Indian Preventive Medicine 1859–1914, Cambridge 1994.

———: *Climates and Constitutions*. Health, Race, Environment and British Imperialism in India, 1600–1850, New Delhi 1999.

———: *Medicine* in an Age of Commerce and Empire: Britain and Its Tropical Colonies, 1660–1830, Oxford 2010.

Hartley, Janet M.: A *Social History* of the Russian Empire, 1650–1825, London 1999.

Harvey, David: The Condition of *Postmodernity*. An Enquiry into the Origins of Cultural Change, Oxford 1989.

Harzig, Christiane, and Dirk Hoerder, with Donna Gabaccia: *What Is Migration History?* Cambridge 2009.

Hastings, Adrian (ed.): A *World History of Christianity*, London 1999.

Haumann, Heiko: A History of *East European Jews*, New York 2002.

Hauner, Milan: *What Is Asia to Us*? Russia's Asian Heartland Yesterday and Today, Boston 1990.

Hausman, William J., et al.: *Global Electrification*. Multinational Enterprise and International Finance in the History of Light and Power, 1878–2007, Cambridge 2008.

Haußig, Hans-Michael. Der *Religionsbegriff* in den Religionen. Studien zum Selbst- und Religionsverständis in Hinduismus, Buddhismus, Judentum und Islam, Bodenheim 1999.

Hawes, C. J.. *Poor Relations*. The Making of a Eurasian Community in British India 1773–1833, Richmond 1996.

Hay, Stephen N.: *Asian Ideas* of East and West. Tagore and his Critics in Japan, China, and India. Cambridge, MA 1970.

Hayami Akira: The *Historical Demography* of Pre-modern Japan, Tokyo 1997.

———: *Population,* Family and Society in Pre-Modern Japan, Folkestone 2009.

Hayami Akira, et al. (eds.): *Economic History of Japan*, 1600–1990, vol. 1: Emergence of Economic Society in Japan, 1600–1859, Oxford 2004.

Hayhoe, Ruth: *China's Universities*, 1895–1995. A Century of Conflict, New York 1996.

Hays, Jo N.: The *Burdens of Disease*. Epidemics and Human Response in Western History, New Brunswick 1998.

Headrick, Daniel R.: The *Tools* of Empire. Technology and European Imperialism in the Nineteenth Century, New York 1981.

———: The *Tentacles* of Progress. Technolopgy Transfer in the Age of Imperialism, 1850–1940, New York 1988.

———: The *Invisible Weapon*. Telecommunications and International Politics, 1851–1945, New York 1991.

———: When *Information* Came of Age. Technologies of Knowledge in the Age of Reason and Revolution, Oxford 2000.

Heathcote, T. A.: The *Military in British India*. The Development of British Land Forces in South Asia, 1600–1947, Manchester 1995.

Heerma van Voss, Lex, and Marcel van der Linden (eds.): *Class* and Other Identities. Gender, Religion and Ethnicity in the Writing of European Labour History, New York 2002.

Heffer, Jean: The *United States* and the Pacific. History of a Frontier, Notre Dame, IN 2002.

Heilbroner, Robert L.: The *Nature and Logic* of Capitalism, New York 1985.

Helmstadter, Richard J. (ed.): *Freedom and Religion* in the Nineteenth Century, Stanford, CA 1997.

Hemming, John: *Amazon Frontier*. The Defeat of the Brazilian Indians, London 1987.

Henare, Amiria J. M.: *Museums*, Anthropology and Imperial Exchange, Cambridge 2005.

Henkin, David M.: The *Postal Age*. The Emergence of Modern Communications in Nineteenth-Century America, Chicago 2006.

Hennessy, Alistair: The *Frontier in Latin American History*, London 1978.

Henning, Joseph M.: *Outposts of Civilization*. Race, Religion, and the Formative Years of American-Japanese Relations, New York 2000.

Hennock, Ernest P.: The *Origin of the Welfare State* in England and Germany, 1850–1914. Social Policies Compared, Cambridge 2007.

Henze, Dietmar: *Enzyklopädie* der Entdecker und Erforscher der Erde, 5 vols., Graz 1978–2004.

Henze, Paul B.: *Layers of Time*. A History of Ethiopia, New York 2000.

Herbert, Christopher: *War of No Pity*. The Indian Mutiny and Victorian Trauma, Princeton 2008.

Herlihy, Patricia: *Odessa*. A History, 1794–1914, Cambridge, MA 1991.

Herr, Richard (ed.): *Themes* in Rural History of the Western World, Ames, IOW. 1993.

Herren, Madeleine: *Hintertüren zur Macht*. Internationalismus und modernisierungsorientierte Außenpolitik in Belgien, der Schweiz und den USA 1865–1914, Munich 2000.

Herrigel, Gary: *Industrial Constructions*. The Sources of German Industrial Power, Cambridge 1996.

Herring, George C.: *From Colony to Superpower*. U.S. Foreign Relations since 1776, Oxford 2008.

Herrmann, David G.: The *Arming of Europe* and the Making of the First World War, Princeton, NJ 1996.

Herzfeld, Michael: *Anthropology*. Theoretical Practice in Culture and Society, Malden, MA 2001.

Heuman, Gad, and Trevor Burnard (eds.): The Routledge *History of Slavery*, London 2011.

Hevia, James L.: *Cherishing Men from Afar*. Qing Guest Ritual and the Macartney Embassy of 1793, Durham, NC 1995.

———: *English Lessons*. The Pedagogy of Imperialism in Nineteenth-Century China, Durham, NC 2003.

Hibbert, Christopher: *Queen Victoria* in Her Letters and Journals, London 1984.

Hight, Eleanor M., and Gary D. Sampson (eds.): *Colonialist Photography*. Imag(in)ing Race and Place, London 2002.

Higman, B. W.: *Slave Populations* of the British Caribbean, 1807–1834, Baltimore, MD 1984.

———: A *Concise History of the Caribbean*, Cambridge 2011.

Higonnet, Patrice: *Paris*. Capital of the World, Cambridge, MA 2002.

Hildebrand, Klaus: *Das vergangene Reich*. Deutsche Außenpolitik von Bismarck bis Hitler, Stuttgart 1995.

———: *No Intervention*. Die Pax Britannica und Preussen 1865/66–1869/70. Eine Untersuchung zur englischen Weltpolitik im 19. Jahrhundert, Munich 1997.

Hildermeier, Manfred: *Bürgertum* und Stadt in Rußland 1760–1870. Rechtliche Lage und soziale Struktur, Cologne 1986.

———: *Geschichte Russlands*. Vom Mittelalter bis zur Oktoberrevolution, Munich 2013.

Hills, Jill: The *Struggle for Control* of Global Communication. The Formative Century, Urbana, IL 2002.

Hilton, Boyd: *A Mad, Bad, and Dangerous People?* England 1783–1846, Oxford 2006.

Hinde, Andrew: *England's Population*. A History Since the Domesday Survey, London 2003.

Hine, Robert V., and John Mack Faragher: The *American West*. A New History, New Haven, CT 2000.

Hinsley, F. Harry: *Power and the Pursuit of Peace*. Theory and Practice in the History of Relations between States, Cambridge 1963.

Hirschhausen, Ulrike von, and Jörn Leonhard (eds.): *Nationalismen* in Europa. West- und Osteuropa im Vergleich, Göttingen 2001.

——— (eds.): *Comparing Empires*. Encounters and Transfers in the Long Nineteenth Century, Göttingen 2011.

Hirst, Leonard F.: The *Conquest of Plague*. A Study of the Evolution of Epidemiology, Oxford 1953.

Hiskett, Mervyn: The Course of *Islam in Africa*, Edinburgh 1994.

Ho Ping-ti: *Studies* on the Population of China, 1368–1953, Cambridge, MA 1959.

Hoare, James E.: *Japan's Treaty Ports* and Foreign Settlements. The Uninvited Guests, 1858–99, London 1994.

Hobsbawm, Eric J.: The *Age of Revolution* (1789–1848), London 1962.

———: *Primitive Rebels*, Manchester 1963.

———: *Industry and Empire*, London 1968.

———: The *Age of Capital* (1848–1875), London 1975.

———: The *Age of Empire* (1875–1914), London 1987.

Hobson, John A.: *Imperialism*. A Study [1902], 3rd ed., London 1988.

Hobson, Rolf: *Maritimer Imperialismus*. Seemachtideologie, seestrategisches Denken und der Tirpitzplan 1875 bis 1914, Munich 2004.

Hochreiter, Walter: Vom *Musentempel* zum Lernort. Zur Sozialgeschichte deutscher Museen 1800–1914, Darmstadt 1994.

Hochschild, Adam: *Bury the Chains*. The British Struggle to Abolish Slavery, London 2005.

Hochstadt, Steve: *Mobility* and Modernity. Migration in Germany, 1820–1989, Ann Arbor, MI 1999.

Hodgson, Marshall G. S.: The *Venture of Islam*. Conscience and History in a World Civilization, 3 vols., Chicago 1974.

Hoensch, Jörg K.: A History of *Modern Hungary*, 1867–1994, 2nd ed., New York 1996

Hoerder, Dirk: *Cultures in Contact*. World Migration in the Second Millenium, Durham, NC 2002.

Hoerder, Dirk, and Leslie Page Moch (eds.): *European Migrants*. Global and Local Perspectives, Boston 1996.

Hoffenberg, Peter H.: *An Empire on Display*. English, Indian, and Australian Exhibitions from the Crystal Palace to the Great War, Berkeley, CA 2001.

Hoffman, Philip T., et al.: *Real Inequality* in Europe since 1500, in: JEH 62 (2002), Spp322–55.

Hofmeister, Burkhard: *Australia* and Its Urban Centres, Berlin 1988.

Hohenberg, Paul M., and Lynn Hollen Lees: The *Making of Urban Europe*, 1000–1950, Cambridge, MA 1985.

Holden, Robert H.: *Armies* Without Nations. Public Violence and State Formation in Central America, 1821–1960, Oxford 2004.

Holt, Thomas C.: The *Problem of Freedom*. Race, Labor, and Politics in Jamaica and Britain, 1832–1938, Baltimore, MD 1992.

Homans, Margaret: *Royal Representations*. Queen Victoria and British Culture, 1837–1876, Chicago 1998.

Homberger, Eric: *Mrs. Astor's New York*. Money and Social Power in a Gilded Age, New Haven, CT 2002.

Hopkins, A. G. (ed.): *Globalization* in World History, London 2002.

—— (ed.): *Global History*. Interactions Between the Universal and the Local. Basingstoke 2006.

Hopkins, Donald R.: *Princes* and Peasants. Smallpox in History, Chicago 1983.

Hoppen, K. Theodore: The *Mid-Victorian Generation*, 1846–1886, Oxford 1998.

Horden, Peregrine, and Nicholas Purcell: The *Corrupting Sea*. A Study of Mediterranean History, Oxford 2000.

Horel, Catherine: Histoire de *Budapest*, Paris 1999.

Hörisch, Jochen: Der *Sinn* und die Sinne. Eine Geschichte der Medien, Frankfurt a.M. 2001.

Horn, Eva, et al. (eds.): *Grenzverletzer*. Von Schmugglern, Spionen und anderen subversiven Gestalten, Berlin 2002.

Horn, Jeff, et al. (eds.): *Reconceptualizing the Industrial Revolution*, Cambridge, MA 2010.

Horowitz, Roger, et al.: *Meat* for the Multitudes. Market Culture in Paris, New York City, and Mexico over the Long Nineteenth Century, in: AHR 109 (2004), pp. 1055–83.

Horton, Mark, and John Middleton: *The Swahili*. The Social Landscape of a Mercantile Society, Oxford 2000.

Hösch, Edgar: Geschichte der *Balkanländer*. Von der Frühzeit bis zur Gegenwart, Munich 1988.

Hounshell, David A.: *From the American System* to Mass Production, 1800–1932. The Development of Manufacturing Technology on the United States, Baltimore, MD 1984.

Hourani, Albert: *Arabic Thought* in the Liberal Age 1798–1939, London 1962.

———: *A History of the Arab Peoples,* New York 1992.

Hourani, Albert, et al. (eds.): The *Modern Middle East.* A Reader, London 1993.

Howard, Michael: *War in European History,* 2nd ed., Oxford 2009.

Howe, Anthony: *Free Trade* and Liberal England, Oxford 1997.

Howe, Christopher: The *Origins* of Japanese Trade Supremacy. Development and Technology in Asia from 1540 to the Pacific War, London 1996.

Howe, Daniel Walker: *What Hath God Wrought.* The Transformation of America, 1815–1848, Oxford 2007.

Howe, Stephen: *Ireland* and Empire. Colonial Legacies in Irish History and Culture, Oxford 2000.

———: *Empire.* A Very Short Introduction, Oxford 2002.

Howland, Douglas, R.: *Translating the West.* Language and Political Reason in Nineteenth-Century Japan, Honolulu 2002.

Hoxie, Frederick E.: A *Final Promise.* The Campaign to Assimilate the Indians, 1880–1920. Lincoln, NE 1984.

Hroch, Miroslav: Das *Europa der Nationen.* Die moderne Nationsbildung im europäischen Vergleich, Göttingen 2005.

Hsiao Kung-chuan: *Rural China.* Imperial Control in the Nineteenth Century, Seattle, WA 1960.

———: *A Modern China* and a New World. K'ang Yu-wei, Reformer and Utopian, 1858–1927, Seattle, WA 1975.

Hsü, Immanuel C. Y.: The Rise of *Modern China,* 6th ed., New York 2000.

Huang, Philip C. C.: The *Peasant Economy* and Social Change in North China, Stanford, CA 1985.

———: The *Peasant Family* and Rural Development in the Yangzi Delta, 1350–1988, Stanford, CA 1990.

Huber, Valeska: The *Unification of the Globe by Disease?* The International Sanitary Conferences on Cholera, 1851–1894, in: HJ 49 (2006), pp. 453–76.

———: *Channelling Mobilities.* Migration and Globalisation in the Suez Canal Region and Beyond, Cambridge 2013.

Hucker, Charles O.: A *Dictionary* of Official Titles in Imperial China, Stanford, CA 1985.

Huenemann, Ralph William: The *Dragon* and the Iron Horse. The Economics of Railroads in China, 1876–1937, Cambridge, MA 1984.

Huerkamp, Claudia: Der *Aufstieg der Ärzte* im 19. Jahrhundert. Vom gelehrten Stand zum professionellen Experten. Das Beispiel Preußens, Göttingen 1985.

Huff, Toby: The Rise of *Early Modern Science.* Islam, China, and the West, 2nd ed., Cambridge 2003.

Huffman, James L.: *Creating a Public.* People and Press in Meiji Japan, Honolulu 1997.

Hughes, Kathryn: *The Victorian Governess,* London 1993.

Hughes, Thomas P.: *Networks of Power.* Electrification in Western Society, 1880–1930, Baltimore, MD 1983.

———: *Human-Built World.* How to Think about Technology and Culture, Chicago 2004.

Hugill, Peter J.: *World Trade* since 1431. Geography, Technology, and Capitalism, Baltimore, MD 1993.

————: *Global Communications* since 1844. Geopolitics and Technology, Baltimore, MD 1999.

Humboldt, Alexander von: *Relation historique* du voyage aux régions équinoxiales du Nouveau Continent, 3 vols., Paris 1814–25.

————: *Essai politique* sur le Royaume de la Nouvelle-Espagne [1808], 2nd ed. 3 vols., Paris 1825–27.

————: *Reise durchs Baltikum* nach Russland und Sibirien 1829, ed. Hanno Beck, 2nd ed., Stuttgart 1984.

————: *Studienausgabe*, ed. Hanno Beck, 7 vols., Darmstadt 1989–93.

————: *Political Essay* on the Island of Cuba. A Critical Edition, ed. Vera M. Kutzinski, and Ottmar Ette, Chicago 2011.

Humboldt, Wilhelm von: *Werke*, ed. Andreas Flitner, 5 vols., Darmstadt 1960–81.

Humphries, Jane: *Childhood and Child Labour* in the British Industrial Revolution, Cambridge 2010.

Hunt, Michael H.: The Making of a *Special Relationship*. The United States and China to 1914, New York 1983.

Hunter, Louis C.: A History of *Industrial Power* in the United States, 1780–1930, vol. 1: Waterpower in the Century of the Steam Engine, Charlottesville, VA 1979.

Hurewitz, Jacob C. (ed.): *Diplomacy* in the Near and Middle East. A Documentary Record, 1535–1956, 2 vols., New York 1956.

Hurt, R. D.: *Indian Agriculture* in America. Prehistory to the Present. Lawrence, KS 1987.

Hurtado, Albert L.: *Indian Survival* on the California Frontier, New Haven, CT 1988.

Huston, James L.: *Securing the Fruits of Labor*. The American Concept of Wealth Distribution, 1765–1900, Baton Rouge, LA 1998.

Hwang, Kyung Moon: *Beyond Birth*. Social Status in the Emergence of Modern Korea, Cambridge, MA 2005.

Hyam, Ronald: *Britain's Imperial Century*, 1815–1914. A Study of Empire and Expansion, 3rd ed., Basingstoke 2002.

Iggers, Georg G., and Q. Edward Wang: A Global History of *Modern Historiography*, Harlow 2008.

Igler, David: *Diseased Goods*. Global Exchanges in the Eastern Pacific Basin, 1770–1850, in: AHR 109 (2004), pp. 693–719.

İhsanoğlu, Ekmeleddin: *Science*, Technology and Learning in the Ottoman Empire. Western Influence, Local Institutions, and the Transfer of Knowledge, Aldershot 2004.

Ikegami, Eiko: *Citizenship* and National Identity in Early Meiji Japan, 1868–1889. A Comparative Assessment, in: IRSH 40, Supplement 3 (1995), pp. 185–221.

————: The *Taming of the Samurai*. Honorific Individualism and the Making of Modern Japan, Cambridge, MA 1995.

Iliffe, John: A Modern History of *Tanganyika*, Cambridge 1979.

————: The *African Poor*. A History, Cambridge 1987.

————: *Famine in Zimbabwe* 1890–1960, Gweru (Simbabwe) 1990.

————: *Africans*. The History of a Continent, Cambridge 1995.

————: *East African Doctors*. A History of the Modern Profession, Cambridge 1998.

Imhof, Arthur E.: Die *Lebenszeit*. Vom aufgeschobenen Tod und von der Kunst des Lebens. Munich 1988.

İnalcık, Halil, and Donald Quataert (eds.): Economic and Social History of the *Ottoman Empire*, 2 vols., Cambridge 1994.

Ingrao, Charles W.: The *Habsburg Monarchy 1618–1815*, 2nd ed., Cambridge 2000.

Inikori, Joseph E.: *Africans* and the Industrial Revolution in England. A Study in International Trade and Economic Development, Cambridge 2002.

Inkster, Ian: *Science* and Technology in History. An Approach to Industrial Development, Basingstoke 1991.

Inwood, Stephen: A History *of London*, London 1998.

Irick, Robert L.: Ch'ing Policy toward the *Coolie Trade*, 1847–1878, San Francisco 1982.

Irigoin, Maria Alejandra: *Gresham on Horseback*. The Monetary Roots of Spanish American Political Fragmentation in the Nineteenth Century, in: EcHR 62 (2009), pp. 551–75.

Iriye, Akira, and Jürgen Osterhammel (general eds.): *A History of the World*, 6 vols., Cambridge, MA 2012ff. See also E. S. Rosenberg (2012).

Irokawa Daikichi: The *Culture of the Meiji Period*, Princeton, NJ 1985.

Irschick, Eugene F.: *Dialogue and History*. Constructing South India; 1795–1895, Berkeley, CA 1994.

Irving, Robert G.: *Indian Summer*. Lutyens, Baker, and Imperial Delhi. New Haven, CT 1981.

Isabella, Maurizio: *Risorgimento in Exile*. Italian Emigrés and the Liberal International in the post-Napoleonic Era, Oxford 2009.

Isenberg, Andrew C.: The *Destruction of the Bison*. An Environmental History, 1750–1920, Cambridge 2000.

Isichei, Elizabeth: A *History* of African Societies to 1870, Cambridge 1997.

Jackson, Carl T.: *Oriental Religions* and American Thought. Nineteenth-Century Explorations, Westport, CT 1981.

Jackson, James Harvey: Migration and Urbanization in the *Ruhr Valley*, 1821–1914, Atlantic Highlands, NJ 1997.

Jackson, Kenneth T.: *Crabgrass Frontier*. The Suburbanization of the United States, New York 1985.

Jackson, R. V.: The *Population History* of Australia, Fitzroy (Victoria) 1988.

Jacobs, Wilbur R.: *On Turner's Trail*. 100 Years of Writing Western History, Lawrence, KS 1994.

Jacobson, Matthew Frye: *Whiteness* of a Different Color. European Immigrants and the Alchemy of Race, Cambridge, MA 1998.

———: *Barbarian Virtues*. The United States Encounters Foreign Peoples at Home and Abroad, 1876–1917, New York 2000.

Jacoby, Karl: *Crimes* against Nature. Squatters, Poachers, Thieves, and the Hidden History of American Conservation, Berkeley, CA 2001.

Jäger, Jens: *Photographie*. Bilder der Neuzeit. Einführung in die Historische Bildforschung, Tübingen 2000.

James, John A., and Mark Thomas (eds.): *Capitalism* in Context, Chicago 1994.

Janich, Nina, and Albrecht Greule (ed.): *Sprachkulturen* in Europa. Ein internationales Handbuch, Tübingen 2002.

Janik, Allan, and Stephen E. Toulmin: *Wittgenstein's Vienna*, New York 1973.

Janku, Andrea: *Nur leere Reden*. Politischer Diskurs und die Shanghaier Presse im China des späten 19. Jahrhunderts, Wiesbaden 2003.

Jannetta, Ann Bowman: The *Vaccinators*. Smallpox, Medical Knowledge, and the "Opening" of Japan, Stanford, CA 2007.

Jansen, Harry S.: *Wrestling with the Angel*. Problems of Definition in Urban Historiography, in: Urban History 23 (1996), pp. 277–99.

Jansen, Marius B.: *China in the Tokugawa World*, Cambridge, MA 1998.

———: *The Making of Modern Japan*, Cambridge, MA 2000.

Janssen, Helmut: Die *Übertragung von Rechtsvorstellungen* auf fremde Kulturen am Beispiel des englischen Kolonialrechts. Ein Beitrag zur Rechtsvergleichung, Tübingen 2000.

Jardin, André: *Alexis de Tocqueville, 1805–1859*, Paris 1984.

Jasanoff, Maya: *Edge of Empire*. Conquest and Collecting in the East, 1750–1850, New York 2005.

Jayawardena, Kumari: *Nobodies* to Somebodies. The Rise of the Colonial Bourgeoisie in Sri Lanka, New York 2002.

Jelavich, Barbara: History of the *Balkans*, 2 vols., Cambridge 1983.

———: *Russia's Balkan Entanglements, 1806–1914*, Cambridge 1991.

Jelavich, Charles, and Barbara Jelavich: The *Establishment* of the Balkan National States, 1804–1920, Seattle, WA 1977.

Jen Yu-wen: The *Taiping* Revolutionary Movement, New Haven, CT 1973.

Jennings, Francis: *Founders of America*, New York 1993.

Jensen, Lionel M.: *Manufacturing Confucianism*. Chinese Traditions and Universal Civilization, Durham, NC 1997.

Jeremy, David J.: *Transatlantic Industrial Revolution*. The Diffusion of Textile Technologies between Britain and America, 1790–1830, Oxford 1981.

——— (ed.): *International Technology Transfer*. Europe, Japan and the USA, 1700–1914, Aldershot 1991.

Jersild, Austin: *Orientalism* and Empire. North Caucasus Mountain Peoples and the Georgian Frontier, 1845–1917, Montréal 2002.

Jeurgens, Charles: *De Haarlemmermeer*. Een studie in planning en beleid 1836–1858, Amsterdam 1991.

Johnson, James H.: *Listening in Paris*. A Cultural History, Berkeley, CA 1995.

Johnson, Linda Cooke: *Shanghai*. From Market Town to Treaty Port, 1074–1858, Stanford, CA 1995.

Johnson, Lonnie R.: *Central Europe*. Enemies, Neighbors, Friends, New York 1996.

Johnson, Paul: The *Birth of the Modern*. World Society 1815–1830, New York 1991.

Johnson, Robert Eugene: *Peasant and Proletarian*. The Working Class of Moscow in the Late Nineteenth Century, Leicester 1979.

Johnston, William: The *Modern Epidemic*. A History of Tuberculosis in Japan, Cambridge, MA 1995.

Joll, James: The *Second International* 1889–1914, London 1974.

———: The *Origins* of the First World War, London 1984.

Jonas, Raymond A.: The *Battle of Adwa*. African Victory in the Age of Empire, Cambridge, MA 2011.

Jones, Charles A.: *International Business* in the Nineteenth Century. The Rise and Fall of a Cosmopolitan Bourgeoisie, Brighton (Sussex) 1987.

Jones, Emrys: *Metropolis*. The World's Great Cities, Oxford 1990.

Jones, Eric L.: The *European Miracle*. Environments, Economies and Geopolitics in the History of Europe and Asia, Cambridge 1981.

———: *Growth Recurring*. Economic Change in World History, Oxford 1988.

Jones, Eric L. et al.: *Coming Full Circle*. An Economic History of the Pacific Rim, Boulder, CO 1993.

Jones, Geoffrey: *Multinationals* and Global Capitalism. From the Nineteenth to the Twenty-First Century, Oxford 2005.

Jones, Geoffrey, and Zeitlin, Jonathan (eds.): The *Oxford Handbook of Business History*, Oxford 2008.

Jones, Howard: *Union in Peril*. The Crisis over British Intervention in the Civil War, Chapel Hill, NC 1992.

Jones, Jacqueline: *The Dispossessed*. America's Underclasses from the Civil War to the Present, New York 1992.

Jones, Kenneth W.: Socio-Religious *Reform Movements* in British India, Cambridge 1990.

Jones, Kristine L.; *Warfare*, Reorganization, and Readaptation at the Margins of Spanish Rule. The Southern Margin (1573–1882), in: Frank Salomon (ed.), The Cambridge History of the Native Peoples of the Americas, vol. 3, part 2, Cambridge 1999, pp. 138–87.

Jones, William C.: The *Great Qing Code*, Oxford 1994.

Jordan, David P.: *Transforming Paris*. The Life and Labors of Baron Haussmann, Chicago 1996.

Jordan, Teresa: *Cowgirls*. Women of the American West, Lincoln, NE 1982.

Jordan, Winthrop D.: *White over Black*. American Attitudes toward the Negro, 1550–1812, New York 1968.

Jordheim, Helge: *Against Periodization*. Koselleck's Theory of Multiple Temporalities, in: HT 51 (2012), pp. 151–71.

Josephson, Jason Ānanda: The *Invention of Religion in Japan*, Chicago 2012.

Jourdan, Annie: *L'Empire de Napoléon*, Paris 2000.

———: *La révolution*, une exception française? Paris 2004.

Judge, Joan: *Print and Politics*. "Shibao" and the Culture of Reform in Late Qing China, Stanford, CA 1996.

Juergens, George: *Joseph Pulitzer* and the New York World, Princeton, NJ 1966.

Jungnickel, Christa, and Russell McCormmach: *Intellectual Mastery* of Nature. Theoretical Physics from Ohm to Einstein, 2 vols., Chicago 1986.

Kaczyńska, Elżbieta: Das größte *Gefängnis* der Welt. Sibirien als Strafkolonie zur Zarenzeit, Frankfurt a.M. 1994.

Kaderas, Christoph: Die *Leishu* der imperialen Bibliothek des Kaisers Qianlong (reg. 1736–96), Wiesbaden 1998.

Kaelble, Hartmut: *Industrialisierung* und soziale Ungleichheit. Europa im 19. Jahrhundert. Eine Bilanz, Göttingen 1983.

———: Der Wandel der *Erwerbsstruktur* in Europa im 19. und 20. Jahrhundert, in: Historical Social Research 22 (1997), pp. 5–28.

——— (ed.): The *European Way*. European Societies during the Nineteenth and Twentieth Centuries, New York 2004.

Kaiwar, Vasant: *Nature*, Property and Polity in Colonial Bombay, in: JPS 27 (2000), pp. 1–49.

Kale, Madhavi: *Fragments of Empire*. Capital, Slavery, and Indian Indentured Labor Migration in the British Caribbean, Philadelphia 1998.

Kalland, Arne, and Brian Moeran: *Japanese Whaling*. End of an Era? London 1992.

Kalyvas, Stathis N.: The *Logic of Violence* in Civil War, Cambridge 2006.

K'ang Yu-wei: The *One-world Philosophy* of K'ang Yu-wei, transl. by Lawrence G. Thompson, London 1958.

Kanwar, Pamela: *Imperial Simla*. The Political Culture of the Raj, Delhi 1990.

Kapitza, Peter (ed.): *Japan in Europa*. Texte und Bilddokumente zur europäischen Japan-kenntnis von Marco Polo bis Wilhelm von Humboldt, 2 vols., Munich 1990.

Kappeler, Andreas: The *Russian Empire*. A Multiethnic History, Harlow 2001.

Karabell, Zachary: *Parting the Desert*. The Creation of the Suez Canal, London 2003.

Karady, Victor: The *Jews of Europe* in the Modern Era. A Socio-historical Outline. Budapest 2004.

Karateke, Hakan T., and Maurus Reinkowski (eds.): *Legitimizing the Order*. The Ottoman Rhetoric of State Power, Leiden 2005.

Karl, Rebecca E.: *Staging the World*. Chinese Nationalism at the Turn of the Twentieth Century, Durham, NC 2002.

Karpat, Kemal H.: *Ottoman Population*, 1830–1914. Demographic and Social Characteristics, Madison, WI 1985.

———: The *Politicization of Islam*. Reconstructing Identity, State, Faith, and Community in the Late Ottoman State, Oxford 2001.

Kasaba, Reşat (ed.): The *Cambridge History of Turkey*, vol. 4: Turkey in the Modern World, Cambridge 2008.

———: *A Moveable Empire*. Ottoman Nomads, Migrants, and Refugees, Seattle, WA 2009.

Kaschuba, Wolfgang: Die *Überwindung* der Distanz. Zeit und Raum in der europäischen Moderne, Frankfurt a.M. 2004.

Kashani-Sabet, Firoozeh: *Frontier Fictions*. Shaping the Iranian Nation, 1804–1946, Princeton, NJ 1999.

Kassir, Samir: Histoire de *Beyrouth*, Paris 2003.

Kato Shuichi: History of *Japanese Literature*, vol. 3: The Modern Years, Tokyo 1991.

Katsu Kokichi: *Musui's Story*. The Autobiography of a Tokugawa Samurai, Tucson, AZ, 1993.

Katz, Jacob: *Out of the Ghetto*. The Social Background of Jewish Emancipation, 1770–1870, Cambridge, MA 1973.

———: From *Prejudice* to Destruction. Antisemitism, 1700–1933, Cambridge, MA 1980.

Kaufmann, Stefan: *Kommunikationstechnik* und Kriegführung 1815–1945. Stufen telemedi-aler Rüstung, Munich 1996.

——— (ed.): *Ordnungen der Landschaft*. Natur und Raum technisch und symbolisch entwer-fen. Würzburg 2002.

Kaukiainen, Yrjö: *Shrinking the World*. Improvements in the Speed of Information Transmis-sion, c. 1820–1870, in: *EREH* 5 (2001), pp. 1–28.

Kaur, Amarjit: *Economic Change* in East Malaysia. Sabah and Sarawak since 1850, Basingstoke 1998.

Kavanagh, Thomas W.: *Comanche Political History*. An Ethnohistorical Perspective, 1706–1875. Lincoln, NE 1996.

Kazemi, Farhad, and John Waterbury (eds.): *Peasants and Politics* in the Modern Middle East, Miami, FL 1991.

Keddie, Nikki R.: *Iran* and the Muslim World. Resistance and Revolution, Basingstoke 1995.

———: *Qajar Iran* and the Rise of Reza Khan, 1796–1925, Costa Mesa, CA 1999.

———: *Modern Iran*. Roots and Results of Revolution, New Haven, CT 2006.

Keegan, Timothy: *Colonial South Africa* and the Origins of the Racial Order, Charlottesville, VA 1996.

Keene, Donald: The *Japanese Discovery of Europe*, 1720–1830, rev. ed., Stanford, CA 1969.

———: *Emperor of Japan*. Meiji and His World, 1852–1912, New York 2002.

Keirstead, Thomas: *Inventing Medieval Japan*. The History and Politics of National Identity, in: Medieval History Journal 1 (1998), pp. 47–71.

Keith, Arthur Berriedale (ed.): *Speeches* and Documents on Indian Policy, 1750–1921, vol. 1, London 1922.

——— (ed.): *Selected Speeches* and Documents on British Colonial Policy, 1763–1917, 2 vols., Oxford 1961.

Kellett, John R.: The *Impact of Railways* on Victorian Cities, London 1969.

Kelly, David, and Anthony Reid (eds.): *Asian Freedoms*. The Idea of Freedom in East and Southeast Asia, Cambridge 1998.

Kennedy, Dane: *Islands of White*. Settler Society and Culture in Kenya and Southern Rhodesia, Durham, NC 1987.

———: The *Magic Mountains*. Hill Stations and the British Raj, Berkeley, CA 1996.

Kennedy, Paul M.: The Rise and Fall of *British Naval Mastery*, London 1983.

———: The *Costs and Benefits* of British Imperialism, 1846–1914, in: P&P 125 (1989), pp. 186–99.

———: The *Rise and Fall* of the Great Powers: Economic Change and Military Conflict from 1500 to 2000, New York 1989.

Kennedy, Roger G.: *Orders from France*. The Americans and the French in a Revolutionary World, 1780–1820, New York 1989.

Kent, Neil: The *Soul of the North*. A Social, Architectural and Cultural History of the Nordic Countries, 1700–1940, London 2000.

Kenwood, A. G., and A. L. Lougheed: The *Growth* of the International Economy 1820–1990, 4th ed., London 1999.

Kern, Stephen: The *Culture* of Time and Space, 1880–1918, Cambridge, MA 1983.

Kerr, Ian J.: *Building* the Railways of the Raj, 1850–1900, Delhi 1997.

Kershaw, Roger: *Monarchy in South-East Asia*. The Faces of Tradition in Transition, London 2001.

Keyssar, Alexander: The *Right to Vote*. The Contested History of Democracy in the United States, New York 2000.

Khater, Akram Fouad: *Inventing Home*. Emigration, Gender, and the Middle Class in Lebanon, 1870–1920, Berkeley, CA 2001.

Khazanov, Anatoly M.: *Nomads* and the Outside World, 2nd ed., Madison, WI 1994.

Khodarkovsky, Michael: *Russia's Steppe Frontier*. The Making of a Colonial Empire, 1500–1800, Bloomington, IN 2002.

Khoury, Philip S., and Joseph Kostiner (eds.): *Tribes* and State Formation in the Middle East, London 1991.

Kiernan, Victor G.: The *Lords of Human Kind*. European Attitudes to the Outside World in the Imperial Age, Harmondsworth 1972.

Kieser, Hans-Lukas: *Der verpasste Friede*. Mission, Ethnie und Staat in den Ostprovinzen der Türkei, 1839–1938, Zurich 2000.

Kim Key-hiuk: The *Last Phase* of the East Asian World Order. Korea, Japan, and the Chinese Empire, 1860–1882, Berkeley, CA 1980.

Kimmel, Michael S.: *Revolution*. A Sociological Interpretation, Cambridge 1990.

Kindleberger, Charles P.: The *Rise of Free Trade* in Western Europe, in: JEH 35 (1975), pp. 20–55.

———: A *Financial History* of Western Europe, London 1984.

Kinealy, Christine: A *Death-Dealing Famine*. The Great Hunger in Ireland, London 1997.

————: The *Great Irish Famine*. Impact, Ideology and Rebellion, Basingstoke 2002.

King, Anthony D.: *Colonial Urban Development*. Culture, Social Power and Environment, London 1976.

————: *Global Cities*. Post-imperialism and the Internationalisation of London, London 1990.

King, Charles: The *Black Sea*. A History, Oxford 2004.

King, Michael: The *Penguin History of New Zealand*, Auckland 2003.

Kingston, Beverley: The *Oxford History of Australia*. vol. 3: 1860–1900: Glad, Confident Morning, Melbourne 1988.

Kingston–Mann, Esther: *In Search of the True West*. Culture, Economics, and Problems of Russian Development, Princeton, NJ 1999.

Kiple, Kenneth F.: The *Caribbean Slave*. A Biological History, Cambridge 1984.

———— (ed.): The Cambridge World History of *Human Disease*, Cambridge 1993.

———— (ed.): The Cambridge World History of *Food*, Cambridge 2000.

————: A *Movable Feast*. Ten Millennia of Food Globalization, Cambridge 2007.

Kirby, David: The *Baltic World, 1772–1993*. Europe's Northern Periphery in an Age of Change, London 1995.

————: A Concise History of *Finland*, Cambridge 2006.

Kirch, Patrick V.: *On the Road* of the Winds. An Archaeological History of the Pacific Islands before European Contact, Berkeley, CA 2000.

Kirimli, Hakan: *National Movements* and National Identity among the Crimean Tatars (1905–1916). Leiden 1996.

Kirsch, Martin: *Monarch* und Parlament im 19. Jahrhundert. Der monarchische Konstitutionalismus als europäischer Verfassungstyp. Frankreich im Vergleich, Göttingen 1999.

Klaits, Joseph, and Michael H. Haltzel (eds.): The *Global Ramifications* of the French Revolution, Cambridge 1994.

Klarén, Peter Flindell: *Peru*. Society and Nationhood in the Andes, New York 2000.

Klein, Bernhard, and Gesa Mackenthun (eds.): *Sea Changes*. Historicizing the Ocean, New York 2004.

Klein, Herbert S.: *African Slavery* in Latin America and the Caribbean, New York 1986.

————: The Atlantic *Slave Trade*, Cambridge 1999.

————: A Concise History of *Bolivia*, Cambridge 2003.

————: A *Population History* of the United States, Cambridge 2004.

Klein, Kerwin Lee: *Frontiers* of Historical Imagination. Narrating the European Conquest of Native America, 1890–1990, Berkeley, CA 1997.

Klein, Martin A.: *Slavery and Colonial Rule* in French West Africa, Cambridge 1998.

Klein, Milton M. (ed.): The *Empire State*. A History of New York, Ithaca, NY 2001.

Klein, Thoralf, and Frank Schumacher (eds.): *Kolonialkriege*. Militärische Gewalt im Zeichen des Imperialismus, Hamburg 2006.

Klier, John D.: *Imperial Russia's Jewish Question*, 1855–1881, Cambridge 1995.

Klier, John D., and Shlomo Lambroza (eds.): *Pogroms*. Anti-Jewish Violence in Modern Russian History, Cambridge 1992.

Klooster, Wim: *Revolutions in the Atlantic World*. A Comparative History, New York 2009.

Knell, Simon et al. (eds.): *National Museums*. New Studies from Around the World, London 2011.

Knight, Alan: The *Mexican Revolution*, 2 vols., Lincoln, NE 1986.

————: The *Peculiarities* of Mexican History. Mexico Compared to Latin America, 1821–1992, in: JLAS, Supplement 24 (1992), pp. 99–144.

Knight, Franklin W. (ed.): The *Slave Societies* of the Caribbean, London 1997.

———: The *Haitian Revolution*, in: AHR 105 (2000), pp. 103–15.

Knight, Franklin W., and Peggy K. Liss (eds.): *Atlantic Port Cities*. Economy, Culture, and Society in the Atlantic World, 1650–1850, Knoxville, TN 1991.

Knöbl, Wolfgang: *Polizei* und Herrschaft im Modernisierungsprozeß. Staatsbildung und innere Sicherheit in Preußen, England und Amerika 1700–1914, Frankfurt a.M. 1998.

Knox, Paul L., and Peter J. Taylor (eds.): *World Cities* in a World-System, Cambridge 1995.

Koch, Tom: *Disease Maps*. Epidemics on the Ground, Chicago 2011.

Kocka, Jürgen: *Arbeitsverhältnisse* und Arbeiterexistenzen. Grundlagen der Klassenbildung im 19. Jahrhundert, Bonn 1990.

———: *Weder Stand noch Klasse*. Unterschichten um 1800, Bonn 1990.

———: Das lange *19. Jahrhundert*. Arbeit, Nation und bürgerliche Gesellschaft, Stuttgart 2002.

———: *Writing the History of Capitalism*, in: Bulletin of the German Historical Institute Washington, no. 47 (fall 2010), pp. 7–24.

Kocka, Jürgen, and Ute Frevert (eds.): *Bürgertum* im 19. Jahrhundert. Deutschland im europäischen Vergleich, 3 vols., Munich 1988.

Kocka, Jürgen, and Claus Offe (ed.): Geschichte und Zukunft der *Arbeit*, Frankfurt a.M. 2000.

Koebner, Richard, and Schmidt, Helmut Dan: *Imperialism*. The Story and Significance of a Political Word, 1840–1960, Cambridge 1964.

Koenig, William J.: The *Burmese Polity*, 1752–1819. Politics, Administration, and Social Organization in the Early Kon-baung Period, Ann Arbor, MI 1990.

Kohlrausch, Martin: Der *Monarch im Skandal*. Die Logik der Massenmedien und die Transformation der wilhelminischen Monarchie, Berlin 2005.

Kolchin, Peter: *Unfree Labor*. American Slavery and Russian Serfdom, Cambridge, MA 1987.

———: *American Slavery*, 1619–1877, London 1993.

———: A *Sphinx* on the American Land. The Nineteenth-Century South in Comparative Perspective, Baton Rouge, LA 2003.

Kolff, Dirk H. A.: *Naukar*, Rajput and Sepoy. The Ethnohistory of the Military Labour Market in Hindustan, 1450–1850, Cambridge 1990.

Köll, Elisabeth: *From Cotton Mill* to Business Empire. The Emergence of Regional Enterprises in Modern China, Cambridge, MA 2003.

Komlos, John: Ein *Überblick* über die Konzeptionen der Industriellen Revolution, in: VSWG 84 (1997), pp. 461–511.

———: The *Industrial Revolution* as the Escape from the Malthusian Trap, in: JEEcH 29 (2000), pp. 307–31.

König, Hans-Joachim: Kleine *Geschichte Lateinamerikas*, Stuttgart 2006,

König, Wolfgang: Geschichte der *Konsumgesellschaft*, Stuttgart 2000.

———: *Wilhelm II*. und die Moderne. Der Kaiser und die technisch-industrielle Welt, Paderborn 2007.

König, Wolfgang, and Wolfhard Weber: *Netzwerke*, Stahl und Strom. 1840 bis 1914, Berlin 1990.

Koning, Niek: The *Failure* of Agrarian Capitalism. Agrarian Politics in the UK, Germany, the Netherlands and the USA, 1846–1919, London 1994.

Konvitz, Josef W.: *Cities and the Sea*. Port City Planning in Early Modern Europe, Baltimore, MD 1978.

————: The *Urban Millenium*. The City-Building Process from the Early Middle Ages to the Present, Carbondale, IL 1985.

Korhonen, Pekka: The *Pacific Age* in World History, in: JWH 7 (1996), pp. 41–70.

Körner, Axel (ed.): *1848*. A European Revolution? International Ideas and National Memories of 1848, Basingstoke 2000.

Kornicki, Peter: The *Book in Japan*. A Cultural History from the Beginnings to the Nineteenth Century, Leiden 1998.

———— (ed.): *Meiji Japan*. Political, Economic and Social History 1868–1912, 4 vols., London 1998.

Kosambi, Meera: *Bombay* in Transition. The Growth and Social Ecology of a Colonial City, 1880–1980, Stockholm 1986.

Koselleck, Reinhart: *Zeitschichten*. Studien zur Historik, Frankfurt a.M. 2000.

————: *Futures Past*. On the Semantics of Historical Time, New York 2004.

Kossok, Manfred: *Ausgewählte Schriften*, 3 vols., Leipzig 2000.

Kostal, Rande W.: A *Jurisprudence of Power*. Victorian Empire and the Rule of Law, Oxford 2006.

Koven, Seth: *Slumming*. Sexual and Social Politics in Victorian London, Princeton, NJ 2004.

Kowner, Rotem (ed.): The Impact of the *Russo-Japanese War*, London 2007.

Kraay, Hendrik, and Thomas Whigham (eds.): *I Die with My Country*. Perspectives on the Paraguayan War, 1864–1870, Lincoln, NE 2004.

Kracauer, Siegfried: *Jacques Offenbach* and the Paris of His Time [1937], New York 2002.

Kramer, Lloyd S.: *Lafayette* in Two Worlds. Public Cultures and Personal Identities in an Age of Revolutions, Chapel Hill, NC 1996.

Kramer, Paul A.: The *Blood of Government*. Race Empire, the United States, and the Philippines, Chapel Hill, NC 2006.

Krauss, Marita: *Herrschaftspraxis* in Bayern und Preußen im 19. Jahrhundert. Ein historischer Vergleich, Frankfurt a.M. 1997.

Krech, Shepard, III: The *Ecological Indian*. Myth and History, New York 1999.

Krech, Shepard, III, et al. (eds.): *Encyclopedia* of World Environmental History, 3 vols., New York 2004.

Kreiner, Josef (ed.): *Der Russisch-Japanische Krieg* (1904/05), Göttingen 2005.

Kreiser, Klaus: *Istanbul*. Ein historisch-literarischer Stadtführer, Munich 2001.

————: *Der osmanische Staat* 1300–1922, Munich 2001.

Kreiser, Klaus, and Christoph K. Neumann: Kleine Geschichte der *Türkei*, Stuttgart 2003.

Krell, Alan: The *Devil's Rope*. A Cultural History of Barbed Wire, London 2002.

Kreuzer, Helmut: *Bohème*. Analyse und Dokumentation der intellektuellen Subkultur vom 19. Jahrhundert bis zur Gegenwart, Stuttgart 1971.

Kriger, Colleen E.: *Pride of Men*. Ironworking in Nineteenth-Century West Central Africa, Portsmouth, NH. 1999.

Kroen, Sheryl: *Politics and Theater*. The Crisis of Legitimacy in Restoration France, 1815–1830, Berkeley, CA 2000.

Krüger, Peter (ed.): *Das europäische Staatensystem* im Wandel. Strukturelle Bedingungen und bewegende Kräfte seit der Frühen Neuzeit, Munich 1996.

Krüger, Peter, and Paul W. Schroeder (eds.): The *Transformation* of European Politics, 1763–1848. Episode or Model in Modern History? Münster 2002.

Kuban, Doğan: *Istanbul*. An Urban History, Istanbul 1996.

Kudlick, Catherine J.: *Cholera* in Post-revolutionary Paris. A Cultural History, Berkeley, CA 1996.

Kuhn, Philip A.: *Origins of the Modern Chinese State*, Stanford, CA 2003.

———: *Chinese among Others*. Emigration in Modern Times, Lanham, MD 2008.

Kuitenbrouwer, Marten: *The Netherlands* and the Rise of Modern Imperialism. Colonies and Foreign Policy, 1870–1902, New York 1991.

Kulke, Hermann, and Dietmar Rothermund: *A History of India*, 5th ed., London 2010.

Kumar, Ann: *Java* and Modern Europe. Ambiguous Encounters, Richmond 1997.

Kumar, Deepak: *Science* and the Raj, 1857–1905, Delhi 1997.

Kumar, Dharma, and Tapan Raychaudhuri (eds.): The *Cambridge Economic History of India*. 2 vols., Cambridge 1982.

Kume Kunitake: The *Iwakura Embassy*, 1871–73. A True Account of the Ambassador Extraordinary & Plenipotententary's Journey of Observation Trough the United States of America and Europe, ed. Graham Healey et al., 5 vols., Matsudo 2002.

Kuran, Timur: *Islam and Mammon*. The Economic Predicaments of Islamism, Princeton, NJ 2004.

Kurzman, Charles (ed.): *Modernist Islam*, 1840–1940. A Sourcebook, Oxford 2002.

Kwan, Man Bun: The *Salt Merchants* of Tianjin. State Making and Civil Society in Late Imperial China, Honolulu 2001.

Kwong, Luke S. K.: The Rise of the *Linear Perspective* on History and Time in Late Qing China, in: P&P 173 (2001), pp. 157–90.

Kynaston, David: The *City* of London, vol. 1: A World of Its Own, 1815–1890, London 1994.

Laak, Dirk van: *Infra-Strukturgeschichte*, in: GG 27 (2001), pp. 367–93.

Labanca, Nicola: *Oltremare*. Storia dell'espansione coloniale italiana, Bologna 2002.

Laband, John: *Kingdom in Crisis*. The Zulu Response to the British Invasion of 1879, Manchester 1992.

La Berge, Ann F.: *Mission and Method*. The Early Ninteenth-Century French Public Health Movement, Cambridge 1992.

Labisch, Alfons: *Homo hygienicus*. Gesundheit und Medizin in der Neuzeit, Frankfurt a.M. 1992.

Labourdette, Jean-Francois: Histoire du *Portugal*, Paris 2000.

Lackner, Michael, et al. (eds.): *New Terms* for New Ideas. Western Knowledge and Lexical Change in Late Imperial China, Leiden 2001.

LaFeber, Walter: The *American Age*. United States Foreign Policy at Home and Abroad since 1750, New York 1989.

Lai, Cheung-chung (ed.): *Adam Smith across Nations*. Translations and Receptions of the Wealth of Nations, Oxford 1999.

Lake, Marilyn, and Henry Reynolds: Drawing the *Global Colour Line*: White Men's Countries and the International Challenge of Racial Equality, Cambridge 2008.

Lamar, Howard R. (ed.): The New *Encyclopedia* of the American West, New Haven, CT 1998.

Lamar, Howard R., and Leonard Thompson (eds.): The *Frontier in History*. North America and Southern Africa Compared, New Haven, CT 1981.

Landes, David S.: *Revolution in Time*. Clocks and the Making of the Modern World, Cambridge, MA 1983.

———: The *Wealth and Poverty* of Nations. Why Some Are So Rich and Some So Poor, New York 1998.

————: The *Unbound Prometheus*. Technological Change and Industrial Development in Western Europe from 1750 to the Present, 2nd ed., Cambridge 2003.

Langewiesche, Dieter (ed.): *Liberalismus im 19. Jahrhundert*. Deutschland im europäischen Vergleich, Göttingen 1988.

————: *Europa* zwischen Restauration und Revolution 1815–1849, 3rd ed., Munich 1993.

————(ed.): *Demokratiebewegung* und Revolution 1847 bis 1849. Internationale Aspekte und europäische Verbindungen, Karlsruhe 1998.

————: *Liberalism in Germany*, Princeton, NJ 2000.

————: *Nation*, Nationalismus, Nationalstaat in Deutschland und Europa, Munich 2000.

————: *Neuzeit*, Neuere Geschichte, in: Richard van Dülmen (ed.), Das Fischer-Lexikon: Geschichte, Frankfurt a.M. 2003, pp. 466–89.

————: Eskalierte die *Kriegsgewalt* im Laufe der Geschichte? in: Jörg Baberowski (ed.), Moderne Zeiten? Krieg, Revolution und Gewalt im 20. Jahrhundert, Göttingen 2006, pp. 12–36.

Langford, Paul: *A Polite and Commercial People*. England 1727–1783, Oxford 1992.

Langfur, Hal: The *Forbidden Lands*. Colonial Identity, Frontier Violence, and the Persistence of Brazil's Eastern Indians, 1750–1830, Stanford, CA 2006.

Langley, Lester D.: *The Americas* in the Age of Revolution, 1750–1850, New Haven, CT 1996.

Lapidus, Ira M.: *Islamic Societies* to the Nineteenth Century. A Global History, Cambridge 2012.

Lappo, Georgij M., and Fritz W. Hönsch: *Urbanisierung Russlands*, Berlin 2000.

Laruelle, Marlène: *Russian Eurasianism*. An Ideology of Empire, Washington, DC 2008.

Lary, Diana: *Chinese Migrations*, Lanham, MD 2012.

Laslett, Peter: *Social Structural Time*. An Attempt at Classifying Types of Social Change by Their Characteristic Paces, in: Tom Schuller, and Michael Young (eds.), The Rhythms of Society, London 1988, pp. 17–36.

Latham, A.J.H.: *Rice*. The Primary Commodity, London 1998.

Lattimore, Owen: *Inner Asian Frontiers* of China, New York 1940.

Lauren, Paul Gordon: *Power and Prejudice*. The Politics and Diplomacy of Racial Discrimination, Boulder, CO 1988.

Laurens, Henry: *L'Expédition d'Égypte*, 1798–1801, Paris 1989.

Lavedan, Pierre: *Histoire de l'urbanisme à Paris*, 2nd ed., Paris 1993.

Lavely, William, and R. Bin Wong: Revising the *Malthusian Narrative*. The Comparative Study of Population Dynamics in Late Imperial China, in: JAS 57 (1998), pp. 714–48.

Law, Robin: The *Oyo Empire*, c. 1600–c. 1836. A West African Imperialism in the Era of the Atlantic Slave Trade, Oxford 1977.

————: *Ouidah*. The Social History of a West African Slaving "Port," 1727–1892, Athens, OH 2004.

Lawton, Richard, and Robert Lee (eds.): *Population and Society* in Western European Port-Cities, c. 1650–1939, Liverpool 2002.

Layton, Susan: *Russian Literature* and Empire. Conquest of the Caucasus from Pushkin to Tolstoy, Cambridge 1994.

Lê Thành Khôi: *Histoire du Viêt Nam* des origines à 1858, Paris 1982.

LeDonne, John P.: The *Russian Empire* and the World 1700–1917. The Geopolitics of Expansion and Containment, New York 1997.

Lee Chong-sik: The Politics of *Korean Nationalism*, Berkeley, CA 1963.

Lee, Erika: *At America's Gates*. Chinese Immigration during the Exclusion Era, 1882–1943, Chapel Hill, NC 2003.

Lee, James Z., and Cameron D. Campbell: *Fate and Fortune* in Rural China. Social Organization and Population Behavior in Liaoning, 1774–1873, Cambridge 1997.

Lee, James Z., and Wang Feng: *One Quarter of Humanity*. Malthusian Mythology and Chinese Realities, 1700–2000, Cambridge, MA 1999.

Lee Ki-baik: A New History of *Korea*, Cambridge, MA 1984.

Leedham-Green, Elisabeth S.: A *Concise History* of the University of Cambridge. Cambridge 1996.

Leerssen, Joep: *National Thought* in Europe. A Cultural History, Amsterdam 2006.

Lees, Andrew: *Cities Perceived*: Urban Society in European and American Thought, 1820–1940, Manchester 1985.

Lees, Andrew, and Lynn Hollen Lees: *Cities* and the Making of Modern Europe, 1750–1914, Cambridge 2007.

Leeuwen, Richard van: *Waqfs* and Urban Structures. The Case of Ottoman Damascus, Boston 1999.

Lemberg, Hans: Zur *Entstehung* des Osteuropabegriffs im 19. Jahrhundert. Vom "Norden" zum "Osten" Europas, in: JGO 33 (1985), pp. 48–91.

Lemon, James T.: *Liberal Dreams* and Nature's Limits. Great Cities of North America Since 1600, Toronto 1996.

Lenger, Friedrich: *Industrielle Revolution* und Nationalstaatsgründung (1849–1870er Jahre), Stuttgart 2003.

———: *European Cities* in the Modern Era, 1850–1914, Leiden 2012.

Leonard, Jane Kate: *Wei Yuan* and China's Rediscovery of the Maritime World, Cambridge, MA 1984.

Leonard, Thomas C.: The *Power of the Press*. The Birth of American Political Reporting, New York 1987.

———: *News for All*. America's Coming-of-Age with the Press, New York 1995.

Leonhard, Jörn: *Liberalismus*. Zur historischen Semantik eines europäischen Deutungsmusters, Munich 2001.

Lepenies, Wolf: *Between Literature and Science*: The Rise of Sociology, Cambridge 1988.

———: *Sainte-Beuve*. Auf der Schwelle zur Moderne, Munich 1997.

Lepetit, Bernard: *Les villes* dans la France moderne (1740–1840), Paris 1988.

Lesaffer, Randall (ed.): *Peace Treaties* and International Law in European History, Cambridge 2004.

Levine, Bruce C.: The *Spirit of 1848*. German Immigrants, Labor Conflict, and the Coming of the Civil War, Urbana, IL 1992.

———: *Half Slave* and Half Free. The Roots of Civil War, rev. ed., New York 2005.

Levy, Jack S.: *War* in the Modern Great Power System, 1495–1975, Lexington, KY 1983.

Levy, Leonard W.: *Emergence of a Free Press*, New York 1985.

Lewis, Bernard: The *Emergence* of Modern Turkey, 2nd ed., Oxford 1968.

Lewis, David Levering: *W.E.B. Du Bois*: Biography of a Race, 1868–1919, New York 1993.

Lewis, Martin W., and Kären E. Wigen: The *Myth of Continents*. A Critique of Metageography, Berkeley, CA 1997.

Leyda, Jay: *Dianying*. An Account of Films and the Film Audience in China, Cambridge, MA 1972.

Li, Lillian M.: *Fighting Famine in North China*. State, Market and Environmental Decline, 1690s–1990s, Stanford, CA 2007.

Li, Lillian M., et al.: *Beijing*. From Imperial Capital to Olympic City, Basingstoke 2007.

Liauzu, Claude: *Histoire des migrations* en Méditteranée occidentale, Paris 1996.

Liauzu, Claude, et al.: *Colonisation*. Droit d'inventaire, Paris 2004.

Licht, Walter: Working for the *Railroad*. The Organization of Work in the Nineteenth-Century, Princeton, NJ 1983.

Lichtenberger, Elisabeth: Die *Stadt*. Von der Polis zur Metropolis, Darmstadt 2002.

———: *Europa*. Geographie, Geschichte, Wirtschaft, Politik, Darmstadt 2005.

Lieberman, Victor (ed.): *Beyond Binary Histories*. Re-imagining Eurasia to c. 1830, Ann Arbor, MI 1999.

———: *Strange Parallels*: Southeast Asia in Global Context, c. 800–1830, 2 vols., Cambridge 2003–2009.

Liebersohn, Harry: *Aristocratic Encounters*. European Travelers and North American Indians, Cambridge 1998.

Liedtke, Rainer, and Stephan Wendehorst (ed.): The *Emancipation* of Catholics, Jews and Protestants. Minorities and the Nation State in Nineteenth-Century Europe, Manchester 1999.

Lieven, Dominic: *Russia* and the Origins of the First World War, London 1983.

———: The *Aristocracy* in Europe, 1815–1914, New York 1993.

———: *Nicholas II*. Twilight of the Empire, New York 1993.

———: *Empire*. The Russian Empire and Its Rivals, London 2000.

——— (ed.): The *Cambridge History of Russia*, vol. 2: Imperial Russia, 1689–1917, Cambridge 2006.

———: *Russia against Napoleon*. The Battle for Europe, 1807–1814, London 2009.

Lieven, Michael: "*Butchering* the Brutes All Over the Place." Total War and Massacre in Zululand, 1879, in: History 84 (1999), pp. 614–32.

Limerick, Patricia Nelson: The *Legacy of Conquest*. The Unbroken Past of the American West, New York 1987.

———: *Something in the Soil*. Legacies and Reckonings in the New West, New York 2000.

Lin Man-houng: *China Upside Down*: Currency, Society, and Ideologies, 1808–1856, Cambridge, MA 2006.

Lincoln, Abraham: *Speeches and Writings*, 2 vols., ed. Don E. Fehrenbacher, New York 1989.

Lincoln, W. Bruce: *Nicholas I*: Emperor and Autocrat of All the Russians, Bloomington, IN 1978.

———: In the *Vanguard of Reform*. Russia's Enlightened Bureaucrats, 1825–1861, DeKalb, IL 1982.

———: The *Great Reforms*. Autocracy, Bureaucracy, and the Politics of Change in Imperial Russia, DeKalb, IL 1990.

Linden, Marcel van der, and Jürgen Rojahn (eds.): The *Formation* of Labour Movements, 1870–1914. An International Perspective, 2 vols., Leiden 1990.

Lindert, Peter H.: *Poor Relief* before the Welfare State. Britain versus the Continent, 1780–1880, in: EREH 2 (1998), pp. 101–40.

———: *Growing Public*. Social Spending and Economic Growth since the Eighteenth Century, vol. 1: The Story, Cambridge 2004.

Lipsey, Richard G., et al.: *Economic Transformations*. General Purpose Technologies and Long-Term Economic Growth, Oxford 2005.

Lipson, Charles: *Standing Guard.* Protecting Foreign Capital in the Nineteenth and Twentieth Centuries, Berkeley, CA 1985.

Lis, Catharina: *Social Change* and the Labouring Poor. Antwerp, 1770–1860, New Haven, CT 1986.

Liss, Peggy: *Atlantic Empires.* The Network of Trade and Revolution, 1713–1826, Baltimore, MD 1983.

Little, Daniel: *Understanding Peasant China.* Case Studies in the Philosophy of Social Science, New Haven, CT 1989.

Litwack, Leon F.: *Been in the Storm so Long.* The Aftermath of Slavery, New York 1979.

Liu Ts'ui-jung et al. (eds.), *Asian Population History*, Oxford 2001.

Livi-Bacci, Massimo: *Population* and Nutrition. An Essay on European Demographic History, Cambridge 1991.

———: A Concise History of *World Population*, 2nd ed., Oxford 1997.

———: The *Population of Europe.* A History, Oxford 1999.

Livingstone, David N.: The *Geographical Tradition*, Oxford 1992.

Livingstone, David N., and Charles W. Withers (eds.): *Geographies* of Nineteenth Century Science, Chicago 2011.

Livois, René de: *Histoire de la presse française*, 2 vols., Lausanne 1965.

Llewellyn-Jones, Rosie: The *Great Uprising in India*, 1857–58. Untold Stories, Indian and British, Woodbridge (Suffolk) 2007.

Locke, Ralph P.: *Musical Exoticism.* Images and Reflections, Cambridge 2009.

Lockwood, Jeffrey A.: *Locust.* The Devastating Rise and Mysterious Disappearance of the Insect that Shaped the American Frontier, New York 2004.

Lockwood, William W., Jr.: The *Economic Development of Japan*, Princeton, NJ 1968.

Logan, William Stewart: *Hanoi.* Biography of a City, Seattle, WA 2000.

Lombard, Denys: *Le carrefour javanais.* Essai d'histoire globale, 3 vols., Paris 1990.

Lombardi, Mary: The *Frontier* in Brazilian History. An Historiographical Essay, in: PHR 44 (1975), pp. 437–57.

Look Lai, Walton: The *Chinese in the West Indies*, 1806–1995. A Documentary History, Kingston, Jamaica 1998.

Lorcin, Patricia: *Imperial Identities.* Stereotyping, Prejudice and Race in Colonial Algeria, London 1995.

Losty, Jeremiah P.: *Calcutta.* City of Palaces. A Survey of the City in the Days of the East India Company 1690–1858, London 1990.

Louis, Wm. Roger (ed.): *Imperialism.* The Robinson and Gallagher Controversy, New York 1976.

——— (ed.): The *Oxford History of the British Empire*, 5 vols., Oxford 1998–99.

Love, Eric T. L.: *Race over Empire.* Racism and U.S. Imperialism, 1865–1900, Chapel Hill, NC 2004.

Lovejoy, Paul E.: *Transformations* in Slavery. A History of Slavery in Africa, 2nd ed., Cambridge 2002.

Low, D. A. (ed.): The *Indian National Congress.* Centenary Hindsights, Delhi 1988.

Lu, David J.: *Japan.* A Documentary History, 2 vols., Armonk, NY 1997.

Lucassen, Jan (ed.): *Global Labour History.* A State of the Art, Berlin 2006.

Lüddeckens, Dorothea: Das *Weltparlament* der Religionen von 1893. Strukturen interreligiöser Begegnung im 19. Jahrhundert, Berlin 2002.

Ludden, David: *Peasant History* in South India, Delhi 1985.

———: An *Agrarian History* of South Asia, Cambridge 1999.

Lufrano, Richard John: *Honorable Merchants*. Commerce and Self-Cultivation in Late Imperial China, Honolulu 1997.

Lukowski, Jerzy T.: *The European Nobility* in the Eighteenth Century, Basingstoke 2003.

Lundgreen, Peter (ed.): *Sozial- und Kulturgeschichte* des Bürgertums. Eine Bilanz des Bielefelder Sonderforschungsbereichs (1986–1997), Göttingen 2000.

Lunn, Jon: *Capital and Labour* on the Rhodesian Railway System, 1888–1947, Basingstoke 1997.

Lustick, Ian: *State-Building Failure* in British Ireland and French Algeria, Berkeley, CA 1985.

Luthi, Jean-Jacques: *La vie quotidienne en Égypte* au temps des khédives, Paris 1998.

Lydon, Ghislaine: *On Trans-Saharan Trails*. Islamic Law, Trade Networks, and Cross-Cultural Exchange in Nineteenth-Century Western Africa, Cambridge 2009.

Lynch, John: *Argentine Dictator*. Juan Manuel de Rosas, 1829–1852, Oxford 1981.

———: The *Spanish American Revolutions*, 1808–1826, 2nd ed., New York 1986.

———: *Caudillos* in Spanish America, 1800–1850, Oxford 1992.

———: *Simón Bolívar*. A Life, New Haven, CT 2006.

———: *New Worlds*. A Religious History of Latin America, New Haven, CT 2012.

Lynch, Martin: *Mining* in World History, London 2002.

Lynn, Martin: *Commerce* and Economic Change in West Africa. The Palm Oil Trade in the Nineteenth Century, Cambridge 1997.

Lyons, Francis S. L.: *Internationalism in Europe*, 1815–1914, Leiden 1963.

Lyons, Martyn: *Readers* and Society in Nineteenth-Century France. Workers, Women, Peasants, Basingstoke 2001.

MacDermott, Joseph P.: A Social History of the *Chinese Book*. Books and Literati Culture in Late Imperial China, Hong Kong 2006.

Macfarlane, Alan: The *Savage Wars of Peace*. England, Japan and the Malthusian Trap, Oxford 1997.

Macfarlane, Alan, and Iris Macfarlane: *Green Gold*. The Empire of Tea, London 2003.

Macintyre, Stuart: A Concise History of *Australia*, Cambridge 1999.

MacKenzie, John M.: The *Empire of Nature*. Hunting, Conservation and British Imperialism, Manchester 1988.

———: *Orientalism*. History, Theory and the Arts, Manchester 1995.

Mackerras, Colin P.: The Rise of the *Peking Opera*, 1770–1870. Social Aspects of the Theatre in Manchu China, Oxford, 1972.

MacLeod, Roy, and Milton Lewis (ed.): *Disease*, Medicine, and Empire. Perspectives on Western Medicine and the Experience of European Expansion, London 1988.

MacPherson, Kerrie L.: A *Wilderness* of Marshes. The Origins of Public Health in Shanghai, 1843–1893, Hong Kong 1987.

MacRaild, Donald M, and David E. Martin: *Labour in British Society*, 1830–1914, Basingstoke 2000.

Maddison, Angus: *Chinese Economic Performance* in the Long Run, Paris 1998.

———: The *World Economy*. A Millennial Perspective, Paris 2001.

———: *Contours* of the World Economy, 1–2030 AD. Essays in Macroeconomic History, Oxford 2007.

Magee, Gary B., and Andrew S. Thompson: *Empire and Globalisation*. Networks of People, Goods and Capital in the British World, c. 1850–1914, Cambridge 2010.

Maheshwari, Shriram: The *Census Administration* under the Raj and After, New Delhi 1996.

Maier, Charles S.: *Consigning the Twentieth Century* to History. Alternative Narratives for the Modern Era, in: AHR 105 (2000), pp. 807–31.

———: *Among Empires*. American Ascendancy and Its Predecessors, Cambridge, MA 2006.

Majluf, Natalia, et al. (eds.): *La recuperación de la memoria*. Perú 1842–1942, Lima 2001.

Major, Andrew J.: *State and Criminal Tribes* in Colonial Punjab. Surveillance, Control and Reclamation of the "Dangerous Classes," in: MAS 33 (1999), pp. 657–88.

Major, John: *Prize Possession*. The United States and the Panama Canal, 1903–1979, Cambridge 1993.

Mak, Geert: *Amsterdam*. A Brief Life of the City, London 2001.

Malanima, Paolo: *Economia preindustriale*. Mille anni, dal IX al XVIII secolo, Milan 1995.

———: *Uomini*, risorse, tecniche nell'economica europea dal X al XIX secolo, Milan 2003.

———: *Pre-Modern European Economy*. One Thousand Years (10th–19th Centuries), Leiden 2009.

Malcolm, Noel: *Bosnia*. A Short History, London 1994.

Malia, Martin: *Russia* under Western Eyes. From the Bronze Horseman to the Lenin Mausoleum, Cambridge, MA 1999.

Mallon, Florencia: *Peasant and Nation*. The Making of Postcolonial Mexico and Peru. Berkeley, CA 1995.

Mandler, Peter: The *Fall and Rise* of the Stately Home, New Haven, CT 1997.

Manela, Erez: The *Wilsonian Moment*. Self-Determination and the International Origins of Anticolonial Nationalism, Oxford 2007.

Mann, Kristin: *Marrying Well*. Marriage, Status and Social Change among the Educated Elite in Colonial Lagos, Cambridge 1985.

Mann, Michael: The *Sources of Social Power*, 4 vols., Cambridge 1986–2013.

Manning, Patrick: *Slavery* and African Life. Occidental, Oriental and African Slave Trades, Cambridge 1990.

———: The *African Diaspora*: A History through Culture, New York 2009.

Manrique, Luis Esteban G.: *De la conquista a la globalización*. Estados, naciones y nacionalismos en América Latina. Madrid 2006.

Mansergh, Nicholas: The *Commonwealth Experience*, 2nd ed., 2 vols., London 1982.

Mantran, Robert: Histoire d'*Istanbul*, Paris 1996.

Mantran, Robert, et al.: *Histoire de l'Empire Ottoman*, Paris 1989.

Marchand, Suzanne L.: *German Orientalism* in the Age of Empire: Religion, Race, and Scholarship, Cambridge 2009.

Mardin, Şerif: The *Genesis of Young Ottoman Thought*. A Study in the Modernization of Turkish Political Ideas, Princeton, NJ 1962.

Marichal, Carlos: A Century of *Debt Crises* in Latin America. From Independence to the Great Depression, 1820–1930, Princeton, NJ 1989.

Markovits, Claude: The *Global World* of Indian Merchants, 1750–1947. Traders of Sind from Bukhara to Panama, Cambridge 2000.

———: *Merchants*, Traders, Entrepreneurs. Indian Business in the Colonial Era, Ranikhet (India) 2008.

Markovits, Claude, et al.: A History of *Modern India*, 1480–1950, London 2002.

——— (eds.): *Society and Circulation*. Mobile People and Itinerant Cultures in South Asia 1750–1950, Delhi 2003.

Marks, Steven G.: *Road to Power*. The Trans-Siberian Railroad and the Colonization of Asian Russia, 1850–1917, Ithaca, NY 1991.

Marks, Steven G.: *How Russia Shaped the Modern World*, Princeton, NJ 2003.

Markus, Andrew: *Australian Race Relations, 1788–1993*, St. Leonards 1994.

Marr, David G.: *Vietnamese Anticolonialism* 1885–1925, Berkeley, CA 1971.

———(ed.): *Reflections from Captivity*. Phan Boi Chau's "Prison Notes" and Ho Chi Minh's "Prison Diary," Athens, OH 1978.

Marrus, Michael R.: *The Unwanted*. Euopean Refugees from the First World War Through the Cold War, 2nd ed., Philadelphia 2002.

Marsden, Ben: *Watt's Perfect Engine*. Steam and the Age of Invention, Cambridge 2002.

Marsden, Ben, and Crosbie Smith: *Engineering Empires*. A Cultural History of Technology in Nineteenth-Century Britain, Basingstoke 2005.

Marshall, Peter J.: The *Making* and Unmaking of Empires. Britain, India, and America c. 1750–1783, Oxford 2005.

Martin, David: *On Secularization*. Towards a Revised General Theory, Aldershot 2005.

Martin, Vanessa: *Islam and Modernism*. The Iranian Revolution of 1906, London 1989.

———: The *Qajar Pact*. Bargaining, Protest and the State in Nineteenth Century Persia, London 2005.

Martin, Virginia: *Law and Custom* in the Steppe. The Kazakhs of the Middle Horde and Russian Colonialism in the Nineteenth Century, Richmond 2001.

Marx, Anthony W.: *Making Race* and Nation. A Comparison of South Africa, the United States, and Brazil, Cambridge 1998.

Marx, Christoph: *Grenzfälle*. Zu Geschichte und Potential des Frontierbegriffs, in: Saeculum 54 (2003), pp. 123–43.

———: *Geschichte Afrikas*. Von 1800 bis zur Gegenwart, Paderborn 2004.

Marx, Karl, and Friedrich Engels: *Collected Works*, 37 vols., New York, 1975–2005.

Mason, R. H. P.: *Japan's First General Election* 1890, Cambridge 1969.

Masuzawa Tomoko: The *Invention* of World Religions, Chicago 2005.

Mathias, Peter: The *First Industrial Nation*. An Economic History of Britain 1700–1914, London 1969.

Matis, Herbert: Das *Industriesystem*. Wirtschaftswachstum und sozialer Wandel im 19. Jahrhundert, Vienna 1988.

Matsusaka, Yoshihisa Tak: The Making of *Japanese Manchuria*, 1904–1932, Cambridge, MA 2001.

Matthew, Colin (ed.): The *Nineteenth Century*. The British Isles, 1815–1901, Oxford 2000.

Matthew, H.C.G.: *Gladstone* 1809–1898, Oxford 1997.

Mawer, Granville Allen: *Ahab's Trade*. The Saga of South Seas Whaling, New York 1999.

May, Henry F.: The *Enlightenment in America*, Oxford 1976.

Mayaud, Jean-Luc, and Lutz Raphael (eds.): *Histoire de l'Europe rurale contemporaine*, Paris 2006.

Mayer, Arno J.: The *Persistence of the Old Regime*. Europe to the Great War, London 1981.

Mayer, Harold M., and Richard C. Wade: *Chicago*. Growth of a Metropolis, Chicago 1970.

Mayer, Henry: *All on Fire*. William Lloyd Garrison and the Abolition of Slavery, New York 1998.

Mayhew, Henry: *London Labour* and the London Poor, 4 vols., London 1861–62.

Maylam, Paul: A *History of the African People of South Africa*, London 1986.

———: *South Africa's Racial Past*. The History and Historiography of Racism, Segregation, and Apartheid, Aldershot 2001.

Maza, Sarah C.: The *Myth of the French Bourgeoisie*. An Essay on the Social Imaginary, 1750–1850, Cambridge, MA 2003.

Mazlish, Bruce: *Civilization* and Its Contents, Stanford, CA 2004.

Mazower, Mark: *The Balkans*, London 2000.

———: *Salonica*, City of Ghosts. Christians, Muslims and Jews, 1430–1950, New York 2004.

McCaffray, Susan Purves, and Michael S. Melancon (eds.): *Russia* in the European Context, 1789–1914. A Member of the Family, New York 2005.

McCalman, Iain (ed.): An Oxford Companion to the *Romantic Age*. British Culture 1776–1832, Oxford 1999.

McCarthy, Justin: *Death and Exile*. The Ethnic Cleansing of Ottoman Muslims, 1821–1922, Princeton, NJ 1995.

McClain, James L.: *Japan*. A Modern History, New York 2002.

McClain, James L., et al. (eds.): *Edo and Paris*. Urban Life and the State in the Early Modern Era, Ithaca, NY 1994.

McClain, James L., and Wakita Osamu (eds.): *Osaka*. The Merchants' Capital of Early Modern Japan, Ithaca, NY 1999.

McClellan, Michael E.: *Performing Empire*: Opera in Colonial Hanoi, in: Journal of Musicological Research 22 (2003), pp. 135–66.

McCord, Edward A.: The *Power of the Gun*. The Emergence of Modern Chinese Warlordism, Berkeley, CA 1993.

McCreery, David J.: The *Sweat* of Their Brow. A History of Work in Latin America, New York 2000.

McCusker, John J., and Russell R. Menard: The *Economy of British America*, 1607–1789, Chapel Hill, NC 1985.

McEvedy, Colin, and Richard Jones: *Atlas* of World Population History, London 1978.

McFadden, Margaret H.: *Golden Cables* of Sympathy. The Transatlantic Sources of Nineteenth-Century Feminism, Lexington, KY 1999.

McKendrick, Neil, et al.: The *Birth of a Consumer Society*. The Commercialization of Eighteenth-Century England, London 1983.

McKeon, Michael: The *Secret History* of Domesticity. Public, Private, and the Division of Knowledge, Baltimore, MD 2005.

McKeown, Adam: *Chinese Migrant Networks* and Cultural Change. Peru, Chicago, Hawaii, 1900–1936, Chicago 2001.

———: *Global Migration*, 1846–1940, in: JWH 15 (2004), pp. 155–89.

———: *Melancholy Order*: Asian Migration and the Globalization of Borders, New York 2008.

———: *Different Transitions*. Comparing China and Europe, 1600–1900, in: JGH 6 (2011), 309–19.

McLeod, Hugh: *Secularisation* in Western Europe 1848–1914, New York 2000.

McLynn, Frank: *1759*. The Year Britain Became Master of the World, London 2004.

McMichael, Philip: *Settlers* and the Agrarian Question. Foundations of Capitalism in Colonial Australia, Cambridge 1984.

McNeill, John R.: *Something New under the Sun*. An Environmental History of the Twentieth-Century World, New York 2000.

———: *Mosquito Empires*. Ecology and War in the Greater Caribbean, 1620–1914, Cambridge 2010.

McNeill, John R., and William H. McNeill: The *Human Web*. A Bird's-Eye View of World History, New York 2003.

McNeill, William H.: *Europe's Steppe Frontier,* 1500–1800, Chicago 1964.

———: The *Shape* of European History, New York 1974.

———: *Plagues* and Peoples, Harmondsworth 1976.

———: The *Pursuit of Power*. Technology, Armed Force, and Society since A.D. 1000, Oxford 1982.

McNeill, William H., et al. (eds.): *Berkshire Encyclopedia* of World History, 5 vols., Great Barrington, MA 2005.

McPherson, James M.: *Battle Cry* of Freedom. The American Civil War, New York 1988.

———: *Abraham Lincoln* and the Second American Revolution, New York 1990.

McShane, Clay, and Joel A. Tarr: *The Horse in the City*. Living Machines in the Nineteenth Century, Baltimore, MD 2007.

Mead, W. R.: A Historical Geography of *Scandinavia*, London 1981.

Mehl, Margaret: *History and the State* in Nineteenth-Century Japan, Basingstoke 1998.

Mehnert, Ute: *Deutschland*, Amerika und die "gelbe Gefahr." Zur Karriere eines Schlagworts in der großen Politik, 1905–1917, Stuttgart 1995.

Mehra, Parshotam: *An "Agreed" Frontier*. Ladakh and India's Northernmost Borders, 1846–1947, Delhi 1992.

Mehrotra, Arvind Krishna (ed.): A History of *Indian Literature in English*, London 2003.

Meinig, Donald W.: The *Shaping of America*. A Geographical Perspective on 500 Years of History, 4 vols., New Haven, CT 1986–2004.

Melinz, Gerhard, and Susan Zimmermann (eds.): *Wien—Prag—Budapest*. Blütezeit der Habsburgermetropolen. Urbanisierung, Kommunalpolitik, gesellschaftliche Konflikte (1867–1918), Vienna 1996.

Melosi, Martin V.: *The Sanitary City*: Environmental Services in Urban America from Colonial Times to the Present. Baltimore, MD 2000.

Mendo Ze, Gervais, et al.: *Le Français* langue africaine. Enjeux et atouts pour la francophonie, Paris 1999.

Merchant, Carolyn: The *Columbia Guide* to American Environmental History, New York 2002.

Merki, Christoph Maria: Der holprige *Siegeszug des Automobils*, 1895–1930. Zur Motorisierung des Straßenverkehrs in Frankreich, Deutschland und der Schweiz, Vienna 2002.

Metcalf, Barbara Daly: *Islamic Revival* in British India. Deoband, 1860–1900, Princeton, NJ 1982.

Metcalf, Thomas R.: *Ideologies of the Raj*, Cambridge 1994.

———: *Imperial Connections*. India in the Indian Ocean Arena; 1860–1920, Berkeley, CA 2007.

Meyer, David R.: *Hong Kong* as a Global Metropolis, Cambridge 2000.

Meyer, James H.: *Immigration*, Return, and the Politics of Citizenship. Russian Muslims in the Ottoman Empire, 1869–1914, in: IJMES 39 (2007), pp. 15–32.

Meyer, Jean, et al.: *Histoire de la France coloniale*. Des origines à 1914, Paris 1991.

Meyer, Michael A.: *Response to Modernity*. A History of the Reform Movement in Judaism, Detroit, MI 1988.

Meyer, Michael C., and William H. Beezley (eds.): The *Oxford History of Mexico*, Oxford 2000.

Meyer, Michael C., and William L. Sherman: The Course of *Mexican History*, 4th ed., New York 1991.

Meyer-Fong, Tobie: *What Remains*: Coming to Terms with Civil War in 19th-Century China, Stanford, CA 2013.

Michael, Franz: The *Taiping Rebellion*, 3 vols., Seattle, WA 1966–71.

Michel, Bernard: Histoire de *Prague*, Paris 1998.

Michie, Ranald C.: The *London Stock Exchange*. A History, Oxford 1999.

———: The *Global Securities Market*. A History, Oxford 2008.

Miers; Suzanne, and Richard L. Roberts (eds.): The *End of Slavery* in Africa, Madison, WI 1988.

Migeod, Heinz-Georg: *Die persische Gesellschaft* unter Nāsiru'd-Dīn Šāh (1848–1896), Berlin 1990.

Mill, John Stuart: *Collected Works*, 33 vols., ed. John M. Robson, Toronto 1965–1991.

Miller, Aleksej I., and Alfred J. Rieber (eds.): *Imperial Rule*, Budapest 2004.

Miller, James R.: *Skyscrapers* Hide the Heavens. A History of Indian-White Relations in Canada, Toronto 1989.

Miller, Joseph C.: The *Significance* of Drought, Disease and Famine in the Agriculturally Marginal Zones of West-Central Africa, in: JAfH 23 (1982), pp. 17–61.

Miller, Rory: *Britain and Latin America* in the 19th and 20th Centuries, Harlow 1993.

Miller, Shawn William: An *Environmental History* of Latin America, Cambridge 2007.

Miller, Stuart Creighton: *"Benevolent Assimilation."* The American Conquest of the Philippines, 1899–1903, New Haven, CT 1982.

Millward, James A.: *Eurasian Crossroads*. A History of Xinjiang, New York 2007.

Millward, Robert: Private and Public *Enterprise* in Europe. Energy, Telecommunications and Transport, 1830–1990, Cambridge 2005.

Milner, Anthony: The *Invention of Politics* in Colonial Malaya, Cambridge 1995.

Milner, Clyde A., et al. (eds.): The Oxford History of the *American West*, New York 1994.

Minami Ryoshin: *Power Revolution* in the Industrialization of Japan, 1885–1940, Tokyo 1987.

Mintz, Sidney W.: *Sweetness* and Power. The Place of Sugar in Modern History, New York 1985.

Mirowski, Philip: *More Heat than Light*. Economics as Social Physics, Physics as Nature's Economics, Cambridge 1989.

Mishra, Girish: An *Economic History* of Modern India, 2nd ed., Delhi 1998.

Mishra, Pankaj: From the *Ruins of Empire*. The Revolt against the West and the Remaking of Asia, London 2012.

Misra, Bankey Bihari.The *Bureaucracy in India*. An Historical Analysis of Development up to 1947, Delhi 1977.

Mitchell, Allan: The Great *Train Race*. Railways and the Franco-German Rivalry 1815–1914, New York 2000.

Mitchell, Brian R.: International Historical Statistics: *Europe,* 1750–1988, 3rd ed., Basingstoke 1992.

———: International Historical Statistics. The *Americas,* 1750–1988, 2nd ed., New York 1993.

———: International Historical Statistics. *Africa*, Asia and Oceania. 1750–1988, New York 1995.

Mitchell, Timothy: The *World as Exhibition*, in: CSSH 31 (1989), pp. 217–36.

———: *Colonising Egypt*, Berkeley, CA 1991.

Moch, Leslie Page: *Moving Europeans*. Migration in Western Europe since 1650, Blooming-ton, IN 1992.

Moe, Nelson: The *View from Vesuvius*. Italian Culture and the Southern Question, Berkeley, CA 2002.

Mokyr, Joel: The *Lever of Riches*. Technological Creativity and Economic Progress, New York 1990.

——: The *Gifts of Athena*, Princeton, NJ 2002.

——: The *Enlightened Economy*. An Economic History of Britain 1700–1850. New Haven, CT 2009.

Mollaret, Henri H., and Jacqueline Brossolet: *Alexandre Yersin*, le vainqeur de la peste, Paris 1985.

Moltke, Helmuth von: *Briefe* über Zustände und Begebenheiten in der Türkei aus den Jahren 1835–1839 [1841], ed. Helmut Arndt, Nördlingen 1987.

Mommsen, Wolfgang J.: *Der europäische Imperialismus*. Aufsätze und Abhandlungen, Göt-tingen 1979.

——: *Theories of Imperialism*, Chicago 1982.

——: *Großmachtstellung* und Weltpolitik. Die Außenpolitik des Deutschen Reiches 1870–1914, Berlin 1993.

——: *Das Ringen um den nationalen Staat*. Die Gründung und der innere Ausbau des Deutschen Reiches unter Otto von Bismarck 1850 bis 1890, Berlin 1993.

——: *Bürgerstolz* und Weltmachtstreben. Deutschland unter Wilhelm II. 1890–1918, Berlin 1995.

——: *1848. Die ungewollte Revolution*. Die revolutionären Bewegungen in Europa 1830–1849, Frankfurt a.M. 1998.

Mommsen, Wolfgang J., and Jürgen Osterhammel (eds.): *Imperialism and After*, London 1986.

Monkkonen, Eric H.: *Police* in Urban America, 1860–1920, Cambridge 1981.

——: *America Becomes Urban*. The Development of US Cities and Towns, 1780–1980, Berkeley, CA 1988.

Monnett, John H.: *Tell Them We Are Going Home*. The Odyssey of the Northern Cheyennes, Norman, OH 2001.

Montanari, Massimo: The *Culture of Food*, Cambridge, MA 1994.

Montel, Nathalie: *Le chantier* du Canal de Suez (1859–1869). Une histoire des pratiques tech-niques, Paris 1998.

Moore, Barrington: *Social Origins* of Dicatorship and Democracy. Lord and Peasant in the Making of the Modern World, Boston 1966.

Moorehead, Caroline: *Dunant's Dream*. War, Switzerland and the History of the Red Cross, London 1998.

Morelli, Federica: *Entre ancien et nouveau régime*. L'histoire politique hispano-américaine du XIXᵉ siècle, in: Annales HSS 59 (2004), pp. 759–81.

Moreman, T. R.: The *Army in India* and the Development of Frontier Warfare, 1849–1947, Basingstoke 1998.

Moretti, Franco: *Atlas of the European Novel, 1800–1900*, London 1998.

Moretti, Franco (ed.), *Il romanzo*, vol. 3: Storia e geografia, Turin 2002.

Morris, Donald R.: The *Washing of the Spears*. A History of the Zulu Nation under Shaka and Its Fall in the Zulu War of 1879, London 1965.

Morris-Suzuki, Tessa: The *Technological Transformation* of Japan. From the Seventeenth to the Twenty–First Century, Cambridge 1994.

Morris-Suzuki, Tessa: *Re-inventing Japan*. Time, Space, Nation, New York 1998.

Morse, Edward S.: *Japanese Homes* and Their Surroundings [1886], Boston 1986.

Mosk, Carl: *Japanese Industrial History*. Technology, Urbanization, and Economic Growth, Armonk, NY 2001.

Mosley, Paul: The *Settler Economies*. Studies in the Economic History of Kenya and Southern Rhodesia 1900–1963, Cambridge 1983.

Mosse, George L.: Toward the *Final Solution*, New York 1978.

Motyl, Alexander J.: *Revolutions*, Nations, Empires. Conceptual Limits and Theoretical Possibilities, New York 1999.

——: *Imperial Ends*. The Decay, Collapse, and Revival of Empires, New York 2001.

Moya, José C.: *Cousins* and Strangers. Spanish Immigrants in Buenos Aires, 1850–1930, Berkeley, CA 1998.

—— (ed.): The *Oxford Handbook of Latin American History*, Oxford 2011.

Mukherjee, S.N.: *Calcutta*. Essays in Urban History, Kalkutta 1993.

Muller, Jerry Z.: *The Mind and the Market*. Capitalism in Western Thought, New York 2003.

Mulligan, William: The *Origins* of the First World War, Cambridge 2010.

Mumford, Lewis: *Technics* and Civilization, New York 1934.

——: The *City in History*. Its Origins, Its Transformations, and Its Prospects, New York 1961.

Münch, Peter: *Stadthygiene* im 19. und 20. Jahrhundert. Die Wasserversorgung, Abwasser und Abfallbeseitigung unter besonderer Berücksichtigung Munichs, Göttingen 1993.

Münkler, Herfried: *Empires*. The Logic of World Domination from Ancient Rome to the United States, Cambridge 2007.

Munn, Nancy D.: The *Cultural Anthropology* of Time. A Critical Essay, in: Annual Review of Anthropology 21 (1992), pp. 93–123.

Munro, J. Forbes: *Maritime Enterprise* and Empire. Sir William Mackinnon and His Business Network, 1823–93, Woodbridge (Suffolk) 2003.

Murphy, Craig N.: *International Organization* and Industrial Change. Global Governance since 1850, Cambridge 1994.

Nadel, Stanley: *Little Germany*. Ethnicity, Religion, and Class in New York City, 1845–80, Urbana, IL 1990.

Nakayama Shigeru: Academic and Scientific *Traditions* in China, Japan and the West, Tokyo 1984.

Nani, Michele: *Ai confini della nazione*. stampa e razzismo nell'Italia di fine ottocento, Rom 2006.

Naquin, Susan: *Peking*. Temples and City Life, 1400–1900, Berkeley, CA 2000.

Naquin, Susan, and Evelyn S. Rawski: *Chinese Society* in the Eighteenth Century, New Haven, CT 1987.

Naquin, Susan, and Yü Chün-fang (eds.): *Pilgrims* and Sacred Sites in China, Berkeley, CA 1992.

Nash, Gary B.: *First City*. Philadelphia and the Forging of Historical Memory, Philadelphia 2002.

Nash, Roderick: *Wilderness and the American Mind*, 3rd ed., New Haven, CT 1982.

Nasson, Bill: The *South African War*, 1899–1902, London 1999.

Navarro Floria, Pedro: *Sarmiento* y la frontera sur argentina y chilena. De tema antropológico a custión social (1837–1856), in: JbLA 37 (2000), pp. 125–47.

Navarro García, Luis (ed.): *Historia de las Américas*, 4 vols., Madrid 1991.

Neal, Larry: The Rise of *Financial Capitalism*. International Capital Markets in the Age of Reason, Cambridge 1990.

Nee, Victor, and Richard Swedberg (eds.): The *Economic Sociology* of Capitalism, Princeton, NJ 2005.

Needell, Jeffrey D.: A *Tropical "belle époque."* Elite Culture and Society in Turn-of-the-Century Rio de Janeiro, Cambridge 1987.

Needham, Joseph: The *Grand Titration*. Science and Society in East and West, London 1969.

Neff, Stephen C.: *War* and the Law of Nations. A General History, Cambridge 2005.

Nelson, Marie C.: *Bitter Bread*. The Famine in Norrbotten, 1867–1868, Stockholm 1988.

Neubach, Helmut: Die *Ausweisungen* von Polen und Juden aus Preußen 1885/86, Wiesbaden 1967.

Newbury, Colin: *Patrons*, Clients and Empire. Chieftaincy and Over-rule in Asia, Africa and the Pacific, Oxford 2003.

Newhall, Beaumont: The *History of Photography*, New York 1982.

Newitt, Malyn: A History of *Mozambique*, London 1996.

Nichols, Roger L.: *American Indians* in U.S. History, Norman, OK 2003.

Nickel, Herbert J.: *Soziale Morphologie* der mexikanischen Hacienda, Wiesbaden 1978.

Nickerson, Thomas, and Owen Chase: The *Loss of the Ship "Essex,"* Sunk by a Whale, ed. Nathaniel Philbrick, New York 2000.

Ninkovich, Frank: *Global Dawn*. The Cultural Foundations of American Internationalism, 1865–1890, Cambridge, MA 2009.

Nipperdey, Thomas: *Deutsche Geschichte 1866–1918*, 2 vols., Munich 1990–92.

———: *Germany* from Napoleon to Bismarck 1800–1866, Princeton, NJ 1996.

Nish, Ian H.: The *Origins* of the Russo-Japanese War, London 1985.

Nishiyama Matsunosuke: *Edo Culture*. Daily Life and Diversions in Urban Japan, 1600–1868, Honolulu 1997.

Nitschke, August, et al. (eds.): *Jahrhundertwende*. Der Aufbruch in die Moderne 1880–1930, 2 vols., Reinbek 1990

Noiriel, Gérard: *Immigration*, antisémitisme et racisme en France (XIXe–XXe siècle). Discours publics, humiliations privées. Paris 2007.

Nolte, Hans-Heinrich: *Weltgeschichte*. Imperien, Religionen und Systeme. 15.–19. Jahrhundert, Vienna 2005.

Nolte, Paul: *1900*. Das Ende des 19. und der Beginn des 20. Jahrhunderts in sozialgeschichtlicher Perspektive, in: GWU 47 (1996), pp. 281–300.

Nolte, Paul: Gibt es noch eine *Einheit* der Neueren Geschichte? in: ZHF 24 (1997), pp. 377–99.

Nordman, Daniel: *Frontières* de France. De l'espace au territoire, Paris 1998.

North, Douglass C.: *Institutions*, Institutional Change, and Economic Performance, Cambridge 1990.

———: *Understanding* the Process of Economic Change, Princeton, NJ 2005.

North, Douglass C., and Robert Paul Thomas: The *Rise* of the Western World. A New Economic History, Cambridge 1973.

North, Michael (ed.): *Kommunikationsrevolutionen*. Die neuen Medien des 16. und 19. Jahrhunderts, 2nd ed., Cologne 2001.

Northrup, David: *Indentured Labour* in the Age of Imperialism, 1834–1922, Cambridge 1995.

———: *Africa's Discovery* of Europe: 1450–1850, New York 2002.

Nouzille, Jean: *Histoire des frontières*. L'Autriche et l'Empire Ottoman, Paris 1991.

Nuckolls, Charles: The *Durbar Incident*, in: MAS 24 (1990), pp. 529–59.

Nugent, Walter: *Crossings*. The Great Transatlantic Migrations, 1870–1914, Bloomington, IN 1992.

———: *Into the West*. The Story of Its People, New York 1999.

Nussbaum, Felicity A. (ed.): The *Global Eighteenth Century,* Baltimore, MD 2003.

Nutini, Hugo G.: The *Wages of Conquest*. The Mexican Aristocracy in the Context of Western Aristocracies, Ann Arbor, MI 1995.

Nwauwa, Apollos O.: *Imperialism*, Academe and Nationalism. Britain and University Education for Africans 1860–1960, London 1997.

Oakes, James: The *Radical and the Republican*. Frederick Douglass, Abraham Lincoln, and the Triumph of Antislavery Politics, New York 2007.

O'Brien, Patrick K. (ed.): *Industrialisation*. Critical Perspectives on the World Economy, 4 vols., London 1998.

———: *Historiographical Traditions* and Modern Imperatives for the Restoration of Global History, in: JGH 1 (2006), pp. 3–39.

O'Brien, Patrick K., and Armand Clesse (eds.): *Two Hegemonies*. Britain 1846–1914 and the United States 1941–2001, Aldershot 2002.

Ó Cadhla, Stiofán: *Civilizing Ireland*. Ordnance Survey 1824–1842. Ethnograpphy, Cartography, Translation, Dublin 2007.

Ochsenwald, William: The *Hijaz Railroad*, Charlottesville, VA 1980.

———: *Religion*, Society and the State in Arabia. The Hijaz under Ottoman Control, 1840–1908, Columbus, OH 1984.

Offen, Karen: *European Feminisms*, 1700–1950. A Political History, Stanford, CA 2000.

Offer, Avner: The *First World War*. An Agrarian Interpretation, Oxford 1989.

———: The *British Empire*, 1870–1914. A Waste of Money, in: EcHR 46 (1993), pp. 215–38.

——— (ed.): *In Pursuit* of the Quality of Life, Oxford 1996.

Ogilvie, Sarah: *Words of the World*. A Global History of the "Oxford English Dictionary," Cambridge 2012.

Ogilvie, Sheilagh C., and Markus Cerman (eds.): European *Proto-Industrialization*. An Introductory Handbook, Cambridge 1996.

Ó Gráda, Cormac: *Ireland*. A New Economic History 1780–1939, Oxford 1995.

———: *Ireland's Great Famine*. Interdisciplinary Perspectives, Dublin 2006.

———: *Famine. A Short History*, Princeton, NJ 2009.

Okey, Robin: The *Habsburg Monarchy* c. 1765–1918. From Enlightenment to Eclipse, Basingstoke 2001.

———: *Taming Balkan Nationalism*. The Habsburg "Civilizing Mission" in Bosnia, 1878–1914, Oxford 2007.

Oldenburg, Veena Talwar: The Making of *Colonial Lucknow*, 1856–1877, Delhi 1984.

Olender, Maurice: The *Languages of Paradise*. Race, Religion, and Philology in the Nineteenth Century, Cambridge, MA 1992.

Oliver, Roland, and Anthony Atmore: *Medieval Africa* 1250–1800, Cambridge 2001.

———: *Africa since 1800*, Cambridge 2005.

Oliver, W. H. (ed.): The *Oxford History of New Zealand*, Oxford 1981.

Olsen, Donald J.: The *City* as a Work of Art. London, Paris, Vienna, New Haven, CT 1986.

Omissi, David E.: The *Sepoy* and the Raj. The Indian Army, 1860–1940, Basingstoke 1994.

O'Rourke, Kevin H., and Jeffrey G. Williamson: *Globalization* and History. The Evolution of a Nineteenth-Century Atlantic Economy, Cambridge, MA 1999.

O'Rourke, Shane: *Warriors* and Peasants. The Don Cossacks in Late Imperial Russia, Basingstoke 2000.

——: The *Cossacks*, Manchester 2007.

Osmani, S. R. (ed.): *Nutrition* and Poverty, Oxford 1992.

Osterhammel, Jürgen: *China* und die Weltgesellschaft. Vom 18. Jahrhundert bis in unsere Zeit, Munich 1989.

——: *Britain and China* 1842–1914, in: Louis, Oxford History of the British Empire, vol. 3 (1999), pp. 146–69.

——: *Geschichtswissenschaft* jenseits des Nationalstaats. Studien zu Beziehungsgeschichte und Zivilisationsvergleich, Göttingen 2001.

——: *Expansion und Imperium*, in: Peter Burschel et al. (eds.), Historische Anstöße. Festschrift für Wolfgang Reinhard, Berlin 2002, pp. 371–92.

——: *Ex-zentrische Geschichte*. Außenansichten europäischer Modernität, in: Jahrbuch des Wissenschaftskollegs zu Berlin 2000/2001, Berlin 2002, pp. 296–318

——: *In Search of a Nineteenth Century*, in: Bulletin of the German Historical Institute (Washington, DC) 32 (Spring 2003), pp. 9–28.

——: *Baylys Moderne*, in: NPL 50 (2005), pp. 7–17.

——: *Colonialism*. A Theoretical Overview, 2nd ed., Princeton, NJ 2005.

——: *Europe*, the "West" and the Civilizing Mission (2005 Annual Lecture at the German Historical Institute), London 2006.

——:*Globalgeschichte*, in: Hans-Jürgen Goertz (ed.), Geschichte. Ein Grundkurs, 3rd ed., Reinbek 2007, pp. 592–610.

——: Die *Entzauberung Asiens*. Europa und die asiatischen Reiche im 18. Jahrhundert, 2nd ed., Munich 2010.

——:*Globalizations*, in: J. H. Bentley, Oxford Handbook of World History (2011), pp. 89–104.

——:*Globale Horizonte* europäischer Kunstmusik, in: GG 38 (2012), pp. 86–132.

Osterhammel, Jürgen, et al. (eds.): *Wege* der Gesellschaftsgeschichte, Göttingen 2006.

Osterhammel, Jürgen, and Niels P. Petersson: *Globalization*. A Short History, Princeton, NJ 2005.

Ostler, Nicholas: *Empires of the Word*. A Language History of the World, London 2005.

Östör, Ákos: *Vessels of Time*. An Essay on Temporal Change and Social Transformation, Delhi 1993.

Otto, Frank: Die *Entstehung eines nationalen Geldes*. Integrationsprozesse der deutschen Währungen im 19. Jahrhundert, Berlin 2002.

Overton, Mark: *Agricultural Revolution* in England. The Transformation of the Agrarian Economy 1500–1850, Cambridge 1996.

Owen, Norman G.: The *Paradox* of Nineteenth–Century Population Growth in Southeast Asia. Evidence from Java and the Philippines, in: JSEAS 18 (1987), pp. 45–57.

Owen, Norman G., et al.: The *Emergence* of Modern Southeast Asia. A New History, Honolulu 2005.

Owen, Roger: The *Middle East* in the World Economy 1800–1914, London 1981.

——: *Lord Cromer*. Victorian Imperialist, Edwardian Proconsul, Oxford 2004.

Özmucur, Süleyman, and Şevket Pamuk: *Real Wages* and Standards of Living in the Ottoman Empire, 1489–1914, in: JEH 62 (2002), pp. 293–321.

Pagden, Anthony: *Lords* of all the World. Ideologies of Empire in Spain, Britain and France c.1500–c.1800, New Haven, CT 1995.

Paine, Sarah C. M.: *Imperial Rivals*. Russia, China and Their Disputed Frontier, 1858–1924, Armonk, NY 1996.

——: The *Sino-Japanese War* of 1894–1895. Perceptions, Power, and Primacy, Cambridge 2003.

Paine, Tom: *Common Sense* [1776], ed. Isaac Kramnick, Harmondsworth 1976.

Palairet, Michael: *Rural Serbia* in the Light of the Census of 1863, in: JEEcH 24 (1995), pp. 41–107.

——: The *Balkan Economies* c. 1800–1914. Evolution without Development, Cambridge 1997.

Palais, James B.: *Politics and Policy* in Traditional Korea, Cambridge, MA 1991.

——: A Search for *Korean Uniqueness*, in: HJAS 55 (1995), pp. 409–25.

Palmer, Robert R.: The Age of the *Democratic Revolution*, 2 vols., Princeton, NJ 1959–64.

Palmié, Stephan, and Francisco A. Scarano (eds.): The *Caribbean*. A History of the Region and Its Peoples, Chicago 2011.

Pamuk, Şevket: The *Ottoman Empire* and European Capitalism, 1820–1913. Trade, Investment and Production, Cambridge 1987.

Panda, Chitta: The Decline of the *Bengal Zamindars*. Midnapore, 1870–1920, Delhi 1996.

Pankhurst, Richard: The *Ethiopians*, Oxford 1998.

Pantzer, Peter (ed.): Die *Iwakura-Mission*. Das Logbuch des Kume Kunitake über den Besuch der japanischen Sondergesandtschaft in Deutschland, Österreich und der Schweiz im Jahre 1873, Munich 2002.

Panzac, Daniel: *Quarantaines* et lazarets. L'Europe et la peste d'orient (XVII^c–XX^c siècles), Aix-en-Provence 1986.

——: *Population et santé* dans l'Empire ottoman (XVIII^c–XX^c siècles), Istanbul 1996.

——: *Les Corsaires barbaresques*. La fin d'une épopée 1800–1820, Paris 1999.

Panzac, Daniel, and André Raymond (eds.): *La France et l'Égypte* à l'époque des vice-rois 1805–1882. Kairo 2002.

Papayanis, Nicholas: *Coachmen* of Nineteenth-Century Paris. Service Workers and Class Consciousness, Baton Rouge, LA 1993.

Papin, Philippe: Histoire de *Hanoi*, Paris 2001.

Paquette, Robert Louis, and Mark W. Smith (eds.): The *Oxford Handbook of Slavery* in the Americas, Oxford 2010.

Parissien, Steven: *Station to Station*, London 1997.

Parker, Linda S.: *Native American Estate*. The Struggle over Indian and Hawaiian Lands, Honolulu 1989.

Parsons, Neil: *King Khama*, Emperor Joe and the Great White Queen. Victorian Britain through African Eyes, Chicago 1998.

Parthasarathi, Prasannan: *Why Europe Grew Rich* and Asia Did Not: Global Economic Divergence, 1600–1850, Cambridge 2011.

Pasquier, Thierry du: *Les baleiniers français* au XIX^c siècle (1814–1868), Grenoble 1982.

Pati, Biswamoy (ed.): The *Great Rebellion* of 1857 in India: Exploring Transgressions, Contests and Diversities, London 2010.

Patriarca, Silvana: *Numbers* and Nationhood. Writing Statistics in Nineteenth-Century Italy, Cambridge 1996.

Patterson, Orlando: *Slavery and Social Death*. A Comparative Study, Cambridge, MA 1982.

Paul, Rodman W.: The *Far West* and the Great Plains in Transition, 1859–1900, New York 1988.

Paulmann, Johannes: *Pomp und Politik*. Monarchenbegegnungen in Europa zwischen Ancien Régime und Erstem Weltkrieg, Paderborn 2000.

Pearson, Michael: The *Indian Ocean*, London 2010.

Peck, Gunther: *Reinventing Free Labor*. Padrones and Immigrant Workers in the North American West, 1880–1930, Cambridge 2000.

Peebles, Patrick: The History of *Sri Lanka*, Westport, CT 2006.

Peers, Douglas M.: Between *Mars* and Mammon. Colonial Armies and the Garrison State in India 1819–1835, London 1995.

———: *Colonial Knowledge* and the Military in India, 1780–1860, in: JICH 33 (2005), pp. 157–80.

Pelzer, Erich (ed.): *Revolution und Klio*. Die Hauptwerke zur Französischen Revolution, Göttingen 2004.

Pennell, C. R.: *Morocco* since 1830. A History, London 2000.

Penny, H. Glenn: *Objects of Culture*. Ethnology and Ethnographic Museums in Imperial Germany, Chapel Hill, NC 2001.

Pepper, Suzanne: *Radicalism* and Education Reform in Twentieth-Century China. The Search for an Ideal Development Model, Cambridge 1996.

Perdue, Peter C.: *China Marches West*. The Qing Conquest of Central Eurasia, Cambridge, MA 2005.

Pérennès, Roger: *Déportés* et forçats de la Commune. De Belleville à Noumea, Nantes 1991.

Perkin, Harold: The *Origins* of Modern English Society, 1780–1880, London 1969.

Perkins, Kenneth J.: A History of *Modern Tunisia*, Cambridge 2004.

Perlin, Frank: The *Invisible City*. Monetary, Administrative and Popular Infrastructures in Asia and Europe, 1500–1900, Aldershot 1993.

Pernau, Margrit: *Bürger mit Turban*. Muslime in Delhi im 19. Jahrhundert, Göttingen 2008.

Petermann, Werner: Die *Geschichte der Ethnologie*, Wuppertal 2004.

Peters, F. E.: *The Hajj*: The Muslim Pilgrimage to Mecca and the Holy Places, Princeton, NJ 1994.

Peterson, Merrill D.: *Lincoln* in American Memory, New York 1994.

Petersson, Niels P.: *Anarchie* und Weltrecht: Das Deutsche Reich und die Institutionen der Weltwirtschaft, 1890–1930, Göttingen 2009.

Pétré-Grenouilleau, Olivier (ed.): *From Slave Trade to Empire*. Europe and the Colonisation of Black Africa 1780s–1880s, London 2004.

Petrow, Stefan: *Policing Morals*. The Metropolitan Police and the Home Office, 1870–1914, Oxford 1994.

Pflanze, Otto: *Bismarck* and the Development of Germany, 3 vols., Princeton, NJ 1990.

Philipp, Thomas, and Guido Schwald (eds.): *Abd-al-Rahman al-Jabarti's History of Egypt*, 3 vols., Stuttgart 1994.

Philippot, Robert: *Les Zemstvos*. Société civile et état bureaucratique dans la Russie tsariste, Paris 1991.

Phillips, Gordon A., and Whiteside, Noel: *Casual Labour*. The Unemployment Question in the Port Transport Industry, 1880–1970, Oxford 1985.

Phillipson, Robert: *Linguistic Imperialism*, Oxford 1992.

Pickering, Mary: *Auguste Comte*. An Intellectual Biography, 3 vols., Cambridge 1993–2009.

Pietrow-Ennker, Bianka: *Wirtschaftsbürger* und Bürgerlichkeit im Königreich Polen. Das Beispiel vom Lodz, dem "Manchester des Ostens," in: GG 31 (2005), pp. 169–202.

Pietschmann, Horst (ed.): *Atlantic History*. History of the Atlantic System 1580–1830, Göttingen 2002.

Pike, David L.: *Subterranean Cities*. The World beneath Paris and London, 1800–1945, Ithaca, NY 2005.

Pilbeam, Pamela M.: The *Middle Classes* in Europe 1789–1914. France, Germany, Italy and Russia, Basingstoke 1990.

————: The *1830 Revolution in France*, Basingstoke 1991.

Pinol, Jean-Luc: *Le Monde des villes* au XIXᵉ siècle, Paris 1991.

Pinol, Jean-Luc, et al.: *La Ville coloniale*, Paris 2012.

Pitts, Jennifer: A *Turn to Empire*. The Rise of Imperial Liberalism in Britain and France, Princeton, NJ 2005.

Plaggenborg, Stefan: *Ordnung und Gewalt*. Kemalismus, Faschismus, Sozialismus, Munich 2012.

Planhol, Xavier de: *Les Fondéments* géographiques de l'histoire de l'Islam, Paris 1968.

————: *Les Nations* du prophète. Manuel géographique de politique musulmane, Paris 1993.

————: *Le Paysage animal*. L'homme et la grande faune: une zoogéographie historique, Paris 2004.

Plato, Alice von: *Präsentierte Geschichte*. Ausstellungskultur und Massenpublikum im Frankreich des 19. Jahrhunderts, Frankfurt a.M. 2001.

Plunz, Richard: A History of *Housing in New York City*. Dwelling Type and Social Change in the American Metropolis, New York 1990.

Pocock, J. G. A.: The *Discovery* of Islands. Essays in British History, Cambridge 2005.

Pohl, Hans: *Aufbruch* zur Weltwirtschaft, 1840–1914. Geschichte der Weltwirtschaft von der Mitte des 19. Jahrhunderts bis zum Ersten Weltkrieg, Stuttgart 1989.

Poignant, Roslyn: *Professional Savages*. Captive Lives and Western Spectacle, New Haven, CT 2004.

Polachek, James M.: The *Inner Opium War*, Cambridge, MA 1992.

Polanyi, Karl: The *Great Transformation* [1944], Boston 1957.

Poliakov, Léon: The *Aryan Myth*. A History of Racist and Nationalist Ideas in Europe, New York 1974.

Pollard, Sidney: *Peaceful Conquest*. The Industrialization of Europe, 1760–1970, Oxford 1981.

Pollock, Sheldon: The *Cosmopolitan Vernacular*, in: JAS 57 (1998), pp. 6–37.

Polunov, Aleksandr J.: *Russia* in the Nineteenth Century. Autocracy, Reform, and Social Change, 1814–1914, Armonk, NY 2005.

Pomeranz, Kenneth: The *Great Divergence*. China, Europe, and the Making of the Modern World Economy, Princeton, NJ 2000.

Pomian, Krzysztof: *Europa* und seine Nationen, Berlin 1990.

————: *Sur l'histoire*, Paris 1999.

Pooley, Colin G. (ed.): *Housing Strategies* in Europe, 1880–1930, Leicester 1992.

Popkin, Jeremy D.: *You Are All Free*. The Haitian Revolution and the Abolition of Slavery, Cambridge 2010.

Port, Michael H.: *Imperial London*. Civil Government Building in London; 1850–1915, New Haven, CT 1995.

Porter, Andrew: *Atlas* of British Overseas Expansion, London 1991.

————: *European Imperialism*, 1860–1914, Basingstoke 1994.

————: *Religion versus Empire?* British Protestant Missionaries and Overseas Expansion, 1700–1914, Manchester 2004.

Porter, Bernard: The *Absent-Minded Imperialists*. Empire, Society and Culture in Britain, Oxford 2004.

———: The *Lion's Share*. A Short History of British Imperialism 1850–2011, 5th ed., London 2012.

Porter, Roy: *London*. A Social History, London 1994.

———: The *Greatest Benefit to Mankind*. A Medical History of Mankind from Antiquity to the Present, London 1999.

Porter, Theodore M., and Dorothy Ross (eds.): The *Modern Social Sciences* (= The Cambridge History of Science, vol. 7), Cambridge 2003.

Pot, Johan Hendrik Jacob van der: *Sinndeutung* und Periodisierung der Geschichte. Eine systematische Übersicht der Theorien und Auffassungen, Leiden 1999.

Potter, David M.: The *Impending Crisis*, 1848–1861, New York 1976.

Potter, Simon J.: *Communication* and Integration. The British and Dominions Press and the British World, c. 1876–1914, in: JICH 31 (2003), pp. 190–206.

———: *News* and the British World. The Emergence of an Imperial Press System, 1876–1922, Oxford 2003.

Pounds, Norman J. G.: An *Historical Geography* of Europe, 1800–1914, Cambridge 1985.

———: *Hearth* and Home. A History of Material Culture, Bloomington, IN 1989.

Prak, Maarten (ed.): *Early Modern Capitalism*. Economic and Social Change in Europe, 1400–1800, London 2001.

Prakash, Gyan (ed.): The *World of the Rural Labourer* in Colonial India, Delhi 1994.

———: *Another Reason*. Science and the Imagination of Modern India, Princeton, NJ 1999.

Prendergast, Christopher: *Paris* and the Nineteenth Century, Oxford 1992.

Price, Don C.: *Russia* and the Roots of the Chinese Revolution, 1896–1911, Cambridge, MA 1974.

Price, Jacob M.: *Economic Function* and the Growth of American Port Towns in the Eighteenth Century, in: Perspectives in American History 8 (1974), pp. 121–86.

Price, Pamela G.: *Acting in Public* versus Forming a Public. Conflict Processing and Political Mobilization in Nineteenth-Century South India, in: South Asia 14 (1991), pp. 91–121.

———: *Kingship* and Political Practice in Colonial India, Cambridge 1996.

Price, Roger: The *French Second Empire*. An Anatomy of Political Power, Cambridge 2001.

———: *People* and Politics in France, 1848–1870, Cambridge 2004.

Pröve, Ralf: *Militär*, Staat und Gesellschaft im 19. Jahrhundert. Munich 2006.

Prucha, Francis Paul: The *Great Father*. The United States Government and the American Indians, 2 vols., Lincoln, NE 1984.

Prude, Jonathan: The Coming of *Industrial Order*. Town and Factory Life in Rural Massachusetts, 1810–1860, Cambridge 1983.

Pückler-Muskau, Hermann Fürst von: *Briefe eines Verstorbenen*, ed. Heinz Ohff, Berlin 1986.

Purdy, Daniel L.: The *Tyranny of Elegance*: Consumer Cosmopolitanism in the Era of Goethe, Baltimore, MD 1998.

Pusey, James Reeve: *China and Charles Darwin*, Cambridge, MA 1983.

Quataert, Donald: *Social Desintegration* and Popular Resistance in the Ottoman Empire, 1881–1908. Reactions to European Economic Penetration, New York 1983.

———: *Ottoman Manufacturing* in the Age of the Industrial Revolution, Cambridge 1993.

——— (ed.): *Consumption Studies* and the History of the Ottoman Empire, New York 2000.

———: The *Ottoman Empire*, 1700–1922, Cambridge 2000.

Rabibhadana, Akin: The Organization of *Thai Society* in the Early Bangkok Period, 1782–1873, Ithaca, NY 1969.

Rabinbach, Anson: The *Human Motor*. Energy, Fatigue, and the Origins of Modernity, New York 1990.

Radkau, Joachim: Das Zeitalter der *Nervosität*. Deutschland zwischen Bismarck und Hitler, Munich 1998.

———: *Nature und Power*: A Global History of the Environment, Cambridge 2008.

Rahikainen, Marjatta: *Centuries of Child Labour*. European Experiences from the Seventeenth to the Twentieth Century, Aldershot 2004.

Raina, Dhruv: *Images and Contexts*. The Historiography of Science and Modernity in India, New Delhi 2003.

Rallu, Jean-Louis: *Les Populations océanniennes* aux XIXe et XXe siècles, Paris 1990.

———: *Population* of the French Overseas Territories in the Pacific, Past, Present and Projected, in: JPH 26 (1991), pp. 169–86.

Ralston, David B.: *Importing the European Army*. The Introduction of European Military Techniques and Institutions into the Extra-European World, 1600–1914, Chicago 1990.

Ramachandran, Ranganathan: *Urbanization* and Urban Systems in India, Delhi 1989.

Rangarajan, Mahesh: *Fencing the Forest*. Conservation and Ecological Change in India's Central Provinces 1860–1914, Delhi 1996.

Ranke, Leopold von: *Aus Werk und Nachlaß*, 4 vols., Munich 1965–75.

———: *Die großen Mächte* [1833], ed. Ulrich Muhlack, Frankfurt a.M. 1995.

Rankin, Mary Backus: *Elite Activism* and Political Transformation in China. Zhejiang Province, 1865–1911, Stanford, CA 1986.

———: *Managed by the People*. Officials, Gentry, and the Foshan Charitable Granary, 1795–1845, in: LIC 15 (1994), pp. 1–52.

Ransom, P.J.G.: The *Archaeology of the Transport Revolution*, 1750–1850, Tadworth (Surrey) 1984.

Raphael, Lutz: *Recht und Ordnung*. Herrschaft durch Verwaltung im 19. Jahrhundert, Frankfurt a.M. 2000.

Rasler, Karen A., and William R. Thompson: *War* and State Making. The Shaping of the Global Powers, Boston 1989.

Rathenau, Walther: *Der Kaiser*. Eine Betrachtung, Berlin 1919.

Ratzel, Friedrich: *Politische Geographie*, Munich 1897.

Raulff, Ulrich: *Der unsichtbare Augenblick*. Zeitkonzepte in der Geschichte, Göttingen 1999.

Ravina, Mark: *Land and Lordship* in Early Modern Japan, Stanford, CA 1999.

———: The *Last Samurai*. The Life and Battles of Saigō Takamori, Hoboken, NJ 2004.

Rawlinson, John L.: *China's Struggle* for Naval Development, Cambridge, MA 1967.

Rawski, Evelyn S.: *Education* and Popular Literacy in Ch'ing China, Ann Arbor, MI 1979.

———: The *Last Emperors*. A Social History of the Qing Imperial Institution, Berkeley, CA 1998.

Raymond, André: *Grandes villes arabes* à l'époque ottomane, Paris 1985.

———: *Le Caire*, Paris 1993.

Read, Donald: The *Power of News*. The History of Reuters, 1849–1989, Oxford 1992.

Reardon-Anderson, James: *Reluctant Pioneers*. China's Expansion Northward, 1644–1937, Stanford, CA 2005.

Reclus, Elisée: *Nouvelle géographie universelle*, 19 vols., Paris 1876–94.

———: *L'homme et la terre*. Histoire contemporaine [1908], 2 vols., Paris 1990.

Reed, Bradly W.: *Talons and Teeth*. County Clerks and Runners in the Qing Dynasty, Stanford, CA 2000.

Reichardt, Rolf: Das *Blut der Freiheit*. Französische Revolution und demokratische Kultur, Frankfurt a.M. 1998.

Reid, Anthony: An *Age of Commerce* in Southeast Asian History, in: MAS 24 (1990), pp. 1–30.

———: *Humans and Forests* in Pre-colonial Southeast Asia, in: Environment and History 1 (1995), pp. 93–110.

——— (ed.): The *Last Stand* of Asian Autonomies. Responses to Modernity in the Diverse States of Southeast Asia and Korea, 1750–1900, Basingstoke 1997.

———: *Charting the Shape* of Early Modern Southeast Asia, Chiang Mai 1999.

——— (ed.): *Sojourners* and Settlers. Histories of Southeast Asia and the Chinese, Honolulu 2001.

Reid, Donald Malcolm: *Whose Pharaohs?* Archaeology, Museums, and Egyptian National Identity from Napoleon to World War I, Berkeley, CA 2002.

Reid, Richard: The *Ganda* on the Lake Victoria. A Nineteenth-Century East African Imperialism, in: JAfH 39 (1998), pp. 349–63.

———: *War in Pre-colonial Eastern Africa,* London 2007.

Reimers, David M.: *Other Immigrants.* The Global Origins of the American People, New York 2005.

Reinhard, Marcel, et al.: *Histoire générale* de la population mondiale, Paris 1968.

Reinhard, Rudolf: Weltwirtschaftliche und politische *Erdkunde*, 6th ed., Breslau 1929.

Reinhard, Wolfgang: Geschichte der europäischen *Expansion*, 4 vols., Stuttgart 1983–90.

———: Geschichte der *Staatsgewalt*. Eine vergleichende Verfassungsgeschichte Europas von den Anfängen bis zur Gegenwart, Munich 1999.

——— (ed.): *Verstaatlichung* der Welt? Europäische Staatsmodelle und außereuropäische Machtprozesse, Munich 1999.

———: *Lebensformen* Europas. Eine historische Kulturanthropologie, Munich 2004.

———: A Short History of *Colonialism*, Manchester 2011.

Reinkowski, Maurus: Die *Dinge der Ordnung*. Eine vergleichende Untersuchung über die osmanische Reformpolitik im 19. Jahrhundert, Munich 2005.

Reiter, Herbert: Politisches *Asyl* im 19. Jahrhundert. Die deutschen politischen Flüchtlinge der Vörmärz und der Revolution von 1848/49 in Europa und den USA, Berlin 1992.

Reps, John W.: The *Making of Urban America*. A History of City Planning in the United States, Princeton, NJ 1965.

Reséndez, Andrés: *Changing National Identities* at the Frontier. Texas and New Mexico, 1800–1850, Cambridge 2005.

Rétif, André: *Pierre Larousse* et son œuvre (1817–1875), Paris 1975.

Reulecke, Jürgen: Geschichte der *Urbanisierung in Deutschland*, 3rd ed., Frankfurt a.M. 1992.

———: Die *Mobilisierung* der "Kräfte und Kapitale." Der Wandel der Lebensverhältnisse im Gefolge von Industrialisierung und Verstädterung, in: idem (ed.): Geschichte des Wohnens, vol. 3: 1800–1918, Das bürgerliche Zeitalter, Stuttgart 1997, pp. 15–144.

Reynolds, Douglas R.: *China*, 1898–1912. The Xinzheng Revolution and Japan, Cambridge, MA 1993.

Rhoads, Edward J. M.: *Manchus and Han*. Ethnic Relations and Political Power in Late Qing and Early Republican China, 1861–1928, Seattle, WA 2000.

Ribbe, Wolfgang: *Geschichte Berlins*, 2 vols., Munich 1987.

Ribeiro, Darcy: *The Americas* and Civilization, New York 1971.

Ribeiro, Darcy, and Gregory Rabassa: *The Brazilian People*. The Formation and Meaning of Brazil, Gainesville, FL 2000.

Rich, Norman: The *Age of Nationalism* and Reform 1850–1890, 2nd ed., New York 1977.

———: *Great Power Diplomacy*, 1814–1914, New York 1992.

Richards, Edward G.: *Mapping Time*. The Calendar and Its History, Oxford 1998.

Richards, Eric: How Did *Poor People* Emigrate from the British Isles to Australia in the Nineteenth Century? in: JBS 32 (1993), pp. 250–79.

Richards, John F.: The *Mughal Empire*, Cambridge 1993.

———: The *Unending Frontier*. An Environmental History of the Early Modern World, Berkeley, CA 2003.

Richardson, Peter: *Chinese Mine Labour* in the Transvaal, London 1982.

Richter, Daniel K.: *Facing East* from Indian Country. A Native History of Early America, Cambridge, MA 2001.

Richthofen, Ferdinand Freiherr von: *China*. Ergebnisse eigener Reisen und darauf begründeter Studien, 5 vols., Berlin 1877–1912.

Rickard, John: *Australia*. A Cultural History, 2nd ed., Harlow 1996.

Ricklefs, M. C.: A *History of Modern Indonesia* since c. 1200, 4th ed., Basingstoke 2008.

Ridley, Jane: *Edwin Lutyens*. His Life, His Wife, His Work, London 2003.

Rieger, Bernhard: *Technology* and the Culture of Modernity in Britain and Germany, 1890–1945, Cambridge 2005.

Riekenberg, Michael: *Ethnische Kriege* in Lateinamerika im 19. Jahrhundert, Stuttgart 1997.

———: *Gewaltsegmente*. Über einen Ausschnitt der Gewalt in Lateinamerika, Leipzig 2003.

Riello, Giorgio, and Patrick O'Brien: The *Future* Is Another Country. Offshore Views of the British Industrial Revolution, in: Journal of Historical Sociology 22 (2009), pp. 1–29.

Riello, Giorgio, and Prasannan Parthasarathi (eds.): The *Spinning World*. A Global History of Cotton Textiles, 1200–1850, Oxford 2009.

Riello, Giorgio, and Tirthankar Roy (eds.): How *India Clothed the World*. The World of South Asian Textiles, 1500–1850, Leiden 2009.

Riesenberger, Dieter: *Für Humanität* in Krieg und Frieden. Das Internationale Rote Kreuz 1863–1977, Göttingen 1992.

Riley, James C.: *Rising Life Expectancy*, Cambridge 2001.

Ringer, Fritz K.: The Decline of the *German Mandarins*. The German Academic Community, 1890–1933, Cambridge, MA 1969.

———: *Education and Society* in Modern Europe, Bloomington, IN 1979.

Rinke, Stefan: *Las revoluciones en América Latina*. Las vías a la independencia, 1760–1830, México 2011.

Rittaud-Hutinet, Jacques: *Le cinéma* des origines. Les frères Lumière et leurs opérateurs, Seyssel 1985.

Ritter, Carl: Die *Erdkunde* im Verhältniß zur Natur und zur Geschichte des Menschen, 19 vols., Berlin 1822–59.

———: *Einleitung* zur allgemeinen und vergleichenden Geographie, Berlin 1852.

Ritter, Gerhard A., and Klaus Tenfelde: *Arbeiter* im Deutschen Kaiserreich 1871 bis 1914, Bonn 1992.

Rivet, Daniel: *Le Maroc* de Lyautey à Mohammed V. Le double visage du protectorat, Paris 1999.

Roach, Joseph: *Cities of the Dead*. Circum-Atlantic Performance, New York 1996.

Robb, Graham: The *Discovery of France*. A Historical Geography from the Revolution to the First World War, New York 2007.

Robb, Peter: *Peasants' Choices?* Indian Agriculture and the Limits of Commercialization in Nineteenth-Century Bihar, in: EcHR 45 (1992), pp. 97–119.

Robbins, Richard G.: *Famine in Russia*, 1891–1892. The Imperial Government Responds to a Crisis, New York 1975.

Roberts, Andrew: *Salisbury*. Victorian Titan, London 1999.

Roberts, J.A.G.: *China to Chinatown*. Chinese Food in the West, London 2002.

Roberts, John M.: *Twentieth Century*. The History of the World, 1901 to 2000, New York 1999.

Robertson, Bruce Carlisle: *Raja Rammohan Ray*. The Father of Modern India, Delhi 1995.

Robinson, David: *Muslim Societies* in African History, Cambridge 2004.

Robinson, Francis: *Islam* and Muslim History in South Asia, New Delhi 2000.

Robinson, Michael E.: *Korea's Twentieth-Century Odyssey*. A Short History, Honolulu 2007.

Roche, Daniel: *Le cheval moteur*, Paris 2008 (= La culture équestre de l'Occident, XVIᵉ–XIXᵉ siècle, vol. 1).

Rock, David: *Argentina*, 1516–1987. From Spanish Colonization to Alfonsín, Berkeley, CA 1987.

Rodger, Nicholas A. M.: The *Command of the Sea*. A Naval History of Britain, 1649–1815, London 2004.

Rodger, Richard (ed.): *European Urban History*. Prospect and Retrospect, Leicester 1993.

Rodgers, Daniel T.: *Atlantic Crossings*. Social Politics in an Progressive Age, Cambridge, MA 1998.

Rodríguez O., Jaime E.: The *Independence of Spanish America*, Cambridge 1998.

———: The *Emancipation* of America, in: AHR 105 (2000), pp. 131–52.

Roediger, David R.: *Working toward Whiteness*. How America's Immigrants Became White, New York 2005.

Rogan, Eugene L.: *Frontiers of the State* in the Late Ottoman Empire. Transjordan, 150–1921, Cambridge 1999.

Rogin, Michael Paul: *Fathers and Children*. Andrew Jackson and the Subjugation of the American Indians, New York 1975.

Rogowski, Ronald: *Commerce and Coalitions*. How Trade Affects Domestic Political Alignments, Princeton, NJ 1989.

Röhl, John C. G.: *Wilhelm II.*, 3 vols., Munich 1993–2008.

Röhl, Wilhelm (ed.): *History of Law in Japan* since 1868, Leiden 2005.

Rohrbough, Malcolm J.: *Aspen*. The History of a Silver Mining Town, 1879–1893, New York 1986.

———: *Days of Gold*. The California Gold Rush and the American Nation, Berkeley, CA 1997.

Rokkan, Stein: *State Formation,* Nation-Building, and Mass Politics in Europe. The Theory of Stein Rokkan, ed. Peter Flora, Oxford 1999.

Romein, Jan: The *Watershed* of Two Eras. Europe in 1900, Middletown, CT 1978.

Rosanvallon, Pierre: *L'état en France* de 1789 à nos jours, Paris 1990.

———: *Le sacre du citoyen*. Histoire du suffrage universel en France, Paris 1992.

———: *La démocratie inachevée*. Histoire de la souveraineté du peuple en France, Paris 2000.

Rosen, Charles: The *Classical Style*. Haydn, Mozart, Beethoven, London 1971.

———: The *Romantic Generation*, Cambridge, MA 1995.

Rosen, George: A *History of Public Health*, New York 1958.

Rosenberg, Charles E.: The *Cholera Years*. The United States in 1832, 1849, and 1866, Chicago 1962.

Rosenberg, Emily S. (ed.): *A World Connecting*, 1870–1945, Cambridge, MA 2012.

Rösener, Werner: The *Peasantry* of Europe, Cambridge, MA 1994.

Rosner, Erhard: *Medizingeschichte Japans*, Leiden 1989.

Ross, Dorothy: The *Origins* of American Social Science, Cambridge 1991.

Ross, Robert: *Status* and Respectability in the Cape Colony, 1750–1870. A Tragedy of Manners, Cambridge 1999.

———: *Clothing*. A Global History, Cambridge 2008.

Ross, Robert, et al. (eds.): The *Cambridge History of South Africa*, vol. 2: 1885–1994, Cambridge 2010.

Ross, Robert, and Gerard J. Telkamp (eds.): *Colonial Cities*, Dordrecht 1985.

Rossabi, Morris: *China and Inner Asia*. From 1368 to the Present Day, London 1975.

Rosselli, John: The *Opera Business* and the Italian Immigrant Community in Latin America, 1820–1939: The Example of Buenos Aires, in: P&P 127 (May 1990), pp. 155–82.

———: *Singers* of Italian Opera. The History of a Profession, Cambridge 1992.

Rostow, Walt W.: The *World Economy*. History and Prospect, Austin, TX 1978.

Rotberg, Robert I.: *The Founder*. Cecil Rhodes and the Pursuit of Power, New York 1988.

Roth, Ralf: Das *Jahrhundert der Eisenbahn*. Die Herrschaft über Raum und Zeit 1800–1914, Ostfildern 2005.

Rothblatt, Sheldon: The *Revolution of the Dons*. Cambridge and Society in Victorian England, London 1968.

Rothblatt, Sheldon, and Björn Wittrock (eds.): The *European and American University* Since 1800. Historical and Sociological Essays, Cambridge 1993.

Rothermund, Dietmar: An *Economic History of India*: From Pre-colonial Times to 1991, 2nd ed., London 1993.

——— (ed.): *Aneignung* und Selbstbehauptung: Antworten auf die europäische Expansion, Munich 1999.

Rothfels, Nigel: *Savages and Beasts*. The Birth of the Modern Zoo, Baltimore, MD 2002.

Rouleau, Bernard: *Paris*. Histoire d'un espace, Paris 1997.

Roussillon, Alain: *Identité et modernité*. Les voyageurs égyptiens au Japon (XIXe–XXe siècle), Arles 2005.

Rowe, William T.: *Hankow*, 2 vols., Stanford, CA 1984–89.

———: *Saving the World*. Chen Hongmou and Elite Consciousness in Eighteenth-Century China, Stanford, CA 2001.

———: *China's Last Empire*. The Great Qing, Cambridge, MA 2009.

Rowley, Charles D.: The *Destruction* of Aboriginal Society, Harmondsworth 1974.

Roy, Tirthankar: The *Economic History of India*, 1857–1947, New Delhi 2000.

———: *India in the World Economy*. From Antiquity to the Present, Cambridge 2012.

Royle, Edward: *Revolutionary Britannia?* Reflections on the Threat of Revolution in Britain, 1789–1848, Manchester 2000.

Rubinger, Richard: *Popular Literacy* in Early Modern Japan, Honolulu 2007.

Rubinstein, William D. (ed.): *Wealth* and the Wealthy in the Modern World, London 1980.

Ruble, Blair A.: *Second Metropolis*. Pragmatic Pluralism in Gilded Age Chicago, Silver Age Moscow and Meiji Osaka, Cambridge 2001.

Ruedy, John: *Modern Algeria*. The Origins and Development of a Nation, 2nd ed. Bloomington, IN 2005.

Rüegg, Walter (ed.): A *History of the University* in Europe, 4 vols., Cambridge 1992–2011.

Ruggiero, Guido de: The History of *Liberalism*, New York 1927.

Rupp, Leila J.: *Worlds of Women*. The Making of an International Women's Movement, Princeton, NJ 1997.

Rusconi, Gian Enrico: *Cavour e Bismarck*. Due leader fra liberalismo e cesarismo, Bologna 2011.

Russell, William Howard: *My Diary in India* [1860], 2 vols., Cambridge 2010.

Rustemeyer, Angela: *Dienstboten* in Petersburg und Moskau 1861–1917. Hintergrund, Alltag, soziale Rolle, Stuttgart 1996.

Rutherford, Susan: The *Prima Donna* and Opera, 1815–1930, Cambridge 2006.

Ruthven, Malise, and Azim Nanji: *Historical Atlas of Islam*, Cambridge, MA 2004.

Ryan, James R.: *Picturing Empire*. Photography and the Visualization of the British Empire, London 1997.

Ryan, Mary P.: *Civic Wars*. Democracy and Public Life in the American City during the Nineteenth Century, Berkeley, CA 1997.

Sabel, Charles F., and Jonathan Zeitlin (eds.): *World of Possibilities*. Flexibility and Mass Production in Western Industrialization, Cambridge 1997.

Sachs, Jeffrey D. (ed.): *Developing Country Debt* and the World Economy, Chicago 1989.

———: *Tropical Underdevelopment*, Cambridge, MA 2001.

Sahlins, Marshall D.: *Anahulu*. The Anthropology of History in the Kingdom of Hawaii. vol. 1: Historical Ethnography, Chicago 1992.

Sahlins, Peter: *Forest Rites*. The War of the Demoiselles in Nineteenth-Century France, Cambridge, MA 1994.

Said, Edward W.: *Orientalism*, London 1978.

Samson, Jim (ed.): The *Cambridge History of Nineteenth-Century Music*, Cambridge 2002.

Sanneh, Lamin: The *Crown* and the Turban. Muslims and West African Pluralism. Boulder, CO 1997.

Sanneh, Lamin: *Abolitionists Abroad*. American Blacks and the Making of Modern West Africa. Cambridge, MA 1999.

Sarasin, Philipp: *Stadt der Bürger*. Bürgerliche Macht und städtische Gesellschaft. Basel 1846–1914, 2nd ed., Göttingen 1997.

Sarmiento, Domingo Faustino: *Facundo: Civilization and Barbarism* [1845], Berkeley, CA 2004.

Sartorius von Waltershausen, August: Die *Entstehung* der Weltwirtschaft. Geschichte des zwischenstaatlichen Wirtschaftslebens vom letzten Viertel des achtzehnten Jahrhunderts bis 1914, Jena 1931.

Sassoon, Donald: The *Culture* of the Europeans. From 1800 to the Present, London 2006.

Satre, Lowell J.: *Chocolate on Trial*. Slavery, Politics, and the Ethics of Business, Athens, OH 2005.

Saunders, David: *Russia* in the Age of Reaction and Reform, 1801–1881, London 1992.

Sayer, Derek: The Coasts of *Bohemia*. A Czech History, Princeton, NJ 1998.

Scaff, Lawrence A.: *Max Weber in America*, Princeton, NJ 2011.

Scarr, Deryck: The History of the *Pacific Islands*. Kingdoms of the Reefs, Basingstoke 1990.

Schalenberg, Marc: *Humboldt auf Reisen?* Die Rezeption des "deutschen Universitätsmodells" in den französischen und britischen Reformdiskursen (1810–1870). Basel 2002.

Schama, Simon: *Patriots* and Liberators. Revolution in the Netherlands, London 1977.

Scheffler, Thomas: *"Fertile Crescent,"* "Orient," "Middle East." The Changing Mental Maps of Southwest Asia, in: European Review of History 10 (2003), pp. 253–72.

Schendel, Willem van, and Henk Schulte Nordholt (eds.): *Time Matters.* Global and Local Time in Asian Societies, Amsterdam 2001.

Scherer, F. M.: *Quarter Notes* and Bank Notes. The Economics of Music Composition in the Eighteenth and Nineteenth Centuries, Princeton, NJ 2004.

Scherzer, Kenneth A.: The *Unbound Community.* Neighborhood Life and Social Structure in New York City, 1830–1875, Durham, NC 1992.

Schieder, Theodor: *Staatensystem* als Vormacht der Welt 1848–1918, Frankfurt a.M. 1977.

———: *Nationalismus* und Nationalstaat. Studien zum nationalen Problem im modernen Europa, Göttingen 1991.

Schilling, Heinz: *Die neue Zeit.* Vom Christenheitseuropa zum Europa der Staaten 1250 bis 1750, Berlin 1999.

Schinz, Alfred: *Cities in China*, Berlin 1989.

Schivelbusch, Wolfgang: The *Railway Journey.* The Industrialization of Time and Space in the Nineteenth Century, Berkeley, CA 1986.

———: *Disenchanted Night.* The Industrialization of Light in the Nineteenth Century, London 1988.

Schleier, Hans: Geschichte der deutschen *Kulturgeschichtsschreibung*, 2 vols., Waltrop 2003.

Schlögel, Karl: *Im Raume* lesen wir die Zeit. Über Zivilisationsgeschichte und Geopolitik, Munich 2003.

Schlör, Joachim: *Night in the Big City*: Paris, Berlin, London 1840–1930, London 1998.

Schmid, André: *Korea* between Empires, 1895–1919, New York 2002.

Schmidt, Manfred G.: *Sozialpolitik* in Deutschland. Historische Entwicklung und internationaler Vergleich, Opladen 1998.

Schmidt, Nelly: *L'Abolition de l'esclavage.* Cinq siècles de combats XVIe–XXe siecle, Paris 2005.

Schmidt, Peer: *Der Guerrillero.* Die Entstehung des Partisanen in der Sattelzeit der Moderne. Eine atlantische Perspektive 1776–1848, in: GG 29 (2003), pp. 161–90.

Schmidt-Glintzer, Helwig: *Geschichte der chinesischen Literatur*, Bern 1990.

———: *Eurasien* als kulturwissenschaftliches Forschungsthema, in: Wolfgang Gantke et al. (eds.): Religionsbegegnung und Kulturaustausch in Asien, Wiesbaden 2002, pp. 185–99.

Schmidt-Nowara, Christopher: *Empire and Antislavery*. Spain, Cuba, and Puerto Rico, 1833–1874, Pittsburgh 1999.

Schmied, Gerhard: *Soziale Zeit.* Umfang, "Geschwindigkeit" und Evolution, Berlin 1985.

Schneer, Jonathan: *London 1900.* The Imperial Metropolis, New Haven, CT 1999.

Schneider, Ronald M.: *Latin American Political History. Patterns and Personalities,* Boulder, CO 2007.

Schölch, Alexander: *Egypt for the Egyptians!* The Socio-political Crisis in Egypt, 1878–1882, London 1981.

Scholte, Jan Aart: *Globalization.* A Critical Introduction, Basingstoke 2000.

Schön, Lennart: An Economic History of *Modern Sweden*, London 2012.

Schoppa, R. Keith: *Xiang Lake.* Nine Centuries of Chinese Life, New Haven, CT 1989.

Schorkowitz, Dittmar: *Staat und Nationalitäten* in Russland. Der Integrationsprozess der Burjaten und Kalmücken, 1822–1925, Stuttgart 2001.

Schorske, Carl E.: *Fin-de-Siècle Vienna.* Politics and Culture, New York 1981.

Schrader, Abby M.: *Languages of the Lash.* Corporal Punishment and Identity in Imperial Russia, DeKalb, IL 2002.

Schreiber, Ulrich: Die *Kunst der Oper*, 3 vols., Frankfurt a.M. 1988–2000.

Schroeder, Paul W.: The Nineteenth-Century *International System*. Changes in the Structure, in: WP 39 (1986), pp. 1–26.

———: The *Transformation* of European Politics 1763–1848, Oxford 1994.

———: *Systems, Stability and Statecraft*: Essays on the International History of Modern Europe, New York 2004.

Schroeter, Daniel J.: *Merchants of Essaouira*. Urban Society and Imperialism in Southwestern Morocco, 1844–1886, Cambridge 1988.

Schudson, Michael: The *Good Citizen*. A History of American Civic Life, New York 1998.

Schularick, Moritz: *Finanzielle Globalisierung* in historischer Perspektive, Tübingen 2006.

Schulin, Ernst: Die *Französische Revolution*, 4th ed., Munich 2004.

Schulte Nordholt, Henk: The *Spell of Power*. A History of Balinese Politics 1650–1940, Leiden 1996.

Schultz, Hans-Dietrich: *Deutschlands "natürliche Grenzen."* "Mittellage" und "Mitteleuropa," in der Diskussion der Geographen seit Beginn des 19. Jahrhunderts, in: GG 15 (1989), pp. 248–91.

———: *Raumkonstrukte* der klassischen deutschsprachigen Geographie des 19./20. Jahrhunderts im Kontext ihrer Zeit, in: GG 28 (2002), pp. 343–77.

Schultz, Kirsten: *Tropical Versailles*. Empire, Monarchy, and the Portuguese Royal Court in Rio de Janeiro, 1808–1821, New York 2001.

Schulze, Hagen: *States, Nations, and Nationalism*. From the Middle Ages to the Present, Oxford 1996.

Schulze, Reinhard: The *Birth of Tradition* and Modernity in 18th and 19th Century Islamic Culture. The Case of Printing, in: Culture and History 16 (1997), pp. 29–72.

Schumpeter, Joseph A.: *Business Cycles*. A Theoretical, Historical and Statistical Analysis of the Capitalist Process, 2 vols., New York 1939.

———: History of *Economic Analysis*, London 1954.

———: The *Economics and Sociology of Capitalism*, ed. Richard Swedberg, Princeton, NJ 1991.

Schwab, Raymond: The *Oriental Renaissance*. Europe's Rediscovery of India and the East, 1680–1880 [1950], New York 1988.

Schwarcz, Vera: The *Chinese Enlightenment*. Intellectuals and the Legacy of the May Fourth Movement of 1919, Berkeley, CA 1986.

Schwentker, Wolfgang: *Max Weber in Japan*. Eine Untersuchung zur Wirkungsgeschichte 1905–1995, Tübingen 1998.

———: Die *Samurai*, Munich 2003.

Scott, Hamish M.: The *Birth* of a Great Power System, 1740–1815, Harlow 2006.

Scott, James C.: *Seeing Like a State*. How Certain Schemes to Improve the Human Condition Have Failed, New Haven, CT 1998.

Scott, Rebecca J.: *Degrees of Freedom*. Louisiana and Cuba after Slavery, Cambridge, MA 2005.

Scott, Tom (ed.): The *Peasantries* of Europe from the Fourteenth to the Eighteenth Centuries, London 1998.

Sdvižkov, Denis: Das *Zeitalter der Intelligenz*. Zur vergleichenden Geschichte der Gebildeten in Europa bis zum Ersten Weltkrieg, Göttingen 2006.

Searing, James F.: *West African Slavery* and Atlantic Commerce. The Senegal River Valley, 1700–1860, Cambridge 1993.

Searle, Geoffrey R: *Morality* and the Market in Victorian Britain, Oxford 1998.

———: *A New England?* Peace and War 1886–1918, Oxford 2004.

Seavoy, Ronald E.: *Famine* in Peasant Societies, New York 1986.

Seely, Robert: The *Russian-Chechen Conflict*, 1800–2000. A Deadly Embrace, London 2001.

Seidensticker, Edward: *Low City*, High City. Tokyo from Edo to the Earthquake, London 1983.

Seidl, Wolf: *Bayern in Griechenland*. Die Geburt des griechischen Nationalstaats und die Regierung König Ottos, Munich 1981.

Semmel, Bernard: *Jamaican Blood* and Victorian Conscience. The Governor Eyre Controversy, Westport, CT 1962.

———: The *Liberal Ideal* and the Demons of Empire. Theories of Imperialism from Adam Smith to Lenin, Baltimore, MD 1993.

Sen, Amartya: The *Argumentative Indian*. Writings on Indian Culture, History and Identity, London 2005.

Sennett, Richard: *Flesh and Stone*. The Body and the City in Western Civilization, New York 1994.

Seton-Watson, Hugh: The *Russian Empire*, 1801–1917, Oxford 1967.

———: *Nations and States*. An Inquiry into the Origins of Nations and the Politics of Nationalism, London 1977.

Shang Keqiang, and Liu Haiyan: *Tianjin* zujie shehui yanjiu [Studies on the society of the concessionary areas of Tianjin], Tianjin 1996.

Shannon, Richard: *Gladstone*, 2 vols., London 1982–99.

Shao Qin: *Culturing Modernity*. The Nantong Model, 1890–1930, Stanford, CA 2003.

Sharma, Arvind (ed.): *Modern Hindu Thought*. The Essential Texts, Oxford 2002.

Shaw, Stanford J.: *Between Old and New*. The Ottoman Empire under Sultan Selim III., 1789–1807, Cambridge, MA 1971.

———: The *Jews of the Ottoman Empire* and the Turkish Republic, New York 1991.

Sheehan, James J.: *Museums* in the German Art World. From the End of the Old Regime to the Rise of Modernism, Oxford 2000.

Shih, Vincent Y. C.: The *Taiping Ideology*. Its Sources, Interpretations, and Influences, Seattle, WA 1967.

Shimazu Naoko: *Japan, Race and Equality*. The Racial Equality Proposal of 1919, London 1998.

Shipps, Jan: *Mormonism*. The Story of a New Religious Tradition, Urbana, IL 1985.

Showalter, Dennis: The *Wars of German Unification*, London 2004.

Siddiqi, Asiya: *Ayesha's World*. A Butcher's Family in Nineteenth-Century Bombay, in: CSSH 43 (2001), pp. 101–29.

Sieder, Reinhard, and Ernst Langthaler (eds.): *Globalgeschichte 1800–2010*, Vienna 2010.

Sieferle, Rolf Peter, et al.: Das *Ende der Fläche*. Zum gesellschaftlichen Stoffwechsel der Industrialisierung, Cologne 2006.

Siegrist, Hannes, et al. (eds.): *Europäische Konsumgeschichte*. Zur Gesellschafts- und Kulturgeschichte des Konsums (18. bis 20. Jahrhundert), Frankfurt a.M. 1997.

Silberman, Bernard S.: *Cages of Reason*. The Rise of the Rational State in France, Japan, the United States, and Great Britain, Chicago 1993.

Silva, K. M. de: A History of *Sri Lanka*, London 1981.

Simey, T. S., and M. B. Simey: *Charles Booth*. Social Scientist, Oxford 1960.

Simmons, I. G.: An *Environmental History* of Great Britain. From 10,000 Years Ago to the Present, Edinburgh 2001.

Simonton, Deborah: A History of *European Women's Work*, 1700 to the Present, London 1998.

Singaravélou, Pierre: *L'École Française d'Extrême-Orient* ou l'institution des marges (1898–1956), Paris 1999.

Singer, James D., and Melvin Small: *Resort to Arms*. International and Civil Wars, 1816–1980, Beverly Hills 1982.

Singh, Chetan: *Natural Premises*. Ecology and Peasant Life in the Western Himalaya, 1800–1950, Delhi 1998.

Sinor, Denis: *Introduction*. The Concept of Inner Asia, in: idem (ed.): The Cambridge History of Early Inner Asia, Cambridge 1990, pp. 1–18.

Skinner, G. William: *Chinese Society* in Thailand. An Analytical History, Ithaca, NY 1957.

——— (ed.): The *City* in Late Imperial China, Stanford, CA 1977.

Skocpol, Theda: *States* and Social Revolutions, Cambridge 1979.

Slatta, Richard W.: *Gauchos* and the Vanishing Frontier, Lincoln, NE 1983.

———: *Cowboys* of the Americas, New Haven, CT 1990.

Slezkine, Yuri: *Arctic Mirrors*. Russia and the Small Peoples of the North, Ithaca, NY 1994.

Slotkin, Richard: *Regeneration* through Violence. The Mythology of the American Frontier, 1600–1860, Middletown, CT 1973.

———: The *Fatal Environment*. The Myth of the Frontier in the Age of Industrialization, 1800–1890, New York 1985.

Smallman-Raynor, Matthew, and Andrew D. Cliff: *War Epidemics*. A Historical Geography of Infectious Diseases in Military Conflict and Civil Strife, 1850–2000, Oxford 2004.

Smelser, Neil J., and Paul B. Baltes (eds.): *International Encyclopedia* of the Social and Behavioral Sciences, 26 vols., Amsterdam 2001.

Smil, Vaclav: *Energy* in World History, Boulder, CO 1994.

———: *Creating the Twentieth Century*. Technical Innovations of 1867–1914 and Their Lasting Impact, Oxford 2005.

———: The *Two Prime Movers* of Globalization. History and Impact of Diesel Engines and Gas Turbines, in: JGH 2 (2007), pp. 373–94.

Smith, Andrew B.: *Pastoralism in Africa*. Origins and Development Ecology, London 1992.

Smith, Anthony D.: *Nationalism* and Modernism, London 1998.

Smith, Crosbie: The *Science of Energy*. A Cultural History of Energy Physics in Victorian Britain, London 1998.

Smith, Crosbie, and M. Norton Wise: *Energy and Empire*. A Biographical Study of Lord Kelvin, Cambridge 1989.

Smith, David A., et al. (eds.): *States* and Sovereignty in the Global Economy, London 1999.

Smith, Jeremy: *Europe and the Americas*. State Formation, Capitalism and Civilizations in Atlantic Modernity, Leiden 2006.

Smith, Joseph: The *Spanish-American War*. Conflict in the Caribbean and the Pacific, 1895–1902, Harlow 1994.

Smith, Mark M.: *Mastered by the Clock*. Time, Slavery and Freedom in the American South, Chapel Hill, NC 1997.

———: *Debating Slavery*. Economy and Society in the Antebellum American South, Cambridge 1998.

Smith, Michael Stephen: The *Emergence* of Modern Business Enterprise in France. 1800–1930, Cambridge, MA 2006.

Smith, Richard J.: *China's Cultural Heritage*. The Qing Dynasty, 1644–1912, 2nd ed., Boulder, CO 1994.

Smith, Thomas C.: The *Agrarian Origins* of Modern Japan, Stanford, CA 1959.

———: *Peasant Time* and Factory Time in Japan, in: P&P 111 (1986), pp. 165–97.

Smith, Woodruff D.: Politics and the *Sciences of Culture* in Germany, 1840–1920, New York 1991.

Snouck Hurgronje, Christiaan: *Mekka* in the Latter Part of the 19th Century. Daily Life, Customs and Learning, Leiden 1931.

Snowden, Frank M.: *Naples* in the Time of Cholera, 1884–1911, Cambridge 1995.

Sohrabi, Nader: *Historicizing Revolutions*. Constitutional Revolutions in the Ottoman Empire; Iran, and Russia, 1905–1908, in: AJS 100 (1995), pp. 1383–1447.

———: *Global Waves*, Local Actors. What the Young Turks Knew about Other Revolutions and Why It Mattered, in: CSSH 44 (2002), pp. 45–79.

———: *Revolution and Constitutionalism* in the Ottoman Empire and Iran, Cambridge 2011.

Solé, Robert: *Le grand voyage de l'obélisque*, Paris 2004.

Sombart, Werner: *Why Is There No Socialism in the United States?* [1906], London 1976.

Somel, Selçuk Akşin: The *Modernization of Public Education* in the Ottoman Empire 1839–1908. Islamization, Autocracy and Discipline, Leiden 2001.

Sondhaus, Lawrence: *Naval Warfare*, 1815–1914, London 2001.

Sorin, Gerald: *A Time for Building*. The Third Migration, 1880–1920, Baltimore, MD 1992.

Soto, Hernando de: The *Mystery of Capital*. Why Capitalism Triumphs in the West and Fails Everywhere Else, New York 2000.

Souçek, Svat: *A History of Inner Asia*, Cambridge 2000.

Spang, Rebecca L.: The Invention of the *Restaurant*. Paris and Modern Gastronomic Culture, Cambridge, MA 2000.

———. *Paradigms* and Paranoia. How Modern Is the French Revolution? in: AHR 108 (2003), pp. 119–47.

Spate, O.H.K.: The *Pacific* since Magellan, 3 vols., London 1979–88.

Speirs, Ronald, and John Breuilly (eds.): *Germany's Two Unifications*. Anticipations, Experiences, Responses, Basingstoke 2005.

Spellman, William M.: *Monarchies*, 1000–2000, London 2001.

Spence, Jonathan: The Search for *Modern China*, New York 1990.

———: *God's Chinese Son*. The Taiping Heavenly Kingdom of Hong Xiuquan, New York 1996.

Spence, Mark David: *Dispossessing the Wilderness*. Indian Removal and the Making of the National Parks, New York 1999.

Sperber, Jonathan: *Bürger*, Bürgertum, Bürgerlichkeit, Bürgerliche Gesellschaft: Studies of the German (Upper) Middle Class and Its Sociocultural World, in: JMH 69 (1997), pp. 271–97.

———: The *European Revolutions*, 1848–1851, 2nd ed., Cambridge 2005.

———: *Revolutionary Europe*, 1780–1850, Harlow 2006.

———: *Europe 1850–1914*. Progress, Participation and Apprehension, Harlow 2009.

———: *Karl Marx*. A Nineteenth-Century Life, New York 2013.

Spiro, Melford E.: *Buddhism* and Society. A Great Tradition and Its Burmese Vicissitudes, 2nd ed., Berkeley, CA 1982.

Stafford, Robert A.: *Scientist of Empire*. Sir Roderick Murchison, Scientific Exploration and Victorian Imperialism, Cambridge 1989.

Staiger, Brunhild et al. (eds.): Das große *China-Lexikon*, Darmstadt 2003.

Stampp, Kenneth M.: *America in 1857*. A Nation on the Brink, New York 1990.

Stanley, Peter: *White Mutiny*. British Military Culture in India, 1825–1875, London 1998.

Staples, John R.: *Cross-Cultural Encounters* on the Ukrainian Steppe. Settling the Molochna Basin, 1783–1861, Toronto 2003.

Starrett, Gregory: *Putting Islam to Work*. Education, Politics, and Religious Transformation in Egypt, Berkeley, CA 1998.

Stearns, Peter N.: The *Industrial Revolution in World History*, Boulder, CO 1993.

Steckel, Richard H., and Roderick Floud (eds.): *Health* and Welfare during Industrialization, Chicago 1997.

Steckel, Richard H., and Jerome C. Rose (eds.): The *Backbone* of History. Health and Nutrition in the Western Hemisphere, Cambridge 2002.

Stedman Jones, Gareth: An *End to Poverty*. A Historical Debate, London 2005.

Stedman Jones, Gareth, and Gregory Claeys (eds.): The *Cambridge History of Nineteenth-Century Political Thought*, Cambridge 2011.

Steinberg, David Joel et al. (eds.): In Search of *Southeast Asia*. A Modern History, Honolulu 1987.

Steinberg, Jonathan: *Bismarck*. A Life, Oxford 2011.

Steinberg, John W., et al. (eds.): The *Russo-Japanese War* in Global Perspective. World War Zero, Leiden 2005.

Steinfeld, Robert J.: The *Invention of Free Labor*. The Employment Relation in English and American Law and Culture, 1350–1870, Chapel Hill, NC 1991.

———: *Coercion*, Contract and Free Labor in the 19th Century, Cambridge 2001.

Steinhardt, Nancy Shatzman: *Chinese Imperial City Planning*, Honolulu 1990.

Stephan, John J.: The *Russian Far East*. A History, Stanford, CA 1994.

Stephanson, Anders: *Manifest Destiny*. American Expansionism and the Empire of Right, 2nd ed., New York 1998.

Stern, Fritz: *Gold and Iron*. Bismarck, Bleichröder, and the Building of the German Empire, New York 1977.

Stevenson, David: 1914–1918. The History of the *First World War*, New York 1995.

Stewart, James Brewer: *Holy Warriors*. The Abolitionists and American Slavery, rev. ed., New York 1997.

Stiegler, Bernd: *Philologie des Auges*. Die photographische Entdeckung der Welt im 19. Jahrhundert, Munich 2001.

Stierle, Karlheinz: Der Mythos von *Paris*. Zeichen und Bewußtsein der Stadt, Munich 1993.

Stietencron, Heinrich von: Der *Hinduismus*, Munich 2001.

Stinchcombe, Arthur L.: *Economic Sociology*, New York 1983.

———: *Stratification* and Organization. Selected Papers, Cambridge, MA 1986.

———: *Sugar Island Slavery* in the Age of Enlightenment. The Political Economy of the Caribbean World, Princeton, NJ 1995.

Stites, Richard: *Serfdom*, Society, and the Arts in Imperial Russia. The Pleasure and the Power, New Haven, CT 2005.

Stöber, Rudolf: *Deutsche Pressegeschichte*. Einführung, Systematik, Glossar, Konstanz 2000.

Stocking, George W.: *Victorian Anthropology*, New York 1987.

———: *After Tylor*. British Social Anthropology, 1888–1951, London 1996.

Stokes, Eric: The *English Utilitarians* and India, Oxford 1959.

Stoler, Ann Laura: *Capitalism* and Confrontation in Sumatra's Plantation Belt, 1870–1979, New Haven, CT 1985.

———: *Sexual Affronts* and Racial Frontiers. European Identities and the Cultural Politics of Exclusion in Colonial Southeast Asia, in: CSSH 34 (1992), pp. 514–51.

Stolleis, Michael, and Yanagihara Masaharu (eds.): *East Asian and European Perspectives* on International Law, Baden-Baden 2004.

Stone, Bailey: *Reinterpreting the French Revolution*. A Global-Historical Perspective, Cambridge 2002.

Stone, Irving: The *Global Export* of Capital from Great Britain, 1865–1914. A Statistical Survey, New York 1999.

Strachan, Hew: The *First World War,* vol. 1: To Arms, Oxford 2001.

Strachey, Lytton: *Eminent* Victorians [1918], definitive ed., London 2002.

Streets, Heather: *Martial Races.* The Military, Race and Masculinity in British Imperial Culture, 1857–1914, Manchester 2004.

Stuchtey, Benedikt (ed.): *Science* across the European Empires 1800–1950, Oxford 2005.

———: *Die europäische Expansion* und ihre Feinde. Kolonialismuskritik vom 18. bis in das 20. Jahrhundert, Munich 2008.

Sugihara Kaoru: *Japan as an Engine* of the Asian International Economy, c. 1880–1936, in: Japan Forum 2 (1989), pp. 127–45.

———: *Japanese Imperialism* in Global Resource History (University of Osaka, Department of Economics, Working Paper 07/04), Osaka 2004.

——— (ed.): *Japan,* China, and the Growth of the Asian International Economy, 1850–1949, Oxford 2005.

Sugiyama Shinya, and Linda Grove (eds.): *Commercial Networks* in Modern Asia, Richmond 2001.

Sullivan, Michael: The *Meeting of Eastern and Western Art*, Berkeley, CA 1989.

Sunderland, Willard: *Taming the Wild Field.* Colonization and Empire on the Russian Steppe, Ithaca, NY 2004.

Sundhaussen, Holm: *Geschichte Serbiens.* 19.–21. Jahrhundert, Vienna 2007.

Suny, Ronald Grigor: *Looking toward Ararat.* Armenia in Modern History, Bloomington, IN 1993.

Surun, Isabelle (ed.): *Les sociétés coloniales* à l'âge des Empires, 1850–1960, Paris 2012.

Sutcliffe, Anthony: Towards the *Planned City.* Germany, Britain, the United States and France 1780–1914, Oxford 1981.

——— (ed.): *Metropolis* 1890–1940, London 1984.

———: *Paris.* An Architectural History, New Haven, CT 1993.

Sutherland, Donald M. G.: The *French Revolution and Empire.* The Quest for a Civic Order, Malden, MA 2003.

Suzuki Shogo: *Civilization and Empire.* China and Japan's Encounter with European International Society, London 2009.

Suzuki Toshio: *Japanese Government Loan Issues* on the London Capital Market, 1870–1913, London 1994.

Swedberg, Richard: *Max Weber* and the Idea of Economic Sociology, Princeton, NJ 1998.

Sweetman, John: The *Oriental Obsession.* Islamic Inspiration in British and American Art and Architecture 1500–1920, Cambridge 1988.

Sylla, Richard, and Gianni Toniolo (eds.): *Patterns* of European Industrialization. The Nineteenth Century, London 1991.

Szreter, Simon, and Graham Mooney: *Urbanization*, Mortality, and the Standard of Living Debate. New Estimates of the Life at Birth in Nineteenth-Century British Cities, in: EcHR 51 (1998), pp. 84–112.

Szücs, Jenö: *Les trois Europe*, Paris 1985.

Szyliowicz, Joseph S.: *Education and Modernization* in the Middle East, Ithaca, NY 1973.

Taguieff, Pierre-André: *Le racisme*, Paris 1997.

Tahtawi, Rifa'a Rafi' al-: *An Imam in Paris*. Al-Tahtawi's Visit to France (1826–31), 2nd ed., London 2011.

Takaki, Ronald T.: *Strangers* from a Different Shore. A History of Asian Americans, New York 1989.

———: A Different *Mirror*. A History of Multicultural America, Boston 1993.

Takebayashi Shirō: Die Entstehung der *Kapitalismustheorie* in der Gründungsphase der deutschen Soziologie, Berlin 2003.

Takenaka Toru: *Wagner-Boom* in Meiji-Japan, in: Archiv für Musikwissenschaft 62 (2005), pp. 13–31.

Tamaki Norio: *Japanese Banking*. A History, 1859–1959, Cambridge 1995.

Tanaka, Stefan: *Japan's Orient*. Rendering Pasts into History, Berkeley, CA 1993.

———: *New Times in Modern Japan*, Princeton, NJ 2004.

Tanner, Albert: *Arbeitsame Patrioten*—wohlanständige Damen. Bürgertum und Bürgerlichkeit in der Schweiz 1830–1914, Zurich 1995.

Tarling, Nicholas (ed.): The *Cambridge History of Southeast Asia*, 2 vols., Cambridge 1992.

———: *Imperialism* in Southeast Asia. "A Fleeting, Passing Phase," London 2001.

———: *Southeast Asia*. A Modern History, Oxford 2001.

Taruskin, Richard: The Oxford History of *Western Music*, 6 vols., Oxford 2005.

Taylor, Jean Gelman: The *Social World of Batavia*. European and Eurasian in Dutch Asia, Madison, WI 1984.

Taylor, Miles: The *Decline* of British Radicalism, 1847–1860. Oxford 1995.

———: The *1848 Revolution* and the British Empire, in: P&P 166 (2000), pp. 146–80.

Taylor, P.J.O. (ed.): A *Companion* to the "Indian Mutiny" of 1857, Delhi 1996.

Tedlow, Richard S.: *New and Improved*. The Story of Mass Marketing in America, New York 1990.

Teich, Mikuláš, and Roy Porter (eds.): The *National Question* in Europe in Historical Context, Cambridge 1993.

——— (eds.): The *Industrial Revolution* in National Context, Cambridge 1996.

Temperley, Howard: *British Antislavery*, 1833–1870, London 1972.

———: *White Dreams*, Black Africa. The Antislavery Expedition to the River Niger, 1841–1842. New Haven, CT 1991.

——— (ed.): *After Slavery*. Emancipation and Its Discontents, London 2000.

Teng Ssu-yü: The *Nien Army* and Their Guerilla Warfare, 1851–1868, Paris 1961.

Teng Ssu-yü, and John K. Fairbank (eds.): *China's Response to the West*, Cambridge, MA 1954.

Tenorio Trillo, Mauricio: *Mexico* at the World's Fairs. Crafting a Modern Nation, Berkeley, CA 1996.

———: *Argucias de la historia*. Siglo XIX, cultura y "América Latina," Mexico City 1999.

Terwiel, Barend J.: A History of *Modern Thailand*, 1767–1942, St. Lucia 1983.

———: *Acceptance* and Rejection. The First Inoculation and Vaccination Campaigns in Thailand, in: Journal of the Siam Society 76 (1988), pp. 183–201.

Teuteberg, Hans Jürgen (ed.): *Durchbruch zum modernen Massenkonsum*, Münster 1987.

Teuteberg, Hans-Jürgen, and Cornelius Neutsch (eds.): *Vom Flügeltelegraphen* zum Internet: Geschichte der modernen Kommunikation, Stuttgart 1998.

Thant Myint-U.: The Making of *Modern Burma*, Cambridge 2001.

Thelin, John R.: A History of *American Higher Education*, Baltimore, MD 2004.

Therborn, Göran: *Globalizations*. Dimensions, Historical Waves, Regional Effects, Normative Governance, in: International Sociology 15 (2000), pp. 151–79.

Theye, Thomas: Der geraubte *Schatten*. Eine Weltreise im Spiegel der ethnographischen Photographie, Munich 1989.

Thom, Martin: *Republics*, Nations and Tribes, London 1995.

Thomas, Nicholas: *Islanders*: The Pacific in the Age of Empire, New Haven, CT 2010.

Thompson, Dorothy: *Queen Victoria*. Gender and Power, London 2001.

Thompson, E. P.: The *Making* of the English Working Class, London 1963.

———: *Time*, Work-Discipline and Industrial Capitalism, in: P&P 38 (1967), pp. 56–97.

Thompson, F.M.L. (ed.): The *Cambridge Social History of Britain*, 3 vols., Cambridge 1990.

Thongchai Winichakul: *Siam Mapped*. A History of the Geo-Body of a Nation, Honolulu 1994.

Thornton, Russell: *American Indian Holocaust* and Survival. A Population History since 1492, Norman, OK 1987.

Tilchin, William N.: *Theodore Roosevelt* and the British Empire. A Study in Presidental Statecraft, Basingstoke 1997.

Tilly, Charles (ed.): The *Formation of National States* in Western Europe, Princeton, NJ 1975.

———: *Coercion*, Capital, and European States, AD 990–1992, Oxford 1992.

———: *European Revolutions*, 1492–1992, Oxford 1993.

———: The Politics of *Collective Violence*, Cambridge 2003.

———: *Democracy*, Cambridge 2007.

Tilly, Charles, and Wim P. Blockmans (eds.): *Cities* and the Rise of States in Europe, A.D. 1000 to 1800, Boulder, CO 1994.

Tilly, Chris, and Charles Tilly: *Work* under Capitalism, Boulder, CO 1998.

Tilly, Louise A., and Joan W. Scott: *Women*, Work and Family, New York 1987.

Tinker, Hugh: A *New System of Slavery*. The Export of Indian Labour Overseas, 1830–1920, London 1974.

Tipton, Frank: The *Rise of Asia*. Economics, Society and Politics in Contemporary Asia, Basingstoke 1998.

Tobler, Hans-Werner: *La revolución mexicana*. Transformación social y cambio político, 1876–1940, México 1994.

Toby, Ronald P.: *State and Diplomacy* in Early Modern Japan. Asia in the Development of the Tokugawa Bakufu, Stanford, CA 1984.

Tocqueville, Alexis de: *Democracy in America* [1835–40], trans. by Gerald E. Bevan, New York 2003.

Todorov, Nikolaj: The *Balkan City*, 1400–1900, Seattle, WA 1983.

Todorova, Maria: *Imagining the Balkans*, Oxford 1997.

Toennes, Achim: Die "Frontier." Versuch einer Fundierung eines Analyse-Konzepts, in: JbLA 35 (1998), pp. 280–300.

Toeplitz, Jerzy: *Geschichte des Films*, 1895–1927, Munich 1975.

Toledano, Ehud R.: *State and Society* in Mid-Nineteenth-Century Egypt, Cambridge 1990.

Tombs, Robert: *France*, 1814–1914, London 1996.

———: The *Paris Commune*, 1871, London 1999.

Tomlinson, B. R.: The *Economy of Modern India*, 1860–1970, Cambridge 1993.

Tone, John L.: *War and Genocide* in Cuba, 1895–1898, Chapel Hill, NC 2006.

Toniolo, Gianni: An *Economic History* of Liberal Italy, 1850–1918, London 1990.

Topik, Steven C.: *Coffee Anyone?* Recent Research on Latin American Coffee Societies, in: HAHR 80 (2000), pp. 225–66.

———: *When Mexico Had the Blues*. A Transatlantic Tale of Bonds, Bankers, and Nationalists, 1862–1910, in: AHR 105 (2000), pp. 714–738.

Topik, Steven C., et al. (eds.): *From Silver to Cocaine*. Latin American Commodity Chains and the Building of the World Economy, 1500–2000, Durham, NC 2006.

Torp, Cornelius: Die *Herausforderung* der Globalisierung. Wirtschaft und Politik in Deutschland 1860–1914, Göttingen 2005.

Torpey, John: The *Invention of the Passport*. Surveillance, Citizenship and the State, Cambridge 2000.

Tortella, Gabriel: *Patterns of Economic Retardation* and Recovery in south-western Europe in the Nineteenth and Twentieth Century, in: EcHR 47 (1994), pp. 1–24.

———: The Development of *Modern Spain*. An Economic History of the Nineteenth and Twentieth Centuries, Cambridge, MA 2000.

Totman, Conrad: *Early Modern Japan*, Berkeley, CA 1993.

———: A *History of Japan*, Oxford 2000.

Touchet, Elisabeth de: *Quand les Français armaient le Japon*. La création de l'arsénal de Yokosuka, 1865–1882, Rennes 2003.

Townshend, Charles: *Making the Peace*. Public Order and Public Security in Modern Britain, Oxford 1993.

Tracy, James D. (ed.): *City Walls*. The Urban Enceinte in Global Perspective, Cambridge 2000.

Trautmann, Thomas R.: *Aryans* and British India, Berkeley, CA 1997.

Trebilcock, Clive: The *Industrialization* of the Continental Powers, 1780–1914, Harlow 1981.

Trentmann, Frank (ed.): *Paradoxes* of Civil Society. New Perspectives in Modern German and British History, New York 2000.

———: *Free Trade Nation*. Commerce, Consumption, and Civil Society in Modern Britain, Oxford 2008.

——— (ed.): The *Oxford Handbook of the History of Consumption,* Oxford 2012.

Trigger, Bruce, and Wilcomb E. Washburn (eds.): The *Cambridge History of the Native Peoples of the Americas*, vol. 1: North America, 2 pts., Cambridge 1996.

Trocki, Carl A.: *Opium and Empire*. Chinese Society in Colonial Singapore, 1800–1910, Ithaca, NY 1991.

———: *Opium*, Empire and the Global Political Economy. A Study of the Asian Opium Trade, 1750–1950, London 1999.

Troeltsch, Ernst: Der *Historismus* und seine Probleme. Das logische Problem der Geschichtsphilosophie, Tübingen 1922 (= Gesammelte Schriften, vol. 3).

Trotha, Trutz von: *Koloniale Herrschaft*. Zur soziologischen Theorie der Staatsentstehung am Beispiel des "Schutzgebietes Togo," Tübingen 1994.

———: Was war der *Kolonialismus*? Einige zusammenfassende Befunde zur Soziologie und Geschichte des Kolonialismus und der Kolonialherrschaft, in: Saeculum 55 (2004), pp. 49–95.

Tsang, Steve Yui-sang: A Modern History of *Hong Kong*, London 2004.

Tseng, Alice Y.: *Imperial Museums* of Meiji Japan. Architecture and the Art of the Nation, Seattle, WA 2008.

Tsunoda Ryusaku, et al. (eds.): *Sources of Japanese Tradition*, vol. 2, 2nd ed., New York 2005.

Tsurumi, E. Patricia: *Women* in the Thread Mills of Meiji Japan, Princeton, NJ 1990.

Tuan Yi-fu: *Space and Place*. The Perspective of Experience, Minneapolis 1977.

Tulard, Jean: *Napoleon*. The Myth of the Saviour, London 1984.

Turfan, Naim: *Rise of the Young Turks*. Politics, the Military and Ottoman Collapse, London 2000.

Turley, David: The Culture of *English Anti-slavery*, 1780–1860, London 1991.

Turnbull, C. Mary: A History of *Singapore*, 1819–1975, Kuala Lumpur 1977.

Turner, B. L., et al. (eds.): *The Earth* as Transformed by Human Action. Global and Regional Changes in the Biosphere over the Past 300 Years, Cambridge 1990.

Turner, Frederick Jackson: The *Frontier* in American History, new ed., Tucson, AZ 1986.

Turner, Michael E., et al.: *Farm Production* in England, 1700–1914, Oxford 2001.

Turrell, Robert Vicat: *Capital and Labour* on the Kimberley Diamond Fields, 1871–1890. Cambridge 1987.

Tutino, John: The *Revolution in Mexican Independence*. Insurgency and the Renegotiation of Property, Production, and Patriarchy in the Bajio, 1800–1855, in: HAHR 78 (1998.), pp. 367–418.

Tyrrell, Ian: *Peripheral Visions*. Californian-Australian Environmental Contacts, c. 1850s–1910, in: JWH 8 (1997), pp. 275–302.

——: *True Gardens* of the Gods. Californian-Australian Environmental Reform, 1860–1930, Berkeley, CA 1999.

——: *Transnational Nation*. United States History in Global Perspective Since 1789, Basingstoke 2007.

——: *Reforming the World*. The Creation of America's Moral Empire, Princeton, NJ 2010.

Ullmann, Hans-Peter: Der deutsche *Steuerstaat*. Geschichte der öffentlichen Finanzen vom 18. Jahrhundert bis heute, Munich 2005.

Umar Al-Naqar: The *Pilgrimage Tradition* in West Africa. An Historical Study with Special Reference to the Nineteenth Century, Khartum 1972.

Unruh, John D.: *The Plains Across*. The Overland Emigrants and the Trans-Mississippi West, 1840–60, Urbana, IL 1979.

Uribe, Victor M.: The *Enigma* of Latin American Independence. Analyses of the Last Ten Years, in: LARR 32 (1997), pp. 236–55.

Uribe-Uran, Victor M.: The *Birth of a Public Sphere* in Latin America during the Age of Revolution, in: CSSH 42 (2000), pp. 425–57.

Urlanis, Boris Z.: *Bilanz* der Kriege. Die Menschenverluste Europas vom 17. Jahrhundert bis zur Gegenwart, Berlin (GDR) 1965.

Utley, Robert M.: The *Indian Frontier* of the American West, 1846–1890, Albuquerque 1984.

——: The Lance and the Shield. The Life and Times of *Sitting Bull*, New York 1993.

Van Young, Eric: The *Other Rebellion*. Popular Violence, Ideology, and the Mexican Struggle for Independence, 1810–1821, Stanford, CA 2001.

Van Zanden, Jan Luiten: *Wages* and the Standard of Living in Europe, 1500–1800, in: EEcH 2 (1999), pp. 175–97.

Van Zanden, Jan Luiten, and Daan Marks: An *Economic History of Indonesia*, London 2012.

Van Zanten, David: *Building Paris*. Architectural Institutions and the Transformation of French Capital, 1830–1870, Cambridge 1994.

Vance, James E.: *Capturing the Horizon*. The Historical Geography of Transportation since the Transportation Revolution of the Sixteenth Century, New York 1986.

————: The *Continuing City*. Urban Morphology in Western Civilization, Baltimore, MD 1990.

Vandervort, Bruce: *Wars* of Imperial Conquest in Africa, 1830–1914, London 1998.

————: *Indian Wars* of Mexico, Canada and the United States, 1812–1900, New York 2006.

Vanthemsche, Guy: *La Belgique et le Congo*. Empreintes d'une colonie, 1885–1980, Brüssel 2007.

Varisco, Daniel Martin: *Reading Orientalism*. Said and the Unsaid, Seattle, WA 2007.

Vaughan, Megan: Creating the *Creole Island*. Slavery in Eighteenth Century Mauritius. Durham, NC 2005.

Vec, Miloš: *Recht und Normierung* in der industriellen Revolution. Neue Strukturen der Normsetzung in Völkerrecht, staatlicher Gesetzgebung und gesellschaftlicher Selbst-normierung, Frankfurt a.M. 2006.

Veenendaal, Augustus J., Jr.: *Railways* in the Netherlands. A Brief History, 1834–1994, Stanford, CA 2001.

Veer, Peter van der (ed.): *Conversion* to Modernities. The Globalization of Christianity, New York 1996.

————: *Imperial Encounters*. Religion and Modernity in India and Britain, Princeton, NJ 2001.

Ven, G.P. van de, et al.: *Leefbar laagland*. Geschiedenis van de waterbehersing en landaanwin-ning in Nederland, Utrecht 1993.

Venturi, Franco: *Roots of Revolution*. A History of the Populist and Socialist Movements in Nineteenth-Century Russia, Chicago 1960.

Verley, Patrick: *L'Échelle du monde*. Essai sur l'industrialisation de l'Occident, Paris 1997.

————: *La Révolution industrielle*, 2nd ed., Paris 1997.

Verrier, Anthony: Francis *Younghusband* and the Great Game, London 1991.

Veysey, Laurence R.: The *Emergence* of the American University, Chicago 1965.

Vierhaus, Rudolf, et al. (eds.): *Frühe Neuzeit*—Frühe Moderne? Forschungen zur Vielschich-tigkeit von Übergangsprozessen, Göttingen 1992.

Vigier, Philippe: *Paris* pendant la monarchie de juillet 1830–1848, Paris 1991.

Viglione, Massimo (ed.): *La Rivoluzione Italiana*. Storia critica del Risorgimento, Rom 2001.

Vincent, David: The Rise of *Mass Literacy*. Reading and Writing in Modern Europe, Cambridge 2000.

Viotti da Costa, Emília: The *Brazilian Empire*. Myths and Histories, Chicago 1985.

Virchow, Rudolf: *Sämtliche Werke*, ed. Christian Andree, Berlin 1992 ff.

Vital, David: *A People Apart*. The Jews in Europe 1789–1939, Oxford 1999.

Vittinghoff, Natascha: Die Anfänge des *Journalismus in China* (1860–1911), Wiesbaden 2002.

Vittinghoff, Natascha, and Michael Lackner (eds.): *Mapping Meanings*. The Field of New Learning in Late Qing China, Leiden 2004.

Vögele, Jörg: *Urban Mortality Change* in England and Germany, 1870–1913, Liverpool 1998.

————: *Sozialgeschichte* städtischer Gesundheitsverhältnisse während der Urbanisierung, Berlin 2001.

Voigt, Johannes: *Geschichte Australiens*, Stuttgart 1988.

Volkov, Shulamit: *Antisemitismus als kultureller Code*, 2nd ed., Munich 2000.

————: Die *Juden* in Deutschland 1780–1918, 2nd ed., Munich 2000.

————: *Germans, Jews, and Anti-Semites*: Trials in Emancipation, Cambridge 2006.

Voll, John Obert: *Islam*. Continuity and Change in the Modern World, 2nd ed., Syracuse, NY 1994.

Volwahsen, Andreas: *Imperial Delhi*. The British Capital of the Indian Empire, Munich 2002.

Vormbaum, Thomas: *Politik und Gesinderecht* im 19. Jahrhundert, Berlin 1980.

Voss, Stuart F.: *Latin America in the Middle Period*, 1750–1929, Wilmington, DE 2002.

Voth, Hans-Joachim: *Time and Work* in England 1750–1830, Oxford 2001.

Vries, Jan de: *European Urbanization*, 1500–1800, Cambridge, MA 1984.

——: The *Industrial Revolution* and the Industrious Revolution, in: JEH 54 (1994), pp. 249–70.

——: The *Industrious Revolution*. Consumer Behavior and the Household Economy, 1650 to the Present, Cambridge 2008.

Vries, Jan de, and Ad van der Woude: The *First Modern Economy*. Success, Failure, and Perseverance of the Dutch Economy, 1500–1815, Cambridge 1997.

Vries, Peer H. H.: *Via Peking* back to Manchester. Britain, the Industrial Revolution, and China, Leiden 2003.

Wade, Richard C.: The *Urban Frontier*. The Rise of Western Cities, 1790–1830, Urbana, IL 1996.

Waechter, Matthias: Die *Erfindung* des amerikanischen Westens. Die Geschichte der Frontier-Debatte, Freiburg i.Br. 1996.

Wagner, Kim A.: The *Marginal Mutiny*. The New Historiography of the Indian Uprising of 1857, in: History Compass 9/19 (2011), pp. 760–66.

Wahrman, Dror: *Imagining the Middle Class*. The Political Representation of Class in Britain, c. 1780–1840, Cambridge 1995.

Wakabayashi, Bob Tadashi: *Anti-foreignism* and Western Learning in Early-Modern Japan. The "New Theses" of 1825, Cambridge, MA 1991.

Waley-Cohen, Joanna: *Exile* in Mid-Qing China. Banishment to Xinjiang, 1758–1820, New Haven, CT 1991.

Walker, Brett L.: The *Early Modern Japanese State* and Ainu Vaccinations. Redefining the Body Politic 1799–1868, in: P&P 163 (1999), pp. 121–60.

Walker, Charles F.: *Smoldering Ashes*. Cuzco and the Creation of Republican Peru, 1780–1840, Durham, NC 1999.

Walker, David R.: *Anxious Nation*. Australia and the Rise of Asia, 1850–1939. St. Lucia 1999.

Wallace, Alfred Russel: The *Wonderful Century*. Its Successes and Its Failures, London 1898.

Waller, Philip (ed.): The *English Urban Landscape*, Oxford 2000.

Wallerstein, Immanuel: The *Modern World-System III*. The Second Era of Great Expansion of the Capitalist World-Economy, 1730–1840s, San Diego 1989.

Walsh, Margaret: The *American West*. Visions and Revisions, Cambridge 2005.

Walter, François: *Les figures paysagères de la nation*. Territoire et paysage en Europe (16e–20e siècle), Paris 2004.

Walter, Michael: "Die *Oper* ist ein Irrenhaus." Sozialgeschichte der Oper im 19. Jahrhundert, Stuttgart 1997.

Walton, John K.: The *English Seaside Resort*. A Social History, 1750–1914, Leicester 1983.

——: *Fish and Chips* and the British Working Class, 1870–1940, Leicester 1992.

Walvin, James: *Fruits of Empire*. Exotic Produce and British Taste, 1660–1800, Basingstoke 1997.

Wang Guanhua: *In Search of Justice*. The 1905–1906 Chinese Anti-American Boycott, Cambridge, MA 2001.

Wang Gungwu: *The Chinese Overseas*. From Earthbound China to the Quest for Autonomy, Cambridge, MA 1997.

—— (ed.): *Global History* and Migrations, Boulder, CO 1997.

Wang Sing-wu: The Organization of *Chinese Emigration*, 1848–1888, San Francisco 1978.

Ward, David: *Poverty*, Ethnicity, and the American City, 1840–1925. Changing Conceptions of the Slum and the Ghetto, Cambridge 1989.

Ward, David, and Olivier Zunz (eds.): The *Landscape of Modernity*. Essays on New York City, 1900–1940, New York 1992.

Ward, J. R.: *Poverty* and Progress in the Caribbean, 1800–1960, London 1985.

———: The *Industrial Revolution* and British Imperialism, 1750–1850, in: EcHR 47(1994), pp. 44–65.

Ward, Kevin: A History of *Global Anglicanism*, Cambridge 2006.

Wasserman, Mark: *Everyday Life* and Politics in Nineteenth-Century Mexico. Men, Women, and War, Albuquerque 2000.

Wasson, Ellis A.: *Aristocracy* and the Modern World. Basingstoke 2006.

Waswo, Ann: *Japanese Landlords*. The Decline of Rural Elite, Berkeley, CA 1977.

Watenpaugh, Keith David: *Being Modern* in the Middle East. Revolution, Nationalism, Colonialism, and the Arab Middle Class, Princeton, NJ 2005.

Waters, Malcom (ed.): *Modernity*. Critical Concepts, 4 vols., New York 1999.

Watkins, Harold: *Time Counts*. The Story of the Calendar, London 1954.

Watt, John R.: The *District Magistrate* in Late Imperial China, New York 1972.

Watts, David: The *West Indies*. Patterns of Development, Culture and Environmental Change since 1492, Cambridge 1987.

Watts, Sheldon: *Epidemics* and History. Disease, Power and Imperialism, New Haven, CT 1997.

Wawro, Geoffrey: The *Austro-Prussian War*. Austria's War with Prussia and Italy in 1866, Cambridge 1996.

———: *Warfare* and Society in Europe, 1792–1914, London 2000.

———: The *Franco-Prussian War*. The German Conquest of France in 1870–1871, Cambridge 2003.

Way, Peter: *Common Labor*. Workers and the Digging of North American Canals, 1780–1860, Baltimore, MD 1993.

Weaver, John C.: The *Great Land Rush* and the Making of the Modern World, 1650–1900, Montréal 2006.

Webb, James L. A.: *Desert Frontier*. Ecological and Economic Change along the Western Sahel, 1600–1850, Madison, WI 1995.

Webb, Walter Prescott: The *Great Frontier*, Austin, TX 1964.

Weber, Adna Ferrin: The *Growth of Cities* in the Nineteenth Century. A Study in Statistics, New York 1899.

Weber, David J.: The *Spanish Frontier* in North America, New Haven, CT 1992.

Weber, Eugen: *Peasants into Frenchmen*. The Modernization of Rural France, 1870–1914, London 1977.

Weber, Max: *General Economic History* [1923], Glencoe, IL 1950.

———: Gesammelte Aufsätze zur *Wissenschaftslehre* [1922], 3rd ed., Tübingen 1968.

———: *Economy and Society*. An Outline of Interpretive Sociology [1922], ed. Guenther Roth and Claus Wittich, 2 vols., Berkeley, CA 1978.

Webster, Anthony: *Gentlemen Capitalists*. British Imperialism in Southeast Asia 1770–1890, London 1998.

Wedewer, Hermann: Eine *Reise nach dem Orient*, Regensburg 1877.

Weeks, Theodore R.: *From Assimilation* to Antisemitism. The "Jewish Question" in Poland, 1850–1914. DeKalb, IL 2006.

Wehler, Hans-Ulrich: *Deutsche Gesellschaftsgeschichte*, 5 vols., Munich 1987–2008.

Weintraub, Stanley: *Victoria*. An Intimate Biography, New York 1987.

——: *Uncrowned King*. The Life of Prince Albert, New York 1997.

Weismann, Itzchak: The *Naqshbandiyya*. Orthodoxy and Activism in a Worldwide Sufi Tradition, London 2007.

Wells, Roger: *Wretched Faces*. Famine in Wartime England 1793–1801, Gloucester 1988.

Welskopp, Thomas: Das *Banner der Brüderlichkeit*. Die deutsche Sozialdemokratie vom Vormärz bis zum Sozialistengesetz, Bonn 2000.

Wendorff, Rudolf: *Zeit und Kultur*. Geschichte des Zeitbewußtseins in Europa, 2nd ed., Opladen 1980.

Wendt, Reinhard: Vom *Kolonialismus* zur Globalisierung. Europa und die Welt seit 1500, Paderborn 2007.

Wenzlhuemer, Roland: *Connecting the Nineteenth-Century World*. The Telegraph and Globalization, Cambridge 2013.

Werdt, Christophe von: *Halyč-Wolhynien*—Rotreußen—Galizien. Im Überlappungsgebiet der Kulturen und Völker, in: JGO 46 (1998), pp. 69–99.

Wernick, Andrew: *Auguste Comte* and the Religion of Humanity. The Post-theistic Program of French Social Theory, Cambridge 2001.

Wesseling, H. L.: *Les guerres coloniales* et la paix armée, 1871–1914. Esquisse pour une étude comparative, in: Histoires d'outre-mer. Mélanges en l'honneur de Jean-Louis Miège, Vol. I, Aix-en-Provence 1992.

——: *Divide and Rule*. The Partition of Africa, 1880–1914, Westport, CT 1996.

——: *Imperialism and Colonialism*. Essays on the History of European Expansion, Westport, CT 1997.

——: The *European Colonial Empires*, 1815–1919, New York 2004.

West, Elliott. The *Contested Plains*. Indians, Goldseekers and the Rush to Colorado. Lawrence, KS 1998.

West, James L., and Jurii A. Petrov (eds.): *Merchant Moscow*. Images of Russia's Vanished Bourgeoisie, Princeton, NJ 1998.

Westney, D. Eleanor: *Imitation* and Innovation. The Transfer of Organizational Patterns to Meiji Japan, Cambridge, MA 1987.

Westad, Odd Arne: *Restless Empire*. China and the World since 1750, London 2012.

Wheeler, Tom: *Mr. Lincoln's T-Mails*. The Untold Story of How Abraham Lincoln Used the Telegraph to Win the Civil War, New York 2006.

Whelan, Yvonne: *Reinventing Modern Dublin*. Streetscape, Iconography and the Politics of Identity, Dublin 2003.

White, Jerry: *London* in the Nineteenth Century. A Human Awful Wonder of God, London 2007.

White, Richard: "It's Your *Misfortune* and None of My Own." A History of the American West, Norman, OK 1991.

——: The *Middle Ground*. Indians, Empires, and Republics in the Great Lakes Region, 1650–1815, Cambridge 1991.

——: *Railroaded*. The Transcontinentals and the Making of Modern America, New York 2011.

Whited, Tamara L.: *Forests* and Peasant Politics in Modern France, New Haven, CT 2000.

Whitehead, Alfred North: *Science* and the Modern World [1925], New York 1967.
Whiteman, Jeremy J.: *Reform*, Revolution and French Global Policy, 1787–1791, Aldershot 2003.
Whitrow, G. J.: *Time in History*. The Evolution of Our General Awareness of Time and Temporal Perspective, Oxford 1988.
Wiebe, Robert H.: *The Search for Order*, 1877–1920, London 1967.
Wigen, Kären: The Making of a *Japanese Periphery*, 1750–1920, Berkeley, CA 1995.
Wilcox, Donald J.: *The Measure of Times Past*. Pre-Newtonian Chronologies and the Rhetoric of Relative Time, Chicago 1987.
Wilentz, Sean: *The Rise of American Democracy*. Jefferson to Lincoln, New York 2005.
Wilkinson, Endymion: *Chinese History*. A New Manual, Cambridge, MA 2013.
Wilkinson, Thomas O.: *The Urbanization of Japanese Labor*, 1868–1955, Amherst, MA 1965.
Will, Pierre-Étienne: *Bureaucratie* et famine en Chine au XVIIIᵉ siècle, Paris 1980.
Will, Pierre-Étienne, and R. Bin Wong: *Nourish the People*. The State Civilian Granary System in China 1650–1850, Ann Arbor, MI 1991.
Williams, Brian G.: *The Crimean Tatars*. The Diaspora Experience and the Forging of a Nation, Leiden 2001.
Williams, Michael: *Americans and Their Forests*. A Historical Geography, Cambridge 1989.
———: *Deforesting the Earth*. From Prehistory to Global Crisis, Chicago 2003.
Williamson, Jeffrey G., and Peter H. Lindert: *American Inequality*. A Macroeconomic History, New York 1980.
Wills, John E., Jr.: *1688: A Global History*, New York 2001.
———: *The World from 1450 to 1700*, Oxford 2009.
Wilson, A. N.: *The Victorians*, London 2002.
Wilson, David M.: *The British Museum*. A History, London 2002.
Wilson, Thomas M., and Hastings Donnan: *Introduction*, in: idem (eds.), Border Identities. Nation and State at International Frontiers, Cambridge 1998., pp. 1–30.
Wimmer, Andreas: *Die komplexe Gesellschaft*. Eine Theorienkritik am Beispiel des indianischen Bauerntums, Berlin 1995.
Winant, Howard: *The World Is a Ghetto*. Race and Democracy since World War II, New York 2001.
Winchester, Simon: *Krakatoa*. The Day the World Exploded, August 27, 1883, New York 2003.
———: A *Crack* in the Edge of the World. The Great American Earthquake of 1906, New York 2005.
Windler, Christian: *La Diplomatie* comme expérience de l'autre. Consuls français au Maghreb (1700–1840), Geneva 2002.
———: *Grenzen vor Ort*, in: Rechtsgeschichte 1 (2002), pp. 122–45.
Winkle, Stefan: *Geißeln* der Menschheit. Kulturgeschichte der Seuchen, Düsseldorf 1997.
Winkler, Heinrich August: *Germany*. The Long Road West, 2 vols., Oxford 2006–7.
———: *Geschichte des Westens*, vol. 1: Von den Anfängen in der Antike bis zum 20. Jahrhundert, Munich 2009.
Winseck, Dwayne R., and Robert M. Pike: *Communication and Empire*. Media, Markets, and Globalization, 1860–1930, Durham, NC 2007.
Winston, Brian: *Media Technology* and Society. A History from the Telegraph to the Internet, London 1998.
Wintle, Michael: *The Image of Europe*. Visualizing Europe in Cartography and Iconography Throughout the Ages, Cambridge 2009.

Wirtschafter, Elise Kimerling: *Structures of Society*. Imperial Russia's "People of Various Ranks," DeKalb, IL 1994.

Wischermann, Clemens, and Anne Nieberding: *Die institutionelle Revolution*. Eine Einführung in die deutsche Wirtschaftsgeschichte des 19. und frühen 20. Jahrhunderts, Stuttgart 2004.

Witte, Els, et al. 2006. *Nouvelle histoire de Belgique*, vol. 1: 1830–1905, Brussels 2006.

Witzler, Beate: *Großstadt und Hygiene*. Kommunale Gesundheitspolitik in der Epoche der Urbanisierung, Stuttgart 1995.

Wobring, Michael: Die *Globalisierung* der Telekommunikation im 19. Jahrhundert. Pläne, Projekte und Kapazitätsausbauten zwischen Wirtschaft und Politik, Frankfurt a.M. 2005.

Woerkens, Martine van: *Le voyageur étranglé*. L'Inde des Thugs, le colonialisme et l'imaginaire, Paris 1995.

Wolmar, Christian: The *Subterranean Railway*. How the London Underground Was Built and How It Changed the City Forever, London 2004.

———: *Fire and Steam*. A New History of the Railways in Britain, London 2007.

Woloch, Isser: The *New Regime*. Transformations of the French Civic Order, 1789–1820s, New York 1994.

——— (ed.): *Revolution* and the Meanings of Freedom in the Nineteenth Century, Stanford, CA 1996.

———: *Napoleon and His Collaborators*. The Making of a Dictatorship; New York 2001.

Wong, R. Bin: *China Transformed*. Historical Change and the Limits of European Experience, Ithaca, NY 1997.

———: *Entre monde et nation*. Les régions braudéliennes en Asie, in: Annales HSS 56 (2001), pp. 5–41.

Wood, Gordon S.: The *Radicalism* of the American Revolution. A History, New York 1992.

———: The *American Revolution*. A History, New York 2002.

———: The Americanization of *Benjamin Franklin*, New York 2004.

———: *Empire of Liberty*. A History of the Early Republic 1789–1815, Oxford 2009.

Woodruff, William: *Impact of Western Man*. A Study of Europe's Role in the World Economy, 1750–1960, London 1966.

Woodside, Alexander B.: *Vietnam* and the Chinese Model. A Comparative Study of Nguyen and Ch'ing Civil Administration in the First Half of the Nineteenth Century, Cambridge, MA 1971.

———: *Lost Modernities*. China, Vietnam, Korea, and the Hazards of World History, Cambridge, MA 2006.

Woolf, Daniel R.: *A Global History of History*, Cambridge 2011.

——— (ed.): The *Oxford History of Historical Writing*, 5 vols., Oxford 2011–12.

Woolf, Stuart J.: A History of *Italy*, 1700–1860. The Social Constraints of Political Space, London 1979.

———: *Napoleon's Integration of Europe*, London 1991.

———: The Construction of a *European World View* in the Revolutionary-Napoleonic Years, in: P&P 137 (1992), pp. 72–101.

Worden, Nigel, et al.: *Cape Town*. The Making of a City, Hilversum 1998.

Wortman, Richard S.: *Scenarios of Power*. Myth and Ceremony in Russian Monarchy. From Peter the Great to the Abdication of Nicholas II (1-vol. ed.), Princeton, NJ 2006.

Woud, Auke van der: *Het lege land*. De ruimtelijke orde van Nederland, 1798–1848, Amsterdam 1987.

Woude, A. M. van der, et al. (eds.): *Urbanization* in History. A Process of Dynamic Interactions, Oxford 1990.

Wray, William C.: *Mitsubishi* and the N.Y.K., 1870–1914. Business Strategy in the Japanese Shipping Industry, Cambridge, MA 1984.

Wright, Denis: *The Persians amongst the English*. Episodes in Anglo-Persian History, London 1985.

Wright, Gavin: The *Origins* of American Industrial Success, 1879–1940, in: AER 80 (1990), pp. 651–68.

Wright, Gwendolyn: The *Politics of Design* in French Colonial Urbanism, Chicago 1991.

Wright, James Leitch: *Creeks* and Seminoles. The Destruction and Regeneration of the Muscogulge People. Lincoln, NE 1986.

Wright, Mary C.: The *Last Stand* of Chinese Conservatism. The T'ung-Chih Restoration, 1862–1874, Stanford, CA 1957.

Wrigley, E. A.: *People*, Cities and Wealth, Oxford 1987.

———: *Poverty*, Progress, and Population, Cambridge 2004.

———: *Energy* and the English Industrial Revolution, Cambridge 2010.

Wrobel, David M.: The *End of American Exceptionalism*. Frontier Anxiety from the Old West to the New Deal, Lawrence, KS 1993.

Wunder, Bernd: Geschichte der *Bürokratie* in Deutschland, Frankfurt a.M. 1988.

Wyatt, David K.: *Thailand*. A Short History, New Haven, CT 1984.

Xiang Lanxin: The *Origins of the Boxer War*. A Multinational Study, London 2003.

Yalland, Zoë: *Boxwallahs*. The British in Cawnpore, 1857–1901, Norwich 1994.

Yamada Keiji (ed.): The *Transfer of Science* and Technology between Europe and Asia, 1780–1880, Kyoto 1994.

Yapp, Malcolm E.: The *Making* of the Modern Near East, 1792–1923, London 1987.

Yelling, J. A.: *Slums* and Slum Clearance in Victorian London, London 1986.

Yen Ching-hwang: *Coolies* and Mandarins. China's Protection of Overseas Chinese during the Late Ch'ing Period (1851–1911), Singapore 1985.

Yergin, Daniel: *The Prize*. The Epic Quest for Oil, Money, and Power, New York 1991.

Yoda Yoshiie: The *Foundations of Japan's Modernization*. A Comparison with China's Path towards Modernization, Leiden 1996.

Yonemoto, Marcia: *Mapping Early Modern Japan*. Space, Place, and Culture in the Tokugawa Period (1603–1868), Berkeley, CA 2003.

Yoshitake Oka: *Five Political Leaders* of Modern Japan, Tokyo 1986.

Young, Alfred F., and Gregory H. Nobles: *Whose American Revolution Was It?* Historians Interpret the Founding, New York 2011.

Young, David C.: The *Modern Olympics*. A Struggle for Revival, Baltimore, MD 1996.

Young, Ernest P.: The *Presidency of Yuan Shih-k'ai*. Liberalism and Dictatorship in Early Republican China, Ann Arbor, MI 1977.

Young, G. M.: *Portrait* of an Age. Victorian England, Oxford 1977.

Young, Michael: The *Metronomic Society*. Natural Rhythms and Human Timetables, London 1988.

Young, Paul: *Globalization and the Great Exhibition*. The Victorian New World Order, Basingstoke 2009.

Young, Robert J. C.: *Postcolonialism*. An Historical Introduction, Oxford 2001.

Zachernuk, Philip S.: *Colonial Subjects*. An African Intelligentsia and Atlantic Ideas, Charlottesville, VA 2000.

Zagorin, Perez: How the Idea of *Religious Toleration* Came to the West. Princeton, NJ 2003.

Zamagni, Vera: The *Economic History of Italy*, 1860–1990, Oxford 1993.

Zamoyski, Adam: *Rites of Peace*. The Fall of Napoleon and the Congress of Vienna, London 2007.

Zastoupil, Lynn, and Martin Moir (eds.): The *Great Indian Education Debate*. Documents Relating to the Orientalist-Anglicist Controversy, 1781–1843, Richmond 1999.

Zeleza, Paul Tiyambe: A Modern *Economic History of Africa*, vol. 1: The Nineteenth Century, Dakar 1993.

Zerubavel, Eviatar: *Time Maps*. Collective Memory and the Social Shape of the Past, Chicago 2003.

Zeuske, Michael: Kleine *Geschichte Kubas*, Munich 2000.

———: *Sklaven* und Sklaverei in den Welten des Atlantiks, Berlin 2006.

Zhang Hailin: *Suzhou* zaoqi chengshi xiandaihua yanjiu [Studies on early urban modernization in Suzhou], Nanjing 1999.

Zimmerman, Andrew: *Anthropology* and Antihumanism in Imperial Germany, Chicago 2001.

Zimmermann, Clemens: Zeit der *Metropolen*, Frankfurt a.M. 1996.

Zimmermann, Ekkart: *Political Violence*, Crises, and Revolutions. Theories and Research, Cambridge, MA 1983.

Zöllner, Reinhard: *Japanische Zeitrechnung*. Ein Handbuch, Munich 2003.

Zürcher, Erik J.: *Turkey*. A Modern History, London 1993.

——— (ed.): *Arming the State*. Military Conscription in the Middle East and Central Asia, London 1999.

Zweig, Stefan: The *World of Yesterday*. An Autobiography, New York 1943.

Series Editors
Sven Beckert and Jeremi Suri

ALSO IN THE SERIES

David Ekbladh, *The Great American Mission: Modernization and the Construction of an American World Order*

Martin Klimke, *The Other Alliance: Student Protest in West Germany and the United States in the Global Sixties*

Andrew Zimmerman, *Alabama in Africa: Booker T. Washington, the German Empire, and the Globalization of the New South*

Ian Tyrell, *Reforming the World: The Creation of America's Moral Empire*

Rachel St. John, *Line in the Sand: A History of the Western U.S.–Mexico Border*

Thomas Borstelmann, *The 1970s: A New Global History from Civil Rights to Economic Inequality*

Donna R. Gabaccia, *Foreign Relations: American Immigration in Global Perspective*